SOLUTIONS TO EXERCISES

ROXY WILSON

UNIVERSITY OF ILLINOIS, URBANA-CHAMPAIGN

TENTH EDITION

CHEMISTRY

THE CENTRAL SCIENCE

BROWN | LeMAY | BURSTEN

PEARSON

Prentice
Hall

Upper Saddle River, New Jersey 07458

Project Manager: Kristen Kaiser
Executive Editor: Nicole Folchetti
Executive Managing Editor: Kathleen Schiaparelli
Assistant Managing Editor: Becca Richter
Production Editor: Rhonda Aversa
Supplement Cover Designer: Elizabeth Wright
Manufacturing Buyer: Alan Fischer

© 2006 Pearson Education, Inc.
Pearson Prentice Hall
Pearson Education, Inc.
Upper Saddle River, NJ 07458

Printed in the United States of America

10 9 8 7 6 5 4 3

ISBN 0-13-146491-4

Pearson Education Ltd., *London*
Pearson Education Australia Pty. Ltd., *Sydney*
Pearson Education Singapore, Pte. Ltd.
Pearson Education North Asia Ltd., *Hong Kong*
Pearson Education Canada, Inc., *Toronto*
Pearson Educación de Mexico, S.A. de C.V.
Pearson Education—Japan, *Tokyo*
Pearson Education Malaysia, Pte. Ltd.

Contents

Introduction

Chemistry: The Central Science, 10th edition, contains nearly 2600 end-of-chapter exercises. Considerable attention has been given to these exercises because one of the best ways for students to master chemistry is by solving problems. Grouping the exercises according to subject matter is intended to aid the student in selecting and recognizing particular types of problems. Within each subject matter group, similar problems are arranged in pairs. This provides the student with an opportunity to reinforce a particular kind of problem. There are also a substantial number of general exercises in each chapter to supplement those grouped by topic. Integrative exercises, which require students to integrate concepts from several chapters, are a continuing feature of the 10th edition. Answers to the odd numbered topical exercises plus selected general and integrative exercises, about 1200 in all, are provided in the text. These appendix answers help to make the text a useful self-contained vehicle for learning.

This manual, **Solutions to Exercises in Chemistry: The Central Science, 10th edition**, was written to enhance the end-of-chapter exercises by provided documented solutions. The manual assists the instructor by saving time spent generating solutions for assigned problem sets and aids the student by offering a convenient independent source to check their understanding of the material. Most solutions have been worked in the same detail as the in-chapter sample exercises to help guide students in their studies.

To reinforce the '*Analyze, Plan, Solve, Check*' problem-solving method used extensively in the text, this strategy has also been incorporated into the Solution Manual. Solutions to most red paired exercises and selected Additional and Integrative exercises feature this four-step approach. We strongly encourage students to master this powerful and totally general method.

When using this manual, keep in mind that the numerical result of any calculation is influenced by the precision of the numbers used in the calculation. In this manual, for example, atomic masses and physical constants are typically expressed to four significant figures, or at least as precisely as the data given in the problem. If students use slightly different values to solve problems, their answers will differ slightly from those listed in the appendix of the text or this manual. This is a normal and a common occurrence when comparing results from different calculations or experiments.

Rounding methods are another source of differences between calculated values. In this manual, when a solution is given in steps, intermediate results will be rounded to the correct number of significant figures; however, unrounded numbers will be used in subsequent calculations. By following this scheme, calculators need not be cleared to re-enter rounded intermediate results in the middle of a calculation sequence. The final answer will appear with the correct number of significant figures. This may result in a small discrepancy in the last significant digit between student-calculated answers and those given in this manual. Variations due to rounding can occur in any analysis of numerical data.

The first step in checking your solution and resolving differences between your answer and the listed value is to look for similarities and differences in problem-solving methods. Ultimately, resolving the small numerical differences described above is less important than understanding the general method for solving a problem. The goal of this manual is to provide a reference for sound and consistent problem-solving methods in addition to accurate answers to text exercises.

Extraordinary efforts have been made to keep this manual as error-free as possible. All exercises were worked and proof-read by at least three chemists to ensure clarity in methods and accuracy in mathematics. The work and advice of Dr. Angela Manders Cannon and Mr. David Shinn have been invaluable to this project. However, in a written work as technically challenging as this manual, typos and errors inevitably creep in. Please help us find and eliminate them. We hope that both instructors and students will find this manual accurate, helpful and instructive.

Roxy B. Wilson
University of Illinois
School of Chemical Sciences
505 S. Mathews Ave., Box 49-1
Urbana, IL 61801
rbwilson@uiuc.edu

1 Introduction: Matter and Measurement

Visualizing Concepts

1.1 *Pure elements* contain only one kind of atom. Atoms can be present singly or as tightly bound groups called molecules. *Compounds* contain two or more kinds of atoms bound tightly into molecules. *Mixtures* contain more than one kind of atom and/or molecule, not bound into discrete particles.

(a) pure element: i, v

(b) mixture of elements: vi

(c) pure compound: iv

(d) mixture of an element and a compound: ii, iii

1.2 After a *physical change*, the identities of the substances involved are the same as their identity before the change. That is, molecules retain their original composition. During a *chemical change*, at least one new substance is produced; rearrangement of atoms into new molecules occurs.

The diagram represents a **chemical change**, because the molecules after the change are different than the molecules before the change.

1.3 (a) time   (b) density   (c) length   (d) area   (e) temperature

(f) volume   (g) temperature

1.4 Measurements (darts) that are close to each other are *precise*. Measurements that are close to the "true value" (the bull's eye) are *accurate*.

(a) Figure ii represents data that are both accurate and precise. The darts are close to the bull's eye and each other.

(b) Figure i represents data that are precise but inaccurate. The darts are near each other but their center point (average value) is far from the bull's eye.

(c) Figure iii represents data that are imprecise but their average value is accurate. The darts are far from each other, but their average value, or geometric center point, is close to the bull's eye.

1.5 (a) 7.5 cm. There are two significant figures in this measurement; the number of cm can be read precisely, but there is some estimating (uncertainty) required to read tenths of a centimeter. Listing two significant figures is consistent with the convention that measured quantities are reported so that there is uncertainty in only the last digit.

1

(b)   140°C. The temperature can be read to the nearest 50°C and estimated to the nearest 5–10°C. Since there is uncertainty in the tens digit, the measurement has two significant figures.

1.6   The determined age of the artifact, 1,900 years, has two significant figures. There is uncertainty in the hundreds place, indicating that the minimum uncertainty in age is 100 years. The 20-year period since the age was determined is not significant relative to the determined age.

1.7   In order to cancel units, the conversion factor must have the unit being canceled opposite the starting position. For example, if the unit cm starts in the numerator, then the conversion factor must have cm in its denominator. However, if the unit cm starts in the denominator, the conversion factor must have cm in the numerator. Ideally, this will lead to the desired units in the appropriate location, numerator or denominator. However, the inverse of the answer can be taken when necessary.

1.8   Given: mi/hr Find: km/s

$$\boxed{\begin{array}{c}\text{Given}\\ \text{mi/hr}\end{array}} \xrightarrow[\;0.62\,\text{mi}\;]{\text{use}\;1\,\text{km}} \boxed{\text{km/hr}} \xrightarrow[\;60\,\text{min}\;]{\text{use}\;1\,\text{hr}} \boxed{\text{km/min}} \xrightarrow[\;60\,\text{min}\;]{\text{use}\;1\,\text{hr}} \boxed{\begin{array}{c}\text{Find}\\ \text{km/s}\end{array}}$$

Classification and Properties of Matter

1.9   (a)   heterogeneous mixture

        (b)   homogeneous mixture (If there are undissolved particles, such as sand or decaying plants, the mixture is heterogeneous.)

        (c)   pure substance

        (d)   homogeneous mixture

1.10   (a)   homogeneous mixture

        (b)   heterogeneous mixture (particles in liquid)

        (c)   pure substance

        (d)   heterogeneous mixture

1.11   (a) S   (b) K   (c) Cl   (d) Cu   (e) Si   (f) N   (g) Ca   (h) He

1.12   (a) C   (b) Na   (c) F   (d) Fe   (e) P   (f) Ar   (g) Ni   (h) Ag

1.13   (a)   lithium   (b)   aluminum   (c)   lead   (d)   sulfur   (e)   bromine

        (f)   tin   (g)   chromium   (h)   zinc

1.14   (a)   cobalt   (b)   iodine   (c)   krypton   (d)   mercury   (e)   arsenic

        (f)   titanium   (g)   potassium   (h)   germanium

1.15 $A(s) \xrightarrow{\;\text{heat}\;} B(s) + C(g)$

When solid carbon is burned in excess oxygen gas, the two elements combine to form a gaseous compound, carbon dioxide. Clearly substance C is this compound. Since C is produced when A is heated in the absence of oxygen (from air), both the carbon and oxygen in C must have been present in A originally. A is, therefore, a compound composed of two or more elements chemically combined. Without more information on the chemical or physical properties of B, we cannot determine absolutely whether it is an element or a compound. However, few if any elements exist as white solids, so B is probably also a compound.

1.16   Before modern instrumentation, the classification of a pure substance as an element was determined by whether it could be broken down into component elements. Scientists subjected the substance to all known chemical means of decomposition, and if the results were negative, the substance was an element. Classification by negative results was somewhat ambiguous, since an effective decomposition technique might exist but not yet have been discovered.

1.17   *Physical properties*: silvery white (color); lustrous; melting point = 649°C; boiling point = 1105°C; density at 20°C = 1.738 g/cm^3; pounded into sheets (malleable); drawn into wires (ductile); good conductor. *Chemical properties*: burns in air to give intense white light; reacts with Cl_2 to produce brittle white solid.

1.18   *Physical properties*: silver-grey (color); melting point = 420°C; hardness = 2.5 Mohs; density = 7.13 g/cm^3 at 25°C. *Chemical properties*: metal; reacts with sulfuric acid to produce hydrogen gas; reacts slowly with oxygen at elevated temperatures to produce ZnO.

1.19   (a)   chemical   (b) physical   (c) physical   (d) chemical   (e) chemical

1.20   (a)   chemical

        (b)   physical

        (c)   physical (The production of H_2O is a chemical change, but its **condensation** is a physical change.)

        (d)   physical (The production of soot is a chemical change, but its **deposition** is a physical change.)

1.21   (a)   Take advantage of the different water solubilities of the two solids. Add water to dissolve the sugar; filter this mixture, collecting the sand on the filter paper and the sugar water in the flask. Evaporate the water from the flask to reproduce solid sugar.

        (b)   Either the melting-point difference or magnetism difference between iron and sulfur can be used to separate these two elements. Heat the mixture until the sulfur melts, then decant (pour off) the liquid sulfur. Or use a magnet to attract the iron particles, leaving the solid sulfur behind.

1.22   Take advantage of differences in physical properties to separate the components of a mixture. First heat the liquid to 100°C to evaporate the water. This is conveniently done in a distillation apparatus (Figure 1.13) so that the water can be collected. After the water is completely evaporated and if there is a residue, measure the physical properties of the residue such as color, density, and melting point. Compare the observed properties of the residue to those of table salt, NaCl. If the properties match, the colorless liquid contained table salt. If the properties don't match, the liquid contained a different dissolved solid. If there is no residue, no dissolved solid is present.

Units and Measurement

1.23 (a) 1×10^{-1} (b) 1×10^{-2} (c) 1×10^{-15} (d) 1×10^{-6} (e) 1×10^{6}

(f) 1×10^{3} (g) 1×10^{-9} (h) 1×10^{-3} (i) 1×10^{-12}

1.24 (a) $6.35 \times 10^{-2}\,L \times \dfrac{1\,mL}{1 \times 10^{-3}\,L} = 63.5\,mL$

(b) $6.5 \times 10^{-6}\,s \times \dfrac{1\,\mu s}{1 \times 10^{-6}\,s} = 6.5\,\mu s$

(c) $9.5 \times 10^{-4}\,m \times \dfrac{1\,mm}{1 \times 10^{-3}\,m} = 0.95\,mm$

(d) $4.23 \times 10^{-9}\,m^3 \times \dfrac{1^3\,mm^3}{(1 \times 10^{-3})^3\,m^3} = 4.23\,mm^3$

$4.23\,mm^3 \times \dfrac{(10^{-1})^3\,cm^3}{1^3\,mm^3} \times \dfrac{1\,mL}{1\,cm^3} \times \dfrac{1 \times 10^{-3}\,L}{1\,mL} \times \dfrac{1\,\mu L}{1 \times 10^{-6}\,L} = 4.23\,\mu L$

(e) $12.5 \times 10^{-8}\,kg \times \dfrac{1 \times 10^{3}\,g}{1\,kg} \times \dfrac{1\,mg}{1 \times 10^{-3}\,g} = 0.125\,mg\,(125\,\mu g)$

(f) $3.5 \times 10^{-10}\,g \times \dfrac{1\,ng}{1 \times 10^{-9}\,g} = 0.35\,ng$

(g) $6.54 \times 10^{9}\,fs \times \dfrac{1 \times 10^{-15}\,s}{1\,fs} \times \dfrac{1\,\mu s}{1 \times 10^{-6}\,s} = 6.54\,\mu s$

1.25 (a) $25.5\,mg \times \dfrac{1 \times 10^{-3}\,g}{1\,mg} = 0.0255\,g\,(2.55 \times 10^{-2}\,g)$

(b) $4.0 \times 10^{-10}\,m \times \dfrac{1\,nm}{1 \times 10^{-9}\,m} = 0.40\,nm$

(c) $0.575\,mm \times \dfrac{1 \times 10^{-3}\,m}{1\,mm} \times \dfrac{1\,\mu m}{1 \times 10^{-6}\,m} = 575\,\mu m$

1.26 (a) $9.5 \times 10^{-2}\,kg \times \dfrac{1 \times 10^{3}\,g}{1\,kg} = 95\,g$

(b) $0.0023\,\mu m \times \dfrac{1 \times 10^{-6}\,m}{1\,\mu m} \times \dfrac{1\,nm}{1 \times 10^{-9}\,m} = 2.3\,nm$

(c) $7.25 \times 10^{-4}\,s \times \dfrac{1\,ms}{1 \times 10^{-3}\,s} = 0.725\,ms$

1.27   (a) $\text{density} = \dfrac{\text{mass}}{\text{volume}} = \dfrac{39.73\,\text{g}}{25.0\,\text{mL}} = 1.59\,\text{g/mL or } 1.59\,\text{g/cm}^3$

(The units cm^3 and mL will be used interchangeably in this manual.)

Carbon tetrachloride, 1.59 g/mL, is more dense than water, 1.00 g/mL; carbon tetrachloride will sink rather than float on water.

     (b) $75.00\,\text{cm}^3 \times 21.45\,\dfrac{\text{g}}{\text{cm}^3} = 1.609 \times 10^3\,\text{g}\ (1.609\,\text{kg})$

     (c) $87.50\,\text{g} \times \dfrac{1\,\text{cm}^3}{1.738\,\text{g}} = 50.3452 = 50.35\,\text{cm}^3 = 50.35\,\text{mL}$

1.28   (a) $\text{volume} = \text{length}^3\ (\text{cm}^3);\ \text{density} = \text{mass/volume}\ (\text{g/cm}^3)$

$\text{volume} = (1.500)^3\,\text{cm}^3 = 3.375\,\text{cm}^3$

$\text{density} = \dfrac{76.31\,\text{g}}{3.375\,\text{cm}^3} = 22.61\,\text{g/cm}^3\ \text{osmium}$

     (b) $65.8\,\text{mL} \times \dfrac{1\,\text{cm}^3}{1\,\text{mL}} \times \dfrac{4.51\,\text{g}}{1\,\text{cm}^3} = 296.758 = 297\,\text{g titanium}$

     (c) $0.1500\,\text{L} \times \dfrac{1\,\text{mL}}{1 \times 10^{-3}\,\text{L}} \times \dfrac{0.8787\,\text{g}}{1\,\text{mL}} = 131.8\,\text{g benzene}$

1.29   (a) $\text{density} = \dfrac{38.5\,\text{g}}{45\,\text{mL}} = 0.86\,\text{g/mL}$

The substance is probably toluene, density = 0.866 g/mL.

     (b) $45.0\,\text{g} \times \dfrac{1\,\text{mL}}{1.114\,\text{g}} = 40.4\,\text{mL ethylene glycol}$

     (c) $(5.00)^3\,\text{cm}^3 \times \dfrac{8.90\,\text{g}}{1\,\text{cm}^3} = 1.11 \times 10^3\,\text{g}\ (1.11\,\text{kg})\ \text{nickel}$

1.30   (a) $\dfrac{21.95\,\text{g}}{25.0\,\text{mL}} = 0.878\,\text{g/mL}$

The tabulated value has four significant figures, while the experimental value has three. The tabulated value rounded to three figures is 0.879. The values agree within one in the last significant figure of the experimental value; the two results agree. The liquid could be benzene.

     (b) $15.0\,\text{g} \times \dfrac{1\,\text{mL}}{0.7781\,\text{g}} = 19.3\,\text{mL cyclohexane}$

     (c) $r = d/2 = 5.0\,\text{cm}/2 = 2.5\,\text{cm}$

$V = 4/3\,\pi\,r^3 = 4/3 \times \pi \times (2.5)^3\,\text{cm}^3 = 65\,\text{cm}^3$

$65.4498\,\text{cm}^3 \times \dfrac{11.34\,\text{g}}{\text{cm}^3} = 7.4 \times 10^2\,\text{g}$

(The answer has two significant figures because the diameter had only two figures.)

Note: This is the first exercise where "intermediate rounding" occurs. In this manual, when a solution is given in steps, the intermediate result will be rounded to the correct number of significant figures. However, the **unrounded** number will be used in subsequent calculations. The final answer will appear with the correct number of significant figures. That is, calculators need not be cleared and new numbers entered in the middle of a calculation sequence. This may result in a small discrepancy in the last significant digit between student-calculated answers and those given in the manual. These variations occur in any analysis of numerical data.

For example, in this exercise the volume of the sphere, 65.4498 cm^3, is rounded to 65 cm^3, but 65.4498 is retained in the subsequent calculation of mass, 7.4×10^2 g. In this case, 65 cm^3 × 11.34 g/cm^3 also yields 7.4×10^2 g. In other exercises, the correctly rounded results of the two methods may not be identical.

1.31   thickness = volume/area

$$\text{volume} = 200\,mg \ \times \ \frac{1 \times 10^{-3}\,g}{1\,mg} \ \times \ \frac{1\,cm^3}{19.32\,g} = 0.01035 = 0.0104\,cm^3$$

$$\text{area} = 2.4\,ft \ \times \ 1.0\,ft \ \times \ \frac{12^2\,in^2}{1\,ft^2} \ \times \ \frac{2.54^2\,cm^2}{in^2} = 2.23 \ \times \ 10^3 = 2.2 \ \times \ 10^3\,cm^2$$

$$\text{thickness} = \frac{0.01035\,cm^3}{2{,}230\,cm^2} \ \times \ \frac{1 \times 10^{-2}\,m}{1\,cm} = 4.6 \ \times \ 10^{-8}\,m$$

$$4.6 \times 10^{-8}\,m \ \times \ \frac{1\,nm}{1 \times 10^{-9}\,m} = 46\,nm\ \text{thick}$$

1.32   Calculate the volume of the rod:

$$2.17\,kg \ \times \ \frac{1000\,g}{1\,kg} \ \times \ \frac{1\,cm^3}{2.33\,g} = 931.3 = 931\,cm^3$$

$$V = \pi\,r^2 h; d = 2r, r = d/2; \quad V = \pi\left(\frac{d}{2}\right)^2 h; \quad d^2 = \frac{4\,V}{\pi\,h}; d = \left(\frac{4\,V}{\pi\,h}\right)^{1/2}$$

$$d = \left(\frac{4\,(931.3)\,cm^3}{\pi\,(16.8)\,cm}\right)^{1/2} = 8.401 = 8.40\,cm$$

1.33   (a)   °C = 5/9 (°F – 32°); 5/9 (62 – 32) = 17°C

(b)   °F = 9/5 (°C) + 32°; 9/5 (216.7) + 32 = 422.1°F

(c)   K = °C + 273.15; 233°C + 273.15 = 506 K

(d)   315 K – 273 = 42°C; 9/5 (42°C) + 32 = 108°F

(e)   °C = 5/9 (°F – 32°); 5/9 (2500 – 32) = 1371°C; 1371°C + 273 = 1644 K
        (assuming 2500 C has 4 sig figs)

1.34   (a)   °C = 5/9 (87°F – 32°) = 31°C

(b)   K = 25°C + 273 = 298 K; °F = 9/5 (25°C) + 32 = 77°F

(c)   °C = 5/9 (175°F – 32°) = 79.444 = 79.4°C

        K = °C + 273.15 = 79.444°C + 273.15 = 352.6 K

(d) °F = 9/5 (755°C) + 32 = 1391°F; K = 755°C + 273.15 = 1028 K

(It could be argued that the result of 9/5 (755) has 3 sig figs, so the final Fahrenheit temperature should have 3 sig figs, 1390°F.)

(e) melting point = –248.6°C + 273.15 = 24.6 K
boiling point = –246.1°C + 273.15 = 27.1 K

Uncertainty In Measurement

1.35 Exact: (c), (d), and (f) (All others depend on measurements and standards that have margins of error, e.g., the length of a week as defined by the earth's rotation.)

1.36 Exact: (b), (e) (The number of students is exact on any given day.)

1.37 (a) 3 (b) 2 (c) 5 (d) 3 (e) 5

1.38 (a) 5 (b) 3 (c) 4 (d) 4 (e) 6

1.39 (a) 1.025×10^2 (b) 6.570×10^5 (c) 8.543×10^{-3}

(d) 2.579×10^{-4} (e) -3.572×10^{-2}

1.40 (a) 7.93×10^3 mi (b) 4.001×10^4 km

1.41 (a) 12.0550 + 9.05 = 21.105 = 21.11 (For addition and subtraction, the minimum number of decimal places, here two, determines decimal places in the result.)

(b) 257.2 – 19.789 = 237.4

(c) $(6.21 \times 10^3)(0.1050) = 652$ (For multiplication and division, the minimum number of significant figures, here three, determines sig figs in the result.)

(d) $0.0577/0.753 = 7.66 \times 10^{-2}$

1.42 (a) $[320.55 - 6104.5/2.3] = -2.3 \times 10^3$ (The intermediate result has two significant figures, so only the thousand and hundred places in the answer are significant.)

(b) $[285.3 \times 10^5 - 0.01200 \times 10^5] \times 2.8954 = 8.260 \times 10^7$ (Since subtraction depends on decimal places, both numbers must have the same exponent to determine decimal places/sig figs. The intermediate result has 1 decimal place and 4 sig figs, so the answer has 4 sig figs.)

(c) $(0.0045 \times 20,000.0)$ + (2813×12) = 3.4×10^4
2 sig figs /0 dec pl   2 sig figs /first 2 digits

(d)   863   ×   [1255   – $(3.45 \times 108)]$ = 7.62×10^5
                                                3 sig figs /0 dec pl
        3 sig figs   ×   0 dec pl/3 sig figs   = 3 sig figs

Dimensional Analysis

1.43 (a) cm → ft: $\dfrac{1\,in}{2.54\,cm} \times \dfrac{1\,ft}{12\,in}$ (b) in^3 → cm^3: $\dfrac{(2.54)^3\,cm^3}{1^3\,in^3}$

1.44   (a) $\dfrac{1.6093\,\text{km}}{1\,\text{mi}}$; when converting miles to kilometers, miles goes in the denominator

                   so that it cancels the original unit, leaving km in the numerator.

  (b) $\dfrac{1\,\text{L}}{1.0567\,\text{qt}}$

1.45   (a) $0.076\,\text{L} \times \dfrac{1000\,\text{mL}}{1\,\text{L}} = 76\,\text{mL}$

  (b) $5.0 \times 10^{-8}\,\text{m} \times \dfrac{1\,\text{nm}}{1 \times 10^{-9}\,\text{m}} = 50.\,\text{nm}$

  (c) $6.88 \times 10^{5}\,\text{ns} \times \dfrac{1 \times 10^{-9}\,\text{s}}{1\,\text{ns}} = 6.88 \times 10^{-4}\,\text{s}$

  (d) $0.50\,\text{lb} \times \dfrac{453.6\,\text{g}}{1\,\text{lb}} = 226.8 = 2.3 \times 10^{2}\,\text{g}$ The data and the result have 2 sig figs.

  (e) $\dfrac{1.55\,\text{kg}}{\text{m}^{3}} \times \dfrac{1000\,\text{g}}{1\,\text{kg}} \times \dfrac{1\,\text{m}^{3}}{(10)^{3}\,\text{dm}^{3}} \times \dfrac{1\,\text{dm}^{3}}{1\,\text{L}} = 1.55\,\text{g/L}$

  (f) $\dfrac{5.850\,\text{gal}}{\text{hr}} \times \dfrac{3.7854\,\text{L}}{1\,\text{gal}} \times \dfrac{1\,\text{hr}}{60\,\text{min}} \times \dfrac{1\,\text{min}}{60\,\text{s}} = 6.151 \times 10^{-3}\,\text{L/s}$

Estimated answer: $6 \times 4 = 24$; $24/60 = 0.4$; $0.4/60 = 0.0066 = 7 \times 10^{-3}$. This agrees with the calculated answer of $6.151 \times 10^{-3}\,\text{L/s}$.

1.46   (a) $\dfrac{2.998 \times 10^{8}\,\text{m}}{\text{s}} \times \dfrac{1\,\text{km}}{1000\,\text{m}} \times \dfrac{60\,\text{s}}{1\,\text{min}} \times \dfrac{60\,\text{min}}{1\,\text{hr}} = 1.079 \times 10^{9}\,\text{km/hr}$

  (b) $1454\,\text{ft} \times \dfrac{1\,\text{yd}}{3\,\text{ft}} \times \dfrac{1\,\text{m}}{1.0936\,\text{yd}} = 443.18 = 443.2\,\text{m}$

  (c) $3{,}666{,}500\,\text{m}^{3} \times \dfrac{1^{3}\,\text{dm}^{3}}{(1 \times 10^{-1})^{3}\,\text{m}^{3}} \times \dfrac{1\,\text{L}}{1\,\text{dm}^{3}} = 3.6665 \times 10^{9}\,\text{L}$

  (d) $\dfrac{232\,\text{mg cholesterol}}{100\,\text{mL blood}} \times \dfrac{1\,\text{mL}}{1 \times 10^{-3}\,\text{L}} \times 5.2\,\text{L} \times \dfrac{1 \times 10^{-3}\,\text{g}}{1\,\text{mg}} = 12\,\text{g cholesterol}$

1.47   (a) $5.00\,\text{days} \times \dfrac{24\,\text{hr}}{1\,\text{day}} \times \dfrac{60\,\text{min}}{1\,\text{hr}} \times \dfrac{60\,\text{s}}{1\,\text{min}} = 4.32 \times 10^{5}\,\text{s}$

  (b) $0.0550\,\text{mi} \times \dfrac{1.6093\,\text{km}}{\text{mi}} \times \dfrac{1000\,\text{m}}{1\,\text{km}} = 88.5\,\text{m}$

  (c) $\dfrac{\$1.89}{\text{gal}} \times \dfrac{1\,\text{gal}}{3.7854\,\text{L}} = \dfrac{\$0.499}{\text{L}}$

  (d) $\dfrac{0.510\,\text{in}}{\text{ms}} \times \dfrac{2.54\,\text{cm}}{1\,\text{in}} \times \dfrac{1 \times 10^{-2}\,\text{m}}{1\,\text{cm}} \times \dfrac{1\,\text{km}}{1000\,\text{m}} \times \dfrac{1\,\text{ms}}{1 \times 10^{-3}\,\text{s}} \times \dfrac{60\,\text{s}}{1\,\text{min}} \times \dfrac{60\,\text{min}}{1\,\text{hr}} = 46.6\,\dfrac{\text{km}}{\text{hr}}$

Estimate: $0.5 \times 2.5 = 1.25$; $1.25 \times 0.01 \approx 0.01$; $0.01 \times 60 \times 60 \approx 36\,\text{km/hr}$

(e) $\dfrac{22.50\,\text{gal}}{\text{min}} \times \dfrac{3.7854\,\text{L}}{\text{gal}} \times \dfrac{1\,\text{min}}{60\,\text{s}} = 1.41953 = 1.420\,\text{L/s}$

Estimate: $20 \times 4 = 80;\ 80/60 \approx 1.3\,\text{L/s}$

(f) $0.02500\,\text{ft}^3 \times \dfrac{12^3\,\text{in}^3}{1\,\text{ft}^3} \times \dfrac{2.54^3\,\text{cm}^3}{1\,\text{in}^3} = 707.9\,\text{cm}^3$

Estimate: $10^3 = 1000;\ 3^3 = 27;\ 1000 \times 27 = 27{,}000;\ 27{,}000/0.04 \approx 700\,\text{cm}^3$

1.48   (a) $0.105\,\text{in} \times \dfrac{2.54\,\text{cm}}{\text{in}} \times \dfrac{1 \times 10^{-2}\,\text{m}}{\text{cm}} \times \dfrac{1\,\text{mm}}{1 \times 10^{-3}\,\text{m}} = 2.667 = 2.67\,\text{mm}$

(b) $0.870\,\text{qt} \times \dfrac{1\,\text{L}}{1.057\,\text{qt}} \times \dfrac{1\,\text{mL}}{1 \times 10^{-3}\,\text{L}} = 823.08 = 823\,\text{mL}$

(c) $\dfrac{8.75\,\mu\text{m}}{\text{s}} \times \dfrac{1 \times 10^{-6}\,\text{m}}{1\,\mu\text{m}} \times \dfrac{1\,\text{km}}{1 \times 10^3\,\text{m}} \times \dfrac{60\,\text{s}}{1\,\text{min}} \times \dfrac{60\,\text{min}}{1\,\text{hr}} = 3.15 \times 10^{-5}\,\text{km/hr}$

(d) $4.733\,\text{yd}^3 \times \dfrac{1\,\text{m}^3}{(1.0936)^3\,\text{yd}^3} = 3.61877 = 3.619\,\text{m}^3$

(e) $\dfrac{\$3.99}{\text{lb}} \times \dfrac{2.205\,\text{lb}}{1\,\text{kg}} = 8.798 = \$8.80\text{/kg}$

(f) $\dfrac{8.75\,\text{lb}}{\text{ft}^3} \times \dfrac{453.59\,\text{g}}{1\,\text{lb}} \times \dfrac{1\,\text{ft}^3}{12^3\,\text{in}^3} \times \dfrac{1\,\text{in}^3}{2.54^3\,\text{cm}^3} \times \dfrac{1\,\text{cm}^3}{1\,\text{mL}} = 0.140\,\text{g/mL}$

1.49   (a) $31\,\text{gal} \times \dfrac{4\,\text{qt}}{1\,\text{gal}} \times \dfrac{1\,\text{L}}{1.057\,\text{qt}} = 1.2 \times 10^2\,\text{L}$

Estimate: $(30 \times 4)/1 \approx 120\,\text{L}$

(b) $\dfrac{6\,\text{mg}}{\text{kg (body)}} \times \dfrac{1\,\text{kg}}{2.205\,\text{lb}} \times 150\,\text{lb} = 4 \times 10^2\,\text{mg}$

Estimate: $6/2 = 3;\ 3 \times 150 = 450\,\text{mg}$

(c) $\dfrac{254\,\text{mi}}{11.2\,\text{gal}} \times \dfrac{1.609\,\text{km}}{1\,\text{mi}} \times \dfrac{1\,\text{gal}}{4\,\text{qt}} \times \dfrac{1.057\,\text{qt}}{1\,\text{L}} = \dfrac{9.64\,\text{km}}{\text{L}}$

Estimate: $250/10 = 25;\ 1.6/4 = 0.4;\ 25 \times 0.4 \times 1 \approx 10\,\text{km/L}$

(d) $\dfrac{50\,\text{cups}}{1\,\text{lb}} \times \dfrac{1\,\text{qt}}{4\,\text{cups}} \times \dfrac{1\,\text{L}}{1.057\,\text{qt}} \times \dfrac{1000\,\text{mL}}{1\,\text{L}} \times \dfrac{1\,\text{lb}}{453.6\,\text{g}} = \dfrac{26\,\text{mL}}{\text{g}}$

Estimate: $50/4 = 12;\ 1000/500 = 2;\ (12 \times 2)/1 \approx 24\,\text{mL/g}$

1.50   (a) $1486\,\text{mi} \times \dfrac{1\,\text{km}}{0.62137\,\text{mi}} \times \dfrac{\text{charge}}{225\,\text{km}} = 10.6\,\text{charges}$

Since charges are integral events, 11 charges are required.

(b) $\dfrac{14\,m}{s}\times\dfrac{1\,km}{1\times10^{3}\,m}\times\dfrac{1\,mi}{1.6093\,km}\times\dfrac{60\,s}{1\,min}\times\dfrac{60\,min}{1\,hr}=31\,mi/hr$

(c) $450\,in^{3}\times\dfrac{(2.54)^{3}\,cm^{3}}{1\,in^{3}}\times\dfrac{1\,mL}{1\,cm^{3}}\times\dfrac{1\times10^{-3}\,L}{1\,mL}=7.37\,L$

(d) $2.4\times10^{5}\,barrels\times\dfrac{42\,gal}{1\,barrel}\times\dfrac{4\,qt}{1\,gal}\times\dfrac{1\,L}{1.057\,qt}=3.8\times10^{7}\,L$

1.51 $12.5\,ft\times15.5\,ft\times8.0\,ft=1580=1.6\times10^{3}\,ft^{3}$ (2 sig figs)

$1550\,ft^{3}\times\dfrac{(1\,yd)^{3}}{(3\,ft)^{3}}\times\dfrac{(1\,m)^{3}}{(1.0936)^{3}\,yd^{3}}\times\dfrac{10^{3}\,dm^{3}}{1\,m^{3}}\times\dfrac{1\,L}{1\,dm^{3}}\times\dfrac{1.19\,g}{L}\times\dfrac{1\,kg}{1000\,g}=52\,kg\,air$

Estimate: $1550/30=50;\ (50\times1)/1\approx50\,kg$

1.52 $9.0\,ft\times14.5\,ft\times18.8\,ft=2453.4=2.5\times10^{3}\,ft^{3}$

$2453.4\,ft^{3}\times\dfrac{(1\,yd)^{3}}{(3\,ft)^{3}}\times\dfrac{(1\,m)^{3}}{(1.094\,yd)^{3}}\times\dfrac{48\,\mu g\,CO}{1\,m^{3}}\times\dfrac{1\times10^{-6}\,g}{1\,\mu g}=3.3\times10^{-3}\,g\,CO$

1.53 Select a common unit for comparison, in this case the cm.

1 in ≈ 2.5 cm, 1 m = 100 cm

57 cm = 57 cm

14 in ≈ 35 cm

1.1 m = 110 cm

The order of length from shortest to longest is 14-in shoe < 57-cm string < 1.1-m pipe.

1.54 Select a common unit for comparison, in this case the kg.

1 kg > 2 lb, 1 L ≈ 1 qt

5 lb potatoes < 2.5 kg

5 kg sugar = 5 kg

1 gal = 4 qt ≈ 4 L. 1 mL H_2O = 1 g H_2O. 1 L = 1000 g, 4 L = 4000 g = 4 kg

The order of mass from lightest to heaviest is 5 lb potatoes < 1 gal water < 5 kg sugar.

1.55 (a) $26.73\,g\,total\times\dfrac{0.90\,g\,Ag}{1\,g\,total}\times\dfrac{1\,tr\,oz}{31.1\,g\,Ag}\times\dfrac{\$1.18}{1\,tr\,oz}=\$0.91$

(b) $\$25.00\times\dfrac{1\,tr\,oz}{\$5.30}\times\dfrac{31.1\,g}{1\,tr\,oz}\times\dfrac{1\,g\,total}{0.90\,g\,Ag}\times\dfrac{1\,coin}{26.73\,g}=6.1\,coins$

Since coins come in integer numbers, 7 coins are required.

1.56 A wire is a very long, thin cylinder of volume, $V=\pi\,r^{2}\,h$, where h is the length of the wire and $\pi\,r^{2}$ is the cross-sectional area of the wire.

1 Matter and Measurement   Solutions to Exercises

Strategy: 1) Calculate total volume of copper in cm^3 from mass and density

2) h (length in cm) $= \dfrac{V}{\pi r^2}$

3) Change cm → ft

$$150\,\text{lb Cu} \times \frac{453.6\,\text{g}}{1\,\text{lb Cu}} \times \frac{1\,\text{cm}^3}{8.94\,\text{g}} = 7610.7 = 7.61 \times 10^3\,\text{cm}^3$$

$$r = d/2 = 8.25\,\text{mm} \times \frac{1\,\text{cm}}{10\,\text{mm}} \times \frac{1}{2} = 0.4125 = 0.413\,\text{cm}$$

$$h = \frac{V}{\pi r^2} = \frac{7610.7\,\text{cm}^3}{\pi(0.4125)^2\,\text{cm}^2} = 1.4237 \times 10^4 = 1.42 \times 10^4\,\text{cm}$$

$$1.4237 \times 10^4\,\text{cm} \times \frac{1\,\text{in}}{2.54\,\text{cm}} \times \frac{1\,\text{ft}}{12\,\text{in}} = 467\,\text{ft}$$

(too difficult to estimate)

Additional Exercises

1.57   *Composition* is the contents of a substance, the kinds of elements that are present and their relative amounts. *Structure* is the arrangement of these contents.

1.58   (a)   A gold coin is probably a *solid solution*. Pure gold (element 79) is too soft and too valuable to be used for coinage, so other metals are added. However, the simple term "gold coin" does not give a specific indication of the other metals in the mixture.

A cup of coffee is a *solution* if there are no suspended solids (coffee grounds). It is a heterogeneous mixture if there are grounds. If cream or sugar is added, the homogeneity of the mixture depends on how thoroughly the components are mixed.

A wood plank is a *heterogeneous mixture* of various cellulose components. The different domains in the mixture are visible as wood grain or knots.

(b)   The ambiguity in each of these examples is that the name of the substance does not provide a complete description of the material. We must rely on mental images, and these vary from person to person.

1.59   (a)   A *hypothesis* is a possible explanation for certain phenomena based on preliminary experimental data. A *theory* may be more general, and has a significant body of experimental evidence to support it; a theory has withstood the test of experimentation.

(b)   A scientific *law* is a summary or statement of natural behavior; it tells how matter behaves. A *theory* is an explanation of natural behavior; it attempts to explain why matter behaves the way it does.

11

1.60   Any sample of vitamin C has the same relative amount of carbon and oxygen; the ratio of oxygen to carbon in the isolated sample is the same as the ratio in synthesized vitamin C.

$$\frac{2.00\,g\,O}{1.50\,g\,C} = \frac{x\,g\,O}{6.35\,g\,C};\quad x = \frac{(2.00\,g\,O)(6.35\,g\,C)}{1.50\,g\,C} = 8.47\,g\,O$$

This calculation assumes the law of constant composition.

1.61   (a)   I.   (22.52 + 22.48 + 22.54)/3 = 22.51

                II.   (22.64 + 22.58 + 22.62)/3 = 22.61

            Based on the average, set I is more accurate. That is, it is closer to the true value of 22.52%.

            (b)   Average deviation = $\sum$ | value – average | /3

                I. | 22.52 – 22.51 | + | 22.48 – 22.51 | + | 22.54 – 22.51 | /3 = 0.02

                II. | 22.64 – 22.61 | + | 22.58 – 22.61 | + | 22.62 – 22.61 | /3 = 0.02

            The two sets display the same precision, even though set I is more accurate.

1.62   (a)   Inappropriate. The circulation of a widely read publication would vary over a year's time, and could simply not be counted to the nearest single subscriber. Probably about four significant figures would be appropriate.

            (b)   Inappropriate. In a county with 5 million people, the population surely fluctuates with moves, births, and deaths during a month. The population cannot be known precisely to the nearest person over this time period. There would be uncertainty in at least hundreds, probably thousands place in the population.

            (c)   Inappropriate. Rainfall can be measured to within 0.02 in., but it is probably not possible to record an entire year's rainfall to the nearest 0.01 in. Further, the variation from year to year is sufficiently large that it does not make much sense to report the annual average to this number of significant figures. Probably two significant figures would be appropriate.

            (d)   Appropriate. The percentage has three significant figures. In a population as large as the United States, the number of people named Brown can surely be counted by census data or otherwise to a precision of three significant figures.

1.63   (a)   volume   (b)   area   (c)   volume   (d)   density

            (e)   time   (f)   length   (g)   temperature

1.64   (a) $\dfrac{m}{s^2}$ (b) $\dfrac{kg \cdot m}{s^2}$ (c) $\dfrac{kg \cdot m}{s^2} \times m = \dfrac{kg \cdot m^2}{s^2}$

            (d) $\dfrac{kg \cdot m}{s^2} \times \dfrac{1}{m^2} = \dfrac{kg}{m \cdot s^2}$ (e) $\dfrac{kg \cdot m^2}{s^2} \times \dfrac{1}{s} = \dfrac{kg \cdot m^2}{s^3}$

1.65   The most dense liquid, Hg, will sink; the least dense, cyclohexane, will float; H_2O will be in the middle.

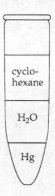

1.66   (a) $$\frac{40 \text{ lb peat}}{14 \times 20 \times 30 \text{ in}^3} \times \frac{1 \text{ in}^3}{(2.54)^3 \text{ cm}^3} \times \frac{453.6 \text{ g}}{1 \text{ lb}} = 0.13 \text{ g/cm}^3 \text{ peat}$$

$$\frac{40 \text{ lb soil}}{1.9 \text{ gal}} \times \frac{1 \text{ gal}}{4 \text{ qt}} \times \frac{1.057 \text{ qt}}{1 \text{ L}} \times \frac{1 \times 10^{-3} \text{ L}}{1 \text{ mL}} \times \frac{1 \text{ mL}}{1 \text{ cm}^3} \times \frac{453.6 \text{ g}}{1 \text{ lb}} = 2.5 \text{ g/cm}^3 \text{ soil}$$

No. Volume must be specified in order to compare mass. The densities tell us that a certain volume of peat moss is "lighter" (weighs less) than the same volume of top soil.

(b)   1 bag peat = $14 \times 20 \times 30 = 8.4 \times 10^3 \text{ in}^3$

$$10. \text{ ft} \times 20. \text{ ft} \times 2.0 \text{ in} \times \frac{12^2 \text{ in}^2}{\text{ft}^2} = 57,600 = 5.8 \times 10^4 \text{ in}^3 \text{ peat needed}$$

$$57,600 \text{ in}^3 \times \frac{1 \text{ bag}}{8.4 \times 10^3 \text{ in}^3} = 6.9 \text{ bags needed} \quad (\text{Buy 7 bags of peat.})$$

1.67   Density is the ratio of mass and volume. For substances with different densities, the greater the density the smaller the volume of substance that will contain a certain mass. Since volume is directly related to diameter ($V = 4/3\ \pi\ r^3 = 1/6\ \pi\ d^3$), the more dense the substance, the smaller the diameter of a ball that contains a certain mass. The order of the sphere diameters is the reverse order of densities: Pb < Ag < Al. Mathematically, assume 10.0 g of material.

Pb: $10.0 \text{ g} \times \dfrac{1 \text{ cm}^3}{11.3 \text{ g}} = 0.88496 = 0.885 \text{ cm}^3; \ d = (6 \ V/\pi)^{1/3} = 1.19 \text{ cm}$

Ag: $10.0 \text{ g} \times \dfrac{1 \text{ cm}^3}{10.5 \text{ g}} = 0.95238 = 0.952 \text{ cm}^3; d = 1.22 \text{ cm}$

Al: $10.0 \text{ g} \times \dfrac{1 \text{ cm}^3}{2.70 \text{ g}} = 3.7037 = 3.70 \text{ cm}^3; d = 1.92 \text{ cm}$

Note that Pb and Ag, with similar densities, have similar diameters; Al, with a much smaller density, has a much larger diameter.

1.68   (a) $23.2 \times 10^9 \text{ lb} \times \dfrac{453.6 \text{ g}}{1 \text{ lb}} = 1.05235 \times 10^{13} = 1.05 \times 10^{13} \text{ g NaOH}$

(b) $1.05235 \times 10^{13} \text{ g} \times \dfrac{1 \text{ cm}^3}{2.130 \text{ g}} \times \dfrac{1 \text{ m}^3}{(100)^3 \text{ cm}^3} \times \dfrac{1 \text{ km}^3}{(1000)^3 \text{ m}^3} = 4.94 \times 10^{-3} \text{ km}^3$

1.69　　(a)　density = (35.66 g – 14.23 g)/4.59 cm^3 = 4.67 g/cm^3

　　　　(b)　$34.5 \, kg \times \dfrac{1000 \, g}{1 \, kg} \times \dfrac{1 \, mL}{13.6 \, g} \times \dfrac{1 \, L}{1000 \, mL} = 2.54 \, L$

　　　　(c)　$V = 4/3 \, \pi \, r^3 = 4/3 \, \pi \, (28.9 \, cm)^3 = 1.0111 \times 10^5 = 1.01 \times 10^5 = 1.01 \times 10^5 \, cm^3$

　　　　　　$1.011 \times 10^5 \, cm^3 \times \dfrac{19.3 \, g}{cm^3} = 1.95 \times 10^6 \, g$

The sphere weighs 1950 kg or 4300 pounds. The student is unlikely to be able to carry the sphere.

1.70　　$0.500 \, L \text{ battery acid} \times \dfrac{1000 \, mL}{L} \times \dfrac{1.28 \, g}{mL} = 640 \, g \text{ battery acid}$

　　　　$640 \, g \text{ battery acid} \times \dfrac{38.1 \, g \text{ sulfuric acid}}{100 \, g \text{ battery acid}} = 243.84 = 244 \, g \text{ sulfuric acid}$

1.71　　mass of toluene = 58.58 g – 32.65 g = 25.93 g

　　　　$\text{volume of toluene} = 25.93 \, g \times \dfrac{1 \, mL}{0.864 \, g} = 30.0116 = 30.0 \, mL$

　　　　volume of solid = 50.00 mL – 30.0116 mL = 19.9884 = 20.0 mL

　　　　$\text{density of solid} = \dfrac{32.65 \, g}{19.9884 \, mL} = 1.63 \, g/mL$

1.72　　There are 209.1 degrees between the freezing and boiling points on the Celsius (C) scale and 100 degrees on the glycol (G) scale. Also, –11.5°C = 0°G. By analogy with °F and °C,

　　　　$^\circ G = \dfrac{100}{209.1} (^\circ C + 11.5) \;\; or \;\; ^\circ C = \dfrac{209.1}{100} (^\circ G) - 11.5$

These equations correctly relate the freezing point (and boiling point) of ethylene glycol on the two scales.

　　　　$\text{f.p. of } H_2O: \;\; ^\circ G = \dfrac{100}{209.1} (0^\circ C + 11.5) = 5.50^\circ G$

1.73　　(a)　The distance is exactly 10 mi (infinite sig figs).

　　　　　　$25 \, min \times \dfrac{1 \, hr}{60 \, min} = 0.41667 = 0.42 \, hr;$

　　　　　　$14 \, s \times \dfrac{1 \, min}{60 \, s} \times \dfrac{1 \, hr}{60 \, min} = 0.00389 = 0.0039 \, hr$

Time can be measured to the nearest second, or 0.0003 hr, so the total time is reported with 4 decimal places and 5 sig figs.

total time = 1 hr + 0.4167 hr + 0.0039 hr = 1.4206 hr

avg. speed = 10 mi/1.4206 hr = 7.0393 mi/hr

　　　　(b)　$\dfrac{1.4206 \, hr}{10 \, mi} \times \dfrac{60 \, min}{1 \, hr} = \dfrac{8.5236 \, min}{mi}$

　　　　　　$\dfrac{1.4206 \, hr}{10 \, mi} \times \dfrac{60 \, min}{1 \, hr} \times \dfrac{60 \, s}{1 \, min} = \dfrac{511.42 \, s}{mi}$

1.74　(a)　$2.4 \times 10^5 \, \text{mi} \times \dfrac{1.609 \, \text{km}}{1 \, \text{mi}} \times \dfrac{1000 \, \text{m}}{1 \, \text{km}} = 3.9 \times 10^8 \, \text{m}$

　　　(b)　$2.4 \times 10^5 \, \text{mi} \times \dfrac{1.609 \, \text{km}}{1 \, \text{mi}} \times \dfrac{1 \, \text{hr}}{2.4 \times 10^3 \, \text{km}} \times \dfrac{60 \, \text{min}}{1 \, \text{hr}} \times \dfrac{60 \, \text{s}}{1 \, \text{min}} = 5.8 \times 10^5 \, \text{s}$

1.75　(a)　$575 \, \text{ft} \times \dfrac{12 \, \text{in}}{1 \, \text{ft}} \times \dfrac{2.54 \, \text{cm}}{1 \, \text{in}} \times \dfrac{10 \, \text{mm}}{1 \, \text{cm}} \times \dfrac{1 \, \text{quarter}}{1.55 \, \text{mm}} = 1.1307 \times 10^5 = 1.13 \times 10^5 \, \text{quarters}$

　　　(b)　$1.1307 \times 10^5 \, \text{quarters} \times \dfrac{5.67 \, \text{g}}{1 \, \text{quarter}} = 6.41 \times 10^5 \, \text{g} \; (641 \, \text{kg})$

　　　(c)　$1.1307 \times 10^5 \, \text{quarters} \times \dfrac{1 \, \text{dollar}}{4 \, \text{quarters}} = \$28,268 = \$2.83 \times 10^4$

　　　(d)　$\$7.2 \times 10^{12} \times \dfrac{1 \, \text{stack}}{\$28,268} = 2.5 \times 10^8 \, \text{stacks}$　(approximately 250 million stacks)

1.76　(a)　$\dfrac{\$2480}{\text{acre} \cdot \text{ft}} \times \dfrac{1 \, \text{acre}}{4840 \, \text{yd}^2} \times \dfrac{3 \, \text{ft}}{1 \, \text{yd}} \times \dfrac{(1.094 \, \text{yd})^3}{(1 \, \text{m})^3} \times \dfrac{(1 \, \text{m})^3}{(10 \, \text{dm})^3} \times \dfrac{(1 \, \text{dm})^3}{1 \, \text{L}} =$

　　　　　$\$2.013 \times 10^{-3}/\text{L}$ or $0.2013 \, \text{¢}/\text{L}$　　$(0.201 \, \text{¢}/\text{L}$ to 3 sig figs$)$

　　　(b)　$\dfrac{\$2480}{\text{acre} \cdot \text{ft}} \times \dfrac{1 \, \text{acre} \cdot \text{ft}}{2 \, \text{households} \cdot \text{year}} \times \dfrac{1 \, \text{year}}{365 \, \text{days}} \times 1 \, \text{household} = \dfrac{\$3.397}{\text{day}} = \dfrac{\$3.40}{\text{day}}$

1.77　$8.0 \, \text{oz} \times \dfrac{1 \, \text{lb}}{16 \, \text{oz}} \times \dfrac{453.6 \, \text{g}}{\text{lb}} \times \dfrac{1 \, \text{cm}^3}{2.70 \, \text{g}} = 84.00 = 84 \, \text{cm}^3$

　　　$\dfrac{84 \, \text{cm}^3}{50 \, \text{ft}^2} \times \dfrac{1^2 \, \text{ft}^2}{12^2 \, \text{in}^2} \times \dfrac{1^2 \, \text{in}^2}{2.54^2 \, \text{cm}^2} \times \dfrac{10 \, \text{mm}}{1 \, \text{cm}} = 0.018 \, \text{mm}$

1.78　$11.86 \, \text{g ethanol} \times \dfrac{1 \, \text{cm}^3}{0.789 \, \text{g ethanol}} = 15.0317 = 15.03 \, \text{cm}^3$, volume of cylinder

　　　$V = \pi r^2 h; \; r = (V/\pi h)^{1/2} = \left[\dfrac{15.0317 \, \text{cm}^3}{\pi \times 15.0 \, \text{cm}} \right]^{1/2} = 0.5648 = 0.565 \, \text{cm}$

　　　$d = 2r = 1.13 \, \text{cm}$

1.79　(a)　Let x = mass of Au in jewelry

　　　　　9.85 - x = mass of Ag in jewelry

　　　　　The total volume of jewelry = volume of Au + volume of Ag

　　　　　$0.675 \, \text{cm}^3 = x \, \text{g} \times \dfrac{1 \, \text{cm}^3}{19.3 \, \text{g}} + (9.85 - x) \text{g} \times \dfrac{1 \, \text{cm}^3}{10.5 \, \text{g}}$

　　　　　$0.675 = \dfrac{x}{19.3} + \dfrac{9.85 - x}{10.5}$　(To solve, multiply both sides by (19.3)(10.5))

　　　　　$0.675 \, (19.3)(10.5) = 10.5 \, x + (9.85 - x)(19.3)$

　　　　　　　$136.79 = 10.5 \, x + 190.105 - 19.3 \, x$

　　　　　　　$-53.315 = -8.8 \, x$

$x = 6.06$ g Au; 9.85 g total – 6.06 g Au = 3.79 g Ag

$$\text{mass \% Au} = \frac{6.06\,\text{g Au}}{9.85\,\text{g jewelry}} \times 100 = 61.5\%\ \text{Au}$$

(b)   24 carats $\times$ 0.615 = 15 carat gold

1.80   A solution can be separated into components by physical means, so separation would be attempted. If the liquid is a solution, the solute could be a solid or a liquid; these two kinds of solutions would be separated differently. Therefore, divide the liquid into several samples and do different tests on each. Try evaporating the solvent from one sample. If a solid remains, the liquid is a solution and the solute is a solid. If the result is negative, try distilling a sample to see if two or more liquids with different boiling points are present. If this result is negative, the liquid is probably a pure substance, but negative results are never entirely conclusive. We might not have tried the appropriate separation technique.

1.81   The separation is successful if two distinct spots are seen on the paper. To quantify the characteristics of the separation, calculate a reference value for each spot that is

$$\frac{\text{distance travelled by spot}}{\text{distance travelled by solvent}}$$

If the values for the two spots are fairly different, the separation is successful. (One could measure the distance between the spots, but this would depend on the length of paper used and be different for each experiment. The values suggested above are independent of the length of paper.)

1.82   The densities are:

carbon tetrachloride (methane, tetrachloro) – 1.5940 g/cm^3

hexane – 0.6603 g/cm^3

benzene – 0.87654 g/cm^3

methylene iodide (methane, diiodo) – 3.3254 g/cm^3

Only methylene iodide will separate the two granular solids. The undesirable solid (2.04 g/cm^3) is less dense than methylene iodide and will float; the desired material is more dense than methylene iodide and will sink. The other three liquids are less dense than both solids and will not produce separation.

1.83   Study (a) is likely to be both precise and accurate, because the errors are carefully controlled. The secondary weight standard will be resistant to chemical and physical changes, the balance is carefully calibrated, and weighings are likely to be made by the same person. The relatively large number of measurements is likely to minimize the effect of random errors on the average value. The accuracy and precision of study (b) depend on the veracity of the participants' responses, which cannot be carefully controlled. It also depends on the definition of "comparable lifestyle." The percentages are not precise, because the broad definition of lifestyle leads to a range of results (scatter). The relatively large number of participants improves the precision and accuracy. In general, controlling errors and maximizing the number of data points in a study improves precision and accuracy.

2 Atoms, Molecules, and Ions

Visualizing Concepts

2.1 (a) The path of the charged particle bends because it is repelled by the negatively charged plate and attracted to the positively charged plate.

(b) Like charges repel and opposite charges attract, so the sign of the electrical charge on the particle is negative.

(c) The greater the magnitude of the charges, the greater the electrostatic repulsion or attraction. As the charge on the plates is increased, the bending will increase.

(d) As the mass of the particle increases and speed stays the same, linear momentum (mv) of the particle increases and bending decreases. (See **A Closer Look**: The Mass Spectrometer.)

2.2 In general, metals occupy the left side of the chart, and nonmetals the right side.

metals: red and green   *nonmetals*: blue and yellow

alkaline earth metal: red   *noble gas*: yellow

2.3 Since the number of electrons (negatively charged particles) does not equal the number of protons (positively charged particles), the particle is an ion. The charge on the ion is 2–.

Atomic number = number of protons = 16. The element is S, sulfur.

Mass number = protons + neutrons = 32

$^{32}_{16}S^{2-}$

2.4 In a solid, particles are close together and their relative positions are fixed. In a liquid, particles are close but moving relative to each other. In a gas, particles are far apart and moving. All ionic compounds are solids because of the strong forces among charged particles. Molecular compounds can exist in any state: solid, liquid, or gas.

Since the molecules in *ii* are far apart, *ii* must be a molecular compound. The particles in *i* are near each other and exist in a regular, ordered arrangement, so *i* is likely to be an ionic compound.

2.5 Formula: IF_5 Name: iodine pentafluoride
Since the compound is composed of elements that are all nonmetals, it is molecular.

2 Atoms, Molecules, and Ions — Solutions to Exercises

2.6 Cations (red spheres) have positive charges; anions (blue spheres) have negative charges. There are twice as many anions as cations, so the formula has the general form CA_2. Only $Ca(NO_3)_2$, calcium nitrate, is consistent with the diagram.

Atomic Theory and the Discovery of Atomic Structure

2.7 Postulate 4 of the atomic theory is the *law of constant composition*. It states that the relative number and kinds of atoms in a compound are constant, regardless of the source. Therefore, 1.0 g of pure water should always contain the same relative amounts of hydrogen and oxygen, no matter where or how the sample is obtained.

2.8 (a) 6.500 g compound – 0.384 g hydrogen = 6.116 g sulfur

(b) *Conservation of mass*

(c) According to postulate 3 of the atomic theory, atoms are neither created nor destroyed during a chemical reaction. If 0.384 g of H are recovered from a compound that contains only H and S, the remaining mass must be sulfur.

2.9 (a)
$$\frac{17.60\,\text{g oxygen}}{30.82\,\text{g nitrogen}} = \frac{0.5711\,\text{g O}}{1\,\text{g N}};\quad 0.5711/0.5711 = 1.0$$

$$\frac{35.20\,\text{g oxygen}}{30.82\,\text{g nitrogen}} = \frac{1.142\,\text{g O}}{1\,\text{g N}};\quad 1.142/0.5711 = 2.0$$

$$\frac{70.40\,\text{g oxygen}}{30.82\,\text{g nitrogen}} = \frac{2.284\,\text{g O}}{1\,\text{g N}};\quad 2.284/0.5711 = 4.0$$

$$\frac{88.00\,\text{g oxygen}}{30.82\,\text{g nitrogen}} = \frac{2.855\,\text{g O}}{1\,\text{g N}};\quad 2.855/0.5711 = 5.0$$

(b) These masses of oxygen per one gram nitrogen are in the ratio of 1:2:4:5 and thus obey the *law of multiple proportions*. Multiple proportions arise because atoms are the indivisible entities combining, as stated in Dalton's theory. Since atoms are indivisible, they must combine in ratios of small whole numbers.

2.10 (a)
1: $\dfrac{3.56\,\text{g fluorine}}{4.75\,\text{g iodine}} = 0.749\,\text{g fluorine/1 g iodine}$

2: $\dfrac{3.43\,\text{g fluorine}}{7.64\,\text{g iodine}} = 0.449\,\text{g fluorine/1 g iodine}$

3: $\dfrac{9.86\,\text{g fluorine}}{9.41\,\text{g iodine}} = 1.05\,\text{g fluorine/1 g iodine}$

(b) To look for integer relationships among these values, divide each one by the smallest.

If the quotients aren't all integers, multiply by a common factor to obtain all integers.

18

1: $0.749/0.449 = 1.67$; $1.67 \times 3 = 5$

2: $0.449/0.449 = 1.00$; $1.00 \times 3 = 3$

3: $1.05/0.449 = 2.34$; $2.34 \times 3 = 7$

The ratio of g fluorine to g iodine in the three compounds is 5:3:7. These are in the ratio of small whole numbers and, therefore, obey the *law of multiple proportions*. This integer ratio indicates that the combining fluorine "units" (atoms) are indivisible entities.

2.11   Evidence that cathode rays were negatively charged particles was (1) that electric and magnetic fields deflected the rays in the same way they would deflect negatively charged particles and (2) that a metal plate exposed to cathode rays acquired a negative charge.

2.12   Since the unknown particle is deflected in the opposite direction from that of a negatively charged beta (β) particle, it is attracted to the ($-$) plate and repelled by the ($+$) plate. The unknown particle is positively charged. The magnitude of the deflection is less than that of the β particle, or electron, so the unknown particle has greater mass than the electron. The unknown is a positively charged particle of greater mass than the electron.

2.13   (a)   If the positive plate were lower than the negative plate, the oil drops "coated" with negatively charged electrons would be attracted to the positively charged plate and would descend much more quickly.

(b)   The more times a measurement is repeated, the better the chance of detecting and compensating for experimental errors. That is, if a quantity is measured five times and four measurements agree but one does not, the measurement that disagrees is probably the result of an error. Also, the four measurements that agree can be averaged to compensate for small random fluctuations. Millikan wanted to demonstrate the validity of his result via its reproducibility.

2.14   (a)   The droplets carry different total charges because there may be 1, 2, 3, or more electrons on the droplet.

(b)   The electronic charge is likely to be the lowest common factor in all the observed charges.

(c)   Assuming this is so, we calculate the apparent electronic charge from each drop as follows:

A: $1.60 \times 10^{-19} / 1 = 1.60 \times 10^{-19}$ C

B: $3.15 \times 10^{-19} / 2 = 1.58 \times 10^{-19}$ C

C: $4.81 \times 10^{-19} / 3 = 1.60 \times 10^{-19}$ C

D: $6.31 \times 10^{-19} / 4 = 1.58 \times 10^{-19}$ C

The reported value is the average of these four values. Since each calculated charge has three significant figures, the average will also have three significant figures.

$(1.60 \times 10^{-19}$ C $+ 1.58 \times 10^{-19}$ C $+ 1.60 \times 10^{-19}$ C $+ 1.58 \times 10^{-19}$ C$) / 4 = 1.59 \times 10^{-19}$ C

2 Atoms, Molecules, and Ions — Solutions to Exercises

Modern View of Atomic Structure; Atomic Weights

2.15   (a) $1.9\,\text{Å} \times \dfrac{1 \times 10^{-10}\,\text{m}}{1\,\text{Å}} \times \dfrac{1\,\text{nm}}{1 \times 10^{-9}\,\text{m}} = 0.19\,\text{nm}$

            $1.9\,\text{Å} \times \dfrac{1 \times 10^{-10}\,\text{m}}{1\,\text{Å}} \times \dfrac{1\,\text{pm}}{1 \times 10^{-12}\,\text{m}} = 1.9 \times 10^{2}$ or $190\,\text{pm}\,(1\,\text{Å} = 100\,\text{pm})$

     (b)   Aligned Kr atoms have **diameters** touching. $d = 2r = 2(1.9\,\text{Å}) = 3.8\,\text{Å}$

            $1.0\,\text{mm} \times \dfrac{1\,\text{m}}{1000\,\text{mm}} \times \dfrac{1\,\text{Å}}{1 \times 10^{-10}\,\text{m}} \times \dfrac{1\,\text{Kr atom}}{3.8\,\text{Å}} = 2.6 \times 10^{6}\,\text{Kr atoms}$

     (c) $V = 4/3\,\pi\,r^{3}.\;\; r = 1.9\,\text{Å} \times \dfrac{1 \times 10^{-10}\,\text{m}}{1\,\text{Å}} \times \dfrac{100\,\text{cm}}{\text{m}} = 1.9 \times 10^{-8}\,\text{cm}$

            $V = (4/3)(\pi)(1.9 \times 10^{-8})^{3}\,\text{cm}^{3} = 2.9 \times 10^{-23}\,\text{cm}^{3}$

2.16   (a) $r = d/2;\; r = \dfrac{2.8 \times 10^{-8}\,\text{cm}}{2} \times \dfrac{1\,\text{Å}}{1 \times 10^{-8}\,\text{cm}} = 1.4\,\text{Å}$

            $r = \dfrac{2.8 \times 10^{-8}\,\text{cm}}{2} \times \dfrac{1\,\text{m}}{100\,\text{cm}} = 1.4 \times 10^{-10}\,\text{m}$

     (b)   Aligned Sn atoms have **diameters** touching. $d = 2.8 \times 10^{-8}\,\text{cm} = 2.8 \times 10^{-10}\,\text{m}$

            $6.0\,\mu\text{m} \times \dfrac{1 \times 10^{-6}\,\text{m}}{1\,\mu\text{m}} \times \dfrac{1\,\text{Sn atom}}{2.8 \times 10^{-10}\,\text{m}} = 2.1 \times 10^{4}\,\text{Sn atoms}$

     (c) $V = 4/3\,\pi\,r^{3};\; r = 1.4 \times 10^{-10}\,\text{m}$

            $V = (4/3)[(\pi(1.4 \times 10^{-10})^{3}]\,\text{m}^{3} = 1.149 \times 10^{-29} = 1.1 \times 10^{-29}\,\text{m}^{3}$

2.17   (a)   proton, neutron, electron

     (b)   proton = +1, neutron = 0, electron = –1

     (c)   The neutron is most massive, the electron least massive. (The neutron and proton have very similar masses).

2.18   (a)   The nucleus has most of the mass **but occupies very little** of the volume of an atom.

     (b)   True

     (c)   The number of electrons in an atom is equal to the number of **protons** in the atom.

     (d)   True

2.19   (a)   *Atomic number* is the number of protons in the nucleus of an atom. *Mass number* is the total number of nuclear particles, protons plus neutrons, in an atom.

     (b)   The mass number can vary without changing the identity of the atom, but the atomic number of every atom of a given element is the same.

2.20   (a) $^{31}_{16}X$ and $^{32}_{16}X$ are isotopes of the same element, because they have identical atomic numbers.

(b)   These are isotopes of the element sulfur, S, atomic number = 16.

2.21   p = protons, n = neutrons, e = electrons

(a) ^{40}Ar has 18 p, 22 n, 18 e   (b) ^{65}Zn has 30 p, 35 n, 30 e

(c) ^{70}Ga has 31 p, 39 n, 31 e   (d) ^{80}Br has 35 p, 45 n, 35 e

(e) ^{184}W has 74 p, 110 n, 74 e   (f) ^{243}Am has 95 p, 148 n, 95 e

2.22   (a) ^{32}P has 15 p, 17 n   (b) ^{51}Cr has 24 p, 27 n

(c) ^{60}Co has 27 p, 33 n   (d) ^{99}Tc has 43 p, 56 n

(e) ^{131}I has 53 p, 78 n   (f) ^{201}Tl has 81 p, 120 n

2.23

Symbol	^{52}Cr	^{55}Mn	^{112}Cd	^{222}Rn	^{207}Pb
Protons	24	25	48	86	82
Neutrons	28	30	64	136	125
Electrons	24	25	48	86	82
Mass no.	52	55	112	222	207

2.24

Symbol	^{121}Sb	^{103}Rh	^{88}Sr	^{127}Te	^{239}Pu
Protons	51	45	38	52	94
Neutrons	70	58	50	75	145
Electrons	51	45	38	52	94
Mass No.	121	103	88	127	239

2.25   (a) $^{196}_{78}Pt$ (b) $^{84}_{36}Kr$ (c) $^{75}_{33}As$ (c) $^{24}_{12}Mg$

2.26   Since the two nuclides are atoms of the same element, by definition they have the same number of protons, 54. They differ in mass number (and mass) because they have different numbers of neutrons. ^{129}Xe has 75 neutrons and ^{130}Xe has 76 neutrons.

2.27   (a) $^{12}_{6}C$

(b)   Atomic weights are really average atomic masses, the sum of the mass of each naturally occurring isotope of an element times its fractional abundance. Each B atom will have the mass of one of the naturally occurring isotopes, while the "atomic weight" is an average value. The naturally occurring isotopes of B, their atomic masses, and relative abundances are:

^{10}B, 10.012937, 19.9%; ^{11}B, 11.009305, 80.1%.

2.28   (a)   12 amu

       (b)   The atomic weight of carbon reported on the front-inside cover of the text is the abundance-weighted average of the atomic masses of the two naturally occurring isotopes of carbon, ^{12}C, and ^{13}C. The mass of a ^{12}C atom is exactly 12 amu, but the atomic weight of 12.011 takes into account the presence of some ^{13}C atoms in every natural sample of the element.

2.29   Atomic weight (average atomic mass) = Σ fractional abundance × mass of isotope

         Atomic weight = 0.6917(62.9296) + 0.3083(64.9278) = 63.5456 = 63.55 amu

2.30   Atomic weight (average atomic mass) = Σ fractional abundance × mass of isotope

         Atomic weight = 0.014(203.97302) + 0.241(205.97444) + 0.221(206.97587) +

                                          0.524(207.97663) = 207.22 = 207 amu

(The result has 0 decimal places and 3 sig figs because the fourth term in the sum has 3 sig figs and 0 decimal places.)

2.31   (a)   Compare Figures 2.4 and 2.13, referring to Solution 2.12. In Thomson's cathode ray experiments and in mass spectrometry a stream of charged particles is passed through a magnetic field. The charged particles are deflected by the magnetic field according to their mass and charge. For a constant magnetic field strength and speed of the particles, the lighter particles experience a greater deflection.

       (b)   The x-axis label (independent variable) is atomic weight and the y-axis label (dependent variable) is signal intensity.

       (c)   Uncharged particles are not deflected in a magnetic field. The effect of the magnetic field on moving, *charged* particles is the basis of their separation by mass.

2.32   (a)   The purpose of the magnet in the mass spectrometer is to change the path of the moving ions. The magnitude of the deflection is inversely related to mass, which is the basis of the discrimination by mass.

       (b)   The atomic weight of Cl, 35.5, is an average atomic mass. It is the average of the masses of two naturally occurring isotopes, weighted by their abundances.

       (c)   The single peak at mass 31 in the mass spectrum of phosphorus indicates that the sample contains a single isotope of P, and the mass of this isotope is 31 amu.

2.33   (a)   Average atomic mass = 0.7899(23.98504) + 0.1000(24.98584) + 0.1101(25.98259)

                         = 24.31 amu

(b)

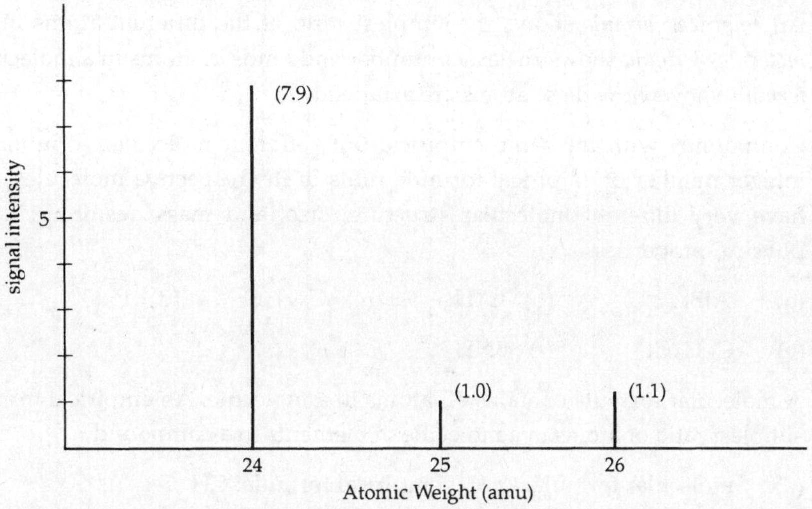

The relative intensities of the peaks in the mass spectrum are the same as the relative abundances of the isotopes. The abundances and peak heights are in the ratio ^{24}Mg: ^{25}Mg: ^{26}Mg as $7.8 : 1.0 : 1.1$.

2.34   (a)   Three peaks: $^{1}H - {}^{1}H$, $^{1}H - {}^{2}H$, $^{2}H - {}^{2}H$

       (b) $^{1}H - {}^{1}H = 2(1.00783) = 2.01566$ amu

              $^{1}H - {}^{2}H = 1.00783 + 2.01410 = 3.02193$ amu

              $^{2}H - {}^{2}H = 2(2.01410) = 4.02820$ amu

              The mass ratios are $1 : 1.49923 : 1.99845$ or $1 : 1.5 : 2$.

       (c) $^{1}H - {}^{1}H$ is largest, because there is the greatest chance that two atoms of the more abundant isotope will combine.

              $^{2}H - {}^{2}H$ is the smallest, because there is the least chance that two atoms of the less abundant isotope will combine.

The Periodic Table; Molecules and Ions

2.35   (a)   Cr (metal)   (b)   He (nonmetal)   (c)   P (nonmetal)   (d)   Zn (metal)

       (e)   Mg (metal)   (f)   Br (nonmetal)   (g)   As (metalloid)

2.36   (a)   sodium (metal)   (b)   titanium (metal)   (c)   gallium (metal)

       (d)   uranium (metal)   (e)   palladium (metal)   (f)   selenium (nonmetal)

       (g)   krypton (nonmetal)

2.37   (a)   K, alkali metals (metal)   (b)   I, halogens (nonmetal)

       (c)   Mg, alkaline earth metals (metal)   (d)   Ar, noble gases (nonmetal)

       (e)   S, chalcogens (nonmetal)

2.38   C, carbon, nonmetal; Si, silicon, metalloid; Ge, germanium, metalloid; Sn, tin, metal; Pb, lead, metal

2.39   An *empirical formula* shows the simplest ratio of the different atoms in a molecule. A *molecular formula* shows the exact number and kinds of atoms in a molecule. A *structural formula* shows how these atoms are arranged.

2.40   Compounds with the same empirical but different molecular formulas differ by the integer number of empirical formula units in the respective molecules. Thus, they can have very different molecular structure, size, and mass, resulting in very different physical properties.

2.41   (a) $AlBr_3$ (b) C_4H_5 (c) C_2H_4O (d) P_2O_5

         (e) C_3H_2Cl (f) BNH_2

2.42   A molecular formula contains all atoms in a molecule. An empirical formula shows the simplest ratio of atoms in a molecule or elements in a compound.

         (a)   molecular formula: C_6H_6; empirical formula: CH

         (b)   molecular formula: $SiCl_4$; empirical formula: $SiCl_4$ (1:4 is the simplest ratio)

         (c)   molecular: B_2H_6; empirical: BH_3

         (d)   molecular: $C_6H_{12}O_6$; empirical: CH_2O

2.43   (a)   6   (b)   6   (c)   12

2.44   (a)   4   (b)   6   (c)   9

2.45   (a) C_2H_6O

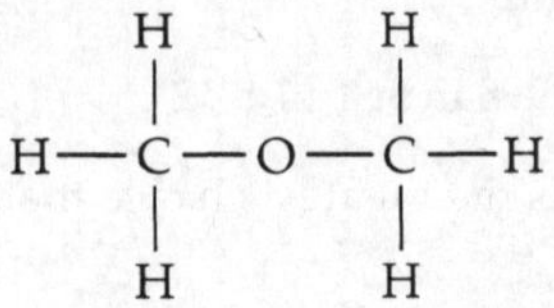

$$\begin{array}{ccccc} & H & & H & \\ & | & & | & \\ H- & C & -O- & C & -H \\ & | & & | & \\ & H & & H & \end{array}$$

(b) C_2H_6O

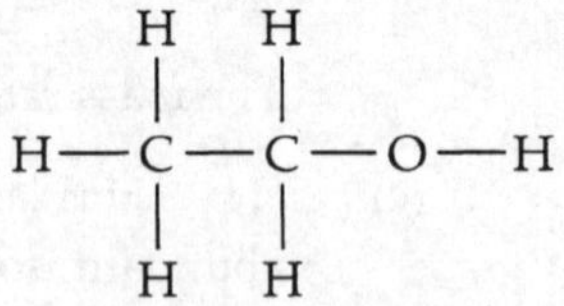

$$\begin{array}{ccccc} & H & H & & \\ & | & | & & \\ H- & C & -C & -O-H \\ & | & | & & \\ & H & H & & \end{array}$$

(c) CH_4O

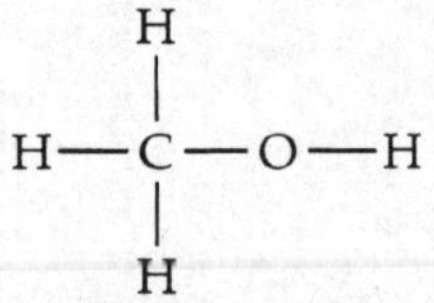

$$\begin{array}{ccc} & H & \\ & | & \\ H- & C & -O-H \\ & | & \\ & H & \end{array}$$

(d) PF_3

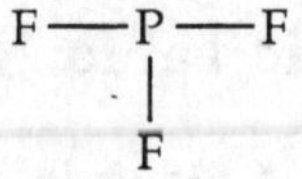

$$\begin{array}{ccc} F- & P & -F \\ & | & \\ & F & \end{array}$$

2.46   (a) C_2H_5Br

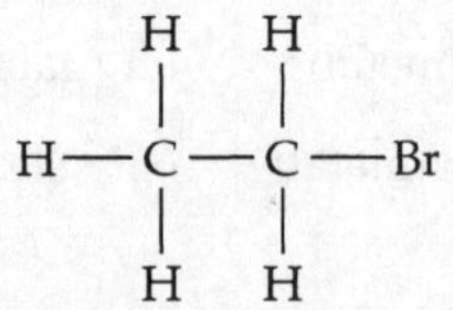

$$\begin{array}{ccc} & H & H \\ & | & | \\ H- & C & -C & -Br \\ & | & | \\ & H & H \end{array}$$

(b) C_2H_7N

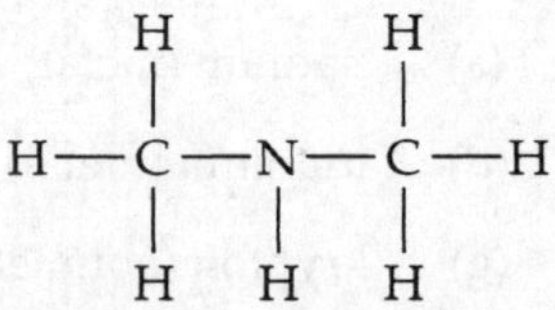

$$\begin{array}{ccccc} & H & & H & \\ & | & & | & \\ H- & C & -N- & C & -H \\ & | & | & | & \\ & H & H & H & \end{array}$$

(c) CH_2Cl_2

$$\begin{array}{ccc} & H & \\ & | & \\ H- & C & -Cl \\ & | & \\ & Cl & \end{array}$$

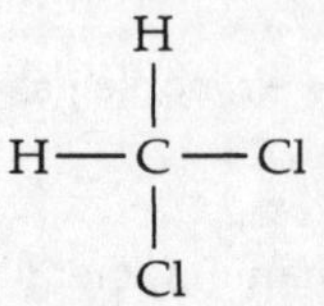

(d) NH_2OH

$$\begin{array}{ccc} H- & N & -O \\ & | & | \\ & H & H \end{array}$$

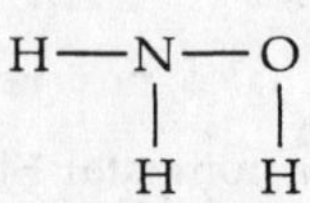

2 Atoms, Molecules, and Ions

Solutions to Exercises

2.47

Symbol	$^{59}Co^{3+}$	$^{80}Se^{2-}$	$^{192}Os^{2+}$	$^{200}Hg^{2+}$
Protons	27	34	76	80
Neutrons	32	46	116	120
Electrons	24	36	74	78
Net Charge	3+	2–	2+	2+

2.48

Symbol	$^{75}As^{3-}$	$^{59}Ni^{2+}$	$^{127}I^{-}$	$^{197}Au^{3+}$
Protons	33	28	53	79
Neutrons	42	31	74	118
Electrons	36	26	54	76
Net Charge	3–	2+	1–	3+

2.49   (a) Mg^{2+} (b) Al^{3+} (c) K^+ (d) S^{2-} (e) F^-

2.50   (a) Sr^{2+} (b) Sc^{2+} or Sc^{3+} (c) P^{3-} (d) I^- (e) Se^{2-}

2.51   (a) GaF_3, gallium(III) fluoride   (b) LiH, lithium hydride

(c) AlI_3, aluminum iodide   (d) K_2S, potassium sulfide

2.52   (a)   AgI   (b) Ag_2S (c) AgF

2.53   (a) $CaBr_2$ (b) K_2CO_3 (c) $Al(C_2H_3O_2)_3$ (d) $(NH_4)_2SO_4$ (e) $Mg_3(PO_4)_2$

2.54   (a) Cu_2S (b) Fe_2O_3 (c) Hg_2CO_3 (d) $Ca_3(AsO_4)_2$ (e) $(NH_4)_2CO_3$

2.55   Molecular (all elements are nonmetals):

(a) B_2H_6 (b) CH_3OH (f) NOCl   (g) NF_3

Ionic (formed by a cation and an anion, usually contains a metal cation):

(c) $LiNO_3$ (d) Sc_2O_3 (e) CsBr   (h) Ag_2SO_4

2.56   Molecular (all elements are nonmetals):

(a) PF_5 (c) SCl   (h) N_2O_4

Ionic (formed from ions, usually contains a metal cation):

(b)   NaI   (d) $Ca(NO_3)_2$ (e) $FeCl_3$ (f) LaP   (g) $CoCO_3$

Naming Inorganic Compounds; Organic Molecules

2.57   (a) ClO_2^- (b) Cl   (c) ClO_3^- (d) ClO_4^- (e) ClO^-

2.58   (a)   selenate   (b) selenide (c) hydrogen selenide (biselenide)

(d)   hydrogen selenite (biselenite)

2.59   (a)   magnesium oxide   (b)   aluminum chloride

       (c)   lithium phosphate   (d)   barium perchlorate

       (e)   copper(II) nitrate (cupric nitrate)   (f)   iron(II) hydroxide (ferrous hydroxide)

       (g)   calcium acetate   (h)   chromium(III) carbonate (chromic carbonate)

       (i)   potassium chromate   (j)   ammonium sulfate

2.60   (a)   lithium oxide   (b)   sodium hypochlorite

       (c)   strontium cyanide   (d)   chromium(III) hydroxide (chromic hydroxide)

       (e)   iron(III) carbonate (ferric carbonate)   (f)   cobalt(II) nitrate (cobaltous nitrate)

       (g)   ammonium sulfite   (h)   sodium dihydrogen phosphate

       (i)   potassium permanganate   (j)   silver dichromate

2.61   (a) $Al(OH)_3$ (b) K_2SO_4 (c) Cu_2O (d) $Zn(NO_3)_2$

       (e) $HgBr_2$ (f) $Fe_2(CO_3)_3$ (g) $NaBrO$

2.62   (a) Na_3PO_4 (b) $Co(NO_3)_2$ (c) $Ba(BrO_3)_2$ (d) $Cu(ClO_4)_2$

       (e) $Mg(HCO_3)_2$ (f) $Cr(C_2H_3O_2)_3$ (g) $K_2Cr_2O_7$

2.63   (a)   bromic acid   (b)   hydrobromic acid   (c)   phosphoric acid

       (d) $HClO$ (e) HIO_3 (f) H_2SO_3

2.64   (a) HBr (b) H_2S (c) HNO_2

       (d)   carbonic acid   (e)   chloric acid   (f)   acetic acid

2.65   (a)   sulfur hexafluoride   (b)   iodine pentafluoride   (c)   xenon trioxide

       (d) N_2O_4 (e) HCN (f) P_4S_6

2.66   (a)   dinitrogen monoxide   (b)   nitrogen monoxide   (c)   nitrogen dioxide

       (d)   dinitrogen pentoxide   (e)   dinitrogen tetroxide

2.67   (a) $ZnCO_3$, ZnO, CO_2 (b) HF, SiO_2, SiF_4, H_2O (c) SO_2, H_2O, H_2SO_3

       (d) PH_3 (e) $HClO_4$, Cd, $Cd(ClO_4)_2$ (f) VBr_3

2.68   (a) $NaHCO_3$ (b) $Ca(ClO)_2$ (c) HCN

       (d) $Mg(OH)_2$ (e) SnF (f) CdS, H_2SO_4, H_2S

2.69   (a)   A hydrocarbon is a compound composed of the elements hydrogen and carbon only.

       (b)

```
        H   H   H   H
        |   |   |   |
   H — C — C — C — C — H
        |   |   |   |
        H   H   H   H
```

       molecular: C_4H_{10}

       empirical: C_2H_5

2.70   *-ane*

(b)   **Hexane** has 6 carbons in its chain.

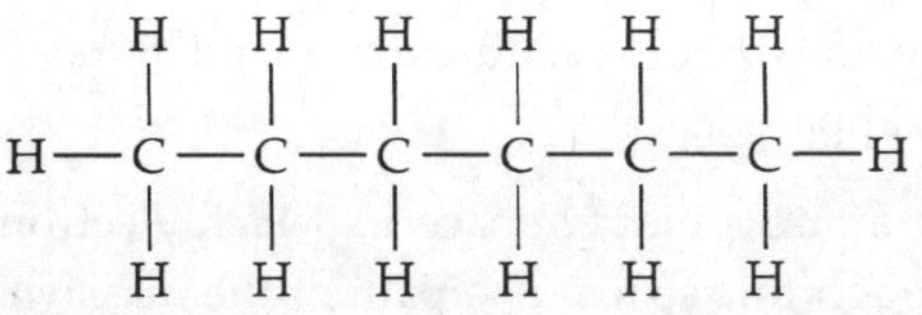

$$
\begin{array}{ccccccccccccc}
 & H & & H & & H & & H & & H & & H & \\
 & | & & | & & | & & | & & | & & | & \\
H- & C & - & C & - & C & - & C & - & C & - & C & -H \\
 & | & & | & & | & & | & & | & & | & \\
 & H & & H & & H & & H & & H & & H &
\end{array}
$$

molecular: C_6H_{14}

empirical: C_3H_7

2.71   (a)   *Functional groups* are groups of specific atoms that are constant from one molecule to the next. For example, the alcohol functional group is an –OH. Whenever a molecule is called an alcohol, it contains the –OH group.

(b)   —OH   (c)

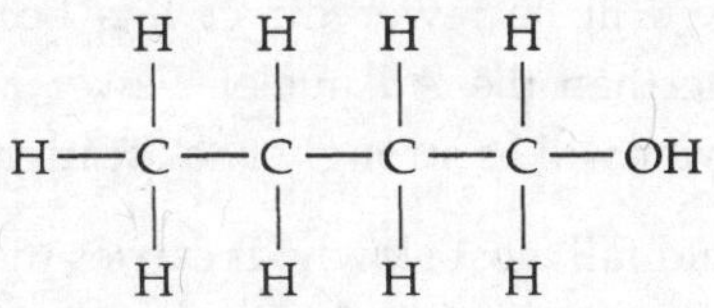

$$
\begin{array}{ccccccccc}
 & H & & H & & H & & H & \\
 & | & & | & & | & & | & \\
H- & C & - & C & - & C & - & C & -OH \\
 & | & & | & & | & & | & \\
 & H & & H & & H & & H &
\end{array}
$$

2.72   (a)   They both have two carbon atoms in their molecular backbone, or chain.

(b)   In 1-propanol one of the H atoms on an outer (terminal) C atom has been replaced by an —OH group.

Additional Exercises

2.73   (a)   Based on data accumulated in the late eighteenth century on how substances react with one another, *Dalton* postulated the atomic theory. Dalton's theory is based on the indivisible atom as the smallest unit of an element that can combine with other elements.

(b)   By determining the effects of electric and magnetic fields on cathode rays, *Thomson* measured the mass-to-charge ratio of the electron. He also proposed the "plum pudding" model of the atom in which most of the space in an atom is occupied by a diffuse positive charge in which the tiny negatively charged electrons are imbedded.

(c)   By observing the rate of fall of oil drops in and out of an electric field, *Millikan* measured the charge of an electron.

(d)   After observing the scattering of alpha particles at large angles when the particles struck gold foil, *Rutherford* postulated the nuclear atom. In Rutherford's atom, most of the mass of the atom is concentrated in a small dense region called the nucleus and the tiny negatively charged electrons are moving through empty space around the nucleus.

2.74   *Radioactivity* is the spontaneous emission of radiation from a substance. Becquerel's discovery showed that atoms could decay, or degrade, *implying* that they are not indivisible. However, it wasn't until Rutherford and others characterized the nature of radioactive emissions, especially the particle nature of α and β rays, that the full significance of the discovery was apparent.

2.75   (a)   Most of the volume of an atom is empty space in which electrons move. Most alpha particles passed through this space. The path of the massive alpha particle would not be significantly altered by interaction with a "puny" electron.

        (b)   Most of the mass of an atom is contained in a very small, dense area called the nucleus. The few alpha particles that hit the massive, positively charged gold nuclei were strongly repelled and essentially deflected back in the direction they came from.

        (c)   The Be nuclei have a much smaller volume and positive charge than the Au nuclei; the charge repulsion between the alpha particles and the Be nuclei will be less, and there will be fewer direct hits because the Be nuclei have an even smaller volume than the Au nuclei. Fewer alpha particles will be scattered in general and fewer will be strongly back scattered.

2.76   (a)   Droplet D would fall most slowly. It carries the most negative charge, so it would be most strongly attracted to the upper (+) plate and most strongly repelled by the lower (–) plate. These electrostatic forces would provide the greatest opposition to gravity.

        (b)   Calculate the lowest common factor.

        A: 3.84×10^{-8} / 2.88×10^{-8} = 1.33; $1.33 \times 3 = 4$

        B: 4.80×10^{-8} / 2.88×10^{-8} = 1.67; $1.67 \times 3 = 5$

        C: 2.88×10^{-8} / 2.88×10^{-8} = 1.00; $1.00 \times 3 = 3$

        D: 8.64×10^{-8} / 2.88×10^{-8} = 3.00; $3.00 \times 3 = 9$

        The total charge on the drops is in the ratio of 4:5:3:9. Divide the total charge on each drop by the appropriate integer and average the four values to get the charge of an electron in warmombs.

        A: 3.84×10^{-8} / 4 = 9.60×10^{-9} wa

        B: 4.80×10^{-8} / 5 = 9.60×10^{-9} wa

        C: 2.88×10^{-8} / 3 = 9.60×10^{-9} wa

        D: 8.64×10^{-8} / 9 = 9.60×10^{-9} wa

        The charge on an electron is 9.60×10^{-9} wa

        (c)   The number of electrons on each drop are the integers calculated in part (b). A has $4\,e^-$, B has $5\,e^-$, C has $3\,e^-$ and D has $9\,e^-$.

        (d) $\dfrac{9.60 \times 10^{-9}\ \text{wa}}{1\,e^-} \times \dfrac{1\,e^-}{1.60 \times 10^{-16}\ \text{C}} = 6.00 \times 10^7\ \text{wa/C}$

2.77   (a) ^{3}He has 2 protons, 1 neutron, and 2 electrons.

(b) ^{3}H has 1 proton, 2 neutrons, and 1 electron.

^{3}He: $2(1.6726231 \times 10^{-24} \text{ g}) + 1.6749286 \times 10^{-24} \text{ g} + 2(9.1093897 \times 10^{-28} \text{ g})$

$= 5.021996 \times 10^{-24} \text{ g}$

^{3}H: $1.6726231 \times 10^{-24} \text{ g} + 2(1.6749286 \times 10^{-24} \text{ g}) + 9.1093897 \times 10^{-28} \text{ g}$

$= 5.023391 \times 10^{-24} \text{ g}$

Tritium, ^{3}H, is more massive.

(c)   The masses of the two particles differ by 0.0014×10^{-24} g. Each particle loses 1 electron to form the +1 ion, so the difference in the masses of the ions is still 1.4×10^{-27}. A mass spectrometer would need precision to 1×10^{-27} g to differentiate ^{3}He$^+$ and ^{3}H.

2.78   (a)   2 protons and 2 neutrons

(b)   the nuclear strong force

(c)   The charge of an α particle is twice the magnitude of the charge of an electron, with the opposite sign. That is, $2 \, (+1.6022 \times 10^{-19}) \text{ C} = +3.2044 \times 10^{-19} \text{ C}$.

(d) $\dfrac{3.2044 \times 10^{-19} \text{ C}}{4.8224 \times 10^4 \text{ g/C}} = 6.6448 \times 10^{-24} \text{ g}$

$6.6448 \times 10^{-24} \text{ g} \times \dfrac{1 \, \text{amu}}{1.66054 \times 10^{-24} \text{ g}} = 4.0016 \, \text{amu}$

(e)   The sum of the particle masses in an α particle is 2(1.0073) amu and 2(1.0087) amu = 4.0320 amu. The actual particle mass, 4.0016 amu, is less than the sum of the masses of the components. The difference is the nuclear binding energy, the energy released when protons and neutrons combine to form a nucleus. Mass and energy are interchangeable according to the Einstein relationship $E = mc^2$.

2.79   (a)   Calculate the mass of a single gold atom, then divide the mass of the cube by the mass of the gold atom.

$\dfrac{197.0 \, \text{amu}}{\text{gold atom}} \times \dfrac{1 \, \text{g}}{6.022 \times 10^{23} \, \text{amu}} = 3.2713 \times 10^{-22} = 3.271 \times 10^{-22} \text{ g/gold atom}$

$\dfrac{19.3 \, \text{g}}{\text{cube}} \times \dfrac{1 \, \text{gold atom}}{3.271 \times 10^{-22} \text{ g}} = 5.90 \times 10^{22} \text{ Au atoms in the cube}$

(b)   The shape of atoms is spherical; spheres cannot be arranged into a cube so that there is no empty space. The question is, how much empty space is there? We can calculate the two limiting cases, no empty space and maximum empty space. The true diameter will be somewhere in this range.

No empty space: volume cube/number of atoms = volume of one atom

$V = 4/3 \pi \, r^3$; $r = (3\pi \, V/4)^{1/3}$; $d = 2r$

$$\text{vol. of cube} = (1.0 \times 1.0 \times 1.0) = \frac{1.0 \, \text{cm}^3}{5.90 \times 10^{22} \, \text{Au atoms}} = 1.695 \times 10^{-23}$$

$$= 1.7 \times 10^{-23} \, \text{cm}^3$$

$r = [\pi \, (1.695 \times 10^{-23} \, \text{cm}^3)/4]^{1/3} = 3.4 \times 10^{-8} \, \text{cm}; \; d = 2r = 6.8 \times 10^{-8} \, \text{cm}$

Maximum empty space: assume atoms are arranged in rows in all three directions so they are touching across their diameters. That is, each atom occupies the volume of a cube, with the atomic diameter as the length of the side of the cube. The number of atoms along one edge of the gold cube is then

$(5.90 \times 10^{22})^{1/3} = 3.893 \times 10^7 = 3.89 \times 10^7$ atoms/1.0 cm.

The diameter of a single atom is 1.0 cm/3.89×10^7 atoms = 2.569×10^{-8}

$$= 2.6 \times 10^{-8} \, \text{cm}.$$

The diameter of a gold atom is between 2.6×10^{-8} cm and 6.8×10^{-8} cm (2.6 – 6.8 Å).

(c)   Some atomic arrangement must be assumed, since none is specified. The solid state is characterized by an orderly arrangement of particles, so it isn't surprising that atomic arrangement is required to calculate the density of a solid. A more detailed discussion of solid-state structure and density appears in Chapter 11.

2.80   (a)   In arrangement A the number of atoms in 1 cm^2 is just the square of the number that fit linearly in 1 cm.

$$1.0 \, \text{cm} \times \frac{1 \, \text{atom}}{4.95 \, \text{Å}} \times \frac{1 \times 10^{10} \, \text{Å}}{1 \, \text{m}} \times \frac{1 \, \text{m}}{100 \, \text{cm}} = 2.02 \times 10^7 = 2.0 \times 10^7 \, \text{atoms/cm}$$

$1.0 \, \text{cm}^2 = (2.02 \times 10^7)^2 = 4.081 \times 10^{14} = 4.1 \times 10^{14} \, \text{atoms/cm}^2$

(b)   In arrangement B, the atoms in the horizontal rows are touching along their diameters, as in arrangement A. The number of Rb atoms in a 1.0 cm row is then 2.0×10^7 Rb atoms. Relative to arrangement A, the vertical rows are offset by 1/2 of an atom. Atoms in a "column" are no longer touching along their vertical diameter. We must calculate the vertical distance occupied by a row of atoms, which is now less than the diameter of one Rb atom.

Consider the triangle shown below. This is an isosceles triangle (equal side lengths, equal interior angles) with a side-length of 2d and an angle of 60°. Drop a bisector to the uppermost angle so that it bisects the opposite side.

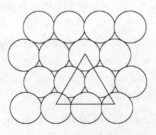

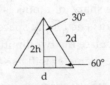

The result is a right triangle with two known side lengths. The length of the unknown side (the angle bisector) is 2h, two times the vertical distance occupied by a row of atoms. Solve for h, the "height" of one row of atoms.

$(2h)^2 + d^2 = (2d)^2; \; 4h^2 = 4d^2 - d^2 = 3d^2; \; h^2 = 3d^2/4$

$h = (3d^2/4)^{1/2} = (3(4.95 \text{ Å})^2/4)^{1/2} = 4.2868 = 4.29 \text{ Å}$

The number of rows of atoms in 1 cm is then

$$1.0 \text{ cm} \times \frac{1 \text{ row}}{4.2868 \text{ Å}} \times \frac{1 \times 10^{10} \text{ Å}}{1 \text{ m}} \times \frac{1 \text{ m}}{100 \text{ cm}} = 2.333 \times 10^7 = 2.3 \times 10^7$$

The number of atoms in a 1.0 cm^2 square area is then

$$\frac{2.020 \times 10^7 \text{ atoms}}{1 \text{ row}} \times 2.333 \times 10^7 \text{ rows} = 4.713 \times 10^{14} = 4.7 \times 10^{14}$$

Note that we have ignored the loss of "1/2" atom at the end of each horizontal row. Out of 2.0×10^7 atoms per row, one atom is not significant.

(c)   The ratio of atoms in arrangement B to arrangement A is then 4.713×10^{14} atoms$/4.081 \times 10^{14} = 1.555 = 1.2:1$. Clearly, arrangement B results in less empty space per unit area or volume. If extended to three dimensions, arrangement B would lead to a greater density for Rb metal.

2.81   (a)   diameter of nucleus $= 1 \times 10^{-4}$ Å; diameter of atom $= 1$ Å

              $V = 4/3 \, \pi \, r^3; \; r = d/2; \; r_n = 0.5 \times 10^{-4} \text{ Å}; \; r_a = 0.5 \text{ Å}$

              volume of nucleus $= 4/3 \, \pi \, (0.5 \times 10^{-4})^3 \text{ Å}^3$

              volume of atom $= 4/3 \, \pi \, (0.5)^3 \text{ Å}^3$

              volume fraction of nucleus $= \dfrac{\text{volume of nucleus}}{\text{volume of atom}} = \dfrac{4/3 \, \pi \, (0.5 \times 10^{-4})^3 \text{ Å}^3}{4/3 \, \pi \, (0.5)^3 \text{ Å}^3} = 1 \times 10^{-12}$

              diameter of atom $= 5$ Å, $r_a = 2.5$ Å

              volume fraction of nucleus $= \dfrac{4/3 \, \pi \, (0.5 \times 10^{-4})^3 \text{ Å}^3}{4/3 \, \pi \, (2.5)^3 \text{ Å}^3} = 8 \times 10^{-15}$

              Depending on the radius of the atom, the volume fraction of the nucleus is between 1×10^{-12} and 8×10^{-15}, that is, between 1 part in 10^{12} and 8 parts in 10^{15}.

     (b)   mass of proton $= 1.0073$ amu

              $1.0073 \text{ amu} \times 1.66054 \times 10^{-24} \text{ g/amu} = 1.6727 \times 10^{-24} \text{ g}$

              diameter $= 1.0 \times 10^{-15}$ m, radius $= 0.50 \times 10^{-15} \text{ m} \times \dfrac{100 \text{ cm}}{1 \text{ m}} = 5.0 \times 10^{-14} \text{ cm}$

              Assuming a proton is a sphere, $V = 4/3 \, \pi \, r^3$.

              density $= \dfrac{\text{g}}{\text{cm}^3} = \dfrac{1.6727 \times 10^{-24} \text{ g}}{4/3 \, \pi \, (5.0 \times 10^{-14})^3 \text{ cm}^3} = 3.2 \times 10^{15} \text{ g/cm}^3$

2.82   (a) $^{16}_{8}\text{O}, \; ^{17}_{8}\text{O}, \; ^{18}_{8}\text{O}$

(b) All isotopes are atoms of the same element, oxygen, with the same atomic number (Z = 8), 8 protons in the nucleus and 8 electrons. Elements with similar electron arrangements have similar chemical properties (Section 2.5). Since the 3 isotopes all have 8 electrons, we expect their electron arrangements to be the same and their chemical properties to be very similar, perhaps identical. Each has a different number of neutrons (8, 9, or 10), a different mass number (A = 16, 17, or 18) and thus a different atomic mass.

2.83 $F = k\, Q_1 Q_2 / d^2$; $k = 9.0 \times 10^9$ N m^2/C^2; $d = 0.53 \times 10^{-10}$ m;

Q (electron) = -1.6×10^{-19} C; Q (proton) = $-$Q (electron) = 1.6×10^{-19} C

$$F = \frac{\dfrac{9.0 \times 10^9 \text{ N m}^2}{\text{C}^2} \times -1.6 \times 10^{-19} \text{ C} \times 1.6 \times 10^{-19} \text{ C}}{(0.53 \times 10^{-10})^2 \text{ m}^2} = 8.202 \times 10^{-8} = 8.2 \times 10^{-8} \text{ N}$$

2.84   (a) The 68.926 amu isotope has a mass number of 69, with 31 protons, 38 neutrons and the symbol $^{69}_{31}$Ga. The 70.925 amu isotope has a mass number of 71, 31 protons, 40 neutrons, and symbol $^{71}_{31}$Ga. (All Ga atoms have 31 protons.)

(b) The average mass of a Ga atom is 69.72 amu. Let x = abundance of the lighter isotope, 1–x = abundance of the heavier isotope. Then x(68.926) + (1–x)(70.925) = 69.72; x = 0.6028 = 0.603, ^{69}Ga = 60.3%, ^{71}Ga = 39.7%.

2.85   (a) There are 24 known isotopes of Ni, from ^{51}Ni to ^{74}Ni.

(b) The five most abundant isotopes are

^{58}Ni, 57.935346 amu, 68.077%

^{60}Ni, 59.930788 amu, 26.223%

^{62}Ni, 61.928346 amu, 3.634%

^{61}Ni, 60.931058 amu, 1.140%

^{64}Ni, 63.927968 amu, 0.926%

Data from *Handbook of Chemistry and Physics*, 74th Ed. [Data may differ slightly in other editions.]

2.86   (a) A Br$_2$ molecule could consist of two atoms of the same isotope or one atom of each of the two different isotopes. This second possibility is twice as likely as the first. Therefore, the second peak (twice as large as peaks 1 and 3) represents a Br$_2$ molecule containing different isotopes. The mass numbers of the two isotopes are determined from the masses of the two smaller peaks. Since 157.836 ≈ 158, the first peak represents a ^{79}Br–^{79}Br molecule. Peak 3, 161.832 ≈ 162, represents a ^{81}Br–^{81}Br molecule. Peak 2 then contains one atom of each isotope, ^{79}Br–^{81}Br, with an approximate mass of 160 amu.

(b) The mass of the lighter isotope is 157.836 amu/2 atoms, or 78.918 amu/atom. For the heavier one, 161.832 amu/2 atoms = 80.916 amu/atom.

(c) The relative size of the three peaks in the mass spectrum of Br$_2$ indicates their relative abundance. The average mass of a Br$_2$ molecule is

$0.2569(157.836) + 0.4999(159.834) + 0.2431(161.832) = 159.79$ amu.

(Each product has four significant figures and two decimal places, so the answer has two decimal places.)

(d) $\dfrac{159.79 \text{ amu}}{\text{avg. Br}_2 \text{ molecule}} \times \dfrac{1\,\text{Br}_2 \text{ molecule}}{2\,\text{Br atoms}} = 79.895$ amu

(e) Let x = the abundance of ^{79}Br, 1 − x = abundance of ^{81}Br. From (b), the masses of the two isotopes are 78.918 amu and 80.916 amu, respectively. From (d), the mass of an average Br atom is 79.895 amu.

$$x(78.918) + (1 - x)(80.916) = 79.895,\ x = 0.5110$$

$$^{79}\text{Br} = 51.10\%,\ ^{81}\text{Br} = 48.90\%$$

2.87　(a)　Five significant figures. ^{1}H$^+$ is a bare proton with mass 1.0073 amu. ^{1}H is a hydrogen atom, with 1 proton and 1 electron. The mass of the electron is 5.486×10^{-4} or 0.0005486 amu. Thus the mass of the electron is significant in the fourth decimal place or fifth significant figure in the mass of ^{1}H.

(b)　Mass of ^{1}H = 1.0073 amu　　(proton)

0.0005486 amu　　(electron)

1.0078 amu　　(We have not rounded up to 1.0079 since 49 < 50 in the final sum.)

$$\text{Mass \% of electron} = \frac{\text{mass of } e^-}{\text{mass of }^1\text{H}} \times 100 = \frac{5.486 \times 10^{-4}\text{ amu}}{1.0078\text{ amu}} \times 100 = 0.05444\%$$

2.88　copper: Cu, 1B (coinage metals, transition metal)

tin: Sn, 4A

zinc: Zn, 2B

phosphorus: P, 5A

lead: Pb, 4A

2.89　(a)　an alkali metal: K　　(b)　an alkaline earth metal: Ca　　(c)　a noble gas: Ar

(d)　a halogen: Br　　(e)　a metalloid: Ge　　(f)　a nonmetal in 1A: H

(g)　a metal that forms a 3+ ion: Al　　(h)　a nonmetal that forms a 2– ion: O

(i)　an element that resembles Al: Ga

2.90　(a)　$^{266}_{106}$Sg has 106 protons, 160 neutrons and 106 electrons

(b)　Sg is in Group 6B (or 6) and immediately below tungsten, W. We expect the chemical properties of Sg to most closely resemble those of W.

2.91　(a)　chlorine gas, Cl$_2$: ii　　(b)　propane, C$_3$H$_8$: v　　(c)　nitrate ion, NO$_3^-$: i

(d)　sulfur trioxide, SO$_3$: iii　　(e)　methylchloride, CH$_3$Cl: iv

2.92   (a)   nickel(II) oxide, 2+   (b)   manganese(IV) oxide, 4+

         (c)   chromium(III) oxide, 3+   (d)   molybdenium(VI) oxide, 6+

2.93   (a) IO_3^- (b) IO_4^- (c) IO (d) HIO (e) HIO_4 or (H_5IO_6)

2.94   (a)   perbromate ion   (b)   selenite ion

         (c) AsO_4^{3-} (d) $HTeO_4^-$

2.95   (a)   sodium chloride   (b)   sodium bicarbonate (or sodium hydrogen carbonate)

         (c)   sodium hypochlorite   (d)   sodium hydroxide

         (e)   ammonium carbonate   (f)   calcium sulfate

2.96   (a)   potassium nitrate   (b)   sodium carbonate   (c)   calcium oxide

         (d)   hydrochloric acid   (e)   magnesium sulfate   (f)   magnesium hydroxide

2.97   (a) $CaS, Ca(HS)_2$ (b) $HBr, HBrO$ (c) $AlN, Al(NO_2)_3$ (d) FeO, Fe_2O_3

         (e) NH_3, NH_4^+ (f) $K_2SO_3, KHSO_3$ (g) $Hg_2Cl_2, HgCl_2$ (h) $HClO_3, HClO_4$

2.98

	Formula	Name	Density, g/mL	Melting Point, °C	Boiling Point, °C
(a)	PF_3	phosphorus trifluoride	3.907	–151.5	–101.5
(b)	$SiCl_4$	silicon tetrachloride	1.483	–70	57.57
(c)	C_2H_6O (C_2H_5OH)	ethanol	0.7893	–117.3	78.5

2.99   (a)   In an alkane, all C atoms have 4 single bonds, so each C in the partial structure needs 2 more bonds. All alkanes are hydrocarbons, so 2 H atoms will bind to each C atom in the ring.

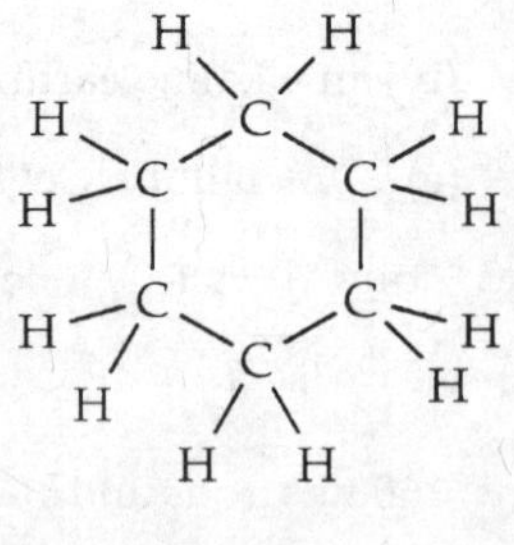

         (b)   The molecular formula of cyclohexane is C_6H_{12}; the molecular formula of n-hexane is C_6H_{14} (see Solution 2.64(d)). Cyclohexane can be thought of as n-hexane in which the two outer (terminal) C atoms are joined to each other. In order to form this C—C bond, each outer C atom must lose 1 H atom. The number of C atoms is unchanged, and each C atom still has 4 single bonds. The resulting molecular formula is $C_6H_{14-2} = C_6H_{12}$.

(c)   On the structure in part (a), replace 1 H atom with an OH group.

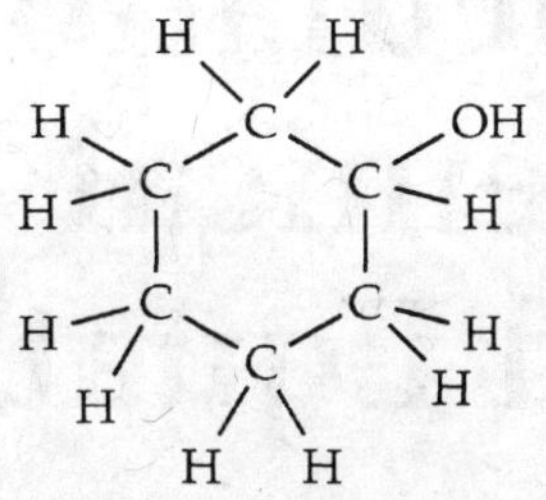

2.100   Elements are arranged in the periodic table by increasing atomic number and so that elements with similar chemical and physical properties form a vertical column or group. By its position in the periodic chart, we know whether an element is a metal, nonmetal, or metalloid, and the common charge of its ion. Members of a group have the same common ionic charge and combine in similar ways with other elements.

3 Stoichiometry: Calculations with Chemical Formulas and Equations

Visualizing Concepts

3.1   Reactant A = red, reactant B = blue

Overall, 4 red A_2 molecules + 4 blue B atoms → 4 A_2B molecules

Since 4 is a common factor, this equation reduces to equation (a).

3.2

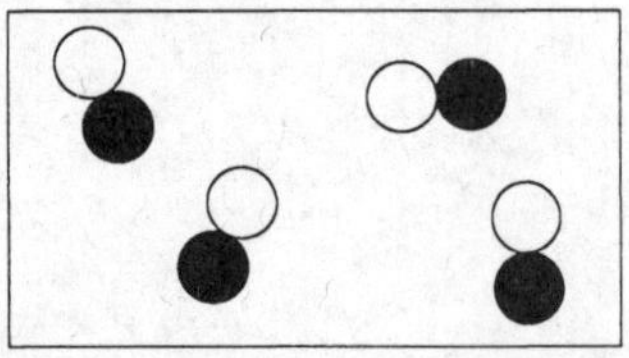

Write the balanced equation for the reaction.

$$2H_2 + CO \rightarrow CH_3OH$$

The combining ratio of H_2: CO is 2:1. If we have 8 H_2 molecules, 4 CO molecules are required for complete reaction. Alternatively, you could examine the atom ratios in the formula of CH_3OH, but the balanced equation is most direct.

3.3   (a)   There are twice as many O atoms as N atoms, so the empirical formula of the original compound is NO_2.

(b)   No, because we have no way of knowing whether the empirical and molecular formulas are the same. NO_2 represents the simplest ratio of atoms in a molecule but not the only possible molecular formula.

3.4   The box contains 4 C atoms and 16 H atoms, so the empirical formula of the hydrocarbon is CH_4.

3.5   (a)   *Analyze.* Given the molecular model, write the molecular formula.

Plan. Use the colors of the atoms (spheres) in the model to determine the number of atoms of each element.

Solve. Observe 2 gray C atoms, 5 white H atoms, 1 blue N atom, 2 red O atoms. $C_2H_5NO_2$

(b)   *Plan.* Follow the method in Sample Exercise 3.9. Calculate formula weight in amu and molar mass in grams.

$$2 \text{ C atoms} = 2(12.0 \text{ amu}) = 24.0 \text{ amu}$$

$$5 \text{ H atoms} = 5(1.0 \text{ amu}) = 5.0 \text{ amu}$$

$$1 \text{ N atoms} = 1(14.0 \text{ amu}) = 14.0 \text{ amu}$$

$$2 \text{ O atoms} = 2(16.0 \text{ amu}) = \underline{32.0 \text{ amu}}$$

$$75.0 \text{ amu}$$

Formula weight = 75.0 amu, molar mass = 75.0 g/mol

(c)   *Plan.* Use the definition of mass % and the results from parts (a) and (b) above to find mass % N in glycine.

Solve. $\text{mass \% N} = \dfrac{\text{g N}}{\text{g C}_2\text{H}_5\text{NO}_2} \times 100$

Assume 1 mol $C_2H_5NO_2$. From the molecular formula of glycine [part (a)], there is 1 mol N/mol glycine.

$$\text{mass \% N} = \frac{1 \times (\text{molar mass N})}{\text{molar mass glycine}} \times 100 = \frac{14.0 \text{ g}}{75.0 \text{ g}} \times 100 = 18.7\%$$

3.6   *Analyze.* Given: 4.0 mol CH_4. Find: mol CO and mol H_2

Plan. Examine the boxes to determine the CH_4:CO mol ratio and CH_4:H_2O mole ratio.

Solve. There are 2 CH_4 molecules in the reactant box and 2 CO molecules in the product box. The mole ratio is 2:2 or 1:1. Therefore, 4.0 mol CH_4 can produce 4.0 mol CO. There are 2 CH_4 molecules in the reactant box and 6 H_2 molecules in the product box. The mole ratio is 2:6 or 1:3. So, 4.0 mol CH_4 can produce 12:0 mol H_2.

Check. Use proportions. 2 mol CH_4/2 mol CO = 4 mol CH_4/4 mol CO;
2 mol CH_4/6 mol H_2 = 4 mol CH_4/12 mol H_2.

3.7   *Analyze.* Given a box diagram and formulas of reactants, draw a box diagram of products.

Plan. Write and balance the chemical equation. Determine combining ratios of elements and decide on limiting reactant. Draw a box diagram of products, containing the correct number of product molecules and only excess reactant.

Solve. $N_2 + 3H_2 \longrightarrow 3NH_3$. $N_2 =$ ⚫⚫ , $NH_3 =$ ○⚫○

Each N atom (1/2 of an N_2 molecule) reacts with 3 H atoms (1.5 H_2 molecules) to form an NH_3 molecule. Eight N atoms (4 N_2 molecules) require 24 H atoms (12 H_2 molecules) for complete reaction. Only 9 H_2 molecules are available, so H_2 is the limiting reactant. Nine H_2 molecules (18 H atoms) determine that 6 NH_3 molecules are produced. One N_2 molecule is in excess.

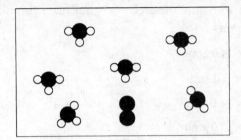

Check. Verify that mass is conserved in your solution, that the number and kinds of atoms are the same in reactant and product diagrams. In this example, there are 8 N atoms and 18 H atoms in both diagrams, so mass is conserved.

3.8   (a) $2NO + O_2 \rightarrow 2NO_2$, $O_2 = $ ⊂⊃ , $NO_2 = $ ⊂●◯

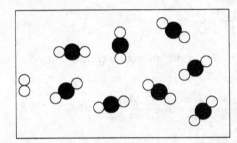

Each NO molecule reacts with 1 O atom (1/2 of an O_2 molecule) to produce 1 NO_2 molecule. Eight NO molecules react with 8 O atoms (4 O_2 molecules) to produce 8 NO_2 molecules. One O_2 molecule doesn't react (is in excess). NO is the limiting reactant.

(b) $\% \text{ yield} = \dfrac{\text{actual yield}}{\text{theoretical yield}} \times 100$; $\text{actual yield} = \dfrac{\% \text{ yield}}{100} \times \text{theoretical yield}$

The theoretical yield from part (a) is 8 NO_2 molecules. If the percent yield is 75%, then $0.75(8) = 6$ NO_2 would appear in the products box.

Balancing Chemical Equations

3.9   (a)   In balancing chemical equations, the law of conservation of mass, that atoms are neither created nor destroyed during the course of a reaction, is observed. This means that the **number** and **kinds** of atoms on both sides of the chemical equation must be the same.

         (b)   Subscripts in chemical formulas should not be changed when balancing equations, because changing the subscript changes the identity of the compound (law of constant composition).

         (c)   gases, (g); liquids, (l); solids, (s); aqueous solutions, (aq)

3.10   (a)   In a CO molecule, there is one O atom bound to C. 2CO indicates that there are **two CO molecules**, each of which contains one C and one O atom. Adding a subscript 2 to CO to form CO_2 means that there are **two O atoms** bound to one C in a CO_2 molecule. The composition of the different molecules, CO_2 and CO, is different and the physical and chemical properties of the two compounds they constitute are very different. The subscript 2 changes molecular composition and thus properties of the compound. The prefix 2 indicates how many molecules (or moles) of the original compound are under consideration.

(b) Yes. There are the same number and kinds of atoms on the reactants side and the products side of the equation.

3.11 (a) $2CO(g) + O_2(g) \rightarrow 2CO_2(g)$

(b) $N_2O_5(g) + H_2O(l) \rightarrow 2HNO_3(aq)$

(c) $CH_4(g) + 4Cl_2(g) \rightarrow CCl_4(l) + 4HCl(g)$

(d) $Al_4C_3(s) + 12H_2O(l) \rightarrow 4Al(OH)_3(s) + 3CH_4(g)$

(e) $2C_5H_{10}O_2(l) + 13O_2(g) \rightarrow 10CO_2(g) + 10H_2O(l)$

(f) $2Fe(OH)_3(s) + 3H_2SO_4(aq) \rightarrow Fe_2(SO_4)_3(aq) + 6H_2O(l)$

(g) $Mg_3N_2(s) + 4H_2SO_4(aq) \rightarrow 3MgSO_4(aq) + (NH_4)_2SO_4(aq)$

3.12 (a) $6Li(s) + N_2(g) \rightarrow 2Li_3N(s)$

(b) $La_2O_3(s) + 3H_2O(l) \rightarrow 2La(OH)_3(aq)$

(c) $2NH_4NO_3(s) \rightarrow 2N_2(g) + O_2(g) + 4H_2O(g)$

(d) $Ca_3P_2(s) + 6H_2O(l) \rightarrow 3Ca(OH)_2(aq) + 2PH_3(g)$

(e) $3Ca(OH)_2(aq) + 2H_3PO_4(aq) \rightarrow Ca_3(PO_4)_2(s) + 6H_2O(l)$

(f) $2AgNO_3(aq) + Na_2SO_4(aq) \rightarrow Ag_2SO_4(s) + 2NaNO_3(aq)$

(g) $4CH_3NH_2(g) + 9O_2(g) \rightarrow 4CO_2(g) + 10H_2O(g) + 2N_2(g)$

3.13 (a) $CaC_2(s) + 2H_2O(l) \rightarrow Ca(OH)_2(aq) + C_2H_2(g)$

(b) $2KClO_3(s) \xrightarrow{\Delta} 2KCl(s) + 3O_2(g)$

(c) $Zn(s) + H_2SO_4(aq) \rightarrow H_2(g) + ZnSO_4(aq)$

(d) $PCl_3(l) + 3H_2O(l) \rightarrow H_3PO_3(aq) + 3HCl(aq)$

(e) $3H_2S(g) + 2Fe(OH)_3(s) \rightarrow Fe_2S_3(s) + 6H_2O(g)$

3.14 (a) $SO_3(g) + H_2O(l) \rightarrow H_2SO_4(aq)$

(b) $B_2S_3(s) + 6H_2O(l) \rightarrow 2H_3BO_3(aq) + 3H_2S(g)$

(c) $4PH_3(g) + 8O_2(g) \rightarrow 6H_2O(g) + P_4O_{10}(s)$

(d) $2Hg(NO_3)_2(s) \xrightarrow{\Delta} 2HgO(s) + 4NO_2(g) + O_2(g)$

(e) $Cu(s) + 2H_2SO_4(aq) \rightarrow CuSO_4(aq) + SO_2(g) + 2H_2O(l)$

Patterns of Chemical Reactivity

3.15   (a)   When a metal reacts with a nonmetal, an ionic compound forms. The combining ratio of the atoms is such that the total positive charge on the metal cation(s) is equal to the total negative charge on the nonmetal anion(s). All ionic compounds are solids. $2\,Na(s) + Br_2(l) \rightarrow 2NaBr(s)$

(b)   The second reactant is oxygen gas from the air, $O_2(g)$. The products are $CO_2(g)$ and $H_2O(l)$. $2C_6H_6(l) + 15O_2(g) \rightarrow 12CO_2(g) + 6H_2O(l)$.

3.16   (a)   Neutral Ca atom loses $2e^-$ to form Ca^{2+}. Neutral O_2 molecule gains $4e^-$ to form $2O^{2-}$. The formula of the product will be CaO, because the cationic and anionic charges are opposite and equal. $2Ca(s) + O_2(g) \rightarrow 2CaO$

(b)   The products are $CO_2(g)$ and $H_2O(l)$. $C_3H_6O(l) + 4O_2(g) \rightarrow 3CO_2(g) + 3H_2O(l)$

3.17   (a) $Mg(s) + Cl_2(g) \rightarrow MgCl_2(s)$

(b) $BaCO_3(s) \xrightarrow{\Delta} BaO(s) + CO_2(g)$

(c) $C_8H_8(l) + 10O_2(g) \rightarrow 8CO_2(g) + 4H_2O(l)$

(d) CH_3OCH_3 is C_2H_6O. $C_2H_6O(l) + 3O_2(g) \rightarrow 2CO_2(g) + 3H_2O(l)$

3.18   (a) $2Al(s) + 3O_2(g) \rightarrow Al_2O_3(s)$

(b) $Cu(OH)_2(s) \xrightarrow{\Delta} CuO(s) + H_2O(g)$

(c) $C_7H_{16}(l) + 11O_2(g) \rightarrow 7CO_2(g) + 8H_2O(l)$

(d) $2C_5H_{12}O(l) + 15O_2(g) \rightarrow 10CO_2(g) + 12H_2O(l)$

3.19   (a) $2Al(s) + 3Cl_2(g) \rightarrow 2AlCl_3(s)$ combination

(b) $C_2H_4(g) + 3O_2(g) \rightarrow 2CO_2(g) + 2H_2O(l)$ combustion

(c) $6Li(s) + N_2(g) \rightarrow 2Li_3N(s)$ combination

(d) $PbCO_3(s) \rightarrow PbO(s) + CO_2(g)$ decomposition

(e) $C_7H_8O_2(l) + 8O_2(g) \rightarrow 7CO_2(g) + 4H_2O(l)$ combustion

3.20   (a) $2C_3H_6(g) + 9O_2(g) \rightarrow 6CO_2(g) + 6H_2O(l)$ combustion

(b) $NH_4NO_3(s) \rightarrow N_2O(g) + 2H_2O(l)$ decomposition

(c) $C_5H_6O(l) + 6O_2(g) \rightarrow 5CO_2(g) + 3H_2O(l)$ combustion

(d) $N_2(g) + 3H_2(g) \rightarrow 2NH_3(g)$ combination

(e) $K_2O(s) + H_2O(l) \rightarrow 2KOH(aq)$ combination

Formula Weights

3.21   *Analyze.* Given molecular formula or name, calculate formula weight.

Plan. If a name is given, write the correct molecular formula. Then, follow the method in Sample Exercise 3.5.   *Solve.*

(a) N_2O_5: $2(14.0) + 5(16.0) = 108.0$ amu

(b) $CuSO_4$: $1(63.6) + 1(32.1) + 4(16.0) = 159.6$ amu

(c) $(NH_4)_3PO_4$: $3(14.0) + 12(1.0) + 1(31.0) + 4(16.0) = 149.0$ amu

(d) $Ca(HCO_3)_2$: $1(40.1) + 2(1.0) + 2(12.0) + 6(16.0) = 162.1$ amu

(e) Al_2S_3: $2(27.0) + 3(32.1) = 150.3$ amu

(f) $Fe_2(SO_4)_3$: $2(55.8) + 3(32.1) + 12(16.0) = 399.9$ amu

(g) Si_2Br_6: $2(28.1) + 6(79.9) = 535.6$ amu

3.22   Formula weight in amu to 1 decimal place.

(a) N_2O: FW $= 2(14.0) + 1(16.0) = 44.0$ amu

(b) $HC_7H_5O_2$: $7(12.0) + 6(1.0) + 2(16.0) = 122.0$ amu

(c) $Mg(OH)_2$: $1(24.3) + 2(16.0) + 2(1.0) = 58.3$ amu

(d) $(NH_2)_2CO$: $2(14.0) + 4(1.0) + 1(12.0) + 1(16.0) = 60.0$ amu

(e) $CH_3CO_2C_5H_{11}$: $7(12.0) + 14(1.0) + 2(16.0) = 130.0$ amu

3.23   *Plan*. Calculate the formula weight (FW), then the mass % oxygen in the compound.
       Solve.

(a) SO_3: FW $= 1(32.1) + 3(16.0) = 80.1$ amu

$$\% \, O = \frac{3(16.0) \, amu}{80.1 \, amu} \times 100 = 59.9\%$$

(b) $CH_3COOCH_3 = C_3H_6O_2$: FW $= 3(12.0) + 6(1.0) + 2(16.0) = 74.0$ amu

$$\% \, O = \frac{2(16.0) \, amu}{74.0 \, amu} \times 100 = 43.2\%$$

(c) $Cr(NO_3)_3$: FW $= 1(52.0) + 3(14.0) + 9(16.0) = 238.0$ amu

$$\% \, O = \frac{9(16.0) \, amu}{238.0 \, amu} \times 100 = 60.5\%$$

(d) Na_2SO_4: FW $= 2(23.0) + 1(32.1) + 4(16.0) = 142.1$ amu

$$\% \, O = \frac{4(16.0) \, amu}{142.1 \, amu} \times 100 = 45.0\%$$

(e) NH_4NO_3: FW $= 2(14.0) + 4(1.0) + 3(16.0) = 80.0$ amu

$$\% \, O = \frac{3(16.0) \, amu}{80.0 \, amu} \times 100 = 60.0\%$$

3.24   (a) C_2H_2: FW $= 2(12.0) + 2(1.0) = 26.0$ amu

$$\% \, C = \frac{2(12.0) \, amu}{26.0 \, amu} \times 100 = 92.3\%$$

(b) $HC_6H_7O_6$: FW = 6(12.0) + 8(1.0) + 6(16.0) = 176.0 amu

$$\% H = \frac{8(1.0)\, amu}{176.0\, amu} \times 100 = 4.5\%$$

(c) $(NH_4)_2SO_4$: FW = 2(14.0) + 8(1.0) + 1(32.1) + 4(16.0) = 132.1 amu

$$\% H = \frac{8(1.0)\, amu}{132.1\, amu} \times 100 = 6.1\%$$

(d) $PtCl_2(NH_3)_2$: FW = 1(195.1) + 2(35.5) + 2(14.0) + 6(1.0) = 300.1 amu

$$\% Pt = \frac{1(195.1)\, amu}{300.1\, amu} \times 100 = 65.01\%$$

(e) $C_{18}H_{24}O_2$: FW = 18(12.0) + 24(1.0) + 2(16.0) = 272.0 amu

$$\% O = \frac{2(16.0)\, amu}{272.0\, amu} \times 100 = 11.8\%$$

(f) $C_{18}H_{27}NO_3$: FW = 18(12.0) + 27(1.0) + 1(14.0) + 3(16.0) = 305.0 amu

$$\% C = \frac{18(12.0)\, amu}{305.0\, amu} \times 100 = 70.8\%$$

3.25   *Plan.* Follow the logic for calculating mass % C given in Sample Exercise 3.6.   *Solve.*

(a) C_7H_6O: FW = 7(12.0) + 6(1.0) + 1(16.0) = 106.0 amu

$$\% C = \frac{7(12.0)\, amu}{106.0\, amu} \times 100 = 79.2\%$$

(b) $C_8H_8O_3$: FW = 8(12.0) + 8(1.0) + 3(16.0) = 152.0 amu

$$\% C = \frac{8(12.0)\, amu}{152.0\, amu} \times 100 = 63.2\%$$

(c) $C_7H_{14}O_2$: FW = 7(12.0) + 14(1.0) + 2(16.0) = 130.0 amu

$$\% C = \frac{7(12.0)\, amu}{130.0\, amu} \times 100 = 64.6\%$$

3.26   (a) CO_2: FW = 1(12.0) + 2(16.0) = 44.0 amu

$$\% C = \frac{12.0\, amu}{44.0\, amu} \times 100 = 27.3\%$$

(b) CH_3OH: FW = 1(12.0) + 4(1.0) + 1(16.0) = 32.0 amu

$$\% C = \frac{12.0\, amu}{32.0\, amu} \times 100 = 37.5\%$$

(c) C_2H_6: FW = 2(12.0) + 6(1.0) = 30.0 amu

$$\% C = \frac{2(12.0)\, amu}{30.0\, amu} \times 100 = 80.0\%$$

(d) $CS(NH_2)_2$: FW = 1(12.0) + 1(32.1) + 2(14.0) + 4(1.0) = 76.1 amu

$$\% C = \frac{12.0\, amu}{76.1\, amu} \times 100 = 15.8\%$$

Avogadro's Number and the Mole

3.27 (a) 6.022×10^{23}. This is the number of objects in a mole of anything.

(b) The formula weight of a substance in amu has the same numerical value as the molar mass expressed in grams.

3.28 (a) <u>exactly</u> 12 g   (b) 6.0221421×10^{23}, Avogadro's number

3.29 *Plan.* Since the mole is a counting unit, use it as a basis of comparison; determine the total moles of atoms in each given quantity.   *Solve.*

23 g Na contains 1 mol of atoms

0.5 mol H_2O contains (3 atoms × 0.5 mol) = 1.5 mol atoms

6.0×10^{23} N_2 molecules contains (2 atoms × 1 mol) = 2 mol atoms

3.30 3.0×10^{23} H_2O_2 molecules contains (4 atoms × 0.5 mol) = 2 mol atoms

32 g O_2 contains (2 atoms × 1 mol) = 2 mol atoms

2.0 mol CH_4 contains (5 atoms × 2 mol) = 10 mol atoms

3.31 *Analyze.* Given: 16 lb/ball; Avogadro's number of balls, 6.022×10^{23} balls. Find: mass in kg of Avogadro's number of balls; compare with mass of Earth.

Plan. balls → mass in lb → mass in kg; mass of balls/mass of Earth

Solve. 6.022×10^{23} balls $\times \dfrac{16\,lb}{ball} \times \dfrac{1\,kg}{2.2046\,lb} = 4.370 \times 10^{24} = 4.4 \times 10^{24}$ kg

$\dfrac{4.370 \times 10^{24}\,kg\,of\,balls}{5.98 \times 10^{24}\,kg\,Earth} = 0.73;$ One mole of shot-put balls weighs 0.73 times as much as Earth.

Check. This mass of balls is reasonable since Avogadro's number is large.

Estimate: 16 lb ≈ 7 kg; $6 \times 10^{23} \times 7 = 4.2 \times 10^{24}$ kg

3.32 292 million = $292 \times 10^6 = 2.92 \times 10^8$ people

$\dfrac{6.022 \times 10^{23}\,¢}{2.92 \times 10^8\,people} \times \dfrac{\$1}{100\,¢} = \dfrac{\$6.022 \times 10^{21}}{2.92 \times 10^8\,people} = \$2.06 \times 10^{13}/person$

$\$7.0\,trillion = \$7.0 \times 10^{12}$ $\dfrac{\$2.06 \times 10^{13}}{\$7.0 \times 10^{12}} = 2.9$

Each person would receive an amount that is 2.9 times the dollar amount of the national debt.

3.33 (a) *Analyze.* Given: 0.773 mol CaH_2. Find: mass in g.

Plan. Use molar mass (g/mol) of CaH_2 to find g CaH_2

Solve. molar mass = 1(40.08) + 2(1.008) = 42.096 = 42.10 g/mol CaH_2

0.773 mol CaH_2 $\times \dfrac{42.096\,g}{1\,mol} = 32.5$ g CaH_2

Check. 3/4(4) = 30 g. The calculated result is reasonable.

(b)   *Analyze.* Given: mass. Find: moles. *Plan.* Use molar mass of $Mg(NO_3)_2$.

Solve. molar mass = $1(24.31) + 2(14.01) + 6(16.00) = 148.33 = 148.3$

$$5.35 \text{ g } Mg(NO_3)_2 \times \frac{1 \text{ mol}}{148.33 \text{ g}} = 0.0361 \text{ mol } Mg(NO_3)_2$$

Check. $5/150 \approx 1/30 = 0.033$ mol

(c)   *Analyze.* Given: moles. Find: molecules. *Plan.* Use Avogadro's number.

$$\text{Solve. } 0.0305 \text{ mol } CH_3OH \times \frac{6.022 \times 10^{23} \text{ molecules}}{1 \text{ mol}} = 1.8367 \times 10^{22}$$
$$= 1.84 \times 10^{22} \; CH_3OH \text{ molecules}$$

Check. $(0.03 \times 6 \times 10^{23}) = 0.18 \times 10^{23} = 1.8 \times 10^{22}$

(d)   *Analyze.* Given: mol C_4H_{10}. Find: C atoms.

Plan. mol $C_4H_{10} \rightarrow$ mol H atoms $\rightarrow$ H atoms

$$\text{Solve. } 0.585 \text{ mol } C_4H_{10} \times \frac{4 \text{ mol C atoms}}{1 \text{ mol } C_4H_{10}} \times \frac{6.022 \times 10^{23} \text{ atoms}}{1 \text{ mol}}$$
$$= 1.41 \times 10^{24} \text{ C atoms}$$

Check. $(0.6 \times 4 \times 6 \times 10^{23}) = 14 \times 10^{23} = 1.4 \times 10^{24}$.

3.34   (a)   molar mass $= 1(137.33) + 2(126.904) = 39.14$ g

$$1.906 \times 10^{-2} \text{ mol } BaI_2 \times \frac{391.14 \text{ g}}{1 \text{ mol}} = 7.455 \text{ g } BaI_2$$

(b)   molar mass $= 1(14.01) + 4(1.008) + 1(35.45) = 53.49$ g/mol

$$48.3 \text{ g } NH_4Cl \times \frac{1 \text{ mol}}{53.49 \text{ g}} = 0.903 \text{ mol } NH_4Cl$$

(c) $$0.05752 \text{ mol } HCHO_2 \times \frac{6.02214 \times 10^{23} \text{ molecules}}{1 \text{ mol}} = 3.464 \times 10^{22} \; HCHO_2 \text{ molecules}$$

(d) $$4.88 \times 10^{-3} \text{ mol } Al(NO_3)_3 \times \frac{9 \text{ mol O}}{1 \text{ mol } Al(NO_3)_3} \times \frac{6.022 \times 10^{23} \text{ O atoms}}{1 \text{ mol}}$$
$$= 2.64 \times 10^{22} \text{ O atoms}$$

3.35   *Analyze/Plan.* See Solution 3.33 for stepwise problem-solving approach.   *Solve.*

(a) $(NH_4)_3PO_4$ molar mass $= 3(14.007) + 12(1.008) + 1(30.974) + 4(16.00) = 149.091$
$$= 149.1 \text{ g/mol}$$

$$2.50 \times 10^{-3} \text{ mol } (NH_4)_3PO_4 \times \frac{149.1 \text{ g } (NH_4)_3PO_4}{1 \text{ mol}} = 0.373 \text{ g } (NH_4)_3PO_4$$

(b) $AlCl_3$ molar mass $= 26.982 + 3(35.453) = 133.341 = 133.34$ g/mol

$$0.2550 \text{ g } AlCl_3 \times \frac{1 \text{ mol}}{133.34 \text{ g } AlCl_3} \times \frac{3 \text{ mol } Cl^-}{1 \text{ mol } AlCl_3} = 5.737 \times 10^{-3} \text{ mol } Cl^-$$

(c) $C_8H_{10}N_4O_2$ molar mass = $8(12.01) + 10(1.008) + 4(14.01) + 2(16.00) = 194.20$

$$= 194.2 \text{ g/mol}$$

$$7.70 \times 10^{20} \text{ molecules} \times \frac{1 \text{ mol}}{6.022 \times 10^{23} \text{ molecules}} \times \frac{194.2 \text{ g } C_8H_{10}N_4O_2}{1 \text{ mol caffeine}}$$

$$= 0.248 \text{ g } C_8H_{10}N_4O_2$$

(d) $\dfrac{0.406 \text{ g cholesterol}}{0.00105 \text{ mol}} = 387 \text{ g cholesterol/mol}$

3.36   (a) $Fe_2(SO_4)_3$ molar mass = $2(55.845) + 3(32.07) + 12(16.00) = 399.900 = 399.9 \text{ g/mol}$

$$0.0714 \text{ mol } Fe_2(SO_4)_3 \times \frac{399.9 \text{ g } Fe_2(SO_4)_3}{1 \text{ mol}} = 28.553 = 28.6 \text{ g } Fe_2(SO_4)_3$$

(b) $(NH_4)_2CO_3$ molar mass = $2(14.007) + 8(1.008) + 12.011 + 3(15.9994) = 96.0872$

$$= 96.087 \text{ g/mol}$$

$$8.776 \text{ g } (NH_4)_2CO_3 \times \frac{1 \text{ mol}}{96.087 \text{ g } (NH_4)_2CO_3} \times \frac{2 \text{ mol } NH_4^+}{1 \text{ mol } (NH_4)_2CO_3} = 0.1827 \text{ mol } NH_4^+$$

(c) $C_9H_8O_4$ molar mass = $9(12.01) + 8(1.008) + 4(16.00) = 180.154 = 180.2 \text{ g/mol}$

$$6.52 \times 10^{21} \text{ molecules} \times \frac{1 \text{ mol}}{6.022 \times 10^{23} \text{ molecules}} \times \frac{180.2 \text{ g } C_9H_8O_4}{1 \text{ mol aspirin}} = 1.95 \text{ g } C_9H_8O_4$$

(d) $\dfrac{15.86 \text{ g Valium}}{0.05570 \text{ mol}} = 284.7 \text{ g Valium/mol}$

3.37   (a) $C_6H_{10}OS_2$ molar mass = $6(12.01) + 10(1.008) + 1(16.00) + 2(32.07) = 162.28$

$$= 162.3 \text{ g/mol}$$

(b)   *Plan.* mg $\rightarrow$ g $\rightarrow$ mol   *Solve.*

$$5.00 \text{ mg allicin} \times \frac{1 \times 10^{-3} \text{ g}}{1 \text{ mg}} \times \frac{1 \text{ mol}}{162.3 \text{ g}} = 3.081 \times 10^{-5} = 3.08 \times 10^{-5} \text{ mol allicin}$$

Check. 5.00 mg is a small mass, so the small answer is reasonable.

$(5 \times 10^{-3})/200 = 2.5 \times 10^{-5}$

(c)   *Plan.* Use mol from part (b) and Avogadro's number to calculate molecules.

$$\textit{Solve. } 3.081 \times 10^{-5} \text{ mol allicin} \times \frac{6.022 \times 10^{23} \text{ molecules}}{\text{mol}} = 1.855 \times 10^{19}$$

$$= 1.86 \times 10^{19} \text{ allicin molecules}$$

Check. $(3 \times 10^{-5})(6 \times 10^{23}) = 18 \times 10^{18} = 1.8 \times 10^{19}$

(d)   *Plan.* Use molecules from part (c) and molecular formula to calculate S atoms.

$$\textit{Solve. } 1.855 \times 10^{19} \text{ allicin molecules} \times \frac{2 \text{ S atoms}}{1 \text{ allicin molecule}} = 3.71 \times 10^{19} \text{ S atoms}$$

Check. Obvious.

3.38   (a) $C_{14}H_{18}N_2O_5$ molar mass = 14(12.01) + 18(1.008) + 2(14.01) + 5(16.00)

$$= 294.30 \text{ g/mol}$$

    (b)   1.00 mg aspartame $\times \dfrac{1\times10^{-3}\text{g}}{1\text{ mg}} \times \dfrac{1\text{ mol}}{294.3\text{ g}} = 3.398\times10^{-6} = 3.40\times10^{-6}$ mol aspartame

    (c) 3.398×10^{-6} mol aspartame $\times \dfrac{6.022\times10^{23}\text{ molecules}}{1\text{ mol}} = 2.046\times10^{18}$

$$= 2.05 \times 10^{18} \text{ aspartame molecules}$$

    (d) 2.046×10^{18} aspartame molecules $\times \dfrac{18\text{ H atoms}}{1\text{ aspartame molecule}} = 3.68\times10^{19}$ H atoms

3.39   (a)   *Analyze.* Given: $C_6H_{12}O_6$, 1.250×10^{21} C atoms. Find: H atoms.

Plan. Use molecular formula to determine number of H atoms that are present with 1.250×10^{21} C atoms.   *Solve.*

$$\dfrac{12\text{ H atoms}}{6\text{ C atoms}} = \dfrac{2\text{ H}}{1\text{ C}} \times 1.250\times10^{21}\text{ C atoms} = 2.500\times10^{21}\text{ H atoms}$$

Check. $(2 \times 1 \times 10^{21}) = 2 \times 10^{21}$

    (b)   *Plan.* Use molecular formula to find the number of glucose molecules that contain 1.250×10^{21} C atoms. *Solve.*

$$\dfrac{1\,C_6H_{12}O_6 \text{ molecule}}{6\text{ C atoms}} \times 1.250\times10^{21}\text{ C atoms} = 2.0833\times10^{20}$$

$$= 2.083 \times 10^{20}\ C_6H_{12}O_6 \text{ molecules}$$

Check. $(12 \times 10^{20}/6) = 2 \times 10^{20}$

    (c)   *Plan.* Use Avogadro's number to change molecules → mol.   *Solve.*

$$2.0833\times10^{20}\ C_6H_{12}O_6 \text{ molecules} \times \dfrac{1\text{ mol}}{6.022\times10^{23}\text{ molecules}}$$

$$= 3.4595\times10^{-4} = 3.460\times10^{-4}\text{ mol } C_6H_{12}O_6$$

Check. $(2 \times 10^{20})/(6 \times 10^{23}) = 0.33 \times 10^{-3} = 3.3 \times 10^{-4}$

    (d)   *Plan.* Use molar mass to change mol → g.   *Solve.*

1 mole of $C_6H_{12}O_6$ weighs 180.0 g (Sample Exercise 3.9)

$$3.4595 \times 10^{-4}\text{ mol } C_6H_{12}O_6 \times \dfrac{180.0\text{ g } C_6H_{12}O_6}{1\text{ mol}} = 0.06227\text{ g } C_6H_{12}O_6$$

Check. $3.5 \times 180 = 630$; $630 \times 10^{-4} = 0.063$

3.40   (a) 7.08×10^{20} H atoms $\times \dfrac{19\text{ C atoms}}{28\text{ H atoms}} = 4.80 \times 10^{20}$ C atoms

    (b) 7.08×10^{20} H atoms $\times \dfrac{1\,C_{19}H_{28}O_2 \text{ molecule}}{28\text{ H atoms}} = 2.529\times10^{19}$

$$= 2.53 \times 10^{19}\ C_{19}H_{28}O_2 \text{ molecules}$$

(c) 2.529×10^{19} $C_{19}H_{28}O_2$ molecules $\times \dfrac{1\,\text{mol}}{6.022 \times 10^{23}\ \text{molecules}} = 4.199 \times 10^{-5}$

$$= 4.20 \times 10^{-5}\ \text{mol}\ C_{19}H_{28}O_2$$

(d) $C_{19}H_{28}O_2$ molar mass $= 19(12.01) + 28(1.008) + 2(16.00) = 288.41 = 288.4$ g/mol

$$4.199 \times 10^{-5}\ \text{mol}\ C_{19}H_{28}O_2 \times \dfrac{288.4\,\text{g}\ C_{19}H_{28}O_2}{1\,\text{mol}} = 0.0121\,\text{g}\ C_{19}H_{28}O_2$$

3.41   *Analyze.* Given: g C_2H_3Cl/L.   Find: mol/L, molecules/L.

Plan. The /L is constant throughout the problem, so we can ignore it. Use molar mass for g $\rightarrow$ mol, Avogadro's number for mol $\rightarrow$ molecules.   *Solve.*

$$\dfrac{2.0 \times 10^{-6}\,\text{g}\ C_2H_3Cl}{1\,\text{L}} \times \dfrac{1\,\text{mol}\ C_2H_3Cl}{62.50\,\text{g}\ C_2H_3Cl} = 3.20 \times 10^{-8} = 3.2 \times 10^{-8}\ \text{mol}\ C_2H_3Cl/L$$

$$\dfrac{3.20 \times 10^{-8}\ \text{mol}\ C_2H_3Cl}{1\,\text{L}} \times \dfrac{6.022 \times 10^{23}\ \text{molecules}}{1\,\text{mol}} = 1.9 \times 10^{16}\ \text{molecules/L}$$

Check. $(200 \times 10^{-8})/60 = 2.5 \times 10^{-8}$ mol

$$(2.5 \times 10^{-8}) \times (6 \times 10^{23}) = 15 \times 10^{15} = 1.5 \times 10^{16}$$

3.42 $25 \times 10^{-6}\,\text{g}\ C_{21}H_{30}O_2 \times \dfrac{1\,\text{mol}\ C_{21}H_{30}O_2}{314.5\,\text{g}\ C_{21}H_{30}O_2} = 7.95 \times 10^{-8} = 8.0 \times 10^{-8}\ \text{mol}\ C_{21}H_{30}O_2$

$$7.95 \times 10^{-8}\ \text{mol}\ C_{21}H_{30}O_2 \times \dfrac{6.022 \times 10^{23}\ \text{molecules}}{1\,\text{mol}} = 4.8 \times 10^{16}\ C_{21}H_{30}O_2\ \text{molecules}$$

Empirical Formulas

3.43   (a)   *Analyze.* Given: moles. Find: empirical formula.

Plan. Find the **simplest ratio of moles** by dividing by the smallest number of moles present.

Solve. 0.0130 mol C / 0.0065 = 2

0.039 mol H / 0.0065 = 6

0.0065 mol O / 0.0065 = 1

The empirical formula is C_2H_6O.

Check. The subscripts are simple integers.

(b)   *Analyze.* Given: grams. Find: empirical formula.

Plan. Calculate the moles of each element present, then the simplest ratio of moles.

Solve. $11.66 \text{ g Fe} \times \dfrac{1 \text{ mol Fe}}{55.85 \text{ g Fe}} = 0.2088 \text{ mol Fe}; \ 0.2088 / 0.2088 = 1$

$5.01 \text{ g O} \times \dfrac{1 \text{ mol O}}{16.00 \text{ g O}} = 0.3131 \text{ mol O}; \ 0.3131 / 0.2088 \approx 1.5$

Multiplying by two, the integer ratio is 2 Fe : 3 O; the empirical formula is Fe_2O_3.

Check. The subscripts are simple integers.

(c) *Analyze.* Given: mass %. Find: empirical formulas.

Plan. Assume 100 g sample, calculate moles of each element, find the simplest ratio of moles.

Solve. $40.0 \text{ g C} \times \dfrac{1 \text{ mol C}}{12.01 \text{ g C}} = 3.33 \text{ mol C}; \ 3.33 / 3.33 = 1$

$6.7 \text{ g H} \times \dfrac{1 \text{ mol H}}{1.008 \text{ mol H}} = 6.65 \text{ mol H}; \ 6.65 / 3.33 \approx 2$

$53.3 \text{ g O} \times \dfrac{1 \text{ mol O}}{16.00 \text{ mol O}} = 3.33 \text{ mol O}; \ 3.33 / 3.33 = 1$

The empirical formula is CH_2O.

Check. The subscripts are simple integers.

3.44 (a) Calculate the simplest ratio of moles.

0.104 mol K / 0.052 = 2

0.052 mol C / 0.052 = 1

0.156 mol O / 0.052 = 3

The empirical formula is K_2CO_3.

(b) Calculate moles of each element present, then the simplest ratio of moles.

$5.28 \text{ g Sn} \times \dfrac{1 \text{ mol Sn}}{118.7 \text{ g Sn}} = 0.04448 \text{ mol Sn}; \ 0.04448 / 0.04448 = 1$

$3.37 \text{ g F} \times \dfrac{1 \text{ mol F}}{19.00 \text{ g FSn}} = 0.1774 \text{ mol F}; 0.1774 / 0.04448 \approx 4$

The integer ratio is 1 Sn : 4 F; the empirical formula is SnF_4.

(c) Assume 100 g sample, calculate moles of each element, find the simplest ratio of moles.

$87.5\% \text{ N} = 87.5 \text{ g N} \times \dfrac{1 \text{ mol N}}{14.01 \text{ g}} = 6.25 \text{ mol N}; \ 6.25 / 6.25 = 1$

$12.5\% \text{ H} = 12.5 \text{ g H} \times \dfrac{1 \text{ mol}}{1.008 \text{ g}} = 12.4 \text{ mol H}; \ 12.4 / 6.25 \approx 2$

The empirical formula is NH_2.

3.45 *Analyze/Plan.* The procedure in all these cases is to assume 100 g of sample, calculate the number of moles of each element present in that 100 g, then obtain the ratio of moles as smallest whole numbers. *Solve.*

(a) $10.4 \text{ g C} \times \dfrac{1 \text{ mol C}}{12.01 \text{ g C}} = 0.866 \text{ mol C}; \; 0.866 / 0.866 = 1$

$27.8 \text{ g S} \times \dfrac{1 \text{ mol S}}{32.07 \text{ g S}} = 0.867 \text{ mol S}; \; 0.867 / 0.866 \approx 1$

$61.7 \text{ g Cl} \times \dfrac{1 \text{ mol Cl}}{35.45 \text{ g Cl}} = 1.74 \text{ mol Cl}; \; 1.74 / 0.866 \approx 2$

The empirical formula is $CSCl_2$.

(b) $21.7 \text{ g C} \times \dfrac{1 \text{ mol C}}{12.01 \text{ g C}} = 1.81 \text{ mol C}; \; 1.81 / 0.600 \approx 3$

$9.6 \text{ g O} \times \dfrac{1 \text{ mol O}}{16.00 \text{ g O}} = 0.600 \text{ mol O}; \; 0.600 / 0.600 = 1$

$68.7 \text{ g F} \times \dfrac{1 \text{ mol F}}{19.00 \text{ g F}} = 3.62 \text{ mol F}; \; 3.62 / 0.600 \approx 6$

The empirical formula is C_3OF_6.

(c) $32.79 \text{ g Na} \times \dfrac{1 \text{ mol Na}}{22.99 \text{ g Na}} = 1.426 \text{ mol Na}; \; 1.426 / 0.4826 \approx 3$

$13.02 \text{ g Al} \times \dfrac{1 \text{ mol Al}}{26.98 \text{ g Al}} = 0.4826 \text{ mol Al}; \; 0.4826 / 0.4826 = 1$

$54.19 \text{ g F} \times \dfrac{1 \text{ mol F}}{19.00 \text{ g F}} = 2.852 \text{ mol F}; \; 2.852 / 0.4826 \approx 6$

The empirical formula is Na_3AlF_6.

3.46   See Solution 3.45 for stepwise problem-solving approach.

(a) $55.3 \text{ g K} \times \dfrac{1 \text{ mol K}}{39.10 \text{ g K}} = 1.414 \text{ mol K}; 1.414/0.4714 \approx 3$

$14.6 \text{ g P} \times \dfrac{1 \text{ mol P}}{30.97 \text{ g P}} = 0.4714 \text{ mol P}; \; 0.4714/0.4714 = 1$

$30.1 \text{ g O} \times \dfrac{1 \text{ mol O}}{16.00 \text{ g O}} = 1.881 \text{ mol O}; \; 1.881/0.4714 \approx 4$

The empirical formula is K_3PO_4.

(b) $24.5 \text{ g Na} \times \dfrac{1 \text{ mol Na}}{22.99 \text{ g Na}} = 1.066 \text{ mol Na}; \; 1.066/0.5304 \approx 2$

$14.9 \text{ g Si} \times \dfrac{1 \text{ mol Si}}{28.09 \text{ Si}} = 0.5304 \text{ mol si}; \; 0.5304/0.5304 = 1$

$60.6 \text{ g F} \times \dfrac{1 \text{ mol F}}{19.00 \text{ g F}} = 3.189 \text{ mol F}; 3.189/0.5304 \approx 6$

The empirical formula is Na_2SiF_6.

(c) $62.1 \, g \, C \times \dfrac{1 \, mol \, C}{12.01 \, g \, C} = 5.17 \, mol \, C; \; 5.17 \, / \, 0.864 \approx 6$

$5.21 \, g \, H \times \dfrac{1 \, mol \, H}{1.008 \, g \, H} = 5.17 \, mol \, O; \; 5.17 \, / \, 0.864 \approx 6$

$12.1 \, g \, N \times \dfrac{1 \, mol \, N}{14.01 \, g \, N} = 0.864 \, mol \, N; \; 0.864 \, / \, 0.864 = 1$

$20.7 \, g \, O \times \dfrac{1 \, mol \, O}{16.00 \, g \, O} = 1.29 \, mol \, O; \; 1.29 \, / \, 0.864 \approx 1.5$

Multiplying by two, the empirical formula is $C_{12}H_{12}N_2O_3$.

3.47   *Analyze.* Given: empirical formula, molar mass. Find: molecular formula.

Plan. Calculate the empirical formula weight (FW); divide FW by molar mass (MM) to calculate the integer that relates the empirical and molecular formulas. Check. If FW/MM is an integer, the result is reasonable. *Solve.*

(a)   FW $CH_2 = 12 + 2(1) = 14$. $\dfrac{MM}{FW} = \dfrac{84}{14} = 6$

The subscripts in the empirical formula are multiplied by 6. The molecular formula is C_6H_{12}.

(b)   FW $NH_2Cl = 14.01 + 2(1.008) + 35.45 = 51.48$. $\dfrac{MM}{FW} = \dfrac{51.5}{51.5} = 1$

The empirical and molecular formulas are NH_2Cl.

3.48   (a) $HCHO_2$ FW $12.01 + 1.008 + 2(16.00) = 45.0$ $\dfrac{MM}{FW} = \dfrac{90.0}{45.0} = 2$

The molecular formula is $H_2C_2O_4$.

(b) C_2H_4O FW $= 2(12) + 4(1) + 16 = 44$. $\dfrac{MM}{FW} = \dfrac{88}{44} = 2$

The molecular formula is $C_4H_8O_2$.

3.49   *Analyze.* Given: mass %, molar mass. Find: molecular formula.

Plan. Use the plan detailed in Solution 3.45 to find an empirical formula from mass % data. Then use the plan detailed in 3.47 to find the molecular formula. Note that some indication of molar mass must be given, or the molecular formula cannot be determined. *Check.* If there is an integer ratio of moles and MM/ FW is an integer, the result is reasonable.   *Solve.*

(a) $92.3 \, g \, C \times \dfrac{1 \, mol \, C}{12.01 \, g \, C} = 7.685 \, mol \, C; \; 7.685/7.639 = 1.006 \approx 1$

$7.7 \, g \, H \times \dfrac{1 \, mol \, H}{1.008 \, g \, H} = 7.639 \, mol \, H; \; 7.639/7.639 = 1$

The empirical formula is CH, FW = 13.

$\dfrac{MM}{FW} = \dfrac{104}{13} = 8$; the molecular formula is C_8H_8.

(b) $49.5 \, \text{g C} \times \dfrac{1 \, \text{mol C}}{12.01 \, \text{g C}} = 4.12 \, \text{mol C}; \quad 4.12/1.03 \approx 4$

$5.15 \, \text{g H} \times \dfrac{1 \, \text{mol H}}{1.008 \, \text{g H}} = 5.11 \, \text{mol H}; \quad 5.11/1.03 \approx 5$

$28.9 \, \text{g N} \times \dfrac{1 \, \text{mol N}}{14.01 \, \text{g N}} = 2.06 \, \text{mol N}; \quad 2.06/1.03 \approx 2$

$16.5 \, \text{g O} \times \dfrac{1 \, \text{mol O}}{16.00 \, \text{g O}} = 1.03 \, \text{mol O}; \quad 1.03/1.03 = 1$

Thus, $C_4H_5N_2O$, FW = 97. If the molar mass is about 195, a factor of 2 gives the molecular formula $C_8H_{10}N_4O_2$.

(c) $35.51 \, \text{g C} \times \dfrac{1 \, \text{mol C}}{12.01 \, \text{g C}} = 2.96 \, \text{mol C}; \; 2.96/0.592 = 5$

$4.77 \, \text{g H} \times \dfrac{1 \, \text{mol H}}{1.008 \, \text{g H}} = 4.73 \, \text{mol H}; \; 4.73/0.592 = 7.99 \approx 8$

$37.85 \, \text{g O} \times \dfrac{1 \, \text{mol O}}{16.00 \, \text{g O}} = 2.37 \, \text{mol O}; \; 2.37/0.592 = 4$

$8.29 \, \text{g N} \times \dfrac{1 \, \text{mol N}}{14.01 \, \text{g N}} = 0.592 \, \text{mol N}; \; 0.592/0.592 = 1$

$13.60 \, \text{g Na} \times \dfrac{1 \, \text{mol Na}}{22.99 \, \text{g Na}} = 0.592 \, \text{mol Na}; \; 0.592/0.592 = 1$

The empirical formula is $C_5H_8O_4NNa$, FW = 169 g. Since the empirical formula weight and molar mass are approximately equal, the empirical and molecular formulas are both $NaC_5H_8O_4N$.

3.50   Assume 100 g in the following problems.

(a) $75.69 \, \text{g C} \times \dfrac{1 \, \text{mol C}}{12.01 \, \text{g C}} = 6.30 \, \text{mol C}; \; 6.30/0.969 = 6.5$

$8.80 \, \text{g H} \times \dfrac{1 \, \text{mol H}}{1.008 \, \text{g H}} = 8.73 \, \text{mol H}; \; 8.73/0.969 = 9.0$

$15.51 \, \text{g O} \times \dfrac{1 \, \text{mol O}}{16.00 \, \text{g O}} = 0.969 \, \text{mol O}; \, 0.969/0.969 = 1$

Multiply by 2 to obtain the integer ratio 13:18:2. The empirical formula is $C_{13}H_{18}O_2$, FW = 206 g. Since the empirical formula weight and the molar mass are equal (206 g), the empirical and molecular formulas are $C_{13}H_{18}O_2$.

(b) $58.55 \, \text{g C} \times \dfrac{1 \, \text{mol C}}{12.01 \, \text{g C}} = 4.875 \, \text{mol C}; \quad 4.875/1.956 \approx 2.5$

$13.81 \, \text{g H} \times \dfrac{1 \, \text{mol H}}{1.008 \, \text{g H}} = 13.700 \, \text{mol H}; \; 13.700/1.956 \approx 7.0$

$$27.40 \text{ g N} \times \frac{1 \text{ mol N}}{14.01 \text{ g N}} = 1.956 \text{ mol N}; \quad 1.956/1.956 = 1.0$$

Multiply by 2 to obtain the integer ratio 5:14:2. The empirical formula is $C_5H_{14}N_2$; FW = 102. Since the empirical formula weight and the molar mass are equal

(102 g), the empirical and molecular formulas are $C_5H_{14}N_2$.

(c) $59.0 \text{ g C} \times \dfrac{1 \text{ mol C}}{12.01 \text{ g C}} = 4.91 \text{ mol C}; \quad 4.91/0.550 \approx 9$

$$7.1 \text{ g H} \times \frac{1 \text{ mol H}}{1.008 \text{ g H}} = 7.04 \text{ mol H}; \quad 7.04/0.550 \approx 13$$

$$26.2 \text{ g O} \times \frac{1 \text{ mol O}}{16.00 \text{ g O}} = 1.64 \text{ mol O}; \quad 1.64/0.550 \approx 3$$

$$7.7 \text{ g N} \times \frac{1 \text{ mol N}}{14.01 \text{ g N}} = 0.550 \text{ mol N}; \quad 0.550/0.550 = 1$$

The empirical formula is $C_9H_{13}O_3N$, FW = 183 amu (or g). Since the molecular weight is approximately 180 amu, the empirical formula and molecular formula are the same, $C_9H_{13}O_3N$.

3.51   (a)   *Analyze.* Given: mg CO_2, mg H_2O Find: empirical formula of hydrocarbon, C_xH_y

Plan. Upon combustion, all $C \rightarrow CO_2$, all $H \rightarrow H_2O$.

mg $CO_2 \rightarrow$ g $CO_2 \rightarrow$ mol C; mg $H_2O \rightarrow$ g H_2O, mol H

Find simplest ratio of moles and empirical formula.   *Solve.*

$$5.86 \times 10^{-3} \text{ g CO}_2 \times \frac{1 \text{ mol CO}_2}{44.01 \text{ g CO}_2} \times \frac{1 \text{ mol C}}{1 \text{ mol CO}_2} = 1.33 \times 10^{-4} \text{ mol C}$$

$$1.37 \times 10^{-3} \text{ g H}_2\text{O} \times \frac{1 \text{ mol H}_2\text{O}}{18.02 \text{ g H}_2\text{O}} \times \frac{2 \text{ mol H}}{1 \text{ mol H}_2\text{O}} = 1.52 \times 10^{-4} \text{ mol H}$$

Dividing both values by 1.33×10^{-4} gives C:H of 1:1.14. This is not "close enough" to be considered 1:1. No obvious multipliers (2, 3, 4) produce an integer ratio. Testing other multipliers (trial and error!), the correct factor seems to be 7. The empirical formula is C_7H_8.

Check. See discussion of C:H ratio above.

(b)   *Analyze.* Given: g of menthol, g CO_2, g H_2O, molar mass. Find: molecular formula.

Plan/Solve. Calculate mol C and mol H in the sample.

$$0.2829 \text{ g CO}_2 \times \frac{1 \text{ mol CO}_2}{44.01 \text{ g CO}_2} \times \frac{1 \text{ mol C}}{1 \text{ mol CO}_2} = 0.0064281 = 0.006428 \text{ mol C}$$

$$0.1159 \text{ g H}_2\text{O} \times \frac{1 \text{ mol H}_2\text{O}}{18.02 \text{ g H}_2\text{O}} \times \frac{2 \text{ mol H}}{1 \text{ mol H}_2\text{O}} = 0.012863 = 0.01286 \text{ mol H}$$

Calculate g C, g H and get g O by subtraction.

$$0.0064281\,\text{mol C}\times\frac{12.01\,\text{g C}}{1\,\text{mol C}}=0.07720\,\text{g C}$$

$$0.012863\,\text{mol H}\times\frac{1.008\,\text{g H}}{1\,\text{mol H}}=0.01297\,\text{g H}$$

mass O = 0.1005 g sample – (0.07720 g C + 0.01297 g H) = 0.01033 g O

Calculate mol O and find integer ratio of mol C: mol H: mol O.

$$0.01033\,\text{g O}\times\frac{1\,\text{mol O}}{16.00\,\text{g O}}=6.456\times10^{-4}\,\text{mol O}$$

Divide moles by 6.456×10^{-4}.

$$\text{C}:\frac{0.006428}{6.456\times10^{-4}}\approx10;\quad \text{H}:\frac{0.01286}{6.456\times10^{-4}}\approx20;\quad \text{O}:\frac{6.456\times10^{-4}}{6.456\times10^{-4}}=1$$

The empirical formula is $C_{10}H_{20}O$.

$$\text{FW}=10(12)+20(1)+16=156;\quad \frac{M}{FW}=\frac{156}{156}=1$$

The molecular formula is the same as the empirical formula, $C_{10}H_{20}O$.

Check. The mass of O wasn't negative or greater than the sample mass; empirical and molecular formulas are reasonable.

3.52 (a) *Plan.* Calculate mol C and mol H, then g C and g H; get g O by subtraction.

Solve.

$$6.32\times10^{-3}\,\text{g CO}_2\times\frac{1\,\text{mol CO}_2}{44.01\,\text{g CO}_2}\times\frac{1\,\text{mol C}}{1\,\text{mol CO}_2}=1.436\times10^{-4}=1.44\times10^{-4}\,\text{mol C}$$

$$2.58\times10^{-3}\,\text{g H}_2\text{O}\times\frac{1\,\text{mol H}_2\text{O}}{18.02\,\text{g H}_2\text{O}}\times\frac{2\,\text{mol H}}{1\,\text{mol H}_2\text{O}}=2.863\times10^{-4}=2.86\times10^{-4}\,\text{mol H}$$

$$1.436\times10^{-4}\,\text{mol C}\times\frac{12.01\,\text{g C}}{1\,\text{mol C}}=1.725\times10^{-3}\,\text{g C}=1.73\,\text{mg C}$$

$$2.863\times10^{-4}\,\text{mol H}\times\frac{1.008\,\text{g H}}{1\,\text{mol H}}=2.886\times10^{-4}\,\text{g H}=0.289\,\text{mg H}$$

mass of O = 2.78 mg sample – (1.725 mg C + 0.289 mg H) = 0.77 mg O

$$0.77\times10^{-3}\,\text{g O}\times\frac{1\,\text{mol O}}{16.00\,\text{g O}}=4.81\times10^{-5}\,\text{mol O}.\ \text{Divide moles by }4.81\times10^{-5}.$$

$$\text{C}:\frac{1.44\times10^{-4}}{4.81\times10^{-5}}\approx3;\quad \text{H}:\frac{2.86\times10^{-4}}{4.81\times10^{-5}}\approx6;\quad \text{O}:\frac{4.81\times10^{-5}}{4.81\times10^{-5}}=1$$

The empirical formula is C_3H_6O.

(b) *Plan.* Calculate mol C and mol H, then g C and g H. In this case, get N by subtraction. *Solve.*

$$14.242 \times 10^{-3} \text{ g CO}_2 \times \frac{1 \text{ mol CO}_2}{44.01 \text{ g CO}_2} \times \frac{1 \text{ mol C}}{1 \text{ mol CO}_2} = 3.2361 \times 10^{-4} \text{ mol C}$$

$$4.083 \times 10^{-3} \text{ g H}_2\text{O} \times \frac{1 \text{ mol H}_2\text{O}}{18.02 \text{ g H}_2\text{O}} \times \frac{2 \text{ mol H}}{1 \text{ mol H}_2\text{O}} = 4.5136 \times 10^{-4} = 4.532 \times 10^{-4} \text{ mol H}$$

$$3.2361 \times 10^{-4} \text{ g mol C} \times \frac{12.01 \text{ g C}}{1 \text{ mol H}} = 3.8866 \times 10^{-3} \text{ g C} = 3.8866 \text{ mg C}$$

$$4.532 \times 10^{-4} \text{ mol H} \times \frac{1.008 \text{ g H}}{1 \text{ mol H}} = 0.45683 \times 10^{-3} \text{ g H} = 0.4568 \text{ mg H}$$

mass of N = 5.250 mg sample – (3.8866 mg C + 0.4568 mg H) = 0.9066

$$= 0.907 \text{ mg N}$$

$$0.9066 \times 10^{-3} \text{ g N} \times \frac{1 \text{ mol N}}{14.01 \text{ g N}} = 6.47 \times 10^{-5} \text{ mol N}. \text{ Divide moles by } 6.47 \times 10^{-5}.$$

$$\text{C:} \quad \frac{3.24 \times 10^{-4}}{6.47 \times 10^{-5}} \approx 5; \quad \text{H:} \quad \frac{4.53 \times 10^{-4}}{6.47 \times 10^{-5}} \approx 7; \quad \text{N:} \quad \frac{6.47 \times 10^{-5}}{6.47 \times 10^{-5}} = 1$$

The empirical formula is C_5H_7N, FW = 81. A molar mass of 160 ± 5 indicates a factor of 2 and a molecular formula of $C_{10}H_{14}N_2$.

3.53   *Analyze.* Given 2.558 g $Na_2CO_3 \bullet xH_2O$, 0.948 g Na_2CO_3. Find: x.

Plan. The reaction involved is $Na_2CO_3 \bullet xH_2O(s) \rightarrow Na_2CO_3(s) + xH_2O(g)$.

Calculate the mass of H_2O lost and then the mole ratio of Na_2CO_3 and H_2O.

Solve. g H_2O lost = 2.558 g sample – 0.948 g Na_2CO_3 = 1.610 g H_2O

$$0.948 \text{ g Na}_2\text{CO}_3 \times \frac{1 \text{ mol Na}_2\text{CO}_3}{106.0 \text{ g Na}_2\text{CO}_3} = 0.00894 \text{ mol Na}_2\text{CO}_3$$

$$1.610 \text{ g H}_2\text{O} \times \frac{1 \text{ mol H}_2\text{O}}{18.02 \text{ g H}_2\text{O}} = 0.08935 \text{ mol H}_2\text{O}$$

The formula is $Na_2CO_3 \bullet \underline{10} H_2O$.

Check. x is an integer.

3.54   The reaction involved is $MgSO_4 \bullet xH_2O(s) \rightarrow MgSO_4(s) + xH_2O(g)$. First, calculate the number of moles of product $MgSO_4$; this is the same as the number of moles of starting hydrate.

$$2.472 \text{ g MgSO}_4 \times \frac{1 \text{ mol MgSO}_4}{120.4 \text{ g MgSO}_4} \times \frac{1 \text{ mol MgSO}_4 \bullet xH_2O}{1 \text{ mol MgSO}_4} = 0.02053 \text{ mol MgSO}_4 \bullet xH_2O$$

Thus, $\dfrac{5.061 \text{ g MgSO}_4 \bullet xH_2O}{0.02053} = 246.5 \text{ g/mol} = \text{FW of MgSO}_4 \bullet xH_2O$

FW of $MgSO_4 \bullet xH_2O$ = FW of $MgSO_4$ + x(FW of H_2O).

246.5 = 120.4 + x(18.02). x = 6.998. The hydrate formula is $MgSO_4 \bullet \underline{7}H_2O$.

Alternatively, we could calculate the number of moles of water represented by weight loss: (5.061 – 2.472) = 2.589 g H_2O lost.

$$2.589 \text{ g } H_2O \times \frac{1 \text{ mol } H_2O}{18.02 \text{ g } H_2O} = 0.1437 \text{ mol } H_2O; \quad \frac{\text{mol } H_2O}{\text{mol } MgSO_4} = \frac{0.1437}{0.02053} = 7.000$$

Again the correct formula is $MgSO_4 \cdot \underline{7}H_2O$.

Calculations Based on Chemical Equations

3.55   The mole ratios implicit in the coefficients of a balanced chemical equation express the fundamental relationship between amounts of reactants and products. If the equation is not balanced, the mole ratios will be incorrect and lead to erroneous calculated amounts of products.

3.56   The **integer coefficients** immediately preceding each molecular formula in a chemical equation give information about relative numbers of moles of reactants and products involved in a reaction.

3.57 $Na_2SiO_3(s) + 8HF(aq) \rightarrow H_2SiF_6(aq) + 2NaF(aq) + 3H_2O(l)$

(a)   *Analyze.* Given: mol Na_2SiO_3. Find: mol HF. *Plan.* Use the mole ratio $8HF:1Na_2SiO_3$ from the balanced equation to relate moles of the two reactants.

Solve.

$$0.300 \text{ mol } Na_2SiO_3 \times \frac{8 \text{ mol } HF}{1 \text{ mol } Na_2SiO_3} = 2.40 \text{ mol } HF$$

Check. Mol HF should be greater than mol Na_2SiO_3.

(b)   *Analyze.* Given: mol HF. Find: g NaF. *Plan.* Use the mole ratio $2NaF:8HF$ to change mol HF to mol NaF, then molar mass to get NaF.   *Solve.*

$$0.500 \text{ mol } HF \times \frac{2 \text{ mol } NaF}{8 \text{ mol } HF} \times \frac{41.99 \text{ g } NaF}{1 \text{ mol } NaF} = 5.25 \text{ g } NaF$$

Check. $(0.5/4) = 0.125; 0.13 \times 42 > 4$ g NaF

(c)   *Analyze.* Given: g HF Find: g Na_2SiO_3.

Plan. g HF $\rightarrow$ mol HF $\left(\dfrac{\text{mol}}{\text{ratio}}\right) \rightarrow$ mol $Na_2SiO_3 \rightarrow$ g Na_2SiO_3

The mole ratio is at the heart of every stoichiometry problem. Molar mass is used to change to and from grams.   *Solve.*

$$0.800 \text{ g } HF \times \frac{1 \text{ mol } HF}{20.01 \text{ g } HF} \times \frac{1 \text{ mol } Na_2SiO_3}{8 \text{ mol } HF} \times \frac{122.1 \text{ g } Na_2SiO_3}{1 \text{ mol } Na_2SiO_3} = 0.610 \text{ g } Na_2SiO_3$$

Check. $0.8 (120/160) < 0.75$ mol

3.58 $C_6H_{12}O_6(aq) \rightarrow 2C_2H_5OH(aq) + 2CO_2(g)$

(a) $0.400 \text{ mol } C_6H_{12}O_6 \times \dfrac{2 \text{ mol } CO_2}{1 \text{ mol } C_6H_{12}O_6} = 0.800 \text{ mol } CO_2$

(b) $7.50 \text{ g } C_2H_5OH \times \dfrac{1 \text{ mol } C_2H_5OH}{46.07 \text{ g } C_2H_5OH} \times \dfrac{1 \text{ mol } C_6H_{12}O_6}{2 \text{ mol } C_2H_5OH} \times \dfrac{180.2 \text{ g } C_6H_{12}O_6}{1 \text{ mol } C_6H_{12}O_6}$

$$= 14.7 \text{ g } C_6H_{12}O_6$$

(c) $7.50 \text{ g C}_2\text{H}_5\text{OH} \times \dfrac{1 \text{ mol C}_2\text{H}_5\text{OH}}{46.07 \text{ g C}_2\text{H}_5\text{OH}} \times \dfrac{2 \text{ mol CO}_2}{2 \text{ mol C}_2\text{H}_5\text{OH}} \times \dfrac{44.01 \text{ g CO}_2}{1 \text{ mol CO}_2} = 7.16 \text{ g CO}_2$

3.59   (a) $\text{Al(OH)}_3(s) + 3\text{HCl}(aq) \rightarrow \text{AlCl}_3(aq) + 3\text{H}_2\text{O}(l)$

(b)   *Analyze.* Given mass of one reactant, find stoichiometric mass of other reactant and products.

Plan. Follow the logic in Sample Exercise 3.16. Calculate mol Al(OH)_3 in 0.500 g $\text{Al(OH}_3)_3$ separately, since it will be used several times.

Solve. $0.500 \text{ g Al(OH)}_3 \times \dfrac{1 \text{ mol Al(OH)}_3}{78.00 \text{ g Al(OH)}_3} = 6.410 \times 10^{-3} = 6.41 \times 10^{-3} \text{ mol Al(OH)}_3$

$6.410 \times 10^{-3} \text{ mol Al(OH)}_3 \times \dfrac{3 \text{ mol HCl}}{1 \text{ mol Al(OH)}_3} \times \dfrac{36.46 \text{ g HCl}}{1 \text{ mol HCl}} = 0.7012 = 0.701 \text{ g HCl}$

(c) $6.410 \times 10^{-3} \text{ molAl(OH)}^3 \times \dfrac{1 \text{ mol HCl}}{1 \text{ mol Al(OH)}_3} \times \dfrac{133.34 \text{ g AlCl}_3}{1 \text{ mol AlCl}_3} = 0.8547$

$= 0.855 \text{ g AlCl}_3$

$6.410 \times 10^{-3} \text{ mol Al(OH)}_3 \times \dfrac{3 \text{ mol H}_2\text{O}}{1 \text{ mol Al(OH)}_3} \times \dfrac{18.02 \text{ g H}_2\text{O}}{1 \text{ mol H}_2\text{O}} = 0.3465 = 0.347 \text{ g H}_2\text{O}$

(d)   Conservation of mass: mass of products = mass of reactants

reactants: Al(OH)_3 + HCl, 0.500 g + 0.701 g = 1.201 g

products: $\text{AlCl}_3 + \text{H}_2\text{O}$, 0.855 g + 0.347 g = 1.202 g

The 0.001 g difference is due to rounding (0.8547 + 0.3465 = 1.2012). This is an excellent *check* of results.

3.60   (a) $\text{Fe}_2\text{O}_3(s) + 3\text{CO}(g) \rightarrow 2\text{Fe}(s) + 3\text{CO}_2(g)$

(b) $0.150 \text{ kg Fe}_2\text{O}_3 \times \dfrac{1000 \text{ g}}{1 \text{ kg}} \times \dfrac{1 \text{ mol Fe}_2\text{O}_3}{159.688 \text{ g Fe}_2\text{O}_3} = 0.9393 = 0.939 \text{ mol Fe}_2\text{O}_3$

$0.9393 \text{ mol Fe}_2\text{O}_3 \times \dfrac{3 \text{ mol CO}}{1 \text{ mol Fe}_2\text{O}_3} \times \dfrac{28.01 \text{ g CO}}{1 \text{ mol CO}} - 78.929 = 78.9 \text{ g CO}$

(c) $0.9393 \text{ mol Fe}_2\text{O}_3 \times \dfrac{2 \text{ mol Fe}}{1 \text{ mol Fe}_2\text{O}_3} \times \dfrac{55.845 \text{ g Fe}}{1 \text{ mol Fe}} = 104.914 = 105 \text{ g Fe}$

$0.9393 \text{ mol Fe}_2\text{O}_3 \times \dfrac{3 \text{ mol CO}_2}{1 \text{ mol Fe}_2\text{O}_3} \times \dfrac{44.01 \text{ g CO}_2}{1 \text{ mol CO}_2} = 124.015 = 124 \text{ g CO}_2$

(d)   reactants: 150 g Fe_2O_3 + 78.9 g CO = 228.9 = 229 g

products: 104.9 g Fe + 124.0 g CO_2 = 228.9 = 229 g

Mass is conserved.

3.61   (a) $\text{Al}_2\text{S}_3(s) + 6\text{H}_2\text{O}(l) \rightarrow 2\text{Al(OH)}_3(s) + 3\text{H}_2\text{S}(g)$

(b)   *Plan.* g A → mol A → mol B → g B. See Solution 3.57 (c).   *Solve.*

$$6.75 \text{ g Al}_2\text{S}_3 \times \frac{1 \text{ mol Al}_2\text{S}_3}{150.2 \text{ g Al}_2\text{S}_3} \times \frac{2 \text{ mol Al(OH)}_3}{1 \text{ mol Al}_2\text{S}_3} \times \frac{78.00 \text{ g Al(OH)}_3}{1 \text{ mol Al(OH)}_3} = 7.01 \text{ g Al(OH)}_3$$

Check. $7 \left(\dfrac{2 \times 78}{150} \right) \approx 7(1) \approx 7 \text{ g Al(OH)}_3$

3.62   (a) $\text{CaH}_2(s) + 2\text{H}_2\text{O}(l) \rightarrow \text{Ca(OH)}_2(aq) + 2\text{H}_2(g)$

(b) $8.500 \text{ g H}_2 \times \dfrac{1 \text{ mol H}_2}{2.016 \text{ g H}_2} \times \dfrac{1 \text{ mol CaH}_2}{2 \text{ mol H}_2} \times \dfrac{42.10 \text{ g CaH}_2}{1 \text{ mol CaH}_2} = 88.75 \text{ g CaH}_2$

3.63   (a)   *Analyze.* Given: mol NaN_3. Find: mol N_2.

Plan. Use mole ratio from balanced equation.   *Solve.*

$$1.50 \text{ mol NaN}_3 \times \frac{3 \text{ mol N}_2}{2 \text{ mol NaN}_3} = 2.25 \text{ mol N}_2$$

Check. The resulting mol N_2 should be greater than mol NaN_3, (the N_2:NaN_3 ratio is > 1), and it is.

(b)   *Analyze.* Given: g N_2 Find: g NaN_3.

Plan. Use molar masses to get from and to grams, mol ratio to relate moles of the two substances.   *Solve.*

$$10.0 \text{ g N}_2 \times \frac{1 \text{ mol N}_2}{28.01 \text{ g N}_2} \times \frac{2 \text{ mol NaN}_3}{3 \text{ mol N}_2} \times \frac{65.01 \text{ g NaN}_3}{1 \text{ mol NaN}_3} = 15.5 \text{ g NaN}_3$$

Check. Mass relations are less intuitive than mole relations. Estimating the ratio of molar masses is sometimes useful. In this case, 65 g NaN_3/28 g $\text{N}_2 \approx 2.25$ Then, $(10 \times 2/3 \times 2.25) \approx 14$ g NaN_3. The calculated result looks reasonable.

(c)   *Analyze.* Given: vol N_2 in ft^3, density N_2 in g/L. Find: g NaN_3.

Plan. First determine how many g N_2 are in 10.0 ft^3, using the density of N_2.

Solve.

$$\frac{1.25 \text{ g}}{1 \text{ L}} \times \frac{1 \text{ L}}{1000 \text{ cm}^3} \times \frac{(2.54)^3 \text{ cm}^3}{1 \text{ in}^3} \times \frac{(12)^3 \text{ in}^3}{1 \text{ ft}^3} \times 10.0 \text{ ft}^3 = 354.0 = 354 \text{ g N}_2$$

$$354.0 \text{ g N}_2 \times \frac{1 \text{ mol N}_2}{28.01 \text{ g N}_2} \times \frac{2 \text{ mol NaN}_3}{3 \text{ mol N}_2} \times \frac{65.01 \text{ g NaN}_3}{1 \text{ mol NaN}_3} = 548 \text{ g NaN}_3$$

Check. 1 ft$^3 \sim 28$ L; 10 ft$^3 \sim 280$ L; 280 L $\times 1.25 \sim 350$ g N_2

Using the ratio of molar masses from part (b), $(350 \times 2/3 \times 2.25) \approx 525$ g NaN_3

3.64 $2\text{C}_8\text{H}_{18}(l) + 25\text{O}_2(g) \rightarrow 16\text{CO}_2(g) + 18\text{H}_2\text{O}(l)$

(a) $1.25 \text{ mol C}_8\text{H}_{18} \times \dfrac{25 \text{ mol O}_2}{2 \text{ mol C}_8\text{H}_{18}} = 15.625 = 15.6 \text{ mol O}_2$

(b) $10.0 \text{ g C}_8\text{H}_{18} \times \dfrac{1 \text{ mol C}_8\text{H}_{18}}{114.2 \text{ g C}_8\text{H}_{18}} \times \dfrac{25 \text{ mol O}_2}{2 \text{ mol C}_8\text{H}_{18}} \times \dfrac{32.00 \text{ g O}_2}{1 \text{ mol O}_2} = 35.0 \text{ g O}_2$

(c) $1.00 \text{ gal } C_8H_{18} \times \dfrac{3.7854 \text{ L}}{1 \text{ gal}} \times \dfrac{1000 \text{ mL}}{1 \text{ L}} \times \dfrac{0.692 \text{ g}}{1 \text{ mL}} = 2619.5 = 2.62 \times 10^3 \text{ g } C_8H_{18}$

$2.6195 \times 10^3 \text{ g } C_8H_{18} \times \dfrac{1 \text{ mol } C_8H_{18}}{114.2 \text{ g } C_8H_{18}} \times \dfrac{25 \text{ mol } O_2}{2 \text{ mol } C_8H_{18}} \times \dfrac{32.00 \text{ g } O_2}{1 \text{ mol } O_2} = 9{,}175.1 \text{ g}$

$$= 9.18 \times 10^3 \text{ g } O_2$$

3.65   (a)   *Analyze.* Given: dimensions of Al foil. Find: mol Al.

Plan.   Dimensions $\rightarrow$ vol $\xrightarrow{\text{density}}$ mass $\xrightarrow[\text{mass}]{\text{molar}}$ mol Al

Solve. $1.00 \text{ cm} \times 1.00 \text{ cm} \times 0.550 \text{ mm} \times \dfrac{1 \text{ cm}}{10 \text{ mm}} = 0.0550 \text{ cm}^3 \text{ Al}$

$0.0550 \text{ cm}^3 \text{ Al} \times \dfrac{2.699 \text{ g Al}}{1 \text{ cm}^3} \times \dfrac{1 \text{ mol Al}}{26.98 \text{ g Al}} = 5.502 \times 10^{-3} = 5.50 \times 10^{-3} \text{ mol Al}$

Check. $2.699 / 26.98 \approx 0.1$; $(0.055 \text{ cm}^3 \times 0.1) = 5.5 \times 10^{-3} \text{ mol Al}$

(b)   *Plan.* Write the balanced equation to get a mole ratio; change mol Al $\rightarrow$ mol $AlBr_3 \rightarrow$ g $AlBr_3$.

Solve. $2Al(s) + 3Br_2(l) \rightarrow 2AlBr_3(s)$

$5.502 \times 10^{-3} \text{ mol Al} \times \dfrac{2 \text{ mol } AlBr_3}{2 \text{ mol Al}} \times \dfrac{266.69 \text{ g } AlBr_3}{1 \text{ mol } AlBr_3} = 1.467 = 1.47 \text{ g } AlBr_3$

Check. $(0.006 \times 1 \times 270) \approx 1.6 \text{ g } AlBr_3$

3.66   (a)   *Plan.* Calculate a "mole ratio" between nitroglycerine and total moles of gas produced. $(12 + 6 + 1 + 10) = 29$ mol gas; 4 mol nitro: 29 total mol gas.   *Solve.*

$2.00 \text{ mL nitro} \times \dfrac{1.592 \text{ g}}{\text{mL}} \times \dfrac{1 \text{ mol nitro}}{227.1 \text{ g nitro}} \times \dfrac{29 \text{ mol gas}}{4 \text{ mol nitro}} = 0.10165 = 0.102 \text{ mol gas}$

(b) $0.10165 \text{ mol gas} \times \dfrac{55 \text{ L}}{\text{mol}} = 5.5906 = 5.6 \text{ L}$

(c) $2.00 \text{ mL nitro} \times \dfrac{1.592 \text{ g}}{\text{mL}} \times \dfrac{1 \text{ mol nitro}}{227.1 \text{ g nitro}} \times \dfrac{6 \text{ mol } N_2}{4 \text{ mol nitro}} \times \dfrac{28.01 \text{ g } N_2}{1 \text{ mol } N_2} = 0.589 \text{ g } N_2$

Limiting Reactants; Theoretical Yields

3.67   (a)   The *limiting reactant* determines the maximum number of product moles resulting from a chemical reaction; any other reactant is an *excess reactant*.

(b)   The limiting reactant regulates the amount of products, because it is completely used up during the reaction; no more product can be made when one of the reactants is unavailable.

3.68   (a)   *Theoretical yield* is the maximum amount of product possible, as predicted by stoichiometry, assuming that the limiting reactant is converted entirely to product.

Actual yield is the amount of product actually obtained, less than or equal to the theoretical yield. *Percent yield* is the ratio of (actual yield to theoretical yield) × 100.

(b)   No reaction is perfect. Not all reactant molecules come together effectively to form products; alternative reaction pathways may produce secondary products and reduce the amount of desired product actually obtained, or it might not be possible to completely isolate the desired product from the reaction mixture. In any case, these factors reduce the actual yield of a reaction.

3.69   (a)   Each bicycle needs 2 wheels, 1 frame, and 1 set of handlebars. A total of 4815 wheels corresponds to 2407.5 pairs of wheels. This is more than the number of frames or handlebars. The 2255 handlebars determine that 2255 bicycles can be produced.

       (b)   2305 frames – 2255 bicycles = 50 frames left over

            2407.5 pairs of wheels – 2255 bicycles = 152.5 pairs of wheels left over 2(152.5) = 305 wheels left over

       (c)   The handlebars are the "limiting reactant" in that they determine the number of bicycles that can be produced.

3.70   (a) $40{,}875 \text{ L beverage} \times \dfrac{1 \text{ bottle}}{0.355 \text{ L}} = 115{,}140.85 = 1.15 \times 10^5$ portions of beverage

            (The uncertainty in 355 mL limits the precision of the number of portions we can reasonably expect to deliver to three significant figures.)

            121,515 bottles; 122,500 caps; 1.15×10^5 bottles can be filled and capped.

       (b)   122,500 caps – 115,141 portions = 7,359 = 7×10^3 caps remain

            121,515 empty bottles – 115,141 portions = 6374 = 6×10^3 bottles remain

            (Uncertainty in the number of portions delivered limits the results to 1 sig fig.)

       (c)   The volume of beverage limits production.

3.71   *Analyze.* Given: 1.85 mol NaOH, 1.00 mol CO_2. Find: mol Na_2CO_3.

       Plan. Amounts of more than one reactant are given, so we must determine which reactant regulates (limits) product. Then apply the appropriate mole ratio from the balanced equation.

       Solve. The mole ratio is 2NaOH:1CO_2, so 1.00 mol CO_2 requires 2.00 mol NaOH for complete reaction. Less than 2.00 mol NaOH are present, so NaOH is the limiting reactant.

       $1.85 \text{ mol NaOH} \times \dfrac{1 \text{ mol } Na_2CO_3}{2 \text{ mol NaOH}} = 0.925 \text{ mol } Na_2CO_3$ can be produced

       The Na_2CO_3:CO_2 ratio is 1:1, so 0.925 mol Na_2CO_3 produced requires 0.925 mol CO_2 consumed. (Alternately, 1.85 mol NaOH × 1 mol CO_2/2 mol NaOH = 0.925 mol CO_2 reacted). 1.00 mol CO_2 initial – 0.925 mol CO_2 reacted = 0.075 mol CO_2 remain.

Check.	$2NaOH(s)$	$+$	$CO_2(g)$	$\rightarrow$	$Na_2CO_3(s)$	$+$	$H_2O(l)$
initial	1.85 mol		1.00 mol		0 mol		
change (reaction)	–1.85 mol		–0.925 mol		+0.925 mol		
final	0 mol		0.075 mol		0.925 mol		

Note that the "change" line (but not necessarily the "final" line) reflects the mole ratios from the balanced equation.

3.72 $0.500 \text{ mol Al(OH)}_3 \times \dfrac{3 \text{ mol H}_2\text{SO}_4}{2 \text{ mol Al(OH)}_3} = 0.750 \text{ mol H}_2\text{SO}_4$ needed for complete reaction

Only 0.500 mol H_2SO_4 available, so H_2SO_4 limits.

$0.500 \text{ mol H}_2\text{SO}_4 \times \dfrac{1 \text{ mol Al}_2(\text{SO}_4)_3}{3 \text{ mol H}_2\text{SO}_4} = 0.1667 = 0.167 \text{ mol Al}_2(\text{SO}_4)_3$ can form

$0.500 \text{ mol H}_2\text{SO}_4 \times \dfrac{2 \text{ mol Al(OH)}_3}{3 \text{ mol H}_2\text{SO}_4} = 0.3333 = 0.333 \text{ mol Al(OH)}_3$ react

$0.500 \text{ mol Al(OH)}_3$ initial – $0.333 \text{ mol react} = 0.167 \text{ mol Al(OH)}_3$ remain

3.73 $3NaHCO_3(aq) + H_3C_6H_5O_7(aq) \rightarrow 3CO_2(g) + 3H_2O(l) + Na_3C_6H_5O_7(aq)$

(a)   *Analyze/Plan.* Abbreviate citric acid as H_3Cit. Follow the approach in Sample Exercise 3.19. *Solve.*

$1.00 \text{ g NaHCO}_3 \times \dfrac{1 \text{ mol NaHCO}_3}{84.01 \text{ g NaHCO}_3} = 1.190 \times 10^{-2} = 1.19 \times 10^{-2} \text{ mol NaHCO}_3$

$1.00 \text{ g H}_3\text{C}_6\text{H}_5\text{O}_7 \times \dfrac{1 \text{ mol H}_3\text{Cit}}{192.1 \text{ g H}_3\text{Cit}} = 5.206 \times 10^{-3} = 5.21 \times 10^{-3} \text{ mol H}_3\text{Cit}$

But $NaHCO_3$ and H_3Cit react in a 3:1 ratio, so 5.21×10^{-3} mol H_3Cit require $3(5.21 \times 10^{-3}) = 1.56 \times 10^{-2}$ mol $NaHCO_3$. We have only 1.19×10^{-2} mol $NaHCO_3$, so $NaHCO_3$ is the limiting reactant.

(b) $1.190 \times 10^{-2} \text{ mol NaHCO}_3 \times \dfrac{3 \text{ mol CO}_2}{3 \text{ mol NaHCO}_3} \times \dfrac{44.01 \text{ g CO}_2}{1 \text{ mol CO}_2} = 0.524 \text{ g CO}_2$

(c) $1.190 \times 10^{-2} \text{ mol NaHCO}_3 \times \dfrac{1 \text{ mol H}_3\text{Cit}}{3 \text{ mol NaHCO}_3} = 3.968 \times 10^{-3}$

$= 3.97 \times 10^{-3} \text{ mol H}_3\text{Cit react}$

$5.206 \times 10^{-3} \text{ mol H}_3\text{Cit} - 3.968 \times 10^{-3} \text{ mol react} = 1.238 \times 10^{-3}$

$= 1.24 \times 10^{-3} \text{ mol H}_3\text{Cit remain}$

$1.238 \times 10^{-3} \text{ mol H}_3\text{Cit} \times \dfrac{192.1 \text{ g H}_3\text{Cit}}{\text{mol H}_3\text{Cit}} = 0.238 \text{ g H}_3\text{Cit remain}$

3.74 $4NH_3(g) + 5O_2(g) \rightarrow 4NO(g) + 6H_2O(g)$

(a)   Follow the approach in Sample Exercise 3.19.

$$1.50 \text{ g NH}_3 \times \frac{1 \text{ mol NH}_3}{17.03 \text{ g NH}_3} = 0.08808 = 0.0881 \text{ mol NH}_3$$

$$2.75 \text{ g O}_2 \times \frac{1 \text{ mol O}_2}{32.00 \text{ g O}_2} = 0.08594 = 0.0859 \text{ mol O}_2$$

$$0.08594 \text{ mol O}_2 \times \frac{4 \text{ mol NH}_3}{5 \text{ mol O}_2} = 0.06875 = 0.0688 \text{ mol NH}_3 \text{ required}$$

More than 0.0688 mol NH_3 is available, so O_2 is the limiting reactant.

(b) $$0.08594 \text{ mol O}_2 \times \frac{4 \text{ mol NO}}{5 \text{ mol O}_2} \times \frac{30.01 \text{ g NO}}{1 \text{ mol NO}} = 2.063 = 2.06 \text{ g NO produced}$$

$$0.08594 \text{ mol O}_2 \times \frac{6 \text{ mol H}_2\text{O}}{5 \text{ mol O}_2} \times \frac{18.02 \text{ g H}_2\text{O}}{1 \text{ mol H}_2\text{O}} = 1.8583 = 1.86 \text{ g H}_2\text{O}$$

(c)   0.08808 mol NH_3 – 0.06875 mol NH_3 reacted = 0.01933 = 0.0193 mol NH_3 remain

$$0.01933 \text{ mol NH}_3 \times \frac{17.03 \text{ g NH}_3}{1 \text{ mol NH}_3} = 0.32919 = 0.329 \text{ g NH}_3 \text{ remain}$$

(d)   mass products = 2.06 g NO + 1.86 g H_2O + 0.329 g NH_3 remaining = 4.25 g products

mass reactants = 1.50 g NH_3 + 2.75 g O_2 = 4.25 g reactants

(For comparison purposes, the mass of excess reactant can be either added to the products, as above, or subtracted from reactants.)

3.75   *Analyze.* Given: initial g Na_2CO_3, g $AgNO_3$. Find: final g Na_2CO_3, $AgNO_3$, Ag_2CO_3, $NaNO_3$

Plan. Write balanced equation; determine limiting reactant; calculate amounts of excess reactant remaining and products, based on limiting reactant.

Solve. $2AgNO_3(aq) + Na_2CO_3(aq) \rightarrow Ag_2CO_3(s) + 2NaNO_3(aq)$

$$3.50 \text{ g Na}_2\text{CO}_3 \times \frac{1 \text{ mol Na}_2\text{CO}_3}{106.0 \text{ g Na}_2\text{CO}_3} = 0.03302 = 0.0330 \text{ mol Na}_2\text{CO}_3$$

$$5.00 \text{ g AgNO}_3 \times \frac{1 \text{ mol AgNO}_3}{169.9 \text{ g AgNO}_3} = 0.02943 = 0.0294 \text{ mol AgNO}_3$$

$$0.02943 \text{ mol AgNO}_3 \times \frac{1 \text{ mol Na}_2\text{CO}_3}{2 \text{ mol AgNO}_3} = 0.01471 = 0.0147 \text{ mol Na}_2\text{CO}_3 \text{ required}$$

$AgNO_3$ is the limiting reactant and Na_2CO_3 is present in excess.

	$2AgNO_3(aq)$ +	$Na_2CO_3(aq)$ $\rightarrow$	$Ag_2CO_3(s)$ +	$2NaNO_3(aq)$
initial	0.0294 mol	0.0330 mol	0 mol	0 mol
reaction	–0.0294 mol	–0.0147 mol	+0.0147 mol	+0.0294 mol
final	0 mol	0.0183 mol	0.0147 mol	0.0294 mol

0.01830 mol $Na_2CO_3 \times 106.0$ g/mol $= 1.940 = 1.94$ g Na_2CO_3

0.01471 mol $Ag_2CO_3 \times 275.8$ g/mol $= 4.057 = 4.06$ g Ag_2CO_3

0.02943 mol $NaNO_3 \times 85.00$ g/mol $= 2.502 = 2.50$ g $NaNO_3$

Check. The initial mass of reactants was 8.50 g, and the final mass of excess reactant and products is 13.50 g; mass is conserved.

3.76   *Plan.* Write balanced equation; determine limiting reactant; calculate amounts of excess reactant remaining and products, based on limiting reactant.

Solve. $H_2SO_4(aq) + Pb(C_2H_3O_2)_2(aq) \rightarrow PbSO_4(s) + 2HC_2H_3O_2(aq)$

$$7.50 \text{ g } H_2SO_4 \times \frac{1 \text{ mol } H_2SO_4}{98.09 \text{ g } H_2SO_4} = 0.07646 = 0.0765 \text{ mol } H_2SO_4$$

$$7.50 \text{ g } Pb(C_2H_3O_2)_2 \times \frac{1 \text{ mol } Pb(C_2H_3O_2)_2}{325.3 \text{ g } Pb(C_2H_3O_2)_2} = 0.023056 = 0.0231 \text{ mol } Pb(C_2H_3O_2)_2$$

1 mol H_2SO_4:1 mol $Pb(C_2H_3O_2)_2$, so $Pb(C_2H_3O_2)_2$ is the limiting reactant.

0 mol $Pb(C_2H_3O_2)_2$, $(0.07646 - 0.023056) = 0.0534$ mol H_2SO_4, 0.0231 mol $PbSO_4$,

$(0.023056 \times 2) = 0.0461$ mol $HC_2H_3O_2$ are present after reaction

0.053405 mol $H_2SO_4 \times 98.09$ g/mol $= 5.2385 = 5.24$ g H_2SO_4

0.023056 mol $PbSO_4 \times 303.3$ g/mol $= 6.9928 = 6.99$ g $PbSO_4$

0.046111 mol $HC_2H_3O_2 \times 60.05$ g/mol $= 2.7690 = 2.77$ g $HC_2H_3O_2$

Check. The initial mass of reactants was 15.00 g; and the final mass of excess reactant and products is 15.00 g; mass is conserved.

3.77   *Analyze.* Given: amounts of two reactants. Find: theoretical yield.

Plan. Determine the limiting reactant and the maximum amount of product it could produce. Then calculate % yield.   *Solve.*

(a) $30.0 \text{ g } C_6H_6 \times \dfrac{1 \text{ mol } C_6H_6}{78.11 \text{ g } C_6H_6} = 0.3841 = 0.384 \text{ mol } C_6H_6$

       $65.0 \text{ g } Br_2 \times \dfrac{1 \text{ mol } Br_2}{159.8 \text{ g } Br_2} = 0.4068 = 0.407 \text{ mol } Br_2$

Since C_6H_6 and Br_2 react in a 1:1 mole ratio, C_6H_6 is the limiting reactant and determines the theoretical yield.

$$0.3841 \text{ mol } C_6H_6 \times \frac{1 \text{ mol } C_6H_5Br}{1 \text{ mol } C_6H_6} \times \frac{157.0 \text{ g } C_6H_5Br}{1 \text{ mol } C_6H_5Br} = 60.30 = 60.3 \text{ g } C_6H_5Br$$

Check. $30/78 \sim 3/8$ mol C_6H_6. $65/160 \sim 3/8$ mol Br_2. Since moles of the two reactants are similar, a precise calculation is needed to determine the limiting reactant. $3/8 \times 160 \approx 60$ g product

(b) $\% \text{ yield} = \dfrac{56.7 \text{ g } C_6H_5Br \text{ actual}}{60.3 \text{ g } C_6H_5Br \text{ theoretical}} \times 100 = 94.0\%$

3 Stoichiometry

Solutions to Exercises

3.78   (a) $C_2H_6 + Cl_2 \rightarrow C_2H_5Cl + HCl$

$$125 \text{ g C}_2\text{H}_6 \times \frac{1 \text{ mol C}_2\text{H}_6}{30.07 \text{ g C}_2\text{H}_6} = 4.157 = 4.16 \text{ mol C}_2\text{H}_6$$

$$255 \text{ g Cl}_2 \times \frac{1 \text{ mol Cl}_2}{70.91 \text{ g Cl}_2} = 3.596 = 3.60 \text{ mol Cl}_2$$

Since the reactants combine in a 1:1 mole ratio, Cl_2 is the limiting reactant. The theoretical yield is:

$$3.596 \text{ mol Cl}_2 \times \frac{1 \text{ mol C}_2\text{H}_5\text{Cl}}{1 \text{ mol Cl}_2} \times \frac{64.51 \text{ g C}_2\text{H}_5\text{Cl}}{1 \text{ mol C}_2\text{H}_5\text{Cl}} = 231.98 = 232 \text{ g C}_2\text{H}_5\text{Cl}$$

(b) $\% \text{ yield} = \dfrac{206 \text{ g C}_2\text{H}_5\text{Cl actual}}{232 \text{ g C}_2\text{H}_5\text{Cl theoretical}} \times 100 = 88.8\%$

3.79   *Analyze.* Given: g of two reactants, % yield. Find: g Li_3N.

Plan. Determine limiting reactant and theoretical yield. Use definition of % yield to calculate actual yield.   *Solve.*

$$5.00 \text{ g Li} \times \frac{1 \text{ mol Li}}{6.941 \text{ g Li}} = 0.7204 = 0.720 \text{ mol Li}$$

$$5.00 \text{ g N}_2 \times \frac{1 \text{ mol N}_2}{28.01 \text{ g N}_2} = 0.1785 = 0.179 \text{ mol N}_2$$

$$0.1785 \text{ mol N}_2 \times \frac{6 \text{ mol Li}}{1 \text{ mol N}_2} = 1.071 = 1.07 \text{ mol Li required}$$

Since there is less than enough Li to react exactly with 0.179 mol N_2, Li is the limiting reactant.

$$0.7204 \text{ mol Li} \times \frac{2 \text{ mol Li}_3\text{N}}{6 \text{ mol Li}} \times \frac{34.83 \text{ g Li}_3\text{N}}{1 \text{ mol Li}_3\text{N}} = 8.363 = 8.36 \text{ g Li}_3\text{N theoretical yield}$$

Check. 5/7 ≈ mol Li; 5/(4 × 7) ≈ mol N_2. There are 1/4 as many mol N_2 as moles Li, but only 1/6 as many moles N_2 are required for exact reaction. N_2 is in excess and Li limits. 0.7 × (36/3) ≈ 8.4 g Li_3N theoretical

$$\% \text{ yield} = \frac{\text{actual}}{\text{theoretical}} \times 100; \quad \frac{\% \text{ yield} \times \text{theoretical}}{100} = \text{actual yield}$$

$$\frac{88.5\%}{100} \times 8.363 \text{ g Li}_3\text{N} = 7.4013 = 7.40 \text{ g Li}_3\text{N actual}$$

3.80 $H_2S(g) + 2NaOH(aq) \rightarrow Na_2S(aq) + 2H_2O(l)$

$$1.50 \text{ g H}_2\text{S} \times \frac{1 \text{ mol H}_2\text{S}}{34.08 \text{ g H}_2\text{S}} = 0.04401 = 0.0440 \text{ mol H}_2\text{S}$$

$$2.00 \text{ g NaOH} \times \frac{1 \text{ mol NaOH}}{40.00 \text{ g NaOH}} = 0.0500 \text{ mol NaOH}$$

By inspection, twice as many mol NaOH as H_2S are needed for exact reaction, but mol NaOH given is less than twice mol H_2S, so NaOH limits.

$$0.0500 \text{ mol NaOH} \times \frac{1 \text{ mol Na}_2\text{S}}{2 \text{ mol NaOH}} \times \frac{78.05 \text{ g Na}_2\text{S}}{1 \text{ mol Na}_2\text{S}} = 1.95125 = 1.95 \text{ g Na}_2\text{S theoretical}$$

$$\frac{92.0\%}{100} \times 1.95125 \text{ g Na}_2\text{S theoretical} = 1.7951 = 1.80 \text{ g Na}_2\text{S actual}$$

Additional Exercises

3.81   (a) $C_4H_8O_2(l) + 5O_2(g) \rightarrow 4CO_2(g) + 4H_2O(l)$

      (b) $Ni(OH)_2(s) \rightarrow NiO(s) + H_2O(g)$

      (c) $Zn(s) + Cl_2(g) \rightarrow ZnCl_2(s)$

3.82   The formulas of the fertilizers are NH_3, NH_4NO_3, $(NH_4)_2SO_4$ and $(NH_2)_2CO$. Qualitatively, the more heavy, non-nitrogen atoms in a molecule, the smaller the mass % of N. By inspection, the mass of NH_3 is dominated by N, so it will have the greatest % N, $(NH_4)_2SO_4$ will have the least. In order of increasing % N:

$(NH_4)_2SO_4 < NH_4NO_3 < (NH_2)_2CO < NH_3$.

Check by calculation:

$(NH_4)_2SO_4$: FW = 2(14.0) + 8(1.0) + 1(32.1) + 4(16.0) = 132.1 amu

% N = [2(14.0)/132.1] × 100 = 21.2%

NH_4NO_3: FW = 2(14.0) + 4(1.0) + 3(16.0) = 80.0 amu

% N = [2(14.0)/80.0] × 100 = 35.0%

$(NH_2)_2CO$: FW = 2(14.0) + 4(1.0) + 1(12.0) + 1(16.0) = 60.0 amu

% N = [2(14.0)/60.0] × 100 = 46.7% N

NH_3: FW = 1(14.0) + 3(1.0) = 17.0

% N = [14.0/17.0] × 100 = 82.4 % N

3.83   (a) $1.25 \text{ carat} \times \dfrac{0.200 \text{ g}}{1 \text{ carat}} \times \dfrac{1 \text{ mol C}}{12.01 \text{ g C}} = 0.020816 = 0.0208 \text{ mol C}$

$$0.020816 \text{ mol C} \times \frac{6.022 \times 10^{23} \text{ C atoms}}{1 \text{ mol C}} = 1.25 \times 10^{22} \text{ C atoms}$$

      (b) $0.500 \text{ g C}_9\text{H}_8\text{O}_4 \times \dfrac{1 \text{ mol C}_9\text{H}_8\text{O}_4}{180.2 \text{ g C}_9\text{H}_8\text{O}_4} = 2.7747 \times 10^{-3} = 2.77 \times 10^{-3} \text{ mol HC}_9\text{H}_7\text{O}_4$

$$0.0027747 \text{ mol C}_9\text{H}_8\text{O}_4 \times \frac{6.022 \times 10^{23} \text{ molecules}}{1 \text{ mol}} = 1.67 \times 10^{21} \text{ HC}_9\text{H}_7\text{O}_4 \text{ molecules}$$

3.84   (a) $\dfrac{5.342 \times 10^{-21} \text{ g}}{1 \text{ molecule penicillin G}} \times \dfrac{6.0221 \times 10^{23} \text{ molecules}}{1 \text{ mol}} = 3217 \text{ g/mol penicillin G}$

      (b)   1.00 g hemoglobin (hem) contains 3.40×10^{-3} g Fe.

$$\frac{1.00 \text{ g hem}}{3.40 \times 10^{-3} \text{ g Fe}} \times \frac{55.85 \text{ g Fe}}{1 \text{ mol Fe}} \times \frac{4 \text{ mol Fe}}{1 \text{ mol hem}} = 6.57 \times 10^4 \text{ g/mol hemoglobin}$$

3.85   (a) 1.000×10^4 Si atoms $\times \dfrac{1\,\text{mol}}{6.022 \times 10^{23}\,\text{atoms}} \times \dfrac{28.0855\,\text{g Si}}{1\,\text{mol Si}} = 4.6638 \times 10^{-19}\,\text{g Si}$

          (b) $4.6638 \times 10^{-19}\,\text{g Si} \times \dfrac{1\,\text{cm}^3\,\text{Si}}{2.3\,\text{g Si}} = 2.03 \times 10^{-19} = 2.0 \times 10^{-19}\,\text{cm}^3$

          (c) $V = l^3;\ l = (V)^{1/3} = (2.03 \times 10^{-19}\,\text{cm}^3)^{1/3} = 5.9 \times 10^{-7}\,\text{cm}\ (= 5.9\,\text{nm})$

3.86   *Plan.* Assume 100 g, calculate mole ratios, empirical formula, then molecular formula from molar mass.   *Solve.*

        $68.2\,\text{g C} \times \dfrac{1\,\text{mol C}}{12.01\,\text{g C}} = 5.68\,\text{mol C};\ 5.68/0.568 \approx 10$

        $6.86\,\text{g H} \times \dfrac{1\,\text{mol H}}{1.008\,\text{g H}} = 6.81\,\text{mol H};\ 6.81/0.568 \approx 12$

        $15.9\,\text{g N} \times \dfrac{1\,\text{mol N}}{14.01\,\text{g N}} = 1.13\,\text{mol N};\ 1.13/0.568 \approx 2$

        $9.08\,\text{g O} \times \dfrac{1\,\text{mol O}}{16.00\,\text{g O}} = 0.568\,\text{mol O};\ 0.568/0.568 = 1$

    The empirical formula is $C_{10}H_{12}N_2O$, FW = 176 amu (or g). Since the molar mass is 176, the empirical and molecular formula are the same, $C_{10}H_{12}N_2O$.

3.87   *Plan.* Assume 1.000 g and get mass O by subtraction.   *Solve.*

          (a) $0.7787\,\text{g C} \times \dfrac{1\,\text{mol C}}{12.01\,\text{g C}} = 0.06484\,\text{mol C}$

                $0.1176\,\text{g H} \times \dfrac{1\,\text{mol H}}{1.008\,\text{g H}} = 0.1167\,\text{mol H}$

                $0.1037\,\text{g O} \times \dfrac{1\,\text{mol C}}{16.00\,\text{g O}} = 0.006481\,\text{mol O}$

    Dividing through by the smallest of these values we obtain $C_{10}H_{18}O$.

          (b)   The formula weight of $C_{10}H_{18}O$ is 154. Thus, the empirical formula is also the molecular formula.

3.88   Since all the C in the vanillin must be present in the CO_2 produced, get g C from g CO_2.

        $2.43\,\text{g CO}_2 \times \dfrac{1\,\text{mol CO}_2}{44.01\,\text{g CO}_2} \times \dfrac{12.01\,\text{g C}}{1\,\text{mol C}} = 0.6631 = 0.663\,\text{g C}$

    Since all the H in vanillin must be present in the H_2O produced, get g H from g H_2O.

        $0.50\,\text{g H}_2\text{O} \times \dfrac{1\,\text{mol H}_2\text{O}}{18.02\,\text{g H}_2\text{O}} \times \dfrac{2\,\text{mol H}}{1\,\text{mol H}_2\text{O}} \times \dfrac{1.008\,\text{g H}}{1\,\text{mol H}} = 0.0559 = 0.056\,\text{g H}$

    Get g O by subtraction. (Since the analysis was performed by combustion, an unspecified amount of O_2 was a reactant, and thus not all the O in the CO_2 and H_2O produced came from vanillin.) 1.05 g vanillin – 0.663 g C – 0.056 g H = 0.331 g O

$$0.6631 \text{ g C} \times \frac{1 \text{ mol C}}{12.01 \text{ g C}} = 0.0552 \text{ mol C}; \ 0.0552 / 0.0207 = 2.67$$

$$0.0559 \text{ g H} \times \frac{1 \text{ mol H}}{1.008 \text{ g H}} = 0.0555 \text{ mol C}; \ 0.0555 / 0.0207 = 2.68$$

$$0.331 \text{ g O} \times \frac{1 \text{ mol O}}{16.00 \text{ g O}} = 0.0207 \text{ mol O}; \ 0.0207 / 0.0207 = 1.00$$

Multiplying the numbers above by **3** to obtain an integer ratio of moles, the empirical formula of vanillin is $C_8H_8O_3$.

3.89   *Plan.* Because different sample sizes were used to analyze the different elements, calculate mass % of each element in the sample.

    i.   Calculate mass % C from g CO_2.

    ii.   Calculate mass % Cl from AgCl.

    iii.   Get mass % H by subtraction.

    iv.   Calculate mole ratios and the empirical formulas.

   Solve.

    i. $3.52 \text{ g CO}_2 \times \dfrac{1 \text{ mol CO}_2}{44.01 \text{ g CO}_2} \times \dfrac{1 \text{ mol C}}{1 \text{ mol CO}_2} \times \dfrac{12.01 \text{ g C}}{1 \text{ mol C}} = 0.9606 = 0.961 \text{ g C}$

$$\frac{0.9606 \text{ g C}}{1.50 \text{ g sample}} \times 100 = 64.04 = 64.0\% \text{ C}$$

    ii. $1.27 \text{ g AgCl} \times \dfrac{1 \text{ mol AgCl}}{143.3 \text{ g AgCl}} \times \dfrac{1 \text{ mol Cl}}{1 \text{ mol AgCl}} \times \dfrac{35.45 \text{ g Cl}}{1 \text{ mol Cl}} = 0.3142 = 0.314 \text{ g Cl}$

$$\frac{0.3142 \text{ g Cl}}{1.00 \text{ g sample}} \times 100 = 31.42 = 31.4\% \text{ Cl}$$

    iii.   % H = 100.0 – (64.04% C + 31.42% Cl) = 4.54 = 4.5% H

    iv.   Assume 100 g sample.

$$64.04 \text{ g C} \times \frac{1 \text{ mol C}}{12.01 \text{ g C}} = 5.33 \text{ mol C}; \ \ 5.33 / 0.886 = 6.02$$

$$31.42 \text{ g Cl} \times \frac{1 \text{ mol Cl}}{35.45 \text{ g Cl}} = 0.886 \text{ mol Cl}; \ 0.886 / 0.886 = 1.00$$

$$4.54 \text{ g H} \times \frac{1 \text{ mol H}}{1.008 \text{ g H}} = 4.50 \text{ mol H}; \ 4.50 / 0.886 = 5.08$$

The empirical formula is probably C_6H_5Cl.

The subscript for H, 5.08, is relatively far from 5.00, but C_6H_5Cl makes chemical sense. More significant figures in the mass data are required for a more accurate mole ratio.

3.90   The mass percentage is determined by the relative number of atoms of the element times the atomic weight, divided by the total formula mass. Thus, the mass percent of bromine in $KBrO_x$ is given by $0.5292 = \dfrac{79.91}{39.10 + 79.91 + x(16.00)}$. Solving for x, we obtain x = 2.00. Thus, the formula is $KBrO_2$.

3.91   (a)   Let AW = the atomic weight of X.

According to the chemical reaction, moles XI_3 reacted = moles XCl_3 produced

$0.5000 \text{ g } XI_3 \times 1 \text{ mol } XI_3 / (AW + 380.71) \text{ g } XI_3$

$$= 0.2360 \text{ g } XCl_3 \times \frac{1 \text{ mol } XCl_3}{(AW + 106.36) \text{ g } XCl_3}$$

$0.5000 (AW + 106.36) = 0.2360 (AW + 380.71)$

$0.5000 \text{ AW} + 53.180 = 0.2360 \text{ AW} + 89.848$

$0.2640 \text{ AW} = 36.67; \text{ AW} = 138.9 \text{ g}$

    (b)   X is lanthanum, La, atomic number 57.

3.92 $C_2H_5OH(l) + 3O_2(g) \rightarrow 2CO_2(g) + 3H_2O(g)$

$C_3H_8(g) + 5O_2(g) \rightarrow 3CO_2(g) + 4H_2O(g)$

$CH_3CH_2COCH_3(l) + 11/2 \, O_2(g) \rightarrow 4CO_2(g) + 4H_2O(l)$

In a combustion reaction, all H in the fuel is transformed to H_2O in the products. The reactant with most mol H/mol fuel will produce the most H_2O. C_3H_8 and $CH_3CH_2COCH_3$ (C_4H_8O) both have 8 mol H/mol fuel, so 1.5 mol of either fuel will produce the same amount of H_2O. 1.5 mol C_2H_5OH will produce less H_2O.

3.93 $O_3(g) + 2NaI(aq) + H_2O(l) \rightarrow O_2(g) + I_2(s) + 2NaOH(aq)$

    (a) $3.8 \times 10^{-5} \text{ mol } O_3 \times \dfrac{2 \text{ mol NaI}}{1 \text{ mol } O_3} = 7.6 \times 10^{-5} \text{ mol NaI}$

    (b) $0.550 \text{ mg } O_3 \times \dfrac{1 \times 10^{-3} \text{ g}}{1 \text{ mg}} \times \dfrac{1 \text{ mol } O_3}{48.00 \text{ g } O_3} \times \dfrac{2 \text{ mol NaI}}{1 \text{ mol } O_3} \times \dfrac{149.9 \text{ g NaI}}{1 \text{ mol NaI}}$

$$= 3.4352 \times 10^{-3} = 3.44 \times 10^{-3} \text{ g NaI} = 3.44 \text{ mg NaI}$$

3.94 $2NaCl(aq) + 2H_2O(l) \rightarrow 2NaOH(aq) + H_2(g) + Cl_2(g)$

Calculate mol Cl_2 and relate to mol H_2, mol NaOH.

$1.5 \times 10^6 \text{ kg} \times \dfrac{1000 \text{ g}}{1 \text{ kg}} \times \dfrac{1 \text{ mol } Cl_2}{70.91 \text{ g } Cl_2} = 2.115 \times 10^7 = 2.1 \times 10^7 \text{ mol } Cl_2$

$2.115 \times 10^7 \text{ mol } Cl_2 \times \dfrac{1 \text{ mol } H_2}{1 \text{ mol } Cl_2} \times \dfrac{2.016 \text{ g } H_2}{1 \text{ mol } H_2} = 4.26 \times 10^7 \text{ g } H_2 = 4.3 \times 10^4 \text{ kg } H_2$

$4.3 \times 10^7 \text{ g} \times \dfrac{1 \text{ metric ton}}{1 \times 10^6 \text{ g (1 Mg)}} = 43 \text{ metric tons } H_2$

$$2.115 \times 10^7 \text{ mol Cl}_2 \times \frac{2 \text{ mol NaOH}}{1 \text{ mol Cl}_2} \times \frac{40.0 \text{ g NaOH}}{1 \text{ mol NaOH}} = 1.69 \times 10^9 = 1.7 \times 10^9 \text{ g NaOH}$$

1.7×10^9 g NaOH $= 1.7 \times 10^6$ kg NaOH $= 1.7 \times 10^3$ metric tons NaOH

3.95 $\quad$ $2C_{57}H_{110}O_6 + 163O_2 \rightarrow 114CO_2 + 110H_2O$

molar mass of fat $= 57(12.01) + 110(1.008) + 6(16.00) = 891.5$

$$1.0 \text{ kg fat} \times \frac{1000 \text{ g}}{1 \text{ kg}} \times \frac{1 \text{ mol fat}}{891.5 \text{ g fat}} \times \frac{110 \text{ mol H}_2\text{O}}{2 \text{ mol fat}} \times \frac{18.02 \text{ g H}_2\text{O}}{1 \text{ mol H}_2\text{O}} \times \frac{1 \text{ kg}}{1000 \text{ g}} = 1.1 \text{ kg H}_2\text{O}$$

3.96 $\quad$ (a) $\quad$ *Plan.* Calculate the total mass of C from g CO and g CO_2. Calculate the mass of H from g H_2O. Calculate mole ratios and the empirical formula. $\quad$ *Solve.*

$$0.467 \text{ g CO} \times \frac{1 \text{ mol CO}}{28.01 \text{ g CO}} \times \frac{1 \text{ mol C}}{1 \text{ mol CO}} \times 12.01 \text{ g C} = 0.200 \text{ g C}$$

$$0.733 \text{ g CO}_2 \times \frac{1 \text{ mol CO}_2}{44.01 \text{ g CO}_2} \times \frac{1 \text{ mol C}}{1 \text{ mol CO}_2} \times 12.01 \text{ g C} = 0.200 \text{ g C}$$

Total mass C is 0.200 g $+ 0.200$ g $= 0.400$ g C.

$$0.450 \text{ g H}_2\text{O} \times \frac{1 \text{ mol H}_2\text{O}}{18.02 \text{ g H}_2\text{O}} \times \frac{2 \text{ mol H}}{1 \text{ mol H}_2\text{O}} \times \frac{1.008 \text{ g H}}{1 \text{ mol H}} = 0.0503 \text{ g H}$$

(Since hydrocarbons contain only the elements C and H, g H can also be obtained by subtraction: 0.450 g sample – 0.400 g C = 0.050 g H.)

$$0.400 \text{ g C} \times \frac{1 \text{ mol C}}{12.01 \text{ g C}} = 0.0333 \text{ mol C}; \ 0.0333 / 0.0333 = 1.0$$

$$0.0503 \text{ g H} \times \frac{1 \text{ mol H}}{1.008 \text{ g H}} = 0.0499 \text{ mol H}; \ 0.0499 / 0.0333 = 1.5$$

Multiplying by a factor of 2, the empirical formula is C_2H_3.

(b) $\quad$ Mass is conserved. Total mass products – mass sample = mass O_2 consumed.

$\quad\quad$ 0.467 g CO + 0.733 g CO_2 + 0.450 g H_2O – 0.450 g sample = 1.200 g O_2 consumed

(c) $\quad$ For complete combustion, 0.467 g CO must be converted to CO_2.

$\quad\quad$ $2CO(g) + O_2(g) \rightarrow 2CO_2(g)$

$$0.467 \text{ g CO} \times \frac{1 \text{ mol CO}}{28.01 \text{ g C}} \times \frac{1 \text{ mol O}_2}{2 \text{ mol CO}} \times \frac{32.00 \text{ g O}_2}{1 \text{ mol O}_2} = 0.267 \text{ g O}_2$$

The total mass of O_2 required for complete combustion is
1.200 g + 0.267 g = 1.467 g O_2.

3.97 $\quad$ $N_2(g) + 3H_2(g) \rightarrow 2NH_3(g)$

Determine the moles of N_2 and H_2 required to form the 2.0 moles of NH_3 present after the reaction has stopped.

$$2.0 \text{ mol NH}_3 \times \frac{3 \text{ mol H}_2}{2 \text{ mol NH}_3} = 3.0 \text{ mol H}_2 \text{ reacted}$$

$$2.0 \, \text{mol NH}_3 \times \frac{1 \, \text{mol N}_2}{2 \, \text{mol NH}_3} = 1 \, \text{mol N}_2 \text{ reacted}$$

mol H_2 initial = 2.0 mol H_2 remain + 3.0 mol H_2 reacted = 5.0 mol H_2

mol N_2 initial = 2.0 mol N_2 remain + 1.0 mol N_2 reacted = 3.0 mol N_2

In tabular form:

	$N_2(g)$	+	$3H_2(g)$	$\rightarrow$	$2NH_3(g)$
initial	3.0 mol		5.0 mol		0 mol
reaction	–1.0 mol		–3.0 mol		+2.0 mol
final	2.0 mol		2.0 mol		2.0 mol

(Tables like this will be extremely useful for solving chemical equilibrium problems in Chapter 15.)

3.98   All of the O_2 is produced from $KClO_3$; get g $KClO_3$ from g O_2. All of the H_2O is produced from $KHCO_3$; get g $KHCO_3$ from g H_2O. The g H_2O produced also reveals the g CO_2 from the decomposition of $NaHCO_3$. The remaining CO_2 (13.2 g CO_2– g CO_2 from $NaHCO_3$) is due to K_2CO_3 and g K_2CO_3 can be derived from it.

$$4.00 \, \text{g O}_2 \times \frac{1 \, \text{mol O}_2}{32.00 \, \text{g O}_2} \times \frac{2 \, \text{mol KClO}_3}{3 \, \text{mol O}_2} \times \frac{122.6 \, \text{g KClO}_3}{1 \, \text{mol KClO}_3} = 10.22 = 10.2 \, \text{g KClO}_3$$

$$1.80 \, \text{g H}_2\text{O} \times \frac{1 \, \text{mol H}_2\text{O}}{18.02 \, \text{g H}_2\text{O}} \times \frac{2 \, \text{mol KHCO}_3}{1 \, \text{mol H}_2\text{O}} \times \frac{100.1 \, \text{g KHCO}_3}{1 \, \text{mol KHCO}_3} = 20.00 = 20.0 \, \text{g KHCO}_3$$

$$1.80 \, \text{g H}_2\text{O} \times \frac{1 \, \text{mol H}_2\text{O}}{18.02 \, \text{g H}_2\text{O}} \times \frac{2 \, \text{mol CO}_2}{1 \, \text{mol H}_2\text{O}} \times \frac{44.01 \, \text{g CO}_2}{1 \, \text{mol CO}_2} = 8.792 = 8.79 \, \text{g CO}_2 \text{ from KHCO}_3$$

13.20 g CO_2 total – 8.792 CO_2 from $KHCO_3$ = 4.408 = 4.41 g CO_2 from K_2CO_3

$$4.408 \, \text{g CO}_2 \times \frac{1 \, \text{mol CO}_2}{44.01 \, \text{g CO}_2} \times \frac{1 \, \text{mol K}_2\text{CO}_3}{1 \, \text{mol CO}_2} \times \frac{138.2 \, \text{g K}_2\text{CO}_3}{1 \, \text{mol K}_2\text{CO}_3} = 13.84 = 13.8 \, \text{g K}_2\text{CO}_3$$

100.0 g mixture – 10.22 g $KClO_3$ – 20.00 g $KHCO_3$ – 13.84 g K_2CO_3 = 56.0 g KCl

3.99   (a) $2C_2H_2(g) + 5O_2(g) \rightarrow 4CO_2(g) + 2H_2O(g)$

(b)   Following the approach in Sample Exercise 3.19,

$$10.0 \, \text{g C}_2\text{H}_2 \times \frac{1 \, \text{mol C}_2\text{H}_2}{26.04 \, \text{g C}_2\text{H}_2} \times \frac{5 \, \text{mol O}_2}{2 \, \text{mol C}_2\text{H}_2} \times \frac{32.00 \, \text{g O}_2}{1 \, \text{mol O}_2} = 30.7 \, \text{g O}_2 \text{ required}$$

Only 10.0 g O_2 are available, so O_2 limits.

(c)   Since O_2 limits, 0.0 g O_2 remain.

Next, calculate the g C_2H_2 consumed and the amounts of CO_2 and H_2O produced by reaction of 10.0 g O_2.

$$10.0 \, \text{g O}_2 \times \frac{1 \, \text{mol O}_2}{32.00 \, \text{g O}_2} \times \frac{2 \, \text{mol C}_2\text{H}_2}{5 \, \text{mol O}_2} \times \frac{26.04 \, \text{g C}_2\text{H}_2}{1 \, \text{mol C}_2\text{H}_2} = 3.26 \, \text{g C}_2\text{H}_2 \text{ consumed}$$

10.0 g C_2H_2 initial – 3.26 g consumed = 6.74 = 6.7 g C_2H_2 remain

$$10.0\,g\,O_2 \times \frac{1\,mol\,O_2}{32.00\,g\,O_2} \times \frac{4\,mol\,CO_2}{5\,mol\,O_2} \times \frac{44.01\,g\,CO_2}{1\,mol\,CO_2} = 11.0\,g\,CO_2 \text{ produced}$$

$$10.0\,g\,O_2 \times \frac{1\,mol\,O_2}{32.00\,g\,O_2} \times \frac{2\,mol\,H_2O}{5\,mol\,O_2} \times \frac{18.02\,g\,H_2O}{1\,mol\,H_2O} = 2.25\,g\,H_2O \text{ produced}$$

3.100　(a)　$1.5 \times 10^5\,g\,C_9H_8O_4 \times \dfrac{1\,mol\,C_9H_8O_4}{180.2\,g\,C_9H_8O_4} \times \dfrac{1\,mol\,C_7H_6O_3}{1\,mol\,C_9H_8O_4} \times \dfrac{138.1\,g\,C_7H_6O_3}{1\,mol\,C_7H_6O_3}$

$$= 1.1496 \times 10^5\,g = 1.1 \times 10^2\,kg\,C_7H_6O_3$$

(b)　If only 80 percent of the acid reacts, then we need $1/0.80 = 1.25$ times as much to obtain the same mass of product: $1.25 \times 1.15 \times 10^2\,kg = 1.4 \times 10^2\,kg\,C_7H_6O_3$

(c)　Calculate the number of moles of each reactant:

$$1.85 \times 10^5\,g\,C_7H_6O_3 \times \frac{1\,mol\,C_7H_6O_3}{138.1\,g\,C_7H_6O_3} = 1.340 \times 10^3 = 1.34 \times 10^3\,mol\,C_7H_6O_3$$

$$1.25 \times 10^5\,g\,C_4H_6O_3 \times \frac{1\,mol\,C_4H_6O_3}{102.1\,g\,C_4H_6O_3} = 1.224 \times 10^3 = 1.22 \times 10^3\,mol\,C_4H_6O_3$$

We see that $C_4H_6O_3$ limits, because equal numbers of moles of the two reactants are consumed in the reaction.

$$1.224 \times 10^3\,mol\,C_4H_6O_3 \times \frac{1\,mol\,C_9H_8O_4}{1\,mol\,C_7H_6O_3} \times \frac{180.2\,g\,C_9H_8O_4}{1\,mol\,C_9H_8O_4} = 2.206 \times 10^5$$

$$= 2.21 \times 10^5\,g\,C_9H_8O_4$$

(d)　$\text{percent yield} = \dfrac{1.82 \times 10^5\,g}{2.206 \times 10^5\,g} \times 100 = 82.5\%$

Integrative Exercises

3.101　*Plan.* Volume cube $\xrightarrow{\text{density}}$ mass $CaCO_3$ → moles $CaCO_3$ → moles O → O atoms

Solve. $(2.005)^3\,in^3 \times \dfrac{(2.54)^3\,cm^3}{1\,in^3} \times \dfrac{2.71\,g\,CaCO_3}{1\,cm^3} \times \dfrac{1\,mol\,CaCO_3}{100.1\,g\,CaCO_3} \times \dfrac{3\,mol\,O}{1\,mol\,CaCO_3}$

$$\times \frac{6.022 \times 10^{23}\,O\,atoms}{1\,mol\,O} = 6.46 \times 10^{24}\,O\,atoms$$

3.102　(a)　*Plan.* volume of Ag cube $\xrightarrow{\text{density}}$ mass of Ag → mol Ag → Ag atoms

Solve. $(1.000)^3\,cm^3\,Ag \times \dfrac{10.49\,g\,Ag}{1\,cm^3\,Ag} \times \dfrac{1\,mol\,Ag}{107.87\,g\,Ag} \times \dfrac{6.022 \times 10^{23}\,atoms}{1\,mol}$

$$= 5.8562 \times 10^{22} = 5.856 \times 10^{22}\,Ag\text{ atoms}$$

(b)　$1.000\,cm^3$ cube volume, 74% is occupied by Ag atoms

$0.7400\,cm^3$ = volume of 5.856×10^{22} Ag atoms

$$\frac{0.7400 \text{ cm}^3}{5.8562 \times 10^{22} \text{ Ag atoms}} = 1.2636 \times 10^{-23} = 1.264 \times 10^{-23} \text{ cm}^3 / \text{Ag atom}$$

Since atomic dimensions are usually given in Å, we will show this conversion.

$$1.264 \times 10^{-23} \text{ cm}^3 \times \frac{(1 \times 10^{-2})^3 \text{ m}^3}{1 \text{ cm}^3} \times \frac{1 \text{ Å}^3}{(1 \times 10^{-10})^3 \text{ m}^3} = 12.64 \text{ Å}^3 / \text{Ag atom}$$

(c) $V = 4/3 \,\pi\, r^3; \; r^3 = 3V/4\pi; \; r = (3V/4\pi)^{1/3}$

$r_A = (3 \times 12.636 \text{ Å}^3 / 4\pi)^{1/3} = 1.4449 = 1.445 \text{ Å}$

3.103   *Analyze.* Given: gasoline = C_8H_{18}, density = 0.69 g/mL, 20.5 mi/gal, 225 mi. Find: kg CO_2.

Plan. Write and balance the equation for the combustion of octane. Change mi → gal octane → mL → g octane. Use stoichiometry to calculate g and kg CO_2 from g octane.

Solve. $2C_8H_{18}(l) + 25O_2(g) \rightarrow 16CO_2(g) + 18H_2O(l)$

$$225 \text{ mi} \times \frac{1 \text{ gal}}{20.5 \text{ mi}} \times \frac{3.7854 \text{ L}}{1 \text{ gal}} \times \frac{1 \text{ mL}}{1 \times 10^{-3} \text{ L}} \times \frac{0.69 \text{ g octane}}{1 \text{ mL}} = 2.8667 \times 10^4 \text{ g}$$

$$= 29 \text{ kg octane}$$

$$2.8667 \times 10^4 \text{ g } C_8H_{18} \times \frac{1 \text{ mol } C_8H_{18}}{114.2 \text{ g } C_8H_{18}} \times \frac{16 \text{ mol } CO_2}{2 \text{ mol } C_8H_{18}} \times \frac{44.01 \text{ g } CO_2}{1 \text{ mol } CO_2} = 8.8382 \times 10^4 \text{ g}$$

$$= 88 \text{ kg } CO_2$$

Check. $\left(\dfrac{225 \times 4 \times 0.7}{20}\right) \times 10^3 = (45 \times 0.7) \times 10^3 = 30 \times 10^3 \text{ g} = 30 \text{ kg octane}$

$\dfrac{44}{114} \approx \dfrac{1}{3}; \; \dfrac{30 \text{ kg} \times 8}{3} \approx 80 \text{ kg } CO_2$

3.104   *Plan.* We can proceed by writing the ratio of masses of Ag to $AgNO_3$, where y is the atomic mass of nitrogen.   *Solve.*

$$\frac{Ag}{AgNO_3} = 0.634985 = \frac{107.8682}{107.8682 + 3(15.9994) + y}$$

Solve for y to obtain y = 14.0088. This is to be compared with the currently accepted value of 14.0067.

3.105   (a) $S(s) + O_2(g) \rightarrow SO_2(g); \; SO_2(g) + CaO(s) \rightarrow CaSO_3(s)$

(b) $\dfrac{2000 \text{ tons coal}}{\text{day}} \times \dfrac{2000 \text{ lb}}{1 \text{ ton}} \times \dfrac{1 \text{ kg}}{2.20 \text{ lb}} \times \dfrac{1000 \text{ g}}{1 \text{ kg}} \times \dfrac{0.025 \text{ g S}}{1 \text{ g coal}} \times \dfrac{1 \text{ mol S}}{32.1 \text{ g S}}$

$\times \dfrac{1 \text{ mol } SO_2}{1 \text{ mol S}} \times \dfrac{1 \text{ mol } CaSO_3}{1 \text{ mol } SO_2} \times \dfrac{120 \text{ g } CaSO_3}{1 \text{ mol } CaSO_3} \times \dfrac{1 \text{ kg } CaSO_3}{1000 \text{ g } CaSO_3}$

$$= 1.7 \times 10^5 \text{ kg } CaSO_3 / \text{day}$$

This corresponds to about 190 tons of $CaSO_3$ per day as a waste product.

3.106   *Analyze.* Given: 2.0 in × 3.0 in boards, 5000 boards, 0.65 mm thick Cu; 8.96 g/cm³ Cu; 85% Cu removed; 97% yield for reaction. Find: mass $Cu(NH_3)_4Cl_2$, mass NH_3.

Plan. vol Cu/board × density → mass Cu/board → 5000 boards × 85% = total Cu removed = actual yield; actual yield/0.97 = theoretical yield Cu.

mass Cu → mol Cu → mol Cu(NH$_3$)$_4$ Cl or NH$_3$ → desired masses.

Solve. $2.0 \, \text{in} \times 3.0 \, \text{in} \times \dfrac{(2.54)^2 \, \text{cm}^2}{\text{in}^2} \times 0.65 \, \text{mm} \times \dfrac{1 \, \text{cm}}{10 \, \text{mm}} = 2.516 = 2.5 \, \text{cm}^3 \, \text{Cu/board}$

$\dfrac{2.516 \, \text{cm}^3 \, \text{Cu}}{\text{board}} \times \dfrac{8.96 \, \text{g}}{\text{cm}^3} \times 5000 \, \text{boards} \times 0.85 \, \text{removed} = 95{,}814 \, \text{g} = 96 \, \text{kg Cu removed}$

$\dfrac{95{,}814 \, \text{g Cu actual yield}}{0.97} = 98{,}777 \, \text{g} = 99 \, \text{kg Cu theoretical}$

$98{,}777 \, \text{g Cu} \times \dfrac{1 \, \text{mol Cu}}{63.546 \, \text{g Cu}} \times \dfrac{1 \, \text{mol Cu(NH}_3)_4\text{Cl}_2}{1 \, \text{mol Cu}} \times \dfrac{202.575 \, \text{g}}{1 \, \text{mol Cu(NH}_3)_4\text{Cl}_2} = 314{,}887 \, \text{g}$

$$= 3.1 \times 10^2 \, \text{kg Cu(NH}_3)_4\text{Cl}_2$$

$98{,}777 \, \text{g Cu} \times \dfrac{1 \, \text{mol Cu}}{63.546 \, \text{g Cu}} \times \dfrac{4 \, \text{mol NH}_3}{1 \, \text{mol Cu}} \times \dfrac{17.03 \, \text{g NH}_3}{\text{mol NH}_3} = 105{,}891 \, \text{g} = 1.1 \times 10^2 \, \text{kg NH}_3$

3.107   (a)   *Plan.* Calculate the kg of air in the room and then the mass of HCN required to produce a dose of 300 mg HCN/kg air.   *Solve.*

$12 \, \text{ft} \times 15 \, \text{ft} \times 8.0 \, \text{ft} = 1440 = 1.4 \times 10^3 \, \text{ft}^3$ of air in the room

$1440 \, \text{ft}^3 \, \text{air} \times \dfrac{(12 \, \text{in})^3}{1 \, \text{ft}^3} \times \dfrac{(2.54 \, \text{cm})^3}{1 \, \text{in}^3} \times \dfrac{0.00118 \, \text{g air}}{1 \, \text{cm}^3 \, \text{air}} \times \dfrac{1 \, \text{kg}}{1000 \, \text{g}} = 48.12 = 48 \, \text{kg air}$

$48.12 \, \text{kg air} \times \dfrac{300 \, \text{mg HCN}}{1 \, \text{kg air}} \times \dfrac{1 \, \text{g}}{1000 \, \text{mg}} = 14.43 = 14 \, \text{g HCN}$

(b) $2\text{NaCN(s)} + \text{H}_2\text{SO}_4\text{(aq)} \rightarrow \text{Na}_2\text{SO}_4\text{(aq)} + 2\text{HCN(g)}$

The question can be restated as: What mass of NaCN is required to produce 14 g of HCN according to the above reaction?

$14.43 \, \text{g HCN} \times \dfrac{1 \, \text{mol HCN}}{27.03 \, \text{g HCN}} \times \dfrac{2 \, \text{mol NaCN}}{2 \, \text{mol HCN}} \times \dfrac{49.01 \, \text{g NaCN}}{1 \, \text{mol NaCN}} = 26.2 = 26 \, \text{g NaCN}$

(c) $12 \, \text{ft} \times 15 \, \text{ft} \times \dfrac{1 \, \text{yd}^2}{9 \, \text{ft}^2} \times \dfrac{30 \, \text{oz}}{1 \, \text{yd}^2} \times \dfrac{1 \, \text{lb}}{16 \, \text{oz}} \times \dfrac{454 \, \text{g}}{1 \, \text{lb}} = 17{,}025$

$$= 1.7 \times 10^4 \, \text{g acrilan in the room}$$

50% of the carpet burns, so the starting amount of CH_2CHCN is $0.50(17{,}025)$

$= 8{,}513 = 8.5 \times 10^3 \, \text{g}$

$8{,}513 \, \text{g CH}_2\text{CHCN} \times \dfrac{50.9 \, \text{g HCN}}{100 \, \text{g CH}_2\text{CHCH}} = 4333 = 4.3 \times 10^3 \, \text{g HCN possible}$

If the actual yield of combustion is 20%, actual g HCN = 4,333(0.20) = 866.6 $= 8.7 \times 10^2 \, \text{g HCN produced}$. From part (a), 14 g of HCN is a lethal dose. The fire produces much more than a lethal dose of HCN.

4 Aqueous Reactions and Solution Stoichiometry

Visualizing Concepts

4.1 *Analyze.* Correlate the formula of the solute with the charged spheres in the diagrams.

Plan. Determine the electrolyte properties of the solute and the relative number of cations, anions, or neutral molecules produced when the solute dissolves.

Solve. Li_2SO_4 is a strong electrolyte, a soluble ionic solid that dissociates into separate Li^+ and SO_4^{2-} when it dissolves in water. There are twice as many Li^+ cations as SO_4^{2-} anions. Diagram (c) represents the aqueous solution of a 2:1 electrolyte.

4.2 Although CH_3OH and HCl are both molecular compounds, HCl is an acid and strong electrolyte. Strong electrolytes exist in solution almost completely as ions, so an aqueous HCl solution conducts electricity. CH_3OH is a nonelectrolyte that exists as neutral molecules in aqueous solution. Since there are no charge carriers, aqueous solutions of nonelectrolytes such as CH_3OH do not conduct electricity.

4.3 *Analyze/Plan.* Correlate the neutral molecules, cations, and anions in the diagrams with the definitions of strong, weak, and nonelectrolytes. *Solve.*

(a) AX is a nonelectrolyte, because no ions form when the molecules dissolve.

(b) AY is a weak electrolyte because a few molecules ionize when they dissolve, but most do not.

(c) AZ is a strong electrolyte because all molecules break up into ions when they dissolve.

4.4 The brightness of the bulb in Figure 4.2 is related to the number of ions per unit volume of solution. If $0.1\ M\ HC_2H_3O_2$ has about the same brightness of $0.001\ M\ HBr$, the two solutions have about the same number of ions. Since $0.1\ M\ HC_2H_3O_2$ has 100 times more solute than $0.001\ M\ HBr$, HBr must be dissociated to a much greater extent than $HC_2H_3O_2$. HBr is one of the few molecular acids that is a strong electrolyte. $HC_2H_3O_2$ is a weak electrolyte; if it were a nonelectrolyte, the bulb in Figure 4.2 wouldn't glow.

4.5 *Analyze.* From the names and/or formulas of 3 substances determine their electrolyte and solubilities properties.

Plan. Determine whether the substance is molecular or ionic. If it is molecular, is it a weak acid or base and thus a weak electrolyte, or a nonelectrolyte? If it is ionic, is it soluble?

73

Solve. Glucose is a molecular compound that is neither a weak acid nor a weak base. It is a nonelectrolyte that dissolves to produce a nonconducting solution; it is solid C. NaOH is ionic; the anion is OH^-. According to Table 4.1, most hydroxides are insoluble, but NaOH is one of the soluble ones. NaOH is a strong electrolyte that dissolves to form a conducting solution; it is solid A. AgBr is ionic; the anion is Br^-. According to Table 4.1, most bromides are soluble, but AgBr is one of the insoluble ones; it is solid B.

Check. We know by elimination that AgBr is solid B, which we verified by solubility rules.

4.6   Certain pairs of ions form precipitates because their attraction is so strong that they cannot be surrounded and separated by solvent molecules. That is, the attraction between solute particles is greater than the stabilization offered by interaction of individual ions with solvent molecules.

4.7   *Analyze.* Given the formulas of some ions, determine whether these ions ever form precipitates in aqueous solution. *Plan.* Use Table 4.1 to determine if the given ions can form precipitates. If not, they will always be spectator ions.   *Solve.*

(a) Cl^- can form precipitates with Ag^+, Hg_2^{2+}, Pb^{2+}.

(b) NO_3^- never forms precipitates, so it is always a spectator.

(c) NH_4^+ never forms precipitates, so it is always a spectator.

(d) S^{2-} usually forms precipitates.

(e) SO_4^{2-} usually forms precipitates.

Check. NH_4^+ is a soluble exception for sulfides, phosphates, and carbonates, which usually form precipitates, so all rules indicate that it is a perpetual spectator.

4.8   Use the difference in reactivities with SO_4^{2-} to identify $Pb^{2+}(aq)$ and $Mg^{2+}(aq)$. Test a portion of each solution with $H_2SO_4(aq)$. $Pb^{2+}(aq)$ is an exception to the soluble sulfates rule, so $Pb(NO_3)_2(aq)$ will form a precipitate, while $Mg(NO_3)_2(aq)$ will not.

4.9   *Analyze.* Given three chemical reactions, match the reaction stoichiometry to the reactant and product particles shown in the diagram. *Plan.* Determine cation:anion stoichometrics of the reactants and products, as well as spectator ions and precipitate in the products.

Solve. Anions are shown as the larger spheres in each container. The first reactant is a 1:2 strong electrolyte. The second reactant is a 2:1 strong electrolyte. The spectator ions are the anion from the first reactant and the cation from the second reactant. The precipitate is a 1:1 solid formed by the cation from the first reactant and the anion from the second reactant. Since the diagram indicates that a precipitate is formed, we can eliminate reaction (b) straight away. The first reactant in reaction (c) is a 1:1 electrolyte, so it cannot be represented by the diagram. Reaction (a) is represented by the diagram.

Check. $BaCl_2$ is a 1:2 electrolyte, Na_2SO_4 is a 2:1 electrolyte, Cl^- (anion from first reactant), and Na^+ (cation from second reactant) are spectators; $BaSO_4$ is a 1:1 precipitate.

4.10   Diagram I shows spectator ions but no precipitate; this corresponds to reaction (b). Diagram II shows spectator ions and a 1:1 precipitate; this corresponds to reaction (c). Diagram III shows only precipitate; this corresponds to reaction (a). The second product in reaction (a) is $H_2O(l)$, which is also the solvent. Solvent molecules are not shown in any of the diagrams.

Electrolytes

4.11   No. Electrolyte solutions conduct electricity because the dissolved ions carry charge through the solution (from one electrode to the other).

4.12   When CH_3OH dissolves, neutral CH_3OH molecules are dispersed throughout the solution. These electrically neutral particles do not carry charge and the solution is nonconducting. When $HC_2H_3O_2$ dissolves, mostly neutral molecules are dispersed throughout the solution. A few of the dissolved molecules ionize to form $H^+(aq)$ and $C_2H_3O_2^-(aq)$. These few ions carry some charge and the solution is weakly conducting.

4.13   Although H_2O molecules are electrically neutral, there is an unequal distribution of electrons throughout the molecule. There are more electrons near O and fewer near H, giving the O end of the molecule a partial negative charge and the H end of the molecule a partial positive charge. Ionic compounds are composed of positively and negatively charged ions. The partially positive ends of H_2O molecules are attracted to the negative ions (anions) in the solid, while the partially negative ends are attracted to the positive ions (cations). Thus, both cations and anions in an ionic solid are surrounded and separated (dissolved) by H_2O molecules.

4.14   Ions are hydrated when they are surrounded by H_2O molecules in aqueous solution.

4.15   *Analyze/Plan.* Given the solute formula, determine the separate ions formed upon dissociation. *Solve.*

        (a) $ZnCl_2(aq) \rightarrow Zn^{2+}(aq) + 2Cl^-(aq)$

        (b) $HNO_3(aq) \rightarrow H^+(aq) + NO_3^-(aq)$

        (c) $(NH_4)_2SO_4(aq) \rightarrow 2NH_4^+(aq) + SO_4^{2-}(aq)$

        (d) $Ca(OH)_2(aq) \rightarrow Ca^{2+}(aq) + 2OH^-(aq)$

4.16   (a) $MgI_2(aq) \rightarrow Mg^{2+}(aq) + 2I^-(aq)$

        (b) $Al(NO_3)_3(aq) \rightarrow Al^{3+}(aq) + 3NO_3^-(aq)$

        (c) $HClO_4(aq) \rightarrow H^+(aq) + ClO_4^-(aq)$

        (d) $KC_2H_3O_2(aq) \rightarrow K^+(aq) + C_2H_3O_2^-(aq)$

4.17   *Analyze/Plan.* Apply the definition of a weak electrolyte to $HCHO_2$.

        Solve. When $HCHO_2$ dissolves in water, neutral $HCHO_2$ molecules, H^+ ions and CHO_2^- ions are all present in the solution. $HCHO_2(aq) \rightleftharpoons H^+(aq) + CHO_2^-(aq)$

4.18   (a)   acetone (nonelectrolyte): $CH_3COCH_3(aq)$ molecules only; hypochlorous acid (weak electrolyte): $HClO(aq)$ molecules, $H^+(aq)$, $ClO^-(aq)$; ammonium chloride (strong electrolyte): $NH_4^+(aq)$, $Cl^-(aq)$

(b) NH_4Cl, 0.2 mol solute particles; HClO, between 0.1 and 0.2 mol particles; CH_3OCH_3, 0.1 mol of solute particles

Precipitation Reactions and Net Ionic Equations

4.19   *Analyze.* Given: formula of compound. Find: solubility.

       Plan. Follow the guidelines in Table 4.1, in light of the anion present in the compound and notable exceptions to the "rules."   *Solve.*

       (a) $NiCl_2$: soluble

       (b) Ag_2S: insoluble

       (c) Cs_3PO_4: soluble (Cs^+ is an alkali metal cation)

       (d) $SrCO_3$: insoluble

       (e) $PbSO_4$: insoluble, Pb^{2+} is an exception to soluble sulfates

4.20   According to Table 4.1:

       (a) $Ni(OH)_2$: insoluble

       (b) $PbBr_2$: insoluble;

       (c) $Ba(NO_3)_2$: soluble

       (d) $AlPO_4$: insoluble

       (e) $AgC_2H_3O_2$: soluble

4.21   *Analyze.* Given: formulas of reactants. Find: balanced equation including precipitates.

       Plan. Follow the logic in Sample Exercise 4.3.

       Solve. In each reaction, the precipitate is in bold type.

       (a) $Na_2CO_3(aq) + 2AgNO_3(aq) \rightarrow \mathbf{Ag_2CO_3(s)} + 2NaNO_3(aq)$

       (b)   No precipitate (all nitrates and most sulfates are soluble).

       (c) $FeSO_4(aq) + Pb(NO_3)_2(aq) \rightarrow \mathbf{PbSO_4(s)} + Fe(NO_3)_2(aq)$

4.22   In each reaction, the precipitate is in bold type.

       (a) $Ni(NO_3)_2(aq) + 2NaOH(aq) \rightarrow \mathbf{Ni(OH)_2(s)} + 2NaNO_3(aq)$

       (b)   No precipitate, and, therefore, no reaction. There is no chemical change to any of the reactant ions.

       (c) $Na_2S(aq) + Cu(C_2H_3O_2)_2(aq) \rightarrow \mathbf{CuS(s)} + 2NaC_2H_3O_2(aq)$

4.23   *Analyze/Plan.* Follow the logic in Sample Exercise 4.4. From the complete ionic equation, identify the ions that don't change during the reaction; these are the spectator ions. *Solve.*

       (a) $2Na^+(aq) + CO_3^{2-}(aq) + Mg^{2+}(aq) + SO_4^{2-}(aq) \rightarrow MgCO_3(s) + 2Na^+(aq) + SO_4^{2-}(aq)$
           Spectators: Na^+, SO_4^{2-}

(b) $Pb^{2+}(aq) + 2NO_3^-(aq) + 2Na^+(aq) + S^{2-}(aq) \rightarrow PbS(s) + 2Na^+(aq) + 2NO_3^-(aq)$
Spectators: Na^+, NO_3^-

(c) $6NH_4^+(aq) + 2PO_4^{3-}(aq) + 3Ca^{2+}(aq) + 6Cl^-(aq) \rightarrow Ca_3(PO_4)_2(s) + 6NH_4^+(aq) +$
$6Cl^-(aq)$ Spectators: NH_4^+, Cl^-

4.24   Spectator ions are those that do not change during reaction.

(a) $2Cr^{3+}(aq) + 3CO_3^{2-}(aq) \rightarrow Cr_2(CO_3)_3(s)$; spectators: NH_4^+, SO_4^{2-}

(b) $Ba^{2+}(aq) + SO_4^{2-}(aq) \rightarrow BaSO_4(s)$; spectators: K^+, NO_3^-

(c) $Fe^{2+}(aq) + 2OH^-(aq) \rightarrow Fe(OH)_2(s)$; spectators: K^+, NO_3^-

4.25   *Analyze.* Given: reactions of unknown with HBr, H_2SO_4, NaOH. Find: The unknown contains a single salt. Is K^+ or Pb^{2+} or Ba^{2+} present?

Plan. Analyze solubility guidelines for Br^-, SO_4^{2-} and OH^- and select the cation that produces a precipitate with each of the anions.

Solve. K^+ forms no precipitates with any of the anions. $BaSO_4$ is insoluble, but $BaCl_2$ and $Ba(OH)_2$ are soluble. Since the unknown forms precipitates with all three anions, it must contain Pb^{2+}.

Check. $PbBr_2$, $PbSO_4$, and $Pb(OH)_2$ are all insoluble according to Table 4.1, so our process of elimination is confirmed by the insolubility of the Pb^{2+} compounds.

4.26 Br^- and NO_3^- can be ruled out because the Ba^{2+} salts are soluble. (Actually all NO_3^- salts are soluble.) CO_3^{2-} forms insoluble salts with the three cations given; it must be the anion in question.

4.27   *Analyze.* Given: three possible salts in an unknown solution react with $Ba(NO_3)_2$ and then NaCl. Find: Can the results identify the unknown salt? Do the three possible unknowns give distinctly different results with $Ba(NO_3)_2$ and NaCl?

Plan. Using Table 4.1, determine whether each of the possible unknowns will form a precipitate with $Ba(NO_3)_2$ and NaCl.

Solve.

Compound	$Ba(NO_3)_2$ result	NaCl result
$AgNO_3(aq)$	no ppt	AgCl ppt
$CaCl_2(aq)$	no ppt	no ppt
$Al_2(SO_4)_3$	$BaSO_4$ ppt	no ppt

This sequence of tests would definitively identify the contents of the bottle, because the results for each compound are unique.

4.28   (a) $Pb(C_2H_3O_2)_2(aq) + Na_2S(aq) \rightarrow PbS(s) + 2NaC_2H_3O_2(aq)$
net ionic: $Pb^{2+}(aq) + S^{2-}(aq) \rightarrow PbS$
$Pb(C_2H_3O_2)_2(aq) + CaCl_2(aq) \rightarrow (PbCl_2)s + Ca(C_2H_3PO)_2(aq)$
net ionic: $Pb^{2+}(aq) + 2Cl^-(aq) \rightarrow (PbCl_2)(s)$

$$Na_2S(aq) + CaCl_2(aq) \rightarrow CaS(aq) + 2NaCl(aq)$$

net ionic: no reaction

(b)   Spectator ions: $Na^+, Ca^{2+}, C_2H_3O_2^-$

Acid-Base Reactions

4.29   *Analyze.* Given: solute and concentration of three solutions. Find: solution with greatest concentration of solvated protons.

        Plan: See Sample Exercise 4.6. Determine whether solutes are strong or weak acids or bases, or nonelectrolytes. For solutions of equal concentration, strong acids will have greatest concentration of solvated protons. Take varying concentration into consideration when evaluating the same class of solutions.

        Solve. LiOH is a strong base, HI is a strong acid, CH_3OH is a molecular compound and nonelectrolyte. The strong acid HI will have the greatest concentration of solvated protons.

        Check. The solution concentrations weren't needed to answer the question.

4.30 $NH_3(aq)$ is a weak base, while KOH and $Ca(OH)_2$ are strong bases. $NH_3(aq)$ is only slightly ionized, so even 0.5 M NH_3 is less basic than 0.1 M KOH. $Ca(OH)_2$ has twice as many OH^- per moles as KOH, so 0.1 M $Ca(OH)_2$ is more basic than 0.1 M KOH. The most basic solution is 0.1 M $Ca(OH)_2$.

4.31   (a)   A *monoprotic acid* has one ionizable (acidic) H and a *diprotic acid* has two.

        (b)   A *strong acid* is completely ionized in aqueous solution, whereas only a fraction of *weak acid* molecules are ionized.

        (c)   An *acid* is an H^+ donor, a substance that increases the concentration of H^+ in aqueous solution. A *base* is an H^+ acceptor and thus increases the concentration of OH^- in aqueous solution.

4.32   (a) NH_3 produces OH^- in aqueous solution by reacting with H_2O (hydrolysis): $NH_3(aq) + H_2O(l) \, f \, NH_4^+(aq) + OH^-(aq)$. The OH^- causes the solution to be basic.

        (b)   The term "weak" refers to the tendency of HF to dissociate into H^+ and F^- in aqueous solution, not its reactivity toward other compounds.

        (c) H_2SO_4 is a **diprotic** acid; it has two ionizable hydrogens. The first hydrogen completely ionizes to form H^+ and HSO_4^-, but HSO_4^- only **partially** ionizes into H^+ and SO_4^{2-} (HSO_4^- is a weak electrolyte). Thus, an aqueous solution of H_2SO_4 contains a mixture of H^+, HSO_4^- and SO_4^{2-}, with the concentration of HSO_4^- greater than the concentration of SO_4^{2-}.

4.33   As a strong acid, $HClO_4$ exists in aqueous solution almost exclusively as $H^+(aq)$ and $ClO_4^-(aq)$; there are almost no neutral $HClO_4$ molecules. As a weak acid, $HClO_2$ exists as a mixture of $H^+(aq)$, $ClO_2^-(aq)$, and $HClO_2(aq)$. There are at least as many neutral $HClO_2$ molecules as anions or cations. For equal solution concentrations, $HClO_4$ will produce a bright light like Figure 4.2(c), while $HClO_2$ will produce a dim light like Figure 4.2(b).

4.34　All soluble ionic hydroxides from Table 4.1 are listed as strong bases in Table 4.2. Insoluble hydroxides like $Cd(OH)_2$ are not listed as strong bases. "Insoluble" means that less than 1% of the base molecules exist as separated ions and are dissolved. Thus, insoluble hydroxide salts produce too few $OH^-(aq)$ to be considered strong bases.

4.35　*Analyze.* Given: chemical formulas. Find: classify as acid, base, salt; strong, weak, or nonelectrolyte.

　Plan. See Table 4.3. Ionic or molecular? Ionic, soluble: OH^-, strong base and strong electrolyte; otherwise, salt, strong electrolyte. Molecular: NH_3, weak base and weak electrolyte; H-first, acid; strong acid (Table 4.2), strong electrolyte; otherwise weak acid and weak electrolyte.　*Solve.*

(a)　HF: acid, mixture of ions and molecules (weak electrolyte)

(b)　CH_3CN: none of the above, entirely molecules (nonelectrolyte)

(c)　$NaClO_4$: salt, entirely ions (strong electrolyte)

(d)　$Ba(OH)_2$: base, entirely ions (strong electrolyte)

4.36　Since the solution does conduct some electricity, but less than an equimolar NaCl solution (a strong electrolyte), the unknown solute must be a weak electrolyte. The weak electrolytes in the list of choices are NH_3 and H_3PO_3; since the solution is acidic, the unknown must be **H_3PO_3**.

4.37　*Analyze.* Given: chemical formulas. Find: electrolyte properties.

　Plan. In order to classify as electrolytes, formulas must be identified as acids, bases, or salts as in Solution 4.35.　*Solve.*

(a)　H_2SO_3: H first, so acid; not in Table 4.2, so weak acid; therefore, weak electrolyte

(b)　C_2H_5OH: not acid, not ionic (no metal cation), contains OH group, but not as anion so not a base; therefore, nonelectrolyte

(c)　NH_3: common weak base; therefore, weak electrolyte

(d)　$KClO_3$: ionic compound, so strong electrolyte

(e)　$Cu(NO_3)_2$: ionic compound, so strong electrolyte

4.38　(a)　$HClO_4$: strong　　(b)　HNO_3: strong　　(c)　NH_4Cl: strong

(d)　CH_3OCH_3: non　　(e)　$CoSO_4$: strong　　(f)　$C_{12}H_{22}O_{11}$: non

4.39　*Plan.* Follow Sample Exercise 4.7.　*Solve.*

(a)　$2HBr(aq) + Ca(OH)_2(aq) \rightarrow CaBr_2(aq) + 2H_2O(l)$

　$H^+(aq) + OH^-(aq) \rightarrow H_2O(l)$

(b)　$Cu(OH)_2(s) + 2HClO_4(aq) \rightarrow Cu(ClO_4)_2(aq) + 2H_2O(l)$

　$Cu(OH)_2(s) + 2H^+(aq) \rightarrow 2H_2O(l) + Cu^{2+}(aq)$

(c)　$Al(OH)_3(s) + 3HNO_3(aq) \rightarrow Al(NO_3)_3(aq) + 3H_2O(l)$

　$Al(OH)_3(s) + 3H^+(aq) \rightarrow 3H_2O(l) + Al^{3+}(aq)$

4.40 (a) $HC_2H_3O_2(aq) + KOH(aq) \rightarrow KC_2H_3O_2(aq) + H_2O(l)$

$HC_2H_3O_2(aq) + OH^-(aq) \rightarrow C_2H_3O_2^-(aq) \, H_2O(l)$

(b) $Cr(OH)_3(s) + 3HNO_3(aq) \rightarrow Cr(NO_3)_3(aq) + 3H_2O(l)$

$Cr(OH)_3(s) + 3H^+(aq) \rightarrow 3H_2O(l) + Cr^{3+}(aq)$

(c) $Ca(OH)_2(aq) + 2HClO(aq) \rightarrow Ca(ClO)_2(aq) + 2H_2O(l)$

$HClO(aq) + OH^-(aq) \rightarrow ClO^-(aq) + H_2O(l)$

4.41 *Analyze.* Given: names of reactants. Find: gaseous products.

Plan. Write correct chemical formulas for the reactants, complete and balance the metathesis reaction, and identify either H_2S or CO_2 products as gases.   *Solve.*

(a) $CdS(s) + H_2SO_4(aq) \rightarrow CdSO_4(aq) + H_2S(g)$

$CdS(s) + 2H^+(aq) \rightarrow H_2S(g) + Cd^{2+}(aq)$

(b) $MgCO_3(s) + 2HClO_4(aq) \rightarrow Mg(ClO_4)_2(aq) + H_2O(l) + CO_2(g)$

$MgCO_3(s) + 2H^+(aq) \rightarrow H_2O(l) + CO_2(g) + Mg^{2+}(aq)$

4.42 (a) $FeO(s) + 2H^+(aq) \rightarrow H_2O(l) + Fe^{2+}(aq)$

(b) $NiO(s) + 2H^+(aq) \rightarrow H_2O(l) + Ni^{2+}(aq)$

4.43 *Analyze.* Given the formulas or names of reactants, write balanced molecular and net ionic equations for the reactions.

Plan. Write correct chemical formulas for all reactants. Predict products of the neutralization reactions by exchanging ion partners. Balance the complete molecular equation, identify spectator ions by recognizing strong electrolytes, write the corresponding net ionic equation (omitting spectators).   *Solve.*

(a) $CaCO_3(s) + 2HNO_3(aq) \rightarrow Ca(NO_3)_2(aq) + H_2O(l) + CO_2(g)$

$2H^+(aq) + CaCO_3(s) \rightarrow H_2O(l) + CO_2(g) + Ca^{2+}(aq)$

(b) $FeS(s) + 2HBr(aq) \rightarrow FeBr_2(aq) + H_2S(g)$

$2H^+(aq) + FeS(s) \rightarrow H_2S(g) + Fe^{2+}(aq)$

4.44 $K_2O(aq) + H_2O(l) \rightarrow 2KOH(aq)$, molecular; $O^{2-}(aq) + H_2O(l) \rightarrow 2OH^-(aq)$, net ionic

base: (H^+ ion acceptor) $O^{2-}(aq)$; acid: (H^+ ion donor) $H_2O(aq)$; spectator: K^+

Oxidation-Reduction Reactions

4.45 (a) In terms of electron transfer, *oxidation* is the loss of electrons by a substance, and *reduction* is the gain of electrons (LEO says GER).

(b) Relative to oxidation numbers, when a substance is oxidized, its oxidation number increases. When a substance is reduced, its oxidation number decreases.

4.46   Oxidation and reduction can only occur together, not separately. When a metal reacts with oxygen, the metal atoms lose electrons and the oxygen atoms gain electrons. Free electrons do not exist under normal conditions. If electrons are lost by one substance they must be gained by another, and vice versa.

4.47   *Analyze.* Given the labeled periodic chart, determine which region is most readily oxidized and which is most readily reduced.

   Plan. Review the definition of oxidation and apply it to the properties of elements in the indicated regions of the chart.

   Solve. Oxidation is loss of electrons. Elements readily oxidized form positive ions; these are metals. Elements not readily oxidized tend to gain electrons and form negative ions; these are nonmetals. Elements in regions A, B, and C are metals, and their ease of oxidation is shown in Table 4.5. Metals in region A, Na, Mg, K, and Ca are most easily oxidized. Elements in region D are nonmetals and are least easily oxidized.

4.48   Elements (metals) from Table 4.5 in region A include Na, Mg, K, and Ca; those from region C are Hg and Au. Let's consider K and Au. Since metals from region A are more readily oxidized than those from region C, K will be oxidized to K^+ and Au^{3+} will be reduced to Au in the redox reaction. (Choose Au^{3+} because it is the Au ion shown in Table 4.5.)

   In a balanced redox reaction, the number of electrons lost must equal the number of electrons gained. Since K loses 1 electron in forming K^+, while Au^{3+} gains 3 electrons when forming Au, 3 K atoms must be oxidized for every 1 Au^{3+} that is reduced. This relationship dictates the coefficients in the balanced redox reaction.

   $$3K(s) + Au^{3+}(aq) \rightleftharpoons Au(s) + 3K^+(aq)$$

4.49   *Analyze.* Given the chemical formula of a substance, determine the oxidation number of a particular element in the substance.

   Plan. Follow the logic in Sample Exercise 4.8. *Solve.*

   (a)   +4   (b) +4   (c) +7   (d) +1   (e) 0   (f) –1 (O_2^{2-} is peroxide ion)

4.50   (a)   +4   (b) +2   (c) +3   (d) –2   (e) +3   (f) +6

4.51   *Analyze.* Given: chemical reaction. Find: element oxidized or reduced. *Plan.* Assign oxidation numbers to all species. The element whose oxidation number becomes more positive is oxidized; the one whose oxidation number decreases is reduced.   *Solve.*

   (a) $Ni \rightarrow Ni^{2+}$, Ni is oxidized; $Cl_2 \rightarrow 2Cl^-$, Cl is reduced

   (b) $Fe^{2+} \rightarrow Fe$, Fe is reduced; $Al \rightarrow Al^{3+}$, Al is oxidized

   (c) $Cl_2 \rightarrow 2Cl$, Cl is reduced; $2I^- \rightarrow I_2$, I is oxidized

   (d) $S^{2-} \rightarrow SO_4^{2-}$(S, +6), S is oxidized; H_2O_2 (O, –1) $\rightarrow H_2O$ (O, –2); O is reduced

4.52   (a)   acid-base reaction

   (b)   oxidation-reduction reaction; Fe is reduced, C is oxidized

(c)   precipitation reaction

(d)   oxidation-reduction reaction; Zn is oxidized, N is reduced

4.53   *Analyze.* Given: reactants. Find: balanced molecular and net ionic equations.

Plan. Metals oxidized by H^+ form cations. Predict products by exchanging cations and balance. The anions are the spectator ions and do not appear in the net ionic equations. *Solve.*

(a) $Mn(s) + H_2SO_4(aq) \rightarrow MnSO_4(aq) + H_2(g)$; $Mn(s) + 2H^+(aq) \rightarrow Mn^{2+}(aq) + H_2(g)$

Products with the metal in a higher oxidation state are possible, depending on reaction conditions and acid concentration.

(b) $2Cr(s) + 6HBr(aq) \rightarrow 2CrBr_3(aq) + 3H_2(g)$; $2Cr(s) + 6H^+(aq) \rightarrow 2Cr^{3+}(aq) + 3H_2(g)$

(c) $Sn(s) + 2HCl(aq) \rightarrow SnCl_2(aq) + H_2(g)$; $Sn(s) + 2H^+(aq) \rightarrow Sn^{2+}(aq) + H_2(g)$

(d) $2Al(s) + 6HCHO_2(aq) \rightarrow 2Al(CHO_2)_3(aq) + 3H_2(g)$;

$2Al(s) + 6HCHO_2(aq) \rightarrow 2Al^{3+}(aq) + 6CHO_2^{-}(aq) + 3H_2(g)$

4.54   (a) $2HCl(aq) + Ni(s) \rightarrow NiCl_2(aq) + H_2(g)$; $Ni(s) + 2H^+(aq) \rightarrow Ni^{2+}(aq) + H_2(g)$

(b) $H_2SO_4(aq) + Fe(s) \rightarrow FeSO_4(aq) + H_2(g)$; $Fe(s) + 2H^+(aq) \rightarrow Fe^{2+}(aq) + H_2(g)$

Products with the metal in a higher oxidation state are possible, depending on reaction conditions and acid concentration.

(c) $2HBr(aq) + Mg(s) \rightarrow MgBr_2(aq) + H_2(g)$; $Mg(s) + 2H^+(aq) \rightarrow Mg^{2+}(aq) + H_2(g)$

(d) $2HC_2H_3O_2(aq) + Zn(s) \rightarrow Zn(C_2H_3O_2)_2(aq) + H_2(g)$;

$Zn(s) + 2HC_2H_3O_2(aq) \rightarrow Zn^{2+}(aq) + 2C_2H_3O_2^{-}(aq) + H_2(g)$

4.55   *Analyze.* Given: a metal and an aqueous solution. Find: balanced equation.

Plan. Use Table 4.5. If the metal is above the aqueous solution, reaction will occur; if the aqueous solution is higher, NR. If reaction occurs, predict products by exchanging cations (a metal ion or H^+), then balance the equation.   *Solve.*

(a) $Fe(s) + Cu(NO_3)_2(aq) \rightarrow Fe(NO_3)_2(aq) + Cu(s)$

(b) $Zn(s) + MgSO_4(aq) \rightarrow NR$

(c) $Sn(s) + 2HBr(aq) \rightarrow SnBr_2(aq) + H_2(g)$

(d) $H_2(g) + NiCl_2(aq) \rightarrow NR$

(e) $2Al(s) + 3CoSO_4(aq) \rightarrow Al_2(SO_4)_3(aq) + 3Co(s)$

4.56   (a) $Mn(s) + NiCl_2(aq) \rightarrow MnCl_2(aq) + Ni(s)$

(b) $Cu(s) + Cr(C_2H_3O_2)(aq) \rightarrow NR$

(c) $2Cr(s) + 3NiSO_4(aq) \rightarrow Cr_2(SO_4)_3(aq) + 3Ni(s)$

(d) $Pt(s) + HBr(aq) \rightarrow NR$

(e) $H_2(g) + CuCl_2(aq) \rightarrow Cu(s) + 2HCl(aq)$

4.57   (a)   i. $Zn(s) + Cd^{2+}(aq) \rightarrow Cd(s) + Zn^{2+}(aq)$

                   ii. $Cd(s) + Ni^{2+}(aq) \rightarrow Ni(s) + Cd^{2+}(aq)$

         (b)   According to Table 4.5, the most active metals are most easily oxidized, and Zn is more active than Ni. Observation (i) indicates that Cd is less active than Zn; observation (ii) indicates that Cd is more active than Ni. Cd is between Zn and Ni on the activity series.

         (c)   Place an iron strip in $CdCl_2(aq)$. If Cd(s) is deposited, Cd is less active than Fe; if there is no reaction, Cd is more active than Fe. Do the same test with Co if Cd is less active than Fe or with Cr if Cd is more active than Fe.

4.58   (a) $Br_2 + 2NaI \rightarrow 2NaBr + I_2$ indicates that Br_2 is more easily reduced than I_2.

                     $Cl_2 + 2NaBr \rightarrow 2NaCl + Br_2$ shows that Cl_2 is more easily reduced than Br_2.

                     The order for ease of reduction is $Cl_2 > Br_2 > I_2$. Conversely, the order for ease of oxidation is $I^- > Br^- > Cl^-$.

         (b)   Since the halogens are nonmetals, they tend to form anions when they react chemically. Nonmetallic character decreases going down a family and so does the tendency to gain electrons during a chemical reaction. Thus, the ease of reduction of the halogen, X_2, decreases going down the family and the ease of oxidation of the halide, X^-, increases going down the family.

         (c) $Cl_2 + 2KI \rightarrow 2KCl + I_2$; $Br_2 + LiCl \rightarrow$ no reaction

Solution Composition; Molarity

4.59   (a)   *Concentration* is an *intensive* property; it is the **ratio** of the amount of solute present in a certain quantity of solvent or solution. This ratio remains constant regardless of how much solution is present.

         (b)   The term *0.50 mol HCl* defines an amount (~18 g) of the pure substance HCl. The term 0.50 *M* HCl is a ratio; it indicates that there are 0.50 mol of HCl solute in 1.0 liter of solution. This same ratio of moles solute to solution volume is present regardless of the volume of solution under consideration.

4.60   (a)   The concentration of the remaining solution is unchanged, assuming the original solution was thoroughly mixed. Molar concentration is a **ratio** of moles solute to liters solution. Although there are fewer moles solute remaining in the flask, there is also less solution volume, so the ratio of moles solute/solution volume remains the same.

         (b)   The second solution is five times as concentrated as the first. An equal volume of the more concentrated solution will contain five times as much solute (five times the number of moles and also five times the mass) as the 0.50 *M* solution. Thus, the mass of solute in the 2.50 *M* solution is $5 \times 4.5\,g = 22.5\,g$.

         Mathematically:

$$\frac{\dfrac{2.50\,\text{mol solute}}{1\,\text{L solution}}}{\dfrac{0.50\,\text{mol solute}}{1\,\text{L solution}}} = \frac{x\,\text{grams solute}}{4.5\,\text{g solute}}$$

$$\frac{2.50\,\text{mol solute}}{0.50\,\text{mol solute}} = \frac{x\,\text{g solute}}{4.5\,\text{g solute}};\ 5.0(4.5\,\text{g solute}) = 23\,\text{g solute}$$

The result has 2 sig figs; 22.5 rounds to 23 g solute.

4.61   *Analyze/Plan.* Follow the logic in Sample Exercises 4.11 and 4.13.   *Solve.*

(a) $M = \dfrac{\text{mol solute}}{\text{L solution}};\ \dfrac{0.0345\,\text{mol NH}_4\text{Cl}}{400\,\text{mL}} \times \dfrac{1000\,\text{mL}}{1\,\text{L}} = 0.0863\,M\,\text{NH}_4\text{Cl}$

Check. $(0.035 \times 0.4) \approx 0.09\,M$

(b) $\text{mol} = M \times \text{L};\ \dfrac{2.20\,\text{mol HNO}_3}{1\,\text{L}} \times 0.0350\,\text{L} = 0.0770\,\text{mol HNO}_3$

Check. $(2 \times 0.035) \approx 0.07\,\text{mol}$

(c) $\text{L} = \dfrac{\text{mol}}{M};\ \dfrac{0.125\,\text{mol KOH}}{1.50\,\text{mol KOH/L}} = 0.0833\,\text{L or }83.3\,\text{mL of }1.50\,M\,\text{KOH}$

Check. $(0.125/1.5)$ is greater than 0.06 and less than 0.12, $\approx 0.08\,M$.

4.62   (a) $M = \dfrac{\text{mol solute}}{\text{L solution}};\ \dfrac{0.145\,\text{mol Na}_2\text{SO}_4}{0.750\,\text{L}} = 0.193\,M\,\text{Na}_2\text{SO}_4$

(b) $\text{mol} = M \times \text{L};\ \dfrac{0.0850\,\text{mol KMnO}_4}{1\,\text{L}} \times 0.125\,\text{L} = 1.06 \times 10^{-2}\,\text{mol KMnO}_4$

(c) $\text{L} = \dfrac{\text{mol}}{M};\ \dfrac{0.255\,\text{mol HCl}}{11.6\,\text{mol HCl/L}} = 2.20 \times 10^{-2}\,\text{L or }22.0\,\text{mL}$

4.63   *Analyze.* Given molarity, M, and volume, L, find mass of Na^+(aq) in the blood.

Plan. Calculate moles Na^+(aq) using the definition of molarity: $M = \dfrac{\text{mol}}{\text{L}};\ \text{mol} = M \times \text{L}.$

Calculate mass Na^+(aq) using the definition moles: $\text{mol} = \text{g/MM/g} = \text{mol} \times \text{MM}.$ (MM is the symbol for molar mass in this manual.)

Solve. $\dfrac{0.135\,\text{mol}}{\text{L}} \times 5.0\,\text{L} \times \dfrac{23.0\,\text{g Na}^+}{\text{mol Na}^+} = 15.525 = 16\,\text{g Na}^+\text{(aq)}$

Check. Since there are more than 0.1 mol/L and we have 5.0 L, there should be more than half a mol (11.5 g) of Na^+. The calculation agrees with this estimate.

4.64   Calculate the mol of Na^+ at the two concentrations; the difference is the mol NaCl required to increase the Na^+ concentration to the desired level.

$\dfrac{0.118\,\text{mol}}{\text{L}} \times 4.6\,\text{L} = 0.5428 = 0.54\,\text{mol Na}^+$

$\dfrac{0.138\,\text{mol}}{\text{L}} \times 4.6\,\text{L} = 0.6348 = 0.63\,\text{mol Na}^+$

(0.6348 − 0.5428) = 0.092 = 0.09 mol NaCl (2 decimal places and 1 sig fig)

$$0.092 \text{ mol NaCl} \times \frac{58.5 \text{ g NaCl}}{\text{mol}} = 5.38 = 5 \text{ g NaCl}$$

4.65   *Plan.* Proceed as in Sample Exercises 4.11 and 4.13.

$$M = \frac{\text{mol}}{\text{L}}; \text{mol} = \frac{\text{g}}{M} \qquad \text{(MM is the symbol for molar mass in this manual.)}$$

Solve.

(a) $$\frac{0.150 \, M \text{ KBr}}{1 \text{ L}} \times 0.250 \text{ L} \times \frac{119.0 \text{ g KBr}}{1 \text{ mol KBr}} = 4.46 \text{ g KBr}$$

   Check. $(0.15 \times 120) \approx 18; 18 \times 0.25 = 18/4 \approx 4.5$ g KBr

(b) $$4.75 \text{ g Ca(NO}_3)_2 \times \frac{1 \text{ mol Ca(NO}_3)_2}{164.1 \text{ g Ca(NO}_3)_2} \times \frac{1}{0.200 \text{ L}} = 0.145 \, M \text{ Ca(NO}_3)_2$$

   Check. $(4.8/0.2) \approx 24; 24/160 = 3/20 \approx 0.15 \, M \text{ Ca(NO}_3)_2$

(c) $$5.00 \text{ g Na}_3\text{PO}_4 \times \frac{1 \text{ mol Na}_3\text{PO}_4}{163.9 \text{ g Na}_3\text{PO}_4} \times \frac{1 \text{ L}}{1.50 \text{ mol Na}_3\text{PO}_4} \times \frac{1000 \text{ mL}}{1 \text{ L}}$$
$$= 20.3 \text{ mL solution}$$

   Check. $[5/(160 \times 1.5)] \approx 5/240 \approx 1/50 \approx 0.02 \text{ L} = 20 \text{ mL}$

4.66 $$M = \frac{\text{mol}}{\text{L}}; \text{mol} = \frac{\text{g}}{\text{MM}} \qquad \text{(MM is the symbol for molar mass in this manual.)}$$

(a) $$\frac{0.360 \text{ mol K}_2\text{Cr}_2\text{O}_7}{1 \text{ L}} \times 50.0 \text{ mL} \times \frac{1 \text{ L}}{1000 \text{ mL}} \times \frac{294.2 \text{ g K}_2\text{Cr}_2\text{O}_7}{1 \text{ mol K}_2\text{Cr}_2\text{O}_7} = 5.30 \text{ g K}_2\text{Cr}_2\text{O}_7$$

(b) $$4.28 \text{ g (NH}_4)_2\text{SO}_4 \times \frac{1 \text{ mol (NH}_4)_2\text{SO}_4}{132.2 \text{ g (NH}_4)_2\text{SO}_4} \times \frac{1}{300. \text{ mL}} \times \frac{1000 \text{ mL}}{1 \text{ L}} = 0.108 \, M \text{ (NH}_4)_2\text{SO}_4$$

(c) $$2.25 \text{ g CuSO}_4 \times \frac{1 \text{ mol CuSO}_4}{159.6 \text{ g CuSO}_4} \times \frac{1 \text{ L}}{0.240 \text{ mol CuSO}_4} \times \frac{1000 \text{ mL}}{1 \text{ L}} = 58.7 \text{ mL solution}$$

4.67   *Analyze.* Given: formula and concentration of each solute. Find: concentration of K^+ in each solution. *Plan.* Note mol K^+/mol solute and compare concentrations or total moles. *Solve.*

(a) $KCl \rightarrow K^+ + Cl^-$; 0.20 M KCl = 0.20 M K^+

   $K_2\text{CrO}_4 \rightarrow \mathbf{2} K^+ + \text{CrO}_4{}^{2-}$; 0.15 M $K_2\text{CrO}_4$ = 0.30 M K^+

   $K_3\text{PO}_4 \rightarrow \mathbf{3} K^+ + \text{PO}_4{}^{3-}$; 0.080 M $K_3\text{PO}_4$ = 0.24 M K^+

   0.15 M $K_2\text{CrO}_4$ has the highest K^+ concentration.

(b) $K_2\text{CrO}_4$: 0.30 M K^+ × 0.0300 L = 0.0090 mol K^+

   $K_3\text{PO}_4$: 0.24 M K^+ × 0.0250 L = 0.0060 mol K^+

   30.0 mL of 0.15 M $K_2\text{CrO}_4$ has more K^+ ions.

4.68   (a) $0.1\ M\ CaCl_2 = 0.2\ M\ Cl^-$; $0.15\ M\ KCl = 0.15\ M\ Cl^-$

             $0.1\ M\ CaCl_2$ has the higher Cl^- concentration.

       (b) $0.1\ M\ KCl$ has a higher Cl^- concentration than $0.080\ M\ LiCl$. Total volume does not affect concentration.

       (c) $0.050\ M\ HCl = 0.050\ M\ Cl^-$; $0.020\ M\ CdCl_2 = 0.040\ M\ Cl^-$

             $0.050\ M\ HCl$ has the higher Cl^- concentration.

4.69   *Analyze.* Given: formula and concentration of each solute. Find: concentration of each species in solution. *Plan.* Decide whether the solute is a strong, weak, or nonelectrolyte, which species are in solution, and concentrations.   *Solve.*

       (a) $0.22\ M\ Na^+$, $0.22\ M\ OH^-$

       (b) $0.16\ M\ Ca^{2+}$, $0.32\ M\ Br^-$

       (c) $0.15\ M$ (CH_3OH is a molecular solute)

       (d)   Mixing two solutions is, in effect, a dilution, Equation [4.35].

             $M_2 = M_2V_1/V_2$, where V_2 is the total solution volume.

$$K^+:\ \frac{0.15\ M \times 0.040\ L}{0.075\ L} = 0.0800 = 0.080\ M$$

             ClO_3^-: concentration ClO_3^- = concentration K^+ = 0.080 M

$$SO_4^{2-}:\ \frac{0.22\ M \times 0.0350\ L}{0.075\ L} = 0.1027 = 0.10\ M\ SO_4^{2-}$$

             Na^+: concentration $Na^+ = 2 \times$ concentration $SO_4^{2-} = 0.21\ M$

4.70   (a) $$H^+:\ \frac{0.130\ M \times 16.0\ mL + 0.600\ M \times 12.0\ mL}{28.0\ mL} = 0.331\ M\ H^+$$

             Cl^-: concentration Cl^- = concentration H^+ = 0.331 M Cl^-

       (b) $$Na^+:\ \frac{2(0.200\ M \times 18.0\ mL)}{33.0\ mL} = 0.218\ M;\ K^+:\ \frac{0.150\ M \times 15.0\ mL}{33.0\ mL} = 0.0682\ M$$

             $$SO_4^{2-}:\ \frac{0.200\ M \times 18.0\ mL}{33.0\ mL} = 0.109\ M;\ Cl^-:\ \frac{0.150\ M \times 15.0\ mL}{33.0\ mL} = 0.0682\ M$$

       (c) $$Na^+:\ \frac{2.38\ g\ NaCl}{0.0500\ L} \times \frac{1\ mol}{58.44\ g} = 0.815\ M;\ Ca^{2+}:\ 0.400\ M$$

             Cl^-: 0.815 M (from NaCl(s)) + 0.800 M (from $CaCl_2$(aq)) = 1.615 M

4.71   *Analyze/Plan.* Follow the logic of Sample Exercise 4.14.

       Solve.

       (a) $$V_1 = M_2V_2/M_1;\ \frac{0.250\ M\ NH_3 \times 100.0\ mL}{14.8\ M\ NH_3} = 1.689 = 1.69\ mL\ 14.8\ M\ NH_3$$

       Check. $250/15 \approx 1.5\ mL$

(b) $M_2 = M_1V_1/V_2;$ $\dfrac{14.8\,M\,NH_3 \times 10.0\,mL}{250\,mL} = 0.592\,M\,NH_3$

Check. $150/250 \approx 0.60\,M$

4.72   (a) $V_1 = M_2V_2/M_1;$ $\dfrac{0.400\,M\,HNO_3 \times 0.350\,mL}{10.0\,M\,HNO_3} = 0.0140\,L = 14.0\,mL\ conc.\ HNO_3$

(b) $M_2 = M_1V_1/V_2;$ $\dfrac{10.0\,M\,HNO_3 \times 25.0\,mL}{500\,mL} = 0.500\,M\,HNO_3$

4.73   (a)   *Plan/Solve.* Follow the logic in Sample Exercise 4.13. The number of moles of sucrose needed is $\dfrac{0.150\,mol}{1\,L} \times 0.125\,L = 0.01875 = 0.0188\,mol$

Weigh out $0.01875\,mol\,C_{12}H_{22}O_{11} \times \dfrac{342.3\,g\,C_{12}H_{22}O_{11}}{1\,mol\,C_{12}H_{22}O_{11}} = 6.42\,g\,C_{12}H_{22}O_{11}$

Add this amount of solid to a 125 mL volumetric flask, dissolve in a small volume of water, and add water to the mark on the neck of the flask. Agitate thoroughly to ensure total mixing.

(b)   *Plan/Solve.* Follow the logic in Sample Exercise 4.14. Calculate the moles of solute present in the final 400.0 mL of $0.100\,M\,C_{12}H_{22}O_{11}$ solution:

moles $C_{12}H_{22}O_{11} = M \times L = \dfrac{0.100\,mol\,C_{12}H_{22}O_{11}}{1\,L} \times 0.4000\,L = 0.0400\,mol\,C_{12}H_{22}O_{11}$

Calculate the volume of 1.50 M glucose solution that would contain 0.04000 mol $C_{12}H_{22}O_{11}$:

$L = moles/M;\ 0.04000\,mol\,C_{12}H_{22}O_{11} \times \dfrac{1\,L}{1.50\,mol\,C_{12}H_{22}O_{11}} = 0.02667 = 0.0267\,L$

$0.02667\,L \times \dfrac{1000\,mL}{1\,L} = 26.7\,mL$

Thoroughly rinse, clean, and fill a 50 mL buret with the 1.50 $M\ C_{12}H_{22}O_{11}$. Dispense 26.7 mL of this solution into a 400 mL volumetric container, add water to the mark, and mix thoroughly. (26.7 mL is a difficult volume to measure with a pipette.)

4.74   (a)   The amount of $AgNO_3$ needed is: $0.150\,M \times 0.2500\,L = 0.0375\,mol\,AgNO_3$

$0.0375\,mol\,AgNO_3 \times \dfrac{169.88\,g\,AgNO_3}{1\,mol\,AgNO_3} = 6.3705 = 6.37\,g\,AgNO_3$

Add this amount of solid to a 250 mL volumetric flask, dissolve in a small amount of water, bring the total volume to exactly 250 mL, and agitate well.

(b)   Dilute the 6.0 $M\ HNO_3$ to prepare 100 mL of 0.50 $M\ HNO_3$. To determine the volume of 6.0 $M\ HNO_3$ needed, calculate the moles HNO_3 present in 100 mL of 0.50 $M\ HNO_3$ and then the volume of 6.0 M solution that contains this number of moles.

$0.100\,L \times 0.50\,M = 0.050\,mol\,HNO_3$ needed;

$$L = \frac{mol}{M}; \quad L\ 6.0\ M\ HNO_3 = \frac{0.050\ mol\ needed}{6.0\ M} = 0.00833\ L = 8.3\ mL$$

Thoroughly clean, rinse, and fill a buret with the 6.0 M HNO_3, taking precautions appropriate for working with a relatively concentrated acid. Dispense 8.3 mL of the 6.0 M acid into a 100 mL volumetric flask, add water to the mark, and mix thoroughly.

4.75   *Analyze.* Given: density of pure acetic acid, volume pure acetic acid, volume new solution. Find: molarity of new solution. *Plan.* Calculate the mass of acetic acid, $HC_2H_3O_2$, present in 20.0 mL of the pure liquid.   *Solve.*

$$20.00\ mL\ acetic\ acid \times \frac{1.049\ g\ acetic\ acid}{1\ mL\ acetic\ acid} = 20.98\ g\ acetic\ acid$$

$$20.98\ g\ HC_2H_3O_2 \times \frac{1\ mol\ HC_2H_3O_2}{60.05\ g\ HC_2H_3O_2} = 0.349375 = 0.3494\ mol\ HC_2H_3O_2$$

$$M = mol/L = \frac{0.349375\ mol\ HC_2H_3O_2}{0.2500\ L\ solution} = 1.39750 = 1.398\ M\ HC_2H_3O_2$$

Check. $(20 \times 1) \approx 20$ g acid; $(20/60) \approx 0.33$ mol acid; $(0.33/0.25 = 0.33 \times 4) \approx 1.33\ M$

4.76 $$50.000\ mL\ glycerol \times \frac{1.2656\ g\ glycerol}{1\ mL\ glycerol} = 63.280\ g\ glycerol$$

$$63.280\ g\ C_3H_8O_3 \times \frac{1\ mol\ C_3H_8O_3}{92.094\ g\ C_3H_8O_3} = 0.687124 = 0.68712\ mol\ C_3H_8O_3$$

$$M = \frac{0.687124\ mol\ C_3H_8O_3}{0.25000\ L\ solution} = 2.7485\ M\ C_3H_8O_3$$

Solution Stoichiometry; Titrations

4.77   *Analyze.* Given: volume and molarity $AgNO_3$. Find: mass NaCl.

Plan. $M \times L = $ mol $AgNO_3 = $ mol Ag^+; balanced equation gives ratio mol NaCl/mol $AgNO_3$; mol NaCl → g NaCl.   *Solve.*

$$\frac{0.100\ mol\ AgNO_3}{1\ L} \times 0.0200\ L = 2.00 \times 10^{-3}\ mol\ AgNO_3(aq)$$

$$AgNO_3(aq) + NaCl(aq) \rightarrow AgCl(s) + NaNO_3(aq)$$

$$mol\ NaCl = mol\ AgNO_3 = 2.00 \times 10^{-3}\ mol\ NaCl$$

$$2.00 \times 10^{-3}\ mol\ NaCl \times \frac{58.44\ g\ NaCl}{1\ mol\ NaCl} = 0.117\ g\ NaCl$$

Check. $(0.1 \times 0.02) = 0.002$ mol; $(0.002 \times 60) \approx 0.12$ g NaCl

4.78   *Plan.* $M \times L = $ mol $Cd(NO_3)_2$; balanced equation → mol ratio → mol NaOH → g NaOH

Solve. $\dfrac{0.500\ mol\ Cd(NO_3)_2}{1\ L} \times 0.02500\ L = 0.0125\ mol\ Cd(NO_3)_2$

$$Cd(NO_3)_2(aq) + 2NaOH(aq) \rightarrow Cd(OH)_2(s) + 2NaNO_3(aq)$$

$$0.0125 \text{ mol } Cd(NO_3)_2 \times \frac{2 \text{ mol NaOH}}{1 \text{ mol } Cd(NO_3)_2} \times \frac{40.00 \text{ g NaOH}}{1 \text{ mol NaOH}} = 1.00 \text{ g NaOH}$$

4.79   (a)   *Analyze.* Given: M and vol base, M acid. Find: vol acid

Plan/Solve. Write the balanced equation for the reaction in question:

$$HClO_4(aq) + NaOH(aq) \rightarrow NaClO_4(aq) + H_2O(l)$$

Calculate the moles of the known substance, in this case NaOH.

$$\text{moles NaOH} = M \times L = \frac{0.0875 \text{ mol NaOH}}{1 \text{ L}} \times 0.0500 \text{ L} = 0.004375 = 0.00438 \text{ mol NaOH}$$

Apply the mole ratio (mol unknown/mol known) from the chemical equation.

$$0.004375 \text{ mol NaOH} \times \frac{1 \text{ mol } HClO_4}{1 \text{ mol NaOH}} = 0.004375 \text{ mol } HClO_4$$

Calculate the desired quantity of unknown, in this case the volume of 0.115 M $HClO_4$ solution.

$$L = \text{mol}/M; \ L = 0.004375 \text{ mol } HClO_4 \times \frac{1 \text{ L}}{0.115 \text{ mol } HClO_4} = 0.0380 \text{ L} = 38.0 \text{ mL}$$

Check. $(0.09 \times 0.045) = 0.0045$ mol; $(0.0045/0.11) \approx 0.040$ L ≈ 40 mL

(b)   Following the logic outlined in part (a):

$$2HCl(aq) + Mg(OH)_2(s) \rightarrow MgCl_2(aq) + 2H_2O(l)$$

$$2.87 \text{ g } Mg(OH)_2 \times \frac{1 \text{ mol } Mg(OH)_2}{58.32 \text{ g } Mg(OH)_2} = 0.049211 = 0.0492 \text{ mol } Mg(OH)_2$$

$$0.0492 \text{ mol } Mg(OH)_2 \times \frac{2 \text{ mol HCl}}{1 \text{ mol } Mg(OH)_2} = 0.0984 \text{ mol HCl}$$

$$L = \text{mol}/M = 0.09840 \text{ mol HCl} \times \frac{1 \text{ L HCl}}{0.128 \text{ mol HCl}} = 0.769 \text{ L} = 769 \text{ mL}$$

(c) $$AgNO_3(aq) + KCl(aq) \rightarrow AgCl(s) + KNO_3(aq)$$

$$785 \text{ mg KCl} \times \frac{1 \times 10^{-3} \text{ g}}{1 \text{ mg}} \times \frac{1 \text{ mol KCl}}{74.55 \text{ g KCl}} \times \frac{1 \text{ mol } AgNO_3}{1 \text{ mol KCl}} = 0.01053 = 0.0105 \text{ mol } AgNO_3$$

$$M = \text{mol/L} = \frac{0.01053 \text{ mol } AgNO_3}{0.0258 \text{ L}} = 0.408 \ M \ AgNO_3$$

(d) $$HCl(aq) + KOH(aq) \rightarrow KCl(aq) + H_2O(l)$$

$$\frac{0.108 \text{ mol HCl}}{1 \text{ L}} \times 0.0453 \text{ L} \times \frac{1 \text{ mol KOH}}{1 \text{ mol HCl}} \times \frac{56.11 \text{ g KOH}}{1 \text{ mol KOH}} = 0.275 \text{ g KOH}$$

4.80   (a) $$2HCl(aq) + Ba(OH)_2(aq) \rightarrow BaCl_2(aq) + 2H_2O(l)$$

$$\frac{0.101 \text{ mol Ba(OH)}_2}{1 \text{ L Ba(OH)}_2} \times 0.0500 \text{ L Ba(OH)}_2 \times \frac{2 \text{ mol HCl}}{1 \text{ mol Ba(OH)}_2}$$

$$\times \frac{1 \text{ L HCl}}{0.120 \text{ mol HCl}} = 0.0842 \text{ L or } 84.2 \text{ mL HCl soln}$$

(b) $\quad H_2SO_4(aq) + 2NaOH(aq) \rightarrow Na_2SO_4(aq) + 2H_2O(l)$

$$0.200 \text{ g NaOH} \times \frac{1 \text{ mol NaOH}}{40.00 \text{ g NaOH}} \times \frac{1 \text{ mol H}_2SO_4}{2 \text{ mol NaOH}} \times \frac{1 \text{ L H}_2SO_4}{0.125 \text{ mol H}_2SO_4}$$

$$= 0.0200 \text{ L or } 20.0 \text{ mL H}_2SO_4 \text{ soln}$$

(c) $\quad BaCl_2(aq) + Na_2SO_4(aq) \rightarrow BaSO_4(s) + 2NaCl(aq)$

$$752 \text{ mg} = 0.752 \text{ g Na}_2SO_4 \times \frac{1 \text{ mol Na}_2SO_4}{142.1 \text{ g Na}_2SO_4} \times \frac{1 \text{ mol BaCl}_2}{1 \text{ mol Na}_2SO_4} \times \frac{1}{0.0558 \text{ L}}$$

$$= 0.0948 \text{ M BaCl}_2$$

(d) $\quad 2HCl(aq) + Ca(OH)_2(aq) \rightarrow CaCl_2(aq) + 2H_2O(l)$

$$0.0427 \text{ L HCl} \times \frac{0.208 \text{ mol HCl}}{1 \text{ L HCl}} \times \frac{1 \text{ mol Ca(OH)}_2}{2 \text{ mol HCl}} \times \frac{74.10 \text{ g Ca(OH)}_2}{1 \text{ mol Ca(OH)}_2}$$

$$= 0.329 \text{ g Ca(OH)}_2$$

4.81　*Analyze/Plan.* See Exercise 4.79(a) for a more detailed approach.　　*Solve.*

$$\frac{6.0 \text{ mol H}_2SO_4}{1 \text{ L}} \times 0.027 \text{ L} \times \frac{2 \text{ mol NaHCO}_3}{1 \text{ mol H}_2SO_4} \times \frac{84.01 \text{ g NaHCO}_3}{1 \text{ mol NaHCO}_3} = 27 \text{ g NaHCO}_3$$

4.82　See Exercise 4.79(a) for a more detailed approach.

$$\frac{0.115 \text{ mol NaOH}}{1 \text{ L}} \times 0.0425 \text{ L} \times \frac{1 \text{ mol HC}_2H_3O_2}{1 \text{ mol NaOH}} \times \frac{60.05 \text{ g HC}_2H_3O_2}{1 \text{ mol HC}_2H_3O_2}$$

$$= 0.29349 = 0.293 \text{ g HC}_2H_3O_2 \text{ in } 3.45 \text{ mL}$$

$$1.00 \text{ qt vinegar} \times \frac{1 \text{ L}}{1.057 \text{ qt}} \times \frac{1000 \text{ mL}}{1 \text{ L}} \times \frac{0.29349 \text{ g HC}_2H_3O_2}{3.45 \text{ mL vinegar}} = 80.5 \text{ g HC}_2H_3O_2/\text{qt}$$

4.83　*Analyze.* Given: M and vol HBr, vol Ca(OH)$_2$. Find: M Ca(OH)$_2$,

g Ca(OH)$_2$/100 mL soln

Plan. Write balanced equation;

$$\text{mol HBr} \xrightarrow[\text{ratio}]{\text{mol}} \text{mol Ca(OH)}_2 \rightarrow M \text{ Ca(OH)}_2 ; \rightarrow \text{g Ca(OH)}_2 / 100 \text{ mL}$$

Solve. The neutralization reaction here is:

$$2HBr(aq) + Ca(OH)_2(aq) \rightarrow CaBr_2(aq) + 2H_2O(l)$$

$$0.0488 \text{ L HBr soln} \times \frac{5.00 \times 10^{-2} \text{ mol HBr}}{1 \text{ L soln}} \times \frac{1 \text{ mol Ca(OH)}_2}{2 \text{ mol HBr}} \times \frac{1}{0.100 \text{ L of Ca(OH)}_2}$$

$$= 1.220 \times 10^{-2} = 1.22 \times 10^{-2} M \text{ Ca(OH)}_2$$

From the molarity of the saturated solution, we can calculate the gram solubility of Ca(OH)$_2$ in 100 mL of H$_2$O.

$$0.100\,\text{L soln} \times \frac{1.220 \times 10^{-2}\,\text{mol Ca(OH)}_2}{1\,\text{L soln}} \times \frac{74.10\,\text{g Ca(OH)}_2}{1\,\text{mol Ca(OH)}_2}$$

$$= 0.0904\,\text{g Ca(OH)}_2 \text{ in } 100\,\text{mL soln}$$

Check. $(0.05 \times 0.05/0.2) = 0.0125\,M$; $(0.1 \times 0.0125 \times 64) \approx 0.085\,\text{g}/100\,\text{mL}$

4.84 The balanced equation for the titration is:

$$Sr(NO_3)_2(aq) + Na_2CrO_4(aq) \rightarrow SrCrO_4(s) + 2NaNO_3(aq)$$

Beginning with a 0.100 L sample, we can do the following conversions:

volume soln $\rightarrow$ g $Sr(NO_3)_2 \rightarrow$ mol $Sr(NO_3)_2 \rightarrow$ mol $Na_2CrO_4 \rightarrow$ vol Na_2CrO_4 soln

$$0.100\,\text{L soln} \times \frac{6.82\,\text{g Sr(NO}_3)_2}{0.500\,\text{L soln}} \times \frac{1\,\text{mol Sr(NO}_3)_2}{211.6\,\text{g Sr(NO}_3)_2} \times \frac{1\,\text{mol Na}_2\text{CrO}_4}{1\,\text{mol Sr(NO}_3)_2}$$

$$\times \frac{1\,\text{L soln}}{0.0335\,\text{mol Na}_2\text{CrO}_4} = 0.192\,\text{L Na}_2\text{CrO}_4 \text{ soln}$$

4.85 (a) $NiSO_4(aq) + 2KOH(aq) \rightarrow Ni(OH)_2(s) + K_2SO_4(aq)$

(b) The precipitate is $Ni(OH)_2$.

(c) *Plan.* Compare mol of each reactant; mol = $M \times$ L

Solve. $0.200\,M\,\text{KOH} \times 0.1000\,\text{L KOH} = 0.0200\,\text{mol KOH}$

$0.150\,M\,\text{NiSO}_4 \times 0.2000\,\text{L KOH} = 0.0300\,\text{mol NiSO}_4$

1 mol $NiSO_4$ requires 2 mol KOH, so 0.0300 mol $NiSO_4$ requires 0.0600 mol KOH. Since only 0.0200 mol KOH is available, KOH is the limiting reactant.

(d) *Plan.* The amount of the limiting reactant (KOH) determines amount of product, in this case $Ni(OH)_2$.

Solve. $0.0200\,\text{mol KOH} \times \frac{1\,\text{mol Ni(OH)}_2}{2\,\text{mol KOH}} \times \frac{92.71\,\text{g Ni(OH)}_2}{1\,\text{mol Ni(OH)}_2} = 0.927\,\text{g Ni(OH)}_2$

(e) *Plan/Solve.* Limiting reactant: OH^-: no excess OH^- remains in solution.

Excess reactant: Ni^{2+}: $M\,Ni^{2+}$ remaining = mol Ni^{2+} remaining/L solution

0.0300 mol Ni^{2+} initial – 0.0100 mol Ni^{2+} reacted = 0.0200 mol Ni^{2+} remaining

0.0200 mol Ni^{2+}/0.3000 L = $0.0667\,M\,Ni^{2+}$(aq)

Spectators: SO_4^{2-}, K^+. These ions do not react, so the only change in their concentration is dilution. The final volume of the solution is 0.3000 L.

$M_2 = M_1V_1/V_2$: $0.200\,M\,K^+ \times 0.1000\,\text{L}/0.3000\,\text{L} = 0.0667\,M\,K^+$(aq)

$0.150\,M\,SO_4^{2-} \times 0.2000\,\text{L}/0.3000\,\text{L} = 0.100\,M\,SO_4^{2-}$(aq)

4.86 (a) $HNO_3(aq) + NaOH(s) \rightarrow NaNO_3(aq) + H_2O(l)$

(b) Determine the limiting reactant, then the identity and concentration of ions remaining in solution. Assume that the $H_2O(l)$ produced by the reaction does **not** increase the total solution volume.

$$12.0 \text{ g NaOH} \times \frac{1 \text{ mol NaOH}}{40.00 \text{ g NaOH}} = 0.300 \text{ mol NaOH}$$

$$0.200 \ M \text{ HNO}_3 \times 0.0750 \text{ L HNO}_3 = 0.0150 \text{ mol HNO}_3.$$

The mol ratio is 1:1, so HNO_3 is the limiting reactant. No excess H^+ remains in solution. The remaining ions are OH^- (excess reactant), Na^+, and NO_3^- (spectators).

OH^-:   0.300 mol OH^- initial – 0.0150 mol OH^- react = 0.285 mol OH^- remain

  0.285 mol OH^- /0.0750 L soln = 3.80 M OH^-(aq)

Na^+:   0.300 mol Na^+ /0.0750 L soln = 4.00 M Na^+(aq)

NO_3^-:   0.0150 mol NO_3^- /0.0750 L = 0.200 M NO_3^-(aq)

(c)   The resulting solution is **basic** because of the large excess of OH^-(aq).

4.87   *Analyze.* Given: mass impure $Mg(OH)_2$; M and vol **excess** HCl; M and vol NaOH.

Find: mass % $Mg(OH)_2$ in sample. *Plan/Solve.* Write balanced equations.

$Mg(OH)_2(s) + 2HCl(aq) \rightarrow MgCl_2(aq) + 2H_2O(l)$

$HCl(aq) + NaOH(aq) \rightarrow NaCl(aq) + H_2O(l)$

Calculate total moles HCl = M HCl × L HCl

$$\frac{0.2050 \text{ mol HCl}}{1 \text{ L soln}} \times 0.1000 \text{ L} = 0.02050 \text{ mol HCl total}$$

mol excess HCl = mol NaOH used = M NaOH × L NaOH

$$\frac{0.1020 \text{ mol NaOH}}{1 \text{ L soln}} \times 0.01985 \text{ L} = 0.0020247 = 0.002025 \text{ mol NaOH}$$

mol HCl reacted with $Mg(OH)_2$ = total mol HCl – excess mol HCl

0.02050 mol total – 0.0020247 mol excess = 0.0184753 = 0.01848 mol HCl reacted

(The result has 5 decimal places and 4 sig. figs.)

Use mol ratio to get mol $Mg(OH)_2$ in sample, then molar mass of $Mg(OH)_2$ to get g pure $Mg(OH)_2$.

$$0.0184753 \text{ mol HCl} \times \frac{1 \text{ mol Mg(OH)}_2}{2 \text{ mol HCl}} \times \frac{58.32 \text{ g Mg(OH)}_2}{1 \text{ mol Mg(OH)}_2} = 0.5387 \text{ Mg(OH)}_2$$

$$\text{mass \% Mg(OH)}_2 = \frac{\text{g Mg(OH)}_2}{\text{g sample}} \times 100 = \frac{0.5388 \text{ g Mg(OH)}_2}{0.5895 \text{ g sample}} \times 100 = 91.40\% \text{ Mg(OH)}_2$$

4.88   *Plan.* $CaCO_3(s) + 2HCl(aq) \rightarrow CaCl_2(aq) + H_2O(l) + CO_2(g)$

  $HCl(aq) + NaOH(aq) \rightarrow NaCl(aq) + H_2O(l)$

total mol HCl – excess mol HCl = mol HCl reacted; mol $CaCO_3$ = (mol HCl)/2;

g $CaCO_3$ = mol $CaCO_3$ × molar mass; mass % = (g $CaCO_3$ /g sample) × 100

Solve:

$$\frac{1.035 \text{ mol HCl}}{1 \text{ L soln}} \times 0.03000 \text{ L} = 0.031050 = 0.03105 \text{ mol HCl total}$$

$$\frac{1.010 \text{ mol NaOH}}{1 \text{ L soln}} \times 0.01156 \text{ L} = 0.011676 = 0.01168 \text{ mol HCl excess}$$

0.031050 total − 0.011676 excess = 0.019374 = 0.01937 mol HCl reacted

$$0.019374 \text{ mol HCl} \times \frac{1 \text{ mol CaCO}_3}{2 \text{ mol HCl}} \times \frac{100.09 \text{ g CaCO}_3}{1 \text{ mol CaCO}_3} = 0.96959 = 0.9696 \text{ g CaCO}_3$$

$$\text{mass \% CaCO}_3 = \frac{\text{g CaCO}_3}{\text{g rock}} \times 100 = \frac{0.96959}{1.248} \times 100 = 77.69\%$$

Additional Exercises

4.89   As soon as the dispersion formed, ion-pairing would begin, and eventually KBr(s) would form. In water, the solute ions stay separated as in solution because they are stabilized by electrostatic attractive forces with the polar water molecules. In a nonpolar solvent like mineral oil, there are no electrostatic attractive forces between the solute ions and the solvent. The oppositely charged ions are attracted to each other, forming KBr solid.

4.90   The precipitate is CdS(s). Na^+(aq) and NO_3^-(aq) are spectator ions and remain in solution. Any excess reactant ions also remain in solution. The net ionic equation is:

Cd^{2+}(aq) + S^{2-}(aq) → CdS(s).

4.91   The two precipitates formed are due to AgCl(s) and $SrSO_4$(s). Since no precipitate forms on addition of hydroxide ion to the remaining solution, the other two possibilities, Ni^{2+} and Mn^{2+}, are absent.

4.92   (a,b)   Expt. 1 No reaction

Expt. 2 $2Ag^+$(aq) + CrO_4^{2-}(aq) → Ag_2CrO_4(s) red precipitate

Expt. 3 No reaction

Expt. 4 $2Ag^+$(aq) + $C_2O_4^{2-}$(aq) → $Ag_2C_2O_4$(s) white precipitate

Expt. 5 Ca^{2+}(aq) + $C_2O_4^{2-}$(aq) → CaC_2O_4(s) white precipitate

Expt. 6 Ag^+(aq) + Cl^-(aq) → AgCl(s) white precipitate

(c)   The silver salts of both ions are insoluble, but many silver salts are insoluble (Expt. 6). The calcium salt of CrO_4^{2-} is soluble (Expt. 3), while the calcium salt of $C_2O_4^{2-}$(aq) is insoluble (Expt. 5). Thus, chromate salts appear more soluble than oxalate salts.

4.93   (a) $Al(OH)_3$(s) + $3H^+$(aq) → Al^{3+}(aq) + $3H_2O$(l)

(b) $Mg(OH)_2$(s) + $2H^+$(aq) → Mg^{2+}(aq) + $2H_2O$(l)

(c) $MgCO_3$(s) + $2H^+$(aq) → Mg^{2+}(aq) + H_2O(l) + CO_2(g)

(d) $NaAl(CO_3)(OH)_2(s) + 4H^+(aq) \rightarrow Na^+(aq) + Al^{3+}(aq) + 3H_2O(l) + CO_2(g)$

(e) $CaCO_3(s) + 2H^+(aq) \rightarrow Ca^{2+}(aq) + H_2O(l) + CO_2(g)$

[In (c), (d) and (e), one could also write the equation for formation of bicarbonate, e.g., $MgCO_3(s) + H^+(aq) \rightarrow Mg^{2+} + HCO_3^-(aq)$.]

4.94   (a) $2H^+(aq) + SO_3^{2-}(aq) \rightarrow H_2SO_3(aq)$; sulfurous acid

(b) $H_2SO_3(aq) \rightarrow H_2O(l) + SO_2(g)$; sulfur dioxide

(c)   The boiling point of $SO_2(g)$ is –10°C. It is a gas at room temperature (23°C) and pressure (1 atm).

(d)   (i) $Na_2SO_3(aq) + 2HCl(aq) \rightarrow 2NaCl(aq) + H_2O(l) + SO_2(g)$

$SO_3^{2-}(aq) + 2H^+(aq) \rightarrow H_2O(l) + SO_2(g)$

(ii) $Ag_2SO_3(s) + 2HCl(aq) \rightarrow 2AgCl(s) + H_2O(l) + SO_2(g)$

$Ag_2SO_3(s) + 2H^+(aq) + 2Cl^-(aq) \rightarrow 2AgCl(s) + H_2O(l) + SO_2(g)$

(iii) $KHSO_3(s) + HCl(aq) \rightarrow KCl(aq) + H_2O(l) + SO_2(g)$

$KHSO_3(s) + H^+(aq) \rightarrow K^+(aq) + H_2O(l) + SO_2(g)$

(iv) $ZnSO_3(aq) + 2HCl(aq) \rightarrow ZnCl_2(aq) + H_2O(l) + SO_2(g)$

$SO_3^{2-}(aq) + 2H^+(aq) \rightarrow H_2O(l) + SO_2(g)$

4.95 $4NH_3(g) + 5O_2(g) \rightarrow 4NO(g) + 6H_2O(g)$.
N = –3   O = 0   N = +2   O = –2

(a)   redox reaction   (b)   N is oxidized, O is reduced

$2NO(g) + O_2(g) \rightarrow 2NO_2(g)$.
N = +2   O = 0   N = +4, O = +2

(a)   redox reaction   (b)   N is oxidized, O is reduced

$3NO_2(g) + H_2O(l) \rightarrow HNO_3(aq) + NO(g)$.
N = +4   N = +5   N = +2

(a)   redox reaction

(b)   N is oxidized ($NO_2 \rightarrow HNO_3$), N is reduced ($NO_2 \rightarrow NO$). A reaction where the same element is both oxidized and reduced is called disproportionation.

4.96   A metal on Table 4.5 is able to displace the metal cations below it from their compounds. That is, zinc will reduce the cations below it to their metals.

(a) $Zn(s) + Na^+(aq) \rightarrow$ no reaction

(b) $Zn(s) + Pb^{2+}(aq) \rightarrow Zn^{2+}(aq) + Pb(s)$

(c) $Zn(s) + Mg^{2+}(aq) \rightarrow$ no reaction

(d) $Zn(s) + Fe^{2+}(aq) \rightarrow Zn^{2+}(aq) + Fe(s)$

(e) $Zn(s) + Cu^{2+}(aq) \rightarrow Zn^{2+}(aq) + Cu(s)$

(f) $Zn(s) + Al^{3+}(aq) \rightarrow$ no reaction

4.97   (a)   A : La_2O_3 Metals often react with the oxygen in air to produce metal oxides.

               B : $La(OH)_3$ When metals react with water (HOH) to form H_2, OH^- remains.

               C : $LaCl_3$ Most chlorides are soluble.

               D : $La_2(SO_4)_3$ Sulfuric acid provides SO_4^{2-} ions.

        (b) $4La(s) + 3O_2(g) \rightarrow 2La_2O_3(s)$

               $2La(s) + 6HOH(l) \rightarrow 2La(OH)_3(s) + 3H_2(g)$

               (There are no spectator ions in either of these reactions.)

               molecular: $La_2O_3(s) + 6HCl(aq) \rightarrow 2LaCl_3(aq) + 3H_2O(l)$

               net ionic: $La_2O_3(s) + 6H^+(aq) \rightarrow 2La^{3+}(aq) + 3H_2O(l)$

               molecular: $La(OH)_3(s) + 3HCl(aq) \rightarrow LaCl_3(aq) + 3H_2O(l)$

               net ionic: $La(OH)_3(s) + 3H^+(aq) \rightarrow La^{3+}(aq) + 3H_2O(l)$

               molecular: $2LaCl_3(aq) + 3H_2SO_4(aq) \rightarrow La_2(SO_4)_3(s) + 6HCl(aq)$

               net ionic: $2La^{3+}(aq) + 3SO_4^{2-}(aq) \rightarrow La_2(SO_4)_3(s)$

        (c)   La metal is oxidized by water to produce $H_2(g)$, so La is definitely above H on the activity series. In fact, since an acid is not required to oxidize La, it is probably one of the more active metals.

4.98   *Plan.* Calculate moles KBr from the two quantities of solution (mol = $M \times$ L), then new molarity (M = mol/L). KBr is nonvolatile, so no solute is lost when the solution is evaporated to reduce the total volume. *Solve.*

        1.00 M KBr × 0.0350 L = 0.0350 mol KBr; 0.600 M KBr × 0.060 L = 0.0360 mol KBr

        0.0350 mol KBr + 0.0360 mol KBr = 0.0710 mol KBr total

        $$\frac{0.0710 \text{ mol KBr}}{0.0500 \text{ L soln}} = 1.42 \, M \text{ KBr}$$

4.99   (a) $0.0400 \text{ L soln} \times \dfrac{0.160 \text{ mol NaCl}}{1 \text{ L soln}} = 6.40 \times 10^{-3} \text{ mol NaCl}$

               $0.0650 \text{ L soln} \times \dfrac{0.150 \text{ mol NaCl}}{1 \text{ L soln}} = 9.75 \times 10^{-3} \text{ mol NaCl}$

               Total moles NaCl = $1.615 \times 10^{-2} = 1.62 \times 10^{-2}$

               Total volume = 0.0400 L + 0.0650 L = 0.1050 L

               $\text{Molarity} = \dfrac{1.615 \times 10^{-2} \text{ mol}}{0.1050 \text{ L}} = 0.154 \, M$

        (b)   Both solutions are the same concentration, 0.750 M, so the resulting solution is 0.750 M.

4.100　(a)　$\dfrac{50\,\text{pg}}{1\,\text{mL}} \times \dfrac{1\times10^{-12}\,\text{g}}{1\,\text{pg}} \times \dfrac{1\times10^3\,\text{mL}}{\text{L}} \times \dfrac{1\,\text{mol Na}}{23.0\,\text{g Na}} = 2.17\times10^{-9} = 2.2\times10^{-9}\,M\,\text{Na}^+$

(b)　$\dfrac{2.17\times10^{-9}\,\text{mol Na}}{1\,\text{L soln}} \times \dfrac{1\,\text{L}}{1\times10^3\,\text{cm}^3} \times \dfrac{6.02\times10^{23}\,\text{Na atom}}{1\,\text{mol Na}}$

$$= 1.3\times10^{12}\,\text{atom or Na}^+\,\text{ions/cm}^3$$

4.101　Na^+ must replace the total positive (+) charge due to Ca^{2+} and Mg^{2+}. Think of this as moles of charge rather than moles of particles.

$$\dfrac{0.010\,\text{mol Ca}^{2+}}{1\,\text{L water}} \times 1.0\times10^3\,\text{L} \times \dfrac{2\,\text{mol + change}}{1\,\text{mol Ca}^{2+}} = 20\,\text{mol of + charge}$$

$$\dfrac{0.0050\,\text{mol Mg}^{2+}}{1\,\text{L water}} \times 1.0\times10^3\,\text{L} \times \dfrac{2\,\text{mol + charge}}{1\,\text{mol Mg}^{2+}} = 10\,\text{mol of + charge}$$

30 moles of + charge must be replaced; 30 mol Na^+ are needed.

4.102　$H_2C_4H_4O_6 + 2OH^-(aq) \rightarrow C_4H_4O_6{}^{2-}(aq) + 2H_2O(l)$

$$0.02262\,\text{L NaOH soln} \times \dfrac{0.2000\,\text{mol NaOH}}{1\,\text{L}} \times \dfrac{1\,\text{mol H}_2\text{C}_4\text{H}_4\text{O}_6}{2\,\text{mol NaOH}} \times \dfrac{1}{0.04000\,\text{L H}_2\text{C}_4\text{H}_4\text{O}_6}$$

$$= 0.05655\,M\,\text{H}_2\text{C}_4\text{H}_4\text{O}_6\,\text{soln}$$

4.103　*Plan.* $\text{mol MnO}_4^- = M \times L \rightarrow \text{mol ratio} \rightarrow \text{mol H}_2\text{O}_2 \rightarrow M\,\text{H}_2\text{O}_2$.　　*Solve.*

$2MnO_4^-(aq) + 5H_2O_2(aq) + 6H^+ \rightarrow 2Mn^{2+}(aq) + 5O_2(aq) + 8H_2O(l)$

$$\dfrac{0.124\,\text{mol MnO}_4^-}{\text{L}} \times 0.0168\,\text{L MnO}_4^- \times \dfrac{5\,\text{mol H}_2\text{O}_2}{2\,\text{mol MnO}_4^-} \times \dfrac{1}{0.0100\,\text{L H}_2\text{O}_2}$$

$$= 0.5208\,\text{mol H}_2\text{O}_2\,/\,\text{L} = 0.521\,M\,\text{H}_2\text{O}_2$$

4.104　mol OH^- from NaOH(aq) + mol OH^- from $\text{Zn(OH)}_2(s)$ = mol H^+ from HBr

$\text{mol H}^+ = M\,\text{HBr} \times \text{L HBr} = 0.500\,M\,\text{HBr} \times 0.400\,\text{L HBr} = 0.200\,\text{mol H}^+$

mol OH^- from NaOH $= M\,\text{NaOH} \times \text{L NaOH} = 0.500\,M\,\text{NaOH} \times 0.0985\,\text{L NaOH}$

$$= 0.04925 = 0.0493\,\text{mol OH}^-$$

mol OH^- from $\text{Zn(OH)}_2(s) = 0.200\,\text{mol H}^+ - 0.04925\,\text{mol OH}^-$ from NaOH $= 0.15075$

$$= 0.151\,\text{mol OH}^-\,\text{from Zn(OH)}_2$$

$$0.15075\,\text{mol OH}^- \times \dfrac{1\,\text{mol Zn(OH)}_2}{2\,\text{mol OH}^-} \times \dfrac{99.41\,\text{g Zn(OH)}_2}{1\,\text{mol Zn(OH)}_2} = 7.49\,\text{g Zn(OH)}_2$$

Integrative Exercises

4.105　*Analyze.* Given the name and mass of a solute and volume of solution, and the name of a second solute, calculate the mass of the second solute needed to make a solution of equal concentration and volume.

Plan. Using definitions of moles and molarity, find a simple ratio for calculating the mass of the second solute.

Solve. $M_1 = M_2$, so $\dfrac{\text{mol(1)}}{\text{L}} = \dfrac{\text{mol(2)}}{\text{L}}$

Since the volume is 1.00 L for both solutions, mol(1) = mol(2); $\dfrac{g(1)}{\text{MM (1)}} = \dfrac{g(2)}{\text{MM (2)}}$

(1) = sodium chlorate = $NaClO_3$, MM = 106.44

(2) = sodium chlorite = $NaClO_2$, MM = 90.44

$\dfrac{1.28\,\text{g}}{106.44} = \dfrac{x\,\text{g}}{90.44}$; $x = \dfrac{1.28(90.44)}{106.44}\,\text{g} = 1.09\,\text{g } NaClO_2$

Check. Since the molar mass of $NaClO_2$ is less than that of $NaClO_3$, a smaller mass of $NaClO_2$ will contain the same number of molecules.

4.106 (a) At the equivalence point of a titration, mol NaOH added = mol H^+ present

$$M_{NaOH} \times L_{NaOH} = \frac{\text{g acid}}{\text{MM acid}}\ \text{(for an acid with 1 acidic hydrogen)}$$

$$\text{MM acid} = \frac{\text{g acid}}{M_{NaOH} \times L_{NaOH}} = \frac{0.2053\,\text{g}}{0.1008\,M \times 0.0150\,\text{L}} = 136\,\text{g/mol}$$

(b) Assume 100 g of acid.

$$70.6\,\text{g C} \times \frac{1\,\text{mol C}}{12.01\,\text{g C}} = 5.88\,\text{mol C; } 5.88\,/\,1.47 \approx 4$$

$$5.89\,\text{g H} \times \frac{1\,\text{mol H}}{1.008\,\text{g H}} = 5.84\,\text{mol H; } 5.84\,/\,1.47 \approx 4$$

$$23.5\,\text{g O} \times \frac{1\,\text{mol O}}{16.00\,\text{g O}} = 1.47\,\text{mol O; } 1.47\,/\,1.47 = 1$$

The empirical formula is C_4H_4O.

$\dfrac{\text{MM}}{\text{FW}}\ \dfrac{136}{68.1} = 2$; the molecular formula is $2 \times$ the empirical formula.

The molecular formula is $C_8H_8O_2$.

4.107 $Ba^{2+}(aq) + SO_4{}^{2-}(aq) \rightarrow BaSO_4(s)$

$$0.2815\,\text{g } BaSO_4 \times \frac{137.3\,\text{g Ba}}{233.4\,\text{g } BaSO_4} = 0.16560 = 0.1656\,\text{g Ba}$$

$$\text{mass \%} = \frac{\text{g Ba}}{\text{g sample}} \times 100 = \frac{0.16560\,\text{g Ba}}{3.455\,\text{g sample}} \times 100 = 4.793\,\%\text{ Ba}$$

4.108 *Plan.* Write balanced equation.

$$\text{mass } H_2SO_4 \text{ soln} \xrightarrow{\text{mass \%}} \text{mass } H_2SO_4 \rightarrow \text{mol } H_2SO_4 \rightarrow \text{mol } Na_2CO_3 \rightarrow \text{mass } Na_2CO_3$$

Solve. $H_2SO_4(aq) + Na_2CO_3(s) \rightarrow Na_2SO_4(aq) + H_2O(l) + CO_2(g)$

$$5.0 \times 10^3 \text{ kg conc. } H_2SO_4 \times \frac{0.950 \text{ kg } H_2SO_4}{1.00 \text{ kg conc. } H_2SO_4} = 4.75 \times 10^3 = 4.8 \times 10^3 \text{ kg } H_2SO_4$$

$$4.75 \times 10^3 \text{ kg } H_2SO_4 \times \frac{1 \times 10^3 \text{ g}}{1 \text{ kg}} \times \frac{1 \text{ mol } H_2SO_4}{98.08 \text{ g } H_2SO_4} \times \frac{1 \text{ mol } Na_2CO_3}{1 \text{ mol } H_2SO_4}$$

$$\times \frac{105.99 \text{ g NaHCO}_3}{1 \text{ mol NaHCO}_3} \times \frac{1 \text{ kg}}{1 \times 10^3 \text{ g}} = 5.133 \times 10^3 = 5.1 \times 10^3 \text{ kg } Na_2CO_3$$

4.109 (a) $Mg(OH)_2(s) + 2HNO_3(aq) \rightarrow Mg(NO_3)_2(aq) + 2H_2O(l)$

(b) $5.53 \text{ g } Mg(OH)_2 \times \dfrac{1 \text{ mol } Mg(OH)_2}{58.32 \text{ g } Mg(OH)_2} = 0.09482 = 0.0948 \text{ mol } Mg(OH)_2$

$0.200 \, M \text{ HNO}_3 \times 0.0250 \text{ L} = 0.00500 \text{ mol HNO}_3$

The 0.00500 mol HNO_3 would neutralize 0.00250 mol $Mg(OH)_2$ and much more $Mg(OH)_2$ is present, so HNO_3 is the limiting reactant.

(c) Since HNO_3 limits, 0 mol HNO_3 is present after reaction.

0.00250 mol $Mg(NO_3)_2$ is produced.

0.09482 mol $Mg(OH)_2$ initial – 0.00250 mol $Mg(OH)_2$ react

$$= 0.0923 \text{ mol } Mg(OH)_2 \text{ remain}$$

4.110 (a) $Na_2SO_4(aq) + Pb(NO_3)_2(s) \rightarrow PbSO_4(s) + 2NaNO_3(aq)$

(b) Calculate mol of each reactant and compare.

$$1.50 \text{ g Pb(NO}_3)_2 \times \frac{1 \text{ mol Pb(NO}_3)_2}{331.2 \text{ g Pb(NO}_3)_2} = 0.004529 = 4.53 \times 10^{-3} \text{ mol Pb(NO}_3)_2$$

$0.100 \, M \text{ Na}_2SO_4 \times 0.125 \text{ L} = 0.0125 \text{ mol Na}_2SO_4$

Since the reactants combine in a 1:1 mol ratio, $Pb(NO_3)_2$ is the limiting reactant.

(c) $Pb(NO_3)_2$ is the limiting reactant, so no Pb^{2+} remains in solution. The remaining ions are: SO_4^{2-} (excess reactant), Na^+ and NO_3^- (spectators).

SO_4^{2-}:   0.0125 mol SO_4^{2-} initial – 0.00453 mol SO_4^{2-} reacted

$$= 0.00797 = 0.0080 \text{ mol } SO_4^{2-} \text{ remain}$$

0.00797 mol SO_4^{2-} / 0.125 L soln = 0.064 M SO_4^{2-}

Na^+:   Since the total volume of solution is the volume of $Na_2SO_4(aq)$ added, the concentration of Na^+ is unchanged.

0.100 M $Na_2SO_4 \times$ (2 mol Na^+ / 1 mol Na_2SO_4) = 0.200 M Na^+

NO_3^-: 4.53×10^{-3} mol $Pb(NO_3)_2 \times$ 2 mol NO_3^- / 1 mol $Pb(NO_3)_2$

$$= 9.06 \times 10^{-3} \text{ mol } NO_3^-$$

9.06×10^{-3} mol NO_3^- / 0.125 L = 0.0725 M NO_3^-

4.111   *Plan.* Cl^- is present in $NaCl$ and $MgCl_2$; using mass %, calculate mass $NaCl$ and $MgCl_2$ in mixture, mol Cl^- in each, then molarity of Cl^- in 0.500 L solution.   *Solve.*

$$7.50 \text{ mixture} \times \frac{0.765 \text{ g NaCl}}{1.00 \text{ g mixture}} \times \frac{1 \text{ mol NaCl}}{58.44 \text{ g NaCl}} \times \frac{1 \text{ mol Cl}^-}{1 \text{ mol NaCl}} = 0.09818 = 0.0982 \text{ mol Cl}^-$$

$$7.50 \text{ mixture} \times \frac{0.065 \text{ g MgCl}_2}{1.00 \text{ g mixture}} \times \frac{1 \text{ mol MgCl}_2}{95.21 \text{ g MgCl}_2} \times \frac{2 \text{ mol Cl}^-}{1 \text{ mol MgCl}} = 0.01024 = 0.010 \text{ mol Cl}^-$$

$$\text{mol Cl}^- = 0.09818 + 0.01024 = 0.10842 = 0.108 \text{ mol Cl}^-; \; M = \frac{0.10842 \text{ mol Cl}^-}{0.5000 \text{ L}} = 0.217 \; M \; Cl^-$$

4.112   *Plan.* $M = \dfrac{\text{mol Br}^-}{\text{L seawater}}$; mg $Br^- \rightarrow$ g $Br^- \rightarrow$ mol Br^-;

1 kg seawater $\rightarrow$ g $\xrightarrow{\text{density}}$ mL water $\rightarrow$ L sea water

Solve. $65 \text{ mg Br}^- \times \dfrac{1 \text{ g Br}^-}{1000 \text{ mg Br}^-} \times \dfrac{1 \text{ mol Br}^-}{79.90 \text{ g Br}^-} = 8.135 \times 10^{-4} = 8.1 \times 10^{-4} \text{ mol Br}^-$

$$1 \text{ kg seawater} \times \frac{1000 \text{ g}}{1 \text{ kg}} \times \frac{1 \text{ mL water}}{1.025 \text{ g water}} \times \frac{1 \text{ L}}{1000 \text{ mL}} = 0.9756 \text{ L}$$

$$M \text{ Br}^- = \frac{8.135 \times 10^{-4} \text{ mol Br}^-}{0.9756 \text{ L seawater}} = 8.3 \times 10^{-4} \; M \; Br^-$$

4.113 $Ag^+(aq) + Cl^-(aq) \rightarrow AgCl(s)$

$$\frac{0.2997 \text{ mol Ag}^+}{1 \text{ L}} \times 0.04258 \text{ L} \times \frac{1 \text{ mol Cl}^-}{1 \text{ mol Ag}^+} \times \frac{35.453 \text{ g Cl}^-}{1 \text{ mol Cl}^-} = 0.45242 = 0.4524 \text{ g Cl}^-$$

$$25.00 \text{ mL seawater} \times \frac{1.025 \text{ g}}{\text{mL}} = 25.625 = 25.63 \text{ g sea water}$$

$$\text{mass \% Cl}^- = \frac{0.45242 \text{ g Cl}^-}{25.625 \text{ g seawater}} \times 100 = 1.766\% \text{ Cl}^-$$

4.114   (a) AsO_4^{3-}; +5

(b) Ag_3PO_4 is silver phosphate; Ag_3AsO_4 is silver arsenate

(c) $0.0250 \text{ L soln} \times \dfrac{0.102 \text{ mol Ag}^+}{1 \text{ L soln}} \times \dfrac{1 \text{ mol Ag}_3\text{AsO}_4}{3 \text{ mol Ag}^+} \times \dfrac{1 \text{ mol As}}{1 \text{ mol Ag}_3\text{AsO}_4} \times \dfrac{74.92 \text{ g As}}{1 \text{ mol As}}$

$$= 0.06368 = 0.0637 \text{ g As}$$

$$\text{mass percent} = \frac{0.06368 \text{ g As}}{1.22 \text{ g sample}} \times 100 = 5.22\% \text{ As}$$

4.115   *Analyze.* Given 10 ppb As, find mass Na_3AsO_4 in 1.00 L of drinking water.

Plan. Use the definition of ppb to calculate g As in 1.0 L of water. Convert g As $\rightarrow$ g Na_3AsO_4 using molar masses. Assume the density of H_2O is 1.00 g/mL.

Solve. $1 \text{ billion} = 1 \times 10^9; 1 \text{ ppb} = \dfrac{1 \text{ g solute}}{1 \times 10^9 \text{ g solution}}$

$$\dfrac{1 \text{ g solute}}{1 \times 10^9 \text{ g solution}} \times \dfrac{1 \text{ g solution}}{1 \text{ mL solution}} \times \dfrac{1 \times 10^3 \text{ mL}}{1 \text{ L solution}} = \dfrac{\text{g As}}{1 \times 10^6 \text{ L H}_2\text{O}}$$

$$10 \text{ ppb As} = \dfrac{10 \text{ g As}}{1 \times 10^6 \text{ L H}_2\text{O}} \times 1 \text{ L H}_2\text{O} = 1.0 \times 10^{-5} \text{ g As/L.}$$

$$1.0 \times 10^{-5} \text{ g As} \times \dfrac{1 \text{ mol As}}{74.92 \text{ g As}} \times \dfrac{1 \text{ mol Na}_3\text{AsO}_4}{1 \text{ mol As}} \times \dfrac{207.89 \text{ g Na}_3\text{AsO}_4}{1 \text{ mol Na}_3\text{AsO}_4}$$

$$= 2.8 \times 10^{-5} \text{ g Na}_3\text{AsO}_4 \text{ in } 1.00 \text{ L H}_2\text{O}$$

4.116   (a)   mol HCl initial – mol NH_3 from air = mol HCl remaining

$$= \text{mol NaOH required for titration}$$

mol NaOH $= 0.0588 \, M \times 0.0131 \text{ L} = 7.703 \times 10^{-4} = 7.70 \times 10^{-4}$ mol NaOH

$$= 7.70 \times 10^{-4} \text{ mol HCl remain}$$

mol HCl initial – mol HCl remaining = mol NH_3 from air

$(0.0105 \text{ M HCl} \times 0.100 \text{ L}) - 7.703 \times 10^{-4}$ mol HCl = mol NH_3

10.5×10^{-4} mol HCl $- 7.703 \times 10^{-4}$ mol HCl $= 2.80 \times 10^{-4} = 2.8 \times 10^{-4}$ mol NH_3

$$2.8 \times 10^{-4} \text{ mol NH}_3 \times \dfrac{17.03 \text{ g NH}_3}{1 \text{ mol NH}_3} = 4.77 \times 10^{-3} = 4.8 \times 10^{-3} \text{ g NH}_3$$

(b)   ppm is defined as molecules of $NH_3/1 \times 10^6$ molecules in air.

Calculate molecules NH_3 from mol NH_3.

$$2.80 \times 10^{-4} \text{ mol NH}_3 \times \dfrac{0.022 \times 20^{23} \text{ molecules}}{1 \text{ mol}} = 1.686 \times 10^{20}$$

$$= 1.7 \times 10^{20} \text{ NH}_3 \text{ molecules}$$

Calculate total volume of air processed, then g air using density, then molecules air using molar mass.

$$\dfrac{10.0 \text{ L}}{1 \text{ min}} \times 10.0 \text{ min} \times \dfrac{1.20 \text{ g air}}{1 \text{ L air}} \times \dfrac{1 \text{ mol air}}{29.0 \text{ g air}} \times \dfrac{6.022 \times 10^{23} \text{ molecules}}{1 \text{ mol}}$$

$$= 2.492 \times 10^{24} = 2.5 \times 10^{24} \text{ air molecules}$$

$$\text{ppm NH}_3 = \dfrac{1.686 \times 10^{20} \text{ NH}_3 \text{ molecules}}{2.492 \times 10^{24} \text{ air molecules}} \times 1 \times 10^6 = 68 \text{ ppm NH}_3$$

(c)   68 ppm > 50 ppm. The manufacturer is **not** in compliance.

5 Thermochemistry

Visualizing Concepts

5.1 The book's potential energy is due to the opposition of gravity by an object of mass m at a distance d above the surface of the earth. Kinetic energy is due to the motion of the book. As the book falls, d decreases and potential energy changes into kinetic energy.

The first law states that the total energy of a system is conserved. At the instant before impact, all potential energy has been converted to kinetic energy, so the book's total kinetic energy is 85 J, assuming no transfer of energy as heat.

5.2 (a) The internal energy, E, of the products is greater than that of the reactants, so the diagram represents an increase in the internal energy of the system.

(b) ΔE for this process is positive, +.

(c) If no work is associated with the process, it is endothermic.

5.3 (a) For an endothermic process, the sign of q is positive; the system gains heat. This is true only for system (iii).

(b) In order for ΔE to be less than 0, there is a net transfer of heat or work from the system to the surroundings. The magnitude of the quantity leaving the system is greater than the magnitude of the quantity entering the system. In system (i), the magnitude of the heat leaving the system is less than the magnitude of the work done on the system. In system (iii), the magnitude of the work done by the system is less than the magnitude of the heat entering the system. None of the systems has $\Delta E < 0$.

(c) In order for ΔE to be greater than 0, there is a net transfer of work or heat to the system from the surroundings. In system (i), the magnitude of the work done on the system is greater than the magnitude of the heat leaving the system. In system (ii), work is done on the system with no change in heat. In system (iii) the magnitude of the heat gained by the system is greater than the magnitude of the work done on the surroundings. $\Delta E > 0$ for all three systems.

5.4 (a). No. This distance traveled to the top of a mountain depends on the path taken by the hiker. Distance is a path function, not a state function.

(b) Yes. Change in elevation depends only on the location of the base camp and the height of the mountain, not on the path to the top. Change in elevation is a state function, not a path function.

5.5 $w = -P\Delta V$. Since ΔV for the process is (–), the sign of w is (+).

At constant pressure, $\Delta E = q + w$. At constant pressure, $\Delta H = q$. If the reaction is endothermic, the signs of ΔH and q are (+). The sign of w is (+), so the sign of ΔE is (+). The internal energy of the system increases during the change. (This situation is described by the diagram (ii) in Exercise 5.3.)

5.6   (a)   The temperature of the system and surroundings will equalize, so the temperature of the hotter system will decrease and the temperature of the colder surroundings will increase. The system loses heat by decreasing its temperature, so the sign of q is (–). The process is exothermic.

          (b)   If neither volume nor pressure of the system changes, $w = 0$ and $\Delta E = q = \Delta H$. The change in internal energy is equal to the change in enthalpy.

5.7   (a) $N_2(g) + O_2(g) \rightarrow 2NO(g)$. Since $\Delta V = 0$, $w = 0$.

          (b) $\Delta H = 90.37$ kJ for production of 1 mol of $NO(g)$. The definition of a formation reaction is one where elements combine to form one mole of a single product. The enthalpy change for such a reaction is the enthalpy of formation.

5.8   (a) $\Delta H_A = \Delta H_B + \Delta H_C$. The net enthalpy change associated with going from the initial state to the final state does not depend on path. The change can be accomplished via reaction A, or via two successive reactions, B then C, with the same net enthalpy change.

          (b) $\Delta H_Z = \Delta H_X + \Delta H_Y$. The diagram indicates that Reaction Z can be written as the sum of reactions X and Y. Hess's Law states that the enthalpy change for the net reaction Z is the sum of the enthalpy changes of the steps X and Y, regardless of whether the reaction actually occurs via this path. $\Delta H_Z = \Delta H_X + \Delta H_Y$ because ΔH is a state function, independent of path.

The Nature of Energy

5.9   An object can possess energy by virtue of its motion or position. Kinetic energy, the energy of motion, depends on the mass of the object and its velocity. Potential energy, stored energy, depends on the position of the object relative to the body with which it interacts.

5.10   (a)   The kinetic energy of the ball **decreases** as it moves higher. As the ball moves higher and opposes gravity, kinetic energy is changed into potential energy.

          (b)   The potential energy of the ball **increases** as it moves higher.

          (c)   The heavier ball would go **half as high** as the tennis ball. At the apex of the trajectory, all initial kinetic energy has been changed into potential energy. The magnitude of the change in potential energy is $m\,g\,\Delta h$, which is equal to the energy initially imparted to the ball. If the same amount of energy is imparted to a ball with twice the mass, m doubles so Δh is half as large.

5.11   (a)   *Analyze.* Given: mass and speed of ball. Find: kinetic energy.

              Plan. Since $1\,J = 1\,kJ \cdot m^2/s^2$, convert g $\rightarrow$ kg to obtain E_k in joules.

Solve. $E_k = 1/2\,mv^2 = 1/2 \times 45\,g \times \dfrac{1\,kg}{1000\,g} \times \left(\dfrac{61\,m}{1\,s}\right)^2 = \dfrac{84\,kg \cdot m^2}{1\,s^2} = 84\,J$

Check. $1/2(45 \times 3600/1000) \approx 1/2(40 \times 4) \approx 80\,J$

(b) $83.72\,J \times \dfrac{1\,cal}{4.184\,J} = 20\,cal$

(c)   As the ball hits the sand, its speed (and hence its kinetic energy) drops to zero. Most of the kinetic energy is transferred to the sand, which deforms when the ball lands. Some energy is released as heat through friction between the ball and the sand.

5.12   (a)   *Plan.* Convert lb → kg, mi/hr → m/s.

               Solve. $950\,lb \times \dfrac{1\,kg}{2.205\,lb} = 430.84 = 431\,kg$

               $\dfrac{68\,mi}{1\,hr} \times \dfrac{1.6093\,km}{1\,mi} \times \dfrac{1000\,m}{1\,km} \times \dfrac{1\,hr}{60\,min} \times \dfrac{1\,min}{60\,sec} = 30.398 = 30\,m/s$

               $E_k = 1/2\,mv^2 = 1/2 \times 430.84\,kg \times (30.398)^2\,m^2/s^2 = 2.0 \times 10^5\,J$

    (b) E_k is proportional to v^2, so if speed decreases by a factor of 2, kinetic energy decreases by a factor of 4.

    (c)   Brakes stop a moving vehicle, so the kinetic energy of the motorcycle is primarily transferred to friction between brakes and wheels, and somewhat to deformation of the tire and friction between the tire and road.

5.13   *Analyze.* Given: heat capacity of water $= 1\,Btu/lb \cdot °F$ Find: J/Btu

      Plan.   heat capacity of water $= 4.184\,J/g \cdot °C$; $\dfrac{J}{g \cdot °C} \rightarrow \dfrac{J}{lb \cdot °F} \rightarrow \dfrac{J}{Btu}$

This strategy requires changing °F to °C. Since this involves the magnitude of a degree on each scale, rather than a specific temperature, the 32 in the temperature relationship is not needed.

100 °C = 180 °F; 5 °C = 9 °F

Solve. $\dfrac{4.184\,J}{g \cdot °C} \times \dfrac{453.6\,g}{lb} \times \dfrac{5\,°C}{9\,°F} \times \dfrac{1\,lb \cdot °F}{1\,Btu} = 1054\,J/Btu$

5.14   (a)   *Analyze.* Given: 1 kwh; 1 watt = 1 J/s; 1 watt $\cdot$ s = 1 J.

               Find: conversion factor for joules and kwh.

               Plan. kwh → wh → ws → J

               Solve. $1\,kwh \times \dfrac{1000\,w}{1\,kw} \times \dfrac{60\,min}{h} \times \dfrac{60\,s}{min} \times \dfrac{1\,J}{1\,w \cdot s} = 3.6 \times 10^6\,J$

               1 kwh = 3.6×10^6 J

    (b)   *Analyze.* Given: 100 watt bulb. Find: heat in kcal radiated by bulb or person in 24 hr.

Plan. 1 watt = 1 J/s; 1 kcal = 4.184 × 10³ J; watt → J/s → J → kcal.   *Solve.*

$$100 \text{ watt} = \frac{100 \text{ J}}{1 \text{ s}} \times \frac{60 \text{ sec}}{\text{min}} \times \frac{60 \text{ min}}{\text{hr}} \times 24 \text{ hr} \times \frac{1 \text{ kcal}}{4.184 \times 10^3 \text{ J}} = 2065 = 2.1 \times 10^3 \text{ kcal}$$

24 hr has 2 sig figs, but 100 watt is ambiguous. The answer to 1 sig fig would be 2 × 10³ kcal.

Check. $(1 \times 10^2 \times 6 \times 10^1 \times 6/10^3) \approx 6^3 \times 10 \approx 2000 \text{ kcal}$

5.15   The energy source of a 100 watt light bulb is electrical current from household wiring. Current passes through and heats a tungsten filament (thin wire) in the bulb. The energy is radiated in the form of heat and visible light.

The energy source for an adult person is food. When a person eats, the food undergoes a complex series of chemical reactions that release the potential energy stored in chemical bonds. Some of this energy is transferred as electrical impulses that trigger muscle action and become kinetic energy. Some is released as heat.

In both cases, the energy must travel through a network (house wiring and lamp or human body) and undergo several changes in form before it is in the correct location and form to accomplish the desired task. In both cases, the energy given off as heat is wasted; it cannot be applied to the tasks of producing light or motion.

5.16   The air gun imparts a certain amount of kinetic energy to the pellet. As the pellet rises against the force of gravity, kinetic energy is changed to potential energy. When all kinetic energy has been transferred to potential energy (or lost as heat through friction) the pellet stops rising and falls to earth. In principle, if enough kinetic energy could be imparted to the pellet, it could escape the force of gravity and move into space. For an air gun and a pellet, this is practically impossible.

5.17   (a)   In thermodynamics, the *system* is the well-defined part of the universe whose energy changes are being studied.

       (b)   A *closed system* can exchange heat but not mass with its surroundings.

5.18   (a)   The system is not closed, because it is exchanging mass with the surroundings. That is, solution flows into and out of the flask.

       (b)   If the system is defined as shown, it can be closed by blocking the flow in and out, but leaving the flask full of solution.

5.19   (a)   *Work* is a force applied over a distance.

       (b)   The amount of work done is the magnitude of the force times the distance over which it is applied. $w = F \times d$.

5.20   (a)   Heat is the energy transferred from a hotter object to a colder object.

       (b)   Heat is transferred from one object (system) to another until the two objects (systems) are at the same temperature.

5.21   (a)   Gravity; work is done because the force of gravity is opposed and the pencil is lifted.

(b)   Mechanical force; work is done because the force of the coiled spring is opposed as the spring is compressed over a distance.

5.22   (a)   Electrostatic attraction; no work is done because the particles are held stationary.

(b)   Magnetic attraction; work is done because the nail is moved a distance.

The First Law of Thermodynamics

5.23   (a)   In any chemical or physical change, energy can be neither created nor destroyed, but it can be changed in form.

(b)   The total *internal energy* (E) of a system is the sum of all the kinetic and potential energies of the system components.

(c)   The internal energy of a system increases when work is done on the system by the surroundings and/or when heat is transferred to the system from the surroundings (the system is heated).

5.24   (a) $\Delta E_{sys} = -\Delta E_{surr}$; $\Delta E_{sys} = q + w$

(b)   The quantities q and w are negative when the system loses heat to the surroundings (it cools), or does work on the surroundings.

5.25   *Analyze.* Given: heat and work. Find: magnitude and sign of ΔE.

Plan. In each case, evaluate q and w in the expression $\Delta E = q + w$. For an exothermic process, q is negative; for an endothermic process, q is positive.   *Solve.*

(a)   q is positive because the system absorbs heat and w is negative because the system does work. $\Delta E = 85 \text{ kJ} - 29 \text{ kJ} = 56 \text{ kJ}$. The process is endothermic.

(b) $\Delta E = 1.50 \text{ kJ} - 657 \text{ J} = 1.50 \text{ kJ} - 0.657 \text{ kJ} = 0.843 = 0.84 \text{ kJ}$. The process is endothermic.

(c)   q is negative because the system releases heat, and w is negative because the system does work. $\Delta E = -57.5 \text{ kJ} - 13.5 \text{ kJ} = -71.0 \text{ kJ}$. The process is exothermic.

5.26   In each case, evaluate q and w in the expression $\Delta E = q + w$. For an exothermic process, q is negative; for an endothermic process, q is positive.

(a)   q is positive and w is negative. $\Delta E = 900 \text{ J} - 422 \text{ J} = 478 \text{ J}$. The process is endothermic.

(b)   q is negative and w is essentially zero. $\Delta E = -3140 \text{ J}$. The process is exothermic.

(c)   q is negative and w is zero. $\Delta E = -8.65 \text{ kJ}$. The process is exothermic.

5.27   *Analyze.* How do the different physical situations (cases) affect the changes to heat and work of the system upon addition of 100 J of energy? *Plan.* Use the definitions of heat and work and the First Law to answer the questions.   *Solve.*

If the piston is allowed to move, case (1), the heated gas will expand and push the piston up, doing work on the surroundings. If the piston is fixed, case (2), most of the electrical energy will be manifested as an increase in heat of the system.

(a)   Since little or no work is done by the system in case (2), the gas will absorb most of the energy as heat; the case (2) gas will have the higher temperature.

(b)   In case (2), $w \approx 0$ and $q \approx 100$ J. In case (1), a significant amount of energy will be used to do work on the surroundings ($-w$), but some will be absorbed as heat ($+q$). (The transfer of electrical energy into work is never completely efficient!)

(c) ΔE is greater for case (2), because the entire 100 J increases the internal energy of the system, rather than a part of the energy doing work on the surroundings.

5.28 $E_{el} = \dfrac{\kappa Q_1 Q_2}{r^2}$ For two oppositely charged particles, the sign of E_{el} is negative; the closer the particles, the greater the magnitude of E_{el}.

(a)   The potential energy becomes less negative as the particles are separated (r increases).

(b) ΔE for the process is positive; the internal energy of the system increases as the oppositely charged particles are separated.

(c)   Work is done on the system to separate the particles so w is positive. We have no direct knowledge of the change in q, except that it cannot be large and negative, because overall $\Delta E = q + w$ is positive.

5.29   (a)   A *state function* is a property of a system that depends only on the physical state (pressure, temperature, etc.) of the system, not on the route used by the system to get to the current state.

(b)   Internal energy and enthalpy **are** state functions; heat **is not** a state function.

(c)   Work **is not** a state function. The amount of work required to move from state A to state B depends on the path or series of processes used to accomplish the change.

5.30   (a)   Independent. Potential energy is a state function.

(b)   Dependent. Some of the energy released could be employed in performing work, as is done in the body when sugar is metabolized; heat is not a state function.

(c)   Dependent. The work accomplished depends on whether the gasoline is used in an engine, burned in an open flame, or in some other manner. Work is not a state function.

Enthalpy

5.31   (a)   Change in enthalpy (ΔH) is usually easier to measure than change in internal energy (ΔE) because, at constant pressure, $\Delta H = q$. The heat flow associated with a process at constant pressure can easily be measured as a change in temperature. Measuring ΔE requires a means to measure both q and w.

(b)   If ΔH is negative, the enthalpy of the system decreases and the process is exothermic.

5.32   (a)   When a process occurs under constant external pressure, the enthalpy change (ΔH) equals the amount of heat transferred. $\Delta H = q_p$.

        (b) $\Delta H = q_p$. If the system absorbs heat, q and ΔH are positive and the enthalpy of the system increases.

5.33   (a) $HC_2H_3O_2(l) + 2O_2(g) \rightarrow 2H_2O(l) + 2CO_2(g)$ $\Delta H = -871.7$ kJ

        (b)   *Analyze.* How are reactants and products arranged on an enthalpy diagram?

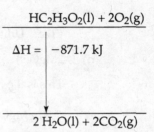

               Plan. The substances (reactants or products, collectively) with higher enthalpy are shown on the upper level, and those with lower enthalpy are shown on the lower level.

               Solve. For this reaction, ΔH is negative, so the products have lower enthalpy and are shown on the lower level; reactants are on the upper level. The arrow points in the direction of reactants to products and is labeled with the value of ΔH.

5.34   (a) $ZnCO_3(s) \rightarrow ZnO(s) + CO_2(g)$

               $\Delta H = 71.5$ kJ

        (b)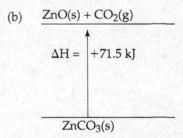

5.35   *Plan.* Consider the sign of ΔH.

        Solve. Since ΔH is negative, the reactants, $2Cl(g)$ have the higher enthalpy.

5.36   *Plan.* Consider the sign of an enthalpy change that would convert one of the substances into the other.   *Solve.*

        (a) $CO_2(s) \rightarrow CO_2(g)$. This change is sublimation, which is endothermic, $+\Delta H$. $CO_2(g)$ has the higher enthalpy.

        (b) $H_2 \rightarrow 2H$. Breaking the H–H bond requires energy, so the process is endothermic, $+\Delta H$. Two moles of H atoms have higher enthalpy.

        (c) $H_2O(g) \rightarrow H_2(g) + 1/2\ O_2(g)$. Decomposing H_2O into its elements requires energy and is endothermic, $+\Delta H$. One mole of $H_2(g)$ and 0.5 mol $O_2(g)$ at 25°C have the higher enthalpy.

        (d) $N_2(g)$ at 100° $\rightarrow N_2(g)$ at 300°. An increase in the temperature of the sample requires that heat is added to the system, $+q$ and $+\Delta H$. $N_2(g)$ at 300° has the higher enthalpy.

5.37   *Analyze/Plan.* Follow the strategy in Sample Exercise 5.4.   *Solve.*

        (a)   Exothermic (ΔH is negative)

(b) $2.4 \, g \, Mg \times \dfrac{1 \, mol \, Mg}{24.305 \, g \, Mg} \times \dfrac{-1204 \, kJ}{2 \, mol \, Mg} = -59 \, kJ$ heat transferred

Check. The units of kJ are correct for heat. The negative sign indicates heat is evolved.

(c) $-96.0 \, kJ \times \dfrac{2 \, mol \, MgO}{-1204 \, kJ} \times \dfrac{40.30 \, g \, MgO}{1 \, mol \, Mg} = 6.43 \, g \, MgO$ produced

Check. Units are correct for mass. $(100 \times 2 \times 40 / 1200) \approx (8000/1200) \approx 6.5 \, g$

(d) $2MgO(s) \rightarrow 2Mg(s) + O_2(g)$ $\Delta H = +1204 \, kJ$

This is the reverse of the reaction given above, so the sign of ΔH is reversed.

$7.50 \, g \, MgO \times \dfrac{1 \, mol \, MgO}{40.30 \, g \, MgO} \times \dfrac{1204 \, kJ}{2 \, mol \, MgO} = +112 \, kJ$ heat absorbed

Check. The units are correct for energy. $(\sim 9000/80) \approx 110 \, kJ)$

5.38   (a)   The reaction is endothermic, so heat is absorbed by the system during the course of reaction.

(b) $45.0 \, g \, CH_3OH \times \dfrac{1 \, mol \, CH_3OH}{32.04 \, g \, CH_3OH} \times \dfrac{90.7 \, kJ}{1 \, mol \, CH_3OH} = 127 \, kJ$ heat transferred (absorbed)

(c) $18.5 \, kJ \times \dfrac{2 \, mol \, H_2}{90.7 \, kJ} \times \dfrac{2.016 \, g \, H_2}{1 \, mol \, H_2} = 0.822 \, g \, H_2$ produced

(d)   The sign of ΔH is reversed for the reverse reaction: $\Delta H = -90.7 \, kJ$

$27.0 \, g \, CO \times \dfrac{1 \, mol \, CO}{28.01 \, g \, CO} \times \dfrac{-90.7 \, kJ}{1 \, mol \, CO} = -87.4 \, kJ$ heat transferred (released)

5.39   *Analyze.* Given: balanced thermochemical equation, various quantities of substances and/or enthalpy. *Plan.* Enthalpy is an extensive property; it is "stoichiometric." Use the mole ratios implicit in the balanced thermochemical equation to solve for the desired quantity. Use molar masses to change mass to moles and vice versa where appropriate. *Solve.*

(a) $0.200 \, mol \, AgCl \times \dfrac{-65.5 \, kJ}{1 \, mol \, AgCl} = -13.1 \, kJ$

Check. Units are correct; sign indicates heat evolved.

(b) $2.50 \, g \, AgCl \times \dfrac{1 \, mol \, AgCl}{143.3 \, g \, AgCl} \times \dfrac{-65.5 \, kJ}{1 \, mol \, AgCl} = -1.14 \, kJ$

Check. Units correct; sign indicates heat evolved.

(c) $0.150 \, mmol \, AgCl \times \dfrac{1 \times 10^{-3} \, mol}{1 \, mmol} \times \dfrac{+65.5 \, kJ}{1 \, mol \, AgCl} = 0.009825 \, kJ = 9.83 \, J$

Check. Units correct; sign of ΔH reversed; sign indicates heat is absorbed during the reverse reaction.

5.40 (a) $0.855 \, \text{mol} \, O_2 \times \dfrac{-89.4 \, \text{kJ}}{3 \, \text{mol} \, O_2} = -25.48 = -25.5 \, \text{kJ}$

(b) $10.75 \, \text{g} \, KCl \times \dfrac{1 \, \text{mol} \, KCl}{74.55 \, \text{g} \, KCl} \times \dfrac{-89.4 \, \text{kJ}}{2 \, \text{mol} \, KCl} = -6.4457 = -6.45 \, \text{kJ}$

(c) Since the sign of ΔH is reversed for the reverse reaction, it seems reasonable that other characteristics would be reversed, as well. If the forward reaction proceeds spontaneously, the reverse reaction is probably not spontaneous. Also, we know from experience that $KCl(s)$ does not spontaneously react with atmospheric $O_2(g)$, even at elevated temperature.

5.41 At constant pressure, $\Delta E = \Delta H - P\Delta V$. In order to calculate ΔE, more information about the conditions of the reaction must be known. For an ideal gas at constant pressure and temperature, $P\Delta V = RT\Delta n$. The values of either P and ΔV or T and Δn must be known to calculate ΔE from ΔH.

5.42 At constant volume ($\Delta V = 0$), $\Delta E = q_v$. According to the definition of enthalpy, $H = E + PV$, so $\Delta H = \Delta E + \Delta(PV)$. For an ideal gas at constant temperature and volume, $\Delta PV = V\Delta P = RT\Delta n$. For this reaction, there are 2 mol of gaseous product and 3 mol of gaseous reactants, so $\Delta n = -1$. Thus $V\Delta P$ or $\Delta(PV)$ is negative. Since $\Delta H = \Delta E + \Delta(PV)$, the negative $\Delta(PV)$ term means that ΔH will be smaller or more negative than ΔE.

5.43 *Analyze/Plan.* q = -79 kJ (heat is given off by the system), w = -18 kJ (work is done by the system). *Solve.*

$\Delta E = q + w = -79 \, \text{kJ} - 18 \, \text{kJ} = -97 \, \text{kJ}$. $\Delta H = q = -79 \, \text{kJ}$ (at constant pressure).

Check. The reaction is exothermic.

5.44 The gas is the system. If 418 J of heat is added, q = +418 J. Work done by the system decreases the overall energy of the system, so w = -107 J.

$\Delta E = q + w = 418 \, \text{J} - 107 \, \text{J} = 311 \, \text{J}$. $\Delta H = q = 418 \, \text{J}$ (at constant pressure).

5.45 *Analyze.* Given: balanced thermochemical equation. *Plan.* Follow the guidelines given in Section 5.4 for evaluating thermochemical equations. *Solve.*

(a) When a chemical equation is reversed, the sign of ΔH is reversed.

$CO_2(g) + 2H_2O(l) \rightarrow CH_3OH(l) + 3/2 \, O_2(g)$ $\qquad \Delta H = 726.5 \, \text{kJ}$

(b) Enthalpy is extensive. If the coefficients in the chemical equation are multiplied by 2 to obtain all integer coefficients, the enthalpy change is also multiplied by 2.

$2CH_3OH(l) + 3O_2(g) \rightarrow 2CO_2(g) + 4H_2O(l)$ $\qquad \Delta H = 2(-726.5) \, \text{kJ} = -1453 \, \text{kJ}$

(c) The exothermic forward reaction is more likely to be thermodynamically favored.

(d) Vaporization (liquid $\rightarrow$ gas) is endothermic. If the product were $H_2O(g)$, the reaction would be more endothermic and would have a smaller negative ΔH. (Depending on temperature, the enthalpy of vaporization for 2 mol H_2O is about +88 kJ, not large enough to cause the overall reaction to be endothermic.)

5.46 (a) $3C_2H_2(g) \rightarrow C_6H_6(l)$ $\qquad \Delta H = -630 \, \text{kJ}$

(b) $C_6H_6(l) \rightarrow 3C_2H_2(g)$ $\Delta H = 630$ kJ

ΔH for the formation of 3 mol of acetylene is 630 kJ. ΔH for the formation of 1 mol of C_2H_2 is then 630 kJ/3 = 210 kJ.

(c) The exothermic reverse reaction is more likely to be thermodynamically favored.

(d)

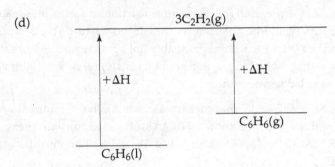

If the reactant is in the higher enthalpy gas phase, the overall ΔH for the reaction has a smaller positive value.

Calorimetry

The specific heat of water to four significant figures, **4.184 J/g • K,** will be used in many of the following exercises; temperature units of K and °C will be used interchangeably.

5.47 (a) J/°C or J/K. Heat capacity is the amount of heat in J required to raise the temperature of an object or a certain amount of a substance 1°C or 1 K. Since the amount is defined, units of amount are not included.

(b) $\dfrac{J}{g \bullet °C}$ or $\dfrac{J}{g \bullet K}$ Specific heat is a particular kind of heat capacity where the amount of substance is 1 g.

(c) To calculate heat capacity from specific heat, the **mass** of the particular piece of copper pipe must be known.

5.48 *Analyze.* Both objects are heated to 100°C. The two hot objects are placed in the same amount of cold water at the same temperature. Object A raises the water temperature more than object B. *Plan.* Apply the definition of heat capacity to heating the water and heating the objects to determine which object has the greater heat capacity.   *Solve.*

(a) Both beakers of water contain the same mass of water, so they both have the same heat capacity. Object A raises the temperature of its water more than object B, so more heat was transferred from object A than from object B. Since both objects were heated to the same temperature initially, object A must have absorbed more heat to reach the 100° temperature. The greater the heat capacity of an object, the greater the heat required to produce a given rise in temperature. Thus, object A has the greater heat capacity.

(b) Since no information about the masses of the objects is given, we cannot compare or determine the specific heats of the objects.

5.49 *Plan.* Manipulate the definition of specific heat to solve for the desired quantity, paying close attention to units. specific heat = q/(m × Δt).   *Solve.*

(a) $\dfrac{4.184\,J}{1\,g\bullet K}$ or $\dfrac{4.184\,J}{1\,g\bullet \,^{\circ}C}$

(b) $\dfrac{4.184\,J}{1\,g\bullet \,^{\circ}C}\times\dfrac{18.02\,g\,H_2O}{1\,mol\,H_2O}=\dfrac{75.40\,J}{mol\bullet \,^{\circ}C}$

(c) $\dfrac{185\,g\,H_2O\times 4.184\,J}{1\,g\bullet \,^{\circ}C}=774\,J/^{\circ}C$

(d) $10.00\,kg\,H_2O\times\dfrac{1000\,g}{1\,kg}\times\dfrac{4.184\,J}{1\,g\bullet \,^{\circ}C}\times\dfrac{1\,kJ}{1000\,J}\times(46.2^{\circ}C-24.6^{\circ}C)=904\,kJ$

Check. $(10\times 4\times 20)\approx 800$ kJ; the units are correct. Note that the conversion factors for kg → g and J → kJ cancel. An equally correct form of specific heat would be kJ/kg $\bullet C^{\circ}$

5.50   (a)   In Table 5.2, Hg(l) has the smallest specific heat, so it will require the smallest amount of energy to heat 50.0 g of the substance 10 k.

(b) $50.0\,g\,Hg(l)\times 10\,K\times\dfrac{0.14\,J}{g\bullet K}=70\,J$

5.51   *Analyze/Plan.* Follow the logic in Sample Exercise 5.5.   *Solve.*

$1.05\,kg\,Fe\times\dfrac{1000\,g}{1\,kg}\times\dfrac{0.450\,J}{g\bullet K}\times(88.5^{\circ}C-25.0^{\circ}C)=3.00\times 10^4\,J\ (or\ 30.0\,kJ)$

5.52 $62.0\,g\ ethylene\ glycol\times\dfrac{2.42\,J}{g\bullet K}\times(40.8^{\circ}C-15.2^{\circ}C)=3.84\times 10^3\,J$

5.53   *Analyze.* Since the temperature of the water increases, the dissolving process is exothermic and the sign of ΔH is negative. The heat lost by the NaOH(s) dissolving equals the heat gained by the solution.

Plan/Solve. Calculate the heat gained by the solution. The temperature change is 47.4 – 23.6 = 23.8°C. The total mass of solution is (100.0 g H_2O + 9.55 g NaOH) = 109.55 = 109.6 g.

$109.55\ solution\times\dfrac{4.184\,J}{1\,g\bullet \,^{\circ}C}\times 23.8^{\circ}C\times\dfrac{1\,kJ}{1000\,J}=10.909=10.9\,kJ$

This is the amount of heat lost when 9.55 g of NaOH dissolves.

The heat loss per mole NaOH is

$\dfrac{-10.909\,kJ}{9.55\,g\,NaOH}\times\dfrac{40.00\,g\,NaOH}{1\,mol\,NaOH}=-45.7\,kJ/mol\quad \Delta H=q_p=-45.7\,kJ/mol\ NaOH$

Check. $(-11/9\times 40)\approx -45$ kJ; the units and sign are correct.

5.54   (a)   Following the logic in Solution 5.53, the dissolving process is endothermic, ΔH is positive. The total mass of the solution is (60.0 g H_2O + 3.88 g NH_4NO_3) = 63.88 = 63.9 g. The temperature change of the solution is 23.0 – 18.4 = 4.6°C. The heat lost by the solution is

$$63.88 \text{ g solution} \times \frac{4.184 \text{ J}}{1 \text{ g} \cdot {}^\circ\text{C}} \times 4.6^\circ\text{C} \times \frac{1 \text{ kJ}}{1000 \text{ J}} = 1.229 = 1.2 \text{ kJ}$$

Thus, 1.2 kJ is absorbed when 3.88 g $NH_4NO_3(s)$ dissolves.

$$\frac{+1.229 \text{ kJ}}{3.88 \text{ } NH_4NO_3} \times \frac{80.05 \text{ g } NH_4NO_3}{1 \text{ mol } NH_4NO_3} = +25.36 = +25 \text{ kJ/mol } NH_4NO_3$$

(b)   This process is endothermic, because the temperature of the surroundings decreases, indicating that heat is absorbed by the system.

5.55   *Analyze/Plan.* Follow the logic in Sample Exercise 5.7.   *Solve.*

$q_{bomb} = -q_{rxn}$; $\Delta T = 30.57^\circ\text{C} - 23.44^\circ\text{C} = 7.13^\circ\text{C}$

$$q_{bomb} = \frac{7.854 \text{ kJ}}{1^\circ\text{C}} \times 7.13^\circ\text{C} = 56.00 = 56.0 \text{ kJ}$$

At constant volume, $q_v = \Delta E$. ΔE and ΔH are very similar.

$$\Delta H_{rxn} \approx \Delta E_{rxn} = q_{rxn} = -q_{bomb} = \frac{-56.0 \text{ kJ}}{2.20 \text{ g } C_6H_4O_2} = -25.454 = -25.5 \text{ kJ/g } C_6H_4O_2$$

$$\Delta H_{rxn} = \frac{-25.454 \text{ kJ}}{1 \text{ g } C_6H_4O_2} \times \frac{108.1 \text{ g } C_6H_4O_2}{1 \text{ mol } C_6H_4O_2} = -2.75 \times 10^3 \text{ kJ/mol } C_6H_4O_2$$

5.56   (a) $C_6H_5OH(s) + 7O_2(g) \rightarrow 6CO_2(g) + 3H_2O(l)$

(b) $q_{bomb} = -q_{rxn}$; $\Delta T = 26.37^\circ\text{C} - 21.36^\circ\text{C} = 5.01^\circ\text{C}$

$$q_{bomb} = \frac{11.66 \text{ kJ}}{1^\circ\text{C}} \times 5.01^\circ\text{C} = 58.417 = 58.4 \text{ kJ}$$

At constant volume, $q_v = \Delta E$. ΔE and ΔH are very similar.

$$\Delta H_{rxn} \approx \Delta E_{rxn} = q_{rxn} = -q_{bomb} = \frac{-58.417 \text{ kJ}}{1.800 \text{ g } C_6H_5OH} = -32.454 = -32.5 \text{ kJ/g } C_6H_5OH$$

$$\Delta H_{rxn} = \frac{-32.454 \text{ kJ}}{1 \text{ g } C_6H_5OH} \times \frac{94.11 \text{ g } C_6H_5OH}{1 \text{ mol } C_6H_5OH} = \frac{-3.054 \times 10^3 \text{ kJ}}{\text{mol } C_6H_5OH}$$

$$= -3.05 \times 10^3 \text{ kJ/mol } C_6H_5OH$$

5.57   *Analyze.* Given: specific heat and mass of glucose, ΔT for calorimeter. Find: heat capacity, C, of calorimeter. *Plan.* All heat from the combustion raises the temperature of the calorimeter. Calculate heat from combustion of glucose, divide by ΔT for calorimeter to get kJ/°C.   *Solve.*

(a) $C_{total} = 2.500 \text{ g glucose} \times \dfrac{15.57 \text{ kJ}}{1 \text{ g glucose}} \times \dfrac{1}{2.70^\circ\text{C}} = 14.42 = 14.4 \text{ kJ/}^\circ\text{C}$

(b)   Qualitatively, assuming the same exact initial conditions in the calorimeter, twice as much glucose produces twice as much heat, which raises the calorimeter temperature by twice as many °C. Quantitatively,

$$5.000 \text{ g glucose} \times \frac{15.57 \text{ kJ}}{1 \text{ g glucose}} \times \frac{1°\text{C}}{14.42 \text{ kJ}} = 5.40°\text{C}$$

Check. Units are correct. ΔT is twice as large as in part (a). The result has 3 sig figs, because the heat capacity of the calorimeter is known to 3 sig figs.

5.58   (a) $C = 1.640 \text{ g } HC_7H_5O_2 \times \dfrac{26.38 \text{ kJ}}{1 \text{ g } HC_7H_5O_2} \times \dfrac{1}{4.95°\text{C}} = 8.740 = 8.74 \text{ kJ/}°\text{C}$

       (b) $\dfrac{8.740 \text{ kJ}}{°\text{C}} \times 4.68°\text{C} \times \dfrac{1}{1.320 \text{ g sample}} = 30.99 = 31.0 \text{ kJ/g sample}$

       (c)   If water is lost from the calorimeter, there is less water to heat, so the same amount of heat (kJ) from a reaction would cause a larger increase in the calorimeter temperature. The calorimeter constant, kJ/°C, would decrease, because °C is in the denominator of the expression.

Hess's Law

5.59   Hess's Law is a consequence of the fact that enthalpy is a state function. Since ΔH is independent of path, we can describe a process by any series of steps that add up to the overall process and ΔH for the process is the sum of the ΔH values for the steps.

5.60   (a)   *Analyze/Plan.* Arrange the reactions so that in the overall sum, B, appears in both reactants and products and can be canceled. This is a general technique for using Hess's Law.   *Solve.*

$$
\begin{array}{ll}
A \rightarrow B & \Delta H = +30 \text{ kJ} \\
\underline{B \rightarrow C} & \underline{\Delta H = +60 \text{ kJ}} \\
A \rightarrow C & \Delta H = +30 \text{ kJ}
\end{array}
$$

       (b)   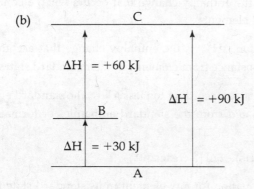

Check. The process of A forming C can be described as A forming B and B forming C.

5.61   *Analyze/Plan.* Follow the logic in Sample Exercise 5.8. Manipulate the equations so that "unwanted" substances can be canceled from reactants and products. Adjust the corresponding sign and magnitude of ΔH.   *Solve.*

$$
\begin{array}{ll}
P_4O_5(s) \rightarrow P_4(s) + 3O_2(g) & \Delta H = 1640.1 \text{ kJ} \\
\underline{P_4(s) + 5O_2(g) \rightarrow P_4O_{10}(s)} & \underline{\Delta H = -2940.1 \text{ kJ}} \\
P_4O_6(s) + 2O_2(g) \rightarrow P_4O_{10}(s) & \Delta H = -1300.0 \text{ kJ}
\end{array}
$$

Check. We have obtained the desired reaction.

5.62 $3H_2(g) + 3/2\,O_2(g) \;\rightarrow\; 3H_2O(g)$ $\Delta H = 3/2(-483.6\text{ kJ})$

                $O_3(g) \;\rightarrow\; 3/2\,O_2(g)$ $\Delta H = 1/2(-284.6\text{ kJ})$

     $3H_2(g) \;+\; O_3(g) \;\rightarrow\; P_4O_{10}(s)$ $\Delta H = -867.7\text{ kJ}$

5.63   *Analyze/Plan.* Follow the logic in Sample Exercise 5.9. Manipulate the equations so that "unwanted" substances can be canceled from reactants and products. Adjust the corresponding sign and magnitude of ΔH.   *Solve.*

         $C_2H_4(g) \;\rightarrow\; 2H_2(g) + 2C(s)$ $\Delta H = -52.3\text{ kJ}$

      $2C(s) + 4F_2(g) \;\rightarrow\; 2CF_4(g)$ $\Delta H = 2(-680\text{ kJ})$

     $2H_2(g) + 2F_2(g) \;\rightarrow\; 4HF(g)$ $\Delta H = 2(-537\text{ kJ})$

     $C_2H_4(g) + 6F_2(g) \rightarrow 2CF_4(g) + 4HF(g)$ $\Delta H = -2.49 \times 10^3\text{ kJ}$

     Check. We have obtained the desired reaction.

5.64 $N_2O(g) \;\rightarrow\; N_2(g) + 1/2\,O_2(g)$ $\Delta H = 1/2\,(-163.2\text{ kJ})$

           $NO_2(g) \;\rightarrow\; NO(g) + 1/2\,O_2(g)$ $\Delta H = 1/2(113.1\text{ kJ})$

     $N_2(g) \;+\; O_2(g) \;\rightarrow\; 2NO(g)$ $\Delta H = 180.7\text{ kJ}$

     $N_2O(g) + NO_2(g) \;\rightarrow 3NO(g)$ $\Delta H = 155.7\text{ kJ}$

Enthalpies of Formation

5.65   (a)   *Standard conditions* for enthalpy changes are usually P = 1 atm and T = 298 K. For the purpose of comparison, standard enthalpy changes, ΔH°, are tabulated for reactions at these conditions.

       (b)   *Enthalpy of formation*, ΔH_f, is the enthalpy change that occurs when a compound is formed from its component elements.

       (c)   Standard enthalpy of formation, ΔH_f°, is the enthalpy change that accompanies formation of one mole of a substance from elements in their standard states.

5.66   (a)   Tables of ΔH_f° are useful because, according to Hess's law, the standard enthalpy of any reaction can be calculated from the standard enthalpies of formation for the reactants and products.

$$\Delta H_{rxn}^\circ = \Sigma \Delta H_f^\circ \,(\text{products}) - \Sigma \Delta H_f^\circ \,(\text{reactants})$$

       (b)   The standard enthalpy of formation for any element in its standard state is zero. Elements in their standard states are the reference point for the enthalpy of formation scale.

       (c) $6C(s) + 6H_2(g) + 3O_2(g) \rightarrow C_6H_{12}O_6(s)$

5.67   (a) $1/2\,N_2(g) + 3/2\,H_2(g) \rightarrow NH_3(g)$ $\Delta H_f^\circ = -46.19\text{ kJ}$

       (b) $1/8\,S_8(s) + O_2(g) \rightarrow SO_2(g)$ $\Delta H_f^\circ = -296.9\text{ kJ}$

       (c) $Rb(s) + 1/2\,Cl_2(g) + 3/2\,O_2(g) \rightarrow RbClO_3(s)$ $\Delta H_f^\circ = -392.4\text{ kJ}$

       (d) $N_2(g) + 2H_2(g) + 3/2\,O_2(g) \rightarrow NH_4NO_3(s)$ $\Delta H_f^\circ = -365.6\text{ kJ}$

5.68 (a) $1/2\, H_2(g) + 1/2\, Br_2(l) \rightarrow HBr(g)$ $\Delta H_f^{\circ} = -36.23\ kJ$

(b) $Ag(s) + 1/2\, N_2(g) + 3/2\, O_2(g) \rightarrow AgNO_3(s)$ $\Delta H_f^{\circ} = -124.4\ kJ$

(c) $2Hg(l) + Cl_2(g) \rightarrow Hg_2Cl_2(s)$ $\Delta H_f^{\circ} = -264.9\ kJ$

(d) $2C(s,\ gr) + 1/2\, O_2(g) + 3H_2(g) \rightarrow C_2H_5OH(l)$ $\Delta H_f^{\circ} = -277.7\ kJ$

5.69 *Plan.* $\Delta H_{rxn}^{\circ} = \Sigma n \Delta H_f^{\circ}\ (products) - \Sigma n \Delta H_f^{\circ}\ (reactants)$. Be careful with coefficients, states, and signs. *Solve.*

$$\Delta H_{rxn}^{\circ} = \Delta H_f^{\circ}\ Al_2O_3(s) + 2\Delta H_f^{\circ}\ Fe(s) - \Delta H_f^{\circ}\ Fe_2O_3 - 2\Delta H_f^{\circ}\ Al(s)$$

$$\Delta H_{rxn}^{\circ} = (-1669.8\ kJ) + 2(0) - (-822.16\ kJ) - 2(0) = -847.6\ kJ$$

5.70 Use heats of formation to calculate ΔH° for the combustion of butane.

$$C_4H_{10}(l) + 13/2\, O_2(g)\ \rightarrow\ 4CO_2(g) + 5H_2O(l)$$

$$\Delta H_{rxn}^{\circ} = 4\Delta H\ CO_2(g) + 5\Delta H_f^{\circ}\ H_2O(l) - \Delta H_f^{\circ}\ C_4H_{10}(l) - 13/2\ \Delta H_f^{\circ}\ O_2(g)$$

$$\Delta H_{rxn}^{\circ} = 4(-393.5\ kJ) + 5(-285.83\ kJ) - (-147.6\ kJ) - 13/2(0) = -2855.6 = -2856\ kJ/mol\ C_4H_{10}$$

$$1.0\ g\ C_4H_{10} \times \frac{1\ mol\ C_4H_{10}}{58.123\ g\ C_4H_{10}} \times \frac{-2855.6\ kJ}{1\ mol\ C_4H_{10}} = -49\ kJ$$

5.71 *Plan.* $\Delta H_{rxn}^{\circ} = \Sigma n \Delta H_f^{\circ}\ (products)\ -\ \Sigma n \Delta H_f^{\circ}\ (reactants\ )$. Be careful with coefficients, states and signs.   *Solve.*

(a) $\Delta H_{rxn}^{\circ} = 2\Delta H_f^{\circ}\ SO_3(g) - 2\Delta H_f^{\circ}\ SO_2(g) - \Delta H_f^{\circ}\ O_2(g)$

$= 2(-395.2\ kJ) - 2(-296.9\ kJ) - 0 = -196.6\ kJ$

(b) $\Delta H_{rxn}^{\circ} = \Delta H_f^{\circ}\ MgO(s) + \Delta H_f^{\circ}\ H_2O(l) - \Delta H_f^{\circ}\ Mg(OH)_2(s)$

$= -601.8\ kJ + (-285.83\ kJ) - (-924.7\ kJ) = 37.1\ kJ$

(c) $\Delta H_{rxn}^{\circ} = \Delta H_f^{\circ}\ CCl_4(l) + 4\Delta H_f^{\circ}\ HCl(g) - \Delta H_f^{\circ}\ CH_4(g) - 4\Delta H_f^{\circ}\ Cl_2(g)$

$= -139.3\ kJ + 4(-92.30\ kJ) - (-74.8\ kJ) - 4(0) = -433.7\ kJ$

(d) $\Delta H_{rxn}^{\circ} = \Delta H_f^{\circ}\ SiO_2(s) + 4\Delta H_f^{\circ}\ HCl(g) - \Delta H_f^{\circ}\ SiCl_4(g) - 2\Delta H_f^{\circ}\ H_2O(l)$

$= -910.9\ kJ + 4(-92.30\ kJ) - (-640.1\ kJ) - 2(-285.83\ kJ) = -68.3\ kJ$

5.72 (a) $\Delta H_{rxn}^{\circ} = 4\Delta H_f^{\circ}\ H_2O(g) + \Delta H_f^{\circ}\ N_2(g) - \Delta H_f^{\circ}\ N_2O_4(g) - 4\Delta H_f^{\circ}\ H_2(g)$

$= 4(-241.82) + 0 - (9.66) - 4(0) = -976.94\ kJ$

(b) $\Delta H_{rxn}^{\circ} = \Delta H_f^{\circ}\ K_2CO_3(s) + \Delta H_f^{\circ}\ H_2O(g) - 2\Delta H_f^{\circ}\ KOH(s) - \Delta H_f^{\circ}\ CO_2(g)$

$= -1150.18\ kJ - 241.82\ kJ - 2(-424.7)\ kJ - (-393.5)\ kJ = -149.1\ kJ$

(c) $\Delta H_{rxn}^{\circ} = 3/8\Delta H_f^{\circ}\ S_8(s) + 2\Delta H_f^{\circ}\ H_2O(g) - \Delta H_f^{\circ}\ SO_2(g) - 2\Delta H_f^{\circ}\ H_2S(g)$

$= 3/8(0) + 2(-241.82) - (-296.9) - 2(-20.17) = -146.4\ kJ$

(d) $\Delta H^{\circ}_{rxn} = 2\Delta H^{\circ}_f\ FeCl_3(s) + 3\Delta H^{\circ}_f\ H_2O(g) - \Delta H^{\circ}_f\ Fe_2O(g) - 6\Delta H^{\circ}_f\ HCl(g)$

           $= 2(-400\ kJ) + 3(-241.82\ kJ) - (-822.16\ kJ) - 6(-92.30\ kJ) = -149.5 = -150\ kJ$

5.73   *Analyze.* Given: combustion reaction, enthalpy of combustion, enthalpies of formation for most reactants and products. Find: enthalpy of formation for acetone.

        Plan. Rearrange the expression for enthalpy of reaction to calculate the desired enthalpy of formation. *Solve.*

        $\Delta H^{\circ}_{rxn} = 3\Delta H^{\circ}_f\ CO_2(g) + 3\Delta H^{\circ}_f\ H_2O(l) - \Delta H^{\circ}_f\ C_3H_6O(l)$

        $-1790\ kJ = 3(-393.5\ kJ) + 3(-285.83\ kJ) - \Delta H^{\circ}_f\ C_3H_6O(l)$

        $\Delta H^{\circ}_f\ C_3H_6O(l) = -248\ kJ$

5.74 $\Delta H^{\circ}_{rxn} = \Delta H^{\circ}_f\ Ca(OH)_2(s) + \Delta H^{\circ}_f\ C_2H_2(g) - 2\Delta H^{\circ}_f\ H_2O(l) - \Delta H^{\circ}_f\ CaC_2(s)$

        $-127.2\ kJ = -986.2\ kJ + 226.7\ kJ - 2(-285.83\ kJ) - \Delta H^{\circ}_f\ CaC_2(s)$

        ΔH°_f for $CaC_2(s) = -60.6\ kJ$

5.75   *Plan.* Use Hess's Law to arrange the given reactions so the overall sum is the formation reaction for $Mg(OH)_2(s)$. Adjust the corresponding ΔH values and calculate ΔH°_f for $Mg(OH)_2(s)$.   *Solve.*

$Mg(s) + 1/2\ O_2(g) \rightarrow MgO(s)$	$\Delta H^{\circ} = 1/2(-1203.6\ kJ)$
$MgO(s) + H_2O(l) \rightarrow Mg(OH)_2(s)$	$\Delta H^{\circ} = -(37.1\ kJ)$
$H_2(g) + 1/2\ O_2(g) \rightarrow H_2O(l)$	$\Delta H^{\circ} = 1/2(-571.7\ kJ)$

        $Mg(s) + O_2(g) + H_2(g) \rightarrow Mg(OH)_2(s)$ $\Delta H^{\circ}_f = -924.8\ kJ$

        Check. The overall reaction is correct.

5.76   (a)

$2B(s) + 3/2\ O_2(g) \rightarrow B_2O_3(s)$	$\Delta H^{\circ} = 1/2(-2509.1\ kJ)$
$3H_2(g) + 3/2\ O_2(g) \rightarrow 3H_2O(l)$	$\Delta H^{\circ} = 3/2(-571.7\ kJ)$
$B_2O_3(s) + 3H_2O(l) \rightarrow B_2H_6(g) + 3O_2(g)$	$\Delta H^{\circ} = -(-2147.5\ kJ)$

        $2B(s) + 3H_2(g) \rightarrow B_2H_6(g)$ $\Delta H^{\circ}_f = +35.4\ kJ$

     (b)   If, like B_2H_6, the combustion of B_5H_9 produces B_2O_3 as the boron-containing product, the heat of combustion of B_5H_9 in addition to data given in part (a) would enable calculation of the heat of formation of B_5H_9.

        The combustion reaction is: $B_5H_9(l) + 6O_2(g) \rightarrow 5/2\ B_2O_3(s) + 9/2\ H_2O(l)$

$5/4\,[4B(s) + 3O_2(g) \rightarrow 2B_2O_3(s)]$	$\Delta H = 5/4(-2509.1\ kJ)$
$9/4\,[2H_2(g) + O_2(g) \rightarrow 2H_2O(l)]$	$\Delta H = 9/4\,(-571.7\ kJ)$
$5/2\ B_2O_3(s) + 9/2\ H_2O(l) \rightarrow B_5H_9(l) + 6O_2(g)$	$\Delta H = -$ (heat of combustion)

        $5B(s) + 9/2\ H_2(g) \rightarrow B_5H_9(l)$ ΔH°_f of $B_5H_9(l)$

        $[\Delta H^{\circ}_f\ B_5H_9(l) = -\,[\text{heat of combustion of } B_5H_9(l)] - 3136.4\ kJ - 1286\ kJ]$

        We need to measure the heat of combustion of $B_5H_9(l)$.

5.77   (a) $C_8H_{18}(l) + 25/2\,O_2(g) \rightarrow 8CO_2(g) + 9H_2O(g)$ $\Delta H^\circ = -5069$ kJ

        (b) $8C(s, gr) + 9H_2(g) \rightarrow C_8H_{18}(l)$ $\Delta H_f^\circ = ?$

        (c)   *Plan.* Follow the logic in Solution 5.73.   *Solve.*

              $\Delta H_{rxn}^\circ = 8\Delta H_f^\circ\,CO_2(g) + 9\Delta H_f^\circ\,H_2O(g) - \Delta H_f^\circ\,C_8H_{18}(l) - 25/2\,\Delta H_f^\circ\,O_2(g)$

              -5069 kJ $= 8(-393.5$ kJ$) + 9(-241.82$ kJ$) - \Delta H_f^\circ\,C_8H_{18}(l) - 25/2(0)$

              $\Delta H_f^\circ\,C_8H_{18}(l) = 8(-393.5$ kJ$) + 9(-241.82$ kJ$) + 5069$ kJ $= -255$ kJ

5.78   (a) $10C(s) + 4H_2(g) \rightarrow C_{10}H_8(s)$ formation

              $C_{10}H_8(s) + 12O_2(g) \rightarrow 10CO_2(g) + 4H_2O(l)$ combustion, $\Delta H^\circ = -5154$ kJ

        (b) $\Delta H_{rxn}^\circ = 10\Delta H_f^\circ\,CO_2(g) + 4\Delta H_f^\circ\,H_2O(l) - \Delta H_f^\circ\,C_{10}H_8 - 12\Delta H_f^\circ\,O_2(g)$

              $-5154 = 10(-393.5$ kJ$) + 4(-285.83$ kJ$) - \Delta H_f^\circ\,C_{10}H_8(s) - 12(0)$

              $\Delta H_f^\circ\,C_{10}H_8(s) = 10(-393.5$ kJ$) + 4(-285.83$ kJ$) + 5154$ kJ $= 76$ kJ

        Check. The result has 0 decimal places because the heat of combustion has 0 decimal places.

Foods and Fuels

5.79   (a)   *Fuel value* is the amount of heat produced when 1 gram of a substance (fuel) is combusted.

        (b)   The fuel value of fats is 9 kcal/g and of carbohydrates is 4 kcal/g. Therefore, 5 g of fat produce 45 kcal, while 9 g of carbohydrates produce 36 kcal; 5 g of fat are a greater energy source.

5.80   (a)   Fats are appropriate for fuel storage because they are insoluble in water (and body fluids) and have a high fuel value.

        (b)   For convenience, assume 100 g of chips.

              $12\text{ g protein} \times \dfrac{17\text{ kJ}}{1\text{ g protein}} \times \dfrac{1\text{ Cal}}{4.184\text{ kJ}} = 48.76 = 49\text{ Cal}$

              $14\text{ g fat} \times \dfrac{38\text{ kJ}}{1\text{ g fat}} \times \dfrac{1\text{ Cal}}{4.184\text{ kJ}} = 127.15 = 130\text{ Cal}$

              $74\text{ g carbohydrates} \times \dfrac{17\text{ kJ}}{1\text{ g carbohydrates}} \times \dfrac{1\text{ Cal}}{4.184\text{ kJ}} = 300.67 = 301\text{ Cal}$

              total Cal $= (48.76 + 127.15 + 300.67) = 476.58 = 480\text{ Cal}$

              $\%\text{ Cal from fat} = \dfrac{127.15\text{ Cal fat}}{476.58\text{ total Cal}} \times 100 = 26.68 = 27\%$

        (Since the conversion from kJ to Cal was common to all three components, we would have determined the same percentage by using kJ.)

(c) $25\,g\,fat \times \dfrac{38\,kJ}{g\,fat} = x\,g\,protein \times \dfrac{17\,kJ}{g\,protein};$ $x = 56\,g\,protein$

5.81   *Plan.* Calculate the Cal (kcal due to each nutritional component of the Campbell's® soup, then sum.   *Solve.*

$$9\,g\,carbohydrates \times \dfrac{17\,kJ}{1\,g\,carbohydrate} = 153\,or\,2 \times 10^2\,kJ$$

$$1\,g\,protein \times \dfrac{17\,kJ}{1\,g\,protein} = 17\,or\,0.2 \times 10^2\,kJ$$

$$7\,g\,fat \times \dfrac{38\,kJ}{1\,g\,fat} = 266\,or\,3 \times 10^2\,kJ$$

total energy = 153 kJ + 17 kJ + 266 kJ = 436 or 4 × 10² kJ

$$436\,kJ \times \dfrac{1\,kcal}{4.184\,kJ} \times \dfrac{1\,Cal}{1\,kcal} = 104\,or\,1 \times 10^2\,Cal/serving$$

Check. 100 Cal/serving is a reasonable result; units are correct. The data and the result have 1 sig fig.

5.82   Calculate the fuel value in a pound of M&M® candies.

$$96\,fat \times \dfrac{38\,kJ}{1\,g\,fat} = 3648\,kJ = 3.6 \times 10^3\,kJ$$

$$320\,g\,carbohydrate \times \dfrac{17\,kJ}{1\,g\,carbohydrate} = 5440\,kJ = 5.4 \times 10^3\,kJ$$

$$21\,g\,protein \times \dfrac{17\,kJ}{1\,g\,protein} = 357\,kJ = 3.6 \times 10^2\,kJ$$

total fuel value = 3648 kJ + 5440 kJ + 357 kJ = 9445 kJ = 9.4 × 10³ kJ/lb

$$\dfrac{9445\,kJ}{lb} \times \dfrac{1\,lb}{453.6\,g} \times \dfrac{42\,g}{serving} = 874.5\,kJ = 8.7 \times 10^2\,kJ/serving$$

$$\dfrac{874.5\,kJ}{serving} \times \dfrac{1\,kcal}{4.184\,kJ} \times \dfrac{1\,Cal}{1\,kcal} = 209.0\,Cal = 2.1 \times 10^2\,Cal/serving$$

Check. 210 Cal is the approximate food value of a candy bar, so the result is reasonable.

5.83   *Plan.* g → mol → kJ → Cal   *Solve.*

$$16.0\,g\,C_6H_{12}O_6 \times \dfrac{1\,mol\,C_6H_{15}O_6}{180.2\,g\,C_6H_{12}O_6} \times \dfrac{2812\,kJ}{mol\,C_6H_{12}O_6} \times \dfrac{1\,Cal}{4.184\,kJ} = 59.7\,Cal$$

Check. 60 Cal is a reasonable result for most of the food value in an apple.

5.84 $$177\,mL \times \dfrac{1.0\,g\,wine}{1\,mL} \times \dfrac{0.106\,g\,ethanol}{1\,g\,wine} \times \dfrac{1\,mol\,ethanol}{46.1\,g\,ethanol} \times \dfrac{1367\,kJ}{1\,mol\,ethanol} \times \dfrac{1\,Cal}{4.184\,kJ}$$

$$= 133 = 1.3 \times 10^2\,Cal$$

Check. A "typical" 6 oz. glass of wine has 150–250 Cal, so this is a reasonable result. Note that alcohol is responsible for most of the food value of wine.

5.85   *Plan.* Use enthalpies of formation to calculate molar heat (enthalpy) of combustion using Hess's Law. Use molar mass to calculate heat of combustion per kg of hydrocarbon.   *Solve.*

Propyne: $C_3H_4(g) + 4O_2(g) \rightarrow 3CO_2(g) + 2H_2(g)$

(a) $\Delta H = 3(-393.5 \text{ kJ}) + 2(-241.82 \text{ kJ}) - (185.4 \text{ kJ}) - 4(0) = -1849.5$
$$= -1850 \text{ kJ/mol } C_3H_4$$

(b) $\dfrac{-1849.5 \text{ kJ}}{1 \text{ mol } C_3H_4} \times \dfrac{1 \text{ mol } C_3H_4}{40.065 \text{ g } C_3H_4} \times \dfrac{1000 \text{ g } C_3H_4}{1 \text{ kg } C_3H_4} = -4.616 \times 10^4 \text{ kJ/kg } C_3H_4$

Propylene: $C_3H_6(g) + 9/2 \, O_2(g) \rightarrow 3CO_2(g) + 3H_2O(g)$

(a) $\Delta H = 3(-393.5 \text{ kJ}) + 3(-241.82 \text{ kJ}) - (20.4 \text{ kJ}) - 9/2(0) = -1926.4$
$$= -1926 \text{ kJ/mol } C_3H_6$$

(b) $\dfrac{-1926.4 \text{ kJ}}{1 \text{ mol } C_3H_6} \times \dfrac{1 \text{ mol } C_3H_6}{42.080 \text{ g } C_3H_6} \times \dfrac{1000 \text{ g } C_3H_6}{1 \text{ kg } C_3H_6} = -4.578 \times 10^4 \text{ kJ/kg } C_3H_6$

Propane: $C_3H_8(g) + 5O_2(g) \rightarrow 3CO_2(g) + 4H_2O(g)$

(a) $\Delta H = 3(-393.5 \text{ kJ}) + 4(-241.82 \text{ kJ}) - (-103.8 \text{ kJ}) - 5(0) = -2044.0$
$$= -2044 \text{ kJ/mol } C_3H_8$$

(b) $\dfrac{-2044.0 \text{ kJ}}{1 \text{ mol } C_3H_8} \times \dfrac{1 \text{ mol } C_3H_8}{44.096 \text{ g } C_3H_8} \times \dfrac{1000 \text{ g } C_3H_8}{1 \text{ kg } C_3H_8} = -4.635 \times 10^4 \text{ kJ/kg } C_3H_8$

(c)   These three substances yield nearly identical quantities of heat per unit mass, but propane is marginally higher than the other two.

5.86 $\Delta H^\circ_{rxn} = \Delta H^\circ_f \, CO_2(g) + 2\Delta H^\circ_f \, H_2O(g) - \Delta H^\circ_f \, CH_4(g) - \Delta H^\circ_f \, O_2(g)$

$= -393.5 \text{ kJ} + 2(-241.82 \text{ kJ}) - (-74.8 \text{ kJ}) - 0 \text{ kJ} = -802.3 \text{ kJ}$

$\Delta H^\circ_{rxn} = \Delta H^\circ_f \, CF_4(g) + 4\Delta H^\circ_f \, HF(g) - \Delta H^\circ_f \, CH_4(g) - 4\Delta H^\circ_f \, F_2(g)$

$= -679.9 \text{ kJ} + 4(-268.61 \text{ kJ}) - (-74.8 \text{ kJ}) - 0 \text{ kJ} = -1679.5 \text{ kJ}$

The second reaction is twice as exothermic as the first. The "fuel values" of hydrocarbons in a fluorine atmosphere are approximately twice those in an oxygen atmosphere. Note that the difference in ΔH° values for the two reactions is in the ΔH°_f for the products, since the ΔH°_f for the reactants is identical.

Additional Exercises

5.87   (a) $mi/hr \rightarrow m/s$

$1050 \, \dfrac{mi}{hr} \times \dfrac{1.6093 \text{ km}}{1 \text{ mi}} \times \dfrac{1000 \text{ m}}{1 \text{ km}} \times \dfrac{1 \text{ hr}}{3600 \text{ s}} = 469.38 = 469.4 \text{ m/s}$

(b)   Find the mass of one N_2 molecule in kg.

$\dfrac{28.0134 \text{ g } N_2}{1 \text{ mol}} \times \dfrac{1 \text{ mol}}{6.022 \times 10^{23} \text{ molecules}} \times \dfrac{1 \text{ kg}}{1000 \text{ g}} = 4.6518 \times 10^{-26}$

$= 4.652 \times 10^{-26} \text{ kg}$

$$E_k = 1/2 \, mv^2 = 1/2 \times 4.6518 \times 10^{-26} \text{ kg} \times (469.38 \text{ m/s})^2$$

$$= 5.1244 \times 10^{-21} \frac{\text{kg} \cdot \text{m}^2}{\text{s}^2} = 5.124 \times 10^{-21} \text{ J}$$

(c) $\dfrac{5.1244 \times 10^{21} \text{ J}}{\text{molecule}} \times \dfrac{6.022 \times 10^{23} \text{ molecules}}{1 \text{ mol}} = 3086 \text{ J/mol} = 3.086 \text{ kJ/mol}$

5.88   (a) $E_p = mgh = 52.0 \text{ kg} \times 9.81 \text{ m/s}^2 \times 10.8 \text{ m} = 5509.3 \text{ J} = 5.51 \text{ kJ}$

       (b) $E_k = 1/2 \, mv^2; \; v = (2E_k/m)^{1/2} = \left(\dfrac{2 \times 5509 \text{ kg} \cdot \text{m}^2/\text{s}^2}{52.0 \text{ kg}} \right)^{1/2} = 14.6 \text{ m/s}$

       (c)   Yes, the diver does work on entering (pushing back) the water in the pool.

5.89   In the process described, one mole of solid CO_2 is converted to one mole of gaseous CO_2. The volume of the gas is much greater than the volume of the solid. Thus the system (that is, the mole of CO_2) must work against atmospheric pressure when it expands. To accomplish this work while maintaining a constant temperature requires the absorption of additional heat beyond that required to increase the internal energy of the CO_2. The remaining energy is turned into work.

5.90   Like the combustion of $H_2(g)$ and $O_2(g)$ described in Section 5.4, the reaction that inflates airbags is spontaneous after initiation. Spontaneous reactions are usually exothermic, $-\Delta H$. The airbag reaction occurs at constant atmospheric pressure, $\Delta H = q_p$; both are likely to be large and negative. When the bag inflates, work is done by the system on the surroundings, so the sign of w is negative.

5.91   Freezing is an exothermic process (the opposite of melting, which is clearly endothermic). When the system, the soft drink, freezes, it releases energy to the surroundings, the can. Some of this energy does the work of splitting the can.

5.92   (a) $q = 0, \, w > 0$ (work done to system), $\Delta E > 0$

       (b)   Since the system (the gas) is losing heat, the sign of q is negative. The changes in state described in cases (a) and (b) are identical and ΔE is the same in both cases. The distribution of energy transferred as either work or heat is different in the two scenarios. In case (b), more work is required to compress the gas because some heat is lost to the surroundings. (The moral of this story is that the more energy lost by the system as heat, the greater the work on the system required to accomplish the desired change.)

5.93 $\Delta E = q + w = +38.95 \text{ kJ} - 2.47 \text{ kJ} = +36.48 \text{ kJ}$

       $\Delta H = q_p = +38.95 \text{ kJ}$

5.94   If a function sometimes depends on path, then it is simply not a state function. Enthalpy is a state function, so ΔH for the two pathways leading to the same change of state pictured in Figure 5.10 must be the same. However, q is not the same for the both. Our conclusion must be that $\Delta H \neq q$ for these pathways. The condition for $\Delta H = q_p$ (other than constant pressure) is that the only possible work on or by the system is pressure-volume work. Clearly, the work being done in this scenario is not pressure-volume work, so $\Delta H \neq q$, even though the two changes occur at constant pressure.

5.95   Find the heat capacity of 1.7×10^3 gal H_2O.

$$C_{H_2O} = 1.7 \times 10^3 \text{ gal } H_2O \times \frac{4 \text{ qt}}{1 \text{ gal}} \times \frac{1 \text{ L}}{1.057 \text{ qt}} \times \frac{1 \times 10^3 \text{ cm}^3}{1 \text{ L}} \times \frac{1 \text{ g}}{1 \text{ cm}^3} \times \frac{4.184 \text{ J}}{1 \text{ g} \bullet {}^\circ C}$$

$$= 2.692 \times 10^7 \text{ J}/{}^\circ C = 2.7 \times 10^4 \text{ kJ}/{}^\circ C; \text{ then,}$$

$$\frac{2.692 \times 10^7 \text{ J}}{1^\circ C} \times \frac{1^\circ C \bullet g}{0.85 \text{ J}} \times \frac{1 \text{ kg}}{1 \times 10^3 \text{ g}} \times \frac{1 \text{ brick}}{1.8 \text{ kg}} = 1.8 \times 10^4 \text{ or } 18,000 \text{ bricks}$$

Check. $(1.7 \times \sim 16 \times 10^6)/(\sim 1.6 \times 10^3) \approx 17 \times 10^3$ bricks; the units are correct.

5.96   (a) $q_{Cu} = \dfrac{0.385 \text{ J}}{g \bullet K} \times 121.0 \text{ g Cu} \times (30.1^\circ C - 100.4^\circ C) = -3274.9 = -3.27 \times 10^3 \text{ J}$

The negative sign indicates the 3.27×10^3 J are lost by the Cu block.

        (b) $q_{H_2O} = \dfrac{4.184 \text{ J}}{g \bullet K} \times 150.0 \text{ g } H_2O \times (30.1^\circ C - 25.1^\circ C) = 3138 = 3.1 \times 10^3 \text{ J}$

The positive sign indicates that 3.14×10^3 J are gained by the H_2O.

        (c)   The difference in the heat lost by the Cu and the heat gained by the water is 3.275 $\times 10^3$ J – 3.138×10^3 J = 0.137×10^3 J = 1×10^2 J. The temperature change of the calorimeter is 5.0°C. The heat capacity of the calorimeter in J/K is

$$0.137 \times 10^3 \text{ J} \times \frac{1}{5.0^\circ C} = 27.4 = 3 \times 10 \text{ J/K}.$$

Since q_{H_2O} is known to one decimal. place, the difference has one decimal place and the result has 1 sig fig.

If the rounded results from (a) and (b) are used,

$$C_{calorimeter} = \frac{0.2 \times 10^3 \text{ J}}{5.0^\circ C} = 4 \times 10 \text{ J/K}.$$

        (d) $q_{H_2O} = 3.275 \times 10^3 \text{ J} = \dfrac{4.184 \text{ J}}{g \bullet K} \times 150.0 \text{ g} \times (\Delta T)$

$\Delta T = 5.22°C$; $T_f = 25.1°C + 5.22°C = 30.3°C$

5.97   (a)   From the mass of benzoic acid that produces a certain temperature change, we can calculate the heat capacity of the calorimeter.

$$\frac{0.235 \text{ g benzoic acid}}{1.642^\circ C \text{ change observed}} \times \frac{26.38 \text{ kJ}}{1 \text{ g benzoic acid}} = 3.7755 = 3.78 \text{ kJ}/{}^\circ C$$

Now we can use this experimentally determined heat capacity with the data for caffeine.

$$\frac{1.525^\circ C \text{ rise}}{0.265 \text{ g caffeine}} \times \frac{3.7755 \text{ kJ}}{1^\circ C} \times \frac{194.2 \text{ g caffeine}}{1 \text{ mol caffeine}} = 4.22 \times 10^3 \text{ kJ/mol caffeine}$$

        (b)   The overall uncertainty is approximately equal to the sum of the uncertainties due to each effect. The uncertainty in the mass measurement is 0.001/0.235 or 0.001/0.265, about 1 part in 235 or 1 part in 265. The uncertainty in the temperature measurements is 0.002/1.642 or 0.002/1.525, about 1 part in 820 or 1 part in 760. Thus the uncertainty in heat of combustion from each measurement is

$$\frac{4220}{235} = 18\,\text{kJ}; \quad \frac{4220}{265} = 16\,\text{kJ}; \quad \frac{4220}{820} = 5\,\text{kJ}; \quad \frac{4220}{760} = 6\,\text{kJ}$$

The sum of these uncertainties is 45 kJ. In fact, the overall uncertainty is less than this because independent errors in measurement do tend to partially cancel.

5.98   *Plan.* Use the heat capacity of H_2O to calculate the energy required to heat the water. Use the enthalpy of combustion of CH_4 to calculate the amount of CH_4 needed to provide this amount of energy, assuming 100% transfer. Assume $H_2O(l)$ is the product of combustion at standard conditions.

$$1.00\,\text{kg}\,H_2O \times \frac{1000\,\text{g}}{1\,\text{kg}} \times \frac{4.184\,\text{J}}{\text{g}\cdot{}^\circ\text{C}} \times (90.0^\circ\text{C} - 25.0^\circ\text{C}) = 2.720 \times 10^5\,\text{J} = 272\,\text{kJ required}$$

$$CH_4(g) + 2O_2(g) \rightarrow CO_2(g) + 2H_2O(l)$$

$$\Delta H_{rxn}^\circ = \Delta H_f^\circ\,CO_2(g) + 2\Delta H_f^\circ\,H_2O(l) - \Delta H_f^\circ\,CH_4(g) - \Delta H_f^\circ\,O_2(g)$$

$$= -393.5\,\text{kJ} + 2(-285.83\,\text{kJ}) - (-74.8\,\text{kJ}) - (0) = -890.4\,\text{kJ}$$

$$2.720 \times 10^2\,\text{kJ} \times \frac{1\,\text{mol}\,CH_4}{-890.4\,\text{kJ}} \times \frac{16.04\,\text{g}\,CH_4}{\text{mol}\,CH_4} = 4.899 = 4.90\,\text{g}\,CH_4$$

5.99   (a)   For comparison, balance the equations so that 1 mole of CH_4 is burned in each.

$$CH_4(g) + O_2(g) \rightarrow C(s) + 2H_2O(l) \qquad \Delta H^\circ = -496.9\,\text{kJ}$$

$$CH_4(g) + 3/2\,O_2(g) \rightarrow CO(g) + 2H_2O(l) \qquad \Delta H^\circ = -607.4\,\text{kJ}$$

$$CH_4(g) + 2O_2(g) \rightarrow CO_2(g) + 2H_2O(l) \qquad \Delta H^\circ = -890.4\,\text{kJ}$$

(b) $$\Delta H_{rxn}^\circ = \Delta H_f^\circ\,C(s) + 2\Delta H_f^\circ\,H_2O(l) - \Delta H_f^\circ\,CH_4(g) - \Delta H_f^\circ\,O_2(g)$$

$$= 0 + 2(-285.83\,\text{kJ}) - (-74.8) - 0 = -496.9\,\text{kJ}$$

$$\Delta H_{rxn}^\circ = \Delta H_f^\circ\,CO(g) + 2\Delta H_f^\circ\,H_2O(l) - \Delta H_f^\circ\,CH_4(g) - 3/2\,\Delta H_f^\circ\,O_2(g)$$

$$= (-110.5\,\text{kJ}) + 2(-285.83\,\text{kJ}) - (-74.8\,\text{kJ}) - 3/2(0) = -607.4\,\text{kJ}$$

$$\Delta H_{rxn}^\circ = \Delta H_f^\circ\,CO_2(g) + 2\Delta H_f^\circ\,H_2O(l) - \Delta H_f^\circ\,CH_4(g) - 2\Delta H_f^\circ\,O_2(g)$$

$$= -393.5\,\text{kJ} + 2(-285.83\,\text{kJ}) - (-74.8\,\text{kJ}) - 2(0) = -890.4\,\text{kJ}$$

(c)   Assuming that $O_2(g)$ is present in excess, the reaction that produces $CO_2(g)$ represents the most negative ΔH per mole of CH_4 burned. More of the potential energy of the reactants is released as heat during the reaction to give products of lower potential energy. The reaction that produces $CO_2(g)$ is the most "downhill" in enthalpy.

5.100   For nitroethane:

$$\frac{1368\,\text{kJ}}{1\,\text{mol}\,C_2H_5NO_2} \times \frac{1\,\text{mol}\,C_2H_5NO_2}{75.072\,\text{g}\,C_2H_5NO_2} \times \frac{1.052\,\text{g}\,C_2H_5NO_2}{1\,\text{cm}^3} = 19.17\,\text{kJ/cm}^3$$

For ethanol:

$$\frac{1367 \text{ kJ}}{1 \text{ mol } C_2H_5OH} \times \frac{1 \text{ mol } C_2H_5OH}{46.069 \text{ g } C_2H_5OH} \times \frac{0.789 \text{ g } C_2H_5OH}{1 \text{ cm}^3} = 23.4 \text{ kJ/cm}^3$$

For methylhydrazine:

$$\frac{1305 \text{ kJ}}{1 \text{ mol } CH_6N_2} \times \frac{1 \text{ mol } CH_6N_2}{46.072 \text{ g } CH_6N_2} \times \frac{0.874 \text{ g } CH_6N_2}{1 \text{ cm}^3} = 24.8 \text{ kJ/cm}^3$$

Thus, **methylhydrazine** would provide the most energy per unit volume, with ethanol a close second.

5.101   (a) $3C_2H_2(g) \rightarrow C_6H_6(l)$

$$\Delta H^{\circ}_{rxn} = \Delta H^{\circ}_f C_6H_6(l) - 3\Delta H^{\circ}_f C_2H_2(g) = 49.0 \text{ kJ} - 3(226.7 \text{ kJ}) = -631.1 \text{ kJ}$$

(b)   Since the reaction is exothermic (ΔH is negative), the product, 1 mole of $C_6H_6(l)$, has less enthalpy than the reactants, 3 moles of $C_2H_2(g)$.

(c)   The fuel value of a substance is the amount of heat (kJ) produced when 1 gram of the substance is burned. Calculate the molar heat of combustion (kJ/mol) and use this to find kJ/g of fuel.

$C_2H_2(g) + 5/2 O_2(g) \rightarrow 2CO_2(g) + H_2O(l)$

$$\Delta H^{\circ}_{rxn} = 2\Delta H^{\circ}_f CO_2(g) + \Delta H^{\circ}_f H_2O(l) - \Delta H^{\circ}_f C_2H_2(g) - 5/2 \Delta H^{\circ}_f O_2(g)$$

$$= 2(-393.5 \text{ kJ}) + (-285.83 \text{ kJ}) - 226.7 \text{ kJ} - 5/2 (0) = -1299.5 \text{ kJ/mol } C_2H_2$$

$$\frac{-1299.5 \text{ kJ}}{1 \text{ mol } C_2H_2} \times \frac{1 \text{ mol } C_2H_2}{26.036 \text{ g } C_2H_2} = 49.912 = 50 \text{ kJ/g}$$

$C_6H_6(g) + 15/2 O_2(g) \rightarrow 6CO_2(g) + 3H_2O(l)$

$$\Delta H^{\circ}_{rxn} = 6\Delta H^{\circ}_f CO_2(g) + 3\Delta H^{\circ}_f H_2O(l) - \Delta H^{\circ}_f C_6H_6(l) - 15/2 \Delta H^{\circ}_f O_2(g)$$

$$= 6(-393.5 \text{ kJ}) + 3(-285.83 \text{ kJ}) - 49.0 \text{ kJ} - 15/2 (0) = -3267.5 \text{ kJ/mol } C_6H_6$$

$$\frac{-3267.5 \text{ kJ}}{1 \text{ mol } C_6H_6} \times \frac{1 \text{ mol } C_6H_6}{78.114 \text{ g } C_6H_6} = 41.830 = 42 \text{ kJ/g}$$

5.102   The reaction for which we want ΔH is:

$$4NH_3(l) + 3O_2(g) \rightarrow 2N_2(g) + 6H_2O(g)$$

Before we can calculate ΔH for this reaction, we must calculate ΔH_f for $NH_3(l)$.

We know that ΔH_f for $NH_3(g)$ is –46.2 kJ, and that for $NH_3(l) \rightarrow NH_3(g)$, $\Delta H = 23.2$ kJ

Thus, $\Delta H_{vap} = \Delta H_f NH_3(g) - \Delta H_f NH_3(l)$.

23.2 kJ = –46.2 kJ – $\Delta H_f NH_3(l)$; $\Delta H_f NH_3(l) = -69.4$ kJ/mol

Then for the overall reaction, the enthalpy change is:

$$\Delta H_{rxn} = 6\Delta H_f H_2O(g) + 2\Delta H_f N_2(g) - 4\Delta H_f NH_3(l) - 3\Delta H_f O_2$$

$$= 6(-241.82 \text{ kJ}) + 2(0) - 4(-69.4 \text{ kJ}) - 3(0) = -1173.3 \text{ kJ}$$

$$\frac{-1173.3\,\text{kJ}}{4\,\text{mol}\,NH_3} \times \frac{1\,\text{mol}\,NH_3}{17.0\,\text{g}\,NH_3} = \frac{0.81\,\text{g}\,NH_3}{1\,\text{cm}^3} \times \frac{1000\,\text{cm}^3}{1\,L} = \frac{1.4\times10^4\,\text{kJ}}{L\,NH_3}$$

(This result has two significant figures because the density is expressed to two figures.)

$2CH_3OH(l) + 3O_2(g) \rightarrow 2CO_2(g) + 4H_2O(g)$

$\Delta H = 2(-393.5\,\text{kJ}) + 4(-241.82\,\text{kJ}) - 2(-239\,\text{kJ}) - 3(0) = -1276\,\text{kJ}$

$$\frac{-1276\,\text{kJ}}{2\,\text{mol}\,CH_3OH} \times \frac{1\,\text{mol}\,CH_3OH}{32.04\,\text{g}\,CH_3OH} \times \frac{0.792\ \text{g}\,CH_3OH}{1\,\text{cm}^3} \times \frac{1000\,\text{cm}^3}{1\,L} = \frac{1.58\times10^4\,\text{kJ}}{L\,CH_3OH}$$

In terms of heat obtained per unit volume of fuel, methanol is a slightly better fuel than liquid ammonia.

5.103   **1,3-butadiene**, C_4H_6, MM = 54.092 g/mol

(a) $C_4H_6(g) + 11/2\,O_2(g) \rightarrow 4CO_2(g) + 3H_2O(l)$

   $\Delta H_{rxn}^{\circ} = 4\Delta H_f^{\circ}\,CO_2(g) + 3\Delta H_f^{\circ}\,H_2O(l) - \Delta H_f^{\circ}\,C_4H_6(g) - 11/2\,\Delta H_f^{\circ}\,O_2(g)$

   $= 4(-393.5\,\text{kJ}) + 3(-285.83\,\text{kJ}) - 111.9\,\text{kJ} + 11/2\,(0) = -2543.4\,\text{kJ/mol}\,C_4H_6$

(b) $\dfrac{-2543.4\,\text{kJ}}{1\,\text{mol}\,C_4H_6} \times \dfrac{1\,\text{mol}\,C_4H_6}{54.092\,\text{g}} = 47.020 \rightarrow 47\,\text{kJ/g}$

(c) $\% H = \dfrac{6(1.008)}{54.092} \times 100 = 11.18\%\,H$

1-butene, C_4H_8, MM = 56.108 g/mol

(a) $C_4H_8(g) + 6O_2(g) \rightarrow 4CO_2(g) + 4H_2O(l)$

   $\Delta H_{rxn}^{\circ} = 4\Delta H_f^{\circ}\,CO_2(g) + 4\Delta H_f^{\circ}\,H_2O(l) - \Delta H_f^{\circ}\,C_4H_8(g) - 6\Delta H_f^{\circ}\,O_2(g)$

   $= 4(-393.5\,\text{kJ}) + 4(-285.83\,\text{kJ}) - 1.2\,\text{kJ} - 6(0) = -2718.5\,\text{kJ/mol}\,C_4H_8$

(b) $\dfrac{-2718.5\,\text{kJ}}{1\,\text{mol}\,C_4H_8} \times \dfrac{1\,\text{mol}\,C_4H_8}{56.108\,\text{g}\,C_4H_8} = 48.451 \rightarrow 48\,\text{kJ/g}$

(c) $\% H = \dfrac{8(1.008)}{56.108} \times 100 = 14.37\%\,H$

n-butane, $C_4H_{10}(g)$, MM = 58.124 g/mol

(a) $C_4H_{10}(g) + 13/2\,O_2(g) \rightarrow 4CO_2(g) + 5H_2O(l)$

   $\Delta H_{rxn}^{\circ} = 4\Delta H_f^{\circ}\,CO_2(g) + 5\Delta H_f^{\circ}\,H_2O(l) - \Delta H_f^{\circ}\,C_4H_{10}(g) - 13/2\,\Delta H_f^{\circ}\,O_2(g)$

   $= 4(-393.5\,\text{kJ}) + 5(-285.83\,\text{kJ}) - (-124.7\,\text{kJ}) - 13/2(0)$

   $= -2878.5\,\text{kJ/mol}\,C_4H_{10}$

(b) $\dfrac{-2878.5\,\text{kJ}}{1\,\text{mol}\,C_4H_{10}} \times \dfrac{1\,\text{mol}\,C_4H_{10}}{58.124\,\text{g}\,C_4H_{10}} = 49.523 \rightarrow 50\,\text{kJ/g}$

(c) $\% H = \dfrac{10(1.008)}{58.124} \times 100 = 17.34\%\,H$

(d) It is certainly true that as the mass % H increases, the fuel value (kJ/g) of the hydrocarbon increases, given the same number of C atoms. A graph of the data in parts (b) and (c) (see below) suggests that mass % H and fuel value are directly proportional when the number of C atoms is constant.

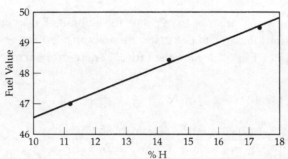

5.104   (a) $C_6H_{12}O_6(s) + 6O_2(g) \rightarrow 6CO_2(g) + 6H_2O(l)$

$\Delta H^\circ_{rxn} = 6\Delta H^\circ_f\, CO_2(g) + 6\Delta H^\circ_f\, H_2O(l) - \Delta H^\circ_f\, C_6H_{12}O_6(s) - 6\Delta H^\circ_f\, O_2(g)$

$= 6(-393.5\ kJ) + 6(-285.83\ kJ) - (-1273\ kJ) - 6(0)$

$= -2803\ kJ/mol\ C_6H_{12}O_6$

$C_{12}H_{22}O_{11}(s) + 12O_2(g) \rightarrow 12CO_2(g) + 11H_2O(l)$

$\Delta H^\circ_{rxn} = 12\Delta H^\circ_f\, CO_2(g) + 11\Delta H^\circ_f\, H_2O(l) - \Delta H^\circ_f\, C_{12}H_{22}O_{11}(s) - 12\Delta H^\circ_f\, O_2(g)$

$= 12(-393.5\ kJ) + 11(-285.83\ kJ) - (-2221\ kJ) - 12(0)$

$= -5645\ kJ/mol\ C_{12}H_{22}O_{11}$

(b) $\dfrac{-2803\ kJ}{1\,mol\ C_6H_{12}O_6} \times \dfrac{1\,mol\ C_6H_{12}O_6}{180.2\ g\ C_6H_{12}O_6} = -\dfrac{15.55\ kJ}{1\,g\ C_6H_{12}O_6} \rightarrow 16\ kJ/g\ C_6H_{12}O_6\ \text{(fuel value)}$

$\dfrac{-5645\ kJ}{1\,mol\ C_{12}H_{22}O_{11}} \times \dfrac{1\,mol\ C_{12}H_{22}O_{11}}{342.3\ g\ C_{12}H_{22}O_{11}} = -\dfrac{16.49\ kJ}{1\,g\ C_{12}H_{22}O_{11}} \rightarrow 16\ kJ/g\ C_{12}H_{22}O_{11}$

(fuel value)

(c)   The average fuel value of carbohydrates (Section 5.8) is 17 kJ/g. These two carbohydrates have fuel values (16 kJ/g), slightly lower but in line with this average. (More complex carbohydrates supply more energy and raise the average value.)

5.105 $\Delta E_p = m\,g\,\Delta h.$ Be careful with units. $1\ J = 1\ kg \bullet m^2/s^2$

$200\ lb \times \dfrac{1\ kg}{2.205\ lb} \times \dfrac{9.81\ m}{s^2} \times \dfrac{45\ ft}{time} \times \dfrac{1\ yd}{3\ ft} \times \dfrac{1\ m}{1.0936\ yd} \times 20\ times$

$= 2.441 \times 10^5\ kg \bullet m^2/s^2 = 2.441 \times 10^5\ J = 2.4 \times 10^2\ kJ$

$1\ Cal = 1\ kcal = 4.184\ kJ$

$2.441 \times 10^2\ kJ \times \dfrac{1\ Cal}{4.184\ kJ} = 58.34 = 58\ Cal$

If all work is used to increase the man's potential energy, 20 rounds of stair-climbing will not compensate for one extra order of 245 Cal fries. In fact, more than 58 Cal of work will be required to climb the stairs, because some energy is required to move limbs and some energy will be lost as heat (see Solution 5.92).

5.106 *Plan.* Use dimensional analysis to calculate the amount of solar energy supplied per m^2 in 1 hr. Use stoichiometry to calculate the amount of plant energy used to produce sucrose per m^2 in 1 hr. Calculate the ratio of energy for sucrose to total solar energy, per m^2 per hr.

Solve. 1 W = 1 J/s, 1 kW = 1 kJ/s

$$\frac{1.0\,kW}{m^2} = \frac{1.0\,kJ/s}{m^2} = \frac{1.0\,kJ}{m^2 \cdot s} \times \frac{60\,s}{1\,min} \times \frac{60\,min}{1\,hr} = \frac{3.6\times10^3\,kJ}{m^2 \cdot hr}$$

$$\frac{5645\,kJ}{mol\,sucrose} \times \frac{1\,mol\,sucrose}{342.3\,g\,sucrose} \times \frac{0.20\,g\,sucrose}{m^2 \cdot hr} = 3.298 = 3.3\,kJ/m^2 \cdot hr$$

for sucrose production

$$\frac{3.298\,kJ\,for\,sucrose}{3.6\times10^3\,kJ\,total\,solar} \times 100 = 0.092\%\ sunlight\ used\ to\ produce\ sucrose$$

5.107 (a) $6CO_2(g) + 6H_2O(l) \rightarrow C_6H_{12}O_6(s) + 6O_2(g)$, $\Delta H° = 2803$ kJ

This is the reverse of the combustion of glucose (Section 5.8 and Solution 5.104), so $\Delta H° = -(-2803)$ kJ = +2803 kJ.

$$\frac{5.5\times10^{16}\,g\,CO_2}{yr} \times \frac{1\,mol\,CO_2}{44.01\,g\,CO_2} \times \frac{2803\,kJ}{6\,mol\,CO_2} = 5.838\times10^{17} = 5.8\times10^{17}\,kJ$$

(b) 1 W = 1 J/s; 1 W · s = 1 J

$$\frac{5.838\times10^{17}\,kJ}{yr} \times \frac{1000\,J}{kJ} \times \frac{1\,yr}{365\,d} \times \frac{1\,d}{24\,hr} \times \frac{1\,hr}{60\,min} \times \frac{1\,min}{60\,s} \times \frac{1\,W \cdot s}{J}$$

$$\times \frac{1\,MW}{1\times10^5\,W} = 1.851\times10^7\,MW = 1.9\times10^7\,MW$$

$$1.9\times10^7\,MW \times \frac{1\,plant}{10^3\,MW} = 1.9\times10^4 = 19,000\ nuclear\ power\ plants$$

Integrative Exercises

5.108 (a) $CH_4(g) + 2O_2(g) \rightarrow CO_2(g) + 2H_2O(l)$

$\Delta H° = \Delta H_f° CO_2(g) + 2\Delta H_f° H_2O(l) - \Delta H_f° CH_4(g) - 2\Delta H_f° O_2(g)$

$= -393.5$ kJ + 2(-285.83 kJ) - (-74.8 kJ) - 2(0) = -890.36 = -890.4 kJ/mol CH_4

$$\frac{-890.36\,kJ}{mol\,CH_4} \times \frac{1000\,J}{1\,kJ} \times \frac{1\,mol}{6.022\times10^{23}\,molecules\,CH_4} = 1.4785\times10^{-18}$$

$$= 1.479\times10^{18}\,J/molecule$$

(b)   1eV = 96.485 kJ/mol

$$8\,keV \times \frac{1000\,eV}{1\,keV} \times \frac{96.485\,kJ}{eV \cdot mol} \times \frac{1\,mol}{6.022 \times 10^{23}} \times \frac{1000\,J}{kJ} = 1.282 \times 10^{-15} = 1 \times 10^{-15}\ J/X\text{-ray}$$

The X-ray has approximately 1000 times more energy than is produced by the combustion of 1 molecule of $CH_4(g)$.

5.109   The situation in Figure 4.3 is a more complex version of that pictured in Exercise 5.28 and discussed in Solution 5.28. An ionic solid is an orderly arrangement of closely spaced ions, oppositely charged particles. The potential energy of any pair of these ions is described as $E_{el} = \kappa\, Q_1 Q_2 / r^2$. Separating these oppositely charged particles leads to an increase in the energy of the system, $+\Delta E$. Work is done to the NaCl by the water molecules. Since both NaCl and water are part of the system, the net amount of work is zero. Since $\Delta E = q + w$, $\Delta E = q$ and both are positive. The dissolving process typically takes place at constant atmospheric pressure, so $\Delta H = q$ and ΔH is also positive.

To verify this conclusion, carry out the dissolution of NaCl in a constant pressure calorimeter. Begin with 1 L of H_2O, record the temperature, add 0.1 mol NaCl, and dissolve completely; record the final temperature. If ΔH is positive, the temperature will increase.

5.110   (a),(b) $Ag^+(aq) + Li(s) \rightarrow Ag(s) + Li^+(aq)$

$$\Delta H^\circ = \Delta H_f^\circ\, Li^+(aq) - \Delta H_f^\circ\, Ag^+(aq)$$

$$= -278.5\ kJ - 105.90\ kJ = -384.4\ kJ$$

$Fe(s) + 2Na^+(aq) \rightarrow Fe^{2+}(aq) + 2Na(s)$

$$\Delta H^\circ = \Delta H_f^\circ\, Fe^{2+}(aq) - 2\Delta H_f^\circ\, Na^+(aq)$$

$$= -87.86\ kJ - 2(-240.1\ kJ) = +392.3\ kJ$$

$2K(s) + 2H_2O(l) \rightarrow 2KOH(aq) + H_2(g)$

$$\Delta H^\circ = 2\Delta H_f^\circ\, KOH(aq) - 2\Delta H_f^\circ\, H_2O(l)$$

$$= 2(-482.4\ kJ) - 2(-285.83\ kJ) = -393.1\ kJ$$

(c)   Exothermic reactions are more likely to be favorable, so the first and third reactions should be favorable and the second reaction should be unfavorable.

(d)   In the activity series of metals, Table 4.5, any metal can be oxidized by the cation of a metal below it on the table.

Ag$^+$ is below Li, so the first reaction will occur.

Na$^+$ is above Fe, so the second reaction will not occur.

H$^+$ (formally in H_2O) is below K, so the third reaction will occur.

These predictions agree with those in part (c).

5.111   (a) $\Delta H^\circ = \Delta H_f^\circ\, NaNO_3(aq) + \Delta H_f^\circ\, H_2O(l) - \Delta H_f^\circ\, HNO_3(aq) - \Delta H_f^\circ\, NaOH(aq)$

$\Delta H^\circ = -446.2 \text{ kJ} - 285.83 \text{ kJ} - (-206.6 \text{ kJ}) - (-469.6 \text{ kJ}) = -55.8 \text{ kJ}$

$\Delta H^\circ = \Delta H^\circ_f \text{ NaCl(aq)} + \Delta H^\circ_f \text{ H}_2\text{O(l)} - \Delta H^\circ_f \text{ HCl(aq)} - \Delta H^\circ_f \text{ NaOH(aq)}$

$\Delta H^\circ = -407.1 \text{ kJ} - 285.83 \text{ kJ} - (-167.2 \text{ kJ}) - (-469.6 \text{ kJ}) = -56.1 \text{ kJ}$

$\Delta H^\circ = \Delta H^\circ_f \text{ NH}_3\text{(aq)} + \Delta H^\circ_f \text{ Na}^+\text{(aq)} + \Delta H^\circ_f \text{ H}_2\text{O(l)} - \Delta H^\circ_f \text{ NH}_4^+\text{(aq)} - \Delta H^\circ_f \text{ NaOH(aq)}$

$= -80.29 \text{ kJ} - 240.1 \text{ kJ} - 285.83 \text{ kJ} - (-132.5 \text{ kJ}) - (-469.6 \text{ kJ}) = -4.1 \text{ kJ}$

(b) $H^+\text{(aq)} + OH^-\text{(aq)} \rightarrow H_2O\text{(l)}$ is the net ionic equation for the first two reactions.

$NH_4^+\text{(aq)} + OH^-\text{(aq)} \rightarrow NH_3\text{(aq)} + H_2O\text{(l)}$

(c) The ΔH° values for the first two reactions are nearly identical, -55.9 kJ and -56.2 kJ. The spectator ions by definition do not change during the course of a reaction, so ΔH° is the enthalpy change for the net ionic equation. Since the first two reactions have the same net ionic equation, it is not surprising that they have the same ΔH°.

(d) Strong acids are more likely than weak acids to donate H^+. The neutralization of the two strong acids is energetically favorable, while the third reaction is not. $NH_4^+\text{(aq)}$ is probably a weak acid.

5.112 (a) mol Cu $= M \times L = 1.00 \, M \times 0.0500 \text{ L} = 0.0500$ mol

g = mol $\times$ / = $0.0500 \times 63.546 = 3.1773 = 3.18$ g Cu

(b) The precipitate is copper(II) hydroxide, $Cu(OH)_2$.

(c) $CuSO_4\text{(aq)} + 2KOH\text{(aq)} \rightarrow Cu(OH)_2\text{(s)} + K_2SO_4\text{(aq)}$, complete

$Cu^{2+}\text{(aq)} + 2OH^-\text{(aq)} \rightarrow Cu(OH)_2\text{(s)}$, net ionic

(d) The temperature of the calorimeter rises, so the reaction is exothermic and the sign of q is negative.

$$q = -6.2^\circ C \times 100 \, g \times \frac{4.184 \, J}{1 \, g \bullet {}^\circ C} = -2.6 \times 10^3 \, J = -2.6 \text{ kJ}$$

The reaction as carried out involves only 0.050 mol of $CuSO_4$ and the stoichiometrically equivalent amount of KOH. On a molar basis,

$$\Delta H = \frac{-2.6 \text{ kJ}}{0.050 \text{ mol}} = -52 \text{ kJ} \text{ for the reaction as written in part (c)}$$

5.113 (a) $AgNO_3\text{(aq)} + NaCl\text{(aq)} \rightarrow NaNO_3\text{(aq)} + AgCl\text{(s)}$

net ionic equation: $Ag^+\text{(aq)} + Cl^-\text{(aq)} \rightarrow AgCl\text{(s)}$

$\Delta H^\circ = \Delta H^\circ_f \text{ AgCl(s)} - \Delta H^\circ_f \text{ Ag}^+\text{(aq)} - \Delta H^\circ_f \text{Cl}^-\text{(aq)}$

$\Delta H^\circ = -127.0 \text{ kJ} - (105.90 \text{ kJ}) - (-167.2 \text{ kJ}) = -65.7 \text{ kJ}$

(b) ΔH° for the complete molecular equation will be the same as ΔH° for the net ionic equation. $Na^+\text{(aq)}$ and $NO_3^-\text{(aq)}$ are spectator ions; they appear on both sides of the chemical equation. Since the overall enthalpy change is the enthalpy of the products minus the enthalpy of the reactants, the contributions of the spectator ions cancel.

(c) $\Delta H^\circ = \Delta H^\circ_f \, NaNO_3(aq) + \Delta H^\circ_f \, AgCl(s) - \Delta H^\circ_f \, AgNO_3(aq) - \Delta H^\circ_f \, NaCl(aq)$

$\Delta H^\circ_f \, AgNO_3(aq) = \Delta H^\circ_f \, NaNO_3(aq) + \Delta H^\circ_f \, AgCl(s) - \Delta H^\circ_f \, NaCl(aq) - \Delta H^\circ$

$\Delta H^\circ_f \, AgNO_3(aq) = -446.2 \, kJ + (-127.0 \, kJ) - (-407.1 \, kJ) - (-65.7 \, kJ)$

$\Delta H^\circ_f \, AgNO_3(aq) = -100.4 \, kJ/mol$

5.114   (a) $21.83 \, g \, CO_2 \times \dfrac{1 \, mol \, CO_2}{44.01 \, g \, CO_2} \times \dfrac{1 \, mol \, C}{1 \, mol \, CO_2} \times \dfrac{12.01 \, g \, C}{1 \, mol \, C} = 5.9572 = 5.957 \, g \, C$

$4.47 \, g \, H_2O \times \dfrac{1 \, mol \, H_2O}{18.02 \, g \, H_2O} \times \dfrac{2 \, mol \, H}{1 \, mol \, H_2O} \times \dfrac{1.008 \, g \, H}{mol \, H} = 0.5001 = 0.500 \, g \, H$

The sample mass is $(5.9572 + 0.5001) = 6.457 \, g$

(b) $5.957 \, g \, C \times \dfrac{1 \, mol \, C}{12.01 \, g \, C} = 0.4960 \, mol \, C; \quad 0.4960/0.496 = 1$

$0.500 \, g \, H \times \dfrac{1 \, mol \, H}{1.008 \, g \, H} = 0.496 \, mol \, H; \quad 0.496/0.496 = 1$

The empirical formula of the hydrocarbon is CH.

(c)   Calculate the ΔH°_f for 6.457 g of the sample.

$6.457 \, g \, sample + O_2(g) \rightarrow 21.83 \, g \, CO_2(g) + 4.47 \, g \, H_2O(g), \, \Delta H^\circ = -311 \, kJ$

$\Delta H^\circ_{comb} = \Delta H^\circ_f \, CO_2(g) + \Delta H^\circ_f \, H_2O(g) - \Delta H^\circ_f \, sample - \Delta H^\circ_f \, O_2(g)$

$\Delta H^\circ_f \, sample = \Delta H^\circ_f \, CO_2(g) + \Delta H^\circ_f \, H_2O(g) - \Delta H^\circ_{comb}$

$\Delta H^\circ_f \, CO_2(g) = 21.83 \, g \, CO_2 \times \dfrac{1 \, mol \, CO_2}{44.01 \, g \, CO_2} \times \dfrac{-393.5 \, kJ}{mol \, CO_2} = -195.185 = -195.2 \, kJ$

$\Delta H^\circ_f \, H_2O(g) = 4.47 \, g \, H_2O \times \dfrac{1 \, mol \, H_2O}{18.02 \, g \, H_2O} \times \dfrac{-241.82 \, kJ}{mol \, H_2O} = 59.985 = -60.0 \, kJ$

$\Delta H^\circ_f \, sample = -195.185 \, kJ - 59.985 \, kJ - (-311 \, kJ) = 55.83 = 56 \, kJ$

$\dfrac{55.83 \, kJ}{6.457 \, g \, sample} \times \dfrac{13.02 \, g}{CH \, unit} = 112.6 = 1.1 \times 10^2 \, kJ/CH \, unit$

(d)   The hydrocarbons in Appendix C with empirical formula CH are C_2H_2 and C_6H_6.

substance	ΔH°_f/mol	ΔH°_f/CH unit
$C_2H_2(g)$	226.7 kJ	113.4 kJ
$C_6H_6(g)$	82.9 kJ	13.8 kJ
$C_6H_6(l)$	49.0 kJ	8.17 kJ
sample		1.1×10^2 kJ

The calculated value of ΔH°_f/ CH unit for the sample is a good match with acetylene, $C_2H_2(g)$.

5.115   (a) $CH_4(g) \rightarrow C(g) + 4H(g)$ (i) reaction given

$CH_4(g) \rightarrow C(s) + 2H_2(g)$ (ii) reverse of formation

The differences are: the state of C in the products; the chemical form, atoms, or diatomic molecules, of H in the products.

(b)   i. $\Delta H^\circ = \Delta H_f^\circ \, C(g) + 4\Delta H_f^\circ \, H(g) - \Delta H_f^\circ \, CH_4(g)$

$= 718.4 \text{ kJ} + 4(217.94) \text{ kJ} - (-74.8) \text{ kJ} = 1665.0 \text{ kJ}$

ii. $\Delta H^\circ = \Delta H_f^\circ \, CH_4 = -(-74.8) \text{ kJ} = 74.8 \text{ kJ}$

The rather large difference in ΔH° values is due to the enthalpy difference between isolated gaseous C atoms and the orderly, bonded array of C atoms in graphite, C(s), as well as the enthalpy difference between isolated H atoms and H_2 molecules. In other words, it is due to the difference in the enthalpy stored in chemical bonds in C(s) and $H_2(g)$ versus the corresponding isolated atoms.

(c) $CH_4(g) + 4F_2(g) \rightarrow CF_4(g) + 4HF(g)$ $\Delta H^\circ = -1679.5 \text{ kJ}$

The ΔH° value for this reaction was calculated in Solution 5.86.

$$3.45 \text{ g } CH_4 \times \frac{1 \text{ mol } CH_4}{16.04 \text{ g } CH_4} \times 0.21509 = 0.215 \text{ mol } CH_4$$

$$1.22 \text{ g } F_2 \times \frac{1 \text{ mol } F_2}{38.00 \text{ g } F_2} = 0.03211 = 0.0321 \text{ mol } F_2$$

There are fewer mol F_2 than CH_4, but 4 mol F_2 are required for every 1 mol of CH_4 reacted, so clearly F_2 is the limiting reactant.

$$0.03211 \text{ mol } F_2 \times \frac{-1679.5 \text{ kJ}}{4 \text{ mol } F_2} = -13.48 = -13.5 \text{ kJ heat evolved}$$

6 Electronic Structure of Atoms

Electronic Structure of Atoms

6.1 (a) Speed is distance traveled per unit time. Measure the distance between the point where the stone is dropped and a second reference point, possibly the shore line. Using a stop watch, measure the elapsed time between when the stone is dropped and when the first wave reaches the second reference point. Be sure to drop the stone exactly at the first reference point. Find the ratio of distance to time.

(b) Measure the distance between two wave crests (or troughs or any analogous points on two adjacent waves). Better yet, measure the distance between two crests (or analogous points) that are several waves apart and divide by the number of waves that separate them.

(c) One way to make this distance measurement is to use a standard. Take a photo of the event and include an object of known dimensions in the picture. Measure the distances described above on the picture. Obtain a conversion factor to actual distances by measuring the object on the picture and comparing the photo dimensions to the true dimensions. Since speed is distance/time, and wavelength is distance, we can calculate frequency by dividing speed by wavelength, $\nu = c/\lambda$.

(d) We can measure frequency of the wave by dropping a cork in the water and counting the number of times per second it moves through a complete cycle of motion.

6.2 (a) No, visible radiation has wavelengths of 4×10^{-7} to 7×10^{-7} m, much shorter than 1 cm.

(b) Energy and wavelength are inversely proportional. Photons of the longer 1 cm radiation have less energy than visible photons.

(c) Radiation of 1 cm, or 1×10^{-2} m, is in the microwave region. The appliance is probably a microwave oven.

6.3 (a) The glowing stove burner is an example of black body radiation, the observational basis for Planck's quantum theory. The wavelengths emitted are related to temperature, with cooler temperatures emitting longer wavelengths and hotter temperatures emitting shorter wavelengths. At the hottest setting, the burner emits orange visible light. At the cooler low setting, the burner emits longer wavelengths out of the visible region, and the burner appears black.

(b) If the burner had a super high setting, the emitted wavelengths would be shorter than those of orange light and the glow color would be more blue. (See Figure 6.4 for color variation with wavelength.)

6.4   (a)   Increase. The rainbow has shorter wavelength blue light on the inside and longer wavelength red light on the outside. (See Figure 6.4.)

       (b)   Decrease. Wavelength and frequency are inversely related. Wavelength increases so frequency decreases going from the inside to the outside of the rainbow.

       (c)   The light from the hydrogen discharge take is not a continuous spectrum, so not all visible wavelengths will be in our "hydrogen discharge rainbow." Starting with the shortest wavelengths, it will be violet followed by blue-violet and blue-green on the inside. Then there will be a gap, and finally a red band. (See the H spectrum in Figure 6.12.)

6.5   (a) $n = 1, n = 4$ (b) $n = 1, n = 2$

       (c)   Wavelength and energy are inversely proportional; the smaller the energy, the longer the wavelength. In order of increasing wavelength (and decreasing energy): (iii) $n = 2$ to $n = 4$ < (iv) $n = 3$ to $n = 1$ < (ii) $n = 3$ to $n = 2$ < (i) $n = 1$ to $n = 2$

6.6   (a) $\psi^2(x)$ will be positive or zero at all values of x, and have two maxima with larger magnitudes than the maximum in $\psi(x)$.

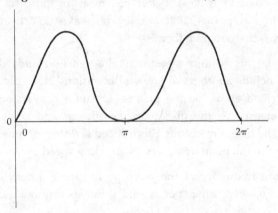

       (b)   The greatest probability of finding the electron is at the two maxima in $\psi^2(x)$ at $x = \pi/2$ and $3\pi/2$.

       (c)   There is zero probability of finding the electron at $x = \pi$. This value is called a node.

6.7   (a)   1

       (b)   p (dumbbell shape, node at the nucleus)

       (c)   The lobes in the contour representation would extend farther along the y axis. A larger principle quantum number (4p vs. 3p) implies a greater average distance from the nucleus for electrons occupying the orbital.

6.8   (a)   In the left-most box, the two electrons cannot have the same spin. The *Pauli principle* states that no two electrons can have the same set of quantum numbers. Since the first three quantum numbers describe an orbital, the fourth quantum number, m_s, must have different values for two electrons in the same orbital; their "spins" must be opposite.

(b)   Flip one of the arrows in the left-most box, so that one points up and the other down.

(c)   Group 6A. The drawing shows three boxes or orbitals at the same energy, so it must represent p orbitals. Since some of these p orbitals are partially filled, they must be the valence orbitals of the element. Elements with four valence electrons in their p orbitals belong to group 6A.

Radiant Energy

6.9   (a)   Meters (m)   (b)   1/seconds (s^{-1})   (c)   meters/second ($m \cdot s^{-1}$ or m/s)

6.10   (a)   Wavelength (λ) and frequency (ν) are inversely proportional; the proportionality constant is the speed of light (c). $\nu = c/\lambda$.

(b)   Light in the 210–230 nm range is in the ultraviolet region of the spectrum. These wavelengths are slightly shorter than the 400 nm short-wavelength boundary of the visible region.

6.11   (a)   True.

(b)   False. The frequency of radiation decreases as the wavelength increases.

(c)   False. Ultraviolet light has shorter wavelengths than visible light. [See Solution 6.10(b).]

(d)   False. Electromagnetic radiation and sound waves travel at different speeds.

6.12   (a)   False. Electromagnetic radiation passes through water. The fact that you can see objects through a glass of water should make this clear.

(b)   True.

(c)   False. Infrared light has lower frequencies than visible light.

(d)   False. A foghorn blast is a form of sound waves, which are not accompanied by oscillating electric and magnetic fields.

6.13   *Analyze/Plan.* Use the electromagnetic spectrum in Figure 6.4 to determine the wavelength of each type of radiation; put them in order from shortest to longest wavelength. *Solve.*

Wavelength of X-rays < ultraviolet < green light < red light < infrared < radio waves

Check. These types of radiation should read from left to right on Figure 6.4

6.14   Wavelength of (a) gamma rays < (d) yellow (visible) light < (e) red (visible) light < (b) 93.1 MHz FM (radio) waves < (c) 680 kHz or 0.680 MHz AM (radio) waves

6.15   *Analyze/Plan.* These questions involve relationships between wavelength, frequency, and the speed of light. Manipulate the equation $\nu = c/\lambda$ to obtain the desired quantities, paying attention to units.   *Solve.*

(a) $\nu = c/\lambda$; $\dfrac{2.998 \times 10^8 \text{ m}}{s} \times \dfrac{1}{955 \text{ μm}} \times \dfrac{1 \text{ μm}}{1 \times 10^{-6} \text{ m}} = 3.14 \times 10^{11} \text{ s}^{-1}$

(b) $\lambda = c/v$; $\dfrac{2.998 \times 10^8 \text{ m}}{\text{s}} \times \dfrac{1 \text{ s}}{5.50 \times 10^{14}} = 5.45 \times 10^{-7} \text{ m}$ (545 nm)

(c)   The radiation in (b) is in the visible region and is "visible" to humans. The $\sim 1 \times 10^{-3}$ m radiation in (a) is in the microwave region near the infrared edge and is not visible.

(d) $50.0 \ \mu s \times \dfrac{1 \text{ s}}{1 \times 10^6 \ \mu s} \times \dfrac{2.998 \times 10^8 \text{ m}}{\text{s}} = 1.50 \times 10^4 \text{ m}$

Check. Confirm that powers of 10 make sense and units are correct.

6.16   (a) $v = c/\lambda$; $\dfrac{2.998 \times 10^8 \text{ m}}{\text{s}} \times \dfrac{1}{10.0 \ \text{Å}} \times \dfrac{1 \ \text{Å}}{1 \times 10^{-10} \text{ m}} = 3.00 \times 10^{17} \text{ s}^{-1}$

(b) $\lambda = c/v$; $\dfrac{2.998 \times 10^8 \text{ m}}{\text{s}} \times \dfrac{1 \text{ s}}{7.6 \times 10^{10}} = 3.94 \times 10^{-3} \text{ m}$

(c)   The 1×10^{-9} m radiation in (a) is X-rays and can be observed by an X-ray detector. Radiation (b) is microwave.

(d) $25.5 \text{ fs} \times \dfrac{1 \times 10^{-15} \text{ s}}{1 \text{ fs}} \times \dfrac{2.998 \times 10^8 \text{ m}}{\text{s}} = 7.64 \times 10^{-6} \text{ m}$ (7.64 μm)

6.17   *Analyze/Plan.* $v = c/\lambda$; change nm $\rightarrow$ m.

Solve. $v = c/\lambda$; $\dfrac{2.998 \times 10^8 \text{ m}}{1 \text{ s}} \times \dfrac{1}{436 \text{ nm}} \times \dfrac{1 \text{ nm}}{1 \times 10^{-9} \text{ m}} = 6.88 \times 10^{14} \text{ s}^{-1}$

The color is blue.

Check. $(3000 \times 10^5 / 500 \times 10^{-9}) = 6 \times 10^{14} \text{ s}^{-1}$; units are correct.

6.18 $v = c/\lambda$; $\dfrac{2.998 \times 10^8 \text{ m}}{1 \text{ s}} \times \dfrac{1}{489 \text{ nm}} \times \dfrac{1 \text{ nm}}{1 \times 10^{-9} \text{ m}} = 6.13 \times 10^{14} \text{ s}^{-1}$

The laser emits visible light; the color is green to blue-green.

Quantized Energy and Photons

6.19   (a)   Quantization means that energy can only be absorbed or emitted in specific amounts or multiples of these amounts. This minimum amount of energy is called a quantum and is equal to a constant times the frequency of the radiation absorbed or emitted. $E = hv$.

(b)   In everyday activities, we deal with macroscopic objects such as our bodies or our cars, which gain and lose total amounts of energy much larger than a single quantum, hv. The gain or loss of the relatively minuscule quantum of energy is unnoticed.

6.20   Planck's original hypothesis was that energy could only be gained or lost in discreet amounts (quanta) with a certain minimum size. The size of the minimum energy change is related to the frequency of the radiation absorbed or emitted, $\Delta E = hv$, and energy changes occur only in multiples of hv.

Einstein postulated that light itself is quantized, that the minimum energy of a photon (a quantum of light) is directly proportional to its frequency, $E = h\nu$. If a photon that strikes a metal surface has less than the threshold energy, no electron is emitted from the surface. If the photon has energy equal to or greater than the threshold energy, an electron is emitted and any excess energy becomes the kinetic energy of the electron.

6.21   *Analyze/Plan.* These questions deal with the relationships between energy, wavelength, and frequency. Use the relationships $E = h\nu = hc/\lambda$ to calculate the desired quantities. Pay attention to units.   *Solve.*

(a) $E = h\nu = hc/\lambda = 6.626 \times 10^{-34} \text{ J} \cdot \text{s} \times \dfrac{2.998 \times 10^8 \text{ m}}{1 \text{ s}} \times \dfrac{1}{438 \text{ nm}} \times \dfrac{1 \text{ nm}}{1 \times 10^{-9} \text{ m}}$

$= 4.54 \times 10^{-19} \text{ J}$

(b) $E = h\nu = 6.626 \times 10^{-34} \text{ J} \cdot \text{s} \times \dfrac{6.75 \times 10^{12}}{1 \text{ s}} = 4.47 \times 10^{-21} \text{ J}$

(c) $\lambda = hc/E = 6.626 \times 10^{-34} \text{ J} \cdot \text{s} \times \dfrac{2.998 \times 10^8 \text{ m}}{1 \text{ s}} \times \dfrac{1}{2.87 \times 10^{-18} \text{ J}} = 6.92 \times 10^{-8} \text{ m}$

$= 69.2 \text{ nm}$

This radiation is in the ultraviolet region.

Check. Units are correct and powers of 10 are reasonable.

6.22   (a) $E = hc/\lambda = 6.626 \times 10^{-34} \text{ J} \cdot \text{s} \times \dfrac{2.998 \times 10^8 \text{ m}}{1 \text{ s}} \times \dfrac{1}{10.8 \text{ mm}} \times \dfrac{1 \text{ mm}}{1 \times 10^{-3} \text{ m}}$

$= 1.84 \times 10^{-23} \text{ J}$

(b) $E = h\nu = 6.626 \times 10^{-34} \text{ J} \cdot \text{s} \times \dfrac{101.1 \times 10^6}{1 \text{ s}} = 6.699 \times 10^{-26} \text{ J}$

(c)   The relationship $\nu = E/h$ requires energy in J/photon. Change kJ/mol to J/photon and divide by h.

$\dfrac{24.7 \text{ kJ}}{\text{mol}} \times \dfrac{1 \times 10^3 \text{ J}}{\text{kJ}} \times \dfrac{1 \text{ mol}}{6.022 \times 10^{23} \text{ photons}} \times \dfrac{1}{6.626 \times 10^{-34} \text{ J} \cdot \text{s}} = 6.19 \times 10^{13} \text{ s}^{-1}$

This radiation is in the infrared region.

6.23   *Analyze/Plan.* Use $E = hc/\lambda$; pay close attention to units.   *Solve.*

(a) $E = hc/\lambda = 6.626 \times 10^{-34} \text{ J} \cdot \text{s} \times \dfrac{2.998 \times 10^8 \text{ m}}{1 \text{ s}} \times \dfrac{1}{3.3 \text{ } \mu\text{m}} \times \dfrac{1 \text{ } \mu\text{m}}{1 \times 10^{-6} \text{ m}}$

$= 6.0 \times 10^{-20} \text{ J}$

$E = hc/\lambda = 6.626 \times 10^{-34} \text{ J} \cdot \text{s} \times \dfrac{2.998 \times 10^8 \text{ m}}{1 \text{ s}} \times \dfrac{1}{0.154 \text{ nm}} \times \dfrac{1 \text{ nm}}{1 \times 10^{-9} \text{ m}}$

$= 1.29 \times 10^{-15} \text{ J}$

Check. $(6.6 \times 3/3.3) \times (10^{-34} \times 10^8/10^{-6}) \approx 6 \times 10^{-20} \text{ J}$

$(6.6 \times 3/0.15) \times (10^{-34} \times 10^8/10^{-9}) \approx 120 \times 10^{-17} \approx 1.2 \times 10^{-15} \text{ J}$

The results are reasonable. We expect the longer wavelength 3.3 μm radiation to have the lower energy.

(b)   The 3.3 μm photon is in the infrared and the 0.154 nm (1.54×10^{-10} m) photon is in the X-ray region; the X-ray photon has the greater energy.

6.24 $E = h\nu$

$$AM : 6.626 \times 10^{-34} \text{ J} \cdot \text{s} \times \frac{1010 \times 10^3}{1 \text{ s}} = 6.69 \times 10^{-28} \text{ J}$$

$$FM : 6.626 \times 10^{-34} \text{ J} \cdot \text{s} \times \frac{98.3 \times 10^6}{1 \text{ s}} = 6.51 \times 10^{-26} \text{ J}$$

The FM photon has about 100 times more energy than the AM photon.

6.25   *Analyze/Plan.* Use $E = hc/\lambda$ to calculate J/photon; Avogadro's number to calculate J/mol; photon/J [the result from part (a)] to calculate photons in 1.00 mJ. Pay attention to units.

Solve.

(a) $E_{photon} = hc/\lambda = \dfrac{6.626 \times 10^{-34} \text{ J} \cdot \text{s}}{325 \times 10^{-9} \text{ m}} \times \dfrac{2.998 \times 10^8 \text{ m}}{\text{s}} = 6.1122 \times 10^{-19}$

$= 6.11 \times 10^{-19}$ J/photon

(b) $\dfrac{6.1122 \times 10^{-19} \text{ J}}{1 \text{ photon}} \times \dfrac{6.022 \times 10^{23} \text{ photons}}{1 \text{ mol}} = 3.68 \times 10^5 \text{ J/mol} = 368 \text{ kJ/mol}$

(c) $\dfrac{1 \text{ photon}}{6.1122 \times 10^{-19} \text{ J}} \times 1.00 \text{ mJ} \times \dfrac{1 \times 10^{-3}}{1 \text{ mJ}} = 1.64 \times 10^{15} \text{ photons}$

Check. Powers of 10 (orders of magnitude) and units are correct.

6.26 $\dfrac{941 \times 10^3 \text{ J}}{\text{mol O}_2} \times \dfrac{1 \text{ mol}}{6.022 \times 10^{23} \text{ photons}} = 1.563 \times 10^{-18} = 1.56 \times 10^{-18} \text{ J/photon}$

$\lambda = hc/E = \dfrac{6.626 \times 10^{-34} \text{ J} \cdot \text{s}}{1.563 \times 10^{-18} \text{ J}} \times \dfrac{2.998 \times 10^8 \text{ m}}{1 \text{ s}} = 1.27 \times 10^{-7} \text{ m} = 127 \text{ nm}$

According to Figure 6.4, this is ultraviolet radiation.

6.27   *Analyze/Plan.* $E = hc/\lambda$ gives J/photon. Use this result with J/s (given) to calculate photons/s. *Solve.*

(a)   The $\sim 1 \times 10^{-6}$ m radiation is infrared but very near the visible edge.

(b) $E_{photon} = hc/\lambda = \dfrac{6.626 \times 10^{-34} \text{ J} \cdot \text{s}}{987 \times 10^{-9} \text{ m}} \times \dfrac{2.998 \times 10^8 \text{ m}}{1 \text{ s}} = 2.0126 \times 10^{-19}$

$= 2.01 \times 10^{-19}$ J/photon

$\dfrac{0.52 \text{ J}}{32 \text{ s}} \times \dfrac{1 \text{ photon}}{2.0126 \times 10^{-19} \text{ J}} = 8.1 \times 10^{16} \text{ photons/s}$

Check. $(7 \times 3/1000) \times (10^{-34} \times 10^8/10^{-9}) \approx 21 \times 10^{-20} \approx 2.1 \times 10^{-19} \text{ J/photon}$

$(0.5/30/2) \times (1/10^{-19}) = 0.008 \times 10^{19} = 8 \times 10^{16} \text{ photons/s}$

Units are correct; powers of 10 are reasonable.

6 Electronic Structure of Atoms   Solutions to Exercises

6.28   (a)   The radiation is microwave.

(b) $E_{photon} = hc/\lambda = \dfrac{6.626 \times 10^{-34} \text{ J}\bullet\text{s}}{3.55 \times 10^{-3} \text{ m}} \times \dfrac{2.998 \times 10^8 \text{ m}}{1 \text{ s}} = 5.5957 \times 10^{-23}$

$= 5.60 \times 10^{-23}$ J/photon

$\dfrac{5.5957 \times 10^{-23} \text{ J}}{1 \text{ photon}} \times \dfrac{3.2 \times 10^8 \text{ photons}}{1 \text{ s}} \times \dfrac{60 \text{ s}}{1 \text{ min}} \times \dfrac{60 \text{ min}}{1 \text{ hr}} = 6.4463 \times 10^{-11}$

$= 6.4 \times 10^{-11}$ J/hr

6.29   *Analyze/Plan.* Use $E = h\nu$ and $\nu = c/\lambda$. Calculate the desired characteristics of the photons.

Compare E_{min} and E_{120} to calculate maximum kinetic energy of the emitted electron. *Solve.*

(a) $E = h\nu = 6.626 \times 10^{-34} \text{ J}\bullet\text{s} \times 1.09 \times 10^{15} \text{ s}^{-1} = 7.22 \times 10^{-19}$ J

(b) $\lambda = c/\nu = \dfrac{2.998 \times 10^8 \text{ m}}{1 \text{ s}} \times \dfrac{1 \text{ s}}{1.09 \times 10^{15}} = 2.75 \times 10^{-7}$ m = 275 nm

(c) $E_{120} = hc/\lambda = 6.626 \times 10^{-34} \text{ J}\bullet\text{s} \times \dfrac{2.998 \times 10^8 \text{ m}}{1 \text{ s}} \times \dfrac{1}{120 \text{ nm}} \times \dfrac{1 \text{ nm}}{1 \times 10^{-9} \text{ m}}$

$= 1.655 \times 10^{-18} = 1.66 \times 10^{-18}$ J

The excess energy of the 120 nm photon is converted into the kinetic energy of the emitted electron.

$E_k = E_{120} - E_{min} = 16.55 \times 10^{-19} \text{ J} - 7.22 \times 10^{-19} \text{ J} = 9.3 \times 10^{-19}$ J/electron

Check. E_{120} must be greater than E_{min} in order for the photon to impart kinetic energy to the emitted electron. Our calculations are consistent with this requirement.

6.30   (a) $\nu = E/h = \dfrac{4.41 \times 10^{-19} \text{ J}}{6.626 \times 10^{-34} \text{ J}\bullet\text{s}} = 6.6556 \times 10^{14} = 6.66 \times 10^{14} \text{ s}^{-1}$

(b) $\lambda = hc/E = \dfrac{6.626 \times 10^{-34} \text{ J}\bullet\text{s}}{4.41 \times 10^{-19} \text{ J}} \times \dfrac{2.998 \times 10^8 \text{ m}}{\text{s}} = 4.50 \times 10^{-7}$ m = 450 nm

(c) $E_{439} = hc/\lambda = \dfrac{6.626 \times 10^{-34} \text{ J}\bullet\text{s}}{439 \times 10^{-9} \text{ m}} \times \dfrac{2.998 \times 10^8 \text{ m}}{\text{s}} = 4.525 \times 10^{-19} = 4.53 \times 10^{-19}$ J

$E_K = E_{439} - E_{min} = 4.525 \times 10^{-19} \text{ J} - 4.41 \times 10^{-19} \text{ J} = 0.115 \times 10^{-19} = 1.1 \times 10^{-20}$ J

(d)   One electron is emitted per photon. Calculate the number of 439 nm photons in 1.00 μJ. The excess energy in each photon will become the kinetic energy of the electron; it cannot be "pooled" to emit additional electrons.

$1.00 \text{ μJ} \times \dfrac{1 \times 10^{-6} \text{ J}}{\text{μJ}} \times \dfrac{1 \text{ photon}}{4.525 \times 10^{-19} \text{ J}} \times \dfrac{1 \text{ e}^-}{1 \text{ photon}} = 2.21 \times 10^{12}$ electrons

Bohr's Model; Matter Waves

6.31   When applied to atoms, the notion of quantized energies means that only certain energies can be gained or lost, only certain values of ΔE are allowed. The allowed values of ΔE are represented by the lines in the emission spectra of excited atoms.

6.32   (a)   According to Bohr theory, when hydrogen emits radiant energy, electrons are moving from a higher allowed energy state to a lower one. Since only certain energy states are allowed, only certain energy changes can occur. These allowed energy changes correspond ($\lambda = hc/\Delta E$) to the wavelengths of the lines in the emission spectrum of hydrogen.

        (b)   When a hydrogen atom changes from the ground state to an excited state, the single electron moves further away from the nucleus, so the atom "expands".

6.33   *Analyze/Plan.* An isolated electron is assigned an energy of zero; the closer the electron comes to the nucleus, the more negative its energy. Thus, as an electron moves closer to the nucleus, the energy of the electron decreases and the excess energy is emitted. Conversely, as an electron moves further from the nucleus, the energy of the electron increases and energy must be absorbed.   *Solve.*

        (a)   As the principle quantum number decreases, the electron moves toward the nucleus and energy is emitted.

        (b)   An increase in the radius of the orbit means the electron moves away from the nucleus; energy is absorbed.

        (c)   An isolated electron is assigned an energy of zero. As the electron moves to the $n = 3$ state closer to the H^+ nucleus, its energy becomes more negative (decreases) and energy is emitted.

6.34   (a)   Absorbed.   (b)   Emitted.   (c)   Absorbed.

6.35   *Analyze/Plan.* Equation 6.5: $E = (-2.18 \times 10^{-18} \text{ J})(1/n^2)$.   *Solve.*

        $E_2 = -2.18 \times 10^{-18} \text{ J}/(2)^2 = -5.45 \times 10^{-19} \text{ J}$

        $E_6 = -2.18 \times 10^{-18} \text{ J}/(6)^2 = -6.0556 \times 10^{-20} = -0.606 \times 10^{-19} \text{ J}$

        $\Delta E = E_6 - E_2 = (-0.606 \times 10^{-19} \text{ J}) - (-5.45 \times 10^{-19} \text{ J}) = 4.844 \times 10^{-19} \text{ J} = 4.84 \times 10^{-19} \text{ J}$

        $\lambda = hc/\Delta E = \dfrac{6.626 \times 10^{-34} \text{ J} \cdot \text{s}}{4.844 \times 10^{-9} \text{ J}} \times \dfrac{2.998 \times 10^8 \text{ m}}{\text{s}} = 4.10 \times 10^{-7} \text{ m} = 410 \text{ nm}$

        The visible range is 400–700 nm, so this line is visible; the observed color is violet.

        Check. We expect E_6 to be a more positive (or less negative) than E_2, and it is. ΔE is positive, which indicates emission. The orders of magnitude make sense and units are correct.

6.36   (a) $\Delta E = -2.18 \times 10^{-18} \text{ J} \left[\dfrac{1}{n_f^2} - \dfrac{1}{n_i^2} \right] = -2.18 \times 10^{-18} \text{ J} (1/1 - 1/16) = -2.044 \times 10^{-18}$

                                                                    $= -2.04 \times 10^{-18} \text{ J}$

$$\nu = E/h = \frac{2.044 \times 10^{-18} \text{ J}}{6.626 \times 10^{-34} \text{ J} \bullet \text{s}} = 3.084 \times 10^{15} = 3.08 \times 10^{15} \text{ s}^{-1}$$

$$\lambda = c/\nu = \frac{2.998 \times 10^8 \text{ m}}{1 \text{ s}} \times \frac{1 \text{ s}}{3.084 \times 10^{15}} = 9.72 \times 10^{-8} \text{ m}$$

Since the sign of ΔE is negative, radiation is emitted.

(b) $\Delta E = -2.18 \times 10^{-18} \text{ J}(1/4 - 1/25) = -4.578 \times 10^{-19} = -4.58 \times 10^{-19} \text{ J}$

$$\nu = \frac{4.578 \times 10^{-19} \text{ J}}{6.626 \times 10^{-34} \text{ J} \bullet \text{s}} = 6.909 \times 10^{14} = 6.91 \times 10^{14} \text{ s}^{-1}; \lambda = \frac{2.998 \times 10^8 \text{ m/s}}{6.909 \times 10^{14}/\text{s}}$$

$\lambda = 4.34 \times 10^{-7}$ m. Visible radiation is emitted.

(c) $\Delta E = -2.18 \times 10^{-18} \text{ J} (1/36 - 1/9) = 1.817 \times 10^{-19} = 1.82 \times 10^{-19} \text{ J}$

$$\nu = \frac{1.817 \times 10^{-19} \text{ J}}{6.626 \times 10^{-34} \text{ J} \bullet \text{s}} = 2.742 \times 10^{14} = 2.74 \times 10^{14} \text{ s}^{-1}; \lambda = \frac{2.998 \times 10^8 \text{ m/s}}{2.742 \times 10^{14}/\text{s}}$$

$\lambda = 1.09 \times 10^{-6}$ m. Radiation is absorbed.

6.37   (a)   Only lines with $n_f = 2$ represent ΔE values and wavelengths that lie in the visible portion of the spectrum. Lines with $n_f = 1$ have larger ΔE values and shorter wavelengths that lie in the ultraviolet. Lines with $n_f > 2$ have smaller ΔE values and lie in the lower energy longer wavelength regions of the electromagnetic spectrum.

(b)   *Analyze/Plan*. Use Equation 6.7 to calculate ΔE, then $\lambda = hc/\Delta E$.   *Solve*.

$$n_i = 3, n_f = 2; \quad \Delta E = -2.18 \times 10^{-18} \text{ J} \left[\frac{1}{n_f^2} - \frac{1}{n_i^2}\right] = -2.18 \times 10^{-18} \text{ J} (1/4 - 1/9)$$

$$\lambda = hc/E = \frac{6.626 \times 10^{-34} \text{ J} \bullet \text{s} \times 2.998 \times 10^8 \text{ m/s}}{-2.18 \times 10^{-18} \text{ J} (1/4 - 1/9)} = 6.56 \times 10^{-7} \text{ m}$$

This is the red line at 656 nm.

$$n_i = 4, n_f = 2; \quad \lambda = hc/E = \frac{6.626 \times 10^{-34} \text{ J} \bullet \text{s} \times 2.998 \times 10^8 \text{ m/s}}{-2.18 \times 10^{-18} \text{ J} (1/4 - 1/16)} = 4.86 \times 10^{-7} \text{ m}$$

This is the blue line at 486 nm.

$$n_i = 5, n_f = 2; \quad \lambda = hc/E = \frac{6.626 \times 10^{-34} \text{ J} \bullet \text{s} \times 2.998 \times 10^8 \text{ m/s}}{-2.18 \times 10^{-18} \text{ J} (1/4 - 1/25)} = 4.34 \times 10^{-7} \text{ m}$$

This is the violet line at 434 nm.

Check. The calculated wavelengths correspond well to three lines in the H emission spectrum in Figure 6.12, so the results are sensible.

6.38   (a)   Transitions with $n_f = 1$ have larger ΔE values and shorter wavelengths than those with $n_f = 2$. These transitions will lie in the ultraviolet region.

(b) $n_i = 2, n_f = 1;$ $\lambda = hc/E = \dfrac{6.626 \times 10^{-34} \text{ J} \cdot \text{s} \times 2.998 \times 10^8 \text{ m/s}}{-2.18 \times 10^{-18} \text{ J} (1/1 - 1/4)} = 1.21 \times 10^{-7} \text{ m}$

$n_i = 3, n_f = 1;$ $\lambda = hc/E = \dfrac{6.626 \times 10^{-34} \text{ J} \cdot \text{s} \times 2.998 \times 10^8 \text{ m/s}}{-2.18 \times 10^{-18} \text{ J} (1/1 - 1/9)} = 1.03 \times 10^{-7} \text{ m}$

$n_i = 4, n_f = 1;$ $\lambda = hc/E = \dfrac{6.626 \times 10^{-34} \text{ J} \cdot \text{s} \times 2.998 \times 10^8 \text{ m/s}}{-2.18 \times 10^{-18} \text{ J} (1/1 - 1/16)} = 0.972 \times 10^{-7} \text{ m}$

6.39   (a) $93.8 \text{ nm} \times \dfrac{1 \times 10^{-9} \text{ m}}{1 \text{ nm}} = 9.38 \times 10^{-8} \text{ m};$ this line is in the ultraviolet region.

(b)   *Analyze/Plan.* Only lines with $n_f = 1$ have a large enough ΔE to lie in the ultraviolet region (see Solutions 6.37 and 6.38). Solve Equation 6.7 for n_i, recalling that ΔE is negative for emission.   *Solve.*

$$\dfrac{-hc}{\lambda} = -2.18 \times 10^{-18} \text{ J} \left[\dfrac{1}{n_f^2} - \dfrac{1}{n_i^2} \right]; \quad \dfrac{hc}{\lambda(2.18 \times 10^{-18} \text{ J})} = \left[1 - \dfrac{1}{n_i^2} \right]$$

$$-\dfrac{1}{n_i^2} = \left[\dfrac{hc}{\lambda(2.18 \times 10^{-18} \text{ J})} - 1 \right]; \quad \dfrac{1}{n_i^2} = \left[1 - \dfrac{hc}{\lambda(2.18 \times 10^{-18} \text{ J})} \right]$$

$$n_i^2 = \left[1 - \dfrac{hc}{\lambda(2.18 \times 10^{-18} \text{ J})} \right]^{-1}; \quad n_i = \left[1 - \dfrac{hc}{\lambda(2.18 \times 10^{-18} \text{ J})} \right]^{-1/2}$$

$$n_i = \left(1 - \dfrac{6.626 \times 10^{-34} \text{ J} \cdot \text{s} \times 2.998 \times 10^8 \text{ m/s}}{9.38 \times 10^{-8} \text{ m} \times 2.18 \times 10^{-18} \text{ J}} \right)^{-1/2} = 6 \ (n \text{ values must be integers})$$

$n_i = 6, n_f = 1$

Check. From Solution 6.38, we know that $n_i > 4$ for $\lambda = 93.8$ nm. The calculated result is close to 6, so the answer is reasonable.

6.40   (a) $2626 \text{ nm} \times \dfrac{1 \times 10^{-9} \text{ m}}{1 \text{ nm}} = 2.626 \times 10^{-6} \text{ m};$ this line is in the infrared.

(b)   Absorption lines with $n_i = 1$ are in the ultraviolet and with $n_i = 2$ are in the visible. Thus, $n_i \geq 3$, but we do not know the exact value of n_i. Calculate the longest wavelength with $n_i = 3$ ($n_f = 4$). If this is less than 2626 nm, $n_i > 3$.

$$\lambda = hc/E = \dfrac{6.626 \times 10^{-34} \text{ J} \cdot \text{s} \times 2.998 \times 10^8 \text{ m/s}}{-2.18 \times 10^{-18} \text{ J} (1/16 - 1/9)} = 1.875 \times 10^{-6} \text{ m}$$

This wavelength is shorter than 2.626×10^{-6} m, so $n_i > 3$; try $n_i = 4$ and solve for n_f as in Solution 6.39.

$$n_f = \left(\dfrac{1}{n_i^2} - \dfrac{hc}{\lambda(2.18 \times 10^{-18} \text{ J})} \right)^{-1/2} = \left(1/16 - \dfrac{6.626 \times 10^{-34} \text{ J} \cdot \text{s} \times 2.998 \times 10^8 \text{ m/s}}{2.626 \times 10^{-6} \text{ m} \times 2.18 \times 10^{-18} \text{ J}} \right)^{-1/2} = 6$$

$n_f = 6, n_i = 4$

6.41   *Analyze/Plan.* $\lambda = \dfrac{h}{mv}$; $1\,J = \dfrac{1\,kg \bullet m^2}{s^2}$; Change mass to kg and velocity to m/s in each case.   *Solve.*

(a) $\dfrac{50\,km}{1\,hr} \times \dfrac{1000\,m}{1\,km} \times \dfrac{1\,hr}{60\,min} \times \dfrac{1\,min}{60\,s} = 13.89 = 14\,m/s$

$\lambda = \dfrac{6.626 \times 10^{-34}\,kg \bullet m^2 \bullet s}{1\,s^2} \times \dfrac{1}{85\,kg} \times \dfrac{1\,s}{13.89\,m} = 5.6 \times 10^{-37}\,m$

(b) $10.0\,g \times \dfrac{1\,kg}{1000\,g} = 0.0100\,kg$

$\lambda = \dfrac{6.626 \times 10^{-34}\,kg \bullet m^2 \bullet s}{1\,s^2} \times \dfrac{1}{0.0100\,kg} \times \dfrac{1\,s}{250\,m} = 2.65 \times 10^{-34}\,m$

(c)   We need to calculate the mass of a single Li atom in kg.

$\dfrac{6.94\,g\,Li}{1\,mol\,Li} \times \dfrac{1\,kg}{1000\,g} \times \dfrac{1\,mol}{6.022 \times 10^{23}\,Li\,atoms} = 1.152 \times 10^{-26} = 1.15 \times 10^{-26}\,kg$

$\lambda = \dfrac{6.626 \times 10^{-34}\,kg \bullet m^2 \bullet s}{1\,s^2} \times \dfrac{1}{1.152 \times 10^{-26}\,kg} \times \dfrac{1\,s}{2.5 \times 10^5\,m} = 2.3 \times 10^{-13}\,m$

6.42 $\lambda = h/mv$; change mass to kg and velocity to m/s

mass of muon $= 206.8 \times 9.1094 \times 10^{-28}\,g \times \dfrac{1\,kg}{1000\,g} = 1.8838 \times 10^{-28} = 1.88 \times 10^{-28}\,kg$

$\lambda = \dfrac{6.626 \times 10^{-34}\,kg \bullet m^2 \bullet s}{1\,s^2} \times \dfrac{1}{1.8838 \times 10^{-28}\,kg} \times \dfrac{1\,s}{8.85 \times 10^3\,m/s} = 3.97 \times 10^{-10}\,m$

$= 3.97\,Å$

6.43   *Analyze/Plan.* Use $v = h/m\lambda$; change wavelength to meters and mass of neutron (back inside cover) to kg.   *Solve.*

$\lambda = 0.955\,Å \times \dfrac{1 \times 10^{-10}\,m}{1\,Å} = 0.955 \times 10^{-10}\,m$; $m = 1.6749 \times 10^{-27}\,kg$

$v = \dfrac{6.626 \times 10^{-34}\,kg \bullet m^2 \bullet s}{1\,s^2} \times \dfrac{1}{1.6749 \times 10^{-27}\,kg} \times \dfrac{1}{0.955 \times 10^{-10}\,m} = 4.14 \times 10^3\,m/s$

Check. $(6.6/1.6/1) \times (10^{-34}/10^{-27}/10^{-10}) \approx 4 \times 10^3\,m/s$

6.44 $m_e = 9.1094 \times 10^{-31}\,kg$ (back inside cover of text)

$\lambda = \dfrac{6.626 \times 10^{-34}\,kg \bullet m^2 \bullet s}{1\,s^2} \times \dfrac{1}{9.1094 \times 10^{-31}\,kg} \times \dfrac{1\,s}{9.38 \times 10^6\,m} = 7.75 \times 10^{-11}\,m$

$7.75 \times 10^{-11}\,m \times \dfrac{1\,Å}{1 \times 10^{-10}\,m} = 0.775\,Å$

Since atomic radii and interatomic distances are on the order of 1–5 Å (Section 2.3), the wavelength of this electron is comparable to the size of atoms.

6.45   *Analyze/Plan.* Use $\Delta x \geq h/4\pi \, m \, \Delta v$, paying attention to appropriate units. Note that the uncertainty in speed of the particle (Δv) is important, rather than the speed itself. *Solve.*

(a) $m = 1.50 \text{ mg} \times \dfrac{1 \text{ g}}{1000 \text{ mg}} \times \dfrac{1 \text{ kg}}{1000 \text{ g}} = 1.50 \times 10^{-6} \text{ kg}; \; \Delta v = 0.01 \text{ m/s}$

$$\Delta x \geq = \dfrac{6.626 \times 10^{-34} \text{ J} \bullet \text{s}}{4\pi (1.50 \times 10^{-6} \text{ kg})(0.01 \text{ m/s})} \geq 3.52 \times 10^{-27} = 4 \times 10^{-27} \text{ m}$$

(b) $m = 1.673 \times 10^{-24} \text{ g} = 1.673 \times 10^{-27} \text{ kg}; \; \Delta v = 0.01 \times 10^4 \text{ m/s}$

$$\Delta x \geq = \dfrac{6.626 \times 10^{-34} \text{ J} \bullet \text{s}}{4\pi (1.673 \times 10^{-27} \text{ kg})(0.01 \times 10^4 \text{ m/s})} \geq 3 \times 10^{-10} \text{ m}$$

Check. The more massive particle in (a) has a much smaller uncertainty in position.

6.46 $\Delta x \geq = h/4\pi m \Delta v$; use masses in kg, Δv in m/s.

(a) $\dfrac{6.626 \times 10^{-34} \text{ J} \bullet \text{s}}{4\pi (9.109 \times 10^{-31} \text{ kg})(0.01 \times 10^5 \text{ m/s})} = 6 \times 10^{-8} \text{ m}$

(b) $\dfrac{6.626 \times 10^{-34} \text{ J} \bullet \text{s}}{4\pi (1.675 \times 10^{-27} \text{ kg})(0.01 \times 10^5 \text{ m/s})} = 3 \times 10^{-11} \text{ m}$

(c)   For particles moving with the same uncertainty in velocity, the more massive neutron has a much smaller uncertainty in position than the lighter electron. In our model of the atom, we know where the massive particles in the nucleus are located, but we cannot know the location of the electrons with any certainty, if we know their speed.

Quantum Mechanics and Atomic Orbitals

6.47   (a)   The uncertainty principle states that there is a limit to how precisely we can simultaneously know the position and momentum (related to energy) of an electron. The Bohr model states that electrons move about the nucleus in precisely circular orbits of known radius; each permitted orbit has an allowed energy associated with it. Thus, according to the Bohr model, we can know the exact distance of an electron from the nucleus and its energy. This violates the uncertainty principle.

(b)   deBroglie stated that electrons demonstrate the properties of both particles and waves, that each particle has a wave associated with it. A wave function is the mathematical description of the matter wave of an electron.

(c)   Although we cannot predict the exact location of an electron in an allowed energy state, we can determine the likelihood or probability of finding an electron at a particular position (or within a particular volume). This statistical knowledge of electron location is called the *probability density* or electron density and is a function of ψ^2, the square of the wave function ψ.

6.48   (a)   The Bohr model states with 100% certainty that the electron in hydrogen can be found 0.53 Å from the nucleus. The quantum mechanical model, taking the wave nature of the electron and the uncertainty principle into account, is a statistical model that states the probability of finding the electron in certain regions around the nucleus. While 0.53 Å might be the radius with highest probability, that probability would always be less than 100%.

        (b)   The equations of classical physics predict the instantaneous position, direction of motion, and speed of a macroscopic particle; they do not take quantum theory or the wave nature of matter into account. For macroscopic particles, these are not significant, but for microscopic particles like electrons, they are crucial. Schrödinger's equation takes these important theories into account to produce a statistical model of electron location given a specific energy.

        (c)   The square of the wave function has the physical significance of an amplitude, or probability. The quantity ψ^2 at a given point in space is the probability of locating the electron within a small volume element around that point at any given instant. The total probability, that is, the sum of ψ^2 over all the space around the nucleus, must equal 1.

6.49   (a)   The possible values of l are $(n - 1)$ to 0. $n = 4, l = 3, 2, 1, 0$
        (b)   The possible values of m_l are $-l$ to $+l$. $l = 2, m_l = -2, -1, 0, 1, 2$

6.50   (a)   For $n = 3$, there are 3 l values (2, 1, 0) and 9 m_l values ($l = 2$; $m_l = -2, -1, 0, 1, 2$; $l = 1, m_l = -1, 0, 1$; $l = 0, m_l = 0$).

        (b)   For $n = 5$, there are 5 l values (4, 3, 2, 1, 0) and 25 m_l values ($l = 4$, $m_l = -4$ to $+4$; $l = 3$, $m_l = -3$ to $+3$; $l = 2$, $m_l = -2$ to $+2$; $l = 1$, $m_l = -1$ to $+1$; $l = 0, = 0$).

        In general, for each principal quantum number n there are n l-values and n^2 m_l-values. For each shell, there are n kinds of orbitals and n^2 total orbitals.

6.51   (a)   3p: $n = 3, l = 1$ (b)   2s: $n = 2, l = 0$

        (c)   4f: $n = 4, l = 3$ (d)   5d: $n = 5, l = 2$

6.52   (a)   2, 1, 1; 2, 1, 0; 2, 1 –1   (b)   5, 2, 2; 5, 2, 1; 5, 2, 0; 5, 2, –1; 5, 2, –2

6.53   Impossible: (a) 1p, only $l = 0$ is possible for $n = 1$; (d) 2d, for $n = 2, l = 1$ or 0, but not 2

6.54   (a)   Permissible, 2p.   (b)   Forbidden, for $l = 0$, m_l can only equal 0.

        (c)   Permissible, 4d.   (d)   Forbidden, for n = 3, the largest l value is 2.

6.55

6.56

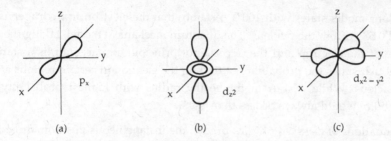

(a)　　　　　　　　　　(b)　　　　　　　　　　(c)

6.57　　(a)　The 1s and 2s orbitals of a hydrogen atom have the same overall spherical shape. The 2s orbital has a larger radial extension and one node, while the 1s orbital has continuous electron density. Since the 2s orbital is "larger," there is greater probability of finding an electron further from the nucleus in the 2s orbital.

　　　　(b)　A single 2p orbital is directional in that its electron density is concentrated along one of the three Cartesian axes of the atom. The $d_{x^2-y^2}$ orbital has electron density along both the x- and y-axes, while the p_x orbital has density only along the x-axis.

　　　　(c)　The average distance of an electron from the nucleus in a 3s orbital is greater than for an electron in a 2s orbital. In general, for the same kind of orbital, the larger the n value, the greater the average distance of an electron from the nucleus of the atom.

　　　　(d)　1s < 2p < 3d < 4f < 6s. In the hydrogen atom, orbitals with the same n value are degenerate and energy increases with increasing n value. Thus, the order of increasing energy is given above.

6.58　　(a)　In an s orbital, there are $(n-1)$ nodes.

　　　　(b)　The $2p_x$ orbital has one node (the yz plane passing through the nucleus of the atom). The 3s orbital has two nodes.

　　　　(c)　Probability density, $\psi^2(r)$, is the probability of finding an electron at a single point, r. The radial probability function, P(r), is the probability of finding an electron at any point that is distance r from the nucleus. Figure 6.18 contains plots of P(r) vs. r for 1s, 2s, and 3s orbitals. The most obvious features of these plots are the radii of maximum probability for the three orbitals, and the number and location of nodes for the three orbitals.

　　　　　　By comparing plots for the three orbitals, we see that as n increases, the number of nodes increases and the radius of maximum probability (orbital size) increases.

　　　　(d)　2s = 2p < 3s < 4d < 5s. In the hydrogen atom, orbitals with the same n value are degenerate and energy increases with increasing n value.

Many-Electron Atoms and Electron Configurations

6.59　　(a)　In the hydrogen atom, orbitals with the same principle quantum number, n, have the same energy; they are degenerate.

　　　　(b)　In a many-electron atom, for a given n-value, orbital energy increases with increasing l-value: s < p < d < f.

144

6.60 (a) The electron with the greater average distance from the nucleus feels a smaller attraction for the nucleus and is higher in energy. Thus the 3p is higher in energy than 3s.

(b) Because it has a larger n value, a 3s electron has a greater average distance from the chlorine nucleus than a 2p electron. The 3s electron experiences a smaller attraction for the nucleus and requires less energy to remove from the chlorine atom.

6.61 (a) $+1/2, -1/2$

(b) Electrons with opposite spins are affected differently by a strong inhomogeneous magnetic field. An apparatus with a strong, inhomogeneous magnetic field, similar to the diagram in Figure 6.26, can be used to distinguish electrons with opposite spins.

(c) The Pauli exclusion principle states that no two electrons can have the same four quantum numbers. Two electrons in a 1s orbital have the same, n, l, and m_l values. They must have different m_s values.

6.62 (a) The Pauli exclusion principle states that no two electrons can have the same four quantum numbers.

(b) An alternate statement of the Pauli exclusion principle is that a single orbital can hold a maximum of two electrons. Thus, the Pauli principle limits the maximum number of electrons in a main shell and its subshells, which determines when a new row of the periodic table begins.

6.63 *Analyze/Plan.* Each subshell has an l-value associated with it. For a particular l-value, permissible m_l-values are $-l$ to $+l$. Each m_l-value represents an orbital, which can hold two electrons.   *Solve.*

(a) 6   (b) 10   (c) 2   (d) 14

6.64 (a) 4   (b) 14   (c) 2   (d) 2

6.65 (a) Each box represents an orbital.

(b) Electron spin is represented by the direction of the half-arrows.

(c) No. The electron configuration of Be is $1s^2 2s^2$. There are no electrons in subshells that have degenerate orbitals, so Hund's rule is not used.

6.66 (a) "Valence electrons" are those beyond the previous noble-gas or core electron configuration.

(b) "Unpaired electrons" are electrons that occupy orbitals singly. That is, when there is only one electron in an orbital, this electron is "unpaired."

(c) A P atom has five valence electrons: $3s^2 3p^3$. Three of them (those in the degenerate 3p orbitals) are unpaired.

6.67 (a) Cs: $[Xe]6s^1$ (b) Ni: $[Ar]4s^2 3d^8$

(c) Se: $[Ar]4s^2 3d^{10} 4p^4$ (d) Cd: $[Kr]5s^2 4d^{10}$

(e)   Ac: $[Rn]7s^26d^1$ (f)   Pb: $[Xe]6s^24f^{14}5d^{10}6p^2$

6.68   (a)   Ga: $[Ar]4s^23d^{10}4p^1$, 1 unpaired electron

         (b)   Ca: $[Ar]4s^2$, 0 unpaired electrons

         (c)   V: $[Ar]4s^23d^3$, 3 unpaired electrons

         (d)   I: $[Kr]5s^24d^{10}5p^5$, 1 unpaired electron

         (e)   Y: $[Kr]5s^24d^1$, 1 unpaired electron

         (f)   Pt: $[Xe]6s^14f^{14}5d^9$, 2 unpaired electrons

         (g)   Lu: $[Xe]6s^24d^{14}5d^1$, 1 unpaired electron

6.69   Mt: $[Rn]7s^25f^{14}6d^7$

6.70   (a) $[Rn]7s^25f^{14}6d^{10}7p^6$

         (b)   None. Element 118 would belong to Group 8A, the Noble Gases. These elements have completely filled sub-shells and orbitals.

         (c)   Group 8A, the Noble Gases

6.71   (a)   Mg   (b)   Al   (c)   Cr   (d)   Te

6.72   (a)   7A (halogens)   (b)   4B   (c)   3A (row 4 and below)

         (d)   the f-block elements Sm and Pu

6.73   (a)   The fifth electron would fill the 2p subshell (same n-value as 2s) before the 3s.

         (b)   The Ne core has filled 2s and 2p subshells. Either the core is [He] or the outer electron configuration should be $3s^23p^3$.

         (c)   The 3p subshell would fill before the 3d because it has the lower l-value and the same n-value.

6.74   Count the total number of electrons to assign the element.

         (a)   N: $[He]2s^22p^3$ (b)   Se: $[Ar]4s^23d^{10}4p^4$ (c)   Rh: $[Kr]5s^24d^7$

Additional Exercises

6.75   (a) $\lambda_A = 1.6 \times 10^{-7}$ m $/ 4.5 = 3.56 \times 10^{-8} = 3.6 \times 10^{-8}$ m

                $\lambda_B = 1.6 \times 10^{-7}$ m $/ 2 = 8.0 \times 10^{-8}$ m

         (b) $\nu = c/\lambda$; $\nu_A = \dfrac{2.998 \times 10^8 \text{ m}}{1 \text{ s}} \times \dfrac{1}{3.56 \times 10^{-8} \text{ m}} = 8.4 \times 10^{15} \text{ s}^{-1}$

                $\nu_B = \dfrac{2.998 \times 10^8 \text{ m}}{1 \text{ s}} \times \dfrac{1}{8.0 \times 10^{-8} \text{ m}} = 3.7 \times 10^{15} \text{ s}^{-1}$

         (c)   A: ultraviolet, B: ultraviolet

6.76   (a)   Elements that emit in the visible: Ba (dark blue), Ca (dark blue), K (dark blue), Na (yellow/orange). (The other wavelengths are in the ultraviolet.)

       (b)   Au: shortest wavelength, highest energy

            Na: longest wavelength, lowest energy

       (c) $\lambda = c/\nu = \dfrac{2.998 \times 10^8 \text{ m/s}}{6.59 \times 10^{14}/\text{s}} \times \dfrac{1 \text{ nm}}{1 \times 10^{-9} \text{ m}} = 455 \text{ nm}, \quad \text{Ba}$

6.77   All electromagnetic radiation travels at the same speed, 2.998×10^8 m/s. Change miles to meters and seconds to some appropriate unit of time.

$$746 \times 10^6 \text{ mi} \times \frac{1.6093 \text{ km}}{1 \text{ mi}} \times \frac{1000 \text{ m}}{1 \text{ km}} \times \frac{1 \text{ s}}{2.998 \times 10^8 \text{ m}} \times \frac{1 \text{ min}}{60 \text{ s}} = 66.7 \text{ min}$$

6.78   (a) $\nu = c/\lambda = \dfrac{2.998 \times 10^8 \text{ m/s}}{320 \text{ nm}} \times \dfrac{1 \text{ nm}}{1 \times 10^{-9} \text{ m}} = 9.37 \times 10^{14} \text{ s}^{-1}$

       (b) $E = hc/\lambda = \dfrac{6.626 \times 10^{-34} \text{ J} \bullet \text{s} \times 2.998 \times 10^8 \text{ m/s}}{3.20 \times 10^{-7} \text{ m}} \times \dfrac{1 \text{ kJ}}{1000 \text{ J}} \times \dfrac{6.022 \times 10^{23} \text{ photons}}{\text{mole}}$

                                           $= 374 \text{ kJ/mol}$

       (c)   UV-B photons have shorter wavelength and higher energy.

       (d)   Yes. The higher energy UV-B photons would be more likely to cause sunburn.

6.79 $E = hc/\lambda \rightarrow$ J/photon; total energy = power × time; photons = total energy / J / photon

$$E = \frac{6.626 \times 10^{-34} \text{ J} \bullet \text{s} \times 2.998 \times 10^8 \text{ m/s}}{780 \times 10^{-9} \text{ m}} = 2.5468 \times 10^{-19} = 2.55 \times 10^{-19} \text{ J/photon}$$

$$0.10 \text{ mW} = \frac{0.10 \times 10^{-3} \text{ J}}{1 \text{ s}} \times 69 \text{ min} \times \frac{60 \text{ s}}{1 \text{ min}} = 0.4140 = 0.41 \text{ J}$$

$$0.4140 \text{ J} \times \frac{1 \text{ photon}}{2.5468 \times 10^{-19} \text{ J}} = 1.626 \times 10^{18} = 1.6 \times 10^{18} \text{ photons}$$

6.80   E/photon = hc/λ

$$E = \frac{6.626 \times 10^{-34} \text{ J} \bullet \text{s} \times 2.998 \times 10^8 \text{ m/s}}{455 \times 10^{-9} \text{ m}} \times \frac{6.022 \times 10^{23} \text{ photons}}{\text{mol}} = 2.63 \times 10^5 \text{ J/mol}$$

                                            $= 263 \text{ kJ/mol}$

6.81 $\dfrac{2.6 \times 10^{-12} \text{ C}}{1 \text{ s}} \times \dfrac{1 \text{ e}^-}{1.602 \times 10^{-19} \text{ C}} \times \dfrac{1 \text{ photon}}{1 \text{ e}^-} = 1.623 \times 10^7 = 1.6 \times 10^7 \text{ photons/s}$

$$\frac{E}{\text{photon}} = hc/\lambda = \frac{6.626 \times 10^{-34} \text{ J} \bullet \text{s}}{630 \text{ nm}} \times \frac{2.998 \times 10^8 \text{ m}}{1 \text{ s}} \times \frac{1 \text{ nm}}{1 \times 10^{-9} \text{ m}} \times \frac{1.623 \times 10^7 \text{ photon}}{\text{s}}$$

                                                        $= 5.1 \times 10^{-12} \text{ J/s}$

6.82   (a) $\dfrac{2.00 \times 10^5 \text{ J}}{\text{mol}} \times \dfrac{1 \text{ mol photons}}{6.022 \times 10^{23} \text{ photons}} = 3.321 \times 10^{-19} = 3.32 \times 10^{-19}$ J/photon

(b) $\lambda = \dfrac{hc}{E} = \dfrac{6.626 \times 10^{-34} \text{ J} \bullet \text{s} \times 2.998 \times 10^8 \text{ m/s}}{3.321 \times 10^{-19} \text{ J}} = 5.98 \times 10^{-7}$ m

(c) 5.98×10^{-7} m = 598 nm is well within the visible portion of the electromagnetic spectrum and corresponds to yellow or yellow-orange light. Red light, with wavelengths near or greater than 700 nm, does not have sufficient energy to initiate electron transfer and darken the film.

6.83   (a) $\nu = c/\lambda; \dfrac{2.998 \times 10^8 \text{ m}}{\text{s}} \times \dfrac{1}{680 \text{ nm}} \times \dfrac{1 \text{ nm}}{1 \times 10^{-9} \text{ m}} = 4.4088 \times 10^{14} = 4.41 \times 10^{14} \text{ s}^{-1}$

(b)   Calculate J/photon using $E = hc/\lambda$; change to kJ/mol.

$E_{photon} = \dfrac{6.626 \times 10^{-34} \text{ J} \bullet \text{s}}{680 \times 10^{-9} \text{ m}} \times \dfrac{2.998 \times 10^8 \text{ m}}{\text{s}} = 2.9213 \times 10^{-19} = 2.92 \times 10^{-19}$ J/photon

$\dfrac{2.9213 \times 10^{-19} \text{ J}}{\text{photon}} \times \dfrac{6.022 \times 10^{23} \text{ photons}}{\text{mol}} \times \dfrac{1 \text{ kJ}}{1000 \text{ J}} = 175.92 = 176$ kJ/mol

(c)   Nothing. The incoming (incident) radiation does not transfer sufficient energy to an electron to overcome the attractive forces holding the electron in the metal.

(d)   For frequencies greater than ν_o, any "extra" energy not needed to remove the electron from the metal becomes the kinetic energy of the ejected electron. The kinetic energy of the electron is directly proportional to this extra energy.

(e)   Let E_{total} be the total energy of an incident photon, E_{min} be the minimum energy required to eject an electron, and E_k be the "extra" energy that becomes the kinetic energy of the ejected electron.

$E_{total} = E_{min} + E_k$, $E_k = E_{total} - E_{min} = h\nu - h\nu_o$, $E_k = h(\nu - \nu_o)$. The slope of the line is the value of h, Planck's constant.

6.84   (a)   Lines with $n_f = 1$ lie in the ultraviolet (see Solution 6.30) and with $n_f = 2$ lie in the visible (see Solution 6.29). Lines with $n_f = 3$ will have smaller ΔE and longer wavelengths and lie in the infrared.

(b)   Use Equation 6.7 to calculate ΔE, then $\lambda = hc/\Delta E$.

$n_i = 4$, $n_f = 3$; $\Delta E = -2.18 \times 10^{-18} \text{ J} \left[\dfrac{1}{n_f^2} - \dfrac{1}{n_i^2} \right] = -2.18 \times 10^{-18}$ J (1/9 − 1/16)

$\lambda = hc/E = \dfrac{6.626 \times 10^{-34} \text{ J} \bullet \text{s} \times 2.998 \times 10^8 \text{ m/s}}{-2.18 \times 10^{-18} \text{ (1/9} - 1/16)} = 1.87 \times 10^{-6}$ m

$n_i = 5$, $n_f = 3$; $\lambda = hc/E = \dfrac{6.626 \times 10^{-34} \text{ J} \bullet \text{s} \times 2.998 \times 10^8 \text{ m/s}}{-2.18 \times 10^{-18} \text{ (1/9} - 1/25)} = 1.28 \times 10^{-6}$ m

$n_i = 6$, $n_f = 3$; $\lambda = hc/E = \dfrac{6.626 \times 10^{-34} \text{ J} \bullet \text{s} \times 2.998 \times 10^8 \text{ m/s}}{-2.18 \times 10^{-18} \text{ (1/9} - 1/36)} = 1.09 \times 10^{-6}$ m

These three wavelengths are all greater than 1 μm or 1×10^{-6} m. They are in the infrared, close to the visible edge (0.7×10^{-6} m).

6.85   (a)   Gaseous atoms of various elements in the sun's atmosphere typically have ground state electron configurations. When these atoms are exposed to radiation from the sun, the electrons change from the ground state to one of several allowed excited states. Atoms absorb the wavelengths of light which correspond to these allowed energy changes. All other wavelengths of solar radiation pass through the atmosphere unchanged. Thus, the dark lines are the wavelengths that correspond to allowed energy changes in atoms of the solar atmosphere. The continuous background is all other wavelengths of solar radiation.

        (b)   The scientist should record the absorption spectrum of pure neon or other elements of interest. The black lines should appear at the same wavelengths regardless of the source of neon.

6.86   (a) He^+ is hydrogen-like because it is a one-electron particle. An He atom has two electrons. The Bohr model is based on the interaction of a single electron with the nucleus, but does not accurately account for additional interactions when two or more electrons are present.

        (b)   Divide each energy by the smallest value to find the integer relationship.

           H: $2.18 \times 10^{-18} / 2.18 \times 10^{-18} = 1; \ Z = 1$

           He^+: $8.72 \times 10^{-18} / 2.18 \times 10^{-18} = 4; \ Z = 2$

           Li^{2+}: $1.96 \times 10^{-17} / 2.18 \times 10^{-18} = 9; \ Z = 3$

        The ground-state energies are in the ratio of 1:4:9, which is also the ratio Z^2, the square of the nuclear charge for each particle.

        The ground state energy for hydrogen-like particles is:

        $E = R_H Z^2$. (By definition, n = 1 for the ground state of a one-electron particle.)

        (c)   Z = 6 for C. $E = -2.18 \times 10^{-18} \text{ J } (6)^2 = -7.85 \times 10^{-17} \text{ J}$

6.87 $\lambda = h/mv; \ v = h/m\lambda. \ \lambda = 0.711 \text{ Å} \times \dfrac{1 \times 10^{-10} \text{ m}}{1 \text{ Å}} = 7.11 \times 10^{-11} \text{ m}; \ m_e = 9.1094 \times 10^{-31} \text{ kg}$

$$v = \dfrac{6.626 \times 10^{-34} \text{ J} \bullet \text{s}}{9.1094 \times 10^{-31} \text{ kg} \times 7.11 \times 10^{-11} \text{ m}} \times \dfrac{1 \text{ kg} \bullet \text{m}^2/\text{s}^2}{1 \text{ J}} = 1.02 \times 10^7 \text{ m/s}$$

6.88   *Plan.* Change keV to J/electron. Calculate v from kinetic energy. $\lambda = h/mv$.   *Solve.*

$$18.6 \text{ keV} \times \dfrac{1000 \text{ eV}}{\text{keV}} \times \dfrac{96.485 \text{ kJ}}{1 \text{ eV} \bullet \text{mol}} \times \dfrac{1000 \text{ J}}{1 \text{ kJ}} \times \dfrac{1 \text{ mol}}{6.022 \times 10^{23} \text{ electrons}}$$

$$= 2.980 \times 10^{-15} = 2.98 \times 10^{-15} \text{ J/electron}$$

$E_k = mv^2/2; \ v^2 = 2E_k/m; \ v = \sqrt{2E_k/m}$

$$v = \left(\frac{2 \times 2.980 \times 10^{-15} \text{ kg} \cdot \text{m}^2/\text{s}^2}{9.1094 \times 10^{-31} \text{ kg}} \right)^{1/2} = 8.089 \times 10^7 = 8.09 \times 10^7 \text{ m/s}$$

$$\lambda = h/mv = \frac{6.626 \times 10^{-34} \text{ J} \cdot \text{s}}{9.1094 \times 10^{-31} \text{ kg} \times 8.089 \times 10^7 \text{ m/s}} \times \frac{1 \text{ kg} \cdot \text{m}^2/\text{s}^2}{1 \text{ J}} = 8.99 \times 10^{-12} \text{ m} = 8.99 \text{ pm}$$

6.89   An *orbit* is an exactly circular path with specified radius for an electron in an allowed energy state. Each allowed orbit is denoted by a principle quantum number n. An *orbital* is a region of space where there is a high probability of finding an electron in an allowed energy state. Orbitals are three-dimensional volumes with characteristic size, shape, and orientation and are denoted by the three quantum numbers n (size), l (shape), and m_l (orientation). An orbital is a statistical prediction while an orbit is a "known" location. The notion of a known orbit violates the *uncertainty principle* (see Solution 6.47).

6.90   (a)   *l*   (b)   *n* and *l*   (c) m_s (d) m_l

6.91   (a)   Probability density, $[\psi(r)]^2$, is the probability of finding an electron at a single point at distance r from the nucleus. The radial probability function, $4\pi r^2$, is the probability of finding an electron at any point on the sphere defined by radius r. $P(r) = 4\pi r^2 [\psi(r)]^2$.

         (b)   The term $4\pi r^2$ explains the differences in plots of the two functions. Plots of the probability density, $[\psi(r)^2]$ for s orbitals shown in Figure 6.21 each have their maximum value at r = 0, with (n – 1) smaller maxima at greater values of r. The plots of radial probability, P(r), for the same s orbitals shown in Figure 6.18 have values of zero at r = 0 and the size of the maxima increases. P(r) is the product of $[\psi(r)]^2$ and $4\pi r^2$. At r = 0, the value of $[\psi(r)]^2$ is finite and large, but the value of $4\pi r^2$ is zero, so the value of P(r) is zero. As r increases, the values of $[\psi(r)]^2$ vary as shown in Figure 6.21, but the values of $4\pi r^2$ increase continuously, leading to the increasing size of P(r) maxima as r increases.

         (c)

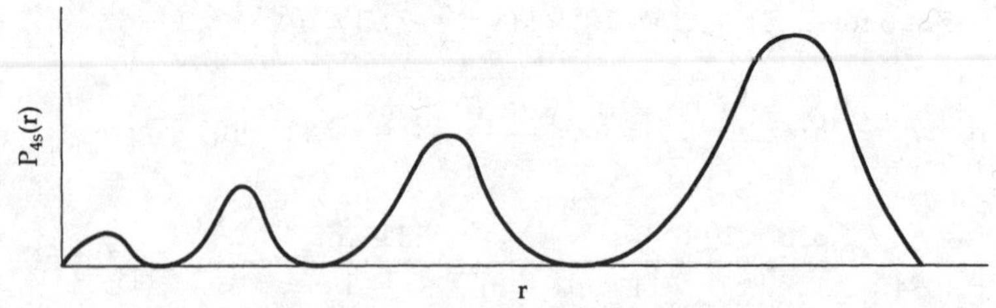

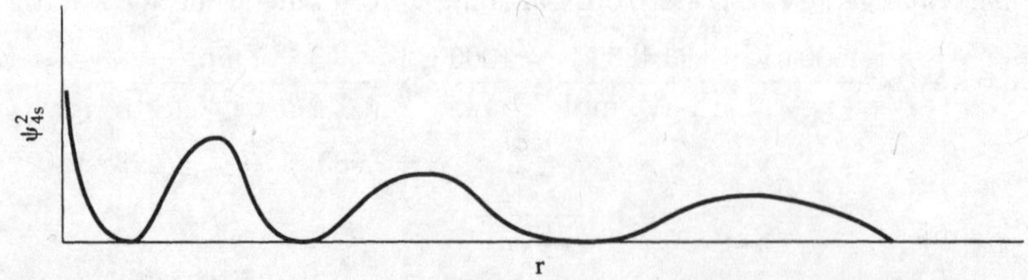

6.92   What the noble gas elements have in common are completed ns and np subshells. Since the Pauli principle limits the number of electrons per orbital to two, this leads to the first three magic numbers, $2(1s^2)$, $10(1s^2 2s^2 2p^6)$, and $18(1s^2 2s^2 2p^6 3s^2 3p^6)$. In the fourth row, $(n - 1)$ d orbitals begin to fill as their energy falls below that of the np orbitals. This leads to the next two magic numbers, $36(1s^2 2s^2 2p^6 3s^2 3p^6 4s^2 3d^{10} 4p^6)$ and $54(1s^2 2s^2 2p^6 3s^2 3p^6 4s^2 3d^{10} 4p^6 5s^2 4d^{10} 5p^6)$. In the sixth row, the energy of the 4f orbitals falls below that of the $(n - 1)$d and np subshells, and it fills. This explains the final magic number, $86(1s^2 2s^2 2p^6 3s^2 3p^6 4s^2 3d^{10} 4p^6 5s^2 4d^{10} 5p^6 6s^2 4f^{14} 5d^{10} 6p^6)$.

6.93   (a)   The p_z orbital has a nodal plane where $z = 0$. This is the xy plane.

        (b)   The d_{xy} orbital has four lobes and two nodal planes, the two planes where $x = 0$ and $y = 0$. These are the yz and xz planes.

        (c)   The $d_{x^2 - y^2}$ has four lobes and two nodal planes, the planes where $x^2 - y^2 = 0$. These are the planes that bisect the x and y axes and contain the z axis.

6.94   (a)   In the absence of a magnetic field, electrons with opposite m_s values have the same energy. Because electrons with opposite spins will have oppositely oriented magnetic fields, only the interaction of the magnetic fields of the electrons with an external magnetic field will cause the energies of the electrons to be different and observable.

        (b)   According to Figure 6.27, the particle with its magnetic field parallel to the external field will have the lower energy. The left electron has its magnetic field oriented parallel to the described magnetic orientation, so it will be lower in energy.

        (c)   Microwave photons used to excite unpaired electrons in the ESR experiment have higher energy than radio wave photons used to excite nuclei in NMR.

6.95   (a)   This is the frequency of microwaves that excite the nuclei from one spin state to the other.

        (b) $\Delta E = h\nu = 6.626 \times 10^{-34} \text{ J} \bullet \text{s} \times \dfrac{450 \times 10^6}{\text{s}} = 2.98 \times 10^{-25} \text{ J}$

        (c)   Since $\Delta E = 0$ in the absence of a magnetic field, it is reasonable to assume that the stronger the external field, the greater ΔE. (In fact, ΔE is directly proportional to field strength). Because ΔE is relatively small [see part (b)], the two spin states are almost equally populated, with a very slight excess in the lower energy state. The stronger the magnetic field, the larger ΔE, the greater number of nuclei in the lower energy spin state. With more nuclei in the lower energy state, more are able to absorb the appropriate radio wave photons and reach the higher energy state. This increases the intensity of the NMR signal, which provides more information and more reliable information than a weak absorption signal.

6.96   If m_s had three allowed values instead of two, each orbital would hold three electrons instead of two. Assuming that the same orbitals are available (that there is no change in the n, l, and m_l values), the number of elements in each of the first four rows would be:

1st row:	1 orbital $\times$ 3 = 3 elements
2nd row:	4 orbitals $\times$ 3 = 12 elements
3rd row:	4 orbitals $\times$ 3 = 12 elements
4th row:	9 orbitals $\times$ 3 = 27 elements

The s-block would be 3 columns wide, the p-block 9 columns wide and the d-block 15 columns wide.

6.97   (a)   Se: $[Ar]4s^2 3d^{10} 4p^4$

(b)   Rh: $[Kr]5s^2 4d^7$

(c)   Si: $[Ne]3s^2 3p^2$

(d)   Hg: $[Xe]6s^2 4f^{14} 5d^{10}$

(e)   Hf: $[Xe]6s^2 4f^{14} 5d^2$

6.98   The core would be the electron configuration of element 118. If no new subshell begins to fill, the condensed electron configuration of element 126 would be similar to those of elements vertically above it on the periodic chart, Pu and Sm. The condensed configuration would be $[118]8s^2 6f^6$. On the other hand, the 5g subshell could begin to fill after 8s, resulting in the condensed configuration $[118]8s^2 5g^6$. Exceptions are also possible (likely).

Integrative Exercises

6.99   We know the wavelength of microwave radiation, the volume of coffee to be heated, and the desired temperature change. Assume the density and heat capacity of coffee are the same as pure water. We need to calculate: (i) the total energy required to heat the coffee and (ii) the energy of a single photon in order to find (iii) the number of photons required.

(i)   From Chapter 5, the heat capacity of liquid water is 4.184 J/g°C.

To find the mass of 200 mL of coffee at 23°C, use the density of water given in Appendix B.

$$200 \text{ mL} \times \frac{0.997 \text{ g}}{1 \text{ mL}} = 199.4 = 199 \text{ g coffee}$$

$$\frac{4.184 \text{ J}}{1 \text{ g} \,°C} \times 199.4 \text{ g} \times (60°C - 23°C) = 3.087 \times 10^4 \text{ J} = 31 \text{ kJ}$$

(ii) $E = hc/\lambda = 6.626 \times 10^{-34} \text{ J} \cdot \text{s} \times \dfrac{2.998 \times 10^8 \text{ m}}{1 \text{ s}} \times \dfrac{1}{0.112 \text{ m}} = \dfrac{1.77 \times 10^{-24} \text{ J}}{1 \text{ photon}}$

(iii) $3.087 \times 10^4 \text{ J} \times \dfrac{1 \text{ photon}}{1.774 \times 10^{-24} \text{ J}} = 1.7 \times 10^{28} \text{ photons}$

(The answer has 2 sig figs because the temperature change, 43°C, has 2 sig figs.)

6.100 $\Delta H^{\circ}_{rxn} = \Delta H^{\circ}_f \, O_2(g) + \Delta H^{\circ}_f \, O(g) - \Delta H^{\circ}_f \, O_3(g)$

         $\Delta H^{\circ}_{rxn} = 0 + 247.5 \text{ kJ} - 142.3 \text{ kJ} = +105.2 \text{ kJ}$

$$\frac{105.2 \text{ kJ}}{\text{mol } O_3} \times \frac{1 \text{ mol } O_3}{6.022 \times 10^{23} \text{ molecules}} \times \frac{1000 \text{ J}}{1 \text{ kJ}} = \frac{1.747 \times 10^{-19} \text{ J}}{O_3 \text{ molecule}}$$

$$\Delta E = hc/\lambda; \; \lambda = \frac{hc}{\Delta E} = \frac{6.626 \times 10^{-34} \text{ J} \bullet \text{s} \times 2.998 \times 10^{8} \text{ m/s}}{1.747 \times 10^{-19} \text{ J}} = 1.137 \times 10^{-6} \text{ m}$$

Radiation with this wavelength is in the infrared portion of the spectrum. (Clearly, processes other than simple photodissociation cause O_3 to absorb ultraviolet radiation.)

6.101   (a)   The electron configuration of Zr is $[Kr]5s^2 4d^2$ and that of Hf is $[Xe]6s^2 4f^{14} 5d^2$. Although Hf has electrons in f orbitals as the rare earth elements do, the 4f subshell in Hf is filled, and the 5d electrons primarily determine the chemical properties of the element. Thus, Hf should be chemically similar to Zr rather than the rare earth elements.

     (b) $ZrCl_4(s) + 4Na(l) \rightarrow Zr(s) + 4NaCl(s)$

         This is an oxidation-reduction reaction; Na is oxidized and Zr is reduced.

     (c) $2ZrO_2(s) + 4Cl_2(g) + 3C(s) \rightarrow 2ZrCl_4(s) + CO_2(g) + 2CO(g)$

$$55.4 \text{ g } ZrO_2 \times \frac{1 \text{ mol } ZrO_2}{123.2 \text{ g } ZrO_2} \times \frac{2 \text{ mol } ZrCl_4}{2 \text{ mol } ZrO_2} \times \frac{233.0 \text{ g } ZrCl_4}{1 \text{ mol } ZrCl_4} = 105 \text{ g } ZrCl_4$$

     (d)   In ionic compounds of the type MCl_4 and MO_2, the metal ions have a 4+ charge, indicating that the neutral atoms have lost four electrons. Zr, $[Kr]5s^2 4d^2$, loses the four electrons beyond its Kr core configuration. Hf, $[Xe]6s^2 4f^{14} 5d^2$, similarly loses its four 6s and 5d electrons, but not electrons from the "complete" 4f subshell.

6.102   (a)   Each oxide ion, O^{2-}, carries a 2- charge. Each metal oxide is a neutral compound, so the metal ion or ions must adopt a total positive charge equal to the total negative charge of the oxide ions in the compound. The table below lists the electron configuration of the neutral metal atom, the positive charge of each metal ion in the oxide, and the corresponding electron configuration of the metal ion.

         i.   K: $[Ar] 4s^1$ 1+ $[Ar]$

         ii.   Ca: $[Ar] 4s^2$ 2+ $[Ar]$

         iii.   Sc: $[Ar] 4s^2 3d^1$ 3+ $[Ar]$

         iv.   Ti: $[Ar] 4s^2 3d^2$ 4+ $[Ar]$

         v.   V: $[Ar] 4s^2 3d^3$ 5+ $[Ar]$

         vi.   Cr: $[Ar] 4s^1 3d^5$ 6+ $[Ar]$

         Each metal atom loses all (valence) electrons beyond the Ar core configuration. In K_2O, Sc_2O_3 and V_2O_5, where the metal ions have odd charges, two metal ions are required to produce a neutral oxide.

(b)   i.   potassium oxide

       ii.   calcium oxide

       iii.   scandium(III) oxide

       iv.   titanium (IV) oxide

       v.   vanadium (V) oxide

       vi.   chromium (VI) oxide

(Roman numerals are required to specify the charges on the transition metal ions, because more than one stable ion may exist.)

(c)   Recall that $\Delta H_f^\circ = 0$ for elements in their standard states. In these reactions, M(s) and $H_2(g)$ are elements in their standard states.

     i. $K_2O(s) + H_2(g) \rightarrow 2K(s) + H_2O(g)$

         $\Delta H^\circ = \Delta H_f^\circ\, H_2O(g) + 2\Delta H_f^\circ\, K(s) - \Delta H\, K_2O(s) - \Delta H_f^\circ\, H_2(g)$

         $\Delta H^\circ = -241.82\,\text{kJ} + 2(0) - (-363.2\,\text{kJ}) - 0 = 121.4\,\text{kJ}$

     ii. $CaO(s) + H_2(g) \rightarrow Ca(s) + H_2O(g)$

         $\Delta H^\circ = \Delta H_f^\circ\, H_2O(g) + \Delta H_f^\circ\, Ca(s) - \Delta H_f^\circ\, CaO(s) - \Delta H_f^\circ\, H_2(g)$

         $\Delta H^\circ = -241.82\,\text{kJ} + 0 - (-635.1\,\text{kJ}) - 0 = 393.3\,\text{kJ}$

     iii. $TiO_2(s) + 2H_2(g) \rightarrow Ti(s) + 2H_2O(g)$

         $\Delta H^\circ = 2\Delta H_f^\circ\, H_2O(g) + \Delta H_f^\circ\, Ti(s) - \Delta H_f^\circ\, TiO_2(s) - 2\Delta H_f^\circ\, H_2(g)$

         $= 2(-241.82) + 0 - (-938.7) - 2(0) = 455.1\,\text{kJ}$

     iv. $V_2O_5(s) + 5H_2(g) \rightarrow 2V(s) + 5H_2O(g)$

         $\Delta H^\circ = 5\Delta H_f^\circ\, H_2O(g) + 2\Delta H_f^\circ\, V(s) - \Delta H_f^\circ\, V_2O_5(s) - 5\Delta H_f^\circ\, H_2(g)$

         $= 5(-241.82) + 2(0) - (-1550.6) - 5(0) = 341.5\,\text{kJ}$

(d) ΔH_f° becomes more negative moving from left to right across this row of the periodic chart. Since Sc lies between Ca and Ti, the median of the two ΔH_f° values is approximately –785 kJ/mol. However, the trend is clearly not linear. Dividing the ΔH_f° values by the positive charge on the pertinent metal ion produces the values –363, –318, –235, and –310. The value between Ca^{2+} (–318) and Ti^{4+} (–235) is Sc^{3+} (–277). Multiplying (–277) by 3, a value of approximately –830 kJ results. A reasonable range of values for ΔH_f° of $Sc_2O_3(s)$ is then –785 to –830 kJ/mol.

6.103   (a)   Bohr's theory was based on the Rutherford "nuclear" model of the atom. That is, Bohr theory assumed a dense positive charge at the center of the atom and a diffuse negative charge (electrons) surrounding it. Bohr's theory then specified the nature of the diffuse negative charge. The prevailing theory before the nuclear model was Thomson's plum pudding or watermelon model, with discrete electrons scattered about a diffuse positive charge cloud. Bohr's theory could not have been based on the Thomson model of the atom.

(b)   DeBroglie's hypothesis is that electrons exhibit both particle and wave properties. Thomson's conclusion that electrons have mass is a particle property, while the nature of cathode rays is a wave property. De Broglie's hypothesis actually rationalizes these two seemingly contradictory observations about the properties of electrons.

6.104   (a) ^{238}U: 92 p, 146 n, 92 e; ^{235}U: 92 p, 143 n, 92 e

In keeping with the definition isotopes, only the number of neutrons is different in the two nuclides. Since the two isotopes have the same number of electrons, they will have the same electron configuration.

(b)   U: $[Rn]7s^2 5f^4$

(c)   From Figure 6.30, the actual electron configuration is $[Rn]7s^2 5f^3 6d^1$. The energies of the 6d and 5f orbitals are very close, and electron configurations of many actinides include 6d electrons.

(d) $^{238}_{92}U \rightarrow {}^{234}_{90}Th + {}^{4}_{2}He$ ^{234}Th has 90 p, 143 n, 90 e. ^{238}U has lost 2 p, 2 n, 2 e.

These are organized into ${}^{4}_{2}He$ shown in the nuclear reaction above.

(e)   From Figure 6.30, the electron configuration of Th is $[Rn]7s^2 6d^2$. This is not really surprising because there are so many rare earth electron configurations that are exceptions to the expected orbital filling order. However, Th is the only rare earth that has two d valence electrons. Furthermore, the configuration of Th is different than that of Ce, the element above it on the periodic chart, so the electron configuration is at least interesting.

7 Periodic Properties of the Elements

Visualizing Concepts

7.1 (a) The light bulb itself represents the nucleus of the atom. The brighter the bulb, the more nuclear charge the electron "sees." A frosted glass lampshade between the bulb and our eyes reduces the brightness of the bulb. The shade is analogous to core electrons in the atom shielding outer electrons (our eyes) from the full nuclear charge (the bare light bulb).

(b) Increasing the wattage of the light bulb mimics moving right along a row of the periodic table. The brighter bulb inside the same shade is analogous to having more protons in the nucleus while the core electron configuration doesn't change.

(c) Moving down a family, both the nuclear charge and the core electron configuration changes. To simulate the addition of core electrons farther from the nucleus, we would add larger frosted glass shades as well as increase the wattage of the bulb to show the increase in Z. The effect of the shade should dominate the increase in wattage, so that the brightness of the light decreases moving down a column.

7.2 The billiard ball has a definite "hard" boundary, while an atom has no definitive edge. According to the quantum mechanical model, we cannot know the exact location of an electron, only the probability of finding it at a particular radius. Thus, there is no exact edge of an atom, only decreasing probability density as atomic radius increases.

Billiard balls can be used to model nonbonding interactions, such as Ne atoms colliding, because there is no penetration of electron clouds. In a bonding interaction, electron clouds do penetrate and the bonding atomic radius is smaller than the nonbonding radius.

7.3 (a) The bonding atomic radius of A, r_A, is $d_1/2$. The distance d_2 is the sum of the bonding atomic radii of A and X, $r_A + r_X$. Since we know that $r_A = d_1/2$, $d_2 = r_X + d_1/2$, $r_X = d_2 - d_1/2$.

(b) The length of the X-X bond is $2r_X$.

$$2r_X = 2(d_2 - d_1/2) = 2d_2 - d_1.$$

7.4

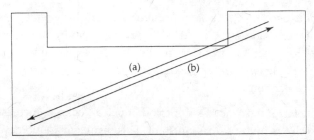

Lines (a) and (b) coincide, but their directions are opposite. Line (a) goes from upper right to lower left, and line (b) from lower left to upper right.

(c)   From the diagram, we observe that the trends in bonding atomic radius (size) and ionization energy are opposite each other. As bonding atomic radius increases, ionization energy decreases, and vice versa.

7.5 $A(g) \rightarrow A^+(g) + e^-$ ionization energy of A

$A(g) + e^- \rightarrow A^-(g)$ electron affinity of A

$$A(g) + A(g) \rightarrow A^+(g) + A^-(g) \qquad \text{ionization energy of A + electron affinity of A}$$

The energy change for the reaction is the ionization energy of A plus the electron affinity of A.

7.6   (a) $X + 2F_2 \rightarrow XF_4$

       (b)   If X is a nonmetal, XF_4 is a molecular compound. If X is a metal, XF_4 is ionic. For an ionic compound with this formula, X would have a charge of 4+, and a much smaller bonding atomic radius than F^-. X in the diagram has about the same bonding radius as F, so it is likely to be a nonmetal.

Periodic Table; Effective Nuclear Charge

7.7   Mendeleev insisted that elements with similar chemical and physical properties be placed within a family or column of the table. Since many elements were as yet undiscovered, Mendeleev left blanks. He predicted properties for the "blanks" based on properties of other elements in the family and on either side.

7.8   (a)   The verification of the existence of many new elements by accurately measuring their atomic weights spurred interest in a classification scheme. Mendeleev (and Meyer) noted that certain chemical and physical properties recur periodically when the elements are arranged by increasing atomic weight. The accurate atomic weights provided a common property on which to base a classification scheme of the elements.

       (b)   Moseley realized that the characteristic X-ray frequencies emitted by each element were related to a unique integer that he assigned to each element. We now know this integer as the atomic number, the number of protons in the nucleus of an atom. In general, atomic weight increases as atomic number increases, but there are a few exceptions. If elements are arranged by increasing atomic number, a few seeming contradictions in the Mendeleev table (the positions of Ar and K or Te and I) are eliminated.

7.9   (a)   *Effective nuclear charge*, Z_{eff}, is a representation of the average electrical field experienced by a single electron. It is the average environment created by the nucleus and the other electrons in the molecule, expressed as a net positive charge at the nucleus. It is approximately the nuclear charge, Z, minus the number of core electrons.

        (b)   Going from left to right across a period, nuclear charge increases while the number of electrons in the core is constant. This results in an increase in Z_{eff}.

7.10   (a)   Electrostatic attraction for the nucleus lowers the energy of an electron, while electron-electron repulsions increase this energy. The concept of effective nuclear charge allows us to model this increase in the energy of an electron as a smaller net attraction to a nucleus with a smaller positive charge, Z_{eff}.

        (b)   In Be (or any element), the 1s electrons are not shielded by any core electrons, so they experience a much greater Z_{eff} than the 2s electrons.

7.11   (a) $Z_{eff} = Z - S$; K: [Ar]$4s^1$; $Z = 19$. In the [Ar] core, there are 18 electrons. If these are 100% effective at screening, $S = 18$. $Z_{eff} = 19 - 18 = 1$.

        (b)   The valence electron in K is a 4s electron. All s electrons have a finite probability of being close to the nucleus and inside the core (Figure 7.3). For this reason, shielding of s electrons is never perfect and calculated values of Z_{eff} reflect this.

7.12   (a) $Z_{eff} = Z - S$; S: [Ne]$3s^2 3p^4$; $Z = 16$. If the 10 core electrons provide perfect shielding and the 6 valence electrons provide no shielding, $S = 10$.

               $Z_{eff} = 16 - 10 = 6$.

        (b)   An estimate of Z_{eff} based on the assumption in part (a) will always be lower than the value based on detailed calculations. Because outer electrons have some probability of being in the core, the core electrons are never 100% effective at shielding; the number of core electrons represents an upper limit for S.

7.13   Krypton has a larger nuclear charge ($Z = 36$) than argon ($Z = 18$). The shielding of electrons in the $n = 3$ shell by the 1s and 2s core electrons in the two atoms is approximately equal, so the $n = 3$ electrons in Kr experience a greater effective nuclear charge and are thus situated closer to the nucleus.

7.14   Mg < P < K < Ti < Rh. The shielding of electrons in the $n = 3$ shell by 1s and 2s core electrons in these elements is approximately equal, so the effective nuclear charge increases as Z increases.

Atomic and Ionic Radii

7.15   Atomic radii are determined by distances between atoms (interatomic distances) in various situations. Bonding radii are calculated from the internuclear separation of two atoms joined by a chemical bond. Nonbonding radii are calculated from the internuclear separation between two gaseous atoms that collide and move apart, but do not bond.

7.16   (a)   Since the quantum mechanical description of the atom does not specify the exact location of electrons, there is no specific distance from the nucleus where the last electron can be found. Rather, the electron density decreases gradually as the distance from the nucleus increases. There is no quantum mechanical "edge" of an atom.

          (b)   When nonbonded atoms touch, it is their electron clouds that interact. These interactions are primarily repulsive because of the negative charges of electrons. Thus, the size of the electron clouds determines the nuclear approach distance of nonbonded atoms.

7.17   The atomic (*metallic*) radius of W is the interatomic W-W distance divided by 2, 2.74 Å/2= 1.37 Å.

7.18   The distance between Ge atoms in solid germanium is two times the bonding atomic radius from Figure 7.6. The Ge-Ge distance is 2×1.22 Å = 2.44 Å.

7.19   From atomic radii, As–I = 1.19 Å + 1.33 Å = 2.52 Å. This is very close to the experimental value of 2.55 Å.

7.20   Bi – I = 2.81 Å = $r_{Bi} + r_I$. From Figure 7.6, r_I = 1.33 Å.

        r_{Bi} = [Bi – I] – r_I = 2.81 Å – 1.33 Å = 1.48 Å.

7.21   (a)   Atomic radii decrease moving from left to right across a row and (b) increase from top to bottom within a group.

          (c)   F < S < P <As. The order is unambiguous according to the trends of increasing atomic radius moving down a column and to the left in a row of the table.

7.22   (a)   The vertical difference in radius is due to a change in principal quantum number of the outer electrons. The horizontal difference in radius is due to the change in electrostatic attraction between the outer electron and a nucleus with one more or one fewer proton. Adding or subtracting a proton has a much smaller radius effect than moving from one principal quantum level to the next.

          (b)   S < Si < Se < Ge. This order is predicted by the trends in increasing atomic radius moving to the left in a row and down a column of the periodic chart, assuming that changes moving down a column are larger [see part (a)]. That is, the order above assumes that the change from S to Se is larger than the change from S to Si. This order is confirmed by the values in Figure 7.5.

7.23   *Plan.* Locate each element on the periodic charge and use trends in radii to predict their order.   *Solve.*

          (a)   Be < Mg < Ca   (b)   Br < Ge < Ga   (c)   Si < Al < Tl

7.24   (a)   K < Rb < Cs   (b)   Te < Sn < ln   (c)   Cl < P < Sr

7.25   (a)   Electrostatic repulsions are reduced by removing an electron from a neutral atom, Z_{eff} increases, and the cation is smaller.

          (b)   The additional electrostatic repulsion produced by adding an electron to a neutral atom causes the electron cloud to expand, so that the radius of the anion is larger than the radius of the neutral atom.

(c)   Going down a column, the n value of the valence electrons increases and they are farther from the nucleus. Valence electrons also experience greater shielding by core electrons. The greater radial extent of the valence electrons outweighs the increase in Z, and the size of particles with like charge increases.

7.26   (a)   As Z stays constant and the number of electrons increases, the electron-electron repulsions increase, the electrons spread apart, and the ions become larger.

$I^- > I > I^+$

(b)   Going down a column, the increasing average distance of the outer electrons from the nucleus causes the size of particles with like charge to increase.

$Ca^{2+} > Mg^{2+} > Be^{2+}$

(c)   Fe: $[Ar]4s^2 3d^6$; Fe^{2+}: $[Ar]3d^6$; Fe^{3+}: $[Ar]3d^5$. The 4s valence electrons in Fe are on average farther from the nucleus than the 3d electrons, so Fe is larger than Fe^{2+}. Since there are five 3d orbitals, in Fe^{2+} at least one orbital must contain a pair of electrons. Removing one electron to form Fe^{3+} significantly reduces repulsion, increasing the nuclear charge experienced by each of the other d electrons and decreasing the size of the ion. $Fe > Fe^{2+} > Fe^{3+}$

7.27   The size of the red sphere decreases on reaction, so it loses one or more electrons and becomes a cation. Metals lose electrons when reacting with nonmetals, so the red sphere represents a metal. The size of the blue sphere increases on reaction, so it gains one or more electrons and becomes an anion. Nonmetals gain electrons when reacting with metals, so the blue sphere represents a nonmetal.

7.28   The order of radii is $Ca > Ca^{2+} > Mg^{2+}$, so the largest sphere is Ca, the intermediate one is Ca^{2+}, and the smallest is Mg^{2+}.

7.29   (a)   An isoelectronic series is a group of atoms or ions that have the same number of electrons, and thus the same electron configuration.

(b)   (i) N^{3-}: Ne   (ii) Ba^{2+}: Xe   (iii) Se^{2-}: Kr   (iv) Bi^{3+}: Hg

7.30   (a) Sr^{2+}, Br^- (b) Y^{3+}, Br^-, Kr   (c) P^{3-}, Ti^{4+} (d) Fe^{3+}, Mn^{2+}

7.31   (a)   Since the electron configurations of the ions in an isoelectronic series are the same, shielding effects do not vary for the different particles. As Z increases, Z_{eff} increases, the valence electrons are more strongly attracted to the nucleus and the size of the particle decreases.

(b)   Because F^-, Ne and Na^+ have the same electron configuration, the 2p electron in the particle with the largest Z experiences the largest effective nuclear charge. A 2p electron in Na^+ experiences the greatest effective nuclear charge.

7.32   (a) $Cl < S < K$ (b) $K^+ < Cl^- < S^{2-}$

(c)   Even though K has the largest Z value, the n-value of the outer electron is larger than the n-value of valence electrons in S and Cl so K atoms are largest. When the 4s electron is removed, K^+ is isoelectronic with Cl^- and S^{2-}. The larger Z value causes the 3p electrons in K^+ to experience the largest effective nuclear charge and K^+ is the smallest ion.

7.33   *Plan.* Use relative location on periodic chart and trends in ionic radii to establish the order.

   Solve. (a) $Se < Se^{2-} < Te^{2-}$ (b) $Co^{3+} < Fe^{3+} < Fe^{2+}$ (c) $Ti^{4+} < Sc^{3+} < Ca$ (d) $Be^{2+} < Na^{+} < Ne$

7.34   (a) Cl^- is larger than Cl because the increase in electron repulsions that accompany addition of an electron causes the electron cloud to expand.

   (b) S^{2-} is larger than O^{2-}, because for particles with like charges, size increases going down a family.

   (c) K^+ is larger than Ca^{2+} because the two ions are isoelectronic and K^+ has the larger Z.

Ionization Energies; Electron Affinities

7.35 $B(g) \rightarrow B^+(g) + 1e^-; \quad B^+(g) \rightarrow B^{2+}(g) + 1e^-; \quad B^{2+}(g) \rightarrow B^{3+}(g) + 1e^-$

7.36   (a) $Sn(g) \rightarrow Sn^+(g) + 1e^-; \quad Sn^+(g) \rightarrow Sn^{2+}(g) + 1e^-$

   (b) $Ti^{3+}(g) \rightarrow Ti^{4+}(g) + 1e^-$

7.37   (a)   According to Coulomb's law, the energy of an electron in an atom is negative, because of the electrostatic attraction of the electron for the nucleus. In order to overcome this attraction, remove the electron and increase its energy; energy must be added to the atom. Ionization energy, ΔE for this process, is positive, regardless of the magnitude of Z or the quantum numbers of the electron.

   (b)   F has a greater first ionization energy than O, because F has a greater Z_{eff} and the outer electrons in both elements are approximately the same distance from the nucleus.

   (c)   The second ionization energy of an element is greater than the first because Z_{eff} is larger for the +1 cation than the neutral atom; more energy is required to overcome the larger Z_{eff}.

7.38   (a)   The effective nuclear charges of Li and Na are similar, but the outer electron in Li has a smaller *n*-value and is closer to the nucleus than the outer electron in Na. More energy is needed to overcome the greater attraction of the Li electron for the nucleus.

   (b)   Sc: $[Ar] 4s^2 3d^1$; Ti: $[Ar] 4s^2 3d^2$. The fourth ionization of titanium involves removing a 4s outer electron, while the fourth ionization of Sc requires removing a 3p electron from the [Ar] core. The effective nuclear charges experienced by the two 4s electrons in Ti are much more similar than the effective nuclear charges of a 4s outer electron and a 3p core electron in Sc. Thus, the difference between the third and fourth ionization energies of Sc is much larger.

   (c)   The electron configuration of Li^+ is $1s^2$ or [He] and that of Be^+ is $[He]2s^1$. Be^+ has one more valence electron to lose while Li^+ has the stable noble gas configuration of He. It requires much more energy to remove a 1s core electron close to the nucleus of Li^+ than a 2s valence electron farther from the nucleus of Be^+.

7.39   (a)   In general, the smaller the atom, the larger its first ionization energy.

       (b)   According to Figure 7.11, He has the largest and Cs the smallest first ionization energy of the nonradioactive elements.

7.40   (a)   Moving from F to I in group 7A, first ionization energies decrease and atomic radii increase. The greater the atomic radius, the smaller the electrostatic attraction of an outer electron for the nucleus and the smaller the ionization energy of the element.

       (b)   First ionization energies increase slightly going from K to Kr and atomic sizes decrease. As valence electrons are drawn closer to the nucleus (atom size decreases), it requires more energy to completely remove them from the atom (first ionization energy increases). Each trend has a discontinuity at Ga, owing to the increased shielding of the 4p electrons by the filled 3d subshell.

7.41   *Plan.* Use periodic trends in first ionization energy.   *Solve.*

       (a)   Ar   (b)   Be   (c)   Co   (d)   S   (e)   Te

7.42   (a)   Mo. As the effective nuclear charge increases in moving from left to right in the fifth row, the energy required to remove an electron increases.

       (b)   N. Valence electrons in N are closer to the nucleus ($n = 2$) and are shielded only by the [He] core, so they experience greater attraction for the nucleus and have a higher ionization energy.

       (c)   Cl. Effective nuclear charge increases moving both right across a row and up a family. Valence electrons in Cl, which is to the right and above Ga, experience the greater Z_{eff} and have the larger first ionization energy.

       (d)   Rn. Pb and Rn are in the same row, so Rn with the larger Z experiences a greater effective nuclear charge and a larger ionization energy.

7.43   *Plan.* Follow the logic of Sample Exercise 7.7.   *Solve.*

       (a) Si^{2+}: $[Ne]3s^2$ (b) Bi^{3+}: $[Xe]6s^2 4f^{14} 5d^{10}$

       (c) Te^{2-}: $[Kr]5s^2 4d^{10} 5p^6$ or $[Xe]$ (d) V^{3+}: $[Ar]3d^2$

       (e) Hg^{2+}: $[Xe]4f^{14} 5d^{10}$ (f) Ni^{2+}: $[Ar]3d^8$

7.44   (a) Mn^{3+}: $[Ar]3d^4$

       (b) Se^{2-}: $[Ar]4s^2 3d^{10} 4p^6 = [Kr]$, noble-gas configuration

       (c) Sc^{3+}: $[Ar]$, noble-gas configuration   (d) Ru^{2+}: $[Kr]4d^6$

       (e) Tl^+: $[Xe]6s^2 4f^{14} 5d^{10}$ (f) Au^+: $[Xe]4f^{14} 5d^{10}$

7.45   *Plan.* Follow the logic in Sample Exercise 7.7. Construct a mental box diagram for the outer electrons to determine how many are unpaired.   *Solve.*

       (a) Mn^{2+}: $[Ar]3d^5$, 5 unpaired electrons   (b) Si^{2-}: $[Ne]3s^2 3p^4$, 2 unpaired electrons

7.46   (a) Cu^{2+}, 1 unpaired electron   (b) Tl^+, 0 unpaired electrons

7.47   Ionization energy : $Se(g)$ $\rightarrow$ $Se^+(g) + 1e^-$

$[Ar]4s^2 3d^{10} 4p^4$ $[Ar]4s^2 3d^{10} 4p^3$

Electron affinity : $Se(g)$ + $1e^- \rightarrow$ $Se^-(g)$

$[Ar]4s^2 3d^{10} 4p^4$ $[Ar]4s^2 3d^{10} 4p^5$

7.48   Argon is a noble gas, with a very stable core electron configuration. This causes the element to resist chemical change. Positive, endothermic, values for ionization energy and electron affinity mean that energy is required to either remove or add electrons. Valence electrons in Ar experience the largest Z_{eff} of any element in the third row, because the nuclear buildup is not accompanied by an increase in screening. This results in a large, positive ionization energy. When an electron is added to Ar, the $n = 3$ electrons become core electrons which screen the extra electrons so effectively that Ar^- has a higher energy than an Ar atom and a free electron. This results in a large positive electron affinity.

7.49 Li + $1e^-$ $\rightarrow$ Li^- ; Be + $1e^-$ $\rightarrow$ Be^-

$[He]2s^1$ $[He]2s^2$ $[He]2s^2$ $[He]2s^2 2p^1$

Adding an electron to Li completes the 2s subshell. The added electron experiences essentially the same effective nuclear charge as the other valence electron, except for the repulsion of pairing electrons in an orbital. There is an overall stabilization; ΔE is negative.

An extra electron in Be would occupy the higher energy 2p subshell. This electron is shielded from the full nuclear charge by the 2s electrons and does not experience a stabilization in energy; ΔE is positive.

7.50   Electron affinity $Br(g)$ + $1e^-$ $\rightarrow$ $Br^-(g)$

$[Ar]4s^2 3d^{10} 4p^5$ $[Ar]4s^2 3d^{10} 4p^6$

When a Br atom gains an electron, the Br^- ion adopts the stable electron configuration of Kr. Since the electron is added to the same 4p subshell as other outer electrons, it experiences essentially the same attraction for the nucleus. Thus, the energy of the Br^- ion is lower than the total energy of a Br atom and an isolated electron, and electron affinity is negative.

Electron affinity : $Kr(g)$ + $1e^-$ $\rightarrow$ $Kr^-(g)$

$[Ar]4s^2 3d^{10} 4p^6$ $[Ar]4s^2 3d^{10} 4p^6 5s^1$

Energy is required to add an electron to a Kr atom; Kr^- has a higher energy than the isolated Kr atom and free electron. In Kr^- the added electron would have to occupy the higher energy 5s orbital; a 5s electron is farther from the nucleus and effectively shielded by the spherical Kr core and is not stabilized by the nucleus.

7.51   Ionization energy of F^-: $F^-(g) \rightarrow F(g) + 1e^-$

Electron affinity of F: $F(g) + 1e^- \rightarrow F^-(g)$

The two processes are the reverse of each other. The energies are equal in magnitude but opposite in sign. $I_1 (F^-) = -E (F)$

7.52

$$Mg^+(g) + 1e^- \rightarrow Mg(g)$$

$$[Ne]\,3s^1 \qquad\qquad\qquad [Ne]\,3s^2$$

This process is the reverse of the first ionization of Mg. The magnitude of the energy change for this process is the same as the magnitude of the first ionization energy of Mg, 738 kJ/mol.

Properties of Metals and Nonmetals

7.53   The smaller the first ionization energy of an element, the greater the metallic character of that element.

7.54   S < Si < Ge < Ca. S is a nonmetal, Si and Ge are metalloids, and Ca is a metal. We expect that electrical conductivity increases as metallic character increases. Since metallic character increases going down a column and to the left in a row, the order of increasing electrical conductivity is as shown above.

7.55   *Analyze/Plan.* Metallic character increases moving down a family and to the left in a period. Use these trends to select the element with greater metallic character.   *Solve.*

(a)   Li   (b) Na   (c) Sn   (d) Al

7.56   (a)   In general, ionization energy decreases as metallic character increases. According to Figure 7.11 the ionization energies of the group 5A elements decrease as atomic weight increases (going down the column). Therefore, metallic character of the group 5A elements increases with increasing atomic weight.

(b)   Ag < Sn < Se < C < F. Nonmetallic character increases going up and to the right on the periodic chart. Vertical trends dominate the relationship between Se and C.

7.57   *Analyze/Plan.* Ionic compounds are formed by combining a metal and a nonmetal; molecular compounds are formed by two or more nonmetals. *Solve.*

Ionic: MgO, Li_2O, Y_2O_3; molecular: SO_2, P_2O_5, N_2O, XeO_3

7.58   (a)   metal oxides: $Li_2O(s) + H_2O(l) \rightarrow 2\,LiOH(aq)$

$$BaO(s) + H_2O(l) \rightarrow Ba(OH)_2(aq)$$

nonmetal oxides: $P_4O_{10}(s) + 6H_2O(l) \rightarrow 4H_3PO_4(aq)$

$$SO_3(g) + H_2O(l) \rightarrow H_2SO_4(aq)$$

(b)   Metals have lower ionization energies than nonmetals, so they tend to form ionic oxides, while nonmetals form molecular oxides. Ionic compounds, in this case oxides, dissociate into ions when they dissolve in water. The reactive oxide ion ends up as hydroxide, separated from the metal ion. Molecular oxides do not ionize upon dissolution, so the oxygen remains bound to the nonmetal.

7.59   (a)   When dissolved in water, an "acidic oxide" produces an acidic (pH < 7) solution. A "basic oxide" dissolved in water produces a basic (pH > 7) solution.

(b)   Oxides of nonmetals are acidic. Example: $SO_3(g) + H_2O(l) \rightarrow H_2SO_4(aq)$. Oxides of metals are basic. Example: CaO (quick lime). $CaO(s) + H_2O(l) \rightarrow Ca(OH)_2(aq)$.

7.60   The more nonmetallic the central atom, the more acidic the oxide. In order of increasing acidity: $CaO < Al_2O_3 < SiO_2 < CO_2 < P_2O_5 < SO_3$

7.61   *Analyze/Plan.* Cl_2O_7 is a molecular compound formed by two nonmetallic elements. More specifically, it is a nonmetallic oxide and acidic.   *Solve.*

(a)   Dichlorineseptoxide

(b)   Elemental chlorine and oxygen are diatomic gases.
$$2Cl_2(g) + 7O_2(g) \rightarrow 2Cl_2O_7(l)$$

(c)   Most nonmetallic oxides we have seen, such as CO_2 and SO_3, are gases. However, oxides with more atoms, such as $P_2O_3(l)$ and $P_2O_5(s)$, exist in other states. A boiling point of 81°C is not totally unexpected for a large molecule like Cl_2O_7.

(d) Cl_2O_7 is an acidic oxide, so it will be more reactive to base, OH^-.
$$Cl_2O_7(l) + 2OH^-(aq) \rightarrow 2ClO_4^-(aq) + H_2O(l)$$

7.62   (a) $XCl_4(l) + 2H_2O(l) \rightarrow XO_2(s) + 4HCl(g)$

The second product is $HCl(g)$.

(b)   If X were a metal, both the oxide and the chloride would be high melting solids. If X were a nonmetal, XO_2 would be a nonmetallic, molecular oxide and probably gaseous, like CO_2, NO_2, and SO_2. Neither of these statements describes the properties of XO_2 and XCl_4, so X is probably a metalloid.

(c)   Use the *Handbook of Chemistry* to find formulas and melting points of oxides, and formulas and boiling points of chlorides of selected metalloids.

metalloid	formula of oxide	m.p. of oxide	formula of chloride	b.p. of chloride
boron	B_2O_3	460°C	BCl_3	12°C
silicon	SiO_2	~1700°C	$SiCl_4$	58°C
germanium	GeO	710°C	$GeCl_2$	decomposes
	GeO_2	~1100°C	$GeCl_4$	84°C
arsenic	As_2O_3	315°C	$AsCl_3$	132°C
	As_2O_5	315°C		

Boron, arsenic, and, by analogy, antimony, do not fit the description of X, because the formulas of their oxides and chlorides are wrong. Silicon and germanium, in the same family, have oxides and chlorides with appropriate formulas. Both SiO_2 and GeO_2 melt above 1000°C, but the boiling point of $SiCl_4$ is much closer to that of XCl_4. Element X is silicon.

7.63   (a) $BaO(s) + H_2O(l) \rightarrow Ba(OH)_2(aq)$

(b) $FeO(s) + 2HClO_4(aq) \rightarrow Fe(ClO_4)_2(aq) + H_2O(l)$

(c) $SO_3(g) + H_2O(l) \rightarrow H_2SO_4(aq)$

(d) $CO_2(g) + 2NaOH(aq) \rightarrow Na_2CO_3(aq) + H_2O(l)$

7.64   (a) $K_2O(s) + H_2O(l) \rightarrow 2KOH(aq)$

       (b) $P_2O_3(l) + 3H_2O(l) \rightarrow 2H_3PO_3(aq)$

       (c) $Cr_2O_3(s) + 6HCl(aq) \rightarrow 2CrCl_3(aq) + 3H_2O(l)$

       (d) $SeO_2(s) + 2KOH(aq) \rightarrow K_2SeO_3(aq) + H_2O(l)$

Group Trends in Metals and Nonmetals

7.65   <u>**Na**</u>   <u>**Mg**</u>

       (a) $[Ne]3s^1$ $[Ne]3s^2$

       (b)   +1   +2

       (c)   +496 kJ/mol   +738 kJ/mol

       (d)   very reactive   reacts with steam, but not $H_2O(l)$

       (e)   1.54 Å   1.30 Å

       (b)   When forming ions, both adopt the stable configuration of Ne, but Na loses one electron and Mg two electrons to achieve this configuration.

       (c),(e) The nuclear charge of Mg (Z = 12) is greater than that of Na, so it requires more energy to remove a valence electron with the same n value from Mg than Na. It also means that the 2s electrons of Mg are held closer to the nucleus, so the atomic radius (e) is smaller than that of Na.

       (d)   Mg is less reactive because it has a filled subshell and it has a higher ionization energy.

7.66   (a)   Rb: $[Kr]5s^1$, r = 2.11 Å   Ag: $[Kr]5s^14d^{10}$, r = 1.53 Å

              The electron configurations both have a [Kr] core and a single 5s electron; Ag has a completed 4d subshell as well. The radii are very different because the 5s electron in Ag experiences a much greater effective nuclear charge. Ag has a much larger Z (47 vs. 37), and although the 4d electrons in Ag shield the 5s electron somewhat, the increased shielding does not compensate for the large increase in Z.

       (b)   Ag is much less reactive (less likely to lose an electron) because its 5s electron experiences a much larger effective nuclear charge and is more difficult to remove.

7.67   (a)   Ca and Mg are both metals; they tend to lose electrons and form cations when they react. Ca is more reactive because it has a lower ionization energy than Mg. The Ca valence electrons in the 4s orbital are less tightly held because they are farther from the nucleus than the 3s valence electrons of Mg.

       (b)   K and Ca are both metals; they tend to lose electrons and form cations when they react. K is more reactive because it has a lower ionization energy. The 4s valence electron in K is less tightly held because it experiences a smaller nuclear charge (Z = 19 for K versus Z = 20 for Ca) with similar shielding effects than the 4s valence electrons of Ca.

7.68   (a)   Cs is much more reactive than Li toward H_2O because its valence electron is less tightly held (greater n value), and Cs is more easily oxidized.

       (b)   The purple flame indicates that the metal is potassium (see Figure 7.23).

       (c) $K_2O_2(s) \ + \ H_2O(l) \ \rightarrow \ H_2O_2(aq) \ + \ K_2O(aq)$

            potassium peroxide   hydrogen peroxide

7.69   (a) $2K(s) \ + Cl_2(g) \ \rightarrow 2KCl(s)$

       (b) $SrO(s) + H_2O(l) \rightarrow Sr(OH)_2(aq)$

       (c) $4Li(s) \ + O_2(g) \ \rightarrow 2Li_2O(s)$

       (d) $2Na(s) + S(l) \ \ \rightarrow Na_2S(s)$

7.70   (a) $2K(s) + 2H_2O(l) \rightarrow 2KOH(aq) + H_2(g)$

       (b) $Ba(s) + 2H_2O(l) \rightarrow Ba(OH)_2(aq) + H_2(g)$

       (c) $6Li(s) + N_2(g) \ \ \rightarrow 2Li_3N(s)$

       (d) $2Mg(s) + O_2(g) \rightarrow 2MgO(s)$

7.71   H: $1s^1$; Li: $[He]2s^1$; F: $[He]2s^22p^5$. Like Li, H has only one valence electron, and its most common oxidation number is +1, which both H and Li adopt after losing the single valence electron. Like F, H needs only one electron to adopt the stable electron configuration of the nearest noble gas. Both H and F can exist in the –1 oxidation state, when they have gained an electron to complete their valence shells.

7.72   (a)   The reactions of the alkali metals with hydrogen and with a halogen are redox reactions. In both classes of reaction, the alkali metal loses electrons and is oxidized. Both hydrogen and the halogen gain electrons and are reduced. The product is an ionic solid, where either hydride ion, H^-, or a halide ion, X^-, is the anion and the alkali metal is the cation.

       (b) $Ca(s) + F_2(g) \rightarrow CaF_2(s)$ $Ca(s) + H_2(g) \rightarrow CaH_2(s)$

            Both products are ionic solids containing Ca^{2+} and the corresponding anion in a 1:2 ratio.

7.73

	F	**Cl**
(a)	$[He]2s^22p^5$	$[Ne]3s^23p^5$
(b)	–1	–1
(c)	1681 kJ/mol	1251 kJ/mol
(d)	reacts exothermically to form HF	reacts slowly to form HCl
(e)	–328 kJ/mol	–349 kJ/mol
(f)	0.71 Å	0.99 Å

       (b)   F and Cl are in the same group, have the same valence electron configuration and common ionic charge.

(c),(f) The $n = 2$ valence electrons in F are closer to the nucleus and more tightly held than the $n = 3$ valence electrons in Cl. Therefore, the ionization energy of F is greater, and the atomic radius is smaller.

(d)   In its reaction with H_2O, F is reduced; it gains an electron. Although the electron affinity, a gas phase single atom property, of F is less negative than that of Cl, the tendency of F to hold its own electrons (high ionization energy) coupled with a relatively large exothermic electron affinity makes it extremely susceptible to reduction and chemical bond formation. Cl is unreactive to water because it is less susceptible to reduction.

(e)   While F has approximately the same Z_{eff} as Cl, its small atomic radius gives rise to large repulsions when an extra electron is added, so the overall electron affinity of F is smaller (less exothermic) than that of Cl.

(f)   The $n = 2$ valence electrons in F are closer to the nucleus so the atomic radius is smaller than that of Cl.

7.74   *Plan.* Predict the physical and chemical properties of At based on the trends in properties in the halogen (7A) family. *Solve.*

(a)   F, at the top of the column, is a gas; I, immediately above At, is a solid; the melting points of the halogens increase going down the column. At is likely to be a solid at room temperature.

(b)   All halogens form ionic compounds with Na; they have the generic formula NaX. The compound formed by At will have the formula NaAt.

7.75   Under ambient conditions, the Group 8A elements are all gases that are extremely unreactive, owing to their stable core electron configurations. Thus, the name "inert gases" seemed appropriate.

In the 1960s, scientists discovered that Xe, which has the lowest ionization energy of the nonradioactive noble gases, would react with substances having a strong tendency to remove electrons, such as PtF_6 or F_2. Thus, the term "inert" no longer described all the Group 8A elements. (Kr also reacts with F_2, but reactions of Ar, Ne, and He are as yet unknown.)

7.76   Xe has a lower ionization energy than Ne. The valence electrons in Xe are much farther from the nucleus than those of Ne ($n = 5$ vs $n = 2$) and much less tightly held by the nucleus; they are more "willing" to be shared than those in Ne. Also, Xe has empty 5d orbitals that can help to accommodate the bonding pairs of electrons, while Ne has all its valence orbitals filled.

7.77   (a) $2O_3(g) \rightarrow 3O_2(g)$

(b) $Xe(g) + F_2(g) \rightarrow XeF_2(g)$

$Xe(g) + 2F_2(g) \rightarrow XeF_4(s)$

$Xe(g) + 3F_2(g) \rightarrow XeF_6(s)$

(c) $S(s) + H_2(g) \rightarrow H_2S(g)$

(d) $2F_2(g) + 2H_2O(l) \rightarrow 4HF(aq) + O_2(g)$

7.78   (a) $Cl_2(g) + H_2O(l) \rightarrow HCl(aq) + HOCl(aq)$

      (b) $Ba(s) + H_2(g) \rightarrow BaH_2(s)$

      (c) $2Li(s) + S(s) \rightarrow Li_2S(s)$

      (d) $Mg(s) + F_2(g) \rightarrow MgF_2(s)$

Additional Exercises

7.79   Up to Z = 82, there are three instances where atomic weights are reversed relative to atomic numbers: Ar and K; Co and Ni; Te and I.

      In each case, the most abundant isotope of the element with the larger atomic number (Z) has one more proton, but fewer neutrons than the element with the smaller atomic number. The smaller number of neutrons causes the element with the larger Z to have a smaller than expected atomic weight.

7.80   (a)   *eka – aluminum* is gallium (Ga), Z = 31.

      (b)   From the *Handbook of Chemistry and Physics*, 74th edition, the physical properties of Ga are:

           atomic weight = 69.92 g/mol   m.p. = 29.78°C

           density = 5.904 g/mL (s)   b.p. = 2403°C

                = 6.095 g/mL (l)   oxide = Ga_2O_3, Ga_2O

          The atomic weight, density, and formula of oxide all agree with Mendeleev's predictions. Gallium does have a low melting point and a high boiling point.

7.81   (a)   4s

      (b)   To a first approximation, s and p valence electrons do not shield each other, so we expect the 4s and 4p electrons in As to experience a similar Z_{eff}. However, since s electrons have a finite probability of being very close to the nucleus (Figure 7.4), they experience less shielding than p electrons with the same n-value. Since $Z_{eff} = Z - S$ and Z is the same for all electrons in As, if S is smaller for 4s than 4p, Z_{eff} will be greater for 4s electrons and they will have a lower energy.

7.82   (a)   P: (+15 – 10) = +5

      (b)   The 3s electrons penetrate the [Ne] core electrons (by analogy to Figure 7.4) and experience less shielding than the 3p electrons. That is, S is greater for 3p electrons, owing to the penetration of the 3s electrons, so Z – S (3p) is less than Z – S(3s).

      (c)   The electron configuration of P is $[Ne]3s^23p^3$. The 3p electrons are the outermost electrons; they experience a smaller Z_{eff} than 3s electrons and thus a smaller attraction for the nucleus, given equal *n*-values. The first electron lost is a 3p electron. Each 3p orbital holds one electron, so there is no preference as to which 3p electron will be lost.

7.83   Close approach by two positively charged nuclei is impossible because of the large electrostatic repulsion between like-charged particles at small distances. The additional space between the nuclei in a molecule like F_2 is occupied by bonding electrons, which are electrostatically stabilized by attraction to both nuclei. The electrons also provide a buffer between the two nuclei.

7.84   (a) $Z_{eff} = Z - S$. According to our simple model, moving from C to N, causes Z to increase by 1 and S to remain the same, so Z_{eff} is greater by 1 for N than for C.

(b)   Our simple model assumed that valence electrons don't screen each other. If there is some mutual screening by valence electrons, S for N would be slightly greater than S for C, because N has one more valence electron than C. The increase in Z_{eff} would be less than the full 1+, as calculated.

(c)   In O, $[He]2s^2 2p^4$, one of the p orbitals is doubly occupied, which significantly increases electron-electron repulsion. This leads to an overall higher energy for the 2p electrons. In our model, this appears as a larger S and a smaller Z_{eff}.

7.85   Atomic size (bonding atomic radius) is strongly correlated to Z_{eff}, which is determined by Z and S. Moving across the representative elements, electrons added to ns or np valence orbitals do not effectively screen each other. The increase in Z is not accompanied by a similar increase in S; Z_{eff} increases and atomic size decreases. Moving across the transition elements, electrons are added to (n–1)d orbitals and become part of the core electrons, which do significantly screen the ns valence electrons. The increase in Z is accompanied by a larger increase in S for the ns valence electrons; Z_{eff} increases more slowly and atomic size decreases more slowly.

7.86   (a)   The estimated distances in the table below are the sum of the radii of the group 5A elements and H from Figure 7.6.

bonded atoms	estimated distance	measured distance
P – H	1.43	1.419
As – H	1.56	1.519
Sb – H	1.75	1.707

In general, the estimated distances are a bit longer than the measured distances. This probably shows a systematic bias in either the estimated radii or in the method of obtaining the measured values.

(b)   The principal quantum number of the outer electrons and thus the average distance of these electrons from the nucleus increases from P (n = 3) to As (n = 4) to Sb (n = 5). This causes the systematic increase in M – H distance.

7.87   Ge – H distance = $r_{Ge} + r_H$ = 1.22 + 0.37 = 1.59Å

Ge – Cl distance = $r_{Ge} + r_{Cl}$ = 1.22 + 0.99 = 2.21 Å

7.88   Y: $[Kr]5s^2 4d^1$, Z = 39   Zr: $[Kr] 5s^2 4d^2$, Z = 40

La: $[Xe]6s^2 5d^1$, Z = 57   Hf: $[Xe] 6s^2 4f^{14} 5d^2$, Z = 72

The completed 4f subshell in Hf leads to a much larger change in Z going from Zr to Hf (72 – 40 = 32) than in going from Y to La (57 – 39 = 18). The 4f electrons in Hf do not completely shield the valence electrons, so there is also a larger increase in Z_{eff}. This larger increase in Z_{eff} going from Zr to Hf leads to a smaller increase in atomic radius than in going from Y to La.

7.89   (a) Hg^{2+}

(b)   No. According to Figure 2.24, Hg is not essential for life, even in trace amounts.

(c) Zn^{2+}: $[Ar]3d^{10}$, r = 0.74 Å; Cd^{2+}: $[Kr]4d^{10}$, r = 0.95 Å; Hg^{2+}: $[Xe]4f^{14}5d^{10}$, r ≈ 1.05 Å.

If there were no 4f electrons in Hg^{2+}, we would expect an ionic radius of around 1.15 Å, an increase of ~0.20 Å due to the increase in principal quantum number of the valence electrons. However, the increase in Z due to filling of the 4f orbitals is not completely offset by shielding. The 5d valence electrons in Hg^{2+} experience a greater than expected Z_{eff} which largely offsets the increase in principal quantum number; ionic radius of Hg^{2+} is smaller than expected. This phenomenon is known as the *lanthanide contraction* and affects the physical properties of elements in the sixth period and beyond. See also Solution 7.88.

(d)   Since the ionic radius of Hg^{2+} is similar to that of Cd^{2+}, Hg^{2+} will be physiologically more similar to Cd^{2+}.

(e)   By common knowledge, and verified with WebElements.com™, both Hg and Hg^{2+} are extremely toxic to humans.

7.90   (a) $2Sr(s) + O_2(g) \rightarrow 2SrO(s)$

(b)   In the Figure, large spheres represent O^{2-} anions and small ones represent Sr^{2+} cations. According to Figure 7.7, anions are larger than their neutral atoms and cations are smaller. The changes in repulsion and Z_{eff} associated with adding or removing electrons from neutral atoms outweigh the increase in principal quantum number between from O to Sr.

(c)   Assume that the corners of the cube are at the centers of the outermost O^{2-} ions, and that the edges pass through the centers of perimeter Sr^{2+} ions. The length of an edge is then $r(O^{2-}) + 2r(Sr^{2+}) + r(O^{2-}) = 2r(O^{2-}) + 2r(Sr^{2+}) = 2(1.40$ Å$) + 2(1.13$Å$) = 5.06$ Å.

(d)   Density is the ratio of mass to volume.

$$d = \frac{mass\ SrO\ in\ cube}{vol\ cube} = \frac{\#\ SrO\ units \times mass\ of\ SrO}{vol\ cube}$$

Calculate the mass of 1 SrO unit in grams and the volume of the cube in cm^3; solve for number of SrO units.

$$\frac{103.62\ g\ SrO}{mol} \times \frac{1\ mol\ SrO}{6.022 \times 10^{23}\ SrO\ units} = 1.7207 \times 10^{-22} = 1.721 \times 10^{-22}\ g/SrO\ unit$$

$$V = (5.06)^3\ Å^3 \times \frac{(1 \times 10^{-8})^3\ cm^3}{Å^3} = 1.2955 \times 10^{-22} = 1.30 \times 10^{-22}\ cm^3$$

$$d = \frac{\text{\# of SrO units} \times 1.7207 \times 10^{-22} \text{ g/SrO unit}}{1.2955 \times 10^{-22} \text{ cm}^3} = 5.10 \text{ g/cm}^3$$

$$\text{\# of SrO units} = 5.10 \text{ g/cm}^3 \times \frac{1.2955 \times 10^{-22} \text{ cm}^3}{1.7207 \times 10^{-22} \text{ g/SrO unit}} = 3.84 \text{ units}$$

Since the number of formula units must be an integer, there are four SrO formula units in the cube. Using average values for ionic radii to estimate the edge length probably leads to the discrepancy.

7.91   C: $1s^2 2s^2 2p^2$. I_1 through I_4 represent loss of the 2p and 2s electrons in the outer shell of the atom. The values of $I_1–I_4$ increase as expected. The nuclear charge is constant, but removing each electron reduces repulsive interactions between the remaining electrons, so effective nuclear charge increases and ionization energy increases. I_5 and I_6 represent loss of the 1s core electrons. These 1s electrons are much closer to the nucleus and experience the full nuclear charge (they are not shielded), so the values of I_5 and I_6 are significantly greater than $I_1–I_4$. I_6 is larger than I_5 because all repulsive interactions have been eliminated.

7.92   (a)   Ca: [Ar] $4s^2$; Ca^{2+}: [Ar]. Zn: [Ar] $4s^2 3d^{10}$; Zn^{2+}: [Ar] $3d^{10}$. In both cases, forming ions involves losing the only electrons in the $n = 4$ shell. The average distance from the nucleus of $n = 3$ electrons is less, so the radii of the ions are smaller.

         (b)   The atomic radius of Ca is greater than that of Zn because Zn has a significantly greater Z and Z_{eff}, which draws the 4s valence electrons closer to the nucleus.

         (c)   In both Ca and Zn atoms, the outermost electrons are in the 4s subshell. The much larger Z_{eff} for Zn causes it to have a much smaller radius. In the +2 ions, both have lost their 4s electrons. The outermost electrons in Zn are in the 3d sublevel and are significantly shielded by the [Ar] core electrons. Thus, the radius of Zn^{2+} is closer to the radius of Ca^{2+} than the radii of the neutral ions.

7.93   The statement is somewhat true, but more accurate if changed to read: "A negative value for the electron affinity of an atom occurs when the outermost electrons only incompletely shield the added electron from the nucleus." This new statement totally explains the negative electron affinity of Br and the positive value for Kr. For Br^-, the electron is added to the 4p subshell and is incompletely shielded by the "other" 4s and 4p electrons. For Kr^-, the electron is added to the 5s subshell, which is effectively shielded by the spherical Kr core.

7.94   O: [He]$2s^2 2p^4$

         O^{2-}: [He]$2s^2 2p^6$ = [Ne]

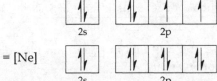

         O^{3-}: [Ne]$3s^1$ The third electron would be added to the 3s orbital, which is farther from the nucleus and more strongly shielded by the [Ne] core. The overall attraction of this 3s electron for the O nucleus is not large enough for O^{3-} to be a stable particle.

7.95   (a)   P: [Ne] $3s^2 3p^3$; S: [Ne] $3s^2 3p^4$. In P, each 3p orbital contains a single electron, while in S one 3p orbital contains a pair of electrons. Removing an electron from S eliminates the need for electron pairing and reduces electrostatic repulsion, so the overall energy required to remove the electron is smaller than in P, even though Z is greater.

    (b)   C: [He] $2s^2 2p^2$; N: [He] $2s^2 2p^3$; O: [He] $2s^2 2p^4$. An electron added to a N atom must be paired in a relatively small 2p orbital, so the additional electron-electron repulsion more than compensates for the increase in Z and the electron affinity is smaller (less exothermic) than that of C. In an O atom, one 2p orbital already contains a pair of electrons, so the additional repulsion from an extra electron is offset by the increase in Z and the electron affinity is greater (more exothermic). Note from Figure 7.12 that the electron affinity of O is only slightly more exothermic than that of C, although the value of Z has increased by 2.

    (c) O^+: [Ne] $2s^2 2p^3$; O^{2+}: [Ne] $2s^2 2p^2$; F^+: [Ne] $2s^2 2p^4$; F^{2+}: [Ne] $2s^2 2p^3$. The decrease in electron-electron repulsion going from F^+ to F^{2+} energetically favors ionization and causes it to be less endothermic than the corresponding process in O, where there is no significant decrease in repulsion.

    (d) Mn^{2+}: [Ar]$3d^5$; Mn^{3+}; [Ar] $3d^4$; Cr^{2+}: [Ar] $3d^4$; Cr^{3+}: [Ar] $3d^3$; Fe^{2+}: [Ar] $3d^6$; Fe^{3+}: [Ar] $3d^5$. The third ionization energy of Mn is expected to be larger than that of Cr because of the larger Z value of Mn. The third ionization energy of Fe is less than that of Mn because going from $3d^6$ to $3d^5$ reduces electron repulsions, making the process less endothermic than predicted by nuclear charge arguments.

7.96   (a)   The group 2B metals have complete $(n–1)$d subshells. An additional electron would occupy an np subshell and be substantially shielded by both ns and $(n–1)$d electrons. Overall this is not a lower energy state than the neutral atom and a free electron.

    (b)   Valence electrons in Group 1B elements experience a relatively large effective nuclear charge due to the buildup in Z with the filling of the $(n–1)$d subshell. Thus, the electron affinities are large and negative. Group 1B elements are exceptions to the usual electron filling order and have the generic electron configuration $n s^1 (n–1)d^{10}$. The additional electron would complete the ns subshell and experience repulsion with the other ns electron. Going down the group, size of the ns subshell increases and repulsion effects decrease. That is, effective nuclear charge is greater going down the group because it is less diminished by repulsion, and electron affinities become more negative.

7.97   (a)   For both H and the alkali metals, the added electron will complete an ns subshell (1s for H and ns for the alkali metals) so shielding and repulsion effects will be similar. For the halogens, the electron is added to an np subshell, so the energy change is likely to be quite different.

    (b)   True. Only He has a smaller estimated "bonding" atomic radius, and no known compounds of He exist. The electron configuration of H is $1s^1$. The single 1s electron experiences no repulsion from other electrons and feels the full unshielded nuclear charge. It is held very close to the nucleus. The outer electrons of all other elements that form compounds are shielded by a spherical inner core of electrons and are less strongly attracted to the nucleus, resulting in larger atomic radii.

(c)   Ionization is the process of removing an electron from an atom. For the alkali metals, the ns electron being removed is effectively shielded by the core electrons, so ionization energies are low. For the halogens, a significant increase in nuclear charge occurs as the np orbitals fill, and this is not offset by an increase in shielding. The relatively large effective nuclear charge experienced by np electrons of the halogens is similar to the unshielded nuclear charge experienced by the H 1s electron. Both H and the halogens have large ionization energies.

7.98   Since Xe reacts with F_2, and O_2 has approximately the same ionization energy as Xe, O_2 will probably react with F_2. Possible products would be O_2F_2, analogous to XeF_2, or OF_2.

$$O_2(g) + F_2(g) \rightarrow O_2F_2(g)$$
$$O_2(g) + 2F_2(g) \rightarrow 2OF_2(g)$$

7.99 $O_2 < Br_2 < K < Mg$. O_2 and Br_2 are (nonpolar) nonmetals. We expect O_2, with the much lower molar mass, to have the lower melting point. This is confirmed by data in Tables 7.6 and 7.7. K and Mg are metallic solids (all metals are solids), with higher melting points than the two nonmetals. Since alkaline earth metals (Mg) are typically harder, more dense and higher melting than alkali metals (K), we expect Mg to have the highest melting point of the group. This is confirmed by data in Tables 7.4 and 7.5.

7.100   Moving one place to the right in a horizontal row of the table, for example, from Li to Be, there is an increase in ionization energy. Moving downward in a given family, for example from Be to Mg, there is usually a decrease in ionization energy. Similarly, atomic size decreases in moving one place to the right and increases in moving downward. Thus, two elements such as Li and Mg that are diagonally related tend to have similar ionization energies and atomic sizes. This in turn gives rise to some similarities in chemical behavior. Note, however, that the valences expected for the elements are not the same. That is, lithium still appears as Li^+, magnesium as Mg^{2+}.

7.101   Fr (1A), Ra (2A), Po (6A), At (7A), Rn (8A)

(a)   Most metallic character: Fr. (Metallic character decreases from left to right in a row.)

(b)   Most nonmetallic character: Rn.

(c)   Largest ionization energy: Rn. (Ionization energy increases from left to right in a row.)

(d)   Smallest ionization energy: Fr.

(e)   Greatest electron affinity: At. (Electron affinity becomes more exothermic from left to right in a row.)

(f)   Largest atomic radius: Fr. (Size decreases from left to right in a row.)

(g)   Appears least like the element above it: Fr. (According to the trends in melting point, Fr may be a liquid or a gas at room temperature, while Cs is a solid. The others should be in the same state as the element above them.)

(h)   Highest melting point: Ra. (The melting points of the group 2A metals are much higher than those of the other groups, even though the values decrease going down the group.)

(i)   Reacts most readily with H_2O: Fr. (It has the lowest ionization energy.)

7.102   (a)   *Plan.* Use qualitative physical (bulk) properties to narrow the range of choices, then match melting point and density to identify the specific element.   *Solve.*

Hardness varies widely in metals and nonmetals, so this information is not too useful. The relatively high density, appearance, and ductility indicate that the element is probably less metallic than copper. Focus on the block of nine main group elements centered around Sn. Pb is not a possibility because it was used as a comparison standard. The melting point of the five elements closest to Pb are:

Tl, 303.5°C; In, 156.1°C; Sn, 232°C; Sb, 630.5°C; Bi, 271.3°C

The best match is In. To confirm this identification, the density of In is 7.3 g/cm^3, also a good match to properties of the unknown element.

(b)   In order to write the correct balanced equation, determine the formula of the oxide product from the mass data, assuming the unknown is In.

5.08 g oxide – 4.20 g In = 0.88 g O

4.20 g In/114.82 g/mol = 0.0366 mol In; 0.0366/0.0366 = 1

0.88 g O/16.00 g/mol = 0.0550 mol O; 0.0550/0.0366 = 1.5

Multiplying by 2 produces an integer ratio of 2 In: 3 O and a formula of In_2O_3. The balanced equation is: $4 In(s) + 3O_2(g) \rightarrow 2 In_2O_3(s)$

(c)   According to Figure 7.2, the element In was discovered between 1843–1886. The investigator who first recorded this data in 1822 could have been the first to discover In.

7.103   Ionic "inorganic" halogen compounds are formed when a metal with low ionization energy and small negative electron affinity combines with a halogen with large ionization energy and large negative electron affinity. That is, it is relatively easy to remove an electron from a metal, and there is only a small energy payback if a metal gains an electron. The opposite is true of a halogen; it is hard to remove an electron and there is a large energy advantage if a halogen gains an electron. Thus, the metal "gives up" an electron to the halogen and an ionic compound is formed. Carbon, on the other hand, is much closer in ionization energy and electron affinity to the halogens. Carbon has a much greater tendency than a metal to keep its own electrons and at least some attraction for the electrons of other elements. Thus, compounds of carbon and the halogens are molecular, rather than ionic.

Integrative Exercises

7.104   (a) $\nu = c/\lambda$; $1\,Hz = 1\,s^{-1}$

Ne: $\nu = \dfrac{2.998 \times 10^{8}\ m/s}{14.610\ \text{Å}} \times \dfrac{1\ \text{Å}}{1 \times 10^{-10}\ m} = 2.052 \times 10^{17}\ s^{-1} = 2.052 \times 10^{17}\ Hz$

Ca : $\nu = \dfrac{2.998 \times 10^{8}\ m/s}{3.358 \times 10^{-10}\ m} = 8.928 \times 10^{17}\ Hz$

Zn : $\nu = \dfrac{2.998 \times 10^{8}\ m/s}{1.435 \times 10^{-10}\ m} = 20.89 \times 10^{17}\ Hz$

Zr : $\nu = \dfrac{2.998 \times 10^{8}\ m/s}{0.786 \times 10^{-10}\ m} = 38.14 \times 10^{17} = 38.1 \times 10^{17}\ Hz$

Sn : $\nu = \dfrac{2.998 \times 10^{8}\ m/s}{0.491 \times 10^{-10}\ m} = 61.06 \times 10^{17} = 61.1 \times 10^{17}\ Hz$

(b)

Element	Z	ν	$\nu^{1/2}$
Ne	10	2.052×10^{17}	4.530×10^{8}
Ca	20	8.928×10^{17}	9.449×10^{8}
Zn	30	20.89×10^{17}	14.45×10^{8}
Zr	40	38.14×10^{17}	19.5×10^{8}
Sn	50	61.06×10^{17}	24.7×10^{8}

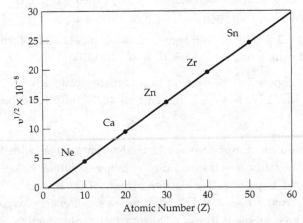

(c)   The plot in part (b) indicates that there is a linear relationship between atomic number and the square root of the frequency of the X-rays emitted by an element. Thus, elements with each integer atomic number should exist. This relationship allowed Moseley to predict the existence of elements that filled "holes" or gaps in the periodic table.

(d)   For Fe, $Z = 26$. From the graph, $\nu^{1/2} = 12.5 \times 10^{8}$, $\nu = 1.56 \times 10^{18}$ Hz.

$\lambda = c/\nu = \dfrac{2.998 \times 10^{8}\ m/s}{1.56 \times 10^{18}\ s^{-1}} \times \dfrac{1\ \text{Å}}{1 \times 10^{-10}\ m} = 1.92\ \text{Å}$

(e) $\lambda = 0.980$ Å $= 0.980 \times 10^{-10}$ m

$$\nu = c/\lambda = \frac{2.998 \times 10^8 \text{ m/s}}{0.980 \times 10^{-10} \text{ m}} = 30.6 \times 10^{17} \text{ Hz; } \nu^{1/2} = 17.5 \times 10^8$$

From the graph, $\nu^{1/2} = 17.5 \times 10^8$, $Z = 36$. The element is krypton, Kr.

7.105   (a)   Li: $[He]2s^1$. Assume that the [He] core is 100% effective at shielding the 2s valence electron $Z_{eff} = Z - S \approx 3 - 2 = 1+$.

       (b)   The first ionization energy represents loss of the 2s electron.

ΔE = energy of free electron (n = ∞) – energy of electron in ground state (n = 2)

$\Delta E = I_1 = [-2.18 \times 10^{-18} \text{ J } (Z^2/\infty^2)] - [-2/18 \times 10^{-18} \text{ J } (Z^2/2^2)]$

$\Delta E = I_1 = 0 + 2.18 \times 10^{-18} \text{ J } (Z^2/2^2)$

For Li, which is not a one-electron particle, let $Z = Z_{eff}$.

$\Delta E \approx 2.18 \times 10^{-18} \text{ J } (+1^2/4) \approx 5.45 \times 10^{-19} \text{ J/atom}$

       (c)   Change the result from part (b) to kJ/mol so it can be compared to the value in

Table 7.4. $5.45 \times 10^{-19} \dfrac{\text{J}}{\text{atom}} \times \dfrac{6.022 \times 10^{23} \text{ atom}}{\text{mol}} \times \dfrac{1 \text{ kJ}}{1000 \text{ J}} = 328$ kJ/mol

The value in Table 7.4 is 520 kJ/mol. This means that our estimate for Z_{eff} was a lower limit, that the [He] core electrons do not perfectly shield the 2s electron from the nuclear charge.

       (d)   From Table 7.4, $I_1 = 520$ kJ/mol.

$$\frac{520 \text{ kJ}}{\text{mol}} \times \frac{1000 \text{ J}}{\text{kJ}} \times \frac{1 \text{ mol}}{6.022 \times 10^{23} \text{ atoms}} = 8.6350 \times 10^{-19} \text{ J/atom}$$

Use the relationship for I_1 and Z_{eff} developed in part (b).

$$Z_{eff}^2 = \frac{4(8.6350 \times 10^{-19} \text{ J})}{2.18 \times 10^{-18} \text{ J}} = 1.5844 = 1.58; Z_{eff} = 1.26$$

This value, $Z_{eff} = 1.26$, based on the experimental ionization energy, is greater than our estimate from part (a), which is consistent with the explanation in part (c).

7.106   (a) $E = hc/\lambda$; 1 nm $= 1 \times 10^{-9}$ m; 58.4 nm $= 58.4 \times 10^{-9}$ m;

1 eV = 96.485 kJ/mol, 1 eV • mol = 96.485 kJ

$$E = \frac{6.626 \times 10^{-34} \text{ J} \bullet \text{s} \times 2.998 \times 10^8 \text{ m/s}}{58.4 \times 10^{-9} \text{ m}} = 3.4015 \times 10^{-18} = 3.40 \times 10^{-18} \text{ J/photon}$$

$$\frac{3.4015 \times 10^{-18} \text{ J}}{\text{photon}} \times \frac{1 \text{ kJ}}{1000 \text{ J}} \times \frac{6.022 \times 10^{23} \text{ photons}}{\text{mol}} \times \frac{1 \text{ eV} \bullet \text{mol}}{96.485 \text{ kJ}} = 21.230 = 21.2 \text{ eV}$$

       (b) $Hg(g) \rightarrow Hg^+(g) + 1e^-$

(c) $I_1 = E_{58.4} - E_K = 21.23 \text{ eV} - 10.75 \text{ eV} = 10.48 = 10.5 \text{ eV}$

$10.48 \text{ eV} \times \dfrac{96.485 \text{ kJ}}{1 \text{ eV} \cdot \text{mol}} = 1.01 \times 10^3 \text{ kJ/mol}$

(d)   From Figure 7.11, iodine (I) appears to have the ionization energy closest to that of Hg, approximately 1000 kJ/mol.

7.107   (a) $Na(g) \rightarrow Na^+ (g) + 1e^-$ (ionization energy of Na)

$Cl(g) + 1e^- \rightarrow Cl^-(g)$ (electron affinity of Cl)

$\overline{Na(g) + C1(g) \rightarrow Na^+(g) + Cl^-(g)}$

(b) $\Delta H = I_1 \text{ (Na)} + E_1(Cl) = +496 \text{ kJ} - 349 \text{ kJ} = +147 \text{ kJ}$, endothermic

(c)   The reaction $2Na(s) + Cl_2(g) \rightarrow 2NaCl(s)$ involves many more steps than the reaction in part (a). One important difference is the production of NaCl(s) versus NaCl(g). The condensation $NaCl(g) \rightarrow NaCl(s)$ is very exothermic and is the step that causes the reaction of the elements in their standard states to be exothermic, while the gas phase reaction is endothermic.

7.108   (a) Mg_3N_2

(b) $Mg_3N_2(s) + 3H_2O(l) \rightarrow 3MgO(s) + 2NH_3(g)$

The driving force is the production of $NH_3(g)$.

(c)   After the second heating, all the Mg is converted to MgO.

Calculate the initial mass Mg.

$0.486 \text{ g MgO} \times \dfrac{24.305 \text{ g Mg}}{40.305 \text{ g MgO}} = 0.293 \text{ g Mg}$

$x = $ g Mg converted to MgO; $y = $ g Mg converted to Mg_3N_2; $x = 0.293 - y$

$\text{g MgO} = x\left(\dfrac{40.305 \text{ g MgO}}{24.305 \text{ g Mg}}\right); \text{g Mg}_3\text{N}_2 = y\left(\dfrac{100.929 \text{ g Mg}_3\text{N}_2}{72.915 \text{ g Mg}}\right)$

$\text{g MgO} + \text{g Mg}_3\text{N}_2 = 0.470$

$(0.293 - y)\left(\dfrac{40.305}{24.305}\right) + y\left(\dfrac{100.929}{72.915}\right) = 0.470$

$(0.293 - y)(1.6583) + y(1.3842) = 0.470$

$-1.6583 y + 1.3842 y = 0.470 - 0.48588$

$-0.2741 y = -0.016$

$y = 0.05794 = 0.058 \text{ g Mg in Mg}_3\text{N}_2$

$\text{g Mg}_3\text{N}_2 = 0.05794 \text{ g Mg} \times \dfrac{100.929 \text{ g Mg}_3\text{N}_2}{72.915 \text{ g Mg}} = 0.0802 = 0.080 \text{ g Mg}_3\text{N}_2$

$\text{mass \% Mg}_3\text{N}_2 = \dfrac{0.0802 \text{ g Mg}_3\text{N}_2}{0.470 \text{ g (MgO + Mg}_3\text{N}_2)} \times 100 = 17\%$

(The final mass % has two sig figs because the mass of Mg obtained from solving simultaneous equations has two sig figs.)

(d) $3Mg(s) + 2NH_3(g) \rightarrow Mg_3N_2(s) + 3H_2(g)$

$$6.3 \text{ g Mg} \times \frac{1 \text{ mol Mg}}{24.305 \text{ g Mg}} = 0.2592 = 0.26 \text{ mol Mg}$$

$$2.57 \text{ g NH}_3 \times \frac{1 \text{ mol NH}_3}{17.031 \text{ g NH}_3} = 0.1509 = 0.16 \text{ mol NH}_3$$

$$0.2592 \text{ mol Mg} \times \frac{2 \text{ mol NH}_3}{3 \text{ mol Mg}} = 0.1728 = 0.17 \text{ mol NH}_3$$

0.26 mol Mg requires more than the available NH_3 so NH_3 is the limiting reactant.

$$0.1509 \text{ mol NH}_3 \times \frac{3 \text{ mol H}_2}{2 \text{ mol NH}_3} \times \frac{2.016 \text{ g H}_2}{\text{mol H}_2} = 0.4563 = 0.46 \text{ g H}_2$$

(e) $\Delta H^\circ_{rxn} = \Delta H^\circ_f \, Mg_3N_2(s) + 3\Delta H^\circ_f \, H_2(g) - 3\Delta H^\circ_f \, Mg(s) - 2\Delta H^\circ_f \, NH_3(g)$

        $= -461.08 \text{ kJ} + 0 - 3(0) - 2(-46.19) = -368.70 \text{ kJ}$

7.109   (a) $r_{Bi} = r_{BiBr_3} - r_{Br} = 2.63 \text{ Å} - 1.14 \text{ Å} = 1.49 \text{ Å}$

        (b) $Bi_2O_3(s) + 6HBr(aq) \rightarrow 2BiBr_3(aq) + 3H_2O(l)$

        (c) Bi_2O_3 is soluble in acid solutions because it act as a base and undergoes acid-base reactions like the one in part (b). It is insoluble in base because it cannot acts as an acid. Thus, Bi_2O_3 is a basic oxide, the oxide of a metal. Based on the properties of its oxide, Bi is characterized as a metal.

        (d)   Bi: $[Xe]6s^24f^{14}5d^{10}6p^3$. Bi has five outer electrons in the 6p and 6s subshells. If all five electrons participate in bonding, compounds such as BiF_5 are possible. Also, Bi has a large enough atomic radius (1.49 Å) and low-energy orbitals available to accommodate more than four pairs of bonding electrons.

        (e)   The high ionization energy and relatively large negative electron affinity of F, coupled with its small atomic radius, make it the most electron withdrawing of the halogens. BiF_5 forms because F has the greatest tendency to attract electrons from Bi. Also, the small atomic radius of F reduces repulsions between neighboring bonded F atoms. The strong electron withdrawing properties of F are also the reason that only F compounds of Xe are known.

7.110   (a) $4KO_2(s) + 2CO_2(g) \rightarrow 2K_2CO_3(s) + 3O_2(g)$

        (b)   K, +1; O, +1/2 (O_2^- is superoxide ion); C, +4; O, -2 $\rightarrow$ K, +1; C, +4; O, -2; O, 0

        (c) $18.0 \text{ g CO}_2 \times \dfrac{1 \text{ mol CO}_2}{44.01 \text{ g CO}_2} \times \dfrac{4 \text{ mol KO}_2}{2 \text{ mol CO}_2} \times \dfrac{71.10 \text{ g KO}_2}{1 \text{ mol KO}_2} = 58.2 \text{ g KO}_2$

        $18.0 \text{ g CO}_2 \times \dfrac{1 \text{ mol CO}_2}{44.01 \text{ g CO}_2} \times \dfrac{3 \text{ mol O}_2}{2 \text{ mol CO}_2} \times \dfrac{32.00 \text{ g O}_2}{1 \text{ mol O}_2} = 19.6 \text{ g O}_2$

8 Basic Concepts of Chemical Bonding

Visualizing Concepts

8.1 *Analyze/Plan.* Count the number of electrons in the Lewis symbol. This corresponds to the 'A'-group number of the family. *Solve.*

(a) Group 14 or 4A

(b) Group 2 or 2A

(c) Group 15 or 5A

(These are the appropriate groups in the s and p blocks, where Lewis symbols are most useful.)

8.2 (a) No. Oppositely charged ions combine (via Coulombic attraction) to form ionic compounds. A_1 and A_2 have like charges and repel each other.

(b) A_1Z_1 has the largest lattice energy. Lattice energy increases as ionic charge increases and decreases as inter-ionic distance increases. Since the magnitude of all charges is 1, this factor doesn't vary among possible ion combinations. A_1 and Z_1 have the smallest radii, which leads to the smallest inter-ionic distance and the largest lattice energy.

(c) A_2Z_2 has the smallest lattice energy, because it has the largest inter-ion distance.

8.3 *Analyze/Plan.* Count the valence electrons in the orbital diagram, take ion charge into account, and find the element with this orbital electron count on the periodic chart. Write the complete electron configuration for the ion. *Solve.*

(a) This ion has seven 3d electrons. Transition metals, or d-block elements, have valence electrons in d-orbitals. Transition metal ions first lose electrons from the 4s orbital, then from 3d if required by the charge. This 2+ ion has lost two electrons from 4s, none from 3d. The transition metal with seven 3d-electrons is cobalt, Co.

(b) The electron configuration of Co is $[Ar]4s^23d^7$. (The configuration of Co^{2+} is $[Ar]3d^7$).

8.4 *Analyze/Plan.* This question is a "reverse" Lewis structure. Count the valence electrons shown in the Lewis structure. For each atom, assume zero formal charge and determine the number of valence electrons an unbound atom has. Name the element. *Solve.*

A: 1 shared e^- pair = 1 valence electron + 3 unshared pairs = 7 valence electrons, F

E: 2 shared pairs = 2 valence electrons + 2 unshared pairs = 6 valence electrons, O

D: 4 shared pairs = 4 valence electrons, C

Q: 3 shared pairs = 3 valence electrons + 1 unshared pair = 5 valence electrons, N

X: 1 shared pair = 1 valence electron, no unshared pairs, H

Z: same as X, H

Check. Count the valence electrons in the Lewis structure. Does the number correspond to the molecular formula CH_2ONF? 12 e^- pair in the Lewis structure. CH_2ONF = 4 + 2 + 6 + 5 + 7 = 24 e^-, 12 e^- pair. The molecular formula we derived matches the Lewis structure.

8.5   *Analyze/Plan.* Since there are no unshared pairs in the molecule, we use single bonds to H to complete the octet of each C atom. For the same pair of bonded atoms, the greater the bond order, the shorter and stronger the bond.   *Solve.*

(a)   Moving from left to right along the molecule, the first C needs two H atoms, the second needs one, the third needs none, and the fourth needs one. The complete molecule is:

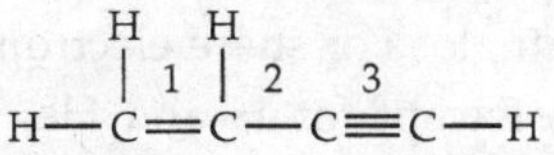

$$H \begin{matrix} H & H \\ | & | \\ C & C \end{matrix}$$

$$\begin{matrix} & H & H & & & \\ & | & | \ ^{1} & | \ ^{2} & & ^{3} \\ H & - C & = C & - C & \equiv C & - H \end{matrix}$$

(b)   In order of increasing bond length: 3 < 1 < 2

(c)   In order of increasing bond enthalpy (strength): 2 < 1 < 3

8.6   (a)   The central atom Xe, is a member of Group 8A, currently known as noble gases. Prior to 1960, this group was known as inert gases, because there were no known compounds of group 8A elements. Noble gas elements have 8 valence electrons, so they satisfy the octet rule without forming chemical bonds. Since forming compounds wasn't necessary for these elements, it was assumed that no compounds of group 8A elements existed.

(b)   There are a total of three resonance structures for the given Lewis structure of XeO_3, each with the single bond in a different Xe–O bonding domain.

(c)   The given Lewis structure does not satisfy the octet rule, because the central Xe atom has more than 8 (12 actually) electrons.

(d)   Below are four possible Lewis structures for XeO_3, with formal charges shown. Several other resonance structures with one or two double bonds can be drawn.

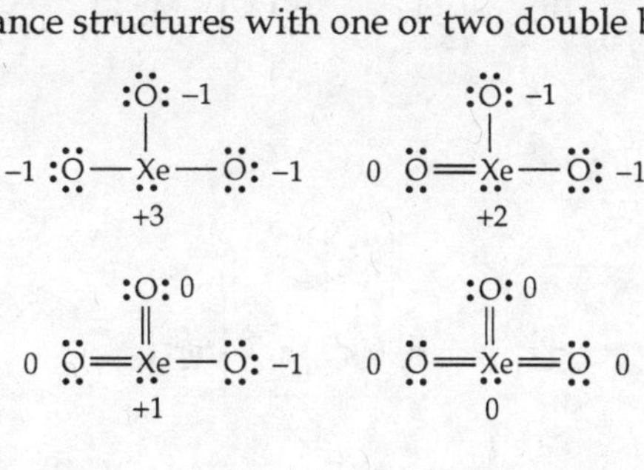

The Lewis structure given in this problem is on the lower left. It neither minimizes formal charge nor obeys the octet rule, so it is not the "best" Lewis structure for XeO_3. According to Section 8.7, the best single Lewis structure is usually the one that obeys the octet rule. This structure is shown on the upper left above. (The structure that minimizes formal charge is on the lower right.) The "best" view of bonding in XeO_3 is a composite of all correct Lewis structures, not any single Lewis structure.

8 Chemical Bonding

Lewis Symbols

8.7   (a)   Valence electrons are those that take part in chemical bonding, those in the outermost electron shell of the atom. This usually means the electrons beyond the core noble-gas configuration of the atom, although it is sometimes only the outer shell electrons.

(b) $N: [He] 2s^2 2p^3$

|_____| A nitrogen atom has 5 valence electrons.

Valence electrons

(c) $1s^2 2s^2 2p^6 \quad 3s^2 3p^2$ The atom (Si) has 4 valence electrons.

|_____||_____|

[Ne]   valence electrons

8.8   (a)   Atoms will gain, lose or share electrons to achieve the nearest noble-gas electron configuration. Except for H and He, this corresponds to eight electrons in the valence shell, thus the term octet rule.

(b) $S: [Ne]3s^2 3p^4$ A sulfur atom has six valence electrons, so it must gain two electrons to achieve an octet.

(c) $1s^2 2s^2 2p^3 = [He]2s^2 2p^3$ The atom (N) has five valence electrons and must gain three electrons to achieve an octet.

8.9 $P: 1s^2 2s^2 2p^6 3s^2 3p^3$. A 3s electron is a valence electron; a 2s (or 1s) electron is a non-valence electron. The 3s valence electron is involved in chemical bonding, while the 2s or 1s non-valence electron is not.

8.10 $Sc: 1s^2 2s^2 2p^6 3s^2 3p^6 4s^2 3d^1 = [Ar]4s^2 3d^1$. Scandium has three (3) valence electrons. These valence electrons are available for chemical bonding, while the core electrons do not participate in bonding.

8.11   (a) $\cdot \overset{\cdot}{Al} \cdot$ (b) $:\overset{\cdot\cdot}{\underset{\cdot\cdot}{Br}}:$ (c) $:\overset{\cdot\cdot}{\underset{\cdot\cdot}{Ar}}:$ (d) $\cdot Sr$

8.12   (a) $K\cdot$ (b) $\cdot \overset{\cdot}{Si} \cdot$ (c) $\left[:\overset{\cdot\cdot}{\underset{\cdot\cdot}{Mg}}: \right]^{2+}$ or Mg^{2+} (d) $\left[:\overset{\cdot\cdot}{\underset{\cdot\cdot}{P}}: \right]^{3-}$ or P^{3-}

Ionic Bonding

8.13 $\overset{\cdot}{Mg}\cdot + \cdot\overset{\cdot\cdot}{\underset{\cdot\cdot}{O}}: \longrightarrow Mg^{2+} + \left[:\overset{\cdot\cdot}{\underset{\cdot\cdot}{O}}: \right]^{2-}$

8.14 $\overset{\cdot}{Mg}\cdot + \cdot\overset{\cdot\cdot}{\underset{\cdot\cdot}{Br}}: + \cdot\overset{\cdot\cdot}{\underset{\cdot\cdot}{Br}}: \longrightarrow Mg^{2+} + 2\left[:\overset{\cdot\cdot}{\underset{\cdot\cdot}{Br}}: \right]^{-}$

8.15   (a) AlF_3 (b) K_2S (c) Y_2O_3 (d) Mg_3N_2

8.16   (a) BaO (b) RbI (c) Li_2S (d) $MgBr_2$

8.17   (a) $Sr^{2+}: [Kr]$, noble-gas configuration

(b) $Ti^{2+}: [Ar]3d^2$

(c) $Se^{2-}: [Ar]4s^2 3d^{10} 4p^6 = [Kr]$, noble-gas configuration

(d) Ni^{2+}: $[Ar]3d^8$

(e) Br^-: $[Ar]4s^23d^{10}4p^6 = [Kr]$, noble-gas configuration

(f) Mn^{3+}: $[Ar]3d^4$

8.18   (a) Zn^{2+}: $[Ar]3d^{10}$

(b) Te^{2-}: $[Kr]5s^24d^{10}5p^6 = [Xe]$, noble-gas configuration

(c) Se^{3+}: $[Ar]4s^23d^{10}4p^1$ (This is a very unlikely ion. A more stable and commonly found ion would be Sc^{3+}.) Sc^{3+}: $[Ar]$, noble-gas configuration

(d) Ru^{2+}: $[Kr]4d^6$

(e) Tl^+: $[Xe]6s^24f^{14}5d^{10}$

(f) Au^+ $[Xe]4f^{14}5d^{10}$

8.19   (a)   *Lattice energy* is the energy required to totally separate one mole of solid ionic compound into its gaseous ions.

(b)   The magnitude of the lattice energy depends on the magnitudes of the charges of the two ions, their radii, and the arrangement of ions in the lattice. The main factor is the charges, because the radii of ions do not vary over a wide range.

8.20   (a)   NaF, 910 kJ/mol; MgO, 3795 kJ/mol

The two factors that affect lattice energies are charge and ionic radii. The Na–F and Mg–O separations are similar (Na^+ is larger than Mg^{2+}, but F^- is smaller than O^{2-}). The charges on Mg^{2+} and O^{2-} are twice those of Na^+ and F^-, so according to Equation 8.4, the lattice energy of MgO is approximately four times that of NaF.

(b) $MgCl_2$, 2326 kJ/mol; $SrCl_2$, 2127 kJ/mol

The two factors that affect lattice energies are charge and ionic radii. The ionic charges are the same in the two compounds. The ionic radius of Mg^{2+} is smaller than that of Sr^{2+} so the Mg–Cl distance is slightly smaller than the Sr–Cl distance. Since lattice energy is inversely proportional to the ion separation, the lattice energy of $MgCl_2$ is slightly larger than that of $SrCl_2$.

8.21   KF, 808 kJ/mol; CaO, 3414 kJ/mol; ScN, 7547 kJ/mol

The sizes of the ions vary as follows: $Sc^{3+} < Ca^{2+} < K^+$ and $F^- < O^{2-} < N^{3-}$. Therefore, the inter-ionic distances are similar. According to Coulomb's law for compounds with similar ionic separations, the lattice energies should be related as the product of the charges of the ions. The lattice energies above are approximately related as (1)(1): (2)(2): (3)(3) or 1:4:9. Slight variations are due to the small differences in ionic separations.

8.22   (a)   According to Equation 8.4, electrostatic attraction increases with increasing charges of the ions and decreases with increasing radius of the ions. Thus, lattice energy (i) increases as the charges of the ions increase and (ii) decreases as the sizes of the ions increase.

(b)   RbBr < NaBr < LiCl < MgO. This order is confirmed by the lattice energies given in Table 8.2. MgO has the highest lattice energy because the ions have 2+ and 2– charges. The other compounds have cations with 1+ charges and anions with 1– charges. They are placed in order of decreasing ionic separation. Rb^+ and Br^- have the largest radii; Na^+ is smaller than Rb^+, Li^+ is smaller than Na^+, and Cl^- is smaller than Br^-.

8.23   Since the ionic charges are the same in the two compounds, the K–Br and Cs–Cl separations must be approximately equal. Since the radii are related as $Cs^+ > K^+$ and $Br^- > Cl^-$, the difference between Cs^+ and K^+ must be approximately equal to the difference between Br^- and Cl^-. This is somewhat surprising, since K^+ and Cs^+ are two rows apart and Cl^- and Br^- are only one row apart.

8.24   (a)   In MgO, the magnitude of the charges on both ions is 2; in $MgCl_2$, the magnitudes of the charges are 2 and 1. Also, the Cl^- ion is larger than the O^{2-} ion, so the charge separation is greater in $MgCl_2$. Thus, the lattice energy of MgO is greater, because the product of the ionic charges is greater and the ion separation is smaller.

         (b)   The ions have 1+ and 1– charges in all three compounds. In NaCl the cationic and anionic radii are smaller than in the other two compounds, so it has the largest lattice energy. In RbBr and CsBr, the anion is the same, but the Cs cation is larger, so CsBr has the smaller lattice energy.

         (c)   In BaO, the magnitude of the charges of both ions is 2; in KF, the magnitudes are 1. Charge considerations alone predict that BaO will have the higher lattice energy. The distance effect is less clear; O^{2-} and F^- are isoelectronic, so F^-, with the larger Z, has a slightly smaller radius. Ba^{2+} is two rows lower on the periodic chart than K^+, but it has a greater positive charge, so the radii are probably similar. In any case, the ionic separations in the two compounds are not very different, and the charge effect dominates.

8.25   Equation 8.4 predicts that as the oppositely charged ions approach each other, the energy of interaction will be large and negative. This more than compensates for the energy required to form Ca^{2+} and O^{2-} from the neutral atoms (see Figure 8.4 for the formation of NaCl).

8.26 $Ca(s) \rightarrow Ca(g)$; $Br_2(l) \rightarrow 2Br(g)$; $Ca(g) \rightarrow Ca^+(g) + 1e^-$;

         $Ca^+(g) \rightarrow Ca^{2+}(g) + 1e^-$; $2Br(g) + 2e^- \rightarrow 2Br^-(g)$, exothermic;

         $Ca^{2+}(g) + 2Br^-(g) \rightarrow CaBr_2(s)$, exothermic

8.27 $RbCl(s) \rightarrow Rb^+(g) + Cl^-(g)$ ΔH (lattice energy) = ?

         By analogy to NaCl, Figure 8.4, the lattice energy is

$$\Delta H_{latt} = -\Delta H_f^\circ \ RbCl(s) + \Delta H_f^\circ \ Rb(g) + \Delta H_f^\circ \ Cl(g) + I_1 \ (Rb) + E \ (Cl)$$
$$= -(-430.5 \ kJ) + 85.8 \ kJ + 121.7 \ kJ + 403 \ kJ + (-349 \ kJ) = +692 \ kJ$$

         This value is smaller than that for NaCl (+788 kJ) because Rb^+ has a larger ionic radius than Na^+. This means that the value of d in the denominator of Equation 8.4 is larger for RbCl, and the potential energy of the electrostatic attraction is smaller.

8.28   By analogy to Figure 8.4:

$$\Delta H_{latt} = -\Delta H_f^\circ\, CaCl_2 + \Delta H_f^\circ\, Ca(g) + 2\Delta H_f^\circ\, Cl(g) + I_1\,(Ca) + I_2\,(Ca) + 2E\,(Cl)$$
$$= -(-795.8\ kJ) + 179.3\ kJ + 2(121.7\ kJ) + 590\ kJ + 1145\ kJ + 2(-349\ kJ) = +2256\ kJ$$

From Table 8.2, the lattice energy of NaCl, +788 kJ/mol, is considerably less than that of CaF_2. The 2+ charge of Ca^{2+} leads to much greater electrostatic attractions and a higher lattice energy.

Covalent Bonding, Electronegativity, and Bond Polarity

8.29   (a)   A *covalent bond* is the bond formed when two atoms share one or more pairs of electrons.

(b)   Any simple compound whose component atoms are nonmetals, such as H_2, SO_2, and CCl_4, are molecular and have covalent bonds between atoms.

(c)   Covalent because it is a gas even below room temperature.

8.30   K and Ar. K is an active metal with one valence electron. It is most likely to achieve an octet by losing this single electron and to participate in ionic bonding. Ar has a stable octet of valence electrons; it is not likely to form chemical bonds of any type.

8.31   *Analyze/Plan.* Follow the logic in Sample Exercise 8.3.   *Solve.*

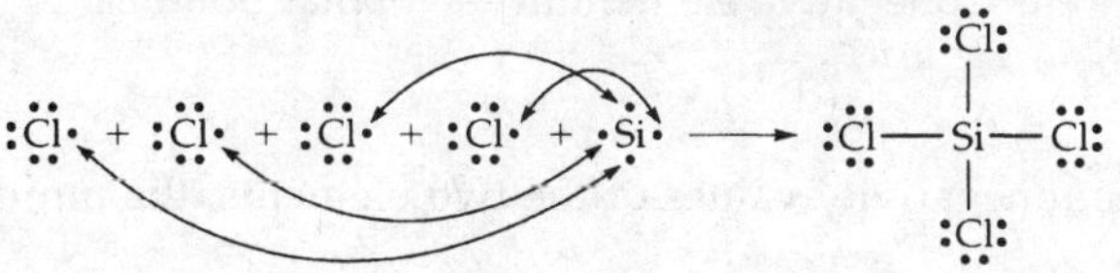

Check. Each pair of shared electrons in $SiCl_4$ is shown as a line; each atom is surrounded by an octet of electrons.

8.32

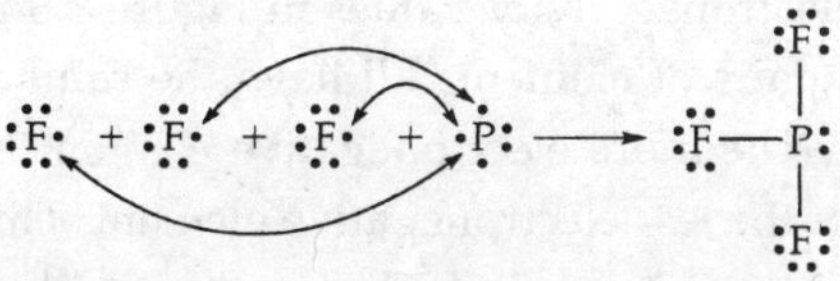

8.33   (a)

(b)   A double bond is required because there are not enough electrons to satisfy the octet rule with single bonds and unshared pairs.

(c)   The greater the number of shared electron pairs between two atoms, the shorter the distance between the atoms. If O_2 has a double bond, the O–O distance will be shorter than the O–O single bond distance.

8.34

The C–S bonds in CS_2 are double bonds, so the C–S distances will be shorter than a C–S single bond distance.

8.35   (a)   Electronegativity is the ability of an atom in a molecule (a bonded atom) to attract electrons to itself.

(b)   The range of electronegativities on the Pauling scale is 0.7–4.0.

(c)   Fluorine, F, is the most electronegative element.

(d)   Cesium, Cs, is the least electronegative element that is not radioactive.

8.36   (a)   The electronegativity of the elements increases going from left to right across a row of the periodic chart.

        (b)   Electronegativity decreases going down a family of the periodic chart.

        (c)   Generally, the trends in electronegativity are the same as those in ionization energy and opposite those in electron affinity. That is, the more positive the ionization energy and the more negative the electron affinity (ignoring a few exceptions), the greater the electronegativity of an element.

8.37   *Plan.* Electronegativity increases going up and to the right in the periodic table.
        Solve.

        (a)   Br   (b) C   (c) P   (d) Be

        Check. The electronegativity values in Figure 8.6 confirm these selections.

8.38   Electronegativity increases going up and to the right in the periodic table.

        (a)   O   (b) Al   (c) Cl   (d) F

8.39   The bonds in (a), (c) and (d) are polar because the atoms involved differ in electronegativity. The more electronegative element in each polar bond is:

        (a) F   (c) O   (d) I

8.40   The more different the electronegativity values of the two elements, the more polar the bond.

        (a)   O–F < C–F < Be–F. This order is clear from the periodic trend.

        (b)   S–Br < C–P < O–Cl. Refer to the electronegativity values in Figure 8.6 to confirm the order of bond polarity. The 3 pairs of elements all have the same positional relationship on the periodic chart. The more electronegative element is one row above and one column to the left of the less electronegative element. This leads us to conclude that ΔEN is similar for the 3 bonds, which is confirmed by values in Figure 8.6. The most polar bond, O–Cl, involves the most electronegative element, O. Generally, the largest electronegativity differences tend to be between row 2 and row 3 elements. The 2 bonds in this exercise involving elements in row 2 and row 3 do have slightly greater ΔEN than the S–Br bond, between elements in rows 3 and 4.

        (c)   C–S < N–O < B–F. You might predict that N–O is least polar since the elements are adjacent on the table. However, the big decrease going from the second row to the third means that the electronegativity of S is not only less than that of O, but essentially the same as that of C. C–S is the least polar.

8.41   *Analyze/Plan.* Q is the charge at either end of the dipole. $Q = \mu/r$. From Table 8.3, the values for HF are $\mu = 1.82$ D and $r = 0.92$ Å. Change Å to m and use the definition of the Debye and the charge of an electron to calculate the charge in units of *e*.   *Solve.*

$$Q = \frac{\mu}{r} = \frac{1.82\,D}{0.92\,\text{Å}} \times \frac{1\,\text{Å}}{1 \times 10^{-10}\,m} \times \frac{3.34 \times 10^{-30}\,C \bullet m}{1\,D} \times \frac{1\,e}{1.60 \times 10^{-19}\,C} = 0.41e$$

Check. The calculated charge on H and F is 0.41 *e*. This can be thought of as the amount of charge "transferred" from H to F. This value is consistent with our idea that HF is a polar covalent molecule; the bonding electron pair is unequally shared, but not totally transferred from H to F.

8.42   (a)   The more electronegative element, Br, will have a stronger attraction for the shared electrons and adopt a partial negative charge.

           (b)   Q is the charge at either end of the dipole.

$$Q = \frac{\mu}{r} = \frac{1.21\,D}{2.49\,\text{Å}} \times \frac{1\,\text{Å}}{1 \times 10^{-10}\,m} \times \frac{3.34 \times 10^{-30}\,C \bullet m}{1\,D} \times \frac{1\,e}{1.60 \times 10^{-19}\,C}$$
$$= 0.1014 = 0.101\,e$$

The charges on I and Br are 0.101 *e*.

8.43   *Analyze/Plan.* Generally, compounds formed by a metal and a nonmetal are described as ionic, while compounds formed from two or more nonmetals are covalent.   *Solve.*

           (a) MnO_2, ionic   (b) P_2S_3, covalent

           (c)   CoO, ionic   (d)   copper(l) sulfide, ionic

           (e)   chlorine trifluoride, covalent   (f)   vanadium(V) fluoride, ionic

8.44   Generally, compounds formed by a metal and a nonmetal are described as ionic, while compounds formed from two or more nonmetals are covalent.

           (a) MnF_3, ionic   (b) CrO_3, ionic

           (c) $AsBr_5$, covalent (As is a metalloid, so the bonding model is not obvious. The structure of the compound, a symmetrical trigonal bipyramid, indicates that it is probably covalent.)

           (d)   sulfur tetrafluoride, covalent

           (e)   molybdenum(IV) chloride, ionic

           (f)   scandium(lll) chloride, ionic

Lewis Structures; Resonance Structures

8.45   *Analyze.* Counting the **correct number of valence electrons** is the foundation of every Lewis structure.   *Plan/Solve.*

           (a)   Count valence electrons: $4 + (4 \times 1) = 8\ e^-$, $4\ e^-$ pairs. Follow the procedure in Sample Exercise 8.6.

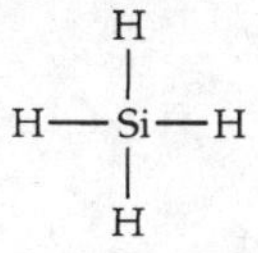

$$
\begin{array}{c}
\text{H} \\
| \\
\text{H} \!-\! \text{Si} \!-\! \text{H} \\
| \\
\text{H}
\end{array}
$$

           (b)   Valence electrons: $4 + 6 = 10\ e^-$, $5\ e^-$ pairs

$$:C \equiv O:$$

187

(c)   Valence electrons: $[6 + (2 \times 7)] = 20\ e^-$, $10\ e^-$ pairs

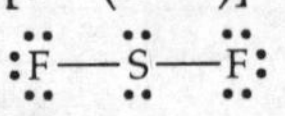

$$:\ddot{F} — \ddot{S} — \ddot{F}:$$

     i.   Place the S atom in the middle and connect each F atom with a single bond; this requires $2\ e^-$ pairs.

    ii.   Complete the octets of the F atoms with nonbonded pairs of electrons; this requires an additional $6\ e^-$ pairs.

   iii.   The remaining $2\ e^-$ pairs complete the octet of the central S atom.

(d) 32 valence e^-, $16\ e^-$ pairs

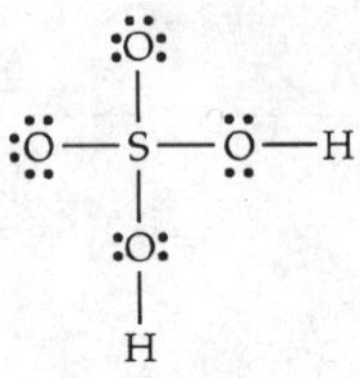

$$
\begin{array}{c}
:\ddot{O}: \\
| \\
:\ddot{O} — S — \ddot{O} — H \\
| \\
:\ddot{O}: \\
| \\
H
\end{array}
$$

(Choose the Lewis structure that obeys the octet rule, Section 8.7.)

(e)   Follow Sample Exercise 8.8. 20 valence e^-, $10\ e^-$ pairs

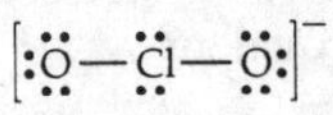

$$\left[:\ddot{O} — \ddot{Cl} — \ddot{O}:\right]^{-}$$

(f) 14 valence e^-, $7\ e^-$ pairs

$$
\begin{array}{c}
H — \ddot{N} — \ddot{O} — H \\
| \\
H
\end{array}
$$

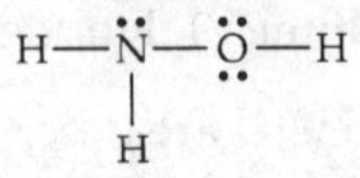

Check. In each molecule, bonding e^- pairs are shown as lines, and each atom is surrounded by an octet of electrons (duet for H).

8.46   (a) $12\ e^-$, $6\ e^-$ pairs   (b) 14 valence e^-, $7\ e^-$ pairs

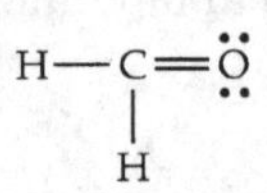

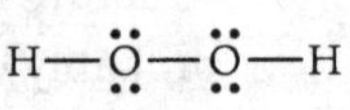

$$
\begin{array}{c}
H — C = \ddot{O} \\
| \\
H
\end{array}
\qquad\qquad\qquad
H — \ddot{O} — \ddot{O} — H
$$

(c) 50 valence e^-, $25\ e^-$ pairs   (d) 26 valence e^-, $13\ e^-$ pairs

$$
\begin{array}{c}
:\ddot{F}: \quad :\ddot{F}: \\
| \quad\quad | \\
:\ddot{F} — C — C — \ddot{F}: \\
| \quad\quad | \\
:\ddot{F}: \quad :\ddot{F}:
\end{array}
\qquad\qquad
\left[
\begin{array}{c}
:\ddot{O} — \ddot{As} — \ddot{O}: \\
| \\
:\ddot{O}:
\end{array}
\right]^{3-}
$$

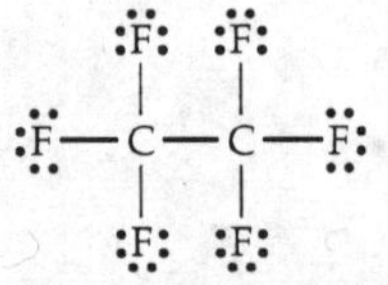

     (Choose the Lewis structure that
     obeys the octet rule, Section 8.7)

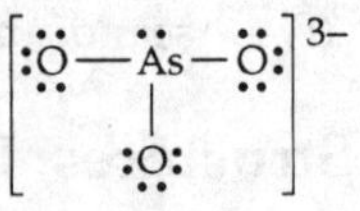

(e) 26 valence e^- $13\ e^-$ pairs   (f) $10\ e^-$, $5\ e^-$ pairs

$$
\begin{array}{c}
H — \ddot{O} — \ddot{S} — \ddot{O} — H \\
| \\
:\ddot{O}:
\end{array}
\qquad\qquad
H — C \equiv C — H
$$

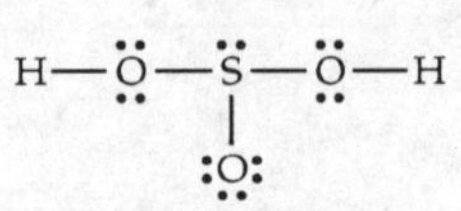

     (Choose the Lewis structure that
     obeys they octet rule, Section 8.7.)

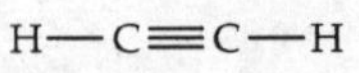

8.47   (a)   *Formal charge* is the charge on each atom in a molecule, assuming all atoms have the same electronegativity.

        (b)   Formal charges are not actual charges. They assume perfect covalency, one extreme for the possible electron distribution in a molecule.

        (c)   The other extreme is represented by oxidation numbers, which assume that the more electronegative element holds all electrons in a bond. The true electron distribution is some composite of the two extremes.

8.48   (a)   26 e$^-$, 13 e$^-$ pairs

               :F—P—F:
                  |
                 :F:

               The octet rule is satisfied for all atoms in the structure.

        (b)   F is more electronegative than P. Assuming F atoms hold all shared electrons, the oxidation number of each F is –1. The oxidation number of P is +3.

        (c)   Assuming perfect sharing, the formal charges on all F and P atoms are 0.

        (d)   The oxidation number on P is +3; the formal charge is 0. These represent extremes in the possible electron distribution, not the best picture. By virtue of their greater electronegativity, the F atoms carry a partial negative charge, and the P atom a partial positive charge.

8.49   *Analyze/Plan.* Draw the correct Lewis structure: count valence electrons in each atom, total valence electrons and electron pairs in the molecule or ion; connect bonded atoms with a line, place the remaining e$^-$ pairs as needed, in nonbonded pairs or multiple bonds, so that each atom is surrounded by an octet (or duet for H). Calculate formal charges: assign electrons to individual atoms [nonbonding e$^-$ + 1/2 (bonding e$^-$)]; formal charge = valence electrons – assigned electrons. Assign oxidation numbers, assuming that the more electronegative element holds all electrons in a bond.

        Solve. Formal charges are shown near the atoms, oxidation numbers (ox. #) are listed below the structures.

        (a)   10 e$^-$, 5 e$^-$ pairs   (b)   32 valence e$^-$, 16 e$^-$ pairs

                                                                          -1
                                                                          :Ö:
               [:N≡O:]$^+$ |
                 0   +1   0 :Cl̈—P—C̈l: 0
                                                                   |+1
               ox. #: N, +3; O, –2   :Cl:
                                                                    0

                                                               ox #: P, +5; Cl, –1; O, –2

(c)   32 valence e⁻, 16 e⁻ pairs   (d)   26 valence e⁻, 13 e⁻ pairs

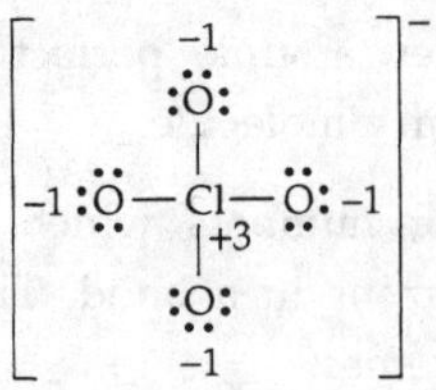

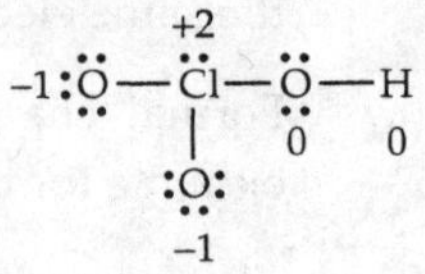

$$\left[\ \underset{-1}{:\ddot{O}:}\ \right]^{-}$$

$$-1\,:\ddot{O}-\underset{+3}{Cl}-\ddot{O}:-1$$

with top and bottom O atoms (−1 each)

$$\underset{-1}{-1\,:\ddot{O}-\underset{+2}{Cl}-\ddot{O}-H}\quad\underset{0\ \ \ 0}{}$$

$$\underset{-1}{:\ddot{O}:}$$

ox. #: Cl, +5; H, +1; O, −2

ox. #: Cl, +7; O, −2

Check. Each atom is surrounded by an octet (or duet) and the sum of the formal charges and oxidation numbers is the charge on the particle.

8.50   Formal charges are given near the atoms, oxidation numbers are listed below the structures.

(a)   18 e⁻, 9 e⁻ pairs   (b)   24 e⁻, 12 e⁻ pairs

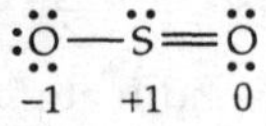

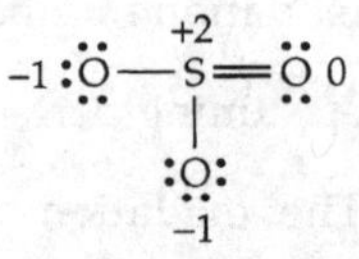

$$\underset{-1}{:\ddot{O}}-\underset{+1}{\ddot{S}}=\underset{0}{\ddot{O}}$$

ox. #: S, +4; O, −2

$$\underset{-1}{-1\,:\ddot{O}}-\underset{+2}{\ddot{S}}=\underset{0}{\ddot{O}}\ 0$$

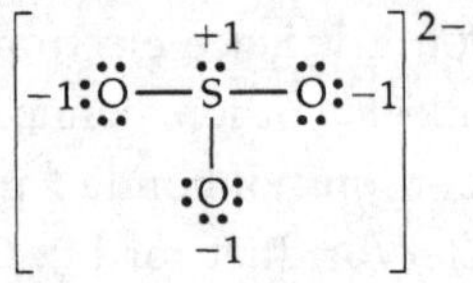

$$\underset{-1}{:\ddot{O}:}$$

ox. #: S, +6; O, −2

(c)   26 e⁻, 13 e⁻ pairs   (d)   32 e⁻, 16 e⁻ pairs

$$\left[\ \underset{-1}{-1\,:\ddot{O}}-\underset{+1}{\ddot{S}}-\underset{-1}{\ddot{O}:}\ \right]^{2-}$$

$$\underset{-1}{:\ddot{O}:}$$

ox. #: S, +4; O, −2

$$\left[\ \underset{-1}{:\ddot{O}:}\ \right]^{2-}$$

$$-1\,:\ddot{O}-\underset{+2}{S}-\ddot{O}:-1$$

$$\underset{-1}{:\ddot{O}:}$$

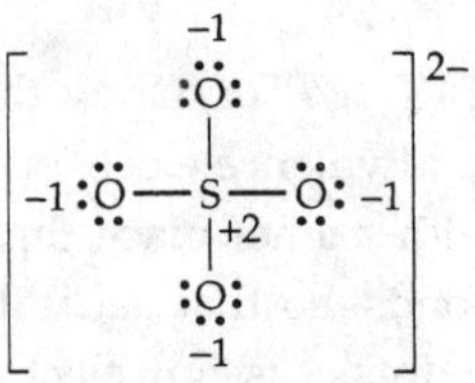

ox. #: S, +6; O, −2

8.51   (a)   *Plan.* Count valence electrons, draw all possible correct Lewis structures, taking note of alternate placements for multiple bonds.   *Solve.*

18 e⁻, 9 e⁻ pairs

$$\left[\ \ddot{O}=\ddot{N}-\ddot{O}:\ \right]^{-}\ \longleftrightarrow\ \left[\ :\ddot{O}-\ddot{N}=\ddot{O}\ \right]^{-}$$

Check. The octet rule is satisfied.

(b)   *Plan.* Isoelectronic species have the same number of valence electrons and the same electron configuration.   *Solve.*

A single O atom has 6 valence electrons, so the neutral ozone molecule O_3 is isoelectronic with NO_2^{-}.

$$\ddot{O}=\ddot{O}-\ddot{O}:\ \longleftrightarrow\ :\ddot{O}-\ddot{O}=\ddot{O}$$

Check. The octet rule is satisfied.

(c)   Since each N–O bond has partial double bond character, the N–O bond length in NO_2^- should be shorter than in species with formal N–O single bonds.

8.52   (a) $16\ e^-$, $8\ e^-$ pairs

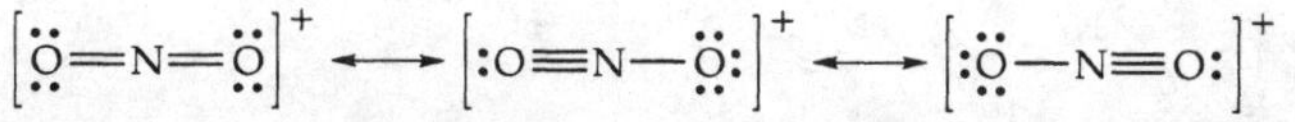

$$\left[\ddot{O}\!=\!N\!=\!\ddot{O} \right]^+ \longleftrightarrow \left[:O\!\equiv\!N\!-\!\ddot{\ddot{O}}: \right]^+ \longleftrightarrow \left[:\ddot{\ddot{O}}\!-\!N\!\equiv\!O: \right]^+$$

(b)   More than one correct Lewis structure can be drawn, so resonance structures are needed to accurately describe the structure.

(c) NO_2^+ has 16 valence electrons. Consider other triatomic molecules involving second-row nonmetallic elements. O_3^{2+} or C_3^{4-} are not "common" (or stable). CO_2 is common and matches the description (as does N_3^-, azide ion).

8.53   *Plan/Solve.* The Lewis structures are as follows:

$5\ e^-$ pairs $8\ e^-$ pairs

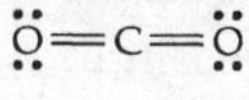

$:C\!\equiv\!O:$ $\ddot{O}\!=\!C\!=\!\ddot{O}$

$12\ e^-$ pairs

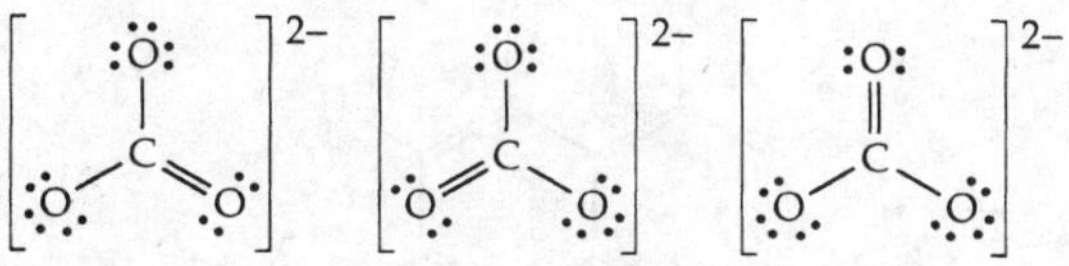

The more pairs of electrons shared by two atoms, the shorter the bond between the atoms. The average number of electron pairs shared by C and O in the three species is 3 for CO, 2 for CO_2, and 1.33 for CO_3^{2-}. This is also the order of increasing bond length: $CO < CO_2 < CO_3^{2-}$.

8.54   The Lewis structures are as follows:

$5\ e^-$ pairs $9\ e^-$ pairs

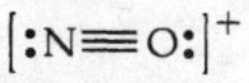

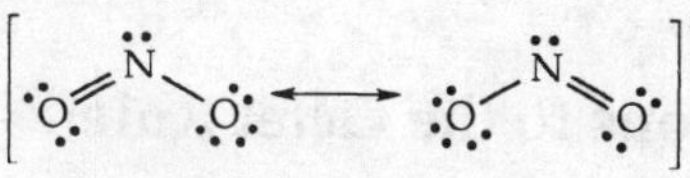

$[:N\!\equiv\!O:]^+$

$12\ e^-$ pairs

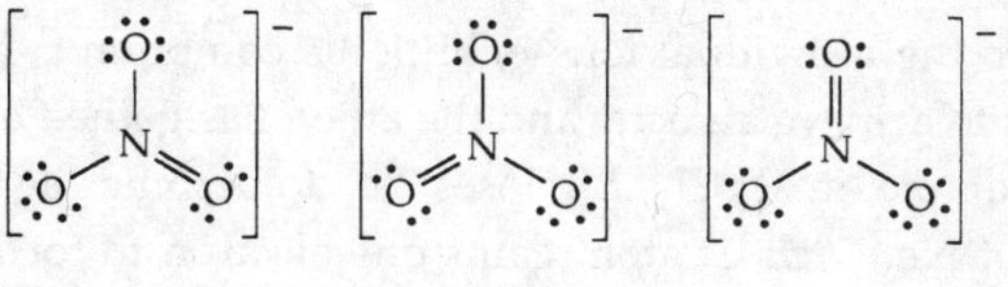

The average number of electron pairs in the N–O bond is 3.0 for NO^+, 1.5 for NO_2^-, and 1.33 for NO_3^-. The more electron pairs shared between two atoms, the shorter the bond. Thus the N–O bond lengths vary in the order $NO^+ < NO_2^- < NO_3^-$.

8.55   (a)   Two equally valid Lewis structures can be drawn for benzene.

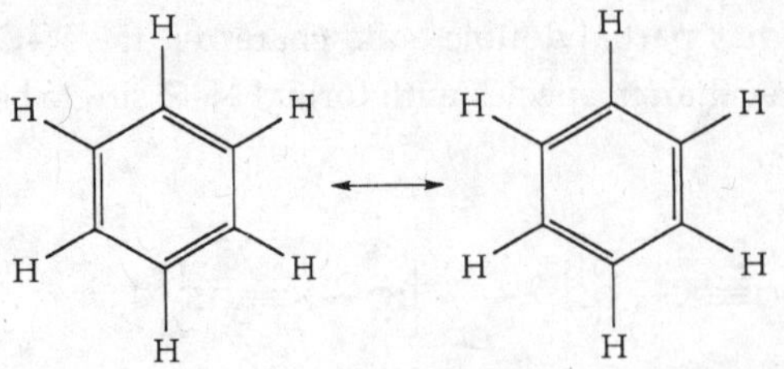

Each structure consists of alternating single and double C–C bonds; a particular bond is single in one structure and double in the other. The concept of resonance dictates that the true description of bonding is some hybrid or blend of the two Lewis structures. The most obvious blend of these two resonance structures is a molecule with six equivalent C–C bonds, each with some but not total double-bond character. If the molecule has six equivalent C–C bonds, the lengths of these bonds should be equal.

(b)   The resonance model described in (a) has six equivalent C–C bonds, each with some double bond character. That is, more than one pair but less than two pairs of electrons is involved in each C–C bond. This model predicts a uniform C–C bond length that is shorter than a single bond but longer than a double bond.

8.56   (a)

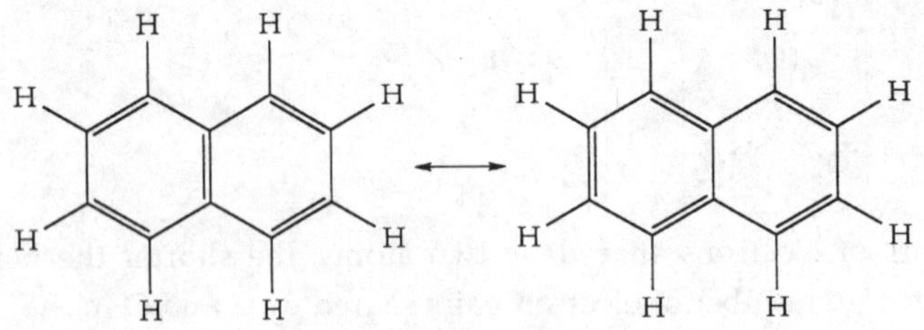

(b)   The resonance model of this molecule has bonds that are neither single nor double, but somewhere in between. This results in bond lengths that are intermediate between C–C single and C=C double bond lengths.

(c)

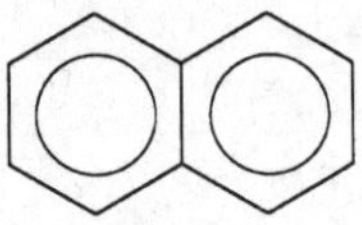

Exceptions to the Octet Rule

8.57   (a)   The octet rule states that atoms will gain, lose, or share electrons until they are surrounded by eight valence electrons.

       (b)   The octet rule applies to the individual ions in an ionic compound. That is, the cation has lost electrons to achieve an octet and the anion has gained electrons to achieve an octet. For example, in $MgCl_2$, Mg loses 2 e^- to become Mg^{2+} with the electron configuration of Ne. Each Cl atom gains one electron to form Cl^- with the electron configuration of Ar.

8.58   Carbon, in group 14, needs to form four single bonds to achieve an octet, as in CH_4. Nitrogen, in group 15, needs to form three, as in NH_3. If G = group number and n = the number of single bonds, G + n = 18 is a general relationship for the representative non-metals.

Check: O as in H_2O (G = 16) + (n = 2 bonds) = 18

8.59   The most common exceptions to the octet rule are molecules with more than eight electrons around one or more atoms, usually the central atom. Examples: SF_6, PF_5

8.60   In the third period, atoms have the space and available orbitals to accommodate extra electrons. Since atomic radius increases going down a family, elements in the third period and beyond are less subject to destabilization from additional electron-electron repulsions. Also, the third shell contains d orbitals that are relatively close in energy to 3s and 3p orbitals (the ones that accommodate the octet) and provide an allowed energy state for the extra electrons.

8.61   (a) $26\ e^-$, $13\ e^-$ pairs

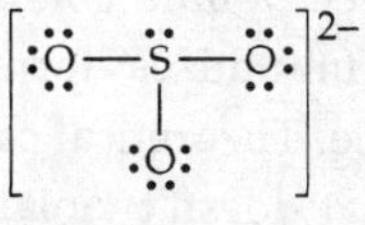

$$\left[:\ddot{O} - \underset{\underset{\displaystyle :\ddot{O}:}{|}}{\overset{..}{S}} - \ddot{O}: \right]^{2-}$$

Other resonance structures with one, two, or three double bonds can be drawn. While a structure with three double bonds minimizes formal charges, all structures with double bonds violate the octet rule. Theoretical calculations show that the single best Lewis structure is the one that doesn't violate the octet rule. Such a structure is shown above.

(b) $6\ e^-$, $3\ e^-$ pairs

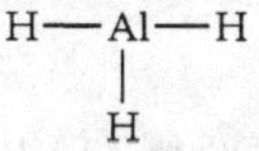

$$H - \underset{\underset{\displaystyle H}{|}}{Al} - H$$

6 electrons around Al; impossible to satisfy octet rule with only 6 valence electrons.

(c) $16\ e^-$, $8\ e^-$ pairs

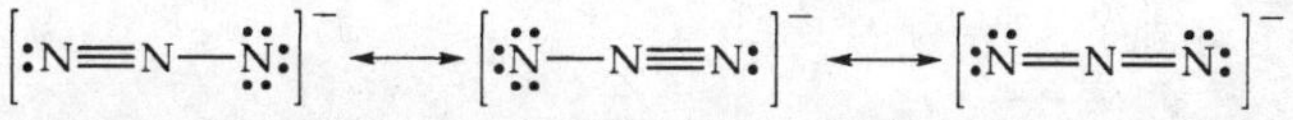

$$\left[:N \equiv N - \ddot{N}: \right]^- \longleftrightarrow \left[:\ddot{N} - N \equiv N: \right]^- \longleftrightarrow \left[:\ddot{N} = N = \ddot{N}: \right]^-$$

3 resonance structures; all obey octet rule.

(d) $20\ e^-$, $10\ e^-$ pairs

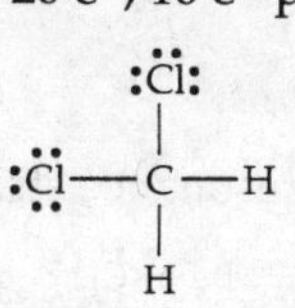

$$\underset{\underset{\displaystyle H}{|}}{:\ddot{Cl} - \underset{}{\overset{\overset{\displaystyle :\ddot{Cl}:}{|}}{C}} - H}$$

Obeys octet rule.

(e) $40\ e^-$, $20\ e^-$ pairs

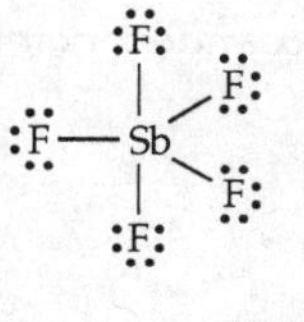

$$:\ddot{F} - \underset{\underset{\displaystyle :\ddot{F}:}{|}}{\overset{\overset{\displaystyle :\ddot{F}:}{|}}{Sb}} \overset{\displaystyle \ddot{F}:}{\underset{\displaystyle \ddot{F}:}{<}}$$

Does not obey octet rule; $10\ e^-$ around central Sb

8.62   (a)   16 e⁻, 8 e⁻ pairs

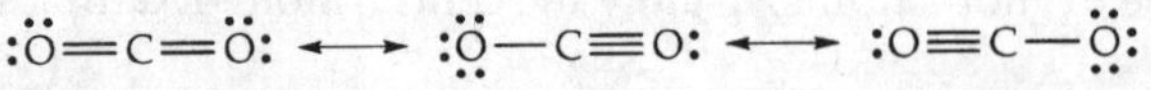

$$:\ddot{O}=C=\ddot{O}: \longleftrightarrow :\ddot{O}-C\equiv O: \longleftrightarrow :O\equiv C-\ddot{O}:$$

Three resonance structures, all obey the octet rule. The left structure minimizes formal charge.

      (b)   26 e⁻, 13 e⁻ pairs

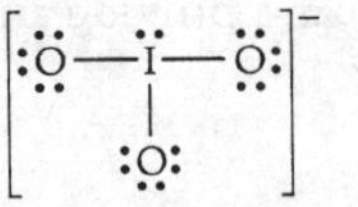

$$\left[:\ddot{O}-\overset{\cdot\cdot}{I}-\ddot{O}:\right]^{-}$$
$$\overset{|}{:\ddot{O}:}$$

Other resonance structures with one, two, or three double bonds can be drawn. While a structure with three double bonds minimizes formal charges, all structures with double bonds violate the octet rule. Theoretical calculations show that the single best Lewis structure is the one that doesn't violate the octet rule. Such a structure is shown above.

      (c)   6 e⁻, 3 e⁻ pairs   (d)   32 e⁻, 16 e⁻ pairs

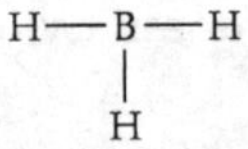

$$H-B-H$$
$$\overset{|}{H}$$

6 electrons around B.

$$\left[\begin{array}{c} :\ddot{F}: \\ | \\ :\ddot{F}-B-\ddot{F}: \\ | \\ :\ddot{F}: \end{array}\right]^{-}$$

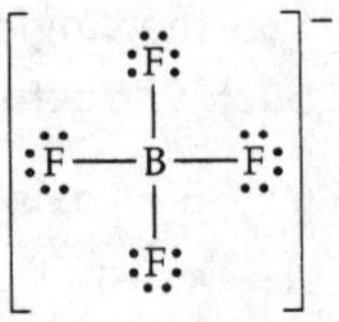

Obeys octet rule.

      (e)   22 e⁻, 11 e⁻ pair

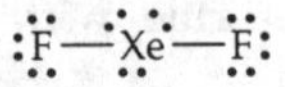

$$:\ddot{F}-\overset{\cdot\cdot}{Xe}-\ddot{F}:$$

Violates the octet rule; 10 e⁻ around central Xe.

8.63   (a)   16 e⁻, 8 e⁻ pairs

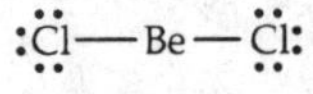

$$:\ddot{Cl}-Be-\ddot{Cl}:$$

This structure violates the octet rule; Be has only 4 e⁻ around it.

      (b) 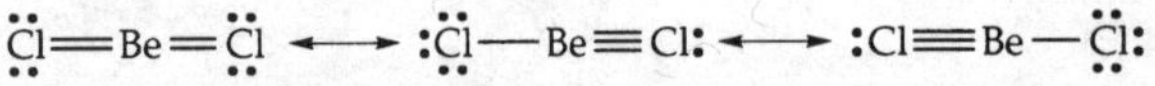 $\ddot{Cl}=Be=\ddot{Cl} \longleftrightarrow :\ddot{Cl}-Be\equiv Cl: \longleftrightarrow :Cl\equiv Be-\ddot{Cl}:$

      (c)   The formal charges on each of the atoms in the four resonance structures are:

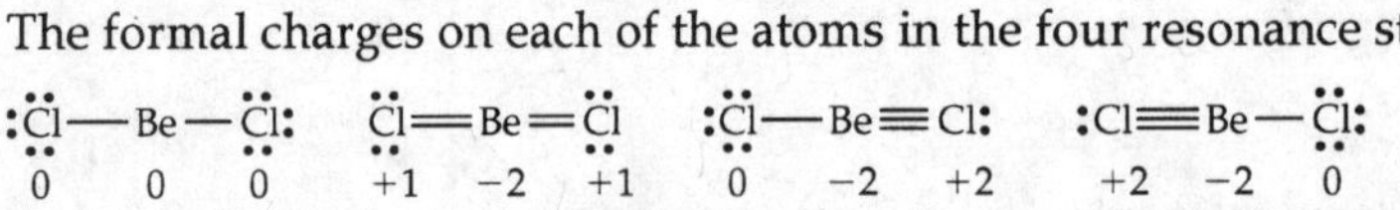

$$\underset{0 \quad\quad 0 \quad\quad 0}{:\ddot{Cl}-Be-\ddot{Cl}:} \quad \underset{+1 \quad -2 \quad +1}{\ddot{Cl}=Be=\ddot{Cl}} \quad \underset{0 \quad -2 \quad +2}{:\ddot{Cl}-Be\equiv Cl:} \quad \underset{+2 \quad -2 \quad 0}{:Cl\equiv Be-\ddot{Cl}:}$$

Formal charges are minimized on the structure that violates the octet rule; this form is probably most important. Note that this is a different conclusion than for molecules that have resonance structures with expanded octets that minimize formal charge.

8.64   (a)   19 e⁻, 9.5 e⁻ pairs, odd electron molecule

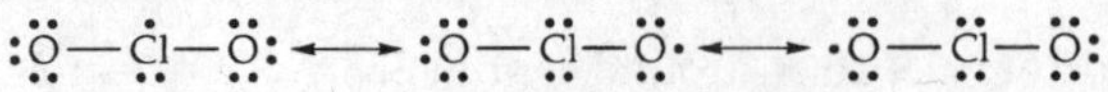

$$:\ddot{O}-\overset{\cdot\cdot}{Cl}-\ddot{O}: \longleftrightarrow :\ddot{O}-\overset{\cdot\cdot}{Cl}-\overset{\cdot}{\ddot{O}}\cdot \longleftrightarrow \cdot\overset{\cdot}{\ddot{O}}-\overset{\cdot\cdot}{Cl}-\ddot{O}:$$

(b) None of the structures satisfies the octet rule. In each structure, one atom has only 7 e⁻ around it. If a molecule has an odd number of electrons in the valence shell, no Lewis structure can satisfy the octet rule.

(c) 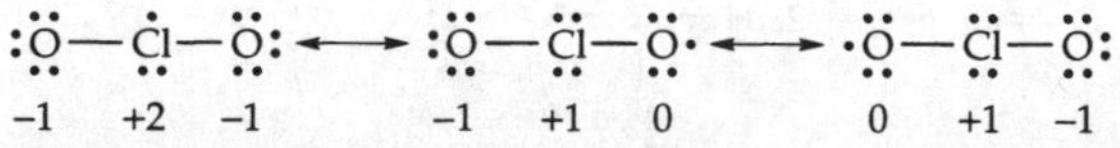 $:\ddot{O}-\ddot{C}l-\ddot{O}: \longleftrightarrow :\ddot{O}-\ddot{C}l-\dot{\ddot{O}} \longleftrightarrow \dot{\ddot{O}}-\ddot{C}l-\ddot{O}:$

       -1   +2   -1   -1   +1   0   0   +1   -1

Formal charge arguments predict that the two resonance structures with the odd electron on O are most important. This contradicts electronegativity arguments, which would predict that the less electronegative atom, Cl, would be more likely to have fewer than 8 e⁻ around it.

Bond Enthalpies

8.65   *Analyze.* Given: structural formulas. Find: enthalpy of reaction.

Plan. Count the number and kinds of bonds that are broken and formed by the reaction. Use bond enthalpies from Table 8.4 and Equation 8.12 to calculate the overall enthalpy of reaction, ΔH.   *Solve.*

(a) ΔH = 2D(O–H) + D(O–O) + 4D(C–H) + D(C=C)

     –2D(O–H) – 2D(O–C) – 4D(C–H) – D(C–C)

ΔH = D(O–O) + D(C=C) – 2D(O–C) – D(C–C)

   = 146 + 614 – 2(358) – 348 = –304 kJ

(b) ΔH = 5D(C–H) + D(C ≡ N) + D(C=C) – 5D(C–H) – D(C ≡ N) – 2D(C–C)

   = D(C=C) – 2D(C–C) = 614 – 2(348) = –82 kJ

(c) ΔH = 6D(N–Cl) – 3D(Cl–Cl) – D(N ≡ N)

   = 6(200) – 3(242) – 941 = –467 kJ

8.66   (a) ΔH = 3D(C–Br) + D(C–H) + D(Cl–Cl) – 3D(C–Br) – (C–Cl) – D(H–Cl)

     = D(C–H) + D(Cl–Cl) – D(C–Cl) – D(H–Cl)

ΔH = 413 + 242 – 328 – 431 = –104 kJ

(b) ΔH = 4D(C–H) + 2D(C–S) + 2D(S–H) + D(C–C) + 2D(H–Br)

     –4D(S–H) – D(C–C) – 2D(C–Br) – 4D(C–H)

   = 2D(C–S) + 2D(H–Br) – 2D(S–H) – 2D(C–Br)

ΔH = 2(259) + 2(366) – 2(339) – 2(276) = 20 kJ

(c) ΔH = 4D(N–H) + D(N–N) + D(Cl–Cl) – 4D(N–H) – 2D(N–Cl)

     = D(N–N) + D(Cl–Cl) – 2D(N–Cl)

ΔH = 163 + 242 – 2(200) = 5 kJ

8.67   *Plan.* Draw structural formulas so bonds can be visualized. Then use Table 8.4 and Equation 8.12.   *Solve.*

(a)

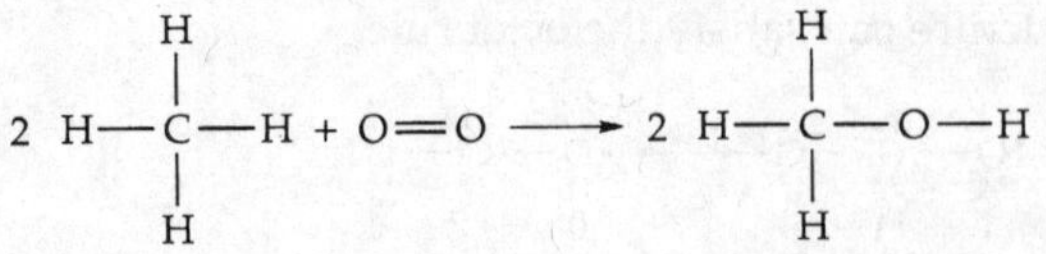

$\Delta H = 8D(C\text{-}H) + D(O\text{=}O) - 6D(C\text{-}H) - 2D(C\text{-}O) - 2D(O\text{-}H)$

$\quad = 2D(C\text{-}H) + D(O\text{=}O) - 2D(C\text{-}O) - 2D(O\text{-}H)$

$\quad = 2(413) + (495) - 2(358) - 2(463) = -321 \text{ kJ}$

(b)   H–H   +   Br–Br   →   2 H–Br

$\Delta H = D(H\text{-}H) + D(Br\text{-}Br) - 2D(H\text{-}Br)$

$\quad = (436) + (193) - 2(366) = -103 \text{ kJ}$

(c) $2\text{ H-O-O-H} \rightarrow 2\text{ H-O-H} + \text{O} = \text{O}$

$\Delta H = 4D(O\text{-}H) + 2D(O\text{-}O) - 4D(O\text{-}H) - D(O\text{=}O)$

$\Delta H = 2D(O\text{-}O) - D(O\text{=}O) = 2(146) - (495) = -203 \text{ kJ}$

8.68   *Plan.* Draw structural formulas so bonds can be visualized.   *Solve.*

(a)

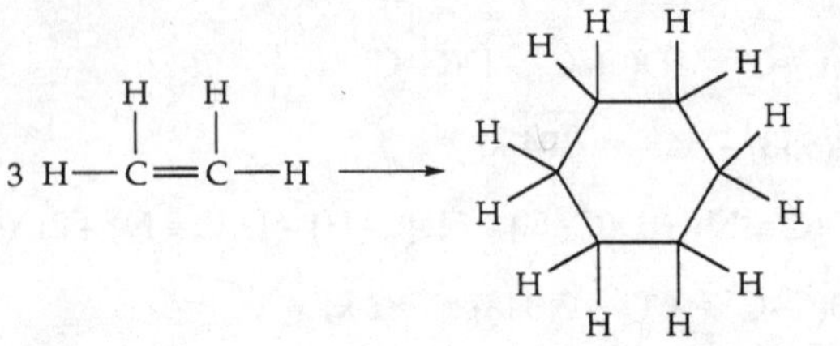

$\Delta H = 12D(C\text{-}H) + 3D(C\text{=}C) - 12D(C\text{-}H) - 6D(C\text{-}C)$

$\quad = 3D(C\text{=}C) - 6D(C\text{-}C) = 3(614) - 6(348) = -246 \text{ kJ}$

(b)

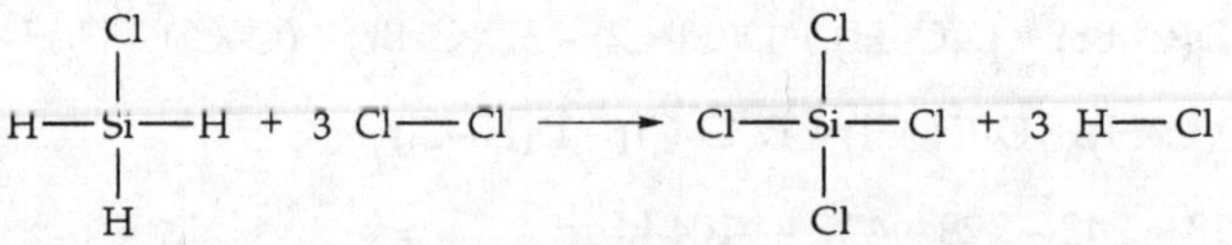

$\Delta H = 3D(Si\text{-}H) + D(Si\text{-}Cl) + 3D(Cl\text{-}Cl) - 4D(Si\text{-}Cl) - 3D(H\text{-}Cl)$

$\quad = 3D(Si\text{-}H) + 3D(Cl\text{-}Cl) - 3D(Si\text{-}Cl) - 3D(H\text{-}Cl)$

$\quad = 3(323) + 3(242) - 3(464) - 3(431) = -990 \text{ kJ}$

(c)   *Plan.* Use bond enthalpies to calculate ΔH for the reaction with $S_8(g)$ as a product. Then,

$$8H_2S(g) \rightarrow 8H_2(g) + S_8(g) \qquad \Delta H$$
$$\underline{S_8(g) \rightarrow S_8(s) \qquad\qquad\qquad -\Delta H_f^{\circ} \text{ for } S_8(g)}$$
$$8H_2S(g) \rightarrow 8H_2(g) + S_8(s) \qquad \Delta H_{rxn} = [\Delta H - \Delta H_f^{\circ} S_8(g)]$$

$$\Delta H = 16D(S\text{-}H) - 8(H\text{-}H) - 8(S\text{-}S)$$

$$= 16(339) - 8(436) - 8(266) = -192 \text{ kJ}$$

$$\Delta H_{rxn} = \Delta H - \Delta H_f^{\circ} S_8(g) = -192 \text{ kJ} - 102.3 \text{ kJ} = -294.3 = -294 \text{ kJ}$$

8.69   *Plan.* Draw structural formulas so bonds can be visualized. Then use Table 8.4 and Equation 8.12. *Solve.*

(a)   :N≡N: + 3 H—H ⟶ 2 H—N̈—H
                                        |
                                        H

$$\Delta H = D(N \equiv N) + 3D(H\text{-}H) - 6(N\text{-}H) = 941 \text{ kJ} + 3(436 \text{ kJ}) - 6(391 \text{ kJ})$$

$$= -97 \text{ kJ}/2 \text{ mol NH}_3; \text{ exothermic}$$

(b)   Plan. Use Eq. 5.31 to calculate ΔH_{rxn} from ΔH_f° values.

$\Delta H_{rxn}^{\circ} = \Sigma n \, \Delta H_f^{\circ} \text{ (products)} - \Sigma n \, \Delta H_f^{\circ} \text{ (reactants)}.$ $\Delta H_f^{\circ} \text{ NH}_3(g) = -46.19 \text{ kJ}.$
Solve.

$$\Delta H_{rxn}^{\circ} = 2 \, \Delta H_f^{\circ} \text{ NH}_3(g) - 3 \, \Delta H_f^{\circ} \text{ H}_2(g) - \Delta H_f^{\circ} \text{ N}_2(g)$$

$$\Delta H_{rxn}^{\circ} = 2(-46.19) - 3(0) - 0 = -92.38 \text{ kJ}/2 \text{ mol NH}_3$$

The ΔH calculated from bond enthalpies is slightly more exothermic (more negative) than that obtained using ΔH_f° values.

8.70   (a)

H   H   H H
 \\   / | |
  C == C   + H—H   ⟶   H—C—C—H
 /   \\ | |
H   H   H H

$$\Delta H = 4D(C\text{-}H) + D(C=C) + D(H\text{-}H) - 6D(C\text{-}H) - D(C\text{-}C)$$

$$= D(C=C) + D(H\text{-}H) - 2D(C\text{-}H) - D(C\text{-}C)$$

$$\Delta H = 614 + 436 - 2(413) - 348 = -124 \text{ kJ}$$

(b) $$\Delta H^{\circ} = \Delta H_f^{\circ} \text{ C}_2\text{H}_6(g) - \Delta H_f^{\circ} \text{ C}_2\text{H}_4(g) - \Delta H_f^{\circ} \text{ H}_2(g)$$

$$= -84.68 - 52.30 - 0 = -136.98 \text{ kJ}$$

The values of ΔH for the reaction differ because the bond enthalpies used in part (a) are average values that can differ from one compound to another. For example, the exact enthalpy of a C–H bond in C_2H_4 is probably not equal to the enthalpy of a C–H bond in C_2H_6. Thus, reaction enthalpies calculated from average bond enthalpies are estimates. On the other hand, standard enthalpies of formation are measured quantities and should lead to accurate reaction enthalpies. The advantage of average bond enthalpies is that they can be used for reactions where no measured enthalpies of formation are available.

8.71   The average Ti–Cl bond enthalpy is just the average of the four values listed. 430 kJ/mol.

8.72   (a)   (i)

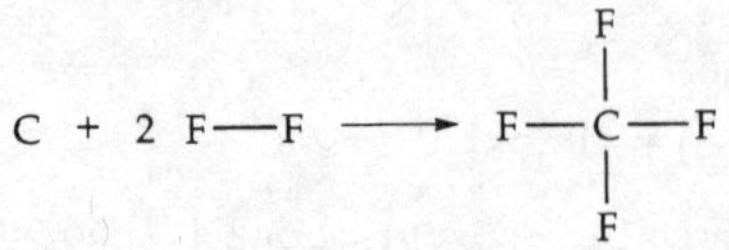

$$C + 2\ F\text{—}F \longrightarrow F\text{—}\underset{\underset{F}{|}}{\overset{\overset{F}{|}}{C}}\text{—}F$$

$$\Delta H = 2D(F\text{-}F) - 4D(C\text{-}F) = 2(155) - 4(485) = -1630 \text{ kJ}$$

       (ii)

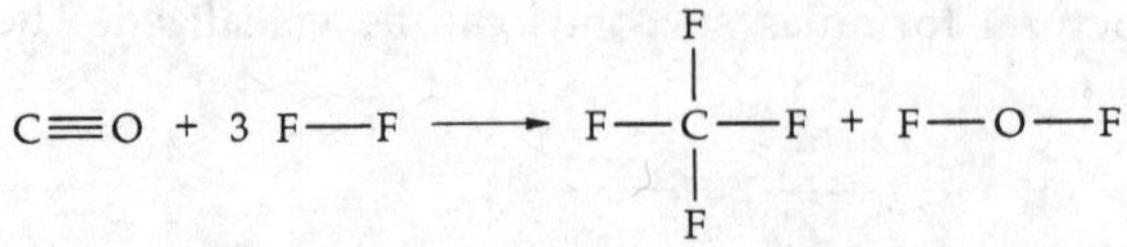

$$C\equiv O + 3\ F\text{—}F \longrightarrow F\text{—}\underset{\underset{F}{|}}{\overset{\overset{F}{|}}{C}}\text{—}F + F\text{—}O\text{—}F$$

$$\Delta H = D(C\equiv O) + 3D(F\text{-}F) - 4D(C\text{-}F) - 2D(D\text{-}F)$$

$$= 1072 + 3(155) - 4(485) - 2(190) = -783 \text{ kJ}$$

       (iii)

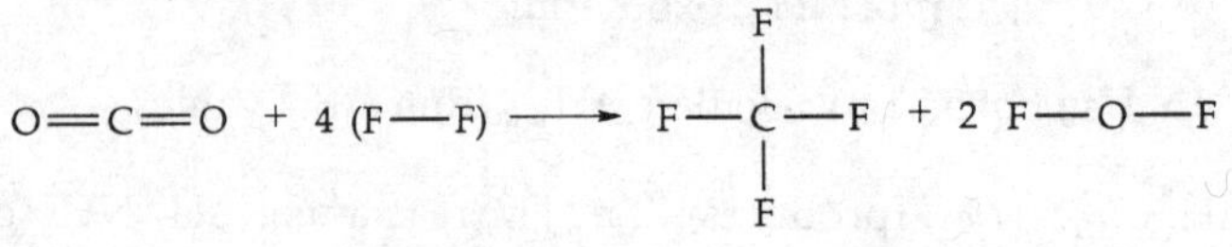

$$O\text{=}C\text{=}O + 4\ (F\text{—}F) \longrightarrow F\text{—}\underset{\underset{F}{|}}{\overset{\overset{F}{|}}{C}}\text{—}F + 2\ F\text{—}O\text{—}F$$

$$\Delta H = 2D(C\text{=}O) + 4D(F\text{-}F) - 4D(C\text{-}F) - 4D(O\text{-}F)$$

$$= 2(799) + 4(155) - 4(485) - 4(190) = -482 \text{ kJ}$$

Reaction (i) is most exothermic.

(b)   The more oxygen atoms bound to carbon, the less exothermic the reaction in this series.

Additional Exercises

8.73   Six nonradioactive elements in the periodic table have Lewis symbols with single dots. Yes, they are in the same family, assuming H is placed with the alkali metals, as it is on the inside cover of the text. This is because the Lewis symbol represents the number of valence electrons of an element, and all elements in the same family have the same number of valence electrons. By definition of a family, all elements with the same Lewis symbol must be in the same family.

8.74   (a)   Lattice energy is proportional to Q_1Q_2/d. For each of these compounds, Q_1Q_2 is the same. The anion H^- is present in each compound, but the ionic radius of the cation increases going from Be to Ba. Thus, the value of d (the cation-anion separation) increases and the ratio Q_1Q_2/d decreases. This is reflected in the decrease in lattice energy going from BeH_2 to BaH_2.

       (b)   Again, Q_1Q_2 for ZnH_2 is the same as that for the other compounds in the series and the anion is H^-. The lattice energy of ZnH_2, 2870 kJ, is closest to that of MgH_2, 2791 kJ. The ionic radius of Zn^{2+} is similar to that of Mg^{2+}.

8.75   (a)

Compound	Lattice Energy (kJ)		Compound	Lattice Energy (kJ)	
NaCl	788		LiCl	834	
NaBr	732	56 kJ	**LiBr**	**779**	55 kJ
Na I	682		Li I	730	

106 kJ (bracket for NaCl–NaBr)   104 kJ (bracket for LiCl–LiBr)

The difference in lattice energy between LiCl and LiI is 104 kJ. The difference between NaCl and NaI is 106 kJ; the difference between NaCl and NaBr is 56 kJ, or 53% of the difference between NaCl and NaI. Applying this relationship to the Li salts, 0.53(104 kJ) = 55 kJ difference between LiCl and LiBr. The approximate lattice energy of LiBr is (834 – 55) kJ = 779 kJ.

(b)

Compound	Lattice Energy (kJ)		Compound	Lattice Energy (kJ)	
NaCl	788		CsCl	657	
NaBr	732		CsBr	**627**	
Na I	682		Cs I	600	

$$106\ kJ \begin{bmatrix} 788 \\ 732 \\ 682 \end{bmatrix} 56\ kJ \qquad 57\ kJ \begin{bmatrix} 657 \\ 627 \\ 600 \end{bmatrix} 30\ kJ$$

By analogy to the Na salts, the difference between lattice energies of CsCl and CsBr should be approximately 53% of the difference between CsCl and CsI. The lattice energy of CsBr is approximately 627 kJ.

(c)

Compound	Lattice Energy (kJ)		Compound	Lattice Energy (kJ)	
MgO	3795		$MgCl_2$	2326	
CaO	3414		$CaCl_2$	**2195**	
SrO	3217		$SrCl_2$	2127	

$$578\ kJ \begin{bmatrix} 3795 \\ 3414 \\ 3217 \end{bmatrix} 381\ kJ \qquad 199\ kJ \begin{bmatrix} 2326 \\ 2195 \\ 2127 \end{bmatrix} 131\ kJ$$

By analogy to the oxides, the difference between the lattice energies of $MgCl_2$ and $CaCl_2$ should be approximately 66% of the difference between $MgCl_2$ and $SrCl_2$. That is, 0.66(199 kJ) = 131 kJ. The lattice energy of $CaCl_2$ is approximately (2326 – 131) kJ = 2195 kJ.

8.76 $$E = \frac{8.99 \times 10^9\ J\bullet m}{C^2} \times \frac{4(1.60 \times 10^{-19}\ C)^2}{(0.99 + 1.40) \times 10^{-10}\ m} = -3.852 \times 10^{-18} = -3.85 \times 10^{-18}\ J$$

On a molar basis: $(-3.852 \times 10^{-18}\ J)(6.022 \times 10^{23}) = -2.319 \times 10^6\ J = -2320\ kJ$

Note that its absolute value is less than the lattice energy, 3414 kJ/mol. The difference represents the added energy of putting all the $Ca^{2+}O^{2-}$ ion pairs together in a three-dimensional array, similar to the one in Figure 8.3.

8.77 $E = Q_1 Q_2 / d; \quad k = 8.99 \times 10^9\ J\bullet m/coul^2$

(a) Na^+, Br^-: $$E = \frac{-8.99 \times 10^9\ J\bullet m}{C^2} \times \frac{(1.60 \times 10^{-19}\ C)^2}{(0.97 + 1.96) \times 10^{-10}\ m} = -7.8547 \times 10^{-19}$$
$$= -7.85 \times 10^{-19}\ J$$

The sign of E is negative because one of the interacting ions is an anion; this is an attractive interaction.

On a molar basis: $-7.855 \times 10^{-19} \times 6.022 \times 10^{23} = -4.73 \times 10^5\ J = -473\ kJ$

(b) Rb^+, Br^-: $$E = \frac{-8.99 \times 10^9\ J\bullet m}{C^2} \times \frac{(1.60 \times 10^{-19}\ C)^2}{(1.47 + 1.96) \times 10^{-10}\ m} = -6.71 \times 10^{-19}\ J$$

On a molar basis: $-4.04 \times 10^5\ J = -404\ kJ$

(c) Sr^{2+}, S^{2-}: $$E = \frac{-8.99 \times 10^9\ J\bullet m}{C^2} \times \frac{(2 \times 1.60 \times 10^{-19}\ C)^2}{(1.13 + 1.84) \times 10^{-10}\ m} = -3.10 \times 10^{-18}\ J$$

On a molar basis: $-1.87 \times 10^6\ J = -1.87 \times 10^3\ kJ$

8.78 (a) A polar molecule has a measurable dipole moment; its centers of positive and negative charge do not coincide. A nonpolar molecule has a zero net dipole moment; its centers of positive and negative charge do coincide.

(b) Yes. If X and Y have different electronegativities, they have different attractions for the electrons in the molecule. The electron density around the more electronegative atom will be greater, producing a charge separation or dipole in the molecule.

(c) $\mu = Qr$. The dipole moment, μ, is the product of the magnitude of the separated charges, Q, and the distance between them, r.

8.79 Molecule (b) H_2S and ion (c) NO_2^- contain polar bonds. The atoms that form the bonds (H–S) and N–O) have different electronegativity values.

8.80 (a) B–O. The most polar bond will be formed by the two elements with the greatest difference in electronegativity. Since electronegativity increases moving right and up on the periodic chart, the possibilities are B–O and Te–O. These two bonds are likely to have similar electronegativitiy differences (3 columns apart vs. 3 rows apart). Values from Figure 8.6 confirm the similarity, and show that B–O is slightly more polar.

(b) Te–I. Both are in the fifth row of the periodic chart and have the two largest covalent radii among this group of elements.

(c) TeI_2. Te needs to participate in two covalent bonds to satisfy the octet rule, and each I atom needs to participate in one bond, so by forming a TeI_2 molecule, the octet rule can be satisfied for all three atoms.

$$:\ddot{\underset{\cdot\cdot}{I}}\!-\!\ddot{Te}\!-\!\ddot{\underset{\cdot\cdot}{I}}: $$

(d) B_2O_3. Although this is probably not a purely ionic compound, it can be understood in terms of gaining and losing electrons to achieve a noble-gas configuration. If each B atom were to lose 3 e^- and each O atom were to gain 2 e^-, charge balance and the octet rule would be satisfied.

P_2O_3. Each P atom needs to share 3 e^- and each O atom 2 e^- to achieve an octet. Although the correct number of electrons seem to be available, a correct Lewis structure is difficult to imagine. In fact, phosphorus (III) oxide exists as P_4O_6 rather than P_2O_3 (Chapter 22).

8.81 (a) $12 + 3 + 15 = 30$ valence e^-, 15 e^- pairs.

Structures with H bound to N and nonbonded electron pairs on C can be drawn, but the structures above minimize formal charges on the atoms.

(b) The resonance structures indicate that triazine will have six equal C–N bond lengths, intermediate between C–N single and C–N double bond lengths. (See

Solutions 8.55 and 8.56.) From Table 8.5, an average C–N length is 1.43 Å, a C=N length is 1.38 Å. The average of these two lengths is 1.405 Å. The C–N bond length in triazine should be in the range 1.40–1.41 Å.

8.82   Use the method detailed in Section 8.5, *A Closer Look*, to estimate partial charges from electronegativity values. From Figure 8.6, the electronegativity of Br is 2.8 and of Cl is 3.0.

Br has $2.8/(3.0 + 2.8) = 0.48$ of the charge of the bonding e^- pair.

Cl has $3.0/(3.0 + 2.8) = 0.52$ of the charge of the bonding e^- pair.

This amounts to $0.52 \times 2e = 1.04e$ on Cl or $0.04e$ more than a neutral Cl atom. This implies a –0.04 charge on Cl and +0.04 charge on Br.

From Figure 7.5, the covalent radius of Br is 1.14 Å and of Cl is 0.99 Å. The Br–Cl separation is 2.13 Å.

$$\mu = Qr = 0.04e \times \frac{1.60 \times 10^{-19}\,C}{e} \times 2.13\,\text{Å} \times \frac{1 \times 10^{-10}\,m}{\text{Å}} \times \frac{1D}{3.34 \times 10^{-30}\,C\bullet m} = 0.41\,D$$

Clearly, this method is approximate. The estimated dipole moment of 0.41 D is within 28% of the measured value of 0.57 D.

8.83 I_3^- has a Lewis structure with an expanded octet of electrons around the central I.

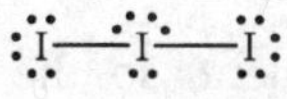

:Ï—Ï—Ï:

F cannot accommodate an expanded octet because it is too small and has no available d orbitals in its valence shell.

8.84   Formal charge (FC) = # valence e^- – (# nonbonding e^- + 1/2 # bonding e^-)

(a)   18 e^-, 9 e^- pairs

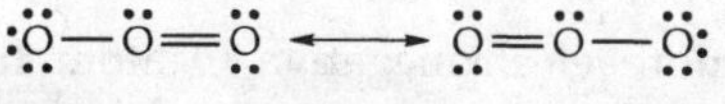

:Ö—Ö=Ö ⟷ Ö=Ö—Ö:

FC for the central O = $6 - [2 + 1/2\,(6)] = +1$

(b)   48 e^-, 24 e^- pairs

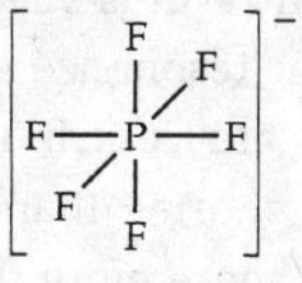

[F₆P]⁻   FC for P = $5 - [0 + 1/2\,(12)] = -1$

The three nonbonded pairs on each F have been omitted.

(c)   17 e^-; 8 e^- pairs, 1 odd e^-

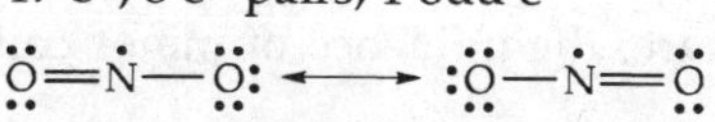

Ö=N—Ö: ⟷ :Ö—N=Ö

The odd electron is probably on N because it is less electronegative than O. Assuming the odd electron is on N, FC for N = $5 - [1 + 1/2\,(6)] = +1$. If the odd electron is on O, FC for N = $5 - [2 + 1/2\,(6)] = 0$.

(d)   28 e$^-$, 14 e$^-$ pairs

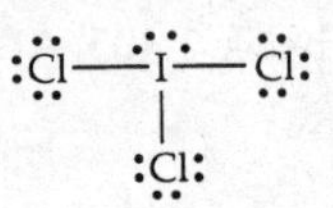

FC for I = 7 − [4 + 1/2 (6)] = 0

(e)   32 e$^-$, 16 e$^-$ pairs

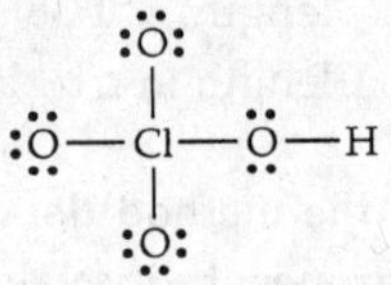

FC for Cl = 7 − [0 + 1/2 (8)] = +3

8.85   (a)   14e$^-$, 7 e$^-$ pairs

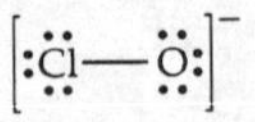

FC on Cl = 7 − [6 + 1/2(2)] = 0

32 e$^-$, 16 e$^-$ pairs

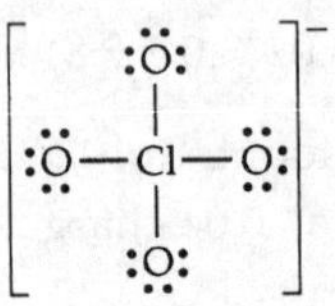

FC on Cl = 7 − [0 + 1/2(8)] = +3

(b)   The oxidation number of Cl is +1 in ClO$^-$ and +7 in ClO$_4^-$.

(c)   The definition of formal charge assumes that all bonding pairs of electrons are equally shared by the two bonded atoms, that all bonds are purely covalent. The definition of oxidation number assumes that the more electronegative element in the bond gets all of the bonding electrons, that the bonds are purely ionic. These two definitions represent the two extremes of how electron density is distributed between bonded atoms.

In ClO$^-$ and ClO$_4^-$, Cl is the less electronegative element, so the oxidation numbers have a higher positive value than the formal charges. The true description of the electron density distribution is somewhere between the extremes indicated by formal charge and oxidation number.

8.86   (a)

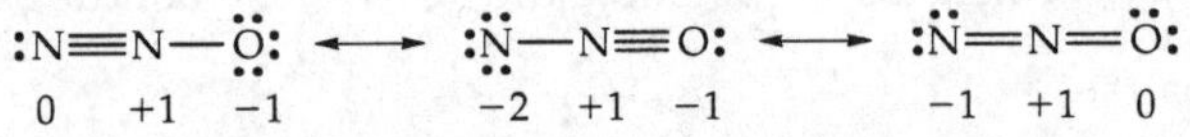

In the leftmost structure, the more electronegative O atom has the negative formal charge, so this structure is likely to be most important.

(b)   In general, the more shared pairs of electrons between two atoms, the shorter the bond, and vice versa. That the N–N bond length in N$_2$O is slightly longer than the typical N≡N indicates that the middle and right resonance structures where the N atoms share less than three electron pairs are contributors to the true structure. That the N–O bond length is slightly shorter than a typical N=O indicates that the middle structure, where N and O share more than two electron pairs, does contribute to the true structure. This physical data indicates that while formal charge can be used to predict which resonance form will be more important to the observed structure, the influence of minor contributors on the true structure cannot be ignored.

8.87   ΔH = 8D(C–H) − D(C–C) − 6D(C–H) − D(H–H)

       = 2D(C–H) − D(C–C) − D(H–H)

       = 2(413) − 348 − 436 = +42 kJ

$\Delta H = 8D(C-H) + 1/2\ D(O=O) - D(C-C) - 6D(C-H) - 2D(O-H)$

$\quad = 2D(C-H) + 1/2\ D(O=O) - D(C-C) - 2D(O-H)$

$\quad = 2(413) + 1/2\ (495) - 348 - 2(463) = -200\ kJ$

The fundamental difference in the two reactions is the formation of 1 mol of H–H bonds versus the formation of 2 mol of O–H bonds. The latter is much more exothermic, so the reaction involving oxygen is more exothermic.

8.88   (a) $\Delta H = 5D(C-H) + D(C-C) + D(C-O) + D(O-H) - 6D(C-H) - 2D(C-O)$

$\quad\quad = D(C-C) + D(O-H) - D(C-H) - D(C-O)$

$\quad\quad = 348\ kJ + 463\ kJ - 413\ kJ - 358\ kJ$

$\Delta H = +40\ kJ$; ethanol has the lower enthalpy

   (b) $\Delta H = 4D(C-H) + D(C-C) + 2D(C-O) - 4D(C-H) - D(C-C) - D(C=O)$

$\quad\quad = 2D(C-O) - D(C=O)$

$\quad\quad = 2(358\ kJ) - 799\ kJ$

$\Delta H = -83\ kJ$; acetaldehyde has the lower enthalpy

   (c) $\Delta H = 8D(C-H) + 4D(C-C) + D(C=C) - 8D(C-H) - 2D(C-C) - 2D(C=C)$

$\quad\quad = 2D(C-C) - D(C=C)$

$\quad\quad = 2(348\ kJ) - 614\ kJ$

$\Delta H = +82\ kJ$; cyclopentene has the lower enthalpy

   (d) $\Delta H = 3D(C-H) + D(C-N) + D(C \equiv N) - 3D(C-H) - D(C-C) - D(C \equiv N)$

$\quad\quad = D(C-N) - D(C-C)$

$\quad\quad = 293\ kJ - 348\ kJ$

$\Delta H = -55\ kJ$; acetonitrile has the lower enthalpy

8.89   (a)

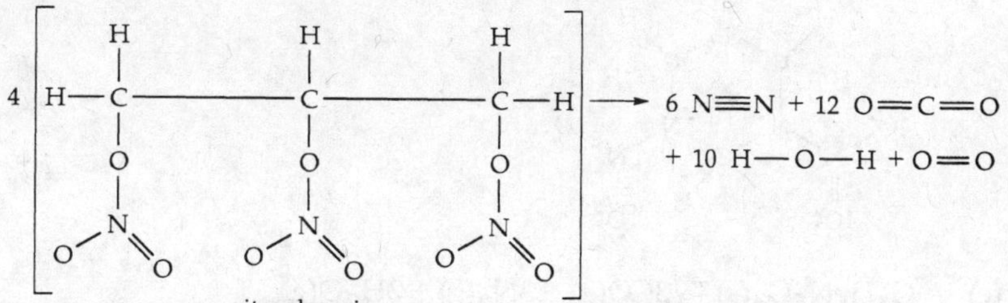

nitroglycerine

$\Delta H = 20D(C-H) + 8D(C-C) + 12D(C-O) + 24D(O-N) + 12D(N=O)$

$\quad\quad\quad - [6D(N \equiv N) + 24D(C=O) + 20D(H-O) + D(O=O)]$

$\Delta H = 20(413) + 8(348) + 12(358) + 24(201) + 12(607)$

$\quad\quad\quad - [6(941) + 24(799) + 20(463) + 495]$

$\quad = -7129\ kJ$

$$1.00 \text{ g } C_3H_5N_3O_9 \times \frac{1 \text{ mol } C_3H_5N_3O_9}{227.1 \text{ g } C_3H_5N_3O_9} \times \frac{-7129 \text{ kJ}}{4 \text{ mol } C_3H_5N_3O_9} = 7.85 \text{ kJ/g } C_3H_5N_3O_9$$

(b) $4C_7H_5N_3O_6(s) \rightarrow 6N_2(g) + 7CO_2(g) + 10H_2O(g) + 21C(s)$

8.90   (a)

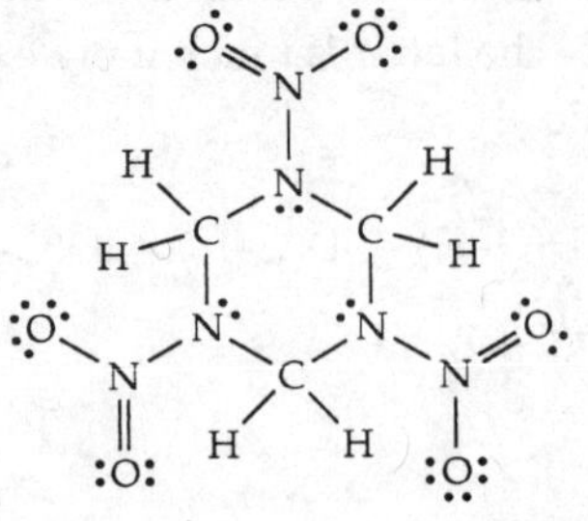

$C_3H_6N_6O_6$ $12 + 6 + 30 + 36 = 84 \text{ e}^-, 42 \text{ e}^- \text{ pairs}$

$42 \text{ e}^- \text{ pairs} - 24 \text{ shared e}^- \text{ pairs } 18 \text{ unshared (lone) e}^- \text{ pairs}$

Use unshared pairs to complete octets on terminal O atoms (15 unshared pairs) and ring N atoms (3 unshared pairs).

(b)   No C=N bonds in the 6-membered ring are possible, because all C octets are complete with 4 bonds to other atoms. N=N are possible, as shown below. There are 8 possibilities involving some combination of N–N and N=N groups [1 with 0 N=N, 3 with 1 N=N, 3 with 2N=N, 1 with 3N=N]. A resonance structure with 1 N=N is shown below.

Each terminal O=N–O group has two possible placements for the N=O. This generates 8 structures with 0 N=N groups (and 3 O = N–O groups), 4 with 1 N=N and 2 O=N–O, 2 with 2 N=N and 1 O=N–O, and 1 with 3 N=N and no O=N–O. This sums to a total of 15 resonance structures (that I can visualize). Can you find others?

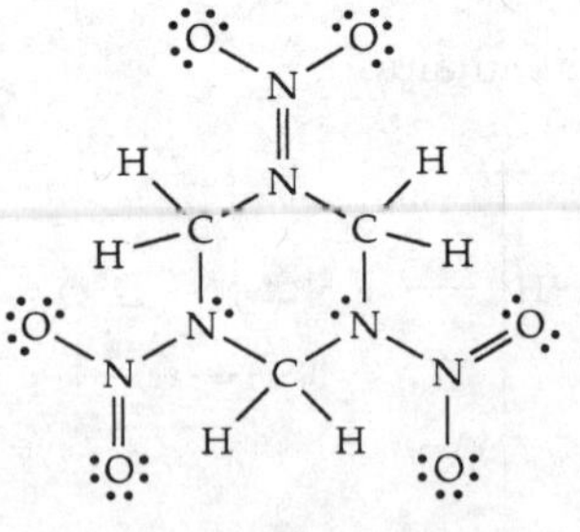

(c) $C_3H_6N_6O_6(s) \rightarrow 3CO(g) + 3N_2(g) + 3H_2O(g)$

(d)   The molecule contains N=O, N=N, C–H, C–N, N–O, and N–N bonds. According to Table 8.4, N–N bonds have the smallest bond enthalpy and are weakest.

(e)   Calculate the enthalpy of decomposition for the resonance structure drawn in part (a).

$\Delta H = 3D(N=O) + 3D(N-O) + 3D(N-N) + 6D(N-C) + 6D(C-H)$

$\quad - 3D(C\equiv O) - 3D(N\equiv N) - 6D(O-H)$

$$= 3(607) + 3(201) + 3(163) + 6(293) + 6(413) - 3(1072) - 3(941) - 6(463)$$

$$= -1668 \text{ kJ/mol } C_3H_6N_6O_6$$

$$5.0 \text{ g } C_3H_6N_6O_6 \times \frac{1 \text{ mol } C_3H_6N_6O_6}{222.1 \text{ g } C_3H_6N_6O_6} \times \frac{-1668 \text{ kJ}}{\text{mol } C_3H_6N_6O_6} = 37.55 = 38 \text{ kJ}$$

While exchanging N=O and N–O bonds has no effect on the enthalpy calculation, structures with N=N and 2 N–O do have different enthalpy of decomposition. For the resonance structure with 3 N=N and 6 N–O bonds instead of 3 N–N, 3 N–O and 3 N=O, $\Delta H = -2121$ kJ/mol. The actual enthalpy of decomposition is probably somewhere between –1668 and –2121 kJ/mol. The enthalpy charge for the decomposition of 5.0 g RDX is then in the range 38–48 kJ.

8.91   (a) $\Delta H = 2D(A=A) - 4D(A–A)$

       (b)   For the reaction to be exothermic ($-\Delta H$), $D(A=A) < 2D(A–A)$.

       (c)   If the reaction is exothermic, the second bond between atoms in A=A must be weaker than the first bond. (The first bond is A–A. If A=A < 2A–A, then the second bond has a smaller bond dissociation enthalpy than the first.)

8.92

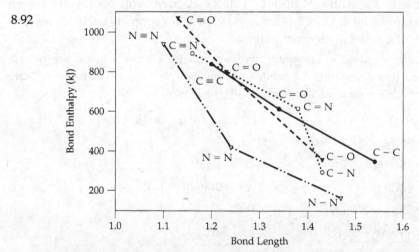

When comparing the same pair of bonded atoms (C–N vs. C=N vs. C≡N), the shorter the bond the greater the bond energy, but the two quantities are not necessarily directly proportional. The plot clearly shows that there are no simple length/strength correlations for single bonds alone, double bonds alone, triple bonds alone, or among different pairs of bonded atoms (all C–C bonds vs. all C–N bonds, etc.).

8.93   (a)   S–N ≈ 1.77 Å (sum of the bonding atomic radii from Figure 7.6).

       (b)   S–O ≈ 1.75 Å (the sum of the bonding atomic radii from Figure 7.6.) Alternatively, half of the S–S distance in S_8 (1.02) plus half of the O–O distance from Table 8.5 (0.74) is 1.76 Å.

       (c)   Owing to the resonance structures for SO_2, we assume that the S–O bond in SO_2 is intermediate between a double and single bond, so the distance of 1.43 Å should be significantly shorter than an S–O single bond distance, 1.75 Å.

(d)   54 e^-, 27 e^- pair

The observed S–O bond distance, 1.48 Å, is similar to that in SO_2, 1.43 Å, which can be described by resonance structures showing both single and double S–O bonds. Thus, S_8O must have resonance structures with both single and double S–O bonds. The structure with the S=O bond has 5 e^- pairs about this S atom. To the extent that this resonance form contributes to the true structure, the S atom bound to O has more than an octet of electrons around it.

Integrative Exercises

8.94   (a) Ti^{2+} : $[Ar]3d^2$; Ca : $[Ar]4s^2$. Yes. The two valence electrons in Ti^{2+} and Ca are in different principle quantum levels and different subshells.

         (b)   According to the Aufbau Principle, valence electrons will occupy the lowest energy empty orbital. Thus, in Ca the 4s is lower in energy than the 3d, while in Ti^{2+}, the 3d is lower in energy than the 4s.

         (c)   Since there is only one 4s orbital, the two valence electrons in Ca are paired. There are five degenerate 3d orbitals, so the two valence electrons in Ti^{2+} are unpaired. Ca has no unpaired electrons, Ti^{2+} has two.

8.95   (a) $Sr(s) \rightarrow Sr(g)$ ΔH_f° Sr(g) [ΔH_{sub}° Sr(s)]

              $Sr(g) \rightarrow Sr^+(g) + 1\ e^-$ I_1 Sr

              $Sr^+(g) \rightarrow Sr^{2+}(g) + 1\ e^-$ I_2 Sr

              $Cl_2(g) \rightarrow 2Cl(g)$ $2\ \Delta H_f^\circ$ Cl(g) [$D(Cl_2)$]

              $2Cl(g) + 2\ e^- \rightarrow 2Cl^-(g)$ $2E_1$ Cl

              $\underline{SrCl_2(s) \rightarrow Sr(s) + Cl_2(g)}$ $\underline{-\Delta H_f^\circ\ SrCl_2}$

              $SrCl_2(s) \rightarrow Sr^{2+}(g) + 2Cl^-(g)$ ΔH_{latt}

         (b) ΔH_f° $SrCl_2(s) = \Delta H_f^\circ$ $Sr(g) + I_1(Sr) + I_2(Sr) + 2\Delta H_f^\circ$ $Cl(g) + 2E(Cl) - \Delta H_{latt}$ $SrCl_2$

              ΔH_f° $SrCl_2(s) = 164.4$ kJ $+ 549$ kJ $+ 1064$ kJ $+ 2(121.7)$ kJ $+ 2(-349)$ kJ $- 2127$ kJ
                           $= -804$ kJ

8.96   The pathway to the formation of K_2O can be written:

         $2K(s) \rightarrow 2K(g)$ $2\Delta H_f^\circ$ K(g)

         $2K(g) \rightarrow 2K^+(g) + 2\ e^-$ $2\ I_1(K)$

         $1/2\ O_2(g) \rightarrow O(g)$ ΔH_f° O(g)

         $O(g) + 1\ e^- \rightarrow O^-(g)$ $E_1(O)$

         $O^-(g) + 1\ e^- \rightarrow O^{2-}(g)$ $E_2(O)$

         $\underline{2K^+(g) + O^{2-}(g) \rightarrow K_2O(s)}$ $\underline{-\Delta H_{latt}\ K_2O(s)}$

         $2K(s) + 1/2\ O_2(g) \rightarrow K_2O(s)$ ΔH_f° $K_2O(s)$

$$\Delta H_f^\circ \; K_2O(s) = 2\Delta H_f^\circ \; K(g) + 2\,I_1(K) + \Delta H_f^\circ \; O(g) + E_1(O) + E_2(O) - \Delta H_{latt} \; K_2O(s)$$

$$E_2(O) = \Delta H_f^\circ \; K_2O(s) + \Delta H_{latt} \; K_2O(s) - 2\Delta H_f^\circ \; K(g) - 2\,I_t(K) - \Delta H_f^\circ \; O(g) - E_1(O)$$

$$E_2(O) = -363.2 \text{ kJ} + 2238 \text{ kJ} - 2(89.99) \text{ kJ} - 2(419) \text{ kJ} - 247.5 \text{ kJ} - (-141) \text{ kJ}$$

$$= +750 \text{ kJ}$$

8.97   (a)   Assume 100 g.

              A: 87.7 g In/114.82   = 0.764 mol In; 0.764/0.384 ≈ 2

                     12.3 g S/32.07   = 0.384 mol S; 0.384/0.384 = 1

              B: 78.2 g In/114.82   = 0.681 mol In; 0.681/0.68 0 ≅ 1

                     21.8 g S/32.07   = 0.680 mol S; 0.680/0.680 = 1

              C: 70.5 g In/114.82   = 0.614 mol In; 0.614/0.614 = 1

                     29.5 g S/32.07   = 0.920 mol S; 0.920/0.614 = 1.5

              A: In_2S; B: InS; C: In_2S_3

    (b)   A: In(I); B: In(II); C: In(III)

    (c)   In(I) : $[Kr]5s^2 4d^{10}$; In(II) : $[Kr]5s^1 4d^{10}$; In(III) : $[Kr]4d^{10}$
            None of these is a noble-gas configuration.

    (d)   The ionic radius of In^{3+} in compound C will be smallest. Removing successive electrons from an atom reduces electron repulsion, increases the effective nuclear charge experienced by the valence electrons and decreases the ionic radius. The higher the charge on a cation, the smaller the radius.

    (e)   Lattice energy is directly related to the charge on the ions and inversely related to the interionic distance. Only the charge and size of the In varies in the three compounds. In(I) in compound A has the smallest charge and the largest ionic radius, so compound A has the smallest lattice energy and the lowest melting point. In(III) in compound C has the greatest charge and the smallest ionic radius, so compound C has the largest lattice energy and highest melting point.

8.98   (a)   Even though Cl has the greater (more negative) electron affinity, F has a much larger ionization energy, so the electronegativity of F is greater.

            F: k(IE–EA) = k(1681 – (–328)) = k(2009)

            Cl: k(IE–EA) = k(1251 – (–349)) = k(1600)

    (b)   Electronegativiy is the ability of an atom in a molecule to attract electrons to itself. It can be thought of as the ability to hold its own electrons (as measured by ionization energy) and the capacity to attract the electrons of other atoms (as measured by electron affinity). Thus, both properties are relevant to the concept of electronegativity.

    (c)   EN = k(IE – EA). For F: 4.0 = k(2009), k = 4.0/2009 = 2.0×10^{-3}

    (d)   Cl: EN = 2.0×10^{-3} (1600) = 3.2

            O: EN = 2.0×10^{-3} (1314 – (–141)) = 2.9

These values do not follow the trend on Figure 8.6. The Pauling scale on the figure shows O to be second only to F in electronegativity, more electronegative than Cl. The simple definition EN = k(IE – EA) that employs thermochemical properties of isolated, gas phase atoms does not take into account the complex bonding environment of molecules.

8.99 (a) Assume 100 g.

$$14.52 \text{ g C} \times \frac{1 \text{ mol}}{12.011 \text{ g C}} = 1.209 \text{ mol C}; 1.209 / 1.209 = 1$$

$$1.83 \text{ g H} \times \frac{1 \text{ mol}}{1.008 \text{ g H}} = 1.816 \text{ mol H}; 1.816 / 1.209 = 1.5$$

$$64.30 \text{ g Cl} \times \frac{1 \text{ mol}}{35.453 \text{ g Cl}} = 1.814 \text{ mol Cl}; 1.814 / 1.209 = 1.5$$

$$19.35 \text{ g O} \times \frac{1 \text{ mol}}{15.9994 \text{ g O}} = 1.209 \text{ mol O}; 1.209 / 1.209 = 1.0$$

Multiplying by 2 to obtain an integer ratio, the empirical formula is $C_2H_3Cl_3O_2$.

(b) The empirical formula weight is 2(12.0) + 3(1.0) + 3(35.5) + 2(16) = 165.5. The empirical formula is the molecular formula.

(c) 44 e^-, 22 e^- pairs

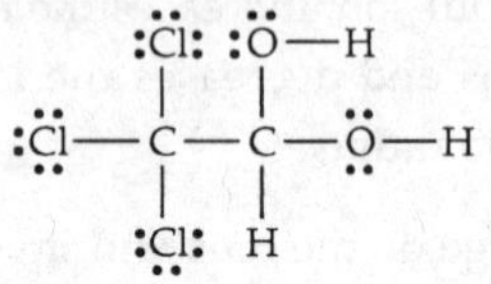

8.100 (a) Assume 100 g.

$$62.04 \text{ g Ba} \times \frac{1 \text{ mol}}{137.33 \text{ g Ba}} = 0.4518 \text{ mol Ba}; 0.4518 / 0.4518 = 1.0$$

$$37.96 \text{ g N} \times \frac{1 \text{ mol}}{14.007 \text{ g N}} = 2.710 \text{ mol N}; 2.710 / 0.4518 = 6.0$$

The empirical formula is BaN_6. Ba has an ionic charge of 2+, so there must be two 1– azide ions to balance the charge. The formula of each azide ion is N_3^-.

(b) 16 e^-, 8 e^- pairs

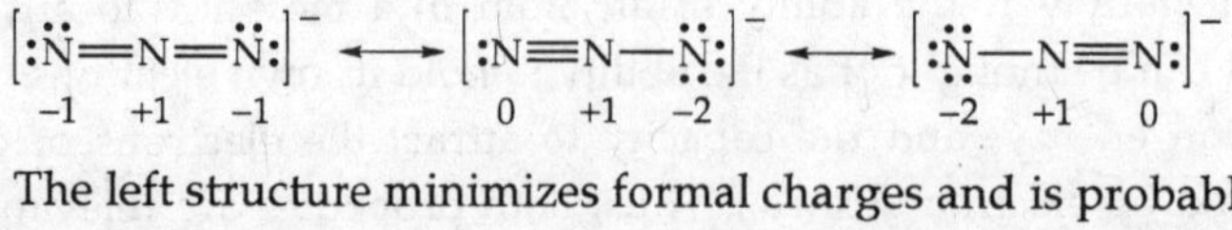

(c) The left structure minimizes formal charges and is probably the main contributor.

(d) The two N–N bond lengths will be equal. The two minor contributors would individually cause unequal N–N distances, but collectively they contribute equally to the lengthening and shortening of each bond. The N–N distance will be approximately 1.24 Å, the average N=N distance.

8.101   (a) C_2H_2: 10 e$^-$, 5 e$^-$ pair N_2: 10 e$^-$, 5 e$^-$ pair

              H—C≡C—H   :N≡N:

        (b) N_2 is an extremely stable, unreactive compound. Under appropriate conditions, it can be either oxidized (Section 22.7) or reduced (Sections 14.7 and 15.1). C_2H_2 is a reactive gas, used in combination with O_2 for welding and as starting material for organic synthesis (Section 25.4).

        (c) $2N_2(g) + 5O_2(g) \rightarrow 2N_2O_5(g)$

             $2C_2H_2(g) + 5O_2(g) \rightarrow 4CO_2(g) + 2H_2O(g)$

        (d) $\Delta H^\circ_{rxn}\,(N_2) = 2\Delta H^\circ_f\,N_2O_5(g) - 2\Delta H^\circ_f\,N_2(g) - 5\Delta H^\circ_f\,O_2(g)$

                    $= 2(11.30) - 2(0) - 5(0) = 22.60$ kJ

          $\Delta H^\circ_{ox} = 11.30$ kJ/mol N_2

          $\Delta H^\circ_{rxn}\,(C_2H_2) = 4\Delta H^\circ_f\,CO_2(g) + 2\Delta H^\circ_f\,H_2O(g) - 2\Delta H^\circ_f\,C_2H_2(g) - 5\Delta H^\circ_f\,O_2(g)$

                        $= 4(-393.5 \text{ kJ}) + 2(-241.82 \text{ kJ}) - 2(226.7 \text{ kJ}) - 5(0)$

                        $= -2511.0$ kJ

          $\Delta H^\circ_{ox}\,(C_2H_2) = -1255.5$ kJ/mol C_2H_2

          The oxidation of C_2H_2 is highly exothermic, which means that the energy state of the combined products is much lower than that of the reactants. The reaction is "downhill" in an energy sense, and occurs readily. The oxidation of N_2 is mildly endothermic (energy of products higher than reactants) and the reaction does not readily occur. This is in agreement with the general reactivities from part (b).

          Referring to bond enthalpies in Table 8.4, when the C–H bonds are taken into account, even more energy is required for bond breaking in the oxidation of C_2H_2 than in the oxidation of N_2. The difference seems to be in the enthalpies of formation of the products. $CO_2(g)$ and $H_2O(g)$ have extremely exothermic ΔH°_f values, which cause the oxidation of C_2H_2 to be energetically favorable. $N_2O_5(g)$ has an endothermic ΔH°_f value, which causes the oxidation of N_2 to be energetically unfavorable.

8.102   (a)   Assume 100 g of compound

          $69.6 \text{ g S} \times \dfrac{1 \text{ mol S}}{32.07 \text{ g}} = 2.17 \text{ mol S}$

          $30.4 \text{ g N} \times \dfrac{1 \text{ mol N}}{14.01 \text{ g}} = 2.17 \text{ mol N}$

          S and N are present in a 1:1 mol ratio, so the empirical formula is SN. The empirical formula weight is 46. MM/FW = 184.3/46 = 4 The molecular formula is S_4N_4.

        (b)   44 e$^-$, 22 e$^-$ pairs. Because of its small radius, N is unlikely to have an expanded octet. Begin with alternating S and N atoms in the ring. Try to satisfy the octet rule with single bonds and lone pairs. At least two double bonds somewhere in the ring are required.

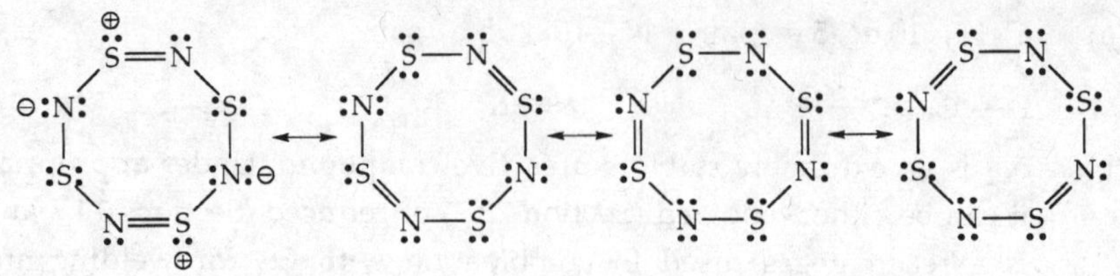

These structures carry formal charges on S and N atoms as shown. Other possibilities include:

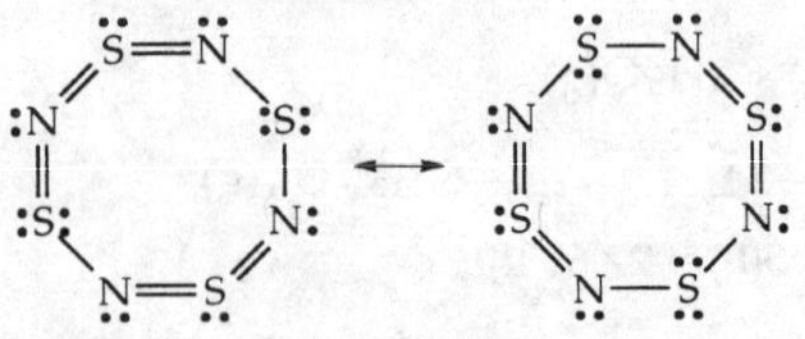

These structures have zero formal charges on all atoms and are likely to contribute to the true structure. Note that the S atoms that are shown with two double bonds are not necessarily linear, because S has an expanded octet. Other resonance structures with four double bonds are.

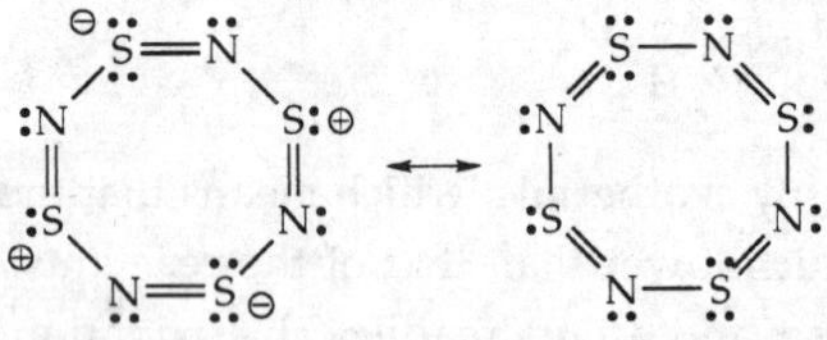

In either resonance structure, the two 'extra' electron pairs can be placed on any pair of S atoms in ring, leading to a total of 10 resonance structures. The sulfur atoms alternately carry formal charges of +1 and –1. Without further structural information, it is not possible to eliminate any of the above structures. Clearly, the S_4N_4 molecule stretches the limits of the Lewis model of chemical bonding.

(c) Each resonance structure has 8 total bonds and more than 8 but less than 16 bonding e⁻ pairs, so an "average" bond will be intermediate between a S–N single and double bond. We estimate an average S–N single bond length to be 1.77 Å (sum of bonding atomic radii from Figure 7.6). We do not have a direct value for a S–N double bond length. Comparing double and single bond lengths for C–C (1.34 Å, 1.54 Å), N–N (1.24 Å, 1.47 Å) and O–O (1.21 Å, 1.48 Å) bonds from Table 8.5, we see that, on average, a double bond is approximately 0.23 Å shorter than a single bond. Applying this difference to the S–N single bond length, we estimate the S–N double bond length as 1.54 Å. Finally, the intermediate S–N bond length in S_4N_4 should be between these two values, approximately 1.60–165 Å. (The measured bond length is 1.62 Å.)

(d) $S_4N_4 \rightarrow 4S(g) + 4N(g)$

$\Delta H = 4\Delta H_f^\circ\ S(g) + 4\Delta H_f^\circ\ N(g) - \Delta H_f^\circ\ S_4N_4$

$\Delta H = 4(222.8\ kJ) + 4(472.7\ kJ) - 480\ kJ = 2302\ kJ$

This energy, 2302 kJ, represents the dissociation of 8 S–N bonds in the molecule; the average dissociation energy of one S–N bond in S_4N_4 is then 2302 kJ/8 bonds = 287.8 kJ.

8.103   (a)   Yes. In the structure shown in the exercise, each P atom needs 1 unshared pair to complete its octet. This is confirmed by noting that only 6 of the 10 valence e^- pairs are bonding pairs.

         (b)   There are six P–P bonds in P_4.

         (c)   The atomization reaction is: $P_4(g) \rightarrow 4P(g)$

$$\Delta H_{atom} = 4\Delta H_f^\circ \, P(g) - \Delta H_f^\circ \, P_4(g)$$

$$= 4(316.4 \text{ kJ}) - 58.9 \text{ kJ} = 1206.7 \text{ kJ}$$

         (d)   Since there are six P–P bonds in P_4, the bond dissociation enthalpy for an P–P bond, $D(P\text{–}P) = 1206.7/6 = 201$ kJ.

         (e)   From Table 8.4, $D(N\text{–}N) = 163$ kJ. Our calculated value for $D(P\text{–}P)$ is 201 kJ, so the P–P bond is stronger than the N–N bond.

8.104   (a)

$HF(g) \rightarrow H(g) + F(g)$	$D(H\text{–}F)$	567 kJ
$H(g) \rightarrow H^+(g) + 1\,e^-$	$I(H)$	1312 kJ
$F(g) + 1\,e^- \rightarrow F^-(g)$	$E(F)$	–328 kJ
$HF(g) \rightarrow H^+(g) + F^-(g)$	ΔH	1551 kJ

         (b) $\Delta H = D(H\text{–}Cl) + I(H) + E(Cl)$

                $\Delta H = 431 \text{ kJ} + 1312 \text{ kJ} + (-349) \text{ kJ} = 1394 \text{ kJ}$

         (c) $\Delta H = D(H\text{–}Br) + I(H) + E(Br)$

                $\Delta H = 366 \text{ kJ} + 1312 \text{ kJ} + (-325) \text{ kJ} = 1353 \text{ kJ}$

8.105   (a) $C_6H_6(g) \rightarrow 6H(g) + 6C(g)$

                $\Delta H^\circ = 6\Delta H_f^\circ \, H(g) + 6\Delta H_f^\circ \, C(g) - \Delta H_f^\circ \, C_6H_6(g)$

                $\Delta H^\circ = 6(217.94) \text{ kJ} + 6(718.4) \text{ kJ} - 82.9 \text{ kJ} = 5535 \text{ kJ}$

         (b) $C_6H_6(g) \rightarrow 6CH(g)$

         (c)

$C_6H_6(g) \rightarrow 6H(g) + 6C(g)$	ΔH°	5535 kJ
$6H(g) + 6C(g) \rightarrow 6CH(g)$	$-6D(C\text{–}H)$	$-6(413)$ kJ
$C_6H_6(g) \rightarrow 6CH(g)$		3057 kJ

            3057 kJ is the energy required to break the six C–C bonds in $C_6H_6(g)$. The average bond dissociation energy for one carbon-carbon bond in $C_6H_6(g)$ is

$$\frac{3057 \text{ kJ}}{6\,C\text{–}C \text{ bonds}} = 509.5 \text{ kJ}.$$

         (d)   The value of 509.5 kJ is between the average value for a C–C single bond (348 kJ) and a C=C double bond (614 kJ). It is somewhat greater than the average of these two values, indicating that the carbon-carbon bond in benzene is a bit stronger than we might expect.

8.106 (a) $Br_2(l) \rightarrow 2Br(g)$ $\Delta H^\circ = 2\Delta H_f^\circ \, Br(g) = 2(111.8) \, kJ = 223.6 \, kJ$

(b) $CCl_4(l) \rightarrow C(g) + 4Cl(g)$

$\Delta H^\circ = \Delta H_f^\circ \, C(g) + 4\Delta H_f^\circ \, Cl(g) - \Delta H_f^\circ \, CCl_4(l)$

$\qquad = 718.4 \, kJ + 4(121.7) \, kJ - (-139.3) \, kJ = 1344.5$

$$\frac{1344.5 \, kJ}{4 \, C - Cl \; bonds} = 336.1 \, kJ$$

(c) $\qquad H_2O_2(l) \rightarrow 2H(g) + 2O(g)$

$\underline{2H(g) + 2O(g) \rightarrow 2OH(g)}$

$\qquad\quad H_2O_2(l) \rightarrow 2OH(g)$

$D(O - O)(l) = 2\Delta H_f^\circ \, H(g) + 2\Delta H_f^\circ \, O(g) - \Delta H_f^\circ \, H_2O_2(l) - 2D(O-H)(g)$

$\qquad\qquad = 2(217.94) \, kJ + 2(247.5) \, kJ - (-187.8) \, kJ - 2(463) \, kJ$

$\qquad\qquad = 193 \, kJ$

(d) The data are listed below.

bond	D gas kJ/mol	D liquid kJ/mol
Br–Br	193	223.6
C–Cl	328	336.1
O–O	146	192.7

Breaking bonds in the liquid requires more energy than breaking bonds in the gas phase. For simple molecules, bond dissociation from the liquid phase can be thought of in two steps:

molecule (l) $\rightarrow$ molecule (g)

molecule (g) $\rightarrow$ atoms (g)

The first step is evaporation or vaporization of the liquid and the second is bond dissociation in the gas phase. Average bond enthalpy in the liquid phase is then the sum of the enthalpy of vaporization for the molecule and the gas phase bond dissociation enthalpies, divided by the number of bonds dissociated. This is greater than the gas phase bond dissociation enthalpy owing to the contribution from the enthalpy of vaporization.

9 Molecular Geometry and Bonding Theories

Visualizing Concepts

9.1   Removing an atom from the equatorial plane of trigonal bipyramid in Figure 9.3 creates a seesaw shape. It might appear that you could also obtain a seesaw by removing two atoms from the square plane of the octahedron. However, one of the B–A–B angles in the seesaw is 120°, so it must be derived from a trigonal bipyramid.

9.2   (a)   120°

      (b)   If the blue balloon expands, the angle between red and green balloons decreases.

      (c)   Nonbonding (lone) electron pairs exert greater repulsive forces than bonding pairs, resulting in compression of adjacent bond angles.

9.3   (a)   Square pyramidal

      (b)   Yes, there is one nonbonding electron domain on A. If there were only five bonding domains, the shape would be trigonal bipyramidal. With five bonding and one nonbonding electron domains, the molecule has octahedral domain geometry.

      (c)   Yes. If the B atoms are halogens, each will have three nonbonding electron pairs; there are five bonding pairs, and A has one nonbonded pair, for a total of [5(3) + 5 + 1] = 21 e$^-$ pairs and 42 electrons in the Lewis structure. If the five halogens contribute 35 e$^-$, A must contribute seven valence electrons. A is also a halogen.

9.4   (a)   4 e$^-$ domains

      (b)   The molecule has a non-zero dipole moment, because the C–H and C–F bond dipoles do not cancel each other.

      (c)   The dipole moment vector bisects the F–C–F and H–C–H angles, with the negative end of the vector toward the F atoms.

9.5   (a)   The reference point for zero energy on the diagram corresponds to a state where the two Cl atoms are separate and not interfacing. This corresponds to an infinite Cl–Cl distance beyond the right extreme of the horizontal axis. The point near the left side of the plot where the curve intersects the x-axis at E = 0 has no special meaning.

      (b)   Energy decreases as atom separation decreases because the valence electrons of one atom come close enough to the other atom to be stabilized by both nuclei instead of just one nucleus.

(c)   The Cl–Cl distance at the energy minimum on the plot is the Cl–Cl bond distance.

(d)   At interatomic separations shorter than the bond distance, the two nuclei begin to repel each other, increasing the overall energy of the system.

9.6   (a)   90° angles are characteristic of hybrids with a d atomic orbital contribution. This pair of orbitals could be sp^3d or sp^3d^2.

(b)   Angles of 109.5° are characteristic of sp^3 hybrid orbitals only.

(c)   Angles of 120° can be formed by sp^2 hybrids or sp^3d hybrids.

9.7   (a)   The ground state electron configuration of Si is $[Ne]3s^23p^2$. The left side of the orbital diagram corresponds to the configuration $3s^13p^3$, so one or more electrons has been promoted.

(b)   The diagram illustrates mixing of a single s and three p atomic orbitals to form sp^3 hybrids.

9.8   (a)   Recall that π bonds require p atomic orbitals, so the maximum hybridization of a C atom involved in a double bond is sp^2 and in a triple bond is sp. There are 6 C atoms in the molecule. Starting on the left, the hybridizations are: sp^2, sp^2, sp^3, sp, sp, sp^3.

(b)   All single bonds are σ bonds. Double and triple bonds each contain 1 σ bond. This molecule has 8 C–H σ bonds and 5 C–C σ bonds, for a total of 13 σ bonds.

(c)   Double bonds have 1 π bond and triple bonds have 2 π bonds. This molecule has a total of 3 π bonds.

9.9   *Analyze/Plan.* σ molecular orbitals (MOs) are symmetric about the internuclear axis, π MOs are not. Bonding MOs have most of their electron density in the area between the nuclei, antibonding MOs have a node between the nuclei.

(a)   (i)   The shape of the molecular orbital (MO) indicates that it is formed by two s atomic orbitals (electron density at each nucleus).

(ii) σ-type MO (symmetric about the internuclear axis, s orbitals can produce only σ overlap).

(iii)   antibonding MO (node between nuclei)

(b)   (i)   Shape indicates this MO is formed by two p atomic orbitals overlapping end-to-end (node near each nucleus).

(ii) σ-type MO (symmetric about internuclear axis)

(iii)   bonding MO (concentration of electron density between nuclei)

(c)   (i)   Shape indicates this MO is formed by two p atomic orbitals overlapping side-to-side (node near each nucleus).

(ii) π-type MO (not symmetric about internuclear axis, side-to-side overlap)

(iii)   antibonding MO (node between nuclei)

9.10   (a)   The diagram has five electrons in MOs formed by 2p atomic orbitals. C has two 2p electrons, so X must have three 2p electrons. X is N.

        (b)   The molecule has an unpaired electron, so it is paramagnetic.

        (c)   Atom X is N, which is more electronegative than C. The atomic orbitals of the more electronegative N are slightly lower in energy than those of C. The lower energy π_{2p} bonding molecular orbitals will have a greater contribution from the lower energy N atomic orbitals. (Higher energy π_{2p}^{*} MOs will have a greater contribution from higher energy C atomic orbitals.)

Molecular Shapes; the VSEPR Model

9.11   (a)   Yes. The bond angle specifies the shape and the bond length tells the size.

        (b)   Yes. This description means that the three terminal atoms point toward the corners of an equilateral triangle and the central atom is in the plane of this triangle. Only 120° bond angles are possible in this arrangement.

9.12   (a)   In a symmetrical tetrahedron, the four bond angles are equal to each other, with values of 109.5°. The H–C–H angles in CH_4 and the O–Cl–O angles in ClO_4^- will have values close to 109.5°.

        (b)   Two, the H–N–H angle and the N–H distance, assuming that NH_3 is a symmetric trigonal pyramid with equal bond distances and angles. In a trigonal pyramidal molecule, there are three bonding and one nonbonding electron domains. Since a nonbonding electron domain compresses bond angles, the H–N–H angles will not be exactly 109.5°, the bond angles in a perfect tetrahedron. The bond angle must be specified. Also, N does not sit in the plane of the H atoms. The distance of N out of the plane is determined by the N–H distances, as well as the H–N–H angles. The N–H distance must also be specified.

9.13   (a)   An *electron domain* is a region in a molecule where electrons are most likely to be found.

        (b)   Each balloon in Figure 9.5 occupies a volume of space. The best arrangement is one where each balloon has its "own" space, where they are as far apart as possible and repulsions are minimized. Electron domains are negatively charged regions, so they also adopt an arrangement where repulsions are minimized.

9.14   (a)   The number of electron domains in a molecule or ion is the number of bonds (double and triple bonds count as one domain) **plus** the number of nonbonding (lone) electron pairs.

        (b)   A *bonding electron domain* is a region between two bonded atoms that contains one or more pairs of bonding electrons. A *nonbonding electron domain* is localized on a single atom and contains one pair of nonbonding electrons (a lone pair).

9.15   *Analyze/Plan.* See Table 9.1.   *Solve.*

        (a)   trigonal planar   (b)   tetrahedral

        (c)   trigonal bipyramidal   (d)   octahedral

9.16   (a)   3 (or 5 if 90° angles are also present)

        (b)   2 (or 5 if 120° angles are also present, 6 if more than one 180° angle is present)

        (c)   4

        (d)   6 (or 5 if 120° angles are also present)

9.17   The electron-domain geometry indicated by VSEPR describes the arrangement of all bonding and nonbonding electron domains. The molecular geometry describes just the atomic positions. H_2O has the Lewis structure given below; there are four electron domains around oxygen so the electron-domain geometry is tetrahedral, but the molecular geometry of the three atoms is bent.

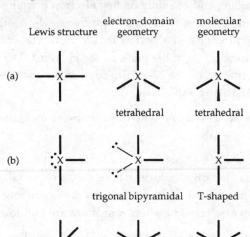

     Lewis structure   electron-domain   molecular geometry
                  geometry

9.18   If the electron-domain geometry is trigonal bipyramidal, there are five total electron domains around the central atom. An AB_3 molecule has three bonding domains, so there must be two nonbonding domains on A.

9.19   *Analyze/Plan.* See Tables 9.2 and 9.3.   *Solve.*

	Lewis structure	electron-domain geometry	molecular geometry
(a)		tetrahedral	tetrahedral
(b)		trigonal bipyramidal	T-shaped
(c)		octahedral	square pyramidal

9.20　(a)

trigonal planar　　trigonal planar

(b)

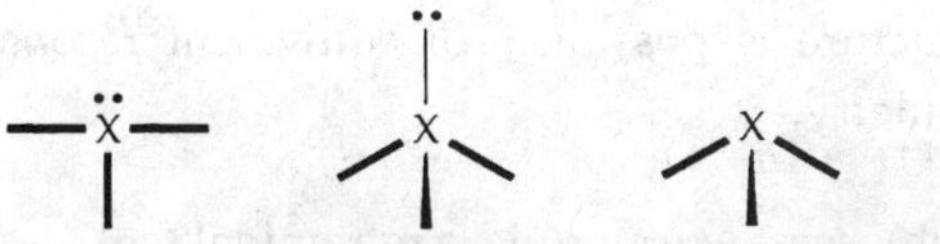

tetrahedral　　trigonal pyramidal

(c)

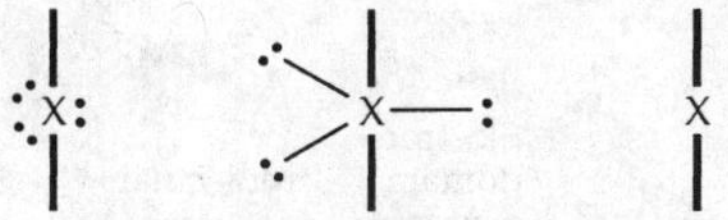

trigonal bipyramidal　　linear

9.21　*Analyze/Plan.* Follow the logic in Sample Exercise 9.1.

Solve. bent (b), linear (l), octahedral (oh), seesaw (ss) square pyramidal (sp), square planar (spl), tetrahedral (td), trigonal bipyramidal (tbp), trigonal planar (tr), trigonal pyramidal (tp), T-shaped (T)

Molecule or ion	Valence electrons	Lewis structure	Electron-domain geometry	Molecular geometry
(a) PF_3	26	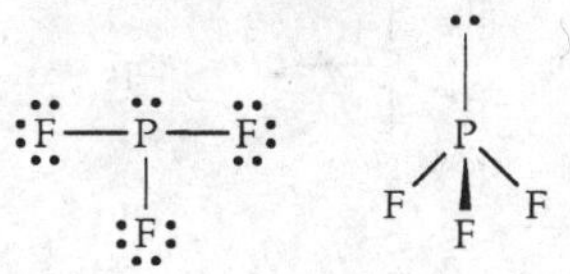	td	tp
(b) CH_3^+	6	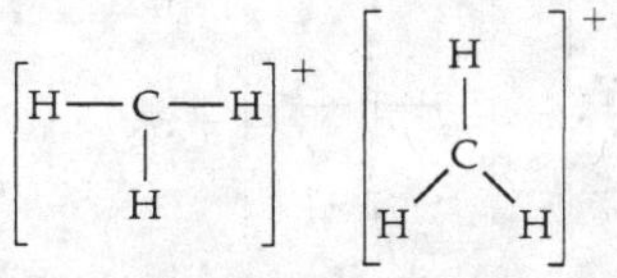	tr	tr
(c) BrF_3	28	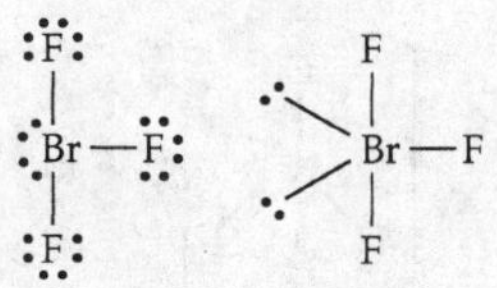	tbp	T
(d) ClO_4^-	32	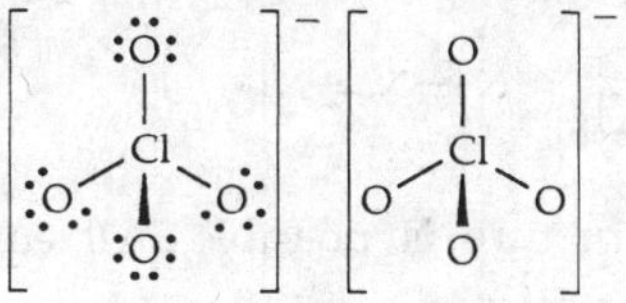	td	td
(e) XeF_2	22	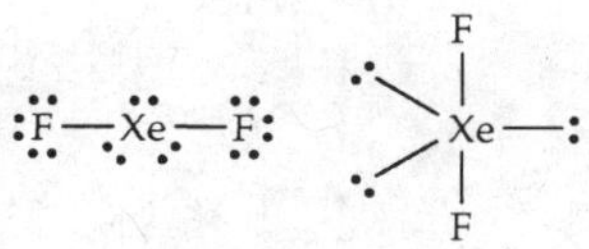	tbp	l

217

(f) BrO_2^- 20   td   b

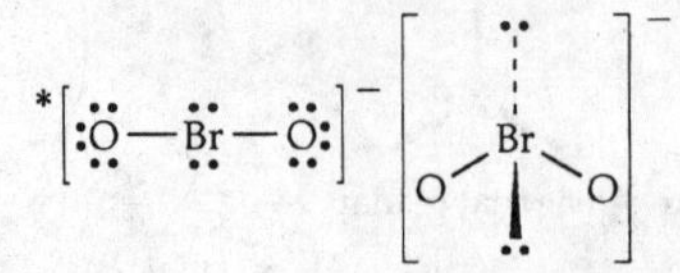

*More than one resonance structure is possible. All equivalent resonance structures predict the same molecular geometry.

9.22   bent (b), linear (l), octahedral (oh), seesaw (ss), square pyramidal (sp),
square planar (spl), tetrahedral (td), trigonal bipyramidal (tbp), trigonal planar (tr),
trigonal pyramidal (tp), T-shaped (T)

Molecule or ion	Valence electrons	Lewis structure	Electron-domain geometry	Molecular geometry
(a) HCN	10		I	I
(b) SO_3^{2-}	26	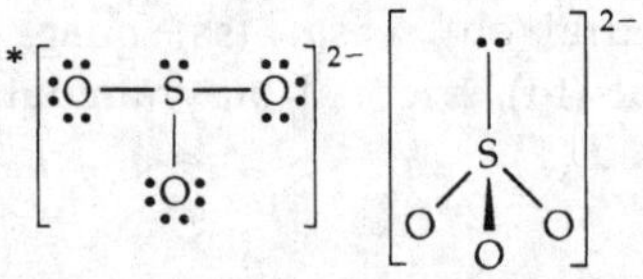	td	tp
(c) SF_4	34	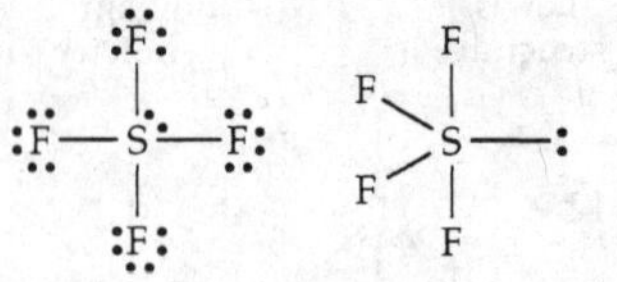	tbp	ss
(d) PF_6^-	48	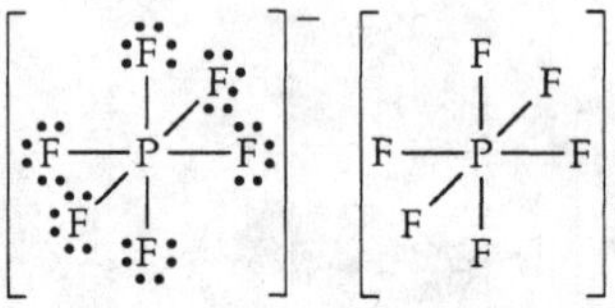	oh	oh
(e) NH_3Cl^+	14	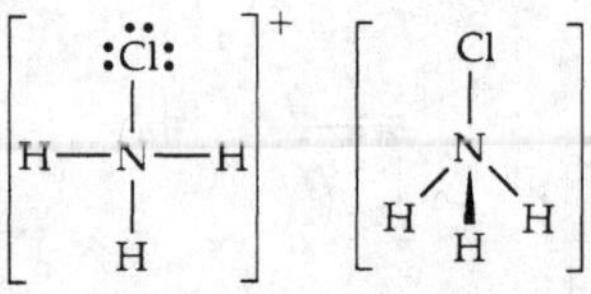	tb	tb
(f) N_3^-	16	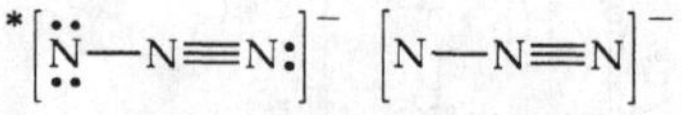	I	I

*More than one resonance structure is possible. All equivalent resonance structures predict the same molecular geometry.

9.23   *Analyze/Plan* Work backwards from molecular geometry, using Tables 9.2 and 9.3. *Solve*.

    (a)   Electron-domain geometries: i, trigonal planar; ii, tetrahedral; iii, trigonal bipyramidal

    (b)   nonbonding electron domains: i, 0; ii, 1; iii, 2

    (c)   N and P. Shape ii has three bonding and one nonbonding electron domains. Li and Al would form ionic compounds with F, so there would be no nonbonding electron domains. Assuming that F always has three nonbonding domains, BF_3 and ClF_3 would have the wrong number of nonbonding domains to produce shape ii.

    (d)   Cl (also Br and I, since they have seven valence electrons). This T-shaped molecular geometry arises from a trigonal bipyramidal electron-domain geometry with two nonbonding domains (Table 9.3). Assuming each F atom has three nonbonding domains and forms only single bonds with A, A must have seven valence electrons to produce these electron-domain and molecular geometries. It must be in or below the third row of the periodic table, so that it can accommodate more than four electron domains.

9.24   (a)   Electron-domain geometries: i, octahedral; ii, tedrahedral; iii, trigonal bipyramial

    (b)   nonbonding electron domains: i, 2; ii, 0; iii, 1

    (c)   S or Se. Shape iii has five electron domains, so A must be in or below the third row of the periodic table. This eliminates Be and C. Assuming each F atom has three nonbonding electron domains and forms only single bonds with A, A must have six valence electrons to produce these electron-domain and molecular geometries.

    (d)   Xe. (See Table 9.3) Assuming F behaves typically, A must be in or below the third row and have eight valence electrons. Only Xe fits this description. (Noble gas elements above Xe have not been shown to form molecules of the type AF_4. See Section 7.8.)

9.25   *Analyze/Plan*. Follow the logic in Sample Exercise 9.3.   *Solve*.

    (a)   1 – 109°, 2 – 109°   (b)   3 – 109°, 4 – 109°

    (c)   5 – 180°   (d)   6 – 120°, 7 – 109°, 8 – 109°

9.26   (a)   1 – 109°, 2 – 120°   (b)   3 – 109°, 4 – 120°

    (c)   5 – 109°, 6 – 109°   (d)   7 – 180°, 8 – 109°

9.27   *Analyze/Plan*. Given the formula of each molecule or ion, draw the correct Lewis structure and use principles of VSEPR to answer the question. *Solve*.

$$\left[H\!-\!\ddot{\underset{\displaystyle ..}{N}}\!-\!H \right]^{-} \qquad H\!-\!\underset{\displaystyle |}{\overset{\displaystyle ..}{N}}\!-\!H \qquad \left[H\!-\!\underset{\displaystyle |}{\overset{\displaystyle H}{N}}\!-\!H \right]^{+}$$
$$\qquad\qquad\qquad\qquad H \qquad\qquad\qquad H$$

Each species has four electron domains around the N atom, but the number of nonbonding domains decreases from two to zero, going from NH_2^- to NH_4^+. Since nonbonding domains exert greater repulsive forces on adjacent domains, the bond angles expand as the number of nonbonding domains decreases.

9.28

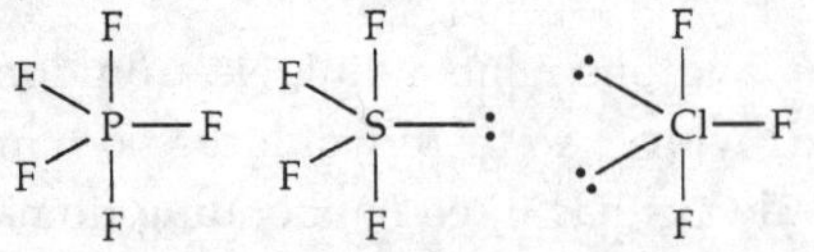

The three nonbonded electron pairs on each F atom have been omitted for clarity.

The F (axial) –A–F (equatorial) angle is largest in PF_5 and smallest in ClF_3. As the number of nonbonding domains in the equatorial plane increases, they push back the axial A–F bonds, decreasing the F (axial) –A–F (equatorial) bond angles.

9.29   *Analyze.* Given: molecular formulas. Find: explain features of molecular geometries.

Plan. Draw the correct Lewis structures for the molecules and use VSEPR to predict and explain observed molecular geometry.   *Solve.*

(a) BrF_4^- 36 e⁻, 18 e⁻ pr

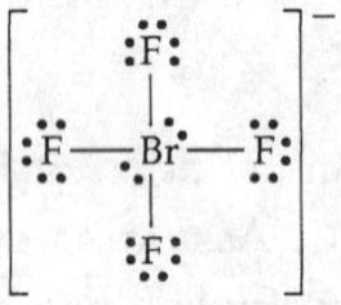

6 e⁻ pairs around Br, octahedral e⁻ domain geometry,
square planar molecular geometry

BF_4^- 32 e⁻, 16 e⁻ pr

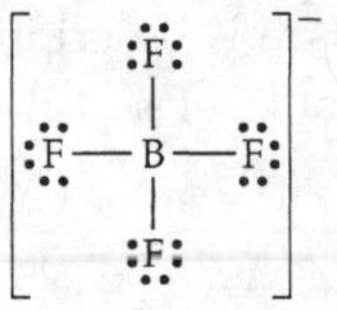

4 e⁻ pairs around B, tetrahedral e⁻ domain geometry,
tetrahedral molecular geometry

The fundamental feature that determines molecular geometry is the number of electron domains around the central atom, and the number of these that are bonding domains. Although BrF_4^- and BF_4^- are both of the form AX_4^-, the central atoms and thus the number of valence electrons in the two ions are different. This leads to different numbers of e⁻ domains about the two central atoms. Even though both ions have four bonding electron domains, the six total domains around Br require octahedral domain geometry and square planar molecular geometry, while the four total domains about B lead to tetrahedral domain and molecular geometry.

(b) CF_4 32 e⁻, 16 e⁻ pr

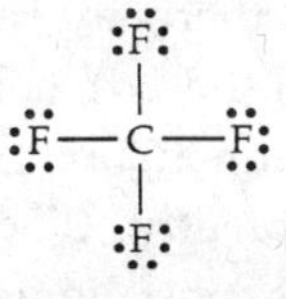

$$\begin{array}{c} :\!\ddot{F}\!: \\ | \\ :\!\ddot{F}\!-\!C\!-\!\ddot{F}\!: \\ | \\ :\!\ddot{F}\!: \end{array}$$

4 e⁻ domains around C, tetrahedral e⁻ domain geometry,
tetrahedral molecular geometry

SF_4 34 e⁻, 17 e⁻ pr

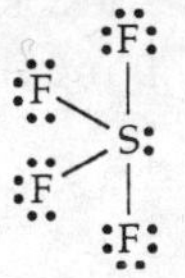

$$\begin{array}{c} :\!\ddot{F}\!: \\ \diagdown \;\; | \\ :\!\ddot{F}\!\diagdown \!S\!: \\ :\!\ddot{F}\!\diagup \;\; | \\ :\!\ddot{F}\!: \end{array}$$

5 e⁻ domains around S, trigonal bipyramidal e⁻ domain geometry,
seesaw molecular geometry

CF_4 will have bond angles closest to the value predicted by VSEPR, because there are no nonbonding e⁻ domains around C. The four bonding domains in CF_4 are equivalent and lead to the balance of repulsions implicit in VSEPR theory. In SF_4, one of the e⁻ domains is nonbonding. A nonbonding domain is surely not equivalent to a bonding domain; we expect it to be more diffuse. That is, nonbonding domains will occupy more space, "push back" the bonding domains, and lead to bond angles that are nonideal.

9.30   (a) ClO_2^- 20 e⁻, 10 e⁻ pr

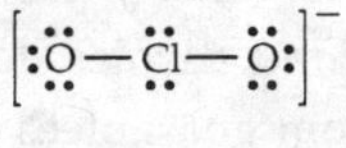

$$\left[:\!\ddot{O}\!-\!\ddot{Cl}\!-\!\ddot{O}\!:\right]^-$$

4 e⁻ domains around Cl, tetrahedral e⁻ domain geometry,
bent molecular geometry bond angle $\lesssim 109.5°$

NO_2^- 18 e⁻, 9 e⁻ pr

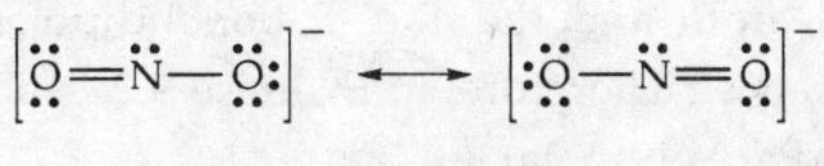

$$\left[\ddot{O}\!=\!\ddot{N}\!-\!\ddot{O}\!:\right]^- \longleftrightarrow \left[:\!\ddot{O}\!-\!\ddot{N}\!=\!\ddot{O}\right]^-$$

3 e⁻ domains about N (both resonance structures), trigonal planar e⁻ domain geometry bent molecular geometry bond angle $\lesssim 120°$

Both molecular geometries are described as 'bent" because both molecules have two nonlinear bonding electron domains. The bond angles (the angle between the two bonding domains) in the two ions are different because the total number of electron domains, and thus the electron domain geometries are different.

(b) XeF_2 22 e⁻, 11 e⁻ pr

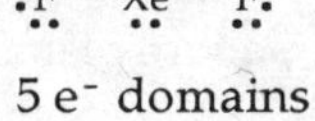

$$:\!\ddot{F}\!-\!\ddot{Xe}\!-\!\ddot{F}\!:$$

5 e⁻ domains around Xe, trigonal bipyramidal e⁻ domain geometry,
linear molecular geometry

The question here really is: why do the three nonbonding domains all occupy the equatorial plane of the trigonal bipyramid? In a tbp, there are several different kinds of repulsions, bonding domain-bonding domain (bd-bd), bonding domain-nonbonding domain (bd-nd), and nonbonding domain-nonbonding domain (nd-nd). Each of these can have 90°, 120°, or 180° geometry. Since nonbonding domains occupy more space, 90° nd-nd repulsions are most significant and least desirable. The various electron domains arrange themselves to minimize these 90° nd-nd interactions. The arrangement shown above has no 90° nd-nd repulsions. An arrangement with one or two nonbonding domains in axial positions would lead to at least two 90° nd-nd repulsions, a less stable situation. (To convince yourself, tabulate the number and kinds of repulsions for each possible tbp arrangement of 2bd's and 3nd's.)

Polarity of Polyatomic Molecules

9.31   See Sample Exercise 9.4(b) for the correct resonance structures and analysis of S–O bond dipoles. According to the electron density model, the net dipole moment vector points along the O–S–O angle bisector with the negative end pointing away from S. the magnitude of this vector is 1.63 D.

9.32   If PH_3 were planar, the PH_3 bond dipoles would cancel, and the molecule would be nonpolar. Since PH_3 is polar, the 3 P–H bond dipoles do not cancel, and the molecule can't be planar.

9.33   (a)   In Exercise 9.23, molecules ii and iii will have nonzero dipole moments. Molecule i has no nonbonding electron pairs on A, and the 3 A–F dipoles are oriented so that the sum of their vectors is zero (the bond dipoles cancel). Molecules ii and iii have nonbonding electron pairs on A and their bond dipoles do not cancel. A nonbonding electron pair (or pairs) on a central atom guarantees at least a small molecular dipole moment, because no bond dipole exactly cancels a nonbonding pair.

         (b) AF_4 molecules will have a zero dipole moment if there are no nonbonding electron pairs on the central atom and the 4 A–F bond dipoles are arranged (symmetrically) so that they cancel. Therefore, in Exercise 9.24, molecules i and ii have zero dipole moments and are nonpolar.

9.34   (a)   For a molecule with polar bonds to be nonpolar, the polar bonds must be (symmetrically) arranged so that the bond dipoles cancel. In most cases, nonbonding e^- domains must be absent from the central atom. Square planar structures may not meet the second condition.

         (b) AB_2: linear e^- domain geometry (edg), linear molecular geometry (mg), trigonal bipyramidal edg, linear mg

                AB_3: trigonal planar edg, trigonal planar mg

                AB_4: tetrahedral edg, tetrahedral mg; octahedral edg, square planar mg

9.35   *Analyze/Plan.* Given molecular formulas, draw correct Lewis structures, determine molecular structure and polarity. *Solve.*

(a)   Nonpolar, in a symmetrical tetrahedral structure (Figure 9.1) the bond dipoles cancel.

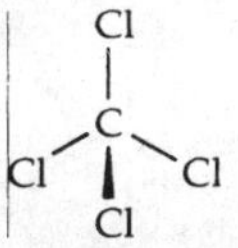

(b)   Polar, there is an unequal charge distribution due to the nonbonded electron pair on N.

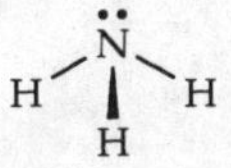

(c)   Polar, there is an unequal charge distribution due to the nonbonded electron pair on S.

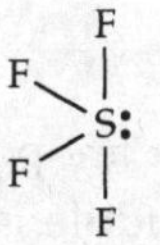

(d)   Nonpolar, the bond dipoles and the nonbonded electron pairs cancel.

(e)   Polar, the C–H and C–Br bond dipoles are not equal and do not cancel.

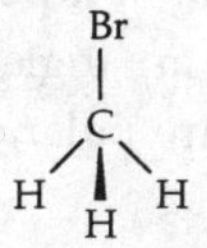

(f)   Nonpolar, in a symmetrical trigonal planar structure, the bond dipoles cancel.

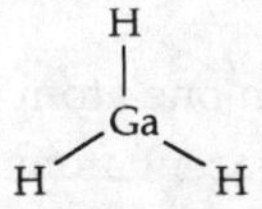

9.36   (a)   Polar, $\Delta EN > 0$

I–F

(b)   Nonpolar, the molecule is linear and the bond dipoles cancel.

S═C═S

(c)   Nonpolar, in a symmetrical trigonal planar structure, the bond dipoles cancel.

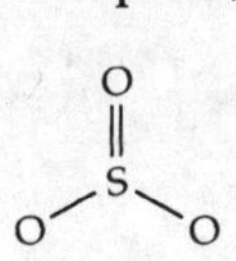

(d)   Polar, there is an unequal charge distribution due to the nonbonded electron pair on P.

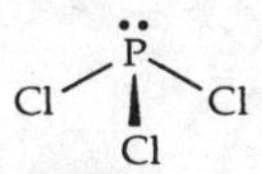

(e)   Nonpolar, symmetrical octahedron

(f)   Polar, square pyramidal molecular geometry, bond dipoles do not cancel.

9.37

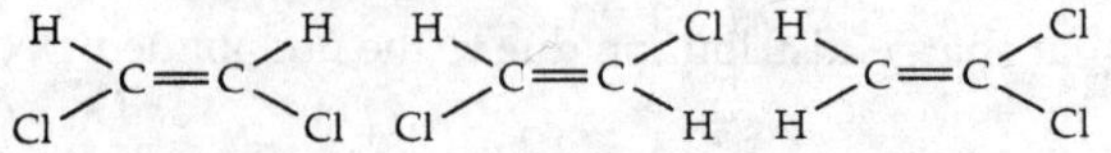

polar   nonpolar   polar

All three isomers are planar. The molecules on the left and right are polar because the C–Cl bond dipoles do not point in opposite directions. In the middle isomer, the C–Cl bonds and dipoles are pointing in opposite directions (as are the C–H bonds), the molecule is nonpolar and has a measured dipole moment of zero.

9.38   Each C–Cl bond is polar. The question is whether the vector sum of the C–Cl bond dipoles in each molecule will be nonzero. In the *ortho* and *meta* isomers, the C–Cl vectors are at 60° and 120° angles, respectively, and their resultant dipole moments are nonzero. In the *para* isomer, the C–Cl vectors are opposite, at an angle of 180°, with a resultant dipole moment of zero. The *ortho* and *meta* isomers are polar, the *para* isomer is nonpolar.

Orbital Overlap; Hybrid Orbitals

9.39   (a)   *Orbital overlap* occurs when a valence atomic orbital on one atom shares the same region of space with a valence atomic orbital on an adjacent atom.

(b)   In valence bond theory, overlap of orbitals allows the two electrons in a chemical bond to mutually occupy the space between the bonded nuclei.

(c)   Valence bond theory is a combination of the atomic orbital concept with the Lewis model of electron pair bonding.

9.40   (a)

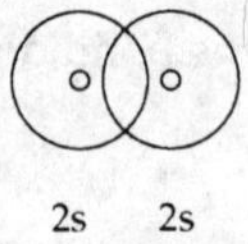

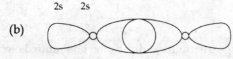

2s   2s

(b)

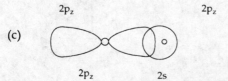

$2p_z$ $2p_z$

(c)

$2p_z$ 2s

9.41   (a)   4 valence e⁻, 2 e⁻ pairs

H—Mg—H

2 bonding e⁻ domains, linear e⁻ domain and molecular geometry

(b)   The Mg atom has two electrons in its 2s orbital; these electrons are paired. Without promotion, Mg has no unpaired electrons available to form bonds with H.

(c)   The linear electron domain geometry in MgH_2 requires sp hybridization.

(d)

9.42   (a)   8 valence e⁻, 4 e⁻ pairs

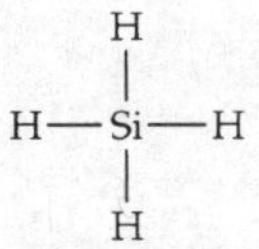

$$\begin{array}{c} H \\ | \\ H-Si-H \\ | \\ H \end{array}$$

4 bonding e⁻ domains, tetrahedral domain and molecular geometry

(b) SiH_4 has four equivalent Si–H bonds. A free Si atom has the valence electron configuration $2s^2 2p^2$; the two 2p electrons are unpaired and available for bonding, but the 2s electrons are paired and unavailable. In order to have four unpaired electrons available for bonding, one of the 2s electrons must be promoted to the empty 2p orbital.

(c)   The tetrahedral electron domain geometry of SiH_4 requires sp^3 hybridization.

(d)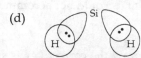

The other two Si–H bonds are in a plane perpendicular to the plane of the page, and pointing up. They form angles of 109.5° with each other and the two Si–H bonds drawn above.

9.43   *Analyze/Plan.* Given electron domain geometry, list the appropriate orbital hybridization and associated bond angles; refer to Table 9.4.   *Solve.*

(a)   sp – 180°   (b) sp^3 – 109°   (c) sp^2 – 120°

(d) sp^3d^2 – 90° and 180°   (e) sp^3d – 90°, 120° and 180°

9.44   (a) sp^2 – 120° angles in a plane

(b) sp^3d – 90°, 120° and 180° bond angles (trigonal bipyramid)

(c) sp^3d^2 – 90° and 180° bond angles (octahedron)

9.45   (a)   B: $[\text{He}]2s^2 2p^1$

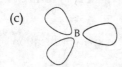

       2s   2p   promote   2s   2p   hybridize sp^2 p

   (b)   The hybrid orbitals are called sp^2.

   (c)

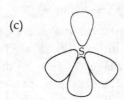

   (d)   A single 2p orbital is unhybridized. It lies perpendicular to the trigonal plane of the sp^2 hybrid orbitals.

9.46   (a)   S: $[\text{Ne}]3s^2 3p^4$

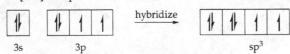

       3s   3p   hybridize sp^3

   (b)   The hybrid orbitals are called sp^3.

   (c)

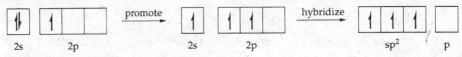

   (d)   The hybrid orbitals formed in (a) would not be appropriate for SF_4. There are five electron domains in SF_4, four bonding and one nonbonding, so five hybrid orbitals are required. A set of four sp^3 hybrid orbitals could not accommodate all the electron pairs around S.

9.47   *Analyze/Plan.* Given the molecular (or ionic) formula, draw the correct Lewis structure and determine the electron domain geometry, which determines hybridization. *Solve.*

   (a)   24 e^-, 12 e^- pairs

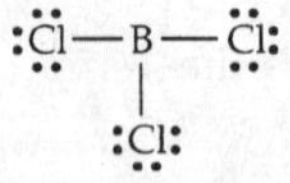

       :C̈l — B — C̈l:
             |
           :C̈l:

       3 e^- pairs around B trigonal, planar e^- domain geometry, sp^2 hybrid orbitals

   (b)   32 e^-, 16 e^- pairs

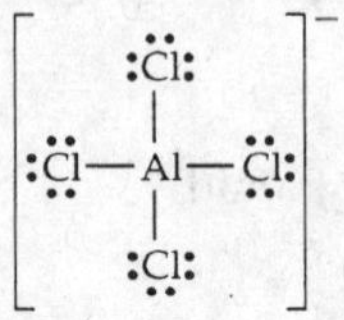

$$
\left[\begin{array}{c} :\ddot{C}l: \\ | \\ :\ddot{C}l - Al - \ddot{C}l: \\ | \\ :\ddot{C}l: \end{array} \right]^-
$$

       4 e^- domains around Al, tetrahedral e^- domain geometry, sp^3 hybrid orbitals

(c)   16 e⁻, 8 e⁻ pairs

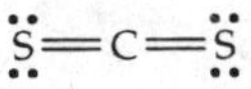

$$:\ddot{S}=C=\ddot{S}:$$

2 e⁻ domains around C, linear e⁻ domain geometry, sp hybrid orbitals

(d)   22 e⁻, 11 e⁻ pairs

$$:\ddot{F}-\dot{\ddot{K}r}-\ddot{F}:$$

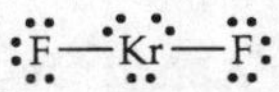

5 e⁻ pairs around Kr, trigonal bipyramidal e⁻ domain geometry, sp³d hybrid orbitals

(e)   48 e⁻, 24 e⁻ pairs

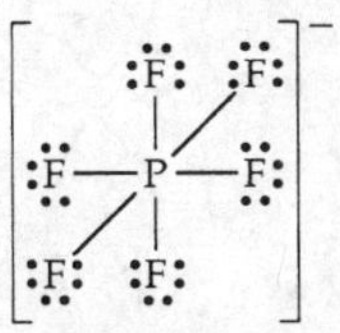

6 e⁻ pairs around P, octahedral e⁻ domain geometry, sp³d² orbitals

9.48   (a)   32 e⁻, 16 e⁻ pairs

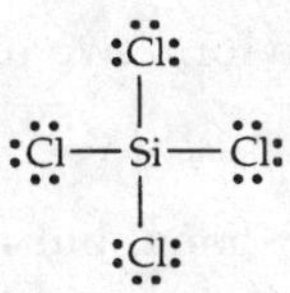

4 e⁻ pairs around Si, tetrahedral e⁻ domain geometry, sp³ hybrid orbitals

(b)   10 e⁻, 5 e⁻ pairs

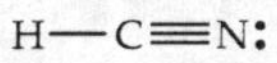

$$H-C\equiv N:$$

2 e⁻ domains around C, linear e⁻ domain geometry, sp hybrid orbitals

(c)   24 e⁻, 12 e⁻ pairs

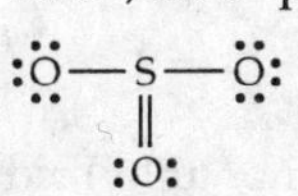

$$:\ddot{O}-S-\ddot{O}:$$

(other resonance structures are possible)

3 e⁻ domains around S, trigonal planar e⁻ domain geometry, sp² hybrid orbitals

(d)   22 e⁻, 11 e⁻ pairs

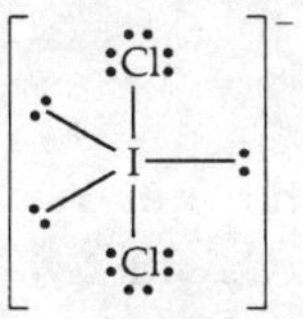

5 e⁻ domains around I, trigonal bipyramidal e⁻ domain geometry, sp³d hybrid orbitals (In a trigonal bipyramid, placing nonbonding e⁻ pairs in the equatorial position minimizes repulsion.)

(e)   36 e⁻, 18 e⁻ pairs

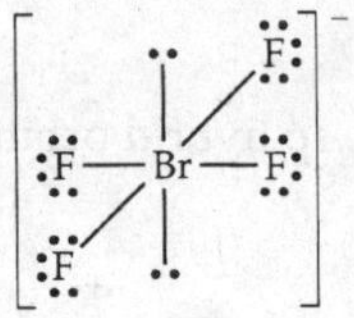

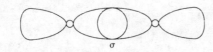

6 e⁻ domains around Br, octahedral e⁻ domain geometry, sp^3d^2 hybrid orbitals

Multiple Bonds

9.49   (a)   (b)

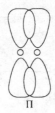

     (c)   A σ bond is generally stronger than a π bond, because there is more extensive orbital overlap.

9.50   (a)   Two unhybridized p orbitals remain, and the atom can form two pi bonds.

     (b)   A triple bond is composed of 1 σ and 2 π bonds.

     (c)   There is free rotation of attached groups around a σ bond, but not around a π bond. Rotation about a π bond would require that the bond be broken; the p orbitals would no longer be in the correct orientation for π overlap. The π overlap that is part of all multiple bonds introduces rigidity into molecules.

9.51   (a)

     (b) sp^3 sp^2

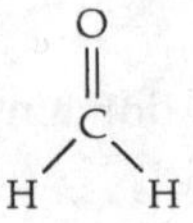

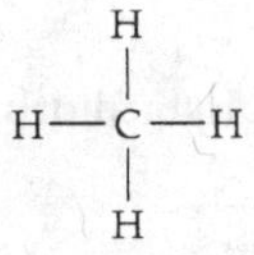

     (c)   The C atom in CH_4 is sp^3 hybridized; there are no unhybridized p orbitals available for the π overlap required by multiple bonds. In CH_2O, the C atom is sp^2 hybridized, with one p atomic orbital available to form the π overlap in the C=O double bond.

9.52   H—N̈—N̈—H   :N≡N:
                  | |
                  H H

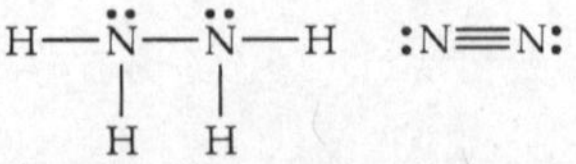

The N atoms in N_2H_4 are sp^3 hybridized; there are no unhybridized p orbitals available for π bonding. In N_2, the N atoms are sp hybridized, with two unhybridized p orbitals on each N atom available to form the two π bonds in the N≡N triple bond.

9.53   *Analyze/Plan.* Single bonds are σ bonds, double bonds consist of 1 σ and 1 π bond. Each bond is formed by a pair of valence electrons.   *Solve.*

    (a) C_3H_6 has $3(4) + 6(1) = 18$ valence electrons

    (b)   8 pairs or 16 total valence electrons form σ bonds

    (c)   1 pair or 2 total valence electrons form π bonds

    (d)   no valence electrons are nonbonding

    (e)   The left and central C atoms are sp^2 hybridized; the right C atom is sp^3 hybridized.

9.54   (a)   The C bound to O has three electron domains and is sp^2 hybridized; the other three C atoms are sp^3 hybridized.

    (b) $C_4H_8O_2$ has $4(4) + 8(1) + 2(6) = 36$ valence electrons.

    (c)   13 pairs or 26 total valence electrons form σ bonds

    (d)   1 pair or 2 total valence electrons form π bonds

    (e)   4 pairs or 8 total valence electrons are nonbonding

9.55   *Analyze/Plan.* Given the correct Lewis structure, analyze the electron domain geometry at each central atom. This determines the hybridization and bond angles at that atom. *Solve.*

    (a)   ~109° about the left most C, sp^3; ~120° about the right-hand C, sp^2

    (b)   The doubly bonded O can be viewed as sp^2, the other as sp^3; the nitrogen is sp^3 with approximately 109° bond angles.

    (c)   9 σ bonds, 1 π bond

9.56   (a)   1, 120°; 2, 120°; 3, 109°

    (b)   1, sp^2; 2, sp^2; 3, sp^3

    (c)   21 σ bonds

9.57   (a)   In a localized π bond, the electron density is concentrated strictly between the two atoms forming the bond. In a delocalized π bond, parallel p orbitals on more than two adjacent atoms overlap and the electron density is spread over all the atoms that contribute p orbitals to the network. There are still two regions of overlap, above and below the σ framework of the molecule.

    (b)   The existence of more than one resonance form is a good indication that a molecule will have delocalized π bonding.

    (c)

$$\left[\ddot{\underset{..}{O}} = \overset{..}{N} - \ddot{\underset{..}{O}} \colon \right]^{-} \longleftrightarrow \left[\colon \ddot{\underset{..}{O}} - \overset{..}{N} = \ddot{\underset{..}{O}} \right]^{-}$$

The existence of more than one resonance form for NO_2 indicates that the π bond is delocalized. From an orbital perspective, the electron-domain geometry around N is trigonal planar, so the hybridization at N is sp^2. This leaves a p orbital on N and one on each O atom perpendicular to the trigonal plane of the molecule, in the correct orientation for delocalized π overlap. Physically, the two N–O bond lengths are equal, indicating that the two N–O bonds are equivalent, rather than one longer single bond and one shorter double bond.

9.58   (a) 24 e^-, 12 e^- pairs   (b)

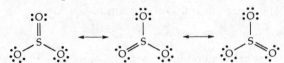

3 electron domains around S, trigonal planar electron-domain geometry, sp^2 hybrid orbitals

(c)   The multiple resonance structures indicate delocalized π bonding. All four atoms lie in the trigonal plane of the sp^2 hybrid orbitals. On each atom there is a p atomic orbital perpendicular to this plane in the correct orientation for π overlap. The resulting delocalized π electron cloud is Y-shaped (the shape of the molecule) and has electron density above and below the plane of the molecule.

Molecular Orbitals

9.59   (a)   Both atomic and molecular orbitals have a characteristic energy and shape (region where there is a high probability of finding an electron). Each atomic or molecular orbital can hold a maximum of two electrons. Atomic orbitals are localized on single atoms and their energies are the result of interactions between the subatomic particles in a single atom. MOs can be delocalized over several or even all the atoms in a molecule and their energies are influenced by interactions between electrons on several atoms.

(b)   There is a net stabilization (lowering in energy) that accompanies bond formation because the bonding electrons in H_2 are strongly attracted to both H nuclei.

(c)   Two

9.60   (a)   In the σ anti-bonding MO, electron density is concentrated away from the nuclei; an electron in this orbital experiences less stabilization by the nucleus than an electron in an isolated atom.

(b)   Yes. The Pauli principle, that no two electrons can have the same four quantum numbers, means that an orbital can hold at most two electrons. (Since n, l, and m_1 are the same for a particular orbital and m_s has only two possible values, an orbital can hold at most two electrons). This is true for atomic and molecular orbitals.

(c)   Four. When AOs combine to form MOs, the total number of orbitals is conserved. Combination of four AOs must result in formation of four MOs. Both can accommodate eight electrons.

9.61   (a)

(b)   There is one electron in H_2^+.

(c)   ☐ σ_{1s}^*

    ⊞↑ σ_{1s}

(d)   Bond order = 1/2 (1-0) = 1/2

(e)   Fall apart. The stability of H_2^+ is due to the lower energy state of the σ bonding molecular orbital relative to the energy of a H 1s atomic orbital. If the single electron in H_2^+ is excited to the σ^*_{1s} orbital, its energy is higher than the energy of an H 1s atomic orbital and H_2^+ will decompose into a hydrogen atom and a hydrogen ion.

$$H_2^+ \xrightarrow{h\nu} H + H^+.$$

9.62   (a)

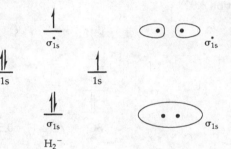

H_2^-

(b)   ⊞↑ σ_{1s}^*

    ⊞↑↓ σ_{1s}

(c)   Bond order = 1/2 (2-1) = 1/2

(d)   If one electron moves from σ_{1s} to σ^*_{1s}, the bond order becomes –1/2. There is a net increase in energy relative to isolated H atoms, so the ion will decompose.

$$H_2^- \xrightarrow{h\nu} H + H^-.$$

9.63   *Analyze/Plan.* In a σ molecular orbital, the electron density is spherically symmetric about the internuclear axis and is concentrated along this axis. In a π MO, the electron density is concentrated above and below the internuclear axis and zero along it. *Solve.*

(a)

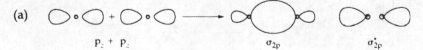

$P_z + P_z$ σ_{2p} σ_{2p}^*

(b)

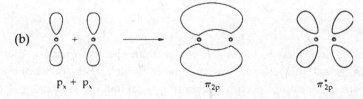

$P_x + P_x$ π_{2p} π_{2p}^*

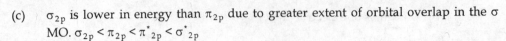

(c) σ_{2p} is lower in energy than π_{2p} due to greater extent of orbital overlap in the σ MO. $\sigma_{2p} < \pi_{2p} < \pi^*_{2p} < \sigma^*_{2p}$

9.64   (a)   Zero

        (b)   The two π_{2p} molecular orbitals are degenerate; they have the same energy, but they have different spatial orientations 90° apart.

        (c)   In the bonding MO the electrons are stabilized by both nuclei. In an antibonding MO, the electrons are directed away from the nuclei, so π_{2p} is lower in energy than π^*_{2p}.

9.65   (a)   When comparing the same two bonded atoms, the greater the bond order, the shorter the bond length and the greater the bond energy. That is, bond order and bond energy are directly related, while bond order and bond length are inversely related. When comparing different bonded nuclei, there are no simple relationships (see Solution 8.92).

        (b)

BO = 1/2(2-2) = 0   BO = 1/2(2-1) = 0.5

Be_2 has a bond order of zero and is not energetically favored over isolated Be atoms; it is not expected to exist. Be_2^+ has a bond order of 0.5 and is slightly lower in energy than isolated Be atoms. It will probably exist under special experimental conditions, but be unstable.

9.66   (a) O_2^{2-} has a bond order of 1.0, while O_2^- has a bond order of 1.5. For the same bonded atoms, the greater the bond order the shorter the bond, so O_2^- has the shorter bond.

        (b)   The two possible orbital energy level diagrams are:

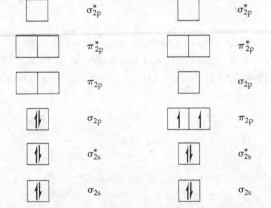

If the σ_{2p} molecular orbital is lower in energy than the π_{2p} orbitals, there are no unpaired electrons, and the molecule is diamagnetic. Switching the order of σ_{2p} and π_{2p} gives one unpaired electron in each degenerate π_{2p} orbital and explains the observed paramagnetism of B_2 (see Figure 9.45).

9.67   (a), (b) Substances with no unpaired electrons are weakly repelled by a magnetic field. This property is called *diamagnetism*.

(c) O_2^{2-}, Be_2^{2+} (see Figure 9.45)

9.68   (a)   Substances with unpaired electrons are attracted into a magnetic field. This property is called *paramagnetism*.

(b)   Weigh the substance normally and in a magnetic field, as shown in Figure 9.46. Paramagnetic substances appear to have a larger mass when weighed in a magnetic field.

(c)   See Figures 9.37 and 9.45. O_2^+, one unpaired electron; N_2^{2-}, two unpaired electrons; Li_2^+, one unpaired electron

9.69

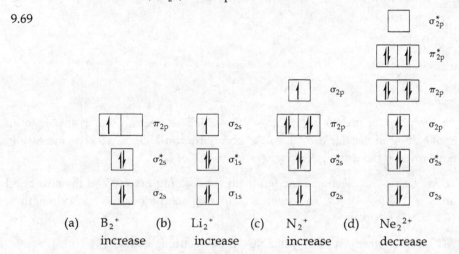

(a) B_2^+ (b) Li_2^+ (c) N_2^+ (d) Ne_2^{2+}
       increase   increase   increase   decrease

Addition of an electron increases bond order if it occupies a bonding MO and decreases stability if it occupies an antibonding MO.

9.70   Determine the number of "valence" (non-core) electrons in each molecule or ion. Use the homonuclear diatomic MO diagram from Figure 9.42 (shown below) to calculate bond order and magnetic properties of each species. The electronegativity difference between heteroatomics increases the energy difference between the 2S AO on one atom and the 2p AO on the other, rendering the "no interaction" MO diagram in Figure 9.42 appropriate.

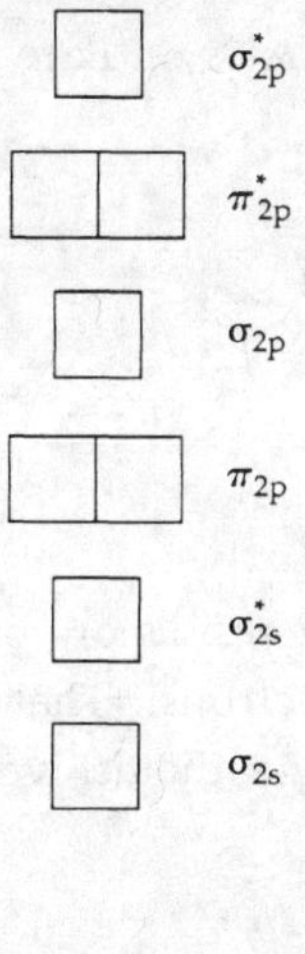

(a) CO^+: 9 e^-, B.O. = (7 – 2) / 2 = 2.5, paramagnetic

(b) NO^-: 12 e^-, B.O. = (8 – 4) / 2 = 2.0, paramagnetic

(c) OF^+: 12 e^-, B.O. = (8 – 4) / 2 = 2.0, paramagnetic

(d) NeF^+: 14 e^-, B.O. = (8 – 6) / 2 = 1.0, diamagnetic

9.71   *Analyze/Plan.* Determine the number of "valence" (non-core) electrons in each molecule or ion. Use the homonuclear diatomic MO diagram from Figure 9.42 (shown below) to calculate bond order and magnetic properties of each species. The electronegativity difference between heteroatomics increases the energy difference between the 2s AO on one atom and the 2p AO on the other, rendering the "no interaction" MO diagram in Figure 9.42 appropriate.   *Solve.*

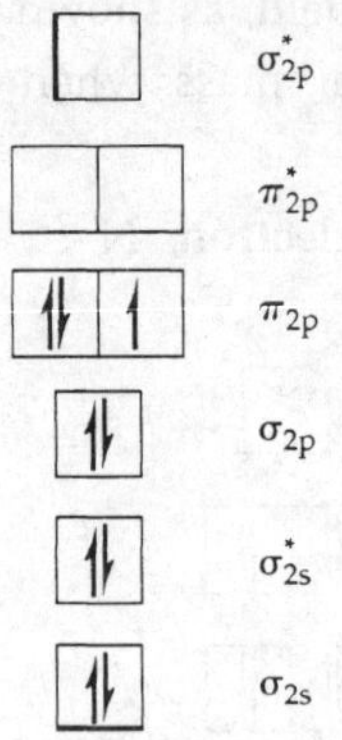

σ_{2p}^*

π_{2p}^*

π_{2p}

σ_{2p}

σ_{2s}^*

σ_{2s}

CN: 9 e$^-$, B.O. = (7 – 2) / 2 = 2.5, paramagnetic

CN$^+$:   8 e$^-$, B.O. = (6 – 2) / 2 = 2.0, paramagnetic

CN$^-$:   10 e$^-$, B.O. = (8 – 2) / 2 = 3.0, diamagnetic

9.72   (a)   The bond order of NO is [1/2 (8 – 3)] = 2.5. The electron that is lost is in an antibonding molecular orbital, so the bond order in NO$^+$ is 3.0. The increase in bond order is the driving force for the formation of NO$^+$.

(b)   To form NO$^-$, an electron is added to an antibonding orbital, and the new bond order is [1/2 (8 – 4)] = 2. The order of increasing bond order and bond strength is: NO$^-$ < NO < NO$^+$.

(c)   NO$^+$ is isoelectronic with N_2, and NO$^-$ is isoelectronic with O_2.

9.73   (a) $3s, 3p_x, 3p_y, 3p_z$ (b) π_{3p} (c)   Two

(d)   If the MO diagram for P_2 is similar to that of N_2, P_2 will have no unpaired electrons and be diamagnetic.

9.74   (a)   I: $5s, 5p_x, 5p_y, 5p_z$; Br: $4s, 4p_x, 4p_y, 4p_z$

(b)   By analogy to F_2, the BO of IBr will be 1.

(c)   I and Br have valence atomic orbitals with different principal quantum numbers. This means that the radial extensions (sizes) of the valence atomic orbital that contribute to the MO are different. The n = 5 valence AOs on I are larger than the n = 4 valence AOs on Br.

(d) σ_{np}^*

(e)   None

Additional Exercises

9.75   (a)   The physical basis of VSEPR is the electrostatic repulsion of like-charged particles, in this case groups or domains of electrons. That is, owing to electrostatic repulsion, electron domains will arrange themselves to be as far apart as possible.

(b)   The σ-bond electrons are localized in the region along the internuclear axes. The positions of the atoms and geometry of the molecule are thus closely tied to the locations of these electron pairs. Because the π-bond electrons are distributed above and below the plane that contains the σ bonds, these electron pairs do not, in effect, influence the geometry of the molecule. Thus, all σ- and π-bond electrons localized between two atoms are located in the same electron domain.

9.76

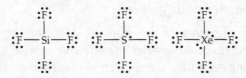

e⁻ domain geometry td   tbp   octahedral (oh)
molecular shape td   seesaw (ss)   square planar (s)

Although there are four bonding electron domains in each molecule, the number of nonbonding domains is different in each case. The bond angles and thus the molecular shape are influenced by the total number of electron domains.

9.77   For any triangle, the law of cosines gives the length of side c as $c^2 = a^2 + b^2 - 2ab\cos\theta$.

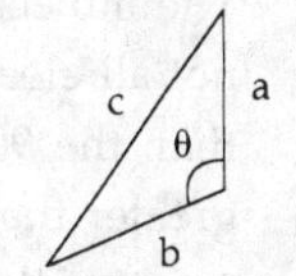

Let the edge length of the cube (uy = vy = vz) = X

The length of the face diagonal (uv) is

$(uv)^2 = (uy)^2 + (vy)^2 - 2(uy)(vy)\cos 90$

$(uv)^2 = X^2 + X^2 - 2(X)(X)\cos 90$

$(uv)^2 = 2X^2; uv = \sqrt{2}X$

The length of the body diagonal (uz) is

$(uz)^2 = (vz^2) + (uv)^2 - 2(vz)(uv)\cos 90$

$(uz)^2 = X^2 + (\sqrt{2}X)^2 - 2(X)(\sqrt{2}X)\cos 90$

$(uz)^2 = 3X^2; uz = \sqrt{3}X$

For calculating the characteristic tetrahedral angle, the appropriate triangle has vertices u, v, and w. Theta, θ, is the angle formed by sides wu and wv and the hypotenuse is side uv.

$wu = wv = uz/2 = \sqrt{3}/2X; uv = \sqrt{2}X$

$(\sqrt{2}X)^2 = (\sqrt{3}/2X)^2 + (\sqrt{3}/2X)^2 - 2(\sqrt{3}/2X)(\sqrt{3}/2)\cos\theta$

$2X^2 = 3/4\,X^2 + 3/4\,X^2 - 3/2\,X^2\cos\theta$

$2X^2 = 3/2\,X^2 - 3/2\,X^2\cos\theta$

$1/2\,X^2 = -3/2\,X^2\cos\theta$

$\cos\theta = -(1/2\,X^2)\,/\,(3/2\,X^2) = -1/3 = -0.3333$

$\theta = 109.47°$

9.78 (a) 40 e^-, 20 e^- pairs

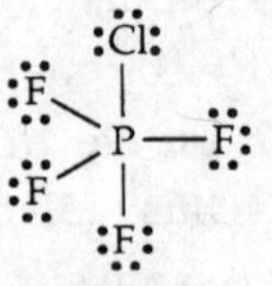

5 e^- domains

trigonal pyramidal electron domain geometry

(b) The greater the electronegativity of the terminal atom, the larger the negative charge centered on the atom, the greater the effective size of the P–X electron domain. A P–F bond will produce a larger electron domain than a P–Cl bond.

(c) The molecular geometry (shape) is also trigonal bipyramidal, because all five electron domains are bonding domains. Because we predicted the P–F electron domain to be larger, the maximum number of three P–F bonds will occupy the equatorial plane of the molecule, minimizing the number of 90° P–F to P–F repulsions. This is the same argument that places a "larger" nonbonding domain in the equatorial position of a molecule like SF_4. The P–Cl bond is then axial, as shown in the Lewis structure.

(d) The molecular geometry is distorted from a perfect trigonal bipyramid because not all electron domains are alike. The 90° P–F to P–F repulsions will be greater than the 90° P–F to P–Cl repulsions, so the F(axial)–P–F angles will be slightly greater than 90°, and the Cl(axial)–P–F angles will be slightly less than 90°. The equatorial F–P–F angles of 120° are distorted little, if at all.

9.79 (a) CO_2, 16 valence e^-

2σ 2π

(b) NCS^-, 16 valence e^-

+ two other resonance structures

2σ 2π (for any of the resonance structures)

(c) H_2CO, 12 valence e^-

3σ, 1π

(d) HCO(OH), 18 valence e^-

4σ, 1π

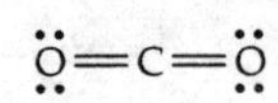

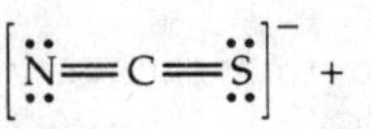

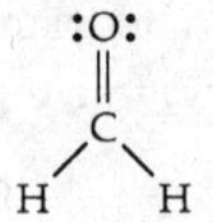

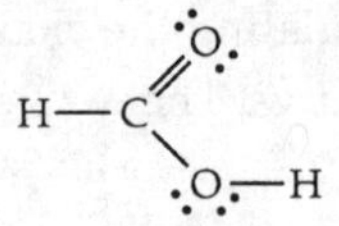

9.80 (a)

$3(4) + 3(6) + 6(1) = 36\ e^-$, $18\ e^-$ pr

(b) There are 11 σ and 1 π bonds.

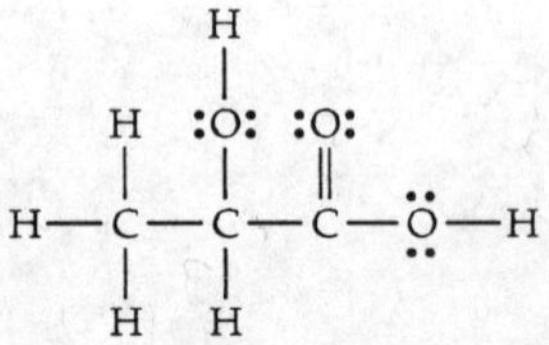

(c)   The C=O on the right-hand C atom is shortest. For the same bonded atoms, in this case C and O, the greater the bond order, the shorter the bond.

(d, e) The right-most C has three e^- domains, so the hybridization is sp^2; bond angles about this C atom are approximately 120°. The middle and left-hand C atoms both have four e^- domains, are sp^3 hybridized, and have bond angles of approximately 109°.

9.81

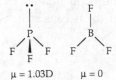

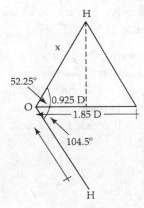

BF$_3$ is a trigonal planar molecule with the central B atom symmetrically surrounded by the three F atoms (Figure 9.13). The individual B–F bond dipoles cancel, and the molecule has a net dipole moment of zero. PF$_3$ has tetrahedral electron-domain geometry with one of the positions in the tetrahedron occupied by a nonbonding electron pair. The individual P–F bond dipoles do not cancel and the presence of a nonbonding electron pair ensures an asymmetrical electron distribution; the molecule is polar.

9.82   The compound on the right has a dipole moment. In the square planar trans structure on the left, all equivalent bond dipoles can be oriented opposite each other, for a net dipole moment of zero.

9.83

(a)   The bond dipoles in H$_2$O lie along the O–H bonds with the positive end at H and the negative end at O. The dipole moment vector of the H$_2$O molecule is the resultant (vector sum) of the two bond dipoles. This vector bisects the H–O–H angle and has a magnitude of 1.85 D with the negative end pointing toward O.

(b)   Since the dipole moment vector bisects the H–O–H bond angle, the angle between one H–O bond and the dipole moment vector is 1/2 the H–O–H bond angle, 52.25°. Dropping a perpendicular line from H to the dipole moment vector creates the right triangle pictured. If x = the magnitude of the O–H bond dipole, x cos (52.25) = 0.925 D. x = 1.51 D.

(c)   The X–H bond dipoles (Table 8.3) and the electronegativity values of X (Figure 8.7) are

	Electronegativity	Bond dipole
F	4.0	1.82
O	3.5	1.51
Cl	3.0	1.08

Since the electronegativity of O is midway between the values for F and Cl, the O–H bond dipole should be approximately midway between the bond dipoles of HF and HCl. The value of the O–H bond dipole calculated in part (b) is consistent with this prediction.

9.84   (a) XeF_6 50 e^-, 25 e^- pairs

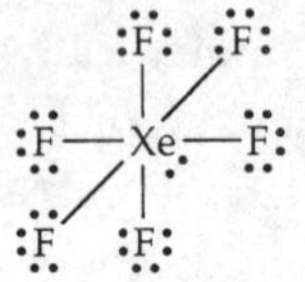

(b)   There are seven electron domains around Xe, and the maximum number of e^- domains in Table 9.3 is six.

(c)   Tie seven balloons together and see what arrangement they adopt (seriously! see Figure 9.5). Alternatively, go to the chemical literature where VSEPR was first proposed and see if there is a preferred orientation for seven e^- domains.

(d)   Since the hybrid orbitals for five e^- domains involve one d orbital and for six pairs, two d orbitals, a reasonable suggestion would be sp^3d^3.

(e)   One of the seven e^- domains is a nonbonded domain. The question is whether it occupies an axial or equatorial position. The equatorial plane of a pentagonal bipyramid has F–Xe–F angles of 72°. Placing the nonbonded domain in the equatorial plane would create severe repulsions between it and the adjacent bonded domains. Thus, the nonbonded domain will reside in the axial position. The molecular structure is a pentagonal pyramid.

9.85

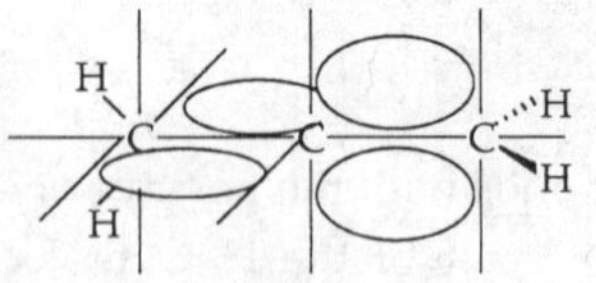

(a)   The molecule is not planar. The CH_2 planes at each end are twisted 90° from one another.

(b)   Allene has no dipole moment.

(c)   The bonding in allene would not be described as delocalized. The π electron clouds of the two adjacent C=C are mutually perpendicular. The mechanism for delocalization of π electrons is mutual overlap of parallel p atomic orbitals on adjacent atoms. If adjacent π electron clouds are mutually perpendicular, there is no overlap and no delocalization of π electrons.

9.86   (a) $16 \, e^-$, $8 \, e^-$ pairs

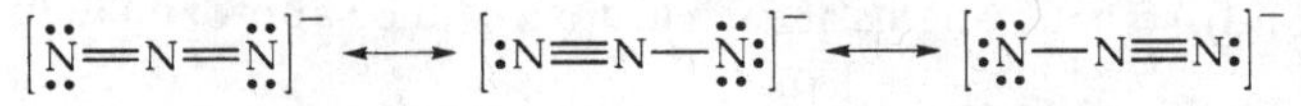

$$\left[\ddot{\text{N}} = \text{N} = \ddot{\text{N}} \right]^- \longleftrightarrow \left[:\text{N} \equiv \text{N} - \ddot{\ddot{\text{N}}}: \right]^- \longleftrightarrow \left[:\ddot{\ddot{\text{N}}} - \text{N} \equiv \text{N}: \right]^-$$

        (b)   The observed bond length of 1.16 Å is intermediate between the values for N=N, 1.24 Å, and N ≡ N, 1.10 Å. This is consistent with the resonance structures, which indicate contribution from formally double and triple bonds to the true bonding picture in N_3^-.

        (c)   In each resonance structure, the central N has two electron domains, so it must be sp hybridized. It is difficult to predict the hybridization of terminal atoms in molecules where there are resonance structures because there are a different number of electron domains around the terminal atoms in each structure. Since the "true" electronic arrangement is a combination of all resonance structures, we will assume that the terminal N–N bonds have some triple bond character and that the terminal N atoms are sp hybridized. (There is no experimental measure of hybridization at terminal atoms, since there are no bond angles to observe.)

        (d)   In each resonance structure, N–N σ bonds are formed by sp hybrids and π bonds are formed by unhybridized p orbitals. Nonbonding e^- pairs can reside in sp hybrids or p atomic orbitals.

        (e)   Recall that electrons in 2s orbitals are on the average closer to the nucleus than electrons in 2p orbitals. Since sp hybrids have greater s orbital character, it is reasonable to expect the radial extension of sp orbitals to be smaller than that of sp^2 or sp^3 orbitals and σ bonds formed by sp orbitals to be slightly shorter than those formed by other hybrid orbitals, assuming the same bonded atoms.

               There are no solitary σ bonds in N_3^-. That is, the two σ bonds in N_3^- are each accompanied by at least one π bond between the bonding pair of atoms. Sigma bonds that are part of a double or triple bond must be shorter so that the p orbitals can overlap enough for the π bond to form. Thus, the observation is not applicable to this molecule. (Comparison of C–H bond lengths in C_2H_2, C_2H_4, C_2H_6 and related molecules would confirm or deny the observation.)

9.87   (a) $\ddot{\text{O}} = \ddot{\text{O}} - \ddot{\ddot{\text{O}}}: \longleftrightarrow :\ddot{\ddot{\text{O}}} - \ddot{\text{O}} = \ddot{\text{O}}$

               To accommodate the π bonding by all 3 O atoms indicated in the resonance structures above, all O atoms are sp^2 hybridized.

        (b)   For the first resonance structure, both sigma bonds are formed by overlap of sp^2 hybrid orbitals, the π bond is formed by overlap of atomic p orbitals, one of the nonbonded pairs on the right terminal O atom is in a p atomic orbital, and the remaining five nonbonded pairs are in sp^2 hybrid orbitals.

        (c)   Only unhybridized p atomic orbitals can be used to form a delocalized π system.

        (d)   The unhybridized p orbital on each O atom is used to form the delocalized π system, and in both resonance structures one nonbonded electron pair resides in a p atomic orbital. The delocalized π system then contains four electrons, two from the π bond and two from the nonbonded pair in the p orbital.

9.88   (a)   Each C atom is surrounded by three electron domains (two single bonds and one double bond), so bond angles at each C atom will be approximately 120°.

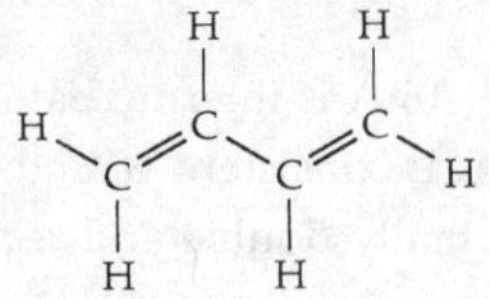

Since there is free rotation around the central C–C single bond, other conformations are possible.

(b)   According to Table 8.5, the average C–C length is 1.54 Å, and the average C=C length is 1.34 Å. While the C=C bonds in butadiene appear "normal," the central C–C is significantly shorter than average. Examination of the bonding in butadiene reveals that each C atom is sp^2 hybridized and the π bonds are formed by the remaining unhybridized 2p orbital on each atom. If the central C–C bond is rotated so that all four C atoms are coplanar, the four 2p orbitals are parallel, and some delocalization of the π electrons occurs.

9.89   (a)   The diagram shows two s atomic orbitals with opposite phases. Because they are spherically symmetric, the interaction of s orbitals can only produce a σ molecular orbital. Because the two orbitals in the diagram have opposite phases, the interaction excludes electron density from the region between the nuclei. The resulting MO has a node between the two nuclei. Relative to Figure 9.42, the MO is a σ_{2s}^*. The principal quantum number designation is arbitrary, because it defines only the size of the pertinent AOs and MOs. Shapes and phases of MOs depend only on these same characteristics of the interacting AOs.

(b)   The diagram shows two p atomic orbitals with oppositely phased lobes pointing at each other. End-to-end overlap produces a σ-type MO; opposite phases mean a node between the nuclei and an antibonding MO. The interaction results in a σ_{2p}^* MO.

(c)   The diagram shows parallel p atomic orbitals with like-phased lobes aligned. Side-to-side overlap produces a π-type MO; overlap of like-phased lobes concentrates electron density between the nuclei and a bonding MO. The interaction results in a π_{2p} MO.

9.90   (a) $C_5H_5^-$, 26 e⁻, 13 e⁻ pair

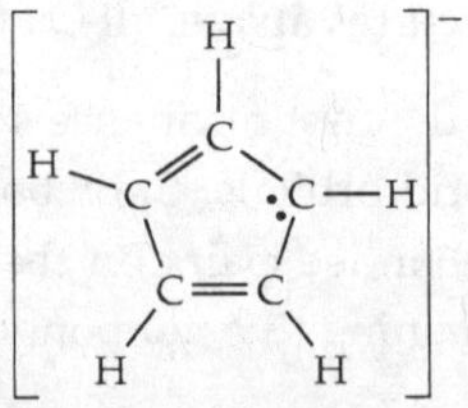

No. According to the single Lewis structure above, the four C atoms involved in double bonds would be sp^2 hybridized, but the C atom with the lone pair would be sp^3 hybridized. Not all C atoms would have the same hybridization.

(b)   Equivalent sp^2 hybridization at all C atoms is consistent with the planar structure of the ion. The VSEPR model applied to the single Lewis structure above does not predict uniform hybridization or planarity of the ion. The VSEPR/Lewis model is consistent with the known structural features of the ion only if the other resonance structures are considered.

(c)   If all C atoms are sp^2 hybridized, as required by the planar structure of the ion, the three hybrid orbitals on each C are used to form the σ framework of the molecule. The unshared pair would then reside in an unhybridized p orbital.

(d)   Yes, there are resonance structures equivalent to the Lewis structure above. It is possible to place the unshared pair on any of the five C atoms, resulting in five equivalent resonance structures (four plus the one above).

(e)   The five resonance structures indicate that there is delocalization of the π electron density in the molecule. The interior circle conveys uniform delocalization over the entire ring, as in benzene.

(f)   In a rigid planar structure with each C atom sp^2 hybridized, the unhybridized p orbitals are aligned parallel, facilitating overlap to form a delocalized π network above and below the plane of the molecule. Since the unshared pair occupies an unhybridized p orbital (part c), it is part of the delocalized π network, along with the four electrons involved in π bonds; the delocalized system contains a total of six electrons.

9.91 σ^*_{2p} ☐

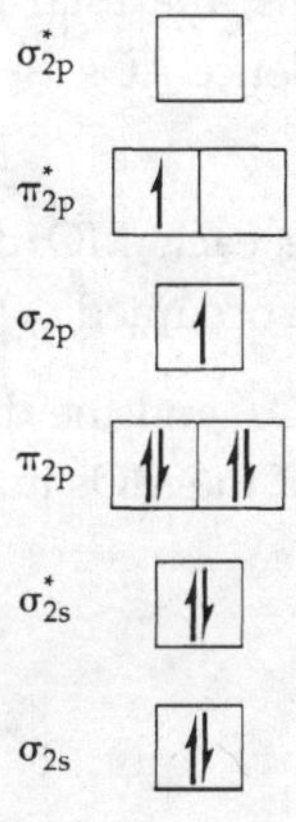

        π^*_{2p} [↑][ ] N_2 in the ground state has a B.O. of 3; in the first excited state (at left) it has a B.O. of 2. Owing to the reduction in bond order, N_2 in the first excited state would be more reactive, have a smaller bond energy and a longer N–N separation.

        σ_{2p} [↑]

        π_{2p} [↑↓][↑↓]

        σ^*_{2s} [↑↓]

        σ_{2s} [↑↓]

9.92   Paramagnetic materials appear to weigh more when the mass measurement is made in the presence of a magnetic field. Air contains $O_2(g)$, which is paramagnetic because of its two unpaired electrons in a σ^*_{2p} molecular orbital (Figure 9.45). In the presence of a magnetic field, mass measurements made in air would be skewed by the paramagnetism of O_2, and lead to inaccurate results.

9.93   We will refer to azo benzene (on the left) as A and hydrazobenzene (on the right) as H.

(a)   A: sp^2; H: sp^3

(b)   A: Each N and C atom has one unhybridized p orbital. H: Each C atom has one unhybridized p orbital, but the N atoms have no unhybridized p orbitals.

(c)   A: 120°; H: 109°

(d)   Since all C and N atoms in A have unhybridized p orbitals, all can participate in delocalized π bonding. The delocalized π system extends over the entire molecule, including both benzene rings and the azo "bridge." In H, the N atoms have no unhybridized p orbitals, so they cannot participate in delocalized π bonding. Each of the benzene rings in H is delocalized, but the network cannot span the N atoms in the bridge.

(e)   This is consistent with the answer to (d). In order for the unhybridized p orbitals in A to overlap, they must be parallel. This requires a planar σ bond framework where all atoms in the molecule are coplanar.

(f)   For a substance to appear colored, it must absorb light in the visible region of the electromagnetic spectrum. This requires a HOMO-LUMO energy gap in the visible region. For organic molecules, the size of the gap is related to the number of conjugated π bonds; the more conjugated π bonds, the smaller the gap and the more likely the molecule is to be colored. Azobenzene has seven conjugated π bonds (π network delocalized over entire molecule) while hydrazobenzene has only three (π network on benzene rings only). Thus, the size of the HOMO-LUMO energy gap is smaller in A, and A is more likely than H to appear colored.

9.94   (a)   H: $1s^1$; F: $[He]2s^2 2p^5$

When molecular orbitals are formed from atomic orbitals, the total number of orbitals is conserved. Since H and F have a total of five valence AOs ($H_{1s} + F_{2s} + 3F_{2p}$), the MO diagram for HF has five MOs.

(b)   H and F have a total of eight valence electrons. Since each MO can hold a maximum of two electrons, four of the five MOs would be occupied.

(c)   No. The MO diagram for NO in Figure 9.48 has eight MOs, and the diagram for HF has only five. The number, type, and energy spacing of the MOs in HF will be different.

9.95   (a)   CO, 10 e⁻, 5 e⁻ pair

:C≡O:

(b)   The bond order for CO, as predicted by the MO diagram in Figure 9.48, is 1/2[8 – 2] = 3.0. A bond order of 3.0 agrees with the triple bond in the Lewis structure.

(c)   Applying the MO diagram in Figure 9.48 to the CO molecule, the highest energy electrons would occupy the π_{2p} MOs. That is, π_{2p} would be the HOMO, highest occupied molecular orbital. If the true HOMO of CO is a σ-type MO, the order of the π_{2p} and σ_{2p} orbitals must be reversed. Figure 9.44 shows how the interaction of the 2s orbitals on one atom and the 2p orbitals on the other atom can affect the relative energies of the resulting MOs. This 2s-2p interaction in CO is significant enough so that the σ_{2p} MO is higher in energy than the π_{2p} MOs, and the σ_{2p} is the HOMO.

(d)   We expect the atomic orbitals of the more electronegative element to have lower energy than those of the less electronegative element. When atoms of the two elements combine, the lower energy atomic orbitals make a greater contribution to the bonding MOs and the higher energy atomic orbitals make a larger contribution to the antibonding orbitals. Thus, the π_{2p} bonding MOs will have a greater contribution from the more electronegative O atom.

Integrative Exercises

9.96   (a)   Assume 100 g of compound

$$2.1\,g\,H \times \frac{1\,mol\,H}{1.008\,g\,H} = 2.1\,mol\,H;\ 2.1/2.1 = 1$$

$$29.8\,g\,N \times \frac{1\,mol\,N}{14.01\,g\,N} = 2.13\,mol\,N;\ 2.13/2.1 \approx 1$$

$$68.1\,g\,O \times \frac{1\,mol\,O}{16.00\,g\,O} = 4.26\,mol\,O;\ 4.26/2.1 \approx 2$$

The empirical formula is HNO_2; formula weight = 47. Since the approximate molar mass is 50, the molecular formula is HNO_2.

(b)   Assume N is central, since it is unusual for O to be central, and part (d) indicates as much. HNO_2: 18 valence e^-

$$\ddot{O}{=}\ddot{N}{-}\ddot{O}{-}H \longleftrightarrow :\ddot{O}{-}\ddot{N}{=}\ddot{O}{-}H$$
$$ {-1} \quad 0 \quad {+1}$$

The second resonance form is a minor contributor due to unfavorable formal charges.

(c)   The electron domain geometry around N is trigonal planar with an O–N–O angle of approximately 120°. If the resonance structure on the right makes a significant contribution to the molecular structure, all four atoms would lie in a plane. If only the left structure contributes, the H could rotate in and out of the molecular plane. The relative contributions of the two resonance structures could be determined by measuring the O–N–O and N–O–H bond angles.

(d)   3 VSEPR e^- domains around N, sp^2 hybridization

(e)   3 σ, 1 π for both structures (or for H bound to N).

9.97   (a) $2SF_4(g) + O_2(g) \rightarrow 2OSF_4(g)$

(b)   40 e^-, 20 e^- pairs

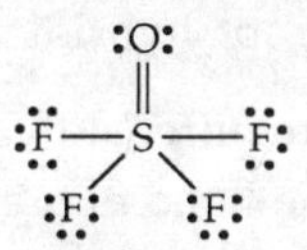

$$:\!O:$$
$$\|$$
$$:\!F{-}S{-}F\!:$$
$$:\!F:\quad :\!F:$$

There must be a double bond drawn between O and S in order for their formal charges to be zero.

(c) $\Delta H = 8D(S{-}F) + D(O{=}O) - 8D(S{-}F) - 2D(S{=}O)$

$\Delta H = D(O{=}O) - 2D(S{=}O) = 495 - 2(523) = -551$ kJ, exothermic

(d) trigonal bipyramidal electron-domain geometry

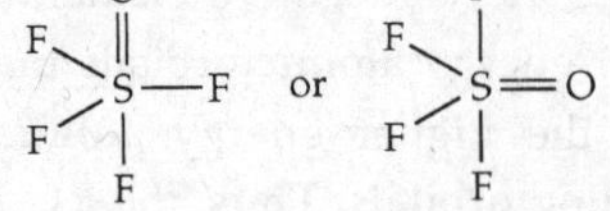

(e) Because F is more electronegative than O, the structure that minimizes S–F repulsions is more likely (see Solution 9.98). That is, the structure with fewer 90° F–S–F angles and more 120°F–S–F angles is favored. The structure on the left, with O in the axial position is more likely. Note that a double bond involving an atom with an expanded octet of electrons, such as the S=O in this molecule, does not have the same geometric implications as a double bond to a first row element.

9.98 (a) PX_3, 26 valence e$^-$, 13$^-$ pairs

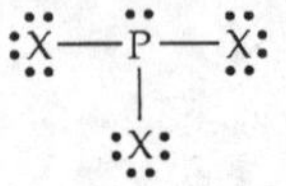

4 electron domains around P, tetrahedral e$^-$ domain geometry, 107° bond angles (less than 109° due to repulsion with nonbonded pair)

(b) As electronegativity increases (I < Br < Cl < F), the X–P–X angles decreases.

(c) For P–I, ΔEN is (2.5 – 2.1) = 0.4 and the bond dipole is small. For P–F, ΔEN is (4.0 – 2.1) = 1.9 and the bond dipole is large. The greater the ΔEN and bond dipole, the larger the magnitude of negative charge centered on X. The more negative charge centered on X, the greater the repulsion between X and the nonbonding electron pair on P and the smaller the bond angle. (Since electrostatic attractions and repulsions vary as $1/r$, the shorter P–F bond may also contribute to this effect.)

(d) $PBrCl_4$, 40 valence electrons, 20 e$^-$ pairs. The molecule will have trigonal bipyramidal electron-domain geometry (similar to PCl_5 in Table 9.3.) Based on the argument in part (c), the P–Br bond will have smaller repulsions with P–Cl bonds than P–Cl bonds have with each other. Therefore, the Br will occupy an axial position in the trigonal bipyramid, so that the more unfavorable P–Cl to P–Cl repulsions can be situated at larger angles in the equatorial plane.

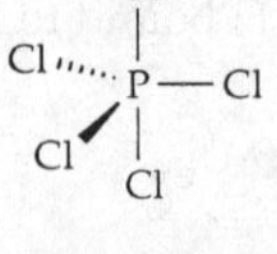

9.99 (a) Three electron domains around each central C atom, sp^2 hybridization

(b) A 180° rotation around the C=C double bond is required to convert the trans isomer into the cis isomer. A 90° rotation around the bond eliminates all overlap of the p orbitals that form the π bond and it is broken.

(c) **average bond enthalpy**

C=C   614 kJ/mol

C–C   348 kJ/mol

The difference in these values, 266 kJ/mol, is the average bond enthalpy of a C–C π bond. This is the amount of energy required to break 1 mol of C–C π bonds. The energy per molecule is

$$266 \text{ kJ/mol} \times \frac{1000 \text{ J}}{1 \text{ kJ}} \times \frac{1 \text{ mol}}{6.022 \times 10^{23} \text{ molecules}} = 4.417 \times 10^{-19}$$

$$= 4.42 \times 10^{-19} \text{ J/molecule}$$

(d) $\lambda = hc/E = \dfrac{6.626 \times 10^{-34} \text{ J} \cdot \text{s} \times 2.998 \times 10^{8} \text{ m/s}}{4.417 \times 10^{-19} \text{ J}} = 4.50 \times 10^{-7} \text{ m} = 450 \text{ nm}$

(e) Yes, 450 nm light is in the visible portion of the spectrum. A cis-trans isomerization in the retinal portion of the large molecule rhodopsin is the first step in a sequence of molecular transformations in the eye that leads to vision. The sequence of events enables the eye to detect visible photons, in other words, to see.

9.100 (a) C ≡ C   839 kJ/mol   (1 σ, 2 π)

  C=C   614 kJ/mol   (1 σ, 1 π)

  C–C   348 kJ/mol   (1 σ)

The contribution from 1 π bond would be (614–348) 266 kJ/mol. From a second π bond, (839 – 614), 225 kJ/mol. An average π bond contribution would be (266 + 225)/2 = 246 kJ/mol.

This is $\dfrac{246 \text{ kJ/π bond}}{348 \text{ kJ/σ bond}} \times 100 = 71\%$ of the average enthalpy of a σ bond.

(b) N ≡ N   941 kJ/mol

  N=N   418 kJ/mol

  N–N   163 kJ/mol

first π = (418 – 163) = 255 kJ/mol

second π = (941 – 418) = 523 kJ/mol

average π bond enthalpy = (255 + 523)/2 = 389 kJ/mol

This is $\dfrac{389 \text{ kJ/π bond}}{163 \text{ kJ/σ bond}} \times 100 = 240\%$ of the average enthalpy of a σ bond.

N–N σ bonds are weaker than C–C σ bonds, while N–N π bonds are stronger than C–C π bonds. The relative energies of C–C σ and π bonds are similar, while N–N π bonds are much stronger than N–N σ bonds.

(c) N_2H_4, 14 valence e⁻, 7 e⁻ pairs

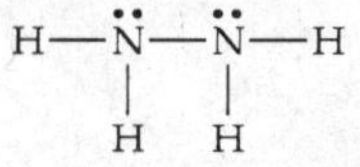

H—N̈—N̈—H
  | |
  H   H

4 electron domains around N, sp³ hybridization

N_2H_2, 12 valence e^-, 6 e^- pairs

$$H-\ddot{N}=\ddot{N}-H$$

3 electron domains around N, sp^2 hybridization

N_2, 10 valence e^-, 5 e^- pairs

$$:N\equiv N:$$

2 electron domains around N, sp hybridization

(d)   In the three types of N–N bonds, each N atom has a nonbonding or lone pair of electrons. The lone pair to bond pair repulsions are minimized going from 109° to 120° to 180° bond angles, making the π bonds stronger relative to the σ bond. In the three types of C–C bonds, no lone-pair to bond-pair repulsions exist, and the σ and π bonds have more similar energies.

9.101

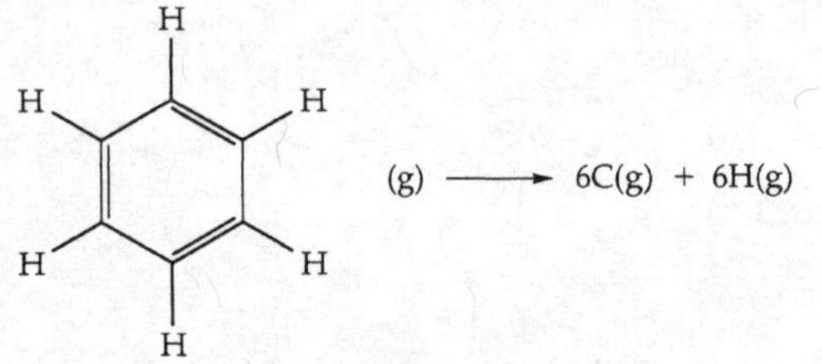

$$(g) \longrightarrow 6C(g) + 6H(g)$$

$\Delta H = 6D(C–H) + 3D(C–C) + 3D(C=C) - 0$
     $= 6(413 \text{ kJ}) + 3(348 \text{ kJ}) + 3(614 \text{ kJ})$
     $= 5364 \text{ kJ}$

(The products are isolated atoms; there is no bond making.)

According to Hess' law:

$\Delta H^{\circ} = 6\Delta H_f^{\circ} \, C(g) + 6\Delta H_f^{\circ} \, H(g) - \Delta H_f^{\circ} \, C_6H_6(g)$
     $= 6(718.4 \text{ kJ}) + 6(217.94 \text{ kJ}) - (82.9 \text{ kJ})$
     $= 5535 \text{ kJ}$

The difference in the two results, 171 kJ/mol C_6H_6 is due to the resonance stabilization in benzene. That is, because the π electrons are delocalized, the molecule has a lower overall energy than that predicted for the presence of 3 localized C–C and C=C bonds. Thus, the amount of energy actually required to decompose 1 mole of $C_6H_6(g)$, represented by the Hess' law calculation, is greater than the sum of the localized bond enthalpies (not taking resonance into account) from the first calculation above.

9.102   (a)   1 eV = 96.485 kJ/mol

$$H_2: 15.4 \, eV \times \frac{96.485 \text{ kJ/mol}}{1 \, eV} = 1486 = 1.49 \times 10^3 \text{ kJ/mol}$$

$$N_2: 15.6 \, eV \times \frac{96.485 \text{ kJ/mol}}{1 \, eV} = 1505 = 1.51 \times 10^3 \text{ kJ/mol}$$

$$O_2: 12.1 \, eV \times \frac{96.485 \text{ kJ/mol}}{1 \, eV} = 1167 = 1.17 \times 10^3 \text{ kJ/mol}$$

$$F_2: 15.7 \, eV \times \frac{96.485 \text{ kJ/mol}}{1 \, eV} = 1515 = 1.52 \times 10^3 \text{ kJ/mol}$$

(b)

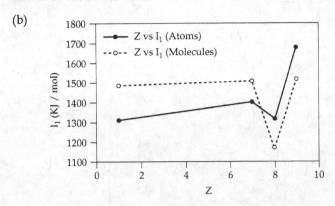

(c)   In general, I_1 for atoms and molecules increases going across a row of the periodic chart. In both cases, there is a discontinuity at oxygen. The details of the trends are different. The deviation at O is larger for the molecules than the atoms, while the increase at F is much greater for the atoms than the molecules.

(d)   According to Figures 9.35 and 9.45, H_2, N_2, and F_2 are diamagnetic and O_2 is paramagnetic. That is, ionization in H_2, N_2, and F_2 has to overcome spin-pairing energy, while ionization of O_2 removes an already unpaired electron. Thus, the ionization energy of O_2 is much less than I_1 for H_2, N_2, and F_2.

Despite differences in bond order, bond length, and the bonding or antibonding nature of the HOMO in H_2, N_2, and F_2, the ionization energies for these molecules are very similar.

9.103   (a) $3d_{z^2}$

        (b)   Ignoring the donut of the d_{z^2} orbital

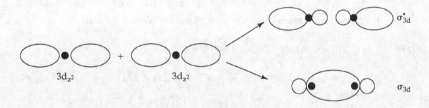

        (c)   A node is generated in σ_{3d}^* because antibonding MOs are formed when AO lobes with opposite phases interact. Electron density is excluded from the internuclear region and a node is formed in the MO.

        (d)   Sc: $[Ar]4s^2 3d^1$ Omitting the core electrons, there are six e^- in the energy level diagram.

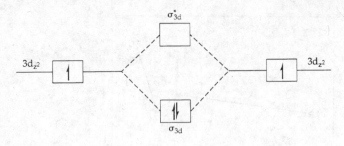

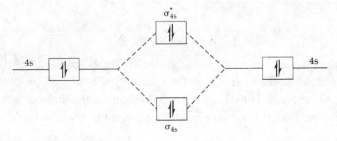

(e)   The bond order in Sc_2 is $1/2 (4 - 2) = 1.0$.

9.104   (a)   The molecular and empirical formulas of the four molecules are:

benzene: molecular, C_6H_6; empirical, CH

napthalene: molecular, $C_{10}H_8$; empirical, C_5H_4

anthracene: molecular, $C_{14}H_{10}$; empirical, C_7H_5

tetracene: molecular, $C_{18}H_{12}$, empirical, C_3H_2

(b)   Yes.   Since the compounds all have different empirical formulas, combustion analysis could in principle be used to distinguish them. In practice, the mass % of C in the four compounds is not very different, so the data would have to be precise to at least 3 decimal places and 4 would be better.

(c) $C_{10}H_8(s) + 12O_2(g) \rightarrow 10CO_2(g) + 4H_2O(l)$

(d) $\Delta H_{comb} = 5D(C=C) + 5D(C-C) + 8D(C-H) + 12D(O=O) - 20D(C=O) - 8D(O-H)$

$= 5(614) + 5(348) + 8(413) + 12(495) - 20(799) - 8(463)$

$= -5630 \text{ kJ/mol } C_{10}H_8$

(e)   Yes.

(f)   For molecules with delocalized π systems, the number of conjugated double bonds is related to the HOMO-LUMO energy gap in the MO diagram. The more conjugated double bonds, the smaller the energy gap. Large energy gaps require higher energy, shorter wavelength radiation to excite electrons from the HOMO to the LUMO. Since tetracene has the most conjugated double bonds of the four molecules (9), it has the smallest HOMO-LUMO gap and can absorb visible light, which causes it to appear colored. The other molecules absorb wavelengths shorter than the visible and appear colorless.

9.105 (a) Parentheses around the atomic weight indicate that the element is radioactive. The nuclei of radioactive elements spontaneously decay to form other, lighter nuclei. The difficulty with studying At is that it decays before a reliable set of physical and chemical properties can be measured.

(b) $1s^2 2s^2 2p^6 3s^2 3p^6 4s^2 3d^{10} 4p^6 5s^2 4d^{10} 5p^6 6s^2 4f^{14} 5d^{10} 6p^5$

(c) AtI is expected to have a slightly polar covalent bond, similar to other interhalogen molecules. Since At and I are adjacent in the halogen family, their electronegativities will be very similar, but not equal. When two atoms with similar electronegativies combine, the result is a covalent molecule with a small net dipole moment.

(d) $22\ e^-$, $11\ e^-$ pairs;

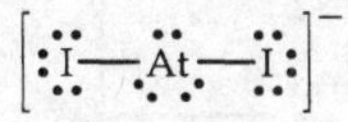

$$\left[\ddot{\underset{\cdot\cdot}{\ddot{I}}} - \ddot{\underset{\cdot\cdot}{At}} - \ddot{\underset{\cdot\cdot}{\ddot{I}}}\ \right]^-$$

trigonal bipyramidal electron-domain geometry

linear molecular geometry

(In a trigonal bipyramid, placing nonbonding electron pairs in the equatorial plane minimizes repulsions.)

(e) By analogy to F_2 in Figure 9.45, the bond order should be 1 and the HOMO is π_{6p}^*.

10 Gases

Visualizing Concepts

10.1   (a) $V_1/T_1 = V_2/T_2$ (Charles' Law)   (b) $P_1V_1 = P_2V_2$ (Boyle's Law)

$V_1/300\ K = V_2/500\ K$ $1\ atm \times V_1 = 2\ atm \times V_2$

$V_2 = 5/3\ V_1$ $V_2 = 1/2\ V_1$

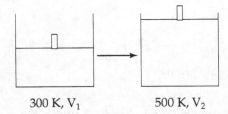
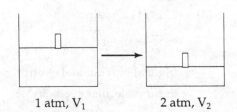

300 K, V_1 500 K, V_2 1 atm, V_1 2 atm, V_2

10.2   At constant temperature and volume, pressure depends on total number of particles (Charles' Law). In order to reduce the pressure by a factor of 2, the number of particles must be reduced by a factor of 2. At the lower pressure, the container would have half as many particles as at the higher pressure.

10.3.   (a)   At constant pressure and temperature, the container volume is directly proportional to the number of particles present (Avogadro's Law). As the reaction proceeds, 3 gas molecules are converted to 2 gas molecules, so the container volume decreases. If the reaction goes to completion, the final volume would be 2/3 of the initial volume.

          (b)   At constant volume, pressure is directly proportional to the number of particles (Charles' Law). Since the number of molecules decreases as the reaction proceeds, the pressure also decreases. At completion, the final pressure would be 2/3 the initial pressure.

10.4   Over time, the gases will mix perfectly. Each bulb will contain 4 blue and 3 red atoms. The "blue" gas has the greater partial pressure after mixing, because it has the greater number of particles (at the same T and V as the "red" gas.)

10.5   (a)   Partial pressure depends on the number of particles of each gas present. Red has the fewest particles, then yellow, then blue. $P_{red} < P_{yellow} < P_{blue}$

          (b) $P_{gas} = \chi_{gas}\ P_t$. Calculate the mole fraction, χ_{gas} = [mol gas / total moles] or [particles gas / total particles]. This is true because Avogadro's number is a counting number, and mole ratios are also particle ratios.

250

χ_{red} = 2 red atoms / 10 total atoms = 0.2; P_{red} = 0.2(0.90 atm) = 0.18 atm

χ_{yellow} = 3 yellow atoms / 10 total atoms = 0.3; P_{yellow} = 0.3(0.90 atm) = 0.27 atm

χ_{blue} = 5 blue atoms / 10 total atoms = 0.5; P_{blue} = 0.5(0.90 atm) = 0.45 atm

10.6

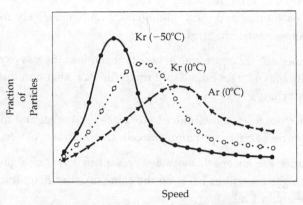

10.7 (a) At constant temperature, the "average" (really root mean square) speed of a collection of gas particles is inversely related to molar mass; the lighter the particle, the faster it moves. Therefore, curve B represents He and curve A represents O_2. Curve B has the higher avg. molecular speed and He is the lighter gas. Curve A has the lower avg. molecular speed and O_2 is the heavier gas.

(b) For the same gas "avg" kinetic energy $(1/2 \, mv^2)$, and therefore "avg" speed (v) is directly related to Kelvin temperature. Curve A is the lower temperature and curve B is the higher temperature.

10.8 (a) Total pressure is directly related to total number of particles (or total mol particles). P(ii) < P(i) = P(iii)

(b) Partial pressure of He is directly related to number of He atoms (yellow) or mol He atoms. P_{He}(iii) < P_{He}(ii) < P_{He}(i)

(c) Density is total mass of gas per unit volume. We can use the atomic or molar masses of He (4) and N_2(28), as relative masses of the particles.

$$mass(i) = 5(4) + 2(28) = 76$$

$$mass(ii) = 3(4) + 1(28) = 40$$

$$mass(iii) = 2(4) + 5(28) = 148$$

Since the container volumes are equal, d(ii) < d(i) < d(iii).

(d) At the same temperature, all gases have the same "avg" kinetic energy. The average kinetic energies of the particles in the three containers are equal.

Gas Characteristics; Pressure

10.9 In the gas phase molecules are far apart, while in the liquid they are touching.

(a) A gas is much less dense than a liquid because most of the volume of a gas is empty space.

(b)   A gas is much more compressible because of the distance between molecules.

(c)   Gaseous molecules are so far apart that there is no barrier to mixing, regardless of the identity of the molecule. All mixtures of gases are homogeneous. Liquid molecules are touching. In order to mix, they must displace one another. Similar molecules displace each other and form homogeneous mixtures. Very dissimilar molecules form heterogeneous mixtures.

10.10   (a)   Because gas molecules are far apart and in constant motion, the gas expands to fill the container. Attractive forces hold liquid molecules together and the volume of the liquid does not change.

(b) H_2O and CCl_4 molecules are too dissimilar to displace each other and mix in the liquid state. All mixtures of gases are homogeneous. (See Solution 10.9 (c)).

(c)   Because gas molecules are far apart, the mass present in 1 mL of a gas is very small. The mass of a gas present in 1 L is on the same order of magnitude as the mass of a liquid present in 1 mL.

10.11   *Analyze.* Given: mass, area. Find: pressure. *Plan.* P=F/A = m × a/A; use this relationship, paying attention to units. *Solve.*

$$1\,Pa = \frac{1\,N}{m^2} = \frac{1\,kg \bullet m}{s^2} \times \frac{1}{m^2} = \frac{1\,kg}{m \bullet s^2} \quad \text{Change mass to kg and area to } m^2.$$

$$P = \frac{m \times a}{A} = \frac{130\,lb}{0.50\,in^2} \times \frac{9.81\,m}{1\,s^2} \times \frac{0.454\,kg}{1\,lb} \times \frac{39.4^2\,in^2}{1\,m^2} = 1.8 \times 10^6 \frac{kg}{m \bullet s^2}$$
$$= 1.8 \times 10^6\,Pa = 1.8 \times 10^3\,kPa$$

Check. $[1.30 \times 10 \times 0.5 \times (40)^2/0.5] \approx (130 \times 16{,}000) \approx 2.0 \times 10^6\,Pa \approx 2.0 \times 10^3\,kPa$. The units are correct.

10.12 $P = m \times a/A$; $1\,Pa = 1\,kg/m \bullet s^2$; $A = 3.0\,cm \times 4.1\,cm \times 4 = 49.2 = 49\,cm^2$

$$\frac{262\,kg}{49.2\,cm^2} \times \frac{9.81\,m}{s^2} \times \frac{(100)^2\,cm^2}{1\,m^2} = 5.224 \times 10^5 \frac{kg}{m \bullet s^2} = 5.2 \times 10^5\,Pa$$

10.13   *Analyze.* Given: 760 mm column of Hg, densities of Hg and H_2O. Find: height of a column of H_2O at same pressure.

Plan. We must develop a relationship between pressure, height of a column of liquid, and density of the liquid. Relationships that might prove useful: P = F/A; F = m × a; m = d × V(density)(volume); V = A × height   *Solve.*

$$P = \frac{F}{A} = \frac{m \times a}{A} = \frac{d \times V \times a}{A} = \frac{d \times A \times h \times a}{A} = d \times h \times a$$

(a) $P_{Hg} = P_{H_2O}$; Using the relationship derived above: $(d \times h \times a)_{H_2O} = (d \times h \times a)_{Hg}$

Since a, the acceleration due to gravity, is equal in both liquids,
$(d \times h)_{H_2O} = (d \times h)_{Hg}$

$1.00\,g/mL \times h_{H_2O} = 13.6\,g/mL \times 760\,mm$

$$h_{H_2O} = \frac{13.6\,g/mL \times 760\,mm}{1.00\,g/mL} = 1.034 \times 10^4 = 1.03 \times 10^4\,mm = 10.3\,m$$

(b)   Pressure due to H_2O:

1 atm = 1.034×10^4 mm H_2O (from part (a))

$$36 \text{ ft } H_2O \times \frac{12 \text{ in}}{1 \text{ ft}} \times \frac{2.54 \text{ cm}}{1 \text{ in}} \times \frac{10 \text{ mm}}{1 \text{ cm}} \times \frac{1 \text{ atm}}{1.034 \times 10^4 \text{ mm}} = 1.061 = 1.1 \text{ atm}$$

$$P_{total} = P_{atm} + P_{H_2O} = 0.95 \text{ atm} + 1.061 \text{ atm} = 2.011 = 2.0 \text{ atm}$$

10.14   Using the relationship derived in Solution 10.13 for two liquids under the influence of gravity, $(d \times h)_{1id} = (d \times h)_{Hg}$. At 752 torr, the height of an Hg barometer is 752 mm.

$$\frac{1.20 \text{ g}}{1 \text{ mL}} \times h_{1id} = \frac{13.6 \text{ g}}{1 \text{ mL}} \times 760 \text{ mm}; \ h_{1id} = \frac{13.6 \text{ g/mL} \times 752 \text{ mm}}{1.20 \text{ g/mL}} = 8.52 \times 10^3 \text{ mm} = 8.52 \text{ m}$$

10.15   (a)   The tube can have any cross-sectional area. (The height of the Hg column in a barometer is independent of the cross-sectional area. See the expression for pressure derived in Solution 10.13.)

         (b)   At equilibrium, the force of gravity per unit area acting on the mercury column at the level of the outside mercury is not equal to the force of gravity per unit area acting on the atmosphere. (F = ma; the acceleration due to gravity is equal for the two substances, but the mass of Hg for a given cross-sectional area is different than the mass of air for this same area.)

         (c)   The column of mercury is held up by the pressure of the atmosphere applied to the exterior pool of mercury.

10.16   The mercury would fill the tube completely; there would be no vacuum at the closed end. This is because atmospheric pressure will support a mercury column higher than 70 cm, while our tube is only 50 cm. No mercury flows from the tube into the dish and no vacuum forms at the top of the tube.

10.17   *Analyze/Plan.* Follow the logic in Sample Exercise 10.1.   *Solve.*

         (a) $265 \text{ torr} \times \dfrac{1 \text{ atm}}{760 \text{ torr}} = 0.349 \text{ atm}$

         (b) $265 \text{ torr} \times \dfrac{1 \text{ mm Hg}}{1 \text{ torr}} = 265 \text{ mm Hg}$

         (c) $265 \text{ torr} \times \dfrac{1.01325 \times 10^5 \text{ Pa}}{760 \text{ torr}} = 3.53 \times 10^4 \text{ Pa}$

         (d) $265 \text{ torr} \times \dfrac{1.01325 \times 10^5 \text{ Pa}}{760 \text{ torr}} \times \dfrac{1 \text{ bar}}{1 \times 10^5 \text{ Pa}} = 0.353 \text{ bar}$

10.18   (a) $0.850 \text{ atm} \times \dfrac{760 \text{ torr}}{1 \text{ atm}} = 646 \text{ torr}$

         (b) $785 \text{ torr} \times \dfrac{101.325 \text{ kPa}}{760 \text{ torr}} = 105 \text{ kPa}$

         (c) $655 \text{ mm Hg} \times \dfrac{1 \text{ atm}}{760 \text{ mm Hg}} = 0.862 \text{ atm}$

(d) $1.323 \times 10^5 \text{ Pa} \times \dfrac{1 \text{ atm}}{1.01325 \times 10^5 \text{ Pa}} = 1.3057 = 1.306 \text{ atm}$

(e) $2.50 \text{ atm} \times \dfrac{1.01325 \times 10^5 \text{ Pa}}{1 \text{ atm}} \times \dfrac{1 \text{ bar}}{1 \times 10^5 \text{ Pa}} = 2.53 \text{ bar}$

10.19   *Analyze/Plan.* Follow the logic in Sample Exercise 10.1.   *Solve.*

(a) $30.45 \text{ in Hg} \times \dfrac{25.4 \text{ mm}}{1 \text{ in}} \times \dfrac{1 \text{ torr}}{1 \text{ mm Hg}} = 773.4 \text{ torr}$

[The result has 4 sig figs because 25.4 mm/in is considered to be an exact number. (Section 1.5)]

(b)   The pressure in Chicago is greater than **standard atmospheric pressure**, 760 torr, so it makes sense to classify this weather system as a "high pressure system."

10.20   (a) $\dfrac{1.63105 \text{ Pa}}{1 \text{ Titan atm}} \times \dfrac{1 \text{ Earth atm}}{101{,}325 \text{ Pa}} = \dfrac{1.60972 \times 10^{-5} \text{ Earth atm}}{1 \text{ Titan atm}}$

(b) $\dfrac{90 \text{ Earth atm}}{1 \text{ Venus atm}} \times \dfrac{101.3 \text{ kPa}}{1 \text{ Earth atm}} = \dfrac{9.1 \times 10^3 \text{ kPa}}{1 \text{ Venus atm}}$

10.21   *Analyze/Plan.* Follow the logic in Sample Exercise 10.2.   *Solve.*

(i)   The Hg level is lower in the open end than the closed end, so the gas pressure is less than atmospheric pressure.

$P_{gas} = 0.985 \text{ atm} - \left(52 \text{ cm} \times \dfrac{1 \text{ atm}}{76 \text{ cm}} \right) = 0.30 \text{ atm}$

(ii)   The Hg level is higher in the open end, so the gas pressure is greater than atmospheric pressure.

$P_{gas} = 0.985 \text{ atm} + \left(67 \text{ mm Hg} \times \dfrac{1 \text{ atm}}{760 \text{ mm Hg}} \right) = 1.073 \text{ atm}$

(iii)   This is a closed-end manometer, so $P_{gas} = h$.

$P_{gas} = 10.3 \text{ cm} \times \dfrac{1 \text{ atm}}{76 \text{ cm}} = 0.136 \text{ atm}$

10.22   (a)   The atmosphere is exerting 10.7 cm = 107 mm Hg (torr) more pressure than the gas.

$P_{gas} = P_{atm} - 107 \text{ torr} = \left(0.977 \text{ atm} \times \dfrac{760 \text{ torr}}{1 \text{ atm}} \right) - 107 \text{ torr} = 636 \text{ torr}$

(b)   The gas is exerting 9.5 mm Hg (torr) more pressure than the atmosphere.

$P_{gas} = P_{atm} + 9.5 \text{ torr} = \left(1.02 \text{ atm} \times \dfrac{760 \text{ torr}}{1 \text{ atm}} \right) + 9.5 \text{ torr} = 785 \text{ torr}$

The Gas Laws

10.23   *Analyze/Plan.* Given certain changes in gas conditions, predict the effect on other conditions. Consider the gas law relationships in Section 10.3.   *Solve.*

(a)   P and V are inversely proportional at constant T. If the volume decreases by a factor of 4, the pressure increases by a factor of 4.

(b)   P and T are directly proportional at constant V. If T decreases by a factor of 2, P also decreases by a factor of 2.

(c)   P and n are directly proportional at constant V and T. If n decreases by a factor of 2, P also decreases by a factor of 2.

10.24   *Analyze.* Given: initial P, V, T. Find: final values of P, V, T for certain changes of condition. *Plan.* Select the appropriate gas law relationships from Section 10.3; solve for final conditions, paying attention to units.   *Solve.*

(a) $P_1V_1 = P_2V_2$; the proportionality holds true for any pressure or volume units.

$P_1 = 735$ torr, $V_1 = 5.22$ L, $P_2 = 1.88$ atm

$$V_2 = \frac{P_1V_1}{P_2} = \frac{735 \text{ torr} \times 5.22 \text{ L}}{1.88 \text{ atm}} \times \frac{1 \text{ atm}}{760 \text{ torr}} = 2.69 \text{ L}$$

Check. As pressure increases, volume should decrease; our result agrees with this.

(b) $V_1/T_1 = V_2/T_2$; T must be in Kelvins for the relationship to be true.

$V_1 = 5.22$ L, $T_1 = 23°C = 296$ K, $T_2 = 165°C = 438$ K

$$V_2 = \frac{V_1T_2}{T_1} = \frac{5.22 \text{ L} \times 438 \text{ K}}{296 \text{ K}} = 7.72 \text{ L}$$

Check. As temperature increases, volume should increase; our result is consistent with this.

10.25   (a)   Avogadro's hypothesis states that equal volumes of gases at the same temperature and pressure contain equal numbers of molecules. Since molecules react in the ratios of small whole numbers, it follows that the volumes of reacting gases (at the same temperature and pressure) are in the ratios of small whole numbers.

(b)   Since the two gases are at the same temperature and pressure, the ratio of the numbers of atoms is the same as the ratio of volumes. There are 1.5 times as many Xe atoms as Ne atoms.

10.26   According to Avogadro's hypothesis, the mole ratios in the chemical equation will be volume ratios for the gases if they are at the same temperature and pressure.

$$N_2(g) + 3H_2(g) \rightarrow 2NH_3(g)$$

The volumes of H_2 and N_2 are in a stoichiometric $\dfrac{3.6 \text{ L}}{1.2 \text{ L}}$ or $\dfrac{3 \text{ vol } H_2}{1 \text{ vol } N_2}$ ratio, so either can be used to determine the volume of $NH_3(g)$ produced.

$$1.2 \text{ L } N_2 \times \frac{2 \text{ mol } NH_3}{1 \text{ mol } N_2} = 2.4 \text{ L } NH_3(g) \text{ produced.}$$

10 Gases

Solutions to Exercises

The Ideal-Gas Equation

(In *Solutions to Exercises*, the symbol for molar mass is MM.)

10.27 (a) PV = nRT; P in atmospheres, V in liters, n in moles, T in kelvins

(b) An ideal gas exhibits pressure, volume, and temperature relationships which are described by the equation PV = nRT. (An ideal gas obeys the ideal-gas equation.)

10.28 (a) STP stands for standard temperature, 0°C (or 273 K), and standard pressure, 1 atm.

(b) $V = \dfrac{nRT}{P}$; $V = 1\,mol \times \dfrac{0.08206\,L \bullet atm}{K \bullet mol} \times \dfrac{273\,K}{1\,atm}$

V = 22.4 L for 1 mole of gas at STP

(c) 25°C + 273 = 298 K

$V = \dfrac{nRT}{P}$; $V = 1\,mol \times \dfrac{0.08206\,L \bullet atm}{K \bullet mol} \times \dfrac{298\,K}{1\,atm}$

V = 24.5 L for 1 mol of gas at 1 atm and 25°C

10.29 *Analyze/Plan.* PV = nRT. At constant volume and temperature, P is directly proportional to n.

Solve. For samples with equal masses of gas, the gas with MM = 30 will have twice as many moles of particles and twice the pressure. Thus, flask A contains the gas with MM = 30 and flask B contains the gas with MM = 60.

10.30 n = g/MM; PV = nRT = gRT/MM; MM = gRT/PV.

2-L flask: MM = 4.8 RT/2.0(X) = 2.4 RT/X

3-L flask: MM = 0.36 RT/3.0 (0.1 X) = 1.2 RT/X

The molar masses of the two gases are not equal. The gas in the 2-L flask has a molar mass that is twice as large as the gas in the 3-L flask.

10.31 *Analyze/Plan.* Follow the strategy for calculations involving many variables given in Section 10.4. *Solve.*

$T = \dfrac{PV}{nR} = 2.00\,atm \times \dfrac{1.00\,L}{0.500\,mol} \times \dfrac{K \bullet mol}{0.08206\,L \bullet atm} = 48.7\,K$

K = 27°C + 273 = 300 K

$n = \dfrac{PV}{RT} = 0.300\,atm \times \dfrac{0.250\,L}{300\,K} \times \dfrac{K \bullet mol}{0.08206\,L \bullet atm} = 3.05 \times 10^{-3}\,mol$

$650\,torr \times \dfrac{1\,atm}{760\,torr} = 0.85526 = 0.855\,atm$

$V = \dfrac{nRT}{P} = 0.333\,mol \times \dfrac{350\,K}{0.85526\,atm} \times \dfrac{0.08206\,L \bullet atm}{K \bullet mol} = 11.2\,L$

585 mL = 0.585 L

$P = \dfrac{nRT}{V} = 0.250\,mol \times \dfrac{295\,K}{0.585\,L} \times \dfrac{0.08206\,L \bullet atm}{K \bullet mol} = 10.3\,atm$

256

P	V	N	T
2.00 atm	1.00 L	0.500 mol	**48.7 K**
0.300 atm	0.250 L	**3.05×10^{-3} mol**	27°C
650 torr	**11.2 L**	0.333 mol	350 K
10.3 atm	585 mL	0.250 mol	295 K

10.32   *Analyze/Plan.* Follow the strategy for calculations involving many variables given in Section 10.4.   *Solve.*

(a)   n = 1.75 mol, P = 0.985 atm, T = –6°C = 267 K

$$V = \frac{nRT}{P} = 1.75 \text{ mol} \times \frac{0.08206 \text{ L} \bullet \text{atm}}{\text{K} \bullet \text{mol}} \times \frac{267 \text{ K}}{0.985 \text{ atm}} = 38.9 \text{ L}$$

(b)   n = 3.33×10^{-3} mol, V = 255 mL = 0.255 L

$$P = 720 \text{ torr} \times \frac{1 \text{ atm}}{760 \text{ torr}} = 0.9474 = 0.947 \text{ atm}$$

$$T = \frac{PV}{nR} = 0.9474 \text{ atm} \times \frac{0.255 \text{ L}}{3.33 \times 10^{-3} \text{ mol}} \times \frac{1 \text{ K} \bullet \text{mol}}{0.08206 \text{ L} \bullet \text{atm}} = 884 \text{ K}$$

(c)   n = 4.67×10^{-2} mol, V = 413 mL = 0.413 L, T = 122°C = 395 K

$$P = \frac{nRT}{V}; = 0.00467 \text{ mol} \times \frac{0.08206 \text{ L} \bullet \text{atm}}{\text{K} \bullet \text{mol}} \times \frac{395 \text{ K}}{0.413 \text{ L}} = 3.67 \text{ atm}$$

(d)   V = 67.5 L, T = 54°C = 327 K,

$$P = 11.25 \text{ kPa} \times \frac{1 \text{ atm}}{101.325 \text{ kPa}} = 0.11103 = 0.1110 \text{ atm}$$

$$n = \frac{PV}{RT} = 0.11103 \text{ atm} \times \frac{\text{K} \bullet \text{mol}}{0.08206 \text{ L} \bullet \text{atm}} \times \frac{67.5 \text{ L}}{327 \text{ K}} = 0.279 \text{ mol}$$

10.33   *Analyze/Plan.* Follow the strategy for calculations involving many variables.   *Solve.*

n = g/MM; PV = nRT; PV = gRT/MM; g = MM×PV/RT

P = 1.0 atm, T = 23°C = 296 K, V = 2.0×10^5 m^3. Change m^3 to L, then calculate grams (or kg).

$$2.0 \times 10^5 \text{ m}^3 \times \frac{10^3 \text{ dm}^3}{1 \text{ m}^3} \times \frac{1 \text{ L}}{1 \text{ dm}^3} = 2.0 \times 10^8 \text{ L H}_2$$

$$g = \frac{2.02 \text{ g H}_2}{1 \text{ mol H}_2} \times \frac{\text{K} \bullet \text{mol}}{0.08206 \text{ L} \bullet \text{atm}} \times \frac{1.0 \text{ atm} \times 2.0 \times 10^8 \text{ L}}{296 \text{ K}} = 1.7 \times 10^7 \text{ g} = 1.7 \times 10^4 \text{ kg H}_2$$

10.34   Find the volume of the tube in cm^3; 1 cm^3 = 1 mL.

r = d/2 = 2.5 cm/2 = 1.25 = 1.3 cm; h = 5.5 m = 5.5×10^2 cm

V = $\pi r^2 h$ = 3.14159 × (1.25 cm)^2 × (5.5×10^2 cm) = 2.700×10^3 cm^3 = 2.7 L

$$PV = \frac{g}{MM}RT; \quad g = \frac{MM \times PV}{RT}; \quad P = 1.78 \text{ torr} \times \frac{1 \text{ atm}}{760 \text{ torr}} = 2.342 \times 10^{-3} = 2.34 \times 10^{-3} \text{ atm}$$

$$g = \frac{20.18\,\text{g Ne}}{1\,\text{mol Ne}} \times \frac{\text{K} \cdot \text{mol}}{0.08206\,\text{L} \cdot \text{atm}} \times \frac{2.342 \times 10^{-3}\,\text{atm} \times 2.700\,\text{L}}{308\,\text{K}} = 5.049 \times 10^{-3} = 5.0 \times 10^{-3}\,\text{g Ne}$$

10.35 *Analyze/Plan.* Follow the strategy for calculations involving many variables. *Solve.*

$$V = 2.50\,\text{L};\ T = 273 + 37°\text{C} = 310\,\text{K};\ P = 735\,\text{torr} \times \frac{1\,\text{atm}}{760\,\text{torr}} = 0.96710 = 0.967\,\text{atm}$$

$PV = nRT,\ n = PV/RT$, number of molecules (#) = $n \times 6.022 \times 10^{23}$

$$\# = \frac{0.9671\,\text{atm} \times 2.50\,\text{L}}{310\,\text{K}} \times \frac{\text{K} \cdot \text{mol}}{0.08206\,\text{L} \cdot \text{atm}} \times \frac{6.022 \times 10^{23}\,\text{molecules}}{\text{mol}}$$

$$= 5.72 \times 10^{22}\,\text{molecules}$$

10.36 $P_{O_3} = 3.0 \times 10^{-3}\,\text{atm};\ T = 250\,\text{K};\ V = 1\,\text{L (exact)}$

$$\#\text{ of }O_3\text{ molecules} = \frac{PV}{RT} \times 6.022 \times 10^{23}$$

$$\# = \frac{3.0 \times 10^{-3}\,\text{atm} \times 1\,\text{L}}{250\,\text{K}} \times \frac{\text{K} \cdot \text{mol}}{0.08206\,\text{L} \cdot \text{atm}} \times \frac{6.022 \times 10^{23}\,\text{molecules}}{\text{mol}}$$

$$= 8.8 \times 10^{19}\ O_3\text{ molecules}$$

10.37 *Analyze/Plan.* Follow the strategy for calculations involving many variables. *Solve.*

(a) $P = \dfrac{nRT}{V};\ n = 0.29\,\text{kg}\ O_2 \times \dfrac{1000\,\text{g}}{1\,\text{kg}} \times \dfrac{1\,\text{mol}\ O_2}{32.00\,\text{g}\ O_2} = 9.0625 = 9.1\,\text{mol};\ V = 2.3\,\text{L};$

$T = 273 + 9°\text{C} = 282\,\text{K}$

$$P = \frac{9.0625\,\text{mol}}{2.3\,\text{L}} \times \frac{0.08206\,\text{L} \cdot \text{atm}}{\text{K} \cdot \text{mol}} \times 282\ \text{K} = 91\,\text{atm}$$

(b) $V = \dfrac{nRT}{P};\ = \dfrac{9.0625\,\text{mol}}{0.95\,\text{atm}} \times \dfrac{0.08206\,\text{L} \cdot \text{atm}}{\text{K} \cdot \text{mol}} \times 299\,\text{K} = 2.3 \times 10^2\,\text{L}$

10.38 (a) $V = 0.250\,\text{L},\ T = 23°\text{C} = 296\ \text{K},\ n = 2.30\,\text{g}\ C_3H_8 \times \dfrac{1\,\text{mol}\ C_3H_8}{44.1\,\text{g}\ C_3H_8} = 0.052154$

$$= 0.0522\,\text{mol}$$

$$P = \frac{nRT}{V} = 0.052154\,\text{mol} \times \frac{0.08206\,\text{L} \cdot \text{atm}}{\text{K} \cdot \text{mol}} \times \frac{296\,\text{K}}{0.250\,\text{L}} = 5.07\,\text{atm}$$

(b) STP = 1.00 atm, 273 K

$$V = \frac{nRT}{P} = 0.052154\,\text{mol} \times \frac{0.08206\,\text{L} \cdot \text{atm}}{\text{K} \cdot \text{mol}} \times \frac{273\,\text{K}}{1.00\,\text{atm}} = 1.1684 = 1.17\,\text{L}$$

(c) $°\text{C} = 5/9\ (°\text{F} - 32°);\ \text{K} = °\text{C} + 273.15 = 5/9\ (130°\text{F} - 32°) + 273.15 = 327.59 = 328\ \text{K}$

$$P = \frac{nRT}{V} = 0.052154\,\text{mol} \times \frac{0.08206\,\text{L} \cdot \text{atm}}{\text{K} \cdot \text{mol}} \times \frac{327.59\,\text{K}}{0.250\,\text{L}} = 5.608 = 5.61\,\text{atm}$$

10.39 *Analyze/Plan.* Follow the strategy for calculations involving many variables. *Solve.*

$$V = 8.70\,\text{L},\ T = 24°\text{C} = 297\,\text{K},\ P = 895\,\text{torr} \times \frac{1\,\text{atm}}{760\,\text{torr}} = 1.1776 = 1.18\,\text{atm} = 5.0 \times 10^{-3}$$

(a) $g = \dfrac{MM \times PV}{RT};\ g = \dfrac{70.91\,\text{g}\ Cl_2}{1\,\text{mol}\ Cl_2} \times \dfrac{\text{K} \cdot \text{mol}}{0.08206\,\text{L} \cdot \text{atm}} \times \dfrac{1.1776\,\text{atm}}{297\,\text{K}} \times 8.70\,\text{L}$

$$= 29.8\,\text{g}\ Cl_2$$

(b) $V_2 = \dfrac{P_1 V_1 T_2}{T_1 P_2} = \dfrac{895 \text{ torr} \times 8.70 \text{ L} \times 273 \text{ K}}{297 \text{ K} \times 760 \text{ torr}} = 9.42 \text{ L}$

(c) $T_2 = \dfrac{P_2 V_2 T_1}{P_1 V_1} = \dfrac{876 \text{ torr} \times 15.00 \text{ L} \times 297 \text{ K}}{895 \text{ torr} \times 8.70 \text{ L}} = 501 \text{ K}$

(d) $P_2 = \dfrac{P_1 V_1 T_2}{V_2 T_1} = \dfrac{895 \text{ torr} \times 8.70 \text{ L} \times 331 \text{ K}}{6.00 \text{ L} \times 297 \text{ K}} = 1.45 \times 10^3 \text{ torr} = 1.90 \text{ atm}$

10.40 $V = 65.0 \text{ L}, T = 23°C = 296 \text{ K}, P = 16{,}500 \text{ kPa} \times \dfrac{1 \text{ atm}}{101.325} = 162.84 = 163 \text{ atm}$

(a) $g = \dfrac{\text{MM} \times PV}{RT}; g = \dfrac{32.0 \text{ g O}_2}{1 \text{ mol O}_2} \times \dfrac{\text{K} \cdot \text{mol}}{0.08206 \text{ L} \cdot \text{atm}} \times \dfrac{162.84 \text{ atm}}{296 \text{ K}} \times 65.0 \text{ L}$

$$= 1.39446 \times 10^4 \text{ g O}_2 = 13.9 \text{ kg O}_2$$

(b) $V_2 = \dfrac{P_1 V_1 T_2}{T_1 P_2} = \dfrac{16{,}500 \text{ kPa} \times 65.0 \text{ L} \times 273 \text{ K}}{296 \text{ K} \times 101.325 \text{ kPa}} = 9.76 \times 10^3 \text{ L}$

(c) $T_2 = \dfrac{P_2 T_1}{P_1} = \dfrac{150.0 \text{ atm} \times 296 \text{ K}}{16{,}500 \text{ kPa}} \times \dfrac{101.325 \text{ kPa}}{1 \text{ atm}} = 272.7 = 2.73 \text{ K}$

(d) $P_2 = \dfrac{P_1 V_1 T_2}{V_2 T_1} = \dfrac{16{,}500 \text{ kPa} \times 65.0 \text{ L} \times 297 \text{ K}}{55.0 \text{ L} \times 296 \text{ K}} = 19{,}566 = 1.96 \times 10^4 \text{ kPa}$

10.41   *Analyze.* Given: mass of cockroach, rate of O_2 consumption, temperature, percent O_2 in air, volume of air. Find: mol O_2 consumed per hour; mol O_2 in 1 quart of air; mol O_2 consumed in 48 hr.

(a)   *Plan/Solve.* V of O_2 consumed = rate of consumption × mass × time. n = PV/RT.

$$5.2 \text{ g} \times 1 \text{ hr} \times \dfrac{0.8 \text{ mL O}_2}{1 \text{ g} \cdot \text{hr}} = 4.16 = 4 \text{ mL O}_2 \text{ consumed}$$

$$n = \dfrac{PV}{RT} = 1 \text{ atm} \times \dfrac{\text{K} \cdot \text{mol}}{0.08206 \text{ L} \cdot \text{atm}} \times \dfrac{0.00416 \text{ L}}{297 \text{ K}} = 1.71 \times 10^{-4} = 2 \times 10^{-4} \text{ mol O}_2$$

(b)   *Plan/Solve.* qt air → L air → L O_2 available. mol O_2 available = PV/RT.
mol O_2/hr (from part (a)) → total mol O_2 consumed. Compare O_2 available and O_2 consumed.

$$1 \text{ qt air} \times \dfrac{0.946 \text{ L}}{1 \text{ qt}} \times 0.21 \text{ O}_2 \text{ in air} = 0.199 \text{ L O}_2 \text{ available}$$

$$n = 1 \text{ atm} \times \dfrac{\text{K} \cdot \text{mol}}{0.08206 \text{ L} \cdot \text{atm}} \times \dfrac{0.199 \text{ L}}{297 \text{ K}} = 8.16 \times 10^{-3} = 8 \times 10^{-3} \text{ mol O}_2 \text{ available}$$

$$\text{roach uses} \dfrac{1.71 \times 10^{-4} \text{ mol}}{1 \text{ hr}} \times 48 \text{ hr} = 8.19 \times 10^{-3} = 8 \times 10^{-3} \text{ mol O}_2 \text{ consumed}$$

Not only does the roach use 20% of the available O_2, it needs all the O_2 in the jar.

10.42   mass $= 1800 \times 10^{-9} \text{ g} = 1.8 \times 10^{-6} \text{ g}; V = 1 \text{ m}^3 = 1 \times 10^3 \text{ L}; T = 273 + 10°\text{ C} = 283 \text{ K}$

(a) $P = \dfrac{g\,RT}{MM \times V}$; $P = \dfrac{1.8 \times 10^{-6}\,g\,Hg \times 1\,mol\,Hg}{200.6\,g\,Hg} \times \dfrac{0.08206\,L \bullet atm}{K \bullet mol} \times \dfrac{283\,K}{1 \times 10^3\,L}$

$$= 2.1 \times 10^{-10}\,atm$$

(b) $\dfrac{1.8 \times 10^{-6}\,g\,Hg}{1\,m^3} \times \dfrac{1\,mol\,Hg}{200.6\,g\,Hg} \times \dfrac{6.022 \times 10^{23}\,Hg\,atoms}{1\,mol\,Hg} = 5.4 \times 10^{15}\,Hg\,atoms/m^3$

(c) $1600\,km^3 \times \dfrac{1000^3\,m^3}{1\,km^3} \times \dfrac{1.8 \times 10^{-6}\,g\,Hg}{1\,m^3} = 2.9 \times 10^6\,g\,Hg/day$

Further Applications of the Ideal-Gas Equation

10.43   (c) $Cl_2(g)$ is the most dense at 1.00 at and 298 K. Gas density is directly proportional to molar mass and pressure, and inversely proportional to temperature (Equation [10.10]). For gas samples at the same conditions, molar mass determines density. Of the three gases listed, Cl_2 has the largest molar mass.

10.44   (c) $CO_2(g)$ is least dense. For gases at the same conditions, density is directly proportional to molar mass, and CO_2 has the smallest molar mass.

10.45   (c) Because the helium atoms are of lower mass than the average air molecule, the helium gas is less dense than air. The balloon thus weighs less than the air displaced by its volume.

10.46   (b) Xe atoms have a higher mass than N_2 molecules. Because both gases at STP have the same number of molecules per unit volume, the Xe gas must be denser.

10.47   *Analyze/Plan.* Conditions (P, V, T) and amounts of gases are given. Rearrange the relationship $PV \times MM = gRT$ to obtain the desired of quantity, paying attention (as always!) to units.   *Solve.*

(a) $d = \dfrac{MM \times P}{RT}$; $MM = 46.0\,g/mol$; $P = 0.970\,atm$, $T = 35°C = 308\,K$

$$d = \dfrac{46.0\,g\,NO_2}{1\,mol} \times \dfrac{K \bullet mol}{0.08206\,L \bullet atm} \times \dfrac{0.970\,atm}{308\,K} = 1.77\,g/L$$

(b) $MM = \dfrac{gRT}{PV} = \dfrac{2.50\,g}{0.875\,L} \times \dfrac{0.08206\,L \bullet atm}{K \bullet mol} \times \dfrac{308\,K}{685\,torr} \times \dfrac{760\,torr}{1\,atm} = 80.1\,g/mol$

10.48   (a) $d = \dfrac{MM \times P}{RT}$; $MM = 146.1\,g/mol$, $T = 28°C = 301\,K$, $P = 678\,torr$

$$d = \dfrac{146.1\,g}{1\,mol} \times \dfrac{K \bullet mol}{0.08206\,L \bullet atm} \times \dfrac{678\,torr}{301\,K} \times \dfrac{1\,atm}{760\,torr} = 5.28\,g/L$$

(b) $MM = \dfrac{dRT}{P} = \dfrac{7.135\,g}{1\,L} \times \dfrac{0.08206\,L \bullet atm}{K \bullet mol} \times \dfrac{285\,K}{743\,torr} \times \dfrac{760\,torr}{1\,atm} = 171\,g/mol$

10.49   *Analyze/Plan.* Given: mass, conditions (P, V, T) of unknown gas. Find: molar mass. $MM = gRT/PV$. *Solve.*

$$MM = \dfrac{gRT}{PV} = \dfrac{1.012\,g}{0.354\,L} \times \dfrac{0.08206\,L \bullet atm}{K \bullet atm} \times \dfrac{372\,K}{742\,torr} \times \dfrac{760\,torr}{1\,atm} = 89.4\,g/mol$$

10.50 $MM = \dfrac{gRT}{PV} = \dfrac{0.846\,g}{0.354\,L} \times \dfrac{0.08206\,L \bullet atm}{K \bullet atm} \times \dfrac{373\,K}{752\,torr} \times \dfrac{760\,torr}{1\,atm} = 73.9\,g/mol$

10.51 *Analyze/Plan.* Follow the logic in Sample Exercise 10.9. *Solve.*

$mol\ O_2 = \dfrac{PV}{RT} = 3.5 \times 10^{-6}\,torr \times \dfrac{1\,atm}{760\,torr} \times \dfrac{K \bullet atm}{0.08206\,L \bullet atm} \times \dfrac{0.382\,L}{300\,K} = 7.146 \times 10^{-11}$

$= 7.1 \times 10^{-11}\,mol\ O_2$

$7.146 \times 10^{-11}\,mol\ O_2 \times \dfrac{2\,mol\ Mg}{1\,mol\ O_2} \times \dfrac{24.3\,g\ Mg}{1\,mol\ Mg} = 3.5 \times 10^{-9}\,g\ Mg$

10.52 $n_{H_2} = \dfrac{P_{H_2}V}{RT}$; $P = 814\,torr \times \dfrac{1\,atm}{760\,torr} = 1.071 = 1.07\,atm$; $T = 273 + 21°C = 294\,K$

$n_{H_2} = 1.071\,atm \times \dfrac{K \bullet mol}{0.08206\ L \bullet atm} \times \dfrac{53.5\,L}{294\,K} = 2.3751 = 2.38\,mol\ H_2$

$2.3751\,mol\ H_2 \times \dfrac{1\,mol\ CaH_2}{2\,mol\ H_2} \times \dfrac{42.10\,g\ CaH_2}{1\,mol\ CaH_2} = 50.0\,g\ CaH_2$

10.53 *Analyze/Plan.* g glucose → mol glucose → mol CO_2 → V CO_2 *Solve.*

$24.5\,g \times \dfrac{1\,mol\ glucose}{180.1\,g} \times \dfrac{6\,mol\ CO_2}{1\,mol\ glucose} = 0.8162 = 0.816\,mol\ CO_2$

$V = \dfrac{nRT}{P} = 0.8162\ mol \times \dfrac{0.08206\,L \bullet atm}{K \bullet mol} \times \dfrac{310\,K}{0.970\,atm} = 21.4\,L\ CO_2$

10.54 Balance the chemical equation.

Then, g C_5H_{12} → mol C_5H_{12} → mol O_2 → V O_2 at given P, T.

$C_5H_{12}(l) + 8O_2(g) \rightarrow 5CO_2(g) + 6H_2O(g)$

$2.50\ g\ C_5H_{12} \times \dfrac{1\,mol\ C_5H_{12}}{72.15\,g\ C_5H_{12}} \times \dfrac{8\,mol\ O_2}{1\,mol\ C_5H_{12}} = 0.2772 = 0.277\,mol\ O_2$

$V = \dfrac{nRT}{P} = 0.2772\,mol\ O_2 \times \dfrac{296\,K}{0.980\,atm} \times \dfrac{0.08206\,L \bullet atm}{K \bullet mol} = 6.87054 = 6.87\,L\ O_2$

According to Avogadro's law, mole ratios are volume ratios for gases at the same pressure and temperature. Use mole ratios from the chemical equation to calculate volumes of $CO_2(g)$ and $H_2O(g)$.

$6.87054\,L\ O_2 \times \dfrac{5\,CO_2}{8\,O_2} = 4.29\,L\ CO_2$; $6.87054\,L\ O_2 \times \dfrac{6\,H_2O}{8\,O_2} = 5.15\,L\ H_2O$

10.55 *Analyze/Plan.* The gas sample is a mixture of $H_2(g)$ and $H_2O(g)$. Find the partial pressure of $H_2(g)$ and then the moles of $H_2(g)$ and Zn(s). *Solve.*

$P_t = 738\,torr = P_{H_2} + P_{H_2O}$

From Appendix B, the vapor pressure of H_2O at 24°C = 22.38 torr

$P_{H_2} = (738\,torr - 22.38\,torr) \times \dfrac{1\,atm}{760\,torr} = 0.9416 = 0.942\,atm$

$$n_{H_2} = \frac{P_{H_2}V}{RT} = 0.9416\,atm \times \frac{K \bullet mol}{0.08206\ L \bullet atm} \times \frac{0.159\,L}{297\,K} = 0.006143 = 0.00614\ mol\ H_2$$

$$0.006143\ mol\ H_2 \times \frac{1\,mol\,Zn}{1\,mol\,H_2} \times \frac{65.39\,g\,Zn}{1\,mol\,Zn} = 0.402\ g\ Zn$$

10.56   The gas sample is a mixture of $C_2H_2(g)$ and $H_2O(g)$. Find the partial pressure of C_2H_2, then moles CaC_2 and C_2H_2.

$$P_t = 726\,torr = P_{C_2H_2} + P_{H_2O}. \quad P_{H_2O} \text{ at } 26°C = 25.21\,torr$$

$$P_{C_2H_2} = (726\,torr - 25.21\,torr) \times \frac{1\,atm}{760\,torr} = 0.9221 = 0.922\ atm$$

$$0.887\ g\ CaC_2 \times \frac{1\,mol\,CaC_2}{64.10\,g} \times \frac{1\,mol\,C_2H_2}{1\,mol\,CaC_2} = 0.013838 = 0.0138\ mol\ C_2H_2$$

$$V = 0.013838\ mol \times \frac{0.08206\,L \bullet atm}{K \bullet mol} \times \frac{299\,K}{0.9221\,atm} = 0.368\ L\ C_2H_2$$

Partial Pressures

10.57   (a)   When the stopcock is opened, the volume occupied by $N_2(g)$ increases from 2.0 L to 5.0 L. At constant T, $P_1V_1 = P_2V_2$. 1.0 atm × 2.0 L = P_2 × 5.0 L; P_2 = 0.40 atm

   (b)   When the gases mix, the volume of $O_2(g)$ increases from 3.0 L to 5.0 L. At constant T, $P_1V_1 = P_2V_2$. 2.0 atm × 3.0 L = P_2 × 5.0 L; P_2 = 1.2 atm

   (c) $P_t = P_{N_2} + P_{O_2} = 0.40\,atm + 1.2\,atm = 1.6\ atm$

10.58   (a)   The partial pressure of gas A is **not affected** by the addition of gas C. The partial pressure of A depends only on moles of A, volume of container, and conditions; none of these factors changes when gas C is added.

   (b)   The total pressure in the vessel **increases** when gas C is added, because the total number of moles of gas increases.

   (c)   The mole fraction of gas B **decreases** when gas C is added. The moles of gas B stay the same, but the total moles increase, so the mole fraction of B (n_B/n_t) decreases.

10.59   *Analyze.* Given: amount, V, T of three gases. Find: P of each gas, total P.

   Plan. P = nRT/V; $P_t = P_1 + P_2 + P_3 + \cdots$ *Solve.*

   (a) $$P_{He} = \frac{nRT}{V} = 0.538\,mol \times \frac{0.08206\,L \bullet atm}{K \bullet atm} \times \frac{298\,K}{7.00\,L} = 1.88\ atm$$

   $$P_{Ne} = \frac{nRT}{V} = 0.315\,mol \times \frac{0.08206\,L \bullet atm}{K \bullet atm} \times \frac{298\,K}{7.00\,L} = 1.10\ atm$$

   $$P_{Ar} = \frac{nRT}{V} = 0.103\,mol \times \frac{0.08206\,L \bullet atm}{K \bullet atm} \times \frac{298\,K}{7.00\,L} = 0.360\ atm$$

   (b) $P_t = 1.88\,atm + 1.10\,atm + 0.360\,atm = 3.34\ atm$

10.60   (a) $2.50 \text{ g CH}_4 \times \dfrac{1 \text{ mol CH}_4}{16.04 \text{ g CH}_4} = 0.15586 = 0.156 \text{ mol CH}_4$

$P_{CH_4} = \dfrac{nRT}{V} = 0.15586 \text{ mol} \times \dfrac{0.08206 \text{ L} \bullet \text{atm}}{\text{K} \bullet \text{mol}} \times \dfrac{288 \text{ K}}{2.00 \text{ L}} = 1.842 = 1.84 \text{ atm}$

$2.50 \text{ g } C_2H_4 \times \dfrac{1 \text{ mol } C_2H_4}{28.05 \text{ g } C_2H_4} = 0.08913 = 0.0891 \text{ mol } C_2H_4$

$P_{C_2H_4} = \dfrac{nRT}{V} = 0.08913 \text{ mol } C_2H_4 \times \dfrac{0.08206 \text{ L} \bullet \text{atm}}{\text{K} \bullet \text{mol}} \times \dfrac{288 \text{ K}}{2.00 \text{ L}} = 1.053 = 1.05 \text{ atm}$

$2.50 \text{ g } C_4H_{10} \times \dfrac{1 \text{ mol } C_4H_{10}}{58.12 \text{ g } C_4H_{10}} = 0.04301 = 0.0430 \text{ mol } C_4H_{10}$

$P_{C_4H_{10}} = \dfrac{nRT}{V} = 0.04301 \text{ mol} \times \dfrac{0.08206 \text{ L} \bullet \text{atm}}{\text{K} \bullet \text{mol}} \times \dfrac{288 \text{ K}}{2.00 \text{ L}} = 0.5083 = 0.508 \text{ atm}$

   (b) $P_t = 1.842 \text{ atm} + 1.053 \text{ atm} + 0.508 \text{ atm} = 3.403 = 3.40 \text{ atm}$

10.61   *Analyze.* Given: mass CO_2 at V, T; pressure of air at same V, T. Find: partial pressure of CO_2 at these conditions, total pressure of gases at V, T.

   Plan. g $CO_2 \rightarrow$ mol $CO_2 \rightarrow P_{CO_2}$ (via P = nRT/V); $P_t = P_{CO_2} + P_{air}$ *Solve.*

$5.50 \text{ g } CO_2 \times \dfrac{1 \text{ mol } CO_2}{44.01 \text{ g } CO_2} = 0.12497 = 0.125 \text{ mol } CO_2 ; T = 273 + 24°C = 297 \text{ K}$

$P_{CO_2} = 0.12497 \text{ mol} \times \dfrac{297 \text{ K}}{10.0 \text{ L}} \times \dfrac{0.08206 \text{ L} \bullet \text{atm}}{\text{K} \bullet \text{mol}} = 0.30458 = 0.305 \text{ atm}$

$P_{air} = 705 \text{ torr} \times \dfrac{1 \text{ atm}}{760 \text{ torr}} = 0.92763 = 0.928 \text{ atm}$

$P_t = P_{CO_2} + P_{air} = 0.30458 + 0.92763 = 1.23221 = 1.232 \text{ atm}$

(Result has 3 decimal places and 4 sig figs.)

10.62 $V \; C_4H_{10}O(l) \xrightarrow{\text{density}} \text{mass } C_4H_{10}O \rightarrow \text{mol } C_4H_{10}O \rightarrow P_{C_4H_{10}O}$ at given V, T.

$P_t = P_{N_2} + P_{O_2} + P_{C_4H_{10}O}; \; T = 273.15 + 35.0°C = 308.15 = 308.2 \text{ K}$

   (a) $4.00 \text{ mL } C_4H_{10}O \times \dfrac{0.7134 \text{ g } C_4H_{10}O}{\text{mL}} \times \dfrac{1 \text{ mol } C_4H_{10}O}{74.12 \text{ g } C_4H_{10}O} = 0.0385 \text{ mol } C_4H_{10}O$

$P = \dfrac{nRT}{V} = 0.03850 \text{ mol} \times \dfrac{308.15}{5.00 \text{ L}} \times \dfrac{0.08206 \text{ L} \bullet \text{atm}}{\text{K} \bullet \text{mol}} = 0.1947 = 0.195 \text{ atm}$

   (b) $P_t = P_{N_2} + P_{O_2} + P_{C_4H_{10}O} = 0.751 \text{ atm} + 0.208 \text{ atm} + 0.195 \text{ atm} = 1.154 \text{ atm}$

10.63   *Analyze/Plan.* The partial pressure of each component is equal to the mole fraction of that gas times the total pressure of the mixture. Find the mole fraction of each component and then its partial pressure.   *Solve.*

$n_t = 0.75 \text{ mol } N_2 + 0.30 \text{ mol } O_2 + 0.15 \text{ mol } CO_2 = 1.20 \text{ mol}$

$$\chi_{N_2} = \frac{0.75}{1.20} = 0.625 = 0.63; \ P_{N_2} = 0.625 \times 1.56 \ atm = 0.98 \ atm$$

$$\chi_{O_2} = \frac{0.30}{1.20} = 0.250 = 0.25; \ P_{O_2} = 0.250 \times 1.56 \ atm = 0.39 \ atm$$

$$\chi_{CO_2} = \frac{0.15}{1.20} = 0.125 = 0.13; \ P_{CO_2} = 0.125 \times 1.56 \ atm = 0.20 \ atm$$

10.64 $\quad n_{N_2} = 10.25 \ g \ N_2 \times \dfrac{1 \ mol}{28.02 \ g} = 0.3658 \ mol; \ n_{H_2} = 2.05 \ g \ H_2 \times \dfrac{1 \ mol}{2.016 \ g} = 1.0169 = 1.02 \ mol$

$\quad n_{NH_3} = 7.63 \ g \ NH_3 \times \dfrac{1 \ mol}{17.03 \ g} = 0.448 \ mol; \ n_t = 0.3658 + 1.0169 + 0.4480 = 1.8307$

$$= 1.83 \ mol$$

$$P_{N_2} = \frac{n_{N_2}}{n_t} \times P_t = \frac{0.3658}{1.8307} \times 2.35 \ atm = 0.470 \ atm$$

$$P_{H_2} = \frac{1.0169}{1.8307} \times 2.35 \ atm = 1.31 \ atm; \ P_{NH_3} = \frac{0.4480}{1.8307} \times 2.35 \ atm = 0.575 \ atm$$

10.65 $\quad$ *Analyze/Plan.* Mole fraction = pressure fraction. Find the desired mole fraction of O_2 and change to mole percent. *Solve.*

$$\chi_{O_2} = \frac{P_{O_2}}{P_t} = \frac{0.21 \ atm}{8.38 \ atm} = 0.025; \ mole \ \% = 0.025 \times 100 = 2.5\%$$

10.66 $\quad$ (a) $\quad n_{O_2} = 5.08 \ g \ O_2 \times \dfrac{1 \ mol}{32.00 \ g} = 0.159 \ mol; \ n_{N_2} = 7.17 \ g \ N_2 \times \dfrac{1 \ mol}{28.02 \ g} = 0.256 \ mol$

$\qquad n_{H_2} = 1.32 \ g \ H_2 \times \dfrac{1 \ mol}{2.016 \ g} = 0.655 \ mol; \ n_t = 0.159 + 0.256 + 0.655 = 1.070 \ mol$

$$\chi_{O_2} = \frac{n_{O_2}}{n_t} = \frac{0.159}{1.07} = 0.149; \ \chi_{N_2} = \frac{n_{N_2}}{n_t} = \frac{0.256}{1.07} = 0.239$$

$$\chi_{H_2} = \frac{0.655}{1.07} = 0.612$$

$\qquad$ (b) $\quad P_{O_2} = n \times \dfrac{RT}{V}; \ P_{O_2} = 0.159 \ mol \times \dfrac{0.08206 \ L \bullet atm}{K \bullet mol} \times \dfrac{288 \ K}{12.40 \ L} = 0.303 \ atm$

$$P_{N_2} = 0.256 \ mol \times \frac{0.08206 \ L \bullet atm}{K \bullet mol} \times \frac{288 \ K}{12.40 \ L} = 0.488 \ atm$$

$$P_{H_2} = 0.655 \ mol \times \frac{0.08206 \ L \bullet atm}{K \bullet mol} \times \frac{288 \ K}{12.40 \ L} = 1.25 \ atm$$

10.67 $\quad$ *Analyze/Plan.* $N_2(g)$ and $O_2(g)$ undergo changes of conditions and are mixed. Calculate the new pressure of each gas and add them to obtain the total pressure of the mixture.

$$P_2 = P_1 V_1 T_2 \ / \ V_2 T_1; \ P_T = P_{N_2} + P_{O_2}. \qquad \textit{Solve.}$$

$$P_{N_2} = \frac{P_1 V_1 T_2}{V_2 T_1} = \frac{4.75 \ atm \times 1.00 \ L \times 293 \ K}{10.0 \ L \times 299 \ K} = 0.46547 = 0.465 \ atm$$

$$P_{O_2} = \frac{P_1 V_1 T_2}{V_2 T_1} = \frac{5.25 \text{ atm} \times 5.00 \text{ L} \times 293 \text{ K}}{10.0 \text{ L} \times 299 \text{ K}} = 2.5723 = 2.57 \text{ atm}$$

$P_t = 0.46547 \text{ atm} + 2.5723 \text{ atm} = 3.0378 = 3.04 \text{ atm}$

10.68   Calculate the pressure of the gas in the second vessel directly from mass and conditions using the ideal-gas equation.

(a) $P_{SO_2} = \dfrac{gRT}{MV} = \dfrac{4.00 \text{ g } SO_2}{64.07 \text{ g } SO_2/\text{mol}} \times \dfrac{0.08206 \text{ L} \cdot \text{atm}}{\text{K} \cdot \text{mol}} \times \dfrac{298 \text{ K}}{10.0 \text{ L}} = 0.15267$

$= 0.153 \text{ atm}$

(b) $P_{N_2} = \dfrac{gRT}{MV} = \dfrac{2.35 \text{ g } N_2}{28.01 \text{ g } N_2/\text{mol}} \times \dfrac{0.08206 \text{ L} \cdot \text{atm}}{\text{K} \cdot \text{mol}} \times \dfrac{298 \text{ K}}{10.0 \text{ L}} = 0.20516$

$= 0.205 \text{ atm}$

(c) $P_t = P_{SO_2} + P_{N_2} = 0.15267 \text{ atm} + 0.20516 \text{ atm} = 0.358 \text{ atm}$

Kinetic-Molecular Theory; Graham's Law

10.69   (a)   Increase in temperature at constant volume, decrease in volume, increase in pressure

(b)   Decrease in temperature

(c)   Increase in volume, decrease in pressure

(d)   Increase in temperature

10.70   (a)   False. The average kinetic energy per molecule in a collection of gas molecules is the same for all gases at the same temperature.

b)   True.

(c)   False. The molecules in a gas sample at a given temperature exhibit a distribution of kinetic energies.

(d)   True.

10.71   The fact that gases are readily compressible supports the assumption that most of the volume of a gas sample is empty space.

10.72   Newton's model provides no explanation of the effect of a change in temperature on the pressure of a gas at constant volume or on the volume of a gas at constant pressure. On the other hand, the assumption that the average kinetic energy of gas molecules increases with increasing temperature explains Charles' Law, that an increase in temperature requires an increase in volume to maintain constant pressure.

10.73   _Analyze/Plan._ We have samples of two different gases at different pressures and temperatures. Compare the two samples by considering the postulates of the kinetic-molecular theory that pertain to the quantities in (a)–(d).   _Solve._

(a) $n \propto P/T$ (V/R is the same for A and B.) Since P is greater and T is smaller for vessel A, it has more molecules.

(b)   Vessel A has more molecules but the molar mass of CO is smaller than the molar mass of SO_2, so we need to calculate the masses. Since volume is not specified, calculate g/L.

$$\frac{g_A}{V} = \frac{MM \times P}{RT} = \frac{28.01\,g\,CO}{1\,mol\,CO} \times \frac{K \bullet mol}{0.08206\,L \bullet atm} \times \frac{1\,atm}{273\,K} = 1.25\,g\,CO/L$$

$$\frac{g_B}{V} = \frac{MM \times P}{RT} = \frac{64.07\,g\,SO_2}{1\,mol\,SO_2} \times \frac{K \bullet mol}{0.08206\,L \bullet atm} \times \frac{0.5\,atm}{293\,K} = 1.33\,g\,SO_2/L$$

Vessel B has more mass.

(c)   Vessel B is at a higher temperature so the average kinetic energy of its molecules is higher.

(d)   The two factors that affect rms speed are temperature and molar mass. The molecules in vessel A have smaller molar mass but are at the lower temperature, so we must calculate the rms speeds.

Mathematically, according to Equation [10.24],

$$\frac{u_A}{u_B} = \sqrt{\frac{T_A/MM_A}{T_B/MM_B}} = \sqrt{\frac{273/28.01}{293/64.07}} = 1.46$$

The ratio is greater than 1; vessel A has the greater rms speed.

10.74   (a)   They have the same number of molecules (equal volumes of gases at the same temperature and pressure contain equal numbers of molecules).

         (b) N_2 is more dense because it has the larger molar mass. Since the volumes of the samples and the number of molecules are equal, the gas with the larger molar mass will have the greater density.

         (c)   The average kinetic energies are equal (statement 5, section 10.7).

         (d) CH_4 will effuse faster. The lighter the gas molecules, the faster they will effuse (Graham's Law).

10.75   (a)   *Plan.* The larger the molar mass, the slower the average speed (at constant temperature).

                 Solve. In order of increasing speed (and decreasing molar mass):
$HBr < NF_3 < SO_2 < CO < Ne$

         (b)   *Plan.* Follow the logic of Sample Exercise 10.14. *Solve.*

$$u = \sqrt{\frac{3RT}{M}} = \left(\frac{3 \times 8.314\,kg \bullet m^2/s^2 \bullet K \bullet mol \times 298\,K}{71.0 \times 10^{-3}\,kg/mol}\right)^{1/2} = 324\,m/s$$

10.76   (a)   *Plan.* The greater the molecular (and molar) mass, the smaller the rms speed of the molecules. Calculate the molar mass of each gas, and place them in decreasing order of mass and increasing order of rms speed. *Solve.*

                 $CO = 28\,g/mol$; $SF_6 = 146\,g/mol$; $H_2S = 34\,g/mol$; $Cl_2 = 71\,g/mol$;
$HBr = 81\,g/mol$. In order of increasing speed (and decreasing molar mass):

                 $SF_6 < HBr < Cl_2 < H_2S < CO$

(b)   *Plan.* Follow the logic of Sample Exercise 10.14. *Solve.*

$$u_{CO} = \sqrt{\frac{3RT}{M}} = \left(\frac{3 \times 8.314 \text{ kg} \cdot \text{m}^2/\text{s}^2 \cdot \text{K} \cdot \text{mol} \times 300 \text{ K}}{28.0 \times 10^{-3} \text{ kg/mol}}\right)^{1/2} = 5.17 \times 10^2 \text{ m/s}$$

$$u_{Cl_2} = \left(\frac{3 \times 8.314 \text{ kg} \cdot \text{m}^2/\text{s}^2 \cdot \text{K} \cdot \text{mol} \times 300 \text{ K}}{70.9 \times 10^{-3} \text{ kg/mol}}\right)^{1/2} = 3.25 \times 10^2 \text{ m/s}$$

As expected, the lighter molecule moves at the greater speed.

10.77   *Plan.* The heavier the molecule, the slower the rate of effusion. Thus, the order for increasing rate of effusion is in the order of decreasing mass. *Solve.*

rate $^2\text{H}^{37}\text{Cl}$ < rate $^1\text{H}^{37}\text{Cl}$ < rate $^2\text{H}^{35}\text{Cl}$ < rate $^1\text{H}^{35}\text{Cl}$

10.78 $\dfrac{\text{rate}^{235}\text{U}}{\text{rate}^{238}\text{U}} = \sqrt{\dfrac{238.05}{235.04}} = \sqrt{1.0128} = 1.0064$

There is a slightly greater rate enhancement for $^{235}\text{U}(g)$ atoms than $^{235}\text{UF}_6(g)$ molecules (1.0043), because ^{235}U is a greater percentage (100%) of the mass of the diffusing particles than in $^{235}\text{UF}_6$ molecules. The masses of the isotopes were taken from *The Handbook of Chemistry and Physics*.

10.79   *Analyze.* Given: relative effusion rates of two gases at same temperature. Find: molecular formula of one of the gases. *Plan.* Use Graham's law to calculate the formula weight of arsenic (III) sulfide, and thus the molecular formula. *Solve.*

$$\frac{\text{rate (sulfide)}}{\text{rate (Ar)}} = \left[\frac{39.9}{\text{MM (sulfide)}}\right]^{1/2} = 0.28$$

MM (sulfide) = $(39.9 / 0.28)^2$ = 510 g/mol (two significant figures)

The empirical formula of arsenic(III) sulfide is As_2S_3, which has a formula mass of 246.1. Twice this is 490 g/mol, close to the value estimated from the effusion experiment. Thus, the formula of the vapor phase molecule is As_4S_6.

10.80   The time required is proportional to the reciprocal of the effusion rate.

$$\frac{\text{rate (X)}}{\text{rate (O}_2)} = \frac{105 \text{ s}}{31 \text{ s}} = \left[\frac{32 \text{ g O}_2}{\text{MM}_x}\right]^{1/2}; \text{ MM}_x = 32 \text{ g O}_2 \times \left[\frac{105}{31}\right]^2 = 370 \text{ g/mol (two sig figs)}$$

Nonideal-Gas Behavior

10.81   (a)   Nonideal gas behavior is observed at very high pressures and/or low temperatures.

            (b)   The real volumes of gas molecules and attractive intermolecular forces between molecules cause gases to behave nonideally.

            (c)   The ratio PV/RT is equal to the number of moles of particles in an ideal-gas sample; this number should be a constant for all pressure, volume, and temperature conditions. If the value of this ratio changes with increasing pressure, the gas sample is not behaving ideally. That is, the gas is not behaving according to the ideal-gas equation.

Negative deviations indicate fewer "effective" particles in the sample, a result of attractive forces among particles. Positive deviations indicate more "effective" particles, a result of the real volume occupied by the particles.

10.82   Ideal-gas behavior is most likely to occur at high temperature and low pressure, so the atmosphere on Mercury is more likely to obey the ideal-gas law. The higher temperature on Mercury means that the kinetic energies of the molecules will be larger relative to intermolecular attractive forces. Further, the gravitational attractive forces on Mercury are lower because the planet has a much smaller mass. This means that for the same column mass of gas (Figure 10.1), atmospheric pressure on Mercury will be lower.

10.83   *Plan.* The constants a and b are part of the correction terms in the van der Waals equation. The smaller the values of a and b, the smaller the corrections and the more ideal the gas. *Solve.*

Ar ($a = 1.34$, $b = 0.0322$) will behave more like an ideal gas than CO_2 ($a = 3.59$, $b = 0.0427$) at high pressures.

10.84   The constant a is a measure of the strength of intermolecular attractions among gas molecules; b is a measure of molecular volume. Both increase with increasing molecular mass and structural complexity.

10.85   *Analyze.* Conditions and amount of $CCl_4(g)$ are given. *Plan.* Use ideal-gas equation and van der Waals equation to calculate pressure of gas at these conditions.   *Solve.*

(a) $P = 1.00 \text{ mol} \times \dfrac{0.08206 \text{ L} \bullet \text{atm}}{\text{K} \bullet \text{mol}} \times \dfrac{313 \text{ K}}{28.0 \text{ L}} = 0.917 \text{ atm}$

(b) $P = \dfrac{nRT}{V - nb} - \dfrac{an^2}{V^2} = \dfrac{1.00 \times 0.08206 \times 313}{28.0 - (1.00 \times 0.1383)} - \dfrac{20.4(1.00)^2}{(28.0)^2} = 0.896 \text{ atm}$

Check. The van der Waals result indicates that the real pressure will be less than the ideal pressure. That is, intermolecular forces reduce the effective number of particles and the real pressure. This is reasonable for 1 mole of gas at relatively low temperature and pressure.

10.86   (a)   At STP, the molar volume $= 1 \text{ mol} \times \dfrac{0.08206 \text{ L} \bullet \text{atm}}{\text{K} \bullet \text{mol}} \times \dfrac{273 \text{ K}}{1 \text{ atm}} = 22.4 \text{ L}$

Dividing the value for b, 0.0322 L/mol, by 4, we obtain 0.0080 L. Thus, the volume of the Ar atoms is $(0.0080/22.4)100 = 0.036\%$ of the total volume.

(b)   At 100 atm pressure (and 0°C, standard temperature) the molar volume is 0.224 L, and the volume of the Ar atoms is 3.6% of the total volume.

Additional Exercises

10.87   A mercury barometer with water trapped in its tip would not read the correct pressure. The standard relationship between height of an Hg column and atmospheric pressure (760 mm Hg = 1 atm) assumes that there is a vacuum in the closed end of the barometer and that gravity is the only downward force on the Hg column. Water at the top of the Hg column would establish a vapor pressure, which would exert additional downward pressure and partially counterbalance the pressure of the atmosphere. The Hg column would read lower than the actual atmospheric pressure.

10.88   Only item (b) is satisfactory. Item (c) would not have supported a column of Hg because it is open at both ends. The atmosphere would exert pressure on the top of the column as well as on the reservoir; the column would only be as high as the reservoir and the height would not change with changing pressure. Item (d) is not tall enough to support a nearly 760 mm Hg column. Items (a) and (e) are inappropriate for the same reason: they don't have a uniform cross-sectional area. The height of the Hg column is a direct measure of atmospheric pressure only if the cross-sectional area is constant over the entire tube.

10.89 $P_1 V_1 = P_2 V_2$; $V_2 = P_1 V_1 / P_2$

$$V_2 = \frac{3.0\,\text{atm} \times 1.0\,\text{mm}^3}{695\,\text{torr}} \times \frac{760\,\text{torr}}{1\,\text{atm}} = 3.3\,\text{mm}^3$$

10.90 $PV = nRT$, $n = PV/RT$. Since RT is constant, n is proportional to PV.

Total available $n = (15.0\,\text{L} \times 1.00 \times 10^2\,\text{atm}) - (15.0\,\text{L} \times 1.00\,\text{atm}) = 1485$

$$= 1.49 \times 10^3\,\text{L} \bullet \text{atm}$$

Each balloon holds $2.00\,\text{L} \times 1.00\,\text{atm} = 2.00\,\text{L} \bullet \text{atm}$

$$\frac{1485\,\text{L} \bullet \text{atm available}}{2.00\,\text{L} \bullet \text{atm/balloon}} = 742.5 = 742\,\text{balloons}$$

(Only 742 balloons can be filled completely, with a bit of He left over.)

10.91 $P = \dfrac{nRT}{V}$; $n = 1.4 \times 10^{-5}$ mol, $V = 0.600$ L, $T = 23°C = 296$ K

$$P = 1.4 \times 10^{-5}\,\text{mol} \times \frac{0.08206\,\text{L} \bullet \text{atm}}{\text{K} \bullet \text{mol}} \times \frac{296\,\text{K}}{0.600\,\text{L}} = 5.7 \times 10^{-4}\,\text{atm} = 0.43\,\text{mm Hg}$$

10.92   (a) $n = \dfrac{PV}{RT} = 3.00\,\text{atm} \times \dfrac{\text{K} \bullet \text{mol}}{0.08206\,\text{L} \bullet \text{atm}} \times \dfrac{110\,\text{L}}{300\,\text{K}} = 13.4\,\text{mol}\ C_3H_8(g)$

(b) $\dfrac{0.590\,\text{g}\ C_3H_8\,(l)}{1\,\text{mL}} \times 110 \times 10^3\,\text{mL} \times \dfrac{1\,\text{mol}\ C_3H_8}{44.094\,\text{g}} = 1.47 \times 10^3\,\text{mol}\ C_3H_8\ (l)$

(c)   Using C_3H_8 in a 110 L container as an example, the ratio of moles liquid to moles gas that can be stored in a certain volume is $\dfrac{1.47 \times 10^3\,\text{mol liquid}}{13.4\,\text{mol gas}} = 110$.

A container with a fixed volume holds many more moles (molecules) of $C_3H_8(l)$ because in the liquid phase the molecules are touching. In the gas phase, the molecules are far apart (statement 2, section 10.7), and many fewer molecules will fit in the container.

10.93   If the air in the room is at STP, the partial pressure of O_2 is 0.2095×1 atm $= 0.2095$ atm. Since the gases in air are perfectly mixed, the volume of O_2 is the volume of the room.

$$V = 10.0\,\text{ft} \times 8.0\,\text{ft} \times 8.0\,\text{ft} \times \frac{(12)^3\,\text{in}^3}{\text{ft}^3} \times \frac{(2.54)^3\,\text{cm}^3}{\text{in}^3} \times \frac{1\,\text{L}}{1000\,\text{cm}^3} = 1.812 \times 10^4$$

$$= 1.8 \times 10^4\,\text{L}$$

$$g = \frac{MM \times PV}{RT} = \frac{32.00 \text{ g } O_2}{\text{mol } O_2} \times \frac{K \bullet mol}{0.08026 \text{ L} \bullet atm} \times \frac{0.2095 \text{ atm} \times 1.812 \times 10^4 \text{ L}}{273 \text{ K}} = 5.4 \times 10^3 \text{ g } O_2$$

10.94   Volume of laboratory $= 54 \text{ m}^2 \times 3.1 \text{ m} \times \frac{1000 \text{ L}}{1 \text{ m}^3} = 1.674 \times 10^5 = 1.7 \times 10^5 \text{ L}$

Calculate the **total** moles of gas in the laboratory at the conditions given.

$$n_t = \frac{PV}{RT} = 1.00 \text{ atm} \times \frac{K \bullet mol}{0.08206 \text{ L} \bullet atm} \times \frac{1.674 \times 10^5 \text{ L}}{297 \text{ K}} = 6.869 \times 10^3 = 6.9 \times 10^3 \text{ mol gas}$$

An $Ni(CO)_4$ concentration of 1 part in 10^9 means 1 mol $Ni(CO)_4$ in 1×10^9 total moles of gas.

$$\frac{x \text{ mol } Ni(CO)_4}{6.869 \times 10^3 \text{ mol gas}} = \frac{1}{10^9} = 6.869 \times 10^{-6} \text{ mol } Ni(CO)_4$$

$$6.869 \times 10^{-6} \text{ mol } Ni(CO)_4 \times \frac{170.74 \text{ g } Ni(CO)_4}{1 \text{ mol } Ni(CO)_4} = 1.2 \times 10^{-3} \text{ g } Ni(CO)_4$$

10.95   It is simplest to calculate the partial pressure of each gas as it expands into the total volume, then sum the partial pressures.

$P_2 = P_1 V_1 / V_2; \; P_{N_2} = 265 \text{ torr} (1.0 \text{ L}/2.5 \text{ L}) = 106 = 1.1 \times 10^2 \text{ torr}$

$P_{Ne} = 800 \text{ torr} (1.0 \text{ L}/2.5 \text{ L}) = 320 = 3.2 \times 10^2 \text{ torr}; \; P_{H_2} = 532 \text{ torr} (0.5 \text{ L}/2.5 \text{ L})$

$$= 106 = 1.1 \times 10^2 \text{ torr}$$

$P_t = P_{N_2} + P_{Ne} + P_{H_2} = (106 + 320 + 106) \text{ torr} = 532 = 5.3 \times 10^2 \text{ torr}$

10.96   (a) $n = \frac{PV}{RT} = 0.980 \text{ atm} \times \frac{K \bullet mol}{0.08206 \text{ L} \bullet atm} \times \frac{0.524 \text{ L}}{347 \text{ K}} = 0.018034 = 0.0180 \text{ mol air}$

$$\text{mol } O_2 = 0.018034 \text{ mol air} \times \frac{0.2095 \text{ mol } O_2}{1 \text{ mol air}} = 0.003778 = 0.00378 \text{ mol } O_2$$

(b) $C_8H_{18}(l) + 25/2 \, O_2(g) \rightarrow 8CO_2(g) + 9H_2O(g)$

(The H_2O produced in an automobile engine is in the gaseous state.)

$$0.003778 \text{ mol } O_2 \times \frac{1 \text{ mol } C_8H_{18}}{12.5 \text{ mol } O_2} \times \frac{114.2 \text{ g } C_8H_{18}}{1 \text{ mol } C_8H_{18}} = 0.0345 \text{ g } C_8H_{18}$$

10.97   (a) $5.00 \text{ g HCl} \times \frac{1 \text{ mol HCl}}{36.46 \text{ g HCl}} = 0.1371 = 0.137 \text{ mol HCl}$

$$5.00 \text{ g } NH_3 \times \frac{1 \text{ mol } NH_3}{17.03 \text{ g } NH_3} = 0.2936 = 0.294 \text{ mol } NH_3$$

The gases react in a 1:1 mole ratio, HCl is the limiting reactant and is completely consumed. $(0.2936 \text{ mol} - 0.1371 \text{ mol}) = 0.1565 = 0.157 \text{ mol } NH_3$ remain in the system. $NH_3(g)$ is the only gas remaining after reaction. $V_t = 4.00 \text{ L}$

(b) $P = \frac{nRT}{V} = 0.1565 \text{ mol} \times \frac{0.08206 \text{ L} \bullet atm}{K \bullet mol} \times \frac{298 \text{ K}}{4.00 \text{ L}} = 0.957 \text{ atm}$

10.98   V and T are the same for He and O_2.

$P_{He} V = n_{He} RT, \quad P_{He} / n_{He} = RT/V; \quad P_{O_2} / n_{O_2} = RT/V$

$$\frac{P_{He}}{n_{He}} = \frac{P_{O_2}}{n_{O_2}} = n_{O_2} = \frac{P_{O_2} \times n_{He}}{P_{He}} ; \; n_{He} = 1.42 \text{ g He} \times \frac{1 \text{ mol He}}{4.003 \text{ g He}} = 0.3547 = 0.355 \text{ mol He}$$

$$n_{O_2} = \frac{158 \text{ torr}}{42.5 \text{ torr}} \times 0.355 \text{ mol} = 1.3188 = 1.32 \text{ mol } O_2 ; \; 1.3188 \text{ mol } O_2 \times \frac{32.00 \text{ g } O_2}{1 \text{ mol } O_2} = 42.2 \text{ g } O_2$$

10.99 $\quad MM_{avg} = \dfrac{dRT}{P} = \dfrac{1.104 \text{ g}}{1 \text{ L}} \times \dfrac{0.08206 \text{ L} \cdot \text{atm}}{\text{K} \cdot \text{mol}} \times \dfrac{300 \text{ K}}{435 \text{ torr}} \times \dfrac{760 \text{ torr}}{1 \text{ atm}} = 47.48 = 47.5 \text{ g/mol}$

χ = mole fraction O_2; $1 - \chi$ = mole fraction Kr

$47.48 \text{ g} = \chi(32.00) + (1 - \chi)(83.80)$

$36.3 = 51.8 \; \chi; \; \chi = 0.701; \; 70.1\% \; O_2$

10.100 Calculate the number of moles of Ar in the vessel:

$n = (339.854 - 337.428)/39.948 = 0.060729 = 0.06073 \text{ mol}$

The total number of moles of the mixed gas is the same (Avogadro's Law). Thus, the average atomic weight is $(339.076 - 337.428)/0.060729 = 27.137 = 27.14$. Let the mole fraction of Ne be χ. Then,

$\chi \; (20.183) + (1 - \chi) \; (39.948) = 27.137; \; 12.811 = 19.765 \; \chi; \; \chi = 0.6482$

Neon is thus 64.82 mole percent of the mixture.

10.101 (a) The quantity $d/P = MM/RT$ should be a constant at all pressures for an ideal gas. It is not, however, because of nonideal behavior. If we graph d/P vs P, the ratio should approach ideal behavior at low P. At $P = 0$, $d/P = 2.2525$. Using this value in the formula $MM = d/P \times RT$, $MM = 50.46 \text{ g/mol}$.

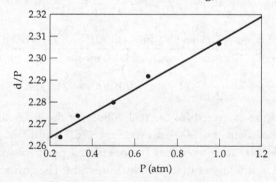

(b) The ratio d/P varies with pressure because of the finite volumes of gas molecules and attractive intermolecular forces.

10.102 (a) The effect of intermolecular attraction becomes more significant as a gas is compressed to a smaller volume at constant temperature. This compression causes the pressure, and thus the number of intermolecular collisions, to increase. Intermolecular attraction causes some of these collisions to be inelastic, which amplifies the deviation from ideal behavior.

(b) The effect of intermolecular attraction becomes less significant as the temperature of a gas is increased at constant volume. When the temperature of a gas is increased at constant volume, the pressure of the gas, the number of intermolecular collisions, and the average kinetic energy of the gas particles increases. This higher average kinetic energy means that a larger fraction of the molecules has sufficient kinetic energy to overcome intermolecular attractions, even though there are more total collisions. This increases the fraction of elastic collisions, and the gas more closely obeys the ideal-gas equation.

10.103 (a) The initial drop in the value of PV/RT is due to attractive forces between molecules. These intermolecular forces cause the molecules to "stick" together and behave as if there are fewer net particles in the sample. At lower pressures, this is the dominant effect. At the same time, the real volume of gas molecules causes the amount of free space in the gas sample to be less than the container volume. Using the container volume to calculate PV/RT gives a value larger than that for an ideal gas (which assumes that the total container volume is free space). At higher pressures, this effect more than compensates for molecular attraction, and PV/RT is greater than 1.

(b) As the temperature of a gas increases, the average kinetic energy of the particles increases. The increased kinetic energy overcomes the attractive forces between molecules and keeps them separate.

10.104 (a) $120.00 \text{ kg N}_2(g) \times \dfrac{1000 \text{ g}}{1 \text{ kg}} \times \dfrac{1 \text{ mol N}_2}{28.0135 \text{ g N}} = 4283.6 \text{ mol N}_2$

$P = \dfrac{nRT}{V} = 4283.6 \text{ mol} \times \dfrac{0.08206 \text{ L} \bullet \text{atm}}{\text{K} \bullet \text{mol}} \times \dfrac{553 \text{ K}}{1100.0 \text{ L}} = 176.72 = 177 \text{ atm}$

(b) According to Equation [10.26], $P = \dfrac{nRT}{V - nb} - \dfrac{n^2 a}{V^2}$

$P = \dfrac{(4283.6 \text{ mol})(0.08206 \text{ L} \bullet \text{atm/K} \bullet \text{mol})(553 \text{ K})}{1100.0 \text{ L} - (4283.6 \text{ mol})(0.0391 \text{ L/mol})} - \dfrac{(4283.6 \text{ mol})^2 (1.39 \text{ L}^2 \bullet \text{atm/mol}^2)}{(1100.0 \text{ L})^2}$

$P = \dfrac{194{,}388 \text{ L} \bullet \text{atm}}{1100.0 \text{ L} - 167.5 \text{ L}} - 21.1 \text{ atm} = 208.5 \text{ atm} - 21.1 \text{ atm} = 187.4 \text{ atm}$

(c) The pressure corrected for the real volume of the N_2 molecules is 208.5 atm, 31.8 atm higher than the ideal pressure of 176.7 atm. The 21.1 atm correction for intermolecular forces reduces the calculated pressure somewhat, but the "real" pressure is still higher than the ideal pressure. The correction for the real volume of molecules dominates. Even though the value of b is small, the number of moles of N_2 is large enough so that the molecular volume correction is larger than the attractive forces correction.

Integrative Exercises

10.105 (a) $MM = \dfrac{gRT}{VP} = \dfrac{1.56 \text{ g}}{1.00 \text{ L}} \times \dfrac{0.08206 \text{ L} \bullet \text{atm}}{\text{K} \bullet \text{mol}} \times \dfrac{323 \text{ K}}{0.984 \text{ atm}} = 42.0 \text{ g/mol}$

Assume 100 g cyclopropane

$100 \text{ g} \times 0.857 \text{ C} = 85.7 \text{ g C} \times \dfrac{1 \text{ mol C}}{12.01 \text{ g}} = \dfrac{7.136 \text{ mol C}}{7.136} = 1 \text{ mol C}$

$$100\,g \times 0.143\,H = 14.3\,g\,H \times \frac{1\,mol\,H}{1.008\,g} = \frac{14.19\,mol\,H}{7.136} = 2\,mol\,H$$

The empirical formula of cyclopropane is CH_2 and the empirical formula weight is $12 + 2 = 14$ g. The ratio of molar mass to empirical formula weight, $42.0\,g/14\,g$, is 3; therefore, there are three empirical formula units in one cyclopropane molecule. The molecular formula is $3 \times (CH_2) = C_3H_6$.

(b) Ar is a monoatomic gas. Cyclopropane molecules are larger and more structurally complex, even though the molar masses of Ar and C_3H_6 are similar. If both gases are at the same relatively low temperature, they have approximately the same average kinetic energy, and the same ability to overcome intermolecular attractions. We expect intermolecular attractions to be more significant for the more complex C_3H_6 molecules, and that C_3H_6 will deviate more from ideal behavior at the conditions listed. This conclusion is supported by the $\underline{a}$ values in Table 10.3. The $\underline{a}$ values for CH_4 and CO_2, more complex molecules than Ar atoms, are larger than the value for Ar. If the pressure is high enough for the volume correction in the van der Waals equation to dominate behavior, the larger C_3H_6 molecules definitely deviate more than Ar atoms from ideal behavior.

10.106   *Plan*. Write the balanced equation for the combustion of methanol. Since amounts of both reactants are given, determine the limiting reactant. Use mole ratios to calculate mol H_2O produced, based on the amount of limiting reactant.   *Solve*.

methanol = $CH_3OH(l)$. $2CH_3OH(l) + 3O_2(g) \rightarrow 2CO_2(g) + 4H_2O(g)$

$$25.0\,mL\,CH_3OH \times \frac{0.850\,g\,CH_3OH}{mL} \times \frac{1\,mol\,CH_3OH}{32.04\,g} = 0.6632 = 0.663\,mol\,CH_3OH$$

$$mol\,O_2 = n = \frac{PV}{RT} = 1.00\,atm \times \frac{12.5\,L}{273\,K} \times \frac{K \cdot mol}{0.08206\,L \cdot atm} = 0.5580 = 0.558\,mol\,O_2$$

$$0.558\,mol\,O_2 \times \frac{2\,mol\,CH_3OH}{3\,mol\,O_2} = 0.372\,mol\,CH_3OH$$

0.558 mol O_2 can react with only 0.372 mol CH_3OH, so O_2 is the limiting reactant. Note that a large volume of $O_2(g)$ is required to completely react with a relatively small volume of $CH_3OH(l)$.

$$0.558\,mol\,O_2 \times \frac{4\,mol\,H_2O}{3\,mol\,O_2} = 0.744\,mol\,H_2O$$

10.107   (a)   *Plan.* Use the ideal-gas law to calculate the moles CO_2 that react.

        Solve. P(reacted) = P(initial) – P(final), at constant V, T. Since both CaO and BaO react with CO_2 in a 1:1 mole ratio, mol CaO + mol BaO = mol CO_2. Use molar masses to calculate % CaO in sample.

$$P(reacted) = 730\,torr - 150\,torr = 580\,torr;\ 580\,torr \times \frac{1\,atm}{760\,torr} = 0.76316 = 0.763\,atm$$

$$n = \frac{PV}{RT} = 0.76316\,atm \times \frac{1.0\,L}{298\,K} \times \frac{K \cdot mol}{0.08206\,L \cdot atm} = 0.03121 = 0.0312\,mol\,CO_2$$

  (b)   *Plan.* Use the stoichiometry of the reaction and definition of moles to calculate the mass and Mass % of CaO.

Solve. $CaO(s) + CO_2(s) \rightarrow CaCO_3(s)$. $BaO(s) + CO_2(g) \rightarrow BaCO_3(s)$

mol CO_2 reacted = mol CaO + mol BaO

Let x = g CaO, 4.00 – x = g BaO

$$0.03121 = \frac{x}{56.08} + \frac{4.00 - x}{153.3}$$

$0.03121(56.08)(153.3) = 153.3x + 56.08(4.00 - x)$

$268.3 = (153.3x - 56.08x) + 224.3$

$43.98 = 97.22x, x = 0.452 = 0.45$ g CaO

$$\frac{0.452 \text{ g CaO}}{4.00 \text{ g sample}} \times 100 = 11.3 = 11\% \text{ CaO}$$

(By strict sig fig rules, the result has 2 sig figs, because 268 – 224 = 44 has 0 decimal places and 2 sig figs.)

10.108 $n = \dfrac{PV}{RT} = 1.00 \text{ atm} \times \dfrac{K \cdot mol}{0.08206 \text{ L} \cdot atm} \times \dfrac{2.7 \times 10^{12} \text{ L}}{273 \text{ K}} = 1.205 \times 10^{11} = 1.2 \times 10^{11} \text{ mol CH}_4$

$CH_4(g) + 2O_2(g) \rightarrow CO_2(g) + 2H_2O(l)$ $\Delta H° = -890.4$ kJ

(At STP, H_2O is in the liquid state.)

$\Delta H_{rxn}^° = \Delta H_f^° CO_2(g) + 2\Delta H_f^° H_2O(l) - \Delta H_f^° CH_4(g) - \Delta H_f^° O_2(g)$

$\Delta H_{rxn}^° = -393.5 \text{ kJ} + 2(-285.83 \text{ kJ}) - (-74.8 \text{ kJ}) - 0 = -890.4$ kJ

$$\frac{-890.4 \text{ kJ}}{1 \text{ mol CH}_4} \times 1.205 \times 10^{11} \text{ mol CH}_4 = -1.073 \times 10^{14} = -1.1 \times 10^{14} \text{ kJ}$$

The negative sign indicates heat evolved by the combustion reaction.

10.109   (a)   *Analyze/Plan.* $AgF + S_8 \xrightarrow{\Delta}$ unknown gas

The gas probably contains the elements S and F in an unknown mole ratio. (Most compounds containing Ag are ionic and therefore solids.) Calculate the molar mass of the gas from its density at the given conditions. Determine the relative amount of F from the data on the reaction of the gas with water to produce HF. *Solve.*

$$MM = \frac{dRT}{P} = \frac{0.803 \text{ g}}{L} \times \frac{0.08206 \text{ L} \cdot atm}{K \cdot mol} \times \frac{305 \text{ K}}{150 \text{ mm}} \times \frac{760 \text{ mm}}{1 \text{ atm}} = 101.83$$

$$= 102 \text{ g/mol}$$

mol F in 480 mL sample: $M \times L$ = mol

0.081 M HF $\times$ 0.080 L = 0.00648 = 0.0065 mol HF = 0.0065 mol F in the sample total mol gas in 480 mL sample: n = PV/RT

$$n = 126 \text{ mm} \times \frac{1 \text{ atm}}{760 \text{ mm}} \times \frac{0.480 \text{ L}}{301 \text{ K}} \times \frac{K \cdot mol}{0.08206 \text{ L} \cdot atm} = 3.222 \times 10^{-3}$$

$$= 3.22 \times 10^{-3} \text{ mol}$$

mole ratio of S and F:

total g gas = 3.222×10^{-3} mol gas $\times$ 101.83 g/mol; = 0.32808 = 0.328 g gas g F = 6.48×10^{-3} mol F $\times$ 18.998 g/mol F = 0.12311 = 0.123 g F

g S = 0.32808 – 0.12311 = 0.20497 = 0.205 g S

mol S = 0.205 g S/32.07 g/mol = 0.0639 mol S

0.00639 mol S/0.00648 mol F = 1:1 mole ratio of S:F

The empirical formula is SF; empirical FW = 32.07 + 19.00 = 51.07;

since MM = 102, the empirical formula is S_2F_2

Check. The empirical and molecular formula weights are in an integer ratio, so the result is reasonable.

(b) 26 valence e⁻, 13 e⁻ pairs

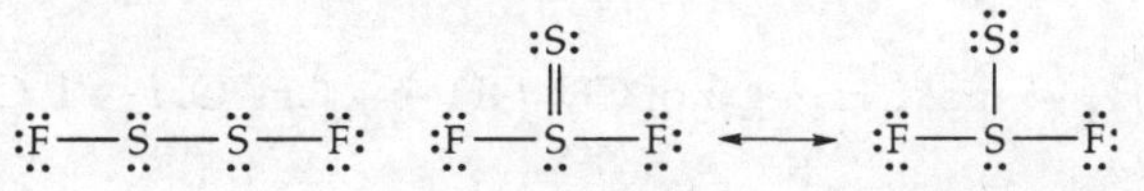

(c) According to VSEPR the electron domain geometry about S in each of the molecules will be tetrahedral, with bond angles of 109° or less. The left molecule will have a structure similar to hydrogen peroxide, H_2O_2; the 4 atoms are not necessarily coplanar, and in fact we expect a dihedral angle of approximately 110°. Since the right molecule has a single central atom, we can describe the molecular geometry as trigonal pyramidal.

From the given bond distances, we expect the single-bond covalent radii to be S = 1.02 Å, F = 0.72 Å. The simple conclusion is that the S–S distances will be ~2.04 Å and the S-F distances ~1.74 Å. In fact, the S–S distance in the left compound is 1.89 Å and in the right it is 1.86 Å. Clearly each of these bonds has some double bond character, as indicated by one of the resonance structures for the right molecule. The actual S-F distances are 1.63 Å (left) and 1.60 Å (right), also shorter than the predicted S-F single bond distance. One possible conclusion is that each of the bonds has some double bond character, that some of the "nonbonding" electron density is incorporated into a delocalized π-bonding network in the molecules.

10.110 (a) 19 e⁻, 9.5 e⁻ pairs

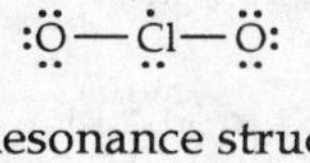

:Ö—Ċl—Ö:

Resonance structures can be drawn with the odd electron on O, but electronegativity considerations predict that it will be on Cl for most of the time.

(b) ClO_2 is very reactive because it is an odd-electron molecule. Adding an electron (reduction) both pairs the odd electron and completes the octet of Cl. Thus, ClO_2 has a strong tendency to gain an electron and be reduced.

(c) ClO_2^-, 20 e⁻, 10 e⁻ pairs

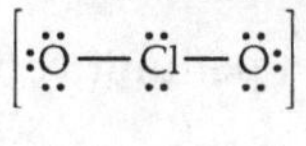

[:Ö—Ċl—Ö:]

(d) 4 e⁻ domains around Cl, O–Cl–O bond angle ~109°

(e)   Calculate mol Cl_2 from ideal-gas equation; determine limiting reactant; mass ClO_2 via mol ratios.

$$\text{mol } Cl_2 = \frac{PV}{RT} = 1.50 \text{ atm} \times \frac{2.00 \text{ L}}{294 \text{ K}} \times \frac{K \cdot mol}{0.08206 \text{ L} \cdot atm} = 0.1243 = 0.124 \text{ mol } Cl_2$$

$$10.0 \text{ g NaClO}_2 \times \frac{1 \text{ mol NaClO}_2}{90.44 \text{ g}} = 0.1106 = 0.111 \text{ mol NaClO}_2$$

2 mol $NaClO_2$ are required for 1 mol Cl_2, so $NaClO_2$ is the limiting reactant. For every 2 mol $NaClO_2$ reacted, 2 mol ClO_2 are produced, so mol ClO_2 = mol $NaClO_2$.

$$0.1106 \text{ mol ClO}_2 \times \frac{67.45 \text{ g ClO}_2}{mol} = 7.46 \text{ g ClO}_2$$

10.111   (a) ft^3 CH_4 → L CH_4 → mol CH_4 → mol CH_3OH → g CH_3OH → L CH_3OH

$$10.7 \times 10^9 \text{ ft } CH_4 \times \frac{1 \text{ yd}^3}{3^3 \text{ ft}^3} \times \frac{1 m^3}{(1.0936)^3 \text{ yd}^3} \times \frac{1 \text{ L}}{1 \times 10^{-3} \text{ m}^3} = 3.03001 \times 10^{11}$$

$$= 3.03 \times 10^{11} \text{ L } CH_4$$

$$n = \frac{PV}{RT} = \frac{3.03 \times 10^{11} \text{ L} \times 1.00 \text{ atm}}{298 \text{ K}} \times \frac{K \cdot mol}{0.08206 \text{ L} \cdot atm} = 1.2391 \times 10^{10}$$

$$= 1.24 \times 10^{10} \text{ mol } CH_4$$

1 mol CH_4 = 1 mol CH_3OH

$$1.2391 \times 10^{10} \text{ mol } CH_3OH \times \frac{32.04 \text{ g } CH_3OH}{mol \, CH_3OH} \times \frac{1 \text{ mL } CH_3OH}{0.791 g} \times \frac{1 L}{1000 \text{ mL}}$$

$$= 5.0189 \times 10^8 = 5.02 \times 10^8 \text{ L } CH_3OH$$

(b) $CH_4(g) + 2O_2(g) \rightarrow CO_2(g) + 2H_2O(l)$ $\Delta H° = -890.4$ kJ (see Solution 10.108)

$$1.2391 \times 10^{10} \text{ mol } CH_4 \times \frac{-890.4 \text{ kJ}}{1 \text{ mol } CH_4} = -1.10 \times 10^{13} \text{ kJ}$$

$CH_3OH(l) + 3/2 \, O_2(g) \rightarrow CO_2(g) + 2H_2O(l)$ $\Delta H° = -726.6$ kJ

$$\Delta H° = \Delta H_f° CO_2(g) + 2\Delta H_f° H_2O(l) - \Delta H_f° CH_2OH(l) - 3/2 \, \Delta H_f° O_2(g)$$

$$= -393.5 \text{ kJ} + 2(-285.83 \text{ kJ}) - (-238.6 \text{ kJ}) - 0 = -726.6 \text{ kJ}$$

$$1.2391 \times 10^{10} \text{ mol } CH_3OH \times \frac{-726.6 \text{ kJ}}{1 \text{ mol } CH_3OH} = -9.00 \times 10^{12} \text{ kJ}$$

(c)   Assume a volume of 1.00 L of each liquid.

$$1.00 \text{ L } CH_4(l) \times \frac{466 \text{ g}}{1 L} \times \frac{1 \text{ mol}}{16.04 g} \times \frac{-890.4 \text{ kJ}}{mol \, CH_4} = -2.59 \times 10^4 \text{ kJ/L } CH_4$$

$$1.00 \text{ L } CH_3OH \times \frac{791 \text{ g}}{1 \text{ L}} \times \frac{1 \text{ mol}}{32.04 g} \times \frac{-726.6 \text{ kJ}}{mol \, CH_3OH} = -1.79 \times 10^4 \text{ kJ/L } CH_3OH$$

Clearly $CH_4(l)$ has the higher enthalpy of combustion per unit volume.

10.112   After reaction, the flask contains $IF_5(g)$ and whichever reactant is in excess. Determine the limiting reactant, which regulates the moles of IF_5 produced and moles of excess reactant.

$$I_2(s) + 5F_2(g) \rightarrow 2\,IF_5(g)$$

$$10.0\,g\,I_2 \times \frac{1\,mol\,I_2}{253.8\,g\,I_2} \times \frac{5\,mol\,F_2}{1\,mol\,I_2} = 0.1970 = 0.197\,mol\,F_2$$

$$10.0\,g\,F_2 \times \frac{1\,mol\,F_2}{38.00\,g\,F_2} = 0.2632 = 0.263\,mol\,F_2\ available$$

I_2 is the limiting reactant; F_2 is in excess.

0.263 mol F_2 available – 0.197 mol F_2 reacted = 0.066 mol F_2 remain.

$$10.0\,g\,I_2 \times \frac{1\,mol\,I_2}{253.8\,g\,I_2} \times \frac{2\,mol\,IF_5}{1\,mol\,I_2} = 0.0788\,mol\,IF_5\ produced$$

(a) $P_{IF_5} = \dfrac{nRT}{V} = 0.0788\,mol \times \dfrac{0.08206\,L \bullet atm}{K \bullet mol} \times \dfrac{398\,K}{5.00\,L} = 0.515\,atm$

(b) $\chi_{IF_5} = \dfrac{mol\,IF_5}{mol\,IF_5 + mol\,F_2} = \dfrac{0.0788}{0.0788 + 0.066} = 0.544$

10.113   (a) $MgCO_3(s) + 2HCl(aq) \rightarrow MgCl_2(aq) + H_2O(l) + CO_2(g)$

               $CaCO_3(s) + 2HCl(aq) \rightarrow CaCl_2(aq) + H_2O(l) + CO_2(g)$

(b) $n = \dfrac{PV}{RT} = 743\,torr \times \dfrac{1\,atm}{760\,torr} \times \dfrac{K \bullet mol}{0.08206\,L \bullet atm} \times \dfrac{1.72\,L}{301\,K}$

$$= 0.06808 = 0.0681\,mol\ CO_2$$

(c)   x = g $MgCO_3$, y = g $CaCO_3$, x + y = 6.53 g

         mol $MgCO_3$ + mol $CaCO_3$ = mol CO_2 total

$$\frac{x}{84.32} + \frac{y}{100.09} = 0.06808;\ y = 6.53 - x$$

$$\frac{x}{84.32} + \frac{6.53 - x}{100.09} = 0.06808$$

100.09x – 84.32x + 84.32(6.53) = 0.06808 (84.32)(100.09)

15.77x + 550.610 = 574.549; x = 1.52 g $MgCO_3$

$$mass\ \%\ MgCO_3 = \frac{1.52\,g\,MgCO_3}{6.53\,g\,sample} \times 100 = 23.3\%$$

[By strict sig fig rules, the answer has 2 sig figs: 15.77x + 551 (3 digits from 6.53) = 575; 575 – 551 = 24 (no decimal places, 2 sig figs) leads to 1.5 g $MgCO_3$.]

11 Intermolecular Forces, Liquids and Solids

Visualizing Concepts

11.1   The diagram best describes a **liquid**. In the diagram, the particles are close together, mostly touching but there is no regular arrangement or order. This rules out a gaseous sample, where the particles are far apart, and a crystalline solid, which has a regular repeating structure in all three directions.

11.2   (a)   Hydrogen bonding; H–F interactions qualify for this narrowly defined interaction.

       (b)   London dispersion forces, the only intermolecular forces between nonpolar F_2 molecules.

       (c)   Ion-dipole forces between Na^+ cation and the negative end of a polar covalent water molecule.

       (d)   Dipole-dipole forces between oppositely charged portions of two polar covalent SO_2 molecules.

11.3   The viscosity of glycerol will be greater than that of 1-propanol. Viscosity is the resistance of a substance to flow. The stronger the intermolecular forces in a liquid, the greater its viscosity. Hydrogen bonding is the predominant force for both molecules. Glycerol has three times as many O–H groups and many more H-bonding interactions than 1-propanol, so it experiences stronger intermolecular forces and greater viscosity. (Both molecules have the same carbon-chain length, so dispersion forces are similar.)

11.4   (a)   385 mm Hg. Find 30°C on the horizontal axis, and follow a vertical line from this point to its intersection with the red vapor pressure curve. Follow a horizontal line from the intersection to the vertical axis and read the vapor pressure.

       (b)   22°C. Reverse the procedure outlined in part (a). Find 300 torr on the vertical axis, follow it to the curve and down to the value on the horizontal axis.

       (c)   47°C. The normal boiling point of a liquid is the temperature at which its vapor pressure is 1 atm, or 760 mm Hg. On this diagram, the vapor pressure curve ends at this point, approximately 47°C.

11.5   The stronger the intermolecular forces, the greater the average kinetic energy required to escape these forces, and the higher the boiling point. $CH_3CH_2CH_2OH$ has hydrogen bonding, by virtue of its –OH group, so it has the higher boiling point. Dispersion forces are similar because molar masses are the same for both molecules.

11.6   (a)   360 K, the normal boiling point; 265 K, normal freezing point. The left-most line is the freezing/melting curve, the right-most line is the condensation/boiling curve. The normal boiling and freezing points are the temperatures of boiling and freezing at 1 atm pressure.

       (b)   The material is solid in the white zone, liquid in the blue zone, and gas in the yellow zone. (i) gas (ii) solid (iii) liquid

11.7   (a)   Nb: $6 \times 1/2 = 3$; O: $12 \times 1/4 = 3$

       (b)   NbO

       (c)   This is primarily an ionic solid, because Nb is a metal and O is a nonmetal. There may be some covalent character to the Nb $\cdots$ O bonds.

11.8   (a)   Clearly, the structure is close-packed. The question is: cubic or hexagonal? Without looking deeper into the layers of oranges, one cannot distinguish whether the layer structure is cubic (ABCABC) or hexagonal (ABABAB) close-packed.

       (b)   CN = 12, regardless of whether the structure is hexagonal or cubic close packed.

       (c)   Molecular. There are no strong bonds between particles.

Kinetic-Molecular Theory

11.9   (a)   solid < liquid < gas

       (b)   gas < liquid < solid

11.10   (a)   In solids, particles are in essentially fixed positions relative to each other, so the average energy of attraction is stronger than average kinetic energy. In liquids, particles are close together but moving relative to each other. The average attractive energy and average kinetic energy are approximately balanced. In gases, particles are far apart and in constant, random motion. Average kinetic energy is much greater than average energy of attraction.

       (b)   As the temperature of a substance is increased, the average kinetic energy of the particles increases. In a collection of particles (molecules), the state is determined by the strength of interparticle forces relative to the average kinetic energy of the particles. As the average kinetic energy increases, more particles are able to overcome intermolecular attractive forces and move to a less ordered state, from solid to liquid to gas.

       (c)   At constant temperature, the average kinetic energy of a collection of particles is constant. Compression brings particles closer together and increases the number of particle-particle collisions. With more collisions, the likelihood of intermolecular attractions causing the particles to coalesce (liquefy) is greater.

11.11   (a)   Gases are more compressible than liquids because there is much empty space between gas particles.

       (b)   The solid and liquid forms of a substance are called *condensed phases* because in both states there is very little space between particles; the volume of the substance is condensed.

(c)   Liquids have greater ability to flow because their average kinetic energy is on the same order of magnitude as the average attractive energy. This is easy to see for the solid and liquid forms of the same substance, where intermolecular attractive forces are the same, but the average kinetic energy of the liquid is greater because its temperature is higher.

11.12   (a)   The average distance between molecules is greater in the liquid state. Density is the ratio of the mass of a substance to the volume it occupies. For the same substance in different states, mass will be the same. The smaller the density, the greater the volume occupied, and the greater the distance between molecules. The liquid at 130° has the lower density (1.08 g/cm^3), so the average distance between molecules is greater.

(b)   As the temperature of a substance increases, the average kinetic energy and speed of the molecules increases. At the melting point the molecules, on average, have enough kinetic energy to break away from the very orderly array that was present in the solid. As the translational motion of the molecules increases, the occupied volume increases and the density decreases. Thus, the solid density, 1.266 g/cm^3 at 15°C, is greater than the liquid density, 1.08 g/cm^3 at 130°C.

Intermolecular Forces

11.13   (a)   London-dispersion forces

(b)   dipole-dipole and London-dispersion forces

(c)   dipole-dipole or in certain cases hydrogen bonding

11.14   (a)   London-dispersion forces

(b)   dipole-dipole and London-dispersion forces

(c)   hydrogen bonding (dominates properties of these small molecules) and London-dispersion forces (less significant)

11.15   (a) Br_2 is a nonpolar covalent molecule, so only London-dispersion forces must be overcome to convert the liquid to a gas.

(b) CH_3OH is a polar covalent molecule that experiences London-dispersion, dipole-dipole, and hydrogen-bonding (O–H bonds) forces. All of these forces must be overcome to convert the liquid to a gas.

(c) H_2S is a polar covalent molecule that experiences London-dispersion and dipole-dipole forces, so these must be overcome to change the liquid into a gas. (H–S bonds do not lead to hydrogen-bonding interactions.)

11.16   (a) CH_3OH experiences hydrogen bonding, but CH_3SH does not.

(b)   Both gases are influenced by London-dispersion forces. The heavier the gas particles, the stronger the London-dispersion forces. The heavier Xe is a liquid at the specified conditions, while the lighter Ar is a gas.

(c)   Both gases are influenced by London-dispersion forces. The larger, diatomic Cl_2 molecules are more polarizable, experience stronger dispersion forces, and have the higher boiling point.

(d)   Acetone and 2-methylpropane are molecules with similar molar masses and London-dispersion forces. Acetone also experiences dipole-dipole forces and has the higher boiling point.

11.17   (a)   *Polarizability* is the ease with which the charge distribution (electron cloud) in a molecule can be distorted to produce a transient dipole.

(b)   Te is most polarizable because its valence electrons are farthest from the nucleus and least tightly held.

(c)   Polarizability increases as molecular size (and thus molecular weight) increases. In order of increasing polarizability: $CH_4 < SiH_4 < SiCl_4 < GeCl_4 < GeBr_4$

(d)   The magnitude of London-dispersion forces and thus the boiling points of molecules increase as polarizability increases. The order of increasing boiling points is the order of increasing polarizability:
$CH_4 < SiH_4 < SiCl_4 < GeCl_4 < GeBr_4$

11.18   (a)   A more polarizable molecule can develop a larger transient dipole, increasing the strength of electrostatic attractions among polarized molecules.

(b)   The noble gases are all monoatomic. Going down the column, the atomic radius and the size of the electron cloud increase. The larger the electron cloud, the more polarizable the atom, the stronger the London-dispersion forces and the higher the boiling point.

(c)   It is generally true that the greater the molecular weight, the stronger the dispersion forces experienced by a molecule. This is true because trends in molecular size and molecular weight are usually parallel.

(d)   It is usually true that as the number of electrons in a molecule increases, the size of the molecule increases. Larger molecules tend to have diffuse electron clouds, which lead to greater polarizability. Thus, the statement that more electrons lead to increased dispersion forces (and greater polarizability) is correct.

11.19   *Analyze/Plan.* For molecules with similar structures, the strength of dispersion forces increases with molecular size (molecular weight and number of electrons in the molecule).

Solve: (a) H_2S (b) CO_2 (c) SiH_4

11.20   For molecules with similar structures, the strength of dispersion forces increases with molecular size (molecular weight and number of electrons in the molecule).

(a) Br_2

(b) $CH_3CH_2CH_2SH$

(c) $CH_3CH_2CH_2Cl$. These two molecules have the same molecular formula and molecular weight (C_3H_7Cl, molecular weight = 78.5 amu), so the shapes of the molecules determine which has the stronger dispersion forces. According to Figure 11.6, the cylindrical (not branched) molecule will have stronger dispersion forces.

11.21   Both hydrocarbons experience dispersion forces. Rod-like butane molecules can contact each other over the length of the molecule, while spherical 2-methylpropane molecules can only touch tangentially. The larger contact surface of butane produces greater polarizability and a higher boiling point.

11.22   Both molecules experience hydrogen bonding through their –OH groups and dispersion forces between their hydrocarbon portions. The position of the –OH group in isopropyl alcohol shields it somewhat from approach by other molecules and slightly decreases the extent of hydrogen bonding. Also, isopropyl alcohol is less rod-like (it has a shorter chain) than propyl alcohol, so dispersion forces are weaker. Since hydrogen bonding and dispersion forces are weaker in isopropyl alcohol, it has the lower boiling point.

11.23   (a)   Molecules with N–H, O–H, and F–H bonds form hydrogen bonds with like molecules.

           (b)   **CH_3NH_2** and **CH_3OH** have N–H and O–H bonds, respectively. (CH_3F has C–F and C–H bonds, but no H–F bonds.)

11.24   (a)   Replacing a hydroxyl hydrogen with a CH_3 group eliminates hydrogen bonding in that part of the molecule. This reduces the strength of intermolecular forces and leads to a (much) lower boiling point.

           (b) $CH_3OCH_2CH_2OCH_3$ is a larger, more polarizable molecule with stronger London-dispersion forces and thus a higher boiling point.

11.25   (a)   HF has the higher boiling point because hydrogen bonding is stronger than dipole-dipole forces.

           (b) $CHBr_3$ has the higher boiling point because it has the higher molar mass, which leads to greater polarizability and stronger dispersion forces.

           (c)   ICl has the higher boiling point because it is a polar molecule. For molecules with similar structures and molar masses, dipole-dipole forces are stronger than dispersion forces.

11.26   (a) C_6H_{14}, dispersion; C_8H_{18}, dispersion. C_8H_{18} has the higher boiling point due to greater molar mass and similar strength of forces.

           (b) C_3H_8, dispersion; CH_3OCH_3, dipole-dipole, and dispersion. CH_3OCH_3 has the higher boiling point due to stronger intermolecular forces and similar molar mass.

           (c)   HOOH, hydrogen bonding, dipole-dipole, and dispersion; HSSH, dipole-dipole, and dispersion. HOOH has the higher boiling point due to the influence of hydrogen bonding (Figure 11.7).

           (d) NH_2NH_2, hydrogen bonding, dipole-dipole, and dispersion; CH_3CH_3, dispersion. NH_2NH_2 has the higher boiling point due to much stronger intermolecular forces.

11.27   (a)   Hydrogen bonding occurs in both water and ice. In water, the molecules are as close together as possible, but in constant motion relative to each other. Hydrogen bonds are constantly broken and new ones formed. In ice, molecules are fixed relative to each other. They adopt an ordered structure that maximizes the number of hydrogen bonding interactions. For one molecule, the two lone pairs interact with H atoms of adjacent molecules and both H atoms interact with lone pairs of two other adjacent molecules. Each molecule participates in four hydrogen bonding interactions oriented tetrahedrally with respect to each other. Maintaining these four interactions per molecule creates a network structure (Figure 11.10) with more open space between molecules than in the liquid state and a corresponding lower density for the solid than the liquid.

(b)   The temperature of a substance is a measure of the average kinetic energy of the sample. In order to increase the kinetic energy of water molecules, strong attractive hydrogen bonds must be broken. A large amount of heat must be added to water to break hydrogen bonds and increase its temperature.

11.28   (a)   In the solid state, NH_3 molecules are arranged so as to form the maximum number of hydrogen bonds. At the melting point, the average kinetic energy of the molecules is large enough so that they are free to move relative to each other. As they move, old hydrogen bonds break and new ones form, but the strict relative order required for maximum hydrogen bonding is no longer present.

(b)   In the liquid state, molecules are moving relative to one another while touching, which makes some hydrogen bonding possible. When molecules achieve enough kinetic energy to vaporize, the distance between them increases beyond the point where hydrogen bonds can form.

Viscosity and Surface Tension

11.29   (a)   Surface tension and viscosity are the result of intermolecular attractive forces or cohesive forces among molecules in a liquid sample. As temperature increases, the number of molecules with sufficient kinetic energy to overcome these attractive forces increases, and viscosity and surface tension decrease.

(b)   Surface tension and viscosity are both directly related to the strength of intermolecular attractive forces. The same attractive forces that cause surface molecules to be difficult to separate cause molecules elsewhere in the sample to resist movement relative to one another. Liquids with high surface tension have intermolecular attractive forces sufficient to produce a high viscosity as well.

11.30   (a)   *Cohesive* forces bind molecules to each other.
            Adhesive forces bind molecules to surfaces.

(b)   The cohesive forces are hydrogen bonding among water molecules and also hydrogen bonding among cellulose molecules in the paper towel. Adhesive forces are any attractive forces between water and cellulose (the paper towel), likely also hydrogen bonding. If adhesive forces between cellulose and water weren't significant, paper towels wouldn't absorb water.

(c)   The shape of a meniscus depends on the strength of the cohesive forces within a liquid relative to the adhesive forces between the walls of the tube and the liquid. Adhesive forces between polar water molecules and silicates in glass (Figure 11.16) are even stronger than cohesive hydrogen-bonding forces among water molecules, so the meniscus is U-shaped (concave-upward).

11.31   (a) $CHBr_3$ has a higher molar mass, is more polarizable, and has stronger dispersion forces, so the surface tension is greater [see Solution 11.25(b)].

         (b)   As temperature increases, the viscosity of the oil decreases because the average kinetic energy of the molecules increases [Solution 11.29(a)].

         (c)   Adhesive forces between polar water and nonpolar car wax are weak, so the large surface tension of water draws the liquid into the shape with the smallest surface area, a sphere.

11.32   (a)   H—N̈—N̈—H   H—Ö—Ö—H   H—Ö—H
                       | |
                      H   H

         (b)   All have bonds (N–H or O–H, respectively) capable of forming hydrogen bonds. Hydrogen bonding is the strongest intermolecular interaction between neutral molecules and leads to very strong cohesive forces in liquids. The stronger the cohesive forces in a liquid, the greater the surface tension.

Changes of State

11.33   (a)   freezing, exothermic

         (b)   evaporation or vaporization, endothermic

         (c)   deposition, exothermic

11.34   (a)   condensation, exothermic

         (b)   sublimation, endothermic

         (c)   vaporization (evaporation), endothermic

         (d)   freezing, exothermic

11.35   The heat energy required to increase the kinetic energy of molecules enough to melt the solid does not produce a large separation of molecules. The specific order is disrupted, but the molecules remain close together. On the other hand, when a liquid is vaporized, the intermolecular forces which maintain close molecular contacts must be overcome. Because molecules are being separated, the energy requirement is higher than for melting.

11.36   (a)   Liquid ethyl chloride at room temperature is far above its boiling point. When the liquid contacts the metal surface, heat sufficient to vaporize the liquid is transferred from the metal to the ethyl chloride, and the heat content of the molecules increases. At constant atmospheric pressure, $\Delta H = q$, so the heat content and the enthalpy content of $C_2H_5Cl(g)$ is higher than that of $C_2H_5Cl(l)$.

         (b)   Attractive intermolecular forces hold the C_2H_5Cl molecules in close contact in the liquid phase. In order to overcome these attractive forces and maintain separation in the gas phase, the enthalpy content of the C_2H_5Cl molecules must increase when they change from the liquid to the gaseous state.

11.37   *Analyze.* The heat required to vaporize 50 g of H_2O equals the heat lost by the cooled water.

Plan. Using the enthalpy of vaporization, calculate the heat required to vaporize 50 g of H_2O in this temperature range. Using the specific heat capacity of water, calculate the mass of water than can be cooled 13°C if this much heat is lost.

Solve. Evaporation of 50 g of water requires:

$$50\,g\,H_2O \times \frac{2.4\,kJ}{1\,g\,H_2O} = 1.2 \times 10^2 \; kJ \; or \; 1.2 \times 10^5 \; J$$

Cooling a certain amount of water by 13°C:

$$1.2 \times 10^5 \; J \times \frac{1\,g \bullet K}{4.184\,J} \times \frac{1}{13°C} = 2206 = 2.2 \times 10^3 \; g\,H_2O$$

Check. The units are correct. A surprisingly large mass of water (2200 g ≈ 2.2 L) can be cooled by this method.

11.38   Energy released when 100 g of H_2O is cooled from 18°C to 0°C:

$$\frac{4.184\,J}{g \bullet K} \times 100\,g\,H_2O \times 18°C = 7.53 \times 10^3 \; J = 7.5 \; kJ$$

Energy released when 100 g of H_2O is frozen (there is no change in temperature during a change of state):

$$\frac{334\,J}{g} \times 100\,g\,H_2O = 3.34 \times 10^4 \; J = 33.4 \; kJ$$

Total energy released = 7.53 kJ + 33.4 kJ = 40.93 = 40.9 kJ

Mass of freon that will absorb 40.9 kJ when vaporized:

$$40.93\,kJ \times \frac{1 \times 10^3 \; J}{1\,kJ} \times \frac{1\,g\,CCl_2F_2}{289\,J} = 142 \; g\,CCl_2F_2$$

11.39   *Analyze/Plan.* Follow the logic in Sample Exercise 11.4. *Solve.*

Heat the solid from –120°C to –114°C (153 K to 159 K), using the specific heat of the solid.

$$75.0\,g\,C_2H_5OH \times \frac{0.97\,J}{g \bullet K} \times 6\,K \times \frac{1\,kJ}{1000\,J} = 0.4365 = 0.4 \; kJ$$

At –114°C (159 K), melt the solid, using its enthalpy of fusion.

$$75.0\,g\,C_2H_5OH \times \frac{1\,mol\,C_2H_5OH}{46.07\,g\,C_2H_5OH} \times \frac{5.02\,kJ}{1\,mol} = 8.172 = 8.17 \; kJ$$

Heat the liquid from –114°C to 78°C (159 K to 351 K), using the specific heat of the liquid.

$$75.0\,g\,C_2H_5OH \times \frac{2.3\,J}{g \bullet K} \times 192\,K \times \frac{1\,kJ}{1000\,J} = 33.12 = 33 \; kJ$$

At 78°C (351 K), vaporize the liquid, using its enthalpy of vaporization.

$$75.0 \text{ g C}_2\text{H}_5\text{OH} \times \frac{1 \text{ mol C}_2\text{H}_5\text{OH}}{46.07 \text{ g C}_2\text{H}_5\text{OH}} \times \frac{38.56 \text{ kJ}}{1 \text{ mol}} = 62.77 = 62.8 \text{ kJ}$$

The total energy required is 0.4365 kJ + 8.172 kJ + 33.12 kJ + 62.77 kJ = 104.50 = 105 kJ. (The result has zero decimal places, from 33 kJ required to heat the liquid.)

Check. The relative energies of the various steps are reasonable; vaporization is the largest.

11.40   Consider the process in steps, using the appropriate thermochemical constant.

Heat the liquid from 5.00°C to 47.6°C (278.00 K to 320.6 K), using the specific heat of the liquid.

$$25.0 \text{ g C}_2\text{Cl}_3\text{F}_3 \times \frac{0.91 \text{ J}}{\text{g} \bullet \text{K}} \times 42.6 \text{ K} \times \frac{1 \text{ kJ}}{1000 \text{ J}} = 0.969 = 0.97 \text{ kJ}$$

Boil the liquid at 47.6°C (320.6 K), using the enthalpy of vaporization.

$$25.0 \text{ g C}_2\text{Cl}_3\text{F}_3 \times \frac{1 \text{ mol C}_2\text{Cl}_3\text{F}_3}{187.4 \text{ g C}_2\text{Cl}_3\text{F}_3} \times \frac{27.49 \text{ kJ}}{\text{mol}} = 3.667 = 3.67 \text{ kJ}$$

Heat the gas from 47.6°C to 82.00°C (320.6 K to 355.00 K), using the specific heat of the gas.

$$25.0 \text{ g C}_2\text{Cl}_3\text{F}_3 \times \frac{0.67 \text{ J}}{\text{g} \bullet \text{K}} \times 34.4 \text{ K} \times \frac{1 \text{ kJ}}{1000 \text{ J}} = 0.576 = 0.58 \text{ kJ}$$

The total energy required is 0.969 kJ + 3.667 kJ + 0.576 kJ = 5.21 kJ.

11.41   (a)   The critical pressure is the pressure required to cause liquefaction at the critical temperature.

         (b)   The critical temperature is the highest temperature at which a gas can be liquefied, regardless of pressure. As the force of attraction between molecules increases, the critical temperature of the compound increases.

         (c)   The temperature of $N_2(l)$ is 77 K. All of the gases in Table 11.5 have critical temperatures higher than 77 K, so all of them can be liquefied at this temperature, given sufficient pressure.

11.42   (a)   According to Solution 11.41(b), the higher the critical temperature, the stronger the intermolecular forces of a substance. Therefore, the strength of intermolecular forces decreases moving from left to right across the series and as molecular weight decreases.

         (b)   The molecules in this series experience London-dispersion forces and, except for CF_4, dipole-dipole forces. We expect the strength of dispersion forces to increase with increasing molecular weight, which agrees with the trends in critical temperature and pressure.

Vapor Pressure and Boiling Point

11.43   (a)   No effect.

        (b)   No effect.

        (c)   Vapor pressure decreases with increasing intermolecular attractive forces because fewer molecules have sufficient kinetic energy to overcome attractive forces and escape to the vapor phase.

        (d)   Vapor pressure increases with increasing temperature because average kinetic energies of molecules increases.

11.44   (a)   The pressure difference on the manometer is 130 mm Hg and the gas in the vessel is essentially 100% molecules of the substance in the vapor phase. When the vessel is evacuated, virtually all air is removed. As the frozen liquid warms, it establishes a vapor pressure of 130 mm Hg. This is the pressure difference on the manometer.

        (b)   The pressure difference is 1 atm. The gas is 130 mm Hg of the molecular vapor and the rest is air. The liquid vaporizes in contact with the atmosphere, so atmospheric pressure is maintained above the liquid, but the equilibrium gas composition reflects the amount of vapor necessary to maintain 130 mm pressure, plus enough air to maintain a total pressure of 1 atm.

        (c)   The pressure difference is 890 mm Hg (1 atm + 130 mm Hg) and the gas is a mixture of 130 mm vapor and 1 atm air. The initial air pressure in the flask is 1 atm and no air is allowed to escape. The gas in the flask is not in equilibrium with the atmosphere and the final pressure in the flask does not equal atmospheric pressure. After a most of the liquid vaporizes, the total gas pressure is the result of 130 mm vapor and 1 atm air.

11.45   (a)   *Analyze/Plan.* Given the molecular formulae of several substances, determine the kind of intermolecular forces present, and rank the strength of these forces. The weaker the forces, the more volatile the substance.   *Solve.*

            $CBr_4 < CHBr_3 < CH_2Br_2 < CH_2Cl_2 < CH_3Cl < CH_4$

            The weaker the intermolecular forces, the higher the vapor pressure, the more volatile the compound. The order of increasing volatility is the order of decreasing strength of intermolecular forces. By analogy to the boiling points of HCl and HBr (Section 11.2), the trend will be dominated by dispersion forces, even though four of the molecules ($CHBr_3$, CH_2Br_2, CH_2Cl_2 and CH_3Cl) are polar. Thus, the order of increasing volatility is the order of decreasing molar mass and decreasing strength of dispersion forces.

        (b) $CH_4 < CH_3Cl < CH_2Cl_2 < CH_2Br_2 < CHBr_3 < CBr_4$

            Boiling point increases as the strength of intermolecular forces increases, so the order of boiling points is the order of increasing strength of forces. This is the order of decreasing volatility and the reverse of the order in part (a).

11.46   Both molecules are pyramidal, with a nonbonding electron pair on the central atom. Even though the molecules are polar covalent, differences in their intermolecular forces and physical properties are likely to be dominated by differences in dispersion forces.

(a) PCl_3

(b) $AsCl_3$

(c)   At the same temperature, the average kinetic energy of molecules of the two substances is equal.

(d)   The strength of dispersion forces increases with increasing molecular weight, so $AsCl_3$ will experience stronger intermolecular forces. This is the basis of predictions in parts (a) and (b) above.

11.47   (a)   The water in the two pans is at the same temperature, the boiling point of water at the atmospheric pressure of the room. During a phase change, the temperature of a system is constant. All energy gained from the surroundings is used to accomplish the transition, in this case to vaporize the liquid water. The pan of water that is boiling vigorously is gaining more energy and the liquid is being vaporized more quickly than in the other pan, but the temperature of the phase change is the same.

(b)   Vapor pressure does not depend on either volume or surface area of the liquid. As long as the containers are at the same temperature, the vapor pressures of water in the two containers are the same.

11.48   (a)   On a humid day, there are more gaseous water molecules in the air and more are recaptured by the surface of the liquid, making evaporation slower.

(b)   At high altitude, atmospheric pressure is lower and water boils at a lower temperature. The eggs must be cooked longer at the lower temperature.

11.49   The boiling point is the temperature at which the vapor pressure of a liquid equals atmospheric pressure.

(a)   The boiling point of diethyl ether at 400 torr is ~17°C, or, at 17°C, the vapor pressure of diethyl ether is 400 torr.

(b)   The vapor pressure of ethyl alcohol at 70°C is approximately 510 torr. Thus, at 70°C ethyl alcohol would boil at an external pressure of 510 torr.

(c)   The vapor pressure of water at 25°C is 23.76 torr, and at 26°C is 25.21 torr. At a pressure of 25 torr, the boiling point of water is much closer to 26°C then 25°C. To two significant figures, the boiling point is 26°C.

(d)   A pressure of 1.2 atm corresponds to $1.2 \text{ atm} \times \dfrac{760 \text{ torr}}{1 \text{ atm}} = 912 = 9.1 \times 10^2$ torr.

According to Appendix B, the vapor pressure of water reaches 912 torr somewhere between 104–106°C. By linear interpolation, the boiling point should be near

$$104°C + \left[\frac{(912-875)\text{torr}}{(938-875)\text{torr}} \times 2°C\right] = 105.2°C = 105°C$$

11.50   (a)   The boiling point of a liquid is the temperature at which its vapor pressure equals atmospheric pressure. According to Appendix B, the vapor pressure of water is 680 torr at approximately 97°C.

        (b)   The temperature at which the vapor pressure of water is 752 mm Hg is almost 100°C, greater than the boiling temperature at 680 mm Hg. The average kinetic energy of the H_2O molecules at the boiling temperature in Chicago is greater than the average kinetic energy of the molecules at the boiling temperature in Reno. A liquid boils when its vapor pressure equals the external pressure acting on the liquid. If the external pressure acting on the liquid molecules is smaller (as is the atmospheric pressure in Reno), a smaller average kinetic energy is required for boiling (bubble formation within the liquid).

Phase Diagrams

11.51   (a)   The *critical point* is the temperature and pressure beyond which the gas and liquid phases are indistinguishable.

        (b)   The gas/liquid line ends at the critical point because at conditions beyond the critical temperature and pressure, there is no distinction between gas and liquid. In experimental terms, a gas cannot be liquefied at temperatures higher than the critical temperature, regardless of pressure.

11.52   (a)   The *triple point* on a phase diagram represents the temperature and pressure at which the gas, liquid, and solid phases are in equilibrium.

        (b)   No. A phase diagram represents a closed system, one where no matter can escape and no substance other than the one under consideration is present; air cannot be present in the system. Even if air is excluded, at 1 atm of external pressure, the triple point of water is inaccessible, regardless of temperature [see Sample Exercise 11.6(b)].

11.53   (a)   The water vapor would condense to form a solid at a pressure of around 4 torr. At higher pressure, perhaps 5 atm or so, the solid would melt to form liquid water. This occurs because the melting point of ice, which is 0°C at 1 atm, decreases with increasing pressure.

        (b)   In thinking about this exercise, keep in mind that the **total** pressure is being maintained at a constant 0.50 atm. That pressure is composed of water vapor pressure and some other pressure, which could come from an inert gas. At 100°C and 0.50 atm, water is in the vapor phase. As it cools, the water vapor will condense to the liquid at the temperature where the vapor pressure of liquid water is 0.50 atm. From Appendix B, we see that condensation occurs at approximately 82°C. Further cooling of the liquid water results in freezing to the solid at approximately 0°C. The freezing point of water increases with decreasing pressure, so at 0.50 atm, the freezing temperature is very slightly above 0°C.

11.54   (a)   Solid CO_2 sublimes to form $CO_2(g)$ at a temperature of about –60°C.

        (b)   Solid CO_2 melts to form $CO_2(l)$ at a temperature of about –50°C. The $CO_2(l)$ boils when the temperature reaches approximately –40°C.

11.55   (a)

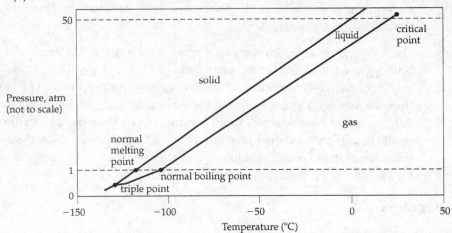

(b)   The solid-liquid line on the phase diagram is normal and the melting point of Xe(s) increases with increasing pressure. This means that Xe(s) is denser than Xe(l).

(c)   Cooling Xe(g) at 100 torr will cause deposition of the solid. A pressure of 100 torr is below the pressure of the triple point, so the gas will change directly to the solid upon cooling.

11.56   (a)

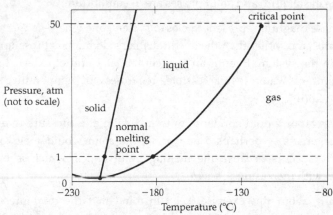

On the diagram above, the triple point is at the dot closest to the x-axis. The normal melting point is at the intersection of the solid-liquid line and the 1-atm dashed line. The normal boiling point is at the intersection of the liquid-gas line and the 1-atm line. The critical point is clearly marked.

(b) $O_2(s)$ will not float on $O_2(l)$. $O_2(s)$ is denser than $O_2(l)$ because the solid-liquid line on the phase diagram is normal. That is, as pressure increases, the melting temperature increases. [Note that the solid-liquid line for O_2 is nearly vertical, indicating a small difference in the densities of $O_2(s)$ and $O_2(l)$].

(c) $O_2(s)$ will melt when heated at a pressure of 1 atm, since this is a much greater pressure than the pressure at the triple point.

Structures of Solids

11.57   In a crystalline solid, the component particles (atoms, ions, or molecules) are arranged in an ordered repeating pattern. In an amorphous solid, there is no orderly structure. Quartz glass (Figure 11.30(b)) is an example of an amorphous solid. Paraffin wax is another example of an amorphous solid. Also, most plastics show no long range order and are amorphous overall, although they can show regions of order (see Section 12.2).

11.58   In amorphous silica (SiO_2) the regular structure of quartz is disrupted; the loose, disordered structure, Figure 11.30(b), has many vacant "pockets" throughout. There are fewer SiO_2 groups per volume in the amorphous solid; the packing is less efficient and less dense.

11.59   Ca: Ca atoms occupy the 8 corners of the cube.

                     8 corners × 1/8 sphere/corner = 1 Ca atom

        O: O atoms occupy the centers of the 6 faces of the cube.

                     6 faces × 1/2 atom/face = 3 O atoms

        Ti: There is 1 Ti atom at the body center of the cube.

        Formula: $CaTiO_3$

11.60   (a)   Ti: 8 corners × 1/8 sphere/corner + [1 center × 1 sphere/center] = 2 Ti atoms

               O: 4 faces × 1/2 sphere/face + [2 interior × 1 sphere/interior] = 4 O atoms

               Formula: TiO_2

        (b)   Rutile is an ionic solid; ion-ion forces among Ti^{4+} cations and O^{2-} anions are quite strong, owing to the magnitudes of the charges, and lead to the ordered structure.

11.61   *Analyze.* Given the cubic unit cell edge length and arrangement of Ir atoms, calculate the atomic radius and the density of the metal. *Plan.* There is space between the atoms along the unit cell edge, but they touch along the face diagonal. Use the geometry of the right equilateral triangle to calculate the atomic radius. From the definition of density and paying attention to units, calculate the density of Ir(s).   *Solve.*

        (a)   The length of the face diagonal of a face-centered cubic unit cell is four times the radius of the atom and $\sqrt{2}$ times the unit cell dimension or edge length, usually designated *a* for cubic unit cells.

$$4r = \sqrt{2}\,a; \quad r = \sqrt{2}\,a/4 = \frac{\sqrt{2} \times 3.833\,\text{Å}}{4} = 1.3552 = 1.355\,\text{Å}$$

        (b)   The density of iridium is the mass of the unit cell contents divided by the unit cell volume. There are 4 Ir atoms in a face-centered cubic unit cell.

$$\frac{4\,\text{Ir atoms}}{(3.833 \times 10^{-8}\,\text{cm})^3} \times \frac{192.22\,\text{g Ir}}{6.022 \times 10^{23}\,\text{Ir atoms}} = 22.67\,\text{g/cm}^3$$

             Check. The units of density are correct. Note that Ir is quite dense.

11.62   (a)   8 corners × 1/8 atom/corner + 6 faces × ½ atom/face = 4 atoms

        (b)   Each aluminum atom is in contact with 12 nearest neighbors, 6 in one plane, 3 above that plane, and 3 below. Its coordination number is thus 12.

(c)   The length of the face diagonal of a face-centered cubic unit cell is four times the radius of the metal and $\sqrt{2}$ times the unit cell dimension (usually designated a for cubic cells).

$$4 \times 1.43\,\text{Å} = \sqrt{2} \times a; \quad a = \frac{4 \times 1.43\,\text{Å}}{\sqrt{2}} = 4.0447 = 4.04\,\text{Å} = 4.04 \times 10^{-8}\,\text{cm}$$

(d)   The density of the metal is the mass of the unit cell contents divided by the volume of the unit cell.

$$\text{density} = \frac{4\,\text{Al atoms}}{(4.0447 \times 10^{-8}\,\text{cm})^3} \times \frac{26.98\,\text{g Al}}{6.022 \times 10^{23}\,\text{Al atoms}} = 2.71\,\text{g/cm}^3$$

11.63   *Analyze.* Given the atomic arrangement, length of the cubic unit cell edge and density of the solid, calculate the atomic weight of the element. *Plan.* If we calculate the mass of a single unit cell, and determine the number of atoms in one unit cell, we can calculate the mass of a single atom and of a mole of atoms.   *Solve.*

The volume of the unit cell is $(2.86 \times 10^{-8}\,\text{cm})^3$. The mass of the unit cell is:

$$\frac{7.92\,\text{g}}{\text{cm}^3} \times \frac{(2.86 \times 10^{-8})^3\,\text{cm}^3}{\text{unit cell}} = 1.853 \times 10^{-22}\,\text{g/unit cell}$$

There are two atoms of the element present in the body-centered cubic unit cell. Thus the atomic weight is:

$$\frac{1.853 \times 10^{-22}\,\text{g}}{\text{unit cell}} \times \frac{1\,\text{unit cell}}{2\,\text{atoms}} \times \frac{6.022 \times 10^{23}\,\text{atoms}}{1\,\text{mol}} = 55.8\,\text{g/mol}$$

Check. The result is a reasonable atomic weight and the units are correct. The element could be iron.

11.64   Avogadro's number is the number of KCl formula units in 74.55 g of KCl.

$$74.55\,\text{g KCl} \times \frac{1\,\text{cm}^3}{1.984\,\text{g}} \times \frac{(1 \times 10^{10}\,\text{pm})^3}{1\,\text{cm}^3} \times \frac{4\,\text{KCl units}}{628^3\,\text{pm}^3} = 6.07 \times 10^{23}\,\text{KCl formula units}$$

11.65   (a)   Each sphere is in contact with 12 nearest neighbors; its coordination number is thus 12.

(b)   Each sphere has a coordination number of six.

(c)   Each sphere has a coordination number of eight.

11.66   (a) Na^+, 6   (b) Zn^{2+}, 4   (c) Ca^{2+}, 8

11.67   *Analyze.* Given the atomic arrangement and density of the solid, calculate the unit cell edge length. *Plan.* Calculate the mass of a single unit cell and then use density to find the volume of a single unit cell. The edge length is the cube-root of the volume of a cubic cell. *Solve.* There are four PbSe units in the unit cell. The unit cell edge is designated a.

$$8.27\,\text{g/cm}^3 = \frac{4\,\text{PbSe units}}{a^3} \times \frac{286.2\,\text{g}}{6.022 \times 10^{23}\,\text{PbSe units}} \times \left(\frac{1\,\text{Å}}{1 \times 10^{-8}\,\text{cm}}\right)^3$$

$a^3 = 229.87\,\text{Å}^3, a = 6.13\,\text{Å}$

11.68   In the face-centered cubic structure, there are four NiO units in the unit cell. Density is the mass of the unit cell contents divided by the unit cell volume (a^3).

$$\text{density} = \frac{4\ \text{NiO units}}{(4.18\ \text{Å})^3} \times \frac{74.69\ \text{g NiO}}{6.022 \times 10^{23}\ \text{NiO units}} \times \left(\frac{1\ \text{Å}}{1 \times 10^{-8}\ \text{cm}}\right)^3 = 6.79\ \text{g/cm}^3$$

11.69   (a)   The U ions in UO_2 are represented by the smaller spheres in Figure 11.42(c). The chemical formula requires twice as many O^{2-} ions as U^{4+} ions. There are eight complete large spheres and four total ($8 \times 1/8 + 6 \times 1/2$) small spheres, so the small ones must represent U^{4+}. (It is probably true that O^{2-} has a physically larger radius than U^{4+}, but the elements' large separation on the periodic chart makes the relative radii difficult to estimate from trends.)

         (b)   According to Figure 11.42(c), there are four UO_2 units in the "fluorite" unit cell.

$$\frac{4\ UO_2\ \text{units}}{(5.468\ \text{Å})^3} \times \frac{270.03\ \text{g}}{6.022 \times 10^{23}\ UO_2\ \text{units}} \times \left(\frac{1\ \text{Å}}{1 \times 10^{-8}\ \text{cm}}\right)^3 = 10.97\ \text{g/cm}^3$$

11.70   (a)   According to Figure 11.42(b), there are 4 HgS units in a unit cell with the zinc blende structure. [4 complete Hg^{2+} ions, $6(1/2) + 8(1/8)$ S^{2-} ions]

$$\text{density} = \frac{4\ \text{HgS units}}{(5.852\ \text{Å})^3} \times \frac{232.655\ \text{g}}{6.022 \times 10^{23}\ \text{HgS units}} \times \left(\frac{1\ \text{Å}}{1 \times 10^{-8}\ \text{cm}}\right)^3 = 7.711\ \text{g/cm}^3$$

         (b)   We expect Se^{2-} to have a larger ionic radius than S^{2-}, since Se is below S in the chalcogen family and both ions have the same charge. Thus, HgSe will occupy a larger volume and the unit cell edge will be longer.

         (c)   For HgSe:

$$\text{density} = \frac{4\ \text{HgSe units}}{(6.085\ \text{Å})^3} \times \frac{279.55\ \text{g HgSe}}{6.022 \times 10^{23}\ \text{HgSe units}} \times \left(\frac{1\ \text{Å}}{1 \times 10^{-8}\ \text{cm}}\right)^3 = 8.241\ \text{g/cm}^3$$

               Even though HgSe has a larger unit cell volume than HgS, it also has a larger molar mass. The mass of Se is more than twice that of S, while the radius of Se^{2-} is only slightly larger than that of S (Figure 7.7). The greater mass of Se accounts for the greater density of HgSe.

Bonding in Solids

11.71   (a)   hydrogen bonding, dipole-dipole forces, London dispersion forces

         (b)   covalent chemical bonds (mainly)

         (c)   ionic bonds (mainly)   (d)   metallic bonds

11.72   (a)   ionic   (b)   metallic

         (c)   ionic (somewhat borderline, could be modeled as ionic with some covalent character to the bonds, in keeping with the high oxidation state of Zr, or as a network solid with ionic character to the bonding, in keeping with the electronegativity difference between Zr and O.)

         (d)   molecular   (e)   molecular   (f)   molecular

11.73   In molecular solids, relatively weak intermolecular forces (hydrogen bonding, dipole-dipole, dispersion) bind the molecules in the lattice, so relatively little energy is required to disrupt these forces. In covalent-network solids, covalent bonds join atoms into an extended network. Melting or deforming a covalent-network solid means breaking these covalent bonds, which requires a large amount of energy.

11.74   (a)   metallic

       (b)   molecular or metallic (physical properties of metals vary widely)

       (c)   covalent-network or ionic   (d)   covalent-network   (e)   ionic

11.75   Because of its relatively high melting point and properties as a conducting solution, the solid must be ionic.

11.76   According to Table 11.7, the solid could be either ionic with low water solubility or network covalent. Due to the extremely high sublimation temperature, it is probably covalent-network.

11.77   (a)   Xe — greater atomic weight, stronger dispersion forces

       (b) SiO_2 — covalent-network lattice versus weak dispersion forces

       (c)   KBr — strong ionic versus weak dispersion forces

       (d) C_6Cl_6 — both are influenced by dispersion forces, C_6Cl_6 has the higher molar mass.

11.78   (a)   HF — hydrogen bonding versus dipole-dipole for HCl

       (b)   C(graphite) — covalent-network bonding versus London dispersion forces for CH_4

       (c)   KBr — ionic versus dispersion forces for nonpolar Br_2

       (d) MgF_2 — higher charge on Mg^{2+} than Li^+

Additional Exercises

11.79   (a)   decrease   (b) increase   (c) increase   (d) increase

       (e)   increase   (f) increase   (g) increase

11.80   (a)   Correct.

       (b)   The lower boiling liquid must experience less total intermolecular forces.

       (c)   If both liquids are structurally similar nonpolar molecules, the lower boiling liquid has a lower molecular weight than the higher boiling liquid.

       (d)   Correct.

       (e)   At their boiling points, both liquids have vapor pressures of 760 mm Hg.

11.81   (a)   The *cis* isomer has stronger dipole-dipole forces; the *trans* isomer is nonpolar. The higher boiling point of the *cis* isomer supports this conclusion.

       (b)   While boiling points are primarily a measure of strength of intermolecular forces, melting points are influenced by crystal packing efficiency as well as intermolecular forces. Since the nonpolar *trans* isomer with weaker intermolecular forces has the higher melting point, it must pack more efficiently.

11.82   (a)   In dibromomethane, CH_2Br_2, the dispersion force contribution will be larger than for CH_2Cl_2, because bromine is more polarizable than the lighter element chlorine. At the same time, the dipole-dipole contribution for CH_2Cl_2 is greater than for CH_2Br_2 because CH_2Cl_2 has a larger dipole moment.

         (b)   Just the opposite comparisons apply to CH_2F_2, which is less polarizable and has a higher dipole moment than CH_2Cl_2.

11.83   When a halogen atom (Cl or Br) is substituted for H in benzene, the molecule becomes polar. These molecules experience dispersion forces similar to those in benzene plus dipole-dipole forces, so they have higher boiling points than benzene. C_6H_5Br has a higher molar mass and is more polarizable than C_6H_5Cl, so it has the higher boiling point. C_6H_5OH experiences hydrogen bonding, the strongest force between neutral molecules, so it has the highest boiling point.

11.84   (a)   Propylamine experiences hydrogen bonding interactions while trimethylamine, with no N–H bonds, does not. Also, the rod-like shape of propylamine (see Solution 11.21) leads to stronger dispersion forces than in pyramidal trimethylamine. The stronger intermolecular forces in propylamine lead to the higher boiling point.

         (b)   Propylamine is most soluble in water by virtue of its –NH_2 group, which can act as both a donor and acceptor in hydrogen bonding. Trimethylamine is much more soluble than the structurally similar isobutane, because the pyramidal $\ddot{N}$ atom can act as a hydrogen bond acceptor. Trimethylamine is less soluble than propylamine because its participation in hydrogen bonding is less extensive.

11.85   The two O–H groups in ethylene glycol are involved in many hydrogen bonding interactions, leading to its high boiling point and viscosity, relative to pentane, which experiences only dispersion forces.

11.86   The more carbon atoms in the hydrocarbon, the longer the chain, the more polarizable the electron cloud, the higher the boiling point. A plot of the number of carbon atoms versus boiling point is shown below. For 8 C atoms, C_8H_{18}, the boiling point is approximately 130°C.

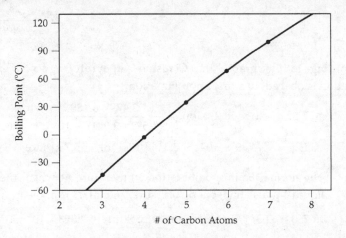

11.87   According to Figure 11.20, as the pressure of a gas above its critical temperature increases beyond critical pressure, the solubility of a solute increases. The solubility of the solute is essentially zero below critical pressure. Supercritical CO_2 at very high pressure dissolves caffeine, and the solution leaves the extractor. The pressure reduction value reduces the pressure of CO_2 enough so that the caffeine becomes insoluble. The solid caffeine is deposited in the separator, and the low pressure CO_2 gas is recycled.

11.88   (a)   Sweat, or salt water, on the surface of the body vaporizes to establish its typical vapor pressure at atmospheric pressure. Since the atmosphere is a totally open system, typical vapor pressure is never reached, and the sweat evaporates continuously. Evaporation is an endothermic process. The heat required to vaporized sweat is absorbed from your body, helping to keep it cool.

(b)   The vacuum pump reduces the pressure of the atmosphere (air + water vapor) above the water. Eventually, atmospheric pressure equals the vapor pressure of water and the water boils. Boiling is an endothermic process, and the temperature drops if the system is not able to absorb heat from the surroundings fast enough. As the temperature of the water decreases, the water freezes. (On a molecular level, the evaporation of water removes the molecules with the highest kinetic energies from the liquid. This decrease in average kinetic energy is what we experience as a temperature decrease.)

11.89   (a)   If the Clausius-Clapeyron equation is obeyed, a graph of ln P vs 1/T(K) should be linear. Here are the data in a form form for graphing.

T(K)	1/T	P(torr)	ln P
280.0	3.571×10^{-3}	32.42	3.479
300.0	3.333×10^{-3}	92.47	4.527
320.0	3.125×10^{-3}	225.1	5.417
330.0	3.030×10^{-3}	334.4	5.812
340.0	2.941×10^{-3}	482.9	6.180

According to the graph, the Clausius-Clapeyron equation is obeyed, to a first approximation.

$$\Delta H_{vap} = -slope \times R; \quad slope = \frac{3.479 - 6.180}{(3.571 - 2.941) \times 10^{-3}} = -\frac{2.701}{0.630 \times 10^{-3}} = -4.29 \times 10^3$$

$$\Delta H_{vap} = -(-4.29 \times 10^3) \times 8.314 \text{ J/K} \cdot \text{mol} = 35.7 \text{ kJ/mol}$$

(b)   The normal boiling point is the temperature at which the vapor pressure of the liquid equals atmospheric pressure, 760 torr. From the graph,

ln 760 = 6.63, 1/T for this vapor pressure = 2.828×10^{-3}; T = 353.6 K

11.90   (a)   The Clausius-Clapeyron equation is $\ln P = \dfrac{-\Delta H_{vap}}{RT} + C$.

For two vapor pressures, P_1 and P_2, measured at corresponding temperatures T_1 and T_2, the relationship is

$$\ln P_1 - \ln P_2 = \left(\frac{-\Delta H_{vap}}{RT_1} + C \right) - \left(\frac{-\Delta H_{vap}}{RT_2} + C \right)$$

$$\ln P_1 - \ln P_2 = \frac{-\Delta H_{vap}}{R}\left(\frac{1}{T_1} - \frac{1}{T_2} \right) + C - C; \quad \ln \frac{P_1}{P_2} = \frac{-\Delta H_{vap}}{R}\left(\frac{1}{T_1} - \frac{1}{T_2} \right)$$

(b) $P_1 = 10.00$ torr, $T_1 = 716$ K; $P_2 = 400.0$ torr, $T_2 = 981$ K

$$\ln \frac{10.00}{400.0} = \frac{-\Delta H_{vap}}{8.314 \text{ J/K} \cdot \text{mol}}\left(\frac{1}{716} - \frac{1}{981} \right)$$

$$-3.6889\,(8.314 \text{ J/K} \cdot \text{mol}) = -\Delta H_{vap}\,(3.773 \times 10^{-4}/\text{K})$$

$$\Delta H_{vap} = 8.129 \times 10^4 = 8.1 \times 10^4 \text{ J/mol} = 81 \text{ kJ/mol}$$

$[(1/T_1) - (1/T_2)]$ has 2 sig figs and so does the result.

(c)   The normal boiling point of a liquid is the temperature at which the vapor pressure of the liquid is 760 torr.

$P_1 = 400.0$ torr, $T_1 = 981$ K; $P_2 = 760$ torr, $T_2 =$ b.p. of potassium

$$\ln\left(\frac{400.0}{760.0} \right) = \frac{-8.129 \times 10^4 \text{ J/mol}}{8.314 \text{ J/K} \cdot \text{mol}}\left(\frac{1}{981 \text{ K}} - \frac{1}{T_2} \right)$$

$$\frac{-0.64185}{-9.7775 \times 10^3} = 1.0194 \times 10^{-3} - \frac{1}{T_2}; \quad \frac{1}{T_2} = 1.0194 \times 10^{-3} - 6.565 \times 10^{-5}$$

$$\frac{1}{T_2} = 9.5375 \times 10^{-4} = 9.5 \times 10^{-4}; \quad T_2 = 1048 = 1.0 \times 10^3 \text{ K } (8 \times 10^{2}{}^{\circ}\text{C})$$

(d) $P_1 =$ VP of K(l) at 100°C, $T_1 = 373$ K; $P_2 = 10.00$ torr, $T_2 = 716$ K

$$\ln \frac{P_1}{10.00 \text{ torr}} = \frac{-8.129 \times 10^4 \text{ J/mol}}{8.314 \text{ J/K} \cdot \text{mol}}\left(\frac{1}{373} - \frac{1}{716} \right)$$

$$\ln \frac{P_1}{10.00 \text{ torr}} = \frac{-8.129 \times 10^4 \text{ J/mol}}{8.314 \text{ J/K} \cdot \text{mol}} \times 1.284 \times 10^{-3} = -12.5543$$

$$\frac{P_1}{10.00 \text{ torr}} = e^{-12.5543} = 3.530 \times 10^{-6}; \quad P_1 = 3.5 \times 10^{-5} \text{ torr}$$

11.91   Physical data for the two compounds from the *Handbook of Chemistry and Physics*:

	MM	dipole moment	boiling point
CH_2Cl_2	85 g/mol	1.60 D	40.0°C
CH_3I	142 g/mol	1.62 D	42.4°C

(a)   The two substances have very similar molecular structures; each is an unsymmetrical tetrahedron with a single central carbon atom and no hydrogen bonding. Since the structures are very similar, the magnitudes of the dipole-

dipole forces should be similar. This is verified by their very similar dipole moments. The heavier compound, CH_3I, will have slightly stronger London dispersion forces. Since the nature and magnitude of the intermolecular forces in the two compounds are nearly the same, it is very difficult to predict which will be more volatile [or which will have the higher boiling point as in part (b)].

(b)   Given the structural similarities discussed in part (a), one would expect the boiling points to be very similar, and they are. Based on its larger molar mass (and dipole-dipole forces being essentially equal) one might predict that CH_3I would have a slightly higher boiling point; this is verified by the known boiling points.

(c)   According to Equation 11.1, $\ln P = \dfrac{-\Delta H_{vap}}{RT} + C$

A plot of $\ln P$ vs. $1/T$ for each compound is linear. Since the order of volatility changes with temperature for the two compounds, the two lines must cross at some temperature; the slopes of the two lines, ΔH_{vap} for the two compounds, and the y-intercepts, C, must be different.

(d)   **CH_2Cl_2**

ln P	T(K)	1/T
2.303	229.9	4.351×10^{-3}
3.689	250.9	3.986×10^{-3}
4.605	266.9	3.747×10^{-3}
5.991	297.3	3.364×10^{-3}

CH_3I

ln P	T(K)	1/T
2.303	227.4	4.398×10^{-3}
3.689	249.0	4.016×10^{-3}
4.605	266.2	3.757×10^{-3}
5.991	298.5	3.350×10^{-3}

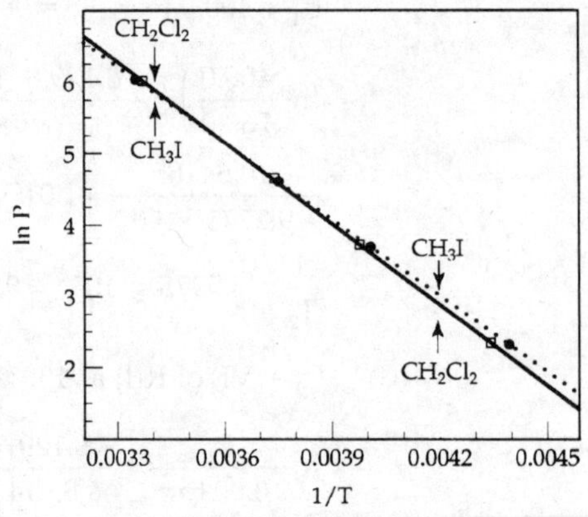

For CH_2Cl_2, $-\Delta H_{vap}/R = \text{slope} = \dfrac{(5.991 - 2.303)}{(3.364 \times 10^{-3} - 4.350 \times 10^{-3})} = \dfrac{-3.688}{0.987 \times 10^{-3}}$

$$= -3.74 \times 10^3 = -\Delta H_{vap}/R$$

$\Delta H_{vap} = 8.314 \,(3.74 \times 10^3) = 3.107 \times 10^4 \text{ J/mol} = 31.1 \text{ kJ/mol}$

For CH_3I, $-\Delta H_{vap}/R = \text{slope} = \dfrac{(5.991 - 2.303)}{(3.350 \times 10^{-3} - 4.398 \times 10^{-3})} = \dfrac{-3.688}{1.048 \times 10^{-3}} = -3.519 \times 10^3$

$$= -\Delta H_{vap}/R$$

$\Delta H_{vap} = 8.314 \,(3.519 \times 10^3) = 2.926 \times 10^4 \text{ J/mol} = 29.3 \text{ kJ/mol}$

11.92   In a metallic solid such as gold, the atoms are held in their very orderly arrangement by metallic bonding, the result of valence electrons delocalized throughout the three-dimensional lattice. A large amount of kinetic energy is required to disrupt this delocalized bonding network and allow the atoms to translate relative to each, so the melting point of Au(s) is quite high. Xe atoms are held in a cubic close-packed arrangement by London-dispersion forces much weaker than metallic bonding. Very little kinetic energy is required for Xe atoms to overcome these forces and melt, so the melting point of Xe is quite low.

11.93   (a)   The face diagonal of a face-centered cubic unit cell has length $a\sqrt{2}$ and also 4 r, where a is the cubic cell dimension and r is the atomic radius.

$$4\,r = a\sqrt{2}; r = (a\sqrt{2})/4 = (4.078\,\text{Å})(\sqrt{2})/4 = 1.44179 = 1.442\,\text{Å}$$

(b)   Density is the mass of the unit cell contents divided by the unit cell volume (a^3). In a face-centered cubic unit cell, there are 4 Au atoms.

$$\text{density} = \frac{4\,\text{Au atoms}}{(4.078\,\text{Å})^3} \times \frac{196.97\,\text{g}}{6.0221 \times 10^{23}\,\text{Au atoms}} \times \left(\frac{1\,\text{Å}}{1 \times 10^{-8}\,\text{cm}}\right)^3 = 19.29\,\text{g/cm}^3$$

11.94   (a)   (i)   8 corners × 1/8 atom/corner = 1 atom

         (ii)   8 corners × 1/8 atom/corner + 1 center × 1 atom/center = 2 atoms

         (iii)   8 corners × 1/8 atom/corner + 6 faces × 1/2 atom/face = 4 atoms

(b)   Fundamentally, $CaCl_2$ must have a different crystal structure than NaCl because the two formulas are different. Na^+ has an ionic radius of 0.97 Å and Ca^{2+}, 0.99 Å; these values are very similar. So, ion size does not prohibit Ca^{2+} from replacing Na^+ in an ionic lattice, but electrostatic effects do. In $CaCl_2$, there are two Cl^- anions for every Ca^{2+} cation.

Use Figure 11.35(b) structure to visualize the contents of one face-centered cubic unit cell of the NaCl structure.

Na^+ (purple): (8 corners × 1/8 Na^+/corner) + (6 faces × 1/2 Na^+/face) = 4 Na^+

Cl^- (green): (12 edges × 1/4 Cl^-/edge) + (1 center × 1 Cl^-/center) = 4 Cl^-

In the NaCl structure, there is one anion site for every cation. $CaCl_2$ requires two anion sites for each cation, so it cannot have the same crystal structure as NaCl. (In fact, the unit cell of $CaCl_2$ has three unequal edge lengths and is not cubic; it is much different than the NaCl unit cell.)

11.95   The most effective diffraction of light by a grating occurs when the wavelength of light and the separation of the slits in the grating are similar. When X-rays are diffracted by a crystal, layers of atoms serve as the "slits." The most effective diffraction occurs when the distances between layers of atoms are similar to the wavelength of the X-rays. Typical interlayer distances in crystals range from 2 Å to 20 Å. Visible light, 400–700 nm or 4,000 to 7,000 Å, is too long to be diffracted effectively by crystals. Molybdenum X-rays of 0.71 Å are on the same order of magnitude as interlayer distances in crystals and are diffracted.

11.96 $n\lambda = 2d \sin\theta$; $n = 1$, $\lambda = 1.54$ Å, $\theta = 14.22°$; calculate d.

$$d = \frac{n\lambda}{2\sin\theta} = \frac{1 \times 1.54\,\text{Å}}{2\sin(14.22)} = 3.1346 = 3.13\,\text{Å}$$

11.97   (a)   Both diamond ($d = 3.5$ g/cm^3) and graphite ($d = 2.3$ g/cm^3) are covalent-network solids with efficient packing arrangements in the solid state; there is relatively little empty space in their respective crystal lattices. Diamond, with bonded C–C distances of 1.54 Å in all directions, is more dense than graphite, with shorter C–C distances within carbon sheets but longer 3.41 Å separations between sheets (Figure 11.41). Buckminsterfullerene has much more empty space, both inside each C$_{60}$ "ball" and between balls, than either diamond or graphite, so its density will be considerably less than 2.3 g/cm^3.

(b)   In a face-centered cubic unit cell, there are 4 complete C$_{60}$ units.

$$\frac{4\,\text{C}_{60}\,\text{units}}{(14.2\,\text{Å})^3} \times \frac{720.66\,\text{g}}{6.022 \times 10^{23}\,\text{C}_{60}\,\text{units}} \times \left(\frac{1\,\text{Å}}{1 \times 10^{-8}\,\text{cm}}\right)^3 = 1.67\,\text{g/cm}^3$$

(1.67 g/cm^3 is the smallest density of the three allotropes, diamond, graphite, and buckminsterfullerene.)

Integrative Exercises

11.98   *Analyze.* Given: mass % of Al, Mg, O; density, unit cell edge length. Find: number of each type of atom. *Plan.* We are not given the type of cubic unit cell, primitive, body centered, face-centered. So we must calculate the number of formula units in the unit cell, using density, cell volume, and formula weight. Begin by determining the empirical formula and formula weight from mass % data. *Solve.* Assume 100 g spinel.

$$37.9\,\text{g Al} \times \frac{1\,\text{mol Al}}{26.98\,\text{g Al}} = 1.405\,\text{mol Al}; 1.405/0.7036 \approx 2$$

$$17.1\,\text{g Mg} \times \frac{1\,\text{mol Mg}}{24.305\,\text{g Mg}} = 0.7036\,\text{mol Mg}; 0.7036/0.7036 = 1$$

$$45.0\,\text{g O} \times \frac{1\,\text{mol O}}{16.00\,\text{g Al}} = 2.813\,\text{mol O}; 2.813/0.7036 \approx 4$$

The empirical formula is Al$_2$MgO$_4$; formula weight = 142.3 g/mol

Calculate the number of formula units per unit cell.

809 pm = 809×10^{-12} m = 8.09×10^{-8} cm; V = $(8.09 \times 10^{-8})^3$ cm^3

$$\frac{3.57\,\text{g}}{\text{cm}^3} \times (8.09 \times 10^{-8})^3\,\text{cm}^3 \times \frac{1\,\text{mol}}{142.3\,\text{g}} \times \frac{6.022 \times 10^{23}\,\text{units}}{\text{mol}} = 7.999 = 8$$

There are 8 formula units per unit cell, for a total of 16 Al atoms, 8 Mg atoms, and 32 O atoms.

[The relationship between density (d), unit cell volume (V), number of formula units (Z), formula weight (FW), and Avogadro's number (N) is a useful one. It can be rearranged to calculate any single variable, knowing values for the others. For densities in g/cm^3 and unit cell volumes in cm^3 the relationship is Z = (N × d × V)/FW.]

11.99   (a)   In Table 11.4, viscosity increases as the length of the carbon chain increases. Longer molecular chains become increasingly entangled, increasing resistance to flow.

        (b)   Whereas viscosity depends on molecular chain length in a critical way, surface tension depends on the strengths of intermolecular interactions between molecules. These dispersion forces do not increase as rapidly with increasing chain length and molecular weight as viscosity does.

        (c)   The –OH group in n-octyl alcohol gives rise to hydrogen bonding among molecules, which increases molecular entanglement and leads to greater viscosity and higher boiling point.

11.100   (a)   24 valence e^-, 12 e^- pairs

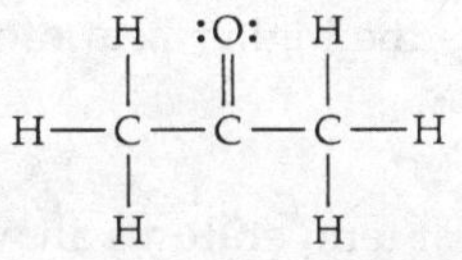

        The geometry around the central C atom is trigonal planar, and around the two terminal C atoms, tetrahedral.

        (b)   Polar. The C=O bond is quite polar and the dipoles in the trigonal plane around the central C atom do not cancel.

        (c)   Dipole-dipole and London-dispersion forces

        (d)   Since the molecular weights of acetone and 1-propanol are similar, the strength of the London-dispersion forces in the two compounds is also similar. The big difference is that 1-propanol has hydrogen bonding, while acetone does not. These relatively strong attractive forces lead to the higher boiling point for 1-propanol.

11.101

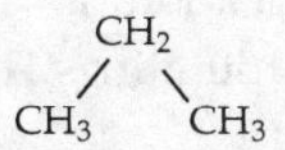

(i) MM = 44 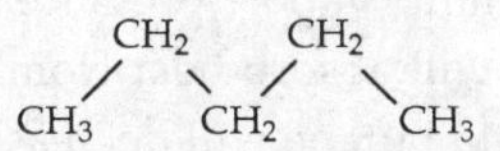   (ii) MM = 72 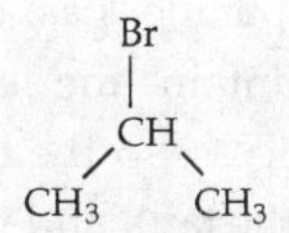   (iii) MM = 123

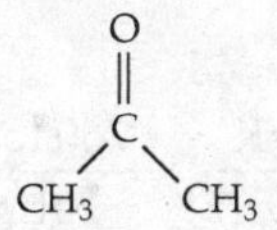

(iv) MM = 58 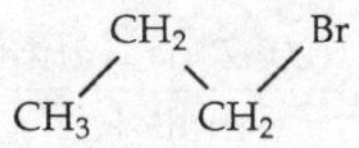   (v) MM = 123 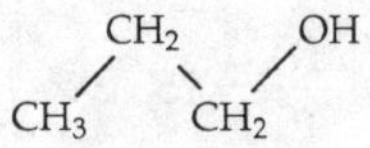   (vi) MM = 60

It is useful to draw the structural formulas because intermolecular forces are determined by the size and shape (structure) of molecules.

        (a)   *Molar mass*: compounds (i) and (ii) have similar rod-like structures; (ii) has a longer rod. The longer chain leads to greater molar mass, stronger London-dispersion forces and higher heat of vaporization.

(b)   *Molecular shape*: compounds (iii) and (v) have the same chemical formula and molar mass but different molecular shapes (they are structural isomers). The more rod-like shape of (v) leads to more contact between molecules, stronger dispersion forces and higher heat of vaporization.

(c)   *Molecular polarity*: rod-like hydrocarbons (i) and (ii) are essentially nonpolar, owing to free rotation about C–C σ bonds, while (iv) is quite polar, owing to the C=O group. (iv) has a smaller molar mass than (ii) but a larger heat of vaporization, which must be due to the presence of dipole-dipole forces in (iv). [Note that (iii) and (iv), with similar shape and molecular polarity, have very similar heats of vaporization.]

(d)   *Hydrogen-bonding interactions*: molecules (v) and (vi) have similar structures, but (vi) has hydrogen bonding and (v) does not. Even though molar mass and thus dispersion forces are larger for (v), (vi) has the higher heat of vaporization. This must be due to hydrogen bonding interactions.

11.102   (a)   In order for butane to be stored as a liquid at temperatures above its boiling point (–5°C), the pressure in the tank must be greater than atmospheric pressure. In terms of the phase diagram of butane, the pressure must be high enough so that, at tank conditions, the butane is "above" the gas-liquid line and in the liquid region of the diagram.

           The pressure of a gas is described by the ideal gas law as P = nRT/V; pressure is directly proportional to moles of gas. The more moles of gas present in the tank the greater the pressure, until sufficient pressure is achieved for the gas to liquify. At the point where liquid and gas are in equilibrium and temperature is constant, liquid will vaporize or condense to maintain the equilibrium vapor pressure. That is, as long as some liquid is present, the gas pressure in the tank will be constant.

   (b)   If butane gas escapes the tank, butane liquid will vaporize (evaporate) to maintain the equilibrium vapor pressure. Vaporization is an endothermic process, so the butane will absorb heat from the surroundings. The temperature of the tank and the liquid butane will decrease.

   (c) $155 \text{ g } C_4H_{10} \times \dfrac{1 \text{ mol } C_4H_{10}}{58.12 \text{ g } C_4H_{10}} \times \dfrac{21.3 \text{ kJ}}{\text{mol}} = 56.8 \text{ kJ}$

           $V = \dfrac{nRT}{P} = 155 \text{ g} \times \dfrac{1 \text{ mol}}{58.12 \text{ g}} \times \dfrac{0.08206 \text{ L} \cdot \text{atm}}{\text{mol} \cdot \text{K}} \times \dfrac{308 \text{ K}}{755 \text{ torr}} \times \dfrac{760 \text{ torr}}{1 \text{ atm}} = 67.851 = 67.9 \text{ L}$

11.103   *Plan*:

   (i)   Using thermochemical data from Appendix B, calculate the energy (enthalpy) required to melt and heat the H_2O.

   (ii)   Using Hess's Law, calculate the enthalpy of combustion, ΔH_{comb}, for C_3H_8.

   (iii)   Solve the stoichiometry problem.

Solve.

(i)   Heat $H_2O(s)$ from $-14.0\text{º}C$ to $0.0\text{º}C$; $2500\,g\,H_2O \times \dfrac{2.092\,J}{g \bullet \text{º}C} \times 14.0\text{º}C = 73.22$

$$= 73.2\,kJ$$

Melt $H_2O(s)$; $2500\,g\,H_2O \times \dfrac{6.008\,kJ}{mol\,H_2O} \times \dfrac{1\,mol\,H_2O}{18.02\,g\,H_2O} = 833.52 = 834\,kJ$

Heat $H_2O(l)$ from $0.0\text{º}C$ to $60.0\text{º}C$; $2500\,g\,H_2O \times \dfrac{4.184\,J}{g \bullet \text{º}C} \times 60.0\text{º}C = 627.6 = 628\,kJ$

Total energy = $73.22\,kJ + 833.52\,kJ + 627.6\,kJ = 1534.34 = 1.534 \times 10^3\,kJ$

(The result has zero decimal places because 628 kJ has zero decimal places.)

(ii) $C_3H_8(g) + 5O_2(g) \rightarrow 3CO_2(g) + 4H_2O(l)$

Assume that one product is $H_2O(l)$, since this leads to a more negative ΔH_{comb} and fewer grams of $C_3H_8(g)$ required.

$\Delta H_{comb} = 3\Delta H_f^\circ\,CO_2(g) + 4\Delta H_f^\circ\,H_2O(l) - \Delta H_f^\circ\,C_3H_8(g) - 5\Delta H_f^\circ\,O_2(g)$

$= 3(-393.5\,kJ) + 4\,(-285.83\,kJ) - (-103.85\,kJ) - 5(0) = -2219.97 = -2220\,kJ$

(iii) $1.53434 \times 10^3\,kJ$ required $\times \dfrac{1\,mol\,C_3H_8}{2219.97\,kJ} \times \dfrac{44.096\,g\,C_3H_8}{1\,mol\,C_3H_8} = 30.48\,g\,C_3H_8$

(1.534×10^3 kJ required has 4 sig figs and so does the result)

11.104 (a)   Low viscosity, low surface tension, and, especially, high thermal conductivity owing to metallic properties.

(b)   Metallic bonding between sodium atoms, which persists in the liquid state, is probably the main inhibitor of movement of atoms relative to one another. As temperature increases, thermal motion of the atoms increases and the liquid expands, weakening bonding interactions relative to thermal energies.

11.105 $P = \dfrac{nRT}{V} = \dfrac{g\,RT}{M\,V}$; $T = 273.15 + 26.0\text{º}C = 299.15 = 299.2\,K$; $V = 5.00\,L$

$g\,C_6H_6(g) = 7.2146 - 5.1493 = 2.0653\,g\,C_6H_6(g)$

$P\,(vapor) = \dfrac{2.0653\,g}{78.11\,g/mol} \times \dfrac{299.15\,K}{5.00\,L} \times \dfrac{0.08206\,L \bullet atm}{K \bullet mol} \times \dfrac{760\,torr}{1\,atm} = 98.660 = 98.7\,torr$

11.106 *Plan.* relative humidity and v.p. of H_2O at $23°\,C \rightarrow P_{H_2O} \rightarrow$ ideal-gas law $\rightarrow$ mol $H_2O(g) \rightarrow H_2O$ molecules.

Solve. r.h. $= (P_{H_2O}$ in air / v.p. of $H_2O) \times 100$

From Appendix B, v.p. of H_2O at $23°C = 21.07$ torr

P_{H_2O} in air = r.h. $\times$ v.p. of $H_2O/100 = 45 \times 21.07$ torr$/100 = 9.4815 = 9.5$ torr

$$n = PV/RT; V = 14\,m \times 9.0\,m \times 8.6\,m \times \frac{1\,dm^3}{(0.1)^3\,m^3} \times \frac{1\,L}{dm^3} = 1.0836 \times 10^6 = 1.1 \times 10^6\,L$$

$$n = 9.4815\,torr \times \frac{1\,atm}{760\,torr} \times \frac{mol \cdot K}{0.08206\,L \cdot atm} \times \frac{1.0836 \times 10^6\,L}{296\,K} = 556.6 = 5.6 \times 10^2\,mol\,H_2O$$

$$556.6\,mol\,H_2O \times \frac{6.022 \times 10^{23}\,molecules}{1\,mol} = 3.4 \times 10^{26}\,H_2O\,molecules$$

11.107　Data are taken from the 74th edition of the *Handbook of Chemistry and Physics*. T_m = melting point, T_b = boiling point

(a)　W: T_m = 3410°C, T_b = 5660 °C; WF_6: T_m = 2.5°C, T_b = 17.5°C

W is a metal, with strong metallic bonding, and very high T_m and T_b. WF_6 is an octahedral, nonpolar molecule. Even though it has high molar mass, the spherical shape of the molecule prevents extensive molecular contacts. The resulting London-dispersion forces are very weak, which leads to the low T_m and T_b.

(b)　SO_2: T_m = -72.7°C, T_b = -10°C; SF_4: T_m = -124°C, T_b = -40°C (sublimes)

Both SO_2 and SF_4 are polar covalent molecules with a nonbonding electron pair on the central S atom. The electron-domain geometry in SO_2 is trigonal planar and the molecule shape is bent. The electron-domain geometry in SF_4 is trigonal bipyramidal and the molecular shape is see-saw. Both are gases at ambient temperature and pressure. SF_4 has higher molar mass but lower melting and boiling points than SO_2. This indicates that dipole-dipole forces are more influential on the properties of these molecules and that SF_4 has a smaller dipole moment than SO_2.

(c)　SiO_2: T_m = 1723°C, T_b = 2230°C, MM = 60 g/mol

$SiCl_4$: T_m = -70°C, T_b = 57.57°C, MM = 170 g/mol

SiO_2 is a covalent-network substance, with high T_m and T_b. Covalent bonds hold SiO_2 units in a rigid lattice, and high energy is required to break these bonds and melt or boil the substance. $SiCl_4$ is a tetrahedral, nonpolar molecule. Weak dispersion forces between the approximately spherical molecules result in predictably low T_m and T_b.

12 Modern Materials

Visualizing Concepts

12.1   *Analyze/Plan.* Given energy diagrams of band structures for two materials, predict which material is a metal. Consider the conductivity of a metal and how it relates to band structure. Review the band structures in Figure 12.2.

       Solve. The band structure of material A is that of a metal. Metals are excellent conductors of electricity and current; they have essentially no barrier to electron flow and a correspondingly small or zero energy band gap. This describes the band structure of material A.

12.2   Material Y, with the smaller band gap, is more suitable for solar energy conversion. According to Section 12.4, semiconductors with small band gaps are most appropriate, because they are able to capture a wider range of solar wavelengths. Semiconductors with larger band gaps capture only higher energy, shorter wavelength light.

12.3   The polymer chains in cartoon (a) are much more tightly wound than those in (b). This indicates that there are stronger intermolecular forces attracting polymer (a) chains to each other. Large empty spaces in polymer (b) suggest only weak attractive forces among chains.

12.4   *Analyze/Plan.* Given diagrams with rod-like (columnar) molecules in different orientations, decide which diagram represents molecules in a liquid crystalline phase. Recall the molecular orientations that define solids, liquid crystals and isotropic liquids.

       Solve. Liquid crystals exhibit molecular order in at least one dimension. Solids display three-dimensional order, and liquids are randomly ordered with close molecular contacts. Diagram (a) has columnar molecules with their long directions oriented parallel to the vertical direction of the box, a one-dimensional order characteristic of a liquid crystalline phase. Diagram (b) has the close but randomly oriented molecules characteristic of an isotropic liquid.

12.5   (a)   Addition polymers are formed from monomers with C–C double or triple bonds, but not by aromatic compounds derived from benzene. Molecule (i), with a terminal C=C, is the only monomer shown that is capable of addition polymerization.

       (b)   Condensation is the combination of two or more molecules to form one larger molecule, in this case a polymer, and a small molecule, usually H_2O or NH_3. Monomers for condensation polymerization contain a carboxyl (–COOH) group and either an alcohol (–OH) or amine (–NH_2) group. Molecule (iii) contains both a carboxyl group and an amine group; it can "condense" with like monomers to form a polymer and NH_3.

(c)   Liquid crystals are typically formed by columnar molecules that contain functional groups capable of strong intermolecular interactions that promote long-range order. Molecule (ii) fits this description (and is the only remaining choice). The –CN group experiences dipole-dipole and London dispersion forces. The rod-like 5 C chain and the planar benzene-like (phenyl) ring both have shapes the encourage strong dispersion forces. Taken together, these intermolecular forces are likely to encourage the long range order required to form a liquid crystal.

12.6   An integrated circuit with 20 billion components has 20,000,000,000 or 20×10^9 components. The log of this number is 10.30. By finding this number on the y-axis of the graph, following the value across to the red line, and reading the corresponding year from the x-axis, we see that the year is ~2015. That is, Moore's law predicts that current chip technology will reach its complexity (and performance) limit in 2015.

Classes of Materials

12.7   *Analyze.* Given: formula of pure substance. Find: metal, semiconductor, insulator.

Plan. Use the periodic table and specific examples as a guideline. Metals are on the left of the periodic chart, the s-block, d-block, and p-block elements up to metalloids. Elemental semiconductors are Si, Ge, and C(graphite). Compound semiconductors are often combinations of Group 3A and 5A elements. Other nonmetals and most ionic and molecular compounds are insulators.   *Solve.*

(a)   GaN—semiconductor; Gp 3A + Gp 5A

(b)   B—insulator; nonmetal, not an elemental semiconductor

(c)   ZnO—semiconductor, like CdS, average 4 e^- per atom

(d)   Pb—metal; p-block metal left of metalloid line

12.8   (a)   InAs—semiconductor, Gp 3A + Gp 5A

(b)   MgO—insulator; ionic compound

(c)   HgS—semiconductor, like CdS, average 4 e^- per atom

(d)   Sn—metal; p-block metal left of metalloid line

12.9   *Analyze.* Given: GaAs. Find: dopant to make n-type semiconductor.

Plan. An n-type semiconductor has extra negative charges. If the dopant replaces a few Ga atoms, it should have more valence electrons than Ga, Group 3A.

Solve. The obvious choice is a Group 4A element, either Ge or Si. Ge would be closer to Ga in bonding atomic radius (Figure 7.5).

12.10   p-type semiconductors have a slight e^- deficit. If the dopant replaces As, Group 5A, it should have fewer than five valence electrons. The dopant will be a 4A element, again probably Si or Ge. Si would be closer to As in bonding atomic radius.

12.11   (a)   False. Semiconductors conduct some electricity, while insulators do not. Semiconductors must have a smaller barrier to electron mobility, a smaller band gap.

       (b)   True. Dopants create either more electrons or more "holes," both of which increase the conductivity of the semiconductor.

       (c)   True. Metals conduct electricity (and heat) because delocalized electrons in the lattice provide a mechanism for charge mobility.

       (d)   True. Metal oxides are ionic substances with essentially localized electrons. There is a large barrier to charge mobility, a band gap so large that metal oxides are insulators.

12.12   (a)   True. 400 kJ/mol is sufficient energy to break many chemical bonds. It is an energy barrier large enough to prohibit any charge mobility.

       (b)   True. The energy difference between valence and conduction can be small as in the case of metals, but the conduction band is always at the higher energy. Otherwise all materials would be conductors.

       (c)   False. Electrons can conduct only if they are in a partially filled band. If electrons in filled bands could conduct, there would be no insulators.

       (d)   False. "Holes" are empty electron sites. They are the absence of an electron where one would normally exist, leading to a region of at least partial positive charge.

12.13   A superconducting material offers no resistance to the flow of electrical current; *superconductivity* is the frictionless flow of electrons. Superconductive materials could transmit electricity with no heat loss and therefore much greater efficiency than current carriers. Because of the Meisner effect, they are also potential materials for magnetically levitated trains.

12.14   A conductive metal such as Ag conducts electricity with a characteristic resistance to the flow of electrons given by Ohm's law, $E = IR$. A superconducting substance such as Nb_3Sn below its transition temperature conducts electricity with no resistance to the flow of electrons. Such a superconductor can transfer energy with no net loss, while a metallic conductor cannot be 100% efficient.

12.15   Below 39 K, MgB_2 conducts electricity with zero resistivity, the definition of a superconductor. Above 39 K, the material is not superconducting. The sharp drop in resistivity of MgB_2 near 39 K is the superconducting transition temperature, T_c.

12.16   (a)   The superconducting transition temperature, T_c, is the temperature at which a material loses all resistance to the flow of electrical current, the temperature below which the material becomes superconducting.

       (b)   The temperature 77 K is significant because that is the temperature of liquid nitrogen, a readily available, inexpensive, and safe coolant. Materials with T_c temperatures above 77 K produce more financially viable devices than materials which must be cooled with liquid helium below 77 K to achieve superconductivity.

12.17 Superconductivity acts on electrons that are conduction electrons. Thus you need to have electrons available to be conductors. Insulators have no electrons available for conduction.

12.18 The phenomenon that superconductors exclude all magnetic fields from their volume is the Meisner effect (Figure 12.4). This can be used to levitate trains, by having either the tracks or the train wheels made from a magnetic material, and the other from a superconductor. The superconductor would need to be cooled below its transition temperature. There are some practical advantages to having the superconductor aboard the train and the magnetic field along the tracks. The train components require less of the costly superconductor and can be cooled more efficiently than the tracks. This is the reverse of the orientation shown in Figure 12.4, where the magnet floats above the cooled superconductor.

Materials for Structure

12.19 *n*-decane does not have a sufficiently high chain length or molecular mass to be considered a polymer.

12.20 Monomers are small molecules with low molecular mass that are joined together to form polymers. They are the repeating units of a polymer. Three (of the many) monomers mentioned in this chapter are

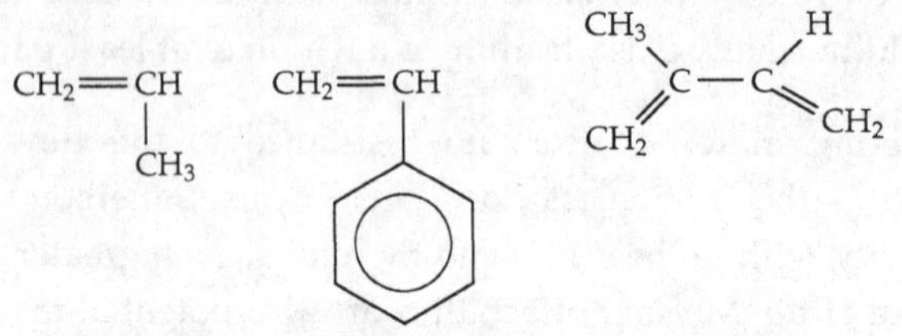

propylene   styrene   isoprene
(propene)   (phenyl ethene)   (2-methyl-1,3-butadiene)

12.21 *Analyze.* Given two types of reactant molecules, we are asked to write a condensation reaction with an ester product. *Plan.* A condensation reaction occurs when two smaller molecules combine to form a larger molecule and a small molecule, often water. Consider the structures of the two reactants and how they could combine to join the larger fragments and split water.   *Solve.*

A carboxylic acid contains the 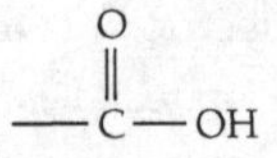   functional group; an alcohol contains the

–OH functional group. These can be arranged to form the 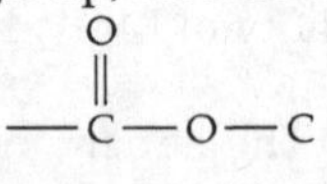   ester functional group and H_2O. Condensation reaction to form an ester:

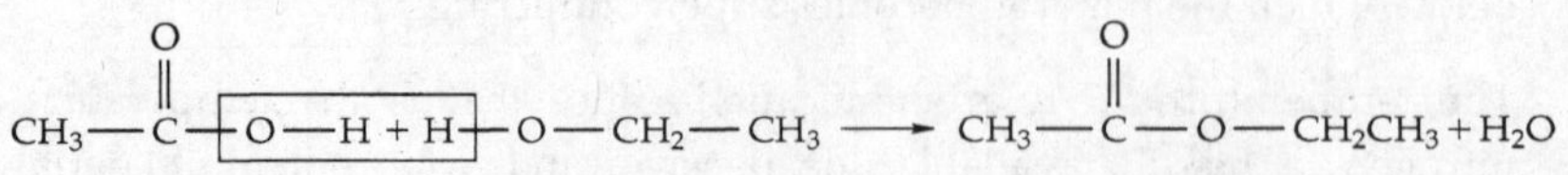

If a dicarboxylic acid (two –COOH groups, usually at opposite ends of the molecule) and a dialcohol (two –OH groups, usually at opposite ends of the molecule) are

combined, there is the potential for propagation of the polymer chain at both ends of both monomers. Polyethylene terephthalate (Table 12.4) is an example of a polyester formed from the monomers ethylene glycol and terephthalic acid.

12.22

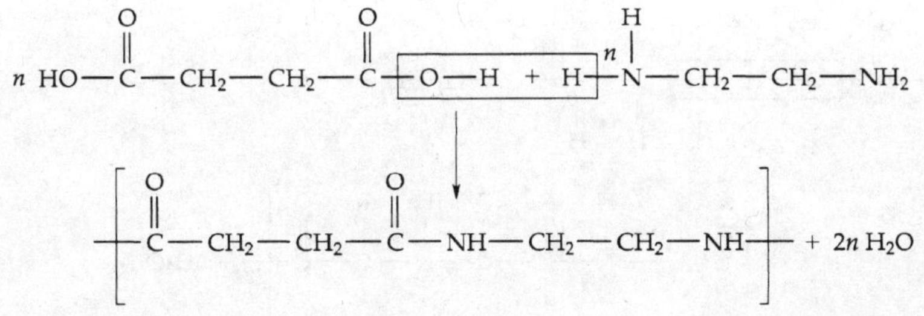

12.23   *Analyze/Plan.* Decide whether the given polymer is an addition or condensation polymer. Select the smallest repeat unit and deconstruct it into the monomer(s) with the specific functional group(s) that would form the stated polymer.   *Solve.*

(a)

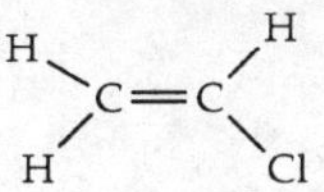

vinyl chloride (chloroethylene or chloroethene)

(b)

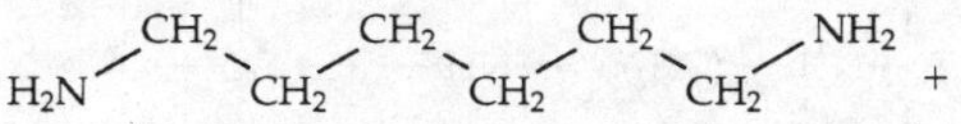

hexanediamine

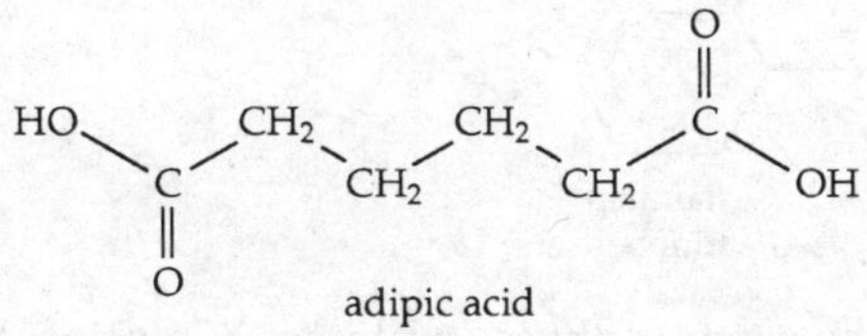

adipic acid

(Formulas given in Equation [12.3].)

(c)

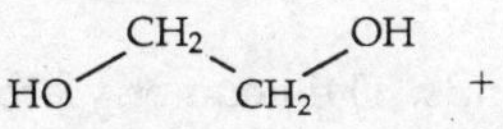

ethylene glycol    terephthalic acid

12.24   (a)   By analogy to polyisoprene, Equation [12.4],

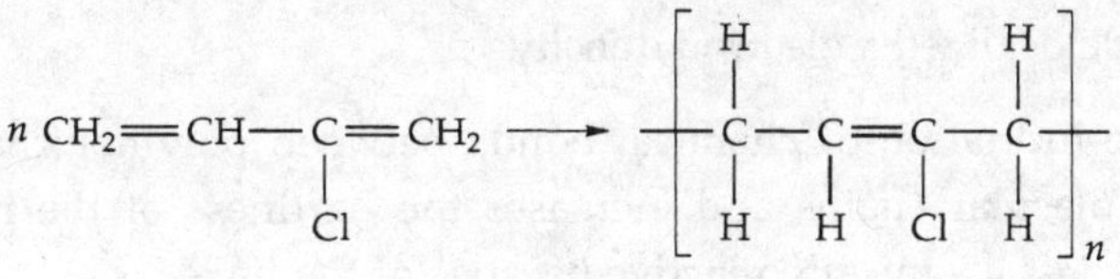

(b) $n\ CH_2{=}CH \longrightarrow$

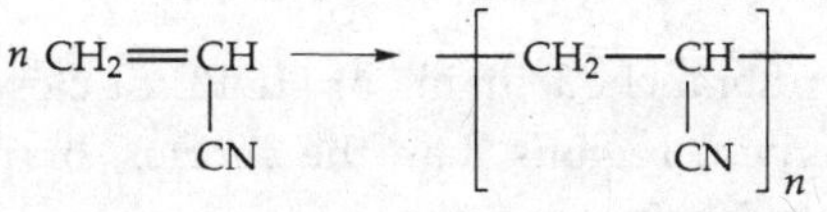

12.25   *Plan/Solve.* When nylon polymers are made, H_2O is produced as the C–N bonds are formed. Reversing this process (adding H_2O across the C–N bond), we see that the monomers used to produce Nomex™ are:

$$HOOC{-}\langle C_6H_4\rangle{-}COOH \quad \text{and} \quad H_2N{-}\langle C_6H_4\rangle{-}NH_2$$

12.26

$$H{-}\underset{\underset{H}{|}}{N}{-}\underset{\underset{H}{|}}{\overset{\overset{R}{|}}{C}}{-}\overset{\overset{O}{\|}}{C}\boxed{-O{-}H + H{-}}\underset{\underset{H}{|}}{N}{-}\underset{\underset{H}{|}}{\overset{\overset{R}{|}}{C}}{-}\overset{\overset{O}{\|}}{C}{-}O{-}H \longrightarrow$$

$$-\underset{\underset{H}{|}}{\overset{\overset{H}{|}}{N}}{-}\overset{\overset{R}{|}}{C}{-}\overset{\overset{O}{\|}}{C}{-}\left[\underset{\underset{H}{|}}{\overset{\overset{H}{|}}{N}}{-}\overset{\overset{R}{|}}{C}{-}\overset{\overset{O}{\|}}{C}\right]_n + nH_2O$$

12.27 *Analyze/Plan.* Given the formula of a monomer, write the equation for condensation polymerization. The monomers are aligned so that the caroboxyl end of one monomer joins the amine end of another molecule.   *Solve.*

$$n\; H_2N{-}(CH_2)_3\overset{\overset{O}{\|}}{C}\boxed{-O{-}H + H{-}}\underset{\underset{H}{|}}{N}{-}(CH_2)_3\overset{\overset{O}{\|}}{C}{-}OH \longrightarrow$$

$$\left[-NH{-}(CH_2)_3\overset{\overset{O}{\|}}{C}{-}NH{-}(CH_2)_3\overset{\overset{O}{\|}}{C}-\right]_n$$

12.28

$$HOOC{-}\langle C_6H_4\rangle{-}COOH \qquad H_2N{-}\langle C_6H_4\rangle{-}NH_2$$

diacid   diamine
terephthalic acid   *p*-diaminobenzene

Note that these monomers are the same as those in Nomex (Solution 12.25) except for the orientation of the functional groups on the benzene rings. Clearly monomer structure strongly impacts polymer structure.

12.29 Most of a polymer backbone is composed of σ bonds. The geometry around individual atoms is tetrahedral with bond angles of 109°, so the polymer is not flat, and there is relatively free rotation around the σ bonds. The flexibility of the molecular chains causes flexibility of the bulk material. Flexibility is enhanced by molecular features that inhibit order, such as branching, and diminished by features that encourage order, such as cross-linking or delocalized π electron density.

Cross-linking is the formation of chemical bonds between polymer chains. It reduces flexibility of the molecular chains and increases the hardness of the material. Cross-linked polymers are less chemically reactive because of the links.

12.30 At the molecular level, the longer, unbranched chains of HDPE fit closer together and have more crystalline (ordered, aligned) regions than the shorter, branched chains of LDPE. Closer packing leads to higher density.

12.31 The function of the material (polymer) determines whether high molecular mass and high degree of crystallinity are desirable properties. If the material will be formed into containers or pipes, the rigidity and structural strength associated with high molecular mass are required. If the polymer will be used as a flexible wrapping or as a garment material, high molecular mass and rigidity are undesirable properties.

12.32 (a)   An elastomer is a polymer material that recovers its shape when released from a distorting force. A typical elastomeric polymer can be stretched to at least twice its original length and return to its original dimensions upon release.

(b)   A thermoplastic material can be shaped and reshaped by application of heat and/or pressure.

(c)   A thermosetting plastic can be shaped once, through chemical reaction in the shape-forming process, but cannot easily be reshaped, due to the presence of chemical bonds that cross-link the polymer chains.

(d)   A plasticizer is a substance of relatively low molar mass added to a polymer material to soften it.

12.33 Ceramics are not readily recyclable because of their extremely high melting points and rigid ionic or covalent-network structures. According to Table 12.2, the melting points of ceramic materials are much higher than those of Al and steel. This makes recycling ceramics technologically difficult and expensive. Crystalline ceramics have rigid, precise three-dimensional structures. If and when these materials can be melted, either covalent or ionic bonds are broken. The precise structures are usually not reformed upon cooling. Recyclable ceramics such as bottle glass are amorphous; there is no exact repeating structure that must be reformed after melting.

12.34 (a)   The object that shatters is ceramic; the one that dents is metal.

(b)   The two materials differ in their behavior because of their different solid state bonding characteristics. Ceramics are formed from inorganic materials linked by ionic or highly polar covalent bonds into three-dimensional bonding networks. During catastrophic failure (dropping 10 feet onto cement), the network structure prevents atoms from sliding over one another and the ceramic shatters. A series of bond ruptures occurs, often along planes in the three-dimensional structure, leading to fragments with sharp edges.

Metallic bonding is characterized by delocalization of loosely held valence electrons among metal atoms. Unlike the covalent or ionic network bonding in ceramics, metallic bonding is multidirectional. This allows metal atoms to slide over each other during deformation, resulting in dents and stress cracks rather than shattering.

12.35 Very small, uniformly sized and shaped particles are required for the production of a strong ceramic object by sintering. During sintering, the small ceramic particles are heated to a high temperature below the melting point of the solid. This high temperature initiates condensation reactions between molecules at the surfaces of the spheres; the spheres are then connected by chemical bonds between atoms in different spheres. The more uniform the particle size and the greater the total surface area of the solid, the more chemical bonds are formed, and the stronger the ceramic object.

12.36   Since Zr and Ti are in the same family, assume that the stoichiometry of the compounds in a sol-gel process will be the same for the two metals.

     i.   Alkoxide formation: oxidation-reduction reaction

$$Zr(s) + 4CH_3CH_2OH(l) \rightarrow Zr(OCH_2CH_3)_4(s) + 2H_2(g)$$
$$\text{alkoxide}$$

     ii.   Sol formation: metathesis reaction

$$Zr(OCH_2CH_3)_4(soln) + 4H_2O(l) \rightarrow \quad Zr(OH)_4(s) \quad + 4CH_3CH_2OH(l)$$
$$\text{"precipitate"} \quad \text{nonelectrolyte}$$
$$\text{sol}$$

          $Zr(OCH_2CH_3)_4(s)$ is dissolved in an alcohol solvent and then reacted with water. In general, reaction with water is called *hydrolysis*. The alkoxide anions $(CH_3CH_2O^-)$ combine with H^+ from H_2O to form the nonelectrolyte $CH_3CH_2OH(l)$, and Zr^{2+} cations combine with OH^- to form the $Zr(OH)_4$ solid. The product $Zr(OH)_4(s)$ is not a traditional coagulated precipitate, but a finely divided evenly dispersed collection of particles called a sol.

     iii.   Gel formation: condensation reaction

$$(OH)_3 Zr{-}O{-}H(s) + H{-}O{-}Zr(OH)_3(s) \rightarrow (HO)_3 Zr{-}O{-}Zr(OH)_3(s) + H_2O(l)$$
$$\text{gel}$$

          Adjusting the acidity of the $Zr(OH)_4$ sol initiates condensation, the splitting-out of $H_2O(l)$ and formation of a zirconium-oxide network solid. The solid remains suspended in the solvent mixture and is called a gel.

     iv.   Processing: physical changes

          The gel is heated to drive off solvent and the resulting solid consists of dry, uniform and finely divided ZrO_2 particles.

12.37   The ceramics are: MgO, soda lime glass, ZrB_2, Al_2O_3, and TaC. The criteria are a combination of chemical formula (with corresponding bonding characteristics) and Knoop values. Ceramics are ionic or covalent-network solids with fairly large hardness values. $CaCO_3$ is ionic, but carbonates are not one of the typical types of ceramics listed in Section 12.1. Ag and Cr are metals.

Hardness alone is not a sufficient criteria for ceramics. The range of hardness values for the ceramics in this group is large; Cr, a metal, lies in the middle of the range, as could other metals. Nonetheless, ceramics as a group are hard materials; hardness is a necessary, but not a sufficient condition for classification as a ceramic.

12.38   By analogy to the ZnS structure, the C atoms form a face-centered cubic array with Si atoms occupying alternate tetrahedral holes in the lattice. This means that the coordination numbers of both Si and C are 4; each Si is bound to four C atoms in a tetrahedral arrangement, and each C is bound to four Si atoms in a tetrahedral arrangement, producing an extended three-dimensional network. ZnS, an ionic solid, sublimes at 1185° and 1 atm pressure and melts at 1850° and 150 atm pressure. The considerably higher melting point of SiC, 2800° at 1 atm, indicates that SiC is probably not a purely ionic solid and that the Si–C bonding network has significant covalent character. This is reasonable, since the electronegativities of Si and C are similar (Figure

8.7). SiC is high-melting because a great deal of chemical energy is stored in the covalent Si–C bonds, and it is hard because the three-dimensional lattice resists any change that would weaken the Si–C bonding network.

12.39   Ceramics are inorganic ionic solids. Their rigid three-dimensional order is the result of strong electrostatic attractions among fully charged ions. Bulk ceramics are hard, because it is difficult to budge an ion from this ionic network. They have extremely high melting points because tremendous thermal energy must be supplied before the ions have sufficient kinetic energy to break away from the network and move relative to each other.

12.40   Correlation and learning vary with each learner. The interpretations below are one of many possibilities, with a lettered statement for each verse of the poem.

(a)   The bulk properties of a substance depend on its component particles. A metal composed of Pb atoms will have the properties of lead, not those of gold. The properties of materials depend on their component atoms and molecules, and the nature of the bonding between particles.

(b)   Silicon and graphite, a form of carbon, are semiconductors because of their ordered covalent-network structure and the resulting bands of energy states. These solid-state structures were determined by X-ray crystallography (Chapter 11).

(c)   Many of the properties of metals, including ductility and conductivity, are due to the delocalized electrons characteristic of metallic bonding. The electrons are delocalized because of the nearly continuous valence and conduction bonds (Figure 12.2) in metals.

(d)   Ceramics are inorganic solids with localized electrons and a large band gap; "no free electrons form a lubricating tide."

(e)   Glass is one of the only recyclable ceramics. It is an amorphous solid, lacking the three-dimensional order of most ceramics. It can be melted and reformed, because its properties do not depend on a perfectly ordered structure.

(f)   Polymers are very long-chain, high molecular weight substances, formed by covalent bonding among one or more types of monomer units. Intermolecular forces between chains are typically weak, so the materials are flexible. Those that can be formed into shapes are called plastics. Rigidity can be increased by cross-linking, chemical bonding between chains. Proteins and nucleic acids, "cross-linked helixes unknown to *Robert Hooke*," are copolymers formed by bonding several different monomers.

(g)   The melting points of all materials depend on their structure.

(h)   Electrical conductivity depends on the size of the band gap in a material. Doping with very small amounts of appropriate impurities increases the conductivity of semiconductors.

(i) Doped semiconductors and ceramic superconductors are examples of nonstoichiometric solids discussed in this chapter. p-type semiconductors have "strange holes" that "wander loose," leading to their semiconductivity.

(j) Semiconductors can emit light if electrons have been excited across the band gap by an applied voltage or light of energy equal to or higher than the band gap. (Figure 12.41)

Nano particles of the semiconductor Cd_3P_2 absorb different colors of visible light and appear different colors, depending on particle size. "Each element absorbs its signature."

(k) Superconductors have no resistance to current flow. They exclude all magnetic fields from their volume (Meisner effect). While the mechanism of superconductivity in ceramics is not well understood, "Diffuse material becomes magnetic when another Field aligns domains."

Materials for Medicine

12.41 Is the neoprene biocompatible: is the surface smooth enough and is the chemical composition appropriate so that there are no inflammatory reactions in the body? Does neoprene meet the physical requirements of a flexible lead: will it remain resistant to degradation by body fluids over a long time period; will it maintain elasticity over the same time period? Can neoprene be prepared in sufficiently pure form (free of trace amounts of monomer, catalyst, etc.) so that it can be classified as medical grade?

12.42 One structural characteristic of polymers that forms effective interfaces with biological systems is the presence of polar functional groups in the polymer backbone or as substituents. Polystyrene is a hydrocarbon; it has no polar functional groups and is a nonpolar substance. Polyurethane has polar carbon–oxygen, carbon–nitrogen, and nitrogen–hydrogen functional groups. The N–H groups mean that it can act as a hydrogen-bond donor as well as an acceptor. In fact, the polyurethane backbone is very similar to the protein backbone shown in this section. We expect polyurethane to be the superior biointerface.

12.43 Current vascular-graft materials cannot be lined with cells similar to those in the native artery. The body detects that the graft is "foreign" and platelets attach to the inside surfaces, causing blood clots. The inside surfaces of the future vascular implants need to accommodate a lining of cells that do not attract or attach to platelets.

12.44 Surface roughness in synthetic heart valves causes hemolysis, the breakdown of red blood cells. The surface of the valve implant was probably not smooth enough.

12.45 In order for skin cells in a culture medium to develop into synthetic skin, a mechanical matrix must be present that holds the cells in contact with one another and allows them to differentiate. The matrix must be mechanically strong, biocompatible, and biodegradable. It probably has polar functional groups that are capable of hydrogen bonding with biomolecules in the tissue cells.

12.46 Polystyrene is an essentially nonpolar hydrocarbon, while polyethyleneterephthalate (PET) contains polar ester groups, as well as nonpolar hydrocarbon portions. PET is more appropriate, because it provides polar ester groups $\left(\!\!\begin{array}{c} O \\ \| \\ C-O-C \end{array}\!\!\right)$ with hydrogen bonding capabilities where the cells can attach. Also, the ester linkages are susceptible to hydrolysis (the reverse of condensation); this renders the synthetic matrix biodegradable when employed in the body.

Materials for Electronics

12.47   Silicon is the semiconductor material of choice for integrated circuits because it is abundant, cheap, nontoxic, can be highly purified, and grows nearly perfect enormous crystals. That is, it has appropriate chemical and physical properties and is cost effective.

12.48   Impurities can be thought of as dopants. Since dopants at concentrations of parts per million (ppm) can change the conductivity of Si, it needs to be more pure than the ppm level to ensure known and constant conductivity. Purity at ppm level would be 99.9999% pure. Purity of 99.999999999% is pure at nearly the parts per trillion level.

12.49   The bonding atomic radius of a Si atom is 1.11 Å (Figure 7.5), so the diameter is 2.22 Å.

$$60 \text{ nm} \times \frac{1 \times 10^{-9} \text{ m}}{1 \text{ nm}} \times \frac{1 \text{ Å}}{1 \times 10^{-10} \text{ m}} \times \frac{1 \text{ Si atom}}{2.22 \text{ Å}} = 2.7 \times 10^{2} \text{ Si atoms.}$$

12.50   "Plastic" semiconductors would be shapable as well as flexible. This would be very desirable for medical applications such as prosthetic hands or feet, where the flexibility would render the device more natural and more durable to constant movements. Flexible semiconductors would also be useful in forming irregularly shaped components for a wide variety of devices.

12.51   *Analyze.* Given: 1.1 eV. Find: wavelength in meters that corresponds to the energy 1.1 eV. *Plan.* Use dimensional analysis to find wavelength.

Solve. 1 eV = 1.602×10^{-19} J (inside-back cover of text); $\lambda = hc/E$

$$\lambda = 6.626 \times 10^{-34} \text{ J} \bullet \text{s} \times \frac{3.00 \times 10^{8} \text{ m}}{\text{s}} \times \frac{1}{1.1 \text{ eV}} \times \frac{1 \text{ eV}}{1.602 \times 10^{-19} \text{ J}} = 1.128 \times 10^{-6}$$

$$= 1.1 \times 10^{-6} \text{ m}$$

Si can absorb energies $\geq$ 1.1 eV, or wavelengths $\leq 1.1 \times 10^{-6}$ m. The range of wavelengths in the solar spectrum at sea level is 3×10^{-6} to 2×10^{-7} m, or 30×10^{-7} to 2×10^{-7} m, a span of 28×10^{-7} m. Si can absorb 11×10^{-7} to 2×10^{-7} m, a span of 9×10^{-7} m.

This represents $\dfrac{9 \times 10^{-7}}{28 \times 10^{-7}} \times 100 = 32\%$ of the wavelengths in the solar spectrum.

According to the diagram, these wavelengths represent much more than 32% of the total flux.

12.52   From Table 12.1, the band gap energy for TiO_2 is 3.0 eV.

$$\lambda = 6.626 \times 10^{-34} \text{ J} \bullet \text{s} \times \frac{3.00 \times 10^8 \text{ m}}{\text{s}} \times \frac{1}{3.0 \text{ eV}} \times \frac{1 \text{ eV}}{1.602 \times 10^{-19} \text{ J}} = 4.136 \times 10^{-7}$$

$$= 4.1 \times 10^{-7} \text{ m}$$

TiO_2 absorbs 4.1×10^{-7} to 2.0×10^{-7}, a span of 2.1×10^{-7} m.

$$\frac{2.1 \times 10^{-7}}{28 \times 10^{-7}} \times 100 = 7.5\% \text{ of the wavelengths in the solar spectrum.}$$

This is a much smaller portion of the spectrum, and of the total flux, than Si absorbs.

Materials for Optics

12.53   Both an ordinary liquid and a nematic liquid crystal phase are fluids; they are converted directly to the solid phase upon cooling. The nematic phase is cloudy and more viscous than an ordinary liquid. Upon heating, the nematic phase is converted to an ordinary liquid.

12.54   In an ordinary liquid, molecules are oriented randomly and their relative orientations are continuously changing. In liquid crystals, the molecules are aligned in at least one dimension. The relative orientations in the other two dimensions may change, but alignment in the oriented direction is maintained.

12.55   In the solid state, there is three-dimensional order; the relative orientation of the molecules is fixed and repeating in all three dimensions. Essentially no translational or rotational motion is allowed. When a substance changes to the nematic liquid-crystalline phase, the molecules remain aligned in one dimension (the long dimension of the molecule). Translational motion is allowed, but rotational motion is restricted. Transformation to the isotropic-liquid phase destroys the one-dimensional order. Free translational and rotational motion result in random molecular orientations that change continuously.

12.56   Reinitzer observed that cholesteryl benzoate has a phase that exhibits properties intermediate between those of the solid and liquid phases. This "liquid-crystalline" phase, formed by melting at 145°C, is opaque, changes color as the temperature increases, and becomes clear at 179°C.

12.57   The presence of polar groups or nonbonded electron pairs leads to relatively strong dipole-dipole interactions between molecules. These are a significant part of the orienting forces necessary for liquid crystal formation.

12.58   Because order is maintained in at least one dimension, the molecules in a liquid-crystalline phase are not totally free to change orientation. This makes the liquid-crystalline phase more resistant to flow, more viscous, than the isotropic liquid.

12.59   In the nematic phase, molecules are aligned in one dimension, the long dimension of the molecule. In a smectic phase (A or C), molecules are aligned in two dimensions. Not only are the long directions of the molecules aligned, but the ends are also aligned. The molecules are organized into layers; the height of the layer is related to the length of the molecule.

12 Modern Materials   Solutions to Exercises

12.60   The "LCD molecule" is long relative to its thickness. It has C=C and C≡N groups that promote rigidity and polarizability along the length of the molecule. The C≡N group also provides dipole-dipole interactions that encourage alignment. Unlike the molecules in Figure 12.34, which contain planar phenyl rings, the LCD molecule contains nonaromatic, nonplanar six-membered rings. These rings could contribute to specific physical properties such as the liquid crystal temperature range that make this molecule particularly functional in LCD displays.

12.61   A nematic phase is composed of sheets of molecules aligned along their lengths, but with no additional order within the sheet or between sheets. A cholesteric phase also contains this kind of sheet, but with some ordering between sheets. In a cholesteric phase, there is a characteristic angle between molecules in one sheet and those in an adjacent sheet. That is, one sheet of molecules is twisted at some characteristic angle relative to the next, producing a "screw" axis perpendicular to the sheets.

12.62   As the temperature of a substance increases, the average kinetic energy of the molecules increases. More molecules have sufficient kinetic energy to overcome intermolecular attractive forces, so overall ordering of the molecules decreases as temperature increases. Melting provides kinetic energy sufficient to disrupt alignment in one dimension in the solid, producing a smectic phase with ordering in two dimensions. Additional heating of the smectic phase provides kinetic energy sufficient to disrupt alignment in another dimension, producing a nematic phase with one-dimensional order.

12.63   *Plan/Solve.* Follow the logic in Solution 12.51.

$$\lambda = hc/E = 6.626 \times 10^{-34} \text{ J} \bullet \text{s} \times \frac{3.00 \times 10^8 \text{ m}}{\text{s}} \times \frac{1}{2.2 \text{ eV}} \times \frac{1 \text{ eV}}{1.602 \times 10^{-19} \text{ J}} = 5.640 \times 10^{-7}$$

$$= 5.6 \times 10^{-7} \text{ m} = 560 \text{ nm}$$

12.64   From Table 12.1, E_g for GaAs (x = 0) is 1.4 and for GaP (x = 1) is 2.2. If E_g varies linearly with x, the band gap for x = 0.5 should be approximately the average of the two extreme values: (1.4 + 2.2)/2 = 1.8 eV.

$$\lambda = hc/E = 6.626 \times 10^{-34} \text{ J} \bullet \text{s} \times \frac{3.00 \times 10^8 \text{ m}}{\text{s}} \times \frac{1}{1.8 \text{ eV}} \times \frac{1 \text{ eV}}{1.602 \times 10^{-19} \text{ J}} = 6.89 \times 10^{-7}$$

$$= 6.9 \times 10^{-7} \text{ m} = 690 \text{ nm}$$

Materials for Nanotechnology

12.65   Continuous energy bands of molecular orbitals require a large number of atoms contributing a large number of atomic orbitals to the molecular orbital scheme. If a solid has dimensions 1–10 nm, nanoscale dimensions, there may not be enough contributing atomic orbitals to produce continuous energy bands of molecular orbitals.

12.66   *Analyze.* Given: 1/16 inch diameter pin, Encyclopedia Britannica, EB (over 20 volumes), line width = 32 atoms. Find: Can we print EB on the head of a pin, using a 32 atom line width?

Plan. There are many ways to approach this problem. Some data about EB must be discovered, via either a trip to the library or the internet, or both. Then, use this data about EB, along with that given in the problem, to test the validity of Feynman's assertion. The exact approach depends on the EB data discovered. It will probably include calculating the area of a pinhead using πr^2, and estimating the diameter of an "average" atom.

Solve. In the exercise, we are given information about the width of a line and the total printing area for the nano printing of EB. Calculate these dimensions explicitly, using any necessary assumptions.

Line width: If each line is 32 atoms wide, what is the diameter of an 'atom'? Use Figure 7.6 to calculate the radius and diameter of an average atom, ignoring H (because it is an outlier) and the noble gases (because they don't typically form bonds). The smallest atom is F and the largest is Rb. The average of their diameters is $[2(0.71) + 2(2.11)]/2 =$ 2.82 Å. (This is a generous estimate; Feynman used 2.5 Å.)

$$\frac{32 \text{ atoms}}{\text{line}} \times \frac{2.82 \text{ Å}}{\text{atom}} \times \frac{1 \times 10^{-8} \text{ cm}}{\text{Å}} = 9.024 \times 10^{-7} = 9.0 \times 10^{-7} \text{ cm}$$

Area of pinhead: πr^2, d = 1/16 inch, r = 1/32 inch

$$\frac{1}{32} \text{ inch} \times 2.54 \text{ cm} = 0.07938 = 0.079 \text{ cm}$$

area = 3.14159 $(0.07938 \text{ cm})^2 = 0.01979 = 0.020 \text{ cm}^2$

The question is, how do we relate the one-dimensional line-width to a two-dimensional area? The smallest area we can print is a square $(9.0 \times 10^{-7} \text{ cm})$ by $(9.0 \times 10^{-7} \text{ cm})$ = $8.1 \times 10^{-13} \text{ cm}^2$; define this area as one pixel. Each character (letter) is a combination of printed lines and white space. Assume that an area 9×9 pixels will be sufficient to print any character, including all lines and white space. This accommodates any character and generalizes our solution to languages other than English This is the method used by dot-matrix printers. A word or a sentence is then a string of these boxes. The number of characters we can print on the pinhead is then

$$\frac{0.020 \text{ cm}^2}{\text{pinhead}} \times \frac{1 \text{ character}}{8.1 \times 10^{-13} \text{ cm}^2} = 2.5 \times 10^{10} \text{ characters.}$$

Is this enough to print EB? Here's where research comes in. According to the EB website, the modern encyclopedia has 32 volumes and 44 million (44×10^6 or 4.4×10^7) words.

$$\frac{2.5 \times 10^{10} \text{ characters}}{4.4 \times 10^7 \text{ words}} = 561 \text{ characters/word.}$$

Since no word, let alone an "average" one, contains 561 characters, it's safe to say that Feynman's assertion is correct, with room to spare.

12.67   (a)   False. As particle size decreases, the band gap increases. The smaller the particle, the fewer AOs that contribute to the MO scheme, the more localized the bonding and the larger the band gap.

(b)   False. The wavelength of emitted light corresponds to the energy of the band gap. As particle size decreases, band gap increases and wavelength decreases $(E = hc/\lambda)$.

12.68   True. Blue light has short wavelengths, corresponding to a relatively large band gap. As particle size decreases, band gap increases and wavelength decreases. We could begin with a semiconductor with a smaller band gap and make it a nanoparticle to increase E_g and decrease wavelength. (Nanoparticle size becomes one more way to tune the properties of semiconductors.)

12.69   *Analyze.* Given: Au, 4 atoms per unit cell, 4.08 Å cell edge, volume of sphere = $4/3 \pi r^3$.
         Find: Au atoms in 20 nm diameter sphere.

   Plan. Relate the number of Au atoms in the volume of 1 cubic unit cell to the number of Au atoms in a 20 nm diameter sphere. Change units to Å (you could just as well have chosen nm as the common unit), calculate the volumes of the unit cell and sphere, and use a ratio to calculate atoms in the sphere.

   Solve. vol of unit cell = $(4.08 \text{ Å})^3$ = 67.9173 = 67.9 Å^3

   20 nm diameter = 10 nm radius; $10 \text{ nm} \times \dfrac{1 \times 10^{-9} \text{ m}}{1 \text{ nm}} \times \dfrac{1 \text{ Å}}{1 \times 10^{-10} \text{ m}} = 100$ Å radius

   (Note that 1 nm = 10 Å.)

   vol. of sphere = $4/3 \times 3.14159 \times (100 \text{ Å})^3$ = 4.18879×10^8 = 4.19×10^6 Å^3

   $\dfrac{4 \text{ Au atoms}}{67.9173 \text{ Å}^3} = \dfrac{x \text{ Au atoms}}{4.18879 \times 10^6 \text{ Å}}$; x = 2.46699×10^6 = 2.47×10^5 Au atoms

12.70   Vol of unit cell = $(4.08 \text{ Å})^3$ = 67.9173 = 67.9 Å^3

   22 nm diameter = 11 nm radius; $11 \text{ nm} \times \dfrac{10 \text{ Å}}{1 \text{ nm}} = 110$ Å radius

   vol of sphere = $4/3 \times 3.14159 \times (110 \text{ Å})^3$ = 5.5753×10^6 = 5.58×10^6 Å^3

   $\dfrac{4 \text{ Au atoms}}{67.9173 \text{ Å}^3} = \dfrac{x \text{ Au atoms}}{5.5753 \times 10^6 \text{ Å}^3}$; x = 3.2836×10^5 = 3.28×10^5 Au atoms

Additional Exercises

12.71   Semiconductors have a filled valence band and an empty conduction bond, separated by a characteristic difference in energy, the band gap, E_g. When a semiconductor is heated, more electrons have sufficient energy to jump the band gap, and conductivity increases. Metals have a partially-filled continuous energy band. Heating a metal increases the average kinetic energy of the metal atoms, usually through increased vibrations within the lattice. The greater vibrational energy of the atoms leads to imperfections in the lattice and discontinuities in the energy band. Thermal vibrations create barriers to electron delocalization and reduce the conductivity of the metal.

12.72   A dipole moment (permanent, partial charge separation) roughly parallel to the long dimension of the molecule would cause the molecules to reorient when an electric field is applied perpendicular to the usual direction of molecular orientation.

12.73

Teflon™ is formed by addition polymerization.

12.74   (a)   polymer   (b) ceramic   (c) ceramic   (d) polymer

         (e)   liquid crystal (an organic molecule with a characteristic long axis and the kinds of functional groups often found in compounds with liquid-crystalline phases (Figure 12.34); not enough repeating units to be a polymer)

12.75   Ceramics are usually three-dimensional network solids, whereas plastics most often consist of large, chain-like molecules (the chain may be branched) held loosely together by relatively weak van der Waals forces. Ceramics are rigid precisely because of the many strong bonding interactions intrinsic to the network. Once a crack forms, atoms near the defect are subject to great stress, and the crack is propagated. They are stable to high temperatures because tremendous kinetic energy (temperature) is required for an atom to break free from the bonding network. On the other hand, plastics are flexible because the molecules themselves are flexible (free rotation around the sigma bonds in the polymer chain), and it is easy for the molecules to move relative to one another (weak intermolecular forces). [However, recall that rigidity of the plastic can be increased in a number of ways, including cross-linking (Figure 12.15), or reinforcement with a second polymer (Figure 12.17).] Plastics are not thermally stable because their largely organic molecules are subject to oxidation and/or bond breaking at high temperatures.

12.76   In a liquid crystal display (Figure 12.37), the molecules must be free to rotate by 90°. The long directions of molecules remain aligned but any attractive forces between the ends of molecules are disrupted. At low Antarctic temperatures, the liquid crystalline phase is closer to its freezing point. The molecules have less kinetic energy due to temperature and the applied voltage may not be sufficient to overcome orienting forces among the ends of molecules. If some or all of the molecules do not rotate when the voltage is applied, the display will not function properly.

12.77   At the temperature where a substance changes from the solid to the liquid-crystalline phase, kinetic energy sufficient to overcome most of the long range order in the solid has been supplied. A few van der Waals forces have sufficient attractive energy to impose the one-dimensional order characteristic of the liquid-crystalline state. Very little additional kinetic energy (and thus a relatively small increase in temperature) is required to overcome these aligning forces and produce an isotropic liquid.

12.78   This phenomenon is similar to supercooling, Section 11.4. When the isotropic liquid is cooled below the liquid crystal-liquid transition temperature, the kinetic energy of the molecules has been decreased enough so that formation of the liquid crystalline phase is energetically favorable. However, the molecules may not be correctly organized so that long range ordering can take place.

12.79

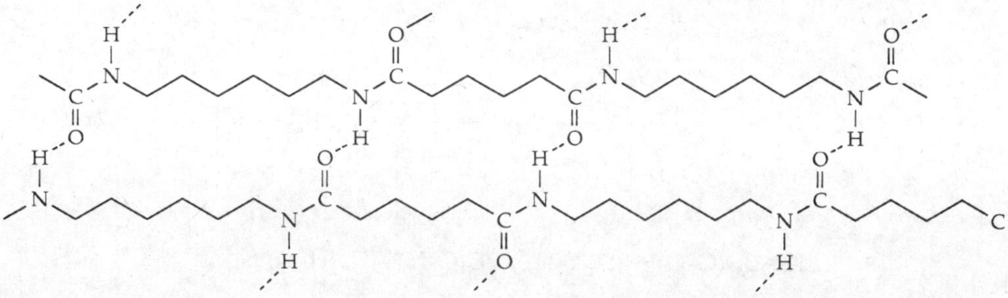

Hydrogen bonding occurs between amide groups of adjacent chains.

12.80

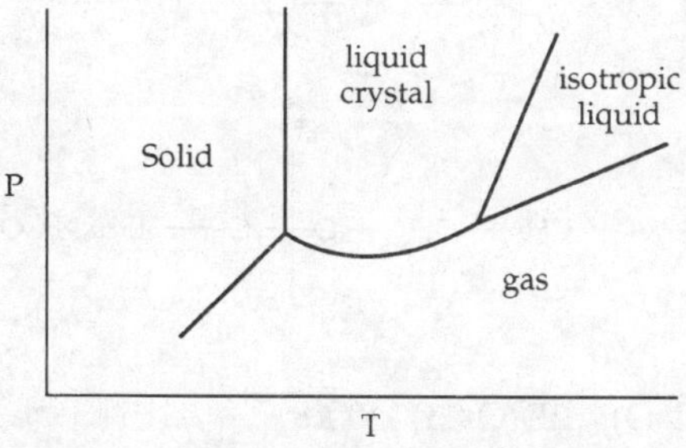

12.81 $TiCl_4(g) + 2SiH_4(g) \rightarrow TiSi_2(s) + 4HCl(g) + 2H_2(g)$

As a ceramic, TiSi2 will have a three dimensional network structure similar to that of Si. [Si(s) has a diamond-like covalent-network structure, Figure 11.41.] At the surface of the thin film there will be Ti atoms and Si atoms with incomplete valences that can and will chemically bond with Si atoms on the surface of the substrate. This kind of bonding would not be possible with a Cu thin film. Strong adherence to the surface is an essential component of thin film performance.

12.82 Solar energy conversion is the process of changing light energy from the sun into useful electrical energy, electricity. Semiconductors are appropriate for this purpose because sunlight excites electrons from the valence to the conduction band, rendering them useful. In order to absorb the full range of solar wavelengths, the semiconductor should have a relatively small band gap, ~50–150 kJ/mol. In order to efficiently translate excited electrons into useful current, the material should have a high conductivity. Practical considerations are, as usual, to be cheap, available, and nontoxic.

Integrative Exercises

12.83 (a)

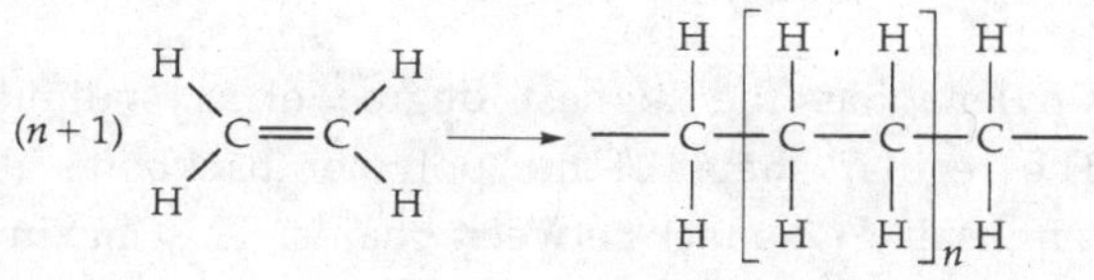

HDPE

$$\Delta H = D(C=C) - 2D(C-C) = 614 - 2(348) = -82 \text{ kJ/mol } C_2H_4$$

(b) $(n+1)$ HOOC—$(CH_2)_6$—COOH + $(n+1)$ H_2N—$(CH_2)_6$—NH_2 $\longrightarrow$

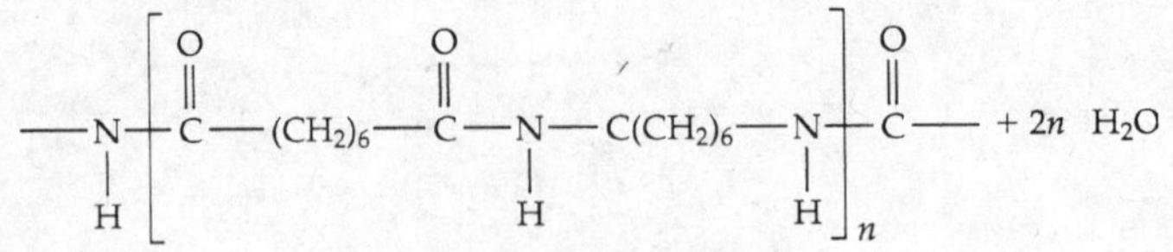

$$—N{\overset{\displaystyle H}{\underset{}{|}}}{\Bigg[}{\overset{\displaystyle O}{\overset{\displaystyle \|}{C}}}—(CH_2)_6—{\overset{\displaystyle O}{\overset{\displaystyle \|}{C}}}—N{\underset{\displaystyle H}{|}}—C(CH_2)_6—N{\underset{\displaystyle H}{\overset{\displaystyle |}{}}}{\Bigg]}_n{\overset{\displaystyle O}{\overset{\displaystyle \|}{C}}}— + 2n\ H_2O$$

Nylon 6,6

$\Delta H = 2D(C{-}O) + 2D(N{-}H) - 2D(C{-}N) - 2D(H{-}O)$

$\Delta H = 2(358) + 2(391) - 2(293) - 2(463) = -14$ kJ/mol

(This is –14 kJ/mol of either reactant.)

(c) $(n+1)$ HOOC—⬡—COOH + $(n+1)$ HO—CH_2—CH_2—OH $\longrightarrow$

$$—O{\Bigg[}{\overset{\displaystyle O}{\overset{\displaystyle \|}{C}}}—⬡—{\overset{\displaystyle O}{\overset{\displaystyle \|}{C}}}—O—CH_2—CH_2—O{\Bigg]}_n{\overset{\displaystyle O}{\overset{\displaystyle \|}{C}}}— + 2n\ H_2O$$

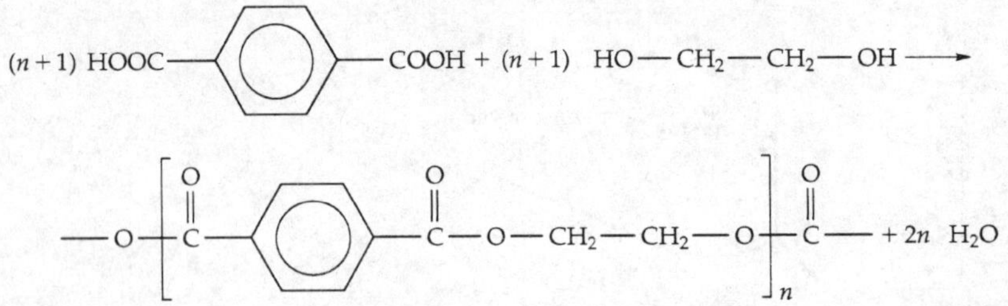

PET

$\Delta H = 2D(C{-}O) + 2D(O{-}H) - 2D(C{-}O) - 2D(O{-}H) = 0$ kJ

12.84   (a) sp^3 hybrid orbitals at C, 109° bond angles around C

(b)

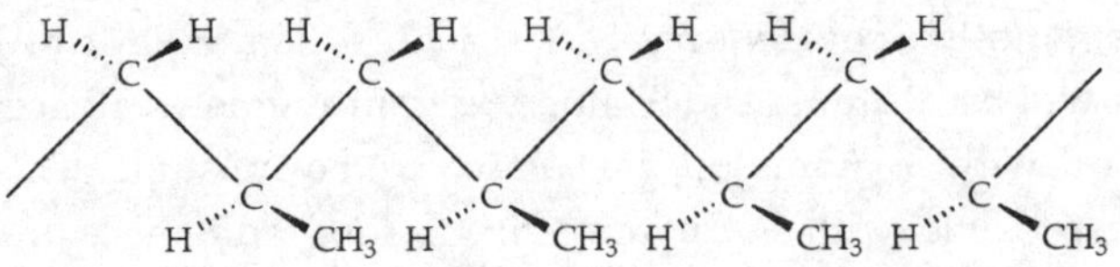

isotactic

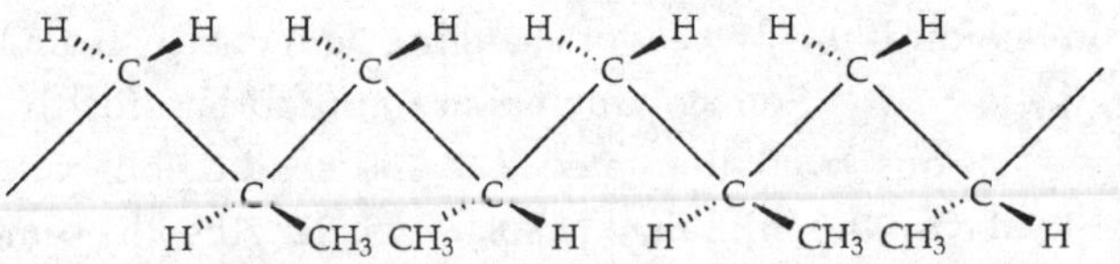

syndiotactic

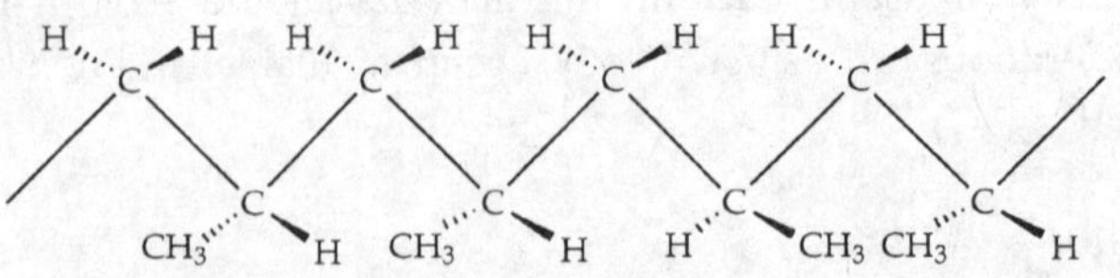

atactic

Isotactic polypropylene has the highest degree of crystallinity and highest melting point. The regular shape of the polymer backbone allows for close, orderly (almost zipper-like) contact between chains. This maximizes dispersion forces between chains and produces higher order (crystallinity) and melting point. Atactic polypropylene has the least order and the lowest melting point.

(c)   Cotton, with $-\!\!-\!\!C\!\!-\!\!-$ groups and polyester, with $-\!\!-\!\!\overset{\displaystyle O}{\overset{\|}{C}}\!\!-\!\!O\!\!-\!\!C$

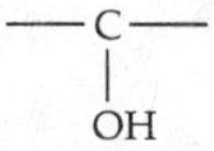

with the C having an OH group below it.

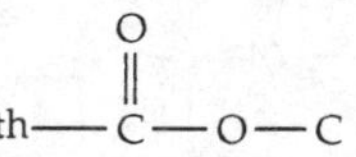

groups, both participate in hydrogen bonding interactions with H_2O molecules. These are strong intermolecular forces that hold the "moisture" at the surface of the fabric next to the skin. Polypropylene has no strong interactions with water, and capillary action "wicks" the moisture away from the skin.

12.85   If the expected oxidation states on Y and Ba are +3 and +2, respectively, the average oxidation state of Cu is +2 1/3. That is, two Cu ions are in the +2 state and one is in the +3 state. Y^{3+} and Ba^{2+} have the stable electron configurations of their nearest noble gases, while Cu has an incomplete d orbital set. Cu(II) is d^9, Cu(III) is d^8 and both have unpaired electrons. Although the mechanism by which copper 3d electrons interact through bridging oxygen atoms to form a superconducting state is still not clear, it is evident that the electronic structure of copper ions is essential to the observed superconductivity.

12.86   (a)   The data (14.99%) has 4 sig figs, so use molar masses to 5 sig figs.

$$\text{mass \% O} = 14.99 = \frac{(8+x)\,15.999}{746.04 + (8+x)\,15.999} \times 100$$

rounded (to show sig figs)

$(8+x)\,15.999$

   $= 0.1499\,[746.04 + (8+x)\,15.999]$

$127.99 + 15.999x$

   $= 0.1499(874.04 + 15.999x)$

$15.999x - 2.398x$

   $= 131.0 - 127.99$

$13.601x = 3.0;\ x = 0.22$

unrounded

$(8+x)\,15.999$

   $= 0.1499\,[746.04 + (8+x)\,15.999]$

$127.992 + 15.999x$

   $= 0.1499\,(874.036 + 15.999x)$

$15.999x - 2.3983x$

   $= 131.018 - 127.992$

$13.6007x = 3.026;\ x = 0.2225$

(b)   Hg and Cu both have more than one stable oxidation state. If different Cu ions (or Hg ions) in the solid lattice have different charges, then the average charge is a noninteger value. Ca and Ba are stable only in the +2 oxidation state; they are unlikely to have noninteger average charge.

(c) Ba^{2+} is largest; Cu^{2+} is smallest. For ions with the same charge, size decreases going up or across the periodic table. In the +2 state, Hg is smaller than Ba. If Hg has an average charge greater than 2+, it will be smaller yet. The same argument is true for Cu and Ca.

12.87   (a)

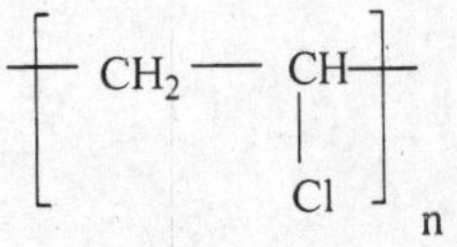

$$C—Cl \quad 328 \text{ kJ/mol} \longleftarrow \text{lowest}$$
$$C—C \quad 348 \text{ kJ/mol}$$
$$C—H \quad 413 \text{ kJ/mol}$$

(b)   C–Cl bonds are weakest, so they are most likely to break upon heating.

(c)   The repeating unit in polyvinyl chloride consists of two C atoms, each in a different environment. Consider the net changes in these two C atoms when the polymer is converted to diamond a high pressure.

Diamond is a covalent-network structure where each C atom is tetrahedrally bound to four other C atoms [Figure 11.41 (a)].

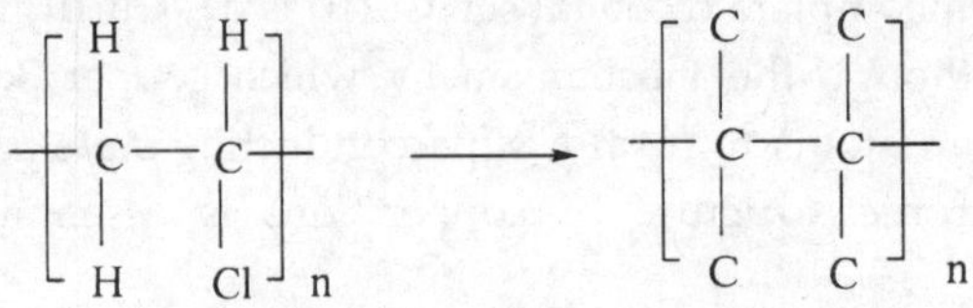

Assume that there is no net change to the C–C bonds in the structure, even though they may be broken and reformed. The net change to the 2-C vinyl chloride unit is then breaking three C–H bonds and one C–Cl bond, and making four C–C bonds.

$$\Delta H = D(C–H) - D(C–I) - 4D(C–) = 3(413) + 328 - 4(348) =$$

523 kJ/vinyl choride unit

12.88   (a)

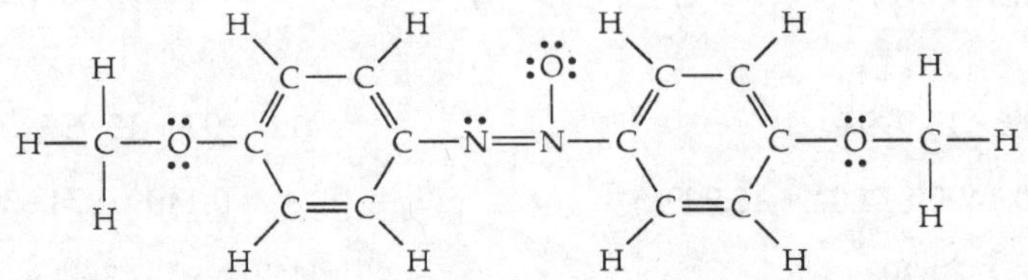

There are several other resonance structures involving alternate placement of the double bonds in the benzene rings.

(b)   Both N atoms are surrounded by 3 VSEPR electron domains, so the hybridization at both atoms is sp^2. The bond angles around the N attached to O will be approximately 120°.

(c)   When the $-OCH_3$ group is replaced by the $-CH_2CH_2CH_2CH_3$ group, a small rather compact group with some polarity is replaced by a larger, more flexible, nonpolar group. Thus, the molecules don't line up as well in the solid, and the melting point and liquid crystal temperature range are lower.

(d)   The density decreases going from solid to nematic liquid crystal to isotropic liquid. In the nematic liquid crystal, most of the long range order of the solid state is lost. The molecules are moving relative to one another with their long axes more or less aligned. The result is more empty space and a lower density than the solid. There is a further small decrease in density when the last degree of order is lost and the substance becomes an isotropic liquid.

12.89   (a)   Follow the logic outlined in Solution 12.69.

vol. of unit cell = $(5.43 \text{ Å})^3 = 160.1030 = 1.60 \times 10^2 \text{ Å}^3$

$$1 \text{ cm}^3 \times \frac{(1)^3 \text{ Å}^3}{(1 \times 10^{-8})^3 \text{ cm}^3} = 1 \times 10^{24} \text{ Å}^3 \text{ (volume of material)}$$

$$\frac{4 \text{ Si atoms}}{160.103 \text{ Å}^3} = \frac{x \text{ Si atoms}}{1 \times 10^{24} \text{ Å}^3}; x = 2.4984 \times 10^{22} = 2.50 \times 10^{22} \text{ Si atoms}$$

(To 1 sig fig, the result is 2×10^{22} Si atoms.)

(b)   1 ppm phosphorus = 1 P atom per 1×10^6 Si atoms

$$\frac{1 \text{ P atom}}{1 \times 10^6 \text{ Si atoms}} = \frac{x \text{ P atoms}}{2.4984 \times 10^{22} \text{ Si Atoms}}; x = 2.4984 \times 10^{22}$$

$$= 2.50 \times 10^{16} \text{ P atoms}$$

$$2.4984 \times 10^{16} \text{ P atom} \times \frac{1 \text{ mol}}{6.022 \times 10^{23} \text{ atoms}} \times \frac{30.97376 \text{ g P}}{\text{mol}} \times \frac{1 \text{ mg}}{1 \times 10^{-3} \text{ g}}$$

$$= 1.29 \times 10^{-3} \text{ mg P} (1.29 \,\mu\text{g})$$

12.90   (a)   According to Table 12.1, the values of E_g for the Group 4A elements with the diamond structure are: C(diamond), 5.5 eV; Si, 1.1 eV; Ge, 0.67 eV; Sn(gray), 0.08 eV. Going down Group 4A, covalent radius and bond length increases, while E_g decreases.

(b)   For the III–V semiconductors, E_g values are: GaP, 2.2 eV; GaAs, 1.43 eV. For the II–VI semiconductors: CdS, 2.4 eV; CdSe, 1.7 eV; CdTe, 1.44 eV. In both sets of mixed semiconductors, holding the element with fewer valence electrons constant, the value of E_g decreases moving down the group of elements with more valence electrons.

(c) I_1 Values from Chapter 7.

element	E_g, eV	I_1, kJ/mol
C(dia)	5.5	1086
Si	1.1	786
Ge	0.67	762
Sn(gray)	0.08	709

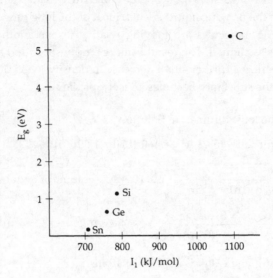

The plot shows a strong relationship between I_1, and E_g; as I_1 increases, E_g increases. I_1 is the energy required to completely remove an electron from a gas-phase atom of the element. E_g is the energy required to excite an electron from the valence band to the conduction band in a bulk sample of the element. I_1 is largely determined by the attraction of an electron for the nucleus. Going down a family, valence electrons are farther from the nucleus and shielded from the full nuclear charge by an increasing electron core, so effective nuclear charge and ionization energy decrease. This relationship carries over to properties of the bulk elements like metallic character and E_g. As electrons are less strongly attracted to nuclei, they are more easily delocalized and the energy required to excite them to the conduction band decreases.

13 Properties of Solutions

Visualizing Concepts

13.1 The energy of the ion-solvent interaction is greater for Li^+ than Na^+. The smaller size of the Li^+ ion means that ion-dipole interactions with polar water molecules are stronger.

13.2 ΔH_3 contains the interaction of a cation with the solvent. In the figure for Exercise 13.1, we see a single Na^+ cation separated from the bulk sample, and solvent molecules interacting with each other as well as the cation. The diagram shows attractive solvent-solvent and ion-solvent interactions, which contribute to an overall negative (–) ΔH. In Figure 13.4, only ΔH_3 is negative.

13.3 The pink solid is hydrated $CoCl_2$, $CoCl_2 \bullet xH_2O$, where x is a specific integer. The waters of hydration are either associated with Co^{2+}, Cl^-, or sit in specific sites in the crystal lattice. When heated in an oven, the water molecules incorporated into the crystal lattice gradually gain kinetic energy and vaporize. The blue solid is anhydrous $CoCl_2$, absent the waters of hydration and with a different solid-state structure than the pink hydrate.

13.4 Diagram (b) is the best representation of a saturated solution. There is some undissolved solid with particles that are close together and ordered, in contact with a solution containing mobile, separated solute particles. As much solute has dissolved as can dissolve, leaving some undissolved solid in contact with the saturated solution.

13.5 Solubility increases in the order Ar, 1.50×10^{-3} M < Kr, 2.79×10^{-3} M < Xe, 5×10^{-5} M, the order of increasing polarizability. As the molar mass of the ideal gas increases, atomic size increases and the electron cloud is less tightly held by the nucleus, causing the cloud to be more polarizable. The greater the polarizability, the stronger the dispersion forces between the gas atoms and water, the more likely the gas atom is to stay dissolved rather than escape the solution, the greater the solubility of the gas.

13.6 Vitamin B_6 is likely to be largely water soluble. The three –OH groups and the $—\ddot{N}—$ can enter into many hydrogen bonding interactions with water. The relatively small molecular size indicates that dispersion forces will not play a large role in intermolecular interactions and the hydrogen bonding will dominate. Vitamin E is likely to be largely fat soluble. The long, rod-like hydrocarbon chain will lead to strong dispersion forces among vitamin E and mostly nonpolar fats. Although vitamin E has one –OH and one $—\ddot{O}—$ group, the long hydrocarbon chain prevents water from surrounding and separating the vitamin E molecules, reducing its water-solubility.

13.7   (a)   Yes, the *molarity* changes with a change in temperature. Molarity is defined as moles solute per unit volume of solution. If solution volume is different, molarity is different.

             (b)   No, *molality* does not change with change in temperature. Molality is defined as moles solute per kilogram of solvent. Even though the volume of solution has changed due to increased kinetic energy, the mass of solute and solvent have not changed, and the molality stays the same.

13.8   Ideally, 0.50 L. If the volume outside the balloon is very large compared to 0.25 L, solvent will flow across the semipermeable membrane until the molarities of the inner and outer solutions are equal, 0.10 M. This requires an "inner" solution volume twice as large as the initial volume, or 0.50 L. (In reality, osmosis across the balloon membrane is not perfect. The solution concentration inside the balloon will be slightly greater than 0.10 M and the volume of the balloon will be slightly less than 0.50 L.)

13.9   A detergent for solubilizing large hydrophobic proteins (or any other large nonpolar solute, such as greasy dirt) needs a hydrophobic part to interact with the solute, and a hydrophilic part to interact with water. In n-octyl glycoside, the eight-carbon n-octyl chain has strong dispersion interactions with the hydrophobic (nonpolar) protein. The –OH groups on the glycoside (sugar) ring form strong hydrogen bonds with water. This causes the glycoside to dissolve, dragging the hydrophobic protein along with it.

13.10   According to Figure 13.18, the solubility of CO at 25°C and 1 atm pressure is approximately 0.96 mM. By Henry's Law, $S_g = k\,P_g$. At the same temperature and pressure, k will be the same, so $S_1/P_1 = S_2/P_2$.

$$\frac{0.96\,\text{m}M}{1\,\text{atm}} = \frac{2.5\,\text{m}M}{x\,\text{atm}}\,; x = \frac{2.5\,\text{m}M \times 1\,\text{atm}}{0.96\,\text{atm}} = 2.6\,\text{atm}$$

The Solution Process

13.11   If the enthalpy released due to solute-solvent attractive forces (ΔH_3) is at least as large as the enthalpy required to separate the solute particles (ΔH_1), the overall enthalpy of solution (ΔH_{soln}) will be either slightly endothermic (owing to $+\Delta H_2$) or exothermic. Even if ΔH_{soln} is slightly endothermic, the increase in disorder due to mixing will cause a significant amount of solute to dissolve. If the magnitude of ΔH_3 is small relative to the magnitude of ΔH_1, ΔH_{soln} will be large and endothermic (energetically unfavorable) and not much solute will dissolve.

13.12   (a)   For the same solute, NaCl, in different solvents, solute-solute interactions (ΔH_1) are the same. Because water experiences hydrogen bonding while benzene has only dispersion forces, solvent-solvent interactions (ΔH_2) are greater for water. On the other hand, solute-solvent interactions (ΔH_3) are much weaker between ionic NaCl and nonpolar benzene than between ionic NaCl and polar water. It is the large difference in ΔH_3 that causes NaCl to be soluble in water but not in benzene.

             (b)   Lattice energy is the main component of ΔH_1, the enthalpy required to separate solute particles. If ΔH_1 is too large, the dissolving process is prohibitively endothermic, and the substance is not very soluble.

(c)   Ion-dipole forces between cations and water molecules and relatively small lattice energies (ion-ion forces between cations and anions) lead to strongly hydrated cations.

13.13   *Analyze/Plan.* Decide whether the solute and solvent in question are ionic, polar covalent, or nonpolar covalent. Draw Lewis structures as needed. Then state the appropriate type of solute-solvent interaction. *Solve.*

(a) CCl_4, nonpolar; benzene, nonpolar; dispersion forces

(b)   methanol, polar with hydrogen bonding; water, polar with hydrogen bonding; hydrogen bonding

(c)   KBr, ionic; water, polar; ion-dipole forces

(d)   HCl, polar; CH_3CN, polar; dipole-dipole forces

13.14   From weakest to strongest solvent-solute interactions:

(b), dispersion forces < (c), hydrogen bonding < (a), ion-dipole

13.15   (a)   Lattice energy is the amount of energy required to completely separate a mole of solid ionic compound into its gaseous ions (Section 8.2). For ionic solutes, this corresponds to ΔH_1 (solute-solute interactions) in Equation [13.1].

(b)   In Equation [13.1], ΔH_3 is always exothermic. Formation of attractive interactions, no matter how weak, always lowers the energy of the system, relative to the energy of the isolated particles.

13.16   Separation of solvent molecules, ΔH_2, will be smallest in this case, because hydrogen bonding is the weakest of the intermolecular forces involved. ΔH_1 involves breaking ionic bonds, and ΔH_3 involves formation of ion-dipole interactions, both stronger forces than hydrogen bonding.

13.17   (a) ΔH_{soln} is determined by the relative magnitudes of the "old" solute-solute (ΔH_1) and solvent-solvent (ΔH_2) interactions and the new solute-solvent interactions (ΔH_3); $\Delta H_{soln} = \Delta H_1 + \Delta H_2 + \Delta H_3$. Since the solute and solvent in this case experience very similar London dispersion forces, the energy required to separate them individually and the energy released when they are mixed are approximately equal.

$\Delta H_1 + \Delta H_2 \approx -\Delta H_3$. Thus, ΔH_{soln} is nearly zero.

(b)   Mixing hexane and heptane produces a homogeneous solution from two pure substances, and the randomness of the system increases. Since no strong intermolecular forces prevent the molecules from mixing, they do so spontaneously due to the increase in disorder.

13.18   KBr is quite soluble in water because of the sizeable increase in disorder of the system (ordered KBr lattice → freely moving hydrated ions) associated with the dissolving process. An increase in disorder or randomness in a process tends to make that process spontaneous.

Saturated Solutions; Factors Affecting Solubility

13.19   (a)   Supersaturated

         (b)   Add a seed crystal. Supersaturated solutions exist because not enough solute molecules are properly aligned for crystallization to occur. A seed crystal provides a nucleus of already aligned molecules, so that ordering of the dissolved particles is more facile.

13.20   (a) $\dfrac{1.22 \text{ mol MnSO}_4 \cdot \text{H}_2\text{O}}{1 \text{ L soln}} \times \dfrac{169.0 \text{ g MnSO}_4 \cdot \text{H}_2\text{O}}{1 \text{ mol}} \times 0.100 \text{ L}$

$$= 20.6 \text{ g MnSO}_4 \cdot \text{H}_2\text{O}/100 \text{ mL}$$

             The 1.22 M solution is unsaturated.

         (b)   Add a known mass, say 5.0 g, of $\text{MnSO}_4 \cdot \text{H}_2\text{O}$, to the unknown solution. If the solid dissolves, the solution is unsaturated. If there is undissolved $\text{MnSO}_4 \cdot \text{H}_2\text{O}$, filter the solution and weigh the solid. If there is less than 5.0 g of solid, some of the added $\text{MnSO}_4 \cdot \text{H}_2\text{O}$ dissolved and the unknown solution is unsaturated. If there is exactly 5.0 g, no additional solid dissolved and the unknown is saturated. If there is more than 5.0 g, excess solute has precipitated and the solution is supersaturated.

13.21   *Analyze/Plan.* On Figure 13.17, find the solubility curve for the appropriate solute. Find the intersection of 40°C and 40 g solute on the graph. If this point is below the solubility curve, more solute can dissolve and the solution is unsaturated. If the intersection is on or above the curve, the solution is saturated.   *Solve.*

         (a)   unsaturated   (b)   saturated   (c)   saturated   (d)   unsaturated

13.22   (a)   at $30^\circ \text{C}, \dfrac{10 \text{ g KClO}_3}{100 \text{ g H}_2\text{O}} \times 250 \text{ g H}_2\text{O} = 25 \text{ g KClO}_3$

         (b) $\dfrac{66 \text{ g Pb(NO}_3)_2}{100 \text{ g H}_2\text{O}} \times 250 \text{ g H}_2\text{O} = 165 = 1.7 \times 10^2 \text{ g Pb(NO}_3)_2$

         (c) $\dfrac{3 \text{ g Ce}_2(\text{SO}_4)_3}{100 \text{ g H}_2\text{O}} \times 250 \text{ g H}_2\text{O} - 7.5 = 8 \text{ g Ce}_2(\text{SO}_4)_3$

13.23   The liquids water and glycerol form homogenous mixtures (solutions), regardless of the relative amounts of the two components. Glycerol has an –OH group on each C atom in the molecule. This structure facilitates strong hydrogen bonding similar to that in water. Like dissolves like and the two liquids are miscible in all proportions.

13.24   Immiscible means that oil and water do not mix homogeneously; they do not dissolve. Many substances are called "oil," but they are typically nonpolar carbon-based molecules with fairly high molecular weights. As such, there are fairly strong dispersion forces among oil molecules. The properties of water are dominated by its strong hydrogen bonding. The dispersion-dipole interactions between water and oil are likely to be weak. Thus, ΔH_1 and ΔH_2 are large and positive, while ΔH_3 is small and negative. The net ΔH_{soln} is large and positive, and mixing does not occur.

13.25   (a)   Dispersion interactions among nonpolar $CH_3(CH_2)_{16}$ –chains dominate the properties of stearic acid. It is more soluble in nonpolar CCl_4 than polar (hydrogen bonding) water, despite the presence of the –COOH group.

        (b)

cyclohexane   dioxane

Dioxane can act as a hydrogen bond acceptor, so it will be more soluble than cyclohexane in water.

13.26   For small *n* values, the dominant interactions among acid molecules will be hydrogen-bonding. As *n* increases, dispersion forces between carbon chains become more important and eventually dominate. Thus, as *n* increases, water solubility decreases and hexane solubility increases.

13.27   *Analyze/Plan.* Hexane is a nonpolar hydrocarbon that experiences dispersion forces with other C_6H_{14} molecules. Solutes that primarily experience dispersion forces will be more soluble in hexane.   *Solve.*

        (a) CCl_4 is more soluble because dispersion forces among nonpolar CCl_4 molecules are similar to dispersion forces in hexane. Ionic bonds in $CaCl_2$ are unlikely to be broken by weak solute-solvent interactions. For $CaCl_2$, ΔH_1 is too large, relative to ΔH_3.

        (b)   Benzene, C_6H_6, is also a nonpolar hydrocarbon and will be more soluble in hexane. Glycerol experiences hydrogen bonding with itself; these solute-solute interactions are less likely to be overcome by weak solute-solvent interactions.

        (c)   Octanoic acid, $CH_3(CH_2)_6COOH$, will be more soluble than acetic acid CH_3COOH. Both solutes experience hydrogen bonding by –COOH groups, but octanoic acid has a long, rod-like hydrocarbon chain with dispersion forces similar to those in hexane, facilitating solubility in hexane.

13.28   *Analyze/Plan.* Water, H_2O, is a polar solvent that forms hydrogen bonds with other H_2O molecules. The more soluble solute in each case will have intermolecular interactions that are most similar to the hydrogen bonding in H_2O.   *Solve.*

        (a)   Glucose, $C_6H_{12}O_6$, is more soluble because it is capable of hydrogen bonding (Figure 13.12). Nonpolar C_6H_{12} is capable only of dispersion interactions and does not have strong intermolecular interactions with polar (hydrogen bonding) H_2O.

        (b)   Ionic sodium propionate, CH_3CH_2COONa, is more soluble. Sodium propionate is a crystalline solid, while propionic acid is a liquid. The increase in disorder or entropy when an ionic solid dissolves leads to significant water solubility, despite the strong ion-ion forces (large ΔH_1) present in the solute (see Solution 13.18).

(c)   HCl is more soluble because it is a strong electrolyte and completely ionized in water. Ionization leads to ion-dipole solute-solvent interactions, and an increase in disorder. CH_3CH_2Cl is a molecular solute capable of relatively weak dipole-dipole solute-solvent interactions and is much less soluble in water.

13.29   (a)   Carbonated beverages are stored with a partial pressure of $CO_2(g)$ greater than 1 atm above the liquid. A sealed container is required to maintain this CO_2 pressure.

(b)   Since the solubility of gases increases with decreasing temperature, some $CO_2(g)$ will remain dissolved in the beverage if it is kept cool.

13.30   Pressure has an effect on O_2 solubility in water because, at constant temperature and volume, pressure is directly related to the amount of O_2 available to dissolve. The greater the partial pressure of O_2 above water, the more O_2 molecules are available for dissolution, and the more molecules that strike the surface of the liquid.

Pressure does not affect the amount or physical properties of NaCl, or ionic solids in general, so it has little influence on the dissolving of NaCl in water.

13.31   *Analyze/Plan.* Follow the logic in Sample Exercise 13.3.   *Solve.*

$S_{He} = 3.7 \times 10^{-4} \, M/atm \times 1.5 \, atm = 5.6 \times 10^{-4} \, M$

$S_{N_2} = 6.0 \times 10^{-4} \, M/atm \times 1.5 \, atm = 9.0 \times 10^{-4} \, M$

13.32 $665 \, torr \times \dfrac{1 \, atm}{760 \, torr} = 0.875 \, atm; \; P_{O_2} = \chi_{O_2}(P_t) = 0.21(0.875 \, atm) = 0.1838 = 0.18 \, atm$

$S_{O_2} = kP_{O_2} = \dfrac{1.38 \times 10^{-3} \, mol}{L \bullet atm} \times 0.1838 \, atm = 2.5 \times 10^{-4} \, M$

Concentrations of Solutions

13.33   *Analyze/Plan.* Follow the logic in Sample Exercise 13.4.   *Solve.*

(a) $mass \, \% = \dfrac{mass \, solute}{total \, mass \, solution} \times 100 = \dfrac{10.6 \, g \, Na_2SO_4}{10.6 \, g \, Na_2SO_4 + 483 \, g \, H_2O} \times 100 = 2.15\%$

(b) $ppm = \dfrac{mass \, solute}{total \, mass \, solution} \times 10^6; \dfrac{2.86 \, g \, Ag}{1 \, ton \, ore} \times \dfrac{1 \, ton}{2000 \, lb} \times \dfrac{1 \, lb}{453.6 \, g} \times 10^6 = 3.15 \, ppm$

13.34   (a) $mass \, \% = \dfrac{mass \, solute}{total \, mass \, solution} \times 100$

$mass \, solute = 0.045 \, mol \, I_2 \times \dfrac{253.8 \, g \, I_2}{1 \, mol \, I_2} = 11.421 = 11 \, g \, I_2$

$mass \, \% \, I_2 = \dfrac{11.421 \, g \, I_2}{11.421 \, g \, I_2 + 115 \, g \, CCl_4} \times 100 = 9.034 = 9.0\% \, I_2$

(b) $ppm = \dfrac{mass \, solute}{total \, mass \, solution} \times 10^6 = \dfrac{0.0079 \, g \, Sr^{2+}}{1 \times 10^3 \, g \, H_2O} \times 10^6 = 7.9 \, ppm \, Sr^{2+}$

13 Properties of Solutions

Solutions to Exercises

13.35 *Analyze/Plan.* Given masses of CH_3OH and H_2O, calculate moles of each component.

(a) Mole fraction CH_3OH = (mol CH_3OH)/(total mol)

(b) mass % CH_3OH = [(g CH_3OH)/(total mass)] × 100

(c) molality CH_3OH = (mol CH_3OH)/(kg H_2O).   *Solve.*

(a) $14.6 \text{ g } CH_3OH \times \dfrac{1 \text{ mol } CH_3OH}{32.04 \text{ g } CH_3OH} = 0.4557 = 0.456 \text{ mol } CH_3OH$

$184 \text{ g } H_2O \times \dfrac{1 \text{ mol } H_2O}{18.02 \text{ g } H_2O} = 10.211 = 10.2 \text{ mol } H_2O$

$\chi_{CH_3OH} = \dfrac{0.4557}{0.4557 + 10.211} = 0.04272 = 0.0427$

(b) $\text{mass \% } CH_3OH = \dfrac{14.6 \text{ g } CH_3OH}{14.6 \text{ g } CH_3OH + 184 \text{ g } H_2O} \times 100 = 7.35\% \ CH_3OH$

(c) $m = \dfrac{0.4557 \text{ mol } CH_3OH}{0.184 \text{ kg } H_2O} = 2.477 = 2.48 \text{ m } CH_3OH$

13.36 (a) $\dfrac{25.5 \text{ g } C_6H_5OH}{94.11 \text{ g/mol}} = 0.2710 = 0.271 \text{ mol } C_6H_5OH$

$\dfrac{495 \text{ g } CH_3CH_2OH}{46.07 \text{ g/mol}} = 10.7445 = 10.7 \text{ mol } CH_3CH_2OH$

$\chi_{C_6H_5OH} = \dfrac{0.2710}{0.2710 + 10.7445} = 0.02460 = 0.0246$

(b) $\text{mass \%} = \dfrac{25.5 \text{ g } C_6H_5OH}{25.5 \text{ g } C_6H_5OH + 495 \text{ g } CH_3CH_2OH} \times 100 = 4.90\% \ C_6H_5OH$

(c) $m = \dfrac{0.2710 \text{ mol } C_6H_5OH}{0.495 \text{ kg } CH_3CH_2OH} = 0.54747 = 0.547 \text{ m } C_6H_5OH$

13.37 *Analyze/Plan.* Given mass solute and volume solution, calculate mol solute, then molarity = mol solute/L solution. Or, for dilution, $M_c \times L_c = M_d \times L_d$ *Solve.*

(a) $M = \dfrac{\text{mol solute}}{\text{L soln}}; \dfrac{0.540 \text{ g } Mg(NO_3)_2}{0.2500 \text{ L soln}} \times \dfrac{1 \text{ mol } Mg(NO_3)_2}{148.3 \text{ g } Mg(NO_3)_2} = 1.46 \times 10^{-2} \ M \ Mg(NO_3)_2$

(b) $\dfrac{22.4 \text{ g } LiClO_4 \cdot 3H_2O}{0.125 \text{ L soln}} \times \dfrac{1 \text{ mol } LiClO_4 \cdot 3H_2O}{160.4 \text{ g } LiClO_4 \cdot 3H_2O} = 1.12 \ M \ LiClO_4 \cdot 3H_2O$

(c) $M_c \times L_c = M_d \times L_d$; $3.50 \ M \ HNO_3 \times 0.0250 \text{ L} = ?M \ HNO_3 \times 0.250 \text{ L}$
250 mL of 0.350 $M \ HNO_3$

13.38 (a) $M = \dfrac{\text{mol solute}}{\text{L soln}}; \dfrac{25.0 \text{ g } Al_2(SO_4)_3}{0.350 \text{ L soln}} \times \dfrac{1 \text{ mol } Al_2(SO_4)_3}{342.2 \text{ g } Al_2(SO_4)_3} = 0.209 \ M \ AL_2(SO_4)_3$

(b) $\dfrac{5.25 \text{ g } Mn(NO_3)_2 \cdot 2H_2O}{0.175 \text{ L soln}} \times \dfrac{1 \text{ mol } Mn(NO_3)_2 \cdot 2H_2O}{215.0 \text{ g } Mn(NO_3)_2 \cdot 2H_2O} = 0.140 \ M \ Mn(NO_3)_2$

(c) $M_c \times L_c = M_d \times L_d$; $9.00 \ M \ H_2SO_4 \times 0.0350 \text{ L} = ?M \ H_2SO_4 \times 0.500 \text{ L}$
500 mL of 0.630 $M \ H_2SO_4$

13.39   *Analyze/Plan.* Follow the logic in Sample Exercise 13.5.   *Solve.*

(a) $m = \dfrac{\text{mol solute}}{\text{kg solvent}}; \dfrac{8.66 \text{ g } C_6H_6}{23.6 \text{ g } CCl_4} \times \dfrac{1 \text{ mol } C_6H_6}{78.11 \text{ g } C_6H_6} \times \dfrac{1000 \text{ g } CCl_4}{1 \text{ kg } CCl_4} = 4.70 \, m \, C_6H_6$

(b)   The density of $H_2O = 0.997$ g/mL $= 0.997$ kg/L.

$\dfrac{4.80 \text{ g NaCl}}{0.350 \text{ L } H_2O} \times \dfrac{1 \text{ mol NaCl}}{58.44 \text{ g NaCl}} \times \dfrac{1 \text{ L } H_2O}{0.997 \text{ kg } H_2O} = 0.235 \, m \, \text{NaCl}$

13.40   (a) $16.0 \text{ mol } H_2O \times \dfrac{18.02 \text{ g } H_2O}{1 \text{ mol } H_2O} = 288.3 \text{ g } H_2O = 0.288 \text{ kg } H_2O$

$m = \dfrac{1.50 \text{ mol KCl}}{0.2883 \text{ kg } H_2O} = 5.2026 = 5.20 \, m \, \text{KCl}$

(b) $m = \dfrac{\text{mol solute}}{\text{kg solute}}; \text{mol } S_8 = m \times \text{kg } C_{10}H_8 = 0.12 \, m \times 0.1000 \text{ kg } C_{10}H_8 = 0.012 \text{ mol}$

$0.012 \text{ mol } S_8 \times \dfrac{256.5 \text{ g } S_8}{1 \text{ mol } S_8} = 3.078 = 3.1 \text{ g } S_8$

13.41   *Analyze/Plan.* Assume 1 L of solution. Density gives the total mass of 1 L of solution. The g H_2SO_4/L are also given in the problem. Mass % = (mass solute/total mass solution) × 100. Calculate mass solvent from mass solution and mass solute. Calculate moles solute and solvent and use the appropriate definitions to calculate mole fraction, molality, and molarity.   *Solve.*

(a) $\dfrac{571.6 \text{ g } H_2SO_4}{1 \text{ L soln}} \times \dfrac{1 \text{ L soln}}{1329 \text{ g soln}} = 0.430098 \text{ g } H_2SO_4/\text{g soln}$

mass percent is thus $0.4301 \times 100 = 43.01\% \, H_2SO_4$

(b)   In a liter of solution there are $1329 - 571.6 = 757.4 = 757 \text{ g } H_2O$.

$\dfrac{571.6 \text{ g } H_2SO_4}{98.09 \text{ g/mol}} = 5.827 \text{ mol } H_2SO_4 \, ; \dfrac{757.4 \text{ g } H_2O}{18.02 \text{ g/mol}} = 42.03 = 42.0 \text{ mol } H_2O$

$\chi_{H_2SO_4} = \dfrac{5.827}{42.03 + 5.827} = 0.122$

(The result has 3 sig figs because 42.0 mol H_2O limits the denominator to 3 sig figs.)

(c) $\text{molality} = \dfrac{5.827 \text{ mol } H_2SO_4}{0.7574 \text{ kg } H_2O} = 7.693 = 7.69 \, m \, H_2SO_4$

(d) $\text{molarity} = \dfrac{5.827 \text{ mol } H_2SO_4}{1 \text{ L soln}} = 5.827 \, M \, H_2SO_4$

13.42   (a) $\text{mass \%} = \dfrac{\text{mass } C_6H_8O_6}{\text{total mass solution}} \times 100;$

$\dfrac{80.5 \text{ g } C_6H_8O_6}{80.5 \text{ g } C_6H_8O_6 + 210 \text{ g } H_2O} \times 100 = 27.71 = 27.7\% \, C_6H_8O_6$

(b) $\text{mol } C_6H_8O_6 = \dfrac{80.5 \text{ g } C_6H_8O_6}{176.1 \text{ g/mol}} = 0.4571 = 0.457 \text{ mol } C_6H_8O_6$

$\text{mol } H_2O = \dfrac{210 \text{ g } H_2O}{18.02 \text{ g/mol}} = 11.654 = 11.7 \text{ mol } H_2O$

$\chi_{C_6H_8O_6} = \dfrac{0.4571 \text{ mol } C_6H_8O_6}{0.4571 \text{ mol } C_6H_8O_6 + 11.654 \text{ mol } H_2O} = 0.0377$

(c) $m = \dfrac{0.4571 \text{ mol } C_6H_8O_6}{0.210 \text{ kg } H_2O} = 2.18 \, m \, C_6H_8O_6$

(d) $M = \dfrac{\text{mol } C_6H_8O_6}{\text{L solution}}; \; 290.5 \text{ g soln} \times \dfrac{1 \text{ mL}}{1.22 \text{ g}} \times \dfrac{1 \text{ L}}{1000 \text{ mL}} = 0.2381 = 0.238 \text{ L}$

$M = \dfrac{0.4571 \text{ mol } C_6H_8O_6}{0.2381 \text{ L soln}} = 1.92 \, M \, C_6H_8O_6$

13.43   *Analyze/Plan.* Given: 98.7 mL of $CH_3CN(l)$, 0.786 g/mL; 22.5 mL CH_3OH, 0.791 g/mL. Use the density and volume of each component to calculate mass and then moles of each component. Use the definitions to calculate mole fraction, molality, and molarity. *Solve.*

(a) $\text{mol } CH_3CN = \dfrac{0.786 \text{ g}}{1 \text{ mL}} \times 98.7 \text{ mL} \times \dfrac{1 \text{ mol } CH_3CN}{41.05 \text{ g } CH_3CN} = 1.8898 = 1.89 \text{ mol}$

$\text{mol } CH_3OH = \dfrac{0.791 \text{ g}}{1 \text{ mL}} \times 22.5 \text{ mL} \times \dfrac{1 \text{ mol } CH_3OH}{32.04 \text{ g } CH_3OH} = 0.5555 = 0.556 \text{ mol}$

$\chi_{CH_3OH} = \dfrac{0.5555 \text{ mol } CH_3OH}{1.8898 \text{ mol } CH_3CN + 0.5555 \text{ mol } CH_3OH} = 0.227$

(b) Assuming CH_3OH is the solute and CH_3CN is the solvent,

$98.7 \text{ mL } CH_3CN \times \dfrac{0.786 \text{ g}}{1 \text{ mL}} \times \dfrac{1 \text{ kg}}{1000 \text{ g}} = 0.07758 = 0.0776 \text{ kg } CH_3CN$

$m_{CH_3OH} = \dfrac{0.5555 \text{ mol } CH_3OH}{0.07758 \text{ kg } CH_3CN} = 7.1602 = 7.16 \, m \, CH_3OH$

(c) The total volume of the solution is 121.2 mL, assuming volumes are additive.

$M = \dfrac{0.5555 \text{ mol } CH_3OH}{0.1212 \text{ L solution}} = 4.58 \, M \, CH_3OH$

13.44   Given: 10.0 g C_4H_4S, 1.065 g/mL; 250.0 mL C_7H_8, 0.867 g/mL

(a) $\text{mol } C_4H_4S = 10.0 \text{ g } C_4H_4S \times \dfrac{1 \text{ mol } C_4H_4S}{84.15 \text{ g } C_4H_4S} = 0.1188 = 0.119 \text{ mol } C_4H_4S$

$\text{mol } C_7H_8 = \dfrac{0.867 \text{ g}}{1 \text{ mL}} \times 250.0 \text{ mL} \times \dfrac{1 \text{ mol } C_7H_8}{92.14 \text{ g } C_7H_8} = 2.352 = 2.35 \text{ mol}$

$\chi_{C_4H_4S} = \dfrac{0.1188 \text{ mol } C_4H_4S}{0.1188 \text{ mol } C_4H_4S + 2.352 \text{ mol } C_7H_8} = 0.04809 = 0.0481$

(b) $m_{C_4H_4S} = \dfrac{mol\, C_4H_4S}{kg\, C_7H_8}$; $250.0\, mL \times \dfrac{0.867\, g}{1\, mL} \times \dfrac{1\, kg}{1000\, g} = 0.2168 = 0.217\, kg\, C_7H_8$

$m_{C_4H_4S} = \dfrac{0.1188\, mol\, C_4H_4S}{0.2168\, kg\, C_7H_8} = 0.548\, m\, C_4H_4S$

(c) $10.0\, g\, C_4H_4S \times \dfrac{1\, mL}{1.065\, g} = 9.390 = 9.39\, mL\, C_4H_4S$;

$V_{soln} = 9.39\, mL\, C_4H_4S + 250.0\, mL\, C_7H_8 = 259.4\, mL$

$M_{C_4H_4S} = \dfrac{0.1188\, mol\, C_4H_4S}{0.2594\, L\, soln} = 0.458\, M\, C_4H_4S$

13.45   *Analyze/Plan.* Given concentration and volume of solution use definitions of the appropriate concentration units to calculate amount of solute; change amount to moles if needed. *Solve.*

(a) $mol = M \times L$; $\dfrac{0.250\, mol\, SrBr_2}{1\, L\, soln} \times 0.600\, L = 0.150\, mol\, SrBr_2$

(b)   Assume that for dilute aqueous solutions, the mass of the solvent is the mass of solution. Use proportions to get mol KCl.

$\dfrac{0.180\, mol\, KCl}{1\, kg\, H_2O} = \dfrac{x\, mol\, KCl}{0.0864\, kg\, H_2O}$; $x = 1.56 \times 10^{-2}\, mol\, KCl$

(c)   Use proportions to get mass of glucose, then change to mol glucose.

$\dfrac{6.45\, g\, C_6H_{12}O_6}{100\, g\, soln} = \dfrac{x\, g\, C_6H_{12}O_6}{124.0\, g\, soln}$; $x = 8.00\, g\, C_6H_{12}O_6$

$8.00\, g\, C_6H_{12}O_6 \times \dfrac{1\, mol\, C_6H_{12}O_6}{180.2\, g\, C_6H_{12}O_6} = 4.44 \times 10^{-2}\, mol\, C_6H_{12}O_6$

13.46   (a) $\dfrac{1.50\, mol\, HNO_3}{1\, L\, soln} \times 0.245\, L = 0.3675 = 0.368\, mol\, HNO_3$

(b)   Assume that for dilute aqueous solutions, the mass of the solvent is the mass of solution.

$\dfrac{1.25\, mol\, NaCl}{1\, kg\, H_2O} \times \dfrac{x\, mol}{50.0 \times 10^{-6}\, kg}$; $x = 6.25 \times 10^{-5}\, mol\, NaCl$

(c) $\dfrac{1.50\, g\, C_{12}H_{22}O_{11}}{100\, g\, soln} = \dfrac{x\, g\, C_{12}H_{22}O_{11}}{124.0\, g\, soln}$; $x = 1.125 = 1.13\, g\, C_{12}H_{22}O_{11}$

$1.125\, g\, C_{12}H_{22}O_{11} \times \dfrac{1\, mol\, C_{12}H_{22}O_{11}}{342.3\, g\, C_{12}H_{22}O_{11}} = 3.287 \times 10^{-3} = 3.29 \times 10^{-3}\, mol\, C_{12}H_{22}O_{11}$

13.47   *Analyze/Plan.* When preparing solution, we must know amount of solute and solvent. Use the appropriate concentration definition to calculate amount of solute. If this amount is in moles, use molar mass to get grams; use mass in grams directly. Amount of solvent can be expressed as total volume or mass of solution. Combine mass solute and solvent to produce the required amount (mass or volume) of solution. *Solve.*

(a) $mol = M \times L$; $\dfrac{1.50 \times 10^{-2}\ mol\ KBr}{1\ L\ soln} \times 0.75\ L \times \dfrac{119.0\ g\ KBr}{1\ mol\ KBr} = 1.3\ g\ KBr$

Weigh out 1.5 g KBr, dissolve in water, dilute with stirring to 0.75 L (750 mL).

(b)   Mass of solution is required, but density is not specified. Use molality to calculate mass fraction, and then the masses of solute and solvent needed for 125 g of solution.

$$\frac{0.180\ mol\ KBr}{1000\ g\ H_2O} \times \frac{119.0\ g\ KBr}{1\ mol\ KBr} = 21.42 = 21.4\ g\ KBr/kg\ H_2O$$

Thus, mass fraction $= \dfrac{21.42\ g\ KBr}{1000 + 21.42} = 0.02097 = 0.0210$

In 125 g of the 0.180 *m* solution, there are

$(125\ g\ soln) \times \dfrac{0.02097\ g\ KBr}{1\ g\ soln} = 2.621 = 2.62\ g\ KBr$

Weigh out 2.62 g KBr, dissolve it in 125 – 2.62 = 122.38 = 122 g H_2O to make exactly 125 g of 0.180 *m* solution.

(c)   Using solution density, calculate the total mass of 1.85 L of solution, and from the mass % of KBr, the mass of KBr required.

$1.85\ L\ soln \times \dfrac{1000\ mL}{1\ L} \times \dfrac{1.10\ g\ soln}{1\ mL} = 2035 = 2.04 \times 10^3\ g\ soln$

0.120 (2035 g soln) = 244.2 = 244 g KBr

Dissolve 244 g KBr in water, dilute with stirring to 1.85 L.

(d)   Calculate moles KBr needed to precipitate 16.0 g AgBr. $AgNO_3$ is present in excess.

$16.0\ g\ AgBr \times \dfrac{1\ mol\ AgBr}{187.8\ g\ AgBr} \times \dfrac{1\ mol\ KBr}{1\ mol\ AgBr} = 0.08520 = 0.0852\ mol\ KBr$

$0.0852\ mol\ KBr \times \dfrac{1\ L\ soln}{0.150\ mol\ KBr} = 0.568\ L\ soln$

Weigh out 0.0852 mol KBr (10.1 g KBr), dissolve it in a small amount of water, and dilute to 0.568 L.

13.48   (a) $\dfrac{0.110\ mol\ (NH_4)_2SO_4}{1\ L\ soln} \times 1.50\ L \times \dfrac{132.2\ g\ (NH_4)_2SO_4}{1\ mol\ (NH_4)_2SO_4} = 21.81 = 21.8\ g\ (NH_4)_2SO_4$

Weigh 21.8 g $(NH_4)_2SO_4$, dissolve in a small amount of water, continue adding water with thorough mixing up to a total solution volume of 1.50 L.

(b)   Determine the mass fraction of Na_2CO_3 in the solution:

$\dfrac{0.65\ mol\ Na_2CO_3}{1000\ g\ H_2O} \times \dfrac{106.0\ g\ Na_2CO_3}{1\ mol\ Na_2CO_3} = 68.9\ g = \dfrac{69\ g\ Na_2CO_3}{1000\ g\ H_2O}$

mass fraction $= \dfrac{68.9\ g\ Na_2CO_3}{1000\ g\ H_2O + 68.9\ g\ Na_2CO_3} = 0.06446 = 0.064$

In 120 g of solution, there are $0.06446(120) = 7.735 = 7.7$ g Na_2CO_3.

Weigh out 7.7 g Na_2CO_3 and dissolve it in $120 - 7.7 = 112.3$ g H_2O to make exactly 120 g of solution.

(112.3 g H_2O/0.997 g H_2O/mL @ 25° = 112.6 mL H_2O)

(c) $1.20 \, L \times \dfrac{1000 \, mL}{1 \, L} \times \dfrac{1.16 \, g}{1 \, mL} = 1392$ g solution; $0.150(1392 \, g \, soln) = 209$ g $Pb(NO_3)_2$

Weigh 209 g $Pb(NO_3)$ and add $(1392 - 209) = 1183$ g H_2O to make exactly ($1392 = 1.39 \times 10^3$) g or 1.20 L of solution.

(1183 g H_2O/0.997 g/mL @ 25°C = 1187 mL H_2O)

(d)   Calculate the mol HCl necessary to neutralize 5.5 g $Ba(OH)_2$.

$Ba(OH)_2(s) + 2HCl(aq) \rightarrow BaCl_2(aq) + 2H_2O(l)$

$$5.5 \, g \, Ba(OH)_2 + \dfrac{1 \, mol \, Ba(OH)_2}{171 \, g \, Ba(OH)_2} \times \dfrac{2 \, mol \, HCl}{1 \, mol \, Ba(OH)_2} = 0.0643 = 0.064 \, mol \, HCl$$

$$M = \dfrac{mol}{L}; \, L = \dfrac{mol}{M} = \dfrac{0.0643 \, mol \, HCl}{0.50 \, M \, HCl} = 0.1287 = 0.13 \, L = 130 \, mL$$

130 mL of 0.50 M HCl are needed.

$M_c \times L_c = M_d \times L_d$; $6.0 \, M \times L_c = 0.50 \, M \times 0.1287 \, L$; $L_c = 0.01072 \, L = 11 \, mL$

Using a pipette, measure exactly 11 mL of 6.0 M HCl and dilute with water to a total volume of 130 mL.

13.49   *Analyze/Plan.* Assume a solution volume of 1.00 L. Calculate the mass of 1.00 L of solution and the mass of HNO_3 in 1.00 L of solution. Mass % = (mass solute/mass solution) × 100.   *Solve.*

$$1.00 \, L \times \dfrac{1000 \, mL}{1 \, L} \times \dfrac{1.42 \, g \, soln}{mL \, soln} = 1.42 \times 10^3 \, g \, soln$$

$$16 \, M = \dfrac{16 \, mol \, HNO_3}{1 \, L \, soln} \times \dfrac{63.02 \, g \, HNO_3}{1 \, mol \, HNO_3} = 1008 = 1.0 \times 10^3 \, g \, HNO_3$$

$$mass \, \% = \dfrac{1008 \, g \, HNO_3}{1.42 \times 10^3 \, g \, soln} \times 100 = 71\% \, HNO_3$$

13.50   *Analyze/Plan.* Assume 1.00 L of solution. Calculate mass of 1 L of solution using density. Calculate mass of NH_3 using mass %, then mol NH_3 in 1.00 L. *Solve.*

$$1.00 \, L \, soln \times \dfrac{1000 \, mL}{1 \, L} \times \dfrac{0.90 \, g \, soln}{1 \, mL \, soln} = 9.0 \times 10^2 \, g \, soln/L$$

$$\dfrac{900 \, g \, soln}{1.00 \, L \, soln} \times \dfrac{28 \, g \, NH_3}{100 \, g \, soln} \times \dfrac{1 \, mol \, NH_3}{17.03 \, g \, NH_3} = 14.80 = 15 \, mol \, NH_3/L \, soln = 15 \, M \, NH_3$$

13.51   *Analyze.* Given: 80.0% Cu, 20.0% Zn by mass; density = 8750 kg/m³. Find: (a) m of Zn (b) M of Zn

    (a)   *Plan.* In the brass alloy, Zn is the solute (lesser component) and Cu is the solvent (greater component). m = mol Zn/kg Cu. Assume 1 m³ → 8750 kg of brass. 80.0% is Cu, 20.0% is Zn. Change g Zn → mol Zn and solve for m.   *Solve.*

$$8750 \text{ kg brass} \times \frac{80 \text{ g Cu}}{100 \text{ g brass}} = 7.00 \times 10^3 \text{ kg Cu}$$

$$8750 \text{ kg brass} - 7000 \text{ kg Cu} = 1750 \text{ kg Zn}$$

$$1750 \text{ kg Zn} \times \frac{1000 \text{ g}}{\text{kg}} \times \frac{1 \text{ mol Zn}}{65.39 \text{ g Zn}} = 26,762.5 = 2.68 \times 10^4 \text{ mol Zn}$$

$$m = \frac{2.676 \times 10^4 \text{ mol Zn}}{7000 \text{ kg Cu}} = 3.82 \, m \text{ Zn}$$

    (b)   *Plan.* M = mol Zn/L brass. Use mol Zn from part (a). Change 1 m³ → L brass and calculate M.   *Solve.*

$$1 \text{ m}^3 \times \frac{(10)^3 \text{ dm}^3}{\text{m}^3} \times \frac{1 \text{ L}}{1 \text{ dm}^3} = 1000 \text{ L}$$

$$M = \frac{2.676 \times 10^4 \text{ mol Zn}}{1000 \text{ L brass}} = 26.76 = 26.8 \, M \text{ Zn}$$

13.52   (a) $\dfrac{0.0750 \text{ mol C}_8\text{H}_{10}\text{N}_4\text{O}_2}{1 \text{ kg CHCl}_3} \times \dfrac{194.2 \text{ g C}_8\text{H}_{10}\text{N}_4\text{O}_2}{1 \text{ mol C}_8\text{H}_{10}\text{N}_4\text{O}_2} = 14.565$

$$= 14.6 \text{ g C}_8\text{H}_{10}\text{N}_4\text{O}_2/\text{kg CHCl}_3$$

$$\frac{14.565 \text{ g C}_8\text{H}_{10}\text{N}_4\text{O}_2}{14.565 \text{ g C}_8\text{H}_{10}\text{N}_4\text{O}_2 + 1000.00 \text{ g CHCl}_3} \times 100 = 1.436 = 1.44\% \text{ C}_8\text{H}_{10}\text{N}_4\text{O}_2 \text{ by mass}$$

    (b) $1000 \text{ g CHCl}_3 \times \dfrac{1 \text{ mol CHCl}_3}{119.4 \text{ CHCl}_3} = 8.375 = 8.38 \text{ mol CHCl}_3$

$$\chi_{\text{C}_8\text{H}_{10}\text{N}_4\text{O}_2} = \frac{0.0750}{0.0750 + 8.375} = 0.00888$$

13.53   *Analyze.* Given: 4.6% CO_2 by volume (in air), 1 atm total pressure. Find: partial pressure and molarity of CO_2 in air.

Plan. 4.6% CO_2 by volume means 4.6 mL of CO_2 could be isolated from 100 mL of air, at the same temperature and pressure. According to Avogadro's Law, equal volumes of gases at the same temperature and pressure contain equal numbers of moles. By inference, the volume ratio of CO_2 to air, 4.6/100 or 0.046, is also the mole ratio. *Solve.*

$$P_{\text{CO}_2} = \chi_{\text{CO}_2} \times P_t = 0.046 \,(1 \text{ atm}) = 0.046 \text{ atm}$$

$$M = \text{mol CO}_2/\text{L air} = n/V. \quad PV = nRT, \quad M = n/V = P/RT$$

$$M_{\text{CO}_2} = \frac{P_{\text{CO}_2}}{RT} = \frac{0.046 \text{ atm}}{310 \text{ K}} \times \frac{\text{K} \cdot \text{mol}}{0.08206 \text{ L} \cdot \text{atm}} = 1.8 \times 10^{-3} \, M$$

13.54   (a)   For gases at the same temperature and pressure, volume % = mol %. The volume and mol % of CO_2 in this breathing air is 4.0%.

       (b) $P_{CO_2} = \chi_{CO_2} \times P_t = 0.040\,(1\,atm) = 0.040\ atm$

$$M_{CO_2} = \frac{P_{CO_2}}{RT} = \frac{0.040\,atm}{310\,K} \times \frac{K \bullet mol}{0.08206\,L \bullet atm} = 1.6 \times 10^{-3}\ M$$

Colligative Properties

13.55   freezing point depression, $\Delta T_f = K_f(m)$; boiling point elevation, $\Delta T_b = K_b(m)$;

          osmotic pressure, $\pi = M\,RT$; vapor pressure lowering, $P_A = \chi_A P_A{}^\circ$

13.56   (a)   decrease   (b)   decrease

       (c)   increase   (d)   increase

13.57   The vapor pressure over the sucrose solution is higher than the vapor pressure over the glucose solution. Since sucrose has a greater molar mass, 10 g of sucrose contains fewer particles than 10 g of glucose. The solution that contains fewer particles, the sucrose solution, will have the higher vapor pressure.

13.58   (a)   An *ideal solution* is a solution that obeys Raoult's Law.

       (b)   *Analyze/Plan.* Calculate the vapor pressure predicted by Raoult's law and compare it to the experimental vapor pressure. Assume ethylene glycol (eg) is the solute. *Solve.*

          $\chi_{H_2O} = \chi_{eg} = 0.500$; $P_A = \chi_A P_A{}^\circ = 0.500(149)\,mm\ Hg = 74.5\,mm\ Hg$

          The experimental vapor pressure (P_A), 67 mm Hg, is less than the value predicted by Raoult's law for an ideal solution. The solution is not ideal.

          Check. An ethylene glycol-water solution has extensive hydrogen bonding, which causes deviation from ideal behavior. We expect the experimental vapor pressure to be less than the ideal value and it is.

13.59   (a)   *Analyze/Plan.* H_2O vapor pressure will be determined by the mole fraction of H_2O in the solution. The vapor pressure of pure H_2O at 338 K (65°C) = 187.5 torr. *Solve.*

$$\frac{22.5\,g\ C_{12}H_{22}O_{11}}{342.3\,g/mol} = 0.06573 = 0.0657\ mol; \quad \frac{200.0\,g\ H_2O}{18.02\,g/mol} = 11.09878 = 11.10\ mol$$

$$P_{H_2O} = \chi_{H_2O}\,P_{H_2O}^{\underline{o}} = \frac{11.09878\ mol\ H_2O}{11.09878 + 0.06573} \times 187.5\ torr = 186.4\ torr$$

       (b)   *Analyze/Plan.* For this problem, it will be convenient to express Raoult's law in terms of the lowering of the vapor pressure of the solvent, ΔP_A.

          $\Delta P_A = P_A{}^\circ - \chi_A P_A{}^\circ = P_A{}^\circ\,(1 - \chi_A)$. $1 - \chi_A = \chi_B$, the mole fraction of the *solute* particles

          $\Delta P_A = \chi_B P_A{}^\circ$; the vapor pressure of the solvent (A) is lowered according to the mole fraction of solute (B) particles present.   *Solve.*

P_{H_2O} at 40°C = 55.3 torr; $\dfrac{340\,g\,H_2O}{18.02\,g/mol} = 18.868 = 18.9\,mol\,H_2O$

$\chi_{C_3H_8O_2} = \dfrac{2.88\,torr}{55.3\,torr} = \dfrac{y\,mol\,C_3H_8O_2}{y\,mol\,C_3H_8O_2 + 18.868\,mol\,H_2O} = 0.05208 = 0.0521$

$0.05208 = \dfrac{y}{y + 18.868}$;$0.05208\,y + 0.98263 = y$; $0.94792\,y = 0.98263$,

$y = 1.0366 = 1.04\,mol\,C_3H_8O_2$

This result has 3 sig figs because (18.9 × 0.0521 = 0.983) has 3 sig figs.

$1.0366\,mol\,C_3H_8O_2 \times \dfrac{76.09\,g\,C_3H_8O_2}{mol\,C_3H_8O_2} = 78.88 = 78.9\,g\,C_3H_8O_2$

13.60   (a) H_2O vapor pressure will be determined by the mole fraction of H_2O in the solution. The vapor pressure of pure H_2O at 343 K (70°C) = 233.7 torr.

$\dfrac{35.0\,g\,C_3H_8O_3}{92.10\,g/mol} = 0.3800 = 0.380\,mol$; $\dfrac{125\,g\,H_2O}{18.02\,g/mol} = 6.937 = 6.94\,mol$

$P_{H_2O} = \dfrac{6.937\,mol\,H_2O}{6.937 + 0.380} \times 233.7\,torr = 221.6 = 222\,torr$

      (b)   Calculate χ_B by vapor pressure lowering; $\chi_B = \Delta P_A / P_A°$ (see Solution 13.59(b)). Given moles solvent, calculate moles solute from the definition of mole fraction.

$\chi_{C_2H_6O_2} = \dfrac{10.0\,torr}{100\,torr} = 0.100$

$\dfrac{1.00 \times 10^3\,g\,C_2H_5OH}{46.07\,g/mol} = 21.71 = 21.7\,mol\,C_2H_5OH$; let $y = mol\,C_2H_6O_2$

$\chi_{C_2H_6O_2} = \dfrac{y\,mol\,C_2H_6O_2}{y\,mol\,C_2H_6O_2 + 21.71\,mol\,C_2H_5OH} = 0.100 = \dfrac{y}{y + 21.71}$

$0.100\,y + 2.171 = y$; $0.900\,y = 2.171$; $y = 2.412 = 2.41\,mol\,C_2H_6O_2$

$2.412\,mol\,C_2H_6O_2 \times \dfrac{62.07\,g}{1\,mol} = 150\,g\,C_2H_6O_2$

13.61   *Analyze/Plan.* At 63.5°C, $P_{H_2O}° = 175\,torr$, $P_{Eth}° = 400\,torr$. Let G = the mass of H_2O and/or C_2H_5OH.   *Solve.*

      (a) $\chi_{Eth} = \dfrac{\dfrac{G}{46.07\,g\,C_2H_5OH}}{\dfrac{G}{46.07\,g\,C_2H_5OH} + \dfrac{G}{18.02\,g\,H_2O}}$

Multiplying top and bottom of the right side of the equation by 1/G gives:

$\chi_{Eth} = \dfrac{1/46.07}{1/46.07 + 1/18.02} = \dfrac{0.02171}{0.02171 + 0.05549} = 0.2812$

      (b) $P_t = P_{Eth} + P_{H_2O}$; $P_{Eth} = \chi_{Eth} \times P_{Eth}°$; $P_{H_2O} = \chi_{H_2O}\,P_{H_2O}°$

$\chi_{Eth} = 0.2812$, $P_{Eth} = 0.2812\,(400\,torr) = 112.48 = 112\,torr$

$$\chi_{H_2O} = 1 - 0.2812 = 0.7188; \quad P_{H_2O} = 0.7188(175\ \text{torr}) = 125.8 = 126\ \text{torr}$$

$$P_t = 112.5\ \text{torr} + 125.8\ \text{torr} = 238.3 = 238\ \text{torr}$$

(c)　　χ_{Eth} in vapor $= \dfrac{P_{Eth}}{P_{total}} = \dfrac{112.5\ \text{torr}}{238.3\ \text{torr}} = 0.4721 = 0.472$

13.62　(a)　Since C_6H_6 and C_7H_8 form an ideal solution, we can use Raoult's Law. Since both components are volatile, both contribute to the total vapor pressure of 35 torr.

$$P_t = P_{C_6H_6} + P_{C_7H_8};\ P_{C_6H_6} = \chi_{C_6H_6} P^{\circ}_{C_6H_6};\ P_{C_7H_8} = \chi_{C_7H_8} P^{\circ}_{C_7H_8}$$

$$\chi_{C_7H_8} = 1 - \chi_{C_6H_6};\ P_T = \chi_{C_6H_6} P^{\circ}_{C_6H_6} + (1 - \chi_{C_6H_6}) P^{\circ}_{C_7H_8}$$

$$35\ \text{torr} = \chi_{C_6H_6}(75\ \text{torr}) + (1 - \chi_{C_6H_6})22\ \text{torr}$$

$$13\ \text{torr} = 53\ \text{torr}\,(\chi_{C_6H_6});\ \chi_{C_6H_6} = \frac{13\ \text{torr}}{53\ \text{torr}} = 0.2453 = 0.25;\ \chi_{C_7H_8} = 0.7547 = 0.75$$

(b)　$P_{C_6H_6} = 0.2453(75\ \text{torr}) = 18.4\ \text{torr};\ P_{C_7H_8} = 0.7547(22\ \text{torr}) = 16.6\ \text{torr}$

In the vapor,　$\chi_{C_6H_6} = \dfrac{P_{C_6H_6}}{P_t} = \dfrac{18.4\ \text{torr}}{35\ \text{torr}} = 0.53;\ \chi_{C_7H_8} = 0.47$

13.63　(a)　Because NaCl is a soluble ionic compound and a strong electrolyte, there are 2 mol dissolved particles for every 1 mol of NaCl solute. $C_6H_{12}O_6$ is a molecular solute, so there is 1 mol of dissolved particles per mol solute. Boiling point elevation is directly related to total moles of dissolved particles; $0.10\ m$ NaCl has more dissolved particles so its boiling point is higher than $0.10\ m\ C_6H_{12}O_6$.

(b)　*Analyze/Plan.* $\Delta T = K_b\ m$; K_b for H_2O is $0.51\ °C/m$ (Table 13.4)　　*Solve.*

$$0.10\ m\ \text{NaCl}:\ \Delta T = \frac{0.51°C}{m} \times 0.20\ m = 0.102\ °C;\ T_b = 100.0 + 0.102 = 100.1°C$$

$$0.10\ m\ C_6H_{12}O_6:\Delta T = \frac{0.51°C}{m} \times 0.10\ m = 0.051\ °C;\ T_b = 100.0 + 0.051 = 100.1°C$$

Check. Because K_b for H_2O is so small, there is little real difference in the boiling points of the two solutions.

(c)　In solutions of strong electrolytes like NaCl, electrostatic attractions between ions lead to ion pairing. Ion pairing reduces the effective number of particles in solution, decreasing the **change** in boiling point. The actual boiling point is then lower than the calculated boiling point for a $0.1\ M$ solution.

13.64　*Analyze/Plan.* ΔT_b depends on mol dissolved particles. Assume 100 g of each solution, calculate mol solute and mol dissolved particles. Glucose and sucrose are molecular solutes, but $NaNO_3$ dissociates into 2 mol particles per mol solute.　*Solve.*

10% by mass means 10 g solute in 100 g solution. If we have 10 g of each solute, the one with the smallest molar mass will have the largest mol solute. The molar masses are: glucose, $180.2\ g/mol$; sucrose, $342.3\ g/mol$; $NaNO_3$, $85.0\ g/mol$. $NaNO_3$ has most mol solute, and twice as many dissolved particles, so it will have the highest boiling point. Sucrose has least mol solute and lowest boiling point. Glucose is intermediate.

In order of increasing boiling point: 10% sucrose < 10% glucose < 10% $NaNO_3$.

13.65   For dilute aqueous solutions such as these, M is essentially equal to m. For the purposes of comparison, assume we can use M.

The more solute particles, the higher the boiling point. Since LiBr and $Zn(NO_3)_2$ are electrolytes, the particle concentrations in these solutions are 0.10 M and 0.15 M, respectively (although ion-ion attractive forces may decrease the real concentrations somewhat). Thus, the order of increasing boiling points is:

0.050 M LiBr < 0.120 M glucose < 0.050 M $Zn(NO_3)_2$

13.66   0.030 m phenol > 0.040 m glycerin = 0.020 m KBr. Phenol is very slightly ionized in water, but not enough to match the number of particles in a 0.040 m glycerin solution. The KBr solution is 0.040 m in particles, so it has the same freezing point as 0.040 m glycerin, which is a nonelectrolyte.

13.67   *Analyze/Plan.* $\Delta T = K\,(m)$; first, calculate the **molality** of each solution. *Solve.*

(a)   0.22 m

(b) $2.45 \text{ mol CHCl}_3 \times \dfrac{119.4 \text{ g CHCl}_3}{\text{mol CHCl}_3} = 292.53 \text{ g} = 0.293 \text{ kg}$;

$\dfrac{0.240 \text{ mol C}_{10}\text{H}_8}{0.29253 \text{ kg CHCl}_3} = 0.8204 = 0.820\ m$

(c) $2.04 \text{ g KBr} \times \dfrac{1 \text{ mol KBr}}{119.0 \text{ g KBr}} \times \dfrac{2 \text{ mol particles}}{1 \text{ mol KBr}} = 0.03429 = 0.0343 \text{ mol particles}$

$4.82 \text{ g C}_6\text{H}_{12}\text{O}_6 \times \dfrac{1 \text{ mol C}_6\text{H}_{12}\text{O}_6}{180.2 \text{ g C}_6\text{H}_{12}\text{O}_6} = 0.02675 = 0.0268 \text{ mol particles}$

$m = \dfrac{(0.03429 + 0.02675) \text{ mol particles}}{0.188 \text{ kg H}_2\text{O}} = 0.32465 = 0.325\ m$

Solve. Then, f.p. = $T_f - K_f(m)$; b.p. = $T_b + K_b(m)$; T in °C

	m	T_f	$-K_f(m)$	f.p.	T_b	$+K_b(m)$	b.p.
(a)	0.22	−114.6	−1.99(0.22) = −0.44	−115.0	78.4	1.22(0.22) = 0.27	78.7
(b)	0.820	−63.5	−4.68(0.820) = −3.84	−67.3	61.2	3.63(0.820) = 2.98	64.2
(c)	0.325	0.0	−1.86(0.325) = −0.604	−0.6	100.0	0.51(0.325) = 0.17	100.2

13.68 $\Delta T = K(m)$; first calculate the **molality** of the solute particles.

(a)   0.40 m

(b) $\dfrac{20.0 \text{ g C}_{10}\text{H}_{22}}{0.455 \text{ kg CHCl}_3} \times \dfrac{1 \text{ mol C}_{10}\text{H}_{22}}{142.3 \text{ g C}_{10}\text{H}_{22}} = 0.3089 = 0.309\ m$

(c) $m = \dfrac{0.45 \text{ mol eg} + 2(0.15) \text{ mol KBr}}{0.150 \text{ kg H}_2\text{O}} = \dfrac{0.75 \text{ mol particles}}{0.150 \text{ kg H}_2\text{O}} = 5.0\ m$

Then, f.p. = $T_f - K_f(m)$; b.p. = $T_b + K_b(m)$; T in °C

	m	T_f	$-K_f(m)$	f.p.	T_b	$+K_b(m)$	b.p.
(a)	0.40	−114.6	−1.99(0.40) = −0.80	−115.4	78.4	1.22(0.40) = 0.49	78.9
(b)	3.09	−63.5	−4.68(3.09) = −14.5	−78.0	61.2	3.63(3.09) = 11.2	72.4
(c)	5.0	0.0	−1.86(5.0) = −9.3	−9.3	100.0	0.51(5.0) = 2.6	102.6

13.69   *Analyze/Plan.* $\pi = M$ RT; T = 25°C + 273 = 298 K; M = mol $C_9H_8O_4$/L soln   *Solve.*

$$M = \frac{44.2 \text{ mg } C_9H_8O_4}{0.358 \text{ L}} \times \frac{1 \text{ g}}{1000 \text{ mg}} \times \frac{1 \text{ mol } C_9H_8O_4}{180.2 \text{ g } C_9H_8O_4} = 6.851 \times 10^{-4} = 6.85 \times 10^{-4} \, M$$

$$\pi = \frac{6.851 \times 10^{-4} \text{ mol}}{L} \times \frac{0.08206 \text{ L} \cdot \text{atm}}{K \cdot \text{mol}} \times 298 \text{ K} = 0.01675 = 0.0168 \text{ atm}$$

13.70 $\pi = MRT$; T = 20°C + 273 = 293 K

$$M \text{ (of ions)} = \frac{\text{mol NaCl} \times 2}{\text{L soln}} = \frac{3.4 \text{ g NaCl}}{1 \text{ L soln}} \times \frac{1 \text{ mol NaCl}}{58.4 \text{ g NaCl}} \times \frac{2 \text{ mol ions}}{1 \text{ mol NaCl}} = 0.116 = 0.12 \, M$$

$$\pi = \frac{0.116 \text{ mol}}{L} \times \frac{0.08206 \text{ L} \cdot \text{atm}}{K \cdot \text{mol}} \times 293 \text{ K} = 2.8 \text{ atm}$$

13.71   *Analyze/Plan.* Follow the logic in Sample Exercise 13.12 to calculate the molar mass of adrenaline based on the boiling point data. Use the structure to obtain the molecular formula and molar mass. Compare the two values.   *Solve.*

$$\Delta T_b = K_b \, m; \quad m = \frac{\Delta T_b}{K_b} = \frac{+0.49}{5.02} = 0.0976 = 0.098 \, m \text{ adrenaline}$$

$$m = \frac{\text{mol adrenaline}}{\text{kg } CCl_4} = \frac{\text{g adrenaline}}{\text{MM adrenaline} \times \text{kg } CCl_4}$$

$$\text{MM adrenaline} = \frac{\text{g adrenaline}}{m \times \text{kg } CCl_4} = \frac{0.64 \text{ g adrenaline}}{0.0976 \, m \times 0.0360 \text{ kg } CCl_4} = 1.8 \times 10^2 \text{ g/mol adrenaline}$$

The molecular formula is $C_9H_{13}NO_3$, MM = 183 g/mol. The values agree to 2 sig figs, the precision of the experimental value.

13.72 $\Delta T_f = 5.5 - 4.1 = 1.4$; $m = \dfrac{\Delta T_f}{K_f} = \dfrac{1.4}{5.12} = 0.273 = 0.27 \, m$

$$\text{MM lauryl alcohol} = \frac{\text{g lauryl alcohol}}{m \times \text{kg } C_6H_6} = \frac{5.00 \text{ g lauryl alcohol}}{0.273 \times 0.100 \text{ kg } C_6H_6}$$
$$= 1.8 \times 10^2 \text{ g/mol lauryl alcohol}$$

13.73   *Analyze/Plan.* Follow the logic in Sample Exercise 13.13.   *Solve.*

$$\pi = MRT; \quad M = \frac{\pi}{RT}; \quad T = 25°C + 273 = 298 \text{ K}$$

$$M = 0.953 \text{ torr} \times \frac{1 \text{ atm}}{760 \text{ torr}} \times \frac{K \cdot \text{mol}}{0.08206 \text{ L} \cdot \text{atm}} \times \frac{1}{298 \text{ K}} = 5.128 \times 10^{-5} = 5.13 \times 10^{-5} \, M$$

$$\text{mol} = M \times L = 5.128 \times 10^{-5} \times 0.210\,L = 1.077 \times 10^{-5} = 1.08 \times 10^{-5}\,\text{mol lysozyme}$$

$$MM = \frac{g}{\text{mol}} = \frac{0.150\,g}{1.077 \times 10^{-5}\,\text{mol}} = 1.39 \times 10^4\,\text{g/mol lysozyme}$$

13.74 $$M = \pi/RT = \frac{0.605\,\text{atm}}{298\,K} \times \frac{\text{mol} \cdot K}{0.08206\,L \cdot \text{atm}} = 0.02474 = 0.0247\,M$$

$$MM = \frac{g}{M \times L} = \frac{2.35\,g}{0.02474\,M \times 0.250\,L} = 380\,\text{g/mol}$$

13.75 (a) *Analyze/Plan.* $i = \pi$ (measured) / π (calculated for a nonelectrolyte);

π (calculated) = M RT. *Solve.*

$$\pi\,(\text{calculated}) = \frac{0.010\,\text{mol}}{L} \times \frac{0.08206\,L \cdot \text{atm}}{\text{mol} \cdot K} \times 298\,K = 0.2445 = 0.24\,\text{atm}$$

$i = 0.674\,\text{atm}/0.2445\,\text{atm} = 2.756 = 2.8$

(b) The van't Hoff factor is the effective number of particles per mole of solute. The closer the measured i value is to a theoretical integer value, the more ideal the solution. Ion-pairing and other interparticle attractive forces reduce the effective number of particles in solution and reduce the measured value of i. The more concentrated the solution, the greater the ion-pairing and the smaller the measured value of i.

13.76 If these were ideal solutions, they would have equal ion concentrations and equal ΔT_f values. Data in Table 13.5 indicates that the van't Hoff factors (i) for both salts are less than the ideal values. For 0.030 m NaCl, i is between 1.87 and 1.94, about 1.92. For 0.020 m K_2SO_4, i is between 2.32 and 2.70, about 2.62. From Equation 13.14,

ΔT_f (measured) = $i \times \Delta T_f$ (calculated for nonelectrolyte)

NaCl: ΔT_f (measured) = 1.92 × 0.030 m × 1.86 °C/m = 0.11 °C

K_2SO_4: ΔT_f (measured) = 2.62 × 0.020 m × 1.86°C/m = 0.097 °C

0.030 m NaCl would have the larger ΔT_f.

The deviations from ideal behavior are due to ion-pairing in the two electrolyte solutions. K_2SO_4 has more extensive ion-pairing and a larger deviation from ideality because of the higher charge on SO_4^{2-} relative to Cl^-.

Colloids

13.77 (a) In the gaseous state, the particles are far apart and intermolecular attractive forces are small. When two gases combine, all terms in Equation [13.1] are essentially zero and the mixture is always homogeneous.

(b) The outline of a light beam passing through a colloid is visible, whereas light passing through a true solution is invisible unless collected on a screen. This is the Tyndall effect. To determine whether Faraday's (or anyone's) apparently homogeneous dispersion is a true solution or a colloid, shine a beam of light on it and see if the light is scattered.

13.78   (a)   Suspensions are classified as solutions or colloids according to the size of the dispersed particles. Solute particles have diameters less than 10 Å. Clearly a protein with a molecular mass of 30,000 amu will be longer than 10 Å. The aqueous suspensions are colloids because of the size of protein molecules.

        (b)   Emulsion. An emulsifying agent is one that aids in the formation of an emulsion. It usually has a polar part and a nonpolar part, to facilitate mixing of immiscible liquids with very different molecular polarities.

13.79   (a)   hydrophobic   (b)   hydrophilic   (c)   hydrophobic

        (d)   hydrophobic (but stabilized by adsorbed charges)

13.80   (a)   When the colloid *particle mass* becomes large enough so that gravitational and interparticle attractive forces are greater than the kinetic energies of the particles, settling and aggregation can occur.

        (b)   *Hydrophobic* colloids do not attract a sheath of water molecules around them and thus tend to aggregate from aqueous solution. They can be stabilized as colloids by adsorbing charges on their surfaces. The charged particles interact with solvent water, stabilizing the colloid.

        (c)   *Charges on colloid particles* can stabilize them against aggregation. Particles carrying like charges repel one another and are thus prevented from aggregating and settling out.

13.81   Proteins form hydrophilic colloids because they carry charges on their surface (Figure 13.28). When electrolytes are added to a suspension of proteins, the dissolved ions form ion pairs with the protein surface charges, effectively neutralizing them. The protein's capacity for ion-dipole interactions with water is diminished and the colloid separates into a protein layer and a water layer.

13.82   (a)   The nonpolar hydrophobic tails of soap particles (the hydrocarbon chain of stearate ions) establish attractive intermolecular dispersion forces with the nonpolar oil molecules, while the charged hydrophilic head of the soap particles interacts with H_2O to keep the oil molecules suspended. (This is the mechanism by which laundry detergents remove greasy dirt from clothes.)

        (b)   Electrolytes from the acid neutralize surface charges of the suspended particles in milk, causing the colloid to coagulate.

Additional Exercises

13.83   The outer periphery of the BHT molecule is mostly hydrocarbon-like groups, such as $-CH_3$. The one $-OH$ group is rather buried inside, and probably does little to enhance solubility in water. Thus, BHT is more likely to be soluble in the nonpolar hydrocarbon hexane, C_6H_{14}, than in polar water.

13.84   In this equilibrium system, molecules move from the surface of the solid into solution, while molecules in solution are deposited on the surface of the solid. As molecules leave the surface of the small particles of powder, the reverse process preferentially deposits other molecules on the surface of a single crystal. Eventually, all molecules that were present in the 50 g of powder are deposited on the surface of a 50 g crystal; this can only happen if the dissolution and deposition processes are ongoing.

13.85   Assume that the density of the solution is 1.00 g/mL.

(a) $4 \text{ ppm O}_2 = \dfrac{4 \text{ mg O}_2}{1 \text{ kg soln}} = \dfrac{4 \times 10^{-3} \text{ g O}_2}{1 \text{ L soln}} \times \dfrac{1 \text{ mol O}_2}{32.0 \text{ g O}_2} = 1.25 \times 10^{-4} = 1 \times 10^{-4} \ M$

(b) $S_{O_2} = kP_{O_2}; \ P_{O_2} = S_{O_2}/k = \dfrac{1.25 \times 10^{-4} \text{ mol}}{L} \times \dfrac{L \bullet atm}{1.71 \times 10^{-3} \text{ mol}} = 0.0731 = 0.07 \text{ atm}$

$0.0731 \text{ atm} \times \dfrac{760 \text{ mm Hg}}{1 \text{ atm}} = 55.6 = 60 \text{ mm Hg}$

13.86   (a) $C_{Rn} = kP_{Rn}; k = C_{Rn}/P_{Rn} = 7.27 \times 10^{-3} \ M/1 \text{ atm} = 7.27 \times 10^{-3} \text{ mol/L} \bullet atm$

(b) $P_{Rn} = \chi_{Rn}P_{total}; \ P_{Rn} = 3.5 \times 10^{-6} \ (32 \text{ atm}) = 1.12 \times 10^{-4} = 1.1 \times 10^{-4} \text{ atm}$

$S_{Rn} = kP_{Rn}; S_{Rn} = \dfrac{7.27 \times 10^{-3} \text{ mol}}{L \bullet atm} \times 1.12 \times 10^{-4} \text{ atm} = 8.1 \times 10^{-7} \ M$

13.87   0.10% by mass means 0.10 g glucose/100 g blood.

(a) $\text{ppm glucose} = \dfrac{\text{g glucose}}{\text{g solution}} \times 10^6 = \dfrac{0.10 \text{ g glucose}}{100 \text{ g blood}} \times 10^6 = 1000 \text{ ppm glucose}$

(b)   *m* = mol glucose/kg solvent. Assume that the mixture of nonglucose components is the 'solvent'.

mass solvent = 100 g blood – 0.10 g glucose = 99.9 g solvent = 0.0999 kg solvent

$\text{mol glucose} = 0.10 \text{ g} \times \dfrac{1 \text{ mol}}{180.2 \text{ g C}_6\text{H}_{12}\text{O}_6} = 5.55 \times 10^{-4} = 5.6 \times 10^{-4} \text{ mol glucose}$

$m = \dfrac{5.55 \times 10^{-4} \text{ mol glucose}}{0.0999 \text{ kg solvent}} = 5.6 \times 10^{-3} \ m \text{ glucose}$

In order to calculate molarity, solution volume must be known. The density of blood is needed to relate mass and volume.

13.88   *Plan.* The definition of ppm is (mass solute/mass solution) $\times 10^6$. Use the ratio of (mass K^+/mass KCl) to calculate (mass K^+/mass solution) $\times 10^6$.   *Solve.*

$260 \text{ ppm KCl} = \dfrac{260 \text{ g KCl}}{1 \times 10^6 \text{ g solution}} \times \dfrac{39.10 \text{ g K}^+}{74.55 \text{ g KCl}} = \times \dfrac{136 \text{ g K}^+}{1 \times 10^6 \text{ g solution}} = 136 \text{ ppm K}^+$

Note that even though 1 mol KCl contains 1 mol K^+, the ppm concentration of KCl and K^+ are not equal. This is because ppm is a mass-based, not mole-based, concentration unit.

13.89   Both solutions have 15 g alcohol per 100 g solution. If the densities are equal, then equal volumes of the two solutions contain equal masses of solute.

(a)   *M* = mol solute/L solution. The molar mass of ethanol, CH_3CH_2OH, is less than the molar mass of propanol, $CH_3CH_2CH_2OH$. In equal volumes of the two solutions with equal masses of solute, the ethanol solution will contain more moles of solute particles and have the greater molarity.

(b) m = mol solute/kg solvent. For equal solution mass and equal solute mass, solvent mass must also be equal. According to part (a), there are more mol solute in the ethanol solution. So it also has the greater molality.

(c) χ = mol solute/total mol. We can express the two ratios as:

$$\chi_E = \frac{\text{mol E}}{\text{mol E} + \text{mol solv}}; \chi_P = \frac{\text{mol P}}{\text{mol P} + \text{mol solv}}$$

We have established that mol E > mol P, and that mol solv is the same in both solutions. The question is, are the mole fraction ratios equal?

No, because mol solv is constant in both denominators. Even though the larger (mol E) appears in both numerator and denominator of χ_E, the value of (mol solv) is not proportionally larger, and the value of χ_E is higher.

13.90   (a) $\dfrac{1.80 \text{ mol LiBr}}{1 \text{ L soln}} \times \dfrac{86.85 \text{ g LiBr}}{1 \text{ mol LiBr}} = 156.3 = 156 \text{ g LiBr}$

     1 L soln = 826 g soln; g CH_3CN = 826 − 156.3 = 669.7 = 670 g CH_3CN

     m LiBr = $\dfrac{1.80 \text{ mol LiBr}}{0.6697 \text{ kg } CH_3CN} = 2.69 \ m$

  (b) $\dfrac{669.7 \text{ g } CH_3CN}{41.05 \text{ g/mol}} = 16.31 = 16.3 \text{ mol } CH_3CN$; $\chi_{LiBr} = \dfrac{1.80}{1.80 + 16.31} = 0.0994$

  (c)   mass % = $\dfrac{669.7 \text{ g } CH_3CN}{826 \text{ g soln}} \times 100 = 81.1\% \ CH_3CN$

13.91   (a) $m = \dfrac{\text{mol Na(s)}}{\text{kg Hg(l)}}$; $1.0 \text{ cm}^3 \text{Na(s)} \times \dfrac{0.97 \text{ g}}{1 \text{ cm}^3} \times \dfrac{1 \text{ mol}}{23.0 \text{ g Na}} = 0.0422 = 0.042 \text{ mol Na}$

     $20.0 \text{ cm}^3 \text{Hg(l)} \times \dfrac{13.6 \text{ g}}{1 \text{ cm}^3} \times \dfrac{1 \text{ kg}}{1000 \text{ g}} = 0.272 \text{ kg Hg(l)}$; $m = \dfrac{0.0422 \text{ mol Na}}{0.272 \text{ kg Hg(l)}} = 0.155 = 0.16 \ m$ Na

  (b) $M = \dfrac{\text{mol Na(s)}}{\text{L soln}} - \dfrac{0.0422 \text{ mol Na}}{0.021 \text{ L soln}} = 2.01 = 2.0 \ M$ Na

  (c)   Clearly, molality and molarity are not the same for this amalgam. Only in the instance that one kg solvent and the mass of one liter solution are nearly equal do the two concentration units have similar values. In this example, one kg Hg has a volume much less than one liter.

13.92   Mole fraction ethyl alcohol, $\chi_{C_2H_5OH} = \dfrac{P_{C_2H_5OH}}{P^\circ_{C_2H_5OH}} = \dfrac{8 \text{ torr}}{100 \text{ torr}} = 0.08$

    $\dfrac{620 \times 10^3 \text{ g } C_{24}H_{50}}{338.6 \text{ g/mol}} = 1.83 \times 10^3 \text{ mol } C_{24}H_{50}$; let y = mol C_2H_5OH

    $\chi_{C_2H_5OH} = 0.08 = \dfrac{y}{y + 1.83 \times 10^3}$; 0.92 y = 146.4; y = 1.6×10^2 mol C_2H_5OH

(Strictly speaking, y should have 1 sig fig because 0.08 has 1 sig fig, but this severely limits the calculation.)

$$1.6 \times 10^2 \text{ mol C}_2\text{H}_5\text{OH} \times \frac{46 \text{ g C}_2\text{H}_5\text{OH}}{1 \text{ mol}} = 7.4 \times 10^3 \text{ g or } 7.4 \text{ kg C}_2\text{H}_5\text{OH}$$

13.93   (a)   The solvent vapor pressure over each solution is determined by the total particle concentrations present in the solutions. When the particle concentrations are equal, the vapor pressures will be equal and equilibrium established. The particle concentration of the nonelectrolyte is just 0.050 M, the ion concentration of the NaCl is $2 \times 0.035 \ M = 0.070 \ M$. Solvent will diffuse from the less concentrated nonelectrolyte solution. The level of the NaCl solution will rise, and the level of the nonelectrolyte solution will fall.

       (b)   Let x = volume of solvent transferred

$$\frac{0.050 \ M \times 30.0 \text{ mL}}{(30.0 - x) \text{ mL}} = \frac{0.070 \ M \times 30.0 \text{ mL}}{(30.0 + x) \text{ mL}}; 1.5(30.0 + x) = 2.1(30.0 - x)$$

$$45 + 1.5 \ x = 63 - 2.1 \ x; \quad 3.6 \ x = 18; \quad x = 5.0 = 5 \text{ mL transferred}$$

The volume in the nonelectrolyte beaker is (30.0 − 5.0) = 25.0 mL; in the NaCl beaker (30.0 + 5.0) = 35.0 mL.

13.94   (a)   0.100 m K$_2$SO$_4$ is 0.300 m in particles. H$_2$O is the solvent.

$$\Delta T_f = K_f m = -1.86(0.300) = -0.558; T_f = 0.0 - 0.558 = -0.558^\circ\text{C} = -0.6^\circ\text{C}$$

       (b) ΔT_f (nonelectrolyte) $= -1.86(0.100) = -0.186; T_f = 0.0 - 0.186 = -0.186^\circ\text{C} = -0.2^\circ\text{C}$

$$T_f \text{ (measured)} = i \times T_f \text{ (nonelectrolyte)}$$

From Table 13.5, i for 0.100 m K$_2$SO$_4$ = 2.32

$$T_f \text{ (measured)} = 2.32(-0.186^\circ\text{C}) = -0.432^\circ\text{C} = -0.4^\circ\text{C}$$

13.95   The compound with the larger i value is the stronger electrolyte.

$$i = \frac{\Delta T_f \text{(measured)}}{\Delta T_f \text{(calculated)}} \quad \text{The idealized value is 3 for both salts.}$$

$$\text{Hg(NO}_3)_2 : \quad m = \frac{10.0 \text{ g Hg(NO}_3)_2}{1.00 \text{ kg H}_2\text{O}} \times \frac{1 \text{ mol Hg(NO}_3)_2}{324.6 \text{ g Hg(NO}_3)_2} = 0.0308 \ m$$

$$\Delta T_f \text{ (nonelectrolyte)} = -1.86(0.0308) = -0.0573^\circ\text{C}$$

$$i = \frac{-0.162^\circ \text{ C}}{-0.0573^\circ \text{ C}} = 2.83$$

$$\text{HgCl}_2: \quad m = \frac{10.0 \text{ g HgCl}_2}{1.00 \text{ kg H}_2\text{O}} \times \frac{1 \text{ mol HgCl}_2}{271.5 \text{ g HgCl}_2} = 0.0368 \ m$$

$$\Delta T_f \text{ (nonelectrolyte)} = -1.86(0.0368) = -0.0685^\circ\text{C}$$

$$i = \frac{-0.0685}{-0.0685} = 1.00$$

With an *i* value of 2.83, $Hg(NO_3)_2$ is almost completely dissociated into ions; with an *i* value of 1.00, the $HgCl_2$ behaves essentially like a nonelectrolyte. Clearly, $Hg(NO_3)_2$ is the stronger electrolyte.

13.96   (a) $K_b = \dfrac{\Delta T_b}{m}$; $\Delta T_b = 47.46°C - 46.30°C = 1.16°C$

$m = \dfrac{\text{mol solute}}{\text{kg } CS_2} = \dfrac{0.250\,\text{mol}}{400.0\,\text{mL } CS_2} \times \dfrac{1\,\text{mL } CS_2}{1.261\,\text{g } CS_2} \times \dfrac{1000\,\text{g}}{1\,\text{kg}} = 0.4956 = 0.496\,m$

$K_b = \dfrac{1.16°C}{0.4956\,m} = 2.34°\,C/m$

(b) $m = \dfrac{\Delta T_b}{K_b} = \dfrac{(47.08 - 46.30)°C}{2.34°\,C/m} = 0.333 = 0.33\,m$

$m = \dfrac{\text{mol unknown}}{\text{kg } CS_2}$; $m \times \text{kg } CS_2 = \dfrac{\text{g unknown}}{\text{MM unknown}}$; $MM = \dfrac{\text{g unknown}}{m \times \text{kg } CS_2}$

$50.0\,\text{mL } CS_2 \times \dfrac{1.261\,\text{g } CS_2}{1\,\text{mL}} \times \dfrac{1\,\text{kg}}{1000\,\text{g}} = 0.06305 = 0.0631\,\text{kg } CS_2$

$MM = \dfrac{5.39\,\text{g unknown}}{0.333\,m \times 0.06305\,\text{kg } CS_2} = 257 = 2.6 \times 10^2\,\text{g/mol}$

13.97   (a)   Assume 1000 g of solution. $1000\,\text{g soln} \times \dfrac{1\,\text{mL}}{1.22\,\text{g}} = 819.7 = 820\,\text{mL}$

$1000\,\text{g soln} \times \dfrac{40.0\,\text{g KSCN}}{100\,\text{g soln}} = 400\,\text{g KSCN}$; $1000\,\text{g soln} - 400\,\text{g KSCN} = 600\,\text{g } H_2O$

$\dfrac{400\,\text{g KSCN}}{97.19\,\text{g/mol}} = 4.116 = 4.12\,\text{mol KSCN}$; $\dfrac{600\,\text{g } H_2O}{18.02\,\text{g/mol}} = 33.30 = 33.3\,\text{mol } H_2O$

$\chi_{KSCN} = \dfrac{4.116}{4.116 + 33.30} = 0.110$; $m = \dfrac{4.116\,\text{mol KSCN}}{0.600\,\text{kg } H_2O} = 6.86\,m$

$M = \dfrac{4.116\,\text{mol KSCN}}{0.8197\,\text{L}} = 5.02\,M$

(b)   If there are 4.12 mol of KSCN, there are 8.24 moles of ions. There are then 33.3 mol H_2O/8.24 mol ions ≈ 4 mol H_2O for each mol of ions, or 4 water molecules for each ion. This is too few water molecules to completely hydrate the anions and cations in the solution.

For a solution that is this concentrated, one would expect significant ion-pairing, because the ions are not completely surrounded and separated by H_2O molecules. Because of ion-pairing, the effective number of particles will be less than that indicated by *m* and *M*, so the observed colligative properties will be significantly different from those predicted by formulas for ideal solutions. The observed freezing point will be higher, the boiling point lower, and the osmotic pressure lower than predicted.

13.98 $M = \dfrac{\pi}{RT} = \dfrac{57.1\,\text{torr}}{298\,\text{K}} \times \dfrac{1\,\text{atm}}{760\,\text{torr}} \times \dfrac{K \cdot \text{mol}}{0.08206\,L \cdot \text{atm}} = 3.072 \times 10^{-3} = 3.07 \times 10^{-3}\,M$

$\dfrac{0.036\,\text{g solute}}{100\,\text{g } H_2O} \times \dfrac{1000\,\text{g } H_2O}{1\,\text{kg } H_2O} = 0.36\,\text{g solute/kg } H_2O$

Assuming molarity and molality are the same in this dilute solution, we can then say 0.36 g solute = 3.072×10^{-3} mol; MM = 117 g/mol. Because the salt is completely ionized, the formula weight of the lithium salt is **twice** this calculated value, or **234 g/mol**. The organic portion, $C_nH_{2n+1}O_2^-$, has a formula weight of 234-7 = 227 g. Subtracting 32 for the oxygens, and 1 to make the formula C_nH_{2n}, we have C_nH_{2n}, MM = 194 g/mol. Since each CH_2 unit has a mass of 14, n ≈ 194/14 ≈ 14. The formula for our salt is $LiC_{14}H_{29}O_2$.

Integrative Exercises

13.99   Since these are very dilute solutions, assume that the density of the solution ≈ the density of H_2O ≈ 1.0 g/mL at 25°C. Then, 100 g solution = 100 g H_2O = 0.100 kg H_2O.

(a) CF_4 : $\dfrac{0.0015 \text{ g } CF_4}{0.100 \text{ kg } H_2O} \times \dfrac{1 \text{ mol } CF_4}{88.00 \text{ g } CF_4} = 1.7 \times 10^{-4} \, m$

        $CClF_3$: $\dfrac{0.009 \text{ g } CClF_3}{0.100 \text{ kg } H_2O} \times \dfrac{1 \text{ mol } CClF_3}{104.46 \text{ g } CClF_3} = 8.6 \times 10^{-4} \, m = 9 \times 10^{-4} \, m$

        CCl_2F_2 : $\dfrac{0.028 \text{ g } CCl_2F_2}{0.100 \text{ kg } H_2O} \times \dfrac{1 \text{ mol } CCl_2F_2}{120.9 \text{ g } CCl_2F_2} = 2.3 \times 10^{-3} \, m$

        $CHClF_2$: $\dfrac{0.30 \text{ g } CHClF_2}{0.100 \text{ kg } H_2O} \times \dfrac{1 \text{ mol } CHClF_2}{86.47 \text{ g } CHClF_2} = 3.5 \times 10^{-2} \, m$

(b) $m = \dfrac{\text{mol solute}}{\text{kg solvent}}; \; M = \dfrac{\text{mol solute}}{\text{L solution}}$

Molality and molarity are numerically similar when kilograms solvent and liters solution are nearly equal. This is true when solutions are dilute, so that the density of the solution is essentially the density of the solvent, and when the density of the solvent is nearly 1 g/mL. That is, for dilute aqueous solutions such as the ones in this problem, $M \approx m$.

(c)   Water is a polar solvent; the solubility of solutes increases as their polarity increases. All the fluorocarbons listed have tetrahedral molecular structures. CF_4, a symmetrical tetrahedron, is nonpolar and has the lowest solubility. As more different atoms are bound to the central carbon, the electron density distribution in the molecule becomes less symmetrical and the molecular polarity increases. The most polar fluorocarbon, $CHClF_2$, has the greatest solubility in H_2O. It may act as a weak hydrogen bond acceptor for water.

(d) $S_g = kP_g$. Assume $M = m$ for $CHClF_2$. $P_g = 1$ atm

        $k = \dfrac{S_g}{P_g} = \dfrac{M}{P_g}; \, k = \dfrac{3.5 \times 10^{-2} \, M}{1.0 \text{ atm}} = 3.5 \times 10^{-2} \text{ mol/L} \bullet \text{atm}$

This value is greater than the Henry's law constant for $N_2(g)$, because $N_2(g)$ is nonpolar and of lower molecular mass than $CHClF_2$. In fact, the Henry's law constant for nonpolar CF_4, 1.7×10^{-4} mol/L $\bullet$ atm is similar to the value for N_2, 6.8×10^{-4} mol L $\bullet$ atm.

13.100 $\dfrac{0.015\,\text{g N}_2}{1\,\text{L blood}} \times \dfrac{1\,\text{mol N}_2}{28.01\,\text{g N}_2} = 5.355 \times 10^{-4} = 5.4 \times 10^{-4}\,\text{mol N}_2/\text{L blood}$

At 100 ft, the partial pressure of N_2 in air is 0.78 (4.0 atm) = 3.12 atm. This is just four times the partial pressure of N_2 at 1.0 atm air pressure. According to Henry's law, $S_g = kP_g$, a 4-fold increase in P_g results n a 4-fold increase in S_g, the solubility of the gas. Thus, the solubility of N_2 at 100 ft is $4(5.355 \times 10^{-4}\,M) = 2.142 \times 10^{-3} = 2.1 \times 10^{-3}\,M$. If the diver suddenly surfaces, the amount of N_2/L blood released is the difference in the solubilities at the two depths: $(2.142 \times 10^{-3}\,\text{mol/L} - 5.355 \times 10^{-4}\,\text{mol/L}) = 1.607 \times 10^{-3} = 1.6 \times 10^{-3}\,\text{mol N}_2/\text{L blood}$.

At surface conditions of 1.0 atm external pressure and 37°C = 310 K,

$$V = \frac{nRT}{P} = 1.607 \times 10^{-3}\,\text{mol} \times \frac{310\,\text{K}}{1.0\,\text{atm}} \times \frac{0.08206\,\text{L} \cdot \text{atm}}{\text{mol} \cdot \text{K}} = 0.041\,\text{L}$$

That is, 41 mL of tiny N_2 bubbles are released from each L of blood.

13.101 The stronger the intermolecular forces, the higher the heat (enthalpy) of vaporization.

(a) None of the substances are capable of hydrogen bonding in the pure liquid, and they have similar molar masses. All intermolecular forces are van der Waals forces, dipole-dipole, and dispersion forces. In decreasing order of strength of forces:

acetone > acetaldehyde > ethylene oxide > cyclopropane

The first three compounds have dipole-dipole and dispersion forces, the last only dispersion forces.

(b) The order of solubility in hexane should be the reverse of the order above. The least polar substance, propane, will be most soluble in hexane. Ethanol, CH_3CH_2OH, is capable of hydrogen bonding with the three polar compounds. Thus, acetaldehyde, acetone, and ethylene oxide should be more soluble than cyclopropane, but without further information we cannot distinguish among the polar molecules.

13.102 For ionic solids, the exothermic part of the solution process is step (3), surrounding the separated ions by solvent molecules. The released energy comes from the attractive interaction of the solvent with the separated ions. In hydrates, one or more water molecules are already associated with the ions, reducing the total energy released during solvation.

13.103 (a)

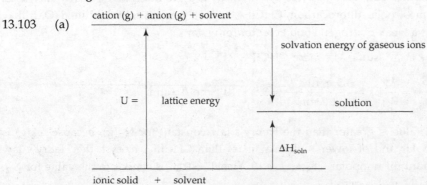

(b)   If the lattice energy (U) of the ionic solid (ion-ion forces) is too large relative to the solvation energy of the gaseous ions (ion-dipole forces), ΔH_{soln} will be too large and positive (endothermic) for solution to occur. This is the case for solutes like NaBr. Lattice energy is inversely related to the distance between ions, so salts with large cations like $(CH_3)_4N^+$ have smaller lattice energies than salts with simple cations like Na^+. The smaller lattice energy of $(CH_4)_3NBr$ causes it to be more soluble in nonaqueous polar solvents. Also, the $-CH_3$ groups in the large cation are capable of dispersion interactions with the $-CH_3$ (or other nonpolar groups) of the solvent molecules. This produces a more negative solvation energy for the salts with large cations.

Overall, for salts with larger cations, U is smaller (less positive), the solvation energy of the gaseous ions is more negative, and ΔH_{soln} is less endothermic. These salts are more soluble in polar nonaqueous solvents.

13.104   (a) $Zn(s) + H_2SO_4(aq) \rightarrow ZnSO_4(aq) + H_2(g)$

$$2.050 \text{ g Zn} \times \frac{1 \text{ mol Zn}}{65.39 \text{ g Zn}} = 0.03135 \text{ mol Zn}$$

$$1.00 \, M \text{ H}_2\text{SO}_4 \times 0.0150 \text{ L} = 0.0150 \text{ mol H}_2\text{SO}_4$$

Since Zn and H_2SO_4 react in a 1:1 mole ratio, H_2SO_4 is the limiting reactant; 0.0150 mol of $H_2(g)$ are produced.

(b) $P = \dfrac{nRT}{V} = \dfrac{0.0150 \text{ mol}}{0.122 \text{ L}} \times \dfrac{0.08206 \text{ L} \bullet \text{atm}}{\text{mol} \bullet \text{K}} \times 298 \text{ K} = 3.0066 = 3.01 \text{ atm}$

(c) $S_{H_2} = kP_{H_2} = \dfrac{7.8 \times 10^{-4} \text{ mol}}{\text{L} \bullet \text{atm}} \times 3.0066 \text{ atm} = 0.002345 = 2.3 \times 10^{-3} \, M$

$$\frac{0.002345 \text{ mol H}_2}{\text{L soln}} \times 0.0150 \text{ L} = 3.518 \times 10^{-5} = 3.5 \times 10^{-5} \text{ mol dissolved H}_2$$

$$\frac{3.5 \times 10^{-5} \text{ mol dissolved H}_2}{0.0150 \text{ mol H}_2 \text{ produced}} \times 100 = 0.23\% \text{ dissolved H}_2$$

This is approximately 2.3 ppt; for every 10,000 H_2 molecules, 23 are dissolved. It was reasonable to ignore dissolved $H_2(g)$ in part (b).

13.105   (a) $\dfrac{1.3 \times 10^{-3} \text{ mol CH}_4}{\text{L soln}} \times 4.0 \text{ L} = 5.2 \times 10^{-3} \text{ mol CH}_4$

$$V = \frac{nRT}{P} = \frac{5.2 \times 10^{-3} \text{ mol} \times 273 \text{ K}}{1.0 \text{ atm}} \times \frac{0.08206 \text{ L} \bullet \text{atm}}{\text{K} \bullet \text{mol}} = 0.12 \text{ L}$$

(b)   All three hydrocarbons are nonpolar; they have zero net dipole moment. In CH_4 and C_2H_6, the C atoms are tetrahedral and all bonds are σ bonds. C_2H_6 has a higher molar mass than CH_4, which leads to stronger dispersion forces and greater water solubility. In C_2H_4, the C atoms are trigonal planar and the π electron cloud is symmetric above and below the plane that contains all the atoms. The π cloud in C_2H_4 is an area of concentrated electron density that experiences attractive forces with the positive ends of H_2O molecules. These forces increase the solubility of C_2H_4 relative to the other hydrocarbons.

(c)   The molecules have similar molar masses. NO is most soluble because it is polar. The triple bond in N_2 is shorter than the double bond in O_2. It is more difficult for H_2O molecules to surround the smaller N_2 molecules, so they are less soluble than O_2 molecules.

(d) H_2S and SO_2 are polar molecules capable of hydrogen bonding with water. Hydrogen bonding is the strongest force between neutral molecules and causes the much greater solubility. H_2S is weakly acidic in water. SO_2 reacts with water to form H_2SO_3, a weak acid. The large solubility of SO_2 is a sure sign that a chemical process has occurred.

(e) N_2 and C_2H_4. N_2 is too small to be easily hydrated, so C_2H_4 is more soluble in H_2O.

   NO (31) and C_2H_6 (30). The structures of these two molecules are very different, yet they have similar solubilities. NO is slightly polar, but too small to be easily hydrated. The larger C_2H_6 is nonpolar, but more polarizable (stronger dispersion forces).

   NO (31) and O_2 (32). The slightly polar NO is more soluble than the slightly larger (longer O=O bond than $N \equiv O$ bond) but nonpolar O_2.

13.106   The process is spontaneous with no significant change in enthalpy, so we suspect that there is an increase in entropy. In general, dilute solutions of the same solute have greater entropy than concentrated ones, because the solute particles are more free to move about the solution. There are a greater number of equivalent environments available to the solute. In the limiting case that the more dilute solution is pure solvent, there is a definite increase in entropy as the concentrated solution is diluted. In Figure 13.23, there may be some entropy decrease as the dilute solution loses solvent, but this is more than offset by the entropy increase that accompanies dilution. There is a net increase in entropy of the system, going from the left to center panels in Figure 13.23.

13.107   (a) $$\Delta T_f = K_f m = K_f \times \frac{mol\, C_7H_6O_2}{kg\, C_6H_6} = K_f \times \frac{g\, C_7H_6O_2}{kg\, C_6H_6 \times M\, C_7H_6O_2}$$

   $$MM = \frac{K_f \times g\, C_7H_6O_2}{\Delta T_f \times kg\, C_6H_6} = \frac{5.12 \times 0.55}{0.360 \times 0.032} = 2444.4 = 2.4 \times 10^2 \, g/mol$$

(b)   The formula weight of $C_7H_6O_2$ is 122 g/mol. The experimental molar mass is twice this value, indicating that benzoic acid is associated into dimers in benzene solution. This is reasonable, since the carboxyl group, –COOH, is capable of strong hydrogen bonding with itself. Many carboxylic acids exist as dimers in solution.

   The structure of benzoic acid dimer in benzene solution is:

13.108 $\chi_{CHCl_3} = \chi_{C_3H_6O} = 0.500$

(a)   For an ideal solution, Raoult's Law is obeyed.

$P_t = P_{CHCl_3} + P_{C_3H_6O}; \quad P_{CHCl_3} = 0.5(300 \text{ torr}) = 150 \text{ torr}$

$P_{C_3H_6O} = 0.5(360 \text{ torr}) = 180 \text{ torr}; \quad P_t = 150 \text{ torr} + 180 \text{ torr} = 330 \text{ torr}$

(b)   The real solution has a lower vapor pressure, 250 torr, than an ideal solution of the same composition, 330 torr. Thus, fewer molecules escape to the vapor phase from the liquid. This means that fewer molecules have sufficient kinetic energy to overcome intermolecular attractions. Clearly, even weak hydrogen bonds such as this one are stronger attractive forces than dipole-dipole or dispersion forces. These hydrogen bonds prevent molecules from escaping to the vapor phase and result in a lower than ideal vapor pressure for the solution. There is essentially no hydrogen bonding in the individual liquids.

(c)   According to Coulomb's law, electrostatic attractive forces lead to an overall lowering of the energy of the system. Thus, when the two liquids mix and hydrogen bonds are formed, the energy of the system is decreased and $\Delta H_{soln} < 0$; the solution process is exothermic.

14 Chemical Kinetics

Visualizing Concepts

14.1   (a)   X is a product, because its concentration increases with time.

         (b)   The average rate of reaction between any two points on the graph is the slope of the line connecting the two points. The average rate is greater between points 1 and 2 than between points 2 and 3 because they are different stages in the overall process. Points 1 and 2 are earlier in the reaction when more reactants are available, so the rate of formation of products is greater. As reactants are used up, the rate of X production decreases, and the average rate between points 2 and 3 is smaller.

14.2   Chemical equation (d), $B \rightarrow 2A$, is consistent with the data. The concentration of A increases with time, and concentration B decreases with time, so B must be a reactant and A must be a product. The ending concentration of A is approximately twice as large as the starting concentration of B, so mole ratio of A:B is 2:1. The reaction is $B \rightarrow 2A$.

14.3   *Analyze/Plan.* Using the relationship rate = $k[A]^x$, determine the value of x that produces a rate law to match the described situation. *Solve.*

         (a) $x = 0$. The rate of reaction does not depend on [A], so the reaction is zero-order in A.

         (b) $x = 2$. When [A] increases by a factor of 3, rate increases by a factor of $(3)^2 = 9$.

         (c) $x = 3$. When [A] increases by a factor of 2, rate increases by a factor of $(2)^3 = 8$.

14.4   *Plan.* For a first-order reaction, a plot of ln[A] vs. time is linear, as shown in the diagram. The slope is –k, and the intercept is $[A]_0$. According to the Arrhenius equation [14.19], k increases with increasing temperature. *Solve.*

         (a)   Lines 1 and 2 have the same slope, and thus the same rate constant, k. These experiments are done at the same temperature. The y-intercepts of the two lines are different; the experiments had different initial concentrations of A.

         (b)   Lines 2 and 3 have the same y-intercept and thus the same starting concentration of A. The slopes of the two lines are different, so their rate constants are different and they occur at different temperatures. Line 3, with the smaller slope and k value will occur at the lower temperature.

14.5   *Analyze.* Given concentrations of reactants and products at two times, as represented in the diagram, find $t_{1/2}$ for this first-order reaction.

         Plan. For a first order reaction, $t_{1/2} = 0.693/k$; $t_{1/2}$ depends only on k. Use equation [14.12] to solve for k. *Solve.*

(a)   Since reactants and products are in the same container, use number of particles as a measure of concentration. The red dots are reactant A, and the blue are product B. $[A]_0 = 8$, $[A]_{30} = 2$, $t = 30$ min.

$$\ln \frac{[A]_t}{[A]_0} = -kt. \quad \ln(2/8) = -k(30 \text{ min}); \quad \frac{-1.3863}{-30\text{min}} = k;$$

$k = 0.046210 = 0.0462 \text{ min}^{-1}$

$t_{1/2} = 0.693/k = 0.693/0.046210 = 15 \text{ min}$

By examination, $[A]_0 = 8$, $[A]_{30} = 2$. After 1 half-life, $[A] = 4$; after a second half-life, $[A] = 2$. Thirty minutes represents exactly 2 half-lives, so $t_{1/2} = 15$ min. [This is more straightforward than the calculation, but a less general method.]

(b)   After 4 half-lives, $[A]_t = [A]_0 \times 1/2 \times 1/2 \times 1/2 \times 1/2 = [A]_0/16$. In general, after n half-lives, $[A] = [A]_0/2^n$.

14.6   On a plot of ln k vs. $1/T$, the slope is $-E_a$ and the y-intercept is ln A, where E_a is activation energy and A is the frequency factor.

(a)   If $E_a(2) > E_a(1)$ and $A_1 = A_2$, the lines will have the same y-intercept, negative slope direction, and the slope of line 2 will be steeper than the slope of line 1.

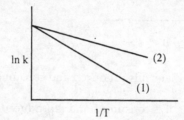

(b)   If $A_2 > A_1$ and $E_a(1) = E_a(2)$, the lines will be parallel with the same negative slopes and different y-intercepts.

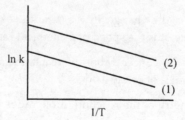

14.7   (a) $NO_2 + F_2 \rightarrow NO_2F + F$

$NO_2 + F \rightarrow NO_2F$

(b) $2NO_2 + F_2 \rightarrow 2NO_2F$

(c)   F is an intermediate, because it is produced and then consumed during the reaction.

(d)   rate $= k[NO_2][F_2]$

14.8   This is the profile of a two-step mechanism, A → B and B → C. There is one intermediate, B. Because there are two energy maxima, there are two transition states. The B → C step is faster, because its activation energy is smaller. The reaction is exothermic because the energy of the products is lower than the energy of the reactants.

14.9   The most likely transition state shows the relative geometry of both reactants and products. It is reasonable to assume that multiple bonds, with greater total bond energy, remain intact at the expense of single bonds. In the black-and-white diagram below, open circles represent the red balls and closed circles represent the blue.

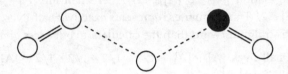

14.10   (a) $A_2 + AB + AC \rightarrow BA_2 + A + AC$

                  $BA_2 + A + AC \rightarrow A_2 + BA_2 + C$

                 _____

                  net: $AB + AC \rightarrow BA_2 + C$

       (b)   A is the intermediate; it is produced and consumed.

       (c) A_2 is the catalyst; it is consumed and reproduced.

Reaction Rates

14.11   (a)   *Reaction rate* is the change in the amount of products or reactants in a given amount of time; it is the speed of a chemical reaction.

       (b)   Rates depend on concentration of reactants, surface area of reactants, temperature and presence of catalyst.

       (c)   The stoichiometry of the reaction (mole ratios of reactants and products) must be known to relate rate of disappearance of reactants to rate of appearance of products.

14.12   (a)   *M/s*

       (b)   The hotter the oven, the faster the cake bakes. Milk sours faster in hot weather than cool weather.

       (c)   The *average rate* is the rate over a period of time, while the *instantaneous rate* is the rate at a particular time.

14.13   *Analyze/Plan.* Given mol A at a series of times in minutes, calculate mol B produced, molarity of A at each time, change in M of A at each 10 min interval, and ΔM A/s. For this reaction, mol B produced equals mol A consumed. M of A or [A] = mol A/0.100 L. The average rate of disappearance of A for each 10 minute interval is

$$-\frac{\Delta[A]}{s} = -\frac{[A]_1 - [A]_0}{10 \text{ min}} \times \frac{1 \text{ min}}{60 \text{ s}}$$

Solve.

Time (min)	Mol A	(a) Mol B	[A]	Δ[A]	(b) Rate $-(\Delta$[A]/s)
0	0.065	0.000	0.65		
10	0.051	0.014	0.51	−0.14	2.3×10^{-4}
20	0.042	0.023	0.42	−0.09	2×10^{-4}
30	0.036	0.029	0.36	−0.06	1×10^{-4}
40	0.031	0.034	0.31	−0.05	0.8×10^{-4}

(c) $\dfrac{\Delta M_B}{\Delta t} = \dfrac{(0.029 - 0.014) \text{ mol}/0.100 \text{ L}}{(30 - 10) \text{ min}} \times \dfrac{1 \text{ min}}{60 \text{ s}} = 1.25 \times 10^{-4} = 1.3 \times 10^{-4} \; M/s$

14.14

Time(s)	Mol A	(a) Mol B	Δ Mol A	(b) Rate $-(\Delta$ mol A/s)
0	0.100	0.000		
40	0.067	0.033	−0.033	8.3×10^{-4}
80	0.045	0.055	−0.022	5.5×10^{-4}
120	0.030	0.070	−0.015	3.8×10^{-4}
160	0.020	0.080	−0.010	2.5×10^{-4}

(c)   The volume of the container must be known to report the rate in units of concentration (mol/L) per time.

14.15   (a)   *Analyze/Plan.* Follow the logic in Sample Exercise 14.1.   *Solve.*

Time (sec)	Time Interval (sec)	Concentration (M)	ΔM	Rate (M/s)
0		0.0165		
2,000	2,000	0.0110	−0.0055	28×10^{-7}
5,000	3,000	0.00591	−0.0051	17×10^{-7}
8,000	3,000	0.00314	−0.00277	9.23×10^{-7}
12,000	4,000	0.00137	−0.00177	4.43×10^{-7}
15,000	3,000	0.00074	−0.00063	2.1×10^{-7}

(b)   From the slopes of
the lines in the fig-
ure at right, the
rates are 12×10^{-7}
M/s at 5000 s,   5.8
$\times 10^{-7} M/s$ at 8000
s.

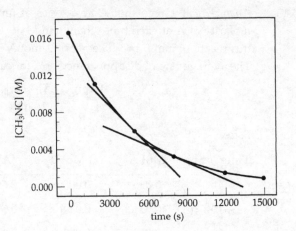

14.16   (a)

Time (min)	Time Interval (min)	Concentration (M)	ΔM	Rate (M/s)
0.0		1.85		
54.0	54.0	1.58	−0.27	8.3×10^{-5}
107.0	53.0	1.36	−0.22	6.9×10^{-5}
215.0	108	1.02	−0.34	5.2×10^{-5}
430.0	215	0.580	−0.44	3.4×10^{-5}

(b)   From the slopes of the lines
in the figure at the right,
the rates are: at 75.0 min,
$4.2 \times 10^{-3} M/min$, or
$7.0 \times 10^{-5} M/s$; at 250 min,
$2.1 \times 10^{-3} M/min$ or
$3.5 \times 10^{-5} M/s$

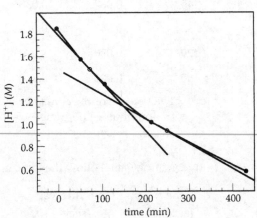

14.17   *Analyze/Plan.* Follow the logic in Sample Exercise 14.3.   *Solve.*

(a) $-\Delta[H_2O_2]/\Delta t = \Delta[H_2]/\Delta t = \Delta[O_2]/\Delta t$

(b) $-\Delta[N_2O]/2\Delta t = \Delta[N_2]/2\Delta t = \Delta[O_2]/\Delta t$

$-\Delta[N_2O]/\Delta t = \Delta[N_2]/\Delta t = 2\Delta[O_2]/\Delta t$

(c) $-\Delta[N_2]/\Delta t = \Delta[NH_3]/2\Delta t; \ -\Delta[H_2]/3\Delta t = \Delta[NH_3]/2\Delta t$

$-2\Delta[N_2]/\Delta t = \Delta[NH_3]/\Delta t; \ -\Delta[H_2]/\Delta t = 3\Delta[NH_3]/2\Delta t$

14.18   (a)   rate = $-\Delta[HBr]/2\Delta t = \Delta[H_2]/\Delta t = \Delta[Br_2]/\Delta t$

        (b)   rate = $-\Delta[SO_2]/2\Delta t = -\Delta[O_2]/\Delta t = \Delta[SO_3]/2\Delta t$

        (c)   rate = $-\Delta[NO]/2\Delta t = -\Delta[H_2]/2\Delta t = \Delta[N_2]/\Delta t = \Delta[H_2O]/2\Delta t$

14.19   *Analyze/Plan.* Use Equation [14.4] to relate the rate of disappearance of reactants to the rate of appearance of products. Use this relationship to calculate desired quantities. *Solve.*

        (a) $\Delta[H_2O]/2\Delta t = -\Delta[H_2]/2\Delta t = -\Delta[O_2]/\Delta t$

              H_2 is burning, $-\Delta[H_2]/\Delta t = 0.85$ mol/s

              O_2 is consumed, $-\Delta[O_2]/\Delta t = -\Delta[H_2]/2\Delta t = 0.85$ mol/s/2 = 0.43 mol/s

              H_2O is produced, $+\Delta[H_2O]/\Delta t = -\Delta[H_2]/\Delta t = 0.85$ mol/s

        (b)   The change in total pressure is the sum of the changes of each partial pressure. NO and Cl_2 are disappearing and NOCl is appearing.

              $-\Delta P_{NO}/\Delta t = -23$ torr/min

              $-\Delta P_{Cl_2}/\Delta t = \Delta P_{NO}/2\Delta t = -12$ torr/min

              $+\Delta P_{NOCl}/\Delta t = -\Delta P_{NO}/\Delta t = +23$ torr/min

              $\Delta P_T/\Delta t = -23$ torr/min $- 12$ torr/min $+ 23$ torr/min $= -12$ torr/min

14.20   (a) $-\Delta[C_2H_4]/\Delta t = \Delta[CO_2]/2\Delta t = \Delta[H_2O]/2\Delta t$

              $-2\Delta[C_2H_4]/\Delta t = \Delta[CO_2]/\Delta t = \Delta[H_2O]/\Delta t$

              C_2H_4 is burning, $-\Delta[C_2H_4]/\Delta t = 0.37$ *M*/s

              CO_2 and H_2O are produced, at twice the rate that C_2H_4 is consumed.

              $\Delta[CO_2]/\Delta t = \Delta[H_2O]/\Delta t = 2(0.37)$ *M*/s = 0.74 *M*/s

        (b)   In this reaction, pressure is a measure of concentration.

              $-\Delta[N_2H_4]/\Delta t = -\Delta[H_2]/\Delta t = \Delta[NH_3]/2\Delta t$

              N_2H_4 is consumed, $-\Delta[N_2H_4]/\Delta t = 63$ torr/hr

              H_2 is consumed, $-\Delta[H_2]/\Delta t = 63$ torr/hr

              NH_3 is produced at twice the rate that N_2H_4 and H_2 are consumed,

              $\Delta[NH_3]/\Delta t = -2\Delta[N_2H_4]/\Delta t = 2(63)$ torr/hr = 126 torr/hr

              $\Delta P_T/\Delta t = (+126$ torr/hr $- 63$ torr/hr $- 63$ torr/hr$) = 0$ torr/hr

Rate Laws

14.21   *Analyze/Plan.* Follow the logic in Sample Exercises 14.4 and 14.5. *Solve.*

        (a)   If [A] is doubled, there will be no change in the rate or the rate constant. The overall rate is unchanged because [A] does not appear in the rate law; the rate constant changes only with a change in temperature.

        (b)   The reaction is zero order in A, second order in B and second order overall.

(c)   Units of $k = \dfrac{M/s}{M^2} = M^{-1} s^{-1}$

14.22   (a)   If [A] doubles, the rate will increase by a factor of four; the rate constant, k, is unchanged. Rate is proportional to $[A]^2$, so when the value of [A] doubles, rate changes by 2^2 or 4. The rate constant, k, is the proportionality constant that does not change (unless the temperature changes).

        (b)   The reaction is second order in A, first order in B, and third order overall.

        (c)   Units of $k = \dfrac{M/s}{M^3} = M^{-2} s^{-1}$

14.23   *Analyze/Plan.* Follow the logic in Sample Exercise 14.4.   *Solve.*

        (a)   rate = $k[N_2O_5]$ = $4.82 \times 10^{-3} s^{-1} [N_2O_5]$

        (b)   rate = $4.82 \times 10^{-3} s^{-1} (0.0240 \, M) = 1.16 \times 10^{-4} \, M/s$

        (c)   rate = $4.82 \times 10^{-3} s^{-1} (0.0480 \, M) = 2.31 \times 10^{-4} \, M/s$

            When the concentration of N_2O_5 doubles, the rate of the reaction doubles.

14.24   (a)   rate = $k[H_2][NO]^2$

        (b)   rate = $(6.0 \times 10^4 \, M^{-2} s^{-1})(0.025 \, M)^2 (0.015 \, M) = 0.56 \, M/s$

        (c)   rate = $(6.0 \times 10^4 \, M^{-2} s^{-1}) (0.10 \, M)^2 (0.010 \, M) = 6.0 \, M/s$

14.25   *Analyze/Plan.* Write the rate law and rearrange to solve for k. Use the given data to calculate k, including units.   *Solve.*

        (a, b)   rate = $k[CH_3Br][OH^-]$; $k = \dfrac{\text{rate}}{[CH_3Br][OH^-]}$

            at 298 K, $k = \dfrac{0.0432 \, M/s}{(5.0 \times 10^{-3} \, M)(0.050 \, M)} = 1.7 \times 10^2 \, M^{-1}s^{-1}$

        (c)   Since the rate law is first order in $[OH^-]$, if $[OH^-]$ is tripled, the rate triples.

14.26   (a, b)   rate = $k[C_2H_5Br][OH^-]$; $k = \dfrac{\text{rate}}{[C_2H_5Br][OH^-]}$

            at 298 K, $k = \dfrac{1.7 \times 10^{-7} \, M/s}{[0.0477 \, M][0.100 \, M]} = 3.6 \times 10^{-5} \, M^{-1}s^{-1}$

        (c)   Adding an equal volume of ethyl alcohol reduces both $[C_2H_5Br]$ and $[OH^-]$ by a factor of two. new rate = (1/2)(1/2) = 1/4 of old rate

14.27   *Analyze/Plan.* Follow the logic in Sample Exercise 14.6.   *Solve.*

        (a)   From the data given, when $[OCl^-]$ doubles, rate doubles. When $[I^-]$ doubles, rate doubles. The reaction is first order in both $[OCl^-]$ and $[I^-]$. rate = $[OCl^-][I^-]$

        (b)   Using the first set of data:

            $k = \dfrac{\text{rate}}{[OCl^-][I^-]} = \dfrac{1.36 \times 10^{-4} \, M/s}{(1.5 \times 10^{-3} \, M)(1.5 \times 10^{-3} \, M)} = 60.444 = 60 \, M^{-1} s^{-1}$

(c) rate $= \dfrac{60.444}{M \bullet s} (2.0 \times 10^{-3}\ M)(5.0 \times 10^{-4}\ M) = 6.0444 \times 10^{-5} = 6.0 \times 10^{-5}\ M/s$

14.28   (a)   From the data given, when [ClO₂] increases by a factor of 3 (experiment 2 to experiment 1), the rate increases by a factor of 9. When [OH⁻] increases by a factor of 3 (experiment 2 to experiment 3), the rate increases by a factor of 3. The reaction is second order in [ClO₂] and first order in [OH⁻]. rate = k[ClO₂]²[OH⁻].

        (b)   Using data from Expt 2:

$$k = \frac{\text{rate}}{[ClO_2]^2 [OH^-]} = \frac{0.00276\ M/s}{(0.020\ M)^2 (0.030\ M)} = 2.3 \times 10^2\ M^{-2} s^{-1}$$

        (c)   rate $= 2.3 \times 10^2\ M^{-2}\ s^{-1} (0.010\ M)^2 (0.025\ M) = 5.75 \times 10^{-4} = 5.8 \times 10^{-4}\ M/s$

14.29   *Analyze/Plan.* Follow the logic in Sample Exercise 14.6 to deduce the rate law. Rearrange the rate law to solve for k and deduce units. Calculate a k value for each set of concentrations and then average the three values.   *Solve.*

        (a)   Doubling [NH₃] while holding [BF₃] constant doubles the rate (experiments 1 and 2). Doubling [BF₃] while holding [NH₃] constant doubles the rate (experiments 4 and 5).

                Thus, the reaction is first order in both BF₃ and NH₃; rate = k[BF₃][NH₃].

        (b)   The reaction is second order overall.

        (c)   From experiment 1: k $= \dfrac{0.2130\ M/s}{(0.250\ M)(0.250\ M)} = 3.41\ M^{-1}\ s^{-1}$

                (Any of the five sets of initial concentrations and rates could be used to calculate the rate constant k. The average of these 5 values is $k_{avg} = 3.408 = 3.41\ M^{-1}s^{-1}$)

        (d)   rate $= 3.41\ M^{-1}s^{-1}(0.100\ M)(0.500\ M) = 0.1704 = 0.170\ M/s$

14.30   *Analyze/Plan.* Follow the logic in Sample Exercise 14.6 to deduce the rate law. Rearrange the rate law to solve for k and deduce units. Calculate a k value for each set of concentrations and then average the three values.   *Solve.*

        (a)   Doubling [NO] while holding [O₂] constant increases the rate by a factor of 4 (experiments 1 and 3). Reducing [O₂] by a factor of 2 while holding [NO] constant reduces the rate by a factor of 2 (experiments 2 and 3). The rate is second order in [NO] and first order in [O₂]. rate = k[NO]²[O₂]

    (b, c)   From experiment 1: $k_1 = \dfrac{1.41 \times 10^{-2}\ M/s}{(0.0126\ M)^2 (0.0125\ M)} = 7105 = 7.11 \times 10^3\ M^{-2}s^{-1}$

                $k_2 = 0.113/(0.0252)^2(0.0250) = 7118 = 7.12 \times 10^3\ M^{-2}\ s^{-1}$

                $k_3 = 5.64 \times 10^{-2}/(0.0252)^2(0.125) = 7105 = 7.11 \times 10^3\ M^{-2}\ s^{-1}$

                $k_{avg} = (7105 + 7118 + 7105)/3 = 7109 = 7.11 \times 10^3\ M^{-2}\ s^{-1}$

        (d)   rate $= 7.109 \times 10^3\ M^{-2}s^{-1} (0.100\ M)^2(0.0200\ M) = 1.422 = 1.42\ M/s$

(e)   The data are given in terms of the disappearance of NO. Use Equation 14.4 to relate the disappearance of NO to the disappearance of O_2.

$-\Delta[NO]/2\Delta t = -[O_2]/\Delta t$

For the concentrations given in part (d), $\Delta[NO]/\Delta t = 1.42\ M/s$.

$\Delta[O_2]/\Delta t = \Delta[NO]/2\Delta t = 1.42\ M/s/2 = 0.711\ M/s$

14.31   *Analyze/Plan.* Follow the logic in Sample Exercise 4.6 to deduce the rate law. Rearrange the rate law to solve for k and deduce units. Calculate a k value for each set of concentrations and then average the three values.   *Solve.*

(a)   Increasing [NO] by a factor of 2.5 while holding $[Br_2]$ constant (experiments 1 and 2) increases the rate by a factor 6.25 or $(2.5)^2$. Increasing $[Br_2]$ by a factor of 2.5 while holding [NO] constant increases the rate by a factor of 2.5. The rate law for the appearance of NOBr is: rate $= \Delta[NOBr]/\Delta t = k[NO]^2[Br_2]$.

(b)   From experiment 1: $k_1 = \dfrac{24\ M/s}{(0.10\ M)^2\,(0.20\ M)} = 1.20 \times 10^4 = 1.2 \times 10^4\ M^{-2}\,s^{-1}$

$k_2 = 150/(0.25)^2(0.20) = 1.20 \times 10^4 = 1.2 \times 10^4\ M^{-2}\,s^{-1}$

$k_3 = 60/(0.10)^2(0.50) = 1.20 \times 10^4 = 1.2 \times 10^4\ M^{-2}\,s^{-1}$

$k_4 = 735/(0.35)^2(0.50) = 1.2 \times 10^4 = 1.2 \times 10^4\ M^{-2}\,s^{-1}$

$k_{avg} = (1.2 \times 10^4 + 1.2 \times 10^4 + 1.2 \times 10^4 + 1.2 \times 10^4)/4 = 1.2 \times 10^4\ M^{-2}\,s^{-1}$

(c)   Use the reaction stoichiometry and Equation 14.4 to relate the designated rates. $\Delta[NOBr]/2\Delta t = -\Delta[Br_2]/\Delta t$; the rate of disappearance of Br_2 is half the rate of appearance of NOBr.

(d)   Note that the data are given in terms of appearance of NOBr.

$$\dfrac{-\Delta[Br_2]}{\Delta t} = \dfrac{k[NO]^2\,[Br_2]}{2} = \dfrac{1.2 \times 10^4}{2\ M^2\ s} \times (0.075\ M)^2 \times (0.25\ M) = 8.4\ M/s$$

14.32   (a)   Increasing $[S_2O_8{}^{2-}]$ by a factor of 1.5 while holding $[I^-]$ constant increases the rate by a factor of 1.5 (Experiments 1 and 2). Doubling $[S_2O_8{}^{2-}]$ and increasing $[I^-]$ by a factor of 1.5 triples the rate ($2 \times 1.5 = 3$, experiments 1 and 3). Thus the reaction is first order in both $[S_2O_8{}^{2-}]$ and $[I^-]$; rate $= k\,[S_2O_8{}^{2-}]\,[I^-]$.

(b) $k = rate/[S_2O_8{}^{2-}]\,[I^-]$

$k_1 = 2.6 \times 10^{-6}\ M/s/(0.018\ M)(0.036\ M) = 4.01 \times 10^{-3} = 4.0 \times 10^{-3}\ M^{-1}s^{-1}$

$k_2 = 3.9 \times 10^{-6}/(0.027)(0.036) = 4.01 \times 10^{-3} = 4.01 \times 10^{-3} = 4.0 \times 10^{-3}\ M^{-1}s^{-1}$

$k_3 = 7.8 \times 10^{-6}/(0.036)(0.054) = 4.01 \times 10^{-3} = 4.01 \times 10^{-3} = 4.0 \times 10^{-3}\ M^{-1}s^{-1}$

$k_4 = 1.4 \times 10^{-5}/(0.050)(0.072) = 3.89 \times 10^{-3} = 3.9 \times 10^{-3}\ M^{-1}s^{-1}$

$k_{avg} = 3.98 \times 10^{-3} = 4.0 \times 10^{-3}\ M^{-1}s^{-1}$

(c) $-\Delta[S_2O_8{}^{2-}]/\Delta t = -\Delta[I^-]/3\Delta t$; the rate of disappearance of $S_2O_8{}^{2-}$ is one-third the rate of disappearance of I^-.

(d)   Note that the data are given in terms of disappearance of $S_2O_8^{2-}$.

$$\frac{-\Delta[I^-]}{\Delta t} = \frac{-3\Delta[S_2O_8^{2-}]}{\Delta t} = 3(3.98 \times 10^{-3}\ M^{-1}s^{-1})(0.015\ M)(0.040\ M) = 7.2 \times 10^{-6}\ M/s$$

Change of Concentration with Time

14.33   (a) $[A]_0$ is the molar concentration of reactant A at time zero, the initial concentration of A. $[A]_t$ is the molar concentration of reactant A at time t. $t_{1/2}$ is the time required to reduce $[A]_0$ by a factor of 2, the time when $[A]_t = [A]_0/2$. k is the rate constant for a particular reaction. k is independent of reactant concentration but varies with reaction temperature.

        (b)   A graph of ln[A] vs time yields a straight line for a first-order reaction.

14.34   (a)   A graph of 1/[A] vs time yields a straight line for a second-order reaction.

        (b)   The half-life of a first-order reaction is independent of $[A]_0$, $t_{1/2} = 0.693/k$. Whereas, the half-life of a second-order reaction does depend on $[A]_0$, $t_{1/2} = 1/k[A]_0$.

14.35   *Analyze/Plan.* The half-life of a first-order reaction depends only on the rate constant, $t_{1/2} = 0.693/k$. Use this relationship to calculate k for a given $t_{1/2}$, and, at a different temperature, $t_{1/2}$ given k. *Solve.*

        (a) $t_{1/2} = 2.3 \times 10^5$ s; $t_{1/2} = 0.693/k$, $k = 0.693/t_{1/2}$

             $k = 0.693/2.3 \times 10^5$ s $= 3.0 \times 10^{-6}$ s^{-1}

        (b) $k = 2.2 \times 10^{-5}$ s^{-1}. $t_{1/2} = 0.693/2.2 \times 10^{-5}$ s^{-1} $= 3.15 \times 10^4 = 3.2 \times 10^4$ s

14.36   *Analyze.* Given rate constants for the decay of two radioisotopes, determine half-lives, decay rates, and amount remaining after three half-lives. *Plan.* Determine reaction order. Based on reaction-order, select the appropriate relationships for (a) rate constant and half-life and (c) rate-constant, time and concentration. In this example, mass is a measure of concentration.

        Solve. Decay of radioiosotopes is a first-order process, since only one species is involved and the decay is not initiated by collision.

        (a)   For a first-order process, $t_{1/2} = 0.693/k$.

             ^{241}Am: $t_{1/2} = 0.693/1.6 \times 10^{-3}$ yr^{-1} $= 433.1 = 4.3 \times 10^2$ yr

             ^{125}I: $t_{1/2} = 0.693/0.011$ day^{-1} $= 63.00 = 63$ days

        (b)   For a given sample size, half of the ^{241}Am sample decays in 433 years, whereas half of the ^{125}I sample decays in 63 days. ^{125}I decays at a much faster rate.

        (c)   For a first order process, $\ln[A]_t - \ln[A]_0 = -kt$. $\ln[A]_t = -kt + \ln[A]_0$.

             $[A]_0 = 1.0$ mg; $t = 3\ t_{1/2}$.

^{241}Am: $t = 3\ t_{1/2} = 3(433.1\ \text{yr}) = 1.299 \times 10^3 = 1.3 \times 10^3\ \text{yr}$

$\ln[\text{Am}]_t = -1.6 \times 10^{-3}\ \text{yr}\ (1.299 \times 10^3 - \ln(1.0) = -2.079 - 0 = -2.08$

$[\text{Am}]_t = 0.125 = 0.13\ \text{mg}$

or, mass ^{241}Am remaining $= 1.0\ \text{mg}/2^3 = 0.125 = 0.13\ \text{mg}$

^{125}I: For the same size starting sample and number of elapsed half-lives, the same mass, 0.13 mg ^{125}I, will remain. (The difference is that the elapsed time of 3 half-lives for ^{125}I is 3(63) = 189 days = 0.52 yr, vs. 433 yr for ^{241}Am.)

14.37   *Analyze/Plan.* Follow the logic in Sample Exercise 14.7. In this reaction, pressure is a measure of concentration. In (a) we are given k, $[A]_0$, t and asked to find $[A]_t$, using Equation [14.13], the integrated form of the first-order rate law. In (b), $[A_t] = 0.1[A_0]$, find t. *Solve.*

(a) $\ln P_t = -kt + \ln P_0;\ P_0 = 375\ \text{torr};\ t = 65\ \text{s}$

        $\ln P_{65} = -4.5 \times 10^{-2}\ \text{s}^{-1}(65) + \ln(375) = -2.925 + 5.927 = 3.002$

        $P_{65} = 20.12 = 20\ \text{torr}$

(b) $P_t = 0.10\ P_0;\ \ln(P_t/P_0) = -kt$

        $\ln(0.10\ P_0/P_0) = -kt,\ \ln(0.10) = -kt;\ -\ln(0.10)/k = t$

        $t = -(-2.303)/4.5 \times 10^{-2}\ \text{s}^{-1} = 51.2 = 51\ \text{s}$

Check. From part (a), the pressure at 65 s is 20 torr, $P_t \sim 0.05\ P_0$. In part (b) we calculate the time where $P_t = 0.10\ P_0$ to be 51 s. This time should be smaller than 65 s, and it is. Data and results in the two parts are consistent.

14.38   (a)   Using Equation [14.13] for a first order reaction: $\ln[A]_t = -kt + \ln[A]_0$

        $2.5\ \text{min} = 150\ \text{s};\ [N_2O_5]_0 = (0.0250\ \text{mol}/2.0\ \text{L}) = 0.0125 = 0.013\ M$

        $\ln[N_2O_5]_{150} = -(6.82 \times 10^{-3}\ \text{s}^{-1})(150\ \text{s}) + \ln(0.0125)$

        $\ln[N_2O_5]_{150} = -1.0230 + (-4.3820) = -5.4050 = -5.41$

        $[N_2O_5]_{150} = 4.494 \times 10^{-3} = 4.5 \times 10^{-3}\ M;\ \text{mol}\ N_2O_5 = 4.494 \times 10^{-3}\ M \times 2.0\ \text{L}$

                                                    $= 9.0 \times 10^{-3}\ \text{mol}$

(b) $[N_2O_5]_t = 0.010\ \text{mol}/2.0\ \text{L} = 0.0050\ M;\ [N_2O_5]_0 = 0.0125\ M$

        $\ln(0.0050) = -(6.82 \times 10^{-3}\ \text{s}^{-1})\ (t) + \ln(0.0125)$

        $t = \dfrac{-[\ln(0.0050) - \ln(0.0125)]}{(6.82 \times 10^{-3}\ \text{s}^{-1})} = 134.35 = 1.3 \times 10^2\ \text{s} \times \dfrac{1\ \text{min}}{60\ \text{s}} = 2.24 = 2.2\ \text{min}$

(c) $t_{1/2} = 0.693/k = 0.693/6.82 \times 10^{-3}\ \text{s}^{-1} = 101.6 = 102\ \text{s}\ \text{or}\ 1.69\ \text{min}$

14.39   *Analyze/Plan.* Given reaction order, various values for t and P_t, find the rate constant for the reaction at this temperature. For a first-order reaction, a graph of lnP vs t is linear with as slope of –k.   *Solve.*

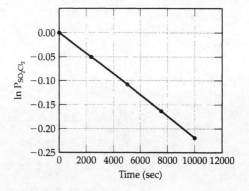

t(s)	$P_{SO_2Cl_2}$	ln $P_{SO_2Cl_2}$
0	1.000	0
2500	0.947	–0.0545
5000	0.895	–0.111
7500	0.848	–0.165
10000	0.803	–0.219

Graph ln $P_{SO_2Cl_2}$ vs. time. (Pressure is a satisfactory unit for a gas, since the concentration in moles/liter is proportional to P.) The graph is linear with slope -2.19×10^{-5} s^{-1} as shown on the figure. The rate constant k = –slope = 2.19×10^{-5} s^{-1}.

14.40

t(s)	P_{CH_2NC}	ln P_{CH_3NC}
0	502	6.219
2000	335	5.814
5000	180	5.193
8000	95.5	4.559
12000	41.7	3.731
15000	22.4	3.109

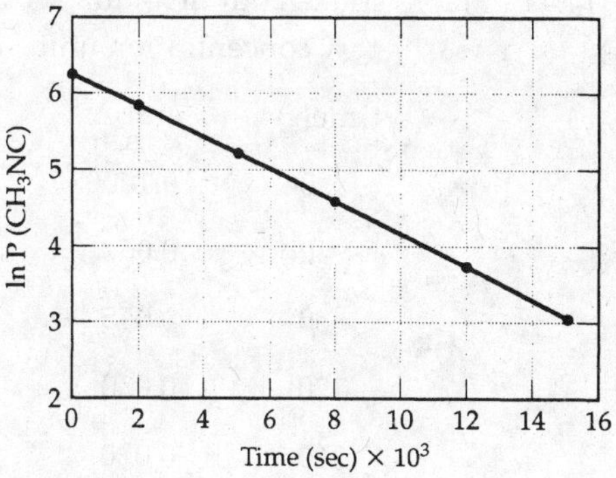

A graph of ln P vs t is linear with a slope of -2.08×10^{-4} s^{-1}. The rate constant k = –slope = 2.08×10^{-4} s^{-1}. Half-life = $t_{1/2}$ = 0.693/k = 3.33×10^{3} s.

14.41   *Analyze/Plan.* Given: mol A, t. Change mol to *M* at various times. Make both first- and second-order plots to see which is linear.   *Solve.*

(a)

time(min)	mol A	[A] (*M*)	ln[A]	1/mol A
0	0.065	0.65	–0.43	1.5
10	0.051	0.51	–0.67	2.0
20	0.042	0.42	–0.87	2.4
30	0.036	0.36	–1.02	2.8
40	0.031	0.31	–1.17	3.2

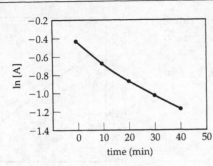
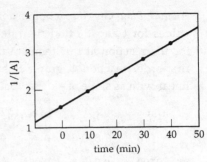

The plot of 1/[A] vs time is linear, so the reaction is second-order in [A].

(b)   For a second-order reaction, a plot of 1/[A] vs. t is linear with slope k.

k = slope = $(3.2 - 2.0)\ M^{-1}\ /\ 30\ min = 0.040\ M^{-1}\ min^{-1}$

(The best fit to the line yields slope = $0.042\ M^{-1}\ min^{-1}$.)

(c) $t_{1/2} = 1/k[A]_0 = 1/(0.040\ M^{-1}\ min^{-1})(0.65\ M) = 38.46 = 38\ min$

(Using the "best-fit" slope, $t_{1/2} = 37\ min$.)

14.42   (a)   Make both first- and second-order plots to see which is linear. Moles is a satisfactory concentration unit, since volume is constant.

time(s)	mol A	ln (mol A)	1/mol A
0	0.1000	–2.303	10.00
40	0.067	–2.70	14.9
80	0.045	–3.10	22.2
120	0.030	–3.51	33.3
160	0.020	–3.91	50.0

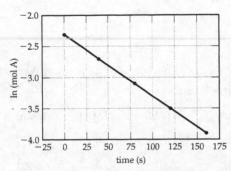
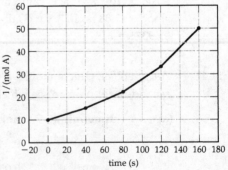

The plot of ln (mol A) vs time is linear, so the reaction is first-order in A.

(b)   k = –slope = – [–3.91 – (–2.70)]/120 = 0.010083 = $0.0101\ s^{-1}$
(The best fit to this line yields the same value for the slope, 0.01006 = $0.0101\ s^{-1}$)

(c) $t_{1/2} = 0.693/k = 0.693/0.010083\ s^{-1} = 68.7\ s$

14.43   *Analyze/Plan.* Follow the logic in Solution 14.41. Make both first and second order plots to see which is linear.   *Solve.*

(a)

time(s)	[NO$_2$](M)	ln[NO$_2$]	1/[NO$_2$]
0.0	0.100	−2.303	10.0
5.0	0.017	−4.08	59
10.0	0.0090	−4.71	110
15.0	0.0062	−5.08	160
20.0	0.0047	−5.36	210

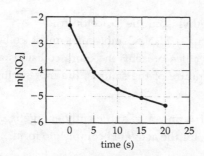   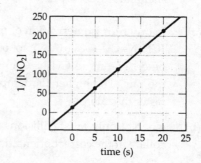

The plot of 1/[NO$_2$] vs time is linear, so the reaction is second order in NO$_2$.

(b)   The slope of the line is (210 − 59) M^{-1} / 15.0 s = 10.07 = 10 $M^{-1}s^{-1}$ = k. (The slope of the best-fit line is 10.02 = 10 $M^{-1}s^{-1}$.)

14.44   (a)   Make both first- and second-order plots to see which is linear.

time(min)	[C$_{12}$H$_{22}$O$_{11}$](M)	ln[C$_{12}$H$_{22}$O$_{11}$]	1/[C$_{12}$H$_{22}$O$_{11}$]
0	0.316	−1.152	3.16
39	0.274	−1.295	3.65
80	0.238	−1.435	4.20
140	0.190	−1.661	5.26
210	0.146	−1.924	6.85

The plot of ln [C$_{12}$H$_{22}$O$_{11}$] is linear, so the reaction is first order in C$_{12}$H$_{22}$O$_{11}$.

(b) $k = -\text{slope} = -[-1.924 - (-1.295)] / 171 \text{ min} = 3.68 \times 10^{-3} \text{ min}^{-1}$

(The slope of the best-fit line is $-3.67 \times 10^{-3} \text{ min}^{-1}$.)

Temperature and Rate

14.45   (a)   The energy of the collision and the orientation of the molecules when they collide determine whether a reaction will occur.

(b)   According to the kinetic-molecular theory (Chapter 10), the higher the temperature, the greater the speed and kinetic energy of the molecules. Therefore, at a higher temperature, there are more total collisions and each collision is more energetic.

14.46   (a)   In order for isomerization of methyl isonitrile to acetonitrile ($CH_3N\equiv C \rightarrow CH_3C\equiv N$) to occur, $CH_3N\equiv C$ molecules must collide with each other. The more $CH_3N\equiv C$ molecules present, the more collisions and the faster the rate. The higher the temperature of the sample, the more collisions and the faster the rate.

(b)   No. Not only must collisions of A and B be sufficiently energetic, A and B must collide in the correct orientation for the activated complex to form.

(c)   The kinetic-molecular theory tells us that at some temperature T, there will be a distribution of molecular speeds and kinetic energies, and that the average kinetic energy of the sample is proportional to temperature. That is, as temperature of the sample increases, the average speed and kinetic energy of the molecules increases. At higher temperatures, there will be more molecular collisions (owing to greater speeds) and more energetic collisions (owing to greater kinetic energies). Overall there will be more collisions that have sufficient energy to form an activated complex, and the reaction rate will be greater.

14.47   *Analyze/Plan.* Given the temperature and energy, use Equation [14.18] to calculate the fraction of Ar atoms that have at least this energy.   *Solve.*

$f = e^{-E_a/RT}$ $E_a = 10.0 \text{ kJ/mol} = 1.00 \times 10^4 \text{ J/mol}$; $T = 400 \text{ K } (127°C)$

$$-E_a/RT = -\frac{1.00 \times 10^4 \text{ J/mol}}{400 \text{ K}} \times \frac{\text{mol} \cdot \text{K}}{8.314 \text{ J}} = -3.0070 = -3.01$$

$f = e^{-3.0070} = 4.9 \times 10^{-2}$

At 400 K, approximately 1 out of 20 molecules has this kinetic energy.

14.48   (a) $f = e^{-E_a/RT}$ $E_a = 160 \text{ kJ/mol} = 1.60 \times 10^5 \text{ J/mol}$, $T = 500 \text{ K}$

$$-E_a/RT = -\frac{1.60 \times 10^5 \text{ J/mol}}{500 \text{ K}} \times \frac{\text{mol} \cdot \text{K}}{8.314 \text{ J}} = -38.489 = -38.5$$

$f = e^{-38.489} = 1.924 \times 10^{-17} = 2 \times 10^{-17}$

(b) $-E_a/RT = -\dfrac{1.60 \times 10^5 \text{ J/mol}}{510 \text{ K}} \times \dfrac{\text{mol} \cdot \text{K}}{8.314 \text{ J}} = -37.735 = -37.7$

$f = e^{-37.735} = 4.093 \times 10^{-17} = 4.09 \times 10^{-17}$

$\dfrac{f \text{ at } 510 \text{ K}}{f \text{ at } 500 \text{ K}} = \dfrac{4.09 \times 10^{-17}}{1.92 \times 10^{-17}} = 2.13$

An increase of 10 K means that 2.13 times more molecules have this energy.

14.49   *Analyze/Plan.* Use the definitions of activation energy ($E_{max} - E_{react}$) and ΔE ($E_{prod} - E_{react}$) to sketch the graph and calculate E_a for the reverse reaction.   *Solve.*

(a)   (b) E_a(reverse) = 73 kJ

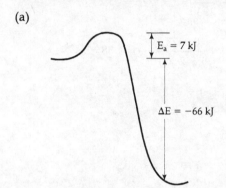

14.50.   *Analyze/Plan.* Use the definitions of activation energy ($E_{max} - E_{react}$) and ΔE ($E_{prod} - E_{react}$) to sketch the graph and calculate E_a for the reverse reaction. *Solve.*

(a)   (b) E_a(reverse) = 18 kJ

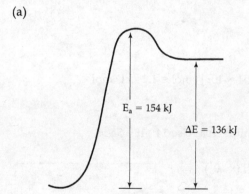

14.51   Assuming all collision factors (A) to be the same, reaction rate depends only on E_a; it is independent of ΔE. Based on the magnitude of E_a, reaction (b) is fastest and reaction (c) is slowest.

14.52 E_a for the reverse reaction is:

(a) 45 – (–25) = 70 kJ   (b) 35 – (–10) = 45 kJ   (c) 55 – 10 = 45 kJ

Based on the magnitude of E_a, the reverse of reactions (b) and (c) occur at the same rate, which is faster than the reverse of reaction (a).

14.53   *Analyze/Plan.* Given k_1, at T_1, calculate k_2 at T_2. Change T to Kelvins, then use the Equation [14.21] to calculate k_2. *Solve.*

$T_1 = 20°C + 273 = 293$ K; $T_2 = 60°C + 273 = 333$ K; $k_1 = 2.75 \times 10^{-2} s^{-1}$

(a) $\ln\left(\dfrac{k_1}{k_2}\right) = \dfrac{E_a}{R}\left(\dfrac{1}{333} - \dfrac{1}{293}\right) = \dfrac{75.5 \times 10^3 \text{ J/mol}}{8.314 \text{ J/mol}}(-4.100 \times 10^{-4})$

$\ln(k_1/k_2) = -3.7229 = -3.7$; $k_1/k_2 = 0.0242 = 0.02$; $k_2 = \dfrac{0.0275 \text{ s}^{-1}}{0.0242} = 1.14 = 1 \text{ s}^{-1}$

(b) $\ln\left(\dfrac{k_1}{k_2}\right) = \dfrac{105 \times 10^3 \text{ J/mol}}{8.314 \text{ J/mol}}\left(\dfrac{1}{333} - \dfrac{1}{293}\right) = -5.1776 = -5.2$

$k_1/k_2 = 5.642 \times 10^{-3} = 6 \times 10^{-3}$; $k_2 = \dfrac{0.0275 \text{ s}^{-1}}{5.642 \times 10^{-3}} = 4.88 = 5 \text{ s}^{-1}$

14.54 $T_1 = 737°C + 273 = 1010$ K, $k_1 = 0.0796 \ M^{-1}s^{-1}$;

$T_2 = 947°C + 273 = 1220$ K, $k_2 = 0.0815 \ M^{-1}s^{-1}$

$\ln\left(\dfrac{k_1}{k_2}\right) = \dfrac{E_a}{R}\left(\dfrac{1}{T_2} - \dfrac{1}{T_1}\right)$

$\ln\left(\dfrac{0.0796}{0.0815}\right) = \dfrac{E_a}{8.314 \text{ J/mol}}\left(\dfrac{1}{1220} - \dfrac{1}{1010}\right)$

$-0.023589 = \dfrac{E_a \ (-1.704 \times 10^{-4})}{8.314 \text{ J/mol}}$

$E_a = \dfrac{8.314 \ (-0.023589) \text{ J/mol}}{(-1.704 \times 10^{-4})} = 1.151 \times 10^3 \text{ J/mol} = 1.15 \text{ kJ/mol}$

14.55   *Analyze/Plan.* Follow the logic in Sample Exercise 14.11. *Solve.*

k	ln k	T(K)	1/T($\times 10^3$)
0.0521	−2.955	288	3.47
0.101	−2.293	298	3.36
0.184	−1.693	308	3.25
0.332	−1.103	318	3.14

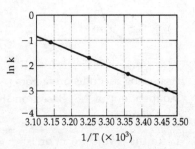

The slope, -5.71×10^3, equals $-E_a/R$. Thus,
$E_a = 5.71 \times 10^3 \times 8.314 \text{ J/mol} = 47.5 \text{ kJ/mol}$.

14.56

k	ln k	T(K)	1/T($\times 10^3$)
0.028	–3.58	600	1.67
0.22	–1.51	650	1.54
1.3	0.26	700	1.43
6.0	1.79	750	1.33
23	3.14	800	1.25

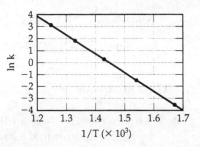

Using the relationship $\ln k = \ln A - E_a/RT$, the slope, $-15.94 \times 10^3 = -16 \times 10^3$, is $-E_a/R$. $E_a = 15.94 \times 10^3 \times 8.314$ J/mol $= 1.3 \times 10^2$ kJ/mol. To calculate A, we will use the rate data at 700 K. From the equation given above, $0.262 = \ln A - 15.94 \times 10^3/700$; $\ln A = 0.262 + 22.771$. $A = 1.0 \times 10^{10}$.

14.57   *Analyze/Plan.* Given E_a, find the ratio of rates for a reaction at two temperatures. Assuming initial concentrations are the same at the two temperatures, the ratio of rates will be the ratio of rate constants, k_1/k_2. Use Equation [14.21] to calculate this ratio.   *Solve.*

$T_1 = 50°C + 273 = 323$ K; $T_2 = 0°C + 273 = 273$ K

$$\ln\left(\frac{k_1}{k_2}\right) = \frac{E_a}{R}\left[\frac{1}{T_2} - \frac{1}{T_1}\right] = \frac{65.7 \text{ kJ/mol}}{8.314 \text{ J/mol}} \times \frac{1000 \text{ J}}{1 \text{ kJ}}\left[\frac{1}{273} - \frac{1}{323}\right]$$

$\ln(k_1/k_2) = 7.902 \times 10^3 (5.670 \times 10^{-4}) = 4.481 = 4.5$; $k_1/k_2 = 88.3 = 9 \times 10^1$

The reaction will occur 90 times faster at 50°C, assuming equal initial concentrations.

14.58   (a) $T_1 = 77°F$; $°C = 5/9$ ($°F - 32$) $= 5/9$ ($77 - 32$) $= 25°C = 298$ K

$T_2 = 59°F$; $°C = 5/9$ ($59-32$) $= 15°C = 288$ K; $k_1/k_2 = 6$

$$\ln\left(\frac{k_1}{k_2}\right) = \frac{E_a}{R}\left[\frac{1}{T_2} - \frac{1}{T_1}\right]; \quad \ln(6) = \frac{E_a}{8.314 \text{ J/mol}}\left[\frac{1}{288} - \frac{1}{298}\right]$$

$$E_a = \frac{\ln(6)(8.314 \text{ J/mol})}{1.165 \times 10^{-4}} = 1.28 \times 10^5 \text{ J} = 1.3 \times 10^2 \text{ kJ/mol}$$

$T_1 = 77°F = 25°C = 298$ K; $T_2 = 41°F = 5°C = 278$ K, $k_1/k_2 = 40$

$$\ln(40) = \frac{E_a}{8.314 \text{ J/mol}}\left[\frac{1}{278} - \frac{1}{296}\right]; \quad E_a = \frac{\ln(40)(8.314 \text{ J/mol})}{2.414 \times 10^{-4}}$$

$E_a = 1.27 \times 10^5 \text{ J} = 1.3 \times 10^2 \text{ kJ/mol}$

The values are amazingly consistent, considering the precision of the data.

(b)   For a first order reaction, $t_{1/2} = 0.693/k$, $k = 0.693/t_{1/2}$

k_1 at 298 K $= 0.693/2.7$ yr $= 0.257 = 0.26$ yr^{-1}

$T_1 = 298$ K, $T_2 = 273 - 15°C = 258$ K

$$\ln\left(\frac{0.257}{k_2}\right) = \frac{1.27 \times 10^5 \text{ J}}{8.314 \text{ J/mol}}\left[\frac{1}{258} - \frac{1}{298}\right] = 7.9497$$

$$0.257/k_2 = e^{7.9497} = 2.835 \times 10^3; k_2 = 0.257/2.835 \times 10^3 = 9.066 \times 10^{-5} = 9.1 \times 10^{-5} \text{ yr}^{-1}$$

$$t_{1/2} = 0.693/k = 0.693/9.066 \times 10^{-5} = 7.64 \times 10^3 \text{ yr} = 7.6 \times 10^3 \text{ yr}$$

Reaction Mechanisms

14.59   (a)   An *elementary reaction* is a process that occurs in a single event; the order is given by the coefficients in the balanced equation for the reaction.

        (b)   A *unimolecular* elementary reaction involves only one reactant molecule; the activated complex is derived from a single molecule. A *bimolecular* elementary reaction involves two reactant molecules in the activated complex and the overall process.

        (c)   A *reaction mechanism* is a series of elementary reactions that describe how an overall reaction occurs and explain the experimentally determined rate law.

14.60   (a)   The *molecularity* of a process indicates the number of molecules that participate as reactants in the process. A unimolecular process has one reactant molecule, a bimolecular process has two reactant molecules and a termolecular process has three reactant molecules.

        (b)   Termolecular processes are rare because it is highly unlikely that three molecules will simultaneously collide with the correct energy and orientation to form an activated complex.

        (c)   An *intermediate* is a substance that is produced and then consumed during a chemical reaction. It does not appear in the balanced equation for the overall reaction.

14.61   *Analyze/Plan.* Elementary reactions occur as a single step, so the molecularity is determined by the number of reactant molecules; the rate law reflects reactant stoichiometry. *Solve.*

        (a)   unimolecular, rate = $k[Cl_2]$

        (b)   bimolecular, rate = $k[OCl^-][H_2O]$

        (c)   bimolecular, rate = $k[NO][Cl_2]$

14.62   (a)   bimolecular, rate = $k[NO]^2$

        (b)   unimolecular, rate = $k[C_3H_6]$

        (c)   unimolecular, rate = $k[SO_3]$

14.63   *Analyze/Plan.* Use the definitions of the terms 'intermediate' and 'exothermic', along with the characteristics of reaction profiles, to answer the questions. *Solve.*

This is a three-step mechanism, A $\rightarrow$ B, B $\rightarrow$ C, and C $\rightarrow$ D.

(a)   There are 2 intermediates, B and C.

(b)   There are 3 energy maxima in the reaction profile, so there are 3 transition states.

(c)   Step C → D has the lowest activation energy, so it is fastest.

(d)   The energy of D is slightly greater than the energy of A, so the overall reaction is endothermic.

14.64   (a)   Two elementary reactions; two energy maxima

        (b)   One intermediate; one energy minimum between reactants and products

        (c)   The second step is rate-limiting; second energy maximum and E_a is larger.

        (d)   Overall reaction is exothermic; energy of products is lower than energy of reactants.

14.65   (a) $H_2(g) + ICl(g) \rightarrow HI(g) + HCl(g)$

                $\underline{HI(g) + ICl(g) \rightarrow I_2(g) + HCl(g)}$

                $H_2(g) + 2ICl(g) \rightarrow I_2(g) + 2HCl(g)$

        (b)   Intermediates are produced and consumed during reaction. HI is the intermediate.

        (c)   Follow the logic in Sample Exercise 14.13.

                First step: rate = $k_1[H_2][ICl]$

                Second step: rate = $k_2[HI][ICl]$

        (d)   The slow step determines the rate law for the overall reaction. If the first step is slow, the observed rate law is: rate = $k[H_2][HCl]$.

14.66   *Analyze/Plan.* Follow the logic in Sample Exercise 14.14.   *Solve.*

        (a)   First step: rate = $k_1[H_2O_2][I^-]$; second step: rate = $k_2[IO^-][H_2O_2]$

        (b) $2H_2O_2(aq) \rightarrow 2H_2O(l) + O_2(g)$

        (c) $IO^-(aq)$ is the intermediate.

        (d)   rate = $k[H_2O_2][I^-]$

14.67   *Analyze/Plan.* Given a proposed mechanism and an observed rate law, determine which step is rate determining.   *Solve.*

        (a)   If the first step is slow, the observed rate law is the rate law for this step.
                rate = $k[NO][Cl_2]$

        (b)   Since the observed rate law is second-order in [NO], the second step must be slow relative to the first step; the second step is rate determining.

14.68   (a)   i. $HBr + O_2 \rightarrow HOOBr$

                ii. $HOOBr + HBr \rightarrow 2HOBr$

                iii. $2HOBr + 2HBr \rightarrow 2H_2O + 2Hr_2$
                        _____
                        $4HBr + O_2 \rightarrow 2H_2O + 2Br_2$

        (b)   The observed rate law is: rate = $k[HBr][O_2]$, the rate law for the first elementary step. The first step must be rate-determining.

        (c)   HOOBr and HOBr are both intermediates; HOOBr is produced in i and consumed in ii and HOBr is produced in ii and consumed in iii.

        (d)   Since the first step is rate-determining, it is possible that neither of the intermediates accumulates enough to be detected. This does not disprove the mechanism, but indicates that steps ii and iii are very fast, relative to step i.

Catalysis

14.69   (a)   A catalyst increases the rate of reaction by decreasing the activation energy, E_a, or increasing the frequency factor A. Lowering the activation energy is more common and more dramatic.

        (b)   A homogeneous catalyst is in the same phase as the reactants; a heterogeneous catalyst is in a different phase and is usually a solid.

14.70   (a)   The smaller the particle size of a solid catalyst, the greater the surface area. The greater the surface area, the more active sites and the greater the increase in reaction rate.

        (b)   Adsorption is the binding of reactants onto the surface of the heterogeneous catalyst. It is usually the first step in the catalyzed reaction.

14.71   (a) $2[NO_2(g) + SO_2(g) \rightarrow NO(g) + SO_3(g)]$
                $2NO(g) + O_2(g) \rightarrow 2NO_2(g)$
                _____
                $2SO_2(g) + O_2(g) \rightarrow 2SO_2(g)$

        (b) $NO_2(g)$ is a catalyst because it is consumed and then reproduced in the reaction sequence. ($NO(g)$ is an intermediate because it is produced and then consumed.)

        (c)   Since NO_2 is in the same state as the other reactants, this is homogeneous catalysis.

14.72   (a) $2[NO(g) + N_2O(g) \rightarrow N_2(g) + NO_2(g)]$
                $2NO_2(g) \rightarrow 2NO(g) + O_2(g)$
                _____
                $2N_2O(g) \rightarrow 2N_2(g) + O_2(g)$

        (b)   An intermediate is produced and then consumed during the course of the reaction. A catalyst is consumed and then reproduced. In other words, the catalyst is present when the reaction sequence begins and after the last step is completed. In this reaction, NO is the catalyst and NO_2 is an intermediate.

(c)   No. The proposed mechanism cannot be ruled out, based on the behavior of NO_2. NO_2 functions as an intermediate; it is produced and then consumed during the reaction. That there is no measurable build-up of NO_2 indicates the first step is slow relative to the second; as soon as NO_2 is produced by the slow first step, it is consumed by the faster second step.

14.73   Use of chemically stable supports such as alumina and silica makes it possible to obtain very large surface areas per unit mass of the precious metal catalyst. This is so because the metal can be deposited in a very thin, even monomolecular, layer on the surface of the support.

14.74   (a)   Catalytic converters are heterogeneous catalysts that adsorb gaseous CO and hydrocarbons and speed up their oxidation to $CO_2(g)$ and $H_2O(g)$. They also adsorb nitrogen oxides, NO_x, and speed up their reduction to $N_2(g)$ and $O_2(g)$. If a catalytic converter is working effectively, the exhaust gas should have very small amounts of the undesirable gases CO, $(NO)_x$ and hydrocarbons.

         (b)   The high temperatures could increase the rate of the desired catalytic reactions given in part (a). It could also increase the rate of undesirable reactions such as corrosion, which decrease the lifetime of the catalytic converter.

         (c)   The rate of flow of exhaust gases over the converter will determine the rate of adsorption of CO, $(NO)_x$ and hydrocarbons onto the catalyst and thus the rate of conversion to desired products. Too fast an exhaust flow leads to less than maximum adsorption. A very slow flow leads to back pressure and potential damage to the exhaust system. Clearly the flow rate must be adjusted to balance chemical and mechanical efficiency of the catalytic converter.

14.75   As illustrated in Figure 14.21, the two C–H bonds that exist on each carbon of the ethylene molecule before adsorption are retained in the process in which a D atom is added to each C (assuming we use D_2 rather than H_2). To put two deuteriums on a single carbon, it is necessary that one of the already existing C–H bonds in ethylene be broken while the molecule is adsorbed, so the H atom moves off as an adsorbed atom, and is replaced by a D. This requires a larger activation energy than simply adsorbing C_2H_4 and adding one D atom to each carbon.

14.76   Just as the π electrons in C_2H_4 are attracted to the surface of a hydrogenation catalyst, the nonbonding electron density on S causes compounds of S to be attracted to these same surfaces. Strong interactions could cause the sulfur compounds to be permanently attached to the surface, blocking active sites and reducing adsorption of alkenes for hydrogenation.

14.77   (a)   Living organisms operate efficiently in a very narrow temperature range; heating to increase reaction rate is not an option. Therefore, the role of enzymes as homogeneous catalysts that speed up desirable reactions without heating and undesirable side-effects is crucial for biological systems.

         (b)   *catalase*: $2H_2O_2 \rightarrow 2H_2O + O_2$; *nitrogenase*: $N_2 \rightarrow 2NH_3$ (nitrogen fixation)

14.78   The individual structure of each enzyme molecule leads to a unique coiling and folding pattern. The resulting shape and electronic properties of the active site in each enzyme leads to its substrate specificity.

14.79   *Analyze/Plan.* Let k = the rate constant for the uncatalyzed reaction,
         k_c = the rate constant for the catalyzed reaction

         According to Equation [14.20], $\ln k = -E_a/RT + \ln A$

         Subtracting $\ln k$ from $\ln k_c$,

         $$\ln k_c - \ln k = -\left[\frac{55 \text{ kJ/mol}}{RT} + \ln A\right] - \left[-\frac{95 \text{ kJ/mol}}{RT} + \ln A\right]. \quad \textit{Solve.}$$

(a) $RT = 8.314 \text{ J/K} \cdot \text{mol} \times 298 \text{ K} \times 1 \text{ kJ}/1000 \text{ J} = 2.478 \text{ kJ/mol}$; $\ln A$ is the same for
         both reactions.

         $$\ln(k_c/k) = \frac{95 \text{ kJ/mol} - 55 \text{ kJ/mol}}{2.478 \text{ kJ/mol}}; \quad k_c/k = 1.024 \times 10^7 = 1 \times 10^7$$

         The catalyzed reaction is approximately 10,000,000 (ten million) times faster at
         25°C.

(b) $RT = 8.314 \text{ J/K} \cdot \text{mol} \times 398 \text{ K} \times 1 \text{ kJ}/1000 \text{ J} = 3.309 \text{ kJ/mol}$

         $$\ln(k_c/k) = \frac{40 \text{ kJ/mol}}{3.309 \text{ kJ/mol}}; \quad k_c/k = 1.778 \times 10^5 = 2 \times 10^5$$

         The catalyzed reaction is 200,000 times faster at 125°C.

14.80   Let k and E_a equal the rate constant and activation energy for the uncatalyzed reaction.
         Let k_c and E_{ac} equal the rate constant and activation energy of the catalyzed reaction. A
         is the same for the uncatalyzed and catalyzed reactions. $k_c/k = 1 \times 10^5$, T = 37°C =
         310 K.

         According to Equation [14.20], $\ln k = E_a/RT + \ln A$. Subtracting $\ln k$ from $\ln k_c$

         $$\ln k_c - \ln k = \left[\frac{-E_{ac}}{RT}\right] + \ln A - \left[\frac{-E_a}{RT}\right] - \ln A$$

         $$\ln(k_c/k) = \frac{E_a - E_{ac}}{RT}; \quad E_a - E_{ac} = RT \ln(k_c/k)$$

         $$E_a - E_{ac} = \frac{8.314 \text{ J}}{K \cdot \text{mol}} \times 310 \text{ K} \times \ln(1 \times 10^5) = 2.966 \times 10^4 \text{ J} = 29.66 \text{ kJ} = 3 \times 10^1 \text{ kJ}$$

         The enzyme must lower the activation energy by 30 kJ in order to achieve a
         1×10^5-fold increase in reaction rate.

Additional Exercises

14.81 $\text{rate} = \dfrac{-\Delta[H_2S]}{\Delta t} = \dfrac{\Delta[Cl^-]}{2\Delta t} = k[H_2S][Cl_2]$

         $\dfrac{-\Delta[H_2S]}{\Delta t} = (3.5 \times 10^{-2} \ M^{-1}s^{-1})(2.0 \times 10^{-4} \ M)(0.050 \ M) = 3.50 \times 10^{-7} = 3.5 \times 10^{-7} \ M/s$

         $\dfrac{\Delta[Cl^-]}{\Delta t} = \dfrac{-2\Delta[H_2S]}{\Delta t} = 2(3.50 \times 10^{-7} \ M/s) = 7.0 \times 10^{-7} \ M/s$

14.82   (a) $\text{rate} = \dfrac{-\Delta[NO]}{2\Delta t} = \dfrac{-\Delta[O_2]}{\Delta t} = \dfrac{9.3 \times 10^{-5} \ M/s}{2} = 4.7 \times 10^{-5} \ M/s$

(b,c)   rate = $k[NO]^2[O_2]$; $k = rate/[NO]^2[O_2]$

$$k = \frac{4.7 \times 10^{-5} \ M/s}{(0.040 \ M)^2 \ (0.035 \ M)} = 0.8393 = 0.84 \ M^{-2} \ s^{-1}$$

(d)   Since the reaction is second order in NO, if the [NO] is increased by a factor of 1.8, the rate would increase by a factor of 1.8^2, or (3.24) = 3.2.

14.83   (a)   rate = $k[I^-][OCl^-]/[OH^-]$

        (b)   Since the reaction is first order in [I^-], tripling [I^-] triples the rate.

        (c)   Since the rate is inversely proportional to [OH^-], doubling [OH^-] cuts the rate in half.

14.84   (a)   The rate increases by a factor of nine when [$C_2O_4^{2-}$] triples (compare experiments 1 and 2). The rate doubles when [$HgCl_2$] doubles (compare experiments 2 and 3). The rate law is apparently: rate = $k[HgCl_2][C_2O_4^{2-}]^2$

        (b) $k = \dfrac{rate}{[HgCl_2][C_2O_4^{2-}]^2}$ Using the data for Experiment 1,

$$k = \frac{(3.2 \times 10^{-5} \ M/s)}{[0.164 \ M][0.15 \ M]^2} = 8.672 \times 10^{-3} = 8.7 \times 10^{-3} \ M^{-2}s^{-1}$$

        (c)   rate = $(8.672 \times 10^{-3} \ M^{-2}s^{-1})(0.050 \ M)(0.10 \ M)^2 = 4.3 \times 10^{-6} \ M/s$

14.85   The units of rate are M/s. The reaction must be second order overall if the units of the rate constant are $M^{-1} \ s^{-1}$. If rate = $k[NO_2]^x$, then the cumulative units of $[NO_2]^x$ must be M^2, and x = 2.

If $[NO_2]_0 = 0.100 \ M$ and $[NO_2]_t = 0.025 \ M$, use the integrated form of the second order rate equation, $\dfrac{1}{[A]_t} = kt + \dfrac{1}{[A]_o}$, Equation [14.14], to solve for t.

$$\frac{1}{0.025 \ M} = 0.63 \ M^{-1}s^{-1} \ (t) + \frac{1}{0.100 \ M}; \ \frac{(40 - 10) \ M^{-1}}{0.63 \ M^{-1}s^{-1}} = t = 47.62 = 48 \ s.$$

14.86   (a) $t_{1/2} = 0.693/k = 0.693/7.0 \times 10^{-4} s^{-1} = 990 = 9.9 \times 10^2 s$

        (b) $k = \dfrac{0.693}{t_{1/2}} = \dfrac{0.693}{56.3 \ min} \times \dfrac{1 \ min}{60 \ s} = 2.05 \times 10^{-4} s^{-1}$

14.87   (a) $k = (8.56 \times 10^{-5} \ M/s)/(0.200 \ M) = 4.28 \times 10^{-4} \ s^{-1}$

        (b)   ln [urea] = $-(4.28 \times 10^{-4}s^{-1} \times 4.00 \times 10^3 \ s) + ln \ (0.500)$

             ln [urea] = $-1.712 - 0.693 = -2.405 = -2.41$; [urea] = $0.0903 = 0.090 \ M$

        (c) $t_{1/2} = 0.693/k = 0.693/4.28 \times 10^{-4} \ s^{-1} = 1.62 \times 10^3 \ s$

14.88   (a)   A = abc, Equation [14.5]. A = 0.605, a = $5.60 \times 10^3 \ cm^{-1} \ M^{-1}$, b = 1.00 cm

$$c = \frac{A}{ab} = \frac{0.605}{(5.60 \times 10^3 \ cm^{-1} M^{-1})(1.00 \ cm)} = 1.080 \times 10^{-4} = 1.08 \times 10^{-4} \ M$$

        (b)   Calculate [c]$_t$ using Beer's law. We calculated [c]$_0$ in part (a). Use Equation [14.13] to calculate k.

$$A_{30} = abc_{30}; \ c_{30} = \frac{A_{30}}{ab} = \frac{0.250}{(5.60 \times 10^3 \ cm^{-1} \ M^{-1})(1.00 \ cm)} = 4.464 \times 10^{-5} \ M$$

$$\ln[c]_t = -kt + \ln[c]_0; \quad \frac{\ln[c]_0 - \ln[c]_t}{t} = k; \quad t = 30 \ min \times \frac{60 \ s}{min} = 1800 \ s$$

$$k = \ln(1.080 \times 10^{-4}) - \ln(4.464 \times 10^{-5}) \ / \ 1800 \ s = 4.910 \times 10^{-4} = 4.91 \times 10^{-4} \ s^{-1}$$

(c)   For a first order reaction, $t_{1/2} = 0.693/k$.

$$t_{1/2} = 0.693/4.910 \times 10^{-4} \ s^{-1} = 1.411 \times 10^3 = 1.41 \times 10^3 \ s = 23.5 \ min$$

(d) $A_t = 0.100$; calculate c_t using Beer's law, then t from the first order integrated rate equation.

$$c_t = \frac{A}{ab} = \frac{0.100}{(5.60 \times 10^3 \ cm^{-1} \ M^{-1})(1.00 \ cm)} = 1.786 \times 10^{-5} = 1.79 \times 10^{-5} \ M$$

$$t = \frac{\ln[c]_0 - \ln[c]_t}{k} = \frac{\ln(1.080 \times 10^{-4}) - \ln(1.786 \times 10^{-5})}{4.910 \times 10^{-4} \ s^{-1}}$$

$$t = 3.666 \times 10^3 = 3.67 \times 10^3 \ s = 61.1 \ min$$

14.89

Time (s)	$[C_5H_6]$ (M)	$\ln[C_5H_6]$	$1/[C_5H_6]$
0	0.0400	–3.219	25.0
50	0.0300	–3.507	33.3
100	0.0240	–3.730	41.7
150	0.0200	–3.912	50.0
200	0.0174	–4.051	57.5

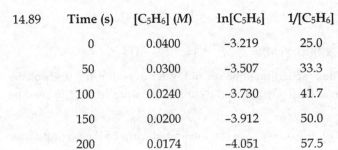

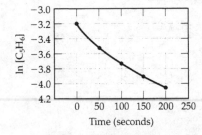

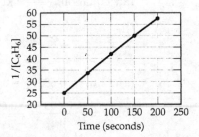

The plot of $1/[C_5H_6]$ vs time is linear and the reaction is second order.

The slope of this line is k. $k = slope = (50.0 - 25.0) \ M^{-1}/(150-0)s = 0.167 \ M^{-1} \ s^{-1}$

(The best-fit slope and k value is $0.163 \ M^{-1} \ s^{-1}$.)

14.90   (a)   No. The value of A, which is related to frequency and effectiveness of collisions, can be different for each reaction and k is proportional to A.

(b)   From Equation [14.21], reactions with different variations of k with respect to temperature have different activation energies, E_a. The fact that k for the two reactions is the same at a certain temperature is coincidental. The reaction with the higher rate at 35°C has the larger activation energy, because it was able to use the increase in energy more effectively.

14.91

ln k	1/T
–24.17	3.33×10^{-3}
–20.72	3.13×10^{-3}
–17.32	2.94×10^{-3}
–15.24	2.82×10^{-3}

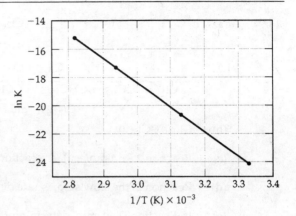

The calculated slope is -1.751×10^4. The activation energy E_a, equals – (slope) × (8.314 J/mol). Thus, $E_a = 1.8 \times 10^4 (8.314) = 1.5 \times 10^5$ J/mol $= 1.5 \times 10^2$ kJ/mol. (The best-fit slope is $-1.76 \times 10^4 = -1.8 \times 10^4$ and the value of E_a is 1.5×10^2 kJ/mol.)

14.92   (a)

$$NO(g) + NO(g) \rightarrow N_2O_2(g)$$

$$N_2O_2(g) + H_2(g) \rightarrow N_2O(g) + H_2O(g)$$

$$\overline{2NO(g) + N_2O_2(g) + H_2(g) \rightarrow N_2O_2(g) + N_2O(g) + H_2O(g)}$$

$$2NO(g) + H_2(g) \rightarrow N_2O(g) + H_2O(g)$$

(b)   First reaction: $- \Delta [NO]/\Delta t = k[NO] [NO] = k[NO]^2$

Second reaction: $- \Delta [H_2]/\Delta t = k[H_2][N_2O_2]$

(c) N_2O_2 is the intermediate: it is produced in the first step and consumed in the second.

(d)   Since $[H_2]$ appears in the rate law, the second step must be slow relative to the first.

14.93   (a)

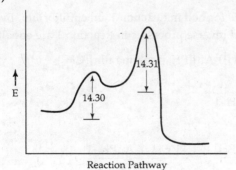

Reaction Pathway

14.30 = E_a for reaction [14.30]

14.31 = E_a for reaction [14.31]

(b)   The fact that Br_2 builds up during the reaction tells us that the appearance of Br_2 (reaction [14.30]) is faster than the disappearance of Br_2 (reaction [14.31]). This is the reason that E_a of [14.31] in the energy profile above is larger than E_a of [14.30].

14.94   (a) $Cl_2(g) \rightleftharpoons 2Cl(g)$

$$Cl(g) + CHCl_3(g) \rightarrow HCl(g) + CCl_3(g)$$
$$Cl(g) + CCl_3(g) \rightarrow CCl_4(g)$$

$$\overline{Cl_2(g) + 2Cl(g) + CHCl_3(g) + CCl_3(g) \rightarrow 2Cl(g) + HCl(g) + CCl_3(g) + CCl_4(g)}$$
$$Cl_2(g) + CHCl_3(g) \rightarrow HCl(g) + CCl_4(g)$$

(b) $Cl(g)$, $CCl_3(g)$

(c)   Reaction 1 - unimolecular, Reaction 2 - bimolecular, Reaction 3 - bimolecular

(d)   Reaction 2, the slow step, is rate determining.

(e)   If Reaction 2 is rate determining, rate = $k_2[CHCl_3][Cl]$. Cl is an intermediate formed in reaction 1, an equilibrium. By definition, the rates of the forward and reverse processes are equal; $k_1 [Cl_2] = k_{-1} [Cl]^2$. Solving for [Cl] in terms of [Cl$_2$],

$$[Cl]^2 = \frac{k_1}{k_{-1}} [Cl_2]; \quad [Cl] = \left(\frac{k_1}{k_{-1}} [Cl_2] \right)^{1/2}$$

Substituting into the overall rate law

$$\text{rate} = k_2 \left(\frac{k_1}{k_{-1}} \right)^{1/2} [CHCl_3][Cl_2]^{1/2} = k[CHCl_3][Cl_2]^{1/2} \text{ (The overall order is 3/2.)}$$

14.95   (a) $(CH_3)_3AuPH_3 \rightarrow C_2H_6 + (CH_3)AuPH_3$

(b) $(CH_3)_3Au$, $(CH_3)Au$ and PH_3 are intermediates.

(c)   Reaction 1 is unimolecular, Reaction 2 is unimolecular, Reaction 3 is bimolecular.

(d)   Reaction 2, the slow one, is rate determining.

(e)   If Reaction 2 is rate determining, rate = $k_2[(CH_3)_3Au]$.

$(CH_3)_3Au$ is an intermediate formed in Reaction 1, an equilibrium. By definition, the rates of the forward and reverse processes in Reaction 1 are equal:

$k_1[(CH_3)_3 AuPH_3] = k_{-1}[(CH_3)_3Au][PH_3]$; solving for [(CH$_3$)$_3$Au],

$$[(CH_3)_3 Au] = \frac{k_1[(CH_3)_3 AuPH_3]}{k_{-1}[PH_3]}$$

Substituting into the rate law

$$\text{rate} = \left(\frac{k_2 k_1}{k_{-1}} \right) \frac{[(CH_3)_3 AuPH_3]}{[PH_3]} = \frac{k[(CH_3)_3 AuPH_3]}{[PH_3]}$$

(f)   The rate is inversely proportional to [PH$_3$], so adding PH$_3$ to the (CH$_3$)$_3$AuPH$_3$ solution would decrease the rate of the reaction.

14.96   *Enzyme*: carbonic anhydrase; *substrate*: carbonic acid (H_2CO_3); *turnover number*: 1×10^7 molecules/s.

14.97   (a)   The fact that the rate doubles with a doubling of the concentration of sugar tells us that the fraction of enzyme tied up in the form of an enzyme-substrate complex is small. A doubling of the substrate concentration leads to a doubling of the concentration of enzyme-substrate complex, because most of the enzyme molecules are available to bind substrates.

         (b)   The behavior of inositol suggests that it acts as a competitor with sucrose for binding at the active sites of the enzyme system. Such a competition results in a lower effective concentration of active sites for binding of sucrose, and thus results in a lower reaction rate.

Integrative Exercises

14.98   *Analyze/Plan.* $2N_2O_5 \rightarrow 4NO_2 + O_2$ rate = $k[N_2O_5] = 1.0 \times 10^{-5}$ s^{-1} $[N_2O_5]$

Use the integrated rate law for a first-order reaction, Equation [14.13], to calculate $k[N_2O_5]$ at 20.0 hr. Build a stoichiometry table to determine mol O_2 produced in 20.0 hr. Assuming that $O_2(g)$ is insoluble in chloroform, calculate the pressure of O_2 in the 10.0 L container. *Solve.*

$$20.0 \text{ hr} \times \frac{60 \text{ min}}{1 \text{ hr}} \times \frac{60 \text{ s}}{1 \text{ min}} \times 7.20 \times 10^4 \text{ s}; \ [N_2O_5]_0 = 0.600 \ M$$

$$\ln[A]_t - \ln[A]_0 = -kt; \ \ln[N_2O_5]_t = -kt + \ln[N_2O_5]_0$$

$$\ln[N_2O_5]_t = -1.0 \times 10^{-5} \text{ s}^{-1} (7.20 \times 10^4 \text{ s}) + \ln(0.600) = -0.720 - 0.511 = -1.231$$

$$[N_2O_5]_t = e^{-1.231} = 0.292 \ M$$

N_2O_5 was present initially as 1.00 L of 0.600 M solution.

mol N_2O_5 = M × L = 0.600 mol N_2O_5 initial, 0.292 mol N_2O_5 at 20.0 hr

	$2N_2O_5$	$\rightarrow$	$4NO_2$	+	O_2
t = 0	0.600 mol		0		0
change	–0.308 mol		0.616 mol		0.154 mol
t = 20 hr	0.292 mol		0.616 mol		0.154 mol

[Note that the reaction stoichiometry is applied to the 'change' line.]

PV = nRT; P = nRT/V; V = 10.0 L, T = 45°C = 318 K, n = 0.154 mol

$$P = 0.154 \text{ mol} \times \frac{318 \text{ K}}{10.0 \text{ L}} \times \frac{0.08206 \text{ L} \bullet \text{atm}}{\text{mol} \bullet \text{K}} = 0.402 \text{ atm}$$

14.99   (a) $\ln k = -E_a/RT + \ln A$; $E_a = 86.8$ kJ/mol = 8.68×10^4 J/mol;

$$T = 35°C + 273 = 308 \text{ K}; \ A = 2.10 \times 10^{11} \ M^{-1} \text{ s}^{-1}$$

$$\ln k = \frac{-8.68 \times 10^4 \text{ J/mol}}{308 \text{ K}} \times \frac{\text{mol} \bullet \text{K}}{8.314 \text{ J}} + \ln(2.10 \times 10^{11} \ M^{-1} \text{ s}^{-1})$$

$$\ln k = -33.8968 + 26.0704 = -7.8264; \ k = 3.99 \times 10^{-4} \ M^{-1} \text{ s}^{-1}$$

(b) $\dfrac{0.335 \text{ g KOH}}{0.250 \text{ L soln}} \times \dfrac{1 \text{ mol KOH}}{56.1 \text{ g KOH}} = 0.02389 = 0.0239 \ M \text{ KOH}$

$\dfrac{1.453 \text{ g } C_2H_5I}{0.250 \text{ L soln}} \times \dfrac{1 \text{ mol } C_2H_5I}{156.0 \text{ g } C_2H_5I} = 0.03726 = 0.0373 \ M \ C_2H_5I$

If equal volumes of the two solutions are mixed, the initial concentrations in the reaction mixture are 0.01194 M KOH and 0.01863 M C_2H_5I. Assuming the reaction is first order in each reactant:

rate = $k[C_2H_5I][OH^-] = 3.99 \times 10^{-4} \ M^{-1}s^{-1} (0.01194 \ M)(0.01863 \ M) = 8.88 \times 10^{-8} \ M/s$

(c)   Since C_2H_5I and OH^- react in a 1 : 1 mole ratio and equal volumes of the solutions are mixed, the reactant with the smaller concentration, KOH, is the limiting reactant.

14.100   (a)   Use an apparatus such as the one pictured in Figure 10.3 (an open-end manometer), a clock, a ruler and a constant temperature bath. Since P = (n/V)RT, $\Delta P/\Delta t$ at constant temperature is an acceptable measure of reaction rate.

Load the flask with HCl(aq) and read the height of the Hg in both arms of the manometer. Quickly add Zn(s) to the flask and record time = 0 when the Zn(s) contacts the acid. Record the height of the Hg in one arm of the manometer at convenient time intervals such as 5 sec. (The decrease in the short arm will be the same as the increase in the tall arm). Calculate the pressure of H_2(g) at each time.

(b)   Keep the amount of Zn(s) constant and vary the concentration of HCl(aq) to determine the reaction order for H^+ and Cl^-. Keep the concentration of HCl(aq) constant and vary the amount of Zn(s) to determine the order for Zn(s). Combine this information to write the rate law.

(c) $-\Delta[H^+]/2\Delta t = \Delta[H_2]/\Delta t; \ -\Delta[H^+]/\Delta t = 2\Delta[H_2]/\Delta t$

$[H_2]$ = mol H_2/L H_2 = n/V; $[H_2]$ = P (in atm)/RT

Then, the rate of disappearance of H^+ is twice the rate of appearance of H_2(g).

(d)   By changing the temperature of the constant temperature bath, measure the rate data at several (at least three) temperatures and calculate the rate constant k at these temperatures. Plot ln k vs 1/T. The slope of the line is $-E_a/R$ and E_a = –slope (R).

(e)   Measure rate data at constant temperature, HCl concentration and mass of Zn(s), varying only the form of the Zn(s). Compare the rate of reaction for metal strips and granules.

14.101   (a)   ln k = $-E_a/RT$ + ln A, Equation [14.20]. E_a = 6.3 kJ/mol = 6.3×10^3 J/mol

T = 100°C + 273 = 373 K

$\ln k = \dfrac{-6.3 \times 10^3 \text{ J/mol}}{8.314 \text{ J/K} \bullet \text{mol} \times 373 \text{ K}} + \ln(6.0 \times 10^8 \ M^{-1}s^{-1})$

ln k = –2.032 + 20.212 = 18.181 = 18.2; k = $7.87 \times 10^7 = 8 \times 10^7 \ M^{-1}s^{-1}$

(b)   NO, 11 valence e⁻, 5.5 e⁻ pair
(Assume the less electronegative N atom will be electron
deficient.)

ONF, 18 valence e⁻, 9 e⁻ pr

$$:\ddot{O}=\ddot{N}-\ddot{F}: \longleftrightarrow \left(:\ddot{\ddot{O}}-\ddot{N}=\ddot{F} \right)$$

$$:\ddot{N}=\ddot{O}:$$

The resonance form on the right is a very
minor contributor to the true bonding picture, due to high formal charges and the
unlikely double bond involving F.

(c)   ONF has trigonal planar electron domain geometry, which leads to a "bent"
structure with a bond angle of approximately 120°.

(d)
$$\begin{bmatrix} O=N \\ \quad\; \diagdown \\ \quad F \overset{\cdot}{\frown} F \end{bmatrix}$$

(e)   The electron deficient NO molecule is attracted to electron-rich F_2, so the driving
force for formation of the transition state is greater than simple random
collisions.

14.102   (a) $\Delta H^{\circ}_{rxn} = 2\Delta H^{\circ}_f\ H_2O(g) + 2\Delta H^{\circ}_f\ Br_2(g) - 4\Delta H^{\circ}_f\ HBr(g) - \Delta H^{\circ}_f\ O_2(g)$

$\Delta H^{\circ}_{rxn} = 2(-241.82) + 2(30.71) - 4(-36.23) - (0) = -277.30$ kJ

(b)   Since the rate of the uncatalyzed reaction is very slow at room temperature, the
magnitude of the activation energy for the rate-determining first step must be
quite large. At room temperature, the reactant molecules have a distribution of
kinetic energies (Chapter 10), but very few molecules even at the high end of the
distribution have sufficient energy to form an activated complex. E_a for this step
must be much greater than 3/2 RT, the average kinetic energy of the sample.

(c)   20 e⁻, 10 e⁻ pr

$$H-\ddot{O}-\ddot{O}-\ddot{B}r:$$

The intermediate resembles hydrogen peroxide, H_2O_2.

14.103   In the lock and key model of enzyme action, the active site is the specific location in the
enzyme where reaction takes place. The precise geometry (size and shape) of the active
site both accommodates and activates the substrate (reactant). Proteins are large biopolymers, with the same structural flexibility as synthetic polymers (Chapter 12). The
three-dimensional shape of the protein in solution, including the geometry of the active
site, is determined by many intermolecular forces of varying strengths.

Changes in temperature change the kinetic energy of the various groups on the enzyme
and their tendency to form intermolecular associations or break free from them. Thus,
changing the temperature changes the overall shape of the protein and specifically the
shape of the active site. At the operating temperature of the enzyme, the competition
between kinetic energy driving groups apart and intermolecular attraction pulling them

together forms an active site that is optimum for a specific substrate. At temperatures above the temperature of maximum activity, sufficient kinetic energy has been imparted so that the forces driving groups apart win the competition, and the three-dimensional structure of the enzyme is destroyed. This is the process of *denaturation*. The activity of the enzyme is destroyed because the active site has collapsed. The protein or enzyme is denatured, because it is no longer capable of its "natural" activity.

14.104 (a)   If the reaction proceeds in a single elementary step, the coefficients in the balanced equation are the reaction orders for the respective reactants.

rate $= k[Ce^{4+}]^2 [Tl^+]$

(b)   If the uncatalyzed reaction occurs in a single step, it is **termolecular**. The activated complex requires collision of three particles with the correct energy and orientation for reaction. The probability of an effective three-particle collision is low and the rate is slow.

(c)   The first step is rate-determining.

(d)   The ability of Mn to adopt every oxidation state from +2 to +7 makes it especially suitable to catalyze this (and many other) reactions.

14.105 (a) $D(Cl-Cl) = 242 \text{ kJ/mol } Cl_2$

$$\frac{242 \text{ kJ}}{\text{mol } Cl_2} \times \frac{1000 \text{ J}}{\text{kJ}} \times \frac{1 \text{ mol}}{6.022 \times 10^{23} \text{ molecules}} = 4.019 \times 10^{-19} = 4.02 \times 10^{-19} \text{ J}$$

$$\lambda = hc/E = \frac{6.626 \times 10^{-34} \text{ J} \cdot \text{s} \times 2.998 \times 10^8 \text{ m/s}}{4.019 \times 10^{-19} \text{ J}} = 4.94 \times 10^{-7} \text{ m}$$

This wavelength, 494 nm, is in the visible portion of the spectrum.

(b)

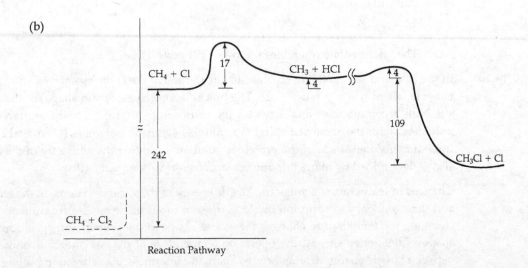

(c)   Since $D(Cl-Cl)$ is 242 kJ/mol, $CH_4(g) + Cl_2(g)$ should be about 242 kJ below the starting point on the diagram. For the reaction $CH_4(g) + Cl_2(g) \rightarrow CH_3(g) + HCl(g) + Cl(g)$, E_a is 242 + 17 = 259 kJ. (From bond dissociation enthalpies, ΔH for the overall reaction $CH_4(g) + Cl_2(g) \rightarrow CH_3Cl(g) + Cl(g)$ is –104 kJ, so the graph above is simply a sketch of the relative energies of some of the steps in the process.)

(d) CH_3, 7 valence e^-, odd electron species

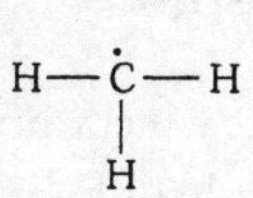

(e)   This sequence is called a chain reaction because $Cl\bullet$ radicals are regenerated in Reaction 4, perpetuating the reaction. Absence of $Cl\bullet$ terminates the reaction, so $Cl\bullet + Cl\bullet \rightarrow Cl_2$ is a termination step.

15 Chemical Equilibrium

Visualizing Concepts

15.1 (a) $k_f > k_r$. According to the Arrhenius equation [14.19], $k = Ae^{-E_a/RT}$. As the magnitude of E_a increases, k decreases. On the energy profile, E_a is the difference in energy between the starting point and the energy at the top of the barrier. Clearly this difference is smaller for the forward reaction, so $k_f > k_r$.

(b) From the Equation [15.5], the equilibrium constant = k_f/k_r. Since $k_f > k_r$, the equilibrium constant for the process shown in the energy profile is greater than 1.

15.2 Yes. The first box is pure reactant A. As the reaction proceeds, some A changes to B. In the fourth and fifth boxes, the relative amounts (concentrations) of A and B are constant. Although the reaction is ongoing the rates of $A \rightarrow B$ and $B \rightarrow A$ are equal, and the relative amounts of A and B are constant.

15.3 *Analyze.* Given box diagram and reaction type, determine whether K > 1 for the equilibrium mixture depicted in the box.

Plan. Assign species in the box to reactants and products. Write an equilibrium expression in terms of concentrations. Find the relationship between numbers of molecules and concentration. Calculate K.

Solve. Let red = A, blue = X, red and blue pairs = AX. (The colors of A and X are arbitrary.) There are 3A, 2B, and 8AX in the box.

M = mol/L. Since moles is a counting unit for particles, mol ratios and particle ratios are equivalent. We can use numbers of particles in place of moles in the molarity formula. V = 1 L, so in this case, [A] = number of A particles.

$$K = \frac{[AX]}{[A][X]}; \quad [AX] = 8/V = 8; [A] = 3/V = 3; [X] = 2/V = 2$$

$$K = \frac{8}{[3][2]} = \frac{8}{6} = 1.33$$

15.4 *Analyze/Plan.* Given that element A = red and element B = blue, evaluate the species in the reactant and product boxes, and write the reaction. Answer the remaining questions based on the balanced equation. *Solve.*

(a) reactants: $4A_2 + 4B$; products: $4A_2B$

balanced equation: $A_2 + B \rightarrow A_2B$

(b) $$K_c = \frac{[A_2B]}{[A_2][B]}$$

(c) $\Delta n = \Sigma n(\text{prod}) - \Sigma n(\text{react}) = 1 - 2 = -1$.

(d) $K_p = K_c(RT)^{\Delta n}$, Equation [15.14].

If you have a balanced equation, calculate Δn. Use Equation [15.14] to calculate K_p from K_c, or vice versa.

15.5   *Analyze.* Given box diagrams, reaction type, and value of K_c, determine whether each reaction mixture is at equilibrium.

Plan. Analyze the contents of each box, express them as concentrations (see Solution 5.3). Write the equilibrium expression, calculate Q for each mixture, and compare it to K_c. If $Q = K$, the mixture is at equilibrium. If $Q < K$, the reaction shifts right (more product). If $Q > K$, the reaction shifts left (more reactant).

Solve. $K_c = \dfrac{[AB]^2}{[A_2][B_2]}$.

For this reaction, $\Delta n = 0$, so the volume terms cancel in the equilibrium expression. In this case, the number of each kind of particle can be used as a representation of moles (see Solution 5.3) and molarity.

(a)   Mixture 1: $1A_2$, $1B_2$, $6AB$; $Q = \dfrac{6^2}{(1)(1)} = 36$

$Q > K_c$, the mixture is not at equilibrium.

Mixture 2: $3A_2$, $2B_2$, $3AB$; $Q = \dfrac{3^2}{(3)(2)} = 1.5$

$Q = K_c$, the mixture is at equilibrium.

Mixture 3: $3A_2$, $3B_2$, $2AB$; $Q = \dfrac{2^2}{(3)(3)} = 0.44$

$Q < K_c$, the mixture is not at equilibrium.

(b)   Mixture 1 proceeds toward reactants.

Mixture 3 proceeds toward products.

15.6   For the reaction $A_2(g) + B(g) \rightleftharpoons A(g) + AB(g)$, $\Delta n = 0$ and $K_p = K_c$. We can evaluate the equilibrium expression in terms of concentration. Also since $\Delta n = 0$, the volume terms in the expression cancel and we can use number of particles as a measure of moles and molarity. The mixture contains $2A$, $4AB$ and $2A_2$.

$K_c = \dfrac{[A][AB]}{[A_2][B]} = \dfrac{(2)(4)}{(2)(B)} = 2; B = 2$

2 B atoms should be added to the diagram.

15.7   *Analyze.* Given the diagram and reaction type, calculate the equilibrium constant K_c.

Plan. Analyze the contents of the cylinder. Express them as concentrations, using number of particles as a measure of moles, and V = 1 L. Write the equilibrium expression in terms of concentration and calculate K_c.   *Solve.*

(a)   The mixture contains $2A_2$, $2B$, $4AB$. $[A_2] = 2$, $[B] = 2$, $[AB] = 4$.

$$K_c = \frac{[AB]^2}{[A_2][B]^2} = \frac{(4)^2}{(2)(2)^2} = 2$$

(b)   A decrease in volume favors the reaction with fewer particles. This reaction has two particles in products and three in reactants, so a decrease in volume favors products. The number of AB (product) molecules will increase.

Note that a change in volume does not change the value of K_c. If V decreases, the number of AB molecules must increase in order to maintain the equilibrium value of K_c.

15.8   If temperature increases, K of an endothermic reaction increases and K of an exothermic reaction decreases. Calculate the value of K for the two temperatures and compare. For this reaction, $\Delta n = 0$ and $K_p = K_c$. We can ignore volume and use number of particles as a measure of moles and molarity. $K_c = [A][AB]/[A_2][B]$.

(1)   300 K, 3A, 5AB, $1A_2$, 1B; $K_c = (3)(5)/(1)(1) = 15$

(2)   500 K, 1A, 3AB, $3A_2$, 3B; $K_c = (1)(3)/(3)(3) = 0.33$

K_c decreases as T increases, so the reaction is exothermic.

Equilibrium; the Equilibrium Constant

15.9   *Analyze/Plan.* Given the forward and reverse rate constants, calculate the equilibrium constant using Equation [15.5]. At equilibrium, the rates of the forward and reverse reactions are equal. Write the rate laws for the forward and reverse reactions and use their equality to answer part (b).   *Solve.*

(a) $K_c = \dfrac{k_f}{k_r}$, Equation [15.5]; $K_c = \dfrac{3.8 \times 10^{-2} \text{ s}^{-1}}{3.1 \times 10^{-1} \text{ s}^{-1}} = 0.12$

For this reaction, $K_p = K_c = 0.12$.

(b) $\text{rate}_f = \text{rate}_r$; $k_f[A] = k_r[B]$

Since $k_f < k_r$, in order for the two rates to be equal, [A] must be greater than [B].

15.10   (a) $K_c = \dfrac{[C][D]}{[A][B]}$; if K_c is large, the numerator of the K_c expression is much greater than the denominator and products will predominate at equilibrium.

(b) $K_c = k_f/k_r$; if K_c is large, k_f is larger than k_r and the forward reaction has the greater rate constant.

15.11   (a)   The *law of mass action* expresses the relationship between the concentrations of reactants and products at equilibrium for any reaction. The law of mass action is a generic equilibrium expression.

$$K_c = \frac{[NOBr_2]}{[NO][Br_2]}$$

(b) The *equilibrium-constant expression* is an algebraic equation where the variables are the equilibrium concentrations of the reactants and products for a specific chemical reaction. The *equilibrium constant* is a number; it is the ratio calculated from the equilibrium expression for a particular chemical reaction. For any reaction, there is an infinite number of sets of equilibrium concentrations, depending on initial concentrations, but there is only one equilibrium constant.

(c) Introduce a known quantity of $NOBr_2(g)$ into a vessel of known volume at constant (known) temperature. After equilibrium has been established, measure the total pressure in the flask. Using an equilibrium table, such as the one in Sample Exercise 15.12, calculate equilibrium pressures and concentrations of $NO(g)$, $Br_2(g)$, and $NOBr_2(g)$ and calculate K_c.

15.12   (a) Yes. The algebraic form of the law of mass action depends only on the coefficients of a chemical equation, not on the reaction mechanism.

       (b) $N_2(g) + 3H_2(g) \rightleftharpoons 2NH_3(g)$. The *Haber process* is the primary industrial method of nitrogen fixation, that is, of converting $N_2(g)$ into usable forms. The major use of $NH_3(g)$ from the Haber process is for fertilizer.

       (c) $K_c = \dfrac{[NH_3]^2}{[N_2][H_2]^3}$

15.13   *Analyze/Plan.* Follow the logic in Sample Exercises 15.1 and 15.5.   *Solve.*

       (a) $K_c = \dfrac{[N_2O][NO_2]}{[NO]^3}$ (b) $K_c = \dfrac{[CS_2][H_2]^4}{[CH_4][H_2S]^2}$

       (c) $K_c = \dfrac{[CO]^4}{[Ni(CO)_4]}$ (d) $K_c = \dfrac{[H^+][F^-]}{[HF]}$

       (e) $K_c = \dfrac{[Ag^+]^2}{[Zn^{2+}]}$

       homogeneous: (a), (b), (d); heterogeneous: (c), (e)

15.14   (a) $K_c = \dfrac{[NO]^2}{[N_2][O_2]}$ (b) $K_c = \dfrac{1}{[Cl_2]^2}$

       (c) $K_c = \dfrac{[C_2H_6]^2[O_2]}{[C_2H_4]^2[H_2O]^2}$ (d) $K_c = \dfrac{[H_2O]}{[H_2]}$

       (e) $K_c = \dfrac{[Cl_2]^2}{[HCl]^4[O_2]}$

       homogeneous: (a), (c); heterogeneous: (b), (d), (e)

15.15   *Analyze.* Given the value of K_c, predict the contents of the equilibrium mixture.

      Plan. If $K_c \gg 1$, products dominate; if $K_c \ll 1$, reactants dominate.   *Solve.*

       (a) mostly reactants ($K_c \ll 1$)

       (b) mostly products ($K_c \gg 1$)

15 Chemical Equilibrium

15.16 (a) equilibrium lies to right, favoring products ($K_c \gg 1$)

(b) equilibrium lies to left, favoring reactants ($K_c \ll 1$)

15.17 *Analyze/Plan.* Follow the logic in Sample Exercise 15.2. *Solve.*

$PCl_3(g) + Cl_2(g) \rightleftharpoons PCl_5(g)$, $K_c = 0.042$. $\Delta n = 1 - 2 = -1$

$K_p = K_c(RT)^{\Delta n} = 0.042(RT)^{-1} = 0.042/RT$

$K_p = \dfrac{0.042}{(0.08206)(500)} = 0.001024 = 1.0 \times 10^{-3}$

15.18 $SO_2(g) + Cl_2(g) \rightleftharpoons SO_2Cl_2(g)$, $K_p = 34.5$. $\Delta n = 1 - 2 = -1$

$K_p = K_c(RT)^{\Delta n}$; $34.5 = K_c(RT)^{-1} = K_c/RT$;

$K_c = 34.5 \, RT = 34.5(0.08206)(303) = 857.81 = 858$

15.19 *Analyze.* Given K_c for a chemical reaction, calculate K_c for the reverse reaction.

Plan. The equilibrium expressions for the reaction and its reverse are the reciprocals of each other, and the values of K_c are also reciprocal. Evaluate which species are favored by examining the magnitude of K_c. *Solve.*

(a) $K_c(\text{forward}) = \dfrac{[NOBr]^2}{[NO]^2[Br_2]} = 1.3 \times 10^{-2}$

$K_c(\text{reverse}) = \dfrac{[NO]^2[Br_2]}{[NOBr]^2} = \dfrac{1}{1.3 \times 10^{-2}} = 76.92 = 77$

(b) $K_c < 1$ when NOBr is the product, and $K_c > 1$ when NOBr is the reactant. At this temperature, the equilibrium favors NO and Br_2.

15.20 (a) $K_p(\text{forward}) = \dfrac{P_{NO_2}^2}{P_{NO}^2 \times P_{O_2}} = 1.48 \times 10^4$

$K_p(\text{reverse}) = \dfrac{P_{NO}^2 \times P_{O_2}}{P_{NO_2}^2} = \dfrac{1}{1.48 \times 10^4} = 6.76 \times 10^{-5}$

(b) $K_p > 1$ when NO_2 is the product, and $K_p < 1$ when NO_2 is the reactant, so the equilibrium favors NO_2 at this temperature.

15.21 *Analyze.* Given K_p for a reaction, calculate K_p for a related reaction.

Plan. The algebraic relationship between the K_p values is the same as the algebraic relationship between equilibrium expressions.

Solve. $K_p = \dfrac{P_{SO_3}}{P_{SO_2} \times P_{O_2}^{1/2}} = 1.85$

(a) $K_p = \dfrac{P_{SO_2} \times P_{O_2}^{1/2}}{P_{SO_3}} = \dfrac{1}{1.85} = 0.541$

(b) $\quad K_p = \dfrac{P_{SO_3}^2}{P_{SO_2}^2 \times P_{O_2}} = (1.85)^2 = 3.4225 = 3.42$

(c) $\quad K_p = K_c(RT)^{\Delta n}; \; \Delta n = 2 - 3 = -1; \; T = 1000 \, K$

$\quad\quad K_p = K_c(RT)^{-1} = K_c/RT; \; K_c = K_p(RT)$

$\quad\quad K_c = 3.4225(0.08206)(1000) = 280.85 = 281$

15.22 $\quad K_p = \dfrac{P_{HCl}^4 \times P_{O_2}}{P_{Cl_2}^2 \times P_{H_2O}^2} = 0.0752$

(a) $\quad K_p = \dfrac{P_{Cl_2}^2 \times P_{H_2O}^2}{P_{HCl}^4 \times P_{O_2}} = \dfrac{1}{0.0752} = 13.298 = 13.3$

(b) $\quad K_p = \dfrac{P_{HCl}^2 \times P_{O_2}^{1/2}}{P_{Cl_2} \times P_{H_2O}} = (0.0752)^{1/2} = 0.2742 = 0.274$

(c) $\quad K_p = K_c(RT)^{\Delta n}; \; \Delta n = 2.5 - 2 = 0.5; \; T = 480°C + 273 = 753 \, K$

$\quad\quad K_p = K_c(RT)^{1/2}, \; K_c = K_p/(RT)^{1/2} = 0.2742/[0.08206 \times 753]^{1/2} = 0.03488 = 0.0349$

15.23 *Analyze/Plan.* Follow the logic in Sample Exercise 15.5.   *Solve.*

$$A(aq) + B(aq) \rightleftharpoons C(aq) \quad\quad\quad\quad K_1 = 1.9 \times 10^{-4}$$

$$C(aq) + D(aq) \rightleftharpoons E(aq) + A(aq) \quad\quad K_2 = 8.5 \times 10^2$$

$$A(aq) + B(aq) + C(aq) + D(aq) \rightleftharpoons A(aq) + C(aq) + E(aq)$$

$$B(aq) + D(aq) \rightleftharpoons E(aq) \quad\quad\quad\quad K_c = K_1 \times K_2 = 0.16$$

$\quad K_c = (1.9 \times 10^{-4})(8.5 \times 10^2) = 0.162 = 0.16$

15.24

$$2NO(g) + Br_2(g) \rightleftharpoons 2NOBr(g) \quad\quad K_1 = 2.0$$

$$N_2(g) + O_2(g) \rightleftharpoons 2NO(g) \quad\quad\quad\quad K_2 = \dfrac{1}{2.1 \times 10^{30}}$$

$$2NO(g) + Br_2(g) + N_2(g) + O_2(g) \rightleftharpoons 2NOBr(g) + 2NO(g)$$

$$N_2(g) + O_2(g) + Br_2(g) \rightleftharpoons 2NOBr(g)$$

$\quad K_c = K_1 \times K_2 = 2.0 \times \dfrac{1}{2.1 \times 10^{30}} = 9.524 \times 10^{-31} = 9.5 \times 10^{-31}$

15.25 *Analyze/Plan.* Follow the logic in Sample Exercise 15.6.   *Solve.*

(a) $\quad K_p = P_{O_2}$

(b) The molar concentration, the ratio of moles of a substance to volume occupied by the substance, is a constant for pure solids and liquids.

15.26   (a) $K_p = 1/ P_{SO_2}$

        (b) Na_2O is a pure solid. The molar concentration, the ratio of moles of a substance to volume occupied by the substance, is a constant for pure solids and liquids.

Calculating Equilibrium Constants

15.27   *Analyze/Plan.* Follow the logic in Sample Exercise 15.8 using concentrations rather than pressures.   *Solve.*

$$K_c = \frac{[H_2][I_2]}{[HI]^2} = \frac{(4.79 \times 10^{-4})(4.79 \times 10^{-4})}{(3.53 \times 10^{-3})^2} = 0.018413 = 0.0184$$

15.28 $[CH_3OH] = \dfrac{0.0406 \text{ mol}}{2.00 \text{ L}} = 0.0203 \text{ M}$

        $[CO] = \dfrac{0.170 \text{ mol CO}}{2.00 \text{ L}} = 0.0850 \text{ M}$

        $[H_2] = \dfrac{0.302 \text{ mol } H_2}{2.00 \text{ L}} = 0.151 \text{ M}$

        $K_c = \dfrac{[CH_3OH]}{[CO][H_2]^2} = \dfrac{0.0203}{(0.0850)(0.151)^2} = 10.4743 = 10.5$

15.29   *Analyze/Plan.* Follow the logic in Sample Exercise 15.8.   *Solve.*

$$2NO(g) + Cl_2(g) \rightleftharpoons 2NOCl(g)$$

$$K_p = \frac{P_{NOCl}^2}{P_{NO}^2 \times P_{Cl_2}} = \frac{(0.28)^2}{(0.095)^2(0.171)} = 50.80 = 51$$

15.30   (a) $K_p = \dfrac{P_{PCl_5}}{P_{PCl_3} \times P_{Cl_2}} = \dfrac{1.30 \text{ atm}}{0.124 \text{ atm} \times 0.157 \text{ atm}} = 66.8$

        (b)   Since $K_p > 1$, products (the numerator of the K_p expression) are favored over reactants (the denominator of the K_p expression).

15.31   *Analyze/Plan.* Follow the logic in Sample Exercise 15.9. Since the container volume is 1.0 L, mol = *M*.   *Solve.*

        (a)   First calculate the change in [NO], 0.10 – 0.062 = 0.038 = 0.04 *M*. From the stoichiometry of the reaction, calculate the changes in the other pressures. Finally, calculate the equilibrium pressures.

	$2NO(g)$ +	$2H_2(g)$ $\rightleftharpoons$	$N_2(g)$ +	$2H_2O(g)$
initial	0.10 *M*	0.050 *M*	0 *M*	0.10 *M*
change	−0.038 *M*	−0.038 *M*	+0.019 *M*	+0.038 *M*
equil.	0.062 *M*	0.012 *M*	0.019 *M*	0.138 *M*

        Strictly speaking, the change in [NO] has one decimal place and thus one sig fig. This limits equilibrium pressures to one decimal place for all but H_2O, and K_c to one sig fig. We compute the extra figures and then round.

(b) $K_c = \dfrac{[N_2][H_2O]^2}{[NO]^2[H_2]^2} = \dfrac{(0.019)(0.138)^2}{(0.062)^2(0.012)^2} = \dfrac{(0.02)(0.14)^2}{(0.06)^2(0.01)^2} = 653.7 = 7 \times 10^2$

15.32 (a)   Calculate the concentrations of $H_2(g)$ and $Br_2(g)$ and the equilibrium concentration of $H_2(g)$. $M = mol/L$.

$[H_2]_{init} = 1.374\ g\ H_2 \times \dfrac{1\ mol\ H_2}{2.0159\ g\ H_2} \times \dfrac{1}{2.00\ L} = 0.34079 = 0.341\ M$

$[Br_2] = 70.31\ g\ Br_2 \times \dfrac{1\ mol\ Br_2}{159.81\ g\ Br_2} \times \dfrac{1}{2.00\ L} = 0.21998 = 0.220\ M$

$[H_2]_{equil} = 0.566\ g\ H_2 \times \dfrac{1\ mol\ H_2}{2.0159\ g\ H_2} \times \dfrac{1}{2.00\ L} = 0.14038 = 0.140\ M$

	$H_2(g)$	$+$	$Br_2(g)$	$\rightleftharpoons$	$2HBr(g)$
initial	0.34079 M		0.21998 M		0
change	–0.20041 M		–0.20041 M		+2(0.20041) M
equil.	0.14038 M		0.01957 M		0.40082 M

The change in H_2 is (0.34079 – 0.14038 = 0.20041 = 0.200). The changes in $[Br_2]$ and $[HBr]$ are set by stoichiometry, resulting in the equilibrium concentrations shown in the table.

(b) $K_c = \dfrac{[HBr]^2}{[H_2][Br_2]} = \dfrac{(0.40082)^2}{(0.14038)(0.01957)} = \dfrac{(0.401)^2}{(0.140)(0.020)} = 58.48 = 58$

The equilibrium concentration of Br_2 has 3 decimal places and 2 sig figs, so the value of K_c has 2 sig figs.

15.33 *Analyze/Plan.* Follow the logic in Sample Exercise 15.9, using partial pressures, rather than concentrations.   *Solve.*

(a) $P = nRT/V$; $P_{CO_2} = 0.2000\ mol \times \dfrac{500\ K}{2.000\ L} \times \dfrac{0.08206\ L \cdot atm}{K \cdot mol} = 4.1030 = 4.10\ atm$

$P_{H_2} = 0.1000\ mol \times \dfrac{500\ K}{2.000\ L} \times \dfrac{0.08206\ L \cdot atm}{K \cdot mol} = 2.0515 = 2.05\ atm$

$P_{H_2O} = 0.1600 \times \dfrac{500\ K}{2.000\ L} \times \dfrac{0.08206\ L \cdot atm}{K \cdot mol} = 3.2824 = 3.28\ atm$

(b)   The change in P_{H_2O} is 3.51 – 3.28 = 0.2276 = 0.23 atm. From the reaction stoichiometry, calculate the change in the other pressures and the equilibrium pressures.

	$CO_2(g)$	$+$	$H_2(g)$	$\rightleftharpoons$	$CO(g)$	$+$	$H_2O(g)$
initial	4.10 atm		2.05 atm		0 atm		3.28 atm
change	–0.23 atm		–0.23 atm		+0.23		+0.23 atm
equil	3.87 atm		1.82 atm		0.23 atm		3.51 atm

(c) $K_p = \dfrac{P_{CO} \times P_{H_2O}}{P_{CO_2} \times P_{H_2}} = \dfrac{(0.23)(3.51)}{(3.87)(1.82)} = 0.1146 = 0.11$

Without intermediate rounding, equilibrium pressures are $P_{H_2O} = 3.51$, $P_{CO} = 0.2276$, $P_{H_2} = 1.8239$, $P_{CO_2} = 3.8754$ and $K_p = 0.1130 = 0.11$, in good agreement with the value above.

15.34   (a)

	$N_2O_4(g)$	$\rightleftharpoons$	$2NO_2(g)$
initial	1.500 atm		1.000 atm
change	+0.244 atm		–0.488 atm
equil	1.744 atm		0.512 atm

The change in P_{NO_2} is $(1.000 - 0.512) = -0.488$ atm, so the change in $P_{N_2O_4}$ is $+(0.488/2) = +0.244$ atm.

(b) $K_p = \dfrac{P_{NO_2}^2}{P_{N_2O_4}} = \dfrac{(0.512)^2}{(1.744)} = 0.1503 = 0.150$

Applications of Equilibrium Constants

15.35   (a)   A reaction quotient is the result of the law of mass action for a general set of concentrations, whereas the equilibrium constant requires equilibrium concentrations.

        (b)   In the direction of more products, to the right.

        (c)   If $Q_c = K_c$, the system is at equilibrium; the concentrations used to calculate Q must be equilibrium concentrations.

15.36   (a)   If the value of Q_c equals the value of K_c, the system is at equilibrium.

        (b)   In the direction of less products (more reactants), to the left.

        (c) $Q_c = 0$ if the concentration of any product is zero.

15.37   *Analyze/Plan.* Follow the logic in Sample Exercise 15.10. We are given molarities, so we calculate Q directly and decide on the direction to equilibrium.   *Solve.*

$K_c = \dfrac{[CO][Cl_2]}{[COCl_2]} = 2.19 \times 10^{-10}$ at 100°C

(a) $Q = \dfrac{(3.3 \times 10^{-6})(6.62 \times 10^{-6})}{(2.00 \times 10^{-3})} = 1.1 \times 10^{-8}; Q > K$

The reaction will proceed left to attain equilibrium.

(b) $Q = \dfrac{(1.1 \times 10^{-7})(2.25 \times 10^{-6})}{(4.50 \times 10^{-2})} = 5.5 \times 10^{-12}; Q < K$

The reaction will proceed right to attain equilibrium.

(c) $Q = \dfrac{(1.48 \times 10^{-6})^2}{(0.0100)} = 2.19 \times 10^{-10}; \; Q = K$

The reaction is at equilibrium.

15.38   Calculate the reaction quotient in each case, compare with

$$K_p = \frac{P_{NH_3}^2}{P_{N_2} \times P_{H_2}^3} = 4.51 \times 10^{-5}$$

(a) $Q = \dfrac{(105)^2}{(35)(495)^3} = 2.6 \times 10^{-6}$

Since $Q < K_p$, the reaction will shift to the right to attain equilibrium.

(b) $Q = \dfrac{(35)^2}{(0)(595)^3} = \infty$

Since $Q > K_p$, reaction must shift to the left to attain equilibrium. There must be **some** N_2 present to attain equilibrium. In this example, the only source of N_2 is the decomposition of NH_3.

(c) $Q = \dfrac{(26)^2}{(42)^3(202)} = 4.52 \times 10^{-5}; \; Q = K_p$ Reaction is at equilibrium.

15.39   *Analyze/Plan.* Follow the logic in Sample Exercise 15.11. We are given concentrations, so write the K_c expression and solve for $[Cl_2]$. Change molarity to partial pressure using the ideal gas equation and the definition of molarity.   *Solve.*

$$K_c = \frac{[SO_2][Cl_2]}{[SO_2Cl_2]}; \; [Cl_2] = \frac{K_c[SO_2Cl_2]}{[SO_2]} = \frac{(0.078)(0.108)}{0.052} = 0.16200 = 0.16 \, M$$

$$PV = nRT, \; P = \frac{n}{V}RT; \; \frac{n}{V} = M; \; P = M\,RT; \; T = 100°C + 273 = 373 \, K$$

$$P_{Cl_2} = \frac{0.16200 \, mol}{L} \times \frac{0.08206 \, L \cdot atm}{mol \cdot K} \times 373 \, K = 4.959 = 5.0 \, atm$$

Check. $K_c = \dfrac{(0.052)(0.162)}{(0.108)} = 0.078$. Our values are self-consistent.

15.40 $K_p = \dfrac{P_{SO_3}^2}{P_{SO_2}^2 \times P_{O_2}}; \; P_{SO_3} = \left(K_p \times P_{SO_2}^2 \times P_{O_2} \right)^{1/2} = [(0.345)(0.165)^2(0.755)]^{1/2} = 0.0842 \, atm$

15.41   *Analyze/Plan.* Follow the logic in Sample Exercise 15.11. In each case, change given masses to molarities solve for the equilibrium molarity of the desired component, and calculate mass of that substance present at equilibrium.   *Solve.*

(a) $K_c = \dfrac{[Br]^2}{[Br_2]} = 1.04 \times 10^{-3}$

$$[Br_2] = \frac{0.245 \, g \, Br_2}{0.200 \, L} \times \frac{1 \, mol \, Br_2}{159.8 \, g \, Br_2} = 0.007666 = 0.00767 \, M$$

$[Br] = (K_c[Br_2])^{1/2} = [(1.04 \times 10^{-3})(0.007666)]^{1/2} = 0.002824 = 0.00282 \, M$

$$\dfrac{0.002824 \text{ mol Br}}{L} \times 0.200 \text{ L} \times \dfrac{79.90 \text{ g Br}}{\text{mol}} = 0.0451 \text{ g Br(g)}$$

Check. $K_c = (0.002824)^2 / (0.007666) = 1.04 \times 10^{-3}$

(b) $K_c = \dfrac{[HI]^2}{[H_2][I_2]} = 55.3; \quad [HI] = (K_c[H_2][I_2])^{1/2}$

$$[H_2] = \dfrac{0.056 \text{ g H}_2}{2.00 \text{ L}} \times \dfrac{1 \text{ mol H}_2}{2.016 \text{ g H}_2} = 0.01389 = 0.014 \, M$$

$$[I_2] = \dfrac{4.36 \text{ g I}_2}{2.00 \text{ L}} \times \dfrac{1 \text{ mol I}_2}{253.8 \text{ g I}_2} = 0.008589 = 0.00859 \, M$$

$[HI] = [(55.3)(0.01389)(0.008589)]^{1/2} = 0.08122 = 0.081 \, M$

$$0.08122 \text{ mol HI} \times 2.00 \text{ L} \times \dfrac{127.9 \text{ g HI}}{\text{mol HI}} = 20.78 = 21 \text{ g HI}$$

Check. $K_c = \dfrac{(0.08122)^2}{(0.01389)(0.008589)} = 55.3$

15.42   (a) $K_c = \dfrac{[I]^2}{[I_2]} = 3.1 \times 10^{-5}$

$$[I] = \dfrac{2.67 \times 10^{-2} \text{ g I}}{10.0 \text{ L}} \times \dfrac{1 \text{ mol I}}{126.9 \text{ g I}} = 2.1040 \times 10^{-5} = 2.10 \times 10^{-5} \, M$$

$$[I_2] = \dfrac{[I]^2}{K_c} = \dfrac{(2.104 \times 10^{-5})^2}{3.1 \times 10^{-5}} = 1.428 \times 10^{-5} = 1.43 \times 10^{-5} \, M$$

$$\dfrac{1.428 \times 10^{-5} \text{ mol I}_2}{L} \times 10.0 \text{ L} \times \dfrac{253.8 \text{ g I}_2}{\text{mol I}_2} = M = 0.0362 \text{ g I}_2$$

Check. $K_c - \dfrac{(2.104 \times 10^{-5})^2}{1.428 \times 10^{-5}} = 3.1 \times 10^{-5}$

(b) $PV = nRT; P = \dfrac{gRT}{MM \, V}$

$$P_{SO_3} = \dfrac{1.57 \text{ g SO}_3}{80.06 \text{ g/mol}} \times \dfrac{0.08206 \text{ L} \cdot \text{atm}}{\text{K} \cdot \text{mol}} \times \dfrac{700 \text{ K}}{2.00 \text{ L}} = 0.5632 = 0.563 \text{ atm}$$

$$P_{O_2} = \dfrac{0.125 \text{ g O}_2}{32.00 \text{ g/mol}} \times \dfrac{0.08206 \text{ L} \cdot \text{atm}}{\text{K} \cdot \text{mol}} \times \dfrac{700 \text{ K}}{2.00 \text{ L}} = 0.1122 = 0.112 \text{ atm}$$

$$K_p = 3.0 \times 10^4 = \dfrac{P_{SO_3}^2}{P_{SO_2}^2 \times P_{O_2}}; P_{SO_2} = \left[P_{SO_3}^2 / (K_p)(P_{O_2})\right]^{1/2}$$

$$P_{SO_2} = [(0.5632)^2 / (3.0 \times 10^4)(0.1122)]^{1/2} = 9.708 \times 10^{-3} = 9.7 \times 10^{-3} \text{ atm}$$

$$g\ SO_2 = \frac{MM\ PV}{RT} = \frac{64.06\ g\ SO_2}{mol\ SO_2} \times \frac{K \cdot mol}{0.08206\ L \cdot atm} \times \frac{9.708 \times 10^{-3}\ atm \times 2.00\ L}{700\ K}$$

$$= 0.02165 = 0.022\ g\ SO_2$$

Check. $K_p = [(9.708 \times 10^{-3})^2/(0.563)^2(0.1122)] = 3.0 \times 10^4$

15.43 *Analyze/Plan.* Follow the logic in Sample Exercise 15.12. Since molarity of NO is given directly, we can construct the equilibrium table straight away. *Solve.*

	$2NO(g)$	$\rightleftharpoons$	$N_2(g)$	$+$	$O_2(g)$	$K_c = \dfrac{[N_2][O_2]}{[NO]^2} = 2.4 \times 10^3$
initial	0.200 *M*		0		0	
change	–2x		+x		+x	
equil.	0.200 – 2x		+x		+x	

$$2.4 \times 10^3 = \frac{x^2}{(0.200 - 2x)^2}; (2.4 \times 10^3)^{1/2} = \frac{x}{0.200 - 2x}$$

$x = (2.4 \times 10^3)^{1/2}(0.200 - 2x); x = 9.798 - 97.98x; 98.98x = 9.798, x = 0.09899 = 0.099\ M$

$[N_2] = [O_2] = 0.099\ M; [NO] = 0.200 - 2(0.09899) = 0.00202 = 0.002\ M$

Check. $K_c = (0.09899)^2/(0.00202)^2 = 2.4 \times 10^3$

15.44 $[Br_2] = [Cl_2] = 0.30\ mol/1.0\ L = 0.30\ M$

	$Br_2(g)$	$+$	$Cl_2(g)$	$\rightleftharpoons$	$2BrCl(g)$	$K_c = \dfrac{[BrCl]^2}{[Br_2][Cl_2]} = 7.0$
initial	0.30 *M*		0.30 *M*		0	
change	–x		–x		+2x	
equil.	(0.30 – x)		(0.30 – x)		+2x	

$$7.0 = \frac{(2x)^2}{(0.30 - x)^2} \quad \text{(We can solve this exactly by taking the square root of both sides.)}$$

$$(7.0)^{1/2} = \frac{2x}{0.30 - x}, 2.646(0.30 - x) = 2x, 0.7937 = 4.646x, x = 0.1709 = 0.17\ M$$

$[BrCl] = 2x = 0.3417 = 0.34\ M; [Br_2] = [Cl_2] = 0.30 - x = 0.1291 = 0.13\ M$

Check. $K_c = (0.3417)^2/(0.1291)^2 = 7.0$

15.45 *Analyze/Plan.* Write the K_p expression, substitute the stated pressure relationship, and solve for P_{Br_2}. *Solve.*

$$K_p = \frac{P_{NO}^2 \times P_{Br_2}}{P_{NOBr}^2}$$

When $P_{NOBr} = P_{NO}$, these terms cancel and $P_{Br_2} = K_p = 0.416$ atm. This is true for all cases where $P_{NOBr} = P_{NO}$.

15.46 $K_c = [NH_3][H_2S] = 1.2 \times 10^{-4}$. Because of the stoichiometry, equilibrium concentrations of H_2S and NH_3 will be equal; call this quantity y. Then, $y^2 = 1.2 \times 10^{-4}$, $y = 0.010954 = 0.011\ M$.

15.47   (a) $CaSO_4(s) \rightleftharpoons Ca^{2+}(aq) + SO_4{}^{2-}(aq)$ $K_c = [Ca^{2+}][SO_4{}^{2-}] = 2.4 \times 10^{-5}$

            At equilibrium, $[Ca^{2+}] = [SO_4{}^{2-}] = x$

            $K_c = 2.4 \times 10^{-5} = x^2;\ x = 4.9 \times 10^{-3}\ M\ Ca^{2+}$ and $SO_4{}^{2-}$

    (b)   A saturated solution of $CaSO_4(aq)$ is $4.9 \times 10^{-3}\ M$.

            3.0 L of this solution contain:

$$\frac{4.9 \times 10^{-3}\ mol}{L} \times 3.0\ L \times \frac{136.14\ g\ CaSO_4}{mol} = 2.001 = 2.0\ g\ CaSO_4$$

            A bit more than 2.0 g $CaSO_4$ is needed in order to have some undissolved $CaSO_4(s)$ in equilibrium with 3.0 L of saturated solution.

15.48   (a)   *Analyze/Plan.* If only $PH_3BCl_3(s)$ is present initially, the equation requires that the equilibrium concentrations of $PH_3(g)$ and $BCl_3(g)$ are equal. Write the K_c expression and solve for $x = [PH_3] = [BCl_3]$. *Solve.*

            $K_c = [PH_3][BCl_3];\ 1.87 \times 10^{-3} = x^2;\ x = 0.043243 = 0.0432\ M\ PH_3$ and BCl_3

    (b)   Since the mole ratios are 1:1:1, mol $PH_3BCl_3(s)$ required = mol PH_3 or BCl_3 produced.

$$\frac{0.043243\ mol\ PH_3}{L} \times 0.500\ L = 0.02162 = 0.0216\ mol\ PH_3 = 0.0216\ mol\ PH_3BCl_3$$

$$0.02162\ mol\ PH_3BCl_3 \times \frac{151.2\ g\ PH_3BCl_3}{1\ mol\ PH_3BCl_3} = 3.269 = 3.27\ g\ PH_3BCl_3$$

            In fact, some $PH_3BCl_3(s)$ must remain for the system to be in equilibrium, so a bit more than 3.27 g PH_3BCl_3 is needed.

15.49   *Analyze/Plan.* Follow the approach in Solution 15.43. Calculate [IBr] from mol IBr and construct the equilibrium table. *Solve.*

[IBr] = 0.500 mol/1.00 L = 0.500 M

Since no I_2 or Br_2 was present initially, the amounts present at equilibrium are produced by the reverse reaction and stoichiometrically equal. Let these amounts equal x. The amount of HBr that reacts is then 2x. Substitute the equilibrium molarities (in terms of x) into the equilibrium expression and solve for x.

            I_2 + Br_2 $\rightleftharpoons$ 2IBr $K_c = \dfrac{[IBr]^2}{[I_2Br_2]} = 280$

	I_2	Br_2	2IBr
initial	0 M	0 M	0.500 M
change	+x M	+x M	−2x M
equil.	x M	x M	(0.500 − 2x) M

$$K_c = 280 = \frac{(0.500 - 2x)^2}{x^2} \; ; \text{taking the square root of both sides}$$

$$16.733 = \frac{0.500 - 2x}{x} \; ; 16.733x + 2x = 0.500; \quad 18.733x = 0.500$$

$$x = 0.02669 = 0.0267 \; M; \; [I_2] = [Br_2] = 0.0267 \; M$$

$$[IBr] = 0.500 - 2(0.02669) = 0.4466 = 0.447 \; M$$

Check. $\dfrac{(0.447)^2}{(0.0267)^2} = 280.$ Our values are self-consistent.

15.50 $CaCrO_4(s) \rightleftharpoons Ca^{2+}(aq) + CrO_4{}^{2-}(aq) \quad K_c = [Ca^{2+}][CrO_4{}^{2-}] = 7.1 \times 10^{-4}$

At equilibrium, $[Ca^{2+}] = [CrO_4{}^{2-}] = x$

$K_c = 7.1 \times 10^{-4} = x^2, \; x = 0.0266 = 0.027 \; M \; Ca^{2+}$ and $CrO_4{}^{2-}$

LeChâtelier's Principle

15.51   *Analyze/Plan.* Follow the logic in Sample Exercise 15.13.   *Solve.*

(a)   Shift equilibrium to the right; more $SO_3(g)$ is formed, the amount of $SO_2(g)$ decreases.

(b)   Heating an exothermic reaction decreases the value of K. More SO_2 and O_2 will form, the amount of SO_3 will decrease.

(c)   Since, $\Delta n = -1$, a change in volume will affect the equilibrium position and favor the side with more moles of gas. The amounts of SO_2 and O_2 increase and the amount of SO_3 decreases; equilibrium shifts to the left.

(d)   No effect. Speeds up the forward and reverse reactions equally.

(e)   No effect. Does not appear in the equilibrium expression.

(f)   Shift equilibrium to the right; amounts of SO_2 and O_2 decrease.

15.52   (a)   increase   (b)   increase   (c)   decrease

(d)   no effect   (e)   no effect   (f)   no effect

15.53   *Analyze/Plan.* Given certain changes to a reaction system, determine the effect on K_p, if any. Only changes in temperature cause changes to the value of K_p.   *Solve.*

(a)   no effect   (b)   no effect   (c)   increase equilibrium constant   (d)   no effect

15.54   (a)   The reaction must be endothermic ($+\Delta H$) if heating increases the fraction of products.

(b)   There must be more moles of gas in the products if increasing the volume of the vessel increases the fraction of products.

15.55   *Analyze/Plan.* Use Hess's Law, $\Delta H° = \Sigma \Delta H_f^{\circ}$ products $- \Sigma \Delta H_f^{\circ}$ reactants, to calculate $\Delta H°$. According to the sign of $\Delta H°$, describe the effect of temperature on the value of K. According to the value of Δn, describe the effect of changes to container volume.

Solve.

(a) $\Delta H° = \Delta H_f° \, NO_2(g) + \Delta H_f° \, N_2O(g) - 3\Delta H_f° \, NO(g)$

$\Delta H° = 33.84 \text{ kJ} + 81.6 \text{ kJ} - 3(90.37 \text{ kJ}) = -155.7 \text{ kJ}$

(b) The reaction is exothermic because it has a negative value of $\Delta H°$. The equilibrium constant will decrease with increasing temperature.

(c) Δn does not equal zero, so a change in volume at constant temperature will affect the fraction of products in the equilibrium mixture. An increase in container volume would favor reactants, while a decrease in volume would favor products.

15.56 (a) $\Delta H° = \Delta H_f° \, CH_3OH(g) - \Delta H_f° \, CO(g) - 2\Delta H_f° \, H_2(g)$

$= -201.2 \text{ kJ} - (-110.5 \text{ kJ}) - 0 \text{ kJ}$

$= -90.7 \text{ kJ}$

(b) The reaction is exothermic; an increase in temperature would decrease the value of K and decrease the yield. A low temperature is needed to maximize yield.

(c) Increasing total pressure would increase the partial pressure of each gas, shifting the equilibrium toward products. The extent of conversion to CH_3OH increases as the total pressure increases.

Additional Exercises

15.57 (a) Since both the forward and reverse processes are elementary steps, we can write the rate laws directly from the chemical equation.

$\text{rate}_f = k_f [CO][Cl_2] = \text{rate}_r = k_r [COCl][Cl]$

$\dfrac{k_f}{k_r} = \dfrac{[COCl][Cl]}{[CO][Cl_2]} = K$

$K_c = \dfrac{k_f}{k_r} = \dfrac{1.4 \times 10^{-28} \, M^{-1} s^{-1}}{9.3 \times 10^{10} \, M^{-1} s^{-1}} = 1.5 \times 10^{-39}$

For a homogeneous equilibrium in the gas phase, we usually write K in terms of partial pressures. In this exercise, concentrations are more convenient because the rate constants are expressed in terms of molarity. For this reaction, the value of K is the same regardless of how it is expressed, because there is no change in the moles of gas in going from reactants to products.

(b) Since the K is quite small, reactants are much more plentiful than products at equilibrium.

15.58 $CH_4(g) + H_2O(g) \rightarrow CO(g) + 3H_2(g)$

$K_p = \dfrac{P_{CO} \times P_{H_2}^3}{P_{CH_4} \times P_{H_2O}} ; \, P = \dfrac{g \, RT}{MM \, V} ; \, T = 1000 \text{ K}$

$P_{CO} = \dfrac{8.62 \text{ g}}{28.01 \text{ g/mol}} \times \dfrac{0.08206 \text{ L} \cdot \text{atm}}{\text{mol} \cdot \text{K}} \cdot \dfrac{1000 \text{ K}}{5.00 \text{ L}} = 5.0507 = 5.05 \text{ atm}$

$$P_{H_2} = \frac{2.60\ g}{2.016\ g/mol} \times \frac{0.08206\ L \bullet atm}{mol \bullet K} \bullet \frac{1000\ K}{5.00\ L} = 21.1663 = 21.2\ atm$$

$$P_{CH_4} = \frac{43.0\ g}{16.04\ g/mol} \times \frac{0.08206\ L \bullet atm}{mol \bullet K} \bullet \frac{1000\ K}{5.00\ L} = 43.9973 = 44.0\ atm$$

$$P_{H_2O} = \frac{48.4\ g}{18.02\ g/mol} \times \frac{0.08206\ L \bullet atm}{mol \bullet K} \bullet \frac{1000\ K}{5.00\ L} = 44.0811 = 44.1\ atm$$

$$K_p = \frac{(5.0507)(21.1663)^3}{(43.9973)(44.0811)} = 24.6949 = 24.7$$

$$K_p = K_c(RT)^{\Delta n}, K_c = K_p/(RT)^{\Delta n}; \Delta n = 4 - 2 = 2$$

$$K_c = (24.6949)/[(0.08206)(1000)]^2 = 3.6673 \times 10^{-3} = 3.67 \times 10^{-3}$$

15.59 $[SO_2Cl_2] = \dfrac{2.00\ mol}{2.00\ L} = 1.00\ M$

The change in $[SOCl_2] = 0.56(1.00\ M) = 0.56\ M$

	$SO_2Cl_2(g)$	$\rightleftharpoons$	$SO_2(g)$	+	$Cl_2(g)$	$K_c = \dfrac{[SO_2][Cl_2]}{[SO_2Cl_2]}$
initial	1.00 M		0		0	
change	–0.56 M		+0.56 M		+0.56 M	
equil.	0.44 M		+0.56 M		+0.56 M	

$$K_c = \frac{(0.56)^2}{0.44} = 0.7127 = 0.71$$

15.60   (a) $H_2(g) + S(s) \rightleftharpoons H_2S(g)$ $K_c = [H_2S]/[H_2]$

   (b)   Calculate the molarities of H_2S and H_2.

$$[H_2S] = \frac{0.46\ g}{34.1\ g/mol} \times \frac{1}{1.0\ L} = 0.01349 = 0.013\ M$$

$$[H_2] = \frac{0.40\ g}{2.02\ g/mol} \times \frac{1}{1.0\ L} = 0.1980 = 0.20\ M$$

$$K_c = 0.01349/0.1980 = 0.06812 = 0.068$$

   (c)   Since S is a pure solid, its concentration doesn't change during the reaction, so [S] does not appear in the equilibrium expression.

15.61   (a) $K_p = \dfrac{P_{Br_2} \times P_{NO}^2}{P_{NOBr}^2}; P = \dfrac{gRT}{PV}; T = 100°C + 273 = 373$

$$P_{Br_2} = \frac{4.19\ g}{159.8\ g/mol} \times \frac{0.08206\ L \bullet atm}{K \bullet mol} \times \frac{373}{5.0\ L} = 0.16051 = 0.161\ atm$$

$$P_{NO} = \frac{3.08\ g}{30.01\ g/mol} \times \frac{0.08206\ L \bullet atm}{K \bullet mol} \times \frac{373}{5.0\ L} = 0.62828 = 0.628\ atm$$

$$P_{NOBr} = \frac{3.22\,g\,NOBr}{109.9\,g/mol} \times \frac{0.08206\,L \bullet atm}{K \bullet mol} \times \frac{373}{5.0\,L} = 0.17936 = 0.179\,atm$$

$$K_p = \frac{(0.16051)\,(0.62828)^2}{(0.17936)^2} = 1.9695 = 1.97 \qquad K_p = K_c(RT)^{\Delta n},\,\Delta n = 3-2 = 1$$

$$K_c = K_p/RT = 1.9695/(0.08206)(373) = 0.064345 = 0.0643$$

(b) $P_t = P_{Br_2} + P_{NO} + P_{NOBr} = 0.16051 + 0.62828 + 0.17936 = 0.96815 = 0.968\,atm$

15.62   (a)

	A(g)	$\rightleftharpoons$	2B(g)
initial	0.55 atm		0
change	−0.19 atm		+0.38 atm
equil.	0.36 atm		0.38 atm

$P_t = P_A + P_B = 0.36\,atm + 0.38\,atm = 0.74\,atm$

(b) $K_p = \dfrac{(P_B)^2}{P_A} = \dfrac{(0.38)^2}{0.36} = 0.4011 = 0.40$

15.63   (a) $K_p = \dfrac{P_{NH_3}^2}{P_{N_2} \times P_{H_2}^3} = 4.34 \times 10^{-3};\,T = 300°C + 273 = 573\,K$

$$P_{NH_3} = \frac{gRT}{MM \times V} = \frac{1.05\,g}{17.03\,g/mol} \times \frac{0.08206\,L \bullet atm}{K \bullet mol} \times \frac{573\,K}{1.00\,L} = 2.899 = 2.90\,atm$$

	$N_2(g) +$	$3H_2(g)$	$\rightleftharpoons$	$2NH_3(g)$
initial	0 atm	0 atm		?
change	x	3x		−2x
equil.	x atm	3x atm		2.899 atm

(Remember, only the change line reflects the stoichiometry of the reaction.)

$$K_p = \frac{(2.899)^2}{(x)\,(3x)^3} = 4.34 \times 10^{-3};\,27\,x^4 = \frac{(2.899)^2}{4.34 \times 10^{-3}};\,x^4 = 71.725$$

$x = 2.910 = 2.91\,atm = P_{N_2};\,P_{H_2} = 3x = 8.730 = 8.73\,atm$

$$g_{N_2} = \frac{MM \times PV}{RT} = \frac{28.02\,g\,N_2}{mol\,N_2} \times \frac{K \bullet mol}{0.08206\,L \bullet atm} \times \frac{2.910\,atm \times 1.00\,L}{573\,K} = 1.73\,g\,N_2$$

$$g_{H_2} = \frac{2.016\,g\,H_2}{mol\,H_2} \times \frac{K \bullet mol}{0.08206\,L \bullet atm} \times \frac{8.730\,atm \times 1.00\,L}{573\,K} = 0.374\,g\,H_2$$

(b)   The initial $P_{NH_3} = 2.899\,atm + 2(2.910\,atm) = 8.719 = 8.72\,atm$

$$g_{NH_3} = \frac{17.03\,g\,NH_3}{mol\,NH_3} \times \frac{K \bullet mol}{0.08206\,L \bullet atm} \times \frac{8.719\,atm \times 1.00\,L}{573\,K} = 3.16\,g\,NH_3$$

(c) $P_t = P_{N_2} + P_{H_2} + P_{NH_3} = 2.910\,atm + 8.730\,atm + 2.899\,atm = 14.54\,atm$

15.64

$$2IBr \rightleftharpoons I_2 + Br_2$$

initial	0.025 atm	0 0
change	−2x	x x
equil.	(0.025 − 2x) atm	x x

$$K_p = 8.5 \times 10^{-3} = \frac{P_{I_2} \times P_{Br_2}}{P_{IBr}^2} = \frac{x^2}{(0.025 - 2x)^2}; \quad \text{Taking the square root of both sides}$$

$$\frac{x}{0.025 - 2x} = (8.5 \times 10^{-3})^{1/2} = 0.0922; \; x = 0.0922(0.025 - 2x)$$

$$x + 0.184x = 0.002305; \; 1.184x = 0.002305; \; x = 0.001947 = 1.9 \times 10^{-3}$$

P_{IBr} at equilibrium = $0.025 - 2(1.947 \times 10^{-3}) = 0.02111 = 0.021$ atm

15.65   (a) $K_p = 0.052$; $K_p = K_c(RT)^{\Delta n}$; $\Delta n = 2 - 0 = 2$; $K_c = K_p/(RT)^2$

             $K_c = 0.052/[0.08206)(333)]^2 = 6.964 \times 10^{-5} = 7.0 \times 10^{-5}$

   (b) PH_3BCl_3 is a solid and its concentration is taken as a constant, C.

         $[BCl_3] = 0.0128$ mol/0.500 L = 0.0256 M

$$PH_3BCl_3 \rightleftharpoons PH_3 + BCl_3$$

initial	C	0 M	0.0256 M
change		+x M	+x M
equil.	C	+x M	(0.0256 + x) M

$$K_c = [PH_3][BCl_3]; \; 6.964 \times 10^{-5} = x(0.0256 + x); \; x^2 + 0.0256x - 6.964 \times 10^{-5} = 0$$

$$x = \frac{-0.0256 \pm [(0.0256)^2 - 4(1)(-6.964 \times 10^{-5})]^{1/2}}{2(1)} = 0.002480 = 2 \times 10^{-3} \; M \; PH_3$$

Check. The numerator in the quadratic has 1 sig fig, which leads to $[PH_3]$ with 1 sig fig. $K_c = (2 \times 10^{-3})(0.0256 + 2 \times 10^{-3}) = 6 \times 10^{-5}$. Using 2 or 3 figures for $[PH_3]$ leads to closer agreement.

15.66 $K_p = P_{NH_3} \times P_{H_2S}$; $P_t = 0.614$ atm

If the equilibrium amounts of NH_3 and H_2S are due solely to the decomposition of $NH_4HS(s)$, the equilibrium pressures of the two gases are equal, and each is 1/2 of the total pressure.

$$P_{NH_3} = P_{H_2S} = 0.614 \text{ atm}/2 = 0.307 \text{ atm}$$

$$K_p = (0.307)^2 = 0.0943$$

15.67   Initial $P_{SO_3} = \dfrac{gRT}{MM \, V} = \dfrac{0.831 \text{ g}}{80.07 \text{ g/mol}} \times \dfrac{0.08206 \text{ L} \cdot \text{atm}}{\text{K} \cdot \text{mol}} \times \dfrac{1100 \text{ K}}{1.00 \text{ L}} = 0.9368 = 0.937$ atm

	$2SO_3$	$\rightleftharpoons$	$2SO_2$	$+$	O_2
initial	0.9368 atm		0		0
change	$-2x$		$+2x$		$+x$
equil.	$0.9368-2x$		$2x$		x
[equil.]	0.2104 atm		0.7264 atm		0.3632 atm

$P_t = (0.9368-2x) + 2x + x; \; 0.9368 + x = 1.300 \text{ atm}; \; x = 1.300 - 0.9368 = 0.3632 = 0.363 \text{ atm}$

$$K_p = \frac{P_{SO_2}^2 \times P_{O_2}}{P_{SO_3}^2} = \frac{(0.7264)^2 (0.3632)}{(0.2104)^2} = 4.3292 = 4.33$$

$K_p = K_p = K_c(RT)^{\Delta n}; \; \Delta n = 3 - 2 = 1; \; K_p = K_c(RT)$

$K_c = K_p/RT = 4.3292/[(0.08206)(1100)] = 0.04796 = 0.0480$

15.68   In general, the reaction quotient is of the form $Q = \dfrac{P_{NOCl}^2}{P_{NO}^2 \times P_{Cl_2}}$.

(a) $Q = \dfrac{(0.11)^2}{(0.15)^2 (0.31)} = 1.7$

$Q > K_p$. Therefore, the reaction will shift toward reactants, to the left, in moving toward equilibrium.

(b) $Q = \dfrac{(0.050)^2}{(0.12)^2 (0.10)} = 1.7$

$Q > K_p$. Therefore, the reaction will shift toward reactants, to the left, in moving toward equilibrium.

(c) $Q = \dfrac{(5.10 \times 10^{-3})^2}{(0.15)^2 (0.20)} = 5.8 \times 10^{-3}$

$Q < K_p$. Therefore, the reaction mixture will shift in the direction of more product, to the right, in moving toward equilibrium.

15.69 $K_c = [CO_2] = 0.0108; \; [CO_2] = \dfrac{g \, CO_2}{44.01 \text{ g/mol}} \times \dfrac{1}{10.0 \text{ L}}$

In each case, calculate $[CO_2]$ and determine the position of the equilibrium.

(a) $[CO_2] = \dfrac{4.25 \text{ g}}{44.01 \text{ g/mol}} \times \dfrac{1}{10.0 \text{ L}} = 9.657 \times 10^{-3} = 9.66 \times 10^{-3} \, M$

$Q = 9.66 \times 10^{-3} > K_c$. The reaction proceeds to the right to achieve equilibrium and the amount of $CaCO_3(s)$ decreases.

(b) $[CO_2] = \dfrac{5.66 \text{ g } CO_2}{44.01 \text{ g/mol}} \times \dfrac{1}{10.0 \text{ L}} = 0.0129 \text{ } M$

Q = 0.0129 > K_c. The reaction proceeds to the left to achieve equilibrium and the amount of $CaCO_3(s)$ increases.

(c)   6.48 g CO_2 means $[CO_2] > 0.0129 \text{ } M$; $Q > 0.0129 > K_c$, the amount of $CaCO_3$ increases.

15.70   (a) $K_p = \dfrac{P_{Ni(CO)_4}}{P_{CO}^4}$

(b)   Increasing the temperature to 200°C favors the reverse process (decomposition of $Ni(CO)_4(g)$ and thus the value of K_p is smaller at the higher temperature. This is the behavior expected from an exothermic reaction (heat is a product).

(c)   At the temperature of the exhaust pipe, the $Ni(CO)_4$ product is a gas and is carried into the atmosphere with other exhaust gases. Thus, equilibrium is never established (we do not have a closed system) and the reaction proceeds to the right as $Ni(CO)_4$ product is removed.

15.71 $K_p = \dfrac{P_{CO_2}}{P_{CO}} = 6.0 \times 10^2$

If P_{CO} is 150 torr, P_{CO_2} can never exceed 760 – 150 = 610 torr. Then Q = 610/150 = 4.1. Since this is far less than K, the reaction will shift in the direction of more product. Reduction will therefore occur.

15.72   (a)

$$CCl_4(g) \rightleftharpoons C(s) + 2Cl_2(g)$$

initial	2.00 atm	0 atm
change	–x atm	+2x atm
equil.	(2.00–x) atm	2x atm

$K_p = 0.76 = \dfrac{P_{Cl_2}^2}{P_{CCl_4}} = \dfrac{(2x)^2}{(2.00-x)}$

$1.52 - 0.76x = 4x^2$; $4x^2 + 0.76x - 1.52 = 0$

Using the quadratic formula, a = 4, b = 0.76, c = –1.52

$x = \dfrac{-0.76 \pm \sqrt{(0.76)^2 - 4(4)(-1.52)}}{2(4)} = \dfrac{-0.76 + 4.99}{8} = 0.5287 = 0.53 \text{ atm}$

Fraction CCl_4 reacted $= \dfrac{x \text{ atm}}{2.00 \text{ atm}} = \dfrac{0.53}{2.00} = 0.264 = 26\%$

(b) $P_{Cl_2} = 2x = 2(0.5287) = 1.06 \text{ atm}$

$P_{CCl_4} = 2.00 - x = 2.00 - 0.5287 = 1.47 \text{ atm}$

15.73   (a) $Q = \dfrac{P_{PCl_5}}{P_{PCl_3} \times P_{Cl_2}} = \dfrac{(0.20)}{(0.50)(0.50)} = 0.80$

0.80 (Q) > 0.0870 (K), the reaction proceeds to the left.

(b)

	$PCl_3(g)$	+	$Cl_2(g)$	$\rightleftharpoons$	$PCl_5(g)$
initial	0.50 atm		0.50 atm		0.20 atm
change	+x atm		+x atm		–x atm
equil.	(0.50 + x) atm		(0.50 + x) atm		(0.20 – x) atm

(Since the reaction proceeds to the left, P_{PCl_5} must decrease and P_{PCl_3} and P_{Cl_2} must increase.)

$K_p = 0.0870 = \dfrac{(0.20 - x)}{(0.50 + x)(0.50 + x)};\quad 0.0870 = \dfrac{(0.20 - x)}{(0.250 + 1.00\,x + x^2)}$

$0.0870(0.250 + 1.00x + x^2) = 0.20 - x;\ -0.17825 + 1.0870x + 0.0870\,x^2 = 0$

$x = \dfrac{-1.0870 \pm \sqrt{(1.0870)^2 - 4(0.0870)(-0.17825)}}{2(0.0870)} = \dfrac{-1.0870 + 1.1152}{0.174} = 0.162$

$P_{PCl_3} = (0.50 + 0.162)\ \text{atm} = 0.662 \qquad P_{Cl_2} = (0.50 + 0.162)\ \text{atm} = 0.662\ \text{atm}$

$P_{PCl_5} = (0.20 - 0.162)\ \text{atm} = 0.038\ \text{atm}$

To two decimal places, the pressures are 0.66, 0.66 and 0.04 atm, respectively. When substituting into the K_p expression, pressures to three decimal places yield a result much closer to 0.0870.

(c)   Increasing the volume of the container favors the process where more moles of gas are produced, so the reverse reaction is favored and the equilibrium shifts to the left; the mole fraction of Cl_2 increases.

(d)   For an exothermic reaction, increasing the temperature decreases the value of K; more reactants and fewer products are present at equilibrium and the mole fraction of Cl_2 increases.

15.74   *Analyze/Plan.* Equilibrium pressures of H_2, I_2, HI $\rightarrow K_p \rightarrow$ equilibrium table $\rightarrow$ new equilibrium pressures.   *Solve.*

$P_{H_2} = P_{I_2} = 0.112\ \text{mol} \times 11.997\ \dfrac{\text{atm}}{\text{mol}} = 1.344 = 1.34\ \text{atm}$

$P_{HI} = 0.775\ \text{mol} \times 11.997\ \dfrac{\text{atm}}{\text{mol}} = 9.298 = 9.30\ \text{atm}$

$$H_2(g) + I_2(g) \rightleftharpoons 2HI(g); \quad K_p = \frac{P_{HI}^2}{P_{H_2} \times P_{I_2}} = \frac{(9.298)^2}{(1.344)^2} = 47.861 = 47.9$$

$$P_{HI} \text{ (added)} = 0.100 \text{ mol} \times \frac{11.997 \text{ atm}}{\text{mol}} = 1.1997 = 1.20 \text{ atm}$$

	$H_2(g)$	+	$I_2(g)$	$\rightleftharpoons$	$2HI(g)$
initial	1.34 atm		1.34 atm		9.30 atm + 1.20 atm
change	+x atm		+x atm		–2x atm
equil.	(1.34+x) atm		(1.34+x) atm		(10.50–2x) atm

$$K_p = 47.86 = \frac{(10.50 - 2x)^2}{(1.34 + x)^2}. \quad \text{Take the square root of both sides :}$$

$$6.918 = \frac{10.50 - 2x}{1.34 + x}; \ 9.270 + 6.918\,x = 10.50 - 2x; \ 8.918\,x = 1.230; \ x = 0.1379 = 0.138$$

$$P_{H_2} = P_{I_2} = 1.34 + 0.138 = 1.48 \text{ atm}; \ P_{HI} = 10.50 - 2(0.138) = 10.22 \text{ atm}$$

Check: $\dfrac{(10.22)^2}{(1.48)^2} = 47.68 = 47.7$

15.75   (a)   Since the volume of the vessel = 1.00 L, mol = *M*. The reaction will proceed to the left to establish equilibrium.

	$A(g)$ +	$2B(g)$	$\rightleftharpoons$	$2C(g)$
initial	0 *M*	0 *M*		1.00 *M*
change	+x *M*	+2x *M*		–2x *M*
equil.	x *M*	2x *M*		(1.00 – 2x) *M*

At equilibrium, [C] = (1.00 – 2x) *M*, [B] = 2x *M*.

  (b)   x must be less than 0.50 *M* (so that [C], 1.00 –2x, is not less than zero).

  (c) $K_c = \dfrac{[C]^2}{[A][B]^2}; \ \dfrac{(1.00 - 2x)^2}{(x)(2x)^2} = 0.25$

$$1.00 - 4x + 4x^2 = 0.25(4x)^3; \ x^3 - 4x^2 + 4x - 1 = 0$$

(d)

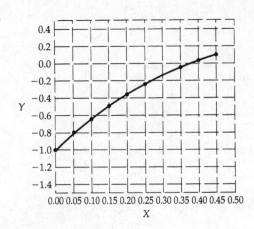

X	Y
0.0	−1.000
0.05	−0.810
0.10	−0.639
0.15	−0.487
0.20	−0.352
0.25	−0.234
0.35	−0.047
0.40	+0.024
0.45	+0.081
~0.383	0.00

(e)   From the plot, $x \approx 0.383\ M$

        $[A] = x = 0.383\ M$; $[B] = 2x = 0.766\ M$

        $[C] = 1.00 - 2x = 0.234\ M$

        Using the K_c expression as a check:

$$K_c = 0.25; \quad \frac{(0.234)^2}{(0.383)(0.766)^2} = 0.24; \quad \text{the estimated values are reasonable.}$$

15.76 $K_p = \dfrac{P_{O_2} \times P_{CO}^2}{P_{CO_2}^2} \approx 1 \times 10^{-13}$; $P_{O_2} = (0.03)(1\ \text{atm}) = 0.03\ \text{atm}$

        $P_{CO} = (0.002)(1\ \text{atm}) = 0.002\ \text{atm}$; $P_{CO_2} = (0.12)(1\ \text{atm}) = 0.12\ \text{atm}$

        $Q = \dfrac{(0.03)(0.002)^2}{(0.12)^2} = 8.3 \times 10^{-6} = 8 \times 10^{-6}$

        Since $Q > K_p$, the system will shift to the left to attain equilibrium. Thus a catalyst that promoted the attainment of equilibrium would result in a lower CO content in the exhaust.

15.77   The patent claim is false. A catalyst does not alter the position of equilibrium in a system, only the rate of approach to the equilibrium condition.

Integrative Exercises

15.78   (a)   (i) $K_c = [Na^+]/[Ag^+]$

              (ii) $K_c = [Hg^{2+}]^3 / [Al^{3+}]^2$

              (iii) $K_c = [Zn^{2+}][H_2] / [H^+]^2$

       (b)   According to Table 4.5, the activity series of the metals, a metal can be oxidized by any metal cation below it on the table.

              (i) Ag^+ is far below Na, so the reaction will proceed to the right and K_c will be large.

(ii) Al^{3+} is above Hg, so the reaction will not proceed to the right and K_c will be small.

(iii) H^+ is below Zn, so the reaction will proceed to the right and K_c will be large.

(c) $K_c < 1$ for this reaction, so Fe^{2+} (and thus Fe) is above Cd on the table. In other words, Cd is below Fe. The value of K_c, 0.06, is small but not extremely small, so Cd will be only a few rows below Fe.

15.79   (a) $AgCl(s) \rightleftharpoons Ag^+(aq) + Cl^-(aq)$

(b) $K_c = [Ag^+][Cl^-]$

(c)   Using thermodynamic data from Appendix C, calculate ΔH for the reaction in part (a).

$$\Delta H° = \Delta H_f° \, Ag^+(aq) + \Delta H_f° \, Cl^-(aq) - \Delta H_f° \, AgCl(s)$$

$$\Delta H° = 105.90 \text{ kJ} - 167.2 \text{ kJ} - (-127.0 \text{ kJ}) = 65.7 \text{ kJ}$$

The reaction is endothermic (heat is a reactant), so the solubility of AgCl(s) in $H_2O(l)$ will increase with increasing temperature.

15.80   (a)   At equilibrium, the forward and reverse reactions occur at **equal** rates.

(b)   One expects the reactants to be favored at equilibrium since they are lower in energy.

(c)   A catalyst lowers the activation energy for both the forward and reverse reactions; the "hill" would be lower.

(d)   Since the activation energy is lowered for both processes, the new rates would be equal and the ratio of the rate constants, k_f/k_r, would remain unchanged.

(e)   Since the reaction is endothermic (the energy of the reactants is lower than that of the products, ΔE is positive), the value of K should increase with increasing temperature.

15.81   Consider the energy profile for an exothermic reaction.

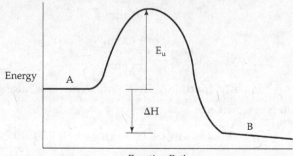

The activation energy in the forward direction, E_{af}, equals E_u, and the activation energy in the reverse reaction, E_{ar}, equals $E_u - \Delta H$. (The same is true for an endothermic reaction because the sign of ΔH is the positive and $E_{ar} < E_{af}$). For the reaction in question,

$$K = \frac{k_f}{k_r} = \frac{A_f e^{-E_{af}/RT}}{A_r e^{-E_{ar}/RT}}$$

Since the ln form of the Arrhenius equation is easier to manipulate, we will consider ln K.

$$\ln K = \ln\left(\frac{k_f}{k_r}\right) = \ln k_f - \ln k_r = \frac{-E_{af}}{RT} + \ln A_f - \left[\frac{-E_{ar}}{RT} + \ln A_r\right]$$

Substituting E_u for E_{af} and $(E_u - \Delta H)$ for E_{ar}

$$\ln K = \frac{-E_u}{RT} + \ln A_f - \left[\frac{-(E_u - \Delta H)}{RT} + \ln A_r\right]; \quad \ln K = \frac{-E_u + (E_u - \Delta H)}{RT} + \ln A_f - \ln A_r$$

$$\ln K = \frac{-\Delta H}{RT} + \ln \frac{A_f}{A_r}$$

For the catalyzed reaction, $E_{cat} < E_u$ and $E_{af} = E_{cat}$, $E_{ar} = E_{cat} - \Delta H$. The catalyst does not change the value of ΔH.

$$\ln K_{cat} = \frac{-E_{cat} + (E_{cat} - \Delta H)}{RT} + \ln A_f - \ln A_r$$

$$\ln K_{cat} = \frac{-\Delta H}{RT} + \frac{\ln A_f}{A_r}$$

Thus, assuming A_f and A_r are not changed by the catalyst, $\ln K = \ln K_{cat}$ and $K = K_{cat}$.

15.82   (a) $P = \dfrac{gRT}{MM\ V} = \dfrac{0.300\,\text{g H}_2\text{S}}{34.08\,\text{g/mol H}_2\text{S}} \times \dfrac{298\,\text{K}}{5.00\,\text{L}} \times \dfrac{0.08206\,\text{L} \cdot \text{atm}}{\text{mol} \cdot \text{K}} = 0.043053$

$$= 0.0431\,\text{atm}$$

(b) $K_p = P_{NH_3} \times P_{H_2S}$. Before solid is added, $Q = P_{NH_3} \times P_{H_2S} = 0 \times 0.0431 = 0$.

Q < K and the reaction will proceed to the right. However, no $NH_4SH(s)$ is present to produce $NH_3(g)$, so the reaction cannot proceed.

(c)

	$NH_4SH(s)$	$\rightleftharpoons$	$NH_3(g)$	+	$H_2S(g)$
initial			0 atm		0.043053 atm
change			+x atm		+x atm
equil.			+x atm		(0.043053+x) atm

Since Q < K initially [part (a)], P_{H_2S} must increase along with P_{NH_3} until equilibrium is established.

$K_p = P_{NH_3} \times P_{H_2S}$; $0.120 = (x)(0.043053 + x)$; $0 = x^2 + 0.043053\,x - 0.120$

Solve for x using the quadratic formula.

$$x = \frac{-0.043053 \pm \sqrt{(0.043053)^2 - 4(1)(-0.120)}}{2(1)}; \quad x = 0.3256 = 0.326\,\text{atm}$$

$P_{NH_3} = 0.326\,\text{atm}$; $P_{H_2S} = (0.043053 + 0.3256)\,\text{atm} = 0.3686 = 0.369\,\text{atm}$

(d) $\quad \chi_{H_2S} = \dfrac{P_{H_2S}}{P_t} = \dfrac{0.3686 \text{ atm}}{(0.3256 + 0.3686) \text{ atm}} = 0.531$

(e) The minimum amount of $NH_4HS(s)$ required is slightly greater than the number of moles NH_3 present at equilibrium. We can calculate the mol NH_3 present at equilibrium using the ideal-gas equation.

$$n_{NH_3} = \dfrac{P_{NH_3} V}{RT} = 0.3256 \text{ atm} \times \dfrac{K \bullet mol}{0.08206 \text{ L} \bullet atm} \times \dfrac{5.00 \text{ L}}{298 \text{ K}} = 0.06657$$

$$= 0.0666 \text{ mol } NH_3$$

$$0.06657 \text{ mol } H_2S \times \dfrac{1 \text{ mol } NH_4SH}{1 \text{ mol } H_2S} \times \dfrac{51.12 \text{ g } NH_4SH}{1 \text{ mol } NH_4SH} = 3.40 \text{ g } NH_4SH$$

The minimum amount is slightly greater than 3.40 g NH_4HS.

15.83 Mole % = pressure %. Since the total pressure is 1 atm, mol %/100 = mol fraction = partial pressure. $K_p = P_{CO}^2 / P_{CO_2}$.

Temp (K)	P_{CO_2} (atm)	P_{CO} (atm)	K_P
1123	0.0623	0.9377	14.1
1223	0.0132	0.9868	73.8
1323	0.0037	0.9963	2.7×10^2
1473	0.0006	0.9994	1.7×10^3 (2×10^3)

Because K grows larger with increasing temperature, the reaction must be endothermic in the forward direction.

15.84 (a) $H_2O(l) \rightleftharpoons H_2O(g)$; $K_p = P_{H_2O}$

(b) At 30°C, the vapor pressure of $H_2O(l)$ is 31.82 torr. $K_P = P_{H_2O} = 31.82$ torr

$K_p = 31.82$ torr $\times 1$ atm/760 torr $= 0.041868 = 0.04187$ atm

(c) From part (b), the value of K_p is the vapor pressure of the liquid at that temperature. By definition, vapor pressure = atmospheric pressure = 1 atm at the normal boiling point. $K_p = 1$ atm

15.85 (a)

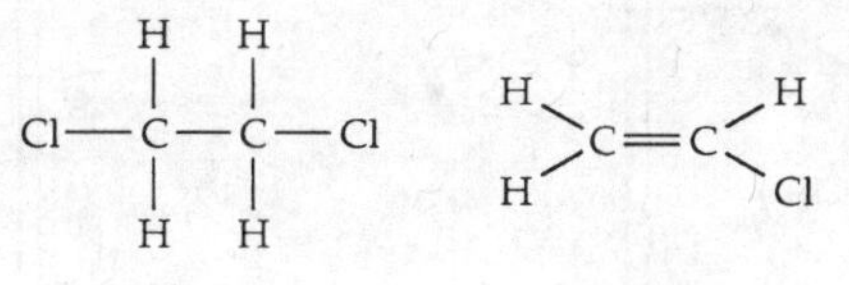

C-C B.O. = 1   C=C B.O. = 2

(b) $\Delta H = D$ (bond breaking) $- D$ (bond making)

E1: $\quad \Delta H = D(C=Cl) - D(C-Cl) - D(C-C) - 2D(C-Cl)$

$\Delta H = 614 + 242 - 348 - 2(328) = -148$ kJ

E2: $\Delta H = D(C\text{–}C) + D(C\text{–}H) + D(C\text{–}Cl) - D(C{=}C) - D(H\text{–}Cl)$

          $= 348 + 413 + 328 - 614 - 431 = 44 \text{ kJ}$

(c)   E1 is exothermic with $\Delta n = -1$. The yield of $C_2H_4Cl_2(g)$ would decrease with increasing temperature and with increasing container volume.

(d)   E2 is endothermic with $\Delta n = 1$. The yield of C_2H_3Cl would increase with increasing temperature and with increasing container volume.

(e)   The boiling points of the reactants and products are: C_2H_4, $-103.7°C$; Cl_2, $-34.6°C$, $C_2H_4Cl_2$, $+83.5°C$; C_2H_3Cl, $-13.4°C$; HCl, $-84.9°C$.

Because the products of E1 and E2 are optimized by different conditions, carry out the two equilibria in separate reactors. Since E1 is exothermic and $\Delta n = -1$, the reactor on the left should be as small and cold as possible to maximize yield of $C_2H_4Cl_2$. At temperatures below 83.5°C, $C_2H_4Cl_2$ will condense to the liquid and it can be easily transferred to the second (right) reactor.

Since E_2 is endothermic, the reactor on the right should be as large and hot as possible to optimize production of C_2H_3Cl. The outlet stream will be a mixture of $C_2H_3Cl(g)$, $C_2H_4Cl_2(g)$, and $HCl(g)$. This mixture could be run through a heat exchanger to condense and subsequently recycle $C_2H_4Cl_2(l)$. Since HCl has a lower boiling point than C_2H_3Cl, it cannot be removed by condensation. The $C_2H_3Cl(g)$ / $HCl(g)$ mixture could be bubbled through a basic aqueous solution such as $NaHCO_3(aq)$ or $NaOH(aq)$ to remove $HCl(g)$, leaving pure $C_2H_3Cl(g)$.

Other details of reactor design such as the use of catalysts to speed up these reactions, the exact costs and benefits of heat exchange, recycling unreacted components, and separation and recovery of products are issues best resolved by chemical engineers.

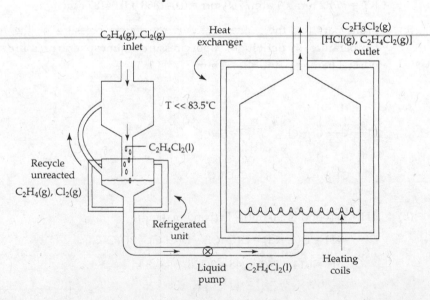

16 Acid-Base Equilibria

Visualizing Concepts

16.1   *Analyze.* From the structures decide which reactant fits the description of a Brønsted-Lowry (B-L) acid, a B-L base, a Lewis acid, and a Lewis base. *Plan.* A B-L acid is an H^+ donor, and a B-L base is an H^+ acceptor. A Lewis acid is an electron pair acceptor and a Lewis base is an electron pair donor.   *Solve.*

   (a)   H–X is a B-L acid, because it loses H^+ during reaction. NH_3 is a B-L base, because it gains H^+ during reaction.

   (b)   By virtue of its unshared electron pair, NH_3 is the electron pair donor and Lewis base. HX is the electron pair acceptor and Lewis acid.

16.2   *Plan.* The stronger the acid, the greater the extent of ionization. The stronger the acid, the weaker its conjugate base. In an acid-base reaction, equilibrium will favor the side with the weaker acid and base.   *Solve.*

   (a)   HY is stronger than HX. Starting with six HY molecules, four are dissociated; of six HX molecules, only two are dissociated. Because it is dissociated to a greater extent, HY is the stronger acid.

   (b)   If HY is the stronger acid, Y^- is the weaker base and X^- is the stronger base.

   (c)   HX and Y^-, the reactants, are the weaker acid and base. Equilibrium lies to the left, and $K_c < 1$.

16.3   *Plan.* Strong acids are completely ionized. The acid that is least ionized is weakest, and has the smallest K_a value. At equal concentrations, the weakest acid has the smallest $[H^+]$ and highest pH.   *Solve.*

   (a)   HY is a strong acid. There are no neutral HY molecules in solution, only H^+ cations and Y^- anions.

   (b)   HX has the smallest K_a value. It has most neutral acid molecules and fewest ions.

   (c)   HX has the fewest H^+ and highest pH.

16.4   *Plan.* The definition of % ionization is $\dfrac{[H^+]}{[HA]_{\text{Initial}}} \times 100$.   *Solve.*

   (a)   Curve C shows the effect of initial concentration on % ionization of a weak acid.

   (b)   The % ionization is inversely related to initial acid concentration; only curve C shows a decrease in % ionization as acid concentration increases.

16.5   *Plan.* The stronger the acid, the greater the extent of ionization. The stronger the acid, the weaker its conjugate base. *Solve.*

(a)   HY has most H^+ and is strongest; HX has the fewest H^+ and is weakest. The order of base strength is the reverse of the order of acid strength. $Y^- < Z^- < X^-$

(b)   The strongest base, X^-, has the largest K_b value.

16.6   *Plan.* Write the molecular formula so we can count the correct number of valence electrons. Use the atom connectivity shown to draw the Lewis structure. *Solve.*

(a)   The molecular formula is $(CH_3)_2NH$, or C_2H_7N. The number of valence electrons is $2(4) + 7 + 5 = 20\ e^-$, $10\ e^-$ pr.

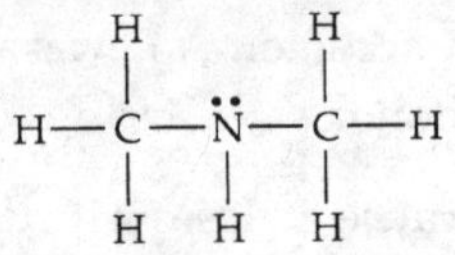

$$\begin{array}{ccccc} & H & & H & \\ & | & & | & \\ H- & C & -\overset{\displaystyle ..}{N}- & C & -H \\ & | & | & | & \\ & H & H & H & \end{array}$$

(b)   The compound is an amine. It is an ammonia molecule where two H atoms have been replaced by CH_3 groups.

16.7   *Plan.* Evaluate the interactions of Na^+ and X^- with H_2O.

Solve. Na^+ does not affect the $[H^+]$ or $[OH^-]$ of an aqueous solution. It is a "negligible" acid in water (which can be thought of as the conjugate acid of the strong base NaOH).

X^- is the conjugate base of HX. It is not a negligible base in water, because we see from the diagram that one X^- has gained an H^+ to form HX. In this solution, H_2O acts as the Brønsted acid, according to the hydrolysis equilibrium:

$$X^-(aq) + H_2O(l) \rightleftharpoons HX(aq) + OH^-(aq).$$

The missing ion is $OH^-(aq)$. According to the equilibrium reaction, the number of HX molecules and OH^- ions are equal. Since there is 1 HX molecule in the diagram, 1 OH^- should be shown.

16.8   *Plan.* Evaluate the molecular structures to determine if the acids are binary acids or oxyacids. Consider the trends in acid strength for both classes of acids. *Solve.*

(a)   The molecules are oxyacids; in both cases, the ionizable H atom is attached to O. The right molecule is a carboxylic acid; the ionizable H is part of a carboxyl,

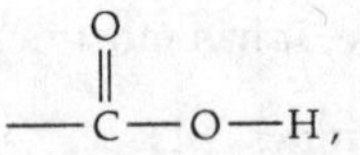

$$\begin{array}{c} O \\ \| \\ -C-O-H \end{array}\ ,\ \text{group.}$$

(b)   Increasing the electronegativity of X increases the strength of both acids. As X becomes more electronegative and attracts more electron density, the O–H bond becomes weaker and more polar. This increases the likelihood of ionization and increases acid strength. An electronegative X group also stabilizes the anionic conjugate bases by delocalizing the negative charge. This causes the ionization equilibrium to favor products, and the values of K_a to increase.

16.9   The carboxyl group in the H-atom group at the "top" of the molecule.

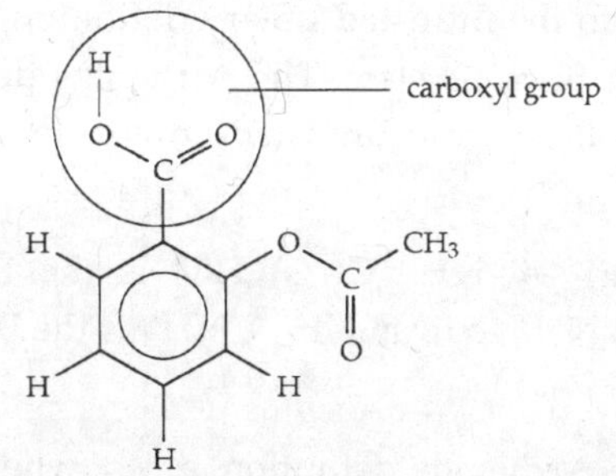

The group on the right of the molecule is <u>not</u> a carboxyl group because it contains no ionizable H.

16.10   (a)   *Plan.* Count valence electrons and draw the correct Lewis structures. Consider the definition of Lewis acids and bases.   *Solve.*

$$PCl_4^- \quad + \quad Cl^- \quad \longrightarrow \quad PCl_5$$
$$32\,e^-,\,16\,e^-\,pr \qquad 8\,e^-,\,4\,e^-\,pr \qquad 40\,e^-,\,20\,e^-\,pr$$

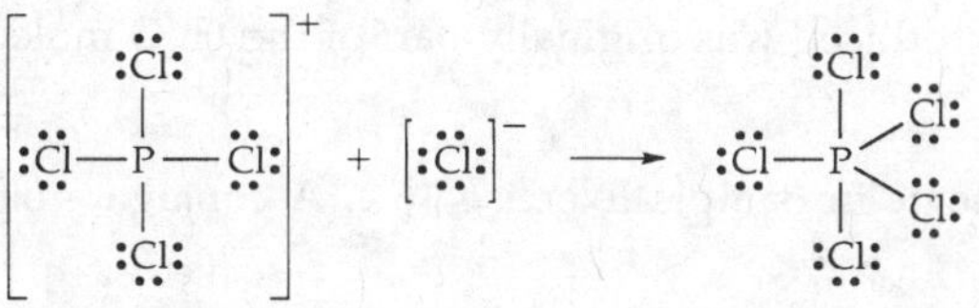

PCl_4^+ accepts an electron pair from Cl^-; PCl_4^+ is the Lewis acid and Cl^- is the Lewis base.

(b)   The hydrated cation is an oxyacid: the ionizable H is attached to O, which is bound to the central cation. As the charge on the cation increases, it attracts more electron density from the O–H bond, which becomes weaker and more polar. The degree of ionization increases and the equilibrium constant (K_a) increases.

Arrhenius and Brønsted-Lowry Acids and Bases

16.11   Solutions of HCl and H_2SO_4 taste sour, turn litmus paper red (are acidic), neutralize solutions of bases, react with active metals to form $H_2(g)$ and conduct electricity. The two solutions have these properties in common because both solutes are strong acids. That is, they both ionize completely in H_2O to form $H^+(aq)$ and an anion. (The first ionization step for H_2SO_4 is complete, but the second is not.) The presence of ions enables the solutions to conduct electricity; the presence of $H^+(aq)$ in excess of $1 \times 10^{-7}\,M$ accounts for all the other listed properties.

16.12   When NaOH dissolves in water, it completely dissociates to form $Na^+(aq)$ and OH^- (aq). CaO is the oxide of a metal; it dissolves in water according to the following process: $CaO(s) + H_2O(l) \rightarrow Ca^{2+}(aq) + 2OH^-(aq)$. Thus, the properties of both solutions are dominated by the presence of $OH^-(aq)$. Both solutions taste bitter, turn litmus paper blue (are basic), neutralize solutions of acids, and conduct electricity.

16.13   (a)   According to the Arrhenius definition, an *acid* when dissolved in water increases $[H^+]$. According to the Brønsted-Lowry definition, an *acid* is capable of donating H^+, regardless of physical state. The Arrhenius definition of an acid is confined to an aqueous solution; the Brønsted-Lowry definition applies to any physical state.

       (b) $HCl(g) + NH_3(g) \rightarrow NH_4^+Cl^-(s)$ HCl is the B-L (Brønsted-Lowry) acid; it donates an H^+ to NH_3 to form NH_4^+. NH_3 is the B-L base; it accepts the H^+ from HCl.

16.14   (a)   According to the Arrhenius definition, a *base* when dissolved in water increases $[OH^-]$. According to the Brønsted-Lowry theory, a *base* is an H^+ acceptor regardless of physical state. A Brønsted-Lowry base is not limited to aqueous solution and need not contain OH^- or produce it in aqueous solution.

       (b) $NH_3(g) + H_2O(l) \rightleftharpoons NH_4^+(aq) + OH^-(aq)$ When NH_3 dissolves in water, it accepts H^+ from H_2O (B-L definition). In doing so, OH^- is produced (Arrhenius definition).

       Note that the OH^- produced was originally part of the H_2O molecule, not part of the NH_3 molecule.

16.15   *Analyze/Plan.* Follow the logic in Sample Exercise 16.1. A conjugate base has one less H^+ than its conjugate acid.   *Solve.*

       (a) IO_3^- (b) NH_3 (c) HPO_4^{2-} (d) $C_7H_5O_2^-$

16.16   A conjugate acid has one more H^+ than its conjugate base.

       (a)   HCN   (b)   OH   (c) $H_2PO_4^-$ (d) $C_2H_5NH_3^+$

16.17   *Analyze/Plan.* Use the definitions of B-L acids and bases, and conjugate acids and bases to make the designations. Evaluate the changes going from reactant to product to inform your choices.   *Solve.*

	B-L acid	+	**B-L base**	$\rightleftharpoons$	**Conjugate acid**	+	**Conjugate base**
(a)	$NH_4^+(aq)$		$CN^-(aq)$		$HCN(aq)$		$NH_3(aq)$
(b)	$H_2O(l)$		$(CH_3)_3N(aq)$		$(CH_3)_3NH^+(aq)$		$OH^-(aq)$
(c)	$HCHO_2(aq)$		$PO_4^{3-}(aq)$		$HPO_4^{2-}(aq)$		$CHO_2^-(aq)$

16.18

	B-L acid	+	**B-L base**	$\rightleftharpoons$	**Conjugate acid**	+	**Conjugate base**
(a)	$HBrO(aq)$		$H_2O(l)$		$H_3O^+(aq)$		$BrO^-(aq)$
(b)	$HSO_4^-(aq)$		$HCO_3^-(aq)$		$H_2CO_3(aq)$		$SO_4^{2-}(aq)$
(c)	$H_3O^+(aq)$		$HSO_3^-(aq)$		$H_2SO_3(aq)$		$H_2O(l)$

16.19   *Analyze/Plan.* Follow the logic in Sample Exercise 16.2.   *Solve.*

       (a)   Acid: $HC_2O_4^-(aq) + H_2O(l) \rightleftharpoons C_2O_4^{2-}(aq) + H_3O^+(aq)$

                 B-L acid   B-L base   conj. base   conj. acid

Base: $HC_2O_4^-(aq) + H_2O(l) \rightleftharpoons H_2C_2O_4(aq) + OH^-(aq)$

            B-L base   B-L acid   conj. acid   conj. Base

(b) $H_2C_2O_4$ is the conjugate acid of $HC_2O_4^-$.

       $C_2O_4^{2-}$ is the conjugate base of $HC_2O_4^-$.

16.20   (a) $H_2C_6H_7O_5^-(aq) + H_2O(l) \rightleftharpoons H_3C_6H_7O_5(aq) + OH^-(aq)$

        (b) $H_2C_6H_7O_5^-(aq) + H_2O(l) \rightleftharpoons HC_6H_7O_5^{2-}(aq) + H_3O^+(aq)$

        (c) $H_3C_6H_7O_5$ is the conjugate acid of $H_2C_6H_7O_5^-$

             $HC_6H_7O_5^{2-}$ is the conjugate base of $H_2C_6H_7O_5^-$

16.21   *Analyze/Plan.* Based on the chemical formula, decide whether the acid is strong, weak, or negligible. Is it one of the known seven strong acids (Section 16.5)? Also check Figure 16.4. Remove a single H and decrease the particle charge by one to write the formula of the conjugate base.   *Solve.*

        (a) HNO_2, weak acid; NO_2^-, weak base

        (b) H_2SO_4, strong acid; HSO_4^-, negligible base

        (c) HPO_4^{2-}, weak acid; PO_4^{3-}, weak base

        (d) CH_4, negligible acid; CH_3^-, strong base

        (e) $CH_3NH_3^+$, weak acid; CH_3NH_2, weak base

16.22   (a) $C_2H_3O_2^-$, weak base; $HC_2H_3O_2$, weak acid

        (b) HCO_3^-, weak base; H_2CO_3, weak acid

        (c) O_2^-, strong base; OH^-, negligible acid

        (d) Cl^-, negligible base; HCl, strong acid

        (e) NH_3, weak base; NH_4^+, weak acid

16.23   *Analyze/Plan.* Given chemical formula, determine strength of acids and bases by checking the known strong acids (Section 16.5). Recall the paradigm "The stronger the acid, the weaker its conjugate base, and vice versa."   *Solve.*

        (a)   HBr. It is one of the seven strong acids (Section 16.5).

        (b) F^-. HCl is a stronger acid than HF, so F^- is the stronger conjugate base.

16.24   (a) HNO_3. It is one of the seven strong acids (Section 16.5). Also, in a series of oxyacids with the same central atom (N), the acid with more O atoms is stronger (Section 16.10).

        (b) NH_3. When NH_3 and H_2O are combined, as in $NH_3(aq)$, NH_3 acts as the B-L base, accepting H^+ from H_2O. NH_3 has the greater tendency to accept H^+. For binary hydrides, base strength increases going to the left across a row of the periodic chart (Section 16.10).

16.25   *Analyze/Plan.* Acid-base equilibria favor formation of the weaker acid and base. Compare the relative strengths of the substances acting as acids on opposite sides of the equation. (Bases can also be compared; the conclusion should be the same.)   *Solve.*

Base	+	Acid	$\rightleftharpoons$	Conjugate acid	+	Conjugate base

(a) $O^{2-}(aq)$ + $H_2O(l)$ $\rightleftharpoons$ $OH^-(aq)$ + $OH^-(aq)$

H_2O is a stronger acid than OH^-, so the equilibrium lies to the right.

(b) $HS^-(aq)$ + $HC_2H_3O_2(aq)$ $\rightleftharpoons$ $H_2S(aq)$ + $C_2H_3O_2^-(aq)$

$HC_2H_3O_2$ is a stronger acid than H_2S, so the equilibrium lies to the right.

(c) $NO_3^-(aq)$ + $H_2O(l)$ $\rightleftharpoons$ $HNO_3(aq)$ + $OH^-(aq)$

HNO_3 is a stronger acid than H_2O (Solution 16.24), so the equilibrium lies to the left.

16.26

Base	+	Acid	$\rightleftharpoons$	Conjugate acid	+	Conjugate base

(a) $OH^-(aq)$ + $NH_4^+(aq)$ $\rightleftharpoons$ $H_2O(l)$ + $NH_3(aq)$

OH^- is a stronger base than NH_3 (Figure 16.4), so the equilibrium lies to the right.

(b) $C_2H_3O_2^-(aq)$ + $H_3O^+(aq)$ $\rightleftharpoons$ $HC_2H_3O_2(aq)$ + $H_2O(l)$

H_3O^+ is a stronger acid than $HC_2H_3O_2$ (Figure 16.4), so the equilibrium lies to the right.

(c) $F^-(aq)$ + $HCO_3^-(aq)$ $\rightleftharpoons$ $HC_2H_3O_2(aq)$ + $H_2O(l)$

CO_3^{2-} is a stronger base than F^-, so the equilibrium lies to the left.

Autoionization of Water

16.27   (a)   *Autoionization* is the ionization of a neutral molecule (in the absence of any other reactant) into an anion and a cation. The equilibrium expression for the autoionization of water is $H_2O(l) \rightleftharpoons H^+(aq) + OH^-(aq)$.

        (b)   Pure water is a poor conductor of electricity because it contains very few ions. Ions, mobile charged particles, are required for the conduction of electricity in liquids.

        (c)   If a solution is acidic, it contains more H^+ than OH^- ($[H^+] > [OH^-]$).

16.28   (a) $H_2O(l) \rightleftharpoons H^+(aq) + OH^-(aq)$

        (b) $K_w = [H^+][OH^-]$. The $[H_2O(l)]$ is omitted because water is a pure liquid. The molarity (mol/L) of pure solids or liquids does not change as equilibrium is established, so it is usually omitted from equilibrium expressions.

        (c)   If a solution is basic, it contains more OH^- than H^+ ($[OH^-] > [H^+]$).

16.29   *Analyze/Plan.* Follow the logic in Sample Exercise 16.5. In pure water at 25°C, $[H^+] = [OH^-] = 1 \times 10^{-7}$ M. If $[H^+] > 1 \times 10^{-7}$ M, the solution is acidic; if $[H^+] < 1 \times 10^{-7}$ M, the solution is basic. *Solve.*

        (a) $[H^+] = \dfrac{K_w}{[OH^-]} = \dfrac{1.0 \times 10^{-14}}{4.5 \times 10^{-4}\ M} = 2.2 \times 10^{-11}\ M < 1 \times 10^{-7}\ M$;   basic

(b) $[H^+] = \dfrac{K_W}{[OH^-]} = \dfrac{1.0 \times 10^{-14}}{8.8 \times 10^{-9}\,M} = 1.1 \times 10^{-6}\,M > 1 \times 10^{-7}\,M$; acidic

(c) $[OH^-] = 100[H^+]$; $K_w = [H^+] \times 100[H^+] = 100[H^+]^2$;

   $[H^+] = (K_w/100)^{1/2} = 1.0 \times 10^{-8}\,M < 1 \times 10^{-7}\,M$; basic

16.30   In pure water at 25°C, $[H^+] = [OH^-] = 1 \times 10^{-7}\,M$. If $[OH^-] > 1 \times 10^{-7}\,M$, the solution is basic; if $[OH^-] < 1 \times 10^{-7}\,M$, the solution is acidic.

(a) $[OH^-] = \dfrac{K_W}{[H^+]} = \dfrac{1.0 \times 10^{-14}}{7.5 \times 10^{-3}\,M} = 1.3 \times 10^{-12}\,M < 1 \times 10^{-7}\,M$; acidic

(b) $[OH^-] = \dfrac{K_w}{[H^+]} = \dfrac{1.0 \times 10^{-14}}{6.5 \times 10^{-10}\,M} = 1.5 \times 10^{-5}\,M > 1 \times 10^{-7}\,M$; basic

(c) $[H^+] = 10[OH^-]$; $K_w = 10[OH^-][OH^-] = 10[OH^-]^2$

   $[OH^-] = (K_w/10)^{1/2} = 3.2 \times 10^{-8}\,M < 1 \times 10^{-7}\,M$; acidic

16.31   *Analyze/Plan.* Follow the logic in Sample Exercise 16.4. Note that the value of the equilibrium constant (in this case, K_w) changes with temperature.   *Solve.*

   At 0°C, $K_w = 1.2 \times 10^{-15} = [H^+][OH^-]$.

   In pure water, $[H^+] = [OH^-]$; $2.4 \times 10^{-14} = [H^+]^2$; $[H^+] = (1.2 \times 10^{-15})^{1/2}$

   $[H^+] = [OH^-] = 3.5 \times 10^{-9}\,M$

16.32 $K_w = [D^+][OD^-]$; for pure D_2O, $[D^+] = [OD^-]$; $8.9 \times 10^{-16} = [D^+]^2$;

   $[D^+] = [OD^-] = 3.0 \times 10^{-8}\,M$

The pH Scale

16.33   *Analyze/Plan.* A change of one pH unit (in either direction) is:

   $\Delta pH = pH_2 - pH_1 = -(\log[H^+]_2 - \log[H^+]_1) = -\log\dfrac{[H^+]_2}{[H^+]_1} = \pm 1$. The antilog of +1 is 10;

   the antilog of –1 is 1×10^{-1}. Thus, a ΔpH of one unit represents an increase or decrease in $[H^+]$ by a factor of 10.   *Solve.*

(a) $\Delta pH = \pm 2.00$ is a change of $10^{2.00}$; $[H^+]$ changes by a factor of 100.

(b) $\Delta pH = \pm 0.5$ is a change of $10^{0.50}$; $[H^+]$ changes by a factor of 3.2.

16.34 $[H^+]_A = 500\,[H^+]_B$ From Solution 16.33, $\Delta pH = -\log\dfrac{[H^+]_B}{[H^+]_A}$

   $\Delta pH = -\log\dfrac{[H^+]_B}{500\,[H^+]_B} = -\log\left(\dfrac{1}{500}\right) = 2.70$

   The pH of solution A is 2.70 pH units lower than the pH of solution B, because $[H^+]_A$ is 500 times greater than $[H^+]_B$. The greater $[H^+]$, the lower the pH of the solution.

16.35   (a) $K_w = [H^+][OH^-]$. If NaOH is added to water, it dissociates into $Na^+(aq)$ and OH^- (aq). This increases $[OH^-]$ and necessarily decreases $[H^+]$. When $[H^+]$ decreases, pH increases.

       (b) $0.0006 \ M = 6 \times 10^{-4} \ M$. On Figure 16.5, this is $[H^+] > 1 \times 10^{-4}$ but $< 1 \times 10^{-3}$. The pH is between 3 and 4, closer to 3. We estimate 3.3. If pH < 7, the solution is acidic.

       (c)   pH = 5.2 is between pH 5 and pH 6 on Figure 16.5, closer to pH = 5. At pH = 6, $[H^+] = 1 \times 10^{-6}$; at pH = 5, $[H^+] = 1 \times 10^{-5} = 10 \times 10^{-6}$. A good estimate is $7 \times 10^{-6} \ M \ H^+$.

By calculation: $[H^+] = 10^{-pOH} = 10^{-5.2} = 6 \times 10^{-6} \ M$

At pH = 5, $[OH^-] = 1 \times 10^{-9}$; at pH = 6, $[OH^-] = 1 \times 10^{-8} = 10 \times 10^{-9}$.

Since pH = 5.2 is closer to pH = 5, we estimate $3 \times 10^{-9} \ M \ OH^-$.

By calculation: pOH = 14.0 – 5.2 = 8.8

$[OH^-] = 10^{-pOH} = 10^{-8.8} = 2 \times 10^{-9} \ M \ OH^-$

16.36   (a) $K_w = [H^+][OH^-]$. If HNO_3 is added to water, it ionizes to form $H^+(aq)$ and NO_3^- (aq). This increases $[H^+]$ and necessarily decreases $[OH^-]$. When $[H^+]$ increases, pH decreases.

       (b)   On Figure 16.5, $1.4 \times 10^{-2} \ M \ OH^-$ is between pH = 12 ($1 \times 10^{-2} \ M \ OH^-$) and pH 13 ($1 \times 10^{-1} \ M \ OH^-$), slightly higher than pH = 12, so we estimate pH = 12.1. If pH > 7, the solution is basic.

       (c)   pH = 6.6 is midway between pH 6 and pH 7 on Figure 16.5.

At pH = 7, $[H^+] = 1 \times 10^{-7}$; at pH = 6, $[H^+] = 1 \times 10^{-6} = 10 \times 10^{-7}$.

A reasonable estimate is $5 \times 10^{-7} \ M \ H^+$. By calculation:

pH = 6.6, $[H^+] = 10^{-pH} = 10^{-6.6} = 3 \times 10^{-7}$

At pH = 6, $[OH^-] = 1 \times 10^{-9}$; at pH = 7, $[OH^-] = 1 \times 10^{-7} = 10 \times 10^{-8}$.

A reasonable estimate is $5 \times 10^{-8} \ M \ OH^-$. By calculation:

pOH = 14.0 – 6.6 = 7.4; $[OH^-] = 10^{-pOH} = 10^{-7.4} = 4 \times 10^{-8} \ M \ OH^-$.

16.37   *Analyze/Plan.* At 25°C, $[H^+][OH^-] = 1 \times 10^{-14}$; pH = pOH = 14. Use these relationships to complete the table. If pH < 7, the solution is acidic; if pH > 7, the solution is basic. *Solve.*

[H$^+$]	[OH$^-$]	pH	pOH	acidic or basic
$7.5 \times 10^{-3} \ M$	$1.3 \times 10^{-12} \ M$	2.12	11.88	acidic
$2.8 \times 10^{-5} \ M$	$3.6 \times 10^{-10} \ M$	4.56	9.44	acidic
$5.6 \times 10^{-9} \ M$	$1.8 \times 10^{-6} \ M$	8.25	5.75	basic
$5.0 \times 10^{-9} \ M$	$2.0 \times 10^{-6} \ M$	8.30	5.70	basic

16.38

pH	pOH	$[H^+]$	$[OH^-]$	acidic or basic
4.75	9.25	$1.8 \times 10^{-5}\ M$	$5.6 \times 10^{-10}\ M$	acidic
2.11	11.89	$7.8 \times 10^{-3}\ M$	$1.3 \times 10^{-12}\ M$	acidic
2.19	11.81	$6.5 \times 10^{-3}\ M$	$1.5 \times 10^{-12}\ M$	acidic
7.93	6.07	$1.2 \times 10^{-8}\ M$	$8.6 \times 10^{-7}\ M$	basic

16.39   *Analyze/Plan.* Given pH and a new value of the equilibrium constant K_w, calculate equilibrium concentrations of H^+(aq) and OH^-(aq). The definition of pH remains pH = $-\log[H^+]$.   *Solve.*

pH = 7.40; $[H^+] = 10^{-pH} = 10^{-7.40} = 4.0 \times 10^{-8}\ M$

$K_w = 2.4 \times 10^{-14} = [H^+][OH^-]$; $[OH^-] = 2.4 \times 10^{-14}\ /\ [H^+]$

$[OH^-] = 2.4 \times 10^{-14}\ /\ 4.0 \times 10^{-8} = 6.0 \times 10^{-7}\ M$, pOH = $-\log(6.0 \times 10^{-7}) = 6.22$

Alternately, pH + pOH = pK_w. At 37°C, pH + pOH = $-\log(2.4 \times 10^{-14})$

pH + pOH = 13.62; pOH = 13.62 – 7.40 = 6.22

$[OH^-] = 10^{-pOH} = 10^{-6.22} = 6.0 \times 10^{-7}\ M$

16.40   The pH ranges from 5.2–5.6; pOH ranges from (14.0–5.2 =) 8.8 to (14.0–5.6 =) 8.4.

$[H^+] = 10^{-pH}$, $[OH^-] = 10^{-pOH}$

$[H^+] = 10^{-5.2} = 6.31 \times 10^{-6} = 6 \times 10^{-6}\ M$; $[H^+] = 10^{-5.6} = 2.51 \times 10^{-6} = 3 \times 10^{-6}\ M$

The range of $[H^+]$ is $6 \times 10^{-6}\ M$ to $3 \times 10^{-6}\ M$.

$[OH^-] = 10^{-8.8} = 1.58 \times 10^{-9} = 2 \times 10^{-9}\ M$; $[OH^-] = 10^{-8.4} = 3.98 \times 10^{-9} = 4 \times 10^{-9}\ M$.

The range of $[OH^-]$ is $2 \times 10^{-9}\ M$ to $4 \times 10^{-9}\ M$.

(The pH has one decimal place, so concentrations are reported to one sig fig.)

Strong Acids and Bases

16.41   (a)   A strong acid is completely ionized in aqueous solution; a strong acid is a strong electrolyte.

(b)   For a strong acid such as HCl, $[H^+]$ = initial acid concentration. $[H^+] = 0.500\ M$

(c)   HCl, HBr, HI

16.42   (a)   A strong base is completely dissociated in aqueous solution; a strong base is a strong electrolyte.

(b) $Sr(OH)_2$ is a soluble strong base.

$Sr(OH)_2(aq)\ \rightarrow\ Sr^{2+}(aq) + 2OH^-(aq)$

0.045 $M\ Sr(OH)_2(aq)$ = 0.090 $M\ OH^-$

(c)   Base strength should not be confused with solubility. Base strength describes the tendency of a dissolved molecule (formula unit for ionic compounds such as $Mg(OH)_2$) to dissociate into cations and hydroxide ions. $Mg(OH)_2$ is a strong base because each $Mg(OH)_2$ unit that dissolves also dissociates into $Mg^{2+}(aq)$ and $OH^-(aq)$. $Mg(OH)_2$ is not very soluble, so relatively few $Mg(OH)_2$ units dissolve when the solid compound is added to water.

16.43   *Analyze/Plan.* Follow the logic in Sample Exercise 16.8. Strong acids are completely ionized, so $[H^+]$ = original acid concentration, and pH = $-\log[H^+]$. For the solutions obtained by dilution, use the "dilution" formula, $M_1 V_1 = M_2 V_2$, to calculate molarity of the acid.   *Solve.*

(a) $8.5 \times 10^{-3}\ M$ HBr = $8.5 \times 10^{-3}\ M$ H^+; pH = $-\log (8.5 \times 10^{-3})$ = 2.07

(b) $\dfrac{1.52\ \text{g } HNO_3}{0.575\ \text{L soln}} \times \dfrac{1\ \text{mol } HNO_3}{63.02\ \text{g } HNO_3}$ = 0.041947 = 0.0419 M HNO_3

       $[H^+]$ = 0.0419 M; pH = $-\log (0.041947)$ = 1.377

(c) $M_c \times V_c = M_d \times V_d$; 0.250 M $\times$ 0.00500 L = ? M $\times$ 0.0500 L

       $M_d = \dfrac{0.250\ M\ \times\ 0.00500\ \text{L}}{0.0500\ \text{L}}$ = 0.0250 M HCl

       $[H^+]$ = 0.0250 M; pH = $-\log (0.0250)$ = 1.602

(d) $[H^+]_{total} = \dfrac{\text{mol } H^+ \text{ from HBr} + \text{mol } H^+ \text{ from HCl}}{\text{total L solution}}$

       $[H^+]_{total} = \dfrac{(0.100\ M\ \text{HBr}\ \times\ 0.0100\ \text{L}) + (0.200\ M\ \times\ 0.0200\ \text{L})}{0.0300\ \text{L}}$

       $[H^+]_{total} = \dfrac{1.00\ \times\ 10^{-3}\ \text{mol } H^+ + 4.00\ \times\ 10^{-3}\ \text{mol } H^+}{0.0300\ \text{L}}$ = 0.1667 = 0.167 M

       pH = $-\log (0.1667\ M)$ = 0.778

16.44   For a strong acid, which is completely ionized, $[H^+]$ = the initial acid concentration.

(a)   0.00835 M HNO_3 = 0.00835 M H^+; pH = $-\log (0.00835)$ = 2.08

(b) $\dfrac{0.525\ \text{g } HClO_4}{2.00\ \text{L soln}} \times \dfrac{1\ \text{mol } HClO_4}{100.5\ \text{g } HClO_4}$ = 2.612×10^{-3} = 2.61×10^{-3} M $HClO_4$

       $[H^+]$ = $2.61 \times 10^{-3}\ M$; pH = $-\log (2.612 \times 10^{-3})$ = 2.583

(c) $M_c \times V_c = M_d \times V_d$; 0.500 L = 500 mL

       1.00 M HCl $\times$ 5.00 mL HCl = M_d HCl $\times$ 500 mL HCl

       M_d HCl = $\dfrac{1.00\ M\ \times\ 5.00\ \text{mL}}{500\ \text{mL}}$ = 1.00×10^{-2} M HCl = 1.00×10^{-2} M H^+

       pH = $-\log (1.00 \times 10^{-2})$ = 2.000

(d) $[H^+]_{total} = \dfrac{\text{mol } H^+ \text{ from HCl} + \text{mol } H^+ \text{ from HI}}{\text{total L solution}}$; $mol = M \times L$

$$[H^+]_{total} = \frac{(0.020\ M\ HCl \times 0.0500\ L) + (0.010\ M\ HI \times 0.150\ L)}{0.200\ L}$$

$$[H^+]_{total} = \frac{1.0 \times 10^{-3}\ mol\ H^+ + 1.50 \times 10^{-3}\ mol\ H^+}{0.200\ L} = 0.0125 = 0.013\ M$$

pH = –log (0.0125) = 1.90

16.45   *Analyze/Plan.* Follow the logic in Sample Exercise 16.9. Strong bases dissociate completely upon dissolving. pOH = –log[OH$^-$]; pH = 14 – pOH.

    (a)   Pay attention to the formula of the base to get [OH$^-$].   *Solve.*

       [OH$^-$] = 2[Sr(OH)$_2$] = 2(1.5 × 10^{-3} M) = 3.0 × 10^{-3} M OH$^-$ (see Exercise 16.42(b))

       pOH = –log (3.0 × 10^{-3}) = 2.52; pH = 14 – pOH = 11.48

    (b)   mol/LiOH = g LiOH/molar mass LiOH. [OH$^-$] = [LiOH].   *Solve.*

$$\frac{2.250\ g\ LiOH}{0.2500\ L\ soln} \times \frac{1\ mol\ LiOH}{23.948\ g\ LiOH} = 0.37581 = 0.3758\ M\ LiOH = [OH^-]$$

       pOH = –log (0.37581) = 0.4250; pH = 14 – pOH = 13.5750

    (c)   Use the dilution formula to get the [NaOH] = [OH$^-$].   *Solve.*

       $M_c \times V_c = M_d \times V_d$; 0.175 M × 0.00100 L = ? M × 2.00 L

$$M_d = \frac{0.0175\ M \times 0.00100\ L}{2.00\ L} = 8.75 \times 10^{-5}\ M\ NaOH = [OH^-]$$

       pOH = –log (8.75 × 10^{-5}) = 4.058; pH = 14 – pOH = 9.942

    (d)   Consider total mol OH$^-$ from KOH and Ca(OH)$_2$, as well as total solution volume. *Solve.*

$$[OH^-]_{total} = \frac{\text{mol } OH^- \text{ from KOH} + \text{mol } OH^- \text{ from Ca(OH)}_2}{\text{total L soln}}$$

$$[OH^-]_{total} = \frac{(0.105\ M \times 0.00500\ L) + 2(9.5 \times 10^{-2} \times 0.0150 L)}{0.0200\ L}$$

$$[OH^-]_{total} = \frac{0.525 \times 10^{-3}\ mol\ OH^- + 2.85 \times 10^{-3}\ mol\ OH^-}{0.0200\ L} = 0.16875 = 0.17\ M$$

       pOH = –log (0.16875) = 0.77; pH = 14 – pOH = 13.23

       (9.5 × 10^{-2} M has 2 sig figs, so the [OH$^-$] has 2 sig figs and pH and pOH have 2 decimal places.)

16.46   For a strong base, which is completely dissociated, [OH$^-$] = the initial base concentration. Then, pOH = –log [OH$^-$] and pH = 14 – pOH.

    (a)   0.012 M KOH = 0.012 M OH$^-$; pOH = –log (0.012) = 1.92; pH = 14 – 1.92 = 12.08

(b) $\dfrac{1.565 \text{ g KOH}}{0.5000 \text{ L}} \times \dfrac{1 \text{ mol KOH}}{56.106 \text{ g KOH}} = 0.055787 = 0.05579 \ M = [OH^-]$

pOH = –log (0.055787) = 1.2535; pH = 14 – pOH = 12.7465

(c) $M_c \times V_c = M_d \times V_d$

$0.250 \ M \ Ca(OH)_2 \times 10.0 \text{ mL} = M_d \ Ca(OH)_2 \times 500 \text{ mL}$

$M_d \ Ca(OH)_2 = \dfrac{0.0105 \ M \ Ca(OH)_2 \times 10.0 \text{ mL}}{500.0 \text{ mL}} = 2.10 \times 10^{-4} \ M \ Ca(OH)_2$

$Ca(OH)_2(aq) \rightarrow Ca^{2+}(aq) + 2OH^-(aq)$

$[OH^-] = 2[Ca(OH)_2] = 2(2.10 \times 10^{-4} \ M) = 4.20 \times 10^{-4} \ M$

pOH = –log (4.20×10^{-4}) = 3.377; pH = 14 – pOH = 10.623

(d) $[OH^-]_{total} = \dfrac{\text{mol } OH^- \text{ from NaOH} + \text{mol } OH^- \text{ from Ba(OH)}_2}{\text{total L solution}}$

$\dfrac{(7.5 \times 10^{-3} \ M \times 0.0400 \text{ L}) + 2(0.015 \ M \times 0.0100 \text{ L})}{0.0500 \text{ L}}$

$[OH^-]_{total} = \dfrac{3.00 \times 10^{-4} \text{ mol } OH^- + 3.0 \times 10^{-4} \text{ mol } OH^-}{0.0500 \text{ L}} = 0.01200 = 0.012 \ M \ OH^-$

pOH = –log (0.0120) = 1.92; pH = 14 – pOH = 12.08

16.47   *Analyze/Plan.* pH $\rightarrow$ pOH $\rightarrow$ $[OH^-]$ = [NaOH].   *Solve.*

pOH = 14 – pH = 14.00 – 11.50 = 2.50

pOH = 2.50 = –log$[OH^-]$; $[OH^-] = 10^{-2.50} = 3.2 \times 10^{-3} \ M$

$[OH^-]$ = [NaOH] = $3.2 \times 10^{-3} \ M$

16.48   pOH = 14 – pH = 14.00 – 12.05 = 1.95

pOH = 1.95 = –log$[OH^-]$; $[OH^-] = 10^{-1.95} = 0.01122 = 1.1 \times 10^{-2} \ M$

$[OH^-] = 2[Ca(OH)_2]$; $[Ca(OH)_2] = [OH^-] / 2 = 0.01122/2 = 5.6 \times 10^{-3} \ M$

16.49   *Analyze/Plan.* NaH(aq) $\rightarrow$ Na^+(aq) + H^-(aq)

H^-(aq) + H_2O(l) $\rightarrow$ H_2(g) + OH^-(aq)

Thus, initial [NaH] = $[OH^-]$; [NaH] = g NaH/[M(NaH) × V].   *Solve.*

$[NaH] = \dfrac{\text{mol NaH}}{\text{L solution}} = 15.00 \text{ g NaH} \times \dfrac{1 \text{ mol NaH}}{24.00 \text{ g NaH}} \times \dfrac{1}{2.50 \text{ L}} = 0.250 \ M$

$[OH^-]$ = 0.250 M; pOH = –log (0.250) = 0.602, pH = 14 – pOH = 13.400

16.50   Upon dissolving, Li_2O dissociates to form Li^+ and O^{2-}. According to Equation 16.22, O^{2-} is completely protonated in aqueous solution.

Li_2O(s) + H_2O(l) $\rightarrow$ $2Li^+$(aq) + $2OH^-$(aq)

Thus, initial $[Li_2O] = [O_2^-]$; $[OH^-] = 2[O^{2-}] = 2[Li_2O]$

$$[Li_2O] = \frac{mol\ Li_2O}{L\ solution} = 2.50\ g\ Li_2O \times \frac{1\,mol\ Li_2O}{29.88\ g\ Li_2O} \times \frac{1}{1.500\ L} = 0.0558 = 0.0558\ M$$

$[OH^-] = 0.11156 = 0.112\ M$; pOH $= 0.9525 = 0.958$ pH $= 14.00 - $ pOH $= 13.0475 = 13.048$

Weak Acids

16.51   *Analyze/Plan.* Remember that $K_{eq} = $ [products]/[reactants]. If $H_2O(l)$ appears in the equilibrium reaction, it will **not** appear in the K_a expression, because it is a pure liquid. *Solve.*

(a) $HBrO_2(aq) \rightleftharpoons H^+(aq) + BrO_2^-(aq)$; $K_a = \dfrac{[H^+][BrO_2^-]}{[HBrO_2]}$

      $HBrO_2(aq) + H_2O(l) \rightleftharpoons H_3O^+(aq) + BrO_2^-(aq)$; $K_a = \dfrac{[H_3O^+][BrO_2^-]}{[HBrO_2]}$

(b) $HC_3H_5O_2(aq) \rightleftharpoons H^+(aq) + C_3H_5O_2^-(aq)$; $K_a = \dfrac{[H^+][C_3H_5O_2^-]}{[HC_3H_5O_2]}$

      $HC_3H_5O_2(aq) + H_2O(l) \rightleftharpoons H_3O^+(aq) + C_3H_5O_2^-(aq)$; $K_a = \dfrac{[H_3O^+][C_3H_5O_2^-]}{[HC_3H_5O_2]}$

16.52   (a) $HC_6H_5O(aq) \rightleftharpoons H^+(aq) + C_6H_5O^-(aq)$; $K_a = \dfrac{[H^+][C_6H_5O^-]}{[HC_6H_5O]}$

         $HC_6H_5O(aq) + H_2O(l) \rightleftharpoons H_3O^+(aq) + C_6H_5O^-(aq)$; $K_a = \dfrac{[H_3O^+][C_6H_5O^-]}{[HC_6H_5O]}$

     (b) $HCO_3^-(aq) \rightleftharpoons H^+(aq) + CO_3^{2-}(aq)$; $K_a = \dfrac{[H^+][CO_3^{2-}]}{[HCO_3^-]}$

         $HCO_3^-(aq) + H_2O(aq) \rightleftharpoons H_3O^+(aq) + CO_3^{2-}(aq)$; $K_a = \dfrac{[H_3O^+][CO_3^{2-}]}{[HCO_3^-]}$

16.53   *Analyze/Plan.* Follow the logic in Sample Exercise 16.10.   *Solve.*

$HC_3H_5O_3(aq)\ f\ H^+(aq) + C_3H_5O_3^-(aq)$; $K_a = \dfrac{[H^+][C_3H_5O_3^-]}{[HC_3H_5O_3]}$

$[H^+] = [C_3H_5O_3^-] = 10^{-2.44} = 3.63 \times 10^{-3} = 3.6 \times 10^{-3}\ M$

$[HC_3H_5O_3] = 0.10 - 3.63 \times 10^{-3} = 0.0964 = 0.096\ M$

$K_a = \dfrac{(3.63 \times 10^{-3})^2}{(0.0964)} = 1.4 \times 10^{-4}$

16.54 $HC_8H_7O_2(aq) \rightleftharpoons H^+(aq) + C_8H_7O_2^-(aq)$; $K_a = \dfrac{[H^+][C_8H_7O_2^-]}{[HC_8H_7O_2]}$

$[H^+] = [C_8H_7O_2^-] = 10^{-2.68} = 2.09 \times 10^{-3} = 2.1 \times 10^{-3}\ M$

$[HC_8H_7O_2] = 0.085 - 2.09 \times 10^{-3} = 0.0829 = 0.083\ M$

$K_a = \dfrac{(2.09 \times 10^{-3})^2}{0.0829} = 5.3 \times 10^{-5}$

16.55 *Analyze/Plan.* Write the equilibrium reaction and the K_a expression. Use % ionization to get equilibrium concentration of $[H^+]$, and by stoichiometry, $[X^-]$ and $[HX]$. Calculate K_a *Solve.*

$[H^+] = 0.110 \times [CH_2ClCOOH]_{initial} = 0.0110\ M$

	$CH_2ClCOOH(aq)$	$\rightleftharpoons$	$H^+(aq)$	$+$	$CH_2ClCOO^-(aq)$
initial	$0.100\ M$		0		0
equil.	$0.089\ M$		$0.0110\ M$		$0.0110\ M$

$K_a = \dfrac{[H^+][CH_2ClCOO^-]}{[CH_2ClCOOH]} = \dfrac{(0.0110)^2}{0.089} = 1.4 \times 10^{-3}$

16.56 $[H^+] = 0.132 \times [BrCH_2COOH]_{initial} = 0.0132\ M$

	$BrCH_2COOH(aq)$	$\rightleftharpoons$	$H^+(aq)$	$+$	$BrCH_2COO^-(aq)$
initial	$0.100\ M$		0		0
equil.	0.087		$0.0132\ M$		$0.0132\ M$

$K_a = \dfrac{[H^+][BrCH_2COO^-]}{[BrCH_2COOH]} = \dfrac{(0.0132)^2}{0.087} = 2.0 \times 10^{-3}$

16.57 *Analyze/Plan.* Write the equilibrium reaction and the K_a expression.

$[H^+] = 10^{-pH} = [C_2H_3O_2^-]\ [HC_2H_3O_2] = x - [H^+]$.

Substitute into the K_a expression and solve for x.   *Solve.*

$[H^+] = 10^{-pH} = 10^{-2.90} = 1.26 \times 10^{-3} = 1.3 \times 10^{-3}\ M$

$K_a = 1.8 \times 10^{-5} = \dfrac{[H^+][C_2H_3O_2^-]}{[HC_2H_3O_2]} = \dfrac{(1.26 \times 10^{-3})^2}{(x - 1.26 \times 10^{-3})}$

$1.8 \times 10^{-5}\ (x - 1.26 \times 10^{-3}) = (1.26 \times 10^{-3})^2;$

$1.8 \times 10^{-5}\ x = 1.585 \times 10^{-6} + 2.266 \times 10^{-8} = 1.608 \times 10^{-6};$

$x = 0.08931 = 0.089\ M\ HC_2H_3O_2$

16.58 $[H^+] = 10^{-pH} = 10^{-3.25} = 5.623 \times 10^{-4} = 5.62 \times 10^{-4}\ M$

$K_a = 6.8 \times 10^{-4} = \dfrac{[H^+][F^-]}{[HF]} = \dfrac{(5.623 \times 10^{-4})^2}{x - 5.623 \times 10^{-4}}$

$6.8 \times 10^{-4}(x - 5.623 \times 10^{-4}) = (5.623 \times 10^{-4})^2;$

$6.8 \times 10^{-4}\ x = 3.824 \times 10^{-7} + 3.162 \times 10^{-7} = 6.986 \times 10^{-7}$

$x = 1.027 \times 10^{-3} = 1.0 \times 10^{-3}\ M\ HF$

$mol = M \times L = 1.027 \times 10^{-3}\ M \times 0.200\ L = 2.055 \times 10^{-4} = 2.1 \times 10^{-4}\ mol\ HF$

16.59   *Analyze/Plan.* Follow the logic in Sample Exercise 16.11. Write K_a, construct the equilibrium table, solve for $x = [H^+]$, then get equilibrium $[C_7H_5O_2^-]$ and $[HC_7H_5O_2]$ by substituting $[H^+]$ for x. *Solve.*

$$HC_7H_5O_2(aq) \quad \rightleftharpoons \quad H^+(aq) \quad + \quad C_7H_5O_2^-(aq)$$

initial	0.050 M	0	0
equil.	(0.050 – x) M	x M	x M

$$K_a = \frac{[H^+][C_7H_5O_2^-]}{[HC_7H_5O_2]} = \frac{x^2}{(0.050-x)} \approx \frac{x^2}{0.050} = 6.3 \times 10^{-5}$$

$x^2 = 0.050 \, (6.3 \times 10^{-5}); \; x = 1.8 \times 10^{-3} \, M = [H^+] = [H_3O^+] = [C_7H_6O_2^-]$

$[HC_7H_5O_2] = 0.050 - 0.0018 = 0.048 \, M$

Check. $\dfrac{1.8 \times 10^{-3} \, M H^+}{0.050 \, M \, HC_7H_5O_2} \times 100 = 3.6\%$ ionization; the assumption is valid

16.60
$$HClO(aq) \quad \rightleftharpoons \quad H^+(aq) \quad + \quad ClO^-(aq)$$

initial	0.0075 M	0	0
equil.	(0.0075 – x) M	x M	x M

$$K_a = \frac{[H^+][ClO^-]}{[HClO]} = \frac{x^2}{(0.0075-x)} \approx \frac{x^2}{0.0075} = 3.0 \times 10^{-5}$$

$x^2 = 0.0075 \, (3.0 \times 10^{-8}); \; x = 1.5 \times 10^{-5} \, M = [H^+] = [H_3O^+] = [ClO^-]$

$[HClO] = 7.5 \times 10^{-3} - 1.5 \times 10^{-6} = 7.485 \times 10^{-3} = 7.5 \times 10^{-3} \, M$

Check. $\dfrac{4.7 \times 10^{-5} \, M H^+}{0.0075 \, M \, HClO} \times 100 = 0.20\%$ ionization; the assumption is valid

16.61   *Analyze/Plan.* Follow the logic in Sample Exercise 16.11.

(a)   *Solve.*
$$HC_3H_5O_2(aq) \quad \rightleftharpoons \quad H^+(aq) \quad + \quad C_3H_5O_2^- \, (aq)$$

initial	0.095 M	0	0
equil	(0.095 – x) M	x M	x M

$$K_a = \frac{[H^+][C_3H_5O_2^-]}{[HC_3H_5O_2]} = \frac{x^2}{(0.095-x)} \approx \frac{x^2}{0.095} = 1.3 \times 10^{-5}$$

$x^2 = 0.095 (1.3 \times 10^{-5}); \; x = 1.111 \times 10^{-3} = 1.1 \times 10^{-3} \, M \, H^+; \, pH = 2.95$

Check. $\dfrac{1.1 \times 10^{-3} \, M H^+}{0.095 \, M \, HC_3H_5O_2} \times 100 = 1.2\%$ ionization; the assumption is valid

(b)   *Solve.*

$$K_a = \frac{[H^+][CrO_4^{2-}]}{[HCrO_4^-]} = \frac{x^2}{(0.100-x)} \approx \frac{x^2}{0.100} = 3.0 \times 10^{-7}$$

$x^2 = 0.100(3.0 \times 10^{-7});\ x = 1.732 \times 10^{-4} = 1.7 \times 10^{-4}\ M\ H^+$

$pH = -\log(1.732 \times 10^{-4}) = 3.7614 = 3.76$

Check. $\dfrac{1.7 \times 10^{-4}\ M\ H^+}{0.100\ M\ HCrO_4^-} \times 100 = 0.17\%$ ionization; the assumption is valid

(c) Follow the logic in Sample Exercise 16.14. $pOH = -\log[OH^-]$. $pH = 14 - pOH$

Solve.

$$C_5H_5N(aq) + H_2O(l) \rightleftharpoons C_5H_5NH^+(aq) + OH^-(aq)$$

initial	0.120 M	0	0
equil	(0.120 – x) M	x M	x M

$K_b = \dfrac{[C_5H_5NH^+][OH^-]}{[C_5H_5N]} = \dfrac{x^2}{(0.120-x)} \approx \dfrac{x^2}{0.120} = 1.7 \times 10^{-9}$

$x^2 = 0.120(1.7 \times 10^{-9});\ x = 1.428 \times 10^{-5} = 1.4 \times 10^{-5}\ M\ OH^-;\ pH = 9.15$

Check. $\dfrac{1.4 \times 10^{-5}\ M\ OH^-}{0.120\ M\ C_5H_5N} \times 100 = 0.011\%$ ionization; the assumption is valid

16.62 (a)

$$HOCl(aq) \rightleftharpoons H^+(aq) + OCl^-(aq)$$

initial	0.125 M	0	0
equil	(0.125 – x) M	x M	x M

$K_a = \dfrac{[H^+][OCl^-]}{[HOCl]} = \dfrac{x^2}{(0.125-x)} \approx \dfrac{x^2}{0.125} = 3.0 \times 10^{-8}$

$x^2 = 0.125\,(3.0 \times 10^{-8});\ x = [H^+] = 6.1 \times 10^{-5}\ M,\ pH = 4.21$

Check. $\dfrac{6.1 \times 10^{-5}\ M\ H^+}{0.125\ M\ HOCl} \times 100 = 0.049\%$ ionization; the assumption is valid

(b) $K_a = \dfrac{[H^+][C_6H_5O^-]}{[C_6H_5OH]} = \dfrac{x^2}{(0.0085-x)} \approx \dfrac{x^2}{0.0085} = 1.3 \times 10^{-10}$

$x^2 = 0.0085\,(1.3 \times 10^{-10});\ x = [H^+] = 1.1 \times 10^{-6}\ M,\ pH = 5.98$

Check. Clearly $1.1 \times 10^{-6}\ M\ H^+$ is small compared to $8.5 \times 10^{-3}\ M\ C_6H_5OH$, and the assumption is valid.

(c)

$$HONH_2(aq) + H_2O(l) \rightleftharpoons HONH_3^+(aq) + OH^-(aq)$$

initial	0.095 M	0	0
equil	(0.095 – x) M	x M	x M

$K_b = \dfrac{[HONH_3^+][OH^-]}{[HONH_2]} = \dfrac{x^2}{(0.095-x)} \approx \dfrac{x^2}{0.095} = 1.1 \times 10^{-8}$

$x^2 = 0.095\,(1.1 \times 10^{-8});\ x = [OH^-] = 3.2 \times 10^{-5}\ M,\ pH = 9.51$

Check. $\dfrac{3.2 \times 10^{-5}\ M\ OH^-}{0.095\ M\ HONH_2} \times 100 = 0.034\%$ ionization; the assumption is valid

16.63 *Analyze/Plan.* $K_a = 10^{-pK_a}$. Follow the logic in Sample Exercise 16.11. *Solve.*

Let $[H^+] = [NC_7H_4SO_3^-] = z$. $K_a = $ antilog $(-2.32) = 4.79 \times 10^{-3} = 4.8 \times 10^{-3}$

$\dfrac{z^2}{0.10 - z} = 4.79 \times 10^{-3}$. Since K_a is relatively large, solve the quadratic.

$z^2 = 4.79 \times 10^{-3}z - 4.79 \times 10^{-4} = 0$

$z = \dfrac{-4.79 \times 10^{-3} \pm \sqrt{(4.79 \times 10^{-3})^2 - 4(1)(-4.79 \times 10^{-4})}}{2(1)} = \dfrac{-4.79 \times 10^{-3} \pm \sqrt{1.937 \times 10^{-3}}}{2}$

$z = 1.96 \times 10^{-2} = 2.0 \times 10^{-2}\ M\ H^+$; pH $= -\log(1.96 \times 10^{-2}) = 1.71$

16.64 Calculate the initial concentration of $HC_9H_7O_4$.

$2\ \text{tablets} \times \dfrac{500\ mg}{\text{tablet}} \times \dfrac{1\ g}{1000\ mg} \times \dfrac{1\ mol\ HC_9H_7O_4}{180.2\ g\ HC_9H_7O_4} = 0.005549 = 0.00555\ mol\ HC_9H_7O_4$

$\dfrac{0.005549\ mol\ HC_9H_7O_4}{0.250\ L} = 0.02220 = 0.0222\ M\ HC_9H_7O_4$

	$HC_9H_7O_4$ (aq) $\rightleftharpoons$	$C_9H_7O_4^-$ +	H^+ (aq)
Initial	0.0222 *M*	0 *M*	0 *M*
equil	(0.0222 − x)	x *M*	x *M*

$K_a = 3.3 \times 10^{-4} = \dfrac{[H^+][C_7H_9O_4^-]}{[HC_7H_9O_4]} = \dfrac{x^2}{(0.0222 - x)}$

Assuming x is small compared to 0.0222,

$x^2 = 0.0222 (3.3 \times 10^{-4})$; x $= [H^+] = 2.7 \times 10^{-3}\ M$

$\dfrac{2.7 \times 10^{-3}\ M\ H^+}{0.0222\ M\ HC_9H_7O_4} \times 100 = 12\%$ ionization; the assumption is not valid

Using the quadratic formula, $x^2 + 3.3 \times 10^{-4}x - 7.325 \times 10^{-6} = 0$

$x = \dfrac{-3.3 \times 10^{-4} \pm \sqrt{(3.3 \times 10^{-4})^2 - 4(1)(-7.325 \times 10^{-6})}}{2(1)} = \dfrac{-3.3 \times 10^{-4} \pm \sqrt{2.941 \times 10^{-5}}}{2}$

$x = 2.547 \times 10^{-3} = 2.5 \times 10^{-3}\ M\ H^+$; pH $= -\log(2.547 \times 10^{-3}) = 2.594 = 2.59$

16.65 *Analyze/Plan.* Follow the logic in Sample Exercise 16.12. *Solve.*

(a)

	HN_3(aq) $\rightleftharpoons$	H^+(aq) +	N_3^- (aq)
initial	0.400 *M*	0	0
equil	(0.400 − x) *M*	x *M*	x *M*

$$K_a = \frac{[H^+][N_3^-]}{[HN_3]} = 1.9 \times 10^{-5}; \; \frac{x^2}{(0.400-x)} = \frac{x^2}{0.400} = 1.9 \times 10^{-5}$$

$$x = 0.00276 = 2.8 \times 10^{-3} \; M = [H^+]; \; \% \text{ ionization} = \frac{2.76 \times 10^{-3}}{0.400} \times 100 = 0.69\%$$

(b) $1.9 \times 10^{-5} = \dfrac{x^2}{0.100}; \; x = 0.00138 = 1.4 \times 10^{-3} \; M \; H^+$

       $\% \text{ ionization} = \dfrac{1.38 \times 10^{-3} \; M \; H^+}{0.100 \; M \; HN_3} \times 100 = 1.4\%$

(c) $1.9 \times 10^{-5} = \dfrac{x^2}{0.0400}; \; x = 8.72 \times 10^{-4} = 8.7 \times 10^{-4} \; M \; H^+$

       $\% \text{ ionization} = \dfrac{8.72 \times 10^{-4} \; M \; H^+}{0.0400 \; M \; HN_3} \times 100 = 2.2\%$

Check. Notice that a tenfold dilution [part (a) versus part (c)] leads to a slightly more than threefold increase in percent ionization.

16.66   (a) $HC_3H_5O_2(aq) \rightleftharpoons H^+(aq) + C_3H_5O_2^-(aq)$

         $K_a = 1.3 \times 10^{-5} = \dfrac{[H^+][C_3H_5O_2^-]}{[HC_3H_5O_2]} = \dfrac{x^2}{0.250-x}$

         $x^2 = 0.250 \, (1.3 \times 10^{-5}); \; x = 1.803 \times 10^{-3} = 1.8 \times 10^{-3} \; M \; H^+$

         $\% \text{ ionization} = \dfrac{1.803 \times 10^{-3} \; M \; H^+}{0.250 \; M \; HC_3H_5O_2} \times 100 = 0.721\%$

   (b) $\dfrac{x^2}{0.0800} \simeq 1.3 \times 10^{-5}; \; x = 1.020 \times 10^{-3} = 1.0 \times 10^{-3} \; M \; H^+$

         $\% \text{ ionization} = \dfrac{1.020 \times 10^{-3} \; M \; H^+}{0.0800 \; M \; HC_3H_5O_2} \times 100 = 1.27\%$

   (c) $\dfrac{x^2}{0.0200} \approx 1.3 \times 10^{-5}; \; x = 5.099 \times 10^{-4} = 5.1 \times 10^{-4} \; M \; H^+$

         $\% \text{ ionization} = \dfrac{5.099 \times 10^{-4} \; M \; H^+}{0.0200 \; M \; HC_3H_5O_2} \times 100 = 2.55\%$

16.67   *Analyze/Plan.* Let the weak acid be HX. $HX(aq) \rightleftharpoons H^+(aq) + X^-(aq)$. Solve the K_a expression symbolically for $[H^+]$ in terms of $[HX]$. Substitute into the formula for % ionization, $([H^+]/[HX]) \times 100$.   *Solve.*

     $K_a = \dfrac{[H^+][X^-]}{[HX]}; \; [H^+] = [X^-] = y; \; K_a = \dfrac{y^2}{[HX]-y};$ assume that % ionization is small

     $K_a = \dfrac{y^2}{[HX]}; \; y = K_a^{1/2} \, [HX]^{1/2}$

$$\% \text{ ionization} = \frac{y}{[HX]} \times 100 = \frac{K_a^{1/2}[HX]^{1/2}}{[HX]} \times 100 = \frac{K_a^{1/2}}{[HX]^{1/2}} \times 100$$

That is, percent ionization varies inversely as the square root of concentration HX.

16.68 $HX(aq) \rightleftharpoons H^+(aq) + X^-(aq); \quad K_a = \dfrac{[H^+][X^-]}{[HX]}$

$[H^+] = [X^-]$; assume the % ionization is small; $K_a = \dfrac{[H^+]^2}{[HX]}$; $[H^+] = K_a^{1/2}[HX]^{1/2}$

$pH = -\log K_a^{1/2}[HX]^{1/2} = -\log K_a^{1/2} - \log[HX]^{1/2}$; $pH = -1/2 \log K_a - 1/2 \log[HX]$

This is the equation of a straight line, where the intercept is $-1/2 \log K_a$, the slope is $-1/2$, and the independent variable is $\log[HX]$.

16.69   Analyze/Plan. Follow the logic in Sample Exercise 16.13. Citric acid is a triprotic acid with three K_a values that do not differ by more than 10^3. We must consider all three steps. Also, $C_6H_5O_7{}^{3-}$ is only produced in step 3. *Solve.*

$H_3C_6H_5O_7(aq) \rightleftharpoons H^+(aq) + H_2C_6H_5O_7{}^-(aq) \qquad K_{a1} = 7.4 \times 10^{-4}$

$H_2C_6H_5O_7{}^-(aq) \rightleftharpoons H^+(aq) + HC_6H_5O_7{}^{2-}(aq) \qquad K_{a2} = 1.7 \times 10^{-5}$

$HC_6H_5O_7{}^{2-}(aq) \rightleftharpoons H^+(aq) + C_5H_5O_7{}^{3-}(aq) \qquad K_{a3} = 4.0 \times 10^{-7}$

To calculate the pH of a 0.050 M solution, assume initially that only the first ionization is important:

	$H_3C_6H_5O_7(aq)$	$\rightleftharpoons$	$H^+(aq)$	+	$H_2C_6H_5O_7{}^-(aq)$
initial	0.050 M		0		0
equil.	$(0.050 - x)\,M$		$x\,M$		$x\,M$

$$K_{a1} = \frac{[H^+][H_2C_6H_5O_7{}^-]}{[H_3C_6H_5O_7]} = \frac{x^2}{(0.050 - x)} = 7.4 \times 10^{-4}$$

$x^2 = (0.050 - x)(7.4 \times 10^{-4})$; $x^2 = (0.050)(7.4 \times 10^{-4})$; $x = 0.00608 = 6.1 \times 10^{-3}\,M$

Since this value for x is rather large in relation to 0.050, a better approximation for x can be obtained by substituting this first estimate into the expression for x^2, then solving again for x:

$x^2 = (0.050 - x)\,(7.4 \times 10^{-4}) = (0.050 - 6.08 \times 10^{-3})\,(7.4 \times 10^{-4})$

$x^2 = 3.2 \times 10^{-5}$; $x = 5.7 \times 10^{-3}\,M$

(This is the same result obtained from the quadratic formula.)

The correction to the value of x, though not large, is significant. Does the second ionization produce a significant additional concentration of H^+?

	$H_2C_6H_5O_7^-$ (aq)	$\rightleftharpoons$	H^+ (aq)	+	$HC_6H_5O_7^{2-}$ (aq)
initial	$5.7 \times 10^{-5}\,M$		$5.7 \times 10^{-3}\,M$		0
equil.	$(5.7 \times 10^{-3} - y)$		$(5.7 \times 10^{-3} + y)$		y

$$K_{a2} = \frac{[H^+][HC_6H_5O_7^{2-}]}{[H_2C_6H_5O_7^-]} = 1.7 \times 10^{-5}; \; \frac{(5.7 \times 10^{-3} + y)(y)}{(5.7 \times 10^{-3} - y)} = 1.7 \times 10^{-5}$$

Assume that y is small relative to 5.7×10^{-3}; that is, that additional ionization of $H_2C_6H_5O_7^-$ is small, then

$$\frac{(5.7 \times 10^{-3})y}{(5.7 \times 10^{-3})} = 1.7 \times 10^{-5}\,M; \; y = 1.7 \times 10^{-5}\,M$$

This value is indeed small compared to $5.7 \times 10^{-3}\,M$; $[H^+]$ and pH are determined by the first ionization step. If we were only interested in pH, we could stop here. However, to calculate $[C_6H_5O_7^{3-}]$, we must consider the third ionization, with adjusted $[H^+] = 5.7 \times 10^{-3} + 1.7 \times 10^{-5} = 5.72 \times 10^{-3}\,M \, (= 5.7 \times 10^{-3})$

	$HC_6H_5O_7^{2-}$	$\rightleftharpoons$	H^+ (aq)	+	$HC_5H_5O_7^{3-}$ (aq)
initial	$1.7 \times 10^{-5}\,M$		$5.72 \times 10^{-3}\,M$		0
equil.	$1.7 \times 10^{-5} - z$		$5.72 \times 10^{-3} + z$		z

$$K_{a3} = \frac{[H^+][C_6H_5O_7^{3-}]}{[HC_6H_5O_7^{2-}]} = \frac{(5.72 \times 10^{-3} + z)(z)}{(1.7 \times 10^{-5} - z)} = 4.0 \times 10^{-7}$$

Assume z is small relative to 5.72×10^{-3}, but not relative to 1.7×10^{-5}.

$(4.0 \times 10^{-7})(1.7 \times 10^{-5} - z) = 5.72 \times 10^{-3}\,z; \; 6.8 \times 10^{-12} - 4.0 \times 10^{-7}\,z = 5.72 \times 10^{-3}\,z;$

$6.8 \times 10^{-12} = 5.72 \times 10^{-3}\,z + 4.0 \times 10^{-7}\,z = 5.72 \times 10^{-3}\,z; \; z = 1.19 \times 10^{-9} = 1.2 \times 10^{-9}\,M$

$[C_6H_5O_7^{3-}] = 1.2 \times 10^{-9}\,M; \; [H^+] = 5.72 \times 10^{-3}\,M + 1.2 \times 10^{-9}\,M = 5.72 \times 10^{-3}\,M$

$pH = -\log(5.72 \times 10^{-3}) = 2.24$

Note that neither the second nor third ionizations contributed significantly to $[H^+]$ and pH.

16.70 $H_2C_4H_4O_6$ (aq) $\rightleftharpoons$ H^+ (aq)   + $HC_4H_4O_6^-$ (aq) $K_{a1} = 1.0 \times 10^{-3}$

        $HC_4H_4O_6^-$ (aq $\rightleftharpoons$ H^+ (aq)   + $C_4H_4O_6^{2-}$ (aq) $K_{a2} = 4.6 \times 10^{-5}$

Begin by calculating the $[H^+]$ from the first ionization. The equilibrium concentrations are $[H^+] = [HC_4H_4O_6^-] = x$, $[H_2C_4H_4O_6] = 0.25 - x$.

$$K_{a1} = \frac{[H^+][HC_4H_4O_6^-]}{[H_2C_4H_4O_8]} = \frac{x^2}{0.25 - x}; \; x^2 + 1.0 \times 10^{-3}\,x - 2.5 \times 10^{-4} = 0$$

Using the quadratic formula, $x = 1.532 \times 10^{-2} = 0.015\,M\,H^+$ from the first ionization. Next calculate the H^+ contribution from the second ionization.

$$\begin{array}{lccc} & HC_4H_4O_6^-(aq) & \rightleftharpoons & H^+(aq) & + & C_4H_4O_6^{2-}(aq) \\ \text{initial} & 0.015 & & 0.015 & & 0 \\ \text{equil.} & (0.015-y) & & (0.015+y) & & y \end{array}$$

$K_{a2} = \dfrac{(0.015+y)\,(y)}{(0.015-y)} = 4.6 \times 10^{-5}$; assuming y is small compared to 0.015,

$y = 4.6 \times 10^{-5}\,M\ HC_4H_4O_6^{2-}(aq)$

This assumption is reasonable, since 4.6×10^{-5} is only 0.3% of 0.015. $[H^+] = 0.015\,M$ (first ionization) $+ 4.6 \times 10^{-5}$ (second ionization). Since 4.6×10^{-5} is 0.3% of $0.015\,M$, it can be safely ignored when calculating total $[H^+]$. Thus, pH = $-\log(0.01532)$ = 1.18148 = 1.181.

Assumptions:

1) The ionization can be treated as a series of steps (valid by Hess' law).

2) The extent of ionization in the second step (y) is small relative to that from the first step (valid for this acid and initial concentration). This assumption was used twice, to calculate the value of y from K_{a2} and to calculate total $[H^+]$ and pH.

Weak Bases

16.71 All Brønsted-Lowry bases contain at least one unshared (lone) pair of electrons to attract H^+.

16.72 Organic amines (neutral molecules with nonbonded pairs on N atoms) and anions that are the conjugate bases of weak acids function as weak bases.

16.73 *Analyze/Plan.* Remember that K_{aq} = [products]/[reactants]. If $H_2O(l)$ appears in the equilibrium reaction, it will not appear in the K_b expression, because it is a pure liquid. *Solve.*

(a) $(CH_3)_2NH(aq) + H_2O(l) \rightleftharpoons (CH_3)_2NH_2^+(aq) + OH^-(aq);\ K_b = \dfrac{[(CH_3)_2NH_2^+][OH^-]}{[(CH_3)_2NH]}$

(b) $CO_3^{2-}(aq) + H_2(l) \rightleftharpoons HCO_3^-(aq) + OH^-(aq);\ K_b = \dfrac{[HCO_3^-][OH^-]}{[CO_3^{2-}]}$

(c) $CHO_2^-(aq) + H_2O(l) \rightleftharpoons HCHO_2(aq) + OH^-(aq);\ K_b = \dfrac{[HCHO_2][OH^-]}{[CHO_2^-]}$

16.74 (a) $C_3H_7NH_2(aq) + H_2O(l) \rightleftharpoons C_3H_7NH_3^+(aq) + OH^-(aq);\ K_b = \dfrac{[C_3H_7NH_3^+][OH^-]}{[C_3H_7NH_2]}$

(b) $HPO_4^{2-}(aq) + H_2O(l) \rightleftharpoons H_2PO_4^-(aq) + OH^-(aq);\ K_b = \dfrac{[H_2PO_4^-][OH^-]}{[HPO_4^{2-}]}$

(c) $C_6H_5CO_2^-(aq) + H_2O(l) \rightleftharpoons C_6H_5CO_2H(aq) + OH^-(aq);\ K_b = \dfrac{[C_6H_5CO_2H][OH^-]}{[C_6H_5CO_2^-]}$

16.75   *Analyze/Plan.* Follow the logic in Sample Exercise 16.14.   *Solve.*

$$C_2H_5NH_2(aq) + H_2O(l) \rightleftharpoons C_2H_5NH_3^+(aq) + OH^-(aq)$$

initial	0.075 M	0	0
equil.	(0.075 – x) M	x M	x M

$$K_b = \frac{[C_2H_5NH_3^+][OH^-]}{[C_2H_5NH_2]} = \frac{(x)(x)}{(0.075-x)} = \frac{x^2}{0.075} = 6.4 \times 10^{-4}$$

$x^2 = 0.075 \,(6.4 \times 10^{-4}); \; x = [OH^-] = 6.9 \times 10^{-3} \, M; \; pH = 11.84$

Check. $\dfrac{6.9 \times 10^{-3} \, M \; OH^-}{0.075 \, M \; C_2H_5NH_2} \times 100 = 9.2\%$ ionization; the assumption is not valid

To obtain a more precise result, the K_b expression is rewritten in standard quadratic form and solved via the quadratic formula.

$$\frac{x^2}{0.075-x} = 6.4 \times 10^{-4}; \; x^2 + 6.4 \times 10^{-4} \, x - 4.8 \times 10^{-5} = 0$$

$$x = \frac{b \pm \sqrt{b^2 - 4ac}}{2a} = \frac{-6.4 \times 10^{-4} \pm \sqrt{(6.4 \times 10^{-4})^2 - 4(1)(-4.8 \times 10^{-5})}}{2}$$

$x = 6.61 \times 10^{-3} = 6.6 \times 10^{-3} \, M \; OH^-; \; pOH = 2.18, \; pH = 14.00 - pOH = 11.82$

Note that the pH values obtained using the two algebraic techniques are very similar.

16.76 $$BrO^-(aq) + H_2O(l) \rightleftharpoons HOBr(aq) + OH^-(aq)$$

initial	1.15 M	0	0
equil.	(1.15 – x) M	x M	x M

$$K_b = \frac{[HOBr][OH^-]}{[BrO^-]} = \frac{x^2}{1.15-x} \approx \frac{x^2}{1.15} = 4.0 \times 10^{-6}$$

$x^2 = 1.15 \,(4.0 \times 10^{-6}); \; x = [OH^-] = 2.14 \times 10^{-3} = 2.1 \times 10^{-3} \, M; \; pH = 11.33$

Check. $\dfrac{2.1 \times 10^{-3} \, M \; OH^-}{1.15 \, M \; C_3H_5O_2^-} \times 100 = 0.19\%$ hydrolysis; the assumption is valid

16.77   *Analyze/Plan.* Given pH and initial concentration of base, calculate all equilibrium concentrations. pH → pOH → $[OH^-]$ at equilibrium. Construct the equilibrium table and calculate other equilibrium concentrations. Substitute into the K_b expression and calculate K_b.   *Solve.*

(a) $[OH^-] = 10^{-pOH}; \; pOH = 14 - pH = 14.00 - 11.33 = 2.67$

$[OH^-] = 10^{-2.67} = 2.138 \times 10^{-3} = 2.1 \times 10^{-3} \, M$

$$C_{10}H_{15}ON(aq) + H_2O(l) \rightleftharpoons C_{10}H_{15}ONH^+(aq) + OH^-(aq)$$

initial	0.035 M	0	0
equil.	0.033 M	$2.1 \times 10^{-3} \, M$	$2.1 \times 10^{-3} \, M$

(b) $K_b = \dfrac{[C_{10}H_{15}ONH^+][OH^-]}{[C_{10}H_{15}ON]} = \dfrac{(2.138 \times 10^{-3})^2}{(0.03286)} = 1.4 \times 10^{-4}$

16.78   (a)   pOH = 14.00 – 9.95 = 4.05; $[OH^-] = 10^{-4.05} = 8.91 \times 10^{-6} = 8.9 \times 10^{-5}\ M$

$$C_{18}H_{21}NO_3(aq) + H_2O(l) \ \rightleftharpoons\ C_{18}H_{21}NO_3H^+(aq)\ +\ OH^-(aq)$$

initial   0.0050 M 0   0

equil. $(0.0050 - 8.9 \times 10^{-5})$ $8.9 \times 10^{-5}\ M$ $8.9 \times 10^{-5}\ M$

$K_b = \dfrac{[C_{18}H_{21}NO_3H^+][OH^-]}{[C_{18}H_{21}NO_3]} = \dfrac{(8.91 \times 10^{-5})^2}{(0.0050 - 8.91 \times 10^{-5})} = 1.62 \times 10^{-5} = 1.6 \times 10^{-6}$

(b) $pK_b = -\log(K_b) = -\log(1.62 \times 10^{-5}) = 5.79$

The K_a – K_b Relationship; Acid-Base Properties of Salts

16.79   (a)   For a conjugate acid/conjugate base pair such as $C_6H_5OH/C_6H_5O^-$, K_b for the conjugate base is always K_w/K_a for the conjugate acid. K_b for the conjugate base can always be calculated from K_a for the conjugate acid, so a separate list of K_b values is not necessary.

       (b) $K_b = K_w/K_a = 1.0 \times 10^{-14} / 1.3 \times 10^{-10} = 7.7 \times 10^{-5}$

       (c) K_b for phenolate (7.7×10^{-5}) > K_b for ammonia (1.8×10^{-5}).

           Phenolate is a stronger base than NH_3.

16.80   (a)   We need K_a for the conjugate acid of $CO_3{}^{2-}$, K_a for $HCO_3{}^-$. K_a for $HCO_3{}^-$ is K_{a2}.

       (b) $K_b = K_w/K_a = 1.0 \times 10^{-14}/5.6 \times 10^{-11} = 1.8 \times 10^{-4}$

       (c) K_b for $CO_3{}^{2-}$ (1.8×10^{-4}) > K_b for NH_3 (1.8×10^{-5}).

           $CO_3{}^{2-}$ is a stronger base than NH_3.

16.81   *Analyze/Plan.* Given K_a, determine relative strengths of the acids and their conjugate bases. The greater the magnitude of K_a, the stronger the acid and the weaker the conjugate base.

       K_b (conjugate base) = K_w/K_a.   *Solve.*

       (a)   Acetic acid is stronger, because it has the larger K_a value.

       (b)   Hypochlorite ion is the stronger base because the weaker acid, hypochlorous acid, has the stronger conjugate base.

       (c) K_b for $C_2H_3O_2{}^- = K_w/K_a$ for $HC_2H_3O_2 = 1.0 \times 10^{-14}/1.8 \times 10^{-5} = 5.6 \times 10^{-10}$

           K_b for $ClO^- = K_w/K_a$ for $HClO = 1 \times 10^{-14}/3.0 \times 10^{-8} = 3.3 \times 10^{-7}$

           Note that K_b for ClO^- is greater than K_b for $C_2H_3O_2{}^-$.

16.82   (a)   Ammonia is the stronger base because it has the larger K_b value.

       (b)   Hydroxylammonium is the stronger acid because the weaker base, hydroxylamine, has the stronger conjugate acid.

(c) K_a for $NH_4^+ = K_w/K_b$ for $NH_3 = 1.0 \times 10^{-14}/1.8 \times 10^{-5} = 5.6 \times 10^{-10}$

K_a for $HONH_3^+ = K_w/K_b$ for $HONH_2 = 1.0 \times 10^{-14}/1.1 \times 10^{-8} = 9.1 \times 10^{-7}$

Note that K_a for $HONH_3^+$ is larger than K_a for NH_4^+.

16.83   *Analyze.* When the solute in an aqueous solution is a salt, evaluate the acid/base properties of the component ions.

(a)   *Plan.* NaCN is a soluble salt and thus a strong electrolyte. When it is dissolved in H_2O, it dissociates completely into Na^+ and CN^-. $[NaCN] = [Na^+] = [CN^-] = 0.10$ M. Na^+ is the conjugate acid of the strong base NaOH and thus does not influence the pH of the solution. CN^-, on the other hand, is the conjugate base of the weak acid HCN and **does** influence the pH of the solution. Like any other weak base, it hydrolyzes water to produce $OH^-(aq)$. Solve the equilibrium problem to determine $[OH^-]$.   *Solve.*

$$CN^-(aq) \; + \; H_2O(l) \; \rightleftharpoons \; HCN(aq) \; + \; OH^-(aq)$$

initial	0.10 M	0	0
equil.	(0.10 − x) M	x M	x M

$$K_b \text{ for } CN^- = \frac{[HCN][OH^-]}{[CN^-]} = \frac{K_w}{K_a \text{ for HCN}} = \frac{1 \times 10^{-14}}{4.9 \times 10^{-10}} = 2.04 \times 10^{-5} = 2.0 \times 10^{-5}$$

$2.04 \times 10^{-5} = \dfrac{(x)(x)}{(0.10 - x)}$; assume the percent of CN^- that hydrolyzes is small

$x^2 = 0.10 \, (2.04 \times 10^{-5})$; $x = [OH^-] = 0.00143 = 1.4 \times 10^{-3} \, M$

pOH = 2.85; pH = 14 − 2.85 = 11.15

(b)   *Plan.* $Na_2CO_3(aq) \rightarrow 2Na^+(aq) + CO_3^{2-}(aq)$

CO_3^{2-} is the conjugate base of HCO_3^- and its hydrolysis reaction will determine the $[OH^-]$ and pH of the solution (see similar explanation for NaCN in part (a)). We will assume the process $HCO_3^-(aq) + H_2O(l) \rightleftharpoons H_2CO_3(aq) + OH^-$ will not add significantly to the $[OH^-]$ in solution because $[HCO_3^- (aq)]$ is so small. Solve the equilibrium problem for $[OH^-]$.   *Solve.*

$$CO_3^{2-}(aq) \; + \; H_2O(l) \; \rightleftharpoons \; HCO_3^-(aq) + OH^-(aq)$$

initial	0.080 M	0	0
equil.	(0.080 − x) M	x	x

$$K_b = \frac{[HCO_3^-][OH^-]}{[CO_3^{2-}]} = \frac{K_w}{K_a \text{ for } HCO_3^-} = \frac{1.0 \times 10^{-14}}{5.6 \times 10^{-11}} = 1.79 \times 10^{-4} = 1.8 \times 10^{-4}$$

$1.8 \times 10^{-4} = \dfrac{x^2}{(0.080 - x)}$; $x^2 = 0.080 \, (1.79 \times 10^{-4})$; $x = 0.00378 = 3.8 \times 10^{-3} \, M \; OH^-$

(Assume x is small compared to 0.080); pOH = 2.42; pH = 14 − 2.42 = 11.58

Check. $\dfrac{3.8 \times 10^{-3}\ M\ OH^-}{0.080\ M\ CO_3^{2-}} \times 100 = 4.75\%$ hydrolysis; the assumption is valid

(c)   *Plan.* For the two salts present, Na^+ and Ca^{2+} are negligible acids. NO_2^- is the conjugate base of HNO_2 and will determine the pH of the solution. *Solve.*

Calculate total $[NO_2^-]$ present initially.

$[NO_2^-]_{total} = [NO_2^-]$ from $NaNO_2 + [NO_2^-]$ from $Ca(NO_2)_2$

$[NO_2^-]_{total} = 0.10\ M + 2(0.20\ M) = 0.50\ M$

The hydrolysis equilibrium is:

$$NO_2^-(aq) + H_2O(l) \rightleftharpoons HNO_2 + OH^-(aq)$$

initial	$0.50\ M$	0	0
equil.	$(0.50 - x)\ M$	$x\ M$	$x\ M$

$$K_b = \frac{[HNO_2][OH^-]}{[NO_2^-]} = \frac{K_w}{K_a\ \text{for}\ HNO_2} = \frac{1.0 \times 10^{-14}}{4.5 \times 10^{-4}} = 2.22 \times 10^{-11} = 2.2 \times 10^{-11}$$

$$2.2 \times 10^{-11} = \frac{x^2}{(0.50 - x)} \approx \frac{x^2}{0.50}\ ; x^2 = 0.50\,(2.22 \times 10^{-11})$$

$x = 3.33 \times 10^{-6} = 3.3 \times 10^{-6}\ M\ OH^-$; pOH = 5.48; pH = 14 − 5.48 = 8.52

16.84   (a)   Proceeding as in Solution 16.83(a):

$$F^-(aq) + H_2O(l) \rightleftharpoons HF(aq) + OH^-(aq)$$

initial	$0.085\ M$	$0\ M$	$0\ M$
equil	$(0.085 - x)\ M$	$x\ M$	$x\ M$

$$K_b\ \text{for}\ F^- = \frac{[HF][OH^-]}{[F^-]} = \frac{K_w}{K_a\ \text{for}\ HF} = \frac{1.0 \times 10^{-14}}{6.8 \times 10^{-4}} = 1.47 \times 10^{-11} = 1.5 \times 10^{-11}$$

$$1.5 \times 10^{-11} = \frac{(x)(x)}{(0.085 - x)}\ ; \text{assume the amount of}\ F^-\ \text{that hydrolyzes is small}$$

$x^2 = 0.085(1.47 \times 10^{-11}); x = [OH^-] = 1.118 \times 10^{-6} = 1.1 \times 10^{-6}\ M$

pOH = 5.95; pH = 14 − 5.95 = 8.05

(b) $Na_2S(aq) \rightarrow S^{2-}(aq) + 2Na^+(aq)$

$S^{2-}(aq) + H_2O(l) \rightleftharpoons HS^-(aq) + OH^-(aq)$

As in part (a) above, $[OH^-] = [HS^-] = x; [S^{2-}] = 0.055\ M$

$$K_b = \frac{[HS^-][OH^-]}{[S^{2-}]} = \frac{K_w}{K_a\ \text{for}\ HS^-} = \frac{1.0 \times 10^{-14}}{1 \times 10^{-19}} = 1 \times 10^5$$

Since $K_b \gg 1$, the equilibrium above lies far to the right and $[OH^-] = [S^{2-}] = 0.055 \, M$. K_b for $HS^- = 1.05 \times 10^{-7}$; $[OH^-]$ produced by further hydrolysis of HS^- amounts to $7.6 \times 10^{-5} \, M$. The second hydrolysis step does not make a significant contribution to the total $[OH^-]$ and pH.

$[OH^-] = 0.055 \, M$; pOH = 1.26, pH = 12.74

(c)   As in Solution 16.83(c), calculate total $[C_2H_3O_2^-]$.

$[C_2H_3O_2^-]_t = [C_2H_3O_2^-]$ from $NaC_2H_3O_2 + [C_2H_3O_2^-]$ from $Ba(C_2H_3O_2)_2$

$[C_2H_3O_2^-]_t = 0.045 \, M + 2(0.055 \, M) = 0.155 \, M$

The hydrolysis equilibrium is $C_2H_3O_2^-(aq) + H_2O(l) \rightleftharpoons HC_2H_3O_2(aq) + OH^-(aq)$

$$K_b = \frac{[HC_2H_3O_2][OH^-]}{[C_2H_3O_2^-]} = \frac{K_w}{K_a \text{ for } HC_2H_3O_2} = \frac{1.0 \times 10^{-14}}{1.8 \times 10^{-5}} = 5.56 \times 10^{-10}$$

$$= 5.6 \times 10^{-10}$$

$[OH^-] = [HC_2H_3O_2] = x$, $[C_2H_3O_2^-] = 0.155 - x$

$$K_b = 5.56 \times 10^{-10} = \frac{x^2}{(0.155 - x)}; \text{ assume x is small compared to } 0.155 \, M$$

$x^2 = 0.155 \, (5.56 \times 10^{-10})$; $x = [OH^-] = 9.280 \times 10^{-6} = 9.3 \times 10^{-6}$

pH = 14 + log (9.280×10^{-6}) = 8.97

16.85   *Analyze/Plan.* Given the formula of a salt, predict whether an aqueous solution will be acidic, basic, or neutral. Evaluate the acid-base properties of both ions and determine the overall effect on solution pH. *Solve.*

(a)   acidic; NH_4^+ is a weak acid, Br^- is negligible.

(b)   acidic; Fe^{3+} is a highly charged metal cation and a Lewis acid; Cl^- is negligible.

(c)   basic; CO_3^{2-} is the conjugate base of HCO_3^-; Na^+ is negligible.

(d)   neutral; both K^+ and ClO_4^- are negligible.

(e)   acidic; $HC_2O_4^-$ is amphoteric, but K_a for the acid dissociation (6.4×10^{-5}) is much greater than K_b for the base hydrolysis $(1.0 \times 10^{-14} / 5.9 \times 10^{-2} = 1.7 \times 10^{-13})$.

16.86   (a)   acidic; Cr^{3+} is a highly charged metal cation and a Lewis acid; Br^- is negligible.

(b)   neutral; both Li^+ and I^- are negligible.

(c)   basic; PO_4^{3-} is the conjugate base of HPO_4^{2-}; K^+ is negligible.

(d)   acidic; $CH_3NH_3^+$ is the conjugate acid of CH_3NH_2; Cl^- is negligible.

(e)   acidic; HSO_4^- is a negligible base, but a fairly strong acid $(K_a = 1.2 \times 10^{-2})$. K^+ is negligible.

16.87   *Plan.* Estimate pH using relative base strength and then calculate to confirm prediction. NaCl is a neutral salt, so it is not the unknown. The unknown is a relatively weak base, because a pH of 8.08 is not very basic. Since F^- is a weaker base than OCl^-, the unknown is probably NaF. Calculate K_b for the unknown from the data provided.

Solve.

$[OH^-] = 10^{-pOH}$; $pOH = 14.00 - pH = 14.00 - 8.08 = 5.92$

$[OH^-] = 10^{-6.92} = 1.202 \times 10^{-6} = 1.2 \times 10^{-6}\ M = [HX]$

$[NaX] = [X^-] = 0.050$ mol salt$/0.500$ L $= 0.10\ M$

$$K_b = \frac{[OH^-][HX]}{[X^-]} = \frac{(1.202 \times 10^{-6})^2}{(0.10 - 1.2 \times 10^{-6})} = \frac{(1.202 \times 10^{-6})^2}{0.10} = 1.4 \times 10^{-11}$$

K_b for $F^- = K_w/K_a$ for HF $= 1.0 \times 10^{-14}/6.8 \times 10^{-4} = 1.5 \times 10^{-11}$

The unknown is NaF.

16.88   *Plan.* Estimate pH of salt solution by evaluating the ions in the salts. Calculate to confirm if necessary. *Solve.*

KBr: salt of strong acid and strong base, neutral solution. The unknown is probably KBr. Check the others to be sure.

NH_4Cl: salt of a weak base and a strong acid, acidic solution

KCN: salt of a strong base and a weak acid, basic solution

K_2CO_3: salt of a strong base and a weak acid (HCO_3^-), basic solution

Only KBr fits the acid-base properties of the unknown.

16.89   *Analyze/Plan.* The solution will be basic because of the hydrolysis of the sorbate anion, $C_6H_7O_2^-$. Calculate the initial molarity of $C_6H_7O_2^-$. Calculate K_b from K_w/K_a. Solve the K_b expression for $[OH^-]$.   *Solve.*

$$\frac{11.25\ g\ KC_6H_7O_2}{1.75\ L} \times \frac{1\ mol\ KC_6H_7O_2}{150.2\ g\ KC_6H_7O_2} = 0.04280 = 0.0428\ M\ KC_6H_7O_2$$

$[C_6H_7O_2^-] = [KC_6H_7O_2] = 0.0428\ M$

	$C_6H_7O_2^-$ (aq)	+ H_2O(l)	$\rightleftharpoons$	$HC_6H_7O_2$(aq)	+	OH^-(aq)
initial	$0.0428\ M$			0		0
equil.	$(0.0428 - x)\ M$			x M		x M

$$K_b = \frac{[HC_6H_7O_2][OH^-]}{[C_6H_7O_2^-]} = \frac{K_w}{K_a\ for\ HC_6H_7O_2} = \frac{1.0 \times 10^{-14}}{1.7 \times 10^{-5}} = 5.88 \times 10^{-10} = 5.9 \times 10^{-10}$$

$$5.88 \times 10^{-10} = \frac{x^2}{0.0428 - x} \approx \frac{x^2}{0.0428};\ x^2 = 0.0428\ (5.88 \times 10^{-10})$$

$x = [OH^-] = 5.018 \times 10^{-6} = 5.0 \times 10^{-6}\ M$; $pOH = 5.30$; $pH = 14 - pOH = 8.70$

16.90   The solution is basic because of the hydrolysis of PO_4^{3-}. The molarity of PO_4^{3-} is

$$\frac{45.0\ g\ Na_3PO_4}{1.00\ L\ soln} \times \frac{1\ mol\ Na_3PO_4}{163.9\ g\ Na_3PO_4} = 0.2746 = 0.275\ M\ PO_4^{3-}$$

$$PO_4^{3-}(aq) + H_2O(l) \rightleftharpoons HPO_4^{2-}(aq) + OH^-(aq)$$

$$K_b = \frac{[HPO_4^{2-}][OH^-]}{[PO_4^{3-}]} = \frac{K_w}{K_a \text{ for } HPO_4^{2-}} = \frac{1.0 \times 10^{-14}}{4.2 \times 10^{-13}} = 0.0238 = 2.4 \times 10^{-2}$$

Ignoring the further hydrolysis of HPO_4^{2-},

$[OH^-] = [HPO_4^{2-}] = x$, $[PO_4^{3-}] = 0.275 - x$

$$2.4 \times 10^{-2} = \frac{x^2}{(0.275 - x)}; x^2 + 2.4 \times 10^{-2}\,x - 0.0065 = 0$$

Since K_b is relatively large, we will not assume x is small compared to 0.275.

$$x = \frac{-0.024 \pm \sqrt{(0.024)^2 - 4(1)(-0.0065)}}{2(1)} = \frac{-0.024 \pm \sqrt{0.0267}}{2(1)}$$

$x = 0.070 \; M \; OH^-$; $pH = 14 + \log(0.070) = 12.84$

Acid-Base Character and Chemical Structure

16.91   (a)   As the electronegativity of the central atom (X) increases, more electron density is withdrawn from the X–O and O–H bonds, respectively. In water, the O–H bond is ionized to a greater extent and the strength of the oxyacid increases.

(b)   As the number of nonprotonated oxygen atoms in the molecule increases, they withdraw electron density from the other bonds in the molecule and the strength of the oxyacid increases.

16.92   (a)   Acid strength increases as the polarity of the X–H bond increases and decreases as the strength of the X–H bond increases.

(b)   Assuming the element, X, is more electronegative than H, as the electronegativity of X increases, the X–H bond becomes more polar and the strength of the acid increases. This trend holds true as electronegativity increases across a row of the periodic chart. However, as electronegativity decreases going down a family, acid strength increases because the strength of the H–X bond decreases, even though the H–X bond becomes less polar.

16.93   (a) HNO_3 is a stronger acid than HNO_2 because it has one more nonprotonated oxygen atom, and thus a higher oxidation number on N.

(b)   For binary hydrides, acid strength increases going down a family, so H_2S is a stronger acid than H_2O.

(c) H_2SO_4 is a stronger acid because H^+ is much more tightly held by the anion HSO_4^-.

(d)   For oxyacids, the greater the electronegativity of the central atom, the stronger the acid, so H_2SO_4 is a stronger acid than H_2SeO_4.

(e) CCl_3COOH is stronger because the electronegative Cl atoms withdraw electron density from other parts of the molecule, which weakens the O–H bond and makes H^+ easier to remove. Also, the electronegative Cl delocalizes negative charge on the carboxylate anion. This stabilizes the conjugate base, favoring products in the ionization equilibrium and increasing K_a.

16.94   (a)   For binary hydrides, acid strength increases going across a row, so HCl is a stronger acid than H_2S.

        (b)   For oxyacids, the more electronegative the central atom, the stronger the acid, so H_3PO_4 is a stronger acid than H_3AsO_4.

        (c) $HBrO_3$ has one more nonprotonated oxygen and a higher oxidation number on Br, so it is a stronger acid than $HBrO_2$.

        (d)   The first dissociation of a polyprotic acid is always stronger because H^+ is more tightly held by an anion, so $H_2C_2O_4$ is a stronger acid than $HC_2O_4^-$.

        (e)   The conjugate base of benzoic acid, $C_7H_5O_2^-$, is stabilized by resonance, while the conjugate base of phenol, $C_6H_5O^-$, is not. $HC_7H_5O_2$ has greater tendency to form its conjugate base and is the stronger acid.

16.95   (a) BrO^- (HClO is the stronger acid due to a more electronegative central atom, so BrO^- is the stronger base.)

        (b) BrO^- ($HBrO_2$ has more nonprotonated O atoms and is the stronger acid, so BrO^- is the stronger base.)

        (c) HPO_4^{2-} (larger negative charge, greater attraction for H^+)

16.96   (a) NO_2^- (HNO_3 is the stronger acid because it has more nonprotonated O atoms, so NO_2^- is the stronger base.)

        (b) PO_4^{3-} (K_a for $HAsO_4^{2-}$ is greater than K_a for HPO_4^{2-}, so K_b for PO_4^{3-} is greater and PO_4^{3-} is the stronger base. Note that P is more electronegative than As and H_3PO_4 is a stronger acid than H_3AsO_4, which could lead to the conclusion that AsO_4^{3-} is the stronger base. As in all cases, the measurement of base strength, K_b, supercedes the prediction. Chemistry is an experimental science.

        (c) CO_3^{2-} (The more negative the anion, the stronger the attraction for H^+.)

16.97   (a)   True.

        (b)   False. In a series of acids that have the same central atom, acid strength increases with the number of nonprotonated oxygen atoms bonded to the central atom.

        (c)   False. H_2Te is a stronger acid than H_2S because the H–Te bond is longer, weaker, and more easily dissociated than the H–S bond.

16.98   (a)   True.

        (b)   False. For oxyacids with the same structure but different central atom, the acid strength **increases** as the electronegativity of the central atom increases.

        (c)   False. HF is a weak acid, weaker than the other hydrogen halides, primarily because the H–F bond energy is exceptionally high.

Lewis Acids and Bases

16.99   Yes. If a substance is an Arrhenius base, it must also be a Brønsted base and a Lewis base. The Arrhenius definition (hydroxide ion) is the most restrictive, the Brønsted (H^+ acceptor) more general and the Lewis (electron pair donor) most general. Since a hydroxide ion is both an H^+ acceptor and an electron pair donor, any substance that fits the narrow Arrhenius definition will fit the broader Brønsted and Lewis definitions.

16.100   No. If a substance is a Lewis acid, it is not necessarily a Brønsted or an Arrhenius acid. The Lewis definition of an acid, an electron pair acceptor, is most general. A Lewis acid does not necessarily fit the more narrow description of a Brønsted or Arrhenius acid. An electron pair acceptor isn't necessarily an H^+ donor, nor must it produce H^+ in aqueous solution. An example is Al^{3+}, which is a Lewis acid, but has no ionizable hydrogen.

16.101   *Analyze/Plan.* Identify each reactant as an electron pair donor (Lewis base) or electron pair acceptor (Lewis acid). Remember that a Brønsted acid is necessarily a Lewis acid, and a Brønsted base is necessarily a Lewis base (Solution 16.89).   *Solve.*

	Lewis Acid	Lewis Base
(a)	$Fe(ClO_4)_3$ or Fe^{3+}	H_2O
(b)	H_2O	CN^-
(c)	BF_3	$(CH_3)_3N$
(d)	HIO	NH_2^-

16.102

	Lewis Acid	Lewis Base
(a)	HNO_2 (or H^+)	OH^-
(b)	$FeBr_3$ (Fe^{3+})	Br^-
(c)	Zn^{2+}	NH_3
(d)	SO_2	H_2O

16.103   (a) Cu^{2+}, higher cation charge

           (b) Fe^{3+}, higher cation charge

           (c) Al^{3+}, smaller cation radius, same charge

16.104   (a) $ZnBr_2$, smaller cation radius, same charge

           (b) $Cu(NO_3)_2$, higher cation charge

           (c) $NiBr_2$, smaller cation radius, same charge

Additional Exercises

16.105   (a) K_w is the equilibrium constant for the reaction of two water molecules to form hydronium ion and hydroxide ion.

           (b) K_a is the equilibrium constant for the reaction of any acid, neutral or ionic, with water to form hydronium ion and the conjugate base of the acid.

           (c)   pOH is the negative log of hydroxide ion concentration; pOH decreases as hydroxide ion concentration increases.

16.106   (a)   Correct.

           (b)   Incorrect. A Brønsted acid must have ionizable hydrogen. Lewis acids are electron pair acceptors, but need not have ionizable hydrogen.

(c)   Correct.

(d)   Incorrect. K^+ is a negligible Lewis acid because it is the conjugate of strong base KOH. Its relatively large ionic radius and low positive charge render it a poor attractor of electron pairs.

(e)   Correct.

16.107   (a)   A higher O_2 concentration displaces protons from Hb, producing a more acidic solution, with lower pH in the lungs than in the tissues.

(b) $[H^+]$ = antilog (-7.4) = 4.0×10^{-8} M. At body temperature, 37°C, K_w = 2.4×10^{-14} (see Solution 16.39). At this temperature, a "neutral" solution has $[H^+]$ = 1.5×10^{-7} and pH 6.81. Even though the frame of reference is a bit different at this temperature, blood at pH 7.4 is slightly basic.

(c)   The equilibrium indicates that a high $[H^+]$ shifts the equilibrium toward the proton-bound form HbH^+, which means a lower concentration of HbO_2 in the blood. Thus the ability of hemoglobin to transport oxygen is impeded.

16.108   Assume T = 25°C. For acid or base solute concentrations less than 1×10^{-6} M, we must consider the autoionization of water as a source of $[OH^-]$ and $[H^+]$.

	$H_2O(l)$	$\rightleftharpoons$	$[H^+]$	+	$[OH^-]$
initial	C		0		2.5×10^{-9} M
equil	C		x		$(x + 2.5 \times 10^{-9})$ M

K_w = 1.0×10^{-14} = $[H^+][OH^-]$ = $(x)(x + 2.5 \times 10^{-9})$; $x^2 + 2.5 \times 10^{-9} x - 1.0 \times 10^{-14} = 0$

From the quadratic formula, $x = \dfrac{-2.5 \times 10^{-9} \pm \sqrt{(2.5 \times 10^{-9})^2 - 4(-1 \times 10^{-14})}}{2}$

$= 9.876 \times 10^{-6} = 9.9 \times 10^{-6}$ M H^+

$[H^+]$ = 9.9×10^{-8} M; $[OH^-]$ = $(9.9 \times 10^{-8} + 2.5 \times 10^{-9})$ = 1.013×10^{-7} = 1.0×10^{-7} M

pH = 7.0054 = 7.01

Check: $[9.876 \times 10^{-8}][1.013 \times 10^{-7}]$ = 1.0×10^{-14}. Our answer makes sense. The very small concentration of OH^- from the solute raises the solution pH to slightly more than 7.

16.109   The solution with the higher pH has the lower $[H^+]$.

(a)   For solutions with equal concentrations, the weaker acid will have a lower $[H^+]$ and higher pH.

(b)   The acid with K_a = 8×10^{-5} is the weaker acid, so it has the higher pH.

(c)   The base with pK_b = 4.5 is the stronger base, has greater $[OH^-]$ and smaller $[H^+]$, so higher pH.

16.110 $K_a = \dfrac{[H^+][C_6H_{11}O_2^-]}{[HC_6H_{11}O_2]}$; $[H^+] = [C_6H_{11}O^{2-}] = 10^{-pH} = 10^{-2.94} = 0.001148 = 1.1 \times 10^{-3}\ M$

$[HC_6H_{11}O_2] = \dfrac{11\text{ g }HC_6H_{11}O_2}{L} \times \dfrac{1\text{ mol }HC_6H_{11}O_2}{116.16\text{ g }HC_6H_{11}O_2} = 0.09470 = 0.095\ M$

$K_a = \dfrac{[H^+][C_6H_{11}O_2^-]}{[HC_6H_{11}O_2]} = \dfrac{(0.001148)^2}{(0.09470 - 0.001148)} = 1.4092 \times 10^{-5} = 1.41 \times 10^{-5}$

16.111 (a) $H_2X \rightarrow H^+ + HX^-$

Assuming HX^- does not ionize, $[H^+] = 0.050\ M$, pH = 1.30

(b) $H_2X \rightarrow 2H^+ + X^-$; $0.050\ M\ H_2X = 0.10\ M\ H^+$; pH = 1.00

(c) The observed pH of a 0.050 M solution of H_2X is only slightly less than 1.30, the pH assuming no ionization of HX^-. HX^- is not completely ionized; H_2X, which is completely ionized, is a stronger acid than HX^-.

(d) Since H_2X is a strong acid, HX^- has no tendency to act like a base. HX^- does act like a weak acid, so a solution of NaHX would be acidic.

16.112 *Analyze/Plan.* Evaluate the acid-base properties of the cation and anion to determine whether a solution of the salt will be acidic, basic, or neutral. *Solve.*

(i) NH_4NO_3: NH_4^+, weak conjugate acid of NH_3; NO_3^-, negligible conjugate base of HNO_3; acidic solution.

(ii) $NaNO_3$: Na^+, negligible conjugate acid of NaOH; NO_3^-, negligible conjugate base of HNO_3; neutral solution.

(iii) $NH_4C_2H_3O_2$: NH_4^+, weak conjugate acid of NH_3, $K_a = K_w/1.8 \times 10^{-5} = 5.6 \times 10^{-10}$; $C_2H_3O_2^-$, weak conjugate base of $HC_2H_3O_2$, $K_b = K_w/1.8 \times 10^{-5} = 5.6 \times 10^{-10}$; neutral solution ($K_a$ for the cation and K_b for the anion are accidentally equal, producing a neutral solution).

(iv) NaF: Na^+, negligible conjugate acid of NaOH; F^-, weak conjugate base of HF, $K_b = K_w/6.8 \times 10^{-4} = 1.5 \times 10^{-11}$; basic solution.

(v) $NaC_2H_3O_2$: Na^+, negligible; $C_2H_3O_2^-$, weak base, $K_b = 5.6 \times 10^{-10}$, basic solution.

In order of increasing acidity and decreasing pH: 0.1 M $NaC_2H_3O_2$ > 0.1 M NaF > 0.1 M $NH_4C_2H_3O_2$ = 0.1 M $NaNO_3$ > 0.1 M NH_4Cl; (v) > (iv) > (iii) ~ (ii) > (i)

(iv) and (v) are both bases, and (v) has the greater K_b value and higher pH. (ii) and (iii) are both neutral and (i) is acidic.

16.113 Considering the stepwise dissociation of H_3PO_4:

$$H_3PO_4(aq) \rightleftharpoons H^+(aq) + H_2PO_4^-(aq)$$

	H_3PO_4	H^+	$H_2PO_4^-$
initial	0.025	0	0
equil.	$(0.025 - x)\ M$	x	x

$$K_{a1} = \frac{[H^+][H_2PO_4^-]}{[H_3PO_4]} = \frac{x^2}{(0.025-x)} = 7.5 \times 10^{-3}; \; x^2 + 7.5 \times 10^{-3}\,x - 1.875 \times 10^{-4} = 0$$

$$x = \frac{-7.5 \times 10^{-3} \pm \sqrt{(7.5 \times 10^{-3})^2 - 4(1)(-1.875 \times 10^{-4})}}{2(1)} = \frac{-7.5 \times 10^{-3} \pm \sqrt{8.06 \times 10^{-4}}}{2}$$

$x = 0.01045 = 0.010 \; M \; H^+$, $0.010 \; M \; H_2PO_4^-$ available for further ionization

	$H_2PO_4^-$ (aq)	$\rightleftharpoons$	H^+ (aq)	+	HPO_4^{2-} (aq)
initial	$0.010 \; M$		$0.010 \; M$		$0 \; M$
equil.	$(0.010-y) \; M$		$(0.010+y) \; M$		$y \; M$

$$K_{a2} = \frac{[H^+][HPO_4^{2-}]}{[H_2PO_4^-]} = \frac{(y)(0.010+y)}{(0.010-y)} = 6.2 \times 10^{-8}$$

Since K_{a2} is very small, assume y is small compared to $0.010 \; M$.

$$6.2 \times 10^{-8} = \frac{0.010\,y}{0.010}; \; y = [HPO_4^{2-}] = 6.2 \times 10^{-8} \; M$$

$[H_2PO_4^-] = (0.010 \; M + 6.2 \times 10^{-8} \; M) = 0.010 \; M$

$[H^+] = (0.010 \; M - 6.2 \times 10^{-8} \; M) = 0.010 \; M$

$[HPO_4^{2-}]$ available for further ionization $= 6.2 \times 10^{-8} \; M$

	HPO_4^{2-}	$\rightleftharpoons$	H^+ (aq)	+	PO_4^{3-} (aq)
initial	$6.2 \times 10^{-8} \; M$		$0.010 \; M$		$0 \; M$
equil.	$(6.2 \times 10^{-8} - z) \; M$		$(0.010 + z) \; M$		$z \; M$

$$K_{a3} = \frac{[H^+][PO_4^{3-}]}{[HPO_4^{2-}]} = \frac{(0.010+z)(z)}{(6.2 \times 10^{-8} - z)} = 4.2 \times 10^{-13}$$

Assuming z is small compared to 6.2×10^{-8} (and 0.010),

$$\frac{(0.010)(z)}{(6.2 \times 10^{-8})} = 4.2 \times 10^{-13}; \; z = 2.6 \times 10^{-18} \; M \; PO_4^{3-}$$

The contribution of z to $[HPO_4^{2-}]$ and $[H^+]$ is negligible. In summary, after all dissociation steps have reached equilibrium:

$[H^+] = 0.010 \; M$, $[H_2PO_4^-] = 0.010 \; M$, $[HPO_4^{2-}] = 6.2 \times 10^{-8} \; M$, $[PO_4^{3-}] = 2.6 \times 10^{-18} \; M$

Note that the first ionization step is the major source of H^+ and the others are important as sources of HPO_4^{2-} and PO_4^{3-}. The $[PO_4^{3-}]$ is very small at equilibrium.

16.114   Call each compound in the neutral form Q.

Then, $Q(aq) + H_2O(l) \rightleftharpoons QH^+(aq) + OH^-$. $K_b = [QH^+][OH^-]/[Q]$

The ratio in question is $[QH^+]/[Q]$, which equals $K_b/[OH^-]$ for each compound.
At pH = 2.5, pOH = 11.5, $[OH^-]$ = antilog $(-11.5) = 3.16 \times 10^{-12} = 3 \times 10^{-12} \; M$. Now calculate $K_b/[OH^-]$ for each compound:

Nicotine $\dfrac{[QH^+]}{[Q]} = 7 \times 10^{-7} / 3.16 \times 10^{-12} = 2 \times 10^5$

Caffeine $\dfrac{[QH^+]}{[Q]} = 4 \times 10^{-14} / 3.16 \times 10^{-12} = 1 \times 10^{-2}$

Strychnine $\dfrac{[QH^+]}{[Q]} = 1 \times 10^{-6} / 3.16 \times 10^{-12} = 3 \times 10^5$

Quinine $\dfrac{[QH^+]}{[Q]} = 1.1 \times 10^{-6} / 3.16 \times 10^{-12} = 3.5 \times 10^5$

For all the compounds except caffeine the protonated form has a much higher concentration than the neutral form. However, for caffeine, a very weak base, the neutral form dominates.

16.115   (a)   Consider the formation of the zwitterion as a series of steps (Hess' law).

$$NH_2-CH_2-COOH + H_2O \rightleftharpoons NH_2-CH_2-COO^- + H_3O^+ \qquad K_a$$

$$NH_2-CH_2-COOH + H_2O \rightleftharpoons {}^+NH_3-CH_2-COOH + OH^- \qquad K_b$$

$$H_3O^+ + OH^- \rightleftharpoons 2H_2O \qquad\qquad\qquad 1/K$$

$$NH_2-CH_2-COOH \rightleftharpoons {}^+NH_3-CH_2-COO^- \qquad \dfrac{K_a \times K_b}{K_w}$$

$$K = \dfrac{K_a \times K_b}{K_w} = \dfrac{(4.3 \times 10^{-3})(6.0 \times 10^{-5})}{1.0 \times 10^{-14}} = 2.6 \times 10^7$$

The large value of K indicates that formation of the zwitterion is favorable. The assumption is that the same NH_2-CH_2-COOH molecule is acting like an acid ($-COOH \rightarrow H^+ + -COO^-$) and a base ($-NH_2 + H^+ \rightarrow NH_3^+$), simultaneously. Glycine is both a stronger acid and a stronger base than water, so the H^+ transfer should be intramolecular, as long as there are no other acids or bases in the solution.

(b)   Since glycine exists as the zwitterion in aqueous solution, the pH is determined by the equilibrium below.

$${}^+NH_3-CH_2-COO^- + H_2O \rightleftharpoons NH_2-CH_2-COO^- + H_3O^+$$

$$K_a = \dfrac{[NH_2-CH_2-COO^-][H_3O^+]}{[{}^+NH_3-CH_2-COO^-]} = \dfrac{K_w}{K_b} = \dfrac{1.0 \times 10^{-14}}{6.0 \times 10^{-5}} = 1.67 \times 10^{-10} = 1.7 \times 10^{-10}$$

$$x = [H_3O^+] = [NH_2-CH_2-COO^-]; \; K_a = 1.67 \times 10^{-10} = \dfrac{(x)(x)}{(0.050-x)} \approx \dfrac{x^2}{0.050}$$

$$x = [H_3O^+] = 2.89 \times 10^{-6} = 2.9 \times 10^{-6} \, M; \; pH = 5.54$$

(c)   In a strongly acidic solution the $-CO_2^-$ function would be protonated, so glycine would exist as $^+H_3NCH_2COOH$. In strongly basic solution the $-NH_3^+$ group would be deprotonated, so glycine would be in the form $H_2NCH_2CO_2^-$.

16.116   The general Lewis structures for these acids and their conjugate bases are shown below.

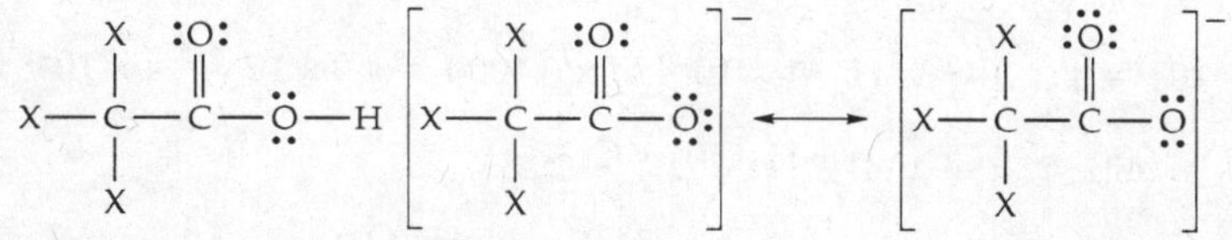

where X = H or Cl.

Replacement of H on the acid by the more electronegative chlorine atoms causes the central carbon to become more positively charged, thus withdrawing more electrons from the attached COOH group, in turn causing the O–H bond to be more polar, so that H^+ is more readily ionized.

For the conjugate base (two resonance structures), the electronegative X atoms delocalize negative charge and stabilize these forms relative to the unsubstituted anions. This favors products in the ionization equilibrium and increases the value of K_a.

To calculate pH proceed as usual, except that the full quadratic formula must be used for all but acetic acid.

Acid	pH
acetic	3.37
chloroacetic	2.51
dichloroacetic	2.09
trichloroacetic	2.0

Integrative Exercises

16.117   At 25°C, $[H^+] = [OH^-] = 1.0 \times 10^{-7} M$

$$\frac{1.0 \times 10^{-7} \text{ mol H}^+}{1 \text{L H}_2\text{O}} \times 0.0010 \text{ L} \times \frac{6.022 \times 10^{23} \text{ H}^+ \text{ ions}}{\text{mol H}^+} = 6.0 \times 10^{13} \text{ H}^+ \text{ ions}$$

16.118   *Analyze.* Given mass % and density of concentrated HCl, calculate volume of concentrated solution required to produce 10.0 L of HCl with pH = 2.05. *Plan.* Calculate molarity of concentrated solution from density and mass %. Calculate molarity of dilute solution from pH. Use the dilution formula to calculate volume (mL) concentrated solution required.   *Solve.*

$$\frac{1.18 \text{ g conc. soln.}}{\text{mL conc. soln.}} \times \frac{36.0 \text{ g HCl}}{100 \text{ g conc. soln.}} \times \frac{1000 \text{ mL}}{1 \text{ L}} \times \frac{1 \text{ mol HCl}}{36.46 \text{ g HCl}} = 11.651 \text{ mol HCl/L}$$

$$= 11.7 \ M \text{ HCl/L}$$

For the dilute HCl solution, $[H^+] = 10^{-pH} = 10^{-2.05} = 8.913 \times 10^{-3} = 8.9 \times 10^{-3} \, M$ HCl

$M_C \times L_C = M_D \times M_D; 11.651 \times L_C = 8.913 \times 10^{-3} \, M \times 10.0 \, L;$

$L_C = 7.650 \times 10^{-3}; 7.650 \times 10^{-3} \, L \times \dfrac{1000 \, mL}{1 \, L} = 7.65 = 7.7 \, mL$ conc. HCl

16.119 $[H^+] = 10^{-pH} = 10^{-2} = 1 \times 10^{-2} \, M \, H^+; 1 \times 10^{-2} \, M \times 0.400 \, L = 4.0 \times 10^{-3} = 4 \times 10^{-3} \, mol \, H^+$

$HCl(aq) + HCO_3^-(aq) \rightarrow Cl^-(aq) + H_2O(l) + CO_2(g)$

$4 \times 10^{-3} \, mol \, H^+ = 4 \times 10^{-3} \, mol \, HCO_3^- \times \dfrac{84.01 \, g \, NaHCO_3}{1 \, mol \, HCO_3^-} = 0.336 = 0.3 \, g \, NaHCO_3$

16.120   *Analyze.* If pH were directly related to CO_2 concentration, this exercise would be simple. Unfortunately, we must solve the equilibrium problem for the diprotic acid H_2CO_3 to calculate $[H^+]$ and pH. We are given ppm CO_2 in the atmosphere at two different times, and the pH that corresponds to one of these CO_2 levels. We are asked to find pH at the other atmospheric CO_2 level.

Plan. Assume all dissolved CO_2 is present as H_2CO_3 (aq) (Sample Exercise 16.13).

pH $\rightarrow$ $[H^+]$ $\rightarrow$ $[H_2CO_3]$. While H_2CO_3 is a diprotic acid, the two K_a values differ by more than 10^3, so we can ignore the second ionization when calculating $[H_2CO_3]$. Change 375 ppm CO_2 to pressure and calculate the Henry's law constant for CO_2. Calculate the dissolved $[CO_2] = [H_2CO_3]$ at 315 ppm, then solve the K_{a1} expression for $[H^+]$ and pH.

(a)   *Solve.* $H_2CO_3(aq) \rightleftharpoons H^+(aq) + HCO_3^-(aq)$

$K_{a1} = 4.3 \times 10^{-7} = \dfrac{[H^+][HCO_3^-]}{[H_2CO_3]}; [H^+] = 10^{-5.4} = 3.98 \times 10^{-6} = 4 \times 10^{-6} \, M$

$[H^+] = [HCO_3^-]; [H_2CO_3] = x - 4 \times 10^{-5}$

$4.3 \times 10^{-7} = \dfrac{(3.98 \times 10^{-6})^2}{(x - 3.98 \times 10^{-6})}; 4.3 \times 10^{-7} x = 1.585 \times 10^{-11} + 1.712 \times 10^{-12}$

$x = 1.756 \times 10^{-11}/4.3 \times 10^{-7} = 4.084 \times 10^{-5} = 4 \times 10^{-5} \, M \, H_2CO_3$

375 ppm = 375 mol CO_2/1×10^6 mol air = 0.000375 mol % CO_2

Because of the properties of gases, mol % = pressure %. $P_{CO_2} = 0.000375$ atm. According to Equation [13.4], $S_{CO_2} = kP_{CO_2}$;

$4.084 \times 10^{-5} \, mol/L = k(3.75 \times 10^{-4} \, atm) \, k = 0.1089 = 0.1 \, mol/L \cdot atm.$

Forty years ago, $S_{CO_2} = 0.1089 \dfrac{mol}{L \cdot atm} \times 3.15 \times 10^{-4} \, atm = 3.4305 \times 10^{-5}$

$= 3.4 \times 10^{-5} \, M$

Now solve K_{a1} for $[H^+]$ at this $[H_2CO_3]$. $[H^+] = x$

We cannot assume x is small, because $[H_2CO_3]$ is so low.

$4.3 \times 10^{-7} = x^2/(3.4305 \times 10^{-5} - x)$; $x^2 + 4.3 \times 10^{-7} x - 1.475 \times 10^{-11} = 0$

$$x = \frac{-4.3 \times 10^{-7} \pm \sqrt{(4.3 \times 10^{-7})^2 - 4(-1.475 \times 10^{-11})}}{2} = \frac{-4.3 \times 10^{-7} + 7.693 \times 10^{-6}}{2}$$

$$= 3.632 \times 10^{-6} = 3.6 \times 10^{-6} \, M \, H^+$$

$[H^+] = 3.6 \times 10^{-6} \, M$, pH = 5.440 = 5.4

(Note that, to the precision that the pH data is reported, the change in atmospheric CO_2 leads to no change in pH.)

(b) From part (a), $[H_2CO_3]$ today = $4.084 \times 10^{-5} \, M$

$$\frac{4.084 \times 10^{-5} \text{ mol } H_2CO_3}{1 \, L} \times 20.0 \, L = 8.168 \times 10^{-4} = 8 \times 10^{-4} \text{ mol } CO_2$$

$$V = \frac{nRT}{P} = 8.168 \times 10^{-5} \text{ mol} \times \frac{298 \, K}{1.0 \text{ atm}} \times \frac{0.08206 \, L \bullet atm}{mol \bullet K} = 0.01997 = 0.02 \, L = 20 \text{ mL}$$

16.121 (a) 24 valence e^-, 12 e^- pairs

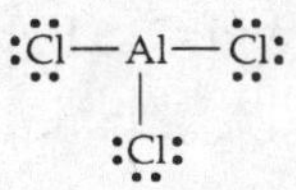

The formal charges on all atoms are zero. Structures with multiple bonds lead to nonzero formal charges. There are three electron domains about Al. The electron-domain geometry and molecular structure are trigonal planar.

(b) The Al atom in $AlCl_3$ has an incomplete octet and is electron deficient. It "needs" to accept another electron pair, to act like a Lewis acid.

(c)

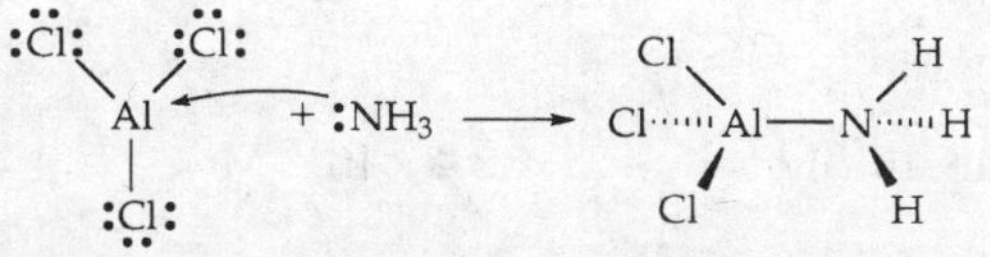

Both the Al and N atoms in the product have tetrahedral geometry.

(d) The Lewis theory is most appropriate. H^+ and $AlCl_3$ are both electron pair acceptors, Lewis acids.

16.122 *Plan.* Use acid ionization equilibrium to calculate the total moles of particles in solution. Use density to calculate kg solvent. From the molality (*m*) of the solution, calculate ΔT_b and T_b. *Solve.*

	HSO_4^-(aq)	$\rightleftharpoons$	H^+(aq)	+	SO_4^{2-}(aq)
Initial	0.10 *M*		0		0
equil.	0.10 – x *M*		x *M*		x *M*
	0.071 *M*		0.029 *M*		0.029 *M*

$K_a = 1.2 \times 10^{-2} = \dfrac{[H^+][SO_4^{2-}]}{[HSO_4^-]} = \dfrac{x^2}{0.10-x}$; K_a is relatively large, so use the quadratic.

$x^2 + 0.012\,x - 0.0012 = 0$; $x = \dfrac{-0.012 \pm \sqrt{(0.012)^2 - 4(1)(-0.0012)}}{2}$; $x = 0.029\ M\ H^+, SO_4^{2-}$

Total ion concentration $= 0.10\ M\ Na^+ + 0.071\ M\ HSO_4^- + 0.029\ M\ H^+ + 0.029\ M\ SO_4^{2-}$

$$= 0.229 = 0.23\ M.$$

Assume 100.0 mL of solution. 1.002 g/mL $\times$ 100.0 mL = 100.2 g solution.

$0.10\ M\ NaHSO_4 \times 0.1000\ L = 0.010\ mol\ NaHSO_4 \times \dfrac{120.1\ g\ NaHSO_4}{mol\ NaHSO_4}$

$$= 1.201 = 12\ g\ NaHSO_4$$

100.2 g soln – 1.201 g $NaHSO_4$ = 99.0 g = 0.099 kg H_2O

$m = \dfrac{mol\ ions}{kg\ H_2O} = \dfrac{0.229\ M \times 0.1000\ L}{0.0990\ kg} = 0.231 = 0.23\ m$ ions

$\Delta T_b = K_b(m) = 0.52°C/m \times (0.23\ m) = +0.12°C$: $T_b = 100.0 + 0.12 = 100.1°C$

16.123   Calculate M of the solution from osmotic pressure, and K_b using the equilibrium expression for the hydrolysis of cocaine. Let Coc = cocaine and CocH$^+$ be the conjugate acid of cocaine.

$\pi = M\,RT$; $M = \pi/RT = \dfrac{52.7\ torr}{288\ K} \times \dfrac{1\ atm}{760\ torr} \times \dfrac{mol \cdot K}{0.08206\ L \cdot atm}$

$$= 0.002934 = 2.93 \times 10^{-3}\ M\ Coc$$

pH = 8.53; pOH = 14 – pH = 5.47; $[OH^-] = 10^{-5.47} = 3.39 \times 10^{-6} = 3.4 \times 10^{-6}\ M$

	Coc(aq) + H$_2$O(l)	$\rightleftharpoons$	CocH$^+$(aq)	+	OH$^-$(aq)
initial	$2.93 \times 10^{-3}\ M$		0		0
equil.	$(2.93 \times 10^{-3} - 3.4 \times 10^{-6})\ M$		$3.4 \times 10^{-6}\ M$		$3.4 \times 10^{-6}\ M$

$K_b = \dfrac{[CocH^+][OH^-]}{[Coc]} = \dfrac{(3.39 \times 10^{-6})^2}{(2.934 \times 10^{-3} - 3.39 \times 10^{-6})} = 3.9 \times 10^{-9}$

Note that % hydrolysis is small in this solution, so "x," $3.4 \times 10^{-6}\ M$, is small compared to $2.93 \times 10^{-3}\ M$ and could be ignored in the denominator of the calculation.

16.124   (a)   rate $= k[IO_3^-][SO_3^{2-}][H^+]$

       (b) $\Delta pH = pH_2 - pH_1 = 3.50 - 5.00 = -1.50$

              $\Delta pH = -\log[H^+]_2 - (-\log[H^+]_1)$; $-\Delta pH = \log[H^+]_2 - \log[H^+]_1$

              $-\Delta pH = \log[H^+]_2 / [H^+]_1$; $[H^+]_2 / [H^+]_1 = 10^{-\Delta pH}$

              $[H^+]_2/[H^+]_1 = 10^{1.50} = 31.6 = 32$. The rate will increase by a factor of 32 if $[H^+]$ increases by a factor of 32. The reaction goes faster at lower pH.

(c)   Since H^+ does not appear in the overall reaction, it is either a catalyst or an intermediate. An intermediate is produced and then consumed during a reaction, so its contribution to the rate law can usually be written in terms of concentrations of other reactants (Sample Exercise 14.15). A catalyst is present at the beginning and end of a reaction and can appear in the rate law if it participates in the rate-determining step (Solution 14.72). This reaction is pH dependent because H^+ is a homogeneous catalyst that participates in the rate-determining step.

16.125   (a)   (i)

$$HCO_3^-(aq) \rightleftharpoons H^+(aq) + CO_3^{2-}(aq) \qquad K_1 = K_{a2} \text{ for } H_2CO_3 = 5.6 \times 10^{-11}$$

$$H^+(aq) + OH^-(aq) \ f \ H_2O(l) \qquad K_2 = 1/K_w = 1 \times 10^{14}$$

$$\overline{HCO_3^-(aq) + OH^-(aq) \rightleftharpoons CO_3^{2-}(aq) + H_2O(l) \qquad K = K_1 \times K_2 = 5.6 \times 10^3}$$

     (ii)

$$NH_4^+(aq) \rightleftharpoons H^+(aq) + NH_3(aq) \qquad K_1 = K_a \text{ for } NH_4^+ = 5.6 \times 10^{-10}$$

$$CO_3^{2-}(aq) + H^+(aq) \rightleftharpoons HCO_3^-(aq) \qquad K_2 = 1/K_{a2} \text{ for } H_2CO_3 = 1.8 \times 10^{10}$$

$$\overline{NH_4^+(aq) + CO_3^{2-}(aq) \rightleftharpoons HCO_3^-(aq) + NH_3(aq) \qquad K = K_1 \times K_2 = 10}$$

(b)   Both (i) and (ii) have K > 1, although K = 10 is not **much** greater than 1. Both could be written with a single arrow. (This is true in general when a strong acid or strong base, $H^+(aq)$ or $OH^-(aq)$, is a reactant.)

16.126   (a)   The structures of the two acids are similar, but lactic acid has an –OH group on the C atom adjacent to the –COOH group. This electronegative substituent withdraws electron density from the –COOH group, and stabilizes its conjugate base, increasing the strength of lactic acid relative to propionic acid. The stronger the acid, the larger the K_a value and the smaller the pK_a.

(b) $pK_a = 3.85$, $K_a = 10^{-3.85} = 1.4 \times 10^{-4}$; $[H^+] = [C_3H_5O_3^-] = x$; $[HC_3H_5O_3] = 0.050 - x$

$$K_a = 1.4 \times 10^{-4} = \frac{[H^+][C_3H_5O_3^-]}{[HC_3H_5O_3]} = \frac{x^2}{0.050-x} \approx \frac{x^2}{0.050}$$

$$x^2 = 0.050(1.4 \times 10^{-4}) = 2.646 \times 10^{-3} = 2.6 \times 10^{-3} \, M \, C_3H_5O_3^-$$

(This represents 5.3% dissociation; solution by the quadratic yields essentially the same result.)

(c)   Strategy: Assume a 100 g sample. Calculate mol Cu in sample. Use mole ratios from formula to calculate mass of O and H not due to H_2O. Subtract masses of Cu, C, H, and O from 100 g to get mass of H_2O. Calculate mol H_2O and X.

Assume a 100 g sample, 22.9 g Cu.

$$22.9 \text{ g Cu} \times \frac{1 \text{ mol Cu}}{63.55 \text{ g Cu}} = 0.3603 = 0.360 \text{ mol Cu}$$

mole ratios of Cu, O, and H (not due to H_2O): 1 Cu:6 O:10 H

$$g \ O = 0.3603 \ mol \ Cu \times \frac{6 \ mol \ O}{1 \ mol \ Cu} \times \frac{16.00 \ g \ O}{1 \ mol \ O} = 34.59 = 34.6 \ g \ O$$

$$g \ H = 0.3603 \ mol \ Cu \times \frac{10 \ mol \ H}{1 \ mol \ Cu} \times \frac{1.008 \ g \ H}{1 \ mol \ H} = 3.632 = 3.63 \ g \ H$$

$$g \ H_2O = 100 \ g \ sample - [22.9 \ g \ Cu + 26.0 \ g \ C + 34.6 \ g \ O + 3.63 \ g \ H]$$

$$= 12.87 = 12.9 \ g \ H_2O$$

$$12.87 \ g \ H_2O \times \frac{1 \ mol \ H_2O}{18.02 \ g \ H_2O} = 0.7142 \ mol \ H_2O/0.3604 = 1.98 \approx 2$$

$X = 2 \ mol \ H_2O$ in the hydrate.

(d)   Compare K_a for $Cu^{2+}(aq)$ and K_b for $C_3H_5O_3^-(aq)$.

The ion with the larger K value will undergo hydrolysis to the greater extend and will determine the pH of the solution.

pK_b for $C_3H_5O_3^- = 14 - pK_a$ for $HC_3H_5O_3 = 14.00 - 3.85 = 10.15$

$K_b = 10^{-10.15} = 7.1 \times 10^{-11}$.

Since K_a for Cu^{2+} (1.0×10^{-8}) is greater than K_b for $C_3H_5O_3^-$ (7.1×10^{-11}) the solution will be slightly acidic.

17 Additional Aspects of Aqueous Equilibria

Visualizing Concepts

17.1 *Analyze.* Given diagrams showing equilibrium mixtures of HX and X⁻ with different compositions, decide which has the highest pH. HX is a weak acid and X⁻ is its conjugate base. *Plan.* Evaluate the contents of the boxes. Use acid-base equilibrium principles to relate [H⁺] to box composition. *Solve.*

Use the following acid ionization equilibrium to describe the mixtures:
$HX(aq) \rightleftharpoons H^+(aq) + X^-(aq)$. Each box has 4 HX molecules, but differing amounts of X⁻ ions. The greater the amount of X⁻ (conjugate base), for the same amount of HX (weak acid), the lower the amount of H⁺ and the higher the pH. The middle box, with most X⁻, has least H⁺ and highest pH.

17.2 (a) According to Figure 16.7, methyl orange is yellow above pH 4.5 and red (really pink) below pH 3.5. The beaker on the left has a pH greater than 4.5, and the one on the right has pH less than 3.5. (By calculation, pH of left beaker = 4.7, pH of right beaker = 2.9.) The right beaker, with lower pH and greater [H⁺], is pure acetic acid. The left beaker contains equal amounts of the weak acid and its conjugate base, acetic acid and acetate ion. Adding the "common-ion" acetate (in the form of sodium acetate) shifts the acid ionization equilibrium to the left, decreases [H⁺], and raises pH.

(b) When small amounts of NaOH are added, the left beaker is better able to maintain its pH. For solutions of the same weak acid, pH depends on the **ratio** of conjugate base to conjugate acid. Small additions of base (or acid) have the least effect when this ratio is close to one. The left beaker is a buffer because it contains a weak conjugate acid-conjugate base pair and resists rapid pH change upon addition of small amounts of strong base or acid.

17.3 *Analyze/Plan.* When strong acid is added to a buffer, it reacts with conjugate base (CB) to produce conjugate acid (CA). [CA] increases and [CB] decreases. The opposite happens when strong base is added to a buffer, [CB] increases and [CA] decreases. Match these situations to the drawings. *Solve.*

The buffer begins with equal concentrations of HX and X⁻.

(a) After addition of strong acid, [HX] will increase and [X⁻] will decrease. Drawing (3) fits this description.

(b) Adding of strong base causes [HX] to decrease and [X⁻] to increase. Drawing (1) matches the description.

(c)   Drawing (2) shows both [HX] and [X$^-$] to be smaller than the initial concentrations shown on the left. This situation cannot be achieved by adding strong acid or strong base to the original buffer.

17.4   *Analyze/Plan.* Consider the reaction $HA + OH^- \rightarrow A^- + H_2O$. What are the major species present in solution at the listed stages of the titration? Which diagram represents these species?   *Solve.*

(a)   *Before addition of NaOH,* the solution is mostly HA. The only A$^-$ is produced by the ionization equilibrium of HA and is too small to appear in the diagram. This situation is shown in diagram (iii), which contains only HA.

(b)   *After addition of NaOH but before the equivalence point,* some, but not all, HA has been converted to A$^-$. The solution contains a mixture of HA and A$^-$; this is shown in diagram (i).

(c)   *At the equivalence point,* all HA has been converted to A$^-$, with no excess HA or OH$^-$ present. This is shown in diagram (iv).

(d)   *After the equivalence point,* the same amount of A$^-$ is present as at the equivalence point, plus some excess OH$^-$. This is diagram (ii).

17.5   *Analyze/Plan.* In each case, the first substance is in the buret, and the second is in the flask. If acid is in the flask, the initial pH is low; with base in the flask, the pH starts high. Strong acids have lower pH than weak acids; strong bases have higher pH than weak bases. Polyprotic acids and bases have more than one "jump" in pH.

(a)   Strong base in flask, pH starts high, ends low as acid is added. Only diagram (ii) fits this description.

(b)   Weak acid in flask, pH starts low, but not extremely low. Diagrams (i), (iii), and (iv) all start at low pH and get higher. Diagram (i) has very low initial pH, and likely has strong acid in the flask. Diagram (iv) has two pH jumps, so it has a polyprotic acid in the flask. Diagram (iii) best fits the profile of adding a strong base to a weak acid.

(c)   Strong acid in the flask, pH starts very low, diagram (i).

(d)   Polyprotic acid, more than one pH jump, diagram (iv).

17.6   *Analyze/Plan.* The product of the ion concentrations in a saturated solution equals K_{sp}. Use numbers of anions and cations as a measure of concentration to calculate relative "K_{sp}" values. Counting cations is not adequate, because excess anions in some of the boxes drive down the cation concentrations. Ion-products must be considered. *Solve.*

AgX: $(4 \, Ag^+)(4X^-) = 16$

AgY: $(1 \, Ag^+)(9 \, Y^-) = 9$

AgZ: $(3 \, Ag^+)(6 \, Y^-) = 18$

AgY has the smallest K_{sp}.

17.7   *Analyze/Plan.* Common anions or cations decrease the solubility of salts. Ions that participate in acid-base or complex ion equilibria increase solubility.   *Solve.*

(a) CO_2^{3-} from $BaCO_3$ reacts with H^+ from HNO_3, causing solubility of $BaCO_3$ to increase with increasing HNO_3 concentration. This behavior matches the right diagram.

(b)   Extra CO_2^{3-} from Na_2CO_3 decreases the solubility of $BaCO_3$. Solubility of $BaCO_3$ decreases as $[Na_2CO_3]$ increases. This behavior matches the left diagram.

(c) $NaNO_3$ has no common ions, nor does it enter into acid-base or complex ion equilibria with Ba^{2+} or CO_3^{2-}; it does not affect the solubility of $BaCO_3$. This behavior is shown in the center diagram.

17.8   A metal hydroxide that is soluble at very low and very high pH's, that is, in strong acid or strong base, is called amphoteric.

Common-Ion Effect

17.9   (a)   The extent of ionization of a weak electrolyte is decreased when a strong electrolyte containing an ion in common with the weak electrolyte is added to it.

(b) $NaNO_2$

17.10   (a)   For a generic weak base B, $K_b = \dfrac{[BH^+][OH^-]}{[B]}$. If an external source of BH^+ such as BH^+Cl^- is added to a solution of B(aq), $[BH^+]$ increases, decreasing $[OH^-]$ and increasing $[B]$, effectively suppressing the ionization (hydrolysis) of B.

(b) NH_4Cl

17.11   *Analyze/Plan.* Given the formula of two substances, determine the effect on pH when one is added to the other. In general, when an acid is added to a solution, pH decreases; when a base is added to a solution, pH increases. Based on its formula, determine whether the substance being added is an acid or a base and predict the change in pH of the solution.   *Solve.*

(a)   pH increases; NO_2^- decreases the ionization of HNO_2 and decreases $[H^+]$.

(b)   pH decreases; $CH_3NH_3^+$ decreases the ionization (hydrolysis) of CH_3NH_2 and decreases $[OH^-]$.

(c)   pH increases; CHO_2^- decreases the ionization of $HCHO_2$ and decreases $[H^+]$.

d)   No change; Br^- is a negligible base and does not affect the 100% ionization of the strong acid HBr.

(e)   pH decreases; the pertinent equilibrium is

$$C_2H_3O_2^-(aq) + H_2O(l) \rightleftharpoons HC_2H_3O_2 + OH^-(aq).$$

HCl reacts with OH^-(aq), decreasing $[OH^-]$ and pH.

17.12 In general, when an acid is added to a solution pH decreases; when a base is added to a solution, pH increases.

(a) pH increases; $C_7H_5O_2^-$ decreases ionization of $HC_7H_5O_2$ and decreases $[H^+]$.

(b) pH decreases; $C_5H_5NH^+$ decreases ionization (hydrolysis) of C_5H_5N and decreases $[OH^-]$.

(c) pH increases; NH_3 reacts with HCl, decreasing $[H^+]$.

(d) pH increases; HCO_3^- decreases ionization of H_2CO_3 and decreases $[H^+]$.

(e) no change; ClO_4^- is a negligible base and Na^+ is a negligible acid.

17.13 *Analyze/Plan.* Follow the logic in Sample Exercise 17.1.

(a)
$$HC_3H_5O_2\,(aq) \;\rightleftharpoons\; H^+\,(aq) \;+\; C_3H_5O_2^-\,(aq)$$

i	$0.085\ M$		$0.060\ M$
c	$-x$	$+x$	$+x$
e	$(0.085-x)\ M$	$+x\ M$	$(0.060+x)\ M$

$$K_a = 1.3 \times 10^{-5} = \frac{[H^+][C_3H_5O_2^-]}{[HC_3H_5O_2]} = \frac{(x)\,(0.060+x)}{(0.085-x)}$$

Assume x is small compared to 0.060 and 0.085.

$$1.3 \times 10^{-5} = \frac{0.060\,x}{0.085}; \; x = 1.8 \times 10^{-5} = [H^+],\ pH = 4.73$$

Check. Since the extent of ionization of a weak acid or base is suppressed by the presence of a conjugate salt, the 5% rule usually holds true in buffer solutions.

(b)
$$(CH_3)_3N(aq) + H_2O(l) \;\rightleftharpoons\; (CH_3)_3NH^+(aq) \;+\; OH^-\,(aq)$$

i	$0.075\ M$	$0.10\ M$	
c	$-x$	$+x$	$+x$
e	$(0.075-x)\ M$	$(0.10+x)\ M$	$+x\ M$

$$K_b = 6.4 \times 10^{-5} = \frac{[OH^-][(CH_3)_3NH^+]}{[(CH_3)_3N]} = \frac{(x)\,(0.10+x)}{(0.075-x)} \approx \frac{0.10\,x}{0.075}$$

$x = 4.8 \times 10^{-5} = [OH^-]$, $pOH = 4.32$, $pH = 14.00 - 4.32 = 9.68$

Check. In a buffer, if [conj. acid] > [conj. base], pH < pK_a of the conj. acid. If [conj. acid] < [conj. base], pH > pK_a of the conj. acid. In this buffer, pK_a of $(CH_3)_3NH^+$ is 9.81. $[(CH_3)_3N]$ and pH = 9.61, less than 9.81.

(c) mol = $M \times$ L; mol $HC_2H_3O_2$ = 0.15 M × 0.0500 L = 7.5×10^{-3} mol

mol $C_2H_3O_2^-$ = 0.20 M × 0.0500 L = 0.010 mol

$$
\begin{array}{cccc}
 & HC_2H_3O_2(aq) & \rightleftharpoons \quad H^+(aq) & + & C_2H_3O_2^-(aq) \\
\end{array}
$$

	$HC_2H_3O_2(aq)$	$\rightleftharpoons$ $H^+(aq)$	+	$C_2H_3O_2^-(aq)$
i	7.5×10^{-3} mol	0		0.010 mol
c	$-x$	$+x$		$+x$
e	$(7.5 \times 10^{-3} - x)$ mol	x mol		$(0.010 + x)$ mol

$[HC_2H_3O_2] = (7.5 \times 10^{-3} - x)$ mol$/0.1000$ L; $[C_2H_3O_2^-] = (0.010 + x)$ mol$/0.1000$ L

$$
K_a = 1.8 \times 10^{-5} = \frac{[H+][C_2H_3O_2^-]}{[HC_2H_3O_2]} = \frac{(x)(0.010+x)/0.1000 \text{ L}}{(0.0075-x)/0.1000 \text{ L}} \approx \frac{x(0.010)}{0.0075}
$$

$x = 1.35 \times 10^{-5} M = 1.4 \times 10^{-5} M\ H^+$; pH $= 4.87$

Check. pK_a for $HC_2H_3O_2 = 4.74$. $[C_2H_3O_2^-] > [HC_2H_3O_2]$, pH of buffer $= 4.87$, greater than 4.74.

17.14   *Analyze/Plan.* Follow the logic in Sample Exercise 17.1.   *Solve.*

(a) $HCHO_2$ is a weak acid, and $NaCHO_2$ contains the common ion CHO_2^-, the conjugate base of $HCHO_2$. Solve the common-ion equilibrium problem.

	$HCHO_2(aq)$	$\rightleftharpoons$ $H^+(aq)$	+	$CHO_2^-(aq)$
i	$0.260\ M$			$0.160\ M$
c	$-x$	$+x$		$+x$
e	$(0.260 - x)\ M$	$+x\ M$		$(0.160 + x)\ M$

$$
K_a = 1.8 \times 10^{-4} = \frac{[H^+][CHO_2^-]}{[HCHO_2]} = \frac{(x)(0.160+x)}{(0.260-x)} \approx \frac{0.160\,x}{0.260}
$$

$x = 2.93 \times 10^{-4} = 2.9 \times 10^{-4} M = [H^+]$, pH $= 3.53$

Check. Since the extent of ionization of a weak acid or base is suppressed by the presence of a conjugate salt, the 5% rule usually holds true in buffer solutions.

(b) C_5H_5N is a weak base, and C_5H_5NHCl contains the common ion $C_5H_5NH^+$, which is the conjugate acid of C_5H_5N. Solve the common ion equilibrium problem.

	$C_5H_5N(aq)$ + $H_2O(l$	$\rightleftharpoons$ $C_5H_5NH^+(aq)$	+	$OH^-(aq)$
i	$0.210\ M$	$0.350\ M$		
c	$-x$	$+x$		$+x$
e	$(0.210 - x)\ M$	$(0.350 + x)\ M$		$+x\ M$

$$
K_b = 1.7 \times 10^{-9} = \frac{[C_5H_5NH^+][OH^-]}{[C_5H_5N]} = \frac{(0.350 + x)(x)}{(0.210-x)} \approx \frac{0.350\,x}{0.210}
$$

$x = 1.02 \times 10^{-9} = 1.0 \times 10^{-9} M = [OH^-]$, pOH $= 8.991$, pH $= 14.00 - 8.991 = 5.01$

Check. In a buffer, if [conj. acid] > [conj. base], pH < pK_a of the conj. acid.
If [conj. acid] < [conj. base], pH > pK_a of the conj. acid. In this buffer, pK_a of
$(CH_3)_3NH^+$ is 9.81. $[(CH_3)_3NH^+] > [(CH_3)_3N]$ and pH = 9.61, less than 9.81.

(c) mol = M × L; mol HF = 0.050 M × 0.125 L = 6.25×10^{-3} = 6.3×10^{-3} mol;

mol F^- = 0.10 M × 0.0500 L = 0.0050 mol

	HF(aq)	$\rightleftharpoons$	H^+(aq)	+	F^-(aq)
i	6.25×10^{-3} mol		0		0.0050 mol
c	$-x$		$+x$		$+x$
e	$(6.25 \times 10^{-3} - x)$ mol		x		$(0.0050 + x)$ mol

[HF] = $(6.25 \times 10^{-3} + x)/0.175$ L; $[F^-]$ = $(0.0050 + x)\, 0.175$ L

Note that the volumes will cancel when substituted into the K_a expression.

$$K_a = 6.8 \times 10^{-4} = \frac{[H^+][F^-]}{[HF]} = \frac{x(0.0050 + x)/0.175}{(6.25\,x - x)/0.175} \approx \frac{x(0.0050)}{0.00625}$$

$x = 8.50 \times 10^{-4}$ = 8.5×10^{-4} M H^+; pH = 3.07

Check. pK_a for HF = 3.17. [HF] > $[F^-]$, pH of buffer = 3.07, less than 3.17.

17.15 *Analyze/Plan.* We are asked to calculate % ionization of (a) a weak acid and (b) a weak
acid in a solution containing a common ion, its conjugate base. Calculate % ionization
as in Sample Exercise 16.12. In part (b), the concentration of the common ion is 0.085 M,
not x, as in part (a).   *Solve.*

	HBu(aq)	$\rightleftharpoons$	H^+ (aq)	+	Bu^-(aq)	$K_a = \dfrac{[H^+][Bu^-]}{[HBu]} = 1.5 \times 10^{-5}$
equil (a)	$0.0075 - x$ M		x M		x M	
equil (b)	$0.0075 - x$ M		x M		$0.085 + x$ M	

(a) $K_a = 1.5 \times 10^{-5} = \dfrac{x^2}{0.0075 - x} \approx \dfrac{x^2}{0.0075}$; $x = [H^+] = 3.354 \times 10^{-4}$ = 3.4×10^{-4} $M\,H^+$

% ionization = $\dfrac{3.4 \times 10^{-4}\ M\,H^+}{0.0075\ M\ HBu} \times 100 = 4.5\%$ ionization

(b) $K_a = 1.5 \times 10^{-5} = \dfrac{(x)(0.085 + x)}{0.0075 - x} \approx \dfrac{0.085\,x}{0.0075}$; $x = 1.3 \times 10^{-6}\ M\,H^+$

% ionization = $\dfrac{1.3 \times 10^{-6}\ M\,H^+}{0.0075\ M\ HBu} \times 100 = 0.018\%$ ionization

Check. Percent ionization is much smaller when the "common ion" is present.

17.16

$$HLac(aq) \rightleftharpoons H^+(aq) + Lac^-(aq) \qquad K_a = \frac{[H^+][Lac^-]}{[HLac]} = 1.4 \times 10^{-4}$$

equil (a) $0.085 - x\ M$ $x\ M$ $x\ M$

equil (b) $0.085 - x\ M$ $x\ M$ $0.050 + x\ M$

(a) $K_a = 1.4 \times 10^{-4} = \dfrac{x^2}{0.085 - x} \approx \dfrac{x^2}{0.085}$; $x = [H^+] = 3.45 \times 10^{-3}\ M = 3.5 \times 10^{-3}\ M\ H^+$

$$\% \text{ ionization} = \frac{3.5 \times 10^{-3}\ M\ H^+}{0.085\ M\ \text{Lac}} \times 100 = 4.1\% \text{ ionization}$$

(b) $K_a = 1.4 \times 10^{-4} = \dfrac{(x)(0.050 + x)}{0.085 - x} \approx \dfrac{0.050\ x}{0.085}$; $x = 2.4 \times 10^{-4}\ M\ H^+$

$$\% \text{ ionization} = \frac{2.4 \times 10^{-4}\ M\ H^+}{0.085\ M\ \text{Lac}} \times 100 = 0.28\% \text{ ionization}$$

Buffers

17.17 $HC_2H_3O_2$ and $NaC_2H_3O_2$ are a weak conjugate acid/conjugate base pair which acts as a buffer because unionized $HC_2H_3O_2$ reacts with added base, while $C_2H_3O_2^-$ combines with added acid, leaving $[H^+]$ relatively unchanged. Although HCl and NaCl are a conjugate acid/conjugate base pair, Cl^- is a negligible base. That is, it has no tendency to combine with added acid to form molecular HCl. Any added acid simply increases $[H^+]$ in an HCl/NaCl mixture. In general, the conjugate bases of strong acids are negligible and mixtures of strong acids and their conjugate salts do not act as buffers.

17.18   NaOH is a strong base and will react with $HC_2H_3O_2$ to form $NaC_2H_3O_2$. As long as $HC_2H_3O_2$ is present in excess, the resulting solution will contain both the conjugate acid $HC_2H_3O_2(aq)$ and the conjugate base $C_2H_3O_2^-(aq)$, the requirements for a buffer.

mmol $= M \times$ mL; mmol $HC_2H_3O_2 = 1.00\ M \times 100$ mL $= 10.0$ mmol

mmol NaOH $= 0.100\ M \times 50$ mL $= 5.0$ mmol

$$HC_2H_3O_2(aq) + NaOH(aq) \rightarrow NaC_2H_3O_2(aq) + H_2O(l)$$

initial   10.0 mmol   5.0 mmol

after rx   5.0 mmol   0   5.0 mmol

Mixing these two solutions has created a buffer by partial neutralization of the weak acid $HC_2H_3O_2$.

17.19   *Analyze/Plan.* Follow the logic in Sample Exercise 17.3. Assume that % ionization is small in these buffers (Solutions 17.15 and 17.16). *Solve.*

(a) $K_a = \dfrac{[H^+][Lac^-]}{[HLac]}; [H^+] = \dfrac{[K_a][HLac]}{[Lac^-]} = \dfrac{1.4 \times 10^{-4}\,(0.12)}{(0.11)}$

$[H^+] = 1.53 \times 10^{-4} = 1.5 \times 10^{-4}; \text{pH} = 3.82$

(b) mol = $M \times$ L; total volume = 85 mL + 95 mL = 180 mL

$[H^+] = \dfrac{K_a[HLac]}{[Lac^-]} = \dfrac{1.4 \times 10^{-4}\,(0.13\,M \times 0.085\,L)/0.180\,L}{(0.15\,M \times 0.095\,L)/0.180\,L} = \dfrac{1.4 \times 10^{-4}\,(0.13 \times 0.085)}{(0.15 \times 0.095)}$

$[H^+] = 1.086 \times 10^{-4} = 1.1\,M\,H^+; \text{pH} = 3.96$

17.20 Assume that % ionization is small in these buffers (Solutions 17.15 and 17.16).

(a) The conjugate acid in this buffer is HCO_3^-, so use K_{a2} for H_2CO_3, 5.6×10^{-11}

$K_a = \dfrac{[H^+][CO_3^{2-}]}{[HCO_3^-]}; [H^+] = \dfrac{K_a[HCO_3^-]}{[CO_3^{2-}]} = \dfrac{5.6 \times 10^{-11}\,(0.120)}{(0.105)}$

$[H^+] = 6.40 \times 10^{-11} = 6.4 \times 10^{-11}\,M; \text{pH} = 10.19$

(b) mol = $M \times$ L; total volume = 140 mL = 0.140 L

$[H^+] = \dfrac{K_a\,(0.20\,M \times 0.065\,L)/0.120\,L}{(0.15\,M \times 0.075\,L)/0.120\,L} = \dfrac{5.6 \times 10^{-11}\,(0.20 \times 0.065)}{(0.15 \times 0.075)}$

$[H^+] = 6.47 \times 10^{-11} = 6.5 \times 10^{-11}\,M; \text{pH} = 10.19$

17.21 (a) *Analyze/Plan.* Follow the logic in Sample Exercises 17.1 and 17.3. As in Sample Exercise 17.1, start by calculating concentrations of the components. *Solve.*

$HC_2H_3O_2(aq) \rightleftharpoons H^+(aq) + C_2H_3O_2^-(aq); K_a = 1.8 \times 10^{-5} = \dfrac{[H^+][C_2H_3O_2^-]}{[HC_2H_3O_2]}$

$[HC_2H_3O_2] = \dfrac{20.0\,g\,HC_2H_3O_2}{2.00\,L\,soln} \times \dfrac{1\,mol\,HC_2H_3O_2}{60.05\,g\,HC_2H_3O_2} = 0.167\,M$

$[C_2H_3O_2^-] = \dfrac{20.0\,g\,NaC_2H_3O_2}{2.00\,L\,soln} \times \dfrac{1\,mol\,NaC_2H_3O_2}{82.04\,g\,NaC_2H_3O_2} = 0.122\,M$

$[H^+] = \dfrac{K_a[HC_2H_3O_2]}{[C_2H_3O_2^-]} = \dfrac{1.8 \times 10^{-5}\,(0.167 - x)}{(0.122 + x)} \approx \dfrac{1.8 \times 10^{-5}\,(0.167)}{(0.122)}$

$[H^+] = 2.4843 \times 10^{-5} = 2.5 \times 10^{-5}\,M, \text{pH} = 4.60$

(b) *Plan.* On the left side of the equation, write all ions present in solution after HCl or NaOH is added to the buffer. Using acid-base properties and relative strengths, decide which ions will combine to form new products. *Solve.*

$Na^+(aq) + C_2H_3O_2^-(aq) + H^+(aq) + Cl^-(aq) \rightarrow HC_2H_3O_2(aq) + Na^+(aq) + Cl^-(aq)$

(c) $HC_2H_3O_2(aq) + Na^+(aq) + OH^-(aq) \rightarrow C_2H_3O_2^-(aq) + H_2O(l) + Na^+(aq)$

17 Additional Aspects of Aqueous Equilibria

17.22 NH_4^+/NH_3 is a basic buffer. Either the hydrolysis of NH_3 or the dissociation of NH_4^+ can be used to determine the pH of the buffer. Using the dissociation of NH_4^+ leads directly to $[H^+]$ and facilitates use of the Henderson-Hasselbach relationship.

(a) $NH_4^+(aq) \rightleftharpoons H^+(aq) + NH_3(aq)$

$$K_a = \frac{K_w}{K_b} = \frac{1.0 \times 10^{-14}}{1.8 \times 10^{-5}} = 5.56 \times 10^{-10} = 5.6 \times 10^{-10}$$

$$[NH_3] = \frac{5.0 \text{ g } NH_3}{2.50 \text{ L soln}} \times \frac{1 \text{ mol } NH_3}{17.0 \text{ g } NH_3} = 0.118 = 0.12 \ M \ NH_3$$

$$[NH_4^+] = \frac{20.0 \text{ g } NH_4Cl}{2.50 \text{ L}} \times \frac{1 \text{ mol } NH_4Cl}{53.50 \text{ g } NH_4Cl} = 0.1495 = 0.15 \ M \ NH_4^+$$

$$K_a = \frac{[H^+][NH_3]}{[NH_4^+]}; [H^+] = \frac{K_a[NH_4^+]}{[NH_3]} = \frac{5.56 \times 10^{-10}(0.1495-x)}{(0.118+x)} \approx \frac{5.56 \times 10^{-10}(0.1495)}{(0.118)}$$

$[H^+] = 7.044 \times 10^{-10} = 7.0 \times 10^{-10} \ M$, pH = 9.15

(b) $NH_3(aq) + H^+(aq) + NO_3^-(aq) \rightarrow NH_4^+(aq) + NO_3^-(aq)$

(c) $NH_4^+(aq) + Cl^-(aq) + K^+(aq) + OH^-(aq) \rightarrow NH_3(aq) + H_2O(l) + Cl^-(aq) + K^+(aq)$

17.23 *Analyze/Plan.* Follow the logic in Sample Exercise 17.4. *Solve.*

In this problem, $[BrO^-]$ is the unknown.

pH = 9.15, $[H^+] = 10^{-9.15} = 7.0795 \times 10^{-10} = 7.1 \times 10^{-10} \ M$

$[HBrO] = 0.050 - 7.1 \times 10^{-10} \approx 0.050 \ M$

$$K_a = 2.5 \times 10^{-9} = \frac{7.0795 \times 10^{-10}[BrO^-]}{0.050}; [BrO] = 0.1766 = 0.18 \ M$$

For 1.00 L, 0.18 mol NaBrO are needed.

17.24 $HC_3H_5O_3(aq) \rightleftharpoons H^+(aq) + C_3H_5O_3^-(aq)$

$$[H^+] = \frac{K_a[HC_3H_5O_3]}{[C_3H_5O_3^-]}; [H^+] = 10^{-4.00} = 1.0 \times 10^{-4}$$

$[HC_3H_5O_3] = 0.150 \ M$; calculate $[C_3H_5O_3^-]$

$$[C_3H_5O_3^-] = \frac{K_a[HC_3H_5O_3]}{[H^+]} = \frac{1.4 \times 10^{-4}(0.150)}{1.0 \times 10^{-4}} = 0.2100 = 0.21 \ M$$

$$\frac{0.210 \text{ mol } NaC_3H_5O_3}{1.00 \text{ L}} \times \frac{112.1 \text{ g } NaC_3H_5O_3}{1 \text{ mol } NaC_3H_5O_3} = 23.54 = 24 \text{ g } NaC_3H_5O_3$$

17.25 *Analyze/Plan.* Follow the logic in Sample Exercise 17.3 and 17.5. *Solve.*

(a) $$K_a = \frac{[H^+][C_2H_3O_2^-]}{[HC_2H_3O_2]}; [H^+] = \frac{K_a[HC_2H_3O_2]}{[C_2H_3O_2^-]}$$

$$[H^+] \approx \frac{1.8 \times 10^{-5}(0.10)}{(0.13)} = 1.385 \times 10^{-5} = 1.4 \times 10^{-5} \ M; pH = 4.86$$

(b)

HC$_2$H$_3$O$_2$(aq)	+	KOH(aq)	→	C$_2$H$_3$O$_2{}^-$(aq) + H$_2$O(l) + K$^+$(aq)
0.10 mol		0.02 mol		0.13 mol
−0.02 mol		−0.02 mol		+0.02 mol
0.08 mol		0 mol		0.15 mol

$$[H^+] = \frac{1.8 \times 10^{-5}\,(0.08\ \text{mol}/0.100\ \text{L})}{(0.15\ \text{mol}/0.100\ \text{L})} = 9.60 \times 10^{-6} = 1 \times 10^{-5}\ M;\ pH = 5.02 = 5.0$$

(c)

C$_2$H$_3$O$_2{}^-$(aq)	+	HNO$_3$(aq)	→	HC$_2$H$_3$O$_2$(aq) + NO$_3{}^-$(aq)
0.13 mol		0.02 mol		0.10 mol
−0.02 mol		−0.02 mol		+0.02 mol
0.11 mol		0 mol		0.12 mol

$$[H^+] = \frac{1.8 \times 10^{-5}\,(0.12\ \text{mol}/0.100\ \text{L})}{(0.11\ \text{mol}/0.100\ \text{L})} = 1.96 \times 10^{-5} = 2.0 \times 10^{-5}\ M;\ pH = 4.71$$

17.26 (a) $$K_a = \frac{[H^+][C_3H_5O_2{}^-]}{[HC_3H_5O_2]};\ [H^+] = \frac{K_a[HC_3H_5O_2]}{[C_3H_5O_2{}^-]}$$

Since this expression contains a ratio of concentrations, we can ignore total volume and work directly with moles.

$$[H^+] = \frac{1.3 \times 10^{-5}\,(0.12 - x)}{(0.10 + x)} \approx \frac{1.3 \times 10^{-5}\,(0.12)}{0.10} = 1.56 \times 10^{-5} = 1.6 \times 10^{-5}\ M,\ pH = 4.81$$

(b)

HC$_3$H$_5$O$_2$(aq)	+	OH$^-$(aq)	→	C$_2$H$_3$O$_2{}^-$(aq) + H$_2$O(l)
0.12 mol		0.01 mol		0.10 mol
−0.01 mol		−0.01 mol		+0.01 mol
0.11 mol		0 mol		0.11 mol

$$[H^+] \approx \frac{1.3 \times 10^{-5}\,(0.11)}{(0.11)} = 1.3 \times 10^{-5}\ M;\ pH = 4.89$$

(c)

C$_3$H$_5$O$_2{}^-$(aq)	+	HI(aq)	→	HC$_3$H$_5$O$_2$(aq) + I$^-$(aq)
0.10 mol		0.01 mol		0.12 mol
−0.01 mol		−0.01 mol		+0.01 mol
0.09 mol		0 mol		0.13 mol

$$[H^+] \approx \frac{1.3 \times 10^{-5}\,(0.13)}{(0.09)} = 1.88 \times 10^{-3} = 2 \times 10^{-3}\ M;\ pH = 4.73 = 4.7$$

17.27 *Analyze/Plan.* Calculate the [conj. base]/[conj. acid] ratio in the H$_2$CO$_3$/HCO$_3{}^-$ blood buffer. Write the acid dissociation equilibrium and K$_a$ expression. Find K$_a$ for H$_2$CO$_3$ in Appendix D. Calculate [H$^+$] from the pH and solve for the ratio. *Solve.*

$$H_2CO_3(aq) \rightleftharpoons H^+(aq) + HCO_3^-(aq) \quad K_a = \frac{[H^+][HCO_3^-]}{[H_2CO_3]}; \frac{[HCO_3^-]}{[H_2CO_3]} = \frac{K_a}{[H^+]}$$

(a)   at pH = 7.4, $[H^+] = 10^{-7.4} = 4.0 \times 10^{-8}$ M; $\frac{[HCO_3^-]}{[H_2CO_3]} = \frac{4.3 \times 10^{-7}}{4.0 \times 10^{-8}} = 11$

(b)   at pH = 7.1, $[H^+] = 7.9 \times 10^{-8}$ M; $\frac{[HCO_3^-]}{[H_2CO_3]} = 5.4$

17.28 $\dfrac{6.5 \text{ g NaH}_2\text{PO}_4}{0.355 \text{ L soln}} \times \dfrac{1 \text{ mol NaH}_2\text{PO}_4}{120 \text{ g NaH}_2\text{PO}_4} = 0.153 = 0.15$ M

$\dfrac{8.0 \text{ g Na}_2\text{HPO}_4}{0.355 \text{ L soln}} \times \dfrac{1 \text{ mol Na}_2\text{HPO}_4}{142 \text{ g Na}_2\text{HPO}_4} = 0.159 = 0.16$ M

Use Equation [17.9] to find the pH of the buffer. K_a for $H_2PO_4^-$ is K_{a2} for H_3PO_4, 6.2×10^{-8}

$$pH = -\log(6.2 \times 10^{-8}) + \log \frac{0.159}{0.153} = 7.2076 + 0.0167 = 7.22$$

17.29   *Analyze.* Given six solutions, decide which two should be used to prepare a pH 3.50 buffer. Calculate the volumes of the two 0.10 M solutions needed to make approximately 1 L of buffer.

Plan. A buffer must contain a conjugate acid/conjugate base (CA/CB) pair. By examining the chemical formulas, decide which pairs of solutions could be used to make a buffer. If there is more than one possible pair, calculate pK_a for the acids. A buffer is most effective when its pH is within 1 pH unit of pK_a for the conjugate acid component. Select the pair with pK_a nearest to 3.50. Use Equation [17.9] to calculate the [CB]/[CA] ratio and the volumes of 0.10 M solutions needed to prepare 1 L of buffer. *Solve.*

There are three CA/CB pairs:

$HCHO_2/NaCHO_2$, $pK_a = 3.74$

$HC_2H_3O_2/NaC_2H_3O_2$, $pK_a = 4.74$

H_3PO_4/NaH_2PO_4, $pK_a = 2.12$

The most appropriate solutions are $HCHO_2/NaCHO_2$, because pK_a for $HCHO_2$ is nearest to 3.50.

$$pH = pKa + \log\frac{[CB]}{[CA]} \quad 3.50 = 3.7447 + \log\frac{[NaCHO_2]}{[HCHO_2]}$$

$$\log\frac{[NaCHO_2]}{[HCHO_2]} = -0.2447, \frac{[NaCHO_2]}{[HCHO_2]} = 0.5692 = 0.57$$

Since we are making a total of 1 L of buffer, let y = vol $NaCHO_2$ and (1 − y) = vol $HCHO_2$.

$$0.5692 = \frac{[NaCHO_2]}{[HCHO_2]} = \frac{(0.1 \, M \times \, y)/1L}{[0.10 \, M \times \, (1-y)]/1 \, L}; \, 0.5692[0.10(1-y)] = 0.10 \, y;$$

0.05692 = 0.15692 y; y = 0.3627 = 0.36 L

360 mL of 0.10 M NaCHO$_2$, 640 mL of 0.10 M HCHO$_2$.

Check. The pH of the buffer is less than pK$_a$ for the conjugate acid, indicating that the amount of CA in the buffer is greater than the amount of CB. This agrees with our result.

17.30 The solutes listed contain three possible conjugate acid/conjugate base (CA/CB) pairs.

These are:

HCHO$_2$/NaCHO$_2$, pK$_a$ = 3.74

HC$_3$H$_5$O$_2$/NaC$_3$H$_5$O$_2$, pK$_a$ = 4.80

H$_3$PO$_4$/NaH$_2$PO$_4$, pK$_a$ = 2.12

For maximum buffer capacity, pK$_a$ should be within 1 pH unit of the buffer. The propionic acid/propionate pair are most appropriate for a buffer with pH 4.80.

$$pH = pK_a + \log \frac{[CB]}{CA}; \; 4.80 = 4.886 + \log \frac{[NaC_3H_5O_2]}{[HC_3H_5O_2]}$$

$$\log \frac{[NaC_3H_5O_2]}{[HC_3H_5O_2]} = -0.0861; \; \frac{[NaC_3H_5O_2]}{[HC_3H_5O_2]} = 0.8202 = 0.82$$

Since we are making a total of 1 L of buffer, let y = vol NaC$_3$H$_5$O$_2$ and (1 – y) = vol HC$_3$H$_5$O$_2$.

$$0.8202 = \frac{[NaC_3H_5O_2]}{[HC_3H_5O_2]} = \frac{(0.1 M \times y)/1.0 \, L}{[0.10 M \times (1-y)]/1.0 \, L} = \frac{0.10 \, y}{0.10 - 0.10 \, y}$$

0.8202(0.10 – 0.10 y) = 0.10 y; 0.08202 = 0.18202 y; y = 0.4506 = 0.45 L

450 mL of 0.10 M NaC$_3$H$_5$O$_2$, 550 mL HC$_3$H$_5$O$_2$

Check. pH (buffer) < pK$_a$ (CA) and the calculated amount of CA in the buffer is greater than the amount of CB.

Acid-Base Titrations

17.31 (a) Curve B. The initial pH is lower and the equivalence point region is steeper.

(b) pH at the approximate equivalence point of curve A = 8.0

pH at the approximate equivalence point of curve B = 7.0

(c) Volume of base required to reach the equivalence point depends only on moles of acid present; it is independent of acid strength. Since acid B requires 40 mL and acid A requires only 30 mL, more moles of acid B are being titrated. For equal volumes of A and B, the concentration of acid B is greater.

17.32 (a) The quantity of base required to reach the equivalence point is the same in the two titrations.

(b)   The pH is higher initially in the titration of a weak acid.

(c)   The pH is higher at the equivalence point in the titration of a weak acid.

(d)   The pH in excess base is essentially the same for the two cases.

(e)   In titrating a weak acid, one needs an indicator that changes at a higher pH than for the strong acid titration. The choice is more critical because the change in pH close to the equivalence point is smaller for the weak acid titration.

17.33   *Analyze.* Given reactants, predict whether pH at the equivalence point of a titration is less than, equal to or greater than 7.

Plan. At the equivalence point of a titration, only product is present in solution; there is no excess of either reactant. Determine the product of each reaction and whether a solution of it is acidic, basic or neutral.   *Solve.*

(a) $NaHCO_3(aq) + NaOH(aq) \rightarrow Na_2CO_3(aq) + H_2O(l)$

At the equivalence point, the major species in solution are Na^+ and CO_3^{2-}. Na^+ is negligible and CO_3^{2-} is the CB of HCO_3^-. The solution is basic, above pH 7.

(b) $NH_3(aq) + HCl(aq) \rightarrow NH_4Cl(aq)$

At the equivalence point, the major species are NH_4^+ and Cl^-. Cl^- is negligible, NH_4^+ is the CA of NH_3. The solution is acidic, below pH 7.

(c) $KOH(aq) + HBr(aq) \rightarrow KBr(aq) + H_2O(l)$

At the equivalence point, the major species are K^+ and Br^-; both are negligible. The solution is at pH 7.

17.34   (a) $HCHO_2(aq) + NaOH(aq) \rightarrow NaCHO_2(aq) + H_2O(l)$

At the equivalence point, the major species are Na^+ and CHO_2^-. Na^+ is negligible and CHO_2^- is the CB of $HCHO_2$. The solution is basic, above pH 7.

(b) $Ca(OH)_2(aq) + 2HClO_4(aq) \rightarrow Ca(ClO_4)_2(aq) + 2H_2O(l)$

At the equivalence point, the major species are Ca^{2+} and ClO_4^-; both are negligible. The solution is at pH 7.

(c) $C_5H_5N(aq) + HNO_3(aq) \rightarrow C_5H_5NH^+NO_3^-(aq)$

At the equivalence point, the major species are $C_5H_5NH^+$ and NO_3^-. NO_3^- is negligible and $C_5H_5NH^+$ is the CA of C_5H_5N. The solution is acidic, below pH 7.

17.35   (a)   HX is weaker. The pH at the equivalence point is determined by the identity and concentration of the conjugate base, X^- or Y^-. The higher the pH at the equivalence point, the stronger the conjugate base (X^-) and the weaker the conjugate acid (HX).

(b)   Phenolphthalein, which changes color in the pH 8-10 range, is perfect for HX and probably appropriate for HY. Bromthymol blue changes from 6–7.5, and thymol blue between from 8–9.5, but these are two-color indicators. One-color indicators such as phenolphthalein are preferred because detection of the color change is more reproducible.

17.36 (a) At the equivalence point, moles HX added = moles B initially present =
0.10 $M \times 0.0300$ L = 0.0030 moles HX added.

(b) $BH^+(aq)$

(c) Both K_a for BH^+ and concentration BH^+ determine pH at the equivalence point.

(d) Because the pH at the equivalence point will be less than 7, methyl red would be more appropriate.

17.37 *Analyze/Plan.* We are asked to calculate the volume of 0.0850 M NaOH required to titrate various acid solutions to their equivalence point. At the equivalence point, moles base added equals moles acid initially present. Solve the stoichiometry problem, recalling that mol = $M \times$ L. In part (c) calculate molarity of HCl from g/L and proceed as outlined above. *Solve.*

(a) 40.0 mL $HNO_3 \times \dfrac{0.0900 \text{ mol } HNO_3}{1000 \text{ mL soln}} \times \dfrac{1 \text{ mol NaOH}}{1 \text{ mol } HNO_3} \times \dfrac{1000 \text{ mL soln}}{0.0850 \text{ mol NaOH}}$

$= 42.353 = 42.4$ mL NaOH soln

(b) 35.0 mL $HC_2H_3O_2 \times \dfrac{0.0850 \, M \, HC_2H_3O_2}{1000 \text{ mL soln}} \times \dfrac{1 \text{ mol NaOH}}{1 \text{ mol } HC_2H_3O_2} \times \dfrac{1000 \text{ mL soln}}{0.0850 \text{ mol NaOH}}$

$= 35.0$ mL NaOH soln

(c) $\dfrac{1.85 \text{ g HCl}}{1 \text{ L soln}} \times \dfrac{1 \text{ mol HCl}}{36.46 \text{ g HCl}} = 0.05074 = 0.0507 \, M$ HCl

50.0 mL HCl $\times \dfrac{0.05074 \text{ mol HCl}}{1000 \text{ mL}} \times \dfrac{1 \text{ mol NaOH}}{1 \text{ mol HCl}} \times \dfrac{1000 \text{ mL soln}}{0.0850 \text{ mol NaOH}}$

$= 29.847 = 29.8$ mL NaOH soln

17.38 (a) 55.0 mL NaOH $\times \dfrac{0.0950 \text{ mol NaOH}}{1000 \text{ mL soln}} \times \dfrac{1 \text{ mol HCl}}{1 \text{ mol NaOH}} \times \dfrac{1000 \text{ mL soln}}{0.105 \text{ mol HCl}}$

$= 49.8$ mL HCl soln

(b) 22.5 mL $NH_3 \times \dfrac{0.118 \text{ mol } NH_3}{1000 \text{ mL soln}} \times \dfrac{1 \text{ mol HCl}}{1 \text{ mol } NH_3} \times \dfrac{1000 \text{ mL soln}}{0.105 \text{ mol HCl}}$

$= 25.3$ mL HCl soln

(c) 125.0 mL $\times \dfrac{1.35 \text{ g NaOH}}{1000 \text{ mL}} \times \dfrac{1 \text{ mol NaOH}}{40.00 \text{ g NaOH}} \times \dfrac{1 \text{ mol HCl}}{1 \text{ mol NaOH}} \times \dfrac{1000 \text{ mL soln}}{0.105 \text{ mol HCl}}$

$= 40.2$ mL HCl soln

17.39 *Analyze/Plan.* Follow the logic in Sample Exercise 17.6 for the titration of a strong acid with a strong base. *Solve.*

moles $H^+ = M_{HBr} \times L_{HBr} = 0.200 \, M \times 0.0200$ L $= 4.00 \times 10^{-3}$ mol

moles $OH^- = M_{NaOH} \times L_{NaOH} = 0.200 \, M \times L_{NaOH}$

	mL_{HBr}	mL_{NaOH}	Total Volume	Moles H^+	Moles OH^-	Molarity Excess Ion	pH
(a)	20.0	15.0	35.0	4.00×10^{-3}	3.00×10^{-3}	$0.0286(H^+)$	1.544
(b)	20.0	19.9	39.9	4.00×10^{-3}	3.98×10^{-3}	$5 \times 10^{-4}(H^+)$	3.3
(c)	20.0	20.0	40.0	4.00×10^{-3}	4.00×10^{-3}	$1 \times 10^{-7}(H^+)$	7.0
(d)	20.0	20.1	40.1	4.00×10^{-3}	4.02×10^{-3}	$5 \times 10^{-4}(H^+)$	10.7
(e)	20.0	35.0	55.0	4.00×10^{-3}	7.00×10^{-3}	$0.0545(OH^-)$	12.737

molarity of excess ion = moles ion / total vol in L

(a) $\dfrac{4.00 \times 10^{-3} \text{ mol H}^+ - 3.00 \times 10^{-3} \text{ mol OH}^-}{0.0350 \text{ L}} = 0.0286 \, M \text{ H}^+$

(b) $\dfrac{4.00 \times 10^{-3} \text{ mol H}^+ - 3.98 \times 10^{-3} \text{ mol OH}^-}{0.0339 \text{ L}} = 5.01 \times 10^{-4} = 5 \times 10^{-4} \, M \text{ H}^+$

(c) equivalence point, mol H^+ = mol OH^-

NaBr does not hydrolyze, so $[H^+] = [OH^-] = 1 \times 10^{-7} \, M$

(d) $\dfrac{4.02 \times 10^{-3} \text{ mol H}^+ - 4.00 \times 10^{-3} \text{ mol OH}^-}{0.041 \text{ L}} = 4.88 \times 10^{-4} = 5 \times 10^{-4} \, M \text{ OH}^-$

(e) $\dfrac{7.00 \times 10^{-3} \text{ mol H}^+ - 4.00 \times 10^{-3} \text{ mol OH}^-}{0.0550 \text{ L}} = 0.054545 = 0.0545 \, M \text{ OH}^-$

17.40   moles $OH^- = M_{KOH} \times L_{KOH} = 0.150 \, M \times 0.0300 \text{ L} = 4.50 \times 10^{-3} \text{ mol}$

moles $H^+ = M_{HClO_4} \times L_{HClO_4} = 0.125 \, M \times L_{HClO_4}$

	mL_{KOH}	mL_{HClO_4}	Total Volume	Moles OH^-	Moles H^+	Molarity Excess Ion	pH
(a)	30.0	30.0	60.0	4.50×10^{-3}	3.75×10^{-3}	$0.0125(OH^-)$	12.10
(b)	30.0	35.0	65.0	4.50×10^{-3}	4.38×10^{-3}	$1.9 \times 10^{-3}(OH^-)$	11.28
(c)	30.0	36.0	66.0	4.50×10^{-3}	4.50×10^{-3}	$1.0 \times 10^{-7}(OH^-)$	7.00
(d)	30.0	37.0	67.0	4.50×10^{-3}	4.63×10^{-3}	$1.9 \times 10^{-3}(H^+)$	2.73
(e)	30.0	40.0	70.0	4.50×10^{-3}	5.00×10^{-3}	$7.1 \times 10^{-3}(H^+)$	2.15

molarity of excess ion $= \dfrac{\text{moles ion}}{\text{total vol in L}}$

(a) $\dfrac{4.50 \times 10^{-3} \text{ mol OH}^- - 3.75 \times 10^{-3} \text{ mol H}^+}{0.0600 \text{ L}} = 0.0125 = 0.013 \, M \text{ OH}^-$

(b) $\dfrac{4.50 \times 10^{-3} \text{ mol OH}^- - 4.38 \times 10^{-3} \text{ mol H}^+}{0.0650 \text{ L}} = 1.92 \times 10^{-3} = 1.9 \times 10^{-3} \, M \text{ OH}^-$

(c)   equivalence point, mol H^+ = mol OH^- M

$KClO_4$ does not hydrolyze, so $[H^+] = [OH^-] = 1 \times 10^{-7}$

(d) $\dfrac{4.50 \times 10^{-3} \text{ mol OH}^- - 4.63 \times 10^{-3} \text{ mol H}^+}{0.0670 \text{ L}} = 1.87 \times 10^{-3} = 1.9 \times 10^{-3} \, M \, H^+$

(e) $\dfrac{4.50 \times 10^{-3} \text{ mol OH}^- - 5.00 \times 10^{-3} \text{ mol H}^+}{0.0700 \text{ L}} = 7.14 \times 10^{-3} = 7.1 \times 10^{-3} \, M \, H^+$

17.41   *Analyze/Plan.* Follow the logic in Sample Exercise 17.7 for the titration of a weak acid with a strong base.   *Solve.*

(a)   At 0 mL, only weak acid, $HC_2H_3O_2$, is present in solution. Using the acid ionization equilibrium

$$HC_2H_3O_2(aq) \quad \rightleftharpoons \quad H^+(aq) \quad + \quad C_2H_3O_2^-(aq)$$

Initial	0.150 M	0	0
equil	0.150 – x M	x M	x M

$$K_a = \frac{[H^+][C_2H_3O_2^-]}{[HC_2H_3O_2]} = 1.8 \times 10^{-5} \text{ (Appendix D)}$$

$$1.8 \times 10^{-5} = \frac{x^2}{(0.150 - x)} \approx \frac{x^2}{0.150}; \, x^2 = 2.7 \times 10^{-6}; \, x = [H^+] = 0.001643$$

$$= 1.6 \times 10^{-3} \, M; \, pH = 2.78$$

(b)–(f)   Calculate the moles of each component after the acid-base reaction takes place.
Moles $HC_2H_3O_2$ originally present = $M \times L = 0.150 \, M \times 0.0350 \, L = 5.25 \times 10^{-3}$ mol.
Moles NaOH added = $M \times L = 0.150 \, M \times y$ mL.

$$NaOH(aq) \quad + \quad HC_2H_3O_2(aq) \rightarrow \quad Na^+C_2H_3O_2 \,(aq) + H_2O(l)$$

(0.150 $M \times$ 0.0175 L) =

(b)	before rx	2.625×10^{-3} mol	5.25×10^{-3} mol	
	after rx	0	2.625×10^{-3} mol	2.63×10^{-3} mol

(0.150 $M \times$ 0.0345 L) =

(c)	before rx	5.175×10^{-3} mol	5.25×10^{-3} mol	
	after rx	0	0.075×10^{-3} mol	5.18×10^{-3} mol

(0.150 $M \times$ 0.0350 L) =

(d)	before rx	5.25×10^{-3} mol	5.25×10^{-3} mol	
	after rx	0	0	5.25×10^{-3} mol

$(0.150\ M \times 0.0355\ L) =$

(e) before rx 5.325×10^{-3} mol 5.25×10^{-3} mol

     after rx   **0.075×10^{-3} mol**   **0**   **5.25×10^{-3} mol**

$(0.150\ M \times 0.0500\ L) =$

(f) before rx 7.50×10^{-3} mol 5.25×10^{-3} mol

     after rx   **2.25×10^{-3} mol**   **0**   **5.25×10^{-3} mol**

Calculate the molarity of each species (M = mol/L) and solve the appropriate equilibrium problem in each part.

(b) $V_T = 35.0$ mL $HC_2H_3O_2$ + 17.5 mL NaOH = 52.5 mL = 0.0525 L

$$[HC_2H_3O_2] = \frac{2.625 \times 10^{-3}\ \text{mol}}{0.0525} = 0.0500\ M$$

$$[C_2H_3O_2^-] = \frac{2.625 \times 10^{-3}\ \text{mol}}{0.0525} = 0.0500\ M$$

$HC_2H_3O_2(aq) \rightleftharpoons H^+(aq) + C_2H_3O_2^-(aq)$

equil $0.0500 - x\ M$ $x\ M$ $0.0500 + x\ M$

$$K_a = \frac{[H^+][C_2H_3O_2^-]}{[HC_2H_3O_2]}; [H^+] = \frac{K_a[HC_2H_3O_2]}{[C_2H_3O_2^-]}$$

$$[H^+] = \frac{1.8 \times 10^{-5}\ (0.0500 - x)}{(0.0500 + x)} = 1.8 \times 10^{-5}\ M\ H^+; pH = 4.74$$

(c) $$[HC_2H_3O_2] = \frac{7.5 \times 10^{-5}\ \text{mol}}{0.0695\ L} = 0.001079 = 1.1 \times 10^{-3}\ M$$

$$[C_2H_3O_2^-] = \frac{5.175 \times 10^{-3}\ \text{mol}}{0.0695\ L} = 0.07446 = 0.074\ M$$

$$[H^+] = \frac{1.8 \times 10^{-5}\ (1.079 \times 10^{-3} - x)}{(0.07446 + x)} \approx 2.6 \times 10^{-7}\ M\ H^+; pH = 6.58$$

(d)   At the equivalence point, only $C_3H_5O_2^-$ is present.

$$[C_2H_3O_2^-] = \frac{5.25 \times 10^{-3}\ \text{mol}}{0.0700\ L} = 0.0750\ M$$

The pertinent equilibrium is the base hydrolysis of $C_2H_3O_2^-$.

     $C_2H_3O_2^-(aq) + H_2O(l) \rightleftharpoons HC_2H_3O_2(aq) + OH^-(aq)$

initial $0.0750\ M$ 0   0

equil $0.0750 - x\ M$ x   x

$$K_b = \frac{K_w}{K_a \text{ for } HC_2H_3O_2} = \frac{1.0 \times 10^{-14}}{1.8 \times 10^{-5}} = 5.56 \times 10^{-10} = 5.6 \times 10^{-10} = \frac{[HC_2H_3O_2][OH^-]}{[C_2H_3O_2^-]}$$

$$5.56 \times 10^{-10} = \frac{x^2}{0.0750 - x}; x^2 \approx 5.56 \times 10^{-10}(0.0750); x = 6.458 \times 10^{-6}$$

$$= 6.5 \times 10^{-6} \ M \ OH^-$$

$$pOH = -\log(6.458 \times 10^{-6}) = 5.19; pH = 14.00 - pOH = 8.81$$

(e) After the equivalence point, the excess strong base determines the pOH and pH. The [OH$^-$] from the hydrolysis of $C_2H_3O_2^-$ is small and can be ignored.

$$[OH^-] = \frac{0.075 \times 10^{-3} \text{ mol}}{0.0705 \text{ L}} = 1.064 \times 10^{-3} = 1.1 \times 10^{-3} \ M; pOH = 2.97$$

$$pH = 14.00 - 2.97 = 11.03$$

(f) $$[OH^-] = \frac{2.25 \times 10^{-3} \text{ mol}}{0.0850 \text{ L}} = 0.0265 \ M \ OH^-; pOH = 1.577; pH = 14.00 - 1.577 = 12.423$$

17.42 (a) Weak base problem: $K_b = 1.8 \times 10^{-5} = \dfrac{[NH_4^+][OH^-]}{[NH_3]}$

At equilibrium, [OH$^-$] = x, [NH$_3$] = (0.030 – x); [NH$_4^+$] = x

$$1.8 \times 10^{-5} = \frac{x^2}{(0.030 - x)} \approx \frac{x^2}{0.030}; x = [OH^-] = 7.348 \times 10^{-4} = 7.3 \times 10^{-4} \ M$$

$$pH = 14.00 - 3.13 = 10.87$$

(b–f) Calculate mol NH$_3$ and mol NH$_4^+$ after the acid-base reaction takes place. 0.030 M NH$_3$ × 0.0300 L = 9.0×10^{-4} mol NH$_3$ present initially.

		NH$_3$(aq)	+	HCl(aq)	→	NH$_4^+$(aq) + Cl$^-$(aq)
				(0.025 M × 0.0100 L) =		
(b)	before rx	9.0×10^{-4} mol		2.5×10^{-4} mol		0 mol
	after rx	6.5×10^{-4} mol		0 mol		2.5×10^{-4} mol
				(0.025 M × 0.0200 L) =		
(c)	before rx	9.0×10^{-4} mol		5.0×10^{-4} mol		0 mol
	after rx	4.0×10^{-4} mol		0 mol		5.0×10^{-4} mol
				(0.025 M × 0.0350 L) =		
(d)	before rx	9.0×10^{-4} mol		8.75×10^{-4} mol		0 mol
	after rx	0.25×10^{-4} mol		0 mol		8.75×10^{-4} mol
				(0.025 M × 0.0360 L) =		
(e)	before rx	9.0×10^{-4} mol		9.0×10^{-4} mol		0 mol
	after rx	0 mol		0 mol		9.0×10^{-4} mol

$$(0.025 \, M \times 0.0370 \, \text{L}) =$$

(f) before rx 9.0×10^{-4} mol 9.25×10^{-4} mol   0 mol

     after rx   **0 mol** $\mathbf{0.25 \times 10^{-4}}$ $\mathbf{9.0 \times 10^{-4}}$ **mol**

                                                   mol

(b) Using the acid dissociation equilibrium for NH_4^+ (so that we calculate $[H^+]$ directly), $NH_4^+(aq) \rightleftharpoons H^+(aq) + NH_3(aq)$

$$K_a = \frac{[H^+][NH_3]}{[NH_4^+]} = \frac{K_w}{K_b \text{ for } NH_3} = \frac{1.0 \times 10^{-14}}{1.8 \times 10^{-5}} = 5.56 \times 10^{-10} = 5.6 \times 10^{-10}$$

$$[NH_3] = \frac{6.5 \times 10^{-4} \text{ mol}}{0.0400 \text{ L}} = 0.01625 \, M; [NH_4^+] = \frac{2.50 \times 10^{-4} \text{ mol}}{0.0400 \text{ L}} = 6.25 \times 10^{-3} \, M$$

$$[H^+] = \frac{5.56 \times 10^{-10} \, [NH_4^+]}{[NH_3]} \approx \frac{5.56 \times 10^{-10} \, (6.25 \times 10^{-3})}{(0.01625)} = 2.14 \times 10^{-10}; pH = 9.67$$

(We will assume $[H^+]$ is small compared to $[NH_3]$ and $[NH_4^+]$.)

(c) $$[NH_3] = \frac{4.0 \times 10^{-4} \text{ mol}}{0.0500 \text{ L}} = 0.0080 \, M; [NH_4^+] = \frac{5.0 \times 10^{-4} \text{ mol}}{0.0500 \text{ L}} = 0.010 M$$

$$[H^+] = \frac{5.56 \times 10^{-10} \, (0.010)}{(0.0080)} = 6.94 \times 10^{-10} = 6.9 \times 10^{-10} \, M; pH = 9.16$$

(d) $$[NH_3] = \frac{0.25 \times 10^{-4} \text{ mol}}{0.0650 \text{ L}} = 3.846 \times 10^{-4} = 4 \times 10^{-4} \, M; [NH_4^+] = \frac{8.75 \times 10^{-4} \text{ mol}}{0.0650 \text{ L}}$$

$$= 0.01346 = 0.013 \, M$$

$$[H^+] = \frac{5.56 \times 10^{-10} \, (0.01346)}{3.846 \times 10^{-4}} = 1.946 \times 10^{-8} = 2 \times 10^{-8} \, M; pH = 7.7$$

(e) At the equivalence point, $[H^+] = [NH_3] = x$

$$[NH_4^+] = \frac{9.0 \times 10^{-4} \, M}{0.0660 \text{ L}} = 0.01364 = 0.014 \, M$$

$$5.56 \times 10^{-10} = \frac{x^2}{0.01364}; x = [H^+] = 2.754 \times 10^{-6} = 2.8 \times 10^{-6}; pH = 5.56$$

(f) Past the equivalence point, $[H^+]$ from the excess HCl determines the pH.

$$[H^+] = \frac{0.25 \times 10^{-4} \text{ mol}}{0.0670 \text{ L}} = 3.731 \times 10^{-4} = 4 \times 10^{-4} \, M; pH = 3.4$$

17.43 *Analyze/Plan.* Calculate the pH at the equivalence point for the titration of several bases with 0.200 M HBr. The volume of 0.200 M HBr required in all cases equals the volume of base and the final volume = $2V_{base}$. The concentration of the salt produced at the equivalence point is $\dfrac{0.200 \, M \times V_{base}}{2V_{base}} = 0.100 \, M$.

In each case, identify the salt present at the equivalence point, determine its acid-base properties (Section 16.9), and solve the pH problem. *Solve.*

(a)   NaOH is a strong base; the salt present at the equivalence point, NaBr, does not affect the pH of the solution. 0.100 M NaBr, pH = 7.00

(b) $HONH_2$ is a weak base, so the salt present at the equivalence point is $HONH_3{}^+Br^-$. This is the salt of a strong acid and a weak base, so it produces an acidic solution.

0.100 M $HONH_3{}^+Br^-$; $HONH_3{}^+(aq)$ $\rightleftharpoons$ $H^+(aq)$ + $HONH_2$

[equil]   0.100 – x   x   x

$$K_a = \frac{[H^+][HONH_2]}{[HONH_3{}^+]} = \frac{K_w}{K_b} = \frac{1.0 \times 10^{-14}}{1.1 \times 10^{-8}} = 9.09 \times 10^{-7} = 9.1 \times 10^{-7}$$

Assume x is small with respect to [salt].

$K_a = x^2 / 0.100$; $x = [H^+] = 3.02 \times 10^{-4} = 3.0 \times 10^{-4} M$, pH = 3.52

(c) $C_6H_5NH_2$ is a weak base and $C_6H_5NH_3{}^+Br^-$ is an acidic salt.

0.100 M $C_6H_5NH_3{}^+Br^-$.   Proceeding as in (b):

$$K_a = \frac{[H^+][C_6H_5NH_2]}{[C_6H_5NH_3{}^+]} = \frac{K_w}{K_b} = 2.33 \times 10^{-5} = 2.3 \times 10^{-5}$$

$[H^+]^2 = 0.100(2.33 \times 10^{-5})$; $[H^+] = 1.52 \times 10^{-3} = 1.5 \times 10^{-3} M$, pH = 2.82

17.44   The volume of NaOH solution required in all cases is

$$V_{base} = \frac{V_{acid} \times M_{acid}}{M_{base}} = \frac{(0.100)\, V_{acid}}{(0.080)} = 1.25\, V_{acid}$$

The total volume at the equivalence point is $V_{base} + V_{acid} = 2.25\, V_{acid}$.

The concentration of the salt at the equivalence point is $\dfrac{M_{acid}\, V_{acid}}{2.25\, V_{acid}} = \dfrac{0.100}{2.25} = 0.0444\ M$

(a)   0.0444 M NaBr, pH = 7.00

(b)   0.0444 M $Na^+\ C_3H_5O_3{}^-$; $C_3H_5O_3{}^-(aq)$ + $H_2O(l)$ $\rightleftharpoons$ $HC_3H_5O_3(aq)$ + OH^- (aq)

$$K_b = \frac{[HC_3H_5O_3][OH^-]}{[C_3H_5O_3{}^-]} = \frac{K_w}{K_a} = \frac{1.0 \times 10^{-14}}{1.4 \times 10^{-4}} = 7.14 \times 10^{-11} = 7.1 \times 10^{-11}$$

$[HC_3H_5O_3] = [OH^-]$; $[C_3H_5O_3{}^-] \approx 0.0444$

$[OH^-]^2 \approx 0.0444(7.14 \times 10^{-11})$; $[OH^-] = 1.78 \times 10^{-6} = 1.8 \times 10^{-6} M$, pOH = 5.75;

pH = 8.25

(c) $CrO_4^{2-}(aq) + H_2O(l) \rightleftharpoons HCrO_4^-(aq) + OH^-(aq)$

$$K_b = \frac{[HCrO_4^-][OH^-]}{[CrO_4^{2-}]} = \frac{K_w}{K_a} = \frac{1.0 \times 10^{-14}}{3.0 \times 10^{-7}} = 3.33 \times 10^{-8} = 3.3 \times 10^{-8}$$

$[OH^-]^2 \approx 0.0444(3.33 \times 10^{-8})$; $[OH^-] = 3.849 \times 10^{-5} = 3.8 \times 10^{-5}$, pH = 9.59

Solubility Equilibria and Factors Affecting Solubility

17.45   (a)   The concentration of undissolved solid does not appear in the solubility product expression because it is constant as long as there is solid present. Concentration is a ratio of moles solid to volume of the solid; solids occupy a specific volume not dependent on the solution volume. As the amount (moles) of solid changes, the volume changes proportionally, so that the ratio of moles solid to volume solid is constant.

   (b)   *Analyze/Plan.* Follow the example in Sample Exercise 17.9.   *Solve.*

   $K_{sp} = [Ag^+][I^-]$; $K_{sp} = [Sr^{2+}][SO_4^{2-}]$; $K_{sp} = [Fe^{2+}][OH^-]^2$; $K_{sp} = [Hg_2^{2+}][Br^-]^2$

17.46   (a)   Solubility is the amount (grams, moles) of solute that will dissolve in a certain volume of solution. Solubility-product constant is an equilibrium constant, the product of the molar concentrations of all the dissolved ions in solution.

   (b) $K_{sp} = [Mn^{2+}][CO_3^{2-}]$; $K_{sp} = [Hg^{2+}][OH^-]^2$; $K_{sp} = [Cu^{2+}]^3[PO_4^{3-}]^2$

17.47   *Analyze/Plan.* Follow the logic in Sample Exercise 17.10.   *Solve.*

   (a) $CaF_2(s) \rightleftharpoons Ca^{2+}(aq) + 2F^-(aq)$; $K_{sp} = [Ca^{2+}][F^-]^2$

   The molar solubility is the moles of CaF_2 that dissolve per liter of solution. Each mole of CaF_2 produces **1** mol $Ca^{2+}(aq)$ and **2** mol $F^-(aq)$.

   $[Ca^{2+}] = 1.24 \times 10^{-3}\,M$; $[F^-] = 2 \times 1.24 \times 10^{-3}\,M = 2.48 \times 10^{-3}\,M$

   $K_{sp} = (1.24 \times 10^{-3})(2.48 \times 10^{-3})^2 = 7.63 \times 10^{-9}$

   (b) $SrF_2(s) \rightleftharpoons Sr^{2+}(aq) + 2F^-(aq)$; $K_{sp} = [Sr^{2+}][F^-]^2$

   Transform the gram solubility to molar solubility.

   $$\frac{1.1 \times 10^{-2}\text{ g } SrF_2}{0.100\text{ L}} \times \frac{1\text{ mol } SrF_2}{125.6\text{ g } SrF_2} = 8.76 \times 10^{-4} = 8.8 \times 10^{-4}\text{ mol } SrF_2/L$$

   $[Sr^{2+}] = 8.76 \times 10^{-4}\,M$; $[F^-] = 2(8.76 \times 10^{-4}\,M)$

   $K_{sp} = (8.76 \times 10^{-4})(2(8.76 \times 10^{-4}))^2 = 2.7 \times 10^{-9}$

   (c) $Ba(IO_3)_2(s) \rightleftharpoons Ba^{2+}(aq) + 2IO_3^-(aq)$; $K_{sp} = [Ba^{2+}][IO_3^-]^2$

   Since 1 mole of dissolved $Ba(IO_3)_2$ produces 1 mole of Ba^{2+}, the molar solubility of $Ba(IO_3)_2 = [Ba^{2+}]$. Let $x = [Ba^{2+}]$; $[IO_3^-] = 2x$

   $K_{sp} = 6.0 \times 10^{-10} = (x)(2x)^2$; $4x^3 = 6.0 \times 10^{-10}$; $x^3 = 1.5 \times 10^{-10}$; $x = 5.3 \times 10^{-4}\,M$

   The molar solubility of $Ba(IO_3)_2$ is 5.3×10^{-4} mol/L.

17.48   (a) $PbBr_2(s) \rightleftharpoons Pb^{2+}(aq) + 2Br^-(aq)$

$K_{sp} = [Pb^{2+}][Br^-]^2; [Pb^{2+}] = 1.0 \times 10^{-2} M, [Br^-] = 2.0 \times 10^{-2} M$

$K_{sp} = (1.0 \times 10^{-2} M)(2.0 \times 10^{-2} M)^2 = 4.0 \times 10^{-6}$

(b) $AgIO_3(s) \rightleftharpoons Ag^+(aq) + IO_3^-(aq); \quad K_{sp} = [Ag^+][IO_3^-]$

$[Ag^+] = [IO_3^-] = \dfrac{0.0490 \text{ g } AgIO_3}{1.00 \text{ L soln}} \times \dfrac{1 \text{ mol } AgIO_3}{282.8 \text{ g } AgIO_3} = 1.733 \times 10^{-4} = 1.73 \times 10^{-4} M$

$K_{sp} = (1.733 \times 10^{-4} M)(1.733 \times 10^{-4} M) = 3.00 \times 10^{-8}$

(c) $Cu(OH)_2(s) \rightleftharpoons Cu^{2+}(aq) + 2OH^-(aq); \quad K_{sp} = [Cu^{2+}][OH^-]^2$

$[Cu^{2+}] = x, [OH^-] = 2x; K_{sp} = 4.8 \times 10^{-20} = (x)(2x)^2$

$4.8 \times 10^{-20} = 4x^3; x = [Cu^{2+}] = 2.290 \times 10^{-7} = 2.3 \times 10^{-7} M$

$\dfrac{2.290 \times 10^{-7} \text{ mol } Cu(OH)_2}{1 \text{ L}} \times \dfrac{97.56 \text{ g } Cu(OH)_2}{1 \text{ mol } Cu(OH)_2} = 2.2 \times 10^{-5} \text{ g } Cu(OH)_2$

However, $[OH^-]$ from $Cu(OH)_2 = 4.58 \times 10^{-7}$ M; this is similar to $[OH^-]$ from the autoionization of water.

$K_w = [H^+][OH^-]; [H^+] = y, [OH^-] = (4.58 \times 10^{-7} + y)$

$1.0 \times 10^{-14} = y(4.58 \times 10^{-7} + y); y^2 + 4.58 \times 10^{-7} y - 1.0 \times 10^{-14}$

$y = \dfrac{-4.58 \times 10^{-7} \pm \sqrt{(4.58 \times 10^{-7})^2 - 4(1)(-1.0 \times 10^{-14})}}{2}; y = 2.09 \times 10^{-8}$

$[OH^-]_{total} = 4.58 \times 10^{-7} M + 0.209 \times 10^{-7} M = 4.79 \times 10^{-7} M$

Recalculating $[Cu^{2+}]$ and thus molar solubility of $Cu(OH)_2(s)$:

$4.8 \times 10^{-20} = x(4.79 \times 10^{-7})^2; x = 2.09 \times 10^{-7} M \text{ } Cu^{2+}$

$\dfrac{2.09 \times 10^{-7} \text{ mol } Cu(OH)_2(s)}{1 \text{ L}} \times \dfrac{97.56 \text{ g } Cu(OH)_2}{1 \text{ mol } Cu(OH)_2} = 2.0 \times 10^{-5} \text{ g } Cu(OH)_2$

Note that the presence of OH^- as a common ion decreases the water solubility of $Cu(OH)_2$.

17.49   *Analyze/Plan.* Given gram solubility of a compound, calculate K_{sp}. Write the dissociation equilibrium and K_{sp} expression. Change gram solubility to molarity of the individual ions, taking the stoichiometry of the compound into account. Calculate K_{sp}. *Solve.*

$CaC_2O_4(s) \rightleftharpoons Ca^{2+}(aq) + C_2O_4^{2-}(aq); \quad K_{sp} = [Ca^{2+}][C_2O_4^{2-}]$

$[Ca^{2+}] = [C_2O_4^{2-}] = \dfrac{0.0061 \text{ g } CaC_2O_4}{1.00 \text{ L soln}} \times \dfrac{1 \text{ mol } CaC_2O_4}{128.1 \text{ g } CaC_2O_4} = 4.76 \times 10^{-5} = 4.8 \times 10^{-5} M$

$K_{sp} = (4.76 \times 10^{-5} M)(4.76 \times 10^{-5} M) = 2.3 \times 10^{-9}$

17.50 $PbI_2(s) \rightleftharpoons Pb^{2+}(aq) + 2I^-(aq)$; $K_{sp} = [Pb^{2+}][I^-]^2$

$$[Pb^{2+}] = \frac{0.54 \text{ g } PbI_2}{1.00 \text{ L soln}} \times \frac{1 \text{ mol } PbI_2}{461.0 \text{ g } PbI_2} = 1.17 \times 10^{-3} = 1.2 \times 10^{-3} \, M$$

$[I^-] = 2[Pb^{2+}]$; $K_{sp} = [Pb^{2+}](2[Pb^{2+}])^2 = 4[Pb^{2+}]^3 = 4(1.17 \times 10^{-3})^3 = 6.4 \times 10^{-9}$

17.51 *Analyze/Plan.* Follow the logic in Sample Exercises 17.11 and 17.12. *Solve.*

(a) $AgBr(s) \rightleftharpoons Ag^+(aq) + Br^-(aq)$; $K_{sp} = [Ag^+][Br^-] = 5.0 \times 10^{-13}$

molar solubility = $x = [Ag^+] = [Br^-]$; $K_{sp} = x^2$

$x = (5.0 \times 10^{-13})^{1/2}$; $x = 7.1 \times 10^{-7}$ mol AgBr/L

(b) Molar solubility = $x = [Br^-]$; $[Ag^+] = 0.030 \, M + x$

$K_{sp} = (0.030 + x)(x) \approx 0.030(x)$

$5.0 \times 10^{-13} = 0.030(x)$; $x = 1.7 \times 10^{-11}$ mol AgBr/L

(c) Molar solubility = $x = [Ag^+]$

There are two sources of Br^-: NaBr(0.10 M) and AgBr($x \, M$)

$K_{sp} = (x)(0.10 + x)$; Assuming x is small compared to 0.10 M

$5.0 \times 10^{-13} = 0.10 (x)$; $x \approx 5.0 \times 10^{-12}$ mol AgBr/L

17.52 $LaF_3(s) \rightleftharpoons La^{3+}(aq) + 3F^-(aq)$; $K_{sp} = [La^{3+}][F^-]^3$

(a) molar solubility = $x = [La^{3+}]$; $[F^-] = 3x$

$K_{sp} = 2 \times 10^{-19} (x)(3x)3$; $2 \times 10^{-19} = 27 x^4$; $x = (7.41 \times 10^{-21})^{1/4}$, $x = 9.28 \times 10^{-6}$

$$= 9 \times 10^{-6} \, M \, La^{3+}$$

$$\frac{9.28 \times 10^{-6} \text{ mol } LaF_3}{1 \text{ L}} \times \frac{195.9 \text{ g } LaF_3}{1 \text{ mol}} = 1.82 \times 10^{-3} = 2 \times 10^{-3} \text{ g } LaF_3/L$$

(b) molar solubility = $x = [La^{3+}]$

There are two sources of F^-: KF(0.010 M) and LaF_3 (3x M)

$K_{sp} = (x)(0.010 + 3x)^3$; assume x is small compared to 0.010 M.

$2 \times 10^{-19} = (0.010)^3 \, x$; $x = 2 \times 10^{-19}/1.0 \times 10^{-6} = 2 \times 10^{-13} \, M \, La^{3+}$

$$\frac{2 \times 10^{-13} \text{ mol } LaF_3}{1 \text{ L}} \times \frac{195.9 \text{ g } LaF_3}{1 \text{ mol}} = 3.92 \times 10^{-11} = 4 \times 10^{-11} \text{ g } LaF_3/L$$

(c) molar solubility = x, $[F^-] = 3x$, $[La^{3+}] = 0.050 \, M + x$

$K_{sp} = (0.050 + x)(3x)^3$; assume x is small compared to 0.050 M.

$2 \times 10^{-19} = (0.050)(27 \, x^3) = 1.35 \, x^3$; $x = (1.48 \times 10^{-19})^{1/3} = 5.29 \times 10^{-7} = 5 \times 10^{-7} \, M$

$$\frac{5.29 \times 10^{-7} \text{ mol } LaF_3}{1 \text{ L}} \times \frac{195.9 \text{ g } LaF_3}{1 \text{ mol}} = 1.04 \times 10^{-4} = 1 \times 10^{-4} \text{ g } LaF_3/L$$

17.53   *Analyze/Plan.* We are asked to calculate the solubility of a slightly-soluble hydroxide salt at various pH values. This is a common ion problem; pH tells us not only $[H^+]$ but also $[OH^-]$, which is an ion common to the salt. Use pH to calculate $[OH^-]$, then proceed as in Sample Exercise 17.12. *Solve.*

$Mn(OH)_2(s) \rightleftharpoons Mn^{2+}(aq) + 2OH^-(aq); \quad K_{sp} = 1.6 \times 10^{-13}$

Since $[OH^-]$ is set by the pH of the solution, the solubility of $Mn(OH)_2$ is just $[Mn^{2+}]$.

(a) $pH = 7.0$, $pOH = 14 - pH = 7.0$, $[OH^-] = 10^{-pOH} = 1.0 \times 10^{-7} M$

$K_{sp} = 1.6 \times 10^{-13} = [Mn^{2+}](1.0 \times 10^{-7})^2; \quad [Mn^{2+}] = \dfrac{1.6 \times 10^{-13}}{1.0 \times 10^{-14}} = 16\, M$

$\dfrac{16\ mol\ Mn(OH)_2}{1\ L} \times \dfrac{88.95\ g\ Mn(OH)_2}{1\ mol\ Mn(OH)_2} = 1423 = 1.4 \times 10^3\ g\ Mn(OH)_2\,/\,L$

Check. Note that the solubility of $Mn(OH)_2$ in pure water is $3.6 \times 10^{-5}\ M$, and the pH of the resulting solution is 9.0. The relatively low pH of a solution buffered to pH 7.0 actually increases the solubility of $Mn(OH)_2$.

(b) $pH = 9.5$, $pOH = 4.5$, $[OH^-] = 3.16 \times 10^{-5} = 3.2 \times 10^{-5}\ M$

$K_{sp} = 1.6 \times 10^{-13} = [Mn^{2+}](3.16 \times 10^{-5})^2; \quad [Mn^{2+}] = \dfrac{1.6 \times 10^{-13}}{1.0 \times 10^{-9}} = 1.6 \times 10^{-4}\ M$

$1.6 \times 10^{-4}\ M\ Mn(OH)_2 \times 88.95\ g/mol = 0.0142 = 0.014\ g/L$

(c) $pH = 11.8$, $pOH = 2.2$, $[OH^-] = 6.31 \times 10^{-3} = 6.3 \times 10^{-3}\ M$

$K_{sp} = 1.6 \times 10^{-13} = [Mn^{2+}](6.31 \times 10^{-3})^2; \quad [Mn^{2+}] = \dfrac{1.6 \times 10^{-13}}{3.98 \times 10^{-5}} = 4.0 \times 10^{-9}\ M$

$4.02 \times 10^{-9}\ M\ Mn(OH)_2 \times 88.95\ g/mol = 3.575 \times 10^{-7} = 3.6 \times 10^{-7}\ g/L$

17.54 $Fe(OH)_2(s) \rightleftharpoons Fe^{2+}(aq) + 2OH^-(aq); \quad K_{sp} = 8.0 \times 10^{-16}$

Since the $[OH^-]$ is set by the pH of the solution, the solubility of $Fe(OH)_2$ is just $[Fe^{2+}]$.

(a) $pH = 7.0$, $pOH = 14 - pH = 7.0$, $[OH^-] = 10^{-pOH} = 1.0 \times 10^{-7} = 1 \times 10^{-7}\ M$

$K_{sp} = 7.9 \times 10^{-16} = [Fe^{2+}](1.0 \times 10^{-7})^2; \quad [Fe^{2+}] = \dfrac{7.9 \times 10^{-16}}{1.0 \times 10^{-14}} = 7.9 \times 10^{-2} = 8 \times 10^{-2}\ M$

Check. In pure water, $[OH^-]$ from $Fe(OH)_2$ is similar to (OH^-) from the autoionization of water, resulting in a cubic equation for $[Fe^{2+}]$. The solubility of $Fe(OH)_2$ at a buffered $pH = 7.0$ is actually greater than the solubility in pure water.

(b) $pH = 10.0$, $pOH = 4.0$, $[OH^-] = 1.0 \times 10^{-4} = 1 \times 10^{-4}\ M$

$K_{sp} = 7.9 \times 10^{-16} = [Fe^{2+}][1.0 \times 10^{-4}]^2; \quad [Fe^{2+}] = \dfrac{7.9 \times 10^{-16}}{1.0 \times 8^{-10}} = 7.9 \times 10^{-8} = 8 \times 10^{-8}\ M$

(c)   pH = 12.0, pOH = 2.0, $[OH^-] = 1.0 \times 10^{-2} = 1 \times 10^{-2} \ M$

$$K_{sp} = 7.9 \times 10^{-16} = [Fe^{2+}][1.0 \times 10^{-2}]^2; [Fe^{2+}] = \frac{7.9 \times 10^{-16}}{1.0 \times 10^{-4}} = 7.9 \times 10^{-12} = 8 \times 10^{-12} \ M$$

17.55   *Analyze/Plan.* Follow the logic in Sample Exercise 17.13.   *Solve.*

If the anion of the salt is the conjugate base of a weak acid, it will combine with H^+, reducing the concentration of the free anion in solution, thereby causing more salt to dissolve. More soluble in acid: (a) $ZnCO_3$, (b) ZnS, (d) AgCN, (e) $Ba_3(PO_4)_2$

17.56   If the anion in the slightly soluble salt is the conjugate base of a strong acid, there will be no reaction.

(a) $MnS(s) + 2H^+(aq) \rightarrow H_2S(aq) + Mn^{2+}(aq)$

(b) $PbF_2(s) + 2H^+(aq) \rightarrow 2HF(aq) + Pb^{2+}(aq)$

(c) $AuCl_3(s) + H^+(aq) \rightarrow$ no reaction

(d) $Hg_2C_2O_4(s) + 2H^+(aq) \rightarrow H_2C_2O_4(aq) + Hg_2^{2+}(aq)$

(e) $CuBr(s) + H^+(aq) \rightarrow$ no reaction

17.57   *Analyze/Plan.* Follow the logic in Sample Exercise 17.14.   *Solve.*

The formation equilibrium is

$$Cu^{2+}(aq) + 4NH_3(aq) \rightleftharpoons Cu(NH_3)_4^{2+}(aq) \quad K_f = \frac{[Cu(NH_3)_4^{2+}]}{[Cu^{2+}][NH_3]^4} = 5 \times 10^{12}$$

Assuming that nearly all the Cu^{2+} is in the form $Cu(NH_3)_4^{2+}$,

$[Cu(NH_3)_4^{2+}] = 1 \times 10^{-3} \ M; [Cu^{2+}] = x; [NH_3] = 0.10 \ M$

$$5 \times 10^{12} = \frac{(1 \times 10^{-3})}{x(0.10)^4}; x = 2 \times 10^{-12} \ M = [Cu^{2+}]$$

17.58 $NiC_2O_4(s) \rightleftharpoons Ni^{2+}(aq) + C_2O_4^{2-}(aq); \quad K_{sp} = [Ni^{2+}][C_2O_4^{2-}] = 4 \times 10^{-10}$

When the salt has just dissolved, $[C_2O_4^{2-}]$ will be 0.020 M. Thus $[Ni^{2+}]$ must be less than $4 \times 10^{-10} / 0.020 = 2 \times 10^{-8} \ M$. To achieve this low $[Ni^{2+}]$ we must complex the Ni^{2+} ion with NH_3: $Ni^{2+}(aq) + 6NH_3(aq)$ f $Ni(NH_3)_6^{2+}(aq)$. Essentially all Ni(II) is in the form of the complex, so $[Ni(NH_3)_6^{2+}] = 0.020$. Find K_f for $Ni(NH_3)_6^{2+}$ in Table 17.1.

$$K_f = \frac{[Ni(NH_3)_6^{2+}]}{[Ni^{2+}][NH_3]^6} = \frac{(0.020)}{(2 \times 10^{-8})[NH_3]^6} = 1.2 \times 10^9; [NH_3]^6 = 8.33 \times 10^{-4};$$

$$[NH_3] = 0.307 = 0.3 \ M$$

17.59   *Analyze/Plan.* We are asked to calculate K_{eq} for a particular reaction, making use of pertinent K_{sp} and K_f values from Appendix D and Table 17.1. Write the dissociation equilibrium for AgI and the formation reaction for $Ag(CN)_2^-$. Use algebra to manipulate these equations and their associated equilibrium constants to obtain the desired reaction and its equilibrium constant. *Solve.*

$$AgI(s) \rightleftharpoons Ag^+(aq) + I^-(aq)$$

$$\underline{Ag^+(aq) + 2CN^-(aq) \rightleftharpoons Ag(CN)_2^-(aq)}$$

$$AgI(s) + 2CN^-(aq) \rightleftharpoons Ag(CN)_2^-(aq) + I^-(aq)$$

$$K = K_{sp} \times K_f = [Ag^+][I^-] \times \frac{[Ag(CN)_2^-]}{[Ag^+][CN^-]^2} = (8.3 \times 10^{-17})(1 \times 10^{21}) = 8 \times 10^4$$

17.60 $Ag_2S(s) \rightleftharpoons 2Ag^+(aq) + S^{2-}(aq)$ K_{sp}

$$S^{2-}(aq) + 2H^+(aq) \rightleftharpoons H_2S(aq) \qquad 1/(K_{a1} \times K_{a2})$$

$$\underline{2[Ag]^+(aq) + 2Cl^-(aq) \rightleftharpoons AgCl_2^-(aq) \qquad K_f^2}$$

$$Ag_2S(s) + 2H^+(aq) + 4Cl^-(aq) \rightleftharpoons 2AgCl_2^-(aq) + H_2S(aq)$$

$$K = \frac{K_{sp} \times K_f^2}{K_{a1} \times K_{a2}} = \frac{(6 \times 10^{-51})(1.1 \times 10^5)^2}{(9.5 \times 10^{-8})(1 \times 10^{-19})} = 7.64 \times 10^{-15} = 8 \times 10^{-15}$$

Precipitation; Qualitative Analysis

17.61   *Analyze/Plan.* Follow the logic in Sample Exercise 17.15. Precipitation conditions: will Q (see Chapter 15) exceed K_{sp} for the compound?   *Solve.*

(a)   In base, Ca^{2+} can form $Ca(OH)_2(s)$.

$Ca(OH)_2(s) \rightleftharpoons Ca^{2+}(aq) + 2OH^-(aq);$ $K_{sp} = [Ca^{2+}][OH^-]^2$

$Q = [Ca^{2+}][OH^-]^2;$ $[Ca^{2+}] = 0.050 \, M;$ $pOH = 14 - 8.0 = 6.0;$ $[OH^-] = 1.0 \times 10^{-6} \, M$

$Q = (0.050)(1.0 \times 10^{-6})^2 = 5.0 \times 10^{-14};$ $K_{sp} = 6.5 \times 10^{-6}$ (Appendix D)

$Q < K_{sp},$ no $Ca(OH)_2$ precipitates.

(b) $Ag_2SO_4(s) \rightleftharpoons 2Ag^+(aq) + SO_4^{2-}(aq);$ $K_{sp} = [Ag+]^2[SO_4^{2-}]$

$$[Ag^+] = \frac{0.050 \, M \times 100 \, mL}{110 \, mL} = 4.545 \times 10^{-2} = 4.5 \times 10^{-2} \, M$$

$$[SO_4^{2-}] = \frac{0.050 \, M \times 10 \, mL}{110 \, mL} = 4.545 \times 10^{-3} = 4.5 \times 10^{-3} \, M$$

$Q = (4.545 \times 10^{-2})^2 (4.545 \times 10^{-3}) = 9.4 \times 10^{-6};$ $K_{sp} = 1.5 \times 10^{-5}$

$Q < K_{sp},$ no Ag_2SO_4 precipitates.

17.62   (a) $Co(OH)_2(s) \rightleftharpoons Co^{2+}(aq) + 2OH^-(aq);$ $K_{sp} = [Co^{2+}][OH^-]^2 = 1.3 \times 10^{-15}$

$pH = 8.5;$ $pOH = 14 - 8.5 = 5.5;$ $[OH^-] = 10^{-5.5} = 3.16 \times 10^{-6} = 3 \times 10^{-6} \, M$

$Q = (0.020)(3.16 \times 10^{-6})^2 = 2 \times 10^{-13};$ $Q > K_{sp},$ $Co(OH)_2$ will precipitate

(b) $AgIO_3(s) \rightleftharpoons Ag^+(aq) + IO_3^-(aq); K_{sp} = [Ag^+][IO_3^-] = 3.1 \times 10^{-8}$

$$[Ag^+] = \frac{0.010 \, M \, Ag^+ \times 0.100 \, L}{0.110 \, L} = 9.09 \times 10^{-3} = 9.1 \times 10^{-3} \, M$$

$$[IO_3^-] = \frac{0.015 \, M \, IO_3^- \times 0.010 \, L}{0.110 \, L} = 1.36 \times 10^{-3} = 1.4 \times 10^{-3} \, M$$

$Q = (9.09 \times 10^{-3})(1.36 \times 10^{-3}) = 1.2 \times 10^{-5}; Q > K_{sp}$, $AgIO_3$ will precipitate

17.63   *Analyze/Plan.* We are asked to calculate pH necessary to precipitate $Mn(OH)_2(s)$ if the resulting Mn^{2+} concentration is $\leq 1 \, \mu g/L$.

$Mn(OH)_2(s) \rightleftharpoons Mn^{2+}(aq) + 2OH^-(aq); K_{sp} = [Mn^{2+}][OH^-]^2 = 1.6 \times 10^{-13}$

At equilibrium, $[Mn^{2+}][OH^-]^2 = 1.6 \times 10^{-13}$. Change concentration $Mn^{2+}(aq)$ to mol/L and solve for $[OH^-]$.   *Solve.*

$$\frac{1 \, \mu g \, Mn^{2+}}{1 \, L} \times \frac{1 \times 10^{-6} \, g}{1 \, \mu g} \times \frac{1 \, mol \, Mn^{2+}}{54.94 \, g \, Mn^{2+}} = 1.82 \times 10^{-8} = 2 \times 10^{-8} \, M \, Mn^{2+}$$

$1.6 \times 10^{-13} = (1.82 \times 10^{-8})[OH^-]^2; [OH^-]^2 = 8.79 \times 10^{-6}; [OH^-] = 2.96 \times 10^{-3} = 3 \times 10^{-3} \, M$

$pOH = 2.53; pH = 14 - 2.53 = 11.47 = 11.5$

17.64 $AgCl(s) \rightleftharpoons Ag^+(aq) + Cl^-(aq); K_{sp} = [Ag^+][Cl^-] = 1.8 \times 10^{-10}$

$$[Ag^+] = \frac{0.10 \, M \times 0.2 \, mL}{10 \, mL} = 2 \times 10^{-3} \, M; \quad [Cl^-] = \frac{1.8 \times 10^{-10}}{2 \times 10^{-3} \, M} = 9 \times 10^{-8} \, M$$

$$\frac{9 \times 10^{-8} \, mol \, Cl^-}{1 \, L} \times \frac{35.45 \, g \, Cl^-}{1 \, mol \, Cl^-} \times 0.010 \, L = 3.19 \times 10^{-8} \, g \, Cl^- = 3 \times 10^{-8} \, g \, Cl^-$$

17.65   *Analyze/Plan.* We are asked which ion will precipitate first from a solution containing $Pb^{2+}(aq)$ and $Ag^+(aq)$ when $I^-(aq)$ is added. Follow the logic in Sample Exercise 17.16. Calculate $[I^-]$ needed to initiate precipitation of each ion. The cation that requires lower $[I^-]$ will precipitate first.   *Solve.*

$$Ag^+: K_{sp} = [Ag^+][I^-]; 8.3 \times 10^{-17} = (2.0 \times 10^{-4})[I^-]; [I^-] = \frac{8.3 \times 10^{-17}}{2.0 \times 10^{-4}} = 4.2 \times 10^{-13} \, M \, I^-$$

$$Pb^{2+}: K_{sp} = [Pb^{2+}][I^-]^2; 7.9 \times 10^{-9} = (1.5 \times 10^{-3})[I^-]^2; [I^-] = \left(\frac{7.9 \times 10^{-9}}{1.5 \times 10^{-3}}\right)^{1/2} = 2.3 \times 10^{-3} \, M \, I^-$$

AgI will precipitate first, at $[I^-] = 4.2 \times 10^{-13} \, M$.

17.66   (a)   Precipitation will begin when $Q = K_{sp}$.

$BaSO_4: K_{sp} = [Ba^{2+}][SO_4^{2-}] = 1.1 \times 10^{-10}$

$1.1 \times 10^{-10} = (0.010)[SO_4^{2-}]; [SO_4^{2-}] = 1.1 \times 10^{-8} \, M$

$SrSO_4$: $K_{sp} = [Sr^{2+}][SO_4^{2-}] = 3.2 \times 10^{-7}$

$3.2 \times 10^{-7} = (0.010)[SO_4^{2-}]$; $[SO_4^{2-}] = 3.2 \times 10^{-5} M$

The $[SO_4^{2-}]$ necessary to begin precipitation is the smaller of the two values, $1.1 \times 10^{-8} M SO_4^{2-}$.

(b) Ba^{2+} precipitates first, because it requires the smaller $[SO_4^{2-}]$.

(c) Sr^{2+} will begin to precipitate when $[SO_4^{2-}]$ in solution (not bound in $BaSO_4$) reaches $3.2 \times 10^{-5} M$.

17.67 *Analyze/Plan.* Use Figure 17.22 and the description of the five qualitative analysis "groups" in Section 17.7 to analyze the given data. *Solve.*

The first two experiments eliminate Group 1 and 2 ions (Figure 17.22). The fact that no insoluble phosphates form in the filtrate from the third experiment rules out Group 4 ions. The ions which might be in the sample are those of Group 3, that is, Al^{3+}, Fe^{3+}, Zn^{2+}, Cr^{3+}, Ni^{2+}, Co^{2+}, or Mn^{2+}, and those of Group 5, NH_4^+, Na^+ or K^+.

17.68 Initial solubility in water rules out CdS and HgO. Formation of a precipitate on addition of HCl indicates the presence of $Pb(NO_3)_2$ (formation of $PbCl_2$). Formation of a precipitate on addition of H_2S at pH 1 probably indicates $Cd(NO_3)_2$ (formation of CdS). (This test can be misleading because enough Pb^{2+} can remain in solution after filtering $PbCl_2$ to lead to visible precipitation of PbS.) Absence of a precipitate on addition of H_2S at pH 8 indicates that $ZnSO_4$ is not present. The yellow flame test indicates presence of Na^+. In summary, $Pb(NO_3)_2$ and Na_2SO_4 are definitely present, $Cd(NO_3)_2$ is probably present, and CdS, HgO and $ZnSO_4$ are definitely absent.

17.69 *Analyze/Plan.* We are asked to devise a procedure to separate various pairs of ions in aqueous solutions. In each case, refer to Figure 17.22 to find a set of conditions where the solubility of the two ions differs. Construct a procedure to generate these conditions. *Solve.*

(a) Cd^{2+} is in Gp. 2, but Zn^{2+} is not. Make the solution acidic using 0.5 M HCl; saturate with H_2S. CdS will precipitate, ZnS will not.

(b) $Cr(OH)_3$ is amphoteric but $Fe(OH)_3$ is not. Add excess base; $Fe(OH)_3(s)$ precipitates, but Cr^{3+} forms the soluble complex $Cr(OH)_4^-$.

(c) Mg^{2+} is a member of Gp. 4, but K^+ is not. Add $(NH_4)_2HPO_4$; Mg^{2+} precipitates as $MgNH_4PO_4$, K^+ remains in solution.

(d) Ag^+ is a member of Gp. 1, but Mn^{2+} is not. Add 6 M HCl, precipitate Ag^+ as AgCl(s).

17.70 (a) Make the solution slightly basic and saturate with H_2S; CdS will precipitate, Na^+ remains in solution.

(b) Make the solution acidic, saturate with H_2S; CuS will precipitate, Mg^{2+} remains in solution.

(c)   Add HCl, $PbCl_2$ precipitates. (It is best to carry out the reaction in an ice-water bath to reduce the solubility of $PbCl_2$.)

(d)   Add dilute HCl; AgCl precipitates, Hg^{2+} remains in solution.

17.71   (a)   Because phosphoric acid is a weak acid, the concentration of free $PO_4{}^{3-}$(aq) in an aqueous phosphate solution is low except in strongly basic media. In less basic media, the solubility product of the phosphates that one wishes to precipitate is not exceeded.

         (b) K_{sp} for those cations in Group 3 is much larger. Thus, to exceed K_{sp} a higher $[S^{2-}]$ is required. This is achieved by making the solution more basic.

         (c)   They should all redissolve in strongly acidic solution, e.g., in 12 M HCl (all the chlorides of Group 3 metals are soluble).

17.72   The addition of $(NH_4)_2HPO_4$ could result in precipitation of salts from metal ions of the other groups. The $(NH_4)_2HPO_4$ will render the solution basic, so metal hydroxides could form as well as insoluble phosphates. It is essential to separate the metal ions of a group from other metal ions before carrying out the specific tests for that group.

Additional Exercises

17.73   The equilibrium of interest is

$$HC_5H_3O_3(aq) \rightleftharpoons H^+(aq) + C_5H_3O_3{}^-(aq); \; K_a = 6.76 \times 10^{-4} = \frac{[H^+][C_5H_3O_3{}^-]}{[HC_5H_3O_3]}$$

Begin by calculating $[HC_5H_3O_3]$ and $[C_5H_3O_3{}^-]$ for each case.

(a) $\dfrac{35.0 \text{ g } HC_5H_3O_3}{0.250 \text{ L soln}} \times \dfrac{1 \text{ mol } HC_5H_3O_3}{112.1 \text{ g } HC_5H_3O_3} = 1.249 = 1.25 \; M \; HC_5H_3O_3$

$\dfrac{30.0 \text{ g } NaC_5H_3O_3}{0.250 \text{ L soln}} \times \dfrac{1 \text{ mol } NaC_5H_3O_3}{134.1 \text{ g } NaC_5H_3O_3} = 0.8949 = 0.895 \; M \; C_5H_3O_3{}^-$

$[H^+] = \dfrac{K_a[HC_5H_3O_3]}{[C_5H_3O_3{}^-]} = \dfrac{6.76 \times 10^{-4}\,(1.249 - x)}{(0.8949 + x)} \approx \dfrac{6.76 \times 10^{-4}(1.249)}{(0.8949)}$

$[H^+] = 9.43 \times 10^{-4} \; M, \; pH = 3.025$

(b)   For dilution, $M_1V_1 = M_2V_2$

$[HC_5H_3O_3] = \dfrac{0.250 \; M \times 30.0 \text{ mL}}{125 \text{ mL}} = 0.0600 \; M$

$[C_5H_3O_3{}^-] = \dfrac{0.220 \; M \times 20.0 \text{ mL}}{125 \text{ mL}} = 0.0352 \; M$

$[H^+] \approx \dfrac{6.76 \times 10^{-4}\,(0.0600)}{0.0352} = 1.15 \times 10^{-3} \; M, \; pH = 2.938$

(yes, $[H^+]$ is < 5% of 0.0352 M)

(c) $0.0850\ M \times 0.500\ L = 0.0425\ mol\ HC_5H_3O_3$

$1.65\ M \times 0.0500\ L = 0.0825\ mol\ NaOH$

	$HC_5H_3O_3(aq)$	$+$	$NaOH(aq)$	$\rightarrow$	$NaC_5H_3O_3(aq) + H_2O(l)$
initial	0.0425 mol		0.0825 mol		
reaction	−0.0425 mol		−0.0425 mol		+0.0425 mol
after	0 mol		0.0400 mol		0.0425 mol

The strong base NaOH dominates the pH; the contribution of $C_5H_3O_3^-$ is negligible. This combination would be "after the equivalence point" of a titration. The total volume is 0.550 L.

$$[OH^-] = \frac{0.0400\ mol}{0.550\ L} = 0.0727\ M;\ pOH = 1.138,\ pH = 12.862$$

17.74 $K_a = \dfrac{[H^+][In^-]}{[HIn]}$; at pH = 4.68, [HIn] = [In^-]; [H^+] = K_a; pH = pK_a = 4.68

17.75   (a) $HA(aq) + B(aq) \rightleftharpoons HB^+(aq) + A^-(aq)$ $K_{eq} = \dfrac{[HB^+][A^-]}{[HA][B]}$

(b)   Note that the solution is slightly basic because B is a stronger base than HA is an acid. (Or, equivalently, that A^- is a stronger base than HB^+ is an acid.) Thus, a little of the A^- is used up in reaction: $A^-(aq) + H_2O(l) \rightleftharpoons HA(aq) + OH^-(aq)$. Since pH is not very far from neutral, it is reasonable to assume that the reaction in part (a) has gone far to the right, and that $[A^-] \approx [HB^+]$ and $[HA] \approx [B]$. Then

$$K_a = \frac{[A^-][H^+]}{[HA]} = 8.0 \times 10^{-5};\ when\ pH = 9.2,\ [H^+] = 6.31 \times 10^{-10} = 6 \times 10^{-10}\ M$$

$$\frac{[A^-]}{[HA]} = 8.0 \times 10^{-5} / 6.31 \times 10^{-10} = 1.268 \times 10^5 = 1 \times 10^5$$

From the assumptions above, $\dfrac{[A^-]}{[HA]} = \dfrac{[HB^+]}{[B]}$, so $K_{eq} \approx \dfrac{[A^-]^2}{[HA]^2} = 1.608 \times 10^{10} = 2 \times 10^{10}$

(c) K_b for the reaction $B(aq) + H_2O(l) \rightleftharpoons BH^+(aq) + OH^-(aq)$ can be calculated by noting that the equilibrium constant for the reaction in part (a) can be written as $K_{eq} = K_a\ (HA) \times K_b\ (B)\ /\ K_w$. (You should prove this to yourself.) Then,

$$K_b\ (B) = \frac{K_{eq} \times K_w}{K_a\ (HA)} = \frac{(1.608 \times 10^{10})(1.0 \times 10^{-14})}{8.0 \times 10^{-5}} = 2.010 = 2$$

$K_b\ (B)$ is larger than $K_a\ (HA)$, as it must be if the solution is basic.

17.76   (a) $K_a = \dfrac{[H^+][CHO_2]}{[HCHO_2]}$; $[H^+] = \dfrac{K_a[HCHO_2]}{[CHO_2^-]}$

Buffer A : $[HCHO_2] = [CHO_2^-] = \dfrac{1.00\ mol}{1.00\ L} = 1.00\ M$

$$[H^+] = \frac{1.8 \times 10^{-4}\,(1.00\ M)}{(1.00\ M)} = 1.8 \times 10^{-4}\ M,\ pH = 3.74$$

Buffer B: $[HCHO_2] = [CHO_2^-] = \dfrac{0.010\ mol}{1.00\ L} = 0.010\ M$

$$[H^+] = \frac{1.8 \times 10^{-4}\,(0.010\ M)}{(0.010\ M)} = 1.8 \times 10^{-4}\ M,\ pH = 3.74$$

The pH of a buffer is determined by the identity of the conjugate acid/conjugate base pair (that is, the relevant K_a value) and the ratio of concentrations of the conjugate acid and conjugate base. The absolute concentrations of the components is not relevant. The pH values of the two buffers are equal because they both contain $HCHO_2$ and $NaCHO_2$ and the $[HCHO_2]$ / $[CHO_2^-]$ ratio is the same in both solutions.

(b)   Buffer capacity is determined by the absolute amount of conjugate acid and conjugate base available to absorb strong acid (H^+) or strong base (OH^-) that is added to the buffer. Buffer A has the greater capacity because it contains the greater absolute concentrations of $HCHO_2$ and CHO_2^-.

(c)   Buffer A:

	CHO_2^-	+	HCl	→	$HCHO_2$	+	Cl^-
	1.00 mol		0.001 mol		1.00 mol		
	0.999 mol		0		1.001 mol		

$$[H^+] = \frac{1.8 \times 10^{-4}\,(1.001)}{(0.999)} = 1.8 \times 10^{-4}\ M,\ pH = 3.74$$

(In a buffer calculation, volumes cancel and we can substitute moles directly into the K_a expression.)

Buffer B:

	CHO_2^-	+	HCl	→	$HCHO_2$	+	Cl^-
	0.010 mol		0.001 mol		0.010 mol		
	0.009 mol		0		0.011 mol		

$$[H^+] = \frac{1.8 \times 10^{-4}\,(0.011)}{(0.009)} = 2.2 \times 10^{-4}\ M,\ pH = 3.66$$

(d)   Buffer A: $1.00\ M$ HCl $\times\ 0.010\ L = 0.010$ mol H^+ added

mol $HCHO_2$ = 1.00 + 0.010 = 1.01 mol

mol CHO_2^- = 1.00 – 0.010 = 0.99 mol

$$[H^+] = \frac{1.8 \times 10^{-4}\,(1.01)}{(0.99)} = 1.8 \times 10^{-4}\ M,\ pH = 3.74$$

Buffer B: mol $HCHO_2 = 0.010 + 0.010 = 0.020$ mol $= 0.020\ M$

mol $CHO_2^- = 0.010 - 0.010 = 0.000$ mol

The solution is no longer a buffer; the only source of CHO_2^- is the dissociation of $HCHO_2$.

$$K_a = \frac{[H^+][CHO_2^-]}{[HCHO_2]} = \frac{x^2}{(0.020 - x)\ M}$$

The extent of ionization is greater than 5%; from the quadratic formula, $x = [H^+] = 1.8 \times 10^{-3}$, pH = 2.74.

(e) Adding 10 mL of 1.00 M HCl to buffer B exceeded its capacity, while the pH of buffer A was unaffected. This is quantitative confirmation that buffer A has a significantly greater capacity than buffer B. In fact, 1.0 L of 1.0 M HCl would be required to exceed the capacity of buffer A. Buffer A, with 100 times more $HCHO_2$ and CHO_2^- has 100 times the capacity of buffer B.

17.77 $\dfrac{0.20\ \text{mol}\ HC_2H_3O_2}{1\ \text{L soln}} \times 0.750\ \text{L} = 0.150 = 0.15\ \text{mol}\ HC_2H_3O_2$

$0.15\ \text{mol}\ HC_2H_3O_2 \times \dfrac{60.05\ \text{g}\ HC_2H_3O_2}{1\ \text{mol}\ HC_2H_3O_2} \times \dfrac{1\ \text{g gl acetic acid}}{0.99\ \text{g}\ HC_2H_3O_2} \times \dfrac{1.00\ \text{mL gl acetic acid}}{1.05\ \text{g gl acetic acid}}$

$= 8.7$ mL glacial acetic acid

At pH 4.50, $[H^+] = 10^{-4.50} = 3.16 \times 10^{-5} = 3.2 \times 10^{-5}\ M$; this is small compared to $0.20\ M\ HC_2H_3O_2$.

$$K_a = \frac{(3.16 \times 10^{-5})[C_2H_3O_2^-]}{0.20} = 1.8 \times 10^{-5}; [C_2H_3O_2^-] = 0.114 = 0.11\ M$$

$\dfrac{0.114\ \text{mol}\ NaC_2H_3O_2}{1\ \text{L soln}} \times 0.750\ \text{L} \times \dfrac{82.03\ \text{g}\ NaC_2H_3O_2}{1\ \text{mol}\ NaC_2H_3O_2} = 7.004 = 7.0\ \text{g}\ NaC_2H_3O_2$

17.78 (a) For a monoprotic acid (one H^+ per mole of acid), at the equivalence point moles OH^- added = moles H^+ originally present

$M_B \times V_B = $ g acid/molar mass

$MM = \dfrac{\text{g acid}}{M_B \times V_B} = \dfrac{0.2140\ \text{g}}{0.0950\ M \times 0.0274\ \text{L}} = 82.21 = 82.2\ \text{g/mol}$

(b) initial mol HA $= \dfrac{0.2140\ \text{g}}{82.21\ \text{g/mol}} = 2.603 \times 10^{-3} = 2.60 \times 10^{-3}$ mol HA

mol OH^- added to pH 6.50 $= 0.0950\ M \times 0.0150\ \text{L} = 1.425 \times 10^{-3}$

$= 1.43 \times 10^{-3}$ mol OH^-

	HA(aq)	+	NaOH(aq)	→	NaA(aq) + H_2O
before rx	2.603×10^{-3} mol		1.425×10^{-3} mol		0
change	-1.425×10^{-3} mol		-1.425×10^{-3} mol		$+1.425 \times 10^{-3}$ mol
after rx	1.178×10^{-3} mol		0		1.425×10^{-3} mol

$$[HA] = \frac{1.178 \times 10^{-3} \text{ mol}}{0.0400 \text{ L}} = 0.02945 = 0.0295 \ M$$

$$[A^-] = \frac{1.425 \times 10^{-3} \text{ mol}}{0.0400 \text{ L}} = 0.03563 = 0.0356 \ M; [H^+] = 10^{-6.50} = 3.162 \times 10^{-7}$$

$$= 3.2 \times 10^{-7} \ M$$

The mixture after reaction (a buffer) can be described by the acid dissociation equilibrium.

	HA(aq)	$\rightleftharpoons$	H$^+$(aq)	+	A$^-$(aq)
initial	0.0295 M		0		0.0356 M
equil	$(0.0295 - 3.2 \times 10^{-7} M)$		3.2×10^{-7} M		$(0.0356 + 3.2 \times 10^{-7})$ M

$$K_a = \frac{[H^+][A^-]}{[HA]} \approx \frac{(3.162 \times 10^{-7})(0.03563)}{(0.02945)} = 3.8 \times 10^{-7}$$

(Although we have carried 3 figures through the calculation to avoid rounding errors, the data dictate an answer with 2 significant figures.)

17.79   At the equivalence point of a titration, moles strong base added equals moles weak acid initially present. $M_B \times V_B$ = mol base added = mol acid initial

At the half-way point, the volume of base is one-half of the volume required to reach the equivalence point, and the moles base delivered equals one-half of the mol acid initially present. This means that one-half of the weak acid HA is converted to the conjugate base A$^-$. If exactly half of the acid reacts, mol HA = mol A$^-$ and [HA] = [A$^-$] at the half-way point.

From Equation [17.9], $pH = pK_a + \log \dfrac{[\text{conj. base}]}{[\text{conj. acid}]} = pK_a + \log \dfrac{[A^-]}{[HA^-]}$.

If [A$^-$]/[HA] = 1, log(1) = 0 and pH = pK$_a$ of the weak acid being titrated.

17.80   (a) $\dfrac{0.4885 \text{ g KHP}}{0.100 \text{ L}} \times \dfrac{1 \text{ mol KHP}}{204.2 \text{ g KHP}} = 0.02392 = 0.0239 \ M \ P^{2-}$ at the equivalence point

The pH at the equivalence point is determined by the hydrolysis of P^{2-}.

$$P^{2-}(aq) + H_2O(l) \rightleftharpoons HP^-(aq) + OH^-(aq)$$

$$K_b = \frac{[HP^-][OH^-]}{[P^{2-}]} = \frac{K_w}{K_a \text{ for } HP^-} = \frac{1.0 \times 10^{-14}}{3.1 \times 10^{-6}} = 3.23 \times 10^{-9} = 3.2 \times 10^{-9}$$

$$3.23 \times 10^{-9} = \frac{x^2}{(0.02392 - x)} \approx \frac{x^2}{0.02392}; X = [OH^-] = 8.8 \times 10^{-6} \ M$$

pH = 14 – 5.06 = 8.94. From Figure 16.7, either phenolphthalein (pH 8.2 – 10.0) or thymol blue (pH 8.0 – 9.6) could be used to detect the equivalence point. Phenolphthalein is usually the indicator of choice because the colorless to pink change is easier to see.

(b) $0.4885 \text{ g KHP} \times \dfrac{1 \text{ mol KHP}}{204.2 \text{ g KHP}} \times \dfrac{1 \text{ mol NaOH}}{1 \text{ mol KHP}} \times \dfrac{1}{0.03855 \text{ L NaOH}} = 0.06206 \, M \text{ NaOH}$

17.81   (a)   Initially, the solution is $0.100 \, M$ in CO_3^{2-}.

$CO_3^{2-}(aq) + H_2O(l) \rightleftharpoons HCO_3^-(aq) + OH^-(aq)$

$K_b = \dfrac{[HCO_3^-][OH^-]}{[CO_3^{2-}]} = \dfrac{K_w}{K_a[HCO_3^-]} = 1.79 \times 10^{-4} = 1.8 \times 10^{-4}$

Proceeding in the usual way for a weak base,

calculate $[OH^-] = 4.23 \times 10^{-3} = 4.2 \times 10^{-3} \, M$, pH = 11.63.

(b)   It will require 40.00 mL of $0.100 \, M$ HCl to reach the first equivalence point, at which point HCO_3^- is the predominant species.

(c)   An additional 40.00 mL are required to react with HCO_3^- to form H_2CO_3, the predominant species at the second equivalence point.

(d)   At the second equivalence point there is a $0.0333 \, M$ solution of H_2CO_3. By the usual procedure for a weak acid:

$H_2CO_3(aq) \rightleftharpoons H^+(aq) + HCO_3^-(aq)$

$K_a = \dfrac{[H^+][HCO_3^-]}{[H_2CO_3]} = 4.3 \times 10^{-7}; \dfrac{(x)^2}{(0.0333 - x)} \approx \dfrac{x^2}{0.0333} \approx 4.3 \times 10^{-7}$

$x = 1.20 \times 10^{-4} = 1.2 \times 10^{-4} \, M \, H^+$; pH = 3.92

17.82   The reaction involved is $HA(aq) + OH^-(aq) \rightleftharpoons A^-(aq) + H_2O(l)$. We thus have 0.080 mol A^- and 0.12 mol HA in a total volume of 1.0 L, so the "initial" molarities of A^- and HA are $0.080 \, M$ and $0.12 \, M$, respectively. The weak acid equilibrium of interest is

$HA(aq) \rightleftharpoons H^+(aq) + A^-(aq)$

(a) $K_a = \dfrac{[H^+][A^-]}{[HA]}$; $[H^+] = 10^{-4.80} = 1.58 \times 10^{-5} = 1.6 \times 10^{-5} \, M$

Assuming $[H^+]$ is small compared to [HA] and $[A^-]$,

$K_a \approx \dfrac{(1.58 \times 10^{-5})(0.080)}{(0.12)} = 1.06 \times 10^{-5} = 1.1 \times 10^{-5}$, $pK_a = 4.98$

(b)   At pH = 5.00, $[H^+] = 1.0 \times 10^{-5} \, M$. Let b = extra moles NaOH.

[HA] = 0.12 – b, $[A^-]$ = 0.080 + b

$1.06 \times 10^{-5} \approx \dfrac{(1.0 \times 10^{-5})(0.080 + b)}{(0.12 - b)}$; $2.06 \times 10^{-5} b = 4.72 \times 10^{-7}$;

b = 0.023 mol NaOH

17.83 Assume that H_3PO_4 will react with NaOH in a stepwise fashion: (This is not unreasonable, since the three K_a values for H_3PO_4 are significantly different.)

$$H_3PO_4(aq) \quad\quad NaOH(aq) \quad \rightarrow \quad H_2PO_4^-(aq) + Na^+(aq) + H_2O(l)$$
$$+$$

before	0.20 mol	0.30 mol	0 mol
after	0 mol	0.10 mol	0.20 mol

$$H_2PO_4^-(aq) \quad\quad NaOH(aq) \quad \rightarrow \quad HPO_4^-(aq) + Na^+(aq) + H_2O(l)$$
$$+$$

before	0.20 mol	0.10 mol	0.25 mol
after	0.10 mol	0	0.35 mol

Thus, after all NaOH has reacted, the resulting 1.00 L solution is a buffer containing 0.10 mol $H_2PO_4^-$ and 0.35 mol HPO_4^{2-}. $H_2PO_4^-(aq) \rightleftharpoons H^+(aq) + HPO_4^{2-}(aq)$

$$K_a = 6.2 \times 10^{-8} = \frac{[HPO_4^{2-}][H^+]}{[H_2PO_4^-]}; [H^+] = \frac{6.2 \times 10^{-8} (0.10\ M)}{0.35\ M} = 1.77 \times 10^{-8} = 1.8 \times 10^{-8}\ M;$$

$$pH = 7.75$$

17.84 The pH of a buffer system is centered around pK_a for the conjugate acid component. For a diprotic acid, two conjugate acid/conjugate base pairs are possible.

$H_2X(aq) \rightleftharpoons H^+(aq) + HX^-(aq); \quad K_{a1} = 2 \times 10^{-2}; \quad pK_{a1} = 1.70$

$HX^-(aq) \rightleftharpoons H^+(aq) + X^{2-}(aq); \quad K_{a2} = 5.0 \times 10^{-7}; \quad pK_{a2} = 6.30$

Clearly HX^- / X^{2-} is the more appropriate combination for preparing a buffer with pH = 6.50. The $[H^+]$ in this buffer = $10^{-6.50} = 3.16 \times 10^{-7} = 3.2 \times 10^{-7}\ M$. Using the K_{a2} expression to calculate the $[X^{2-}] / [HX^-]$ ratio:

$$K_{a2} = \frac{[H^+][X^{2-}]}{[HX^-]}; \frac{K_{a2}}{[H^+]} = \frac{[X^{2-}]}{[HX^-]} = \frac{5.0 \times 10^{-7}}{3.16 \times 10^{-7}} = 1.58 = 1.6$$

Since X^{2-} and HX^- are present in the same solution, the ratio of concentrations is also a ratio of moles.

$$\frac{[X^{2-}]}{[HX^-]} = \left(\frac{mol\ X^{2-}/L\ soln}{mol\ HX^-/L\ soln} \right) = \frac{mol\ X^{2-}}{mol\ HX^-} = 1.58;\ mol\ X^{2-} = (1.58)\ mol\ HX^-$$

In the 1.0 L of 1.0 M H_2X, there is 1.0 mol of X^{2-} containing material.

Thus, mol HX^- + 1.58 (mol HX^-) = 1.0 mol. 2.58 (mol HX^-) = 1.0;

mol HX^- = 1.0 / 2.58 = 0.39 mol HX^-; mol X^{2-} = 1.0 – 0.39 = 0.61 mol X^{2-}.

Thus enough 1.0 M NaOH must be added to produce 0.39 mol HX^- and 0.61 mol X^{2-}.

Considering the neutralization in a step-wise fashion (see discussion of titrations of polyprotic acids in Section 17.3).

	$H_2X(aq)$	$+$	$NaOH(aq)$	$\rightarrow$	$HX^-(aq) + H_2O(l)$
before	1.0 mol		1 mol		0
after	0		0		1.0 mol

	$HX^-(aq)$	$+$	$NaOH(aq)$	$\rightarrow$	$X^{2-}(aq) + H_2O(l)$
before	1.0				0.61
change	–0.61		–0.61		+0.61
after	0.39		0		0.61

Starting with 1.0 mol of H_2X, 1.0 mol of NaOH is added to completely convert it to 1.0 mol of HX^-. Of that 1.0 mol of HX^-, 0.61 mol must be converted to 0.61 mol X^{2-}. The total moles of NaOH added is (1.00 + 0.61) = 1.61 mol NaOH.

$$L\ NaOH = \frac{mol\ NaOH}{M\ NaOH} = \frac{1.61\ mol}{1.0\ M} = 1.6\ L\ of\ 1.0\ M\ NaOH$$

17.85 $C_3H_5O_3^-$ will be formed by reaction of $HC_3H_5O_3$ with NaOH.

0.1000 M × 0.02500 L = 2.500×10^{-3} mol $HC_3H_5O_3$; b = mol NaOH needed

	$HC_3H_5O_3$	$+$	$NaOH$	$\rightarrow$	$C_3H_5O_3^- + H_2O + Na^+$
initial	2.500×10^{-3} mol		b mol		
rx	–b mol		–b mol		+b mol
after rx	$(2.500 \times 10^3 - b)$ mol		0		b mol

$$K_a = \frac{[H^+][C_3H_5O_3^-]}{[HC_3H_5O_3]}; K_a = 1.4 \times 10^{-4}; [H^+] = 10^{-pH} = 10^{-3.75} = 1.778 \times 10^{-4} = 1.8 \times 10^{-4}\ M$$

Since solution volume is the same for $HC_3H_5O_3$ and $C_3H_5O_3^-$, we can use moles in the equation for $[H^+]$.

$$K_a - 1.4 \times 10^{-4} - \frac{1.778 \times 10^{-4}\ (b)}{(2.500 \times 10^{-3} - b)}; 0.7874\ (2.500 \times 10^{-3}\ b) = b, 1.969 \times 10^{-3} = 1.7874\ b,$$

$b = 1.10 \times 10^{-3} = 1.1 \times 10^{-3}$ mol OH^-

(The precision of K_a dictates that the result has 2 sig figs.)

Substituting this result into the K_a expression gives $[H^+] = 1.8 \times 10^{-4}$. This checks and confirms our result. Calculate volume NaOH required from M = mol/L.

$$1.10 \times 10^{-3}\ mol\ OH^- \times \frac{1\ L}{1.000\ mol} \times \frac{1\ \mu L}{1 \times 10^{-6}\ L} = 1.1 \times 10^3\ \mu L\ (1.1\ mL)$$

17.86 (a) $H^+(aq) + HCO_3^-(aq) \rightleftharpoons H_2CO_3(aq) \rightleftharpoons H_2O(l) + CO_2(g)$

A person breathing normally exhales $CO_2(g)$. Rapid breathing causes excess $CO_2(g)$ to be removed from the blood. By LeChatelier's principle, this causes both equilibria above to shift right, reducing $[H^+]$ in the blood and raising blood pH.

(b) Breathing in a paper bag traps the exhaled CO_2; the gas in the bag contains more CO_2 than ambient air. When a person inhales gas from the bag, a greater amount (partial pressure) of $CO_2(g)$ in the lungs shifts the equilibria left, increasing $[H^+]$ and lowering blood pH.

17.87 (a) CdS: 8.0×10^{-28}; CuS: 6×10^{-37} CdS has greater molar solubility.

(b) $PbCO_3$: 7.4×10^{-14}; $BaCrO_4$: 2.1×10^{-10} $BaCrO_4$ has greater molar solubility.

(c) Since the stoichiometry of the two complexes is not the same, K_{sp} values can't be compared directly; molar solubilities must be calculated from K_{sp} values.

$Ni(OH)_2$: $K_{sp} = 6.0 \times 10^{-16} = [Ni^{2+}][OH^-]^2$; $[Ni^{2+}] = x$, $[OH^-] = 2x$

$6.0 \times 10^{-16} = (x)(2x)^2 = 4x^3$; $x = 5.3 \times 10^{-6}\ M\ Ni^{2+}$

Note that $[OH^-]$ from the autoionization of water is less than 1% of $[OH^-]$ from $Ni(OH)_2$ and can be neglected.

$NiCO_3$: $K_{sp} = 1.3 \times 10^{-7} = [Ni^{2+}][CO_3^{2-}]$; $[Ni^{2+}] = [CO_3^{2-}] = x$

$1.3 \times 10^{-7} = x^2$; $x = 3.6 \times 10^{-4}\ M\ Ni^{2+}$

$NiCO_3$ has greater molar solubility than $Ni(OH)_2$, but the values are much closer than expected from inspection of K_{sp} values alone.

(d) Again, molar solubilities must be calculated for comparison.

Ag_2SO_4: $K_{sp} = 1.5 \times 10^{-5} = [Ag^+]^2[SO_4^{2-}]$; $[SO_4^{2-}] = x$, $[Ag^+] = 2x$

$1.5 \times 10^{-5} = (2x)^2(x) = 4x^3$; $x = 1.6 \times 10^{-2}\ M\ SO_4^{2-}$

AgI: $K_{sp} = 8.3 \times 10^{-17} = [Ag^+][I^-]$; $[Ag^+] = [I^-] = x$

$8.3 \times 10^{-17} = x^2$; $x = 9.1 \times 10^{-9}\ M\ Ag^+$

Ag_2SO_4 has greater molar solubility than AgI.

17.88 pH $= 10.38$; pOH $= 14.00 - 10.38 = 3.62$; $[OH^-] = 10^{-3.62}$

$[OH^-] = 2.40 \times 10^{-4} = 2.4 \times 10^{-4}\ M$; $[Mg^{2+}] = 0.5[OH^-] = 1.20 \times 10^{-4} = 1.2 \times 10^{-4}\ M$

$K_{sp} = [Mg^{2+}][OH^-]^2 \approx (1.20 \times 10^{-4})(2.40 \times 10^{-4})^2 \approx 6.9 \times 10^{-12}$

17.89 $Ca(OH)_2(aq) + 2HCl(aq) \rightarrow CaCl_2(aq) + 2H_2O$

mmol HCl $= M \times$ mL $= 0.0983\ M \times 11.23$ mL $= 1.1039 = 1.10$ mmol HCl

mmol $Ca(OH)_2$ = mmol HCl/2 = 1.1039/2 = 0.55195 = 0.552 mmol $Ca(OH)_2$

$[Ca^{2+}] = \dfrac{0.55195\ \text{mmol}}{50.00\ \text{mL}} = 0.01104 = 0.0110\ M$

$[OH^-] = 2[Ca^{2+}] = 0.02208 = 0.0221\ M$

$K_{sp} = [Ca^{2+}][OH^-]^2 = (0.01104)(0.02208)2 = 5.38 \times 10^{-6}$

The value in Appendix D is 6.5×10^{-6}, a difference of 17%. Since a change in temperature does change the value of an equilibrium constant, the solution may not have been kept at $25\,^{\circ}$C. It is also possible that experimental errors led to the difference in K_{sp} values.

17.90 $K_{sp} = [Ba^{2+}][MnO_4^-]^2 = 2.5 \times 10^{-10}$

$[MnO_4^-]^2 = 2.5 \times 10^{-10} / 2.0 \times 10^{-8} = 0.0125; [MnO_4^-] = \sqrt{0.0125} = 0.11\ M$

17.91 $[Ca^{2+}][CO_3^{2-}] = 4.5 \times 10^{-9}; [Fe^{2+}][CO_3^{2-}] = 2.1 \times 10^{-11}$

Since $[CO_3^{2-}]$ is the same for both equilibria:

$[CO_3^{2-}] = \dfrac{4.5 \times 10^{-9}}{[Ca^{2+}]} = \dfrac{2.1 \times 10^{-11}}{[Fe^{2+}]}$; rearranging $\dfrac{[Ca^{2+}]}{[Fe^{2+}]} = \dfrac{4.5 \times 10^{-9}}{2.1 \times 10^{-11}} = 214 = 2.1 \times 10^2$

17.92 $PbSO_4(s) \rightleftharpoons Pb^{2+}(aq) + SO_4^{2-}(aq);\quad K_{sp} = 6.3 \times 10^{-7} = [Pb^{2+}][SO_4^{2-}]$

$SrSO_4(s) \rightleftharpoons Sr^{2+}(aq) + SO_4^{2-}(aq);\quad K_{sp} = 3.2 \times 10^{-7} = [Sr^{2+}][SO_4^{2-}]$

Let $x = [Pb^{2+}],\ y = [Sr^{2+}],\ x + y = [SO_4^{2-}]$

$\dfrac{x(x+y)}{y(x+y)} = \dfrac{6.3 \times 10^{-7}}{3.2 \times 10^{-7}}; \dfrac{x}{y} = 1.9688 = 2.0; x = 1.969\ y = 2.0\ y$

$y(1.969\ y+y) = 3.2 \times 10^{-7}; 2.969\ y^2 = 3.2 \times 10^{-7}; y = 3.283 \times 10^{-4} = 3.3 \times 10^{-4}$

$x = 1.969\ y; x = 1.969(3.283 \times 10^{-4}) = 6.464 \times 10^{-4} = 6.5 \times 10^{-4}$

$[Pb^{2+}] = 6.5 \times 10^{-4}\ M, [Sr^{2+}] = 3.3 \times 10^{-4}\ M, [SO_4^{2-}] = (3.283 + 6.464) \times 10^{-4} = 9.7 \times 10^{-4}\ M$

17.93 $MgC_2O_4(s) \rightleftharpoons Mg^{2+}(aq) + C_2O_4^{2-}(aq)$

$K_{sp} = [Mg^{2+}][C_2O_4^{2-}] = 8.6 \times 10^{-5}$

If $[Mg^{2+}]$ is to be $3.0 \times 10^{-2}\ M, [C_2O_4^{2-}] = 8.6 \times 10^{-5}/3.0 \times 10^{-2} = 2.87 \times 10^{-3} = 2.9 \times 10^{-3}\ M$

The oxalate ion undergoes hydrolysis:

$C_2O_4^{2-}(aq) + H_2O(l) \rightleftharpoons HC_2O_4^-(aq) + OH^-(aq)$

$K_b = \dfrac{[HC_2O_4^-][OH^-]}{[C_2O_4^{2-}]} = 1.0 \times 10^{-14}/6.4 \times 10^{-5} = 1.56 \times 10^{-10} = 1.6 \times 10^{-10}$

$[Mg^{2+}] = 3.0 \times 10^{-2}\ M, [C_2O_4^{2-}] = 2.87 \times 10^{-3} = 2.9 \times 10^{-3}\ M$

$[HC_2O_4^-] = (3.0 \times 10^{-2} - 2.87 \times 10^{-3})\ M = 2.71 \times 10^{-2} = 2.7 \times 10^{-2}\ M$

$[OH^-] = 1.56 \times 10^{-10} \times \dfrac{[C_2O_4^{2-}]}{[HC_2O_4^-]} = 1.56 \times 10^{-10} \times \dfrac{(2.87 \times 10^{-3})}{(2.71 \times 10^{-2})} = 1.652 \times 10^{-11}$

$[OH^-] = 1.7 \times 10^{-11}; pOH = 10.78, pH = 3.22$

17.94 The student failed to account for the hydrolysis of the AsO_4^{3-} ion. If there were no hydrolysis, $[Mg^{2+}]$ would indeed be 1.5 times that of $[AsO_4^{3-}]$. However, as the reaction $AsO_4^{3-}(aq) + H_2O(l) \rightleftharpoons HAsO_4^{2-}(aq) + OH^-(aq)$ proceeds, the ion product $[Mg^{2+}]^3[AsO_4^{3-}]^2$ falls below the value for K_{sp}. More $Mg_3(AsO_4)_2$ dissolves, more hydrolysis occurs, and so on, until an equilibrium is reached. At this point $[Mg^{2+}]$ in solution is much greater than 1.5 times free $[AsO_4^{3-}]$. However, it is exactly 1.5 times the **total** concentration of all arsenic-containing species. That is,

$$[Mg^{2+}] = 1.5\,([AsO_4^{3-}] + [HAsO_4^{2-}] + [H_2AsO_4^-] + [H_3AsO_4])$$

17.95

$$Zn(OH)_2(s) \rightleftharpoons Zn^{2+}(aq) + 2OH^-(aq) \qquad K_{sp} = 3.0 \times 10^{-16}$$

$$Zn^{2+}(aq) + 4OH^-(aq) \rightleftharpoons Zn(OH)_4^{2-}(aq) \qquad K_f = 4.6 \times 10^{17}$$

$$Zn(OH)_2(s) + 2OH^-(aq) \rightleftharpoons Zn(OH)_4^{2-}(aq) \qquad K = K_{sp} \times K_f = 138 = 1.4 \times 10^2$$

$$K = 138 = 1.4 \times 10^2 = \frac{[Zn(OH)_4^{2-}]}{[OH^-]^2}$$

If 0.015 mol $Zn(OH)_2$ dissolves, 0.015 mol $Zn(OH)_4^{2-}$ should be present at equilibrium.

$$[OH^-]^2 = \frac{(0.015)}{138}; [OH^-] = 1.043 \times 10^{-2}\ M\ [OH^-] \geq 1.0 \times 10^{-2}\ M \text{ or } pH \geq 12.02$$

Integrative Exercises

17.96 (a) Complete ionic:

$$H^+(aq) + Cl^-(aq) + Na^+(aq) + CHO_2^-(aq) \rightarrow HCHO_2(aq) + Na^+(aq) + Cl^-(aq)$$

Na^+ and Cl^- are spectator ions.

Net ionic: $H^+(aq) + CHO_2^-(aq) \rightleftharpoons HCHO_2(aq)$

(b) The net ionic equation in part (a) is the reverse of the dissociation of $HCHO_2$.

$$K = \frac{1}{K_a} = \frac{1}{1.8 \times 10^{-4}} = 5.55 \times 10^3 = 5.6 \times 10^3$$

(c) For Na^+ and Cl^-, this is just a dilution problem.

$M_1V_1 = M_2V_2$; V_2 is 50.0 mL + 50.0 mL = 100.0 mL

$$Cl^-: \frac{0.15\ M \times 50.0\ mL}{100.0\ mL} = 0.075\ M;\ Na^+: \frac{0.15\ M \times 50.0\ mL}{100.0\ mL} = 0.075\ M$$

H^+ and CHO_2^- react to form $HCHO_2$. Since $K \gg 1$, the reaction essentially goes to completion.

$0.15\ M \times 0.0500\ mL = 7.5 \times 10^{-3}$ mol H^+

$\underline{0.15\ M \times 0.0500\ mL = 7.5 \times 10^{-3}\ mol\ CHO_2^-}$

$= 7.5 \times 10^{-3}$ mol $HCHO_2$

Solve the weak acid problem to determine $[H^+]$, $[CHO_2^-]$ and $[HCHO_2]$ at equilibrium.

$$K_a = \frac{[H^+][CHO_2^-]}{[HCHO_2]}; [H^+] = [CHO_2^-] = x \, M; [HCHO_2] = \frac{(7.5 \times 10^{-3} - x) \, \text{mol}}{0.100 \, \text{L}}$$
$$= (0.075 - x) \, M$$

$$1.8 \times 10^{-4} = \frac{x^2}{(0.075 - x)} \approx \frac{x^2}{0.075}; x = 3.7 \times 10^{-3} \, M H^+ \text{ and } HCHO_2^-$$

$$[HCHO_2] = (0.075 - 0.0037) = 0.071 \, M$$

$$\frac{[H^+]}{[HNO_2]} \times 100 = \frac{3.7 \times 10^{-3}}{0.075} \times 100 = 4.9\% \text{ dissociation}$$

In summary:

$$[Na^+] = [Cl^-] = 0.075 \, M, [HCHO_2] = 0.071 \, M, [H^+] = [CHO_2^-] = 0.0037 \, M$$

17.97   (a)   For a monoprotic acid (one H^+ per mole of acid), at the equivalence point

moles OH^- added = moles H^+ originally present

$M_B \times V_B$ = g acid/molar mass

$$MM = \frac{\text{g acid}}{M_B \times V_B} = \frac{0.1044 \, \text{g}}{0.0500 \, M \times 0.02210 \, \text{L}} = 94.48 = 94.5 \, \text{g/mol}$$

(b)   11.05 mL is exactly half-way to the equivalence point (22.10 mL). When half of the unknown acid is neutralized, $[HA] = [A^-]$, $[H^+] = K_a$ and $pH = pK_a$.

$$K_a = 10^{-4.89} = 1.3 \times 10^{-5}$$

(c)   From Appendix D, Table D.1, acids with K_a values close to 1.3×10^{-5} are

name	K_a	formula	molar mass
propionic	1.3×10^{-5}	$HC_3H_5O_2$	74.1
butanoic	1.5×10^{-5}	$HC_4H_7O_2$	88.1
acetic	1.8×10^{-5}	$HC_2H_3O_2$	60.1
hydroazoic	1.9×10^{-5}	HN_3	43.0

Of these, butanoic has the closest match for K_a and molar mass, but the agreement is not good.

17.98 $n = \dfrac{PV}{RT} = 735 \, \text{torr} \times \dfrac{1 \, \text{atm}}{760 \, \text{torr}} \times \dfrac{7.5 \, \text{L}}{295 \, \text{K}} \times \dfrac{K \cdot \text{mol}}{0.08206 \, \text{L} \cdot \text{atm}} = 0.300 = 0.30 \, \text{mol NH}_3$

$0.40 \, M \times 0.50 \, \text{L} = 0.20 \, \text{mol HCl}$

	HCl(aq)	+	NH$_3$(g)	→	NH$_4^+$(aq)	+	Cl$^-$(aq)
before	0.20 mol		0.30 mol				
after	0		0.10 mol		0.20 mol		0.20 mol

The solution will be a buffer because of the substantial concentrations of NH$_3$ and NH$_4^+$ present. Use K$_a$ for NH$_4^+$ to describe the equilibrium.

$$NH_4^+(aq) \rightleftharpoons NH_3(aq) + H^+(aq)$$

equil. 0.20 – x   0.10 + x   x

$$K_a = \frac{1.0 \times 10^{-14}}{1.8 \times 10^{-5}} = 5.56 \times 10^{-10} = 5.6 \times 10^{-10}; K_a = \frac{[NH_3][H^+]}{[NH_4^+]}; [H^+] = \frac{K_a[NH_4^+]}{[NH_3]}$$

Since this expression contains a ratio of concentrations, volume will cancel and we can substitute moles directly. Assume x is small compared to 0.10 and 0.20.

$$[H^+] = \frac{5.56 \times 10^{-10} (0.20)}{(0.10)} = 1.111 \times 10^{-9} = 1.1 \times 10^{-9} \, M, pH = 8.95$$

17.99   Calculate the initial M of aspirin in the stomach and solve the equilibrium problem to find equilibrium concentrations of C$_8$H$_7$O$_2$COOH and C$_8$H$_7$O$_2$COO$^-$. At pH = 2, [H$^+$] = 1 × 10^{-2}.

$$\frac{325 \text{ mg}}{\text{tablet}} \times 2 \text{ tablets} \times \frac{1 \text{ g}}{1000 \text{ mg}} \times \frac{1 \text{ mol C}_8\text{H}_7\text{O}_2\text{COOH}}{180.2 \text{ g C}_8\text{H}_7\text{O}_2\text{COOH}} \times \frac{1}{1 \text{ L}} = 3.61 \times 10^{-3} = 4 \times 10^{-3} \, M$$

	C$_8$H$_7$O$_2$COOH(aq)	⇌	C$_8$H$_7$O$_2$COO$^-$	+	H$^+$(aq)
initial	3.61 × 10^{-3} M		0		1 × 10^{-2} M
equil	(3.61 × 10^{-3} – x) M		x M		(1 × 10^{-2} + x) M

$$K_a = 3 \times 10^{-5} = \frac{[H^+][C_8H_7O_2COO^-]}{[C_8H_7O_2COOH]} = \frac{(0.01+x)(x)}{(3.61 \times 10^{-3} - x)} \approx \frac{0.01 x}{3.61 \times 10^{-3}}$$

$$x = [C_8H_7O_2COO^-] = 1.08 \times 10^{-5} = 1 \times 10^{-5} \, M$$

$$\% \text{ ionization} = \frac{1.08 \times 10^{-5} \, M \, C_8H_7O_2COO^-}{3.61 \times 10^{-3} \, M \, C_8H_7O_2COOH} \times 100 = 0.3\%$$

(% ionization is small, so the assumption was valid.)

% aspirin molecules = 100.0% – 0.3% = 99.7% molecules

17.100   According to Equation [13.4], S$_g$ = kP$_g$

$$S_{CO_2} = 3.1 \times 10^2 \frac{\text{mol}}{\text{L} \cdot \text{atm}} \times 1.10 \text{ atm} = 0.0341 = \frac{0.034 \text{ mol}}{\text{L}} = 0.034 \, M \, CO_2$$

$CO_2(g) + H_2O(l) \rightarrow H_2CO_3(aq)$; $0.0341\ M\ CO_2 = 0.0341\ M\ H_2CO_3$

Consider the stepwise dissociation of $H_2CO_3(aq)$.

	$H_2CO_3(aq)$	$\rightleftharpoons$	$H^+(aq)$	$+$	$HCO_3^-(aq)$
initial	$0.0341\ M$		0		0
equil.	$(0.0341\text{-}x)\ M$		x		x

$$K_{a1} = \frac{[H^+][HCO_3^-]}{[H_2CO_3]} = \frac{x^2}{(0.0341-x)} \approx \frac{x^2}{0.0341} \approx 4.3 \times 10^{-7}$$

$x^2 = 1.47 \times 10^{-8}$; $x = 1.2 \times 10^{-4}\ M\ H^+$; $pH = 3.92$

$K_{a2} = 5.6 \times 10^{-11}$; assume the second ionization does not contribute significantly to $[H^+]$.

17.101 $\pi = MRT, M = \dfrac{\pi}{RT} = \dfrac{21\ \text{torr}}{298\ K} \times \dfrac{1\ \text{atm}}{760\ \text{torr}} \times \dfrac{K \bullet \text{mol}}{0.08206\ L \bullet \text{atm}} = 1.13 \times 10^{-3} = 1.1 \times 10^{-3}\ M$

$SrSO_4(s) \rightleftharpoons Sr^{2+}(aq) + SO_4^{2-}(aq)$; $K_{sp} = [Sr^{2+}][SO_4^{2-}]$

The total particle concentration is $1.13 \times 10^{-3}\ M$. Each mole of $SrSO_4$ that dissolves produces 2 mol of ions, so $[Sr^{2+}] = [SO_4^{2-}] = 1.13 \times 10^{-3}\ M/2 = 5.65 \times 10^{-4} = 5.7 \times 10^{-4}\ M$.

$K_{sp} = (5.65 \times 10^{-4})^2 = 3.2 \times 10^{-7}$

17.102   For very dilute aqueous solutions, assume the solution density is 1 g/mL.

$$ppb = \frac{g\ solute}{10^9\ g\ solution} = \frac{1 \times 10^{-6}\ g\ solute}{1 \times 10^3\ g\ solution} = \frac{\mu g\ solute}{L\ solution}$$

(a) $K_{sp} = [Ag^+][Cl^-] = 1.8 \times 10^{-10}$; $[Ag^+] = (1.8 \times 10^{-10})^{1/2} = 1.34 \times 10^{-5} = 1.3 \times 10^{-5}\ M$

$$\frac{1.34 \times 10^{-5}\ mol\ Ag^+}{L} \times \frac{107.9\ g\ Ag^+}{1\ mol\ Ag^+} \times \frac{1\ \mu g}{1 \times 10^{-6}\ g} = \frac{1.4 \times 10^3\ \mu g\ Ag^+}{L}$$

$$= 1.4 \times 10^3\ ppb = 1.4\ ppm$$

(b) $K_{sp} = [Ag^+][Br^-] = 5.0 \times 10^{-13}$; $[Ag^+] = (5.0 \times 10^{-13})^{1/2} = 7.07 \times 10^{-7} = 7.1 \times 10^{-7}\ M$

$$\frac{7.07 \times 10^{-7}\ mol\ Ag^+}{L} \times \frac{107.9\ g\ Ag^+}{1\ mol\ Ag^+} \times \frac{1\ \mu g}{1 \times 10^{-6}\ g} = 76\ ppb$$

(c) $K_{sp} = [Ag^+][I^-] = 8.3 \times 10^{-17}$; $[Ag^+] = (8.3 \times 10^{-17})^{1/2} = 9.11 \times 10^{-9} = 9.1 \times 10^{-9}\ M$

$$\frac{9.11 \times 10^{-9}\ mol\ Ag^+}{L} \times \frac{107.9\ g\ Ag^+}{1\ mol\ Ag^+} \times \frac{1\ \mu g}{1 \times 10^{-6}\ g} = 0.98\ ppb$$

$AgBr(s)$ would maintain $[Ag^+]$ in the correct range.

17.103   To determine precipitation conditions, we must know K_{sp} for $CaF_2(s)$ and calculate Q under the specified conditions. $K_{sp} = 3.9 \times 10^{-11} = [Ca^{2+}][F^-]^2$

$[Ca^{2+}]$ and $[F^-]$: The term 1 ppb means 1 part per billion or 1 g solute per billion g solution. Assuming that the density of this very dilute solution is the density of water:

$$1 \text{ ppb} = \frac{1 \text{ g solute}}{1 \times 10^9 \text{ g solution}} \times \frac{1 \text{ g solution}}{1 \text{ mL solution}} \times \frac{1 \times 10^3 \text{ mL}}{1 \text{ L}} = \frac{1 \times 10^{-6} \text{ g solute}}{1 \text{ L solution}}$$

$$\frac{1 \times 10^{-6} \text{ g solute}}{1 \text{ L solution}} \times \frac{1 \mu g}{1 \times 10^{-6} \text{ g}} = 1 \mu g / 1 \text{ L}$$

$$8 \text{ ppb } Ca^{2+} \times \frac{1 \mu g}{1 \text{ L}} = \frac{8 \mu g \, Ca^{2+}}{1 \text{ L}} = \frac{8 \times 10^{-6} \text{ g } Ca^{2+}}{1 \text{ L}} \times \frac{1 \text{ mol } Ca^{2+}}{40 \text{ g}} = 2 \times 10^{-7} \, M \, Ca^{2+}$$

$$1 \text{ ppb } F^- \times \frac{1 \mu g}{1 \text{ L}} = \frac{1 \mu g \, F^-}{1 \text{ L}} = \frac{1 \times 10^{-6} \text{ g } F^-}{1 \text{ L}} \times \frac{1 \text{ mol } F^-}{19.0 \text{ g}} = 5 \times 10^{-8} \, M \, F^-$$

$Q = [Ca^{2+}][F^-]^2 = (2 \times 10^{-7})(5 \times 10^{-8})^2 = 5 \times 10^{-22}$

$5 \times 10^{-22} < 3.9 \times 10^{-11}$, $Q < K_{sp}$, no CaF_2 will precipitate

18 Chemistry of the Environment

Visualizing Concepts

18.1   *Analyze.* Given that one mole of an ideal gas at 1 atm and 298 K occupies 22.4 L, is the volume of one mole of ideal gas in the middle of the stratosphere greater than, equal to, or less than 22.4 L?

Plan. Consider the relationship between pressure, temperature, and volume of an ideal gas. Use Figure 18.1 to estimate the pressure and temperature in the middle of the stratosphere, and compare the two sets of temperature and pressure.

Solve. According to the ideal gas law, PV = nRT, so V = nRT/P. Since n and R are constant for this exercise, V is proportional to T/P.

(a)   The stratosphere ranges from 10 to 50 km, so the middle is at approximately 30 km. At this altitude, T ≈ 230 K, P ≈ 40 torr (from Figure 18.1). Since we are comparing T/P ratios, either atm or torr can be used as pressure units; we will use torr.

At sea level: T/P = 298 K/760 torr = 0.39

At 30 km: T/P = 230 K/40 torr = 5.75

The proportionality constant (T/P) is much greater at 30 km than sea level, so the volume of 1 mol of an ideal gas is greater at this altitude. The decrease in temperature at 30 km is more than offset by the substantial decrease in pressure.

(b)   Volume is proportional to T/P, not simply T. The relative volumes of one mole of an ideal gas at 50 km and 85 km depend on the temperature and pressure at the two altitudes. From Figure 18.1,

50 km: T ≈ 270 K, P ≈ 20 torr, T/P = 270 K/20 torr = 13.5

85 km: T ≈ 190 K, P < 0.01 torr, T/P = 190 K/0.01 torr = 19,000

Again, the slightly lower temperature at 85 km is more than offset by a much lower pressure. One mole of an ideal gas will occupy a much larger volume at 85 km than 50 km.

18.2   Molecules in the upper atmosphere tend to have multiple bonds because they have sufficiently high bond dissociation enthalpies (Table 8.4) to survive the incoming high energy radiation from the sun. According to Table 8.4, for the same two bonded atoms, multiple bonds have higher bond dissociation enthalpies than single bonds. Molecules with single bonds are likely to undergo photodissociation in the presence of the high energy, short wavelength solar radiation present in the upper atmosphere.

18.3   Ozone concentration varies with altitude because conditions favorable to ozone formation (and unfavorable to decomposition) vary with altitude. Formation and persistence of O_3 require the presence of O atoms, O_2 molecules, and energy carriers (M^*, usually N_2 or O_2). Above 60 km, there are too few O_2 molecules for significant O_3 formation. Below 30 km, there are too few O atoms. Between 30 km and 60 km, O_3 concentration varies depending on the concentrations of O, O_2 and M^*.

18.4   *Analyze.* Given granite, marble, bronze, and other solid materials, what observations and measurements indicate whether the material is appropriate for an outdoor sculpture? If the material changes (erodes) over time, what chemical processes are responsible?

   Plan. An appropriate material resists chemical and physical changes when exposed to environmental conditions. An inappropriate material undergoes chemical reactions with substances in the troposphere, degrading the structural strength of the material and the sculpture.   *Solve.*

   (a)   The appearance and mass of the material upon environmental exposure are both indicators of chemical and physical changes. If the appearance and mass of the material are unchanged after a period of time, the material is well-suited for the sculpture because it is inert to chemical and physical changes. Changes in the color or texture of the material's surface indicate that a chemical reaction has occurred, because a different substance with different properties has formed. A decrease in mass indicates that some of the material has been lost, either by chemical reaction or physical change. An increase in mass indicates corrosion. If the mass of the material is unchanged, it is probably inert to chemical and physical environmental changes and suitable for sculpture.

   (b)   The two main chemical processes that lead to erosion are reaction with acid rain and corrosion or air oxidation, which is encouraged by acid conditions (see Section 20.8).

   Acid rain is primarily H_2SO_3 and/or H_2SO_4, which reacts directly with carbonate minerals such as marble and limestone. Acidic conditions created by acid rain encourage corrosion of metals such as iron, steel, and bronze. Corrosion produces metal oxides which may or may not cling to the surface of the material. If the oxides are washed away, the material will lose mass after corrosion. Physical erosion due to the effects of wind and rain on soft materials such as sandstone also causes mass to decrease.

18.5 $CO_2(g)$ dissolves in seawater to form $H_2CO_3(aq)$. The basic pH of the ocean encourages ionization of $H_2CO_3(aq)$ fo form $HCO_3^-(aq)$ and $CO_3^{2-}(aq)$. Under the correct conditions, carbon is removed from the ocean as $CaCO_3(s)$ (sea shells, coral, chalk cliffs). As carbon is removed, more $CO_2(g)$ dissolves to maintain the balance of complex and interacting acid-base and precipitation equilibria.

18.6   *Analyze/Plan.* Explain how an ion-exchange column "softens" water. See the Closer Look box on "Water Softening" in Section 18.6.

Solve. The plastic beads in an ion-exchange column contain covalently bound anionic groups such as R–COO⁻ and R–SO₃⁻. These groups have Na^+ cations associated with them for charge balance. When "hard" water containing Ca^{2+} and other divalent cations passes over the beads, the 2+ cations are attracted to the anionic groups and Na^+ is displaced. The higher charge on the divalent cations leads to greater electrostatic attractions, which promote the cation exchange. The "soft" water that comes out of the column contains two Na^+ ions in place of each divalent cation, mostly Ca^{2+} and Mg^{2+}, that remains in the column associated with the plastic beads.

18.7   The guiding principle of green chemistry is that "an ounce of prevention is worth a pound of cure." Processes should be designed to minimize or eliminate solvents and waste, generate nontoxic waste, be energy efficient, employ renewable starting materials, and take advantage of catalysts that enable the use of safe and common reagents.

18.8   Some of the missing CO_2 is absorbed by "land plants" (vegetation other than trees) and incorporated into the soil. Soil is the largest land-based carbon reservoir. The amount of carbon-storing capacity of soil is affected by erosion, soil fertility, and other complex factors. For more details, search the internet for "carbon budget."

Earth's Atmosphere

18.9   (a)   The temperature profile of the atmosphere (Figure 18.1) is the basis of its division into regions. The center of each peak or trough in the temperature profile corresponds to a new region.

(b)   Troposphere, 0–12 km; stratosphere, 12–50 km; mesosphere, 50–85 km; thermosphere, 85–110 km.

18.10   (a)   Boundaries between regions of the atmosphere are at maxima and minima (peaks and valleys) in the atmospheric temperature profile. For example, in the troposphere, temperature decreases with altitude, while in the stratosphere, it increases with altitude. The temperature minimum is the tropopause boundary.

(b)   Atmospheric pressure in the troposphere ranges from 1.0 atm to 0.4 atm, while pressure in the stratosphere ranges from 0.4 atm to 0.001 atm. Gas density (g/L) is directly proportional to pressure. The much lower density of the stratosphere means it has the smaller mass, despite having a larger volume than the troposphere.

18.11   *Analyze/Plan.* Given O_3 concentration in ppm, calculate partial pressure. Use the definition of ppm to get mol fraction O_3. For gases mole fraction = pressure fraction;

$P_{O_3} = \chi_{O_3} \cdot P_{atm}$

$$0.441 \text{ ppm } O_3 = \frac{0.441 \text{ mol } O_3}{1 \times 10^6 \text{ mol air}} = 4.41 \times 10^{-7} = \chi_{O_3}$$

Solve. $P_{O_3} = \chi_{O_3} \cdot P_{atm} = 4.41 \times 10^{-7} (0.67 \text{ atm}) = 2.955 \times 10^{-7} = 3.0 \times 10^{-7} \text{ atm}$

18.12 $P_{Ar} = \chi_{Ar} \cdot P_{atm}$; $P_{Ar} = 0.00934 (98.6 \text{ kPa}) = 0.921 \text{ kPa}$; $0.921 \text{ kPa} \times \dfrac{760 \text{ torr}}{101.325 \text{ kPa}} = 6.91 \text{ torr}$

$$P_{CO_2} = \chi_{CO_2} \bullet P_{atm}; \; P_{CO_2} = 0.000375\,(98.6\text{ kPa}) = 0.0370\text{ kPa}; \; 0.0350\text{ kPa} \times \frac{760\text{ torr}}{101.325\text{ kPa}}$$
$$= 0.277\text{ torr}$$

18.13   *Analyze/Plan.*   Given CO concentration in ppm, calculate number of CO molecules in 1.0 L air at given conditions. $\text{ppm CO} \rightarrow \chi_{O_3} \rightarrow \text{atm CO} \rightarrow \text{mol CO} \rightarrow \text{molecules CO}$. Use the ideal gas law to change atm CO to mol CO, then Avogadro's number to get molecules.   *Solve.*

$$3.4\text{ ppm CO} = \frac{3.4\text{ mol CO}}{1 \times 10^6\text{ mol air}} = 3.4 \times 10^{-6} = \chi_{CO}$$

$$P_{CO} = \chi_{CO} \bullet P_{atm} = 3.4 \times 10^{-6} \times 755\text{ torr} \times \frac{1\text{ atm}}{760\text{ torr}} = 3.378 \times 10^{-6} = 3.4 \times 10^{-6}\text{ atm}$$

$$n_{CO} = \frac{P_{CO}V}{RT} = \frac{3.378 \times 10^{-6}\text{ atm} \times 1.0\text{ L}}{295\text{ K}} \times \frac{K \bullet mol}{0.08206\text{ L} \bullet \text{atm}} = 1.395 \times 10^{-7} = 1.4 \times 10^{-7}\text{ mol CO}$$

$$1.395 \times 10^{-7}\text{ mol CO} \times \frac{6.022 \times 10^{23}\text{ molecules}}{\text{mol}} = 8.402 \times 10^{16} = 8.4 \times 10^{16}\text{ molecules CO}$$

18.14   (a) $\text{ppm Ne} = \text{mol Ne}/1 \times 10^6\text{ mol air}; \; \chi_{Ne} = 1.818 \times 10^{-5}\text{ mol Ne}/\text{mol air}$

$$\frac{1.818 \times 10^{-5}\text{ mol Ne}}{1\text{ mol air}} = \frac{x\text{ mol Ne}}{1 \times 10^6\text{ mol air}}; \; x = 18.18\text{ ppm Ne}$$

(b) $P_{Ne} = \chi_{Ne} \bullet P_{atm} = 1.818 \times 10^{-5} \times 743\text{ torr} \times \dfrac{1\text{ atm}}{760\text{ torr}} = 1.7773 \times 10^{-5}$

$$= 1.78 \times 10^{-5}\text{ atm}$$

$$T = 300°C + 273 = 573\text{ K}$$

$$\frac{n_{Ne}}{V} = \frac{P_{Ne}}{RT} = \frac{1.7773 \times 10^{-5}\text{ atm}}{573\text{ K}} \times \frac{K \bullet mol}{0.08206\text{ L} \bullet \text{atm}} = 3.7799 \times 10^{-7} = 3.78 \times 10^{-7}\text{ mol}/\text{L}$$

$$\frac{3.7799 \times 10^{-7}\text{ mol Ne}}{\text{L}} \times \frac{6.022 \times 10^{23}\text{ atoms}}{\text{mol}} = 2.2763 \times 10^{17}$$
$$= 2.28 \times 10^{17}\text{ Ne atoms}/\text{L}$$

The Upper Atmosphere; Ozone

18.15   *Analyze/Plan.* Given bond dissociation energy in kJ/mol, calculate the wavelength of a single photon that will rupture a C–Br bond. $\text{kJ/mol} \rightarrow \text{J/molecule}$. $\lambda = hc/E$. ($\lambda = hc/E$ describes the energy/wavelength relationship of a single photon.)   *Solve.*

$$\frac{210 \times 10^3\text{ J}}{1\text{ mol}} \times \frac{1\text{ mol}}{6.022 \times 10^{23}\text{ molecules}} = 3.487 \times 10^{-19} = 3.49 \times 10^{-19}\text{ J/molecule}$$

$\lambda = c/\nu$ We also have that $E = h\nu$, so $\nu = E/h$. Thus,

$$\lambda = \frac{hc}{E} = \frac{(6.626 \times 10^{-34}\text{ J} \bullet \text{sec})(3.00 \times 10^8\text{ m/sec})}{3.487 \times 10^{-19}\text{ J}} = 5.70 \times 10^{-7}\text{ m} = 570\text{ nm}$$

18.16 $\dfrac{339 \times 10^3 \text{ J}}{1 \text{ mol}} \times \dfrac{1 \text{ mol}}{6.022 \times 10^{23} \text{ molecules}} = 5.6294 \times 10^{-19} = 5.63 \times 10^{-19}$ J/molecule

$\lambda = \dfrac{hc}{E} = \dfrac{(6.626 \times 10^{-34} \text{ J} \cdot \text{sec})(3.00 \times 10^8 \text{ m/sec})}{5.6294 \times 10^{-19} \text{ J}} = 3.53 \times 10^{-7}$ m $= 353$ nm

$\dfrac{293 \times 10^3 \text{ J}}{1 \text{ mol}} \times \dfrac{1 \text{ mol}}{6.022 \times 10^{23} \text{ molecules}} = 4.8655 \times 10^{-19} = 4.87 \times 10^{-19}$ J/molecule

$\lambda = \dfrac{(6.626 \times 10^{-34} \text{ J} \cdot \text{sec})(3.00 \times 10^8 \text{ m/sec})}{4.8655 \times 10^{-19} \text{ J}} = 4.09 \times 10^{-7}$ m $= 409$ nm

Photons of wavelengths longer than 409 nm cannot cause rupture of the C–Cl bond in either CF_3Cl or CCl_4. Photons with wavelengths between 409 and 353 nm can cause C–Cl bond rupture in CCl_4, but not in CF_3Cl.

18.17   (a)   *Photodissociation* is cleavage of the $O{=}O$ bond such that two neutral O atoms are produced: $O_2(g) \rightarrow 2O(g)$

         Photoionization is absorption of a photon with sufficient energy to eject an electron from an O_2 molecule: $O_2(g) + h\nu \rightarrow O_2{}^+ + e^-$

     (b)   Photoionization of O_2 requires 1205 kJ/mol. Photodissociation requires only 495 kJ/mol. At lower elevations, solar radiation with wavelengths corresponding to 1205 kJ/mol or shorter has already been absorbed, while the longer wavelength radiation has passed through relatively well. Below 90 km, the increased concentration of O_2 and the availability of longer wavelength radiation cause the photodissociation process to dominate.

18.18   Photodissociation of N_2 is relatively unimportant compared to photodissociation of O_2 for two reasons. The bond dissociation energy of N_2, 941 kJ/mol, is much higher than that of O_2, 495 kJ/mol. Photons with a wavelength short enough to photodissociate N_2 are not as abundant as the ultraviolet photons that lead to photodissociation of O_2. Also, N_2 does not absorb these photons as readily as O_2 so even if a short-wavelength photon is available, it may not be absorbed by an N_2 molecule.

18.19   A *hydrofluorocarbon* is a compound that contains hydrogen, fluorine, and carbon; it contains hydrogen in place of chlorine. HFCs are potentially less harmful than CFCs because photodissociation does not produce Cl atoms, which catalyze the destruction of ozone.

18.20   32 e⁻, 16 e⁻ pr

CFC–11, $CFCl_3$, contains C–Cl bonds that can be cleaved by UV light in the stratosphere to produce Cl atoms. It is chlorine in atomic form that catalyzes the destruction of stratospheric ozone. CFC–11 is chemically inert and resists decomposition in the troposphere, so that it eventually reaches the stratosphere in molecular form.

18.21   (a)   In order to catalyze ozone depletion, the halogen must be present as single halogen atoms. These halogen atoms are produced in the stratosphere by photo-dissociation of a carbon-halogen bond. According to Table 8.4, the C–F average bond dissociation energy is 485 kJ/mol, while that of C–Cl is 328 kJ/mol. The C–F bond requires more energy for dissociation and is not readily cleaved by the available wavelengths of UV light.

   (b)   Chlorine is present as chlorine atoms and chlorine oxide molecules, Cl and ClO.

18.22   Yes. Assuming $CFBr_3$ reaches the stratosphere intact, it contains C–Br bonds that are even more susceptible to cleavage by UV light than C–Cl bonds. According to Table 8.4, the average C–Br bond dissociation energy is 273 kJ/mol, compared to 328 kJ/mol for C–Cl bonds. Once in atomic form, Br atoms catalyze the destruction of ozone by a mechanism similar to that of Cl atoms.

Chemistry of the Troposphere

18.23   (a)   Methane, CH_4, arises from decomposition of organic matter by certain microorganisms; it also escapes from underground gas deposits.

   (b) SO_2 is released in volcanic gases, and also is produced by bacterial action on decomposing vegetable and animal matter.

   (c)   Nitric oxide, NO, results from oxidation of decomposing organic matter, and is formed in lightning flashes.

   (d)   CO is a possible product of some vegetable matter decay.

18.24   Rainwater is naturally acidic due to the presence of $CO_2(g)$ in the atmosphere. All oxides of nonmetals produce acidic solutions when dissolved in water. Even in the absence of polluting gases such as SO_2, SO_3, NO, and NO_2, CO_2 causes rainwater to be acidic. The important equilibria are:

$$CO_2(g) + H_2O(l) \; \rightleftharpoons \; H_2CO_3(aq) \; \rightleftharpoons \; H^+(aq) + HCO_3{}^-(aq).$$

18.25   (a)   Acid rain is primarily $H_2SO_4(aq)$.

$$H_2SO_4(aq) + CaCO_3(s) \rightarrow CaSO_4(s) + H_2O(l) + CO_2(g)$$

   (b)   The $CaSO_4(s)$ would be much less reactive with acidic solution, since it would require a strongly acidic solution to shift the relevant equilibrium to the right.

$$CaSO_4(s) + 2H^+(aq) \; \rightleftharpoons \; Ca^{2+}(aq) + 2HSO_4{}^-(aq)$$

   Note, however, that $CaSO_4(s)$ is brittle and easily dislodged; it provides none of the structural strength of limestone.

18.26   (a) $Fe(s) + 2H^+(aq) \rightarrow Fe^{2-}(aq) + H_2(g)$

   (b)   No. Silver is a "noble" metal. It is relatively resistant to oxidation, and much more resistant than iron. In Table 4.5, The Activity Series of Metals in Aqueous Solution, Ag is much, much lower than Fe and it is below hydrogen, while Fe is above hydrogen. This means that Fe is susceptible to oxidation by acid, while Ag is not.

18.27 *Analyze/Plan.* Given wavelength of a photon, place it in the electromagnetic spectrum, calculate its energy in kJ/mol, and compare it to an average bond dissociation energy. Use Figure 6.4; $E(J/photon) = hc/\lambda$. $J/photon \rightarrow kJ/mol$. *Solve.*

(a) Ultraviolet (Figure 6.4)

(b) $E_{photon} = hc/\lambda = \dfrac{6.626 \times 10^{-34} \text{ J} \bullet \text{s} \times 3.00 \times 10^{8} \text{ m/s}}{335 \times 10^{-9} \text{ m}} = 5.934 \times 10^{-19}$

$= 5.93 \times 10^{-19} \text{ J/photon}$

$\dfrac{5.934 \times 10^{-19} \text{ J}}{1 \text{ photon}} \times \dfrac{6.022 \times 10^{23} \text{ photons}}{1 \text{ mol}} \times \dfrac{1 \text{ kJ}}{1000 \text{ J}} = 357 \text{ kJ/mol}$

(c) The average C–H bond energy from Table 8.4 is 413 kJ/mol. The energy calculated in part (b), 357 kJ/mol, is the energy required to break 1 mol of C–H bonds in formaldehyde, CH_2O. The C–H bond energy in CH_2O must be less than the "average" C–H bond energy.

(d)
$$H\!-\!\overset{\displaystyle :\!O\!:}{\overset{\|}{C}}\!-\!H + h\nu \longrightarrow H\!-\!\overset{\displaystyle :\!O\!:}{\overset{\|}{C}}\!\cdot + H$$

18.28 (a) Visible (Figure 6.4)

(b) $E_{photon} = hc/\lambda = \dfrac{6.626 \times 10^{-34} \text{ J} \bullet \text{s} \times 3.00 \times 10^{8} \text{ m/s}}{420 \times 10^{-9} \text{ m}} = 4.733 \times 10^{-19}$

$= 4.73 \times 10^{-19} \text{ J/photon}$

$\dfrac{4.733 \times 10^{-19} \text{ J}}{1 \text{ photon}} \times \dfrac{6.022 \times 10^{23} \text{ photons}}{1 \text{ mol}} \times \dfrac{1 \text{ kJ}}{1000 \text{ J}} = 285 \text{ kJ/mol}$

(c) $\ddot{O}\!=\!\ddot{N}\!-\!\ddot{O}\!: + h\nu \longrightarrow \ddot{O}\!=\!\ddot{N}\!\cdot + :\ddot{O}\!\cdot$

18.29 Most of the energy entering the atmosphere from the sun is in the form of visible radiation, while most of the energy leaving the earth is in the form of infrared radiation. CO_2 is transparent to the incoming visible radiation, but absorbs the outgoing infrared radiation.

18.30 (a) A *greenhouse gas* absorbs energy in the 10,000–30,000 nm or infrared region. It absorbs wavelengths of radiation emitted by earth and returns it as heat. A non-greenhouse gas is transparent to radiation in this wavelength range.

(b) Ar(g) is monatomic, while $CH_4(g)$ contains 4 C–H bonds. Infrared radiation has insufficient energy to cause electron transitions or bond cleavage; but it has an appropriate amount of energy to cause molecular deformations, bond stretching, and angle bending. Monatomic gases such as Ar cannot "use" infrared radiation and are transparent to it.

The World Ocean

18.31 *Analyze/Plan.* Given salinity and density, calculate molarity. A salinity of 5.6 denotes that there are 5.6 g of dry salt per kg of water. *Solve.*

$\dfrac{5.6 \text{ g NaCl}}{1 \text{ kg soln}} \times \dfrac{1.03 \text{ kg soln}}{1 \text{ L soln}} \times \dfrac{1 \text{ mol NaCl}}{58.44 \text{ g NaCl}} \times \dfrac{1 \text{ mol Na}^+}{1 \text{ mol NaCl}} = 0.0987 = 0.099 \text{ M Na}^+$

18.32   If the phosphorous is present as phosphate, there is a 1:1 ratio between the molarity of phosphorus and molarity of phosphate. Thus, we can calculate the molarity based on the given mass of P.

$$\frac{0.07 \text{ g P}}{1 \times 10^6 \text{ g H}_2\text{O}} \times \frac{1 \text{ mol P}}{31 \text{ g P}} \times \frac{1 \text{ mol PO}_4^{3-}}{1 \text{ mol P}} \times \frac{1 \times 10^3 \text{ g H}_2\text{O}}{1 \text{ L H}_2\text{O}} = 2.26 \times 10^{-6} = 2 \times 10^{-6} \text{ } M \text{ PO}_4^{3-}$$

18.33   *Analyze/Plan.* g $Mg(OH)_2 \rightarrow$ mol $Mg(OH)_2 \rightarrow$ mol ratio $\rightarrow$ mol CaO $\rightarrow$ g CaO.   *Solve.*

$$1000 \text{ lb Mg(OH)}_2 \times \frac{453.6 \text{ g}}{\text{lb}} \times \frac{1 \text{ mol Mg(OH)}_2}{58.33 \text{ g Mg(OH)}_2} \times \frac{1 \text{ mol CaO}}{1 \text{ mol Mg(OH)}_2} \times \frac{56.08 \text{ g CaO}}{1 \text{ mol CaO}}$$

$$= 4.361 \times 10^5 \text{ g CaO}$$

18.34   0.05 ppb Au = 0.05 g Au/1×10^9 g seawater

$$\$1{,}000{,}000 \times \frac{1 \text{ oz Au}}{\$400} \times \frac{1 \text{ lb}}{16 \text{ oz}} \times \frac{453.6 \text{ g}}{1 \text{ lb}} = 7.0875 \times 10^4 \text{ g} = 7.09 \times 10^4 \text{ g Au needed}$$

$$7.0875 \times 10^4 \text{ g Au} \times \frac{1 \times 10^9 \text{ g seawater}}{0.05 \text{ g Au}} \times \frac{1 \text{ mL seawater}}{1.03 \text{ g seawater}} \times \frac{1 \text{ L}}{1000 \text{ mL}} = 1.3762 \times 10^{12}$$

$$= 1 \times 10^{12} \text{ L seawater}$$

1×10^{12} L seawater are needed if the process is 100% efficient; since it is only 50% efficient, twice as much seawater is needed.

$1.3762 \times 10^{12} \times 2 = 2.7524 \times 10^{12} = 3 \times 10^{12}$ L seawater

Note that the 1 sig fig in 0.05 ppb Au limits the precision of the calculation.

18.35   *Analyze/Plan.* Given temperature and the concentration difference between the two solutions, (ΔM = 0.22 – 0.01 = 0.21 M), calculate the minimum pressure for reverse osmosis. Use the relationship $\pi = MRT$ from Section 13.5. This is the pressure required to halt osmosis from the more dilute (0.01 M) to the more concentrated (0.22 M) solution. Slightly more pressure will initiate reverse osmosis.   *Solve.*

$$\pi = \Delta MRT = \frac{0.21 \text{ mol}}{\text{L}} \times \frac{0.08206 \text{ L} \bullet \text{atm}}{\text{mol} \bullet \text{K}} \times 298 \text{ K} = 5.135 = 5.1 \text{ atm}$$

The minimum pressure required to initiate reverse osmosis is greater than 5.1 atm.

18.36   Calculate the total ion concentration of sea water by summing the molarities given in Table 18.6. Then use $\pi = \Delta MRT$ to calculate pressure.

$M_{\text{total}} = 0.55 + 0.47 + 0.028 + 0.054 + 0.010 + 0.010 + 2.3 \times 10^{-3} + 8.3 \times 10^{-4}$

$\quad\quad + 4.3 \times 10^{-4} + 9.1 \times 10^{-5} + 7.0 \times 10^{-5} = 1.1257 = 1.13 \text{ } M$

$$\pi = \frac{(1.1257 - 0.02) \text{ mol}}{\text{L}} \times \frac{0.08206 \text{ L} \times \text{atm}}{\text{mol} \bullet \text{K}} \times 305 \text{ K} = 27.674 = 27.7 \text{ atm}$$

Check. The largest numbers in the molarity sum have 2 decimal places, so M_{total} has 2 decimal places and 3 sig figs. ΔM also has 2 decimal places and 3 sig figs so the calculated pressure has 3 sig figs. Units are correct.

18 Chemistry of the Environment   Solutions to Exercises

Fresh Water

18.37   *Analyze/Plan.* Under aerobic conditions, excess oxygen is present and decomposition leads to oxidized products, the element in its maximum oxidation state combined with oxygen. Under anaerobic conditions, little or no oxygen is present so decomposition leads to reduced products, the element in its minimum oxidation state combined with hydrogen.   *Solve.*

   (a) $CO_2, HCO_3^-, H_2O, SO_4^{2-}, NO_3^-, HPO_4^{2-}, H_2PO_4^-$.

   (b) $CH_4(g), H_2S(g), NH_3(g), PH_3(g)$

18.38   (a)   Decomposition of organic matter by aerobic bacteria depletes dissolved O_2. A low dissolved oxygen concentration indicates the presence of organic pollutants.

   (b)   According to Section 13.3, the solubility of $O_2(g)$ (or any gas) in water decreases with increasing temperature.

18.39   *Analyze/Plan.* Given the balanced equation, calculate the amount of one reactant required to react exactly with a certain amount of the other reactants. Solve the stoichiometry problem. $g\ C_{18}H_{29}O_3S^- \rightarrow mol \rightarrow mol\ ratio \rightarrow mol\ O_2 \rightarrow g\ O_2$.   *Solve.*

$$1.0\ g\ C_{18}H_{29}O_3S^- \times \frac{1\ mol\ C_{18}H_{29}O_3S^-}{325\ g\ C_{18}H_{29}O_3S^-} \times \frac{51\ mol\ O_2}{2\ mol\ C_{18}H_{29}O_3S^-} \times \frac{32.0\ g\ O_2}{1\ mol\ O_2} = 2.5\ g\ O_2$$

Notice that the mass of O_2 required is 2.5 times greater than the mass of biodegradable material.

18.40 $120,000\ persons \times \dfrac{59\ g\ O_2}{1\ person} \times \dfrac{1\times10^6\ g\ H_2O}{9\ g\ O_2} \times \dfrac{1\ L\ H_2O}{1\times10^3\ g\ H_2O} = 7.9\times10^8 = 8\times10^8\ L\ H_2O$

18.41   *Analyze/Plan.* Slaked lime is $Ca(OH)_2(s)$. The reaction is metathesis.   *Solve.*

$Mg^{2+}(aq) + Ca(OH)_2(s) \rightarrow Mg(OH)_2(s) + Ca^{2+}(aq)$

The excess $Ca^{2+}(aq)$ is removed as $CaCO_3$ by naturally occurring bicarbonate or added Na_2CO_3.

18.42   (a) $Ca^{2+}, Mg^{2+}, Fe^{2+}$

   (b)   Divalent cations (ions with 2+ charges) contribute to water hardness. These ions react with soap to form scum on surfaces or leave undesirable deposits on surfaces, particularly inside pipes, upon heating.

18.43   *Analyze/Plan.* Given $[Ca^{2+}]$ and $[HCO_3^-]$ calculate mole $Ca(OH)_2$ and Na_2CO_3 needed to remove the Ca^{2+} and HCO_3^-. Consider the chemical equations and reaction stoichiometry in the stepwise process.   *Solve.*

$Ca(OH)_2$ is added to remove Ca^{2+} as $CaCO_3(s)$, and Na_2CO_3 removes the remaining Ca^{2+}.

   $Ca^{2+}(aq) + 2HCO_3^-(aq) + [Ca^{2+}(aq) + 2OH^-(aq)] \rightarrow 2CaCO_3(s) + 2H_2O(l)$.

One mole $Ca(OH)_2$ is needed for each 2 moles of $HCO_3^-(aq)$ present.

$$\frac{7.0\times10^{-4}\ mol\ HCO_3^-}{L} \times \frac{1\ mol\ Ca(OH)_2}{2\ mol\ HCO_3^-} \times 1.200\times10^3\ L\ H_2O = 0.42\ mol\ Ca(OH)_2$$

$$1.200 \times 10^3 \text{ L H}_2\text{O} \times \frac{5.0 \times 10^{-4} \text{ mol Ca}^{2+}}{\text{L}} = 0.60 \text{ mol Ca}^{2+}\text{(aq) total}$$

0.42 mol Ca(OH)$_2$ removes 0.42 mol of the 0.60 mol Ca^{2+}(aq) in the sample. This leaves 0.18 mol Ca^{2+}(aq) to be removed by Na$_2$CO$_3$.

$$\text{Ca}^{2+}\text{(aq)} + \text{Na}_2\text{CO}_3\text{(aq)} \rightarrow \text{CaCO}_3\text{(s)} + 2\text{Na}^+\text{(aq)}$$

0.18 mol of Na$_2$CO$_3$ is needed to remove the remaining Ca^{2+}(aq).

18.44   Ca(OH)$_2$ is added to remove Ca^{2+} as CaCO$_3$(s), and Na$_2$CO$_3$ removes the remaining Ca^{2+}.

$$\text{Ca}^{+2}\text{(aq)} + 2\text{HCO}_3^-\text{(aq)} + [\text{Ca}^{2+}\text{(aq)} + 2\text{OH}^-\text{(aq)}] \rightarrow 2\text{CaCO}_3\text{(s)} + 2\text{H}_2\text{O(l)}$$

One mole Ca(OH)$_2$ is needed for each 2 moles of HCO$_3^-$(aq) present.

$$5.0 \times 10^7 \text{ L H}_2\text{O} \times \frac{1.7 \times 10^{-3} \text{ mol HCO}_3^-}{1 \text{ L H}_2\text{O}} \times \frac{1 \text{ mol Ca(OH)}_2}{2 \text{ mol HCO}_3^-} \times \frac{74 \text{ g Ca(OH)}_2}{1 \text{ mol Ca(OH)}_2}$$
$$= 3.1 \times 10^6 \text{ g Ca(OH)}_2$$

Half of the native HCO$_3^-$ precipitates the added Ca^{2+} so this operation reduces the Ca^{2+} concentration from 5.7×10^{-3} M to $(5.7 \times 10^{-3} - 8.5 \times 10^{-4})$ M = 4.85×10^{-3} = 4.9×10^{-3} M. Next we must add sufficient Na$_2$CO$_3$ to further reduce [Ca^{2+}] to 1.1×10^{-3} M (20% of the original [Ca^{2+}]). We thus need to reduce [Ca^{2+}] by $(4.85 \times 10^{-3} - 1.1 \times 10^{-3})$ M = 3.75×10^{-3} = 3.8×10^{-3} M

$$\text{Ca}^{2+}\text{(aq)} + \text{CO}_3^{-2}\text{(aq)} \rightarrow \text{CaCO}_3\text{(s)}.$$

$$5.0 \times 10^7 \text{ L H}_2\text{O} \times \frac{3.75 \times 10^{-3} \text{ mol Ca}^{2+}}{1 \text{ L H}_2\text{O}} \times \frac{1 \text{ mol Na}_2\text{CO}_3}{1 \text{ mol Ca}^{2+}} \times \frac{106 \text{ g Na}_2\text{CO}_3}{1 \text{ mol Na}_2\text{CO}_3}$$
$$= 2.0 \times 10^7 \text{ g Na}_2\text{CO}_3$$

18.45 $4\text{FeSO}_4\text{(aq)} + \text{O}_2\text{(aq)} + 2\text{H}_2\text{O(l)} \rightarrow 4\text{Fe}^{3+}\text{(aq)} + 4\text{OH}^-\text{(aq)} + 4\text{SO}_4^{2-}\text{(aq)}$

SO$_4^{2-}$ is a spectator, so the net ionic equation is

$4\text{Fe}^{2+}\text{(aq)} + \text{O}_2\text{(aq)} + 2\text{H}_2\text{O(l)} \rightarrow 4\text{Fe}^{3+}\text{(aq)} + 4\text{OH}^-\text{(aq)}$.

$\text{Fe}^{3+}\text{(aq)} + 3\text{HCO}_3^-\text{(aq)} \rightarrow \text{Fe(OH)}_3\text{(s)} + 3\text{CO}_2\text{(g)}$

In this reaction, Fe^{3+} acts as a Lewis acid, and HCO$_3^-$ acts as a Lewis base.

18.46   Al$_2$(SO$_4$)$_3$ is a typical coagulant in municipal water purification. It reacts with OH$^-$ in a slightly basic solution to form a gelatinous precipitate that occludes very small particles and bacteria. The precipitate settles slowly and is removed by sand filtration.

Properties of Al$_2$(SO$_4$)$_3$ and other useful coagulants are:

• They react with low concentrations of OH$^-$(aq). That is, K$_{sp}$ of the hydroxide precipitate is very small. The capacity to form a hydroxide precipitates means that no extra salts must be added to form the precipitate. Also, the [OH$^-$] can be easily adjusted by Ca(OH)$_2$ and other reagents that are part of the purification process.

• The hydroxide precipitate is composed of very small, evenly dispersed particles that do not settle quickly. This is required to remove very small bacteria and viruses from all parts of the liquid, not just the sites of solid formation.

Green Chemistry

18.47 The fewer steps in a process, the less waste (solvents as well as unusable by-products) is generated. It is probably true that a process with fewer steps requires less energy at the site of the process, and it is certainly true that the less waste the process generates, the less energy is required to clean or dispose of the waste.

18.48 Catalysts increase the rate of a reaction by lowering activation energy, E_a. For an uncatalyzed reaction that requires extreme temperatures and pressures to generate product at a viable rate, finding a suitable catalyst reduces the required temperature and/or pressure, which reduces the amount of energy used to run the process. A catalyst can also increase rate of production, which would reduce the net time and thus energy required to generate a certain amount of product.

18.49 (a)

(b)
- It is better to prevent waste than to treat it. The alternative process eliminates production of 3-chlorobenzoic acid by-product, chlorine-containing waste that must be treated.

- Produce as little, nontoxic waste as possible. The by-product of the alternative process is nontoxic water. The low molar mass of water means that a small amount of "waste" is generated.

- Chemical processes should be efficient. The alternative process is catalyzed, which could mean that the process will be more energy efficient than the Baeyer-Villiger reaction (see Solution 18.44).

- Raw materials should be renewable. The catalyst can be recovered from the reaction mixture and reused. We don't have information about solvents or other auxiliary substances.

18.50
- In either solvent, the reaction is catalyzed, which usually leads to decreased processing temperatures and times, and greater energy efficiency.

- $scCO_2$ is the preferred solvent. It achieves maximum conversion much faster than CH_2Cl_2 solvent. $scCO_2$ reduces processing time, temperature, and energy requirements. It also means fewer unwanted by-products to be separated and processed. While use of $scCO_2$ increases the amount of a greenhouse gas released to the environment, it eliminates use of CH_2Cl_2, which is implicated in stratospheric ozone depletion. Use of $scCO_2$ rather than CH_2Cl_2 is a good green trade-off.

Additional Exercises

18.51 (a) *Acid rain* is rain with a larger $[H^+]$ and thus a lower pH than expected. The additional H^+ is produced by the dissolution of sulfur and nitrogen oxides such as $SO_3(g)$ and $NO_2(g)$ in rain droplets to form sulfuric and nitric acid, $H_2SO_4(aq)$ and $HNO_3(aq)$.

(b) A *greenhouse gas* absorbs infrared or "heat" radiation emitted from the earth's surface and serves to maintain a relatively constant temperature on the surface. A significant increase in the amount of atmospheric CO_2 (from burning fossil fuels and other sources) could cause a corresponding increase in the average surface temperature and drastically change the global climate.

(c) *Photochemical smog* is an unpleasant collection of atmospheric pollutants initiated by photochemical dissociation of NO_2 to form NO and O atoms. The major components are $NO(g)$, $NO_2(g)$, $CO(g)$, and unburned hydrocarbons, all produced by automobile engines, and $O_3(g)$, ozone.

(d) *Ozone depletion* is the reduction of O_3 concentration in the stratosphere, most notably over Antarctica. It is caused by reactions between O_3 and Cl atoms originating from chlorofluorocarbons (CFC's), CF_xCl_{4-x}. Depletion of the ozone layer allows damaging ultraviolet radiation disruptive to the plant and animal life in our ecosystem to reach earth.

18.52 MM_{avg} at the surface = $83.8(0.17) + 16.0(0.38) + 32.0(0.45) = 34.73 = 35$ g/mol.

Next, calculate the percentage composition at 200 km. The fractions can be "normalized" by saying that the 0.45 fraction of O_2 is converted into **two** 0.45 fractions of O atoms, then dividing by the total fractions, $0.17 + 0.38 + 0.45 + 0.45 = 1.45$:

$$MM_{avg} = \frac{83.8(0.17) + 16.0(0.38) + 16.0(0.90)}{1.45} = 23.95 = 24 \text{ g/mol}$$

18.53   Stratospheric ozone is formed and destroyed in a cycle of chemical reactions. The decomposition of O_3 to O_2 and O produces oxygen atoms, an essential ingredient for the production of ozone. While single O_3 molecules exist for only a few seconds, new O_3 molecules are constantly reformed. This cyclic process ensures a finite concentration of O_3 in the stratosphere available to absorb ultraviolet radiation. (This explanation assumes that the cycle is not disrupted by outside agents such as CFCs.)

18.54

$$
\begin{aligned}
&2[Cl(g) + O_3(g) \rightarrow ClO(g) + O_2(g)] &&[18.7]\\
&2Cl(g) + 2O_3(g) \rightarrow 2ClO(g) + 2O_2(g)\\
&\underline{2ClO(g) \qquad\qquad \rightarrow O_2(g) + 2Cl(g)} &&[18.9]\\
&2Cl(g) + 2O_3(g) + 2ClO(g) \rightarrow 2ClO(g) + 3O_2(g) + 2Cl(g)\\
&\qquad\qquad 2O_3(g) \xrightarrow{\ Cl\ } 3O_2(g) &&[18.10]
\end{aligned}
$$

Note that $Cl(g)$ fits the definition of a catalyst in this reaction.

18.55   (a) The production of Cl atoms in the stratosphere is the result of the photodissociation of a C—H bond in the chlorofluorocarbon molecule.

$$CF_2Cl_2(g) \xrightarrow{\ h\nu\ } CF_2Cl(g) + Cl(g)$$

According to Table 8.4, the bond dissociation energy of a C—Br bond is 276 kJ/mol, while the value for a C—H bond is 328 kJ/mol. Photodissociation of $CBrF_3$ to form Br atoms requires less energy than the production of Cl atoms and should occur readily in the stratosphere.

(b) $CBrF_3(g) \xrightarrow{h\nu} CF_3(g) + Br(g)$

$Br(g) + O_3(g) \rightarrow BrO(g) + O_2(g)$

Also, under certain conditions

$BrO(g) + BrO(g) \rightarrow Br_2O_2(g)$

$Br_2O_2(g) + h\nu \rightarrow O_2(g) + 2Br(g)$

18.56   (a) $\cdot \ddot{O}{-}H$

(b) HNO_3 is a major component in acid rain.

(c)   While it removes CO, the reaction produces NO_2. The photodissociation of NO_2 to form O atoms is the first step in the formation of tropospheric ozone and photochemical smog.

(d)   Again, NO_2 is the initiator of photochemical smog. Also, methoxyl radical, OCH_3, is a reactive species capable of initiating other undesirable reactions.

18.57   From section 18.4:

$N_2(g) + O_2(g) \rightleftharpoons 2NO(g)$ $\Delta H = +180.8$ kJ   [18.11]

$2 NO(g) + O_2(g) \rightleftharpoons 2NO_2(g)$ $\Delta H = -113.1$ kJ   [18.12]

In an endothermic reaction, heat is a reactant. As the temperature of the reaction increases, the addition of heat favors formation of products and the value of K increases. The reverse is true for exothermic reactions; as temperature increases, the value of K decreases. Thus, K for reaction [18.11], which is endothermic, increases with increasing temperature and K for reaction [18.12], which is exothermic, decreases with increasing temperature.

18.58   Oxygen is present in the atmosphere to the extent of 209,000 parts per million. If CO binds 210 times more effectively than O_2, then the **effective** concentration of CO is 210×125 ppm $= 26,250 = 26,300$ ppm. The fraction of carboxyhemoglobin in the blood leaving the lungs is thus $\dfrac{26,250}{26,250 + 209,000} = 0.112$. Thus, 11.2 percent of the blood is in the form of carboxyhemoglobin, 88.8 percent as the O_2-bound oxyhemoglobin.

18.59   (a) $CH_4(g) + 2O_2(g) \rightarrow CO_2(g) + 2H_2O(g)$

(b) $2CH_4(g) + 3O_2(g) \rightarrow 2CO(g) + 4H_2O(g)$

(c)   vol $CH_4 \rightarrow$ vol $O_2 \rightarrow$ volume air $(\chi_{O_2} = 0.20948)$

Equal volumes of gases at the same temperature and pressure contain equal numbers of moles (Avogadro's law). If 2 moles of O_2 are required for 1 mole of CH_4, 2.0 L of pure O_2 are needed to burn 1.0 L of CH_4.

vol $O_2 = \chi_{O_2} \times$ vol$_{air} = \dfrac{\text{vol } O_2}{\chi_{O_2}} = \dfrac{2.0 \text{ L}}{0.20948} = 9.5$ L air

18.60   (a)   According to Section 13.3, the solubility of gases in water decreases with increasing temperature. Thus, the solubility of $CO_2(g)$ in the ocean would decrease if the temperature of the ocean increased.

(b)   If the solubility of $CO_2(g)$ in the ocean decreased because of global warming, more $CO_2(g)$ would be released into the atmosphere, perpetuating a cycle of increasing temperature and concomitant release of $CO_2(g)$ from the ocean.

18.61   Most of the 390 watts/m^2 radiated from Earth's surface is in the infrared region of the spectrum. Tropospheric gases, particularly $H_2O(g)$ and $CO_2(g)$, absorb much of this radiation and prevent it from escaping into space (Figure 18.11). The energy absorbed by these so-called "greenhouse gases" warms the atmosphere close to Earth's surface and makes the planet livable.

18.62   Given 169 watts/m^2 at 10% efficiency, find the land area needed to produce 55 watts/m^2.

169 watts/m^2 (0.10) = 16.9 watts/m^2 solar energy possible with current technology.

$$\frac{55 \text{ watts}}{m^2} \times \frac{1 \, m^2}{16.9 \text{ watts}} = 3.254 = 3.3$$

Solar energy must be harvested from 3.3 times the land area of New York City to supply its energy needs. According to *Wikipedia*, http://en.wikipedia.org/wiki/New_York_City, the land area of New York City is 831 km^2, which is 831×10^6 m^2 or 2.05×10^5 acres. The area needed for solar energy harvesting would then be 2.70×10^3 km^2, 2.70×10^9 m^2 or 6.68×10^5 acres.

18.63   (a) $NO(g) + h\nu \rightarrow N(g) + O(g)$

               (b) $NO(g) + h\nu \rightarrow NO^+(g) + e^-$

               (c) $NO(g) + O_3(g) \rightarrow NO_2(g) + O_2(g)$

               (d) $3NO_2(g) + H_2O(l) \rightarrow 2HNO_3(aq) + NO(g)$

18.64   (a) $CO_3{}^{2-}$ is a relatively strong Brønsted base and produces OH^- in aqueous solution according to the hydrolysis reaction:

$$CO_3{}^{2-}(aq) + H_2O(l) \rightleftharpoons HCO_3{}^-(aq) + OH^-(aq), \quad K_b = 1.8 \times 10^{-4}$$

If $[OH^-(aq)]$ is sufficient to exceed K_{sp} for $Mg(OH)_2$, the solid will precipitate.

               (b) $$\frac{125 \text{ mg } Mg^{2+}}{1 \text{ kg soln}} \times \frac{1 g \, Mg^{2+}}{1000 \text{ mg } Mg^{2+}} \times \frac{1.00 \text{ kg soln}}{1.00 \text{ L soln}} \times \frac{1 \text{ mol } Mg^{2+}}{24.305 \text{ g } Mg^{2+}} = 5.143 \times 10^{-3}$$

$$= 5.14 \times 10^{-3} \, M \, Mg^{2+}$$

$$\frac{4.0 \text{ g } Na_2CO_3}{1.0 \text{ L soln}} \times \frac{1 \text{ mol } CO_3{}^{2-}}{106.0 \text{ g } Na_2CO_3} = 0.03774 = 0.038 \, M \, CO_3{}^{2-}$$

$$K_b = 1.8 \times 10^{-4} = \frac{[HCO_3{}^-][OH^-]}{[CO_3{}^{2-}]} \approx \frac{x^2}{0.03774}; \, x = [OH^-] = 2.606 \times 10^{-3}$$

$$= 2.6 \times 10^{-3} \, M$$

(This represents 6.9% hydrolysis, but the result will not be significantly different using the quadratic formula.)

$Q = [Mg^{2+}][OH^-]^2 = (5.143 \times 10^{-3})(2.606 \times 10^{-3})^2 = 3.5 \times 10^{-8}$

K_{sp} for $Mg(OH)_2 = 1.6 \times 10^{-12}$; $Q > K_{sp}$, so $Mg(OH)_2$ will precipitate.

18.65   Because NO has an odd electron, like $Cl(g)$, it could act as a catalyst for decomposition of ozone in the stratosphere. The increased destruction of ozone by NO would result in less absorption of short wavelength UV radiation now being screened out primarily by the ozone. Radiation in this wavelength range is known to be harmful to humans; it causes skin cancer. There is evidence that many plants don't tolerate it very well either, though more research is needed to test this idea.

In Chapter 22 the oxidation of NO to NO_2 by oxygen is described. On dissolving in water, NO_2 disproportionates into $NO_3^-(aq)$ and $NO(g)$. Thus, over time the NO in the troposphere will be converted into NO_3^-, which is in turn incorporated into soils.

18.66   *Plan.* Calculate the volume of air above Los Angeles and the volume of pure O_3 that would be present at the 85 ppb level. For gases at the same temperature and pressure, volume fractions equal mole fractions.   *Solve.*

$$V_{air} = 4000 \text{ mi}^2 \times \frac{(1.6093)^2 \text{ km}^2}{\text{mi}^2} \times \frac{(1000)^2 \text{ m}^2}{1 \text{ km}^2} \times 10 \text{ m} \times \frac{1 L}{1 \times 10^{-3} \text{ m}^3} = 1.036 \times 10^{14}$$

$$= 1.0 \times 10^{14} \text{ L air}$$

$$85 \text{ ppb } O_3 = \frac{85 \text{ mol } O_3}{1 \times 10^9 \text{ mol air}} = 8.5 \times 10^{-8} = \chi_{O_3}$$

$$V \text{ (pure } O_3) = 8.5 \times 10^{-8} (1.036 \times 10^{14} \text{ L air}) = 8.805 \times 10^6 = 8.8 \times 10^6 \text{ L } O_3$$

Values for P and T are required to calculate mol O_3 from volume O_3, using the ideal-gas law. Since these are not specified in the exercise, we will make a reasonable assumption for a sunny April day in Los Angeles. The city is near sea level and temperatures are moderate throughout the year, so $P = 1$ atm and $T = 25°C$ (78°F) are reasonable values.

$PV = nRT, n = PV/RT$

$$n = 1.000 \text{ atm} \times \frac{8.805 \times 10^6 \text{ L}}{298 \text{ K}} \times \frac{K \cdot mol}{0.08206 \text{ L} \cdot atm} = 3.601 \times 10^5 = 3.6 \times 10^5 \text{ mol } O_3$$

Check. Using known conditions to make reasonable estimates and assumptions is a valuable skill for problem solving. Knowing when assumptions are required is an important step in the learning process.

Integrative Exercises

18.67   (a) $0.019 \text{ ppm } NO_2 = \dfrac{0.019 \text{ mol } NO_2}{1 \times 10^6 \text{ mol air}} = 1.9 \times 10^{-8} = \chi_{NO_2}$

$P_{NO_2} = \chi_{NO_2} \cdot P_{atm} = 1.9 \times 10^{-8} (755 \text{ torr}) = 1.4345 \times 10^{-5} = 1.4 \times 10^{-5} \text{ torr}$

(b) $n = \dfrac{PV}{RT}$; molecules $= n \times \dfrac{6.022 \times 10^{23} \text{ molecules}}{mol} = \dfrac{PV}{RT} \times \dfrac{6.022 \times 10^{23} \text{ molecules}}{mol}$

$V = 15 \text{ ft} \times 14 \text{ ft} \times 8 \text{ ft} \times \dfrac{12^3 \text{ in}^3}{ft^3} \times \dfrac{2.54^3 \text{ cm}^3}{in^3} \times \dfrac{1 L}{1000 \text{ cm}^3} = 4.757 \times 10^4 = 5 \times 10^4 \text{ L}$

$$1.4345 \times 10^{-5} \text{ torr} \times \frac{1 \text{ atm}}{760 \text{ torr}} \times \frac{4.757 \times 10^4 \text{ L}}{293 \text{ K}} \times \frac{\text{K} \cdot \text{mol}}{0.08206 \text{ L} \cdot \text{atm}}$$

$$\times \frac{6.022 \times 10^{23} \text{ molecules}}{\text{mol}} = 2.249 \times 10^{19} = 2 \times 10^{19} \text{ molecule}$$

18.68   (a) $8{,}376{,}726 \text{ tons coal} \times \dfrac{83 \text{ ton C}}{100 \text{ ton coal}} \times \dfrac{44.01 \text{ ton CO}_2}{12.01 \text{ ton C}} = 2.5 \times 10^7 \text{ ton CO}_2$

           $8{,}376{,}726 \text{ tons coal} \times \dfrac{2.5 \text{ ton S}}{100 \text{ ton coal}} \times \dfrac{64.07 \text{ ton SO}_2}{32.07 \text{ ton S}} = 4.2 \times 10^5 \text{ ton SO}_2$

   (b) $CaO(s) + SO_2(g) \rightarrow CaSO_3(s)$

           $4.18 \times 10^5 \text{ ton SO}_2 \times \dfrac{55 \text{ ton SO}_2 \text{ removed}}{100 \text{ ton SO}_2 \text{ produced}} \times \dfrac{120.15 \text{ ton CaSO}_3}{64.07 \text{ ton SO}_2}$

                                     $= 4.3 \times 10^5 \text{ ton CaSO}_3$

18.69   *Coarse sand* is removed by coarse sand filtration. *Finely divided particles* and some *bacteria* are removed by precipitation with aluminum hydroxide. Remaining *harmful bacteria* are removed by ozonation. *Trihalomethanes* are removed by either aeration or activated carbon filtration; use of activated carbon might be preferred because it does not involve release of TCMs into the atmosphere. *Dissolved organic substances* are oxidized (and rendered less harmful, but not removed) by both aeration and ozonation. Dissolved *nitrates* and *phosphates* are not removed by any of these processes, but are rendered less harmful by adequate aeration.

18.70   (a) $H - \ddot{O} - H \longrightarrow H\cdot \; + \; \cdot\ddot{O} - H$

   (b) $\Delta H = 2D(O-H) - D(O-H) = D(O-H) = 463 \text{ kJ/mol}$

           $\dfrac{463 \text{ kJ}}{\text{mol H}_2\text{O}} \times \dfrac{1 \text{ mol H}_2\text{O}}{6.022 \times 10^{23} \text{ molecules}} \times \dfrac{1000 \text{ J}}{\text{kJ}} = 7.688 \times 10^{-19}$

                                     $= 7.69 \times 10^{-19} \text{ J/H}_2\text{O molecule}$

           $\lambda = \dfrac{hc}{\Delta E} = \dfrac{6.626 \times 10^{-34} \text{ J} \cdot \text{sec} \times 2.998 \times 10^8 \text{ m/s}}{7.688 \times 10^{-19} \text{ J}} = 2.58 \times 10^{-7} \text{ m} = 258 \text{ nm}$

       This wavelength is in the UV region of the spectrum, close to the visible.

   (c) $OH(g) + O_3(g) \rightarrow HO_2(g) + O_2(g)$

                      $HO_2(g) + O(g) \rightarrow OH(g) + O_2(g)$

       $\overline{OH(g) + O_3(g) + HO_2(g) + O(g) \;\; \rightarrow \;\; HO_2(g) + 2O_2(g) + OH(g)}$

                      $O_3(g) + O(g) \rightarrow 2O_2(g)$

       OH(g) is the catalyst is this overall reaction, another pathway for the destruction of ozone.

18.71   According to Equation [14.12], $\ln([A]_t / [A]_o) = -kt.$ $[A]_t = 0.10 [A]_o$

       $\ln(0.10 [A]_o / [A]_o) = \ln(0.10) = -(2 \times 10^{-6} \text{ s}^{-1}) t$

       $t = -\ln(0.10) / 2 \times 10^{-6} \text{ s}^{-1} = 1.151 \times 10^6 \text{ s}$

$$1.151 \times 10^6 \text{ s} \times \frac{1 \text{ min}}{60 \text{ s}} \times \frac{1 \text{ hr}}{60 \text{ min}} \times \frac{1 \text{ day}}{24 \text{ hr}} = 13.3 \text{ days} (1 \times 10 \text{ days})$$

The value of the rate constant limits the result to 1 sig fig. This implies that there is minimum uncertainty of $\pm$ 1 in the tens place of our answer. Realistically, the remediation could take anywhere from 1 to 20 days.

18.72   (i) $ClO(g) + O_3(g) \rightarrow ClO_2(g) + O_2(g)$

$$\Delta H_i = \Delta H_f^\circ ClO_2(g) + \Delta H_f^\circ O_2(g) - \Delta H_f^\circ ClO(g) - \Delta H_f^\circ O_3(g)$$

$$\Delta H_i = 102 + 0 - 101 - (142.3) = -141 \text{ kJ}$$

(ii) $ClO_2(g) + O(g) \rightarrow ClO(g) + O_2(g)$

$$\Delta H_{ii} = \Delta H_f^\circ ClO(g) + \Delta H_f^\circ O_2(g) - \Delta H_f^\circ ClO_2(g) + \Delta H_f^\circ O(g)$$

$$\Delta H_{ii} = 101 + 0 - 102 - (247.5) = -249 \text{ kJ}$$

(overall) $ClO(g) + O_3(g) + ClO_2(g) + O(g) \rightarrow ClO_2(g) + O_2(g) + ClO(g) + O_2(g)$

$$O_3(g) + O(g) \rightarrow 2O_2(g)$$

$$\Delta H = \Delta H_i + \Delta H_{ii} = -141 \text{ kJ} + (-249) \text{ kJ} = -390 \text{ kJ}$$

Because the enthalpies of both (i) and (ii) are distinctly exothermic, it is possible that the $ClO - ClO_2$ pair could be a catalyst for the destruction of ozone.

18.73   (a)   Assume the density of water at 20°C is the same as at 25°C.

$$1.00 \text{ gal} \times \frac{4 \text{ qt}}{1 \text{ gal}} \times \frac{1 \text{ L}}{1.057 \text{ qt}} \times \frac{1000 \text{ mL}}{1 \text{ L}} \times \frac{0.99707 \text{ g H}_2\text{O}}{1 \text{ mL}} = 3773$$

$$= 3.77 \times 10^3 \text{ g H}_2\text{O}$$

The $H_2O(l)$ must be heated from 20°C to 100°C and then vaporized at 100°C.

$$3.773 \times 10^3 \text{ g H}_2\text{O} \times \frac{4.184 \text{ J}}{\text{g °C}} \times 80 \text{ °C} \times \frac{1 \text{ kJ}}{1000 \text{ J}} = 1263 = 1.3 \times 10^3 \text{ kJ}$$

$$3.773 \times 10^3 \text{ g H}_2\text{O} \times \frac{1 \text{ mol H}_2\text{O}}{18.02 \text{ g H}_2\text{O}} \times \frac{40.67 \text{ kJ}}{\text{mol H}_2\text{O}} = 8516 = 8.52 \times 10^3 \text{ kJ}$$

energy = 1263 kJ + 8516 kJ = 9779 = 9.8×10^3 kJ/gal H_2O

(b)   According to Solution 5.14, 1 kwh = 3.6×10^6 J.

$$\frac{9779 \text{ kJ}}{\text{gal H}_2\text{O}} \times \frac{1000 \text{ J}}{\text{kJ}} \times \frac{1 \text{ kwh}}{3.6 \times 10^6 \text{ J}} \times \frac{\$0.085}{\text{kwh}} = \$0.23/\text{gal}$$

(c) $\frac{\$0.23}{\$1.26} \times 100 = 18\%$ of the total cost is energy

18.74   (a)   A rate constant of $M^{-1}s^{-1}$ is indicative of a reaction that is second order overall. For the reaction given, the rate law is probably rate = $k[O][O_3]$. (Although rate = $k[O]^2$ or $k[O_3]^2$ are possibilities, it is difficult to envision a mechanism consistent with either one that would result in two molecules of O_2 being produced.)

(b)   Yes. Most atmospheric processes are initiated by collision. One could imagine an activated complex of four O atoms collapsing to form two O_2 molecules. Also, the rate constant is large, which is less likely for a multistep process. The reaction is analogous to the destruction of O_3 by Cl atoms (Equation 18.7), which is also second order with a large rate constant.

(c)   According to the Arrhenius equation, $k = Ae^{-Ea/RT}$. Thus, the larger the value of k, the smaller the activation energy, E_a. The value of the rate constant for this reaction is large, so the activation energy is small.

(d) $\Delta H^{\circ}_f = 2\Delta H^{\circ}_f\ O_2(g) - \Delta H^{\circ}_f\ O(g) - \Delta H^{\circ}_f\ O_3(g)$

$\Delta H^{\circ}_f = 0 - 247.5\ kJ - 142.3\ kJ = -389.8\ kJ$

The reaction is exothermic, so energy is released; the reaction would raise the temperature of the stratosphere.

18.75   (a) $17\ e^-, 8.5\ e^-$ pairs

$$\ddot{O}\!\!=\!\!\dot{N}\!\!-\!\!\ddot{\underset{..}{O}}: \longleftrightarrow :\ddot{\underset{..}{O}}\!\!-\!\!\dot{N}\!\!=\!\!\ddot{O}$$

Owing to its lower electronegativity, N is more likely to be electron deficient and to accommodate the odd electron.

(b)   The fact that NO_2 is an electron deficient molecule indicates that it will be highly reactive. Dimerization results in formation of a N—N single bond which completes the octet of both N atoms. NO_2 and N_2O_4 exist in equilibrium in a closed system. The reaction is exothermic, Equation [22.65]. In an urban environment, NO_2 is produced from hot automobile combustion. At these temperatures, equilibrium favors the monomer because the reaction is exothermic.

(c) $2NO_2(g) + 4CO(g) \rightarrow N_2(g) + 4CO_2(g)$

$NO_2(g) + CO(g) \rightarrow NO(g) + CO_2(g)$

NO_2 is an oxidizing agent and CO is a reducing agent, so we expect products to contain N in a more reduced form, NO or N_2, and C in a more oxidized form, CO_2.

(d)   No. Because it is an odd-electron molecule, NO_2 is very reactive. We expect it to undergo chemical reactions or photodissociate before it can migrate to the stratosphere. The expected half-life of an NO_2 molecule is short.

18.76   Calculate $[H_2SO_4]$ required to produce a solution with pH = 3.5. From the volume of rainfall, calculate the amount of H_2SO_4 present.

$[H^+] = 10^{-3.5} = 3.16 \times 10^{-4} = 3 \times 10^{-4}\ M$

Assume initially that both ionization steps are complete.

$$\underset{x\ M}{H_2SO_4(aq)} \rightleftharpoons \underset{x\ M}{H^+(aq)} + HSO_4^-(aq)$$

$$\underset{+2x\ M}{HSO_4^-(aq)} \rightleftharpoons H^+(aq) + \underset{x\ M}{SO_4^{2-}(aq)}$$

Since $[HSO_4^-]$ at equilibrium is small but finite, let $[HSO_4^-] = y$.

$$HSO_4^-(aq) \rightleftharpoons H^+(aq) + SO_4^{2-}(aq) \qquad K_a = 0.012$$

equil $\qquad\qquad y\,M \qquad\qquad (2x-y)\,M \qquad (x-y)\,M$

$$K_a = 0.012 = \frac{[H^+][SO_4^{2-}]}{[HSO_4^-]} = \frac{(2x-y)(x-y)}{y}$$

But we know that $[H^+]$ at equilibrium $= 3.16 \times 10^{-4}\,M$.

$2x - y = 3.16 \times 10^{-4}$; $y = 2x - 3.16 \times 10^{-4}$; $(x - y) = [x - (2x - 3.16 \times 10^{-4})] = 3.16 \times 10^{-4} - x$

$$K_a = 0.012 = \frac{(3.16 \times 10^{-4})(3.16 \times 10^{-4} - x)}{2x - 3.16 \times 10^{-4}}$$

$(0.012)(2x - 3.16 \times 10^{-4}) = 1.00 \times 10^{-7} - 3.16 \times 10^{-4}x$;

$0.024x - 3.795 \times 10^{-6} = 1.00 \times 10^{-7} - 3.16 \times 10^{-4}x$;

$0.024316x = 3.895 \times 10^{-6}$; $x = 1.60 \times 10^{-4} = 2 \times 10^{-4}\,M\,H_2SO_4$

Check. This result is reasonable, since it is just slightly greater than $[H^+]/2$. The amount of HSO_4^- at equilibrium, $y = 4.1 \times 10^{-6}\,M$.

$$\frac{[H^+][SO_4^{2-}]}{[HSO_4^-]} = \frac{(3.16 \times 10^{-4})(1.60 \times 10^{-4} - 4.1 \times 10^{-6})}{4.1 \times 10^{-6}} = 0.012$$

The calculated results are reasonable and self-consistent.

Now proceed to find the volume of rainfall and corresponding mass of H_2SO_4 if $[H_2SO_4] = 1.60 \times 10^{-4} = 2 \times 10^{-4}\,M$.

$$V = 1.0\,in \times 1500\,mi^2 \times \frac{5280^2\,ft^2}{mi^2} \times \frac{12^2\,in^2}{ft^2} \times \frac{2.54^3\,cm^3}{in^3} \times \frac{1\,L}{1000\,cm^3} = 9.868 \times 10^{10}$$

$$= 9.9 \times 10^{10}$$

$$\frac{1.60 \times 10^{-4}\,mol\,H_2SO_4}{1\,L\,rainfall} \times 9.868 \times 10^{10}\,L \times \frac{98.1g\,H_2SO_4}{1\,mol\,H_2SO_4} \times \frac{1\,kg}{1000\,g} = 1.55 \times 10^6$$

$$= 2 \times 10^6\,kg\,H_2SO_4$$

18.77   (a)   According to Table 18.1, the mole fraction of CO_2 in air is 0.000375.

$$P_{CO_2} = \chi_{CO_2} \cdot P_{atm} = 0.000375\,(1.00\,atm) = 3.75 \times 10^{-4}\,atm$$

$$C_{CO_2} = kP_{CO_2} = 3.1 \times 10^{-2}\,M/atm \times 3.75 \times 10^{-4}\,atm = 1.16 \times 10^{-5} = 1.2 \times 10^{-5}\,M$$

(b) H_2CO_3 is a weak acid, so the $[H^+]$ is regulated by the equilibria:

$$H_2CO_3(aq) \rightleftharpoons H^+(aq) + HCO_3^-(aq)\ K_{a1} = 4.3 \times 10^{-7}$$

$$HCO_3^-(aq) \rightleftharpoons H^+(aq) + CO_3^{2-}(aq)\ \ K_{a2} = 5.6 \times 10^{-11}$$

Since the value of K_{a2} is small compared to K_{a1}, we will assume that most of the $H^+(aq)$ is produced by the first dissociation.

$$K_{a1} = 4.3 \times 10^{-7} = \frac{[H^+][HCO_3^-]}{[H_2CO_3]}; [H^+] = [HCO_3^-] = x, [H_2CO_3] = 1.2 \times 10^{-5} - x$$

Since K_{a1} and $[H_2CO_3]$ have similar values, we cannot assume x is small compared to 1.2×10^{-5}.

$$4.3 \times 10^{-7} = \frac{x^2}{(1.2 \times 10^{-5} - x)}; 5.00 \times 10^{-12} - 4.3 \times 10^{-7} x = x^2$$

$$0 = x^2 + 4.3 \times 10^{-7} - 5.00 \times 10^{-12}$$

$$x = \frac{-4.3 \times 10^{-7} \pm \sqrt{(4.3 \times 10^{-7})^2 - 4(1)(-5.00 \times 10^{-12})}}{2(1)}$$

$$x = \frac{-4.3 \times 10^{-7} \pm \sqrt{1.85 \times 10^{-13} + 2.00 \times 10^{-11}}}{2} = \frac{-4.3 \times 10^{-7} \pm 4.49 \times 10^{-6}}{2}$$

The negative result is meaningless; $x = 2.03 \times 10^{-6} = 2.0 \times 10^{-6}$ M H^+; pH = 5.69

Since this $[H^+]$ is quite small, the $[H^+]$ from the autoionization of water might be significant. Calculation shows that for $[H^+] = 2.0 \times 10^{-6}$ M from H_2CO_3, $[H^+]$ from $H_2O = 5.2 \times 10^{-9}$ M, which we can ignore.

18.78   (a) $Al(OH)_3(s) \rightleftharpoons Al^{3+}(aq) + 3OH^-(aq)$ $K_{sp} = 1.3 \times 10^{-33} = [Al^{3+}][OH^-]^3$

This is a precipitation conditions problem. At what $[OH^-]$ (we can get pH from $[OH^-]$) will $Q = 1.3 \times 10^{-33}$, the requirement for the onset of precipitation?

$Q = 1.3 \times 10^{-33} = [Al^{3+}][OH^-]^3$. Find the molar concentration of $Al_2(SO_4)_3$ and thus $[Al^{3+}]$.

$$\frac{5.0 \text{ lb } Al_2(SO_4)_3}{2000 \text{ gal } H_2O} \times \frac{453.6 \text{ g}}{1 \text{ lb}} \times \frac{1 \text{ mol } Al_2(SO_4)_3}{342.2 \text{ g } Al_2(SO_4)_3} \times \frac{1 \text{ gal}}{4 \text{ qt}} \times \frac{1 \text{ qt}}{0.946 \text{ L}}$$

$$= 8.758 \times 10^{-4} \ M \ Al_2(SO_4)_3 = 1.752 \times 10^{-3} = 1.8 \times 10^{-3} \ M \ Al^{3+}$$

$Q = 1.3 \times 10^{-33} = (1.752 \times 10^{-3})[OH^-]^3; [OH^-]^3 = 7.42 \times 10^{-31}$

$[OH^-] = 9.054 \times 10^{-11} = 9.1 \times 10^{-11}$ M; pOH = 10.04; pH = 14 − 10.04 = 3.96

(b) $CaO(s) + H_2O(l) \rightarrow Ca^{2+}(aq) + 2OH^-(aq)$; $[OH^-] = 9.054 \times 10^{-11}$ mol/L

$$\text{mol } OH^- = \frac{9.054 \times 10^{-11} \text{ mol}}{1 \text{ L}} \times 2000 \text{ gal} \times \frac{4 \text{ qt}}{1 \text{ gal}} \times \frac{0.946 \text{ L}}{1 \text{ qt}} = 6.852 \times 10^{-7}$$

$$= 6.9 \times 10^{-7} \text{ mol } OH^-$$

$$6.852 \times 10^{-7} \text{ mol } OH^- \times \frac{1 \text{ mol } CaO}{2 \text{ mol } OH^-} \times \frac{56.1 \text{ g } CaO}{1 \text{ mol } CaO} \times \frac{1 \text{ lb}}{453.6 \text{ g}} = 4.2 \times 10^{-8} \text{ lb } CaO$$

This is a **very** small amount of CaO, about 20 μg.

19 Chemical Thermodynamics

Visualizing Concepts

19.1 (a)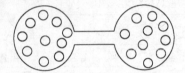

(b) ΔS is positive, because the disorder of the system increases. Each gas has greater motional freedom as it expands into the second bulb, and there are many more possible arrangements for the mixed gases.

By definition, ideal gases experience no attractive or repulsive intermolecular interactions, so ΔH for the mixing of ideal gases is zero, assuming heat exchange only between the two bulbs.

(c) The process is irreversible. It is inconceivable that the gases would reseparate.

(d) The entropy change of the surroundings is related to ΔH for the system. Since we are mixing ideal gases and $\Delta H = 0$, ΔH_{surr} is also zero, assuming heat exchange only between the two bulbs.

19.2 (a) The process depicted is a change of state from a solid to a gas. ΔS increases because of the greater motional freedom of the particles. ΔH increases because both melting and boiling are endothermic processes. Since $\Delta G = \Delta H - T\Delta S$, and both ΔH and ΔS are positive, the sign of ΔG depends on temperature. This is true for all phase changes. If the temperature of the system is greater than the boiling point of the substance, the process is spontaneous and ΔG is negative. If the temperature is lower than the boiling point, the process is not spontaneous and ΔG is positive.

(b) If the process is spontaneous, the second law states that $\Delta S_{univ} \geq 0$. Since ΔS_{sys} increases, ΔS_{surr} must decrease. If the change occurs via a reversible pathway, $\Delta S_{univ} = 0$ and $\Delta S_{surr} = -\Delta S_{sys}$. If the pathway is irreversible, the magnitude of ΔS_{sys} is greater than the magnitude of ΔS_{surr}, but the sign of ΔS_{surr} is still negative.

19.3 In the depicted reaction, both reactants and products are in the gas phase (they are far apart and randomly placed). There are twice as many molecules (or moles) of gas in the products, so ΔS is positive for this reaction.

19.4 (a) At 300 K, $\Delta H = T\Delta S$. Since $\Delta G = \Delta H - T\Delta S$, $\Delta G = 0$ at this point. When $\Delta G = 0$, the system is at equilibrium.

(b)   The reaction is spontaneous when ΔG is negative. This condition is met when $T\Delta S > \Delta H$. From the diagram, $T\Delta S > \Delta H$ when $T > 300$ K. The reaction is spontaneous at temperatures above 300 K.

19.5   (a)   *Analyze.* The boxes depict three different mixtures of reactants and products for the reaction $A_2 + B_2 \rightleftharpoons 2AB$.

Plan. Box 1 is an equilibrium mixture. By definition, $\Delta G = 0$ for box 1. Calculate K and $\Delta G°$ for the reaction from box 1. Boxes 2 and 3 are nonequilibrium mixtures. Calculate Q and ΔG for boxes 2 and 3.

Solve. $K = \dfrac{[AB]^2}{[A][B]}$. Use number of molecules as a measure of concentration.

Box 1: $K = \dfrac{(3)^2}{(3)(3)} = 1$.

$\Delta G = \Delta G° + RT \ln K$; $0 = \Delta G° - RT \ln(1)$, $0 = \Delta G° - 0$; $\Delta G° = 0$

Box 2: $Q = \dfrac{(1)^2}{(4)(4)} = \dfrac{1}{16} = 0.0625 = 0.06$

$\Delta G = \Delta G° + RT \ln Q = 0 - RT \ln(0.0625) = 2.77\ RT = 3RT$

Box 3: $Q = \dfrac{(7)^2}{(1)(1)} = \dfrac{46}{1} = 49$

$\Delta G = \Delta G° + RT \ln Q = 0 - RT \ln(49) = -3.89\ RT = -4RT$

(b)   The magnitudes of ΔG (ignoring sign) are: box 1, 0; box 2, 2.8 RT; box 3, 3.9 RT. The order of increasing *magnitude* of ΔG is: box 1 < box 2 < box 3.

The signs on ΔG indicate in which direction the reaction is spontaneous. The mixture in box 2 will react spontaneously in the forward direction, toward products. The mixture in box 3 will react spontaneously in the reverse direction, toward reactants. The driving force for the reverse reaction in box 3, the *magnitude* of ΔG, is greater than the driving force for the forward reaction in box 2.

19.6   (a)   The minimum in the plot is the equilibrium position of the reaction, where $\Delta G = 0$.

(b)   X is the difference in free energy between reactant and products in their standard states, $\Delta G°$.

Spontaneous Processes

19.7   *Analyze/Plan.* Follow the logic in Sample Exercise 19.1.   *Solve.*

(a)   Nonspontaneous; $-5°C$ is below the melting point of ice, so melting does not happen without continuous intervention.

(b)   Spontaneous; sugar is soluble in water, and even more soluble in hot coffee.

(c)   Spontaneous; N_2 molecules are stable relative to isolated N atoms.

(d)   Spontaneous; the filings organize in a magnetic field without intervention.

(e)   Nonspontaneous; CO_2 and H_2O are in contact continuously at atmospheric conditions in nature and do not form CH_4 and O_2.

19.8   (a)   Spontaneous; a gas, in this case perfume vapor, expands to fill its container, the room.

(b)   Nonspontaneous; a mixture cannot be separated without outside intervention.

(c)   Nonspontaneous; an inflated balloon doesn't burst without external stress, such as a pin prick, a squeeze, or adding more gas.

(d)   Spontaneous; see Figure 8.2.

(e)   Spontaneous; the very polar HCl molecules readily dissolve in water to form concentrated HCl(aq).

19.9   (a) $NH_4NO_3(s)$ dissolves in water, as in a chemical cold pack. Naphthalene (mothballs) sublimes at room temperature.

(b)   Melting of a solid is spontaneous above its melting point but nonspontaneous below its melting point.

19.10   Berthelot's suggestion is incorrect. Some examples of nonexothermic spontaneous processes are expansion of certain pressurized gases, dissolving of one liquid in another, and dissolving of many salts in water.

19.11   *Analyze/Plan.* Define the system and surroundings. Use the appropriate definition to answer the specific questions.   *Solve.*

(a)   Water is the system. Heat must be added to the system to evaporate the water. The process is endothermic.

(b)   At 1 atm, the reaction is spontaneous at temperatures above 100°C.

(c)   At 1 atm, the reaction is nonspontaneous at temperatures below 100°C.

(d)   The two phases are in equilibrium at 100°C.

19.12   (a)   Exothermic. If melting requires heat and is endothermic, freezing must be exothermic.

(b)   At 1 atm (indicated by the term "normal" freezing point), the freezing of 1-propanol is spontaneous at temperatures below –127°C.

(c)   At 1 atm, the freezing of 1-propanol is nonspontaneous at temperatures above –127°C.

(d)   At 1 atm and –127°C, the normal freezing point of 1-propanol, the solid and liquid phases are in equilibrium. That is, at the freezing point, 1-propanol molecules escape to the liquid phase at the same rate as liquid 1-propanol solidifies, assuming no heat is exchanged between 1-propanol and the surroundings.

19.13   *Analyze/Plan.* Define the system and surroundings. Use the appropriate definition to answer the specific questions.   *Solve.*

(a)   For a *reversible* process, the forward and reverse changes occur by the same path. In a reversible process, both the system and the surroundings are restored to their original condition by exactly reversing the change. A reversible change produces the maximum amount of work.

(b)   If a system is returned to its original state via a reversible path, the surroundings are also returned to their original state. That is, there is no net change in the surroundings.

(c)   The vaporization of water to steam is reversible if it occurs at the boiling temperature of water for a specified external (atmospheric) pressure, and only if the needed heat is added infinitely slowly.

19.14.   (a)   A process is *irreversible* if the system cannot be returned to its original state by the same path that the forward process took place.

(b)   Since the system returned to its initial state via a different path (different q_r and w_r than q_f and w_f), there is a net change in the surroundings.

(c)   The condensation of a liquid will be irreversible if it occurs at any temperature other than the boiling point of the liquid, at a specified pressure.

19.15.   No. ΔE is a state function. $\Delta E = q + w$; q and w are not state functions. Their values do depend on path, but their sum, ΔE, does not.

19.16   (a) $\Delta E (1 \rightarrow 2) = -\Delta E (2 \rightarrow 1)$

(b)   We can say nothing about the values of q and w because we have no information about the paths.

(c)   If the changes of state are reversible, the two paths are the same and $w (1 \rightarrow 2) = -w (2 \rightarrow 1)$. This is the maximum realizable work from this system.

19.17   *Analyze/Plan.* Define the system and surroundings. Use the appropriate definition to answer the specific questions.   *Solve.*

(a)   An ice cube can melt reversibly at the conditions of temperature and pressure where the solid and liquid are in equilibrium. At 1 atm external pressure, the normal melting point of water is 0°C.

(b)   We know that melting is a process that increases the energy of the system, even though there is no change in temperature. ΔE is not zero for the process.

19.18   (a)   The detonation of an explosive is definitely spontaneous, once it is initiated.

(b)   The quantity q is related to ΔH. Since the detonation is highly exothermic, q is large and negative.

     If only PV-work is done and P is constant, $\Delta H = q$. Although these conditions probably do not apply to a detonation, we can still predict the sign of q, based on ΔH, if not its exact magnitude.

(c)   The sign (and magnitude) of w depend on the path of the process, the exact details of how the detonation is carried out. It seems clear, however, that work will be done by the system on the surroundings in almost all circumstances (buildings collapse, earth and air are moved), so the sign of w is probably negative.

(d) $\Delta E = q + w$. If q and w are both negative, then the sign of ΔE is negative, regardless of the magnitudes of q and w.

Entropy and the Second Law of Thermodynamics

19.19   (a)   For a process that occurs at constant temperature, an isothermal process, $\Delta S = q_{rev}/T$. Here q_{rev} is the heat that would be transferred if the process were reversible. Since ΔS is a state function, it is independent of path, so ΔS for the reversible path must equal ΔS for any path.

        (b)   No. ΔS is a state function, so it is independent of path.

19.20   (a)   When a liquid freezes, the entropy of the system decreases.

        (b) ΔS is negative.

        (c)   Entropy being a state function means that ΔS is independent of the path of the process. ΔS defined in terms of a reversible path must equal ΔS for any path.

19.21   (a) $CH_3OH(l) \rightarrow CH_3OH(g)$, entropy increases, more mol gas in products, greater motional freedom.

        (b) $\Delta S = \dfrac{\Delta H}{T} = \dfrac{71.8\,kJ}{mol\,CH_3OH(l)} \times 1.00\,mol\,CH_3OH(l) \times \dfrac{1}{(273.15+64.7)K} \times \dfrac{1000\,J}{1\,kJ} = 213\,J/K$

19.22   (a) $Cs(l) \rightarrow Cs(s)$, ΔS is negative

        (b) $\Delta H = 15.0\,g\,Cs \times \dfrac{1\,mol\,Cs}{132.9\,g\,Cs} \times \dfrac{2.09\,kJ}{mol\,Cs} = 0.2359 = 0.236\,kJ$

                $\Delta S = \dfrac{\Delta H}{T} = 0.2359\,kJ \times \dfrac{1000\,J}{1\,kJ} \times \dfrac{1}{(273.15+28.4)K} = 0.782\,J/K$

19.23   (a)   For a spontaneous process, the entropy of the universe increases; for a reversible process, the entropy of the universe does not change.

        (b)   In a reversible process, $\Delta S_{system} + \Delta S_{surroundings} = 0$. If ΔS_{system} is positive, $\Delta S_{surroundings}$ must be negative.

        (c)   Since $\Delta S_{universe}$ must be positive for a spontaneous process, $\Delta S_{surroundings}$ must be greater than –42 J/K.

19.24   (a)   For a spontaneous process, $\Delta S_{universe} > 0$. For a reversible process, $\Delta S_{universe} = 0$.

        (b) $\Delta S_{surroundings}$ is positive and greater than the magnitude of the decrease in ΔS_{system}.

        (c) $\Delta S_{system} = 78\,J/K$.

19.25   *Analyze.* Calculate ΔS for the isothermal expansion of 0.100 mol He from 2.00 L to 5.00 L at 27°C.

        Plan. Use the relationship $\Delta S_{sys} = nR\,ln(V_2/V)$.

Solve. $\Delta S_{sys} = 0.100 \, (8.314 \, \text{J/mol·K})(\ln [5.00 \, \text{L}/2.00 \, \text{L}]) = 0.762 \, \text{J/K}$.

Check. We expect ΔS to be positive when the motional freedom of a gas increases, and our calculation agrees with this prediction.

19.26　According to Boyle's law, $P_1 V_1 = P_2 V_2$ at constant n and T.

0.900 atm $\times \, V_1 = 3.00$ atm $\times \, V_2$; $V_2/V_1 = 0.900$ atm$/3.00$ atm $= 0.300$

$\Delta S_{sys} = nR \ln (V_2/V_1) = 0.500 \, \text{mol} \, (8.314 \, \text{J/mol·K})(\ln 0.300) = -5.0049 = -5.00 \, \text{J/K}$

Check. An increase in pressure results in a decrease in volume at constant T, so we expect ΔS to be negative, and it is.

The Molecular Interpretation of Entropy

19.27　(a)　The higher the temperature, the broader the distribution of molecular speeds and kinetic energies available to the particles. This wider range of accessible kinetic energies leads to more microstates for the system.

　　　　(b)　An increase in volume generates more possible positions for the particles and leads to more microstates for the system.

　　　　(c)　Going from solid to liquid to gas, particles have greater translational motion, which increases the number of positions available to the particles and the number of microstates for the system.

19.28　(a)　ΔH_{vap} for H_2O at 25°C = 44.02 kJ/mol; at 100°C = 40.67 kJ/mol

$$\Delta S = \frac{q_{rev}}{T} = \frac{44.02 \, \text{kJ}}{\text{mol}} \times \frac{1000 \, \text{J}}{\text{kJ}} \times \frac{1}{298 \, \text{K}} = 148 \, \text{J/mol·K}$$

$$\Delta S = \frac{q_{rev}}{T} = \frac{40.67 \, \text{kJ}}{\text{mol}} \times \frac{1000 \, \text{J}}{\text{kJ}} \times \frac{1}{373 \, \text{K}} = 109 \, \text{J/mol·K}$$

　　　　(b)　At both temperatures, the liquid $\rightarrow$ gas phase transition is accompanied by an increase in entropy, as expected. That the magnitude of the increase is greater at the lower temperature requires some explanation.

　　　　In the liquid state, there are significant hydrogen bonding interactions between H_2O molecules. This reduces the number of possible molecular positions and the number of microstates. Liquid water at 100° has sufficient kinetic energy to have broken many hydrogen bonds, so the number of microstates for $H_2O(l)$ at 100° is greater than the number of microstates for $H_2O(l)$ at 25°C. The difference in the number of microstates upon vaporization at 100°C is smaller, and the magnitude of ΔS is smaller.

19.29　*Analyze/Plan.* Consider the conditions that lead to an increase in entropy: more mol gas in products than reactants, increase in volume of sample and, therefore, number of possible arrangements, more motional freedom of molecules, etc.　*Solve.*

　　　　(a)　More gaseous particles means more possible arrangements and greater disorder; ΔS is positive.

(b) ΔS is positive for Exercise 19.8 (a) and (c). Both processes represent an increase in volume and possible arrangements for the sample. [In (e), even though HCl(aq) is a mixture, there are fewer moles of gas in the product, so ΔS is not positive.]

19.30   (a)   Solids are much more ordered than gases, so ΔS is negative.

        (b) ΔS is positive for Exercise 19.7 (a), (b), and (e). [At room temperature and 1 atm pressure, H_2O is a liquid, so there are more moles of gas in the products in part (e) and $\Delta S > 0$.]

19.31   *Analyze/Plan.* Consider the conditions that lead to an increase in entropy: more mol gas in products than reactants, increase in volume of sample and, therefore, number of possible arrangements, more motional freedom of molecules, etc.   *Solve.*

        (a)   S increases; translational motion is greater in the liquid than the solid.

        (b)   S decreases; volume and translational motion decrease going from the gas to the liquid.

        (c)   S increases; volume and translational motion are greater in the gas than the solid.

19.32   (a)   When temperature increases, the range of accessible molecular speeds and kinetic energies increases. This produces more microstates and an increase in entropy.

        (b)   When the volume of a gas increases (even at constant T), there are more possible positions for the particles, more microstates, and greater entropy.

        (c)   When a solid dissolves in water, there are both more possible positions for the particles (ions or molecules) and more motional freedom. The number of microstates and entropy increases.

19.33   (a)   The entropy of a pure crystalline substance at absolute zero is zero.

        (b)   In *translational* motion, the entire molecule moves in a single direction; in *rotational* motion, the molecule rotates or spins around a fixed axis. *Vibrational* motion is reciprocating motion. The bonds within a molecule stretch and bend, but the average position of the atoms does not change.

        (c)

19.34   (a)   Since CO_2 has more than one atom, the thermal energy can be distributed as translational, vibrational, or rotational motion.

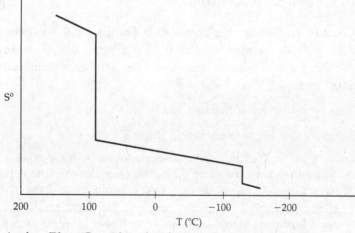

(b)   According to Boltzmann's Law, S = k lnW. The number of microstates, W, is directly proportional to entropy, S. Thus, if the number of microstates for a system increases, the entropy of the system increases.

19.35   *Analyze/Plan.* Consider the physical changes that occur when a substance is heated. How do these changes affect the entropy of the substance?

(a)   Both 1 and 2 represent changes in entropy at constant temperature; these are phase changes. Since 1 happens at lower temperature, it represents melting (fusion), and 2 represents vaporization.

(b)   The substance changes from solid to liquid in 1, from liquid to gas in 2. The larger volume and greater motional freedom of the gas phase causes ΔS for vaporization to (always) be larger than ΔS for fusion.

19.36   Melting = –126.5°C; boiling = 97.4°C.

19.37   *Analyze/Plan.* Consider the factors that lead to higher entropy: more mol gas in products than reactants, increase in volume of sample and, therefore, number of possible arrangements, more motional freedom of molecules, etc.   *Solve.*

(a)   Ar(g) (gases have higher entropy due primarily to much larger volume)

(b)   He(g) at 1.5 atm (larger volume and more motional freedom)

(c)   1 mol of Ne(g) in 15.0 L (larger volume provides more motional freedom)

(d) CO_2(g) (more motional freedom)

19.38   (a)   1 mol of $As_4(g)$ at 300°C, 0.01 atm (As_4 has more massive atoms in a comparable system at the same temperature.)

          (b)   1 mol $H_2O(g)$ at 100°C, 1 atm (larger volume occupied by $H_2O(g)$)

          (c)   0.5 mol $CH_4(g)$ at 298 K, 20-L volume (more complex molecule, more rotational and vibrational degrees of freedom)

          (d)   100 g of $Na_2SO_4(aq)$ at 30°C (more motional freedom in aqueous solution)

19.39   *Analyze/Plan.* Consider the markers of an increase in entropy for a chemical reaction: liquids or solutions formed from solids, gases formed from either solids or liquids, increase in moles gas during reaction.   *Solve.*

          (a) ΔS negative (moles of gas decrease)

          (b) ΔS positive (gas produced, increased disorder)

          (c) ΔS negative (moles of gas decrease)

          (d) ΔS positive (moles of gas increase)

19.40   (a) $Fe(l) \rightarrow Fe(s)$; ΔS is negative (less motional freedom)

          (b) $2Li(s) + Cl_2(g) \rightarrow 2LiCl$; ΔS is negative (moles of gas decrease)

          (c) $Zn(s) + 2HCl(aq) \rightarrow ZnCl_2(aq) + H_2(g)$; ΔS is positive (moles of gas increase)

          (d) $AgNO_3(aq) + KBr(aq) \rightarrow AgBr(s) + KNO_3(aq)$; ΔS is negative (less motional freedom)

Entropy Changes in Chemical Reactions

19.41   *Analyze/Plan.* Given two molecules in the same state, predict which will have the higher molar entropy. In general, for molecules in the same state, the more atoms in the molecule, the more degrees of freedom, the greater the number of microstates and the higher the standard entropy, $S°$.

          (a) $C_2H_6(g)$ has more degrees of freedom and larger $S°$.

          (b) $CO_2(g)$ has more degrees of freedom and larger $S°$.

19.42   Propylene will have a higher $S°$ at 25°C. At this temperature, both are gases, so there are no lattice effects. Since they have the same molecular formula, only the details of their structures are different. In propylene, there is free rotation around the C—C single bond, while in cyclopropane the 3-membered ring severely limits rotation. The greater motional freedom of the propylene molecule leads to a higher absolute entropy.

19.43   *Analyze/Plan.* Consider the conditions that lead to an increase in entropy: more mol gas in products than reactants, increase in volume of sample and, therefore, number of possible arrangements, more motional freedom of molecules, etc. *Solve.*

          (a) $Sc(s)$, 34.6 J/mol•K; $Sc(g)$, 174.7 J/mol•K. In general, the gas phase of a substance has a larger $S°$ than the solid phase because of the greater volume and motional freedom of the molecules.

(b) $NH_3(g)$, 192.5 J/mol•K; $NH_3(aq)$, 111.3 J/mol•K. Molecules in the gas phase have more motional freedom than molecules in solution.

(c) 1 mol of $P_4(g)$, 280 J/K; 2 mol of $P_2(g)$, 2(218.1) = 436.2 J/K. More particles have greater motional energy (more available microstates).

(d) C(diamond), 2.43 J/mol•K; C(graphite) 5.69 J/mol•K. Diamond is a network covalent solid with each C atom tetrahedrally bound to four other C atoms. Graphite consists of sheets of fused planar 6-membered rings with each C atom bound in a trigonal planar arrangement to three other C atoms. The internal entropy in graphite is greater because there is translational freedom among the planar sheets of C atoms while there is very little vibrational freedom within the network covalent diamond lattice.

19.44   (a) CuO(s), 42.59 J/mol•K; $Cu_2O(s)$, 92.36 J/mol•K. Molecules in the solid state have only vibrational motion available to them. The more complex Cu_2O molecule has more vibrational degrees of freedom and a larger standard entropy.

(b) 1 mol $N_2O_4(g)$, 304.3 J/K; 2 mol $NO_2(g)$, 2(240.45) = 480.90 J/K. More particles have a greater number of arrangements.

(c) $CH_3OH(g)$, 237.6 J/mol•K; $CH_3OH(l)$, 126.8 J/mol•K. Molecules in the gas phase occupy a larger volume and have more motional freedom than molecules in the liquid state.

(d) 1 mol $PbCO_3(s)$, 131.0 J/K; 1 mol PbO(s) + 1 mol $CO_2(g)$, (68.70 + 213.6) = 282.3 J/K. The second member of the pair has more total particles and half of them are in the gas phase for greater total motional freedom. Note that 1 mol of $PbCO_3(s)$ has greater entropy than 1 mol of PbO(s), because of the additional ways to store energy in the more complex CO_3^{2-} anion.

19.45   For elements with similar structures, the heavier the atoms, the lower the vibrational frequencies at a given temperature. This means that more vibrations can be accessed at a particular temperature resulting in a greater absolute entropy for the heavier elements.

19.46   (a) C(diamond), $S°$ = 2.43 J/mol•K; C(graphite), $S°$ = 5.69 J/mol•K. Diamond is a network covalent solid with each C atom tetrahedrally bound to four other C atoms. Graphite consists of sheets of fused planar 6-membered rings with each C atom bound in a trigonal planar arrangement to three other C atoms. The internal entropy in graphite is greater because there is translational freedom among the planar sheets of C atoms, while there is very little translational or vibrational freedom within the covalent-network diamond lattice.

(b) $S°$ for buckminsterfullerene will be ≥ 10 J/mol•K. $S°$ for graphite is twice $S°$ for diamond, and $S°$ for the fullerene should be higher than that of graphite. The 60-atom "bucky" balls have more flexibility than graphite sheets. Also, the balls have translational freedom in three dimensions, while graphite sheets have it in only two directions. Because of the ball structure, there is more empty space in the fullerene lattice than in graphite or diamond; essentially, 60 C-atoms in fullerene occupy a larger volume than 60 C-atoms in graphite or diamond. Thus, the fullerene has additional "molecular" complexity, more degrees of translational freedom, and occupies a larger volume, all features that point to a higher absolute entropy.

19.47   *Analyze/Plan.* Follow the logic in Sample Exercise 19.5.   *Solve.*

(a) $\Delta S° = S° \, C_2H_6(g) - S° \, C_2H_4(g) - S° \, H_2(g)$

      $= 229.5 - 219.4 - 130.58 = -120.5 \, J/K$

$\Delta S°$ is negative because there are fewer moles of gas in the products.

(b) $\Delta S° = 2S° \, NO_2(g) - \Delta S° \, N_2O_4(g) = 2(240.45) - 304.3 = +176.6 \, J/K$

$\Delta S°$ is positive because there are more moles of gas in the products.

(c) $\Delta S° = \Delta S° \, BeO(s) + \Delta S° \, H_2O(g) - \Delta S° \, Be(OH)_2(s)$

      $= 13.77 + 188.83 - 50.21 = +152.39 \, J/K$

$\Delta S°$ is positive because the product contains more total particles and more moles of gas.

(d) $\Delta S° = 2S° \, CO_2(g) + 4S° \, H_2O(g) - 2S° \, CH_3OH(g) - 3S° \, O_2(g)$

      $= 2(213.6) + 4(188.83) - 2(237.6) - 3(205.0) = +92.3 \, J/K$

$\Delta S°$ is positive because the product contains more total particles and more moles of gas.

19.48   (a) $\Delta S° = 2S° \, NH_3(g) - S° \, N_2H_4(g) - S° \, H_2(g)$

      $= 2(192.5) - 238.5 - 130.58 = +15.9 \, J/K$

$\Delta S°$ is small because there are the same number of moles of gas in the products as in reactants. The slight increase is due to the relatively small $S°$ value of $H_2(g)$, which has fewer degrees of freedom than molecules with more than two atoms.

(b) $\Delta S° = 2S° \, AlCl_3(s) - 2S° \, Al(s) - 3S° \, Cl_2(g)$

      $= 2(109.3) - 2(28.32) - 3(222.96) = -506.9 \, J/K$

$\Delta S°$ is negative because the products contain fewer (no) moles of gas.

(c) $\Delta S° = S° \, MgCl_2(s) + 2S°H_2O(l) - S° \, Mg(OH)_2(s) - 2S° \, HCl(g)$

      $= 89.6 + 2(69.91) - 63.24 - 2(186.69) = -207.2 \, J/K$

$\Delta S°$ is negative because the products contain fewer (no) moles of gas.

(d) $\Delta S° = S° \, C_2H_6(g) + S° \, H_2(g) - 2S° \, CH_4(g)$

      $= 229.5 + 130.58 - 2(186.3) = -12.5 \, J/K$

$\Delta S°$ is very small because there are the same number of moles of gas in the products and reactants. The slight decrease is related to the relatively small $S°$ value for $H_2(g)$, which has fewer degrees of freedom than molecules with more than two atoms.

19 Chemical Thermodynamics   **Solutions to Exercises**

Gibbs Free Energy

19.49   (a) $\Delta G = \Delta H - T\Delta S$

(b)   If ΔG is positive, the process is nonspontaneous, but the reverse process is spontaneous.

(c)   There is no relationship between ΔG and rate of reaction. A spontaneous reaction, one with a $-\Delta G$, may occur at a very slow rate. For example: $2H_2(g) + O_2(g) \rightarrow 2H_2O(g)$, $\Delta G = -457$ kJ is very slow if not initiated by a spark.

19.50   (a)   The *standard* free energy change, $\Delta G°$, represents the free energy change for the process when all reactants and products are in their standard states. When any or all reactants or products are not in their standard states, the free energy is represented simply as ΔG. The value for ΔG thus depends on the specific states of all reactants and products.

(b)   When $\Delta G = 0$, the system is at equilibrium.

(c)   The sign and magnitude of ΔG give no information about rate; we cannot predict whether the reaction will occur rapidly.

19.51   *Analyze/Plan.* Consider the definitions of $\Delta H°$, $\Delta S°$ and $\Delta G°$, along with sign conventions. $\Delta G° = \Delta H° - T\Delta S°$.   *Solve.*

(a) $\Delta H°$ is negative; the reaction is exothermic.

(b) $\Delta S°$ is negative; the reaction leads to decrease in disorder (increase in order) of the system.

(c) $\Delta G° = \Delta H° - T\Delta S° = -35.4$ kJ $- 298$ K $(-0.0855$ kJ/K$) = -9.921 = -9.9$ kJ

(d)   At 298 K, $\Delta G°$ is negative. If all reactants and products are present in their standard states, the reaction is spontaneous (in the forward direction) at this temperature.

19.52   (a) $\Delta H°$ is negative; the reaction is exothermic.

(b) $\Delta S°$ is positive; the reaction leads to an increase in disorder.

(c) $\Delta G° = \Delta H° - T\Delta S° = -19.5$ kJ $- 298$ K $(0.0427$ kJ/K$) = -32.225 = -32.2$ kJ

(d)   At 298 K, $\Delta G°$ is negative. If all reactants and products are present in their standard states, the reaction is spontaneous (in the forward direction) at this temperature.

19.53   *Analyze/Plan.* Follow the logic in Sample Exercise 19.7. Calculate $\Delta H°$ according to Equation [5.31], $\Delta S°$ by Equation [19.8] and $\Delta G°$ by Equation [19.13]. Then use $\Delta H°$ and $\Delta S°$ to calculate $\Delta G°$ using Equation [19.20], $\Delta G° = \Delta H° - T\Delta S°$.   *Solve.*

(a) $\Delta H° = 2(-268.61) - [0 + 0] = -537.22$ kJ

$\Delta S° = 2(173.51) - [130.58 + 202.7] = 13.74 = 13.7$ J/K

$\Delta G° = 2(-270.70) - [0 + 0] = -541.40$ kJ

$\Delta G° = -537.22$ kJ $- 298(0.01374)$ kJ $= -541.31$ kJ

(b) $\Delta H° = -106.7 - [0 + 2(0)] = -106.7$ kJ

$\Delta S° = 309.4 - [5.69 + 2(222.96)] = -142.21 = -142.2$ J/K

$\Delta G° = -64.0 - [0 + 2(0)] = -64.0$ kJ

$\Delta G° = -106.7$ kJ $- 298(-0.14221)$ kJ $= -64.3$ kJ

(c) $\Delta H° = 2(-542.2) - [2(-288.07) + 0] = -508.26 = -508.3$ kJ

$\Delta S° = 2(325) - [2(311.7) + 205.0] = -178.4 = -178$ J/K

$\Delta G° = 2(-502.5) - [2(-269.6) + 0] = -465.8$ kJ

$\Delta G° = -508.26$ kJ $- 298(-0.1784)$ kJ $= -455.097 = -455.1$ kJ

(The discrepancy in $\Delta G°$ values is due to experimental uncertainties in the tabulated thermodynamic data.)

(d) $\Delta H° = -84.68 + 2(-241.82) - [2(-201.2) + 0] = -165.92 = -165.9$ kJ

$\Delta S° = 229.5 + 2(188.83) - [2(237.6) + 130.58] = 1.38 = 1.4$ J/K

$\Delta G° = -32.89 + 2(-228.57) - [2(-161.9) + 0] = -166.23 = -166.2$ kJ

$\Delta G° = -165.92$ kJ $- 298(0.00138)$ kJ $= -166.33 = -166.3$ kJ

19.54   (a) $\Delta H° = -305.3 - [0 + 0] = -305.3$ kJ

$\Delta S° = 97.65 - [29.9 + 222.96] = -155.21 = -155.2$ J/K

$\Delta G° = -259.0 - [0 + 0] = -259.0$ kJ

$\Delta G° = -305.3$ kJ $- 298(-0.15521)$ kJ $= -259.047 = -259.0$ kJ

(b) $\Delta H° = -635.5 + (-393.5) - (-1207.1) = 178.1$ kJ

$\Delta S° = 39.75 + 213.6 - (92.88) = 160.47 = 160.5$ J/K

$\Delta G° = -604.17 + (-394.4) - (-1128.76) = 130.19 = 130.2$ kJ

$\Delta G° = 178.1$ kJ $- 298(0.16047)$ kJ $= 130.28 = 130.3$ kJ

(c) $\Delta H° = 4(-1288.3) - [-2940.1 + 6(-285.83)] = -498.12 = -498.1$ kJ

$\Delta S° = 4(158.2) - [228.9 + 6(69.91)] = -15.56 = -15.6$ J/K

$\Delta G° = 4(-1142.6) - [-2675.2 + 6(-237.13)] = -472.42 = -472.4$ kJ

$\Delta G° = -498.12$ kJ $- 298(-0.01556)$ kJ $= -493.48 = -493.5$ kJ

(The discrepancy in $\Delta G°$ values is due to experimental uncertainties in the tabulated thermodynamic data.)

(d) $\Delta H° = 2(-393.5) + 4(-285.83) - [2(-238.6) + 3(0)] = -1453.1$ kJ

$\Delta S° = 2(213.6) + 4(69.91) - [2(126.8) + 3(205.0)] = -161.76 = -161.8$ J/K

$\Delta G° = 2(-394.4) + 4(-237.13) - [2(-166.23) + 3(0)] = -1404.86 = -1404.9$ kJ

$\Delta G° = -1453.2$ kJ $- 298(-0.16176)$ kJ $= -1404.996 = -1405.0$ kJ

19.55   *Analyze/Plan.* Follow the logic in Sample Exercise 19.6.   *Solve.*

(a) $\Delta G° = 2\Delta G° \ SO_3(g) - [2\Delta G° \ SO_2(g) + \Delta G° \ O_2(g)]$

$= 2(-370.4) - [2(-300.4) + 0] = -140.0$ kJ, spontaneous

(b) $\Delta G° = 3\Delta G° \; NO(g) - [\Delta G° \; NO_2(g) + \Delta G° \; N_2O(g)]$

   $= 3(86.71) - [51.84 + 103.59] = +104.70 \; kJ$, nonspontaneous

(c) $\Delta G° = 4\Delta G° \; FeCl_3(s) + 3\Delta G° \; O_2(g) - [6\Delta G° \; Cl_2(g) + 2\Delta G° \; Fe_2O_3(s)]$

   $= 4(-334) + 3(0) - [6(0) + 2(-740.98)] = +146 \; kJ$, nonspontaneous

(d) $\Delta G° = \Delta G° \; S(s) + 2\Delta G° \; H_2O(g) - [\Delta G° \; SO_2(g) + 2\Delta G° \; H_2(g)]$

   $= 0 + 2(-228.57) - [(-300.4) + 2(0)] = -156.7 \; kJ$, spontaneous

19.56   (a) $\Delta G° = 2\Delta G° \; HCl(g) - [\Delta G° \; H_2(g) + \Delta G° \; Cl_2(g)]$

   $= 2(-95.27 \; kJ) - 0 - 0 = -190.5 \; kJ$, spontaneous

(b) $\Delta G° = \Delta G° \; MgO(s) + 2\Delta G° \; HCl(g) - [\Delta G° \; MgCl_2(s) + \Delta G° \; H_2O(l)]$

   $= -569.6 + 2(-95.27) - [-592.1 + (-237.13)] = + 69.1 \; kJ$, nonspontaneous

(c) $\Delta G° = \Delta G° \; N_2H_4(g) + \Delta G° \; H_2(g) - 2\Delta G° \; NH_3(g)$

   $= 159.4 + 0 - 2(-16.66) = +192.7 \; kJ$, nonspontaneous

(d) $\Delta G° = 2\Delta G° \; NO(g) + \Delta G° \; Cl_2(g) - 2\Delta G° \; NOCl(g)$

   $= 2(86.71) + 0 - 2(66.3) = +40.8 \; kJ$, nonspontaneous

19.57   *Analyze/Plan.* Follow the logic in Sample Exercise 19.8(a).   *Solve.*

(a) $C_6H_{12}(l) + 9O_2(g) \rightarrow 6CO_2(g) + 6H_2O(l)$

(b)   Because there are fewer moles of gas in the products, $\Delta S°$ is negative, which makes $-T\Delta S$ positive. $\Delta G°$ is less negative (more positive) than $\Delta H°$.

19.58   (a) $\Delta G°$ should be less negative than $\Delta H°$. Products contain fewer moles of gas, so $\Delta S°$ is negative. $\Delta G° = \Delta H° - T\Delta S°$; $-T\Delta S°$ is positive so $\Delta G°$ is less negative than $\Delta H°$.

(b)   We can estimate $\Delta S°$ using a similar reaction and then use $\Delta G° = \Delta H° - T\Delta S°$ (estimate) to get a ballpark figure. There are no sulfite salts listed in Appendix C, so use a reaction such as $CO_2(g) + CaO(s) \rightarrow CaCO_3(s)$ or $CO_2(g) + BaO(s) \rightarrow BaCO_3(s)$. Or calculate both $\Delta S°$ values and use the average as your estimate.

19.59   *Analyze/Plan.* Based on the signs of ΔH and ΔS for a particular reaction, assign a category from Table 19.4 to each reaction.   *Solve.*

(a) ΔG is negative at low temperatures, positive at high temperatures. That is, the reaction proceeds in the forward direction spontaneously at lower temperatures but spontaneously reverses at higher temperatures.

(b) ΔG is positive at all temperatures. The reaction is nonspontaneous in the forward direction at all temperatures.

(c) ΔG is positive at low temperatures, negative at high temperatures. That is, the reaction will proceed spontaneously in the forward direction at high temperature.

19.60 $\Delta G° = \Delta H° - T\Delta S°$

(a) $\Delta G° = -844$ kJ $- 298$ K$(-0.165$ kJ/K$) = -795$ kJ, spontaneous

(b) $\Delta G° = +572$ kJ $- 298$ K$(0.179$ kJ/K$) = +519$ kJ, nonspontaneous

To be spontaneous, ΔG must be negative ($\Delta G < 0$).

Thus, $\Delta H° - T\Delta S° < 0$; $\Delta H° < T\Delta S°$; $T > \Delta H°/\Delta S°$; $T > \dfrac{572 \text{ kJ}}{0.179 \text{ kJ/K}} = 3.20 \times 10^3$ K

19.61   *Analyze/Plan.* We are told that the reaction is spontaneous and endothermic, and asked to estimate the sign and magnitude of ΔS. If a reaction is spontaneous, $\Delta G < 0$. Use this information with Equation [19.20] to solve the problem.   *Solve.*

At 450 K, $\Delta G < 0$; $\Delta G = \Delta H - T\Delta S < 0$

34.5 kJ $- 450$ K $(\Delta S) < 0$; 34.5 kJ < 450 K (ΔS); $\Delta S > 34.5$ kJ/450 K

$\Delta S > 0.0767$ kJ/K or $\Delta S > 76.7$ J/K

19.62   At $-25°$C or 248 K, $\Delta G > 0$. $\Delta G = \Delta H - T\Delta S > 0$

$\Delta H - 248$ K $(95$ J/K$) > 0$; $\Delta H > +2.4 \times 10^4$ J; $\Delta H > +24$ kJ

19.63   *Analyze/Plan.* Use Equation [19.20] to calculate T when $\Delta G = 0$. This is similar to calculating the temperature of a phase trasition in Sample Exercise 19.9. Use Table 19.4 to determine whether the reaction is spontaneous or non-spontaneous above this temperature.   *Solve.*

(a) $\Delta G = \Delta H - T\Delta S$; $0 = -32$ kJ $- T(-98$ J/K$)$; 32×10^3 J $= T(98$ J/K$)$

$T = 32 \times 10^3$ J$/(98$ J/K$) = 326.5 = 330$ K

(b)   Nonspontaneous. The sign of ΔS is negative, so as T increases, ΔG becomes more positive.

19.64 ΔG is negative when $T\Delta S > \Delta H$ or $T > \Delta H/\Delta S$.

$\Delta H° = \Delta H°$ CH$_4$(g) $+ \Delta H°$ CO$_2$(g) $- \Delta H°$ CH$_3$COOH(l)

$\quad = -74.8 + (-393.5) - (-487.0) = +18.7$ kJ

$\Delta S° = S°$ CH$_4$(g) $+ S°$ CO$_2$(g) $- S°$ CH$_3$COOH(l) $= +186.3 + 213.6 - 159.8 = +240.1$ J/K

$T > \dfrac{18.7 \text{ kJ}}{0.2401 \text{ kJ/K}} = 77.9$ K

The reaction is spontaneous above 77.9 K ($-195°$C).

19.65   *Analyze/Plan.* Given a chemical equation and thermodynamic data (values of $\Delta H_f°, \Delta G_f°$ and $S°$) for reactants and products, predict the variation of $\Delta G°$ with temperature and calculate $\Delta G°$ at 800 K and 1000 K. Use Equations [5.31] and [19.8] to calculate $\Delta H°$ and $\Delta S°$, respectively; use these values to calculate $\Delta G°$ at various temperatures, using Equation [19.20]. The signs of $\Delta H°$ and $\Delta S°$ determine the variation of $\Delta G°$ with temperature.   *Solve.*

(a)   Calculate $\Delta H°$ and $\Delta S°$ to determine the sign of $T\Delta S°$.

$\Delta H° = 3\Delta H°$ NO(g) $- \Delta H°$ NO$_2$(g) $- \Delta H°$ N$_2$O(g)

$\quad = 3(90.37) - 33.84 - 81.6 = 155.7$ kJ

$\Delta S° = 3S° \, NO(g) - S° \, NO_2(g) - S° \, N_2O(g)$

$= 3(210.62) - 240.45 - 220.0 = 171.4 \, J/K$

$\Delta G° = \Delta H° - T\Delta S°$. Since $\Delta S°$ is positive, $-T\Delta S°$ becomes more negative as T increases and $\Delta G°$ becomes more negative.

(b) $\Delta G° = \Delta H° - T\Delta S° = 155.7 \, kJ - (800 \, K)(0.1714 \, kJ/K)$

$\Delta G° = 155.7 \, kJ - 137 \, kJ = 19 \, kJ$

Since $\Delta G°$ is positive at 800 K, the reaction is not spontaneous at this temperature.

(c) $\Delta G° = 155.7 \, kJ - (1000 \, K)(0.1714 \, kJ/K) = 155.7 \, kJ - 171.4 \, kJ = -15.7 \, kJ$

$\Delta G°$ is negative at 1000 K and the reaction is spontaneous at this temperature.

19.66 (a) $\Delta H° = \Delta H_f° \, CH_3OH(g) - \Delta H_f° \, CH_4(g) - 1/2 \, \Delta H_f° \, O_2(g)$

$= -201.2 - (-74.8) - 0 = -126.4 \, kJ$

$\Delta S° = S° \, CH_3OH(g) - S° \, CH_4(g) - 1/2 \, S° \, O_2(g)$

$= 237.6 - 186.3 - 1/2(205.0) = -51.2 \, J/K = -0.0512 \, kJ/K$

(b) $\Delta G° = \Delta H° - T\Delta S°$. $-T\Delta S°$ is positive, so $\Delta G°$ becomes more positive as temperature increases.

(c) $\Delta G° = \Delta H° - T\Delta S° = -126.4 \, kJ - 298 \, K(-0.0512 \, kJ/K) = -111.1 \, kJ$

The reaction is spontaneous at 298 K because $\Delta G°$ is negative at this temperature. In this case, $\Delta G°$ could have been calculated from $\Delta G_f°$ values in Appendix C, since these values are tabulated at 298 K.

(d) The reaction is at equilibrium when $\Delta G° = 0$.

$\Delta G° = \Delta H° - T\Delta S° = 0$. $\Delta H° = T\Delta S°$, $T = \Delta H°/\Delta S°$

$T = -126.4 \, kJ/-0.0512 \, kJ/K = 2469 = 2470 \, K$.

This temperature is so high that the reactants and products are likely to decompose. At standard conditions, equilibrium is functionally unattainable for this reaction.

19.67 *Analyze/Plan.* Follow the logic in Sample Exercise 19.9.   *Solve.*

(a) $\Delta S_{vap}° = \Delta H_{vap}°/T_b$; $T_b = \Delta H_{vap}°/\Delta S_{vap}°$

$\Delta H_{vap}° = \Delta H° \, C_6H_6(g) - \Delta H° \, C_6H_6(l) = 82.9 - 49.0 = 33.9 \, kJ$

$\Delta S_{vap}° = S° \, C_6H_6(g) - S° \, C_6H_6(l) = 269.2 - 172.8 = 96.4 \, J/K$

$T_b = 33.9 \times 10^3 \, J/96.4 \, J/K = 351.66 = 352 \, K = 79°C$

(b) From the *Handbook of Chemistry and Physics*, 74th Edition, $T_b = 80.1°C$. The values are remarkably close; the small difference is due to deviation from ideal behavior by $C_6H_6(g)$ and experimental uncertainty in the boiling point measurement and the thermodynamic data.

19.68 (a) As in Sample Exercise 19.9, $T_{sub} = \Delta H_{sub}°/\Delta S_{sub}°$

Use Data from Appendix C to calculate $\Delta H_{sub}°$ and $\Delta S_{sub}°$ for $I_2(s)$.

$$I_2(s) \rightarrow I_2(l) \text{ melting}$$
$$I_2(l) \rightarrow I_2(g) \text{ boiling}$$

$$\overline{I_2(s) \rightarrow I_2(g) \text{ sublimation}}$$

$$\Delta H^{\circ}_{sub} = \Delta H^{\circ}_f I_2(g) - \Delta H^{\circ}_f I_2(s) = 62.25 - 0 = 62.25 \text{ kJ}$$

$$\Delta S^{\circ}_{sub} = S^{\circ} I_2(g) - S^{\circ} I_2(s) = 260.57 - 116.73 = 143.84 \text{ J/K} = 0.14384 \text{ kJ/K}$$

$$T_{sub} = \frac{\Delta H^{\circ}_{sub}}{\Delta S^{\circ}_{sub}} = \frac{62.25 \text{ kJ}}{0.14384 \text{ kJ/K}} = 432.8 \text{ K} = 159.6^{\circ}\text{C}$$

(b) T_m for $I_2(s) = 386.85 \text{ K} = 113.7^{\circ}\text{C}$; $T_b = 457.4 \text{ K} = 184.3^{\circ}\text{C}$

(from WebElements™, 2005)

(c)   The boiling point of I_2 is closer to the sublimation temperature. Both boiling and sublimation begin with molecules in a condensed phase (little space between molecules) and end in the gas phase (large intermolecular distances). Separation of the molecules is the main phenomenon that determines both ΔH and ΔS, so it is not surprising that the ratio of $\Delta H / \Delta S$ is similar for sublimation and boiling.

19.69   *Analyze/Plan.* We are asked to write a balanced equation for the combustion of acetylene, calculate ΔH° for this reaction and calculate maximum useful work possible by the system. Combustion is combination with O_2 to produce CO_2 and H_2O. Calculate ΔH° using data from Appendix C and Equation 5.31. The maximum obtainable work is ΔG (Equation [19.19]), which can be calculated from data in Appendix C and Equation [19.13].   *Solve.*

(a) $C_2H_2(g) + 5/2 \, O_2(g) \rightarrow 2CO_2(g) + H_2O(l)$

(b) $\Delta H^{\circ} = 2\Delta H^{\circ} \, CO_2(g) + \Delta H^{\circ} \, H_2O(l) - \Delta H^{\circ} \, C_2H_2(g) - 5/2\Delta H^{\circ} \, O_2(g)$

        $= 2(-393.5) - 285.83 - 226.7 - 5/2(0) = -1299.5$ kJ produced/mol C_2H_2 burned

(c) $w_{max} = \Delta G^{\circ} = 2\Delta G^{\circ} \, CO_2(g) + \Delta G^{\circ} \, H_2O(l) - \Delta G^{\circ} \, C_2H_2(g) - 5/2 \, \Delta G^{\circ} \, O_2(g)$

        $= 2(-394.4) - 237.13 - 209.2 - 5/2(0) = -1235.1$ kJ

The negative sign indicates that the system does work on the surroundings; the system can accomplish a maximum of 1235.1 kJ of work on its surroundings.

19.70   (a) $C_2H_4(g) + 3O_2(g) \rightarrow 2CO_2(g) + 2H_2O(l)$

        $\Delta H^{\circ} = 2\Delta H^{\circ}_f \, CO_2(g) + 2\Delta H^{\circ}_f \, H_2O(l) - \Delta H^{\circ}_f \, C_2H_4(g) - 3\Delta H^{\circ}_f \, O_2(g)$

          $= 2(-393.5) + 2(-285.83) - 52.30 - 3(0) = -1410.96 = -1411.0$ kJ/mol C_2H_4 burned

(b) $w_{max} = \Delta G^{\circ} = 2\Delta G^{\circ}_f \, CO_2(g) + 2\Delta H^{\circ}_f \, H_2O(l) - \Delta G^{\circ}_f \, C_2H_4(g) - 3\Delta G^{\circ}_f \, O_2(g)$

          $= 2(-394.4) + 2(-237.13) - 68.11 - 3(0) = -1331.2$ kJ

The system can accomplish at most 1331.2 kJ of work per mole of C_2H_4 on the surroundings.

19 Chemical Thermodynamics

Free Energy and Equilibrium

19.71   *Analyze/Plan.* We are given a chemical reaction and asked to predict the effect of the partial pressure of $O_2(g)$ on the value of ΔG for the system. Consider the relationship $\Delta G = \Delta G° + RT \ln Q$ where Q is the reaction quotient.   *Solve.*

    (a) $O_2(g)$ appears in the denominator of Q for this reaction. An increase in pressure of O_2 decreases Q and ΔG becomes smaller or more negative. Increasing the concentration of a reactant increases the tendency for a reaction to occur.

    (b) $O_2(g)$ appears in the numerator of Q for this reaction. Increasing the pressure of O_2 increases Q and ΔG becomes more positive. Increasing the concentration of a product decreases the tendency for the reaction to occur.

    (c) $O_2(g)$ appears in the numerator of Q for this reaction. An increase in pressure of O_2 increases Q and ΔG becomes more positive. Since pressure of O_2 is raised to the third power in Q, an increase in pressure of O_2 will have the largest effect on ΔG for this reaction.

19.72   Consider the relationship $\Delta G = \Delta G° + RT \ln Q$, where Q is the reaction quotient.

    (a) $H_2(g)$ appears in the denominator of Q for this reaction. An increase in pressure of H_2 decreases Q and ΔG becomes smaller or more negative. Increasing the concentration of a reactant increases the tendency for a reaction to occur.

    (b) $H_2(g)$ appears in the numerator of Q for this reaction. Increasing the pressure of H_2 increases Q and ΔG becomes more positive. Increasing the concentration of a product decreases the tendency for the reaction to occur.

    (c) $H_2(g)$ appears in the denominator of Q for this reaction. An increase in pressure of H_2 decreases Q and ΔG becomes smaller or more negative.

19.73   *Analyze/Plan.* Given a chemical reaction, we are asked to calculate $\Delta G°$ from Appendix C data, and ΔG for a given set of initial conditions. Use Equation [19.13] to calculate $\Delta G°$, and Equation [19.21] to calculate ΔG. Follow the logic in Sample Exercise 19.10 when calculating ΔG.   *Solve.*

    (a) $\Delta G° = \Delta G°\ N_2O_4(g) - 2\Delta G°\ NO_2(g) = 98.28 - 2(51.84) = -5.40\ kJ$

    (b) $\Delta G = \Delta G° + RT \ln P_{N_2O_4} / P_{NO_2}^2$

$$= -5.40\ kJ + \frac{8.314 \times 10^{-3}\ kJ}{K \bullet mol} \times 298\ K \times \ln[1.60/(0.40)^2] = 0.3048 = 0.3\ kJ$$

19.74   (a) $\Delta G° = 2\Delta G°\ HF(g) - [\Delta G°\ H_2(g) + \Delta G°\ F_2(g)] = 2(-270.70) - [0 + 0] = -541.40\ kJ$

    (b) $\Delta G = \Delta G° + RT \ln P_{HF}^2 / P_{H_2} \times P_{F_2}$

$$= -541.40 + \frac{8.314 \times 10^{-3}\ kJ}{K \bullet mol} \times 298\ K\ \ln[(0.36)^2/8.0 \times 4.5] = -555.34 = -555\ kJ$$

19.75   *Analyze/Plan.* Given a chemical reaction, we are asked to calculate K using $\Delta G_f°$ data from Appendix C. Follow the logic in Sample Exercise 19.13. $\Delta G° = -RT \ln K$, Equation [19.22]; $\ln K = -\Delta G°/RT$ *Solve.*

(a) $\Delta G° = 2\Delta G° \, HI(g) - \Delta G° \, H_2(g) - \Delta G° \, I_2(g)$

$= 2(1.30) - 0 - 19.37 = -16.77 \text{ kJ}$

$$\ln K = \frac{-(-16.77 \text{ kJ}) \times 10^3 \text{ J/kJ}}{8.314 \text{ J/K} \times 298 \text{ K}} = 6.76876 = 6.769; \quad K = 870$$

(b) $\Delta G° = \Delta G° \, C_2H_4(g) + \Delta G° \, H_2O(g) - \Delta G° \, C_2H_5OH(g)$

$= 68.11 - 228.57 - (-168.5) = 8.04 = 8.0 \text{ kJ}$

$$\ln K = \frac{-(8.04 \text{ kJ}) \times 10^3 \text{ J/kJ}}{8.314 \text{ J/K} \times 298 \text{ K}} = -3.24511 = -3.25; K = 0.039$$

(c) $\Delta G° = \Delta G° \, C_6H_6(g) - 3\Delta G° \, C_2H_2(g) = 129.7 - 3(209.2) = -497.9 \text{ kJ}$

$$\ln K = \frac{-\Delta G°}{RT} = \frac{-(-497.9 \text{ kJ}) \times 10^3 \text{ J/KJ}}{8.314 \text{ J/K} \times 298 \text{ K}} = 200.963 = 201.0; K = 2 \times 10^{87}$$

19.76 $\Delta G° = -RT \ln K; \ln K = -\Delta G° / RT;$ at 298 K, RT = 2.4776 = 2.478 kJ

(a) $\Delta G° = \Delta G° \, NaOH(s) + \Delta G° \, CO_2(g) - \Delta G° \, NaHCO_3(s)$

$= -379.5 + (-394.4) - (-851.8) = +77.9 \text{ kJ}$

$$\ln K = \frac{-\Delta G°}{RT} = \frac{-77.9 \text{ kJ}}{2.478 \text{ kJ}} = -31.442 = -31.4; \quad K = 2 \times 10^{-14}$$

$K = P_{CO_2} = 2 \times 10^{-14}$

(b) $\Delta G° = 2\Delta G° \, HCl(g) + \Delta G° \, Br_2(g) - 2\Delta G° \, HBr(g) - \Delta G° \, Cl_2(g)$

$= 2(-95.27) + 3.14 - 2(-53.22) - 0 = -80.96 \text{ kJ}$

$$\ln K = \frac{-(-80.96)}{2.4776} = +32.68; \quad K = 1.6 \times 10^{14}$$

$$K = \frac{P_{HCl}^2 \times P_{Br_2}}{P_{HBr}^2 \times P_{Cl_2}} = 1.6 \times 10^{14}$$

(c)   From Exercise 19.55(a), $\Delta G°$ at 298 K = -140.0 kJ.

$$\ln K = \frac{-\Delta G°}{RT} = \frac{-(-140.0)}{2.4776} = 56.51; \quad K = 3.5 \times 10^{24}$$

$$K = \frac{P_{SO_3}^2}{P_{SO_2}^2 \times P_{O_2}} = 3.5 \times 10^{24}$$

19.77   *Analyze/Plan.* Given a chemical reaction and thermodynamic data in Appendix C, calculate the equilibrium pressure of $CO_2(g)$ at two temperatures. $K = P_{CO_2}$. Calculate $\Delta G°$ at the two temperatures using $\Delta G° = \Delta H° - T\Delta S°$ and then calculate K and P_{CO_2}. *Solve.*

$\Delta H° = \Delta H° \, BaO(s) + \Delta H° \, CO_2(g) - \Delta H° \, BaCO_3(s)$

$= -553.5 + -393.5 - (-1216.3) = +269.3 \text{ kJ}$

$\Delta S° = S° \, BaO(s) + S° \, CO_2(g) - S° \, BaCO_3(s)$

$= 70.42 + 213.6 - 112.1 = 171.92 \text{ J/K} = 0.1719 \text{ kJ/K}$

(a) ΔG at 298 K = 269.3 kJ – 298 K (0.17192 kJ/K) = 218.07 = 218.1 kJ

$$\ln K = \frac{-\Delta G^{\circ}}{RT} = \frac{-218.07 \times 10^3 \text{ J}}{8.314 \text{ J/K} \times 298 \text{ K}} = -88.017 = -88.02$$

$K = 6.0 \times 10^{-39}; \quad P_{CO_2} = 6.0 \times 10^{-39} \text{ atm}$

(b) ΔG at 1100 K = 269.3 kJ – 1100 K (0.17192 kJ) = 80.19 = +80.2 kJ

$$\ln K = \frac{-\Delta G^{\circ}}{RT} = \frac{-80.19 \times 10^3 \text{ J}}{8.314 \text{ J/K} \times 1100 \text{ K}} = -8.768 = -8.77$$

$K = 1.6 \times 10^{-4}; \quad P_{CO_2} = 1.6 \times 10^{-4} \text{ atm}$

19.78 $K = P_{CO_2}$. Calculate ΔG° at the two temperatures using $\Delta G^{\circ} = \Delta H^{\circ} - T\Delta S^{\circ}$ and then calculate K and P_{CO_2}.

$\Delta H^{\circ} = \Delta H^{\circ} \text{ PbO(s)} + \Delta H^{\circ} \text{ CO}_2(g) - \Delta H^{\circ} \text{ PbCO}_3(s)$

       = –217.3 – 393.5 + 699.1 = 88.3 kJ

$\Delta S^{\circ} = S^{\circ} \text{ PbO(s)} + S^{\circ} \text{ CO}_2(g) - S^{\circ} \text{ PbCO}_3(s)$

       = 68.70 + 213.6 – 131.0 = 151.3 J/K or 0.1513 kJ/K

(a) $\Delta G^{\circ} = \Delta H^{\circ} - T\Delta S^{\circ}$. At 393 K, ΔG° = 88.3 kJ – 393 K(0.1513 kJ/K) = 28.84

                                                 = 28.8 kJ

$$\ln K = \frac{-\Delta G^{\circ}}{RT} = \frac{-28.84 \times 10^3 \text{ J}}{8.314 \text{ J/K} \times 393 \text{ K}} = -8.82631 = -8.83$$

$K = P_{CO_2} = 1.5 \times 10^{-4} \text{ atm}$

(b) $\Delta G^{\circ} = \Delta H^{\circ} - T\Delta S^{\circ}$. At 753 K, ΔG° = 88.3 kJ – 753 K (0.1513 kJ) = –25.629

                                                 = –25.6 kJ

$$\ln K = \frac{-(-25.629 \times 10^3 \text{ J})}{8.314 \text{ J/K} \times 753 \text{ K}} = 4.0938 = 4.09; \; K = P_{CO_2} = 60 \text{ atm}$$

19.79   *Analyze/Plan.* Given an acid dissociation equilibrium and the corresponding K_a value, calculate ΔG° and ΔG for a given set of concentrations. Use Equation [19.22] to calculate ΔG° and Equation [19.21] to calculate ΔG.   *Solve.*

(a) $HNO_2(aq) \;\rightleftharpoons\; H^+(aq) + NO_2^-(aq)$

(b) $\Delta G^{\circ} = -RT \ln K_a = -(8.314 \times 10^{-3})(298) \ln (4.5 \times 10^{-4}) = 19.0928 = 19.1$ kJ

(c) $\Delta G = 0$ at equilibrium

(d) $\Delta G = \Delta G^{\circ} + RT \ln Q$

          $= 19.09 \text{ kJ} + (8.314 \times 10^{-3})(298) \ln \dfrac{(5.0 \times 10^{-2})(6.0 \times 10^{-4})}{0.20} = -2.72 = -3 \text{ kJ}$

19.80   (a) $CH_3NH_2(aq) + H_2O(l) \;\rightleftharpoons\; CH_3NH_3^+(aq) + OH^-(aq)$

          (b) $\Delta G^{\circ} = -RT \ln K_b = -(8.314 \times 10^{-3})(298) \ln (4.4 \times 10^{-4}) = 19.148 = 19.1$ kJ

(c) $\Delta G = 0$ at equilibrium

(d) $\Delta G = \Delta G° + RT \ln Q$; $[OH^-] = 1 \times 10^{-14} / 1.5 \times 10^{-8} = 6.7 \times 10^{-7}$

$$= 19.148 + (8.314 \times 10^{-3})(298) \ln \frac{(5.5 \times 10^{-4})(6.67 \times 10^{-7})}{0.120} = -29.43 = -29 \text{ kJ}$$

Additional Exercises

19.81   (a)   False. The essential question is whether the reaction proceeds far to the right before arriving at equilibrium. The position of equilibrium, which is the essential aspect, is not only dependent on ΔH but on the entropy change as well.

          (b)   True.

          (c)   True.

          (d)   False. **Non**spontaneous processes in general require that work be done to force them to proceed. Spontaneous processes occur without application of work.

          (e)   False. Such a process **might** be spontaneous, but would not necessarily be so. Spontaneous processes are those that are exothermic and/or that lead to increased disorder in the system.

19.82

Process	ΔH	ΔS
(a)	+	+
(b)	–	–
(c)	+	+
(d)	+	+
(e)	–	+

19.83   There is no inconsistency. The second law states that in any spontaneous process there is an increase in the entropy of the universe. While there may be a decrease in entropy of the system, as in the present case, this decrease is more than offset by an increase in entropy of the surroundings.

19.84   If $NH_4NO_3(s)$ dissolves spontaneously in water, $\Delta G = \Delta H - T\Delta S$. If ΔG is negative and ΔH is positive, the sign of ΔS must be positive. Furthermore, $T\Delta S > \Delta H$ at room temperature.

19.85   At the normal boiling point of a liquid, $\Delta G = 0$ and $\Delta H_{vap} = T\Delta S_{vap}$; $T = \Delta H_{vap}/\Delta S_{vap}$. By Trouton's rule, $\Delta S_{vap} = 88$ J/mol•K. The process of vaporization is:

          (a) $Br_2(l) \rightleftharpoons Br_2(g)$

$$\Delta H_{vap} = \Delta H_f^° \, Br_2(g) - \Delta H_f^° \, Br_2(l) = 30.71 \text{ kJ} - 0 = 30.71 \text{ kJ}$$

$$T_b = \frac{\Delta H_{vap}}{\Delta S_{vap}} = \frac{30.71 \text{ kJ}}{88 \text{ J/mol} \cdot \text{K}} \times \frac{1000 \text{ J}}{\text{kJ}} = 349 = 3.5 \times 10^2 \text{ K}$$

          (b)   According to WebElements™ 2005, the normal boiling pont of $Br_2(l)$ is 332 K. Trouton's rule provides a good "ballpark" estimate.

19.86   (a)   Formation reactions are the synthesis of 1 mole of compound from elements in their standard states.

$$1/2 \, N_2(g) + 3/2 \, H_2(g) \rightarrow NH_3(g)$$

$$C(s) + 2Cl_2(g) \rightarrow CCl_4(l)$$

$$K(s) + 1/2 \, N_2(g) + 3/2 \, O_2(g) \rightarrow KNO_3(s)$$

In each of these formation reactions, there are fewer moles of gas in the products than the reactants, so we expect $\Delta S°$ to be negative. If $\Delta G_f° = \Delta H_f° - T\Delta S_f°$ and $\Delta S_f°$ is negative, $-T\Delta S_f°$ is positive and $\Delta G_f°$ is more positive than $\Delta H_f°$.

(b) $C(s) + 1/2 \, O_2(g) \rightarrow CO(g)$

In this reaction, there are more moles of gas in products, $\Delta S_f°$ is positive, $-T\Delta S_f°$ is negative and $\Delta G_f°$ is more negative than $\Delta H_f°$.

19.87   (a)   (i) $2RbCl(s) + 3O_2(g) \rightarrow 2RbClO_3(s)$

$\Delta H° = 2\Delta H° \, RbClO_3(s) - 3\Delta H° \, O_2(g) - 2\Delta H° \, RbCl(s)$

$= 2(-392.4) - 3(0) - 2(-430.5) = +76.2 \text{ kJ}$

$\Delta S° = 2(152) - 3(205.0) - 2(92) = -495 \text{ J/K} = -0.495 \text{ kJ/K}$

$\Delta G° = 2(-292.0) - 3(0) - 2(-412.0) = +240.0 \text{ kJ}$

(ii) $C_2H_2(g) + 4Cl_2(g) \rightarrow 2CCl_4(l) + H_2(g)$

$\Delta H° = 2\Delta H° \, CCl_4(l) + \Delta H° \, H_2(g) - \Delta H° \, C_2H_2(g) - 4\Delta H° \, Cl_2(g)$

$= 2(-139.3) + 0 - (226.7) - 4(0) = -505.3 \text{ kJ}$

$\Delta S° = 2(214.4) + 130.58 - (200.8) - 4(222.96) = -533.3 \text{ J/K} = -0.5333 \text{ kJ/K}$

$\Delta G° = 2(-68.6) + 0 - (209.2) - 4(0) = -346.4 \text{ kJ}$

(iii) $TiCl_4(l) + 2H_2O(l) \rightarrow TiO_2(s) + 4HCl(aq)$

$\Delta H° = \Delta H° \, TiO_2(s) + 4\Delta H° \, HCl(aq) - \Delta H° \, TiCl_4(l) - 2\Delta H° \, H_2O(l)$

$= -944.7 + 4(-167.2) - (-804.2) - 2(-285.83) = -237.6 \text{ kJ}$

$\Delta S° = 50.29 + 4(56.5) - 221.9 - 2(69.91) = -85.43 \text{ J/K} = -0.0854 \text{ kJ/K}$

$\Delta G° = -889.4 + 4(-131.2) - (-728.1) - 2(-237.13) = -211.8 \text{ kJ}$

(b)   (i) $\Delta G°$ is (+), nonspontaneous

(ii) $\Delta G°$ is (−), spontaneous

(iii) $\Delta G°$ is (−), spontaneous

(c)   In each case the manner in which free energy change varies with temperature depends mainly on ΔS: $\Delta G = \Delta H - T\Delta S$. When ΔS is substantially positive, ΔG becomes more negative as temperature increases. When ΔS is substantially negative, ΔG becomes more positive as temperature increases.

(i)　　ΔS° is negative, ΔG° becomes more positive with increasing temperature.

(ii)　　ΔS° is negative, ΔG° becomes more positive with increasing temperature. (The reaction will become nonspontaneous at some temperature.)

(iii)　　ΔS° is negative, ΔG° becomes more positive with increasing temperature. (The reaction will become nonspontaneous at some temperature.)

19.88　　$\Delta G = \Delta G° + RT \ln Q$

(a)　　$Q = \dfrac{P_{NH_3}^2}{P_{N_2} \times P_{H_2}^3} = \dfrac{(1.2)^2}{(2.6)(5.9)^3} = 2.697 \times 10^{-3} = 2.7 \times 10^{-3}$

$\Delta G° = 2\Delta G° \, NH_3(g) - \Delta G° \, N_2(g) - 3\Delta G° \, H_2(g)$

$= 2(-16.66) - 0 - 3(0) = -33.32 \text{ kJ}$

$\Delta G = -33.32 \text{ kJ} + \dfrac{8.314 \times 10^{-3} \text{ kJ}}{K \bullet mol} \times 298 \text{ K} \times \ln(2.69 \times 10^{-3})$

$\Delta G = -33.32 - 14.66 = -47.98 = -48.0 \text{ kJ}$

(b)　　$Q = \dfrac{P_{N_2}^3 \times P_{H_2O}^4}{P_{N_2H_4}^2 \times P_{NO_2}^2} = \dfrac{(0.5)^3(0.3)^4}{(5.0 \times 10^{-2})^2(5.0 \times 10^{-2})^2} = 162 = 2 \times 10^2$

$\Delta G° = 3\Delta G° \, N_2(g) + 4\Delta G° \, H_2O(g) - 2\Delta G° \, N_2H_4(g) - 2\Delta G° \, NO_2(g)$

$= 3(0) + 4(-228.57) - 2(159.4) - 2(51.84) = -1336.8 \text{ kJ}$

$\Delta G = -1336.8 \text{ kJ} + 2.478 \ln 162 = -1324.2 = -1.32 \times 10^3 \text{ kJ}$

(c)　　$Q = \dfrac{P_{N_2} \times P_{H_2}^2}{P_{N_2H_4}} = \dfrac{(1.5)(2.5)^2}{0.5} = 18.75 = 2 \times 10^1$

$\Delta G° = \Delta G° \, N_2(g) + 2\Delta G° \, H_2(g) - \Delta G° \, N_2H_4(g)$

$= 0 + 2(0) - 159.4 = -159.4 \text{ kJ}$

$\Delta G = -159.4 \text{ kJ} + 2.478 \ln 18.75 = -152.1 = -152 \text{ kJ}$

19.89

Reaction	(a) Sign of ΔH°	(a) Sign of ΔS°	(b) K > 1?	(c) Variation in K as Temp. Increases
(i)	–	–	yes	decrease
(ii)	+	+	no	increase
(iii)	+	+	no	increase
(iv)	+	+	no	increase

(a)　　Note that at a particular temperature, positive ΔH° leads to a smaller value of K, while positive ΔS° increases the value of K.

19.90　　(a)　　$K = \dfrac{\chi_{CH_3COOH}}{\chi_{CH_3OH}P_{CO}}$

$\Delta G° = -RT \ln K; \ln K = -\Delta G/RT$

$\Delta G° = \Delta G° \ CH_3COOH(l) - \Delta G° \ CH_3OH(l) - \Delta G° \ CO(g)$

$= -392.4 - (-166.23) - (-137.2) = -89.0 \ kJ$

$\ln K = \dfrac{-(-89.0 \ kJ)}{(8.314 \times 10^{-3} \ kJ/K)(298 \ K)} = 35.922 = 35.9; \ K = 4 \times 10^{15}$

(b) $\Delta H° = \Delta H° \ CH_3COOH(l) - \Delta H° \ CH_3OH(l) - \Delta H° \ CO(g)$

$= -487.0 - (-238.6) - (-110.5) = -137.9 \ kJ$

The reaction is exothermic, so the value of K will decrease with increasing temperature, and the mole fraction of CH_3COOH will also decrease. Elevated temperatures must be used to increase the speed of the reaction. Thermodynamics cannot predict the rate at which a reaction reaches equilibrium.

(c) $\Delta G° = -RT \ln K; \ K = 1, \ln K = 0, \Delta G° = 0$

$\Delta G° = \Delta H° - T\Delta S°$; when $\Delta G° = 0$, $\Delta H° = T\Delta S°$

$\Delta S° = S° \ CH_3COOH(l) - S° \ CH_3OH(l) - S° \ CO(g)$

$= 159.8 - 126.8 - 197.9 = -164.9 \ J/K = -0.1649 \ kJ/K$

$-137.9 \ kJ = T(-0.1649 \ kJ/K), \ T = 836.3 \ K$

The equilibrium favors products up to 836 K or 563°C, so the elevated temperatures to increase the rate of reaction can be safely employed.

19.91   (a)   First calculate $\Delta G°$ for each reaction:

For $C_6H_{12}O_6(s) + 6O_2(g) \ \rightleftharpoons \ 6CO_2(g) + 6H_2O(l)$ (A)

$\Delta G° = 6(-237.13) + 6(-394.4) - (-910.4) + 6(0) = -2878.8 \ kJ$

For $C_6H_{12}O_6(s) \ \rightleftharpoons \ 2C_2H_5OH(l) + 2CO_2(g)$ (B)

$\Delta G° = 2(-394.4) + 2(-174.8) - (-910.4) = -228.0 \ kJ$

For (A), $\ln K = 2879 \times 10^3/(8.314)(298) = 1162; \ K = 5 \times 10^{504}$

For (B), $\ln K = 228 \times 10^3/(8.314)(298) = 92.026 = 92.0; \ K = 9 \times 10^{39}$

(b)   Both these values for K are unimaginably large. However, K for reaction (A) is larger, because $\Delta G°$ is more negative. The magnitude of the work that can be accomplished by coupling a reaction to its surroundings is measured by ΔG. According to the calculations above, considerably more work can in principle be obtained from reaction (A), because $\Delta G°$ is more negative.

19.92   (a) $\Delta G° = -RT \ln K$ (Equation [19.22]); $\ln K = -\Delta G°/RT$

Use $\Delta G° = \Delta H° - T\Delta S°$ to get $\Delta G°$ at the two temperatures. Calculate $\Delta H°$ and $\Delta S°$ using data in Appendix C.

$2CH_4(g) \rightarrow C_2H_6(g) + H_2(g)$

$\Delta H° = \Delta H° \ C_2H_6(g) + \Delta H° \ H_2(g) - 2\Delta H° \ CH_4(g) = -84.68 + 0 - 2(-74.8) = 64.92$

$= 64.9 \ kJ$

$\Delta S° = S° C_2H_6(g) + S° H_2(g) - 2S° CH_4(g) = 229.5 + 130.58 - 2(186.3) = -12.52$

$$= -12.5 \text{ J/K}$$

at 298 K, $\Delta G = 64.92 \text{ kJ} - 298 \text{ K}(-12.52 \times 10^{-3} \text{ kJ/K}) = 68.65 = 68.7 \text{ kJ}$

$$\ln K = \frac{-68.65 \text{ kJ}}{(8.314 \times 10^{-3} \text{ kJ/K})(298 \text{ K})} = -27.709 = -27.7, K = 9.25 \times 10^{-13} = 9 \times 10^{-13}$$

at 773 K, $\Delta G = 64.9 \text{ kJ} - 773 \text{ K}(-12.52 \times 10^{-3} \text{ J/K}) = 74.598 = 74.6 \text{ kJ}$

$$\ln K = \frac{-74.598 \text{ kJ}}{(8.314 \times 10^{-3} \text{ kJ/K})(773 \text{ K})} = -11.607 = -11.6, \ K = 9.1 \times 10^{-6}$$

Because the reaction is endothermic, the value of K increases with an increase in temperature.

$2CH_4(g) + 1/2 \ O_2(g) \rightarrow C_2H_6(g) + H_2O(g)$

$\Delta H° = \Delta H° C_2H_6(g) + \Delta H° H_2O(g) - 2\Delta H° CH_4(g) - 1/2 \ \Delta H° O_2(g)$

$$= -84.68 + (-241.82) - 2(-74.8) - 1/2 \ (0) = -176.9 \text{ kJ}$$

$\Delta S° = S° C_2H_6(g) + S° H_2O(g) - 2S° CH_4(g) - 1/2 \ S° O_2(g)$

$$= 229.5 + 188.83 - 2(186.3) - 1/2 \ (205.0) = -56.77 = -56.8 \text{ J/K}$$

at 298 K, $\Delta G = -176.9 \text{ kJ} - 298 \text{ K}(-56.77 \times 10^{-3} \text{ kJ/K}) = -159.98 = -160.0 \text{ kJ}$

$$\ln K = \frac{-(-159.98 \text{ kJ})}{(8.314 \times 10^{-3} \text{ kJ/K})(298 \text{ K})} = 64.571 = 64.57; \ K = 1.1 \times 10^{28}$$

at 773 K, $\Delta G = -176.9 \text{ kJ} - 773 \text{ K} (-56.77 \times 10^{-3} \text{ kJ/K}) = -133.02 = -133.0 \text{ kJ}$

$$\ln K = \frac{-(-133.02 \text{ kJ})}{(8.314 \times 10^{-3} \text{ kJ/K})(773 \text{ K})} = 20.698 = 20.70; \ K = 9.750 \times 10^{8} = 9.8 \times 10^{8}$$

Because this reaction is exothermic, the value of K decreases with increasing temperature.

(b)   The difference in $\Delta G°$ for the two reactions is primarily enthalpic; the first reaction is endothermic and the second exothermic. Both reactions have $-\Delta S°$, which inhibits spontaneity.

(c)   This is an example of coupling a useful but nonspontaneous reaction with a spontaneous one to spontaneously produce a desired product.

$$2CH_4(g) \rightarrow C_2H_6(g) + H_2(g) \qquad \Delta G°_{298} = +68.7 \text{ kJ, nonspontaneous}$$
$$H_2(g) + 1/2 \ O_2(g) \rightarrow H_2O(g) \qquad \Delta G°_{298} = -228.57 \text{ kJ, spontaneous}$$
$$2CH_4(g) + 1/2 \ O_2(g) \rightarrow C_2H_6(g) + H_2O(g) \qquad \Delta G°_{298} = -159.9 \text{ kJ, spontaneous}$$

(d) $CH_4(g) + 2O_2(g) \rightarrow CO_2(g) + 2H_2O(g)$

19.93 $\Delta G°$ for the metabolism of glucose is:

$6\Delta G° CO_2(g) + 6\Delta G° H_2O(l) - \Delta G° C_6H_{12}O_6(s) - 6\Delta G° O_2(g)$

$\Delta G° = 6(-394.4) + 6(-237.13) - (-910.4) + 6(0) = -2878.8 \text{ kJ}$

moles ATP $= -2878.8 \text{ kJ} \times 1 \text{ mol ATP} / (-30.5 \text{ kJ}) = 94.4 \text{ mol ATP} / \text{ mol glucose}$

Note that this calculation is done at standard conditions, not metabolic conditions. A more accurate answer would be obtained using ΔG values that reflect actual concentration, partial pressure, and pH in a cell.

19.94   (a)   The equilibrium of interest here can be written as:

$$K^+ \text{ (plasma)} \rightleftharpoons K^+ \text{ (muscle)}$$

Since an aqueous solution is involved in both cases, assume that the equilibrium constant for the above process is exactly 1, that is, $\Delta G° = 0$. However, ΔG is not zero because the concentrations are not the same on both sides of the membrane. Use Equation [19.21] to calculate ΔG:

$$\Delta G = \Delta G° + RT \ln \frac{[K^+ \text{ (muscle)}]}{[K^+ \text{ (plasma)}]}$$

$$= 0 + (8.314)(310) \ln \frac{(0.15)}{(5.0 \times 10^{-3})} = 8766 \text{ J} = 8.8 \text{ kJ}$$

(b)   Note that ΔG is positive. This means that work must be done on the system (blood plasma plus muscle cells) to move the K^+ ions "uphill," as it were. The minimum amount of work possible is given by the value for ΔG. This value represents the minimum amount of work required to transfer one mole of K^+ ions from the blood plasma at 5×10^{-3} M to muscle cell fluids at 0.15 M, assuming constancy of concentrations. In practice, a larger than minimum amount of work is required.

19.95   (a)   To obtain $\Delta H°$ from the equilibrium constant data, graph $\ln K$ at various temperatures vs $1/T$, being sure to employ absolute temperature. The slope of the linear relationship that should result is $-\Delta H°/R$; thus, $\Delta H°$ is easily calculated.

(b)   Use $\Delta G° = \Delta H° - T\Delta S°$ and $\Delta G° = -RT \ln K$. Substituting the second expression into the first, we obtain

$$-RT \ln K = \Delta H° - T\Delta S°; \quad \ln K = \frac{-\Delta H°}{RT} - \frac{-\Delta S°}{R} = \frac{-\Delta H°}{RT} + \frac{\Delta S°}{R}$$

Thus, the constant in the equation given in the exercise is $\Delta S°/R$.

19.96 $S = k \ln W$ (Equation [19.5]), $k = R/N$, $W \propto V^m$

$\Delta S = S_2 - S_1; S_1 = k \ln W_1, S_2 = k \ln W_2$

$\Delta S = k \ln W_2 - k \ln W_1; W_2 = cV_2^m; W_1 = cV_1^m$

(The number of particles, m, is the same in both states.)

$\Delta S = k \ln cV_2^m - k \ln cV_1^m; \ln a^b = b \ln a$

$\Delta S = k\, m \ln cV_2 - k\, m \ln cV_1; \ln a - \ln b = \ln (a/b)$

$$\Delta S = k\, m \ln\left(\frac{cV_2}{cV_1}\right) = k\, m \ln\left(\frac{V_2}{V_1}\right) = \frac{R}{N} m \ln\left(\frac{V_2}{V_1}\right)$$

$$\frac{m}{N} = \frac{\text{particles}}{6.022 \times 10^{23}} = n(\text{mol}); \quad \Delta S = nR \ln\left(\frac{V_2}{V_1}\right)$$

19 Chemical Thermodynamics Solutions to Exercises

19.97 Absolute entropy is a fundamental property of matter at a specified set of conditions, that is, a state. In order to lower the entropy of the fuel, either the structure of the molecules or the conditions (temperature, pressure, amount) must be changed. Any of these changes would require energy, which would reduce the amount of energy available to drive the car, not increase it.

Integrative Exercises

19.98 (a) At the boiling point, vaporization is a reversible process, so $\Delta S_{vap}^{\circ} = \Delta H_{vap}^{\circ}/T$.

acetone: $\Delta S_{vap}^{\circ} = \Delta H_{vap}^{\circ}/T = (29.1\,kJ/mol)\,/\,329.25\,K = 88.4\,J/mol\bullet K$

dimethyl ether: $\Delta S_{vap}^{\circ} = (21.5\,kJ/mol)\,/\,248.35\,K = 86.6\,J/mol\bullet K$

ethanol: $\Delta S_{vap}^{\circ} = (38.6\,kJ/mol)\,/\,351.6\,K = 110\,J/mol\bullet K$

octane: $\Delta S_{vap}^{\circ} = (34.4\,kJ/mol)\,/\,398.75\,K = 86.3\,J/mol\bullet K$

pyridine: $\Delta S_{vap}^{\circ} = (35.1\,kJ/mol)\,/\,388.45\,K = 90.4\,J/mol\bullet K$

(b) Ethanol is the only liquid listed that doesn't follow *Trouton's rule* and it is also the only substance that exhibits hydrogen bonding in the pure liquid. Hydrogen bonding leads to more ordering in the liquid state and a greater than usual increase in entropy upon vaporization. The rule appears to hold for liquids with London dispersion forces (octane) and ordinary dipole-dipole forces (acetone, dimethyl ether, pyridine), but not for those with hydrogen bonding.

(c) Owing to strong hydrogen bonding interactions, water probably does not obey Trouton's rule.

From Appendix B, ΔH_{vap}° at 100°C = 40.67 kJ/mol.

$\Delta S_{vap}^{\circ} = (40.67\,kJ/mol)\,/\,373.15\,K = 109.0\,J/mol\bullet K$

(d) Use $\Delta S_{vap}^{\circ} = 88\,J/mol\bullet K$, the middle of the range for Trouton's rule, to estimate ΔH_{vap}° for chlorobenzene.

$\Delta H_{vap}^{\circ} = \Delta S_{vap}^{\circ} \times T = 88\,J/mol\bullet K \times 404.95\,K = 36\,kJ/mol$

19.99 (a) Polymerization is the process of joining many small molecules (monomers) into a few very large molecules (polymers). Polyethylene in particular can have extremely high molecular weights. In general, reducing the number of particles in a system reduces entropy, so ΔS_{poly} is expected to be negative.

(b) $\Delta G_{poly} = \Delta H_{poly} - T\Delta S_{poly}$. If the polymerization of ethylene is spontaneous, ΔG_{poly} is negative. If ΔS_{poly} is negative, $-T\Delta S_{poly}$ is positive, so ΔH_{poly} must be negative for ΔG_{poly} to be negative. The enthalpy of polymerization must be exothermic.

(c) According to Equation [12.1], polymerization of ethylene requires breaking one C=C and forming 2C—C per monomer (1C—C between the C-atoms of the monomer and $2 \times 1/2$ C—C to two other monomers).

$$\Delta H = D(C=C) - 2D(C-C) = 614 - 2(348) = -82 \text{ kJ/mol } C_2H_4$$

$$\frac{-82 \text{ kJ}}{\text{mol } C_2H_4} \times \frac{1 \text{ mol}}{6.022 \times 10^{23} \text{ molecules}} \times \frac{1000 \text{ J}}{1 \text{ kJ}} = 1.36 \times 10^{-19} \text{ J/}C_2H_4 \text{ monomer}$$

(d) The products of a condensation polymerization are the polymer and a small molecule, typically H_2O; there is usually one small molecule formed per monomer unit. Unlike addition polymerization, the total number of particles is not reduced. A condensation polymer does impose more order on the monomer or monomers than an addition polymer. If there is a single monomer, it has different functional groups at the two ends and only one end can react to join the polymer, so orientation is required. If there are two different monomers, as in nylon, the monomers alternate in the polymer, so only the correct monomer can react to join the polymer. In terms of structure, the condensation polymer imposes more order on the monomer(s) than an addition polymer. But, condensation polymerization does not lead to a reduction in the number of particles in the system, so ΔS_{poly} will be less negative than for addition polymerization.

19.100   The activated complex in Figure 14.13 is a single "particle" or entity that contains four atoms. It is formed from an atom A and a triatomic molecule, ABC, that must collide with exactly the correct energy and orientation to form the single entity. There are many fewer degrees of freedom for the activated complex than the separate reactant particles, so the *entropy of activation* is negative.

19.101   (a) $O_2(g) \xrightarrow{h\nu} 2O(g)$; S increases because there are more moles of gas in the products.

          (b) $O_2(g) + O(g) \to O_3(g)$, S decreases because there are fewer moles of gas in the products.

          (c)   S increases as the gas molecules diffuse into the larger volume of the stratosphere; there are more possible positions and therefore more motional freedom.

          (d) $NaCl(aq) \to NaCl(s) + H_2O(l)$; ΔS decreases as the mixture (seawater, greater disorder) is separated into pure substances (fewer possible arrangements, more order).

19.102   (a)   16 e⁻, 8 e⁻ pairs. The C-S bond order is approximately 2.

               $\overset{..}{\underset{..}{S}} = C = \overset{..}{\underset{..}{S}}$

          (b)   2 e⁻ domains around C, linear e⁻ domain geometry, linear molecular structure

          (c) $CS_2(l) + 3O_2(g) \to CO_2(g) + 2SO_2(g)$

          (d) $\Delta H° = \Delta H° \, CO_2(g) + 2\Delta H° \, SO_2(g) - \Delta H° \, CS_2(l) - 3 \, \Delta H \, \Delta H° \, O_2(g)$

               $= -393.5 + 2(-296.9) - (89.7) - 3(0) = -1077.0 \text{ kJ}$

            $\Delta G° = \Delta G° \, CO_2(g) + 2\Delta G° \, SO_2(g) - \Delta G° \, CS_2(l) - 3 \, \Delta G° \, O_2(g)$

               $= -394.4 + 2(-300.4) - (65.3) - 3(0) = -1060.5 \text{ kJ}$

            The reaction is exothermic ($-\Delta H°$) and spontaneous ($-\Delta G°$) at 298 K.

(e)   vaporization: $CS_2(l) \rightarrow CS_2(g)$

$$\Delta G_{vap}^\circ = \Delta H_{vap}^\circ - T\Delta S_{vap}^\circ ; \quad \Delta S_{vap}^\circ = (\Delta H_{vap}^\circ - \Delta G_{vap}^\circ)/T$$

$$\Delta G_{vap}^\circ = \Delta G^\circ \, CS_2(g) - \Delta G^\circ \, CS_2(l) = 67.2 - 65.3 = 1.9 \text{ kJ}$$

$$\Delta H_{vap}^\circ = \Delta H^\circ \, CS_2(g) - \Delta H^\circ \, CS_2(l) = 117.4 - 89.7 = 27.7 \text{ kJ}$$

$$\Delta S_{vap}^\circ = (27.7 - 1.9) \text{ kJ}/298 \text{ K} = 0.086577 = 0.0866 \text{ kJ/K} = 86.6 \text{ J/K}$$

ΔS_{vap} is always positive, because the gas phase occupies a greater volume, has more motional freedom and a larger absolute entropy than the liquid.

(f)   At the boiling point, $\Delta G = 0$ and $\Delta H_{vap} = T_b \Delta S_{vap}$.

$$T_b = \Delta H_{vap}/\Delta S_{vap} = 27.7 \text{ kJ}/0.086577 \text{ kJ/K} = 319.9 = 320 \text{ K}$$

$T_b = 320 \text{ K} = 47°C$. CS_2 is a liquid at 298 K, 1 atm

19.103   (a) $Ag(s) + 1/2 \, N_2(g) + 3/2 \, O_2(g) \rightarrow AgNO_3(s)$; S decreases because there are fewer moles of gas in the product.

(b) $\Delta G_f^\circ = \Delta H_f^\circ - T\Delta S_f^\circ ; \quad \Delta S_f^\circ = (\Delta G_f^\circ - \Delta H_f^\circ)/(-T) = (\Delta H_f^\circ - \Delta G_f^\circ)/T$

$$\Delta S_f^\circ = -124.4 \text{ kJ} - (-33.4 \text{ kJ})/298 \text{ K} = -0.305 \text{ kJ/K} = -305 \text{ J/K}$$

ΔS_f° is relatively large and negative, as anticipated from part (a).

(c)   Dissolving of $AgNO_3$ can be expressed as

$AgNO_3(s) \rightarrow AgNO_3 \text{ (aq, 1 m)}$

$\Delta H° = \Delta H° \, AgNO_3(aq) - \Delta H° \, AgNO_3(s) = -101.7 - (-124.4) = +22.7 \text{ kJ}$

$\Delta H° = \Delta H° \, MgSO_4(aq) - \Delta H° \, MgSO_4(s) = -1374.8 - (-1283.7) = -91.1 \text{ kJ}$

Dissolving $AgNO_3(s)$ is endothermic ($+\Delta H°$), but dissolving $MgSO_4(s)$ is exothermic ($-\Delta H°$).

(d) $AgNO_3$: $\Delta G° = \Delta G_f^\circ \, AgNO_3(aq) - \Delta G_f^\circ \, AgNO_3(s) = -34.2 - (-33.4) = -0.8 \text{ kJ}$

$\Delta S° = (\Delta H° - \Delta G°) / T = [22.7 \text{ kJ} - (-0.8 \text{ kJ})] / 298 \text{ K} = 0.0789 \text{ kJ/K} = 78.9 \text{ J/K}$

$MgSO_4$: $\Delta G° = \Delta G_f^\circ \, MgSO_4(aq) - \Delta G_f^\circ \, MgSO_4(s) = -1198.4 - (-1169.6) = -28.8 \text{ kJ}$

$\Delta S° = (\Delta H° - \Delta G°) / T = [-91.1 \text{ kJ} - (-28.8 \text{ kJ})] / 298 \text{ K} = -0.209 \text{ kJ/K} = -209 \text{ J/K}$

(e)   In general, we expect dissolving a crystalline solid to be accompanied by an increase in positional disorder and an increase in entropy; this is the case for $AgNO_3$ ($\Delta S° = + 78.9$ J/K). However, for dissolving $MgSO_4(s)$, there is a substantial decrease in entropy ($\Delta S = -209$ J/K). According to Section 13.5, ion-pairing is a significant phenomenon in electrolyte solutions, particularly in concentrated solutions where the charges of the ions are greater than 1. According to Table 13.5, a 0.1 m $MgSO_4$ solution has a van't Hoff factor of 1.21. That is, for each mole of $MgSO_4$ that dissolves, there are only 1.21 moles of "particles" in solution instead of 2 moles of particles. For a 1 m solution, the factor is even smaller. Also, the exothermic enthalpy of mixing indicates substantial interactions between solute and solvent. Substantial ion-pairing coupled with ion-dipole interactions with H_2O molecules lead to a decrease in entropy for $MgSO_4(aq)$ relative to $MgSO_4(s)$.

19.104   (a) $K = P_{NO_2}^2 / P_{N_2O_4}$

Assume equal amounts means equal number of moles. For gases, $P = n(RT/V)$. In an equilibrium mixture, RT/V is a constant, so moles of gas are directly proportional to partial pressure. Gases with equal partial pressures will have equal moles of gas present. The condition $P_{NO_2} = P_{N_2O_4}$ leads to the expression $K = P_{NO_2}$. The value of K then depends on P_t for the mixture. For any particular value of P_t, the condition of equal moles of the two gases can be achieved at some temperature. For example, $P_{NO_2} = P_{N_2O_4} = 1.0$ atm, $P_t = 2.0$ atm.

$$K = \frac{(1.0)^2}{1.0} = 1.0; \ \ln K = 0; \ \Delta G^\circ = 0 = \Delta H^\circ - T\Delta S^\circ; \ T = \Delta H^\circ / \Delta S^\circ$$

$$\Delta H^\circ = 2\Delta H^\circ \ NO_2(g) - \Delta H^\circ \ N_2O_4(g) = 2(33.84) - 9.66 = +58.02 \text{ kJ}$$

$$\Delta S^\circ = 2S^\circ \ NO_2(g) - S^\circ \ N_2O_4(g) = 2(240.45) - 304.3 = 0.1766 \text{ kJ/K}$$

$$T = \frac{58.02 \text{ kJ}}{0.1766 \text{ kJ/K}} = 328.5 \text{ K or } 55.5^\circ C$$

     (b) $P_t = 1.00$ atm; $P_{N_2O_4} = x$, $P_{NO_2} = 2x$; $x + 2x = 1.00$ atm

$$x = P_{N_2O_4} = 0.3333 = 0.333 \text{ atm}; \ P_{NO_2} = 0.6667 = 0.667 \text{ atm}$$

$$K = \frac{(0.6667)^2}{0.3333} = 1.334 = 1.33; \ \Delta G^\circ = -RT \ln K = \Delta H^\circ - T\Delta S^\circ$$

$$-(8.314 \times 10^{-3} \text{ kJ/K})(\ln 1.334) \ T = 58.02 \text{ kJ} - (0.1766 \text{ kJ/K}) \ T$$

$$(-0.00239 \text{ kJ/K}) \ T + (0.1766 \text{ kJ/K}) \ T = 58.02 \text{ kJ}$$

$$(0.1742 \text{ kJ/K}) \ T = 58.02 \text{ kJ}; \ T = 333.0 \text{ K}$$

     (c) $P_t = 10.00$ atm; $x + 2x = 10.00$ atm

$$x = P_{N_2O_4} = 3.3333 = 3.333 \text{ atm}; \ P_{NO_2} = 6.6667 = 6.667 \text{ atm}$$

$$K = \frac{(6.6667)^2}{3.3333} = 13.334 = 13.33; \ -RT \ln K = \Delta H^\circ - T\Delta S^\circ$$

$$-(8.314 \times 10^{-3} \text{ kJ/K})(\ln 13.334) \ T = 58.02 \text{ kJ} - (0.1766 \text{ kJ/K}) \ T$$

$$(-0.02154 \text{ kJ/K}) \ T + (0.1766 \text{ kJ/K}) \ T = 58.02 \text{ kJ}$$

$$(0.15506 \text{ kJ/K}) \ T = 58.02 \text{ kJ}; \ T = 374.2 \text{ K}$$

     (d)   The reaction is endothermic, so an increase in the value of K as calculated in parts (b) and (c) should be accompanied by an increase in T.

19.105   (a) $\Delta G^\circ = 3\Delta G_f^\circ \ S(s) + 2\Delta G_f^\circ \ H_2O(g) - \Delta G_f^\circ \ SO_2(g) - 2\Delta G_f^\circ \ H_2S(g)$

$$= 3(0) + 2(-228.57) - (-300.4) - 2(-33.01) = -90.72 = -90.7 \text{ kJ}$$

$$\ln K = \frac{-\Delta G^{\circ}}{RT} = \frac{-(-90.72 \text{ kJ})}{(8.314 \times 10^{-3} \text{ kJ/K})(298 \text{ K})} = 36.6165 = 36.6; \quad K = 7.99 \times 10^{15}$$

$$= 8 \times 10^{15}$$

(b)   The reaction is highly spontaneous at 298 K and feasible in principle. However, use of $H_2S(g)$ produces a severe safety hazard for workers and the surrounding community.

(c) $P_{H_2O} = \dfrac{25 \text{ torr}}{760 \text{ torr/atm}} = 0.033 \text{ atm}$

$$K = \frac{P_{H_2O}^2}{P_{SO_2} \times P_{H_2S}^2}; \quad P_{SO_2} = P_{H_2S} = x \text{ atm}$$

$$K = 7.99 \times 10^{15} = \frac{(0.033)^2}{x(x)^2}; \quad x^3 = \frac{(0.033)^2}{7.99 \times 10^{15}}$$

$$x = 5 \times 10^{-7} \text{ atm}$$

(d) $\Delta H^{\circ} = 3\Delta H_f^{\circ} \text{ S}(s) + 2\Delta H_f^{\circ} \text{ H}_2O(g) - \Delta H_f^{\circ} \text{ SO}_2(g) - 2\Delta H_f^{\circ} \text{ H}_2S(g)$

$$= 3(0) + 2(-241.82) - (-296.9) - 2(-20.17) = -146.4 \text{ kJ}$$

$\Delta S^{\circ} = 3S^{\circ} \text{ S}(s) + 2S^{\circ} \text{ H}_2O(g) - S^{\circ} \text{ SO}_2(g) - 2S^{\circ} \text{ H}_2S(g)$

$$= 3(31.88) + 2(188.83) - 248.5 - 2(205.6) = -186.4 \text{ J/K}$$

The reaction is exothermic $(-\Delta H)$, so the value of K_{eq} will decrease with increasing temperature. The negative ΔS° value means that the reaction will become nonspontaneous at some higher temperature. The process will be less effective at elevated temperatures.

19.106   (a)   When the rubber band is stretched, the molecules become more ordered, so the entropy of the system decreases, ΔS_{sys} is negative.

(b) $\Delta S_{sys} = q_{rev}/T$. Since ΔS_{sys} is negative, q_{rev} is negative and heat is evolved by the system.

20 Electrochemistry

Visualizing Concepts

20.1 In a Brønsted-Lowry acid-base reaction, H^+ is transferred from the acid to the base. In a redox reaction, the substance being oxidized (the reductant) loses electrons and the substance being reduced (the oxidant) gains electrons. Furthermore, the number of electrons gained and lost must be equal. The concept of electron transfer from reductant to oxidant is clearly applicable to redox reactions. (The path of the transfer may or may not be direct, but ultimately electrons are transferred during redox reactions.)

20.2 (a) If a Zn(s) strip was placed in a $CdSO_4$(aq) solution, Cd(s) would form on the strip. Although E°_{red} for Zn^{2+}(aq), –0.763 V, and Cd^{2+}(aq), –0.403 V, are both negative, the value for Cd^{2+} is larger (less negative), so it will be the reduced species in the redox reaction.

(b) If a Cu(s) strip was placed in a $AgNO_3$(aq) solution, Ag(s) would form on the strip. Although E°_{red} for Cu^{2+}(aq), 0.337 V, and Ag^+(aq), 0.799 V, are both positive, the species with the larger E°_{red} value, Ag^+, will be reduced in the reaction.

20.3 Ni^{2+}(aq) + 2e⁻ → Ni(s), E°_{red} = –0.28 V, cathode

Fe^{2+}(aq) + 2e⁻ → Fe(s), E°_{red} = –0.44 V, anode

Ni^{2+}(aq) has the larger E°_{red}, so it will be reduced in the redox reaction. Reduction occurs at the cathode, so Ni^{2+}(aq) and Ni(s) will be in the cathode compartment and Fe^{2+}(aq) and Fe(s) will be in the anode compartment. The voltmeter will read:

$E^\circ_{cell} = E^\circ_{red}$(cathode) - E°_{red}(anode) = –0.28 V – (–044 V) = 0.16 V

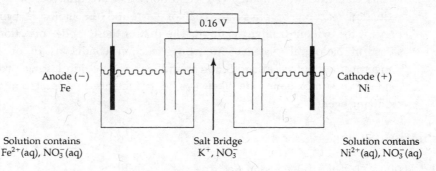

20.4   The species with the largest E°_{red} is easiest to reduce, while the species with the smallest, most negative E°_{red} is easiest to oxidize.

    (a)   The species easiest to oxidize is at the bottom of Figure 20.14.

    (b)   The species easiest to reduce is at the top of Figure 20.14.

20.5 $A(aq) + B(aq) \rightarrow A^-(aq) + B^+(aq)$

    (a)   A gains electrons; it is being reduced. B loses electrons; it is being oxidized.

    (b)   Reduction occurs at the cathode; oxidation occurs at the anode.

        $A(aq) + 1e^- \rightarrow A^-(aq)$ occurs at the cathode.

        $B(aq) \rightarrow B^+(aq) + 1e^-$ occurs at the anode.

    (c)   In a voltaic cell, the anode is at higher potential energy than the cathode. The anode reaction, $B(aq) \rightarrow B^+(aq) + 1e^-$, is higher in potential energy.

20.6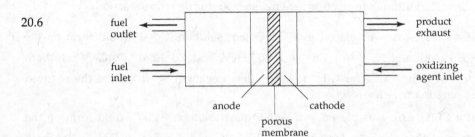

The main difference between a fuel cell and a battery is that a fuel cell is not self-contained. That is, there is a continuous supply of fuel (reductant) and oxidant to the cell, and continuous exhaust of products. The fuel cell produces electrical current as long as reactants are supplied. It never goes "dead."

20.7   Zinc, $E^\circ_{red} = -0.763\,V$, is more easily oxidized than iron, $E^\circ_{red} = -0.440\,V$. If conditions are favorable for oxidation, zinc will be preferentially oxidized, preventing iron from corroding. The protection lasts until all the Zn coating has reacted.

20.8   Unintended oxidation reactions in the body lead to unwanted health effects, just as unwanted oxidation of metals leads to corrosion. Antioxidants probably have modes of action similar to anti-corrosion agents. They can preferentially react with oxidizing agents (cathodic protection), create conditions that are unfavorable to the oxidation-reduction reaction, or physically coat or surround the molecule being oxidized to prevent the oxidant from attacking it. The first of these modes of action is likely to be safest in biological systems. Adjusting reaction conditions in our body can be dangerous, and physical protection is unlikely to provide lasting protection against oxidation. Anti-oxidants are likely to be reductants that preferentially react with oxidizing agents.

Oxidation States

20.9   (a)   *Oxidation* is the loss of electrons.

(b)   The electrons appear on the products side (right side) of an oxidation half-reaction.

(c)   The *oxidant* is the reactant that is reduced; it gains the electrons that are lost by the substance being oxidized.

(d)   An *oxidizing agent* is the substance that promotes oxidation. That is, it gains electrons that are lost by the substance being oxidized. It is the same as the oxidant.

20.10   (a)   *Reduction* is the gain of electrons.

(b)   The electrons appear on the reactants side (left side) of a reduction half-reaction.

(c)   The *reductant* is the reactant that is oxidized; it provides the electrons that are gained by the substance being reduced.

(d)   A *reducing agent* is the substance that promotes reduction. It donates the electrons gained by the substance that is reduced. It is the same as the reductant.

20.11   (a)   True.

(b)   False. Fe^{3+} is reduced to Fe^{2+}, so it is the oxidizing agent, and Co^{2+} is the reducing agent.

(c)   True.

20.12   (a)   False. If something is reduced, it gains electrons.

(b)   True.

(c)   True. Oxidation can be thought of as a gain of oxygen atoms. Looking forward, this view will be useful for organic reactions, Chapter 25.

20.13   *Analyze/Plan.* Given a chemical equation, we are asked to indicate which elements undergo a change in oxidation number and the magnitude of the change. Assign oxidation numbers according to the rules given in Section 4.4. Note the changes and report the magnitudes.   *Solve.*

(a)   I is reduced from +5 to 0; C is oxidized from +2 to +4.

(b)   Hg is reduced from +2 to 0; N is oxidized from –2 to 0.

(c)   N is reduced from +5 to +2; S is oxidized from –2 to 0.

(d)   Cl is reduced from +4 to +3; O is oxidized from –1 to 0.

20.14   (a)   No oxidation-reduction

(b)   I is oxidized from –1 to +5; Cl is reduced from +1 to –1.

(c)   S is oxidized from +4 to +6; N is reduced from +5 to +2.

(d)   S is reduced from +6 to +4; Br is oxidized from –1 to 0.

Balancing Oxidation-Reduction Reactions

20.15   *Analyze/Plan.* Write the balanced chemical equation and assign oxidation numbers. The substance oxidized is the reductant and the substance reduced is the oxidant.   *Solve.*

(a) $TiCl_4(g) + 2Mg(l) \rightarrow Ti(s) + 2MgCl_2(l)$

(b) $Mg(l)$ is oxidized; $TiCl_4(g)$ is reduced.

(c) $Mg(l)$ is the reductant; $TiCl_4(g)$ is the oxidant.

20.16   (a) $2N_2H_4(g) + N_2O_4(g) \rightarrow 3N_2(g) + 4H_2O(g)$

(b) $N_2H_4(g)$ is oxidized; $N_2O_4(g)$ is reduced.

(c) $N_2O_4(g)$ serves as the oxidizing agent; it is itself reduced. $N_2H_4(g)$ serves as the reducing agent; it is itself oxidized.

20.17   *Analyze/Plan.* Follow the logic in Sample Exercises 20.2 and 20.3. If the half-reaction occurs in basic solution, balance as in acid, then add OH^- to each side.   *Solve.*

(a) $Sn^{2+}(aq) \rightarrow Sn^{4+}(aq) + 2e^-$, oxidation

(b) $TiO_2(s) + 4H^+(aq) + 2e^- \rightarrow Ti^{2+}(aq) + 2H_2O(l)$, reduction

(c) $ClO_3^-(aq) + 6H^+(aq) + 6e^- \rightarrow Cl^-(aq) + 3H_2O(l)$, reduction

(d) $4OH^-(aq) \rightarrow O_2(g) + 2H_2O(l) + 4e^-$, oxidation

(e) $SO_3^{2-}(aq) + 2OH^-(aq) \rightarrow SO_4^{2-}(aq) + H_2O(l) + 2e^-$, oxidation

(f) $N_2(g) + 8H^+(aq) + 6e^- \rightarrow 2NH_4^+(aq)$, reduction

(g) $N_2(g) + 6H_2O + 2e^- \rightarrow 2NH_3(g) + OH^-(aq)$, reduction

20.18   (a) $Mo^{3+}(aq) + 3e^- \rightarrow Mo(s)$, reduction

(b) $H_2SO_3(aq) + H_2O(l) \rightarrow SO_4^{2-}(aq) + 4H^+(aq) + 2e^-$, oxidation

(c) $NO_3^-(aq) + 4H^+(aq) + 3e^- \rightarrow NO(g) + 2H_2O(l)$, reduction

(d) $Mn^{2+}(aq) + 4OH^-(aq) \rightarrow MnO_2(s) + 2H_2O(l) + 2e^-$, oxidation

(e) $Cr(OH)_3(s) + 5OH^-(aq) \rightarrow CrO_4^{2-}(aq) + 4H_2O(l) + 3e^-$, oxidation

(f) $O_2(g) + 4H^+(aq) + 4e^- \rightarrow 2H_2O(l)$, reduction

(g) $O_2(g) + 2H_2O(l) + 4e^- \rightarrow 4OH^-(aq)$, reduction

20.19   *Analyze/Plan.* Follow the logic in Sample Exercises 20.2 and 20.3 to balance the given equations. Use the method in Sample Exercise 20.1 to identify oxidizing and reducing agents.   *Solve.*

(a) $Cr_2O_7^{2-}(aq) + I^-(aq) + 8H^+ \rightarrow 2Cr^{3+}(aq) + IO_3^-(aq) + 4H_2O(l)$

oxidizing agent, $Cr_2O_7^{2-}$; reducing agent, I^-

(b)   The half-reactions are:

$4[MnO_4^-(aq) + 8H^+(aq) + 5e^- \rightarrow Mn^{2+}(aq) + 4H_2O(l)]$

$5[CH_3OH(aq) + H_2O(l) \rightarrow HCO_2H(aq) + 4H^+(aq) + 4e^-]$

$\overline{4MnO_4^-(aq) + 5CH_3OH(aq) + 12H^+(aq) \rightarrow 4Mn^{2+}(aq) + 5HCO_2H(aq) + 11H_2O(l)}$

oxidizing agent, MnO_4^-; reducing agent, CH_3OH

(c)
$$I_2(s) + 6H_2O(l) \rightarrow 2IO_3^-(aq) + 12H^+(aq) + 10e^-$$
$$\underline{5[OCl^-(aq) + 2H^+(aq) + 2e^- \rightarrow Cl^-(aq) + H_2O(l)]}$$
$$I_2(s) + 5OCl^-(aq) + H_2O(l) \rightarrow 2IO_3^-(aq) + 5Cl^-(aq) + 2H^+(aq)]$$

oxidizing agent, OCl^-; reducing agent, I_2

(d)
$$As_2O_3(s) + 5H_2O(l) \rightarrow 2H_3AsO_4(aq) + 4H^+(aq) + 4e^-$$
$$\underline{2NO_3^-(aq) + 6H^+(aq) + 4e^- \rightarrow N_2O_3(aq) + 3H_2O(l)}$$
$$As_2O_3(s) + 2NO_3^-(aq) + 2H_2O(l) + 2H^+(aq) \rightarrow 2H_3AsO_4(aq) + N_2O_3(aq)$$

oxidizing agent, NO_3^-; reducing agent, As_2O_3

(e)
$$2[MnO_4^-(aq) + 2H_2O(l) + 3e^- \rightarrow MnO_2(s) + 4OH^-]$$
$$\underline{Br^-(aq) + 6OH^-(aq) \rightarrow BrO_3^-(aq) + 3H_2O(l) + 6e^-}$$
$$2MnO_4^-(aq) + Br^-(aq) + H_2O(l) \rightarrow 2MnO_2(s) + BrO_3^-(aq) + 2OH^-(aq)$$

oxidizing agent, MnO_4^-; reducing agent, Br^-

(f) $Pb(OH)_4^{2-}(aq) + ClO^-(aq) \rightarrow PbO_2(s) + Cl^-(aq) + 2OH^-(aq) + H_2O(l)$

oxidizing agent, ClO^-; reducing agent, $Pb(OH)_4^{2-}$

20.20   (a)
$$3[NO_2^-(aq) + H_2O(l) \rightarrow NO_3^-(aq) + 2H^+(aq) + 2e^-]$$
$$\underline{Cr_2O_7^{2-}(aq) + 14H^+(aq) + 6e^- \rightarrow 2Cr^{3+}(aq) + 7H_2O(l)}$$
Net: $3NO_2^-(aq) + Cr_2O_7^{2-}(aq) + 8H^+(aq) \rightarrow 3NO_3^-(aq) + 2Cr^{3+}(aq) + 4H_2O(l)$

oxidizing agent, $Cr_2O_7^{2-}$; reducing agent, NO_2^-

(b)
$$4[As(s) + 3H_2O(l) \rightarrow H_3AsO_3(aq) + 3H^+(aq) + 3e^-]$$
$$\underline{3[ClO_3^-(aq) + 5H^+(aq) + 4e^- \rightarrow HClO(aq) + 2H_2O(l)]}$$
$$4As(s) + 3ClO_3^-(aq) + 6H_2O(l) + 3H^+(aq) \rightarrow 4H_3AsO_3(aq) + 3HClO(aq)$$

oxidizing agent, ClO_3^-; reducing agent, As

(c)
$$2[Cr_2O_7^{2-}(aq) + 14H^+(aq) + 6e^- \rightarrow 2Cr^{3+}(aq) + 7H_2O(l)]$$
$$\underline{3[CH_3OH(aq) + H_2O(l) \rightarrow HCO_2H(aq) + 4H^+(aq) + 4e^-]}$$
Net: $2Cr_2O_7^{2-}(aq) + 3CH_3OH(aq) + 16H^+(aq) \rightarrow 4Cr^{3+}(aq) + 3HCO_2H(aq) + 11H_2O(l)$

oxidizing agent, $Cr_2O_7^{2-}$; reducing agent, CH_3OH

(d)
$$2[MnO_4^-(aq) + 8H^+(aq) + 5e^- \rightarrow Mn^{2+}(aq) + 4H_2O(l)]$$
$$\underline{5[2Cl^-(aq) \rightarrow Cl_2(aq) + 2e^-]}$$
Net: $2MnO_4^-(aq) + 10Cl^-(aq) + 16H^+(aq) \rightarrow 2Mn^{2+}(aq) + 5Cl_2(g) + 8H_2O(l)$

oxidizing agent, MnO_4^-; reducing agent, Cl^-

(e) $H_2O_2(aq) + 2e^- \rightarrow O_2(g) + 2H^+(aq)$

Since the reaction is in base, the H^+ can be "neutralized" by adding $2OH^-$ to each side of the equation to give $H_2O_2(aq) + 2OH^-(aq) \rightarrow O_2(g) + 2H_2O(l) + 2e^-$. The other half reaction is $2[ClO_2(aq) + e^- \rightarrow ClO_2^-(aq)]$.

Net: $H_2O_2(aq) + 2ClO_2(aq) + 2OH^-(aq) \rightarrow O_2(g) + 2ClO_2^-(aq) + 2H_2O(l)$

oxidizing agent, ClO_2; reducing agent, H_2O_2

(f) $\qquad 4[H_2O_2(aq) + 2OH^-(aq) \rightarrow O_2(g) + 2H_2O(l) + 2e^-]$

$\qquad\quad \underline{Cl_2O_7(aq) + 3H_2O(l) + 8e^- \rightarrow 2ClO_2^-(aq) + 6OH^-(aq)}$

Net: $Cl_2O_7(aq) + 4H_2O_2(aq) + 2OH^-(aq) \rightarrow 2ClO_2^-(aq) + 4O_2(g) + 5H_2O(l)$

oxidizing agent, Cl_2O_7; reducing agent, H_2O_2

Voltaic Cells

20.21 (a) The reaction $Cu^{2+}(aq) + Zn(s) \rightarrow Cu(s) + Zn^{2+}(aq)$ is occurring in both figures. In Figure 20.3, the reactants are in contact, and the concentrations of the ions in solution aren't specified. In Figure 20.4 the oxidation half-reaction and reduction half-reaction are occurring in separate compartments, joined by a porous connector. The concentrations of the two solutions are initially 1.0 M. In Figure 20.4, electrical current is isolated and flows through the voltmeter. In Figure 20.3, the flow of electrons cannot be isolated or utilized.

(b) In the cathode compartment of the voltaic cell in Figure 20.5, Cu^{2+} cations are reduced to Cu atoms, decreasing the number of positively charged particles in the compartment. Na^+ cations are drawn into the compartment to maintain charge balance as Cu^{2+} ions are removed.

20.22 (a) The porous glass dish in Figure 20.4 provides a mechanism by which ions not directly involved in the redox reaction can migrate into the anode and cathode compartments to maintain charge neutrality of the solutions. Ionic conduction within the cell, through the glass disk, completes the cell circuit.

(b) In the anode compartment of Figure 20.5, Zn atoms are oxidized to Zn^{2+} cations, increasing the number of positively charged particles in the compartment. NO_3^- anions migrate into the compartment to maintain charge balance as Zn^{2+} ions are produced.

20.23 *Analyze/Plan.* Follow the logic in Sample Exercise 20.4. *Solve.*

(a) Fe(s) is oxidized, $Ag^+(aq)$ is reduced.

(b) $Ag^+(aq) + 1e^- \rightarrow Ag(s)$; $Fe(s) \rightarrow Fe^{2+}(aq) + 2e^-$

(c) Fe(s) is the anode, Ag(s) is the cathode.

(d) Fe(s) is negative; Ag(s) is positive.

(e) Electrons flow from the Fe(–) electrode toward the Ag(+) electrode.

(f) Cations migrate toward the Ag(s) cathode; anions migrate toward the Fe(s) anode.

20.24 (a) Al(s) is oxidized, $Ni^{2+}(aq)$ is reduced.

(b) $Al(s) \rightarrow Al^{3+}(aq) + 3e^-$; $Ni^{2+}(aq) + 2e^- \rightarrow Ni(s)$

(c) Al(s) is the anode; Ni(s) is the cathode.

(d)   Al(s) is negative (–); Ni(s) is positive (+).

(e)   Electrons flow from the Al(–) electrode toward the Ni(+) electrode.

(f)   Cations migrate toward the Ni(s) cathode; anions migrate toward the Al(s) anode.

Cell EMF under Standard Conditions

20.25   (a)   *Electromotive force*, emf, is the driving force that causes electrons to flow through the external circuit of a voltaic cell. It is the potential energy difference between an electron at the anode and an electron at the cathode.

           (b)   One *volt* is the potential energy difference required to impart 1 J of energy to a charge of 1 coulomb. 1 V = 1 J/C.

           (c)   *Cell potential*, E_{cell}, is the emf of an electrochemical cell.

20.26   (a)   In a voltaic cell, the anode has the higher potential energy for electrons. To achieve a lower potential energy, electrons flow from the anode to the cathode.

           (b)   The units of electrical potential are volts. A potential of one volt imparts one joule of energy to one coulomb of charge.

           (c)   A *standard* cell potential describes the potential of an electrochemical cell where all components are present at standard conditions: elements in their standard states, gases at 1 atm pressure and 1 *M* aqueous solutions.

20.27   (a) $2H^+(aq) + 2e^- \rightarrow H_2(g)$

           (b)   A *standard* hydrogen electrode is a hydrogen electrode where the components are at standard conditions, 1 *M* $H^+(aq)$ and $H_2(g)$ at 1 atm.

           (c)   The platinum foil in an SHE serves as an inert electron carrier and a solid reaction surface.

20.28   (a) $H_2(g) \rightarrow 2H^+(aq) + 2e^-$

           (b)   The platinum electrode serves as a reaction surface; the greater the surface area, the more H_2 or H^+ that can be adsorbed onto the surface to facilitate the flow of electrons.

           (c)

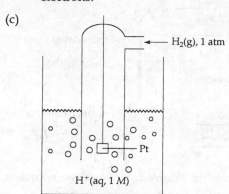

20.29   (a)   A *standard reduction potential* is the relative potential of a reduction half-reaction measured at standard conditions, 1 M aqueous solution and 1 atm gas pressure.

(b) $E^\circ_{red} = 0\,V$ for a standard hydrogen electrode.

(c)   The reduction of $Ag^+(aq)$ to $Ag(s)$ is much more energetically favorable, because it has a substantially more positive E°_{red} (0.799 V) than the reduction of $Sn^{2+}(aq)$ to $Sn(s)$ (–0.136 V).

20.30   (a)   It is not possible to measure the standard reduction potential of a single half-reaction because each voltaic cell consists of two half-reactions and only the potential of a complete cell can be measured.

(b)   The standard reduction potential of a half-reaction is determined by combining it with a reference half-reaction of known potential and measuring the cell potential. Assuming the half-reaction of interest is the reduction half-reaction:

$E^\circ_{cell} = E^\circ_{red}(cathode) - E^\circ_{red}(anode) = E^\circ_{red}(unknown) - E^\circ_{red}(reference);$

$E^\circ_{red}(unknown) = E^\circ_{cell} + E^\circ_{red}(reference).$

(c) $Cd^{2+}(aq) + 2e^- \rightarrow Cd(s)$ E° = –0.403 V

$Ca^{2+}(aq) + 2e^- \rightarrow Ca(s)$ E° = –2.87 V

The reduction of $Ca^{2+}(aq)$ to $Ca(s)$ is the more energetically unfavorable reduction because it has a more negative E° value.

20.31   *Analyze/Plan.* Follow the logic in Sample Exercise 20.5.   *Solve.*

(a)   The two half-reactions are:

$Tl^{3+}(aq) + 2e^- \rightarrow Tl^+(aq)$ cathode $E^\circ_{red} = ?$

$2[Cr^{2+}(aq) \rightarrow Cr^{3+}(aq) + e^-]$ anode $E^\circ_{red} = -0.41\,V$

(b) $E^\circ_{cell} = E^\circ_{red}(cathode) - E^\circ_{red}(anode); 1.19\,V = E^\circ_{red} - (-0.41\,V);$

$E^\circ_{red} = 1.19\,V - 0.41\,V = 0.78\,V$

(c)

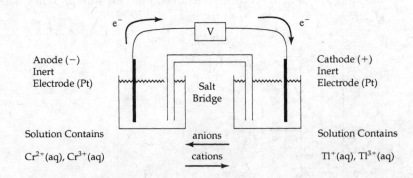

Note that because $Cr^{2+}(aq)$ is readily oxidized, it would be necessary to keep oxygen out of the left-hand cell compartment.

20.32   (a) $PdCl_4^{2-}(aq) + 2e^- \rightarrow Pd(s) + 4Cl^-$ cathode $E_{red}^{\circ} = ?$

$Cd(s) \rightarrow Cd^{2+}(aq) + 2e^-$ anode $E_{red}^{\circ} = -0.403 \, V$

(b) $E_{cell}^{\circ} = E_{red}^{\circ} \, (cathode) - E_{red}^{\circ} \, (anode); 1.03 \, V = E_{red}^{\circ} - (-0.403 \, V);$

$E_{red}^{\circ} = 1.03 \, V - 0.403 = 0.63 \, V$

(c)

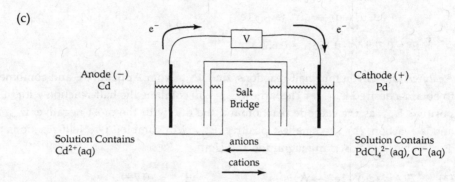

Anode (−)   Cathode (+)
Cd   Pd

Solution Contains   anions   Solution Contains
$Cd^{2+}(aq)$ ⟵ $PdCl_4^{2-}(aq), Cl^-(aq)$
                       cations

20.33   *Analyze/Plan.* Follow the logic in Sample Exercise 20.6.   *Solve.*

(a) $Cl_2(g) \rightarrow 2Cl^-(aq) + 2e^-$ $E_{red}^{\circ} = 1.359 \, V$

$I_2(s) + 2e^- \rightarrow 2I^-(aq)$ $E_{red}^{\circ} = 0.536 \, V$

$E° = 1.359 \, V - 0.536 \, V = 0.823 \, V$

(b) $Ni(s) \rightarrow Ni^{2+}(aq) + 2e^-$ $E_{red}^{\circ} = -0.28 \, V$

$2[Ce^{4+}(aq) + 1e^- \rightarrow Ce^{3+}(aq)]$ $E_{red}^{\circ} = 1.61 \, V$

$E° = 1.61 \, V - (-0.28 \, V) = 1.89 \, V$

(c) $Fe(s) \rightarrow Fe^{2+}(aq) + 2e^-$ $E_{red}^{\circ} = -0.440 \, V$

$2[Fe^{3+}(aq) + 1e^- \rightarrow Fe^{2+}(aq)]$ $E_{red}^{\circ} = 0.771 \, V$

$E° = 0.771 \, V - (-0.440 \, V) = 1.211 \, V$

(d) $3[Ca(s) \rightarrow Ca^{2+}(aq) + 2e^-]$ $E_{red}^{\circ} = -2.87 \, V$

$2[Al^{3+}(aq) + 3e^- \rightarrow Al(s)]$ $E_{red}^{\circ} = -1.66 \, V$

$E° = -1.66 \, V - (-2.87 \, V)] = 1.21 \, V$

20.34   (a) $F_2(g) + 2e^- \rightarrow 2F^-(aq)$ $E_{red}^{\circ} = 2.87 \, V$

$H_2(g) \rightarrow 2H^+(aq) + 2e^-$ $E_{red}^{\circ} = 0.00 \, V$

$E° = 2.87 \, V - 0.00 \, V = 2.87 \, V$

(b) $Cu(s) \rightarrow Cu^{2+}(aq) + 2e^-$ $E_{red}^{\circ} = 0.337 \, V$

$Ba^{2+}(aq) + 2e^- \rightarrow Ba(s)$ $E_{red}^{\circ} = -2.90 \, V$

$E° = -2.90 \, V - (0.337 \, V) = -3.24 \, V$

(c) $Fe^{2+}(aq) + 2e^- \rightarrow Fe(s)$ $E^\circ_{red} = -0.440\ V$

         $2[Fe^{2+}(aq) \rightarrow Fe^{3+}(aq) + 1e^-]$ $E^\circ_{red} = 0.771\ V$

       $E^\circ = -0.440\ V - 0.771\ V = -1.211\ V$

(d) $Hg_2^{2+}(aq) + 2e^- \rightarrow 2Hg(l)$ $E^\circ_{red} = 0.789\ V$

         $2[Cu^+(aq) \rightarrow Cu^{2+}(aq) + 1e^-]$ $E^\circ_{red} = 0.153\ V$

       $E^\circ = 0.789\ V - 0.153\ V = 0.636\ V$

20.35   *Analyze/Plan.* Given four half-reactions, find E°_{red} from Appendix E and combine them to obtain a desired E_{cell}. (a) The largest E_{cell} will combine the half-reaction with the most positive E°_{red} as the cathode reaction and the one with the most negative E°_{red} as the anode reaction. (b) The smallest positive E°_{cell} will combine two half-reactions whose E°_{red} values are closest in magnitude **and** sign.   *Solve.*

(a) $3[Ag^+(aq) + 1e^- \rightarrow Ag(s)]$ $E^\circ_{red} = 0.799$

         $Cr(s) \rightarrow Cr^{3+}(aq) + 3e^-$ $E^\circ_{red} = -0.74$

       $3Ag^+(aq) + Cr(s) \rightarrow 3Ag(s) + Cr^{3+}(aq)$ $E^\circ = 0.799 - (-0.74) = 1.54\ V$

(b)   Two of the combinations have essentially equal E° values.

         $2[Ag^+(aq) + 1e^- \rightarrow Ag(s)]$ $E^\circ_{red} = 0.799\ V$

           $Cu(s) \rightarrow Cu^{2+}(aq) + 2e^-$ $E^\circ_{red} = 0.337\ V$

       $2Ag^+(aq) + Cu(s) \rightarrow 2Ag(s) + Cu^{2+}(aq)$ $E^\circ = 0.799\ V - 0.337\ V = 0.462\ V$

         $3[Ni^{2+}(aq) + 2e^- \rightarrow Ni(s)]$ $E^\circ_{red} = -0.28\ V$

           $2[Cr(s) \rightarrow Cr^{3+}(aq) + 3e^-]$ $E^\circ_{red} = -0.74\ V$

       $3Ni^{2+}(aq) + 2Cr(s) \rightarrow 3Ni(s) + 2Cr^{3+}(aq)$ $E^\circ = -0.28\ V - (-0.74\ V) = 0.46\ V$

20.36   (a) $2[Au(s) + 4Br^-(aq) \rightarrow AuBr_4^-(aq) + 3e^-]$ $E^\circ_{red} = -0.858\ V$

           $3[2e^- + IO^-(aq) + H_2O(l) \rightarrow I^-(aq) + 2OH^-(aq)]$ $E^\circ_{red} = 0.49\ V$

     $2Au(s) + 8Br^-(aq) + 3IO^-(aq) + 3H_2O(l) \rightarrow 2AuBr_4^-(aq) + 3I^-(aq) + 6OH^-(aq)$

                                  $E^\circ = 0.49 - (-0.858) = 1.35\ V$

(b) $2[Eu^{2+}(aq) \rightarrow Eu^{3+}(aq) + 1e^-]$ $E^\circ_{red} = -0.43\ V$

       $Sn^{2+}(aq) + 2e^- \rightarrow Sn(s)$ $E^\circ_{red} = -0.14\ V$

   $2Eu^{2+}(aq) + Sn^{2+}(aq) \rightarrow 2Eu^{3+}(aq) + Sn(s)$ $E^\circ = -0.14 - (-0.43) = 0.29\ V$

20.37   *Analyze/Plan.* Given the description of a voltaic cell, answer questions about this cell. Combine ideas in Sample Exercises 20.4 and 20.7. The reduction half-reactions are:

       $Cu^{2+}(aq) + 2e^- \rightarrow Cu(s)$ $E^\circ = 0.337\ V$

       $Sn^{2+}(aq) + 2e^- \rightarrow Sn(s)$ $E^\circ = -0.136\ V$

Solve.

(a) It is evident that Cu^{2+} is more readily reduced. Therefore, Cu serves as the cathode, Sn as the anode.

(b) The copper electrode gains mass as Cu is plated out, the Sn electrode loses mass as Sn is oxidized.

(c) The overall cell reaction is $Cu^{2+}(aq) + Sn(s) \rightarrow Cu(s) + Sn^{2+}(aq)$

(d) $E° = 0.337 \text{ V} - (-0.136 \text{ V}) = 0.473 \text{ V}$

20.38 (a) The two half-reactions are:

$$Pb^{2+}(aq) + 2e^- \rightarrow Pb(s) \qquad E° = -0.126 \text{ V}$$
$$Cl_2(g) + 2e^- \rightarrow 2Cl^-(aq) \qquad E° = 1.359 \text{ V}$$

Because $E°$ for the reduction of Cl_2 is higher, the reduction of Cl_2 occurs at the Pt cathode. The Pb electrode is the anode.

(b) The Pb anode loses mass as $Pb^{2+}(aq)$ is produced.

(c) $Cl_2(g) + Pb(s) \rightarrow Pb^{2+}(aq) + 2Cl^-(aq)$

(d) $E° = 1.359 \text{ V} - (-0.126 \text{ V}) = 1.485 \text{ V}$

Strengths of Oxidizing and Reducing Agents

20.39 *Analyze/Plan.* Follow the logic in Sample Exercise 20.8. In each case, choose the half-reaction with the more positive reduction potential and with the given substance on the left. *Solve.*

(a) $Cl_2(g)$ (1.359 V vs. 1.065 V)

(b) $Ni^{2+}(aq)$ (−0.28V vs. −0.403 V)

(c) $BrO_3^-(aq)$ (1.52 V vs. 1.195 V)

(d) $O_3(g)$ (2.07 V vs. 1.776 V)

20.40 The more readily a substance is oxidized, the stronger it is as a reducing agent. In each case choose the half-reaction with the more negative reduction potential and the given substance on the right.

(a) Mg(s) (−2.37 V vs. −0.440 V)

(b) Ca(s) (−2.87 V vs. −1.66 V)

(c) H_2(g, acidic) (0.000 V vs. 0.141 V)

(d) $H_2C_2O_4(aq)$ (−0.49 V vs. 0.17 V)

20.41 *Analyze/Plan.* If the substance is on the left of a reduction half-reaction, it will be an oxidant; if it is on the right, it will be a reductant. The sign and magnitude of the $E°_{red}$ determines whether it is strong or weak. *Solve.*

(a) $Cl_2(aq)$: strong oxidant (on the left, $E°_{red} = 1.359 \text{ V}$)

(b) MnO_4^-(aq, acidic): strong oxidant (on the left, $E°_{red} = 1.51 \text{ V}$)

(c)   Ba(s): strong reductant (on the right, $E_{red}^\circ = -2.90\text{ V}$)

(d)   Zn(s): reductant (on the right, $E_{red}^\circ = -0.763\text{ V}$)

20.42   If the substance is on the left of a reduction half-reaction, it will be an oxidant; if it is on the right, it will be a reductant. The sign and magnitude of the E_{red}° determine whether it is strong or weak.

(a)   Na(s): strong reductant (on the right, $E_{red}^\circ = -2.71\text{ V}$)

(b) $O_3(g)$: strong oxidant (on the left, $E_{red}^\circ = 2.07\text{ V}$)

(c) $Ce^{3+}(aq)$: very weak reductant (on the right, $E_{red}^\circ = 1.61\text{ V}$)

(d) $Sn^{2+}(aq)$: reductant (on the right, $E_{red}^\circ = 0.154\text{ V}$) **or** weak oxidant (on the left, -0.136 V)

20.43   *Analyze/Plan.* Follow the logic in Sample Exercise 20.8.   *Solve.*

(a)   Arranged in order of increasing strength as oxidizing agents (and increasing reduction potential):

$Cu^{2+}(aq) < O_2(g) < Cr_2O_7^{2-}(aq) < Cl_2(g) < H_2O_2(aq)$

(b)   Arranged in order of increasing strength as reducing agents (and decreasing reduction potential):

$H_2O_2(aq) < I^-(aq) < Sn^{2+}(aq) < Zn(s) < Al(s)$

20.44   (a)   The strongest oxidizing agent is the species most readily reduced, as evidenced by a large, positive reduction potential. That species is H_2O_2. The weakest oxidizing agent is the species that least readily accepts an electron. We expect that it will be very difficult to reduce Zn(s); indeed, Zn(s) acts as a comparatively strong **reducing** agent. No potential is listed for reduction of Zn(s), but we can safely assume that it is less readily reduced than any of the other species present.

(b)   The strongest reducing agent is the species most easily oxidized (the largest negative reduction potential). Zn, $E_{red}^\circ = -0.76\text{ V}$, is the strongest reducing agent and F^-, $E_{red}^\circ = 2.87\text{ V}$, is the weakest.

20.45   *Analyze/Plan.* In order to reduce Eu^{3+} to Eu^{2+}, we need an oxidizing agent, one of the reduced species from Table 20.1 or Appendix E. It must have a greater tendency to be oxidized than Eu^{3+} has to be reduced. That is, E_{red}° must be more negative than -0.43 V. *Solve.*

Any of the **reduced** species in Table 20.1 or Appendix E from a half-reaction with a reduction potential more negative than -0.43 V will reduce Eu^{3+} to Eu^{2+}. From the list of possible reductants in the exercise, Al and $H_2C_2O_4$ will reduce Eu^{3+} to Eu^{2+}.

Free Energy and Redox Reactions

20.46   Any oxidized species from Table 20.1 or Appendix E with a reduction potential greater than 0.59 V will oxidize RuO_4^{2-} to RuO_4^-. From the list of possible oxidants in the exercise, $Cr_2O_7^{2-}(aq)$ and $ClO^-(aq)$ will oxidize RuO_4^{2-} to RuO_4^-.

20.47 *Analyze/Plan.* In each reaction, $Fe^{2+} \rightarrow Fe^{3+}$ will be the oxidation half-reaction and one of the other given half-reactions will be the reduction half-reaction. Follow the logic in Sample Exercise 20.10 to calculate $E°$ and $\Delta G°$ for each reaction. *Solve.*

(a) $2Fe^{2+}(aq) + S_2O_6^{2-}(aq) + 4H^+(aq) \rightarrow 2Fe^{3+}(aq) + 2H_2SO_3(aq)$

$E° = 0.60\ V - 0.77\ V = -0.17\ V$

$2Fe^{2+}(aq) + N_2O(aq) + 2H^+(aq) \rightarrow 2Fe^{3+}(aq) + N_2(g) + H_2O(l)$

$E° = -1.77\ V - 0.77\ V = -2.54\ V$

$Fe^{2+}(aq) + VO_2^+(aq) + 2H^+(aq) \rightarrow Fe^{3+}(aq) + VO^{2+}(aq) + H_2O(l)$

$E° = 1.00\ V - 0.77\ V = +0.23\ V$

(b) $\Delta G° = -nFE°$ For the first reaction,

$$\Delta G° = -2\ mol \times \frac{96{,}500\ J}{1\ V \cdot mol} \times (-0.17\ V) = 3.281 \times 10^4 = 3.3 \times 10^4\ J\ \text{or}\ 33\ kJ$$

For the second reaction, $\Delta G° = -2(96{,}500)(-2.54) = 4.902 \times 10^5 = 4.90 \times 10^2\ kJ$

For the third reaction, $\Delta G° = -1(96{,}500)(0.23) = -2.22 \times 10^4\ J = -22\ kJ$

(c) $\Delta G° = -RT\ \ln K;\ \ln K = -\Delta G°/RT;\ K = e^{-\Delta G°/RT}$

For the first reaction,

$$\ln K = \frac{-3.281 \times 10^4\ J}{(8.314\ J/mol \cdot K)(298\ K)} = -13.243 = -13;\ K = e^{-13.2428} = 1.78 \times 10^{-6} = 2 \times 10^{-6}$$

[Convert ln to log; the number of decimal places in the log is the number of sig figs in the result.]

For the second reaction,

$$\ln K = \frac{-4.902 \times 10^5\ J}{8.314\ J/mol \cdot K \times 298\ K} = -197.86 = -198;\ K = e^{-198} = 1.23 \times 10^{-86} = 1 \times 10^{-86}$$

For the third reaction,

$$\ln K = \frac{-(-2.22 \times 10^4\ J)}{8.314\ J/mol \cdot K \times 298\ K} = 8.958 = 9.0;\ K = e^{9.0} = 7.77 \times 10^3 = 8 \times 10^3$$

Check. The equilibrium constants calculated here are indicators of equilibrium position, but are not particularly precise numerical values.

20.48 (a)

$$2I^-(aq) \rightarrow I_2(s) + 2e^- \qquad E^°_{red} = 0.536\ V$$

$$\underline{Hg_2^{2+}(aq) + 2e^- \rightarrow 2Hg(l) \qquad E^°_{red} = 0.789\ V}$$

$$2I^-(aq) + Hg_2^{2+}(aq) \rightarrow I_2(s) + 2Hg(l) \qquad E^° = 0.789 - 0.536 = 0.253\ V$$

$$\Delta G° = -nFE° = -2\ mol\ e- \times \frac{96.5\ kJ}{V \cdot mol\ e^-} \times 0.253\ V = -48.829 = -48.8\ kJ$$

$$\ln K = \frac{-(-4.8829 \times 10^4\ J)}{(8.314\ J/mol \cdot K)(298\ K)} = 19.708 = 19.7;\ K = e^{19.7} = 3.61 \times 10^8 = 3.6 \times 10^8$$

(b) $3[Cu^+(aq) \rightarrow Cu^{2+}(aq) + 1e^-]$ $E^\circ_{red} = 0.153$ V

$$\frac{NO_3^-(aq) + 4H^+(aq) + 3e^- \rightarrow NO(g) + H_2O(l) \qquad E^\circ_{red} = 0.96 \text{ V}}{3Cu^+(aq) + NO_3^-(aq) + 4H^+(aq) \rightarrow 3Cu^{2+}(aq) + NO(g) + 2H_2O(l)}$$

$E^\circ = 0.96 - 0.153 = 0.81$ V; $\Delta G^\circ = -3(96.5)(0.81) = -2.345 \times 10^2$ kJ $= -2.3 \times 10^5$ J

$$\ln K = \frac{-(-2.345 \times 10^5 \text{ J})}{(8.314 \text{ J/mol} \cdot \text{K})(298 \text{ K})} = 94.65 = 95; \quad K = e^{95} = 1.3 \times 10^{41} = 10^{41}$$

(c) $2[Cr(OH)_3(s) + 5OH^-(aq) \rightarrow CrO_4^{2-}(aq) + 4H_2O(l) + 3e^-]$ $E^\circ_{red} = -0.13$ V

$$\frac{3[ClO^-(aq) + H_2O(l) + 2e^- \rightarrow Cl^-(aq) + 2OH^-(aq)] \qquad\qquad E^\circ_{red} = 0.89 \text{ V}}{2Cr(OH)_3(s) + 3ClO^-(aq) + 4OH^-(aq) \rightarrow 2CrO_4^{2-}(aq) + 3Cl^-(aq) + 5H_2O(l)}$$

$E^\circ = 0.89 - (-0.13) = 1.02$ V; $\Delta G^\circ = -6(96.5)(1.02) = -590.58$ kJ $= -5.91 \times 10^5$ J

$$\ln K = \frac{-(-5.9058 \times 10^5 \text{ J})}{(8.314 \text{ J/mol} \cdot \text{K})(298 \text{ K})} = 238.37 = 238; \quad K = 3.3 \times 10^{103} = 10^{103}$$

This is an unimaginably large number.

20.49   *Analyze/Plan.* Given K, calculate ΔG° and E°. Reverse the logic in Sample Exercise 20.10. According to Equation [19.22], $\Delta G^\circ = -RT \ln K$. According to Equation [20.12], $\Delta G^\circ = -nFE^\circ$, $E^\circ = -\Delta G^\circ/nF$.   *Solve.*

$K = 1.5 \times 10^{-4}$

$\Delta G^\circ = -RT \ln K = -(8.314 \text{ J/mol} \cdot \text{K})(298) \ln(1.5 \times 10^{-4}) = 2.181 \times 10^4$ J $= 21.8$ kJ

$E^\circ = -\Delta G^\circ/nF$; $n = 2$; $F = 96.5$ kJ/mol e^-

$$E^\circ = \frac{-21.81 \text{ kJ}}{2 \text{ mol } e^- \times 96.5 \text{ kJ/V} \cdot \text{mol } e^-} = -0.113 \text{ V}$$

Check. The unit of ΔG° is actually kJ/mol, which means kJ per 'mole of reaction', or for the reaction as written. Since we don't have a specific reaction, we interpret the unit as referring to the overall reaction.

20.50 $K = 3.7 \times 10^6$; $\Delta G^\circ = -RT \ln K$; $E^\circ = -\Delta G^\circ/nF$; $n = 1$; $T = 298$ K

$\Delta G^\circ = -8.314 \text{ J/mol} \cdot \text{K} \times 298 \text{ K} \times \ln(3.7 \times 10^6) = -3.747 \times 10^4$ J $= -37.5$ kJ

$$E^\circ = -\Delta G^\circ/nF = \frac{-(-37.47 \text{ kJ})}{1e^- \times 96.5 \text{ kJ/V} \cdot \text{mol } e^-} = 0.388 \text{ V}$$

20.51   *Analyze.* Given E°_{red} values for half reactions, calculate the value of K for a given redox reaction.

Plan. Combine the relationships involving E°, ΔG° and K to get a direct relationship between E° and K. For each reaction, calculate E° from E°_{red}, then apply the relationship to calculate K.

Solve. $\Delta G° = -nFE°$, $\Delta G° = -RT \ln K$; $\ln K = 2.303 \log K$

$$-nFE^{\circ} = -RT \ln K, \quad E^{\circ} = \frac{RT}{nF} \ln K = \frac{2.303\,RT}{nF} \log K$$

From Equation [20.15] and [20.16], $2.303\,RT/F = 0.0592$.

$$E^{\circ} = \frac{0.0592}{n} \log K; \quad \log K = \frac{nE^{\circ}}{0.0592}; \quad K = 10^{\log K}$$

(a) $E° = -0.28 - (-0.440) = 0.16$ V, $n = 2$ $(Ni^{2+} + 2e^- \rightarrow Ni)$

$$\log K = \frac{2(0.16)}{0.0592} = 5.4054 = 5.4; \quad K = 2.54 \times 10^5 = 3 \times 10^5$$

(b) $E° = 0 - (-0.277) = 0.277$ V; $n = 2$ $(2H^+ + 2e^- \rightarrow H_2)$

$$\log K = \frac{2(0.277)}{0.0592} = 9.358 = 9.36; \quad K = 2.3 \times 10^9$$

(c) $E° = 1.51 - 1.065 = 0.445 = 0.45$ V; $n = 10$ $(2MnO_4^- + 10e^- \rightarrow 2Mn^{+2})$

$$\log K = \frac{10(0.445)}{0.0592} = 75.169 \approx 75; \quad K = 1.5 \times 10^{75} = 10^{75}$$

Check. Note that small differences in E° values lead to large changes in the magnitude of K. Sig fig rules limit precision of K values; using log instead of ln leads to more sig figs in the K value. This result is strictly numerical and does not indicate any greater precision in the data.

20.52 $E^{\circ} = \dfrac{0.0592\ V}{n} \log K; \quad \log K = \dfrac{nE^{\circ}}{0.0592\ V}.$ See Solution 20.51 for a more complete explanation.

(a) $E° = 1.00$ V $- 0.799$ V $= 0.201 = 0.20$ V; $n = 1$ $(VO_2^+ + 1e^- \rightarrow VO^{2+})$

$$\log K = \frac{1(0.210\ V)}{0.0592\ V} = 3.3953 = 3.40; \quad K = 2.48 \times 10^3 = 2.5 \times 10^3$$

Note: The correctly balanced chemical equation for this reaction is

$VO_2^+(aq) + 2H^+(aq) + Ag(s) \rightarrow 2\,VO^{2+}(aq) + H_2O(l) + Ag^+(aq)$, for which $n = 1$.

(b) $E° = 1.61$ V $- 0.32$ V $= 1.29$ V; $n = 3$ $(3Ce^{4+} + 3e^- \rightarrow 3Ce^{3+})$

$$\log K = \frac{3(1.29)}{0.0592} = 65.372 = 65.4; \quad K = 2.35 \times 10^{65} = 2 \times 10^{65}$$

(c) $E° = 0.36$ V $- (-0.23$ V$) = 0.59$ V; $n = 4$ $(4Fe(CN)_6^{3-} + 4e^- \rightarrow 4Fe(CN)_6^{4-})$

$$\log K = \frac{4(0.59)}{0.0592} = 39.865 = 40; \quad K = 7.3 \times 10^{39} = 10^{40}$$

20.53   *Analyze/Plan.* $E^{\circ} = \dfrac{0.0592\ V}{n} \log K$. See Solution 20.51 for a more complete development.

$\log K = \dfrac{nE^{\circ}}{0.0592\ V}.$ *Solve.*

(a) $\log K = \dfrac{1(0.177\ V)}{0.0592\ V} = 2.9899 = 2.99; \quad K = 9.8 \times 10^2$

(b) $\log K = \dfrac{2(0.177 \text{ V})}{0.0592 \text{ V}} = 5.9797 = 5.98; \; K = 9.5 \times 10^5$

(c) $\log K = \dfrac{3(0.177 \text{ V})}{0.0592 \text{ V}} = 8.9696 = 8.97; \; K = 9.32 \times 10^8 = 9.3 \times 10^8$

20.54 $E^{\circ} = \dfrac{0.0592 \text{ V}}{n} \log K; \; n = \dfrac{0.0592 \text{ V}}{E^{\circ}} \log K$. See Solution 20.51 for a more complete development.

$$n = \dfrac{0.0592 \text{ V}}{0.17 \text{ V}} \log (5.5 \times 10^5); \, n = 2$$

Cell EMF under Nonstandard Conditions

20.55   (a)   The *Nernst equation* is applicable when the components of an electrochemical cell are at nonstandard conditions.

        (b) $Q = 1$ if all reactants and products are at standard conditions.

        (c)   If concentration of reactants increases, Q decreases, and E increases.

20.56   (a)   No. As the spontaneous chemical reaction of the voltaic cell proceeds, the concentrations of products increase and the concentrations of reactants decrease, so standard conditions are not maintained.

        (b)   Yes. The Nernst equation is applicable to cell EMF at nonstandard conditions, so it must be applicable at temperatures other than 298 K. There are two terms in the Nernst Equation. First, values of E° at temperatures other than 298 K are required. Then, in the form of Equation [20.14], there is a variable for T in the second term. In the short-hand form of Equation [20.16], the value 0.0592 assumes 298 K. A different coefficient would apply to cells at temperatures other than 298 K.

        (c)   If concentration of products increases, Q increases, and E decreases.

20.57   *Analyze/Plan.* Given a circumstance, determine its effect on cell emf. Each circumstance changes the value of Q. An increase in Q reduces emf; a decrease in Q increases emf. *Solve.*

$$\text{Zn(s)} + 2\text{H}^+\text{(aq)} \rightarrow \text{Zn}^{2+}\text{(aq)} + \text{H}_2\text{(g)}; \; E = E^{\circ} - \dfrac{0.0592}{n} \log Q; \; Q = \dfrac{[\text{Zn}^{2+}]P_{\text{H}_2}}{[\text{H}^+]^2}$$

        (a) P_{H_2} increases, Q increases, E decreases

        (b) $[\text{Zn}^{2+}]$ increases, Q increases, E decreases

        (c) $[\text{H}^+]$ decreases, Q increases, E decreases

        (d)   No effect; does not appear in the Nernst equation

20.58 $\text{Al(s)} + 3\text{Ag}^+\text{(aq)} \rightarrow \text{Al}^{3+}\text{(aq)} + 3\text{Ag(s)}; \; E = E^{\circ} - \dfrac{0.0592}{n} \log Q; \; Q = \dfrac{[\text{Al}^{3+}]}{[\text{Ag}^+]^3}$

        Any change that causes the reaction to be less spontaneous (that causes Q to increase and ultimately shifts the equilibrium to the left) will result in a less positive value for E.

(a)   Increases E by decreasing $[Al^{3+}]$ on the right side of the equation, which decreases Q.

(b)   No effect; the "concentrations" of pure solids and liquids do not influence the value of K for a heterogeneous equilibrium.

(c)   No effect; the concentration of Ag^+ and the value of Q are unchanged.

(d)   Decreases E; forming $AgCl(s)$ decreases the concentration of Ag^+, which increases Q.

20.59   *Analyze/Plan.* Follow the logic in Sample Exercise 20.11.   *Solve.*

(a)
$$Ni^{2+}(aq) + 2e^- \rightarrow Ni(s) \qquad\qquad E^\circ_{red} = -0.28\ V$$
$$\underline{Zn(s) \rightarrow Zn^{2+}(aq) + 2e^- \qquad\qquad E^\circ_{red} = -0.763\ V}$$
$$Ni^{2+}(aq) + Zn(s) \rightarrow Ni(s) + Zn^{2+}(aq) \qquad E^\circ = -0.28 - (-0.763) = 0.483 = 0.48\ V$$

(b)
$$E = E^\circ - \frac{0.0592}{n} \log \frac{[Zn^{2+}]}{[Ni^{2+}]};\ n = 2$$

$$E = 0.483 - \frac{0.0592}{2} \log \frac{(0.100)}{(3.00)} = 0.483 - \frac{0.0592}{2} \log (0.0333)$$

$$E = 0.483 - \frac{0.0592\,(-1.477)}{2} = 0.483 + 0.0437 = 0.527 = 0.53\ V$$

(c)
$$E = 0.483 - \frac{0.0592}{2} \log \frac{(0.900)}{(0.200)} = 0.483 - 0.0193 = 0.464 = 0.46\ V$$

20.60   (a)
$$3[Ce^{4+}(aq) + 1e^- \rightarrow Ce^{3+}(aq)] \qquad\qquad E^\circ_{red} = 1.61\ V$$
$$\underline{Cr(s) \rightarrow Cr^{3+}(aq) + 3e^- \qquad\qquad E^\circ_{red} = -0.74\ V}$$
$$3Ce^{4+}(aq) + Cr(s) \rightarrow 3Ce^{3+}(aq) + Cr^{3+}(aq) \qquad E^\circ = 1.61 - (-0.74) = 2.35\ V$$

(b)
$$E = E^\circ - \frac{0.0592}{n} \log \frac{[Ce^{3+}]^3[Cr^{3+}]}{[Ce^{4+}]^3};\ n = 3$$

$$E = 2.35 - \frac{0.0592}{3} \log \frac{(0.010)^3(0.010)}{(2.0)^3} = 2.35 - \frac{0.0592}{3} \log(1.250 \times 10^{-9})$$

$$E = 2.35 - \frac{0.0592\,(-8.903)}{3} = 2.35 + 0.176 = 2.53\ V$$

(c)
$$E = 2.35 - \frac{0.0592}{3} \log \frac{(0.85)^3(1.2)}{(0.35)^3} = 2.35 - 0.0244 = 2.33\ V$$

20.61   *Analyze/Plan.* Follow the logic in Sample Exercise 20.11.   *Solve.*

(a)
$$4[Fe^{2+}(aq) \rightarrow Fe^{3+}(aq) + 1e^-] \qquad E^\circ_{red} = 0.771\ V$$
$$\underline{O_2(g) + 4H^+(aq) + 4e^- \rightarrow 2H_2O(l) \qquad E^\circ_{red} = 1.23\ V}$$
$$4Fe^{2+}(aq) + O_2(g) + 4H^+(aq) \rightarrow 4Fe^{3+}(aq) + 2H_2O(l) \quad E^\circ = 1.23 - 0.771 = 0.459 = 0.46\ V$$

(b)

$$E = E^{\circ} - \frac{0.0592}{n}\log\frac{[Fe^{3+}]^4}{[Fe^{2+}]^4[H^+]^4 P_{O_2}}; \ n = 4, \ [H^+] = 10^{-3.50} = 3.2 \times 10^{-4} \ M$$

$$E = 0.459 \ V - \frac{0.0592}{4}\log\frac{(0.010)^4}{(1.3)^4(3.2\times10^{-4})^4(0.50)} = 0.459 - \frac{0.0592}{4}\log(7.0\times10^5)$$

$$E = 0.459 - \frac{0.0592}{4}(5.845) = 0.459 - 0.0865 = 0.3725 = 0.37 \ V$$

20.62   (a) $2[Fe^{3+}(aq) + 1e^- \rightarrow Fe^{2+}(aq)]$ $E^{\circ}_{red} = 0.771 \ V$

$$\underline{\qquad\qquad H_2(g) \rightarrow 2H^+(aq) + 2e^- \qquad\qquad E^{\circ}_{red} = 0.000 \ V \qquad}$$

$2Fe^{3+}(aq) + H_2(g) \rightarrow 2Fe^{2+}(aq) + 2H^+(aq)$ $E^{\circ} = 0.771 - 0.000 = 0.771 \ V$

(b)

$$E = E^{\circ} - \frac{0.0592}{n}\log\frac{[Fe^{2+}]^2[H^+]^2}{[Fe^{3+}]^2 P_{H_2}}; \ [H^+] = 10^{-pH} = 1.0 \times 10^{-5}, \ n = 2$$

$$E = 0.771 - \frac{0.0592}{2}\log\frac{(0.0010)^2(1.0\times10^{-5})^2}{(1.50)^2(0.50)} = 0.771 - \frac{0.0592}{2}\log(8.9\times10^{-17})$$

$$E = 0.771 - \frac{0.0592(-16.05)}{2} = 0.771 + 0.4751 = 1.246 \ V$$

20.63   *Analyze/Plan.* We are given a concentration cell with Zn electrodes. Use the definition of a concentration cell in Section 20.6 to answer the stated questions. Use Equation [20.16] to calculate the cell emf. For a concentration cell, Q = [dilute]/[concentrated].   *Solve.*

(a)   The compartment with the more dilute solution will be the anode. That is, the compartment with $[Zn^{2+}] = 1.00 \times 10^{-2} \ M$ is the anode.

(b)   Since the oxidation half-reaction is the opposite of the reduction half-reaction, E° is zero.

(c) $E = E^{\circ} - \dfrac{0.0592}{n}\log Q; \ Q = [Zn^{2+}, \text{dilute}]/[Zn^{2+}, \text{conc.}]$

$$E = 0 - \frac{0.0592}{2}\log\frac{(1.00 \times 10^{-2})}{(1.8)} = 0.0668 \ V$$

(d)   In the anode compartment, $Zn(s) \rightarrow Zn^{2+}(aq)$, so $[Zn^{2+}]$ increases from 1.00×10^{-2} M. In the cathode compartment, $Zn^{2+}(aq) \rightarrow Zn(s)$, so $[Zn^{2+}]$ decreases from 1.8 M.

20.64   (a)   The compartment with $0.0150 \ M \ Cl^-$ (aq) is the cathode.

(b)   E° = 0 V

(c) $E = E^{\circ} - \dfrac{0.0592}{n}\log Q; \ Q = [Cl^-, \text{dilute}]/[Cl^-, \text{conc.}]$

$$E = 0 - \frac{0.0592}{1}\log\frac{(0.0150)}{(2.55)} = -0.13204 = -0.1320 \ V$$

(d)   In the anode compartment, $[Cl^-]$ will decrease from 2.55 M. In the cathode, $[Cl^-]$ will increase from 0.0150 M.

20.65   *Analyze/Plan.* Follow the logic in Sample Exercise 20.12.   *Solve.*

$$E = E° - \frac{0.0592}{2} \log \frac{[P_{H_2}][Zn^{2+}]}{[H^+]^2}; E° = 0.0\ V - (-0.763\ V) = 0.763\ V$$

$$0.684 = 0.763 - \frac{0.0592}{2} \times (\log[P_{H_2}][Zn^{2+}] - 2\log[H^+])$$

$$= 0.763 - \frac{0.0592}{2} \times (-0.5686 - 2\log[H^+])$$

$$0.684 = 0.763 + 0.0168 + 0.0592\log[H^+]; \log[H^+] = \frac{0.684 - 0.0168 - 0.763}{0.0592}$$

$$\log[H^+] = -1.6188 = -1.6; [H^+] = 0.0241 = 0.02\ M; pH = 1.6$$

20.66   (a) $E° = -0.136\ V - (-0.126\ V) = -0.010\ V; n = 2$

$$0.22 = -0.010 - \frac{0.0592}{2}\log\frac{[Pb^{2+}]}{[Sn^{2+}]} = -0.010 - \frac{0.0592}{2}\log\frac{[Pb^{2+}]}{1.00}$$

$$\log[Pb^{2+}] = \frac{-0.23(2)}{0.0592} = -7.770 = -7.8; [Pb^{2+}] = 1.7 \times 10^{-8} = 2 \times 10^{-8}\ M$$

(b)   For $PbSO_4(s)$, $K_{sp} = [Pb^{2+}][SO_4^{2-}] = (1.0)(1.7 \times 10^{-8}) = 1.7 \times 10^{-8}$

Batteries and Fuel Cells

20.67   (a)   The emf of a battery decreases as it is used. This happens because the concentrations of products increase and the concentrations of reactants decrease. According to the Nernst equation, these changes increase Q and decrease E_{cell}.

(b)   The major difference between AA- and D-size batteries is the amount of reactants present. The additional reactants in a D-size battery enable it to provide power for a longer time.

20.68   First, H_2O is a reactant in the cathodic half-reaction, so it must be present in some form. Additionally, liquid water enhances mobility of the hydroxide ion in the alkaline battery. OH^- is produced in the cathode compartment and consumed in the anode compartment. It must be available at all points where Zn(s) is being oxidized. If the Zn(s) near the separator is mostly reacted, OH^- must diffuse through the gel until it reaches fresh Zn(s). A small amount of $H_2O(l)$ mobilizes OH^- so that redox can continue until reactants throughout the battery are depleted.

20.69   *Analyze/Plan.* Given mass of a reactant (Pb), calculate mass of product (PbO_2). This is a stoichiometry problem; we need the balanced equation for the chemical reaction that occurs in the lead-acid battery. Then, g Pb → mol Pb → mol PbO_2 → g PbO_2.   *Solve.*

The overall cell reaction (Equation [20.19]) is:

$$Pb(s) + PbO_2(s) + 2H^+(aq) + 2HSO_4^-(aq) \rightarrow 2PbSO_4(s) + 2H_2O(l)$$

$$402\ g\ Pb \times \frac{1\ mol\ Pb}{207.2\ g\ Pb} \times \frac{1\ mol\ PbO_2}{1\ mol\ Pb} \times \frac{239.2\ g\ PbO_2}{1\ mol\ PbO_2} = 464\ g\ PbO_2$$

20.70   The overall cell reaction is:

$$2MnO_2(s) + Zn(s) + 2H_2O(l) \rightarrow 2MnO(OH)(s) + Zn(OH)_2(s)$$

$$12.6 \text{ g Zn} \times \frac{1 \text{ mol Zn}}{65.39 \text{ g Zn}} \times \frac{2 \text{ mol MnO}_2}{1 \text{ mol Zn}} \times \frac{86.94 \text{ g MnO}_2}{1 \text{ mol MnO}_2} = 33.5 \text{ g MnO}_2 \text{ reduced}$$

20.71   *Analyze/Plan.* We are given a redox reaction and asked to write half-reactions, calculate E°, and indicate whether Li(s) is the anode or cathode. Determine which reactant is oxidized and which is reduced. Separate into half-reactions, find E°_{red} for the half-reactions from Appendix E and calculate E°.   *Solve.*

(a)   Li(s) is oxidized at the anode.

(b) $Ag_2CrO_4(s) + 2e^- \rightarrow 2Ag(s) + CrO_4^{2-}(aq)$ $E^{\circ}_{red} = 0.446 \text{ V}$

   $\underline{2[Li(s) \rightarrow Li^+(aq) + 1e^-]}$ $E^{\circ}_{red} = -3.05 \text{ V}$

   $Ag_2CrO_4(s) + 2Li(s) \rightarrow 2Ag(s) + CrO_4^-(aq) + 2Li^+(aq)$

   E° = 0.446 V – (–3.05 V) = 3.496 = 3.50 V

(c)   The emf of the battery, 3.5 V, is exactly the standard cell potential calculated in part (b).

(d)   For this battery at ambient conditions, E ≈ E°, so log Q ≈ 0. This makes sense because all reactants and products in the battery are solids and thus present in their standard states. Assuming that E° is relatively constant with temperature, the value of the second term in the Nernst equation is ≈ 0 at 37°C, and E ≈ 3.5 V.

20.72   (a) $HgO(s) + Zn(s) \rightarrow Hg(l) + ZnO(s)$

(b) $E^{\circ}_{cell} = E^{\circ}_{red} \text{ (cathode)} - E^{\circ}_{red} \text{ (anode)}$

   $E^{\circ}_{red} \text{ (anode)} = E^{\circ}_{red} - E^{\circ}_{cell} = 0.098 - 1.35 = -1.25 \text{ V}$

(c) E°_{red} is different from $Zn^{2+}(aq) + 2e^- \rightarrow Zn(s)$ (–0.76 V) because in the battery the process happens in the presence of base and Zn^{2+} is stabilized as ZnO(s). Stabilization of a reactant in a half-reaction decreases the driving force, so E°_{red} is more negative.

20.73   *Analyze/Plan.* (a) Consider the function of Zn in an alkaline battery. What effect would it have on the redox reaction and cell emf if Cd replaces Zn? (b) Both batteries contain Ni. What is the difference in environmental impact between Cd and the metal hydride? *Solve.*

(a) E°_{red} for Cd (–0.40 V) is less negative than E°_{red} for Zn (–0.76 V), so E_{cell} will have a smaller (less positive) value.

(b)   NiMH batteries use an alloy such as $ZrNi_2$ as the anode material. This eliminates the use and concomitant disposal problems associated with Cd, a toxic heavy metal.

20.74   (a)   The alkali metal Li has much greater metallic character than Zn, Cd, Pb or Ni. The reduction potential for Li is thus more negative, leading to greater overall cell emf for the battery. Also, Li is less dense than the other metals, so greater total energy for a battery can be achieved for a given total mass of material. One disadvantage is that Li is very reactive and the cell reactions are difficult to control.

(b)   Li has a much smaller molar mass (6.94 g/mol) than Ni (58.69 g/mol). A Li-ion battery can have many more charge-carrying particles than a Ni-based battery with the same mass. That is, Li-ion batteries have a greater *energy density* than Ni-based batteries.

20.75   The main advantage of a H_2-O_2 fuel cell over an alkaline battery is that the fuel cell is not a closed system. Fuel, H_2, and oxidant, O_2 are continuously supplied to the fuel cell, so that it can produce electrical current for a time limited only by the amount of available fuel. An alkaline battery contains a finite amount of reactant and produces current only until the reactants are spent, or reach equilibrium.

Alkaline batteries are much more convenient, because they are self-contained. Fuel cells require a means to acquire and store volatile and explosive $H_2(g)$. Disposal of spent alkaline batteries, which contain zinc and manganese solids, is much more problematic. H_2-O_2 fuel cells produce only $H_2O(l)$, which is not a disposal problem.

20.76   No. The fuel in a fuel cell must be fluid, either gas or liquid. Because fuel must be continuously supplied to the fuel cell, it must be capable of flow; the fuel cannot be solid.

Corrosion

20.77   *Analyze/Plan.* (a) Decide which reactant is oxidized and which is reduced. Write the balanced half-reactions and assign the appropriate one as anode and cathode. (b) Write the balanced half-reaction for $Fe^{2+}(aq) \rightarrow Fe_2O_3 \cdot 3H_2O$. Use the reduction half-reaction from part (a) to obtain the overall reaction.   *Solve.*

(a)   anode: $Fe(s) \rightarrow Fe^{2+}(aq) + 2e^-$

cathode: $O_2(g) + 4H^+(aq) + 4e^- \rightarrow 2H_2O(l)$

(b) $2Fe^{2+}(aq) + 6H_2O(l) \rightarrow Fe_2O_3 \cdot 3H_2O(s) + 6H^+(aq) + 2e^-$

$O_2(g) + 4H^+(aq) + 4e^- \rightarrow 2H_2O(l)$

(Multiply the oxidation half-reaction by two to balance electrons and obtain the overall balanced reaction.)

20.78   (a)   Calculate E°_{cell} for the given reactants at standard conditions.

$$O_2(g) + 4H^+(aq) + 4\,e^- \rightarrow 2H_2O(l) \qquad E^{\circ}_{red} = 1.23\text{ V}$$

$$\underline{\qquad 2[Cu(s) \rightarrow Cu^{2+}(aq) + 2e^-] \qquad E^{\circ}_{red} = 0.337\text{ V}}$$

$$2Cu(s) + O_2(g) + 4H^+(aq) \rightarrow 2Cu^{2+}(aq) + 2H_2O(l) \quad E^{\circ} = 1.23 - 0.337 = 0.89\text{ V}$$

At standard conditions with $O_2(g)$ and $H^+(aq)$ present, the oxidation of $Cu(s)$ has a positive E° value and is spontaneous. $Cu(s)$ will oxidize (corrode) in air in the presence of acid.

(b) Fe^{2+} has a more negative reduction potential (–0.440 V) than Cu^{2+} (+0.337 V), so Fe(s) is more readily oxidized than Cu(s). If the two metals are in contact, Fe(s) would act as a sacrificial anode and oxidize (corrode) in preference to Cu(s); this would weaken the iron support skeleton of the statue. The teflon spacers prevent contact between the two metals and insure that the iron skeleton doesn't corrode when the Cu(s) skin comes in contact with atmospheric $O_2(g)$ and $H^+(aq)$.

20.79   (a)   A "sacrificial anode" is a metal that is oxidized in preference to another when the two metals are coupled in an electrochemical cell; the sacrificial anode has a more negative E°_{red} than the other metal. In this case, Mg acts as a sacrificial anode because it is oxidized in preference to the pipe metal; it is sacrificed to preserve the pipe.

(b) E°_{red} for Mg^{2+} is –2.37 V, more negative than most metals present in pipes, including Fe $(E^\circ_{red} = -0.44$ V) and Zn $(E^\circ_{red} = -0.763$ V).

20.80   No. To afford cathodic protection, a metal must be more difficult to reduce (have a more negative reduction potential) than Fe^{2+}. E°_{red} Co^{2+} = –0.28 V, E°_{red} Fe^{2+} = –0.44 V.

20.81   *Analyze/Plan.* Given the materials brass, composed of Zn and Cu, and galvanized steel, determine the possible spontaneous redox reactions that could occur when the materials come in contact. Calculate E° values for these reactions.

Solve. The main metallic component of steel is Fe. Galvanized steel is steel plated with Zn. The three metals in question are Fe, Zn, and Cu; their E°_{red} values are shown below.

E°_{red} Fe^{2+}(aq) = –0.440 V

E°_{red} Zn^{2+}(aq) = –0.763 V

E°_{red} Cu^{2+}(aq) = 0.337 V

Zn, with the most negative E°_{red} value, can act as a sacrificial anode for either Fe or Cu. That is, Zn(s) will be preferentially oxidized when in contact with Fe(s) or Cu(s). For environmental corrosion, the oxidizing agent is usually $O_2(g)$ in acidic solution, E°_{red} = 1.23 V. The pertinent reactions and their E° values are:

$2Zn(s) + O_2(g) + 4H^+(aq) \rightarrow 2Zn^{2+}(aq) + 2H_2O(l)$

E° = 1.23 V – (–0.763 V) = 1.99 V

$2Fe(s) + O_2(g) + 4H^+(aq) \rightarrow 2Fe^{2+}(aq) + 2H_2O(l)$

E° = 1.23 V – (–0.440 V) = 1.67 V

$2Cu(s) + O_2(g) + 4H^+(aq) \rightarrow 2Cu^{2+}(aq) + 2H_2O(l)$

E° = 1.23 V – (0.337 V) = 0.893 V

Note, however, that Fe has a more negative E°_{red} than Cu so when the two are in contact Fe acts as the sacrificial anode, and corrosion (of Fe) occurs preferentially. This is verified by the larger E° value for the corrosion of Fe, 1.67 V, relative to the corrosion of Cu, 0.893 V. When the three metals Zn, Fe, and Cu are in contact, oxidation of Zn will happen first, followed by oxidation of Fe, and finally Cu.

20.82 The principal metallic component of steel is Fe. E_{red}° for Fe, –0.763 V, is more negative than that of Cu, 0.337 V. When the two are in contact, Fe acts as the sacrificial anode and corrodes (oxidizes) preferentially in the presence of $O_2(g)$.

$2Fe(s) + O_2(g) + 4H^+(aq) \rightarrow 2Fe^{2+}(aq) + 2H_2O(l)$

$E^{\circ} = 1.23\ V - (-0.440\ V) = 1.67\ V$

$2Cu(s) + O_2(g) + 4H^+(aq) \rightarrow 2Cu^{2+}(aq) + 2H_2O(l)$

$E^{\circ} = 1.23\ V - (0.337\ V) = 0.893\ V$

Both reactions are spontaneous, but the corrosion of Fe has the larger E° value and happens preferentially.

Electrolysis; Electrical Work

20.83 (a) *Electrolysis* is an electrochemical process driven by an outside energy source.

(b) Electrolysis reactions are, by definition, nonspontaneous.

(c) $2Cl^-(l) \rightarrow Cl_2(g) + 2e^-$

20.84 (a) An *electrolytic cell* is the vessel in which electrolysis occurs. It consists of a power source and two electrodes in a molten salt or aqueous solution.

(b) It is the cathode. In an electrolysis cell, as in a voltaic cell, electrons are consumed (via reduction) at the cathode. Electrons flow from the negative terminal of the voltage source and then to the cathode.

(c) A small amount of $H_2SO_4(aq)$ present during the electrolysis of water acts as a change carrier, or supporting electrolyte. This facilitates transfer of electrons through the solution and at the electrodes, speeding up the reaction. (Considering $H^+(aq)$ as the substance reduced at the cathode changes the details of the half-reactions, but not the overall E° for the electrolysis. $SO_4^{2-}(aq)$ cannot be oxidized.)

20.85 *Analyze/Plan.* Follow the logic in Sample Exercise 20.14, paying close attention to units. Coulombs = **amps** ·s; since this is a 3e⁻ reduction, each mole of Cr(s) requires 3 Faradays. *Solve.*

(a) $7.60\ A \times 2.00\ d \times \dfrac{24\ hr}{1\ d} \times \dfrac{60\ min}{1\ hr} \times \dfrac{60\ s}{1\ min} \times \dfrac{1\ C}{1\ amp \bullet s} \times \dfrac{1\ F}{96,500\ C}$

$\times \dfrac{1\ mol\ Cr}{3\ F} \times \dfrac{52.00\ g\ Cr}{1\ mol\ Cr} = 236\ g\ Cr(s)$

(b) $0.250\ mol\ Cr \times \dfrac{3\ F}{1\ mol\ Cr} \times \dfrac{96,500\ C}{F} \times \dfrac{1\ amp \bullet s}{1\ C} \times \dfrac{1}{8.00\ hr} \times \dfrac{1\ hr}{60\ min} \times \dfrac{1\ min}{60\ s}$

$= 2.51\ A$

20.86 Coulombs = **amps** ·s; since this is a 2e⁻ reduction, each mole of Mg(s) requires 2 Faradays.

(a) $5.25 \text{ A} \times 2.50 \text{ d} \times \dfrac{24 \text{ hr}}{1 \text{ d}} \times \dfrac{60 \text{ min}}{1 \text{ hr}} \times \dfrac{60 \text{ s}}{1 \text{ min}} \times \dfrac{1 \text{ C}}{1 \text{ amp} \cdot \text{s}} \times \dfrac{1 \text{ F}}{96,500 \text{ C}}$

$\times \dfrac{1 \text{ mol Mg}}{2 \text{ F}} \times \dfrac{24.31 \text{ g Mg}}{1 \text{ mol Mg}} = 143 \text{ g Mg}$

(b) $10.00 \text{ g Mg} \times \dfrac{1 \text{ mol Mg}}{24.31 \text{ g Mg}} \times \dfrac{2 \text{ F}}{1 \text{ mol Mg}} \times \dfrac{96,500 \text{ C}}{\text{F}} \times \dfrac{1 \text{ amp} \cdot \text{s}}{\text{C}} \times \dfrac{1 \text{ min}}{60 \text{ s}} \times \dfrac{1}{3.50 \text{ A}}$

$= 378 \text{ min}$

20.87   *Analyze/Plan.* Given a spontaneous chemical reaction, calculate the maximum possible work for a given amount of reactant at standard conditions. Separate the equation into half-reactions and calculate cell emf. Use Equation [20.19], $w_{max} = -nFE$, to calculate maximum work. At standard conditions, $E = E°$.   *Solve.*

$I_2(s) + 2e^- \rightarrow 2I^-(aq)$ $E^{°}_{red} = 0.536 \text{ V}$

$\underline{Sn(s) \rightarrow Sn^{2+}(aq) + 2e^-}$ $\underline{E^{°}_{red} = -0.136 \text{ V}}$

$I_2(s) + Sn(s) \rightarrow 2I^-(aq) + Sn^{2+}(aq)$ $E° = 0.536 - (-0.136) = 0.672 \text{ V}$

$w_{max} = -2(96.5)(0.672) = -129.7 \doteq -130 \text{ kJ/mol Sn}$

$\dfrac{-129.7 \text{ kJ}}{\text{mol Sn(s)}} \times \dfrac{1 \text{ mol Sn}}{118.71 \text{ g Sn}} \times 75.0 \text{ g Sn} \times \dfrac{1000 \text{ J}}{\text{kJ}} = -8.19 \times 10^4 \text{ J}$

Check. The (–) sign indicates that work is done by the cell.

20.88   For this cell at standard conditions, $E° = 1.10 \text{ V}$.

$w_{max} = \Delta G° = -nFE° = -2(96.5)(1.10) = -212.3 \doteq -212 \text{ kJ/mol Cu}$

$50.0 \text{ g Cu} \times \dfrac{1 \text{ mol Cu}}{63.55 \text{ g Cu}} \times \dfrac{-212.3 \text{ kJ}}{\text{mol Cu}} = -167 \text{ kJ} = -1.67 \times 10^5 \text{ J}$

20.89   *Analyze/Plan.* Follow the logic in Sample Exercise 20.15, paying close attention to units.

Solve.

(a) $7.5 \times 10^4 \text{ A} \times 24 \text{ hr} \times \dfrac{3600 \text{ s}}{1 \text{ hr}} \times \dfrac{1 \text{ C}}{1 \text{ amp} \cdot \text{s}} \times \dfrac{1 \text{ F}}{96,500 \text{ C}} \times \dfrac{1 \text{ mol Li}}{1 \text{ F}}$

$\times \dfrac{6.94 \text{ g Li}}{1 \text{ mol Li}} \times 0.85 = 3.961 \times 10^5 = 4.0 \times 10^5 \text{ g Li}$

(b)   If the cell is 85% efficient, $\dfrac{96,500 \text{ C}}{\text{F}} \times \dfrac{1 \text{ F}}{0.85 \text{ mol}} = 1.135 \times 10^5$

$= 1.1 \times 10^5 \text{ C/mol Li required}$

$\text{Energy} = 7.5 \text{ V} \times \dfrac{1.135 \times 10^5 \text{ C}}{\text{mol Li}} \times \dfrac{1 \text{ J}}{1 \text{ C} \cdot \text{V}} \times \dfrac{1 \text{ kWh}}{3.6 \times 10^6 \text{ J}} = 0.24 \text{ kWh/mol Li}$

20.90   (a) $6.5 \times 10^3 \text{ A} \times 48 \text{ hr} \times \dfrac{3600 \text{ s}}{1 \text{ hr}} \times \dfrac{1 \text{ C}}{1 \text{ amp} \cdot \text{s}} \times \dfrac{1 \text{ F}}{96,500 \text{ C}} \times \dfrac{1 \text{ mol Ca}}{2 \text{ F}}$

$\times \dfrac{40.08 \text{ g Ca}}{1 \text{ mol Ca}} \times 0.68 = 1.586 \times 10^5 = 1.6 \times 10^5 \text{ g Ca}$

(b) If the cell is 68% efficient, $\frac{96{,}500\ \text{C}}{\text{F}} \times \frac{2\ \text{F}}{0.68\ \text{mol Ca}} = 2.838 \times 10^5$

$= 2.8 \times 10^5$ C/mol Ca required

$$\text{Energy} = 5.00\ \text{V} \times \frac{2.838 \times 10^5\ \text{C}}{\text{mol Ca}} \times \frac{1\ \text{J}}{\text{C} \bullet \text{V}} \times \frac{1\ \text{kWh}}{3.6 \times 10^6\ \text{J}} = 0.3942 = 0.39\ \text{kWh}$$

Additional Exercises

20.91 (a)

$$\begin{aligned} &Ni^+(aq) + 1e^- \rightarrow Ni(s) \\ &Ni^+(aq) \rightarrow Ni^{2+}(aq) + 1e^- \\ \hline &2Ni^+(aq) \rightarrow Ni(s) + Ni^{2+}(aq) \end{aligned}$$

(b)

$$\begin{aligned} &MnO_4^{2-}(aq) + 4H^+(aq) + 2e^- \rightarrow MnO_2(s) + 2H_2O(l) \\ &2[MnO_4^{2-}(aq) \rightarrow MnO_4^-(aq) + 1e^-] \\ \hline &3MnO_4^{2-}(aq) + 4H^+(aq) \rightarrow 2MnO_4^-(aq) + MnO_2(s) + 2H_2O(l) \end{aligned}$$

(c)

$$\begin{aligned} &H_2SO_3(aq) + 4H^+(aq) + 4e^- \rightarrow S(s) + 3H_2O(l) \\ &2[H_2SO_3(aq) + H_2O(l) \rightarrow HSO_4^-(aq) + 3H^+(aq) + 2e^-] \\ \hline &3H_2SO_3(aq) \rightarrow S(s) + 2HSO_4^-(aq) + 2H^+(aq) + H_2O(l) \end{aligned}$$

(d)

$$\begin{aligned} &Cl_2(aq) + 2H_2O(l) \rightarrow 2ClO^-(aq) + 4H^+(aq) + 2e^- \\ &4OH^-(aq) \qquad + 4OH^-(aq) \\ \hline &Cl_2(aq) + 4OH^-(aq) \rightarrow 2ClO^-(aq) + 2H_2O(l) + 2e^- \\ &Cl_2(aq) + 2e^- \rightarrow 2Cl^-(aq) \\ \hline &1/2[2Cl_2(aq) + 4OH^-(aq) \rightarrow 2Cl^-(aq) + 2ClO^-(aq) + 2H_2O(l)] \\ &Cl_2(aq) + 2OH^-(aq) \rightarrow Cl^-(aq) + ClO^-(aq) + H_2O(l) \end{aligned}$$

20.92 (a)

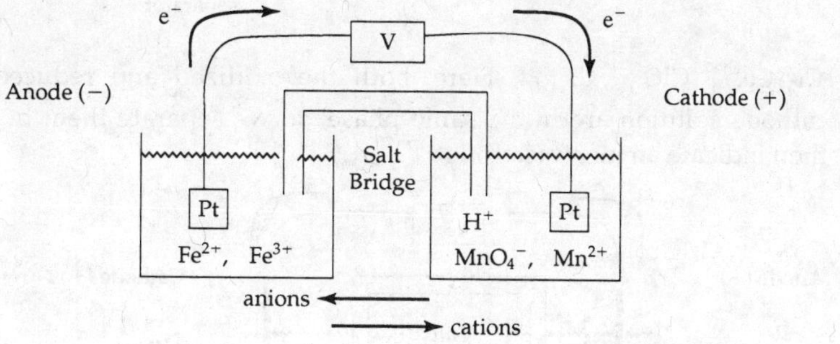

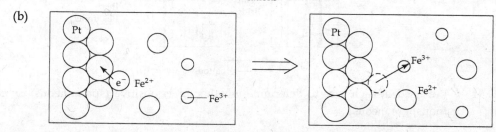

(b)

(c) $MnO_4^-(aq) + 8H^+(aq) + 5e^- \rightarrow Mn^{2+}(aq) + 4H_2O(l)$ $E_{red}^\circ = 1.51$ V

$$5[Fe^{2+}(aq) \rightarrow Fe^{3+}(aq) + 1e^-] \qquad E_{red}^\circ = 0.771 \text{ V}$$

$E^\circ = 1.51$ V $-$ 0.771 V $= 0.74$ V

(d)

$$E = E^\circ - \frac{0.0592}{5} \log \frac{[Fe^{3+}]^5[Mn^{2+}]}{[Fe^{2+}]^5[MnO_4^-][H^+]^8} ; \text{pH} = 0.0, [H^+] = 1.0$$

$$E = 0.74 \text{ V} - \frac{0.0592}{5} \log \frac{(2.5 \times 10^{-4})^5 (0.010)}{(0.10)^5 (1.50)(1.0)^8}; Q = 6.510 \times 10^{-16} = 6.5 \times 10^{-16}$$

$$E = 0.74 \text{ V} - \frac{0.0592(-15.1864)}{5} = 0.74 \text{ V} + 0.18 \text{ V} = 0.92 \text{ V}$$

20.93   (a)

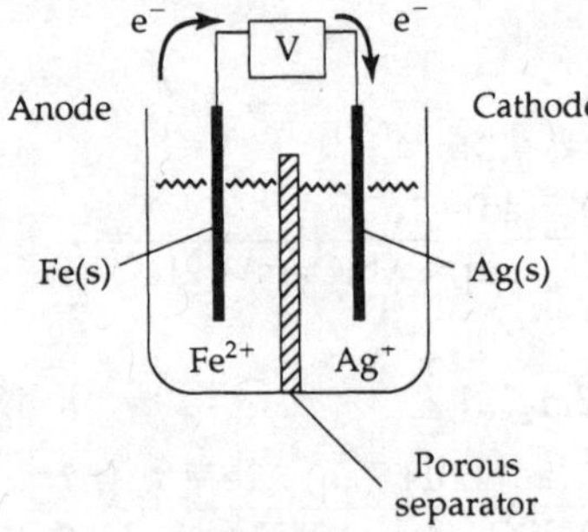

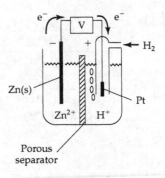

$$Fe(s) \rightarrow Fe^{2+}(aq) + 2e^-$$
$$\underline{2Ag^+(aq) + 2e^- \rightarrow 2Ag(s)}$$
$$Fe(s) + 2Ag^+(aq) \rightarrow Fe^{2+}(aq) + 2Ag(s)$$

(b)

$$Zn(s) \rightarrow Zn^{2+}(s) + 2e^-$$
$$\underline{2H^+(aq) + 2e^- \rightarrow H_2(g)}$$
$$Zn(s) + 2H^+(aq) \rightarrow Zn^{2+}(aq) + H_2(g)$$

(c) $Cu | Cu^{2+} || ClO_3^-, Cl^- | Pt$ Here, both the oxidized and reduced forms of the cathode solution are in the same phase, so we separate them by a comma, and then indicate an inert electrode.

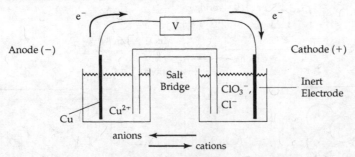

20.94   We need in each case to determine whether E° is positive (spontaneous) or negative (nonspontaneous).

(a)

$$I_2(s) + 2e^- \rightarrow 2I^-(aq) \qquad E^{\circ}_{red} = 0.536 \text{ V}$$

$$\underline{Sn(s) \rightarrow Sn^{2+}(aq) + 2e^- \qquad E^{\circ}_{red} = -0.136}$$

$$Sn(s) + I_2(s) \rightarrow Sn^{2+}(aq) + 2I^-(aq) \quad E^{\circ} = 0.536 - (-0.136) = 0.672 \text{ V, spontaneous}$$

(b)

$$Ni^{2+}(aq) + 2e^- \rightarrow Ni(s) \qquad E^{\circ}_{red} = -0.28 \text{ V}$$

$$\underline{2I^-(aq) \rightarrow I_2(s) + 2e^- \qquad E^{\circ}_{red} = 0.536 \text{ V}}$$

$$Ni^{2+}(aq) + 2I^-(aq) \rightarrow Ni(s) + I_2(s) \quad E^{\circ} = -0.28 - 0.536 = -0.82 \text{ V, nonspontaneous}$$

(c)

$$2[Ce^{4+}(aq) + 1e^- \rightarrow Ce^{3+}(aq)] \qquad\qquad E^{\circ}_{red} = 1.61 \text{ V}$$

$$\underline{H_2O_2(aq) \rightarrow O_2(g) + 2H^+(aq) + 2e^- \qquad E^{\circ}_{red} = 0.68 \text{ V}}$$

$$2Ce^{4+}(aq) + H_2O_2(aq) \rightarrow 2Ce^{3+}(aq) + O_2(g) + 2H^+(aq) \quad E^{\circ} = 1.61 - 0.68 = 0.93 \text{ V, spontaneous}$$

(d)

$$Cu^{2+}(aq) + 2e^- \rightarrow Cu(s) \qquad E^{\circ}_{red} = 0.337 \text{ V}$$

$$\underline{Sn^{2+}(aq) \rightarrow Sn^{4+}(aq) + 2e^- \qquad E^{\circ}_{red} = 0.154 \text{ V}}$$

$$Cu^{2+}(aq) + Sn^{2+}(aq) \rightarrow Cu(s) + Sn^{4+}(aq) \quad E^{\circ} = 0.337 - 1.54 = 0.183 \text{ V, spontaneous}$$

20.95   (a)   The reduction potential for $O_2(g)$ in the presence of acid is 1.23 V. $O_2(g)$ cannot oxidize $Au(s)$ to $Au^+(aq)$ or $Au^{3+}(aq)$, even in the presence of acid.

(b)   The possible oxidizing agents need a reduction potential greater than 1.50 V. These include $Co^{3+}(aq)$, $F_2(g)$, $H_2O_2(aq)$, and $O_3(g)$. Marginal oxidizing agents (those with reduction potential near 1.50 V) from Appendix E are $BrO_3^-(aq)$, $Ce^{4+}(aq)$, $HClO(aq)$, $MnO_4^-(aq)$, and $PbO_2(s)$.

(c) $4Au(s) + 8NaCN(aq) + 2H_2O(l) + O_2(g) \rightarrow 4Na[Au(CN)_2](aq) + 4NaOH(aq)$

$$Au(s) + 2CN^-(aq) \rightarrow [Au(CN)_2]^- + 1e^-$$

$$O_2(g) + 2H_2O(l) + 4e^- \rightarrow 4OH^-(aq)$$

$Au(s)$ is being oxidized and $O_2(g)$ is being reduced.

(d)

$$2[Na[Au(CN)_2](aq) + 1e^- \rightarrow Au(s) + 2CN^-(aq) + Na^+(aq)]$$

$$\underline{Zn(s) \rightarrow Zn^{2+}(aq) + 2e^-}$$

$$2Na[Au(CN)_2](aq) + Zn(s) \rightarrow 2Au(s) + Zn^{2+}(aq) + 2Na^+(aq) + 4CN^-(aq)$$

$Zn(s)$ is being oxidized and $[Au(CN)_2]^-(aq)$ is being reduced. While $OH^-(aq)$ is not included in the redox reaction above, its presence in the reaction mixture probably causes $Zn(OH)_2(s)$ to form as the product. This increases the driving force (and E°) for the overall reaction.

20.96   (a)

$$2[Ag^+(aq) + 1e^- \rightarrow Ag(s)] \qquad E^{\circ}_{red} = 0.80 \text{ V}$$

$$\underline{Ni(s) \rightarrow Ni^{2+}(aq) + 2e^- \qquad E^{\circ}_{red} = -0.28 \text{ V}}$$

$$2Ag^+(aq) + Ni(s) \rightarrow 2Ag(s) + Ni^{2+}(aq) \quad E^{\circ} = 0.80 - (-0.28) = 1.08 \text{ V}$$

(b)   As the reaction proceeds, $Ni^{2+}(aq)$ is produced, so $[Ni^{2+}]$ increases as the cell operates.

(c) $E = E^{\circ} - \dfrac{0.0592}{n} \log K; \ 1.12 = 1.08 - \dfrac{0.0592}{2} \log \dfrac{[Ni^{2+}]}{[Ag^+]^2}$

$$-\frac{0.04(2)}{0.0592} = \log(0.0100) - \log[Ag^+]^2; \; \log[Ag^+]^2 = \log(0.0100) + \frac{0.04(2)}{0.0592}$$

$\log[Ag^+]^2 = -2.000 + 1.351 = -0.649; \; [Ag^+]^2 = 0.255 \; M; \; [Ag^+] = 0.474 = 0.5 \; M$
(Strictly speaking, $[E - E°]$ having only one sig fig leads (after several steps) to the answer having only one sig fig. This is not a very precise or useful result.)

20.97   (a)

$$I_2(s) + 2e^- \rightarrow 2I^-(aq) \qquad\qquad E^°_{red} = 0.536 \; V$$

$$\underline{2[Cu(s) \rightarrow Cu^+(aq) + 1\,e^-] \qquad E^°_{red} = 0.521 \; V}$$

$$I_2(s) + 2Cu(s) \rightarrow 2Cu^+(aq) + 2\,I^-(aq) \quad E° = 0.536 - 0.521 = 0.015 \; V$$

$$E = E° - \frac{0.0592}{n}\log Q = 0.015 - \frac{0.0592}{2}\log [Cu^+]^2 \, [I^-]^2$$

$$E = +0.015 - \frac{0.0592}{2}\log (2.5)^2(3.5)^2 = +0.015 - 0.056 = -0.041 \; V$$

(b)   Since the cell potential is negative at these concentration conditions, the cell would be spontaneous in the opposite direction and the inert electrode in the I_2/I^- compartment would be the anode; $Cu(s)$ would be the cathode.

(c)   No. At standard conditions the cell reaction is as written in part (a) and $Cu(s)$ is the anode.

(d) $E = 0, \; +0.015 = \dfrac{0.0592}{2}\log (1.4)^2 \, [I^-]^2; \; \dfrac{2(0.015)}{0.0592} = \log (1.4)^2 + 2 \log [I^-];$

$\log[I^-] = 0.107 = 0.11; \; [I^-] = 10^{0.107} = 1.28 = 1.3 \; M \; I^-$

20.98   Both $E°$ and K are related to $\Delta G°$. (See Solution 20.51.)

$\Delta G° = -nFE°; \; \Delta G° = -RT \ln K$

$-nFE° = -RT \ln K, \; E° = \dfrac{RT}{nF}\ln K$

In terms of base 10 logs, $\ln K = 2.303 \log K$.

$E° = \dfrac{2.303 \, RT}{nF}\log K$

From the development of the Nernst equation,

$\dfrac{2.303 \, RT}{F} = 0.0592, \; E° = \dfrac{0.0592}{n}\log K$

20.99   Use the relationship developed in Solution 20.98 to calculate K from $E°$. Use data from Appendix E to calculate $E°$ for the disproportionation.

$$Cu^+(aq) + 1e^- \rightarrow Cu(s) \qquad\qquad E^°_{red} = 0.521 \; V$$

$$\underline{Cu^+(aq) \rightarrow Cu^{2+}(aq) + 1e^- \qquad E^°_{red} = 0.153 \; V}$$

$$2Cu^+(aq) \rightarrow Cu(s) + Cu^{2+}(aq) \qquad E° = 0.521 \; V - 0.153 \; V = 0.368 \; V$$

$$E° = \frac{0.0592}{n}\log K, \log K = \frac{nE°}{0.0592} = \frac{1 \times 0.368}{0.0592} = 6.216 = 6.22$$

$K = 10^{6.216} = 1.6 \times 10^6$

20.100   (a)   False. For standard cell potentials derived from E_{red}° values listed in Appendix E, the maximum E_{cell}° value is 5.92 V. This reaction involves $F_2(g)$, the element with the largest positive E_{red}°, and Li, the element with the most negative E_{red}°. Other species stable in aqueous solution will have E_{red}° values between these two limiting values, so other possible E_{cell}° values would be less than 5.92 V. Aqueous E_{cell}° values are not likely to exceed 5.92 V.

       (b)   False. The effect of concentration of voltage is given by the Nernst equation; $E = E^{\circ} - \dfrac{0.0592}{n} \log Q$, where Q is the reaction quotient which incorporates concentrations. Clearly E is not directly proportional to either individual concentrations or Q.

       (c)   True. Oxidizing agents are themselves reduced. The strength of an oxidizing agent is measured by the magnitude of its reduction potential, E_{red}°.

20.101   (a)   In discharge: $Cd(s) + 2NiO(OH)(s) + 2H_2O(l) \rightarrow Cd(OH)_2(s) + 2Ni(OH)_2(s)$
             In charging, the reverse reaction occurs.

       (b) $E^{\circ} = 0.49\ V - (-0.76\ V) = 1.25\ V$

       (c)   The 1.25 V calculated in part (b) is the standard cell potential, E°. The concentrations of reactants and products inside the battery are adjusted so that the cell output is greater than E°. Note that most of the reactants and products are pure solids or liquids, which do not appear in the Q expression. It must be $[OH^-]$ that is other than 1.0 M, producing an emf of 1.30 rather than 1.25.

       (d) $E^{\circ} = \dfrac{0.0592}{n} \log K; \log k = \dfrac{nE^{\circ}}{0.0592}$

             $\log K = \dfrac{2 \times 1.25}{0.0592} = 42.23 = 42.2; K = 1.7 \times 10^{42} = 2 \times 10^{42}$

20.102   The ship's hull should be made negative. By keeping an excess of electrons in the metal of the ship, the tendency for iron to undergo oxidation, with release of electrons, is diminished. The ship, as a negatively charged "electrode," becomes the site of reduction, rather than oxidation, in an electrolytic process.

20.103   It is well established that corrosion occurs most readily when the metal surface is in contact with water. Thus, moisture is a requirement for corrosion. Corrosion also occurs more readily in acid solution, because O_2 has a more positive reduction potential in the presence of $H^+(aq)$. SO_2 and its oxidation products dissolve in water to produce acidic solutions, which encourage corrosion. The anodic and cathodic reactions for the corrosion of Ni are:

$$Ni(s) \rightarrow Ni^{2+}(aq) + 2e^- \qquad E_{red}^{\circ} = -0.28\ V$$
$$O_2(g) + 4H^+(aq) + 4e^- \rightarrow 2H_2O(l) \qquad E_{red}^{\circ} = \ \ 1.23\ V$$

Nickel(ll) oxide, $NiO(s)$, can form by the dry air oxidation of Ni. This NiO coating serves to protect against further corrosion. However, NiO dissolves in acidic solutions such as those produced by SO_2 or SO_3, according to the reaction:
$NiO(s) + 2H^+(aq) \rightarrow Ni^{2+}(aq) + H_2O(l)$.   This exposes Ni(s) to further wet corrosion.

20.104   A battery is a voltaic cell, so the cathode compartment contains the positive terminal of the battery. In the alkaline, Ni–Cd and NiMH batteries, $OH^-(aq)$ is produced at the cathode. The wire that turns the indicator pink is in contact with $OH^-(aq)$, so the rightmost wire is connected to the positive terminal of the battery. [The battery could not be a lead-acid battery because $OH^-(aq)$ is not present in either compartment, so neither wire would turn the indicator pink.]

20.105   (a)   Total volume of Cr = 2.5×10^{-4} m $\times$ 0.32 m^2 = 8.0×10^{-5} m^3

$$mol\ Cr = 8.0 \times 10^{-5}\ m^3\ Cr \times \frac{100^3\ cm^3}{1\ m^3} \times \frac{7.20\ g\ Cr}{1\ cm^3} \times \frac{1\ mol\ Cr}{52.0\ g\ Cr} = 11.077$$

$$= 11\ mol\ Cr$$

The electrode reaction is:

$$CrO_4^{2-}(aq) + 4H_2O(l) + 6e^- \rightarrow Cr(s) + 8OH^-\ (aq)$$

$$Coulombs\ required = 11.077\ mol\ Cr \times \frac{6\ F}{1\ mol\ Cr} \times \frac{96,500\ C}{1\ F} = 6.41 \times 10^6$$

$$= 6.4 \times 10^6\ C$$

(b) $6.41 \times 10^6\ C \times \dfrac{1\ amp \bullet s}{1\ C} \times \dfrac{1}{10.0\ s} = 6.4 \times 10^5\ amp$

(c)   If the cell is 65 efficient, $(6.41 \times 10^6 / 0.65) = 9.867 \times 10^6 = 9.9 \times 10^6$ C are required to plate the bumper.

$$6.0\ V \times 9.867 \times 10^6\ C \times \frac{1\ J}{1\ C \bullet V} \times \frac{1\ kWh}{3.6 \times 10^6\ J} = 16.445 = 16\ kWh$$

20.106   (a)   The work obtainable is given by the product of the voltage, which has units of J/C, times the number of Coulombs of electricity produced:

$$w_{max} = 300\ amp \bullet hr \times \frac{3600\ s}{1\ hr} \times \frac{1\ C}{1\ amp \bullet s} \times \frac{6\ J}{1\ C} \times \frac{1\ kWh}{3.6 \times 10^6\ J} = 1.8\ kWh \approx 2\ kWh$$

(b)   This maximum amount of work is never realized because some of the electrical energy is dissipated in overcoming the internal resistance of the battery, because the cell voltage does not remain constant as the reaction proceeds, and because the systems to which the electrical energy is delivered are not capable of completely converting electrical energy into work.

20.107   (a) $7 \times 10^8\ mol\ H_2 \times \dfrac{2\ F}{1\ mol\ H_2} \times \dfrac{96,500\ C}{1\ F} = 1.35 \times 10^{14} = 1 \times 10^{14}\ C$

(b)

$$2H_2O(l) \rightarrow O_2(g) + 4H^+(aq) + 4e^- \qquad E_{red}^\circ = 1.23\ V$$
$$\underline{2[2H^+(aq) + 2e^- \rightarrow H_2(g)] \qquad\qquad E_{red}^\circ = 0\ V}$$
$$2H_2O(l) \rightarrow O_2(g) + 2H_2(g) \qquad\qquad E^\circ = 0.00 - 1.23 = -1.23\ V$$

$P_t = 300\ atm = P_{O_2} + P_{H_2}$. Since $H_2(g)$ and $O_2(g)$ are generated in a 2:1 mole ratio,

$P_{H_2} = 200\ atm$ and $P_{O_2} = 100\ atm$.

$$E = E^\circ - \frac{0.0592}{4} \log (P_{O_2} \times P_{H_2}^2) = -1.23 \text{ V} - \frac{0.0592}{4} \log [100 \times (200)^2]$$

$$E = -1.23 \text{ V} - 0.100 \text{ V} = -1.33 \text{ V}; \ E_{min} = 1.33 \text{ V}$$

(c) $\text{Energy} = nFE = 2(7 \times 10^8 \text{ mol}) (1.33 \text{ V}) \dfrac{96{,}500 \text{ J}}{\text{V} \bullet \text{mol}} = 1.80 \times 10^{14} = 2 \times 10^{14} \text{ J}$

(d) $1.80 \times 10^{14} \text{ J} \times \dfrac{1 \text{ kWh}}{3.6 \times 10^6 \text{ J}} \times \dfrac{\$0.85}{\text{kWh}} = \$4.24 \times 10^7 = \$4 \times 10^7$

It would cost more than $40 million for the electricity alone.

Integrative Exercises

20.108 $\begin{array}{ll} 2[NO_3^-(aq) + 4H^+(aq) + 3e^- \rightarrow NO(g) + 2H_2O(l)] & E_{red}^\circ = 0.96 \text{ V} \\ 3[Cu(s) \rightarrow Cu^{2+}(aq) + 2e^-] & E_{red}^\circ = 0.34 \text{ V} \end{array}$

$\overline{3Cu(s) + 2NO_3^-(aq) + 8H^+(aq) \rightarrow 3Cu^{2+}(aq) + 2NO(g) + 4H_2O(l)}$

$\qquad\qquad\qquad\qquad\qquad\qquad\qquad\qquad\qquad\qquad E^\circ = 0.96 - 0.34 = 0.62 \text{ V}$

$\begin{array}{ll} 2H^+(aq) + 2e^- \rightarrow H_2(g) & E_{red}^\circ = 0 \text{ V} \\ Cu(s) \rightarrow Cu^{2+}(aq) + 2e^- & E_{red}^\circ = 0.34 \text{ V} \end{array}$

$\overline{Cu(s) + 2H^+(aq) \rightarrow Cu^{2+}(aq) + H_2(g) \qquad\qquad E^\circ = 0 - 0.34 = -0.34 \text{ V}}$

The overall cell potential for the oxidation of Cu(s) by HNO_3 is positive and the reaction is spontaneous. The cell potential for the oxidation of Cu(s) by HCl is negative and the reaction is nonspontaneous. Note that in the reaction with HNO_3, it is NO_3^- that is reduced, not H^+; Cl^- from HCl cannot be further reduced.

20.109 $N_2(g) + 3H_2(g) \rightarrow 2NH_3(g)$

(a) The oxidation number of $H_2(g)$ and $N_2(g)$ is 0. The oxidation number of N in NH_3 is –3, H in NH_3 is +1. H_2 is being oxidized and N_2 is being reduced.

(b) Calculate ΔG° from ΔG_f° values in Appendix C. Use $\Delta G^\circ = -RT \ln K$ to calculate K.

$$\Delta G^\circ = 2\Delta G_f^\circ \ NH_3(g) - \Delta G_f^\circ \ N_2(g) - 3\Delta G_f^\circ \ H_2(g)$$

$$\Delta G^\circ = 2(-16.66 \text{ kJ}) - 0 - 3(0) = -33.32 \text{ kJ}$$

$$\Delta G^\circ = -RT\ln K, \ \ln K = \frac{-\Delta G^\circ}{RT} = \frac{-(-33.32 \times 10^3 \text{ J})}{(8.314 \text{ J/mol} \bullet \text{K})(298 \text{ K})} = 13.4487 = 13.45$$

$$K = e^{13.4487} = 6.9 \times 10^5$$

(c) $\Delta G^\circ = -nFE^\circ, \ E^\circ = \dfrac{-\Delta G^\circ}{nF}$

n = ? 2 N atoms change from 0 to –3, or 6 H atoms change from 0 to +1. Either way, n = 6.

$$E^\circ = \frac{-(-33.32 \text{ kJ})}{6 \times 96.5 \text{ kJ/V}} = 0.05755 \text{ V}$$

20.110   The redox reaction is: $2Ag^+(aq) + H_2(g) \rightarrow 2Ag(s) + 2H^+(aq)$.   n = 2 for this reaction.

$$E_{cell}^{\circ} = E_{red}^{\circ} \text{ cathode} - E_{red}^{\circ} \text{ anode} = 0.799 \text{ V} - 0 \text{ V} = 0.799$$

$$E = E^{\circ} - \frac{0.0592}{n} \log \frac{[H^+]^2}{[Ag^+]^2 \, P_{H_2}}$$

$[H^+]$ in the cell is held essentially constant by the benzoate buffer.

$$C_6H_5COOH(aq) \rightleftharpoons H^+(aq) + C_6H_5COO^-(aq) \quad K_a = ?$$

$$K_a = \frac{[H^+][C_6H_5COO^-]}{[C_6H_5COOH]}; [H^+] = \frac{K_a[C_6H_5COOH]}{[C_6H_5COO^-]} = \frac{0.10 \text{ M}}{0.050 \text{ M}} \times K_a = 2K_a$$

Solve the Nernst expression for $[H^+]$ and calculate K_a and pK_a as shown above.

$$1.030 \text{ V} = 0.799 \text{ V} - \frac{0.0592}{n} \log \frac{[H^+]^2}{(1.00)^2(1.00)}$$

$$0.231 \times \frac{2}{0.0592} = -\log [H^+]^2 = -2 \log[H^+]$$

$$\frac{0.231}{0.0592} = -\log [H^+] = pH; pH = 3.9020 = 3.90; [H^+] = 10^{-3.902} = 1.253 \times 10^{-4} = 1.3 \times 10^{-4};$$

$[H^+] = 2K_a$, $K_a = [H^+]/2 = 6.265 \times 10^{-5} = 6.3 \times 10^{-5}$; $pK_a = 4.20$

Check. According to Appendix D, K_a for benzoic acid is 6.3×10^{-5}.

20.111   (a)   The oxidation potential of A is equal in magnitude but opposite in sign to the reduction potential of A^+.

          (b)   Li(s) has the highest oxidation potential, Au(s) the lowest.

          (c)   The relationship is reasonable because both oxidation potential and ionization energy describe removing electrons from a substance. Ionization energy is a property of gas phase atoms or ions, while oxidation potential is a property of the bulk material.

20.112   (a)

$$NO_3^-(aq) + 4H^+(aq) + 3e^- \rightarrow NO(g) + 2H_2O(l) \quad E_{red}^{\circ} = 0.96 \text{ V}$$

$$\underline{Au(s) \rightarrow Au^{3+}(aq) + 3e^- \quad\quad\quad\quad\quad E_{red}^{\circ} = 1.498 \text{ V}}$$

$$Au(s) + NO_3^-(aq) + 4H^+(aq) \rightarrow Au^{3+}(aq) + NO(g) + 2H_2O(l)$$

$E^{\circ} = 0.96 - 1.498 = -0.54$ V; E° is negative, the reaction is not spontaneous.

          (b)

$$3[2H^+(aq) + 2e^- \rightarrow H_2(g)] \quad\quad\quad\quad E_{red}^{\circ} = 0.000 \text{ V}$$

$$\underline{2[Au(s) + 4Cl^-(aq) \rightarrow AuCl_4^-(aq) + 3e^-] \quad E_{red}^{\circ} = 1.002 \text{ V}}$$

$$2Au(s) + 6H^+(aq) + 8Cl^-(aq) \rightarrow 2AuCl_4^-(aq) + 3H_2(g)$$

$E^{\circ} = 0.000 - 1.002 = -1.002$ V; E° is negative, the reaction is not spontaneous.

          (c)

$$NO_3^-(aq) + 4H^+(aq) + 3e^- \rightarrow NO(g) + 2H_2O(l) \quad E_{red}^{\circ} = 0.96 \text{ V}$$

$$\underline{Au(s) + 4Cl^-(aq) \rightarrow AuCl_4^-(aq) + 3e^- \quad\quad E_{red}^{\circ} = 1.002 \text{ V}}$$

$$Au(s) + NO_3^-(aq) + 4Cl^-(aq) + 4H^+(aq) \rightarrow AuCl_4^-(aq) + NO(g) + 2H_2O(l)$$

$E^{\circ} = 0.96 - 1.002 = -0.04$; E° is small but negative, the process is not spontaneous.

(d) $E = E° - \dfrac{0.0592}{3} \log \dfrac{[AuCl_4^-] P_{NO}}{[NO_3^-][Cl^-]^4[H^+]^4}$

If $[H^+]$, $[Cl^-]$ and $[NO_3^-]$ are much greater than 1.0 M, the log term is negative and the correction to $E°$ is positive. If the correction term is greater than 0.042 V, the value of E is positive and the reaction at nonstandard conditions is spontaneous.

20.113   (a)

$$Ag^+(aq) + e^- \rightarrow Ag(s) \qquad\qquad E°_{red} = 0.799 \text{ V}$$
$$\underline{Fe^{2+}(aq) \rightarrow Fe^{3+}(aq) + 1e^- \qquad E°_{red} = 0.771 \text{ V}}$$
$$Ag^+(aq) + Fe^{2+}(aq) \rightarrow Ag(s) + Fe^{3+}(aq) \quad E° = 0.799 \text{ V} - 0.771 \text{ V} = 0.028 \text{ V}$$

(b) $Ag^+(aq)$ is reduced at the cathode and $Fe^{2+}(aq)$ is oxidized at the anode.

(c) $\Delta G° = -nFE° = -(1)(96.5)(0.028) = -2.7 \text{ kJ}$

$\Delta S° = S° \, Ag(s) + S° \, Fe^{3+}(aq) - S°Ag^+(aq) - S° \, Fe^{2+}(aq)$

      $= 42.55 \text{ J} + 293.3 \text{ J} - 73.93 \text{ J} - 113.4 \text{ J} = 148.5 \text{ J}$

$\Delta G° = \Delta H° - T\Delta S°$ Since $\Delta S°$ is positive, $\Delta G°$ will become more negative and $E°$ will become more positive as temperature is increased.

20.114   (a) $\Delta H° = 2\Delta H° \, H_2O(l) - 2\Delta H° \, H_2(g) - \Delta H° \, O_2(g) = 2(-285.83) - 2(0) - 0 = -571.66 \text{ kJ}$

$\Delta S° = 2S° \, H_2O(l) - 2S° \, H_2(g) - \Delta S° \, O_2(g)$

      $= 2(69.91) - 2(130.58) - (205.0) = -326.34 \text{ J}$

(b)   Since $\Delta S°$ is negative, $-T\Delta S$ is positive and the value of ΔG will become more positive as T increases. The reaction will become nonspontaneous at a fairly low temperature, because the magnitude of $\Delta S°$ is large.

(c) $\Delta G = w_{max}$. The larger the negative value of ΔG, the more work the system is capable of doing on the surroundings. As the magnitude of ΔG decreases with increasing temperature, the usefulness of H_2 as a fuel decreases.

(d)   The combustion method increases the temperature of the system, which quickly decreases the magnitude of the work that can be done by the system. Even if the effect of temperature on this reaction could be controlled, only about 40% of the energy from any combustion can be converted to electrical energy, so combustion is intrinsically less efficient than direct production of electrical energy via a fuel cell.

20.115   First balance the equation:

$4CyFe^{2+}(aq) + O_2(g) + 4H^+(aq) \rightarrow 4CyFe^{3+}(aq) + 2H_2O(l);\ E = +0.60 \text{ V}; n = 4$

(a)   From Equation [20.11] we can calculate ΔG for the process under the conditions specified for the measured potential E:

$$\Delta G = -nFE = -(4 \text{ mol } e^-) \times \dfrac{96.5 \text{ kJ}}{1 \text{ V} \bullet \text{mol } e^-}(0.60 \text{ V}) = -231.6 = -232 \text{ kJ}$$

(b)   The moles of ATP synthesized per mole of O_2 is given by:

$$\frac{231.6 \text{ kJ}}{O_2 \text{ molecule}} \times \frac{1 \text{ mol ATP formed}}{37.7 \text{ kJ}} = \text{approximately 6 mol ATP/mol } O_2$$

20.116 $\text{AgSCN(s)} + e^- \rightarrow \text{Ag(s)} + \text{SCN}^-\text{(aq)} \qquad E^\circ_{red} = 0.0895 \text{ V}$

$$\frac{\text{Ag(s)} \rightarrow \text{Ag}^+\text{(aq)} + e^- \qquad E^\circ_{red} = 0.799 \text{ V}}{\text{AgSCN(s)} \rightarrow \text{Ag}^+\text{(aq)} + \text{SCN}^-\text{(aq)} \quad E^\circ = 0.0895 - 0.799 = -0.710 \text{ V}}$$

$$E^\circ = \frac{0.0592}{n} \log K_{sp}; \; \log K_{sp} = \frac{(-0.710)\,(1)}{0.0592} = -11.993 = -12.0$$

$$K_{sp} = 10^{-11.993} = 1.02 \times 10^{-12} = 1 \times 10^{-12}$$

20.117   The reaction can be written as a sum of the steps:

$$\text{Pb}^{2+}\text{(aq)} + 2e^- \rightarrow \text{Pb(s)} \qquad\qquad E^\circ_{red} = -0.126 \text{ V}$$

$$\frac{\text{PbS(s)} \rightarrow \text{Pb}^{2+}\text{(aq)} + \text{S}^{2-}\text{(aq)} \qquad \text{"}E^\circ\text{"} = ?}{\text{PbS(s)} + 2e^- \rightarrow \text{Pb(s)} + \text{S}^{2-}\text{(aq)} \qquad E^\circ_{red} = ?}$$

"E°" for the second step can be calculated from K_{sp}.

$$E^\circ = \frac{0.0592}{n} \log K_{sp} = \frac{0.0592}{2} \log (8.0 \times 10^{-28}) = \frac{0.0592}{2} (-27.10) = -0.802 \text{ V}$$

E° for the half-reaction = $-0.126 \text{ V} + (-0.802 \text{ V}) = -0.928 \text{ V}$

Calculating an imaginary E° for a nonredox process like step 2 may be a disturbing idea. Alternatively, one could calculate K for step 1 (5.4×10^{-5}), K for the reaction in question (K = $K_1 \times K_{sp} = 4.4 \times 10^{-32}$) and then E° for the half-reaction. The result is the same.

20.118   (a)

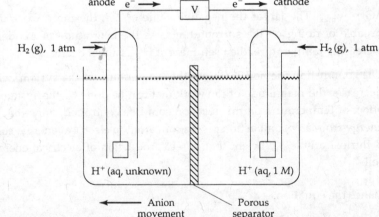

(b) $2H^+\text{(aq, 1 }M\text{)} + 2e^- \rightarrow H_2\text{(g)} \qquad\qquad E^\circ_{red} = 0$

$$\frac{H_2\text{(g)} \rightarrow 2H^+\text{(aq, 1 }M\text{)} + 2e^- \quad E^\circ_{red} = 0}{2H^+\text{(aq, 1 }M\text{)} + H_2\text{(g)} \rightarrow 2H^+\text{(aq, 1 }M\text{)} + H_2\text{(g)} \quad E^\circ = 0}$$

(c)   At standard conditions, $[H^+] = 1 \, M$, pH = 0

(d) $E = E° - \dfrac{0.0592}{2} \log \dfrac{[H^+ \text{ (unknown)}]^2 \, P_{H_2}}{[H^+ \text{ (1 } M\text{)}]^2 \, P_{H_2}}$

            $E = 0 - \dfrac{0.0592}{2} \log [H^+]^2 = \dfrac{0.0592}{2} \times 2(-\log [H^+]) = 0.0592(\text{pH}) = 0.0592(5.0) = 0.30 \text{ V}$

(e) E_{cell} changes $0.0592(0.01) = 0.000592 = 0.0006$ V for each 0.01 pH unit. The voltmeter would have to be precise to at least 0.001 V to detect a change of 0.01 pH units.

20.119   The two half-reactions in the electrolysis of $H_2O(l)$ are:

$2[2H_2O(l) + 2e^- \rightarrow H_2(g) + 2OH^-]$

$\underline{2H_2O(l) \rightarrow O_2(g) + 4H^+ + 4e^-}$

$2H_2O(l) \rightarrow 2H_2(g) + O_2(g)$

4 mol e^-/2 mol $H_2(g)$ or 2 mol e^-/mol $H_2(g)$

Using partial pressures and the ideal-gas law, calculate the mol $H_2(g)$ produced, and the current required to do so.

$P_t = P_{H_2} + P_{H_2O}$. From Appendix B, P_{H_2O} at 25.5 °C is approximately 24.5 torr.

$P_{H_2} = 768 \text{ torr} - 24.5 \text{ torr} = 743.5 = 744 \text{ torr}$

$n = PV/RT = \dfrac{(743.5/760) \text{ atm} \times 0.0123 \text{ L}}{298.5 \text{ K} \times 0.08206 \text{ L} \bullet \text{atm/mol} \bullet \text{K}} = 4.912 \times 10^{-4} = 4.91 \times 10^{-4} \text{ mol } H_2$

$4.912 \times 10^{-4} \text{ mol } H_2 \times \dfrac{2 \text{ mol } e^-}{\text{mol } H_2} \times \dfrac{96,500 \text{ C}}{1 \text{ mol } e^-} \times \dfrac{1 \text{ amp} \bullet \text{s}}{1 \text{ C}} \times \dfrac{1 \text{ min}}{60 \text{ s}} \times \dfrac{1}{2.00 \text{ min}} = 0.790 \text{ amp}$

21 Nuclear Chemistry

Nuclear Chemistry

21.1 *Analyze.* Given the name and mass number of a nuclide, decide if it lies within the belt of stability. If not, suggest a process that moves it toward the belt.

Plan. Calculate the number of protons and neutrons in each nuclide. Locate this point on Figure 21.2. If the point is above the belt, β-decay increases protons and decreases neutrons, decreasing the neutron-to-proton ratio. If the point is below the belt, either positron emission or neutron capture decreases protons and increases neutrons, increasing the neutron-to-proton ratio. *Solve.*

(a) ^{24}Ne: 10 p, 14 n, just above the belt of stability. Reduce the neutron-to proton ratio via β-decay.

(b) ^{32}Cl: 17 p, 15 n, just below the belt of stability. Increase the neutron-to-proton ratio via positron emission or orbital electron capture.

(c) ^{108}Sn: 50 p, 58 n, just below the belt of stability. Increase the neutron-to-proton ratio via positron emission or orbital electron capture.

(d) ^{216}Po: 84 p, 132 n, just beyond the belt of stability. Nuclei with atomic numbers $\geq$ 84 tend to decay via alpha emission, which decreases both protons and neutrons.

21.2 *Analyze/Plan.* From the diagram, determine the atomic number (number of protons) and mass number (number of protons plus neutrons) of the two nuclides involved. Based on the relationship between the two nuclides, decide whether the reaction is α or β-decay, positron emission or electron capture. Complete the nuclear reaction, balancing atomic numbers and mass numbers.

Solve. The two nuclides in the diagram are $^{109}_{46}$Pd and $^{109}_{47}$Ag so the second product is a β-particle. The balanced reaction is:

$$^{109}_{46}\text{Pd} \rightarrow {}^{109}_{47}\text{Ag} + {}^{0}_{-1}\text{e}$$

Check. Atomic number and mass number balance.

21.3 *Analyze/Plan.* Determine the number of protons and neutrons present in the two heavy nuclides in the reaction. Draw a graph with appropriate limits and plot the two points. Draw an arrow from reactant to product. *Solve.*

Bi: 83 p, 128 n; Tl: 81 p, 126 n.

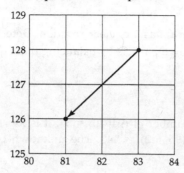

Check. An α-particle has 2 p and 2 n. The diagram shows a decrease in 2 p and 2 n for the reaction.

21.4   *Analyze/Plan.* Write the balanced equation for the decay. Nuclear decay is a first-order process; use appropriate relationships for first-order processes to determine $t_{1/2}$, k and remaining ^{88}Mo after 12 minutes.   *Solve.*

(a) $t_{1/2}$ is the time required for half of the original nuclide to decay. Relative to the graph, this is the time when the amount of ^{88}Mo is reduced from 1.0 to 0.5. This time is 7 minutes.

(b)   For a first-order process, $t_{1/2} = 0.693/k$ or $k = 0.693/t_{1/2}$.

      $k = 0.693/7$ min $= 0.0990 = 0.1$ min^{-1}

(c)   From the graph, the fraction of ^{88}Mo remaining after 12 min is $0.3/1.0 = 0.3$

      Check. $\ln(N_t/N_o) = -kt = -(0.099)(12) = -1.188$; $N_t/N_o = e^{-1.188} = 0.30$.

(d) $^{88}_{42}\text{Mo} \rightarrow {}^{88}_{41}\text{Nb} + {}^{0}_{1}\text{e}$

21.5   (a)   The difference in mass between a nuclide and its component nucleons is the mass defect. It corresponds to the energy required to separate the nuclide into individual nucleons, according to the relationship $E = \Delta mc^2$.

(b)   On Figure 21.13, ^{56}Fe has the largest binding energy per nucleon. ^{100}Ru is near this maximum and has more total nucleons than ^{56}Fe, so we expect the mass defect for ^{100}Ru to be significant.

(c)   *Plan.* Calculate the total mass of the separate nucleons. Subtract the mass of the nuclide to get Δm, the mass defect. Convert Δm to binding energy, divide by 100 to get the binding energy per nucleon.

      Δm = mass of individual protons and neutrons − mass of nuclide

      Δm = 44(1.0072765 amu) + 56(1.0086649 amu) − 99.90422 amu

      Δm = 0.9011804 = 0.90118 amu

$$\Delta E = (2.9979246 \times 10^8 \text{ m/s})^2 \times 0.9011804\,\text{amu} \times \frac{1\,\text{g}}{6.022 \times 10^{23}\,\text{amu}} \times \frac{1\,\text{kg}}{1000\,\text{g}}$$
$$= 1.34497 \times 10^{-10} \text{ J}$$

$$1.34497 \times 10^{-10} \, \frac{J}{\text{nuclide}} \times \frac{1 \, \text{nuclide}}{100 \, \text{nucleons}} = 1.34497 \times 10^{-12} \, \text{J/nucleon}$$

21.6   *Analyze/Plan.* Express the particles in the diagram as a nuclear reaction. Determine the mass number and atomic number of the unknown particle by balancing these quantities in the nuclear reaction.   *Solve.*

(a) $^{239}_{94}\text{Pu} + ^{1}_{0}\text{n} \rightarrow ^{95}_{40}\text{Zr} + ? + 2\,^{1}_{0}\text{n}$

The unknown particle has an atomic number of (94–40) = 54; it is Xe. The mass number of the nuclide is [(239 + 1) – (95 + 2)] = 143. The unknown particle is $^{143}_{54}\text{Xe}$.

(b) ^{95}Zr: 40 p, 55 n is stable. ^{143}Xe: 54 p, 89 n is above the belt of stability and is not stable; it will probably undergo β-decay.

Radioactivity

21.7   *Analyze/Plan.* Given various nuclide descriptions, determine the number of protons and neutrons in each nuclide. The left superscript is the mass number, protons plus neutrons. If there is a left subscript, it is the atomic number, the number of protons. Protons can always be determined from chemical symbol; all isotopes of the same element have the same number of protons. A number following the element name, as in part (c) is the mass number.   *Solve.*

p = protons, n = neutrons, e = electrons; number of protons = atomic number;
number of neutrons = mass number – atomic number

(a) $^{55}_{25}\text{Mn}$: 25p, 30n   (b) ^{201}Hg: 80p, 121n   (c) ^{39}K: 19p, 20n

21.8   p = protons, n = neutrons, e = electrons; number of protons = atomic number;
number of neutrons = mass number – atomic number

(a) $^{126}_{55}\text{Cs}$: 55p, 71n   (b) ^{119}Sn: 50p, 69n   (c) ^{141}Ba: 56p, 85n

21.9   *Analyze/Plan.* See definitions in Section 21.1. In each case, the left superscript is mass number, the left subscript is related to atomic number.   *Solve.*

(a) $^{1}_{1}\text{p}$ or $^{1}_{1}\text{H}$ (b) $^{0}_{1}\text{e}$ (c) $^{0}_{-1}\beta$ or $^{0}_{-1}\text{e}$

21.10   (a) $^{1}_{0}\text{n}$ (b) $^{0}_{-1}\text{e}$ or $^{0}_{-1}\beta$ (c) $^{4}_{2}\text{He}$ or $^{4}_{2}\alpha$

21.11   *Analyze/Plan.* Follow the logic in Sample Exercises 21.1 and 21.2. Pay attention to definitions of decay particles and conservation of mass and charge.   *Solve.*

(a) $^{214}_{83}\text{Bi} \rightarrow ^{214}_{84}\text{Po} + ^{0}_{-1}\text{e}$ (b) $^{195}_{79}\text{Au} + ^{0}_{-1}\text{e (orbital electron)} \rightarrow ^{195}_{78}\text{Pt}$

(c) $^{38}_{19}\text{K} \rightarrow ^{38}_{18}\text{Ar} + ^{0}_{1}\text{e}$ (d) $^{242}_{94}\text{Pu} \rightarrow ^{238}_{92}\text{U} + ^{4}_{2}\text{He}$

21.12   (a) $^{141}_{60}\text{Nd} + ^{0}_{-1}\text{e (orbital electron)} \rightarrow ^{141}_{59}\text{Pr}$ (b) $^{201}_{79}\text{Au} \rightarrow ^{201}_{80}\text{Hg} + ^{0}_{-1}\beta$

(c) $^{81}_{34}\text{Se} \rightarrow ^{81}_{35}\text{Br} + ^{0}_{-1}\beta$ (d) $^{83}_{38}\text{Sr} \rightarrow ^{83}_{37}\text{Rb} + ^{0}_{1}\text{e}$

21.13   *Analyze/Plan.* Using definitions of the decay processes and conservation of mass number and atomic number, work backwards to the reactants in the nuclear reactions.

Solve.

(a) $^{211}_{82}Pb \rightarrow {}^{211}_{83}Bi + {}^{0}_{-1}\beta$ (b) $^{50}_{25}Mn \rightarrow {}^{50}_{24}Cr + {}^{0}_{1}e$

(c) $^{179}_{74}W + {}^{0}_{-1}e \rightarrow {}^{179}_{73}Ta$ (d) $^{230}_{90}Th \rightarrow {}^{226}_{88}Ra + {}^{4}_{2}He$

21.14   (a) $^{24}_{11}Na \rightarrow {}^{24}_{12}Mg + {}^{0}_{-1}e$; a β particle is produced

(b) $^{188}_{80}Hg \rightarrow {}^{188}_{79}Au + {}^{0}_{1}e$; a positron is produced

(c) $^{122}_{53}I \rightarrow {}^{122}_{54}Xe + {}^{0}_{-1}e$; a β particle is produced

(d) $^{242}_{94}Pu \rightarrow {}^{238}_{92}U + {}^{4}_{2}He$; an α particle is produced

21.15   *Analyze/Plan.* Given the starting and ending nuclides in a nuclear decay sequence, we are asked to determine the number of alpha and beta emissions. Use the total change in A and Z, along with definitions of alpha and beta decay, to answer the question.   *Solve.*

The total mass number change is (235–207) = 28. Since each α particle emission decreases the mass number by four, whereas emission of a β particle does not correspond to a mass change, there are 7 α particle emissions. The change in atomic number in the series is 10. Each α particle results in an atomic number lower by two. The 7 α particle emissions alone would cause a decrease of 14 in atomic number. Each β particle emission raises the atomic number by one. To obtain the observed lowering of 10 in the series, there must be 4 β emissions.

21.16   This decay series represents a change of (232–208 =) 24 mass units. Since only alpha emissions change the nuclear mass, and each changes the mass by four, there must be a total of 6 α emissions. Each alpha emission causes a decrease of two in atomic number.

Therefore, the 6 alpha emissions, by themselves, would cause a decrease in atomic number of 12. The series as a whole involves a decrease of 8 in atomic number. Thus, there must be a total of 4 β emissions, each of which increases atomic number by one. Overall, there are 6 α emissions and 4 β emissions.

Nuclear Stability

21.17   *Analyze/Plan.* Follow the logic in sample Exercise 21.3, paying attention to the guidelines for neutron-to-proton ratio.   *Solve.*

(a) $^{8}_{5}B$ - low neutron/proton ratio, positron emission (for low atomic numbers, positron emission is more common than orbital electron capture)

(b) $^{68}_{29}Cu$ - high neutron/proton ratio, beta emission

(c) $^{241}_{93}Np$ - high neutron/proton ratio, beta emission

(Even though ^{241}Np has an atomic number ≥ 84, the most common decay pathway for nuclides with neutron/proton ratios higher than the isotope listed on the periodic chart is beta decay.)

(d) $^{39}_{17}Cl$ - high neutron/proton ratio, beta emission

21.18   (a) $^{66}_{32}$Ge - low neutron/proton ratio, positron emission

(b) $^{105}_{45}$Rh - high neutron/proton ratio, beta emission

(c) $^{137}_{53}$I - high neutron/proton ratio, beta emission

(d) $^{133}_{58}$Ce - low neutron/proton ratio, positron emission

21.19   *Analyze/Plan.* Use the criteria listed in Table 21.3.   *Solve.*

(a)   Stable: $^{39}_{19}$K   odd proton, even neutron more abundant than odd proton, odd neutron; 20 neutrons is a magic number.

(b)   Stable: $^{209}_{83}$Bi   odd proton, even neutron more abundant than odd proton, odd neutron; 126 neutrons is a magic number.

(c)   Stable: $^{25}_{12}$Mg   even though $^{24}_{10}$Ne is an even proton, even neutron nuclide, it has a very high neutron/proton ratio and lies outside the band of stability.

21.20   Use criteria listed in Table 21.3.

(a) $^{112}_{48}$Cd   even, even more abundant

(b) $^{27}_{13}$Al   odd proton, even neutron more abundant

(c) $^{106}_{46}$Pd   even, even more abundant

(d) $^{128}_{54}$Xe   even proton, even neutron much more abundant than odd proton; odd neutron

21.21   *Analyze/Plan.* For each nuclide, determine the number of protons and neutrons and decide if they are magic numbers.   *Solve.*

(a) $^{4}_{2}$He   (b) $^{40}_{20}$Ca   (c) $^{208}_{82}$Pb

(d) $^{58}_{28}$Ni has a magic number of protons, but not neutrons.

21.22 $^{112}_{50}$Sn has a magic number of protons and even numbers of protons and neutrons, good indications of nuclear stability. $^{112}_{49}$In has no magic numbers and odd numbers of protons and neutrons, indicators of nuclear instability or radioactivity.

21.23   *Analyze/Plan.* For each nuclide, determine the number of protons and neutrons and find the location on Figure 21.2. Rationalize the location based on magic numbers, neutron-to-proton ratio and Z value. Predict radioactivity (nonstability of nucleus).   *Solve.*

Radioactive: $^{14}_{8}$O — low neutron/proton ratio, $^{115}_{52}$Te — low neutron/proton ratio;

$^{208}_{84}$Po — atomic number $\geq$ 84

Stable: $^{32}_{16}$S, $^{78}_{34}$Se — even proton, even neutron, stable neutron/proton ratio

21.24   The criterion employed in judging whether the nucleus is likely to be radioactive is the position of the nucleus on the plot shown in Figure 21.2. If the neutron/proton ratio is too high or low, or if the atomic number exceeds 83, the nucleus will be radioactive.

Radioactive: $^{58}_{29}$Cu — odd proton, odd neutron, low neutron/proton ratio

^{206}Po — high atomic number

Stable: $^{62}_{28}$Ni — even proton, even neutron, stable neutron/proton ratio

$^{108}_{47}$Ag — stable neutron/proton ratio, (one of 5 stable odd proton/odd neutron nuclides)

^{184}W — even proton, even neutron, stable neutron/proton ratio

Nuclear Transmutations

21.25   Protons and alpha particles are positively charged and must be moving very fast to overcome electrostatic forces which would repel them from the target nucleus. Neutrons are electrically neutral and not repelled by the nucleus.

21.26   A major difference is that the charge on the nitrogen nucleus, +7, is much smaller than on the gold nucleus, +79. Thus, the alpha particle could more easily penetrate the coulomb barrier (that is, the repulsive energy barrier due to like charges) to make contact with the nitrogen nucleus than the gold nucleus. Rutherford used alpha particles that were being emitted from some radioactive source. He did not have access to machines that can accelerate particles to very high energy. It would be necessary to do just that to observe reaction of an alpha particle with a gold nucleus.

21.27   *Analyze/Plan.* Determine A and Z for the missing particle by conservation principles. Find the appropriate symbol for the particle.   *Solve.*

(a) $^{32}_{16}$S + $^{1}_{0}$n → $^{1}_{1}$p + $^{32}_{15}$P   (b) $^{7}_{4}$Be + $^{0}_{-1}$e (orbital electron) → $^{7}_{3}$Li

(c) $^{187}_{75}$Re → $^{187}_{76}$Os + $^{0}_{-1}$e   (d) $^{98}_{42}$Mo + $^{2}_{1}$H → $^{1}_{0}$n + $^{99}_{43}$Tc

(e) $^{235}_{92}$U + $^{1}_{0}$n → $^{135}_{54}$Xe + $^{99}_{38}$Sr + 2 $^{1}_{0}$n

21.28   (a) $^{252}_{98}$Cf + $^{10}_{5}$B → 3 $^{1}_{0}$n + $^{259}_{103}$Lr   (b) $^{2}_{1}$H + $^{3}_{2}$He → $^{4}_{2}$He + $^{1}_{1}$H

(c) $^{1}_{1}$H + $^{11}_{5}$B → 3 $^{4}_{2}$He   (d) $^{122}_{53}$I → $^{122}_{54}$Xe + $^{0}_{-1}$e

(e) $^{59}_{26}$Fe → $^{0}_{-1}$e + $^{59}_{27}$Co

21.29   *Analyze/Plan.* Follow the logic in Sample Exercise 21.5, paying attention to conservation of A and Z.   *Solve.*

(a) $^{238}_{92}$U + $^{1}_{0}$n → $^{239}_{92}$U + $^{0}_{0}\gamma$ (b) $^{14}_{7}$N + $^{1}_{1}$H → $^{11}_{6}$C + $^{4}_{2}$He

(c) $^{18}_{8}$O + $^{1}_{0}$n → $^{19}_{9}$F + $^{0}_{-1}$e

21.30   (a) $^{238}_{92}$U + $^{4}_{2}$He → $^{241}_{94}$Pu + $^{1}_{0}$n   (b) $^{14}_{7}$N + $^{4}_{2}$He → $^{17}_{8}$O + $^{1}_{1}$H

(c) $^{56}_{26}$Fe + $^{4}_{2}$He → $^{60}_{29}$Cu + $^{0}_{-1}$e

Rates of Radioactive Decay

21.31 Chemical reactions do not affect the character of atomic nuclei. The energy changes involved in chemical reactions are much too small to allow us to alter nuclear properties via chemical processes. Therefore, the nuclei that are formed in a nuclear reaction will continue to be radioactive regardless of any chemical changes we bring to bear. However, we can hope to use chemical means to separate radioactive substances, or remove them from foods or a portion of the environment.

21.32 The suggestion is not reasonable. The energies of nuclear states are very large relative to ordinary temperatures. Thus, merely changing the temperature by less than 100 K would not be expected to significantly affect the behavior of nuclei with regard to nuclear decay rates.

21.33 *Analyze/Plan.* Follow the logic in Sample Exercise 21.6. *Solve.*

After 12.3 yr, one half-life, there are $(1/2)48.0 = 24.0$ mg. 49.2 yr is exactly four half-lives. There are then $(48.0)(1/2)^4 = 3.0$ mg tritium remaining.

21.34 Calculate the decay constant, k, and then $t_{1/2}$.

$$k = \frac{-1}{t} \ln \frac{N_t}{N_o} = \frac{-1}{5.2\,\text{min}} \ln \frac{0.250\,\text{g}}{1.000\,\text{g}} = 0.2666 = 0.27\,\text{min}^{-1}$$

Using Equation [21.20], $t_{1/2} = 0.693/k = 0.693/0.02666\,\text{min}^{-1} = 2.599 = 2.6$ min

21.35 *Analyze/Plan.* Follow the logic in Sample Exercise 21.7. In this case, we are given initial sample mass as well as mass at time t, so we can proceed directly to calculate k (Equation [21.20]) and then t (Equation [21.19]). *Solve.*

$$k = 0.693 / t_{1/2} = 0.693/27.8\,\text{d} = 0.02493 = 0.0249\,\text{d}^{-1}$$

$$t = \frac{-1}{k} \ln \frac{N_t}{N_o} = \frac{-1}{0.02493\,\text{d}^{-1}} \ln \frac{1.50}{5.75} = 53.9\,\text{d}$$

21.36 $k = 0.693 / t_{1/2} = 0.693/5.26\,\text{yr} = 0.1317 = 0.132\,\text{yr}^{-1}$; $N_t/N_o = 0.75$

$$t = \frac{-1}{k} \ln \frac{N_t}{N_o} = -(1/0.1317\,\text{yr}^{-1}) \ln (0.75) = 2.18\,\text{yr}$$

2.18 yr = 26.2 mo = 797 d. The source would have been replaced sometime in the fall of 2007, probably in October.

21.37 (a) *Analyze/Plan.* $^{226}_{88}\text{Ra} \rightarrow \, ^{222}_{86}\text{Rn} + \, ^{4}_{2}\text{He}$

1 α particle is produced for each ^{226}Ra that decays. Calculate the mass of ^{226}Ra remaining after 1.0 min, calculate by subtraction the mass that has decayed, and use Avogadro's number to get the number of $^{4}_{2}$He particles. *Solve.*

Calculate k in min^{-1}. $1600\,\text{yr} \times \frac{365\,\text{d}}{1\,\text{yr}} \times \frac{24\,\text{hr}}{1\,\text{d}} \times \frac{60\,\text{min}}{1\,\text{hr}} = 8.410 \times 10^8\,\text{min}$

$$k = \frac{0.693}{t_{1/2}} = \frac{0.693}{8.410 \times 10^8\,\text{min}} = 8.241 \times 10^{-10}\,\text{min}^{-1}$$

$$\ln \frac{N_t}{N_o} = -kt = (-8.241 \times 10^{-10} \text{ min}^{-1})(1.0 \text{ min}) = -8.241 \times 10^{-10} = -8.2 \times 10^{-10}$$

$$\frac{N_t}{N_o} = e^{-8.2 \times 10^{-10}} = (1.000 - 8.2 \times 10^{-10}); \text{ (don't round here!)}$$

$[e^{-8.2 \times 10^{-10}}$ is a number very close to 1. In this calculation, it is conveneint to express the number as $(1 - 8.2 \times 10^{-10})]$.

$N_t = 5.0 \times 10^{-3}$ g $(1.00 - 8.2 \times 10^{-10})$ The amount that decays is $N_o - N_t$:

$$5.0 \times 10^{-3} \text{ g} - [5.0 \times 10^{-3} (1.00 - 8.2 \times 10^{-10})] = 5.0 \times 10^{-3} \text{ g} (8.2 \times 10^{-10})$$

$$= \sim 4.1 \times 10^{-12} \text{ g Ra}$$

(In terms of sig figs, $[N_o - N_t]$ is a very small number, found by subtracting two numbers known only to two sig figs and one decimal place. At best, we can express the result as an order of magnitude.)

$$[N_o - N_t] = 4.1 \times 10^{-12} \text{ g Ra} \times \frac{1 \text{ mol Ra}}{226 \text{ g Ra}} \times \frac{6.022 \times 10^{23} \text{ Ra atoms}}{1 \text{ mol Ra}} \times \frac{1 \, {}^4_2\text{He}}{1 \text{ Ra atom}}$$

$$= 1.1 \times 10^{10} = \sim 10^{10} \; \alpha \text{ particles emitted in 1 min}$$

(b)   *Plan.* The result from (a) is disintegrations/min. Change this to dis/s and apply the definition $1 \text{ Ci} = 2.7 \times 10^{10}$ dis/s.

$$1.1 \times 10^{10} \frac{\text{dis}}{\text{min}} \times \frac{1 \text{ min}}{60 \text{ s}} \times \frac{1 \text{ Ci}}{3.7 \times 10^{10} \text{ dis/s}} \times \frac{1000 \text{ mCi}}{\text{Ci}} = 4.945 = \sim 5 \text{ mCi}$$

Realistically, the activity is best expressed as between 0.01 and 0.001 Ci.

21.38   (a)   Proceeding as in Solution 21.37, calculate k in s^{-1}.

$$5.26 \text{ yr} \times \frac{365 \text{ d}}{1 \text{ yr}} \times \frac{24 \text{ hr}}{1 \text{ d}} \times \frac{3600 \text{ sec}}{1 \text{ hr}} = 1.659 \times 10^8 = 1.66 \times 10^8 \text{ s}$$

$$k = \frac{0.693}{t_{1/2}} = \frac{0.693}{1.659 \times 10^8} = 4.178 \times 10^{-9} = 4.18 \times 10^{-9} \text{ s}^{-1}$$

$$\ln \frac{N_t}{N_o} = -kt = -(4.178 \times 10^{-9} \text{ s}^{-1})(45.5 \text{ s}) = -1.901 \times 10^{-7} = -1.90 \times 10^{-7}$$

$$\frac{N_t}{N_o} = e^{-1.90 \times 10^{-7}} = (1.000 - 1.90 \times 10^{-7}); N_t = 2.44 \times 10^{-3} \text{ g} (1.000 - 1.90 \times 10^{-7})$$

$[e^{-1.90 \times 10^{-7}}$ is a number very close to 1. In this calculation, it is conveneint to express the number as $(1 - 1.90 \times 10^{-7})]$.

The amount that decays is $N_o - N_t$:

$$2.44 \times 10^{-3} \text{ g} - [2.44 \times 10^{-3} \text{ g} (1.000 - 1.90 \times 10^{-7})] = 2.44 \times 10^{-3} \text{ g} (1.90 \times 10^{-7})$$

$$= 4.638 \times 10^{-10} = \sim 5 \times 10^{-10} \text{ g Co}$$

(In terms of sig figs, $[N_o - N_t]$ is a very small number, found by subtracting two numbers known only to three sig figs and two decimal place. At best, we can express the result as an order of magnitude.)

$$N_o - N_t = 4.638 \times 10^{-10} \text{ g Co} \times \frac{1 \text{ mol Co}}{60 \text{ g Co}} \times \frac{6.022 \times 10^{23} \text{ Co atoms}}{1 \text{ mol Co}} \times \frac{1\beta}{1 \text{ Co atom}}$$

$$= 4.655 \times 10^{12} = \sim 5 \times 10^{12}$$

Between 10^{11} and 10^{12} β particles are emitted in 45.5 seconds

(b) $$\frac{4.655 \times 10^{12} \text{ dis}}{45.5 \text{ s}} \times \frac{1 \text{ Bq}}{1 \text{ dis/s}} = 1.02 \times 10^{11} = \sim 10^{11} \text{ Bq}$$

The activity of the sample is approximately 10^{11} Bq.

21.39   *Analyze/Plan.* Follow the logic in Sample Exercise 21.7.   *Solve.*

$$t = \frac{-1}{k} \ln \frac{N_t}{N_o}; k = 0.693/5715 \text{ yr} = 1.213 \times 10^{-4} \text{ yr}^{-1}$$

$$t = \frac{-1}{1.213 \times 10^{-4} \text{ yr}^{-1}} \ln \frac{24.9}{32.5} = 2.20 \times 10^3 \text{ yr}$$

21.40   Calculate k in yr^{-1} and solve Equation 21.19 for t. $N_o = 15.2/\text{min/g}$, $N_t = 8.9/\text{min/g}$

$$k = 0.693/t_{1/2} = 0.693/5715 \text{ yr} = 1.213 \times 10^{-4} = 1.21 \times 10^{-4} \text{ yr}^{-1}$$

$$t = \frac{-1}{k} \ln \frac{N_t}{N_o} = \frac{-1}{1.213 \times 10^{-4} \text{ yr}^{-1}} \ln \frac{8.9}{15.2} = 4.414 \times 10^3 = 4.4 \times 10^3 \text{ yr}$$

21.41   *Analyze/Plan.* Follow the procedure outlined in Sample Exercise 21.7. The original quantity of ^{238}U is 50.0 mg plus the amount that gave rise to 14.0 mg of ^{206}Pb. This amount is 14.0(238/206) = 16.2 mg.   *Solve.*

$$k = 0.693/4.5 \times 10^9 \text{ yr} = 1.54 \times 10^{-10} = 1.5 \times 10^{-10} \text{ yr}^{-1}$$

$$t = \frac{-1}{k} \ln \frac{N_t}{N_o} = \frac{-1}{1.54 \times 10^{-10} \text{ yr}^{-1}} \ln \frac{50.0}{66.2} = 1.8 \times 10^9 \text{ yr}$$

21.42 $k = 0.693/1.27 \times 10^9 \text{ yr} = 5.457 \times 10^{-10} = 5.46 \times 10^{-10} \text{ yr}^{-1}$

If the mass of ^{40}Ar is 3.6 times that of ^{40}K, then the original mass of ^{40}K must have been 3.6 + 1 = 4.6 times that now present.

$$t = \frac{-1}{5.457 \times 10^{-10} \text{ yr}^{-1}} \times \ln \frac{1}{(4.6)} = 2.8 \times 10^9 \text{ yr}$$

Energy Changes

21.43   *Analyze/Plan.* Given an energy change, find the corresponding change in mass. Use Equation [21.22], $E = mc^2$.   *Solve.*

$\Delta E = c^2 \Delta m$; $\Delta m = \Delta E/c^2$; $1 \text{ J} = \text{kg} \cdot \text{m}^2/\text{s}^2$

$$\Delta m = \frac{393.5 \times 10^3 \, kg \bullet m^2/s^2}{(2.9979 \times 10^8 \, m/s)^2} \times \frac{1000 \, g}{1 \, kg} = 4.378 \times 10^{-9} \, g$$

21.44 $\Delta E = c^2 \Delta m = (3.0 \times 10^8 \, m/s)^2 \times 0.1 \, mg \times \dfrac{1 \, g}{1000 \, mg} \times \dfrac{1 \, kg}{1000 \, g} \times \dfrac{1 \, kJ}{1000 \, J} = 9 \times 10^6 \, kJ$

21.45   *Analyze/Plan.* Given the mass of a ^{23}Na nucleus, find the energy required to separate the nucleus into protons and neutrons. This corresponds to the binding energy of the nucleus. Calculate the total mass of the separate particles and subtract the mass of the nucleus. Convert the difference to energy using Equation [21.22]. Use Avogadro's number to calculate energy per mole of nuclei.   *Solve.*

Δm = mass of individual protons and neutrons – mass of nucleus

Δm = 11(1.0072765 amu) + 12(1.0086649 amu) – 22.983733 amu = 0.2002873

$$= 0.200287 \, amu$$

$$\Delta E = (2.9979246 \times 10^8 \, m/s)^2 \times 0.2002873 \, amu \times \frac{1 \, g}{6.0221421 \times 10^{23} \, amu} \times \frac{1 \, kg}{1 \times 10^3 \, g}$$

$$= 2.989123 \times 10^{-11} = 2.98912 \times 10^{-11} \, J/\,^{23}Na \text{ nucleus required}$$

$$2.989123 \times 10^{-11} \, \frac{J}{nucleus} \times \frac{6.0221421 \times 10^{23} \, atoms}{mol} = 1.80009 \times 10^{13} \, J/mol \,^{23}Na$$

21.46 Δm = mass of individual protons and neutrons – mass of nucleus

Δm = 10(1.0072765 amu) + 11(1.0086649 amu) – 20.98846 amu = 0.1796189 = 0.17962 amu

$$\Delta E = (2.9979246 \times 10^8 \, m/s)^2 \times 0.1796189 \, amu \times \frac{1 \, g}{6.0221421 \times 10^{23} \, amu} \times \frac{1 \, kg}{1000 \, g}$$

$$= 2.680664 \times 10^{-11} = 2.6807 \times 10^{-11} \, J/\,^{21}Ne \text{ nucleus required}$$

$$2.680664 \times 10^{-11} \, \frac{J}{nucleus} \times \frac{6.0221421 \times 10^{23} \, nuclei}{mol}$$

$$= 1.6143 \times 10^{13} \, J/mol \,^{21}Ne \text{ binding energy}$$

21.47   *Analyze/Plan.* In each case, calculate the mass defect (Δm), total nuclear binding energy and then binding energy per nucleon.   *Solve.*

(a) Δm = 6(1.0072765) + 6(1.0086649) – 11.996708 = 0.0989404 = 0.098940 amu

$$\Delta E = 0.0989404 \, amu \times \frac{1 \, g}{6.0221421 \times 10^{23} \, amu} \times \frac{1 \, kg}{1000 \, g} \times \frac{8.987551 \times 10^{16} \, m^2}{s^2}$$

$$= 1.476604 \times 10^{-11} = 1.4766 \times 10^{-11} \, J$$

binding energy/nucleon = $1.476604 \times 10^{-11} \, J/12 = 1.2305 \times 10^{-12} \, J/nucleon$

(b) $\Delta m = 17(1.0072765) + 20(1.0086649) - 36.956576 = 0.3404225 = 0.340423$ amu

$$\Delta E = 0.3404225 \text{ amu} \times \frac{1 \text{g}}{6.0221421 \times 10^{23} \text{ amu}} \times \frac{1 \text{kg}}{1000 \text{g}} \times \frac{8.987551 \times 10^{16} \text{ m}^2}{\text{s}^2}$$

$$= 5.080525 \times 10^{-11} = 5.08053 \times 10^{-11} \text{ J}$$

binding energy/ nucleon $= 5.080525 \times 10^{-11}$ J $/ 37 = 1.37312 \times 10^{-12}$ J/nucleon

(c)   Calculate the nuclear mass by subtracting the electron mass from the atomic mass. 136.905812 amu $- 56(5.485799 \times 10^{-4}$ amu$) = 136.875092$ amu

$\Delta m = 56(1.0072765) + 81(1.0086649) - 136.875092 = 1.2342489 = 1.234249$ amu

$$\Delta E = 1.2342489 \text{ amu} \times \frac{1 \text{g}}{6.0221421 \times 10^{23} \text{ amu}} \times \frac{1 \text{kg}}{1000 \text{g}} \times \frac{8.987551 \times 10^{16} \text{ m}^2}{\text{s}^2}$$

$$= 1.842014 \times 10^{-10} \text{ J}$$

binding energy/nucleon $= 1.842014 \times 10^{-10}$ J $/ 137 = 1.344536 \times 10^{-12}$ J/nucleon

21.48   In each case, calculate the mass defect, total nuclear binding energy and then binding energy per nucleon.

(a) $\Delta m = 7(1.0072765) + 7(1.0086649) - 13.999234 = 0.1123558 = 0.112356$ amu

$$\Delta E = 0.1123558 \text{ amu} \times \frac{1 \text{g}}{6.0221421 \times 10^{23} \text{ amu}} \times \frac{1 \text{kg}}{1000 \text{g}} \times \frac{8.987551 \times 10^{16} \text{ m}^2}{\text{s}^2}$$

$$= 1.676817 \times 10^{-11} = 1.67682 \times 10^{-11} \text{ J}$$

binding energy/nucleon $= 1.676817 \times 10^{-11}$ J $/ 14 = 1.19773 \times 10^{-12}$ J/nucleon

(b) $\Delta m = 22(1.0072765) + 26(1.0086649) - 47.935878 = 0.4494924 = 0.449492$ amu

$$\Delta E = 0.4494924 \text{ amu} \times \frac{1 \text{g}}{6.0221421 \times 10^{23} \text{ amu}} \times \frac{1 \text{kg}}{1000 \text{g}} \times \frac{8.987551 \times 10^{16} \text{ m}^2}{\text{s}^2}$$

$$= 6.708304 \times 10^{-11} = 6.70830 \times 10^{-11} \text{ J}$$

binding energy/nucleon $= 6.708304 \times 10^{-11}$ J $/ 48 = 1.39756 \times 10^{-12}$ J/nucleon

(c)   Calculate the nuclear mass by subtracting the electron mass from the atomic mass. $200.970277 - 80(5.485799 \times 10^{-4}$ amu$) = 200.926391$ amu

$\Delta m = 80(1.0072765) + 121(1.0086649) - 200.926391 = 1.7041819 = 1.704182$ amu

$$\Delta E = 1.7041819 \text{ amu} \times \frac{1 \text{g}}{6.0221421 \times 10^{23} \text{ amu}} \times \frac{1 \text{kg}}{1000 \text{g}} \times \frac{8.987551 \times 10^{16} \text{ m}^2}{\text{s}^2}$$

$$= 2.543351 \times 10^{-10} \text{ J}$$

binding energy/nucleon $= 2.543351 \times 10^{-10}$ J$/201 = 1.265348 \times 10^{-12}$ J/nucleon

21.49   *Analyze/Plan.* Use Equation [21.22] to calculate the mass equivalence of the solar radiation.   *Solve.*

(a) $$\frac{1.07 \times 10^{16}\,\text{kJ}}{1\,\text{min}} \times \frac{60\,\text{min}}{1\,\text{hr}} \times \frac{24\,\text{hr}}{1\,\text{day}} = 1.541 \times 10^{19}\,\frac{\text{kJ}}{\text{day}} = 1.54 \times 10^{22}\,\text{J/day}$$

$$\Delta m = \frac{1.541 \times 10^{22}\,\text{kg} \cdot \text{m}^2/\text{s}^2/\text{d}}{(2.998 \times 10^8\,\text{m/s})^2} = 1.714 \times 10^5 = 1.71 \times 10^5\,\text{kg/d}$$

(b) *Analyze/Plan.* Calculate the mass change in the given nuclear reaction, then a conversion factor for g ^{235}U to mass equivalent. *Solve.*

$\Delta m = 140.8833 + 91.9021 + 2(1.0086649) - 234.9935 = -0.19077 = -0.1908$ amu

Converting from atoms to moles and amu to grams, it requires 1.000 mol or 235.0 g ^{235}U to produce energy equivalent to a change in mass of 0.1908 g. 0.10% of 1.714×10^5 kg is 1.714×10^2 kg = 1.714×10^5 g

$$1.714 \times 10^5\,\text{g} \times \frac{235.0\,\text{g}\ ^{235}\text{U}}{0.1908\,\text{g}} = 2.111 \times 10^8 = 2.1 \times 10^8\,\text{g}\ ^{235}\text{U}$$

(This is about 230 tons of ^{235}U **per day**.)

21.50 The calculated Δm is for one group of single nuclides involved in a reaction, labeled ΔM/'atomic reaction'. Multiplying by Avogadro's number changes the quantity to 'mol of reaction'. Since energy is released, the sign of ΔE is negative.

(a) $\Delta m = 4.00260 + 1.0086649 - 3.01605 - 2.01410 = -0.0188851 = -0.01889$ amu

$$\Delta E = \frac{-0.0188851\,\text{amu}}{\text{'atomic reaction'}} \times \frac{1\,\text{g}}{6.022 \times 10^{23}\,\text{amu}} \times \frac{6.022 \times 10^{23}\,\text{'atomic reaction'}}{\text{mol of reaction}}$$
$$\times \frac{1\,\text{kg}}{10^3\,\text{g}} \times (2.99792458 \times 10^8\,\text{m/sec})^2 = -1.697 \times 10^{12}\,\text{J/mol}$$

(b) $\Delta m = 3.01605 + 1.0086649 - 2(2.01410) = -3.4851 \times 10^{-3} = -3.49 \times 10^{-3}$ amu

$\Delta E = -3.13 \times 10^{11}$ J/mol

(c) $\Delta m = 4.00260 + 1.00782 - 3.01605 - 2.01410 = -1.973 \times 10^{-2}$ amu

$\Delta E = -1.773 \times 10^{12}$ J/mol

21.51 We can use Figure 21.13 to see that the binding energy per nucleon (which gives rise to the mass defect) is greatest for nuclei of mass numbers around 50. Thus (a) $^{59}_{27}$Co should possess the greatest mass defect per nucleon.

21.52 In a fission reactor absorption of neutrons causes a single heavy nucleus to undergo fission, producing two medium mass nuclei. These are seen in Figure 21.13 to have a larger total mass defect than the starting single heavier nucleus, so energy is released.

Effects and Uses of Radioisotopes

21.53 The ^{59}Fe would be incorporated into the diet component, which in turn is fed to the rabbits. After a time blood samples could be removed from the animals, the red blood cells separated, and the radioactivity of the sample measured. If the iron in the dietary compound has been incorporated into blood hemoglobin, the blood cell sample should show beta emission. Samples could be taken at various times to determine the rate of iron uptake, rate of loss of the iron from the blood, and so forth.

21.54   (a)   Add ^{36}Cl to water as a chloride salt. Then dissolve ordinary CCl_3COOH. After a time, distill the volatile materials away from the salt; CCl_3COOH is volatile, and will distill with water. Count radioactivity in the volatile material. If chlorine exchange has occurred, there will be radioactivity.

         (b)   Prepare a saturated solution of $BaCl_2$ containing a small amount of solid $BaCl_2$. Add to this solution solid $BaCl_2$ containing ^{36}Cl. If the solid-solution equilibrium is dynamic, some of the ^{36}Cl in the solid will find itself in solution as chloride ion. After allowing some time for equilibrium to become established, filter the solution, measure radioactivity in the solution that is separated from the solid. If there were no dynamic equilibrium, the ^{36}Cl$^-$ would remain in the added solid, since the solution is already saturated before the addition of more solid.

         (c)   Utilize ^{36}Cl in soils of various pH values; grow plants for a given period of time. Remove plants, and directly measure radioactivity in samples from stems, leaves, and so forth, or reduce the volume of plant sample by some form of digestion and evaporation of solution to give a dry residue that can be counted.

21.55   (a)   Control rods control neutron flux so that there are enough neutrons to sustain the chain reaction but not so many that the core overheats.

         (b)   A moderator slows neutrons so that they are more easily captured by fissioning nuclei.

21.56   (a)   In a chain reaction, one neutron initiates a nuclear transformation that produces more than one neutron. The product neutrons initiate more transformations, so that the reaction is self-sustaining.

         (b)   Critical mass is the mass of fissionable material required to sustain a chain reaction so that only one product neutron is effective at initiating a new transformation.

21.57   *Analyze/Plan.* Use conservation of A and Z to complete the equations, keeping in mind the symbols and definitions of various decay products.   *Solve.*

         (a) $^{235}_{92}U + ^{1}_{0}n \rightarrow ^{160}_{62}Sm + ^{72}_{30}Zn + 4\,^{1}_{0}n$

         (b) $^{239}_{94}Pu + ^{1}_{0}n \rightarrow ^{144}_{58}Ce + ^{94}_{36}Kr + 2\,^{1}_{0}n$

21.58   (a) $^{2}_{1}H + ^{2}_{1}H \rightarrow ^{3}_{2}He + 4\,^{1}_{0}n$

         (b) $^{233}_{92}U + ^{1}_{0}n \rightarrow ^{133}_{51}Sb + ^{98}_{41}Nb + 3\,^{1}_{0}n$

21.59   The extremely high temperature is required to overcome the electrostatic charge repulsions between the nuclei so that they come together to react.

21.60   (a)   If the spent fuel rods are more radioactive than the original rods, the products of fission must lie outside the belt of stability and be radioactive themselves.

(b)   The heavy (Z > 83) nucleus has a high neutron/proton ratio. The lighter radioactive fission products, (e.g., barium-142 and krypton-91) also have high neutron/proton ratios, since only 2 or 3 free neutrons are produced during fission. The preferred decay mode to reduce the neutron/proton ratio is β decay, which has the effect of converting a neutron into a proton. Both barium-142 (86 n, 56 p) and krypton-91 (55 n, 36 p) undergo β decay.

21.61   •OH is a free radical; it contains an unpaired (free) electron, which makes it an extremely reactive species. (As an odd electron molecule, it violates the octet rule.) It can react with almost any particle (atom, molecule, ion) to acquire an electron and become OH^-. This often starts a disruptive chain of reactions, each producing a different free radical.

Hydroxide ion, OH^-, on the other hand, will be readily neutralized in the buffered cell environment. Its most common reaction is ubiquitous and innocuous:

$H^+ + OH^- \rightarrow H_2O$. The acid-base reactions of OH^- are usually much less disruptive to the organism than the chain of redox reactions initiated by •OH radical.

21.62 $$\left[H-\ddot{\underset{\underset{H}{|}}{O}}\cdot \right]^{+} + H-\ddot{\underset{\underset{H}{|}}{O}}: \longrightarrow \left[H-\underset{\underset{H}{|}}{\ddot{O}}-H \right]^{+} + \cdot\ddot{O}-H$$

H_2O^+ is a free radical because it contains seven valence electrons. Around the central O atom there are two bonding and one nonbonding electron pairs and a single unpaired electron (3(2) + 1 = 7 valence electrons).

21.63   *Analyze/Plan.* Use definitions of the various radiation units and conversion factors to calculate the specified quantities. Pay particular attention to units.   *Solve.*

(a)   1 Ci = 3.7×10^{10} disintegrations(dis)/s; 1 Bq = 1 dis/s

$$8.7 \text{ mCi} \times \frac{1 \text{ Ci}}{1000 \text{ mCi}} \times \frac{3.7 \times 10^{10} \text{ dis/s}}{\text{Ci}} = 3.22 \times 10^8 = 3.2 \times 10^8 \text{ dis/s} = 3.2 \times 10^8 \text{ Bq}$$

(b)   1 rad = 1×10^{-2} J/kg; 1 Gy = 1 J/kg = 100 rad. From part (a), the activity of the source is 3.2×10^8 dis/s.

$$3.22 \times 10^8 \text{ dis/s} \times 2.0 \text{ s} \times 0.65 \times \frac{9.12 \times 10^{-13} \text{ J}}{\text{dis}} \times \frac{1}{0.250 \text{ kg}} = 1.53 \times 10^{-3} = 1.5 \times 10^{-3} \text{ J/kg}$$

$$1.5 \times 10^{-3} \text{ J/kg} \times \frac{1 \text{ rad}}{1 \times 10^{-2} \text{ J/kg}} \times \frac{1000 \text{ mrad}}{\text{rad}} = 1.5 \times 10^2 \text{ mrad}$$

$$1.5 \times 10^{-3} \text{ J/kg} \times \frac{1 \text{ Gy}}{1 \text{ J/kg}} = 1.5 \times 10^{-3} \text{ Gy}$$

(c)   rem = rad (RBE); Sv = Gy (RBE) , where 1 Sv = 100 rem

mrem = 1.53×10^2 mrad (9.5) = 1.45×10^3 = 1.5×10^3 mrem (or 1.5 rem)

Sv = 1.53×10^{-3} Gy (9.5) = 1.45×10^{-2} = 1.5×10^{-2} Sv

21.64   (a) $1 \text{ Ci} = 3.7 \times 10^{10} \text{ dis/s}; 1 \text{ Bq} = 1 \text{ dis/s}$

$$21 \text{ mCi} \times \frac{1 \text{ Ci}}{1000 \text{ mCi}} \times 3.7 \times 10^{10} \text{ dis/s} = 7.77 \times 10^8 = 7.8 \times 10^8 \text{ dis/s} = 7.8 \times 10^8 \text{ Bq}$$

(b) $1 \text{ Gy} = 1 \text{ J/kg}; 1 \text{ Gy} = 100 \text{ rad}$

$$7.77 \times 10^8 \text{ dis/s} \times 116 \text{ s} \times 0.065 \times \frac{8.75 \times 10^{-14} \text{ J}}{\text{dis}} \times \frac{1}{65 \text{ kg}} = 7.887 \times 10^{-6} = 7.9 \times 10^{-6} \text{ J/kg}$$

$$7.9 \times 10^{-6} \text{ J/kg} \times \frac{1 \text{ Gy}}{1 \text{ J/kg}} = 7.9 \times 10^{-6} \text{ Gy}; 7.9 \times 10^{-6} \text{ Gy} \times \frac{100 \text{ rad}}{1 \text{ Gy}} = 7.9 \times 10^{-4} \text{ rad}$$

(c)   rem = rad (RBE); Sv = Gy (RBE)

$$7.9 \times 10^{-4} \text{ rad } (1.0) = 7.9 \times 10^{-4} \text{ rem} \times \frac{1000 \text{ mrem}}{1 \text{ rem}} = 0.79 \text{ mrem}$$

$$7.9 \times 10^{-6} \text{ Gy } (1.0) = 7.9 \times 10^{-6} \text{ Sv}$$

(d)   From Figure 21.23, the average annual background radiation is 360 mrem, or about 1 mrem/day. This 0.79 mrem exposure is less than the average background radiation for a day.

Additional Exercises

21.65 $^{222}_{86}\text{Rn} \rightarrow \text{X} + 3\,^4_2\text{He} + 2\,^0_{-1}\beta$

This corresponds to a reduction in mass number of $(3 \times 4 =)$ 12 and a reduction in atomic number of $(3 \times 2 - 2) = 4$. The stable nucleus is $^{210}_{82}\text{Pb}$. (This is part of the sequence in Figure [21.4].)

21.66   (a) $^1_0\text{n} \rightarrow ^1_1\text{p} + ^0_{-1}\text{e (or } ^0_{-1}\beta)$

The other product of neutron decay is a β particle (with the mass and charge of an electron).

(b)   Neutrons in atomic nuclei do not decay at this rate because they are stabilized by the strong forces among subatomic particles in a nucleus. Evidence for strong forces in the nucleus includes nuclear binding energies and the coexistence of like-charged protons in the very small volume of the nucleus.

21.67   The most massive radionuclides will have the highest neutron/proton ratios. Thus, they are most likely to decay by a process that lowers this ratio, beta emission. The least massive nuclides, on the other hand, will decay by a process that increases the neutron/proton ratio, positron emission, or orbital electron capture.

21.68   (a) $^{36}_{17}\text{Cl} \rightarrow ^{36}_{18}\text{Ar} + ^0_{-1}\text{e}$

(b)   According to Table 21.3, nuclei with even numbers of both protons and neutrons, or an even number of one kind of nucleon, are more stable. ^{35}Cl and ^{37}Cl both have an odd number of protons but an even number of neutrons. ^{36}Cl has an odd number of protons and neutrons (17 p, 19 n), so it is less stable than the other two isotopes. Also, ^{37}Cl has 20 neutrons, a nuclear closed shell.

21.69   (a) $^{6}_{3}\text{Li} \rightarrow {}^{56}_{28}\text{Ni} + {}^{62}_{31}\text{Ga}$

(b) $^{40}_{20}\text{Ca} + {}^{248}_{96}\text{Cm} \rightarrow {}^{147}_{62}\text{Sm} + {}^{141}_{54}\text{Xe}$

(c) $^{88}_{38}\text{Sr} + {}^{84}_{36}\text{Kr} \rightarrow {}^{116}_{46}\text{Pd} + {}^{56}_{28}\text{Ni}$

(d) $^{40}_{20}\text{Ca} + {}^{238}_{92}\text{U} \rightarrow {}^{70}_{30}\text{Zn} + 4\,{}^{1}_{0}\text{n} + 2\,{}^{102}_{41}\text{Nb}$

21.70

Time (hr)	N_t (dis/min)	$\ln N_t$
0	180	5.193
2.5	130	4.868
5.0	104	4.644
7.5	77	4.34
10.0	59	4.08
12.5	46	3.83
17.5	24	3.18

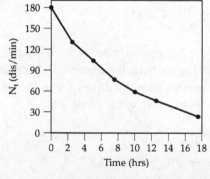 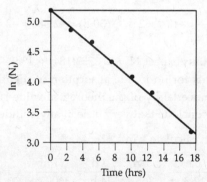

The plot on the left is a graph of activity (disintegrations per minute) vs. time. Choose $t_{1/2}$ at the time where $N_t = 1/2\,N_o = 90$ dis/min. $t_{1/2} \approx 6.0$ hr.

Rearrange Equation [21.19] to obtain the linear relationship shown on the right.

$\ln(N_t / N_o) = -kt; \ln N_t - \ln N_o = -kt; \ln N_t = -kt + \ln N_o$

The slope of this line $= -k = -0.11;\ t_{1/2} = 0.693/0.11 = 6.3$ hr.

21.71 $1 \times 10^{-6}\ \text{curie} \times \dfrac{3.7 \times 10^{10}\ \text{dis/s}}{\text{curie}} = 3.7 \times 10^4\ \text{dis/s}$

rate $= 3.7 \times 10^4\ \text{nuclei/s} = kN$

$k = \dfrac{0.693}{t_{1/2}} = \dfrac{0.693}{28.8\ \text{yr}} \times \dfrac{1\ \text{yr}}{365 \times 24 \times 3600\ \text{sec}} = 7.630 \times 10^{-10} = 7.63 \times 10^{-10}\ \text{s}^{-1}$

3.7×10^4 nuclei/s $= (7.63 \times 10^{-10}/s)$ N; N $= 4.849 \times 10^{13} = 4.8 \times 10^{13}$ nuclei

$$\text{mass } ^{90}\text{Sr} = 4.849 \times 10^{13} \text{ nuclei} \times \frac{90 \text{ g Sr}}{6.022 \times 10^{23} \text{ nuclei}} = 7.2 \times 10^{-9} \text{ g Sr}$$

21.72   First calculate k in s^{-1}

$$k = \frac{0.693}{2.4 \times 10^4 \text{ yr}} \times \frac{1 \text{ yr}}{365 \times 24 \times 3600 \text{ s}} = 9.16 \times 10^{-13} = 9.2 \times 10^{-13} \text{ s}^{-1}$$

Now calculate N:

$$N = 0.173 \text{ g Pu} \times \frac{1 \text{ mol Pu}}{239 \text{ g Pu}} \times \frac{6.022 \times 10^{23} \text{ Pu atoms}}{1 \text{ mol Pu}} = 4.36 \times 10^{20} = 4.4 \times 10^{20} \text{ Pu atoms}$$

rate $= (9.16 \times 10^{-13} \text{ s}^{-1})(4.36 \times 10^{20} \text{ Pu atoms}) = 3.99 \times 10^8 = 4.0 \times 10^8$ dis/s

21.73   The C—OH bond of the acid and the O—H bond of the alcohol break in this reaction. Initially, ^{18}O is present in the C—^{18}OH group of the alcohol. In order for ^{18}O to end up in the ester, the ^{18}O—H bond of the alcohol must break. This requires that the C—OH bond in the acid also breaks. The unlabeled O from the acid ends up in the H_2O product.

21.74   Assume that no depletion of iodide from the water due to plant uptake has occurred. Then the activity after 30 days would be:

$k = 0.693/t_{1/2} = 0.693/8.04 \text{ d} = 0.0862 \text{ d}^{-1}$

$$\ln \frac{N_t}{N_o} = -(0.0862 \text{ d}^{-1})(30 \text{ d}) = -2.586 = -2.6; \quad \frac{N_t}{N_o} = 0.0753 = 0.08$$

We thus expect $N_t = 0.0753(184) = 13.9 = 1 \times 10^1$ counts/min. Although the theoretical counts for no iodide absorption is slightly higher than the observed 13.5 counts/min, the uncertainty of the theoretical value is greater than the difference in the two values. We conclude that very little, if any, iodide is absorbed by the plant.

21.75   First, calculate k in s^{-1}

$$k = \frac{0.693}{12.3 \text{ yr}} \times \frac{1 \text{ yr}}{365 \text{ d}} \times \frac{1 \text{ d}}{24 \text{ hr}} \times \frac{1 \text{ hr}}{3600 \text{ sec}} = 1.7866 \times 10^{-9} = 1.79 \times 10^{-9} \text{ s}^{-1}$$

From Equation [21.18], $1.50 \times 10^3 \text{ s}^{-1} = (1.7866 \times 10^{-9} \text{ s}^{-1})(N)$;

$N = 8.396 \times 10^{11} = 8.40 \times 10^{11}$. In 26.00 g of water, there are

$$26.00 \text{ g } H_2O \times \frac{1 \text{ mol } H_2O}{18.02 \text{ g } H_2O} \times \frac{6.022 \times 10^{23} \text{ } H_2O}{1 \text{ mol } H_2O} \times \frac{2 \text{ H}}{1 H_2O} = 1.738 \times 10^{24} \text{ H atoms}$$

The mole fraction of $^3_1$H atoms in the sample is thus

$8.396 \times 10^{11}/1.738 \times 10^{24} = 4.831 \times 10^{-13} = 4.83 \times 10^{-13}$

21.76   Because of the relationship $\Delta E = \Delta mc^2$, the mass defect (Δm) is directly related to the binding energy (ΔE) of the nucleus.

^{7}Be: 4p, 3n; 4(1.0072765) + 3(1.0086649) = 7.05510 amu

Total mass defect = 7.0551 – 7.0147 = 0.0404 amu

0.0404 amu/7 nucleons = 5.77×10^{-3} amu/nucleon

$$\Delta E = \Delta m \times c^2 = \frac{5.77 \times 10^{-3} \text{ amu}}{\text{nucleon}} \times \frac{1g}{6.022 \times 10^{23} \text{ amu}} \times \frac{1 \text{ kg}}{1 \times 10^3 \text{ g}} \times \frac{8.988 \times 10^{16} \text{ m}^2}{\text{sec}^2}$$

$$= \frac{5.77 \times 10^{-3} \text{ amu}}{\text{nucleon}} \times \frac{1.4925 \times 10^{-10} \text{ J}}{1 \text{amu}} = 8.612 \times 10^{-13} = 8.61 \times 10^{-13} \text{ J/nucleon}$$

^{9}Be: 4p, 5n; 4(1.0072765) + 5(1.0086649) = 9.07243 amu

Total mass defect = 9.0724 – 9.0100 = 0.06243 = 0.0624 amu

0.0624 amu/9 nucleons = $6.937 \times 10^{-3} = 6.94 \times 10^{-3}$ amu/nucleon

6.937×10^{-3} amu/nucleon $\times 1.4925 \times 10^{-10}$ J/amu = $1.035 \times 10^{-12} = 1.04 \times 10^{-12}$ J/nucleon

^{10}Be: 4p, 6n; 4(1.0072765) + 6(1.0086649) = 10.0811 amu

Total mass defect = 10.0811 – 10.0113 = 0.0698 amu

0.0698 amu/10 nucleons = 6.98×10^{-3} amu/nucleon

6.98×10^{-3} amu/nucleon $\times 1.4925 \times 10^{-10}$ J/amu = $1.042 \times 10^{-12} = 1.04 \times 10^{-12}$ J/nucleon

The binding energies/nucleon for ^{9}Be and ^{10}Be are very similar; that for ^{10}Be is slightly higher.

21.77   (a) $\Delta m = \Delta E/c^2$; $\Delta m = \dfrac{3.9 \times 10^{26} \text{ J/s}}{(3.00 \times 10^8 \text{ m/s})^2} \times \dfrac{1 \text{ kg} \bullet \text{m}^2/\text{s}^2}{1 \text{ J}} = 4.3 \times 10^9 \text{ kg/s}$

The rate of mass loss is 4.3×10^9 kg/s.

(b)   The mass loss arises from fusion reactions that produce more stable nuclei from less stable ones, e.g., Equations [21.26-21.29].

21.78 $1000 \text{ Mwatts} \times \dfrac{1 \times 10^6 \text{ watts}}{1 \text{ Mwatt}} \times \dfrac{1 \text{ J}}{1 \text{ watt} \bullet \text{s}} \times \dfrac{1 \, ^{235}\text{U atom}}{3 \times 10^{-11} \text{ J}} \times \dfrac{1 \text{ mol U}}{6.02 \times 10^{23} \text{ atoms}}$

$\times \dfrac{235 \text{ g U}}{1 \text{ mol}} \times \dfrac{3600 \text{ s}}{1 \text{ hr}} \times \dfrac{24 \text{ hr}}{1 \text{ d}} \times \dfrac{365 \text{ d}}{1 \text{ yr}} \times \dfrac{100}{40} \text{(efficiency)} = 1.03 \times 10^6 = 1 \times 10^6 \text{ g U / yr}$

21.79 $2 \times 10^{-12} \text{ curies} \times \dfrac{3.7 \times 10^{10} \text{ dis/s}}{1 \text{curie}} = 7.4 \times 10^{-2} = 7 \times 10^{-2} \text{ dis/s}$

$\dfrac{7.4 \times 10^{-2} \text{ dis/s}}{75 \text{ kg}} \times \dfrac{8 \times 10^{-13} \text{ J}}{\text{dis}} \times \dfrac{1 \text{ rad}}{1 \times 10^{-2} \text{ J/g}} \times \dfrac{3600 \text{ s}}{\text{hr}} \times \dfrac{24 \text{ hr}}{1 \text{ d}}$

$\times \dfrac{365 \text{ d}}{1 \text{ yr}} = 2.49 \times 10^{-6} = 2 \times 10^{-6} \text{ rad/yr}$

Recall that there are 10 rem/rad for alpha particles.

$\dfrac{2.49 \times 10^{-6} \text{ rad}}{1 \text{ yr}} \times \dfrac{10 \text{ rem}}{1 \text{ rad}} = 2.49 \times 10^{-5} = 2 \times 10^{-5} \text{ rem/yr}$

Integrative Exercises

21.80　Calculate the molar mass of $NaClO_4$ that contains 25.6% ^{36}Cl. Atomic mass of the enhanced Cl is $0.256(36.0) + 0.744(35.453) = 35.593 = 35.6$. The molar mass of $NaClO_4$ is then $(22.99 + 35.593 + 64.00) = 122.58 = 122.6$. Calculate N, the number of ^{36}Cl nuclei, the value of k in s^{-1}, and the activity in dis/s.

$$36.9\,mg\;NaClO_4 \times \frac{1g}{1000\,mg} \times \frac{1\,mol\;NaClO_4}{122.58\,g\;NaClO_4} \times \frac{1\,mol\;Cl}{1\,mol\;NaClO_4} \times \frac{6.022 \times 10^{23}\;Cl\,atoms}{mol\;Cl}$$

$$\times \frac{25.6\;^{36}Cl\,atoms}{100\;Cl\,atoms} = 4.641 \times 10^{19} = 4.64 \times 10^{19}\;^{36}Cl\;atoms$$

$$k = 0.693/t_{1/2} = \frac{0.693}{3.0 \times 10^5\;yr} \times \frac{1\,yr}{365 \times 24 \times 3600\,s} = 7.32 \times 10^{-14} = 7.3 \times 10^{-14}\;s^{-1}$$

$$rate = kN = (7.32 \times 10^{-14}\;s^{-1})(4.641 \times 10^{19}\;nuclei) = 3.40 \times 10^6 = 3.4 \times 10^6\;dis/s$$

21.81　Calculate the amount of energy produced by the nuclear fusion reaction, the enthalpy of combustion, $\Delta H°$, of C_3H_8, and then the mass of C_3H_8 required.

Δm for the reaction　　$4\,^1_1H \rightarrow\,^4_2He + 2\,^0_1e$　　is:

$4(1.00782) - 4.00260\;amu - 2(5.4858 \times 10^{-4}\;amu) = 0.027583 = 0.02758\;amu$

$$\Delta E = \Delta mc^2 = 0.027583\,amu \times \frac{1g}{6.02214 \times 10^{23}\,amu} \times \frac{1\,kg}{1000\,g} \times (2.9979246 \times 10^8\;m/s)^2$$

$$= 4.11654 \times 10^{-12} = 4.117 \times 10^{-12}\;J/4\;^1H\;nuclei$$

$$1.0\,g\;^1H \times \frac{1\;^1H\,nucleus}{1.00782\,amu} \times \frac{6.02214 \times 10^{23}\,amu}{g} \times \frac{4.11654 \times 10^{-12}\;J}{4\;^1H\;nuclei}$$

$$= 6.1495 \times 10^{11}\;J = 6.1 \times 10^8\;kJ\;produced\;by\;the\;fusion\;of\;1.0\;g\;^1H.$$

$C_3H_8(g) + 5O_2(g) \rightarrow 3CO_2(g) + 4H_2O(g)$

$\Delta H° = 3(-393.5\;kJ) + 4(-241.82\;kJ) - (-103.85) - (0) = -2043.9\;kJ$

$$6.1495 \times 10^8\;kJ \times \frac{1\,mol\;C_3H_8(g)}{2043.9\;kJ} \times \frac{44.094\,g\;C_3H_8}{mol\;C_3H_8} = 1.327 \times 10^7\;g = 1.3 \times 10^4\;kg\;C_3H_8$$

13,000 kg $C_3H_8(g)$ would have to be burned to produce the same amount of energy as fusion of 1.0 g ^1H.

21.82　(a)　$0.18\;Ci \times \dfrac{3.7 \times 10^{10}\;dis/s}{Ci} \times \dfrac{3600\,s}{hr} \times \dfrac{24\,hr}{d} \times 235\,d = 1.35 \times 10^{17} = 1.4 \times 10^{17}\;\alpha\;particles$

(b)　$P = nRT/V = 1.35 \times 10^{17}\;He\,atoms \times \dfrac{1\,mol\;He}{6.022 \times 10^{23}\,atoms} \times \dfrac{295\;K}{0.0150\;L} \times \dfrac{0.08206\;L \cdot atm}{K \cdot mol}$

$$= 3.62 \times 10^{-4} = 3.6 \times 10^{-4}\;atm = 0.28\;torr$$

21.83　Calculate N_t in dis/min/g C from 1.5×10^{-2} dis/0.788 g $CaCO_3$. $N_o = 15.3$ dis/min/g C. Calculate k from $t_{1/2}$, calculate t from $\ln (N_t / N_o) = -kt$.

$C(s) + O_2(g) \rightarrow CO_2(g) + Ca(OH_2)(aq) \rightarrow CaCO_3(s) + H_2O(l)$

1 C atom → 1 $CaCO_3$ molecule

$$\frac{1.5 \times 10^{-2} \text{ Bq}}{0.788 \text{ g CaCO}_3} \times \frac{1 \text{ dis/s}}{1 \text{ Bq}} \times \frac{60 \text{ s}}{1 \text{ min}} \times \frac{100.1 \text{ g CaCO}_3}{12.01 \text{ g C}} = 9.52 = 9.5 \text{ dis/min/g C}$$

$$k = 0.693/t_{1/2} = 0.693/5.715 \times 10^3 \text{ yr} = 1.213 \times 10^{-4} = 1.21 \times 10^{-4} \text{ yr}^{-1}$$

$$t = -\frac{1}{k} \ln \frac{N_t}{N_o} = \frac{-1}{1.213 \times 10^{-4} \text{ yr}^{-1}} \ln \frac{9.52 \text{ dis/min/g C}}{15.3 \text{ dis/min/g C}} = 3.91 \times 10^3 \text{ yr}$$

21.84   Determine the wavelengths of the photons by first calculating the energy equivalent of the mass of an electron or positron. (Since **two** photons are formed by annihilation of **two** particles of equal mass, we need to calculate the energy equivalent of just one particle.) The mass of an electron is 9.109×10^{-31} kg.

$$\Delta E = (9.109 \times 10^{-31} \text{ kg}) \times (2.998 \times 10^8 \text{ m/s})^2 = 8.187 \times 10^{-14} \text{ J}$$

Also, $\Delta E = h\nu$; $\Delta E = hc/\lambda$; $\lambda = hc/\Delta E$

$$\lambda = \frac{(6.626 \times 10^{-34} \text{ J} \cdot \text{s})(2.998 \times 10^8 \text{ m/s})}{8.187 \times 10^{-14} \text{ J}} = 2.426 \times 10^{-12} \text{ m} = 2.426 \times 10^{-3} \text{ nm}$$

This is a very short wavelength indeed; it lies at the short wavelength end of the range of observed gamma ray wavelengths (see Figure 6.4).

21.85   (a) $Ba(NO_3)_2(aq) + Na_2SO_4(aq) \rightarrow BaSO_4(s) + 2NaNO_3(aq)$

(b) $1.25 \text{ mmol Ba}^{2+} + 1.25 \text{ mmol SO}_4^{2-} \rightarrow 1.25 \text{ mmol BaSO}_4$

Neither reactant is in excess, so the activity of the filtrate is due entirely to $[SO_4^{2-}]$ from dissociation of $BaSO_4(s)$. Calculate $[SO_4^{2-}]$ in the filtrate by comparing the activity of the filtrate to the activity of the reactant.

$$\frac{0.050 \text{ } M \text{ SO}_4^{2-}}{1.22 \times 10^6 \text{ Bq/mL}} = \frac{x \text{ } M \text{ filtrate}}{250 \text{ Bq/mL}}$$

$[SO_4^{2-}]$ in the filtrate $= 1.0246 \times 10^{-5} = 1.0 \times 10^{-5} \text{ } M$

$K_{sp} = [Ba^{2+}][SO_4^{2-}]$; $[SO_4^{2-}] = [Ba^{2+}]$

$K_{sp} = (1.0246 \times 10^{-5})^2 = 1.0498 \times 10^{-10} = 1.0 \times 10^{-10}$

22 Chemistry of the Nonmetals

Chemistry of Nonmetals

22.1 C_2H_4, the structure on the left, is the stable compound. Carbon, with a relatively small covalent radius owing to its location in the second row of the periodic chart, is able to closely approach other atoms. This close approach enables significant π overlap, so carbon can form strong multiple bonds to satisfy the octet rule. Silicon, in the third row of the periodic table, has a covalent radius too large for significant π overlap. Si does not form stable multiple bonds and Si_2H_4 is unstable.

22.2   (a)   Acid-base (Brønsted)

       (b)   Charges on species from left to right in the reaction: 0, 0, 1+, 1–

       (c) $NH_3(aq) + H_2O(l) \rightleftharpoons NH_4^+(aq) + OH^-(aq)$

22.3   *Analyze.* The structure is a trigonal bipyramid where one of the five positions about the central atom is occupied by a lone pair, often called a see-saw.

Plan A: Count the valence electrons in each molecule, draw a correct Lewis structure, and count the electron domains about the central atom.

Plan B: Molecules (a)–(d) each contain four F atoms bound to a central atom through a single bond (F is unlikely to form multiple bonds because of its high electronegativity). This represents 16 electron pairs; the fifth position is occupied by a lone pair, for a total of 17 e^- pairs. A valence e^- count for (a)–(d) will tell us which molecules are likely to have the designated structure. Molecule (e), $HClO_4$, is not exactly of the type AX_4, so a Lewis structure will be required.   *Solve.*

(a) XeF_4 36 e^-, 16 e^- pairs. Plan B predicts that this molecule will **not** adopt the see-saw structure.

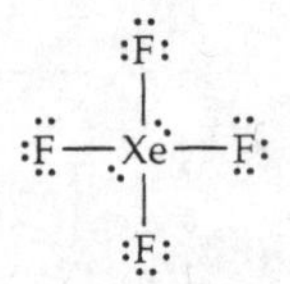

                              6 e^- domains about the Xe
                              octahedral domain geometry
                              square planar structure

(b) BrF_4^- 34 e^-, 17 e^- pairs; structure will be see-saw.

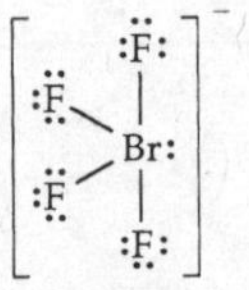

                              5 e^- domains about Br trigonal
                              bipyramidal domain geometry
                              see-saw structure

(c) SiF_4 32 e^-, 16 e^- pairs; structure **will not** be see-saw

4 e^- domains about Si
tetrahedral domain geometry and structure

(d) $TeCl_4$ 34 e^-, 17 e^- pairs; structure **will be** see-saw

5 e^- domains about Te
trigonal bipyramidal domain geometry
see-saw structure

(e) $HClO_4$ 32 e^-, 16 e^- pairs; no prediction

($HClO_4$ is an oxyacid, so H is bound to O, not Cl.
Other Lewis structures that optimize formal charges
are possible; structure predictions are the same.)

22.4   *Analyze.* Equation [22.33] is: $O_3 \rightarrow O_2 + O$ $\Delta H = 105$ kJ

Plan. Use $\Delta H°$ and the dissociation energy barrier to draw the profile.

Solve. $\Delta H°$ tells us that the products are 105 kJ higher in energy than the reactants. The dissociation energy barrier is the total energy difference between reactants and the activated complex (the top of the peak).

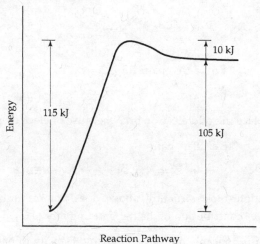

22.5   *Analyze.* Given: space-filling models of molecules containing nitrogen and oxygen atoms.

Find: molecular formulas and Lewis structures.

Plan. Nitrogen atoms are blue, and oxygen atoms are red. Count the number of spheres of each color to determine the molecular formula. From each molecular formula, count the valence electrons (N = 5, O = 6) and draw a correct Lewis structure. Resonance structures are likely.

Solve.

(a) N_2O_5 40 valence electrons, 20 e⁻ pairs

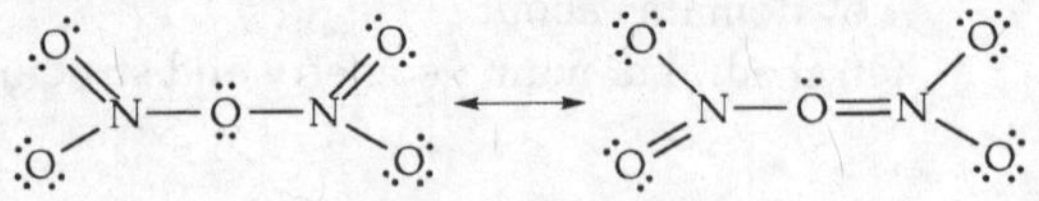

Many other resonance structures are possible. Those with double bonds to the central oxygen (like the right-hand structure above) do not minimize formal charge and are less significant in the net bonding model.

(b) N_2O_4 34 e⁻, 17 e⁻ pairs

Other equivalent resonance structures with different arrangement of the double bonds are possible.

(c) NO_2 17 e⁻, 8.5 e⁻ pairs

We place the odd electron on N because of electronegativity arguments.

(d) N_2O_3 28 e⁻, 14 e⁻ pairs

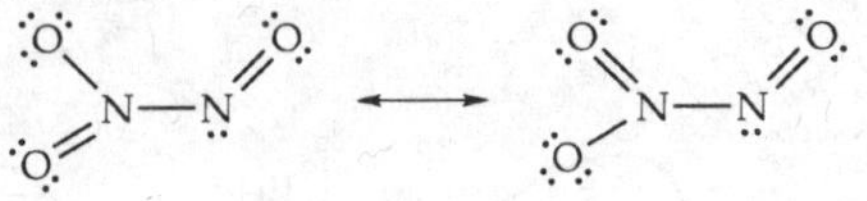

(e) NO   11 e⁻, 5.5 e⁻ pairs

We place the odd electron on N because of electronegativity arguments.

(f) N_2O 16 e⁻, 8 e⁻ pairs

The right-most structure above does not minimize formal charge and makes smaller contribution to the net bonding model.

22.6   The graph is applicable only to (c) density. Density depends on both atomic mass and volume (radius). Both increase going down a family, but atomic mass increases to a greater extent. Density, the ratio of mass to volume, increases going down the family; this trend is consistent with the data in the figure.

According to periodic trends, (a) electronegativity and (b) first ionization energy both decrease rather than increase going down the family. According to Table 22.5 both (d) X—X single bond enthalpy and (e) electron affinity are somewhat erratic, with the trends decreasing from S to Po, and anomalous values for the properties of O, probably owing to its small covalent radius.

22.7   (a)   Atomic radius increases moving downward in a group because the principal quantum number (n) of the valence electrons increases. As n increases, the average distance of an electron from the nucleus increases and so does atomic radius.

        (b)   Anionic radii are greater than atomic radii because of increased electrostatic repulsions among electrons. Additional electrons in the same principle quantum level lead to additional electrostatic repulsion. This increases the energy of the electrons, and their average distance from the nucleus; the anionic radii are thus greater than the atomic radii.

        (c)   The anion that is the strongest base in water is the conjugate base of the weakest conjugate acid. The conjugate acids are OH^-, SH^-, and SeH^-. According to trends in binary hydrides, the acid with the longest X—H bond will be the most readily ionized and the strongest acid. SeH^- is thus the strongest acid and OH^- the weakest. Therefore O^{2-} is the strongest base in water.

22.8   *Analyze/Plan.* Evaluate the graph, describe the trend in data, recall the general trend for each of the properties listed, and use details of the data to discriminate between possibilities.

        Solve. The general trend is an increase in value moving from left to right across the period, with a small discontinuity at S. Considering just this overall feature, both (a) first ionization energy and (c) electronegativity increase moving from left to right, so these are possibilities. (b) Atomic radius decreases, and can be eliminated. Since Si is a solid and Cl and Ar are gases at room temperature, melting points must decrease across the row; (d) melting point can be eliminated. According to data in Tables 22.2, 22.5, 22.7, and 22.8, (e) X-X single bond enthalpies show no consistent trend. Furthermore, there is no known Ar-Ar single bond, so no value for this property can be known; (e) can be eliminated.

        Now let's examine trends in (a) first ionization energy and (c) electronegativity more closely. From electronegativity values in Chapter 8, we see a continuous increase with no discontinuity at S, and no value for Ar. Values in (a) first ionization energy from Chapter 7 do match the pattern in the figure. The slightly lower value of I_1 for S is due to a decrease in repulsion by removing an electron from a fully occupied orbital. In summary, only (a) first ionization energy fits the property depicted in the graph.

22.9   White phosphorus consists of tetrahedral P_4 molecules. This molecular geometry requires al P—P—P bond angles to be much smaller than the idealized 109°. This angular strain is relieved when P_4 reacts with another substance, which it does readily. The chains in red phosphorus require much less extreme bond angles, and there is less need to relieve steric repulsion by chemical reaction.

22.10   *Analyze/Plan.* The structure shown is a diatomic molecule or ion, depending on the value of n. Each species has 10 valence electrons and 5 electron pairs.

        Solve.

(a)   Only second row elements are possible, because of the small covalent radius required for multiple bonding. Likely candidates are CO, N_2, NO^+, CN^-, and C_2^{2-}.

$:C\equiv O:$ $:N\equiv N:$ $[:N\equiv O:]^+$ $[:C\equiv N:]^-$ $[:C\equiv C:]^{2-}$

(b)   Since C_2^{2-} has the highest negative charge, it is likely to be the strongest H^+ acceptor and strongest Brønsted base. This is confirmed in Section 22.9 under "Carbides."

Periodic Trends and Chemical Reactions

22.11   *Analyze/Plan.* Use the color-coded periodic chart on the front-inside cover of the text to classify the given elements.   *Solve.*

Metals: (b) Sr, (c) Mn, (e) Rh;   nonmetals: (a) P, (d) Se, (f) Kr;   metalloids: none

22.12   Metals: (a) Re, (d) Zr, (f) Ga   nonmetals: (c) Ar   metalloid: (b) As, (e) Te

22.13   *Analyze/Plan.* Follow the logic in Sample Exercise 22.1.   *Solve.*

(a)   O   (b)   Br   (c)   Ba

(d)   O   (e)   Co

22.14   (a)   Cl   (b)   K

(c)   K in the gas phase (lowest ionization energy), Li in aqueous solution (most positive E° value)

(d)   Ne; Ne and Ar are difficult to compare to the other elements because they do not form compounds and their radii are not measured in the same way as other elements. However, Ne is several rows to the right of C and surely has a smaller atomic radius. The next smallest is C.

(e)   C

22.15   *Analyze/Plan.* Use the position of the specified elements on the periodic chart, periodic trends, and the arguments in Sample Exercise 22.1 to explain the observations.   *Solve.*

(a)   Nitrogen is too small to accommodate five fluorine atoms about it. The P and As atoms are larger. Furthermore, P and As have available 3d and 4d orbitals, respectively, to form hybrid orbitals that can accommodate more than an octet of electrons about the central atom.

(b)   Si does not readily form π bonds, which would be necessary to satisfy the octet rule for both atoms in SiO.

(c)   A reducing agent is a substance that readily loses electrons. As has a lower electronegativity than N; that is, it more readily gives up electrons to an acceptor and is more easily oxidized.

22.16   (a)   Nitrogen is a highly electronegative element. In HNO_3 it is in its highest oxidation state, +5, and thus is more readily reduced than phosphorus, which forms stable P—O bonds.

(b)   The difference between the third row element and the second lies in the smaller size of C as compared with Si, and the fact that Si has 3d orbitals available to form an sp^3d^2 hybrid set that can accommodate more than an octet of electrons.

(c)   Two of the carbon compounds, C_2H_4 and C_2H_2, contain C—C π bonds. Si does not readily form π bonds (to itself or other atoms), so Si_2H_4 and Si_2H_2 are not known as stable compounds.

22.17   *Analyze/Plan.* Follow the logic in Sample Exercise 22.2.   *Solve.*

(a) $Mg_3N_2(s) + 6H_2O(l) \rightarrow 2NH_3(g) + 3Mg(OH)_2(s)$

     Because $H_2O(l)$ is a reactant, the state of NH_3 in the products could be expressed as $NH_3(aq)$.

(b) $2C_3H_7OH(l) + 9O_2(g) \rightarrow 6CO_2(g) + 8H_2O(l)$

(c) $MnO_2(s) + C(s) \xrightarrow{\Delta} CO(g) + MnO(s)$ or

     $MnO_2(s) + 2C(s) \xrightarrow{\Delta} 2CO(g) + Mn(s)$ or

     $MnO_2(s) + C(s) \xrightarrow{\Delta} CO_2(g) + Mn(s)$

(d) $AlP(s) + 3H_2O(l) \rightarrow PH_3(g) + Al(OH)_3(s)$

(e) $Na_2S(s) + 2HCl(aq) \rightarrow H_2S(g) + 2NaCl(aq)$

22.18   (a) $NaOCH_3(s) + H_2O(l) \rightarrow NaOH(aq) + CH_3OH(aq)$

       (b) $CuO(s) + 2HNO_3(aq) \rightarrow Cu(NO_3)_2(aq) + H_2O(l)$

       (c) $WO_3(s) + 3H_2(g) \rightarrow W(s) + 3H_2O(g)$

       (d) $4NH_2OH(l) + O_2(g) \rightarrow 6H_2O(l) + 2N_2(g)$

       (e) $Al_4C_3(s) + 12H_2O(l) \rightarrow 4Al(OH)_3(s) + 3CH_4(g)$

Hydrogen, the Noble Gases, and the Halogens

22.19   *Analyze/Plan.* Use information on the isotopes of hydrogen in Section 22.2 to list their symbols, names, and relative abundances.   *Solve.*

a) $_1^1H$ - protium; $_1^2H$ - deuterium; $_1^3H$ - tritium

b)   The order of abundance is proteum > deuterium > tritium.

22.20   Tritium is radioactive. $_1^3H \rightarrow {}_2^3He + {}_{-1}^0e$

22.21   *Analyze/Plan.* Consider the electron configuration of hydrogen and the Group 1A elements. *Solve.*

Like other elements in group 1A, hydrogen has only one valence electron and its most common oxidation number is +1.

22.22   Halogens are nonmetals with high electronegativity that need one additional valence electron to have a complete octet. They typically form negative ions with a 1- charge, homonuclear diatomic molecules, interhalogen compounds, or are the more electronegative atom in a polar covalent bond. Hydrogen also needs one additional valence electron to have a complete duet. In compounds (c) NaH and (d) H_2, hydrogen is acting like a halogen in two of the bonding situations described above.

(c)   NaH is ionic, with H acting as hydride anion, H$^-$.

$NaH(s) + H_2O(l) \rightarrow H_2(g) + NaOH(aq)$

$NaF(s) + H_2O(l) \rightarrow HF(aq) + NaOH(aq)$

(d)   Hydrogen forms the nonpolar homonuclear diatomic molecule H$_2$.

$2Li(s) + H_2(g) \rightarrow 2LiH(s)$

$2Li(s) + F_2(g) \rightarrow 2LiF(s)$

22.23   *Analyze/Plan.* Use information on the descriptive chemistry of hydrogen in Section 22.2 to formulate the required equations. Steam is H$_2$O(g).   *Solve.*

(a) $Mg(s) + 2H^+(aq) \rightarrow Mg^{2+}(aq) + H_2(g)$

(b) $C(s) + H_2O(g) \xrightarrow{1000°C} CO(g) + H_2(g)$

(c) $CH_4(g) + H_2O(g) \xrightarrow{1100°C} CO(g) + 3H_2(g)$

22.24   (a)   Electrolysis of brine; reaction of carbon with steam; reaction of methane with steam; by-product in petroleum refining

(b)   Synthesis of ammonia; synthesis of methanol; reducing agent; hydrogenation of unsaturated vegetable oils

22.25   *Analyze/Plan.* Use information on the descriptive chemistry of hydrogen given in Section 22.2 to complete and balance the equations.   *Solve.*

(a) $NaH(s) + H_2O(l) \rightarrow NaOH(aq) + H_2(g)$

(b) $Fe(s) + H_2SO_4(aq) \rightarrow Fe^{2+}(aq) + H_2(g) + SO_4^{2-}(aq)$

(c) $H_2(g) + Br_2(g) \rightarrow 2HBr(g)$

(d) $2Na(l) + H_2(g) \rightarrow 2NaH(s)$

(e) $PbO(s) + H_2(g) \xrightarrow{\Delta} Pb(s) + H_2O(g)$

22.26   (a) $2Al(s) + 6H^+(aq) \rightarrow 2Al^{3+}(aq) + 3H_2(g)$

(b) $Mg(s) + H_2O(g) \rightarrow MgO(s) + H_2(g)$

(c) $MnO_2(s) + H_2(g) \rightarrow MnO(s) + H_2O(g)$

(d) $CaH_2(s) + 2H_2O(l) \rightarrow Ca(OH)_2(aq) + 2H_2(g)$

22.27   *Analyze/Plan.* If the element bound to H is a nonmetal, the hydride is molecular. If H is bound to a metal with integer stoichiometry, the hydride is ionic; with noninteger stoichiometry, the hydride is metallic.   *Solve.*

(a)   ionic (metal hydride)

(b)   molecular (nonmetal hydride)

(c)   metallic (nonstoichiometric transition metal hydride)

22.28   (a)   molecular   (b)   ionic   (c)   metallic

22.29   *Analyze/Plan.* Consider the periodic properties of Xe and Ar.   *Solve.*

Xenon is larger, and can more readily accommodate an expanded octet. More important is the lower ionization energy of xenon; because the valence electrons are a greater average distance from the nucleus, they are more readily promoted to a state in which the Xe atom can form bonds with fluorine.

22.30   Xe(l), an atomic liquid, is capable only of London dispersion forces. All polar covalent molecules experience both dipole-dipole and dispersion forces, so dispersion forces are the most important intermolecular interactions between Xe(l) and polar covalent solutes. This is particularly true for CH_3I. The iodine atom has a large, very polarizable electron cloud, which makes dispersion forces the dominant type of intermolecular interaction for this molecule.

22.31   *Analyze/Plan.* Follow the rules for assigning oxidation numbers in Section 4.4 and the logic in Sample Exercise 4.8.   *Solve.*

    (a) ClO_3^-, +5   (b)   HI, –1   (c) ICl_3; I, +3; Cl, –1

    (d)   NaOCl, +1   (e) $HClO_4$, +7   (f) XeF_4, +4; F, –1

22.32   (a) $Ca(OBr)_2$, +1   (b) $HBrO_3$, +5   (c) XeO_3; Xe, +6

            (d) ClO_4^-, +7   (e) HIO_2, +3   (f) IF_5; I, +5; F, –1

22.33   *Analyze/Plan.* Review the nomenclature rules and ion names in Section 2.8.   *Solve.*

    (a)   iron(III) chlorate   (b)   chlorous acid

    (c)   xenon hexafluoride   (d)   bromine pentafluoride

    (e)   xenon oxide tetrafluoride   (f)   iodic acid

22.34   (a)   potassium chlorate   (b)   calcium iodate

            (c)   aluminum chloride   (d)   bromic acid

            (e)   paraperiodic acid   (f)   xenon tetrafluoride

22.35   *Analyze/Plan.* Consider intermolecular forces and periodic properties, including oxidizing power, of the listed substances.   *Solve.*

    (a)   Van der Waals intermolecular attractive forces increase with increasing numbers of electrons in the atoms.

    (b) F_2 reacts with water: $F_2(g) + H_2O(l) \rightarrow 2HF(aq) + 1/2\ O_2(g)$. That is, fluorine is too strong an oxidizing agent to exist in water.

    (c)   HF has extensive hydrogen bonding.

    (d)   Oxidizing power is related to electronegativity. Electronegativity decreases in the order given.

22.36   (a)   The more electronegative the central atom, the greater the extent to which it withdraws charge from oxygen, in turn making the O—H bond more polar, and enhancing ionization of H^+.

(b)   HF reacts with the silica which is a major component of glass:

$$6HF(aq) + SiO_2(s) \rightarrow SiF_6^{2-}(aq) + 2H_2O(l) + 2H^+(aq)$$

(c)   Iodide is oxidized by sulfuric acid, as shown in Figure 22.12.

(d)   The major factor is size; there is not room about Br for the three chlorides plus the two unshared electron pairs that would occupy the bromine valence shell orbitals.

22.37   Vehicle fuels produce energy via combustion reactions. The reaction $H_2(g) + 1/2\ O_2(g) \rightarrow H_2O(g)$ is very exothermic, producing 242 kJ per mole of H_2 burned. The only product of combustion is H_2O, a nonpollutant (but like CO_2, a greenhouse gas).

22.38   Cracking is a process where higher molecular weight hydrocarbons are broken down into smaller molecules (Section 25.3). One model reaction is

$$CH_4(g) \xrightarrow{\text{catalyst}} C(s) + 2H_2(g)\ \Delta H \sim 75\ kJ.$$

An analogous reaction can be written for "cracking" ethanol, where a third product is $O_2(g)$. The overall process is endothermic and requires energy. The question remains, what energy source drives the cracking? If energy to drive the cracking can be provided by some nonpetroleum fuel (such as grain-based ethanol) or power source (such as wind, geothermal, etc.), this could provide a satisfactory basis for a hydrogen economy.

Additionally, cracking would produce C(s) or soot and other C-based pollutants, which would have to be treated. This is probably easier at the power-plant level than the individual car level, but not insignificant.

22.39   If the lifetime of perchlorate anion, ClO_4^-, in soils and water is decades, the ion must be extremely stable (unreactive) in aqueous solutions and aerobic environments; it is not easily oxidized by O_2. Although chlorine is in a very high oxidation state in ClO_4^-, it is not readily reduced, because the ion has a stable, symmetric structure that protects it against reactions. The anion is a symmetrical tetrahedron with several plausible Lewis structures when expanded octets about Cl are considered.

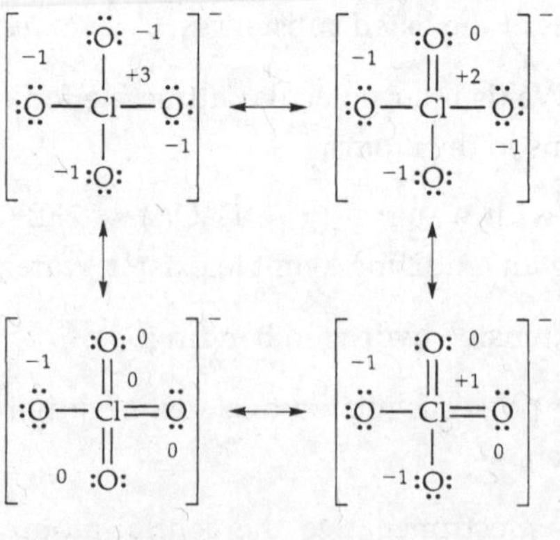

(minimum formal charges)

More structures with alternate locations of the single and double bonds can be drawn. Resonance stabilization could certainly contribute to the ion's high stability.

22.40   Perchlorate salts are highly soluble, so the reaction is probably not a precipitation reaction. Perchloric acid, $HClO_4$, is a strong acid and ClO_4^- is a negligible base in water so the decomposition is probably not an acid-base reaction. Because oxyanions of the halogens are known to participate in oxidation-reduction reactions, the microorganisms probably destroy ClO_4^- via a redox reaction or series of reactions. While several lower oxidation states are accessible to Cl in ClO_4^-, intermediate species ClO_3^-, ClO_2^-, ClO^-, and Cl_2 are quite reactive, so the logical fate of the perchlorate anion is $Cl^-(aq)$ and $O_2(g)$ or $H_2O(l)$ with some oxidized organic compound.

Oxygen and the Group 6A Elements

22.41   *Analyze/Plan.* Consider the industrial uses of oxygen and ozone given in Section 22.5. *Solve.*

     (a)   As an oxidizing agent in steel-making; to bleach pulp and paper; in oxyacetylene torches; in medicine to assist in breathing

     (b)   Synthesis of pharmaceuticals, lubricants, and other organic compounds where $C=C$ bonds are cleaved; in water treatment

22.42

     Ozone has two resonance forms (Section 8.7); the molecular structure is bent, with an $O-O-O$ bond angle of approximately 120°. The π bond in ozone is delocalized over the entire molecule; neither individual $O-O$ bond is a full double bond, so the observed $O-O$ distance of 1.28 Å is greater than the 1.21 Å distance in O_2, which has a full $O-O$ double bond.

22.43   *Analyze/Plan.* Use information on the descriptive chemistry of oxygen given in Section 22.5 to complete and balance the equations.   *Solve.*

     (a) $2HgO(s) \xrightarrow{\Delta} 2Hg(l) + O_2(g)$

     (b) $2Cu(NO_3)_2(s) \xrightarrow{\Delta} 2CuO(s) + 4NO_2(g) + O_2(g)$

     (c) $PbS(s) + 4O_3(g) \rightarrow PbSO_4(s) + 4O_2(g)$

     (d) $2ZnS(s) + 3O_2(g) \xrightarrow{\Delta} 2ZnO(s) + 2SO_2(g)$

     (e) $2K_2O_2(s) + 2CO_2(g) \rightarrow 2K_2CO_3(s) + O_2(g)$

22.44
     (a) $CaO(s) + H_2O(l) \rightarrow Ca^{2+}(aq) + 2OH^-(aq)$

     (b) $Al_2O_3(s) + 6H^+(aq) \rightarrow 2Al^{3+}(aq) + 3H_2O(l)$

     (c) $Na_2O_2(s) + 2H_2O(l) \rightarrow 2Na^+(aq) + 2OH^-(aq) + H_2O_2(aq)$

     (d) $N_2O_3(g) + H_2O(l) \rightarrow 2HNO_2(aq)$

     (e) $2KO_2(s) + 2H_2O(l) \rightarrow 2K^+(aq) + 2OH^-(aq) + O_2(g) + H_2O_2(aq)$

     (f) $NO(g) + O_3(g) \rightarrow NO_2(g) + O_2(g)$

22.45   *Analyze/Plan.* Oxides of metals are bases, oxides of nonmetals are acids, oxides that act as both acids and bases are amphoteric and oxides that act as neither acids nor bases are neutral.   *Solve.*

(a)   acidic (oxide of a nonmetal)

(b)   acidic (oxide of a nonmetal)

(c)   basic (oxide of a metal)

(d)   amphoteric

22.46   (a) Mn_2O_7 (higher oxidation state of Mn)

         (b) SnO_2 (higher oxidation state of Sn)

         (c) SO_3 (higher oxidation state of S)

         (d) SO_2 (more nonmetallic character of S)

         (e) Ga_2O_3 (more nonmetallic character of Ga)

         (f) SO_2 (more nonmetallic character of S)

22.47   *Analyze/Plan.* Follow the rules for assigning oxidation numbers in Section 4.4 and the logic in Sample Exercise 4.8.   *Solve.*

         (a) H_2SeO_3, +4   (b) $KHSO_3$, +4   (c) $H_2\mathbf{Te}$, –2

         (d) CS_2, –2   (e) $CaSO_4$, +6

         Oxygen (a group 6A element) is in the –2 oxidation state in compounds (a), (b), and (e).

22.48   (a) SeO_2, +4   (b) $Na_2S_2O_3$, +2   (c) SF_6, +6

         (d) H_2S, –2   (e) H_2SO_4, +6

         Oxygen (a group 6A element) is in the –2 oxidation state in compounds (a), (b), and (e).

22.49   *Analyze/Plan.* The half-reaction for oxidation in all these cases is:

$H_2S(aq) \rightarrow S(s) + 2H^+ + 2e^-$ (The product could be written as $S_8(s)$, but this is not necessary. In fact it is not necessarily the case that S_8 would be formed, rather than some other allotropic form of the element.) Combine this half-reaction with the given reductions to write complete equations. The reduction in (c) happens only in acid solution. The reactants in (d) are acids, so the medium is acidic.   *Solve.*

         (a) $2Fe^{3+}(aq) + H_2S(aq) \rightarrow 2Fe^{2+}(aq) + S(s) + 2H^+(aq)$

         (b) $Br_2(l) + H_2S(aq) \rightarrow 2Br^-(aq) + S(s) + 2H^+(aq)$

         (c) $2MnO_4^-(aq) + 6H^+(aq) + 5H_2S(aq) \rightarrow 2Mn^{2+}(aq) + 5S(s) + 8H_2O(l)$

         (d) $2NO_3^-(aq) + H_2S(aq) + 2H^+(aq) \rightarrow 2NO_2(aq) + S(s) + 2H_2O(l)$

22.50   An aqueous solution of SO_2 contains H_2SO_3 and is acidic. Use H_2SO_3 as the reducing agent and balance assuming acid conditions.

         (a)

$$2[MnO_4^-(aq) + 8H^+(aq) + 5e^- \rightarrow Mn^{2+}(aq) + 4H_2O(l)]$$

$$\underline{5[H_2SO_3(aq) + H_2O(l) \rightarrow SO_4^{2-}(aq) + 4H^+(aq) + 2e^-]}$$

$$2MnO_4^-(aq) + 5H_2SO_3(aq) \rightarrow 2MnSO_4(aq) + 3SO_4^{2-}(aq) + 3H_2O(l) + 4H^+(aq)$$

(b)
$$Cr_2O_7{}^{2-}(aq) + 14H^+(aq) + 6e^- \rightarrow 2Cr^{3+}(aq) + 7H_2O(l)$$

$$\frac{3[H_2SO_3(aq) + H_2O(l) \rightarrow SO_4{}^{2-}(aq) + 4H^+(aq) + 2e^-]}{Cr_2O_7{}^{2-}(aq) + 3H_2SO_3(aq) + 2H^+(aq) \rightarrow 2Cr^{3+}(aq) + 3SO_4{}^{2-}(aq) + 4H_2O(l)}$$

(c)
$$Hg_2{}^{2+}(aq) + 2e^- \rightarrow 2Hg(l)$$

$$\frac{H_2SO_3(aq) + H_2O(l) \rightarrow SO_4{}^{2-}(aq) + 4H^+(aq) + 2e^-}{Hg_2{}^{2+}(aq) + H_2SO_3(aq) + H_2O(l) \rightarrow 2Hg(l) + SO_4{}^{2-}(aq) + 4H^+(aq)}$$

22.51   *Analyze/Plan.* For each substance, count valence electrons, draw the correct Lewis structure, and apply the rules of VSEPR to decide electron domain geometry and geometric structure.   *Solve.*

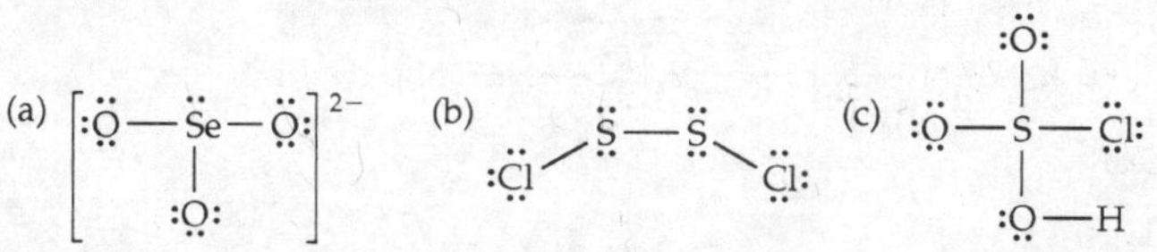

(a)

trigonal pyramidal

(b)

bent (free rotation around S-S bond)

(c)

tetrahedral

22.52 SF_4, 34 e$^-$ $SF_5{}^-$, 42 e$^-$

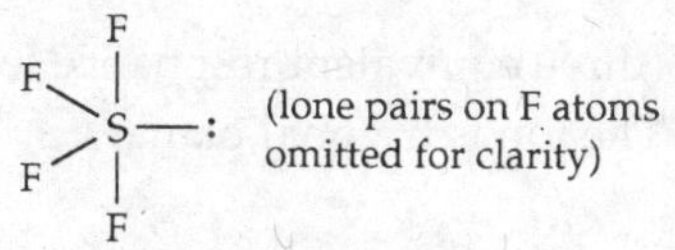

(lone pairs on F atoms omitted for clarity)

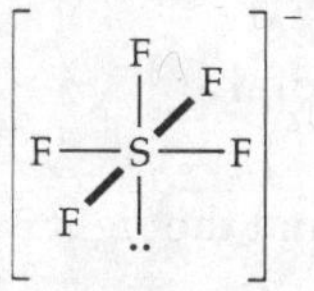

trigonal bipyramidal
electron pair geometry

octahedral electron
pair geometry

see-saw molecular
geometry

square pyramidal molecular
geometry

22.53   *Analyze/Plan.* Use information on the descriptive chemistry of sulfur given in Section 22.6 to complete and balance the equations.   *Solve.*

(a) $SO_2(s) + H_2O(l) \rightarrow H_2SO_3(aq) \rightleftharpoons H^+(aq) + HSO_3{}^-(aq)$

(b) $ZnS(s) + 2HCl(aq) \rightarrow ZnCl_2(aq) + H_2S(g)$

(c) $8SO_3{}^{2-}(aq) + S_8(s) \rightarrow 8S_2O_3{}^{2-}(aq)$

(d) $SO_3(aq) + H_2SO_4(l) \rightarrow H_2S_2O_7(l)$

22.54   (a) $Al_2Se_3(s) + 6H^+(aq) \rightarrow 2Al^{3+}(aq) + 3H_2Se(g)$

(b) $Cl_2(aq) + S_2O_3{}^{2-}(aq) + H_2O(l) \rightarrow 2Cl^-(aq) + S(s) + SO_4{}^{2-}(aq) + 2H^+(aq)$

Nitrogen and the Group 5A Elements

22.55   *Analyze/Plan.* Follow the rules for assigning oxidation numbers in Section 4.4 and the logic in Sample Exercise 4.8.   *Solve.*

(a) $NaNO_2$, +3   (b) NH_3, –3   (c) N_2O, +1

(d) $NaCN$, –3   (e) HNO_3, +5   (f) NO_2, +4

22.56　(a)　HNO_3, +5　　　(b)　N_2H_4, −2　　　(c)　KCN, −3

　　　　(d)　$NaNO_3$, +5　　(e)　NH_4Cl, −3　　(f)　Li_3N, −3

22.57　*Analyze/Plan.* For each substance, count valence electrons, draw the correct Lewis structure, and apply the rules of VSEPR to decide electron domain geometry and geometric structure.　*Solve.*

　　　(a)　$:\ddot{O}{=}\ddot{N}{-}\ddot{O}{-}H \longleftrightarrow :\ddot{O}{-}\ddot{N}{=}\ddot{O}{-}H$

　　　　　The molecule is bent around the central oxygen and nitrogen atoms; the four atoms need not lie in a plane. The right-most form does not minimize formal charges and is less important in the actual bonding model.

　　　(b)　$\left[:\ddot{N}{=}N{=}\ddot{N}:\right]^{-} \longleftrightarrow \left[:N{\equiv}N{-}\ddot{\ddot{N}}:\right]^{-} \longleftrightarrow \left[:\ddot{\ddot{N}}{-}N{\equiv}N:\right]^{-}$

　　　　The molecule is linear.

　　　(c)　$\left[\begin{array}{c} H \quad H \\ | \quad\; | \\ H{-}N{-}N: \\ | \quad\; | \\ H \quad H \end{array}\right]^{+}$　　(d)　$\left[\begin{array}{c} :\ddot{O}: \\ | \\ :\ddot{O}{-}N{=}\ddot{O} \end{array}\right]^{-}$

　　　　The geometry is tetrahedral　　　(three equivalent resonance forms)
　　　　around the left nitrogen,　　　　The ion is trigonal planar.
　　　　trigonal pyramidal around the right.

22.58　(a)　$\left[\begin{array}{c} H \\ | \\ H{-}N{-}H \\ | \\ H \end{array}\right]^{+}$

　　　　　tetrahedral

　　　(b)　$:\ddot{O}{-}\overset{\overset{\displaystyle :O:}{||}}{N}{-}\ddot{O}{-}H \longleftrightarrow \ddot{O}{=}\overset{\overset{\displaystyle :\ddot{O}:}{|}}{N}{-}\ddot{O}{-}H \longleftrightarrow :\ddot{O}{-}\overset{\overset{\displaystyle :\ddot{O}:}{|}}{N}{=}O{-}H$

　　　　The geometry around nitrogen is trigonal planar, but the hydrogen atom is not required to lie in this plane. The third resonance form makes a much smaller contribution to the structure than the first two.

　　　(c)　$:\ddot{N}{=}N{=}\ddot{O}: \longleftrightarrow :N{\equiv}N{-}\ddot{O}: \longleftrightarrow :\ddot{N}{-}N{\equiv}O:$

　　　　The molecule is linear. Again, the third resonance form makes less contribution to the structure because of the high formal charges involved.

　　　(d)　$\ddot{O}{=}\ddot{N}{-}\ddot{O}: \longleftrightarrow :\ddot{O}{-}\ddot{N}{=}\ddot{O}:$

　　　　The molecule is bent (nonlinear).

22.59　*Analyze/Plan.* Use information on the descriptive chemistry of nitrogen given in Section 22.7 to complete and balance the equations.　*Solve.*

(a) $Mg_3N_2(s) + 6H_2O(l) \rightarrow 2NH_3(g) + 3Mg(OH)_2(s)$

Because $H_2O(l)$ is a reactant, the state of NH_3 in the products could be expressed as $NH_3(aq)$.

(b) $2NO(g) + O_2(g) \rightarrow 2NO_2(g)$

(c) $N_2O_5(g) + H_2O(l) \rightarrow 2H^+(aq) + 2NO_3^-(aq)$

(d) $NH_3(aq) + H^+(aq) \rightarrow NH_4^+(aq)$

(e) $N_2H_4(l) + O_2(g) \rightarrow N_2(g) + 2H_2O(g)$

22.60   (a) $4Zn(s) + 2NO_3^-(aq) + 10H^+(aq) \rightarrow 4Zn^{2+}(aq) + N_2O(g) + 5H_2O(l)$

(b) $4NO_3^-(aq) + S(s) + 4H^+(aq) \rightarrow 4NO_2(g) + SO_2(g) + 2H_2O(l)$

(or $6NO_3^-(aq) + S(s) + 4H^+(aq) \rightarrow 6NO_2(g) + SO_4^{2-}(aq) + 2H_2O(l)$

(c) $2NO_3^-(aq) + 3SO_2(g) + 2H_2O(l) \rightarrow 2NO(g) + 3SO_4^{2-}(aq) + 4H^+(aq)$

(d) $N_2H_4(g) + 5F_2(g) \rightarrow 2NF_3(g) + 4HF(g)$

(e) $4CrO_4^{2-}(aq) + 3N_2H_4(aq) + 4H_2O(l) \rightarrow 4Cr(OH)_4^-(aq) + 4OH^-(aq) + 3N_2(g)$

22.61   *Analyze/Plan.* Follow the method for writing balanced half-reactions given in Section 20.2 and Sample Exercises 20.2 and 20.3. Find standard reduction potentials in Figure 22.28.

Solve. For nonadjacent half-reactions on Figure 22.28, use the relationship $\Delta G° = -nFE°$. Because $\Delta G°$ is a state function, $\Delta G°$ for the overall process is the sum of the $\Delta G°$ values for the steps. Then, $\Delta G° = \Delta G_1° + \Delta G_2°$; $-nFE° = -n_1FE_1° - n_2FE_2°$; $n = n_1 + n_2$;

$$E° = \frac{n_1E_1° + N_2E_2°}{n_1 + n_2}$$

(a) $HNO_2(aq) + H_2O(l) \rightarrow NO_3^-(aq) + 3H^+(aq) + 2e^-$, $E_{red}° = 0.96$ V

The steps are: (1) $NO_3^- \rightarrow NO_2$, $1e^-$, 0.79 V;

(2) $NO_2 \rightarrow HNO_2$, $1e^-$, 1.12 V

$E_{red}° = [(1)\ 0.79\ V + (1)\ 1.12\ V)/(1+1) = 0.955 = 0.96$ V

(b) $N_2(g) + H_2O(l) \rightarrow N_2O(g) + 2H^+(aq) + 2e^-$, $E_{red}° = 1.77$ V

$E_{red}°$ can be read directly from Figure 22.28.

22.62 $E_{red}°$ for both half-reactions can be read directly from Figure 22.28.

(a) $NO_3^-(aq) + 4H^+(aq) + 3e^- \rightarrow NO(g) + 2H_2O(l)$ $E_{red}° = +0.96$ V

(b) $HNO_2(aq) \rightarrow NO_2(g) + H^+(aq) + 1e^-$ $E_{red}° = 1.12$ V

22.63   *Analyze/Plan.* Follow the rules for assigning oxidation numbers in Section 4.4 and the logic in Sample Exercise 4.8.   *Solve.*

(a) H_3PO_3, +3   (b) $H_4P_2O_7$, +5   (c) $SbCl_3$, +3

(d) Mg_3As_2, −3   (e) P_2O_5, +5

22.64   (a) H_3PO_4, +5   (b) H_3AsO_3, +3   (c) Sb_2S_3, +3

        (d) $Ca(H_2PO_4)_2$, +5   (e) K_3P, –3

22.65   *Analyze/Plan.* Consider the structures of the compounds of interest when explaining the observations.   *Solve.*

        (a)   Phosphorus is a larger atom and can more easily accommodate five surrounding atoms and an expanded octet of electrons than nitrogen can. Also, P has energetically "available" 3d orbitals which participate in the bonding, but nitrogen does not.

        (b)   Only one of the three hydrogens in H_3PO_2 is bonded to oxygen. The other two are bonded directly to phosphorus and are not easily ionized because the P—H bond is not very polar.

        (c) PH_3 is a weaker base than H_2O (PH_4^+ is a stronger acid than H_3O^+). Any attempt to add H^+ to PH_3 in the presence of H_2O merely causes protonation of H_2O.

        (d)   White phosphorus consists of P_4 molecules, with P—P—P bond angles of 60°. Each P atom has four VSEPR pairs of electrons, so the predicted electron pair geometry is tetrahedral and the preferred bond angle is 109°. Because of the severely strained bond angles in P_4 molecules, white phosphorus is highly reactive.

22.66   (a)   Only two of the hydrogens in H_3PO_3 are bound to oxygen. The third is attached directly to phosphorus, and not readily ionized, because the H—P bond is not very polar.

        (b)   The smaller, more electronegative nitrogen withdraws more electron density from the O—H bond, making it more polar and more likely to ionize.

        (c)   Phosphate rock consists of $Ca_3(PO_4)_2$, which is only slightly soluble in water. The phosphorus is unavailable for plant use.

        (d) N_2 can form stable π bonds to complete the octet of both N atoms. Because phosphorus atoms are larger than nitrogen atoms, they do not form stable π bonds with themselves and must form σ bonds with several other phosphorus atoms (producing P_4 tetrahedral or sheet structures) to complete their octets.

        (e)   In solution Na_3PO_4 is completely dissociated into Na^+ and PO_4^{3-}. PO_4^{3-}, the conjugate base of the very weak acid HPO_4^{2-}, has a K_b of 2.4×10^{-2} and produces a considerable amount of OH^- by hydrolysis of H_2O.

22.67   *Analyze/Plan.* Use information on the descriptive chemistry of phosphorus given in Section 22.8 to complete and balance the equations.   *Solve.*

        (a) $2Ca_3(PO_4)_2(s) + 6SiO_2(s) + 10C(s) \xrightarrow{\Delta} P_4(g) + 6CaSiO_3(l) + 10CO(g)$

        (b) $PBr_3(l) + 3H_2O(l) \rightarrow H_3PO_3(aq) + 3HBr(aq)$

        (c) $4PBr_3(g) + 6H_2(g) \rightarrow P_4(g) + 12HBr(g)$

22.68   (a) $PCl_5(l) + 4H_2O(l) \rightarrow H_3PO_4(aq) + 5HCl(aq)$

        (b) $2H_3PO_4(aq) \xrightarrow{\Delta} H_4P_2O_7(aq) + H_2O(l)$

        (c) $P_4O_{10}(s) + 6H_2O(l) \rightarrow 4H_3PO_4(aq)$

Carbon, the Other Group 4A Elements, and Boron

22.69   *Analyze/Plan.* Review the nomenclature rules and ion names in Section 2.8.   *Solve.*
   (a)   HCN   (b) $Ni(CO)_4$ (c) $Ba(HCO_3)_2$ (d) CaC_2

22.70   (a) H_2CO_3 (b)   NaCN   (c) $KHCO_3$ (d) C_2H_2

22.71   *Analyze/Plan.* Use information on the descriptive chemistry of carbon given in Section 22.9 to complete and balance the equations.   *Solve.*

   (a) $ZnCO_3(s) \xrightarrow{\Delta} ZnO(s) + CO_2(g)$

   (b) $BaC_2(s) + 2H_2O(l) \rightarrow Ba^{2+}(aq) + 2OH^-(aq) + C_2H_2(g)$

   (c) $2C_2H_2(g) + 5O_2(g) \rightarrow 4CO_2(g) + 2H_2O(g)$

   (d) $CS_2(g) + 3O_2(g) \rightarrow CO_2(g) + 2SO_2(g)$

   (e) $Ca(CN)_2(s) + 2HBr(aq) \rightarrow CaBr_2(aq) + 2HCN(aq)$

22.72   (a) $CO_2(g) + OH^-(aq) \rightarrow HCO_3^-(aq)$

   (b) $NaHCO_3(s) + H^+(aq) \rightarrow Na^+(aq) + H_2O(l) + CO_2(g)$

   (c) $2CaO(s) + 5C(s) \xrightarrow{\Delta} 2CaC_2(s) + CO_2(g)$

   (d) $C(s) + H_2O(g) \xrightarrow{\Delta} H_2(g) + CO(g)$

   (e) $CuO(s) + CO(g) \rightarrow Cu(s) + CO_2(g)$

22.73   *Analyze/Plan.* Use information on the descriptive chemistry of carbon given in Section 22.9 to complete and balance the equations.   *Solve.*

   (a) $2CH_4(g) + 2NH_3(g) + 3O_2(g) \xrightarrow[\text{cat}]{800°C} 2HCN(g) + 6H_2O(g)$

   (b) $NaHCO_3(s) + H^+(aq) \rightarrow CO_2(g) + H_2O(l) + Na^+(aq)$

   (c) $2BaCO_3(s) + O_2(g) + 2SO_2(g) \rightarrow 2BaSO_4(s) + 2CO_2(g)$

22.74   (a) $2Mg(s) + CO_2(g) \rightarrow 2MgO(s) + C(s)$

   (b) $6CO_2(g) + 6H_2O(l) \xrightarrow{h\nu} C_6H_{12}O_6(aq) + 6O_2(g)$

   (c) $CO_3^{2-}(aq) + H_2O(l) \rightarrow HCO_3^-(aq) + OH^-(aq)$

22.75   *Analyze/Plan.* Follow the rules for assigning oxidation numbers in Section 4.4 and the logic in Sample Exercise 4.8.   *Solve.*

   (a) H_3BO_3, +3   (b) $SiBr_4$, +4   (c) $PbCl_2$, +2
   (d) $Na_2B_4O_7 \cdot 10H_2O$, +3   (e) B_2O_3, +3

22.76   (a) SiO_2, +4   (b) $GeCl_4$, +4   (c) $NaBH_4$, +3
   (d) $SnCl_2$, +2   (e) B_2H_6, +3

22.77   *Analyze/Plan.* Consider periodic trends within a family, particularly metallic character, as well as descriptive chemistry in Sections 22.9 and 22.10.   *Solve.*

   (a)   Lead; the most metallic element has the lowest first ionization energy.

(b)   Carbon, Si, and Ge; these are the nonmetal and metalloids in group 4A. They form compounds ranging from XH_4 (–4) to XO_2 (+4). The metals Sn and Pb are not found in negative oxidation states.

(c)   Silicon; silicates are the main component of sand.

22.78   (a)   carbon   (b)   lead   (c)   germanium

22.79   *Analyze/Plan.* Consider the structural chemistry of silicates discussed in Section 22.10 and shown in Figures 22.47-22.48.   *Solve.*

(a)   Tetrahedral

(b)   Metasilicic acid will probably adopt the single-strand silicate chain structure shown in Figure 22.48(a). The empirical formula shows 3 O and 2 H atoms per Si atom. The chain has the same Si to O ratio as metasilicic acid. Furthermore, in the chain structure, there are two terminal (not bridging) O atoms on each Si. These can accommodate the 2 H atoms associated with each Si atom of the acid. The sheet structure does not fulfill these requirements.

22.80   (a)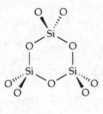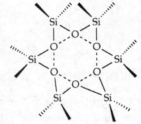

(b) $Si_3O_9^{6-}$ $Si_6O_{18}^{12-}$

22.81   (a)   Diborane (Figure 22.50 and below) has bridging H atoms linking the two B atoms. The structure of ethane shown below has the C atoms bound directly, with no bridging atoms.

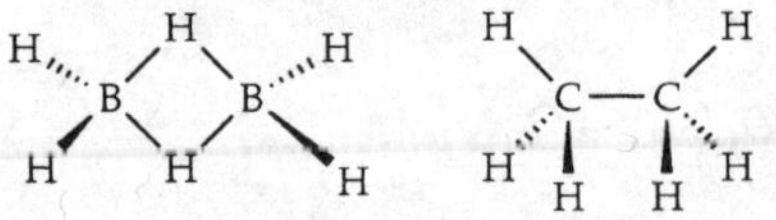

(b) B_2H_6 is an electron deficient molecule. It has 12 valence electrons, while C_2H_6 has 14 valence electrons. The 6 valence electron pairs in B_2H_6 are all involved in B—H sigma bonding, so the only way to satisfy the octet rule at B is to have the bridging H atoms shown in Figure 22.50.

(c)   A hydride ion, H^-, has two electrons while an H atom has one. The term *hydridic* indicates that the H atoms in B_2H_6 have more than the usual amount of electron density for a covalently bound H atom.

22.82   (a) $B_2H_6(g) + 6H_2O(l) \rightarrow 2H_3BO_3(aq) + 6H_2(g)$

(b) $4H_3BO_3(s) \xrightarrow{\Delta} H_2B_4O_7(s) + 5H_2O(g)$

(c) $B_2O_3(s) + 3H_2O(l) \rightarrow 2H_3BO_3(aq)$

Additional Exercises

22.83   (a)   A *reducing agent* agent is an electron-rich substance in a low oxidation state that loses electrons to the reactant being reduced in a redox reaction.

   (b)   *Allotropes* are different structural forms of the same element. They are composed of atoms of a single element bound into different structures. For example, graphite, diamond, and buckey balls are all allotropes of carbon.

   (c)   *Disproportionation* is an oxidation-reduction process where the same element is both oxidized and reduced.

   (d)   *Interhalogen* is a compound formed from atoms of two or more halogens.

   (e)   An *acidic anhydride* is a neutral molecule that is the oxide of a nonmetal. It reacts with water to produce an acid.

   (f)   A *condensation reaction* is the combination of two molecules to form a large molecule and a small one such as H_2O or HCl.

22.84   (a) $10.0 \text{ lb FeTi} \times \dfrac{453.6 \text{ g}}{1 \text{ lb}} \times \dfrac{1 \text{ mol FeTi}}{103.7 \text{ g FeTi}} \times \dfrac{1 \text{ mol } H_2}{1 \text{ mol FeTi}} \times \dfrac{2.016 \text{ g H}}{1 \text{ mol } H_2} = 88.18 = 88.2 \text{ g } H_2$

   (b) $V = \dfrac{88.18 \text{ g } H_2}{2.016 \text{ g/mol } H_2} \times \dfrac{0.08206 \text{ L} \bullet \text{atm}}{\text{mol} \bullet \text{K}} \times \dfrac{273 \text{ K}}{1 \text{ atm}} = 979.9 = 980 \text{ L}$

22.85   (a)   React an ionic nitride with D_2O, e.g.,

   $Mg_3N_2(s) + 6D_2O(l) \rightarrow 2ND_3(aq) + 3Mg(OD)_2(s)$

   (b)   React SO_3 with D_2O: $SO_3(g) + D_2O(l) \rightleftharpoons D_2SO_4(aq)$

   (c)   React Na_2O with D_2O: $Na_2O(s) + D_2O(l) \rightarrow 2NaOD(aq)$

   (d)   Dissolve $N_2O_5(g)$ in D_2O: $N_2O_5(g) + D_2O(l) \rightarrow 2DNO_3(aq)$

   (e)   React CaC_2 with D_2O: $CaC_2(s) + 2D_2O(l) \rightarrow Ca^{2+}(aq) + 2OD^-(aq) + C_2D_2(g)$

   (f)   Add NaCN to the D_2SO_4 solution prepared in (b):

   $NaCN(s) + D^+(aq) \overset{\Delta}{\rightarrow} DCN(aq) + Na^+(aq)$

   The DCN can be removed as gas from the reaction.

22.86 $BrO_3^-(aq) + XeF_2(aq) + H_2O(l) \rightarrow Xe(g) + 2HF(aq) + BrO_4^-(aq)$

22.87   Substances that will burn in O_2: SiH_4, CO, Mg.

   The others, SiO_2, CO_2, and CaO, have Si, C, and Ca in maximum oxidation states, so O_2 cannot act as an oxidizing agent.

22.88   (a) $SO_2(g) + H_2O(l) \rightleftharpoons H_2SO_3(aq)$

   (b) $Cl_2O_7(g) + H_2O(l) \rightleftharpoons 2HClO_4(aq)$

   (c) $Na_2O_2(s) + 2H_2O \rightarrow H_2O_2(aq) + 2NaOH(aq)$

   (d) $BaC_2(s) + 2H_2O(l) \rightarrow Ba^{2+}(aq) + 2OH^-(aq) + C_2H_2(g)$

(e) $2RbO_2(s) + 2H_2O(l) \rightarrow 2Rb^+(aq) + 2OH^-(aq) + O_2(g) + H_2O_2(aq)$

(f) $Mg_3N_2(s) + 6H_2O(l) \rightarrow 3Mg(OH)_2(s) + 2NH_3(g)$

(g) $NaH(s) + H_2O \rightarrow NaOH(aq) + H_2(g)$

22.89   (a) $H_2SO_4 - H_2O \rightarrow SO_3$

      (b) $2HClO_3 - H_2O \rightarrow Cl_2O_5$

      (c) $2HNO_2 - H_2O \rightarrow N_2O_3$

      (d) $H_2CO_3 - H_2O \rightarrow CO_2$

      (e) $2H_3PO_4 - 3H_2O \rightarrow P_2O_5$

22.90 $S_8(s) + 8Fe(s) \rightarrow 8FeS(s)$

      $S_8(s) + 16F_2(g) \rightarrow 8SF_4(g)$ or $S_8(s) + 24F_2(g) \rightarrow 8SF_6(g)$

      $S_8(s) + 8O_2(g) \rightarrow 8SO_2(g)$

      $S_8(s) + 8H_2(g) \rightarrow 8H_2S(g)$

Sulfur acts as an oxidizing agent in reactions with Fe or H_2 and as a reducing agent in reactions with O_2 or F_2. Incidentally, these reactions are often written using the symbol S rather than S_8 for sulfur.

22.91

$$S(g) + O_2(g) \rightarrow SO_2(g) \qquad \Delta H = -296.9 \text{ kJ} \qquad (1)$$
$$SO_2(g) + 1/2\, O_2(g) \rightarrow SO_3(g) \qquad \Delta H = -98.3 \text{ kJ} \qquad (2)$$
$$\underline{SO_3(g) + H_2O(l) \rightarrow H_2SO_4(aq) \qquad \Delta H = -130 \text{ kJ} \qquad (3)}$$
$$S(g) + 3/2\, O_2(g) + H_2O(l) \rightarrow H_2SO_4(aq) \qquad \Delta H = -525 \text{ kJ}$$

$$1 \text{ ton } H_2SO_4 \times \frac{2000 \text{ lb}}{\text{ton}} \times \frac{453.6 \text{ g}}{1 \text{ lb}} \times \frac{1 \text{ mol } H_2SO_4}{98.09 \text{ g}} \times \frac{-525 \text{ kJ}}{\text{mol } H_2SO_4}$$

$$= -4.86 \times 10^6 \text{ kJ of heat/ton } H_2SO_4$$

22.92   (a) $PO_4^{3-}, +5; NO_3^-, +5$

      (b)   The Lewis structure for NO_4^{3-} would be.

The formal charge on N is +1 and on each O atom is -1. The four electronegative oxygen atoms withdraw electron density, leaving the nitrogen deficient. Since N can form a maximum of four bonds, it cannot form a π bond with one or more of the O atoms to regain electron density, as the P atom in PO_4^{3-} does. Also, the short N—O distance would lead to a tight tetrahedron of O atoms subject to steric repulsion.

22.93   (a)   Although P_4, P_4O_6 and P_4O_{10} all have four P atoms in a tetrahedral arrangement, the bonding **between** P atoms and **by** P atoms is not the same in the three molecules. In P_4, the 4 P atoms are bound only to each other by P—P single bonds and strained bond angles of approximately 60°. In the two oxides, the 4 P atoms are directly bound to oxygen atoms, not to each other. Bonding by P atoms in P_4O_6 and P_4O_{10} is very similar. Each contains the P_4O_6 cage, formed by four P_3O_3 rings which share a P—O—P edge. Phosphorus bonding to oxygen maintains the overall P_4 tetrahedron but allows the P atoms to move away from each other so that the angle strain is relieved relative to molecular P_4. The P—O—P and O—P—O angles in both oxides are near the ideal 109°. In P_4O_6, each P is bound to 3 O atoms and has a lone pair completing its octet. In P_4O_{10}, the lone pair is replaced by a terminal O atom and each P is bound to 3 bridging and 1 terminal O atom.

(b)

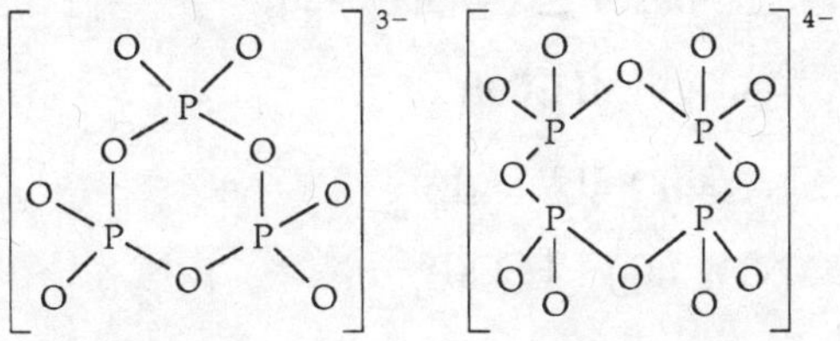

In both structures there are unshared pairs on all oxygens to give octets and the geometry around each P is approximately tetrahedral.

22.94　(a)

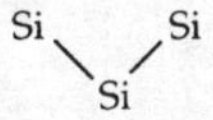

To complete their octets, the two terminal Si atoms each require three H atoms and the central Si requires two, for a total of 8 H atoms. The molecular formula is Si_3H_8.

(b)　　$Si_3H_8 + 5O_2 \rightarrow 3SiO_2 + 4H_2O$

22.95　$GeO_2(s) + C(s) \xrightarrow{\Delta} Ge(l) + CO_2(g)$

$Ge(l) + 2Cl_2(g) \rightarrow GeCl_4(l)$

$GeCl_4(l) + 2H_2O(l) \rightarrow GeO_2(s) + 4HCl(g)$

$GeO_2(s) + 2H_2(g) \rightarrow Ge(s) + 2H_2O(l)$

22.96　(a)　　　　$2[5e^- + MnO_4^-(aq) + 8H^+(aq) \rightarrow Mn^{2+}(aq) + 4H_2O(l)]$

$$\frac{5[H_2O_2(aq) \rightarrow O_2(g) + 2H^+(aq) + 2e^-]}{2MnO_4^-(aq) + 5H_2O_2(aq) + 6H^+(aq) \rightarrow 2Mn^{2+}(aq) + 5O_2(g) + 8H_2O(l)}$$

(b)　　　　　$2[Fe^{2+}(aq) \rightarrow Fe^{3+}(aq) + e^-]$

$$\frac{H_2O_2(aq) + 2H^+(aq) + 2e^- \rightarrow 2H_2O(l)}{2Fe^{2+}(aq) + H_2O_2(aq) + 2H^+(aq) \rightarrow 2Fe^{3+}(aq) + 2H_2O(l)}$$

(c)　　　　　$2I^-(aq) \rightarrow I_2(s) + 2e^-$

$$\frac{H_2O_2(aq) + 2H^+(aq) + 2e^- \rightarrow 2H_2O(l)}{2\,I^-(aq) + H_2O_2(aq) + 2H^+(aq) \rightarrow I_2(s) + 2H_2O(l)}$$

(d)
$$Cu(s) \rightarrow Cu^{2+}(aq) + 2e^-$$

$$\frac{H_2O_2(aq) + 2H^+(aq) + 2e^- \rightarrow 2H_2O(l)}{Cu(s) + H_2O_2(aq) + 2H^+(aq) \rightarrow Cu^{2+}(aq) + 2H_2O(l)}$$

(e)
$$2I^-(aq) \rightarrow I_2(s) + 2e^-$$

$$\frac{O_3(g) + H_2O(l) + 2e^- \rightarrow O_2(g) + 2OH^-(aq)}{2I^-(aq) + O_3(g) + H_2O(l) \rightarrow O_2(g) + I_2(s) + 2OH^-(aq)}$$

22.97 Assume that the reactions occur in acidic solution. The half-reaction for reduction of H_2O_2 is in all cases $H_2O_2(aq) + 2H^+(aq) + 2e^- \rightarrow 2H_2O(aq)$.

(a) $N_2H_4(aq) + 2H_2O_2(aq) \rightarrow N_2(g) + 4H_2O(l)$

(b) $SO_2(g) + H_2O_2(aq) \rightarrow SO_4^{2-}(aq) + 2H^+(aq)$

(c) $NO_2^-(aq) + H_2O_2(aq) \rightarrow NO_3^-(aq) + H_2O(l)$

(d) $H_2S(g) + H_2O_2(aq) \rightarrow S(s) + 2H_2O(l)$

(e)
$$2H^+(aq) + H_2O_2(aq) + 2e^- \rightarrow 2H_2O(l)$$

$$\frac{2[Fe^{2+}(aq) \rightarrow Fe^{3+}(aq) + e^-]}{2Fe^{2+}(aq) + H_2O_2(aq) + 2H^+(aq) \rightarrow 2Fe^{3+}(aq) + 2H_2O(l)}$$

22.98 (a) $Li_3N(s) + 3H_2O(l) \rightarrow 3Li^+(aq) + 3OH^-(aq) + NH_3(aq)$

(b) $NH_3(aq) + H_2O(l) \rightleftharpoons NH_4^+(aq) + OH^-(aq)$

(c) $3NO_2(g) + H_2O(l) \rightarrow NO(g) + 2H^+(aq) + 2NO_3^-(aq)$

(d) $2NO_2(g) \rightleftharpoons N_2O_4(g)$

(e) $4NH_3(g) + 5O_2(g) \xrightarrow{\text{catalyst}} 4NO(g) + 6H_2O(g)$

(f) $2CO(g) + O_2(g) \rightarrow 2CO_2(g)$

(g) $H_2CO_3(aq) \xrightarrow{\Delta} H_2O(g) + CO_2(g)$

(h) $Ni(s) + CO(g) \rightarrow NiO(s) + C(s)$

(i) $CS_2(g) + O_2(g) \rightarrow CO_2(g) + S_2(g)$

(j) $CaO(s) + SO_2(g) \rightarrow CaSO_3(s)$

(k) $2Na(s) + 2H_2O(l) \rightarrow 2NaOH(aq) + H_2(g)$

(l) $CH_4(g) + H_2O(g) \xrightarrow{\Delta} CO(g) + 3H_2(g)$

(m) $LiH(s) + H_2O(l) \rightarrow LiOH(aq) + H_2(g)$

(n) $Fe_2O_3(s) + 3H_2(g) \rightarrow 2Fe(s) + 3H_2O(g)$

22 Chemistry of the Nonmetals Solutions to Exercises

Integrative Exercises

22.99 From Appendix C, we need only ΔH_f° for $F(g)$, so that we can estimate ΔH for the process:

$$F_2(g) \rightarrow F(g) + F(g); \qquad \Delta H^\circ = 160 \text{ kJ}$$
$$\underline{XeF_2(g) \rightarrow Xe(g) + F_2(g) \qquad -\Delta H_f^\circ = 109 \text{ kJ}}$$
$$XeF_2(g) \rightarrow Xe(g) + 2F(g) \qquad \Delta H^\circ = 269 \text{ kJ}$$

The average Xe—F bond enthalpy is thus $269/2 = 134$ kJ. Similarly,

$$XeF_4(g) \rightarrow Xe(g) + 2F_2(g) \qquad -\Delta H_f^\circ = 218 \text{ kJ}$$
$$\underline{2F_2(g) \rightarrow 4F(g) \qquad\qquad \Delta H^\circ = 320 \text{ kJ}}$$
$$XeF_4(g) \rightarrow Xe(g) + 4F(g) \qquad \Delta H^\circ = 538 \text{ kJ}$$

Average Xe—F bond energy = $538/4 = 134$ kJ

$$XeF_6(g) \rightarrow Xe(g) + 3F_2(g) \qquad -\Delta H_f^\circ = 298 \text{ kJ}$$
$$\underline{3F_2(g) \rightarrow 6F(g) \qquad\qquad \Delta H^\circ = 480 \text{ kJ}}$$
$$XeF_6(g) \rightarrow Xe(g) + 6F(g) \qquad \Delta H^\circ = 778 \text{ kJ}$$

Average Xe—F bond energy = $778/6 = 130$ kJ

The average bond enthalpies are: XeF_2, 134 kJ; XeF_4, 134 kJ; XeF_6, 130 kJ. They are remarkably constant in the series.

22.100 (a) $H_2(g) + 1/2 \, O_2(g) \rightarrow H_2O(l); \Delta H = -285.83$ kJ

$CH_4(g) + 2O_2(g) \rightarrow CO_2(g) + 2H_2O(l)$

$\Delta H = 2(-285.83) - 393.5 - (-74.8) = -890.4$ kJ

(b) for H_2: $\dfrac{-285.83 \text{ kJ}}{1 \text{ mol } H_2} \times \dfrac{1 \text{ mol } H_2}{2.0159 \text{ g } H_2} = -141.79 \text{ kJ/g } H_2$

for CH_4: $\dfrac{-890.4 \text{ kJ}}{1 \text{ mol } CH_4} \times \dfrac{1 \text{ mol } CH_4}{16.043 \text{ g } CH_4} = -55.50 \text{ kJ/g } CH_4$

(c) Find the number of moles of gas that occupy 1 m^3 at STP:

$$n = \frac{1 \text{ atm } \times 1 \text{ m}^3}{273 \text{ K}} \times \frac{1 \text{ K} \cdot \text{mol}}{0.08206 \text{ L} \cdot \text{atm}} \times \left[\frac{100 \text{ cm}}{1 \text{ m}}\right]^3 \times \frac{1 \text{ L}}{10^3 \text{ cm}^3} = 44.64 \text{ mol}$$

for H_2: $\dfrac{-285.83 \text{ kJ}}{1 \text{ mol } H_2} \times \dfrac{44.64 \text{ mol } H_2}{1 \text{ m}^3 \, H_2} = 1.276 \times 10^4 \text{ kJ/m}^3 \, H_2$

for CH_4: $\dfrac{-890.4 \text{ kJ}}{1 \text{ mol } CH_4} \times \dfrac{44.64 \text{ mol } CH_4}{1 \text{ m}^3 \, CH_4} = 3.975 \times 10^4 \text{ kJ/m}^3 \, CH_4$

22.101 First calculate the molar solubility of Cl_2 in water.

$$n = \frac{1\,atm\,(0.310\,L)}{\frac{0.08206\,L \cdot atm}{1\,mol \cdot K} \times 273\,K} = 0.01384 = 0.0138\,mol\,Cl_2$$

$$M = \frac{0.01384\,mol}{0.100\,L} = 0.1384 = 0.138M$$

$[Cl^-] = [HOCl] = [H^+]$ Let this quantity = x. Then, $\dfrac{x^3}{(0.1384 - x)} = 4.7 \times 10^{-4}$

Assuming that x is small compared with 0.1384:

$x^3 = (0.1384)(4.7 \times 10^{-4}) = 6.504 \times 10^{-5}$; $x = 0.0402 = 0.040\,M$

We can correct the denominator using this value, to get a better estimate of x:

$$\frac{x^3}{0.1384 - 0.0402} = 4.7 \times 10^{-4};\ x = 0.0359 = 0.036\,M$$

One more round of approximation gives x = 0.0364 = 0.036 M. This is the equilibrium concentration of HClO.

22.102 (a) $2NH_4ClO_4(s) \xrightarrow{\Delta} N_2(g) + 2HCl(g) + 3H_2O(g) + 5/2\,O_2(g)$

$NH_4ClO_4(s) \xrightarrow{\Delta} 1/2\,N_2(g) + HCl(g) + 3/2\,H_2O(g) + 5/4\,O_2(g)$

(b) $\Delta H° = \Sigma\,\Delta H_f^\circ\,prod - \Sigma\,\Delta H\,react$

$\Delta H° = \Delta H_f^\circ\,HCl(g) + 3/2\,\Delta H_f^\circ\,H_2O(g) + 1/2\,\Delta H_f^\circ\,N_2(g) + 5/4\,\Delta H_f^\circ\,O_2(g) - \Delta H_f^\circ\,NH_4ClO_4\Delta H°$

$= -92.30\,kJ + 3/2(-241.82\,kJ) + 1/2\,(0\,kJ) + 5/4\,(0\,kJ) - (-295.8\,kJ)$

$= -159.2\,kJ/mol\,NH_4ClO_4$

(c) The aluminum reacts exothermically with $O_2(g)$ and HCl(g) produced in the decomposition, providing additional heat and thrust.

22.103 (a) $N_2H_4(g) + O_2(g) \rightarrow N_2(g) + 2H_2O(l)$

(b) $\Delta H° = \Delta H_f^\circ\,N_2(g) + 2\Delta H_f^\circ\,H_2O(l) - \Delta H_f^\circ\,N_2H_4(aq) - \Delta H_f^\circ\,O_2(g)$

$= 0 + 2(-285.83) - 95.40 - 0 = -667.06\,kJ$

(c) $\dfrac{9.1\,g\,O_2}{1 \times 10^6\,g\,H_2O} \times \dfrac{1.0\,g\,H_2O}{1\,mL\,H_2O} \times \dfrac{1000\,mL}{1L} \times 3.0 \times 10^4\,L = 273 = 2.7 \times 10^2\,g\,O_2$

$2.73 \times 10^2\,g\,O_2 \times \dfrac{1\,mol\,O_2}{32.00\,g\,O_2} \times \dfrac{1\,mol\,N_2H_4}{1\,mol\,O_2} \times \dfrac{32.05\,g\,N_2H_4}{1\,mol\,N_2H_4} = 2.7 \times 10^2\,g\,N_2H_4$

22.104 (a) $SO_2(g) + 2H_2S(aq) \rightarrow 3S(s) + 2H_2O(g)$ or, if we assume S_8 is the product,

$8SO_2(g) + 16H_2S(aq) \rightarrow 3S_8(s) + 16H_2O(g)$.

(b) Assume that all S in the coal becomes SO_2 upon combustion, so that

1 mol S (coal) = 1 mol SO_2.

$$2000 \text{ lb coal} \times \frac{0.035 \text{ lb S}}{1 \text{ lb coal}} \times \frac{453.6 \text{ g S}}{1 \text{ lb S}} \times \frac{1 \text{ mol S (coal)}}{32.07 \text{ g S}} \times \frac{1 \text{ mol SO}_2}{1 \text{ mol S (coal)}} \times \frac{2 \text{ mol H}_2\text{S}}{1 \text{ mol SO}_2}$$

$$= 1.98 \times 10^3 = 2.0 \times 10^3 \text{ mol H}_2\text{S}$$

$$V = \frac{1.98 \times 10^3 \text{ mol } (0.08206 \text{ L} \cdot \text{atm/mol} \cdot \text{K})(300 \text{ K})}{(740/760) \text{ atm}} = 5.01 \times 10^4 = 5.0 \times 10^4 \text{ L}$$

(c) $1.98 \times 10^3 \text{ mol H}_2\text{S} \times \dfrac{3 \text{ mol S}}{2 \text{ mol H}_2\text{S}} \times \dfrac{32.07 \text{ g S}}{1 \text{ mol S}} = 9.5 \times 10^4 \text{ g S}$

This is about 210 lb S per ton of coal combusted. (However, two-thirds of this comes from the H₂S, which presumably at some point was also obtained from coal.)

22.105   *Plan.* vol air → kg air → g H₂S → g FeS. Use the ideal-gas equation to change volume of air to mass of air, (assuming 1.00 atm, 298 K and an average molar mass (MM) for air of 29.0 g/mol. Use (20 mg H₂S/kg) air to find the mass of H₂S in the given mass of air. *Solve.*

$$V_{air} = 2.7 \text{ m} \times 4.3 \text{ m} \times 4.3 \text{ m} \times \frac{(100)^3 \text{ cm}^3}{1 \text{ m}^3} \times \frac{1 \text{ L}}{1000 \text{ cm}^3} = 4.9923 \times 10^4 = 5.0 \times 10^4 \text{ L}$$

$$g_{air} = \frac{PV \text{ MM}}{RT}; \text{ assume } P = 1.00 \text{ atm}, T = 298 \text{ K}, MM_{air} = 29.0 \text{ g/mol}$$

$$g_{air} = \frac{1.00 \text{ atm} \times 4.9923 \times 10^4 \text{ L} \times 29.0 \text{ g/mol}}{298 \text{ K}} \times \frac{\text{K} \cdot \text{mol}}{0.08206 \text{ L} \cdot \text{atm}} = 59,204 = 5.9 \times 10^4 \text{ g air}$$

$$5.9203 \times 10^4 \text{ g air} \times \frac{1 \text{ kg}}{1000 \text{ g}} \times \frac{20 \text{ mg H}_2\text{S}}{1 \text{ kg air}} \times \frac{1 \text{ g}}{1000 \text{ mg}} = 1.184 = 1.2 \text{ g H}_2\text{S}$$

$$\text{FeS(s)} + 2\text{HCl(aq)} \rightarrow \text{FeCl}_2\text{(aq)} + \text{H}_2\text{S(g)}$$

$$1.184 \text{ g H}_2 \times \frac{1 \text{ mol H}_2}{34.08 \text{ g H}_2\text{S}} \times \frac{1 \text{ mol FeS}}{1 \text{ mol H}_2\text{S}} \times \frac{87.91 \text{ g FeS}}{1 \text{ mol FeS}} = 3.054 = 3.1 \text{ g FeS}$$

22.106   The reactions can be written as follows:

$$\text{H}_2(g) + \text{X(std state)} \rightarrow \text{H}_2\text{X}(g) \qquad \Delta H_f^\circ$$
$$2\text{H}(g) \rightarrow \text{H}_2(g) \qquad \Delta H_f^\circ(\text{H}-\text{H})$$
$$\underline{\text{X}(g) \rightarrow \text{X(std state)} \qquad \Delta H_3}$$
$$\text{Add}: 2\text{H}(g) + \text{X}(g) \rightarrow \text{H}_2\text{X}(g) \qquad \Delta H = \Delta H_f^\circ + \Delta H_f^\circ (\text{H}-\text{H}) + \Delta H_3$$

These are all the necessary ΔH values. Thus,

Compound	ΔH	D H−X
H₂O	ΔH = −242 kJ − 436 kJ − 248 kJ = −926 kJ	463 kJ
H₂S	ΔH = −20 kJ − 436 kJ − 277 kJ = −733 kJ	367 kJ
H₂Se	ΔH = +30 kJ − 436 kJ − 227 kJ = −633 kJ	316 kJ
H₂Te	ΔH = +100 kJ − 436 kJ − 197 kJ = −533 kJ	266 kJ

The average H—X bond energy in each case is just half of ΔH. The H—X bond energy decreases steadily in the series. The origin of this effect is probably the increasing size of the orbital from X with which the hydrogen 1s orbital must overlap.

22.107   (a)   MnSi: more than one element, so not metallic; high melting, so not molecular; insoluble in water, so not ionic; therefore covalent network.

       (b) $MnSi(s) + HF(aq) \rightarrow SiH_4(g) + MnF_4(s)$

           Reduction of Mn(IV) to Mn(II) is unlikely, because F^- is an extremely weak reducing agent. E°_{red} for $F_2(g) + 2\,e^- \rightarrow 2F^-(aq) = 2.87$ V

22.108 $N_2H_5^+(aq) \rightarrow N_2(g) + 5H^+(aq) + 4e^-$ $E^\circ_{red} = -0.23$ V

       Reduction of the metal should occur when E°_{red} of the metal ion is more positive than about -0.15 V. This is the case for (b) Sn^{2+} (marginal), (c) Cu^{2+} and (d) Ag^+.

22.109 $(CH_3)_2N_2H_2(g) + 2N_2O_4(g) \rightarrow 2CO_2(g) + 3N_2(g) + 4H_2O(g)$

$$4.0 \text{ tons } (CH_3)_2N_2H_2 \times \frac{2000 \text{ lb}}{1 \text{ ton}} \times \frac{453.6 \text{ g}}{1 \text{ lb}} \times \frac{1 \text{ mol } (CH_3)_2N_2H_2}{60.10 \text{ g } (CH_3)_2N_2H_2}$$

$$\times \frac{2 \text{ mol } N_2O_4}{1 \text{ mol } (CH_3)_2N_2H_2} \times \frac{92.02 \text{ g } N_2O_4}{1 \text{ mol } N_2O_4} \times \frac{1 \text{ lb}}{453.6 \text{ g}} \times \frac{1 \text{ ton}}{2000 \text{ lb}} = 12 \text{ tons } N_2O_4$$

22.110   First write the balanced equation to give the number of moles of gaseous products per mole of hydrazine.

       (A) $(CH_3)_2NNH_2 + 2N_2O_4 \rightarrow 3N_2(g) + 4H_2O(g) + 2CO_2(g)$

       (B) $(CH_3)HNNH_2 + 5/4\,N_2O_4 \rightarrow 9/4\,N_2(g) + 3H_2O(g) + CO_2(g)$

       In case (A) there are nine moles gas per one mole $(CH_3)_2NNH_2$ plus two moles N_2O_4. The total mass of reactants is $60 + 2(92) = 244$ g. Thus, there are

$$\frac{9 \text{ mol gas}}{244 \text{ g reactants}} = \frac{0.0369 \text{ mol gas}}{1 \text{ g reactants}}$$

       In case (B) there are 6.25 moles of gaseous product per one mole $(CH_3)HNNH_2$ plus 1.25 moles N_2O_4. The total mass of this amount of reactants is $46.0 + 1.25(92.0) = 161$ g.

$$\frac{6.25 \text{ mol gas}}{161 \text{ g reactants}} = \frac{0.0388 \text{ mol gas}}{1 \text{ g reactants}}$$

       Thus the methylhydrazine (B) has marginally greater thrust.

22.111   (a) $HOOC-CH_2-COOH \xrightarrow{P_2O_5} C_3O_2 + 2H_2O$

       (b)   24 valence e^-, 12 e^- pair $\ddot{O}\!=\!C\!=\!C\!=\!C\!=\!\ddot{O}$

(c)   C=O, about 1.23 Å; C=C, 1.34 Å or less. Since consecutive C=C bonds require sp hybrid orbitals on C (as in allene, C_3H_4), we might expect the orbital overlap requirements of this bonding arrangement to require smaller than usual C=C distances.

(d)   The product has the formula $C_3H_4O_2$.

   28 valence e⁻, 14 e⁻ pr

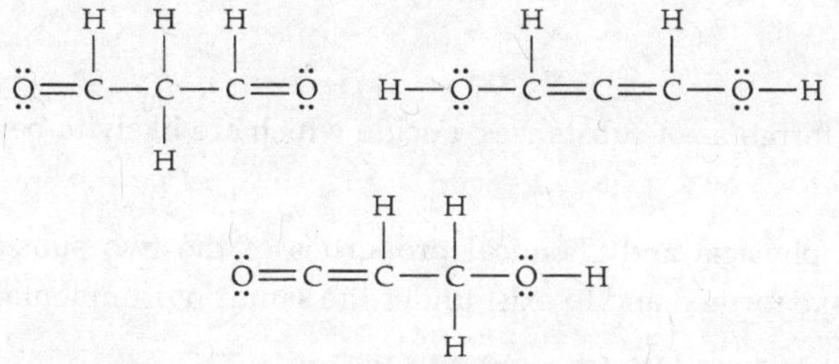

Two possibilities are shown above. The O=C=C group in the structure on the right is uncommon and less likely than the two symmetrical structures.

22.112   BN has the same number of valence electrons per formula unit as carbon. (Three from B, five from N, for an average of four per atom.) To the extent that we can neglect the difference in nuclear charges between B and N, we can think of BN as carbon-like. Indeed, BN takes on the same structural forms as carbon. However, because the B—N bonds are somewhat polar, BN is in fact even harder than diamond.

23 Metals and Metallurgy

Visualizing Concepts

23.1 *Analyze.* Given the formulas of substances, decide which are likely to be found together in nature.

Plan. Consider the physical and chemical properties of the two substances, whether they are likely to have formed and to exist under the same environmental conditions.

Solve. Most likely to be found together: (c) Al_2O_3 and Fe_2O_3.

The compounds have the same stoichiometry, both metals are in high oxidation states, so they could have been formed in the same environment. Both compounds are inert solids that would survive environmental degradation.

Reasons to eliminate other choices:

(a) Al^{3+} in highly oxidized state, Cu^+ in reduced state, unlikely to be produced by similar environmental conditions

(b) Second most likely pair, ionic radii similar. CaS could be water soluble, which would render this an unlikely pair to exist together.

(c) Na^+ has much larger ionic radius than Ag^+, unlikely to share same crystal lattice. NaCl is water soluble, AgCl is not. Unlikely to survive same environmental conditions.

(d) KO_2 is potassium superoxide, extremely reactive and unlikely to exist in nature.

23.2 The diagram indicates that the roasting of ZnS is exothermic. The roasting reaction, once under way, will increase the temperature of the oven. The thermodynamic characteristics of the reaction do not affect its rate. Although heating will decrease the value of the equilibrium constant for an exothermic reaction, it is required so that the roasting reaction occurs at a practical rate.

23.3 (a)

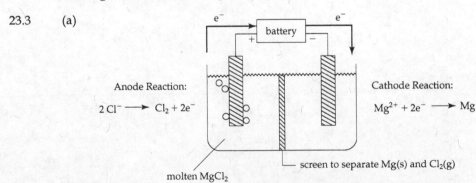

Anode Reaction:

$$2\,Cl^- \longrightarrow Cl_2 + 2e^-$$

Cathode Reaction:

$$Mg^{2+} + 2e^- \longrightarrow Mg$$

screen to separate Mg(s) and Cl_2(g)

molten $MgCl_2$

(b) $MgCl_2(l) \rightarrow Mg(s) + Cl_2(g)$ overall

              $2Cl^- \rightarrow Cl_2(g) + 2e^-$ anode (oxidation)

      $Mg^{2+} + 2e^- \rightarrow Mg(s)$ cathode (reduction)

(c)   Magnesium is an active metal. It must be separated from the $Cl_2(g)$ that is also formed by electrolysis (see screen in diagram) or $MgCl_2$ will form. The $Mg(s)$ also should not come in contact with air (O_2) or moisture (H_2O).

23.4   (a)   Cr: $[Ar]4s^1 3d^5$; Cd: $[Ar]4s^2 3d^{10}$

According to the molecular orbital model of metallic bonding, valence orbitals from each atom of a metallic solid combine to form a metal-metal bonding band and a metal-metal anti-bonding band. Chromium atoms have valence orbitals that are exactly half-filled. This means that the bonding band is full, and the anti-bonding band is empty; Cr has the strongest metal-metal bonding of the transition metals because it has maximum occupancy of the bonding band and minimum occupancy of the anti-bonding bond. Cadmium atoms have filled valence orbitals, resulting in filled bonding and anti-bonding bands. Metal-metal bonding in Cd is much weaker than that in Cr. The strength of metal-metal bonding in Cr causes it to be much harder than Cd.

(b)   The chemical equation that represents formation of gaseous elements is M(s) → M(g), where M(s) represents the metal in its standard state and M(g) is the gaseous metal. In order for this change to occur, metal-metal bonds in the solid must be broken as the atoms move apart to enter the gas phase. The stronger the metal-metal bonding of the element, the greater the magnitude of $\Delta H_f°$. The metals shown in the diagram have valence orbitals that are less than half full and the number of valence electrons increases from K to V. The metal-metal bonding band contains progressively more electrons, the strength of metallic bonding increases, and $\Delta H_f°$ increases.

23.5   The close-packed forms, both cubic and hexagonal, maximize the number of metal-metal contacts at 12. This maximizes the orbital interactions among metal atoms and increases delocalization. Substances with delocalized bonding have a special stability (Section 9.6) relative to those with localized bonding. The most common solid-state structure for metals is one of the close-packed forms because they result in maximum delocalization in metallic bonding.

23.6   Periodic properties are explained in terms of effective nuclear charge, Z_{eff}. Moving from left to right in a period, Z_{eff} increases because the increase in Z is not offset by a significant increase in shielding. Increasing Z_{eff} leads to increasing ionization energy and electronegativity, but decreasing atomic radius.

The chart shows a general decrease in magnitude of the property from K to Ge, so the property must be atomic radius.

23.7   (a)   Sc: $[Ar]4s^2 3d^1$; Sc^{3+}: $[Ar]$

Scandium has a total of 3 valence electrons; loss of all 3 electrons leads to the stable noble gas configuration of Ar.

(b) The oxidation state of Cr in CrO_4^{2-} is +6; that of Mn in MnO_4^- is +7. If we think of the bonding between the metal and oxygen as totally ionic, Cr(VI) and Mn(VII) have lost all their valence electrons and have the electron configuration of Ar. There are no unpaired electrons associated with either metal center, and the ions are diamagnetic.

Assuming perfectly covalent bonding between the metals and oxygen, we can draw the Lewis structures below.

CrO_4^{2-}, valence electrons: 6 from Cr + 24 from O + 2 from charge =
32 e^-, 16 e^- pairs

MnO_4^{2-}, valence electrons: 7 from Mn + 24 from O + 1 from charge =
32 e^-, 16 e^- pairs

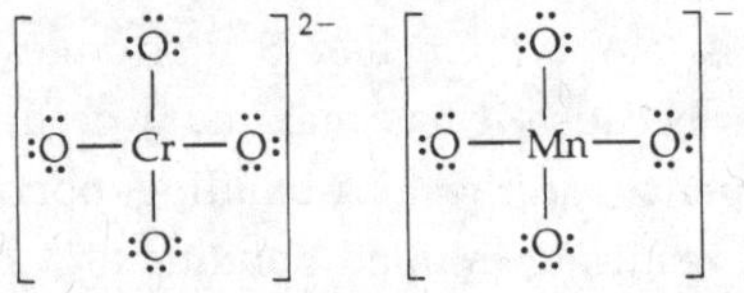

In both Lewis structures, all valence electrons are paired; both metals are in the diamagnetic state. Note that both extreme models predict a diamagnetic state for the metal centers.

(c) "Inorganic" ions are those not based on carbon. SO_4^{2-} and ClO_4^- have the same formula type and charge as CrO_4^{2-} and MnO_4^-. S has 6 valence electrons like Cr, and Cl has 7 like Mn. Both SO_4^- and ClO_4^- are tetrahedral, so we predict that CrO_4^{2-} and MnO_4^- are also tetrahedral. The Lewis structures in part (b) support this prediction.

23.8 (a) $Ni(s) + H_2SO_4(aq) \rightarrow NiSO_4(aq) + H_2(g)$

(b) $Ni(s) + Br_2(l) \rightarrow NiBr_2(s)$

Metallurgy

23.9 *Analyze/Plan.* Use Table 23.1 and other information in Section 23.1 to find important natural sources of Al and Fe. Use the rules for assigning oxidation numbers in Section 4.4 to determine the oxidation state of the metal in each natural source. *Solve.*

The important sources of iron are **hematite** (Fe_2O_3) and **magnetite** (Fe_3O_4). The major source of aluminum is **bauxite** ($Al_2O_3 \cdot xH_2O$). In ores, iron is present as the +3 ion, or in both the +2 and +3 states, as in magnetite. Aluminum is always present in the +3 oxidation state.

23.10 (a) +4

(b) $MnO_2(s) + 4H^+(aq) + 2e^- \rightarrow Mn^{2+}(aq) + 2H_2O(l)$ $\quad E^{\circ}_{red} = 1.23$ V

$Mn^{2+}(aq) + 2e^- \rightarrow Mn(s)$ $\quad E^{\circ}_{red} = -1.18$ V

Standard reduction potentials indicate that a very strong reducing agent is required to reduce the ore to Mn(s), at least if $Mn^{2+}(aq)$ is an intermediate product. According to Appendix E, only Group I and Group II metals (Li, Na,

Mg, etc.) are strong enough to reduce Mn^{2+} to Mn. In practice, $MnO_2(s)$ is reduced by coke in blast furnaces, into which MnO_2 is added to incorporate Mn into steel.

23.11   An ore consists of a little bit of the stuff we want, (chalcopyrite, $CuFeS_2$) and lots of other junk (gangue).

23.12   (a)   *Calcination* is heating an ore to decompose the mineral of interest into a simple solid and volatile compound. Calcination usually produces a metal oxide and a gas that is a nonmetal oxide.

(b)   *Leaching* is dissolving the mineral of interest to remove it from an ore. The solvent is usually water or an aqueous solution of acid, base or salt.

(c)   *Smelting* is heating an ore, often in a reducing atmosphere, to a very high temperature so that two immiscible liquid layers form. The layers are usually the molten metal or metals of interest and slag.

(d)   *Slag* is the unwanted layer of the smelting process. It contains molten silicate, aluminate, phosphate or fluoride compounds.

23.13   *Analyze/Plan.* Use principles of writing and balancing chemical equations from Chapter 3 to complete and balance the given reactions. The Δ above each arrow indicates that the reactions take place at elevated temperature. Information in Section 23.2 on *pyrometallurgy* will probably be useful.   *Solve.*

(a) $Cr_2O_3(s) + 6Na(l) \rightarrow 2Cr(s) + 3Na_2O(s)$

(b) $PbCO_3(s) \xrightarrow{\Delta} PbO(s) + CO_2(g)$

(c) $2CdS(s) + 3O_2(g) \xrightarrow{\Delta} 2CdO(s) + 2SO_2(g)$

(d) $ZnO(s) + CO(g) \xrightarrow{\Delta} Zn(l) + CO_2(g)$

23.14   (a) $2PbS(s) + 3O_2(s) \xrightarrow{\Delta} 2PbO(s) + 2SO_2(g)$

(b) $CoCO_3(s) \xrightarrow{\Delta} CoO(s) + CO_2(g)$

(c) $WO_3(s) + 3H_2(g) \xrightarrow{\Delta} W(s) + 3H_2O(g)$

(d) $VCl_3(g) + 3K(l) \rightarrow V(s) + 3KCl(s)$

(e) $3BaO(s) + P_2O_5(l) \rightarrow Ba_3(PO_4)_2(l)$

23.15   *Analyze/Plan.* Use information on *pyrometallurgy* in Section 23.2, along with principles of writing and balancing equations to provide the requested information.   *Solve.*

(a) $SO_3(g)$

(b) $CO(g)$ provides a reducing environment for the transformation of Pb^{2+} to Pb.

(c) $PbSO_4(s) \rightarrow PbO(s) + SO_3(g)$

$PbO(s) + CO(g) \rightarrow Pb(s) + CO_2(g)$

23.16   (a) $CoO(s) + CO_2(g)$

(b) $CO(g)$

(c) $CoCO_3(s) \xrightarrow{\Delta} CoO(s) + CO_2(g)$; $CoO(s) + CO(g) \rightarrow Co(s) + CO_2(g)$

23.17   *Analyze/Plan.* Use information on *pyrometallurgy* in Section 23.2, along with principles of writing and balancing equations to provide the requested information.   *Solve.*

The major reducing agent is CO, formed by partial oxidation of the coke (C) with which the furnace is charged.

$Fe_2O_3(s) + 3CO(g) \rightarrow 2Fe(l) + 3CO_2(g)$

$Fe_3O_4(s) + 4CO(g) \rightarrow 3Fe(l) + 4CO_2(g)$

23.18 $FeO(s) + H_2(g) \rightarrow Fe(s) + H_2O(g)$

$FeO(s) + CO(g) \rightarrow Fe(s) + CO_2(g)$

$Fe_2O_3(s) + 3H_2(g) \rightarrow 2Fe(s) + 3H_2O(g)$

$Fe_2O_3(s) + 3CO(g) \rightarrow 2Fe(s) + 3CO_2(g)$

23.19   *Analyze/Plan.* Use information on *pyrometallurgy* in Section 23.2, along with principles of writing and balancing equations to provide the requested information.   *Solve.*

(a)   *Air* serves primarily to oxidize coke (C) to CO, the main reducing agent in the blast furnace. This exothermic reaction also provides heat for the furnace.

$2C(s) + O_2(g) \rightarrow 2CO(g)$ $\Delta H = -221$ kJ

(b)   *Limestone*, $CaCO_3$, is the source of basic oxide for slag formation.

$CaCO_3(s) \xrightarrow{\Delta} CaO(s) + CO_2(g)$; $CaO(l) + SiO_2(l) \rightarrow CaSiO_3(l)$

(c)   *Coke* is the fuel for the blast furnace, and the source of CO, the major reducing agent in the furnace.

$2C(s) + O_2(g) \rightarrow 2CO(g)$; $4CO(g) + Fe_3O_4(s) \rightarrow 4CO_2(g) + 3Fe(l)$

(d)   *Water* acts as a source of hydrogen, and as a means of controlling temperature. (See Equation [23.8].) $C(s) + H_2O(g) \rightarrow CO(g) + H_2(g)$ $\Delta H = +131$ kJ

23.20   (a)   In the *converter*, oxidation of C, Si and metals by O_2 are exothermic reactions that raise the temperature.

(b) $2C(s) + O_2(g) \rightarrow 2CO(g)$; $S(s) + O_2(g) \rightarrow SO_2(g)$; $Si(s) + O_2(g) \rightarrow SiO_2(l)$

23.21   *Analyze/Plan.* Use information on the *electrometallurgy* of Cu as a model for describing how electrometallurgy can be employed to purify pure Co. Compare the ease of oxidation and reduction of cobalt with that of water.   *Solve.*

Cobalt could be purified by constructing an electrolysis cell in which the crude metal was the anode and a thin sheet of pure cobalt was the cathode. The electrolysis solution is aqueous with a soluble cobalt salt such as $CoSO_4 \cdot 7H_2O$ serving as the electrolyte. (Other soluble salts with anions that do not participate in the cell reactions could be used.) Anode reaction: $Co(s) \rightarrow Co^{2+}(aq) + 2e^-$; cathode reaction: $Co^{2+}(aq) + 2e^- \rightarrow Co(s)$. Although $E°$ for reduction of $Co^{2+}(aq)$ is slightly negative (-0.277 V), it is less than the standard reduction potential for $H_2O(l)$, -0.83 V.

23.22 $SnO_2(s) + C(s) \xrightarrow{\Delta} Sn(l) + CO_2(g)$

           $Sn(s) \rightarrow Sn^{2+}(aq) + 2e^-$ (anode)

           $Sn^{2+}(aq) + 2e^- \rightarrow Sn(s)$ (cathode)

Metals and Alloys

23.23   *Analyze/Plan.* Compare the bonding characteristics of metallic sodium and ionic sodium chloride and use them to explain the difference in malleability. *Solve.*

        Sodium is metallic; each atom is bonded to many nearest neighbor atoms by metallic bonding involving just one electron per atom, and delocalized over the entire three-dimensional structure. When sodium metal is distorted, each atom continues to have bonding interactions with many nearest neighbors. In NaCl the ionic forces are strong, and the arrangement of ions in the solid is very regular. When subjected to physical stress, the three-dimensional lattice tends to cleave along the very regular lattice planes, rather than undergo the large distortions characteristic of metals.

23.24   Since silicon has the same crystal structure as diamond, it is a covalent-network solid with its bonding electrons localized between Si atoms. Since there is no significant delocalization in Si, it is not likely to have the metallic properties of malleability, ductility and high electrical and thermal conductivity. It is likely to be hard and high-melting like other covalent-network solids.

23.25   *Analyze/Plan.* Apply the description of the electron-sea model of metallic bonding given in Section 23.5 to the conductivity of metals.   *Solve.*

        In the electron-sea model for metallic bonding, valence electrons move about the three-dimensional metallic lattice, while the metal atoms maintain regular lattice positions.

        Under the influence of an applied potential the electrons can move throughout the structure, giving rise to high electrical conductivity. The mobility of the electrons facilitates the transfer of kinetic energy and leads to high thermal conductivity.

23.26   (a)   Cr: $[Ar]4s^1 3d^5$, Z = 24; Se: $[Ar]4s^2 3d^{10} 4p^4$, Z = 34

              Both elements have the [Ar] core configuration and both have six valence electrons. The orbital locations of the six valence electrons are different in the two elements, because Se has more total electrons.

        (b)   Different Z and Z_{eff} for the two elements, and the different orbital locations of the valence electrons, are the main factors that lead to the differences in properties. In Cr, the 4s and 3d electrons are the valence electrons. Its Z and Z_{eff} are smaller than those of Se and it is not likely to gain enough electrons to achieve a noble-gas configuration. Thus, Cr loses electrons when it forms ions, acting like a metal. Se is in the same row of the periodic table as Cr, but its 3d subshell is filled, so its valence electrons are in 4s and 4p. Because Se has a larger Z and Z_{eff}, it is more likely to hold its own valence electrons and gain other electrons when it forms ions. That Se needs only two additional electrons to achieve the noble-gas configuration of Kr is also a driving force for it gaining electrons when it forms ions, acting like a nonmetal.

23.27　*Analyze/Plan.* Consider trends in atomic mass and volume of the elements listed to explain the variation in density. *Solve.*

Moving left to right in the period, atomic mass and Z_{eff} increase. The increase in Z_{eff} leads to smaller bonding atomic radii and thus atomic volume. Mass increases, volume decreases, and density increases in the series.

The variation in densities reflects shorter metal-metal bond distances. These shorter distances suggest that the extent of metal-metal bonding increases in the series. This is consistent with greater occupancy of the bonding band as the number of valence electrons increases up to 6.

23.28　Moving across the fifth period from Y to Mo, the melting points of the metals increase. The number of valence electrons also increases, from 3 for Y to 6 for Mo. More valence electrons (up to 6) mean increased occupancy of the bonding molecular orbital band, and increased strength of metallic bonding. Melting requires that atoms are moving relative to each other. Stronger metallic bonding requires more energy to break bonds and mobilize atoms, resulting in higher melting points from Y to Mo.

23.29　*Analyze/Plan.* Consider the definition of ductility, as well as the discussion of band theory in Chapter 12. *Solve.*

Ductility is the property related to the ease with which a solid can be drawn into a wire. Basically, the softer the solid the more ductile it is. The more rigid the solid, the less ductile it is. For metals, ductility decreases as the number of bonding electrons per atom increases, producing a stiffer lattice less susceptible to distortion.

(a)　K is more ductile. Cr, with 6 valence electrons, has a filled bonding band, strong metal-metal interactions, and a rigid lattice. This predicts high hardness and low ductility.

(b)　Zn is more ductile. Si is a covalent-network solid with all valence electrons localized in bonds between Si atoms. Covalent-network substances are high-melting, hard, and not particularly ductile.

23.30　According to band theory, an *insulator* has a completely filled valence band and a large energy gap between the valence band and the nearest empty band; electrons are localized within the lattice. A *conductor* must have a partially filled energy band; a small excitation will promote electrons to previously empty levels within the band and allow them to move freely throughout the lattice, giving rise to the property of conduction. A *semiconductor* has a filled valence band, but the gap between the filled and empty bands is small enough to jump to the empty conduction band. The presence of an impurity may also place an electron in an otherwise empty band (producing an n-type semiconductor), or create a vacancy in an otherwise full band (producing a p-type semiconductor), providing a mechanism for conduction.

23.31　*Analyze/Plan.* Recall the diamond and closest-packed structures described in Section 11.7. Use these structures to draw conclusions about Sn–Sn distance and electrical conductivity in the two allotropes. *Solve.*

White tin, with a characteristic metallic structure, is expected to be more metallic in character. The electrical conductivity of the white allotropic form is higher because the valence electrons are shared with 12 nearest neighbors rather than being localized in four bonds to nearest neighbors as in gray tin. The Sn–Sn distance should be longer in white tin; there are only four valence electrons from each atom, and 12 nearest neighbors. The **average** tin–tin bond order can, therefore, be only about 1/3, whereas in gray tin the bond order is one. Gray tin, with the higher bond order, has a shorter Sn–Sn distance, 2.81 Å. The bond length in white tin, with the lower bond order, is 3.02 Å.

23.32. Electrical conductivity is related to the extent of valence electron delocalization in the material.

In the hexagonal close-packed structure of titanium, each Ti atom has twelve nearest neighbors. The four valence electrons of a Ti atom are delocalized over bonding interactions with twelve neighbors. In the diamond structure of silicon, each Si atom has four nearest neighbors and four valence electrons. These four valence electrons are essentially localized in four covalent (sigma) bonds to the four nearest neighbors. The much more extensive electron delocalization in Ti leads to its significantly greater electrical conductivity.

23.33 *Analyze/Plan.* Use information in Section 23.6 to define *alloy*, and compare the various types of alloys.   *Solve.*

An *alloy* contains atoms of more than one element and has the properties of a metal. *Solution alloys* are homogeneous mixtures with different kinds of atoms dispersed randomly and uniformly. In *heterogeneous alloys* the components (elements or compounds) are not evenly dispersed and their properties depend not only on composition but methods of preparation. In an *intermetallic compound* the component elements have interacted to form a compound substance, for example, Cu_3As. As with more familiar compounds, these are homogeneous and have definite composition and properties.

23.34 Substitutional and interstitial alloys are both solution alloys. In a *substitutional* alloy, the atoms of the "solute" take positions normally occupied by the "solvent." Substitutional alloys tend to form when solute and solvent atoms are of comparable size and have similar bonding characteristics. In an *interstitial* alloy, the atoms of the "solute" occupy the holes or interstitial positions between "solvent" atoms. Solute atoms are necessarily much smaller than solvent atoms.

Transition Metals

23.35 *Analyze/Plan.* Consider the definitions of the properties listed (Chapter 7 and Chapter 23) and whether they refer to single, isolated atoms or bulk material.   *Solve.*

Of the properties listed, (b) the first ionization energy and (f) electron affinity are characteristic of isolated atoms. Electrical conductivity (a), atomic radius (c), melting point (d), and heat of vaporization (e) are properties of the bulk metal. Although it seems that atomic radius would be a property of isolated atoms, it can only be measured in bulk samples.

23.36 (b) NiCo alloy and (c) W will have metallic properties. The lattices of these substances are composed of neutral metal atoms. Delocalization of valence electrons via the MO model of metallic bonding produces metallic properties.

(d) Ge is a metalloid, not a metal. (a) $TiCl_4$ is an ionic compound and (e) Hg_2^{2+} is a metal ion. In ions and ionic compounds, electrons are localized on the individual ions, precluding metallic properties.

23.37   *Analyze/Plan.* Examine the electron configurations, Z and Z_{eff}, of the two elements to account for their similar atomic radii.

Solve. Zr: $[Kr]5s^2 4d^2$, Z = 40; Hf: $[Xe]6s^2 4f^{14} 5d^2$, Z = 72

Moving down a family of the periodic chart, atomic size increases because the valence electrons are in a higher principle quantum level (and thus further from the nucleus) and are more effectively shielded from the nuclear charge by a larger core electron cloud. However, the build-up in Z that accompanies the filling of the 4f orbitals causes the valence electrons in Hf to experience a much greater relative nuclear charge than those in La, its neighbor to the left. This increase in Z offsets the usual effect of the increase in *n* value of the valence electrons and the radii of Zr and Hf atoms are similar.

23.38   *Analyze/Plan.* Define lanthanide contraction (Section 23.7). Based on the definition, list properties related to atomic radius.   *Solve.*

The *lanthanide contraction* is the name given to the decrease in atomic size due to the build-up in effective nuclear charge as we move through the lanthanides (elements 58–71) and beyond them. This effect offsets the expected increase in atomic size going from the second to the third transition series. The lanthanide contraction affects size-related properties such as ionization energy, electron affinity, and density.

23.39   *Analyze/Plan.* Use Figure 23.20 to determine the highest oxidation state of each metal. Write formulas of the metal fluorides, given that fluoride ion is F^-.   *Solve.*

(a) ScF_3 (b) CoF (c) ZnF_2

(d) MoF_6 (The oxidation states of Mo are similar to those of Cr.)

23.40   (a) CdO (b) TiO (c) Nb_2O_5 (d) NiO_2

23.41   *Analyze/Plan.* Consider the electron configurations of Cr and Al to rationize observed oxidation states.   *Solve.*

Chromium, $[Ar]4s^1 3d^5$, has six valence-shell electrons, some or all of which can be involved in bonding, leading to multiple stable oxidation states. By contrast, aluminum, $[Ne]3s^2 3p^1$, has only three valence electrons which are all lost or shared during bonding, producing the +3 state exclusively.

23.42   V: $[Ar]4s^2 3d^3$, (V^{2+}: $[Ar]3d^3$); Sc: $[Ar]4s^2 3d^1$; (Sc^{2+}: $[Ar]3d^1$)

V has a slightly larger Z (23) and Z_{eff} than Sc (Z > 21), so the 3d electrons in V are more tightly held than those in Sc. Also, losing the lone 3d electron in Sc^{2+} leads to the stable noble-gas configuration of [Ar] for Sc^{3+}.

23.43   *Analyze/Plan.* Write electron configurations for the neutral elements and their positive ions recalling that valence electrons are last in order of descending *n*-value.   *Solve.*

(a) Cr^{3+}: $[Ar]3d^3$ (b) Au^{3+}: $[Xe]4f^{14} 5d^8$ (c) Ru^{2+}: $[Kr]4d^6$

(d) Cu^+: $[Ar]3d^{10}$ (e) Mn^{4+}: $[Ar]3d^3$ (f) Ir^+: $[Xe]4f^{14} 5d^8$

23.44   (a) Ti^{2+}: $[Ar]3d^2$ (b) Co^{3+}: $[Ar]3d^6$ (c) Pd^{2+}: $[Kr]4d^8$

        (d) Mo^{3+}: $[Kr]4d^3$ (e) Ru^{3+}: $[Kr]4d^5$ (f) Ni^{4+}: $[Ar]3d^6$

23.45   *Analyze/Plan.* Oxidation is loss of electrons. Which periodic trend determines how tightly a valence electron is held in a particular atom or ion?   *Solve.*

        Ease of oxidation decreases from left to right across a period (owing to increasing effective nuclear charge); Ti^{2+} should be more easily oxidized than Ni^{2+}.

23.46   The stronger reducing agent is more easily oxidized; Cr^{2+} is more easily oxidized (see Solution 23.45), so it is the stronger reducing agent.

23.47   *Analyze/Plan.* Consider Equation [23.26] regarding the oxidation states of iron. *Solve.* Fe^{2+} is a reducing agent that is readily oxidized to Fe^{3+} in the presence of O_2 from air.

23.48   (a)   Chromate ion, $CrO_4{}^{2-}$, is bright yellow. Dichromate, $Cr_2O_7{}^{2-}$, is orange.

        (b) $Cr_2O_7{}^{2-}$ is more stable in acid solution than $CrO_4{}^{2-}$.

        (c)   Their interconversion in solution involves the acid-base equilibrium

$$2CrO_4{}^{2-}(aq) + 2H^+(aq) \rightleftharpoons Cr_2O_7{}^{2-}(aq) + H_2O(l).$$

23.49   *Analyze/Plan.* Consider information on the descriptive chemistry of iron in Section 23.8. *Solve.*

        (a) $Fe(s) + 2HCl(aq) \rightarrow FeCl_2(aq) + H_2(g)$

        (b) $Fe(s) + 4HNO_3(aq) \rightarrow Fe(NO_3)_3(aq) + NO(g) + 2H_2O(l)$

        (See net ionic equation, Equation [23.28].) In concentrated nitric acid, the reaction can produce $NO_2(g)$ according to the reaction:

        $Fe(s) + 6HNO_3(aq) \rightarrow Fe(NO_3)_3(aq) + 3NO_2(g) + 3H_2O(l)$

23.50   (a) $MnO_2(s) + 4HCl(aq) \rightarrow MnCl_2(aq) + Cl_2(g) + 2H_2O(l)$

        (b)   Yes. MnO_2 is the oxidizing agent; HCl is the reducing agent.

23.51   *Analyze/Plan.* Consider the definitions of paramagnetic and diamagnetic.   *Solve.*

        The unpaired electrons in a *paramagnetic* material cause it to be weakly attracted into a magnetic field. A *diamagnetic* material, where all electrons are paired, is very weakly repelled by a magnetic field.

23.52   (a)   *Ferromagnetic* materials can form "permanent" magnets, whereas *paramagnetic* materials cannot.

        (b)   In order for a substance to be ferromagnetic, the magnetic moments on sites throughout the lattice must interact with one another. That is, the sites must be physically close and overlap of orbitals must enable the individual magnetic sites to couple forming a much larger magnetic moment throughout the solid. Because of these interactions, the sustained existence of the magnetic moment does not require application of an external magnetic field.

        (c)   No. Other metals (Co and Ni, for example), alloys and some oxides (CrO_2) are also ferromagnetic.

Additional Exercises

23.53 $PbS(s) + O_2(g) \rightarrow Pb(l) + SO_2(g)$

Regardless of the metal of interest, $SO_2(g)$ is a product of roasting sulfide ores. In an oxygen rich environment, $SO_2(g)$ is oxidized to $SO_3(g)$, which dissolves in $H_2O(l)$ to form sulfuric acid, $H_2SO_4(aq)$. Because of its corrosive nature, $SO_2(g)$ is a dangerous environmental pollutant (Section 18.4) and cannot be freely released into the atmosphere. A sulfuric acid plant near a roasting plant would provide a means for disposing of $SO_2(g)$ that would also generate a profit.

23.54 Al^{3+}, Mg^{2+}, and Na^+ all have large negative reduction potentials (Al, Mg, and Na are very active metals). A substance with a more negative reduction potential would have to be used to chemically reduce them. All such substances are more expensive and difficult to obtain than Al, Mg, and Na. Electrolysis is thus the most cost-efficient way to reduce Al^{3+}, Mg^{2+}, and Na^+ to their metallic states.

23.55 $CO(g)$: $Pb(s)$; $H_2(g)$: $Fe(s)$; $Zn(s)$: $Au(s)$

23.56   (a) $2VCl_3(s) + O_2(g) \rightarrow 2VOCl_3(s)$

         (b) $Nb_2O_5(s) + 5H_2(g) \rightarrow 2Nb(s) + 5H_2O(l)$

         (c) $2Fe^{3+}(aq) + Zn(s) \rightarrow 2Fe^{2+}(aq) + Zn^{2+}(aq)$

         (d) $NbCl_5(s) + 3H_2O(l) \rightarrow HNbO_3(s) + 5HCl(aq)$

23.57   (a) $NiO(s) + 2H^+(aq) \rightarrow Ni^{2+}(aq) + H_2O(l)$

         (b)   The simple answer is that the solid is subjected to acid hydrolysis:

         $CuCo_2S_4(s) + 8H^+(aq) \rightarrow Cu^{2+}(aq) + 2Co^{3+}(aq) + 4H_2S(g)$

         However, in the absence of a strong complexing ligand, Co^{3+} is not stable in water. It oxidizes water according to the following reaction:

         $4Co^{3+}(aq) + 2H_2O(l) \rightarrow 4Co^{2+}(aq) + O_2(g) + 4H^+(aq)$

         (c) $TiO_2(s) + C(s) + 2Cl_2(g) \rightarrow TiCl_4(g) + CO_2(g)$

         (d)   In this reaction O_2 is reduced and sulfide is oxidized. Writing the sulfur product as S_8, the balanced equation is:

         $8ZnS(s) + 4O_2(g) + 16H^+(aq) \rightarrow 8Zn^{2+}(aq) + S_8(s) + 8H_2O(l)$

23.58   Because selenium and tellurium are both nonmetals, we expect them to be difficult to oxidize. Thus, both Se and Te are likely to accumulate as the free elements in the so-called anode slime, along with noble metals that are not oxidized.

23.59   All transition metals have the generic electron configuration $ns^2(n–1)d^x$. Regardless of the number of d electrons, each transition metal has 2 ns valence electrons that are the first electrons lost when metal ions are formed. Thus, almost every transition metal has a stable +2 oxidation state.

After the 2 ns electrons are lost, a varying number of $(n–1)d$ electrons can be lost, depending on the identity of the transition metal. The availability of different numbers of d electrons leads to a wide variety of accessible oxidation states for the transition metals.

23.60   Assuming that SO_2 and N_2 are the nonmetallic products, two half-reactions can be written:

$$5[MoS_2(s) + 7H_2O(l) \rightarrow MoO_3(s) + 2SO_2(g) + 14H^+(aq) + 14e^-]$$

$$\underline{7[12H^+(aq) + 2NO_3^-(aq) + 10e^- \rightarrow N_2(g) + 6H_2O(l)]}$$

$$5MoS_2(s) + 14H^+(aq) + 14NO_3^-(aq) \rightarrow 5MoO_3(s) + 10SO_2(g) + 7N_2(g) + 7H_2O(l)$$

$$MoO_3(s) + 2NH_3(aq) + H_2O(l) \rightarrow (NH_4)_2MoO_4(s)$$

$$(NH_4)_2MoO_4(s) \xrightarrow{\Delta} 2NH_3(g) + H_2O(g) + MoO_3(s)$$

$$MoO_3(s) + 3H_2(g) \xrightarrow{\Delta} Mo(s) + 3H_2O(g)$$

23.61   (a)   The very low boiling point (130°C) of OsO_4 indicates that it is a covalent compound. An ionic compound would have a much higher boiling point. For a covalent molecule, we can apply VSEPR to predict that OsO_4 will be tetrahedral. Any molecule with four attached groups will be tetrahedral as long as there are no lone pairs on the central atom. Metal atoms like Os are unlikely to have lone pairs, so a tetrahedral structure for OsO_4 is likely.

(b)   The oxidation state of Os in OsO_4 is +8. According to Figure 23.20, +8 is not a common oxidation state for Fe. Also, in a reversal of the typical trend for representative elements, the electronegativity of Fe is less than that of Os (see Figure 8.6). The electronegativity difference between Fe and O is greater than that between Os and O, so the iron oxides are more likely to be ionic.

23.62   Antimony is a metalloid with 5 valence electrons ($5s^2 5p^3$). It is not a semiconductor like graphite, Si and Ge, group 4A elements with 4 valence electrons. As a poor electrical conductor (insulator), it must have a large band gap between the valence band (bonding band) and the conduction band (antibonding band). Valence electrons are localized in covalent Sb–Sb bonds. Niobium is a metal with 5 valence electrons ($5s^2 4d^3$). As a good electrical conductor, it has a very small or zero band gap. Its valence electrons are delocalized over several (more than 5) Nb–Nb close contacts. The extent of valence electron delocalization determines the electrical conductivity of a solid.

23.63   The metallic properties of malleability, ductility, and high electrical and thermal conductivity are results of the delocalization of valence electrons throughout the lattice. Delocalization occurs because metal atom valence orbitals of nearest-neighbor atoms interact to produce nearly continuous molecular orbital energy bands. When C atoms are introduced into the metal lattice, their valence orbitals do not have the same energies as metal orbitals, and their interaction is different. This causes a discontinuity in the band structure and limits delocalization of electrons. The properties of the carbon-infused metal begin to resemble those of a covalent-network lattice with localized electrons (Solution 23.24). The substance is harder and less conductive than the pure metal.

23.64   The equilibrium of interest is $[ZnL_4] \rightleftharpoons Zn^{2+}(aq) + 4L$ $K = 1/K_f$
Since $Zn(H_2O)_4^{2+}$ is $Zn^{2+}(aq)$, its reduction potential is –0.763 V. As the stability (K_f) of the complexes increases, K decreases. Since E° is directly proportional to log K (Equation [20.16]), E° values for the complexes will become more negative as K_f increases.

23.65   (a) Nb^{5+}: [Ar]; diamagnetic, no unpaired electrons

        (b) Cr^{2+}: $[Ar]3d^4$ paramagnetic, unpaired electrons

        (c) Cu^+: $[Ar]3d^{10}$; diamagnetic, no unpaired electrons

        (d) Ru^{8+}: [Ar]; diamagnetic, no unpaired electrons

        (e) Ni^{2+}: $[Ar]3d^8$; paramagnetic, unpaired electrons

23.66   In a ferromagnetic solid, the magnetic centers are coupled such that the spins of all unpaired electrons are parallel. As the temperature of the solid increases, the average kinetic energy of the atoms increases until the energy of motion overcomes the force aligning the electron spins. The substance becomes paramagnetic; it still has unpaired electrons, but their spins are no longer aligned.

23.67   (a) $Mn(s) + 2HNO_3(aq) \rightarrow Mn(NO_3)_2(aq) + H_2(g)$

        (b) $Mn(NO_3)_2(s) \overset{\Delta}{\rightarrow} MnO_2(s) + 2NO_2(g)$

        (c) $3MnO_2(s) \overset{\Delta}{\rightarrow} Mn_3O_4(s) + O_2(g)$

        (d) $2MnCl_2(s) + 9F_2(g) \rightarrow 2MnF_3(s) + 4ClF_3(g)$

23.68   (a)   Nothing. As noted in Section 20.8, a basic environment (OH^-) inhibits oxidation of Fe^{2+} to Fe^{3+}, even in the presence of $O_2(g)$.

        (b) $Cu(NO_3)_2(aq) + 2KOH(aq) \rightarrow Cu(OH)_2(s) + 2KNO_3(aq)$

             $Cu(OH)_2(s)$ precipitates. Cu^{2+} forms a soluble complex ion with $NH_3(aq)$, but not $OH^-(aq)$.

        (c)   The color of the solution changes from orange ($Cr_2O_7{}^{2-}$) to yellow ($CrO_4{}^{2-}$). The equilibrium is

             $Cr_2O_7{}^{2-}(aq) + H_2O(l) \rightleftharpoons 2CrO_4{}^{2-}(aq) + 2H^+(aq)$

             As $OH^-(aq)$ is added, it reacts with and removes $H^+(aq)$ from solution, shifting the equilibrium to the right in favor of the yellow $CrO_4{}^{2-}$.

23.69   (a) $2NiS(s) + 3O_2(g) \rightarrow 2NiO(s) + 2SO_2(g)$

        (b) $2C(s) + O_2(g) \rightarrow 2CO(g)$; $C(s) + H_2O(g) \rightarrow CO(g) + H_2(g)$

             $NiO(s) + CO(g) \rightarrow Ni(s) + CO_2(g)$; $NiO(s) + H_2(g) \rightarrow Ni(s) + H_2O(g)$

        (c) $Ni(s) + 2HCl(aq) \rightarrow NiCl_2(aq) + H_2(g)$

        (d) $NiCl_2(aq) + 2NaOH(aq) \rightarrow Ni(OH)_2(s) + 2NaCl(aq)$

        (e) $Ni(OH)_2(s) \overset{\Delta}{\rightarrow} NiO(s) + H_2O(g)$

23.70   (a)   insulator   (b)   semiconductor   (c)   metallic conductor

        (d)   metallic conductor   (e)   insulator   (f)   metallic conductor

Integrative Exercises

23.71 (a)   Calculate mass Cu_2S and FeS, then mass SO_2 from each.

3.3×10^6 kg sample $\times 0.27 = 8.91 \times 10^5 = 8.9 \times 10^5$ kg $= 8.9 \times 10^8$ g Cu_2S

3.3×10^6 kg sample $\times 0.13 = 4.29 \times 10^5 = 4.3 \times 10^5$ kg $= 4.3 \times 10^8$ g FeS

8.91×10^8 g $Cu_2S \times \dfrac{1 \text{ mol } Cu_2S}{159.1 \text{ g } Cu_2S} \times \dfrac{1 \text{ mol } SO_2}{1 \text{ mol } Cu_2S} \times \dfrac{64.07 \text{ g } SO_2}{1 \text{ mol } SO_2} = 3.588 \times 10^8$

$= 3.6 \times 10^8$ g SO_2

4.29×10^8 g FeS $\times \dfrac{1 \text{ mol FeS}}{87.9 \text{ g FeS}} \times \dfrac{1 \text{ mol } SO_2}{1 \text{ mol FeS}} \times \dfrac{64.07 \text{ g } SO_2}{1 \text{ mol } SO_2} = 3.127 \times 10^8$ g

$= 3.1 \times 10^8$ g SO_2

g $SO_2 = 3.588 \times 10^8 + 3.127 \times 10^8 = 6.715 \times 10^8 = 6.7 \times 10^8$ g SO_2

(b)   Calculate mol Cu, mol Fe and mole ratio Cu: Fe.

8.91×10^8 g $Cu_2S \times \dfrac{1 \text{ mol } Cu_2S}{159.1 \text{ g } Cu_2S} \times \dfrac{2 \text{ mol Cu}}{1 \text{ mol } Cu_2S} = 1.12 \times 10^7 = 1.1 \times 10^7$ mol Cu

4.29×10^8 g FeS $\times \dfrac{1 \text{ mol FeS}}{87.9 \text{ g FeS}} \times \dfrac{1 \text{ mol Fe}}{1 \text{ mol FeS}} = 4.88 \times 10^6 = 4.9 \times 10^6$ mol Fe

1.12×10^7 mol Cu$/4.88 \times 10^6$ mol Fe $= 2.3$ mol Cu/mol Fe

(c)   The oxidizing environment of the converter is likely to produce CuO and Fe_2O_3.

(d) $Cu_2S(s) + 2O_2(g) \rightarrow 2CuO(s) + SO_2(g)$

$4FeS(s) + 7O_2(g) \rightarrow 2Fe_2O_3(s) + 4SO_2(g)$

23.72 Recall from the discussion in Chapter 13 that like substances tend to be soluble in one another, whereas unlike substances do not. Molten metal consists of atoms that continue to be bound to one another by metallic bonding, even though the substance is liquid. In a slag, on the other hand, the attractive forces are those between ions. The slag phase is a highly polar, ionic medium, whereas the metallic phase is nonpolar, and the attractive interactions are due to metallic bond formation. There is little driving force for materials with such different characteristics to dissolve in one another.

23.73 The first equation indicates that one mole Ni^{2+} is formed from passage of two moles of electrons, and the second equation indicates the same thing. Thus, the simple ratio (1 mol Ni^{2+}/2F).

67 A $\times 11.0$ hr $\times \dfrac{3600 \text{ s}}{1 \text{ hr}} \times \dfrac{1 \text{ C}}{1 \text{ A} \cdot \text{s}} \times \dfrac{1 \text{ F}}{96,500 \text{ C}} \times \dfrac{1 \text{ mol } Ni^{2+}}{2 \text{ F}} \times \dfrac{58.7 \text{ g } Ni^{2+}}{1 \text{ mol } Ni^{2+}}$

$\times \dfrac{0.90 \text{ g Ni actual}}{1.00 \text{ g Ni theoretical}} = 7.3 \times 10^2$ g Ni^{2+}(aq)

23.74 (a) $\Delta G° = \Delta H° - T\Delta S°$ (assume $\Delta H°$ and $S°$ are constant with changes in temperature)

$Si(s) + 2MnO(s) \rightarrow SiO_2(s) + 2Mn(s)$

$$\Delta H° = \Delta H_f^° \; SiO_2(s) + 2\Delta H_f^° \; Mn(s) - 2\Delta H_f^° \; MnO(s) - \Delta H_f^° \; Si(s)$$

$$\Delta H° = -910.9 + 2(0) - 2(-385.2) + 0 = -140.5 \; kJ$$

$$\Delta S° = S° \; SiO_2(s) + 2S° \; Mn(s) - 2S° \; MnO(s) - S° \; Si(s)$$

$$= 41.84 + 2(32.0) - 2(59.7) - 18.7 = -32.26 = -32.3 \; J/K$$

$$\Delta G° = -140.5 \; kJ - 1473 \; K(-0.03226 \; kJ/K) = -93.0 \; kJ$$

(b)   At 1473 K, the reactants and products are all solids, so they are in their standard states. Since $\Delta G°$ is negative at this temperature, the reaction should be spontaneous and thus feasible.

23.75   (a)   According to Section 20.8, the reduction of O_2 during oxidation of Fe(s) to Fe_2O_3 requires H^+. Above pH 9, iron does not corrode. At the high temperature of the converter, it is unlikely to find H_2O or H^+ in contact with the molten Fe. Also, the basic slag (CaO(l)) that is present to remove phosphorus will keep the environment basic rather than acidic. Thus, the H^+ necessary for oxidation of Fe in air is not present in the converter.

(b) $C + O_2(g) \rightarrow CO_2(g)$

$$S + O_2(g) \rightarrow SO_2(g)$$

$$P + O_2(g) \rightarrow P_2O_5(l); \; P_2O_5(l) + 3CaO(l) \rightarrow Ca_3(PO_4)(l)$$

$$Si + O_2(g) \rightarrow SiO_2$$

$$M + O_2(g) \rightarrow M_xO_y(l); \; M_xO_y + SiO_2 \rightarrow silicates$$

CO_2 and SO_2 escape as gases. P_2O_5 reacts with CaO(l) to form $Ca_3(PO_4)_2(l)$, which is removed with the basic slag layer. SiO_2 and metal oxides can combine to form other silicates; SiO_2, M_xO_y, and complex silicates are all removed with the basic slag layer.

23.76 $2[Cu^+(aq) + 1e^- \rightarrow Cu(s)]$ $E_{red}^° = 0.521 \; V$

         $Cu(s) \rightarrow Cu^{2+}(aq) + 2e^-$ $E_{red}^° = 0.337 \; V$

         $2Cu^+(aq) \rightarrow Cu^{2+}(aq) + Cu(s)$ $E° = (0.521 \; V - 0.337 \; V) = 0.184 \; V$

Rearrange Equation [20.16] for equilibrium conditions. At equilibrium, Q = K and E = 0;

$$\log K = \frac{nE°}{0.0592}; \; n = 2 \quad \log K = \frac{2(0.184)}{0.0592} = 6.2162 = 6.22; \; K = 1.6 \times 10^6$$

23.77 $\Delta G° = -RT \; lnK; \; \Delta G° = \Delta H° - T\Delta S°$

Calculate $\Delta H°$ and $\Delta S°$ using data from Appendix C, assuming $\Delta H°$ and $\Delta S°$ remain constant with changing temperature. Then calculate $\Delta G°$ and K at the two temperatures.

$$\Delta H° = 2\Delta H_f \; CO(g) - \Delta H_f^° \; C(s) - \Delta H_f^° \; CO_2(g)$$

$$\Delta H° = 2(-110.5) - 0 - (-393.5) = +172.5 \; kJ$$

$$\Delta S° = 2S° \; CO(g) - S° \; C(s) - S° \; CO_2(g)$$

$$= 2(197.9) - 5.69 - 213.6 = +176.5 \; J/K = 0.1765 \; kJ/K$$

$$\Delta G^{\circ}_{298} = 172.5 \text{ kJ} - 298 \text{ K}(0.1765 \text{ kJ/K}) = +119.9 \text{ kJ}$$

$$\ln K = \frac{\Delta G^{\circ}}{-RT} = \frac{119.9 \text{ kJ}}{-(8.314 \times 10^{-3} \text{ kJ/K})(298 \text{ K})} = -48.3942 = -48.39; K = 9.6 \times 10^{-22}$$

$$\Delta G^{\circ}_{2000} = 172.5 \text{ kJ} - 2000 \text{ K}(0.1765 \text{ kJ/K}) = -180.5 \text{ kJ}$$

$$\ln K = \frac{-180.5}{-(8.314 \times 10^{-3} \text{ kJ/K})(2000 \text{ K})} = 10.8552 = 10.86; K = 5.18 \times 10^{4}$$

(log K has 3 decimal places, so K has 3 sig figs.)

23.78   (a)   The standard reduction potential for $H_2O(l)$ is much greater than that of $Mg^{2+}(aq)(-0.83 \text{ V} \text{ vs. } -2.37 \text{ V})$. In aqueous solution, $H_2O(l)$ would be preferentially reduced and no $Mg(s)$ would be obtained.

(b) $$97,000 \text{ A} \times 24 \text{ hr} \times \frac{3600 \text{ s}}{1 \text{ hr}} \times \frac{1 \text{ C}}{1 \text{ A} \cdot \text{s}} \times \frac{1 \text{ F}}{96,500 \text{ C}} \times \frac{1 \text{ mol Mg}}{2 \text{ F}} \times \frac{24.31 \text{ g Mg}}{1 \text{ mol Mg}} \times 0.96$$

$$= 1.0 \times 10^{6} \text{ g Mg} = 1.0 \times 10^{3} \text{ kg Mg}$$

23.79   (a)   The very low melting and boiling points for VF_5 indicate that it is molecular rather than ionic, and that the intermolecular forces are probably weak London-dispersion forces. In order for the molecule to experience only London-dispersion forces, it must be nonpolar covalent, which requires the symmetrical trigonal bipyramidal structure shown below. PF_5 also has this structure.

(b) $$VCl_3(s) + 3HF(g) \xrightarrow{\Delta} VF_3(s) + 3HCl(g)$$

(c)   V(V) has a relatively small covalent radius. F is the smallest and most electronegative halogen. The steric repulsions associated with placing five larger halogens around the small V(V) central atom would be substantial. Also, the extreme electron attracting nature of F might be required to coax V into the +5 oxidation state.

23.80   Calculate the mass of $Zn(s)$ that will be deposited.

$$2.0 \text{ m} \times 80 \text{ m} \times \frac{(100)^2 \text{ cm}^2}{1 \text{ m}^2} \times 0.49 \text{ mm} \times \frac{1 \text{ cm}}{10 \text{ mm}} \times \frac{7.1 \text{ g}}{\text{cm}^3} \times 2 \text{ sides}$$

$$= 1.113 \times 10^{6} = 1.1 \times 10^{6} \text{ g Zn}$$

$$1.113 \times 10^{6} \text{ g Zn} \times \frac{1 \text{ mol Zn}}{65.39 \text{ g Zn}} \times \frac{2 \text{ F}}{0.90 \text{ mol Zn}} \times \frac{96,500 \text{ C}}{\text{F}} = 3.651 \times 10^{9} = 3.7 \times 10^{9} \text{ C}$$

(2 F/0.90 mol Zn takes the 90% efficiency into account.)

$$3.651 \times 10^{9} \text{ C} \times 3.5 \text{ V} \times \frac{1 \text{ J}}{\text{C} \cdot \text{V}} \times \frac{1 \text{ kWh}}{3.6 \times 10^{6} \text{ J}} = 3,550 = 3.6 \times 10^{3} \text{ kWh}$$

$$3.550 \times 10^{3} \text{ kWh} \times \frac{\$0.082}{1 \text{ kWh}} = \$291.06 \rightarrow \$291$$

23.81   (a)

$$Ag_2S(s) \rightleftharpoons 2Ag^+(aq) + S^{2-}(aq) \qquad K_{sp}$$

$$2[Ag^+(aq) + 2CN^-(aq) \rightleftharpoons Ag(CN)_2^-] \qquad K_f^2$$

$$\overline{Ag_2S(s) + 4CN^-(aq) \rightleftharpoons 2Ag(CN)_2^-(aq) + S^{2-}(aq)}$$

$$K = K_{sp} \times K_f^2 = [Ag^+]^2[S^{2-}] \times \frac{[Ag(CN)_2^-]^2}{[Ag^+]^2[CN^-]^4} = (6 \times 10^{-51})(1 \times 10^{21})^2 = 6 \times 10^{-9}$$

(b)   The equilibrium constant for the cyanidation of Ag_2S, 6×10^{-9}, is much less than one and favors the presence of reactants rather than products. The process is not practical.

(c)

$$AgCl(s) \rightleftharpoons Ag^+(aq) + Cl^-(aq) \qquad K_{sp}$$

$$Ag^+(aq) + 2CN^-(aq) \rightleftharpoons Ag(CN)_2^-(aq) \qquad K_f$$

$$\overline{AgCl(s) + 2CN^-(aq) \rightleftharpoons Ag(CN)_2^-(aq) + Cl^-(aq)}$$

$$K = K_{sp} \times K_f = [Ag^+][Cl^-] \times \frac{[Ag(CN)_2^-]}{[Ag^+][CN^-]^2} = (1.8 \times 10^{-10})(1 \times 10^{21}) = 2 \times 10^{11}$$

Since $K \gg 1$ for this process, it is potentially useful for recovering silver from horn silver. However, the magnitude of K says nothing about the rate of reaction. The reaction could be slow and require heat, a catalyst, or both to be practical.

23.82   (a) $M(s) \rightarrow M(g)$. The process of atomization is essentially breaking the "metallic bonds" in the solid metal and separating the particles into isolated gas-phase atoms. This requires relocalizing electrons from the solid lattice onto the individual metal atoms.

(b) ΔH_{atom} is the difference between the energy of a mole of gaseous metal atoms, isolated from one another, and a mole of the metal, with all its metal-metal bonding. The difference will be smaller if: 1) the gaseous atoms have some special stability relative to other metallic elements or 2) the metal-metal bonding in the solid is weaker.

The data indicate that Cr and Mn, in the middle of the first transition series, and Cu at the end, have smaller ΔH_{atom} than their neighbors. The electron configurations for these elements are: Cr, $[Ar]4s^13d^5$ (exception); Mn, $[Ar]4s^23d^5$; Cu, $[Ar]4s^13d^{10}$. The gaseous atoms of each of these elements have special stability due to either full or half-full subshells. Assuming relatively constant metal-metal bond strength, the special stability of the gaseous atoms reduces ΔH_{atom} for these elements, relative to their neighbors.

The lower values of ΔH_{atom} for Fe, Co, and Ni relative to the elements around V (after taking account of the variations in stability of the gaseous atoms) is likely due to decreasing metal-metal bond strength. Moving to the right across the transition series from the middle onward, effective nuclear charge increases, the radial extension of the d-orbitals decreases, and the strength of metallic bonding decreases. This is somewhat of a trend.

24 Chemistry of Coordination Compounds

Visualizing Concepts

24.1   *Analyze.* Given the formula of a coordination compound, determine the coordination geometry, coordination number, and oxidation state of the metal.

Plan. From the formula, determine the identity of the ligands and the number of coordination sites they occupy. From the total coordination number, decide on a likely geometry. Use ligand and overall complex charges to calculate the oxidation number of the metal.

Solve.

(a) The ligands are $2Cl^-$, one coordination site each, and en, ethylenediamine, two coordination sites, for a coordination number of 4. This coordination number has two possible geometries, tetrahedral and square planar. Pt is one of the metals known to adopt square planar geometry when CN = 4.

(b) CN = 4, coordination geometry = square planar

(c) $Pt(en)Cl_2$ is a neutral compound, the en ligand is neutral, and the $2Cl^-$ ligands are each –1, so the oxidation state of Pt must be +2, Pt(II).

24.2   *Analyze.* Given a ball-and-stick figure of a ligand, write the Lewis structure and answer questions about the ligand.

Plan. Assume that each atom in the Lewis structure obeys the octet rule. Complete each octet with unshared electron pairs or multiple bonds, depending on the bond angles in the ball-and-stick model. Black = C, blue = N, red = O, gray = H.

There is a second resonance structure with the double bond drawn to the second O atom.

Check. Write the molecular formula, count the valence electron pairs and see if it matches your structure. $[C_4H_9N_2O_2]^-$ (16 + 9 + 10 + 12 + 1) = 48 valence e^-, 24 e^- pair Our Lewis structure also has 24 e^- pairs.

(a)   Donor atoms have unshared electron pairs. The potential donors in this structure are the two N and two O atoms.

The ligand is tridentate. (Even though there are four possible donor atoms, the structure would be strained if all four were bound to one metal center. It is likely that only one of the two O atoms binds to the same metal as the two N atoms.)

(b)   An octahedral complex has 6 coordination sites. A single ligand has only 4 possible donors, so two ligands are needed. From a steric perspective, the likely donors would be the 2 N atoms and 1 of the carbonyl oxygen atoms. The chelate bite of a carboxyl group is relatively small and would require an O—M—O angle of less than 90°.

24.3   *Analyze.* Given a ball-and-stick structure, name the complex ion, which has a 1– charge.

Plan. Write the chemical formula of the complex ion, determine the oxidation state of the metal, and name the complex.

Solve. $[Pt(NH_3)Cl_3]^-$. Oxidation numbers: $[Pt + 0 + 3(-1) = -1$, $Pt = +2$, $Pt(II)$

Arrange the ligands alphabetically, followed y the metal. Since the complex is an anion, add the suffix -ate, then the oxidation state of the metal: aminotrichloroplatinate(II)

24.4   *Analyze.* Given 5 structures, visualize which are identical to (1) and which are geometric isomers of (1).

Plan. There are two possible ways to arrange MA_3X_3. The first has bond angles of 90° between all similar ligands; this is structure (1). The second has one 180° angle between similar ligands. Visualize which description fits each of the five structures.

Solve. (1) has all 90° angles between similar ligands.

(2) has a 180° angle between similar ligands (see the blue ligands in the equatorial plane of the octahedron)

(3) has all 90° angles between similar ligands

(4) has all 90° angles between similar ligands

(5) has a 180° angle between similar ligands (see the blue axial ligands)

Structures (3) and (4) are identical to (1); (2) and (5) are geometric isomers.

24.5   *Analyze.* Given four structures, decide which are chiral.

Plan. Chiral molecules have nonsuperimposable mirror images. Draw the mirror image of each molecule and visualize whether it can be rotated into the original molecule. If so, the complex is not chiral. If the original orientation cannot be regenerated by rotation, the complex is chiral. *Solve.*

(1)   (1) mirror

The two orientations are not superimposable and molecule (1) is chiral.

The two orientations are superimposible. Rotate the right-most structure 90° counterclockwise about the B-M-B axis to align the G's; the bidentate ligands then also overlap. Molecule (2) is not chiral.

The two orientations are not superimposable and molecule (3) is chiral.

The two orientations are not superimposable and molecule (4) is chiral.

24.6   *Analyze.* Given the visible colors of two solutions, determine the colors of light absorbed by each solution.

Plan. Apparent color is transmitted or reflected light, absorbed color is basically the complement of apparent color. Use the color wheel in Figure 24.24 to obtain the complementary absorbed color for the solutions.

Solve. The left solution appears yellow-orange, so it absorbs blue-violet. The right solution appears blue-green (cyan), so it absorbs orange-red.

24.7   *Analyze.* Fit the crystal field splitting diagram to the complex description in each part.

Plan. Determine the number of d-electrons in each transition metal. On the splitting diagrams match the d-orbital splitting patterns to complex geometry and electron pairing to the definition of high-spin and low-spin.

Solve. Octahedral complexes have the 3 lower, 2 higher splitting pattern, while tetrahedral complexes have the opposite 2 lower, 3 higher pattern. Low spin complexes favor electron pairing because of large d-orbital splitting. High-spin complexes have maximum occupancy because of small orbital splitting.

(a) Fe^{3+}, 5d-electrons; weak field: spins unpaired; octahedral: 3 lower, 2 higher d-splitting ∴ diagram (4)

(b) Fe^{3+}, 5d-electrons; strong field: spins paired; octahedral: 3 lower, 2 higher d-splitting ∴ diagram (1)

(c) Fe^{3+}, 5d-electrons; tetrahedral: 2 lower, 3 higher d-splitting ∴ diagram (3)

(d) Ni^{2+}, 8 d-electrons; tetrahedral: 2 lower, 3 higher d-splitting ∴ diagram (2)

Check. Diagram (2) was the remaining choice for (d) and it fits the description.

24 Coordination Compounds   Solutions to Exercises

24.8   *Analyze/Plan.* Given the linear diagram and axial labels, answer the questions and predict crystal field splitting. Orbitals with lobes nearest ligand charges (or partial charges) will be highest in energy; orbitals with lobes away from charges are lowest in energy.

Solve. d_{z^2} has lobes nearest the charges. $d_{x^2-y^2}$ and d_{xy} have lobes in the xy-plane farthest from the charges. d_{xz} and d_{yz} point between the respective axes and are intermediate in energy.

Introduction to Metal Complexes

24.9   (a)   A *metal complex* consists of a central metal ion bonded to a number of surrounding molecules or ions. The number of bonds formed by the central metal ion is the *coordination number*. The surrounding molecules or ions are the *ligands*.

       (b)   A Lewis acid is an electron pair acceptor and a Lewis base is an electron pair donor. All ligands have at least one unshared pair of valence electrons. Metal ions have empty valence orbitals (d, s, or p) that can accommodate donated electron pairs. Ligands act as electron pair donors, or Lewis bases, and metal ions act as electron pair acceptors, or Lewis acids, via their empty valence orbitals.

24.10   (a)   In Werner's theory, *primary valence* is the charge of the metal cation at the center of the complex. *Secondary valence* is the number of atoms bound or coordinated to the central metal ion. The modern terms for these concepts are oxidation state and coordination number, respectively. (Note that "'oxidation state" is a broader term than ionic charge, but Werner's complexes contain metal ions where cation charge and oxidation state are equal.)

       (b)   Ligands are the Lewis base in metal-ligand interactions [see Solution 24.9(b)]. As such, they must possess at least one unshared electron pair. NH_3 has an unshared electron pair but BH_3, with less than 8 electrons about B, has no unshared electron pair and cannot act as a ligand. In fact, BH_3 acts as a Lewis acid, an electron pair acceptor, because it is electron-deficient.

24.11   *Analyze/Plan.* Follow the logic in Sample Exercises 24.1 and 24.2.   *Solve.*

       (a)   This compound is electrically neutral, and the NH_3 ligands carry no charge, so the charge on Ni must balance the –2 charge of the 2 Br^- ions. The charge and oxidation state of Ni is +2.

       (b)   Since there are 6 NH_3 molecules in the complex, the likely coordination number is 6. In some cases Br^- acts as a ligand, so the coordination number could be other than 6.

       (c)   Assuming that the 6 NH_3 molecules are the ligands, 2 Br^- ions are not coordinated to the Ni^{2+}, so 2 mol AgBr(s) will precipitate. (If one or both of the Br^- act as a ligand, the mol AgBr(s) would be different.)

24.12   (a)   Yes. There are 6 possible ligands, 3 H_2O molecules and 3 Cl^- ions. Any Cl^- ions that are not coordinated to the metal will form $AgCl(s)$ precipitate when the complex is treated with $AgNO_3(aq)$. Absence of $AgCl(s)$ would mean all Cl^- ions were ligands and a coordination number of 6. One mole of $AgCl(s)$ per mole of complex would mean one uncoordinated ligand, and so on. This assumes that all 3 H_2O molecules act as ligands. In fact, they could serve as water of hydration (Section 13.1, A Closer Look: Hydrates). Reaction with $AgNO_3$ gives no information about the nature of H_2O molecules.

        (b)   Yes. Conductivity is directly related to the number of ions in a solution. The lower the conductivity, the more Cl^- ions that act as ligands. Conductivity measurements on a set of standard solutions with various moles of ions per mole of complex would provide a comparative method for quantitative determination of the number of free and bound Cl^- ions.

24.13   *Analyze/Plan.* Count the number of donor atoms in each complex, taking the identity of polydentate ligands into account. Follow the logic in Sample Exercise 24.2 to obtain oxidation numbers of the metals.

        (a)   Coordination number = 4, oxidation number = +2

        (b)   5, +4   (c)   6, +3   (d)   5, +2

        (e)   6, +3   (f)   4, +2

24.14   (a)   Coordination number = 6, oxidation number = +2

        (b)   4, +2   (c)   6, +1   (d)   6, +3

        (e)   6, +3   (f)   5, +2

24.15   *Analyze/Plan.* Given the formula of a coordination compound, determine the number and kinds of donor atoms. The ligands are enclosed in the square brackets. Decide which atom in the ligand has an unshared electron pair it is likely to donate.   *Solve.*

        (a)   4 Cl^- (b)   4 Cl^-, 1 O^{2-} (c)   4 N, 2 Cl^-

        (d)   5 C. In CN^-, both C and N have an unshared electron pair. C is less electronegative and more likely to donate its unshared pair.

        (e)   6 O. $C_2O_4{}^{2-}$ is a bidentate ligand; each ion is bound through 2 O atoms for a total of 6 O donor atoms.

        (f)   4 N. en is a bidentate ligand bound through 2 N atoms.

24.16   (a)   6 C (see Solution 24.15(d))

        (b)   4 N

        (c)   5 C, 1 Br. In CO, both C and O have an unshared electron pair. C is less electronegative and more likely to donate its unshared pair. (This is analogous to the situation in CN^-).

        (d)   4 N, 2 O. en is a bidentate ligand bound through N, for a total of 4 N donors. $C_2O_4{}^{2-}$ is bidentate with 2 O donors.

(e)   6 N. When thiocyanate is written "NCS," it is bound through N. This makes a total of 6 N donors.

(f)   4 N, 1 I. bipy is bidentate bound through N, for a total of 4 N donors.

Polydendate Ligands; Nomenclature

24.17   (a)   A monodendate ligand binds to a metal in through one atom; a bidendate ligand binds through two atoms.

         (b)   If a bidentate ligand occupies two coordination sites, three bidentate ligands fill the coordination sphere of a six-coordinate complex.

         (c)   A tridentate ligand has at least three atoms with unshared electron pairs in the correct orientation to simultanously bind one or more metal ions.

24.18   (a)   2 coordination sites, 2 N donor atoms

         (b)   2 coordination sites, 2 N donor atoms

         (c)   2 coordination sites, 2 O donor atoms (Although there are four potential O donor atoms in $C_2O_4{}^{2-}$, it is geometrically impossible for more than two of these to be bound to a single metal ion.)

         (d)   4 coordination sites, 4 N donor atoms

         (e)   6 coordination sites, 2 N and 4 O donor atoms

24.19   *Analyze/Plan.* Given the formula of a coordination compound, determine the number of coordination sites occupied by the polydentate ligand. The coordination number of the complexes is either 4 or 6. Note the number of monodentate ligands and determine the number of coordination sites occupied by the polydentate ligands.   *Solve.*

         (a)   *ortho*-phenanthroline, *o*-phen, is bidentate

         (b)   oxalate, $C_2O_4{}^{2-}$, is bidentate

         (c)   ethylenediaminetetraacetate, EDTA, is pentadentate

         (d)   ethylenediamine, en, is bidentate

24.20   (a)   4   (b)   4   (c)   6   (d)   6

24.21   (a)   The term *chelate effect* means there is a special stability associated with formation of a metal complex containing a polydentate (chelate) ligand relative to a complex containing only monodentate ligands.

         (b)   When a single chelating ligand replaces two or more monodendate ligands, the number of free molecules in the system increases and the entropy of the system increases. Chemical reactions with $+\Delta S$ tend to be spontaneous, have negative ΔG, and large positive values of K.

         (c)   Polydentate ligands can be used to bind metal ions and prevent them from undergoing unwanted chemical reactions without removing them from solution. The polydentate ligand thus hides or *sequesters* the metal ion.

24.22   (a)   Monodentate; py has only one N donor atom.

         (b)   K for this reaction will be less than one. Two free pyridine molecules are replaced by one free bipy molecule. There are more moles of particles in the reactants than products, so ΔS is predicted to be negative. Processes with a net decrease in entropy are usually nonspontaneous, have positive ΔG, and values of K less than one. This equilibrium is likely to be spontaneous in the reverse direction.

24.23   *Analyze/Plan.* Given the name of a coordination compound, write the chemical formula. Refer to Table 24.2 to find ligand formulas. Place the metal complex (metal ion + ligands) inside square brackets and the counter ion (if there is one) outside the brackets. *Solve.*

         (a) $[Cr(NH_3)_6](NO_3)_3$ (b) $[Co(NH_3)_4CO_3]_2SO_4$ (c) $[Pt(en)_2Cl_2]Br_2$

         (d) $K[V(H_2O)_2Br_4]$ (e) $[Zn(en)_2][HgI_4]$

24.24   (a) $[Mn(H_2O)_5I](ClO_4)_2$ (b) $[Ru(bipy)_3](NO_3)_2$

         (c) $[Rh(o\text{-}phen)_2Cl_2]_2SO_4$ (d) $Na[Cr(NH_3)_2Br_4]$

         (e) $[Co(en)_3]_4[Fe(ox)_3]_3$

24.25   *Analyze/Plan.* Follow the logic in Sample Exercise 24.4, paying attention to naming rules in Section 24.3.   *Solve.*

         (a)   tetraamminedichlororhodium(III) chloride

         (b)   potassium hexachlorotitanate(IV)

         (c)   tetrachlorooxomolybdenum(VI)

         (d)   tetraaqua(oxalato)platinum(IV) bromide

24.26   (a)   dichloroethylenediamminecadmium(II)

         (b)   potassium hexacyanomanganate(II)

         (c)   pentaamminecarbonatochromium(III) chloride

         (d)   tetraamminediaquairidium(III) nitrate

Isomerism

24.27   *Analyze/Plan.* Consider the definitions of the various types of isomerism, and which of the complexes could exhibit isomerism of the specified type.   *Solve.*

         (a)

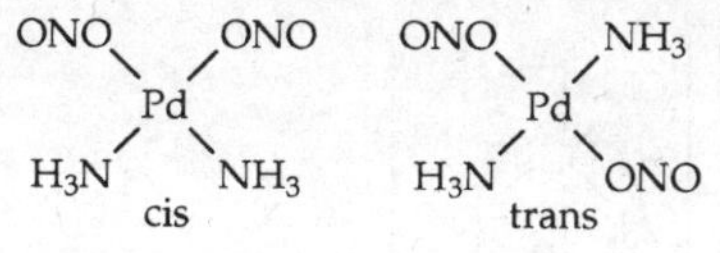

         (b) $[Pd(NH_3)_2(ONO)_2]$, $[Pd(NH_3)_2(NO_2)_2]$

         (c)

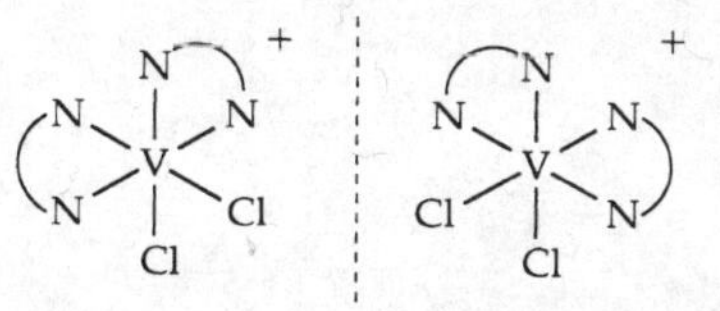

         (d) $[Co(NH_3)_4Br_2]Cl$, $[Co(NH_3)_4BrCl]Br$

24.28 (a)

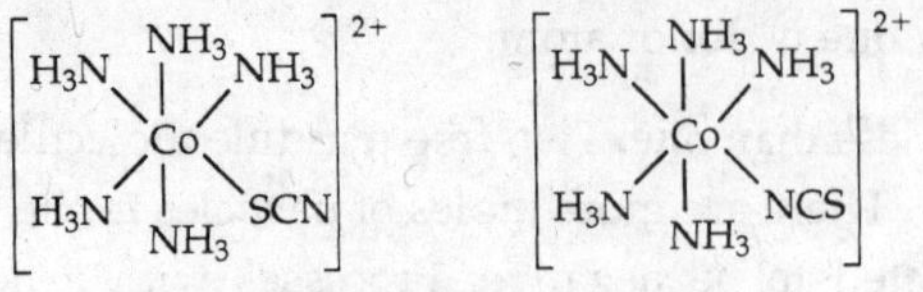

coordination sphere isomerism

(b)

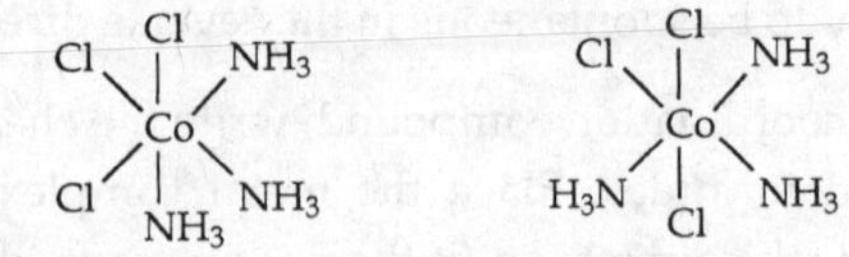

(c)

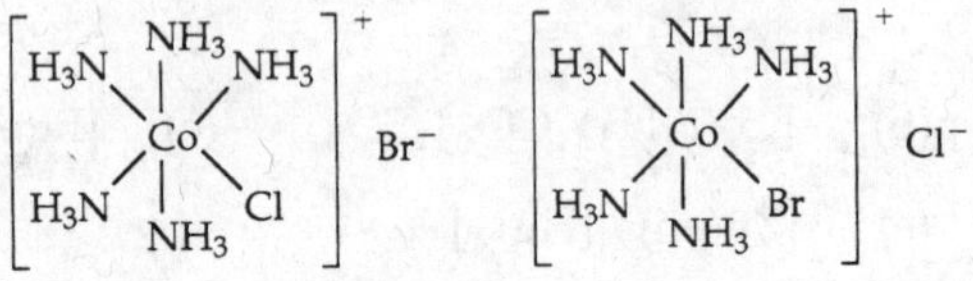

coordination sphere isomerism

24.29 Yes. A tetrahedral complex of the form MA_2B_2 would have neither structural nor stereoisomers. For a tetrahedral complex, no differences in connectivity are possible for a single central atom, so the terms *cis* and *trans* do not apply. No optical isomers with tetrahedral geometry are possible because M is not bound to four different groups. The complex must be square planar with *cis* and *trans* geometric isomers.

24.30 Two geometric isomers are possible for an octahedral MA_3B_3 complex (see below). All other arrangements, including mirror images, can be rotated into these two structures. Neither isomer is optically active.

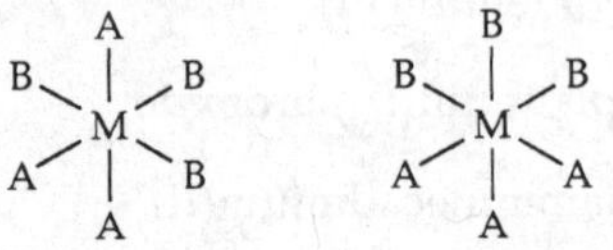

24.31 *Analyze/Plan.* Follow the logic in Sample Exercises 24.5 and 24.6.   *Solve.*

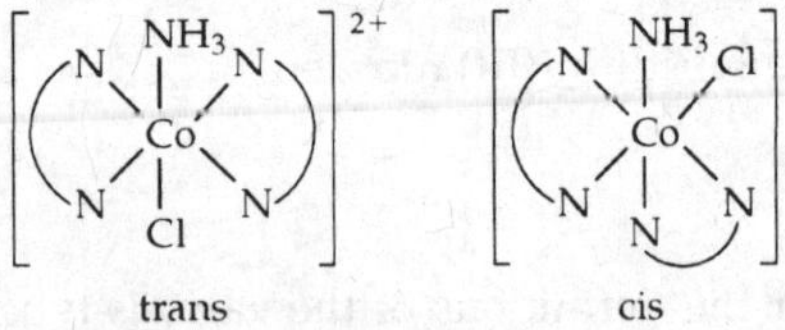

trans   cis

The cis isomer is chiral.

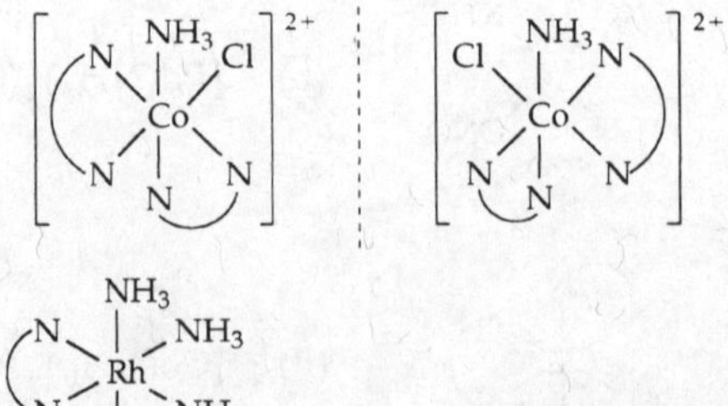

24.32

The symbol N⌒N represents the bidentate ligand (bipy).

There are no optical isomers. The mirror image of each structural isomer can be superimposed on the structure above by a 180° rotation.

24.33 *Analyze/Plan.* Follow the logic in Sample Exercise 24.5 and 24.6. *Solve.*

(a)

(b)

cis trans

(c)

cis cis

optical isomers

trans

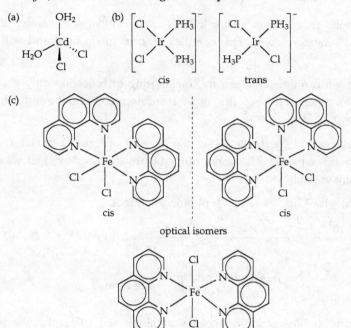

(The three isomeric complex ions in part (c) each have a 1+ charge.)

24.34 (a)

(b)

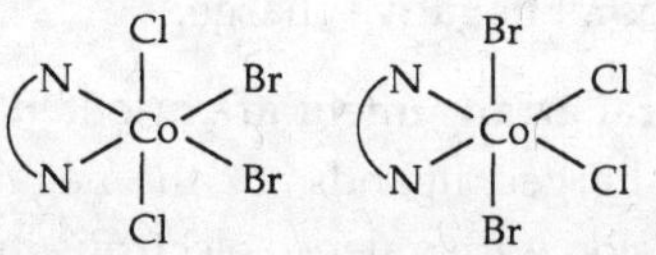

(c)

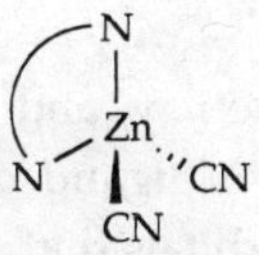

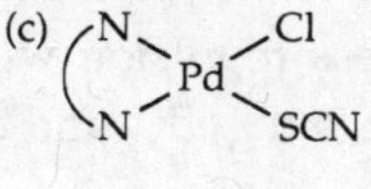

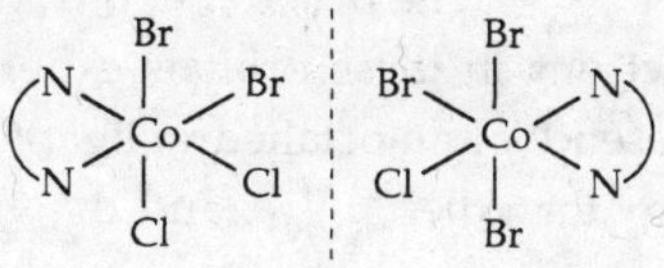

Color, Magnetism; Crystal-Field Theory

24.35 (a) Visible light has wavelengths between 400 and 700 nm.

(b) *Complementary* colors are opposite each other on a color wheel such as Figure 24.25.

(c) A colored metal complex absorbs visible light of its complementary color. For example, a red complex absorbs green light.

(d) E(J/photon) = hv = hc/λ. Change J/photon to kJ/mol.

$$E = \frac{6.626 \times 10^{-34} \text{ J} \bullet \text{s}}{610 \text{ nm}} \times \frac{3.00 \times 10^8 \text{ m}}{\text{s}} \times \frac{1 \text{nm}}{1 \times 10^{-9} \text{ m}} = 3.259 \times 10^{-19} = 3.26 \times 10^{-19} \text{ J}$$

$$\frac{3.259 \times 10^{-19} \text{ J}}{\text{photon}} \times \frac{1 \text{ kJ}}{1000 \text{ J}} \times \frac{6.022 \times 10^{23} \text{ photons}}{\text{mol}} = 196 \text{ kJ/mol}$$

24.36   (a)   Yes. A complex that absorbs visible light of one wavelength or color will appear as the complementary color. This complex absorbs green light and will appear red.

         (b)   No. A solution can appear green by transmitting or reflecting only green light (the situation stated in the exercise) **or** by absorbing red light, the complementary color of green.

         (c)   A visible absorption spectrum shows the amount of light absorbed at a given wavelength. It is a plot of absorbance (dependent variable, y-axis) vs. wavelength (independent variable, x-axis).

         (d) $E(\text{J/photon}) = h\nu = hc/\lambda$. Change J/photon to kJ/mol.

$$E = \frac{6.626 \times 10^{-34} \text{ J} \bullet \text{s}}{530 \text{ nm}} \times \frac{3.00 \times 10^8 \text{ m}}{\text{s}} \times \frac{1 \text{nm}}{1 \times 10^{-9} \text{ m}} = 3.751 \times 10^{-19} = 3.75 \times 10^{-19} \text{ J}$$

$$3.751 \times 10^{-19} = \frac{1 \text{ kJ}}{1000 \text{ J}} \times \frac{6.022 \times 10^{23} \text{ photons}}{\text{mol}} = 226 \text{ kJ/mol}$$

24.37   Most of the electrostatic interaction between a metal ion and a ligand is the attractive interaction between a positively charged metal cation and the full negative charge of an anionic ligand or the partial negative charge of a polar covalent ligand. Whether the interaction is ion-ion or ion-dipole, the ligand is strongly attracted to the metal center and can be modeled as a point negative charge.

24.38   Six ligands in an octahedral arrangement are oriented along the x, y, and z axes of the metal. These negatively charged ligands (or the negative end of ligand dipoles) have greater electrostatic repulsion with valence electrons in metal orbitals that also lie along these axes, the d_{z^2}, and $d_{x^2-y^2}$. The d_{xy}, d_{xz} and d_{yz} metal orbitals point between the x, y, and z axes, and electrons in these orbitals experience less repulsion with ligand electrons. Thus, in the presence of an octahedral ligand field, the d_{xy}, d_{xz} and d_{xy} metal orbitals are lower in energy than the $d_{x^2-y^2}$. and d_{z^2}.

24.39   (a)

         (b)   The magnitude of Δ and the energy of the d-d transition for a d^1 complex are equal.

         (c)

$$\frac{6.626 \times 10^{-34} \text{ J} \bullet \text{s}}{590 \text{ nm}} \times \frac{3.00 \times 10^8 \text{ m}}{\text{s}} \times \frac{1 \text{ nm}}{1 \times 10^{-9} \text{ m}} \times \frac{1 \text{ kJ}}{1000 \text{ J}} \times \frac{6.022 \times 10^{23} \text{ photons}}{\text{mol}}$$

$$= 203 \text{ kJ/mol}$$

24.40   **(a)** $\Delta E = hc/\lambda = \dfrac{6.626 \times 10^{-34}\ \text{J}\cdot\text{s} \times 2.998 \times 10^8\ \text{m/s}}{500 \times 10^{-9}\ \text{m}} = 3.973 \times 10^{-19} = 3.97 \times 10^{-19}\ \text{J/photon}$

$\Delta = 3.973 \times 10^{-19}\ \text{J/photon} \times \dfrac{6.022 \times 10^{23}\ \text{photons}}{1\ \text{mol}} \times \dfrac{1\ \text{kJ}}{1000\ \text{J}} = 239.25 = 239\ \text{kJ/mol}$

(b)   The *spectrochemical* series is an ordering of ligands according to their ability to increase the energy gap Δ. If H_2O is replaced by NH_3 in the complex, the magnitude of Δ would increase because NH_3 is higher in the spectrochemical series and creates a stronger ligand field.

24.41   *Analyze/Plan.* Consider the relationship between the color of a complex, the wavelength of absorbed light, and the position of a ligand in the spectrochemical series.   *Solve.*

Cyanide is a strong field ligand. The d-d electronic transitions occur at relatively high energy, because Δ is large. A yellow color corresponds to absorption of a photon in the violet region of the visible spectrum, between 430 and 400 nm. H_2O is a weaker field ligand than CN^-. The blue or green colors of aqua complexes correspond to absorptions in the region of 620 nm. Clearly, this is a region of lower energy photons than those with characteristic wavelengths in the 430 to 400 nm region. These are very general and imprecise comparisons. Other factors are involved, including whether the complex is high spin or low spin.

24.42   The ions absorb the complement of the color they appear. Green $[Ni(H_2O)_6]^{2+}$ absorbs red light, 650-800 nm. Purple $[Ni(NH_3)_6]^{2+}$ absorbs yellow light, 560–580 nm. Thus, $[Ni(NH_3)_6]^{2+}$ absorbs light with the shorter wavelength. This agrees with the spectrochemical series, which indicates that H_2O will produce a smaller d-orbital splitting (Δ) than NH_3. Thus, $[Ni(H_2O)_6]^{2+}$ should absorb light with a smaller energy and longer wavelength.

24.43   *Analyze/Plan.* Determine the charge on the metal ion, subtract it from the row number (3-12) of the transition metal, and the remainder is the number of d-electrons.   *Solve.*

    (a) Ti^{3+}, d^1 **(b)** Co^{3+}, d^6 **(c)** Ru^{3+}, d^5

    (d) Mo^{5+}, d^1 **(e)** Re^{3+}, d^4

24.44   **(a)** Fe^{3+}, d^5 **(b)** Mn^{2+}, d^5 **(c)** Ag^+, d^{10}

    (d) Cr^{3+}, d^3 **(e)** Sr^{2+}, d^0

24.45   *Analyze/Plan.* Follow the logic in Sample Exercise 24.9.   *Solve.*

    (a) Mn: $[Ar]4s^2 3d^5$ **(b)** Ru: $[Kr]5s^1 4d^7$ **(c)** Rh: $[Kr]5s^1 4d^8$

    Mn^{2+}: $[Ar]3d^5$ Ru^{2+}: $[Kr]4d^6$ Rh^{2+}: $[Kr]4d^7$

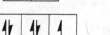

    1 unpaired electron   0 unpaired electrons   1 unpaired electron

24.46 (a) Ru: $[Kr]5s^1 4d^7$ (b) Mo: $[Kr]5s^1 4d^5$ (c) Co: $[Ar]4s^2 3d^7$

Ru^{3+}: $[Kr]4d^5$ Mo^{3+}: $[Kr]3d^3$ Co^{3+}: $[Ar]3d^6$

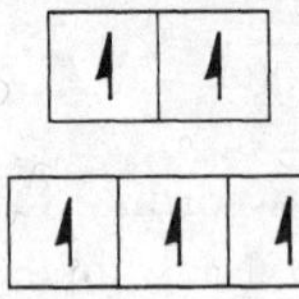

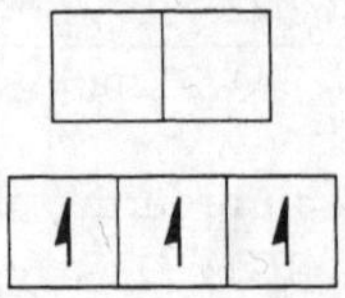

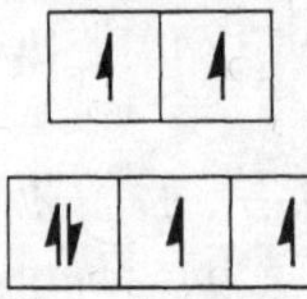

5 unpaired electrons   3 unpaired electrons   4 unpaired electrons

24.47 *Analyze/Plan.* All complexes in this exercise are six-coordinate octahedral. Use the definitions of high-spin and low-spin along with the orbital diagram from Sample Exercise 24.9 to place electrons for the various complexes. *Solve.*

(a) 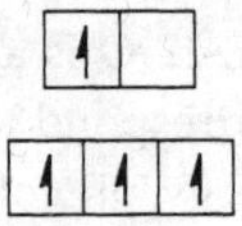 d^4, high spin

(b) 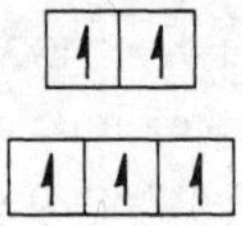 d^5, high spin

(c) 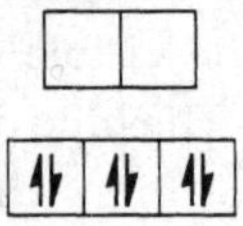 d^6, low spin

(d) 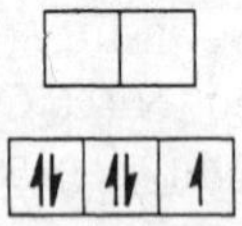 d^5, low spin

(e) 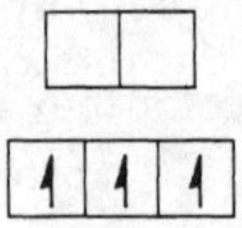 d^3

(f) 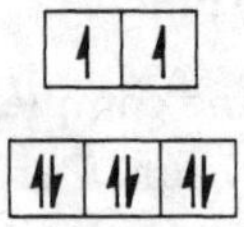 d^8

24.48 (a) 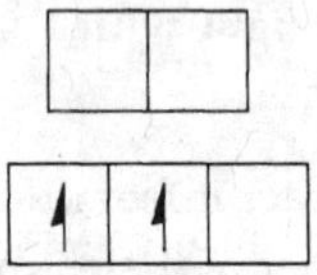 d^2

(b) 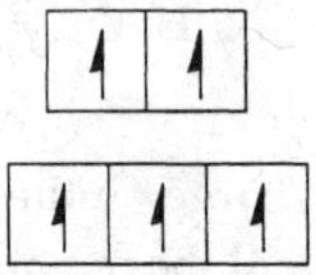 d^5, high spin

(c) 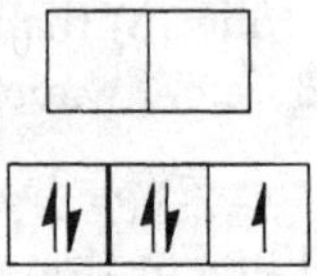 d^5, low spin

(d) 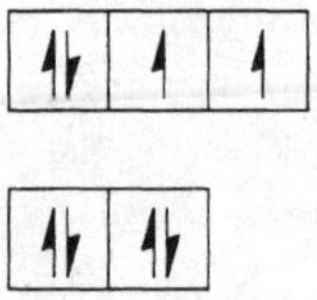 d^8

(e) 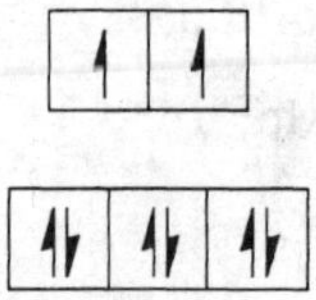 d^8

(f) 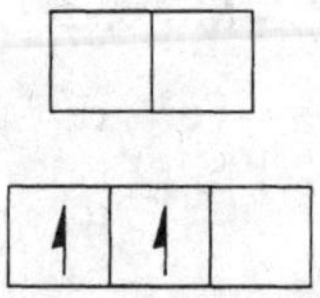 d^2

24.49 *Analyze/Plan.* Follow the ideas but reverse the logic in Sample Exercise 24.9. *Solve.*

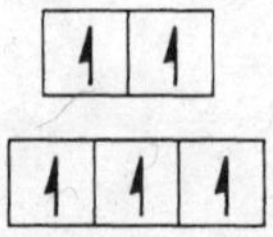

high spin

24.50

$[Fe(CN)_6]^{3-}$
low spin

$[Fe(NCS)_6]^{3-}$
high spin

Both complexes contain Fe^{3+}, a d^5 ion. CN^-, a strong field ligand, produces such a large Δ that the splitting energy is greater than the pairing energy, and the complex is low spin. NCS^- produces a smaller Δ, so it is energetically favorable for d-electrons to be unpaired in the higher energy d-orbitals. NCS^- is a much weaker-field ligand than CN^-. It is probably weaker than NH_3 and near H_2O in the spectrochemical series.

Additional Exercises

24.51 $[Pt(NH_3)_6]Cl_4$; $[Pt(NH_3)_4Cl_2]Cl_2$; $[Pt(NH_3)_3Cl_3]Cl$; $[Pt(NH_3)_2Cl_4]$; $K[Pt(NH_3)Cl_5]$

24.52   (a)

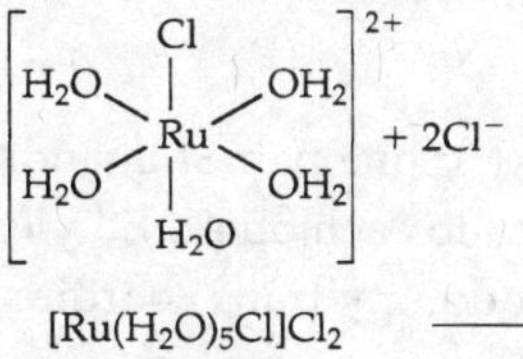

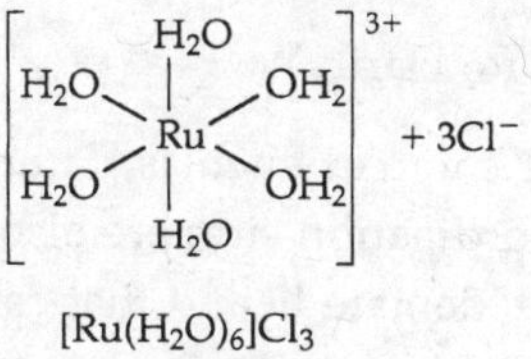

$[Ru(H_2O)_5Cl]Cl_2$ ⟶ $[Ru(H_2O)_6]Cl_3$

24.53   (a)

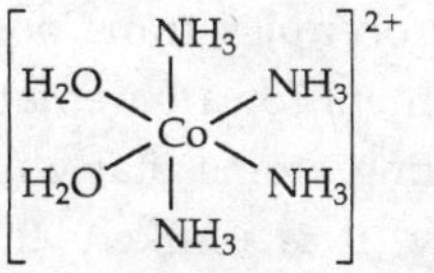

octahedral   octahedral

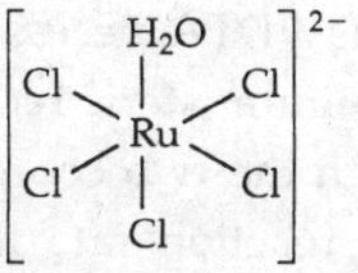

(c)

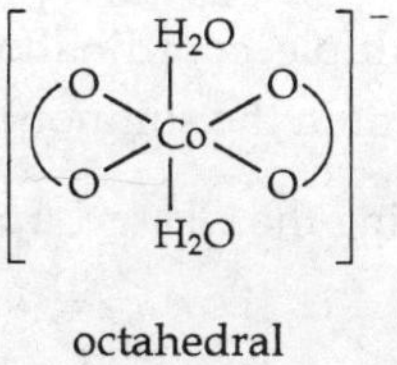

octahedral   octahedral

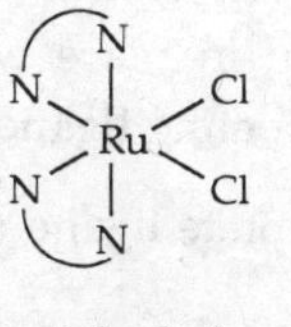

24.54   (a)   [24.53(a)] *cis*-tetraamminediaquacobalt(II) nitrate

              [24.53(b)] sodium aquapentachlororuthenate(III)

              [24.53(c)] ammonium *trans*-diaquabisoxalatocobaltate(III)

              [24.53(d)] *cis*-dichlorobisethylenediamineruthenium(II)

   (b)   Only the complex in 24.53(d) is optically active. The mirror images of (a)-(c) can be superimposed on the original structure. The chelating ligands in (d) prevent its mirror images (enantiomers) from being superimposable.

24.55   (a)   Valence electrons: $2P + 6C + 16H = 10 + 24 + 16 = 50\ e^-$, $25\ e^-$ pr

$$H-\overset{\overset{\displaystyle H}{|}}{\underset{\underset{\displaystyle H}{|}}{C}}-\overset{\cdot\cdot}{P}-\overset{\overset{\displaystyle H}{|}}{\underset{\underset{\displaystyle H-\overset{\overset{\displaystyle H}{|}}{\underset{\underset{\displaystyle H}{|}}{C}}-H}{|}}{C}}-\overset{\overset{\displaystyle H}{|}}{\underset{\underset{\displaystyle H-\overset{\overset{\displaystyle H}{|}}{\underset{\underset{\displaystyle H}{|}}{C}}-H}{|}}{C}}-\overset{\cdot\cdot}{P}-\overset{\overset{\displaystyle H}{|}}{\underset{\underset{\displaystyle H}{|}}{C}}-H$$

(b)   Both CO and dmpe are neutral molecules, so the oxidation state of Mo must be zero.

(c)   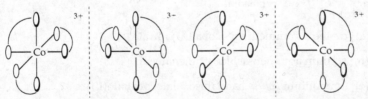 $C\equiv O$ coordinates through C, because it is less electronegative than O, and a better electron pair donor. The molecule has only a single isomer. The dmpe ligand cannot span *trans* positions, so there are no geometric isomers. The mirror image of the structure above is easily superimposable, so there are no optical isomers.

(H atoms omitted for clarity)

24.56   (a)   In a square planar complex such as [Pt(en)Cl$_2$], if one pair of ligands is trans, the remaining two coordination sites are also trans to each other. Ethylenediamine is a relatively short bidentate ligand that cannot occupy trans coordination sites, so the trans isomer is unknown.

(b)   A polydentate ligand such as EDTA necessarily occupies trans positions in an octahedral complex. The minimum steric requirement for a bidentate ligand is a medium-length chain between the two coordinating atoms that will occupy the trans positions. In terms of reaction rate theory, it is unlikely that a flexible bidentate ligand will be in exactly the right orientation to coordinate trans. The polydentate ligand has a much better chance of occupying trans positions, because it locks the metal ion in place with multiple coordination sites (and shields the metal ion from competing ligands present in the solution).

24.57   We will represent the end of the bidentate ligand containing the CF$_3$ group by a shaded oval, the other end by an open oval:

24.58   (a)   Hemoglobin is the iron-containing protein that transports O$_2$ in human blood.

(b)   Chlorophylls are magnesium-containing porphyrins in plants. They are the key components in the conversion of solar energy into chemical energy that can be used by living organisms.

(c)   Siderophores are iron-binding compounds or ligands produced by a microorganism. They compete on a molecular level for iron in the medium outside the organism and carry needed iron into the cells of the organism.

24.59   (a) $AgCl(s) + 2NH_3(aq) \rightarrow [Ag(NH_3)_2]^+(aq) + Cl^-(aq)$

         (b) $[Cr(en)_2Cl_2]Cl(aq) + 2H_2O(l) \rightarrow [Cr(en_2)(H_2O)_2]^{3+}(aq) + 3Cl^-(aq)$

              green   brown-orange

             $3Ag^+(aq) + 3Cl^-(aq) \rightarrow 3AgCl(s)$

             $[Cr(en)_2(H_2O)_2]^{3+}$ and $3NO_3^-$ are spectator ions in the second reaction.

         (c) $Zn(NO_3)_2(aq) + 2NaOH(aq) \rightarrow Zn(OH)_2(s) + 2NaNO_3(aq)$

             $Zn(OH)_2(s) + 2NaOH(aq) \rightarrow [Zn(OH)_4]^{2-}(aq) + 2Na^+(aq)$

         (d) $Co^{2+}(aq) + 4Cl^-(aq) \rightarrow [CoCl_4]^{2-}(aq)$

24.60   (a)   pentacarbonyliron(0)

         (b)   Since CO is a neutral molecule, the oxidation state of iron must be zero.

         (c) $[Fe(CO)_4CN]^-$ has two geometric isomers. In a trigonal bipyramid, the axial and equatorial positions are not equivalent and not superimposable. One isomer has CN in an axial position and the other has it in an equatorial position.

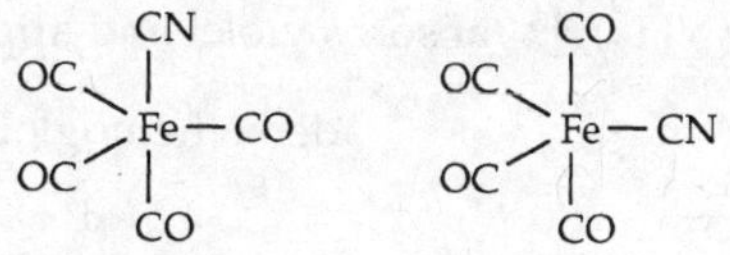

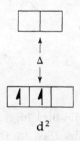

24.61   (a)   left shoe

         (c)   wood screw

         (e)   a typical golf club

24.62   (a)

                          Δ

                          d^2

         (b)   These complexes are colored because the crystal-field splitting energy, Δ, is in the visible portion of the electromagnetic spectrum. Visible light with $\lambda = hc/\Delta$ is absorbed, promoting one of the d-electrons into a higher energy d-orbital. The remaining wavelengths of visible light are reflected or transmitted; the combination of these wavelengths is the color we see.

         (c) $[V(H_2O)_6]^{3+}$ will absorb light with higher energy. H_2O is in the middle of the spectrochemical series, and causes a larger Δ than F^-, a weak-field ligand. Since Δ and λ are inversely related, larger Δ corresponds to higher energy and shorter λ.

24.63   (a)   Formally, the two Ru centers have different oxidation states; one is +2 and the other is +3.

(b) Ru^{2+}, d^6 Ru^{3+}, d^5

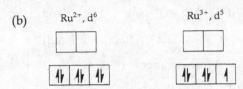

(c)   There is extensive bonding-electron delocalization in the isolated pyrazine molecule. When pyrazine acts as a bridging ligand, its delocalized molecular orbitals provide a pathway for delocalization of the "odd" d-electron in the Creutz-Taube ion. The two metal ions appear equivalent because the odd d-electron is delocalized across the pyrazine bridge.

24.64   According to the spectrochemical series, the order of increasing Δ for the ligands is $Cl^- < H_2O < NH_3$. (The tetrahedral Cl^- complex will have an even smaller Δ than an octahedral one.) The smaller the value of Δ, the longer the wavelength of visible light absorbed. The color of light absorbed is the complement of the observed color. A blue complex absorbs orange light (580–650 nm), a pink complex absorbs green light (490–560 nm) and a yellow complex absorbs violet light (400–430 nm). Since $[CoCl_4]^{2-}$ absorbs the longest wavelength, it appears blue. $[Co(H_2O)_6]^{2+}$ absorbs green and appears pink, and $[Co(NH_3)_6]^{3+}$ absorbs violet and appears yellow.

24.65   oxyhemoglobin   deoxyhemoglobin

             $Fe^{2+} : d^6$ $Fe^{2+} : d^6$

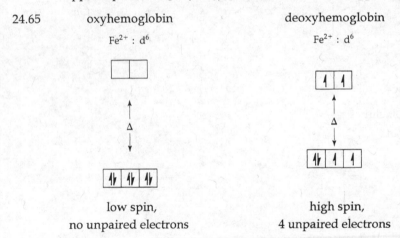

             low spin,   high spin,

    no unpaired electrons   4 unpaired electrons

In general, the crystal field splitting, Δ, is greater in low spin than high spin complexes. The energy, Δ, corresponds to the wavelength of light absorbed by the complex and determines its color. Since oxyhemoglobin absorbs higher energy, shorter wavelength light, longer wavelengths remain and the sample appears red. Deoxyhemoglobin absorbs lower energy (orange-red) light, and the sample appears blue.

24.66   (a) $[FeF_6]^{4-}$. Both complexes contain the same metal ion, Fe^{2+}; F^- is a weak-field ligand that imposes a smaller Δ and longer λ for the complex ion.

        (b) $[V(H_2O)_6]^{2+}$. Both complexes contain the same ligand, H_2O. V^{2+} has a lower charge, so the interaction with the ligand will produce a weaker field, a smaller Δ, and a longer absorbed wavelength.

        (c) $[CoCl_4]^{2-}$. Both complexes contain the same metal ion, Co^{2+}; Cl^- is a weak-field ligand that imposes a smaller Δ and a longer λ for the complex ion.

24.67   (a)   The term *isoelectronic* means that the three ions have the same number of valence electrons and the same electron configuration.

(b)   In each ion, the metal is in its maximum oxidation state and has a d^0 electron configuration. That is, the metal ions have no *d*-electrons, so there should be no *d-d* transitions.

(c)   A *ligand-metal charge transfer* transition occurs when an electron in a filled ligand orbital is excited to an empty d-orbital of the metal.

(d)   Absorption of 565 nm yellow light by MnO_4^- causes the compound to appear violet, the complementary color. CrO_4^{2-} appears yellow, so it is absorbing violet light of approximately 420 nm. The wavelength of the LMCT transition for chromate, 420 nm, is shorter than the wavelength of LCMT transition in permanganate, 565 nm. This means that there is a larger energy difference between filled ligand and empty metal orbitals in chromate than in permanganate.

(e)   Yes. A white compound indicates that no visible light is absorbed. Going left on the periodic chart from Mn to Cr, the absorbed wavelength got shorter and the energy difference between ligand and metal orbitals increased. The 420 nm absorption by CrO_4^- is at the short wavelength edge of the visible spectrum. It is not surprising that the ion containing V, further left on the chart, absorbs at a still shorter wavelength in the ultraviolet region and that VO_4^{3-} appears white.

24.68   Application of pressure would result in shorter metal ionoxide distances. This would have the effect of increasing the ligand-electron repulsions, and would result in a larger splitting in the d-orbital energies. Thus, application of pressure should result in a shift in the absorption to a higher energy and shorter wavelength.

24.69   (a)

$$\begin{bmatrix} & & CO & & \\ NC & \diagdown & | & \diagup & CN \\ & & Fe & & \\ NC & \diagup & | & \diagdown & CN \\ & & CO & & \end{bmatrix}^{2-}$$

(b)   sodium dicarbonyltetracyanoferrate(II)

(c)   +2, 6 d-electrons

(d)   We expect the complex to be low spin. Cyanide (and carbonyl) are high on the spectrochemical series, which means the complex will have a large Δ splitting characteristic of low spin complexes.

24.70

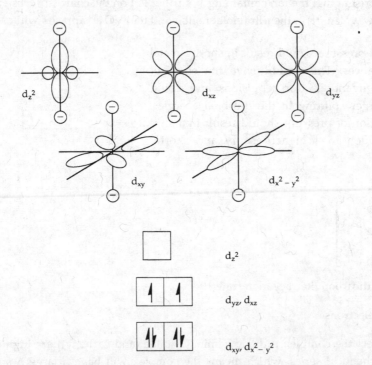

24.71 (a) Only one   (b) Two

(c) Four; two are geometric, the other two are stereoisomers of each of these.

24.72

For a d^6 metal ion in a strong ligand field, there would be two unpaired electrons.

Integrative Exercises

24.73 In a complex ion, the transition metal is an electron pair acceptor, a Lewis acid; the ligand is an electron pair donor, a Lewis base. In carbonic anhydrase, the Zn^{2+} ion withdraws electron density from the O atom of water. The electronegative oxygen atom compensates by withdrawing electron-density from the O—H bond. The O—H bond is polarized and H becomes more ionizable, more acidic than in the bulk solvent. This is similar to the effect of an electronegative central atom in an oxyacid such as H_2SO_4.

24.74 (a) Both compounds have the same general formulation, so Co is in the same (+3) oxidation state in both complexes.

(b) Cobalt(III) complexes are generally inert; that is, they do not rapidly exchange ligands inside the coordination sphere. Therefore, the ions that form precipitates in these two cases are probably outside the coordination sphere. The dark violet compound A forms a precipitate with $BaCl_2(aq)$ but not $AgNO_3(aq)$, so it has SO_4^{2-} outside the coordination sphere and coordinated Br^-, $[Co(NH_3)_5Br]SO_4$. The red-violet compound B forms a precipitate with $AgNO_3(aq)$ but not $BaCl_2(aq)$ so it has Br^- outside the coordination sphere and coordinated SO_4^{2-}, $[Co(NH_3)_5SO_4]Br$.

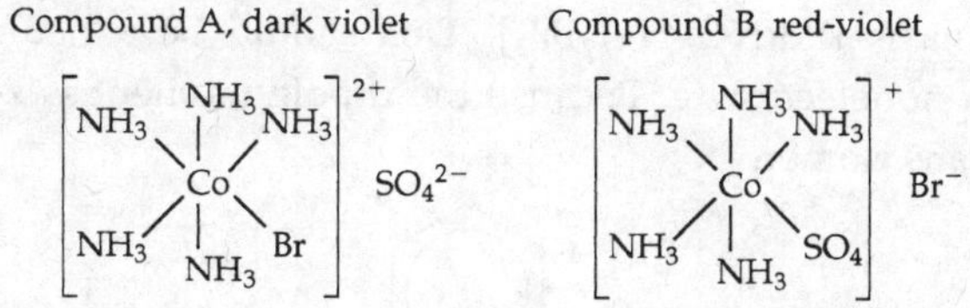

Compound A, dark violet   Compound B, red-violet

(c) Compounds A and B have the same formula but different properties (color, chemical reactivity), so they are isomers. They vary by which ion is inside the coordination sphere, so they are *coordination sphere isomers*.

(d) Compound A is an ionic sulfate and compound B is an ionic bromide, so both are strong electrolytes. According to the solubility rules in Table 4.1, both should be water-soluble.

24.75 Determine the empirical formula of the complex, assuming the remaining mass is due to oxygen, and a 100 g sample.

$$10.0\ g\ Mn \times \frac{1\ mol\ Mn}{54.94\ g\ Mn} = 0.1820\ mol\ Mn;\ 0.182\ /\ 0.182 = 1$$

$$28.6\ g\ K \times \frac{1\ mol\ K}{39.10\ g\ K} = 0.7315\ mol\ K;\ 0.732\ /\ 0.182 = 4$$

$$8.8\ g\ C \times \frac{1\ mol\ C}{12.0\ g\ C} = 0.7327\ mol\ C;\ 0.733\ /\ 0.182 = 4$$

$$29.2\ g\ Br \times \frac{1\ mol\ Br}{79.904\ g\ Br} = 0.3654\ mol\ Br;\ 0.365\ /\ 0.182 = 2$$

$$23.4\ g\ O \times \frac{1\ mol\ O}{16.00\ g\ O} = 1.463\ mol\ O;\ 1.46\ /\ 0.182 = 8$$

There are 2 C and 4 O per oxalate ion, for a total of two oxalate ligands in the complex. To match the conductivity of $K_4[Fe(CN)_6]$, the oxalate and bromide ions must be in the coordination sphere of the complex anion. Thus, the compound is $K_4[Mn(ox)_2Br_2]$.

24.76　First determine the empirical formula, assuming that the remaining mass of complex is Pd.

$$37.6 \text{ g Br} \times \frac{1 \text{ mol Br}}{79.904 \text{ g Br}} = 0.4706 \text{ mol Br}; 0.4706 / 0.2361 = 2$$

$$28.3 \text{ g C} \times \frac{1 \text{ mol C}}{12.01 \text{ g C}} = 2.356 \text{ mol C}; 2.356 / 0.2361 = 10$$

$$6.60 \text{ g N} \times \frac{1 \text{ mol N}}{14.01 \text{ g N}} = 0.4711 \text{ mol N}; 0.4711 / 0.2361 = 2$$

$$2.37 \text{ g H} \times \frac{1 \text{ mol H}}{1.008 \text{ g H}} = 2.351 \text{ mol H}; 2.351 / 0.2361 = 10$$

$$25.13 \text{ g Pd} \times \frac{1 \text{ mol Pd}}{106.42 \text{ g Pd}} = 0.2361 \text{ mol Pd}; 0.2361 / 0.2361 = 1$$

The chemical formula is $[Pd(NC_5H_5)_2Br_2]$. This should be a neutral square-planar complex of Pd(II), a nonelectrolyte. Because the dipole moment is zero, we can infer that it must be the trans isomer.

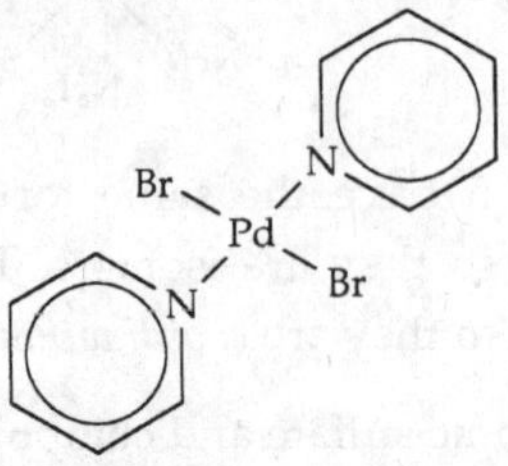

24.77　(a)　The reaction that occurs increases the conductivity of the solution by producing a greater number of charged particles, particles with higher charges, or both. It is likely that H_2O from the bulk solvent exchanges with a coordinated Br according to the reaction below. This reaction would convert the 1:1 electrolyte, $[Co(NH_3)_4Br_2]Br$, to a 1:2 electrolyte, $[Co(NH_3)_3(H_2O)Br]Br_2$.

　　　　(b)　$[Co(NH_3)_4Br_2]^+(aq) + H_2O(l) \rightarrow [Co(NH_3)_4(H_2O)Br]^{2+}(aq) + Br^-(aq)$

　　　　(c)　Before the exchange reaction, there is one mole of free Br^- per mole of complex. mol $Br^- =$ mol Ag^+

　　　　　　$M =$ mol/L; L $AgNO_3 =$ mol $AgNO_3 / M$ $AgNO_3$

$$\frac{3.87 \text{ g complex}}{0.500 \text{ L soln}} \times \frac{1 \text{ mol complex}}{366.77 \text{ g complex}} \times 0.02500 \text{ L soln used} =$$

$$5.276 \times 10^{-4} = 5.28 \times 10^{-4} \text{ mol complex}$$

$$5.276 \times 10^{-4} \text{ mol complex} \times \frac{1 \text{ mol Br}^-}{1 \text{ mol complex}} \times \frac{1 \text{mol Ag}^+}{1 \text{ mol Br}^-} \times \frac{1 \text{ L Ag}^+(\text{aq})}{0.0100 \text{ mol Ag}^+(\text{aq})}$$

$$= 0.05276 \text{ L} = 52.8 \text{ mL AgNO}_3(\text{aq})$$

(d)   After the exchange reaction, there are 2 mol free Br^- per mol of complex. Since M $AgNO_3(aq)$ and volume of complex solution are the same for the second experiment, the titration after conductivity changes will require twice the volume calculated in part (c), 105.52 = 106 mL of 0.0100 M $AgNO_3(aq)$.

24.78   Calculate the concentration of Mg^{2+} alone, and then the concentration of Ca^{2+} by difference. $M \times L = \text{mol}$

$$\frac{0.0104 \text{ mol EDTA}}{1 \text{ L}} \times 0.0187\text{L} \times \frac{1 \text{ mol Mg}^{2+}}{1 \text{ mol EDTA}} \times \frac{24.31 \text{ g Mg}^{2+}}{1 \text{ mol Mg}^{2+}} \times \frac{1000 \text{ mg}}{\text{g}}$$

$$\times \frac{1}{0.100 \text{ L H}_2\text{O}} = 47.28 = 47.3 \text{ mg Mg}^{2+}/\text{L}$$

$$0.0104 \text{ } M \text{ EDTA} \times 0.0315 \text{ L} = \text{mol} \,(\text{Ca}^{2+} + \text{Mg}^{2+})$$

$$\frac{0.0104 \text{ } M \text{ EDTA} \times 0.0187 \text{ L} = \text{mol Mg}^{2+}}{0.0104 \text{ } M \text{ EDTA} \times 0.0128 \text{ L} = \text{mol Ca}^{2+}}$$

$$0.0104 \text{ M EDTA} \times 0.0128 \text{ L} \times \frac{1 \text{ mol Ca}^{2+}}{1\text{mol EDTA}} \times \frac{40.08 \text{ g Ca}^{2+}}{1 \text{ mol Ca}^{2+}} \times \frac{1000 \text{ mg}}{\text{g}} \times \frac{1}{0.100 \text{ L H}_2\text{O}}$$

$$= 53.35 = 53.4 \text{ mg Ca}^{2+}/\text{L}$$

24.79   Use Hess' law to calculate $\Delta G°$ for the desired equilibrium. Then $\Delta G° = -RT\ln K$ to calculate K.

$$\begin{array}{ll} \text{Hb} + \text{CO} \rightarrow \text{HbCO} & \Delta G° = -80 \text{ kJ} \\ \underline{\text{HbO}_2 \rightarrow \text{Hb} + \text{O}_2} & \Delta G° = 70 \text{ kJ} \\ \text{HbO}_2 + \text{Hb} + \text{CO} \rightarrow \text{HbCO} + \text{Hb} + \text{O}_2 & \\ \text{HbO}_2 + \text{CO} \rightarrow \text{HbCO} + \text{O}_2 & \Delta G° = -10 \text{ kJ} \end{array}$$

$$\Delta G° = -RT\ln K, \ln K = \frac{-\Delta G°}{RT} = \frac{-(-10 \text{ kJ})}{8.314 \text{ J/K} \bullet \text{mol} \times 298 \text{ K}} \times \frac{1000 \text{ J}}{\text{kJ}} = 4.036 = 4.04$$

$$K = e^{4.04} = 56.61 = 57$$

24.80   (a) $$[\text{Cd}(\text{CH}_3\text{NH}_2)_4]^{2+} \rightleftharpoons \text{Cd}^{2+}(\text{aq}) + 4\text{CH}_3\text{NH}_2(\text{aq}) \qquad \Delta G° = 37.2 \text{ kJ}$$

$$\underline{\text{Cd}^{2+}(\text{aq}) + 2\text{en}(\text{aq}) \rightleftharpoons [\text{Cd}(\text{en})_2]^{2+}(\text{aq})} \qquad\qquad\qquad \Delta G° = -60.7 \text{ kJ}$$

$$\text{Cd}(\text{CH}_3\text{NH}_2)_4]^{2+} + 2\text{en}(\text{aq}) \rightleftharpoons [\text{Cd}(\text{en})_2]^{2+}(\text{aq}) + 4\text{CH}_3\text{NH}_2(\text{aq}) \qquad \Delta G° = -23.5 \text{ kJ}$$

$$\Delta G° = -RT\ln K; -23.5 \text{ kJ} = -2.35 \times 10^4 \text{ J}$$

$$-2.35 \times 10^4 \text{ J} = \frac{-8.314 \text{ J}}{\text{K} \bullet \text{mol}} \times 298 \text{ K} \times \ln K; \ln K = 9.485, K = 1.32 \times 10^4$$

(b) The magnitude of K is large, so the reaction favors products. The bidentate chelating ligand en will spontaneously replace the monodentate ligand CH_3NH_2. This is an illustration of the chelate effect.

(c) Using the stepwise construction from part (a),

$\Delta H° = 57.3 \text{ kJ} - 56.5 \text{ kJ} = 0.8 \text{ kJ}$

$\Delta S° = 67.3 \text{ J/K} + 14.1 \text{ J/K} = 81.4 \text{ J/K}$

$-T\Delta S = -298 \text{ K} \times 81.4 \text{ J/K} = -2.43 \times 10^4 \text{ J} = -24.3 \text{ kJ}$

The chelate effect is mainly the result of entropy. The reaction is spontaneous due to the increase in the number of free particles and corresponding increase in entropy going from reactants to products. The enthalpic contribution is essentially zero because the bonding interactions of the two ligands are very similar and the reaction is not "downhill" in enthalpy.

(d) $\Delta H°$ will be very small and negative. When NH_3 replaces H_2O in a complex (Closer Look Box), the tighter bonding of the NH_3 ligand causes a substantial negative $\Delta H°$ for the substitution reaction. When a bidentate amine ligand replaces a monodentate amine ligand of similar bond strength, $\Delta H°$ is very small and either positive (part (c)) or negative (Closer Look Box). In the case of NH_3 replacing CH_3NH_2, the bonding characteristics are very similar. The presence of CH_3 groups in CH_3NH_2 produces some steric hindrance in $[Cd(CH_3NH_2)_4]^{2+}$. This complex is at a slightly higher energy than $[Cd(NH_3)_4]^{2+}$, which experiences no steric hindrance, so $\Delta H°$ will have a negative sign but a very small magnitude. Relief of steric hindrance leads to a very small negative $\Delta H°$ for the substitution reaction.

24.81 $\quad \dfrac{182 \times 10^3 \text{ J}}{1 \text{ mol}} \times \dfrac{1 \text{ mol}}{6.022 \times 10^{23} \text{ molecules}} = 3.022 \times 10^{-19} = 3.02 \times 10^{-19} \text{ J/photon}$

$\Delta E = h\nu = 3.02 \times 10^{-19} \text{ J}; \nu = \Delta E / h$

$\nu = 3.022 \times 10^{-19} \text{ J} / 6.626 \times 10^{-34} \text{ J} \cdot \text{s} = 4.561 \times 10^{14} = 4.56 \times 10^{14} \text{s}^{-1}$

$\lambda = \dfrac{2.998 \times 10^8 \text{ m/s}}{4.561 \times 10^{14} \text{ s}^{-1}} = 6.57 \times 10^{-7} \text{ m} = 657 \text{ nm}$

We expect that this complex will absorb in the visible, at around 660 nm. It will thus exhibit a blue-green color (Figure 24.24).

24.82 $\quad$ The process can be written:

$$H_2(g) + 2e \rightarrow 2H^+(aq) \qquad\qquad E_{red}° = 0.0 \text{ V}$$

$$Cu(s) \rightarrow Cu^{2+} + 2e^- \qquad\qquad E_{red}° = 0.337 \text{ V}$$

$$\underline{Cu^{2+}(aq) + 4NH_3(aq) \rightarrow [Cu(NH_3)_4]^{2+} (aq) \qquad\qquad "E_f°" = ?}$$

$$H_2(g) + Cu(s) + 4NH_3(aq) \rightarrow 2H^+(aq) + [Cu(NH_3)_4]^{2+}(aq) \qquad E = 0.08 \text{ V}$$

$$E = E° - RT \ln K; \ K = \frac{[H^+]^2[Cu(NH_3)_4^{2+}]}{P_{H_2}[NH_3]^4}$$

$P_{H_2} = 1 \text{ atm}, \ [H^+] = 1 \ M, \ [NH_3] = 1 \ M, \ [Cu(NH_3)_4]^{2+} = 1 \ M, \ Q = 1$

$E = E° - RT \ln(1); \ E = E° - RT(0); \ E = E° = 0.08 \text{ V}$

Since we know E° values for two steps and the overall reaction, we can calculate "E°" for the formation reaction and then K_f, using $E° = \frac{0.0592}{n} \log K_f$ for the step.

$E_{cell} = 0.08 \text{ V} = 0.0 \text{ V} - 0.337 \text{ V} + "E_f°" \quad "E_f°" = 0.08 \text{ V} + 0.337 \text{ V} = 0.417 \text{ V} = 0.42 \text{ V}$

$"E_f°" = \frac{0.0592}{n} \log K_f; \ \log K_f = \frac{n(E_f°)}{0.0592} = \frac{2(0.417)}{0.0592} = 14.0878 = 14$

$K_f = 10^{14.0878} = 1.2 \times 10^{14} = 10^{14}$

24.83 (a) The units of the rate constant and the rate dependence on the identity of the second ligand show that the reaction is second order. Therefore, the rate-determining step cannot be a dissociation of water, since that would be independent of the concentration and identity of the incoming ligand. The alternative mechanism, a bimolecular association of the incoming ligand with the complex, is indicated.

(b) The relative values of rate constant are a reflection of the kinetic basicities of the three ligands: Pyridine $>$ SCN$^-$ $>$ CH$_3$CN.

(c) Ru(III) is a d^5 ion. In a low-spin d^5 octahedral complex, there is one unpaired electron.

25 The Chemistry of Life: Organic and Biological Chemistry

Visualizing Concepts

25.1 *Analyze/Plan.* Given structural formulas, specify which molecules are unsaturated. Consider the definition of unsaturated and apply it to the molecules in the exercise. *Solve.*

Unsaturated molecules contain one or more multiple bonds. Saturated molecules contain only single bonds. Molecules (c) and (d) are unsaturated.

25.2 *Analyze/Plan.* Given structural formulas, decide which molecule will undergo addition. Consider which functional groups are present in the molecules, and which are most susceptible to addition. *Solve.*

Addition reactions are characteristic of alkenes. Molecule (c), an alkene, will readily undergo addition.

Molecule (a) is an aromatic hydrocarbon, which does not typically undergo addition because the delocalized electron cloud is too difficult to disrupt. Molecules (b) and (d) contain carbonyl groups (actually carboxylic acid groups) that do not typically undergo addition, except under special conditions.

25.3 *Analyze/plan.* Given structural formulas, predict which molecule will have the highest boiling point. Boiling point is determined by strength of intermolecular forces; for neutral molecules with similar molar masses, the strongest intermolecular force is hydrogen bonding. The molecule that experiences hydrogen bonding will have the highest boiling point. *Solve.*

Only $F-H, O-H$, and $N-H$ bonds fit the strict definition of hydrogen bonding. Molecule (b), an alcohol, forms hydrogen bonds with like molecules; it has the highest boiling point.

[Molecules (a) and (d) have dipole-dipole and dispersion forces. The greater molar mass of (d) probably means it has stronger dispersion forces and a slightly higher boiling point than molecule (a). Molecule (c) has only dispersion forces and the lowest boiling point. The probable order of strength of forces and boiling points is: (b) > (d) > (a) > (c).]

25.4 *Analyze.* Given structural formulas, decide which molecules are capable of isomerism, and what type. *Plan.* Analyze each molecule for possible structural, geometric, and optical isomers/enantiomers. *Solve.*

670

(a) $C_5H_{11}NO_2$. Structural, geometric and optical. There are many ways to arrange the atoms in molecules with this empirical formula, so there are many structural isomers. There is one point of unsaturation in the given molecule, the C=O group; structural isomers with their point of unsaturation at a C=C group could have geometric isomers as well.   The C atom to which the $-NH_3^+$ group is bound is a chiral center, so there are enantiomers. All amino acids except glycine have two possible enantiomers.

(b) $C_7H_5O_2Cl$. The most obvious isomers for this aromatic compound are ortho, meta, and para geometric isomers. Because the molecule has several points of unsaturation, the number of structural isomers is limited, but there are a few possibilities with two triple bonds. Switching the $-OH$ and $-Cl$ groups also generates a structural isomer. There are no chiral centers, so no optical isomers.

(c) C_5H_{10}. There are many structural isomers for this empirical formula. The straight-chain alkene shown also has geometric (cis-trans) isomers.

(d) C_3H_8. There are no other structural, geometric, or optical isomers for this molecule.

25.5 *Analyze/Plan.* Given ball-and-stick models, select the molecule that fits the description given. From the models, decide the type of molecule or functional group represented.

Solve. Molecule (i) is a sugar, (ii) is an organic base and a component of nucleic acids, (iii) is an amino acid, and (iv) is an alcohol.

(a) Molecule (i) is a disaccharide composed of galactose (left) and glucose (right); it can be hydrolyzed to form a solution containing glucose. Since it is the only sugar molecule depicted, it was not necessary to know the exact structure of glucose to answer the question.

(b) Amino acids form zwitterions, so the choice is molecule (iii).

(c) Molecule (ii) is an organic base present in DNA (again, the only possible choice).

(d) Molecule (iv) because alcohols react with carboxylic acids to form esthers.

25.6 *Analyze/Plan.* Follow the logic in Sample Exercise 25.1 to name each compound. Decide which structures are the same compound.   *Solve.*

(a) 2,2,4-trimethylpentane   (b) 3-ethyl-2-methylpentane

(c) 2,3,4-trimethylpentane   (d) 2,3,4-trimethylpentane

Structures (c) and (d) are the same molecule.

Introduction to Organic Compounds; Hydrocarbons

25.7 *Analyze/Plan.* Given a condensed structural formula, determine the bond angles and hybridization about each carbon atom in the molecule. Visualize the number of electron domains about each carbon. State the bond angle and hybridization based on electron domain geometry.   *Solve.*

H H O
| | ‖
H—C—C—C—H
  3| 2| 1
   H H

C2 and C3 both have tetrahedral electron domain geometry, 109° bond angles and sp³ hybridization. C1 has trigonal planar electron domain geometry, 120° bond angles and sp² hybridization.

25.8

        H H H H OH O
        | | | | | ‖
N≡C—C—C—C=C—C—C—H
     7 6| 5| 4 3 2| 1
        H H   H

(a) C2, C5 and C6 have sp³ hybridization (4 e⁻ domains around C)

(b) C7 has sp hybridization (2 e⁻ domains around C)

(c) C1, C3, and C4 have sp² hybridization (3 e⁻ domains around C)

25.9 Carbon (of course), hydrogen, oxygen, nitrogen, sulfur, phosphorus, chlorine, (and other halogens). According to periodic trends and Figure 8.6, oxygen, nitrogen, fluorine, and chlorine are more electronegative than carbon. Sulfur has the same electronegativity as carbon.

25.10 From Table 8.4, the bond enthalpies in kJ/mol are: C—H, 413; C—C, 348; C—O, 358; C—Cl, 328. The bond enthalpes indicate that C—H bonds are most difficult to break, and C—Cl bonds least difficult. However, they do not explain the reactivity of C—O bonds, or stability of C—C bonds.

The reactivity of molecules containing C—O and C—Cl bonds is a result of their unequal charge distribution, which attracts reactants that are either electron deficient (electrophilic) or electron rich (nucleophilic).

25.11 (a) A *straight-chain hydrocarbon* has all carbon atoms connected in a continuous chain; no carbon atom is bound to more than two other carbon atoms. A *branched-chain hydrocarbon* has a branch; at least one carbon atom is bound to three or more carbon atoms.

(b) An *alkane* is a complete molecule composed of carbon and hydrogen in which all bonds are single (sigma) bonds. An *alkyl group* is a substituent formed by removing a hydrogen atom from an alkane.

(c) Alkanes are said to be *saturated* because they contain only single bonds. Multiple bonds that enable addition of H_2 or other substances are absent. The bonding capacity of each carbon atom is fulfilled with single bonds to C or H.

25.12 All the classifications listed are hydrocarbons; they contain only the elements hydrogen and carbon.

(a) *Alkanes* are hydrocarbons that contain only single bonds.

(b) *Cycloalkanes* contain at least one ring of three or more carbon atoms joined by single bonds. Because it is a type of alkane, all bonds in a cycloalkane are single bonds.

(c) *Alkenes* contain at least one C=C double bond.

(d) *Alkynes* contain at least one C≡C triple bond.

25 Organic and Biological Chemistry   Solutions to Exercises

(e)   A *saturated hydrocarbon* contains only single bonds. Alkanes and cycloalkanes fit this definition.

(f)   An *aromatic hydrocarbon* contains one or more planar, six-membered rings of carbon atoms with delocalized π-bonding throughout the ring.

25.13   *Analyze/Plan.* Consider the definition of the stated classification and apply it to a compound containing five C atoms.   *Solve.*

(a) $CH_3CH_2CH_2CH_2CH_3$, C_5H_{12}

(b)

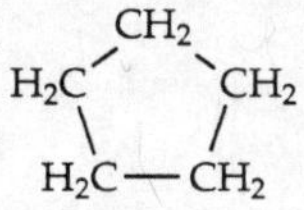

(c) $CH_2=CHCH_2CH_2CH_3$, C_5H_{10}

(d) $HC\equiv CCH_2CH_2CH_3$, C_5H_8 saturated: (a), (b); unsaturated: (c), (d)

25.14   cycloalkane, H_2C-CH_2 , C_6H_{12}, saturated

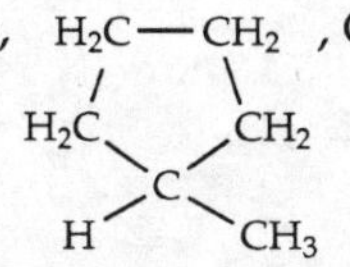

cycloalkene, $HC=CH$, C_6H_{10}, unsaturated

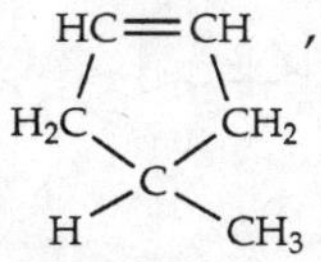

alkyne, $CH_3-CH_2-C\equiv C-CH_2-CH_3$, C_6H_{10}, unsaturated

aromatic hydrocarbon, C_6H_6, unsaturated

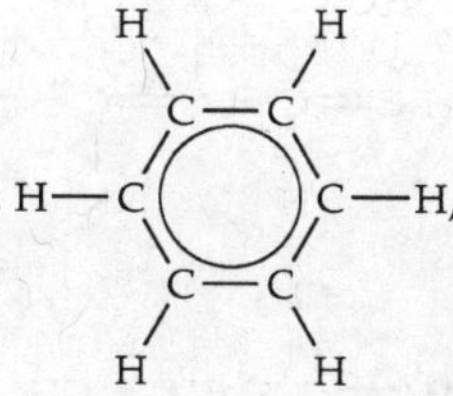

25.15   *Analyze/Plan.* The general formula of an alkane is C_nH_{2n+2}. For an alkene, with 2 fewer H atoms, the general formula is C_nH_{2n}.   *Solve.*

A dialkene has one more $C=C$ and thus two fewer H atoms than an alkene. The general formula is C_nH_{2n-2}.

25.16 C_nH_{2n-2}

673

25.17 *Analyze/Plan.* Follow the logic in Sample Exercise 25.3.   *Solve.*

$CH_3 - CH_2 - CH_2 - CH = CH_2$
pentene

$CH_3 - CH_2 - CH = CH - CH_3$
2-pentene

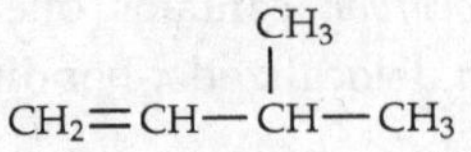

$$\underset{\text{3-methyl-1-butene}}{CH_2 = CH - \underset{\underset{CH_3}{|}}{CH} - CH_3}$$

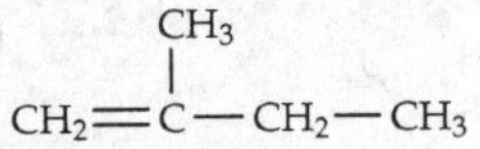

$$\underset{\text{2-methyl-1-butene}}{CH_2 = \underset{\overset{CH_3}{|}}{C} - CH_2 - CH_3}$$

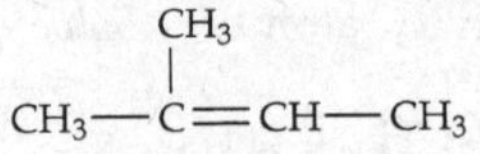

$$\underset{\text{2-methyl-2-butene}}{CH_3 - \underset{\overset{CH_3}{|}}{C} = CH - CH_3}$$

25.18

$CH_3CH_2CH_2CH_2C \equiv CH$

$CH_3CH_2CH_2C \equiv CCH_3$

$CH_3CH_2C \equiv CCH_2CH_3$

$$CH_3\underset{\underset{CH_3}{|}}{CH}CH_2C \equiv CH$$

$$CH_3CH_2\underset{\overset{CH_3}{|}}{C}HC \equiv CH$$

$$CH_3\underset{\underset{CH_3}{|}}{CH}C \equiv CCH_3$$

$$CH_3CH_2\underset{\overset{H}{|}}{C} = \underset{\overset{H}{|}}{C} - CH = CH_2$$

$$CH_3CH_2\underset{\underset{H}{|}}{\overset{\overset{H}{|}}{C}} = C - CH = CH_2$$

$$CH_3\underset{\overset{H}{|}}{C} = \underset{\overset{H}{|}}{C} - CH_2CH = CH_2$$

$$CH_3\underset{\underset{H}{|}}{\overset{\overset{H}{|}}{C}} = C - CH_2CH = CH_2$$

$CH_2 = CHCH_2CH_2CH = CH_2$

$$CH_3\underset{\overset{H}{|}}{C} = \underset{\overset{H}{|}}{C} - \underset{\overset{H}{|}}{C} = \underset{\overset{H}{|}}{C} - CH_3$$

$$CH_3\underset{\overset{H}{|}}{C} = \underset{\underset{H}{|}}{\overset{\overset{H}{|}}{C}} - \underset{\underset{H}{|}}{C} = C - CH_3$$

$$CH_3\underset{\underset{H}{|}}{\overset{\overset{H}{|}}{C}} = C - \underset{\underset{H}{|}}{C} = C - CH_3$$

$$CH_2 = CHCH\underset{\overset{CH_3}{|}}{}CH = CH_2$$

$$CH_2 = \underset{\overset{CH_3}{|}}{C}CH_2CH = CH_2$$

$$CH_2 = CH\underset{\overset{CH_3}{|}}{C} = CH$$

$$CH_2 = CH\underset{\overset{CH_3\ H}{|\ \ |}}{C} = CCH_3$$

$$CH_2 = \underset{\underset{H}{|}}{\overset{\overset{CH_3\ H}{|}}{C}}C = CCH_3$$

$$CH_2 = \underset{\underset{H\ H}{|\ \ |}}{\overset{\overset{CH_3}{|}}{C}}C = CCH_3$$

$$CH_2 = CHCH = \underset{\overset{CH_3}{|}}{C}CH_3$$

$$CH_2 = CH\underset{\overset{CH_2CH_3}{|}}{C} = CH_2$$

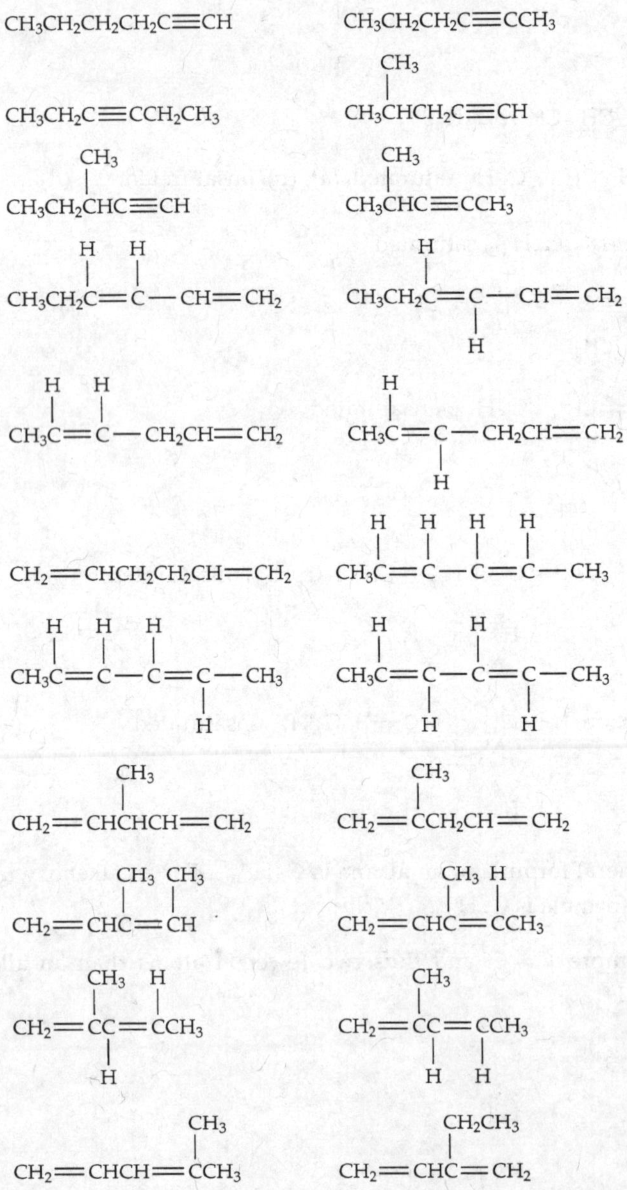

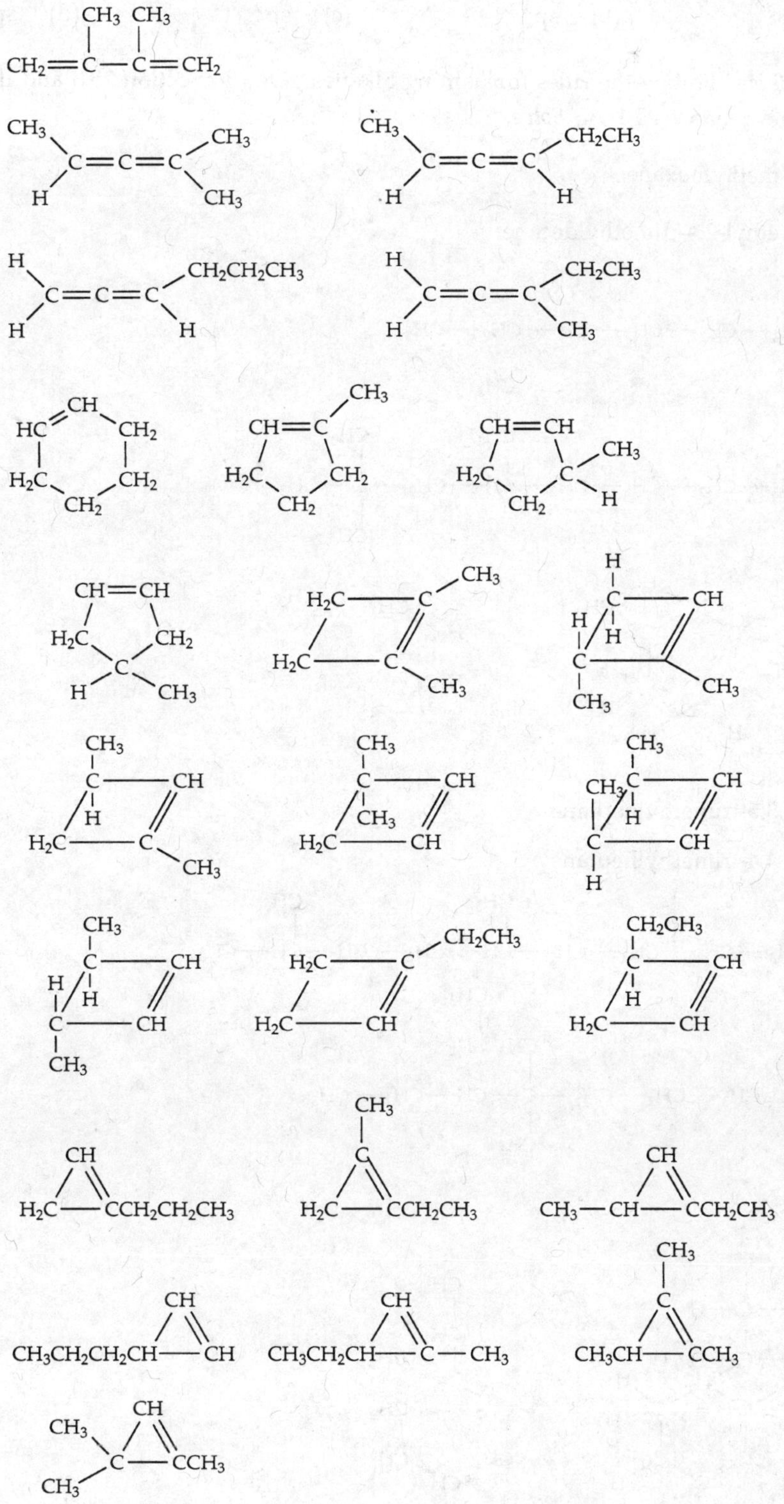

$CH_2=\overset{\overset{\displaystyle CH_3}{|}}{C}-\overset{\overset{\displaystyle CH_3}{|}}{C}=CH_2$

$\overset{\displaystyle CH_3}{\underset{\displaystyle H}{}}C=C=C\overset{\displaystyle CH_3}{\underset{\displaystyle CH_3}{}}$

$\overset{\displaystyle \cdot CH_3}{\underset{\displaystyle H}{}}C=C=C\overset{\displaystyle CH_2CH_3}{\underset{\displaystyle H}{}}$

$\overset{\displaystyle H}{\underset{\displaystyle H}{}}C=C=C\overset{\displaystyle CH_2CH_2CH_3}{\underset{\displaystyle H}{}}$

$\overset{\displaystyle H}{\underset{\displaystyle H}{}}C=C=C\overset{\displaystyle CH_2CH_3}{\underset{\displaystyle CH_3}{}}$

25.19   (a)   109°   (b)   120°   (c)   180°

25.20   (a) sp^3 (b) sp^2 (c) sp^2 (d) sp

25.21   *Analyze/Plan.* Follow the rules for naming alkanes given in Section 25.3 and illustrated in Sample Exercise 25.1.   *Solve.*

     (a)   2-methylhexane

     (b)   4-ethyl-2,4-dimethyldecane

     (c)

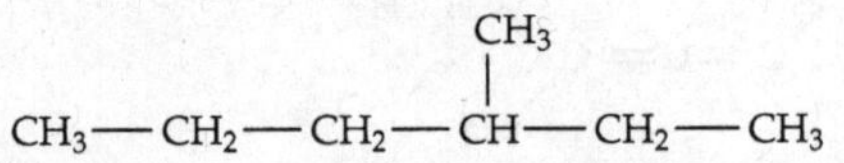

$$CH_3-CH_2-CH_2-\underset{\underset{\displaystyle CH_3}{|}}{CH}-CH_2-CH_3$$

     (d) $CH_3-CH_2-CH_2-CH_2-\underset{\underset{\displaystyle CH_2}{\underset{|}{\overset{|}{\overset{\displaystyle CH_3}{}}}}}{CH}-CH_2-\underset{\underset{\displaystyle CH_3}{|}}{\overset{\overset{\displaystyle CH_3}{|}}{C}}-CH_3$

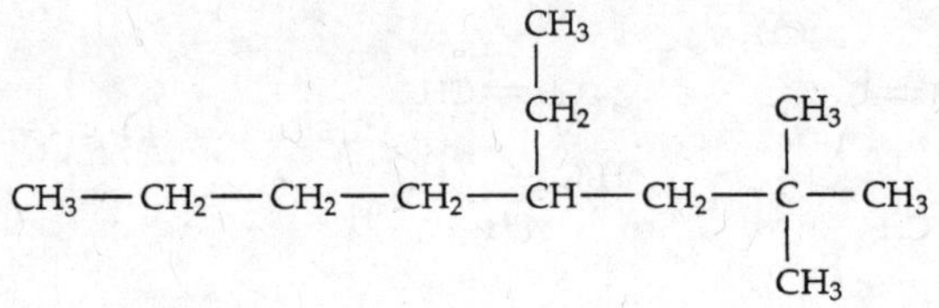

     (e)

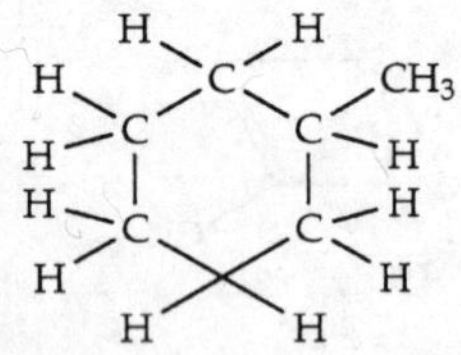

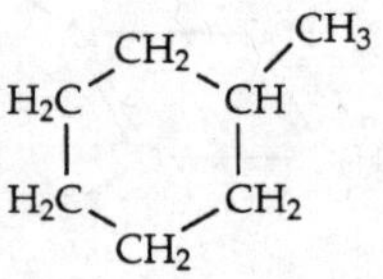

25.22   (a)   3,3,5-trimethylheptane

     (b)   3,4,4-trimethylheptane

     (c) $CH_3-CH_2-CH_2-CH_2-\underset{\underset{\displaystyle CH_3}{|}}{\overset{\overset{\displaystyle CH_3}{|}}{CH}}-CH_2-CH_2-\underset{\underset{\displaystyle CH_3}{|}}{CH}-CH_3$

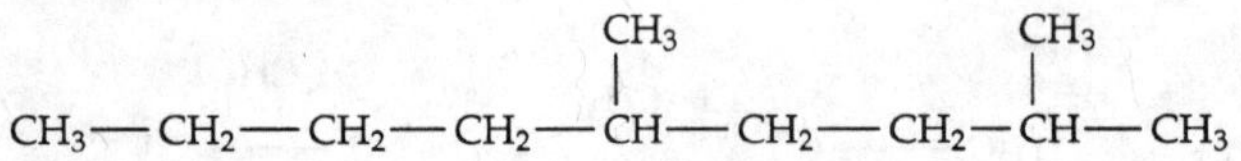

     (d) $CH_3-CH_2-CH_2-\underset{\underset{\displaystyle CH_3}{|}}{\overset{\overset{\displaystyle CH_3}{|}}{C}}-\underset{\underset{\displaystyle CH_3}{}}{\overset{\overset{\displaystyle CH_2}{|}}{CH}}-CH_2-CH_3$

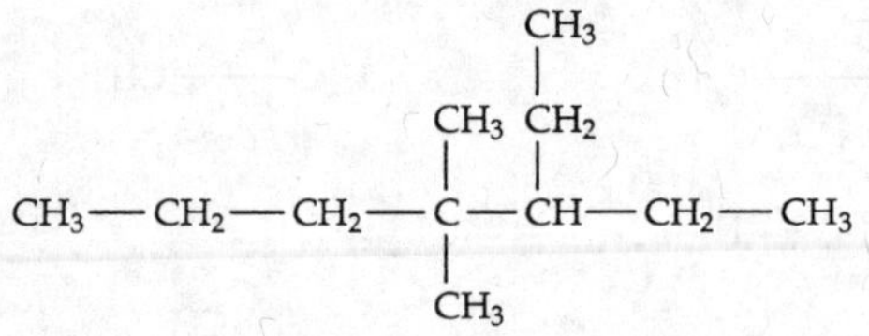

     (e)

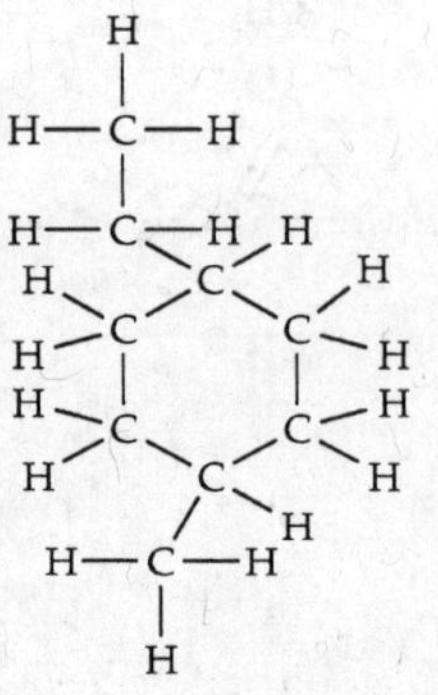

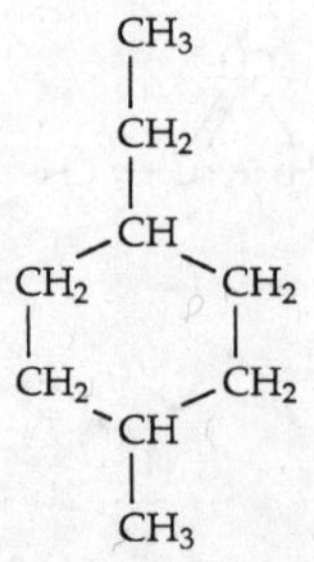

25.23   *Analyze/Plan.* Follow the logic in Sample Exercises 25.1 and 25.4.   *Solve.*

     (a)   2,3-dimethylheptane

     (b)   *cis*-6-methyl-3-octene

     (c)   *para*-dibromobenzene

     (d)   4,4-dimethyl-1-hexyne

     (e)   methylcyclobutane

25.24   (a)   1,4-dichlorocyclohexane

     (b)   3-chloro-1-propyne

     (c)   *trans*-2-hexene

     (d)   1-chloro-2-methyl-2-phenyl-butane or (1-chloro-2-methyl)-2-butylbenzene

     (e)   *cis*-5-chloro-1,3-pentadiene

25.25   Each doubly bound carbon atom in an alkene has two unique sites for substitution. These sites cannot be interconverted because rotation about the double bond is restricted; *geometric isomerism* results. In an alkane, carbon forms only single bonds, so the three remaining sites are interchangeable by rotation about the single bond. Although there is also restricted rotation around the triple bond of an alkyne, there is only one additional bonding site on a triply bound carbon, so no isomerism results.

25.26   Butene is an alkene, C_4H_8. There are two possible placements for the double bond:

$$CH_2=CHCH_2CH_3 \quad \text{or} \quad CH_3CH=CHCH_3$$
$$\text{1-butene} \qquad\qquad\qquad \text{2-butene}$$

These two compounds are *structural isomers*. For 2-butene, there are two different, noninterchangeable ways to construct the carbon skeleton (owing to the absence of free rotation around the double bond). These two compounds are *geometric isomers*.

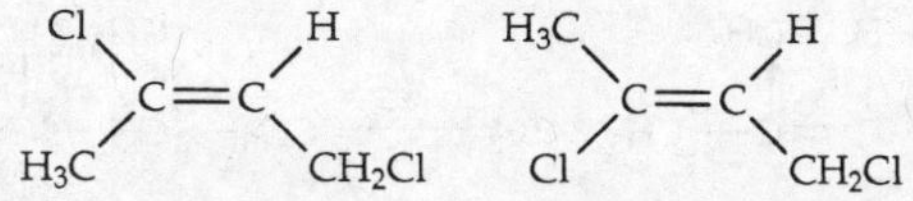

cis-2-butene   *trans*-2-butene

25.27   *Analyze/Plan.* In order for geometrical isomerism to be possible, the molecule must be an alkene with two different groups bound to each of the alkene C atoms.   *Solve.*

     (a)

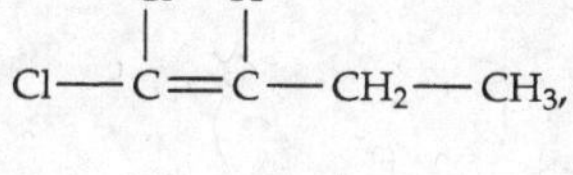

     (b)

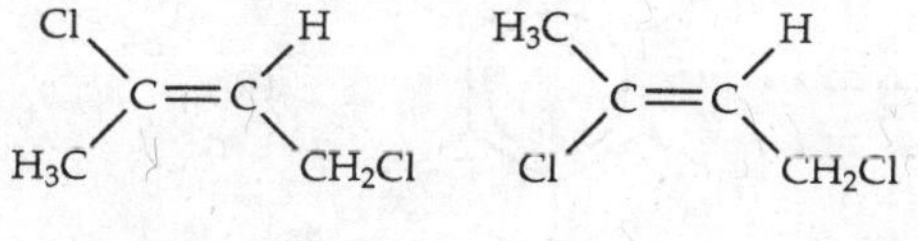

     (c)   no, not an alkene

     (d)   no, not an alkene

25.28

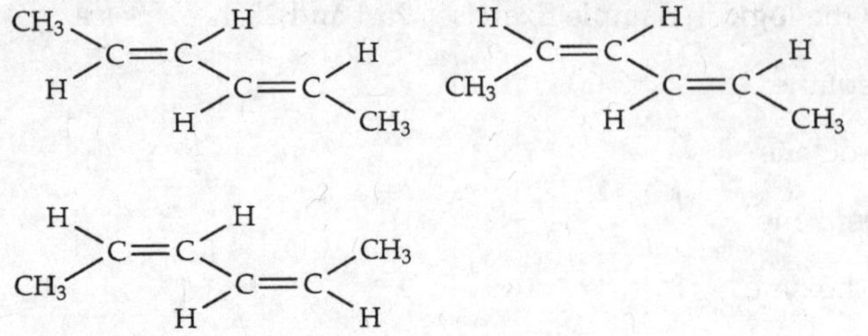

25.29 Assuming that each component retains its effective octane number in the mixture (and this isn't always the case), we obtain: octane number = 0.35(0) + 0.65(100) = 65.

25.30 Octane number can be increased by increasing the fraction of branched-chain alkanes or aromatics, since these have high octane numbers. This can be done by cracking. The octane number also can be increased by adding an anti-knock agent such as tetraethyl lead, $Pb(C_2H_5)_4$ (no longer legal); methyl t-butyl ether (MTBE); or an alcohol, methanol, or ethanol.

Reactions of Hydrocarbons

25.31 **(a)** An *addition reaction* is the addition of some reagent to the two atoms that form a multiple bond. In a *substitution reaction*, one atom or group of atoms replaces (substitutes for) another atom or group of atoms. In an addition reaction, two atoms and a multiple bond on the target molecule are altered; in a substitution reaction, the environment of one atom in the target molecule changes. Alkenes typically undergo addition, while aromatic hydrocarbons usually undergo substitution.

(b) *Plan.* Consider the general form of addition across a double bond. The π bond is broken and one new substituent (in this case two Br atoms) adds to each of the C atoms involved in the π bond. *Solve.*

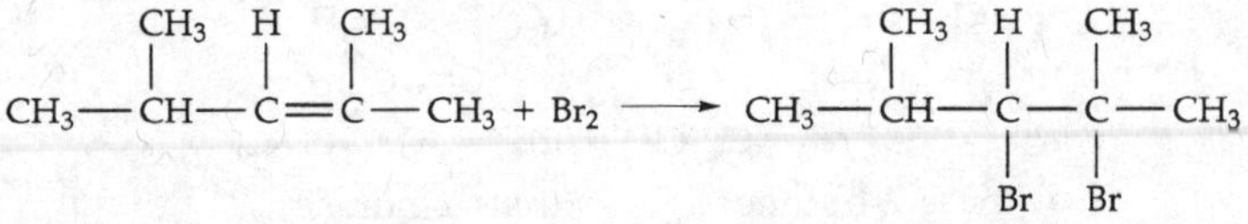

(c) *Plan.* Consider the general form of a *substitution* reaction. A Cl atom will replace one of the H atoms on the benzene ring. In the target molecule, all H atoms are equivalent, so no choice of position is required. *Solve.*

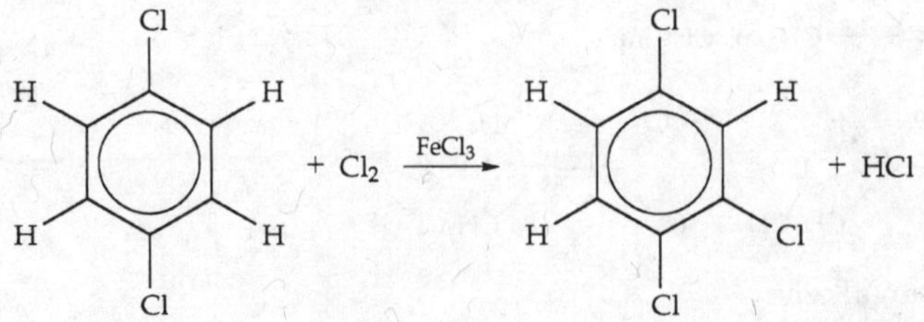

25.32 (a)

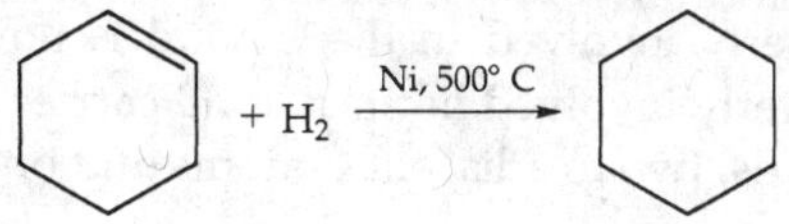

(cyclohexene) $+ H_2 \xrightarrow{\text{Ni, 500°C}}$ (cyclohexane)

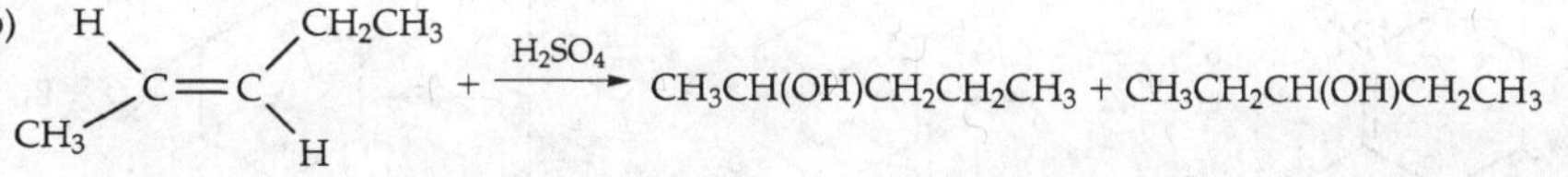

(b)

$$\begin{array}{c} H \\ \diagup \\ CH_3 \end{array} C = C \begin{array}{c} CH_2CH_3 \\ \diagdown \\ H \end{array} + \xrightarrow{H_2SO_4} CH_3CH(OH)CH_2CH_2CH_3 + CH_3CH_2CH(OH)CH_2CH_3$$

(c)

(benzene) $+ CH_3\overset{Cl}{\underset{}{CH}}CH_3 \xrightarrow{AlCl_3}$ (benzene ring)—$\overset{CH_3}{\underset{}{CHCH_3}}$ $+ HCl$

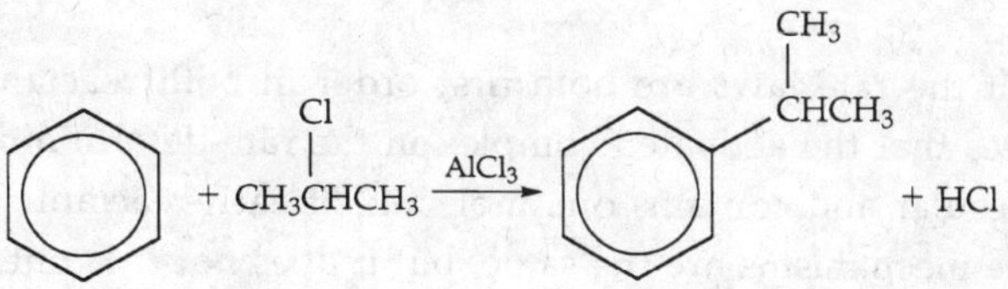

25.33 (a) *Plan.* Consider the structures of cyclopropane, cyclopentane, and cyclohexane. *Solve.*

The small 60° C—C—C angles in the cyclopropane ring cause strain that provides a driving force for reactions that result in ring opening. There is no comparable strain in the five- or six-membered rings.

(b) *Plan.* First form an alkyl halide: $C_2H_4(g) + HBr(g) \rightarrow CH_3CH_2Br(l)$; then carry out a Friedel-Crafts reaction. *Solve.*

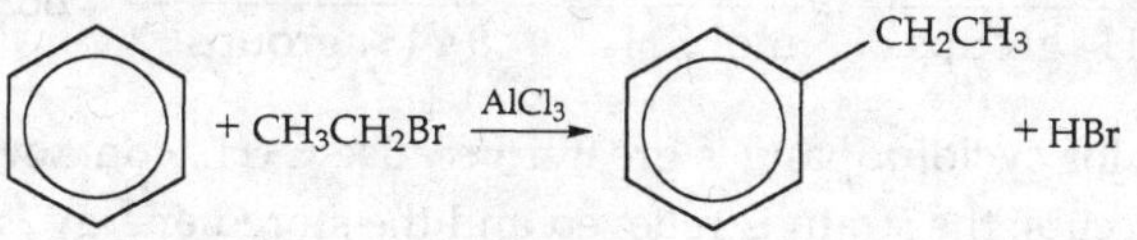

(benzene) $+ CH_3CH_2Br \xrightarrow{AlCl_3}$ (benzene ring)—CH_2CH_3 $+ HBr$

25.34 (a) The reaction of Br_2 with an alkene to form a colorless halogenated alkane is an addition reaction. Aromatic hydrocarbons do not readily undergo addition reactions, because their π-electrons are stabilized by delocalization.

(b) *Plan.* Use a Friedel-Crafts reaction to substitute a —CH_2CH_3 onto benzene. Do a second substitution reaction to get *para*-bromoethylbenzene. *Solve.*

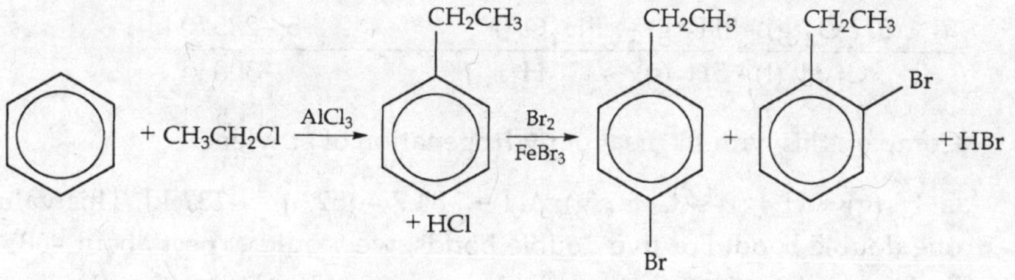

(benzene) $+ CH_3CH_2Cl \xrightarrow{AlCl_3}$ (ethylbenzene) $+ HCl$ $\xrightarrow[FeBr_3]{Br_2}$ (para-bromoethylbenzene) $+$ (ortho-bromoethylbenzene) $+ HBr$

It appears that ortho, meta, and para geometric isomers of bromoethylbenzene would be possible. However, because of electronic effects beyond the scope of this chapter, the ethyl group favors formation of ortho and para isomers, but not the meta. The ortho and para products must be separated by distillation or some other technique.

25.35 The partially positive end of the hydrogen halide, $\overset{\delta^+ \quad \delta^-}{H - X}$, is attached to the π electron cloud of the alkene cyclohexene. The electrons that formed the π bond in cyclohexene form a sigma bond to the H atom of HX, leaving a halide ion, X^-. The intermediate is a carbocation; one of the C atoms formerly involved in the π bond is now bound to a second H atom. The other C atom formerly involved in the π bond carries a full positive charge and forms only three sigma bonds, two to adjacent C atoms and one to H.

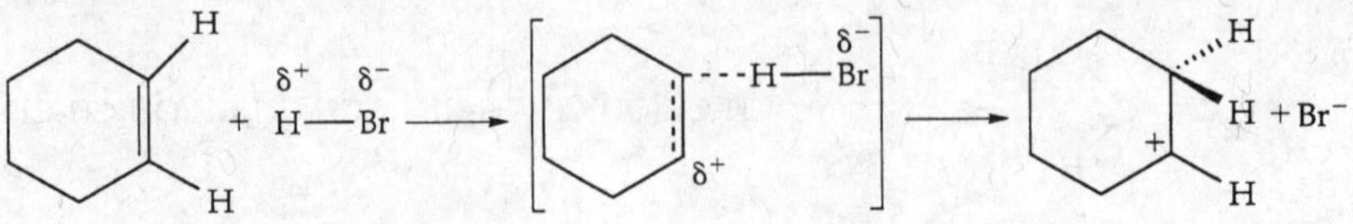

25.36   Not necessarily. That the rate laws are both first order in both reactants and second order overall indicates that the activated complex in the rate-determining step in each mechanism is bimolecular and contains one molecule of each reactant. This is usually an indication that the mechanisms are the same, but it does not rule out the possibility of different fast steps, or a different order of elementary steps.

25.37   *Analyze/Plan.* Both combustion reactions produce CO_2 and H_2O:

$$C_3H_6(g) + 9/2\,O_2(g) \rightarrow 3CO_2(g) + 3H_2O(l)$$

$$C_5H_{10}(g) + 15/2\,O_2(g) \rightarrow 5CO_2(g) + 5H_2O(l)$$

Thus, we can calculate the ΔH_{comb} / CH_2 group for each compound.   *Solve.*

$$\frac{\Delta H_{comb}}{CH_2 \text{ group}} = \frac{2089 \text{ kJ/mol } C_3H_6}{3 \text{ CH}_2 \text{ groups}} = \frac{696.3 \text{ kJ}}{\text{mol } CH_2}; \frac{3317 \text{ kJ/mol } C_5H_{10}}{5 \text{ CH}_2 \text{ groups}} = 663.4 \text{ kJ/mol } CH_2$$

$\Delta H_{comb}/CH_2$ group for cyclopropane is greater because C_3H_6 contains a strained ring. When combustion occurs, the strain is relieved and the stored energy is released during the reaction.

25.38

	ΔH
$C_{10}H_8(l) + 12O_2(g) \rightarrow 10CO_2(g) + 4H_2O(l)$	-5157 kJ
$-[C_{10}H_{18}(l) + 29/2\,O_2(g) \rightarrow 10CO_2(g) + 9H_2O(l)$	$-(-6286)$ kJ
$C_{10}H_8(l) + 5H_2O(l) \rightarrow C_{10}H_{18}(l) + 5/2\,O_2(g)$	$+1129$ kJ
$5/2\,O_2(g) + 5H_2(g) \rightarrow 5H_2O(l)$	$5(-285.8)$ kJ
$C_{10}H_8(l) + 5H_2(g) \rightarrow C_{10}H_{18}(l)$	-300 kJ

Compare this with the heat of hydrogenation of ethylene:

$C_2H_4(g) + H_2(g) \rightarrow C_2H_6(g)$; $\Delta H = -84.7 - (52.3) = -137$ kJ. This value applies to just one double bond. For five double bonds, we would expect about -685 kJ. The fact that hydrogenation of napthalene yields only -300 kJ indicates that the overall energy of the napthalene molecule is lower than expected for five isolated double bonds and that there must be some special stability associated with the aromatic system in this molecule.

Functional Groups and Chirality

25.39   (a)   ketone   (b)   carboxylic acid   (c)   alcohol

(d)   ester   (e)   amide   (f)   amine

25.40   (a) $-C\equiv C-$, alkyne; $-\overset{\overset{\displaystyle O}{\|}}{C}-H$, aldehyde

(b)   , aromatic hydrocarbon (phenol group); $-Cl$, halogen; $-C=C-$, alkene; $-COOH$, carboxylic acid

(c)   , cycloalkane; $-Cl$, halogen; $\overset{\overset{\displaystyle O}{\|}}{C}$, ketone

(d) $-C=C-$, alkene; $-\overset{\overset{\displaystyle O}{\|}}{C}-O-$, ester

(e) $\overset{|}{N}$, amine; $-\overset{\overset{\displaystyle O}{\|}}{C}-N\big\langle$, amide

25.41   *Analyze/Plan.* Given the name of a molecule, write the structural formula of an isomer that contains a specified functional group. Consider the definition of isomer, write the molecular formula of the given molecule, draw the structural formula of a molecule with the same formula that contains the specified functional group.   *Solve.*

(a)   The formula of acetone is C_3H_6O. An aldehyde contains the group $-C\overset{\displaystyle \diagup O}{\diagdown H}$

An aldehyde that is an isomer of acetone is propionaldehyde (or propanal):

$$H-\overset{\overset{\displaystyle H}{|}}{\underset{\underset{\displaystyle H}{|}}{C}}-\overset{\overset{\displaystyle H}{|}}{\underset{\underset{\displaystyle H}{|}}{C}}-C\overset{\diagup O}{\diagdown H}$$

(b)   The formula of 1-propanol is C_3H_8O. An ether contains the group $-O-$. An ether that is an isomer of 1-propanol is ethylmethyl ether:

$$H-\overset{\overset{\displaystyle H}{|}}{\underset{\underset{\displaystyle H}{|}}{C}}-\overset{\overset{\displaystyle H}{|}}{\underset{\underset{\displaystyle H}{|}}{C}}-O-\overset{\overset{\displaystyle H}{|}}{\underset{\underset{\displaystyle H}{|}}{C}}-H$$

25.42   (a) C_4H_8O, $\overset{\displaystyle O}{H_2C\diagup \ \diagdown CH_2}$
             $\underset{\displaystyle H_2C-CH_2}{}$

(b) $CH_3CH_2\overset{\overset{\displaystyle O}{\|}}{C}CH_3$, $CH_3CH_2CH_2\overset{\overset{\displaystyle O}{\|}}{C}-H$

$CH_2=CH_2CH_2CH_2OH$, $CH_3CH=CHCH_2OH$,   (cis and trans)

$CH_2=CHCH(OH)CH_3$ (enantiomers)

(Structures with the $-OH$ group attached to an alkene carbon atom are not included. These molecules are called "vinyl alcohols" and are not the major form at equilibrium.)

25 Organic and Biological Chemistry   Solutions to Exercises

25.43   *Analyze/Plan.* Count the number of C atoms in each chain, including the carboxyl C atom. Name the chain and the acid.   *Solve.*

(a)   methanoic acid

(b)   butanoic acid

(c)   3-methylpentanoic acid

25.44   (a)

$$CH_3CH_2\overset{\overset{\displaystyle O}{\|}}{C}—H$$

(b)

$$CH_3CH_2CH_2\overset{\overset{\displaystyle O}{\|}}{C}CH_3$$

(c)

$$CH_3\underset{\underset{\displaystyle CH_3}{|}}{\overset{\overset{\displaystyle O}{\|}}{C}}HCCH_3$$

(d)

$$CH_3CH_2\underset{\underset{\displaystyle CH_3}{|}}{\overset{\overset{\displaystyle O}{\|}}{C}}HC—H$$

25.45   *Analyze/Plan.* In a condensation reaction between an alcohol and a carboxylic acid, the alcohol loses its —OH hydrogen atom and the acid loses its —OH group. The alkyl group from the acid is attached to the carbonyl group and the alkyl group from alcohol is attached to the ether oxygen of the ester. The name of the ester is the alkyl group from the alcohol plus the alkyl group from the acid plus the suffix *-oate*.   *Solve.*

(a)

$$CH_3CH_2O—\overset{\overset{\displaystyle O}{\|}}{C}—\bigcirc$$

ethylbenzoate

(b)

$$CH_3\overset{\overset{\displaystyle H}{|}}{N}—\overset{\overset{\displaystyle O}{\|}}{C}CH_3$$

N-methylethanamide
or N-methylacetamide

(c)

$$\bigcirc—O—\overset{\overset{\displaystyle O}{\|}}{C}CH_3$$

phenylacetate

25.46   (a)

$$CH_3CH_2CH_2\overset{\overset{\displaystyle O}{\|}}{C}—O—CH_3$$

methylbutanoate

(b)

$$\bigcirc—\overset{\overset{\displaystyle O}{\|}}{C}—O—\underset{\underset{\displaystyle CH_3}{|}}{\overset{\overset{\displaystyle CH_3}{|}}{C}}—H$$

2-propylbenzoate

(c)

$$CH_3CH_2\underset{\underset{\displaystyle CH}{|}}{\overset{\overset{\displaystyle O}{\|}}{C}}—N—CH_3$$

N, N-dimethylpropanamide

25.47   *Analyze/Plan.* Follow the logic in Sample Exercise 25.6.   *Solve.*

(a)

$$CH_3CH_2\overset{\overset{\displaystyle O}{\|}}{C}—O—CH_3 + NaOH \longrightarrow \left[CH_3CH_2C\overset{\diagup O}{\diagdown O}\right]^- + Na^+ + CH_3OH$$

(b)

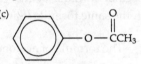

$$CH_3\overset{\overset{\displaystyle O}{\|}}{C}—O—\bigcirc + 2NaOH \longrightarrow \left[CH_3C\overset{\diagup O}{\diagdown O}\right]^- + 2Na^+ + \left[\overset{O}{\bigcirc}\right]^-$$

682

25.48 (a)

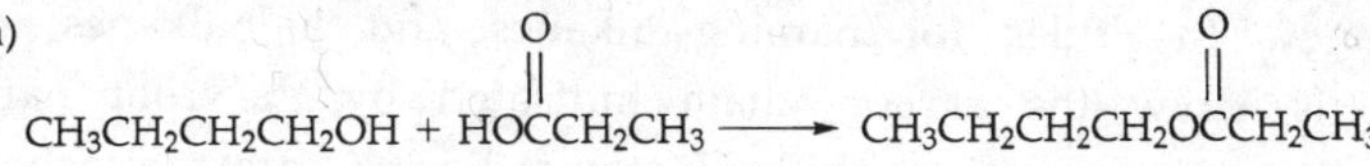

$$CH_3CH_2CH_2CH_2OH + HOCCH_2CH_3 \longrightarrow CH_3CH_2CH_2CH_2OCCH_2CH_3$$

  1-butanol   propionic acid   butyl proprionate
                     (propanoic acid)

(b)

$$CH_3OC\text{—}\langle \rangle + NaOH \longrightarrow [\langle \rangle\text{—}C\langle^O_O]^- + Na^+ + CH_3OH$$

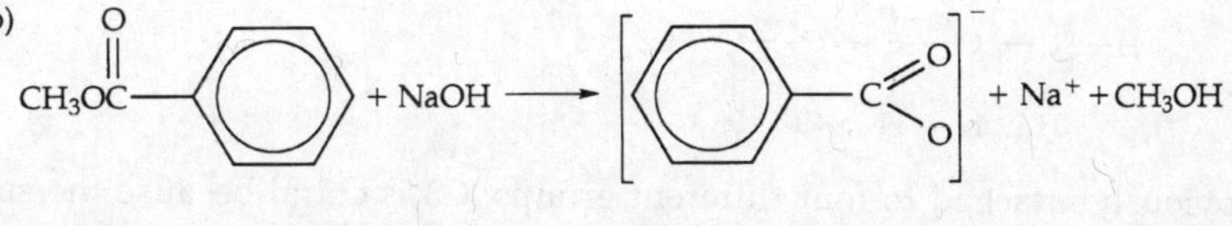

25.49 Yes, we expect acetic acid to be a strongly hydrogen-bonded substance. The carboxyl group has both —OH and —C=O groups that participate in hydrogen bonding. The boiling point of acetic acid is higher than that of water (118°C vs. 100°C), indicating that hydrogen-bonding in acetic acid is even stronger than that of water (Figure 11.10). The melting points, 16.7°C for acetic acid and 0°C for water, show a similar trend.

25.50 $2\ CH_3COOH(l) \longrightarrow CH_3COCH_3(l) + H_2O(l)$

$$CH_3C\text{—}\boxed{OH + H}\text{—}O\text{—}CCH_3 \longrightarrow \underset{CH_3}{C}\text{—}O\text{—}\underset{CH_3}{C} + H_2O$$

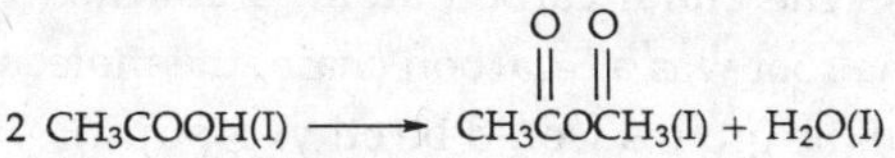

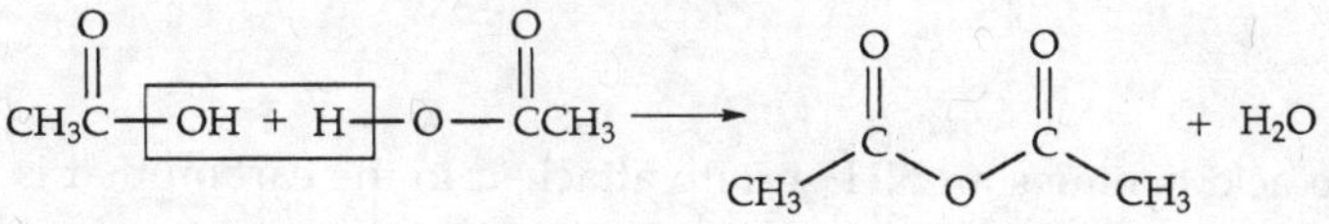

25.51 *Analyze/Plan.* Follow the logic in Sample Exercise 25.2, incorporating functional group information from Table 25.4.   *Solve.*

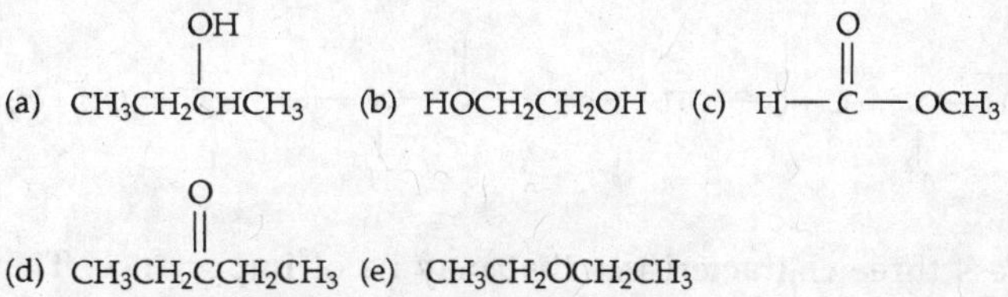

(a) $CH_3CH_2\overset{OH}{\underset{|}{C}}HCH_3$ (b) $HOCH_2CH_2OH$ (c) $H\text{—}\overset{O}{\underset{||}{C}}\text{—}OCH_3$

(d) $CH_3CH_2\overset{O}{\underset{||}{C}}CH_2CH_3$ (e) $CH_3CH_2OCH_2CH_3$

25.52 (a)

$$CH_3\text{—}\overset{Cl}{\underset{Cl}{C}}\text{—}CH_2\text{—}\overset{O}{\underset{||}{C}}\text{—}H$$

(b)

$$CH_3\text{—}\overset{O}{\underset{||}{C}}\text{—}\langle \rangle$$

(c)

$$Br\text{—}\langle \rangle\text{—}\overset{O}{\underset{||}{C}}\text{—}OH$$

(d)

$$CH_3\diagdown_O\diagup^{CH_2}\diagdown_C\diagup_{CH_3}$$

(e)

$$\langle \rangle\text{—}\overset{O}{\underset{||}{C}}\text{—}N\diagup^{CH_3}\diagdown_{CH_3}$$

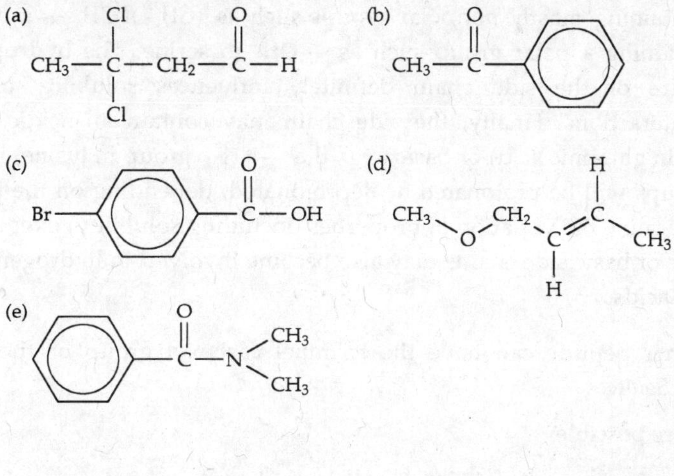

25.53   *Analyze/Plan.* Review the rules for naming alkanes and haloalkanes; draw the structures. That is, draw the carbon chain indicated by the root name, place substituents, fill remaining positions with H atoms. Each C atom attached to four different groups is chiral.   *Solve.*

$$
\begin{array}{ccccc}
H & H & CH_3 & Br & H \\
| & | & | & | & | \\
H-C & -C & -\overset{*}{C} & -\overset{*}{C} & -C-H \\
| & | & | & | & | \\
H & H & H & Cl & H
\end{array}
$$

* chiral C atoms

C2 is obviously attached to four different groups. C3 is chiral because the substituents on C2 render the C1-C2 group different than the C4-C5 group.

25.54
$$
\begin{array}{cccccc}
H & H & H & Cl & H & H \\
| & | & | & | & | & | \\
H-C & -C & -C & -\overset{*}{C} & -C & -C-H \\
| & | & | & | & | & | \\
H & H & H & CH_3 & H & H
\end{array}
$$

Yes, the molecule has optical isomers. The chiral carbon atom is attached to chloro, methyl, ethyl, and propyl groups. (If the root was a 5-carbon chain, the molecule would not have optical isomers because two of the groups would be ethyl groups.)

Proteins

25.55   (a)   An α-amino acid contains an NH_2 group attached to the carbon that is bound to the carbon of the carboxylic acid function.

(b)   In forming a protein, amino acids undergo a condensation reaction between the amino group and carboxylic acid:

$$
\begin{array}{c}
O \\
\parallel \\
-C-\boxed{OH + H}-N-CH_2C-OH \\
\qquad\qquad | \\
\qquad\qquad H
\end{array}
\longrightarrow
\begin{array}{c}
O \qquad\qquad O \\
\parallel \qquad\qquad \parallel \\
-C-N-CH_2-C-OH + H_2O \\
\quad | \\
\quad H
\end{array}
$$

25.56   The side chains possess three characteristics that may be of importance. They may be bulky (e.g., the phenyl group in phenylalanine) and thus impose restraints on where and how the amino acid can undergo reaction. Secondly, the side chain will be either hydrophobic, containing mostly nonpolar groups such as $(CH_3)_2CH-$ in valine, or hydrophilic, containing a polar group such as $-OH$ in serine. The hydrophobic or hydrophilic nature of the side chain definitely influences solubility and other intermolecular interactions. Finally, the side chain may contain an acidic (e.g., the $-COOH$ group in glutamic acid) or basic (e.g., the $-NH_2$ group in lysine) functional group. These groups will be protonated or deprotonated, depending on the pH of the solution, and determine the variation of properties (including solubility) over a range of pH values. Acidic or basic side chains may also become involved in hydrogen-bonding with other amino acids.

25.57   *Analyze/Plan.* Either peptide can have the terminal carboxyl group or the terminal amino group.   *Solve.*

Two dipeptides are possible:

$$H_3\overset{+}{N}-CH_2-\underset{\underset{H}{|}}{\overset{\overset{H_3\overset{+}{N}}{|}}{C}}-\overset{\overset{O}{||}}{C}-\underset{\underset{\underset{\underset{NH_3^+}{|}}{(CH_2)_4}}{|}}{\overset{\overset{H}{|}}{N}}-\underset{|}{\overset{\overset{H}{|}}{C}}-\overset{\overset{O}{||}}{C}-O^-$$

seryllysine

$$H_3\overset{+}{N}-(CH_2)_4-\underset{\underset{H}{|}}{\overset{\overset{H_3\overset{+}{N}}{|}}{C}}-\overset{\overset{O}{||}}{C}-\overset{\overset{H}{|}}{N}-\underset{\underset{\underset{OH}{|}}{CH_2}}{\overset{\overset{H}{|}}{C}}-\overset{\overset{O}{||}}{C}-O^-$$

lysylserine

25.58

$$H_3\overset{+}{N}-\underset{\underset{\underset{\underset{CH_3}{|}}{S}}{\underset{(CH_2)_2}{|}}}{\overset{\overset{H}{|}}{C}}-\overset{\overset{O}{||}}{C}-O^- + H_3\overset{+}{N}-\underset{\underset{H}{|}}{\overset{\overset{H}{|}}{C}}-\overset{\overset{O}{||}}{C}-O^- \longrightarrow H_3\overset{+}{N}-\underset{\underset{\underset{\underset{CH_3}{|}}{S}}{\underset{(CH_2)_2}{|}}}{\overset{\overset{H}{|}}{C}}-\overset{\overset{O}{||}}{C}-\overset{\overset{H}{|}}{N}-\underset{\underset{H}{|}}{\overset{\overset{H}{|}}{C}}-\overset{\overset{O}{||}}{C}-O^-$$

methionine   glycine   methionylglycine

25.59   *Analyze/Plan.* Follow the logic in Sample Exercise 25.7.   *Solve.*

(a)

$$H_3\overset{+}{N}-\underset{\underset{\underset{\underset{CH_3}{|}}{CH_2}}{\underset{CH-CH_3}{|}}}{\overset{\overset{H}{|}}{C}}-\overset{\overset{O}{||}}{C}-\overset{\overset{H}{|}}{N}-\underset{\underset{CH_3}{|}}{\overset{\overset{H}{|}}{C}}-\overset{\overset{O}{||}}{C}-\overset{\overset{H}{|}}{N}-\underset{\underset{\underset{SH}{|}}{CH_2}}{\overset{\overset{H}{|}}{C}}-\overset{\overset{O}{||}}{C}-O^-$$

Ile-Ala-Cys

(b)   Eight: Ser-Ser-Ser; Ser-Ser-Phe; Ser-Phe-Ser; Phe-Ser-Ser; Ser-Phe-Phe;

Phe-Ser-Phe; Phe-Phe-Ser; Phe-Phe-Phe

25.60   (a)   Valine, serine, glutamic acid

(b)   Six (assuming the tripeptid contains all three amino acids):

Gly-Ser-Glu; Gly-Glu-Ser; Ser-Gly-Glu; Ser-Glu-Gly; Glu-Ser-Gly; Glu-Gly-Ser

25.61   The *primary structure* of a protein refers to the sequence of amino acids in the chain. Along any particular section of the protein chain the configuration may be helical, may be an open chain, or arranged in some other way. This is called the *secondary structure*. The overall shape of the protein molecule is determined by the way the segments of the protein chain fold together, or pack. The interactions which determine the overall shape are referred to as the *tertiary structure*.

25.62   It is quite evident from Figure 25.26 that the hydrogen bonds between an NH group along the chain and the unshared electron pairs of a carbonyl group further along are responsible for maintaining the helix. Indeed, the pitch and general shape of the helix are determined by what specific interactions produce a good hydrogen-bonding arrangement.

25 Organic and Biological Chemistry   Solutions to Exercises

Carbohydrates

25.63   (a)   *Carbohydrates*, or sugars, are composed of carbon, hydrogen, and oxygen. From a chemical viewpoint, they are polyhydroxyaldehydes or ketones. Carbohydrates are primarily derived from plants and are a major food source for animals.

   (b)   A *monosaccharide* is a simple sugar molecule that cannot be decomposed into smaller sugar molecules by (acid) hydrolysis.

   (c)   A *disaccharide* is a carbohydrate composed of two simple sugar units. Hydrolysis breaks the disaccharides into two monosaccharides.

25.64   Glucose exists in solution as a cyclic structure in which the aldehyde function on carbon 1 reacts with the OH group of carbon 5 to form what is called a hemiacetal, Figure 25.29. Carbon atom 1 carries an OH group in the hemiacetal form; in α-glucose this OH group is on the opposite side of the ring as the CH_2OH group on carbon atom 5. In the β (beta) form the OH group on carbon 1 is on the same side of the ring as the CH_2OH group on carbon 5.

The condensation product of two glucose units looks like this:

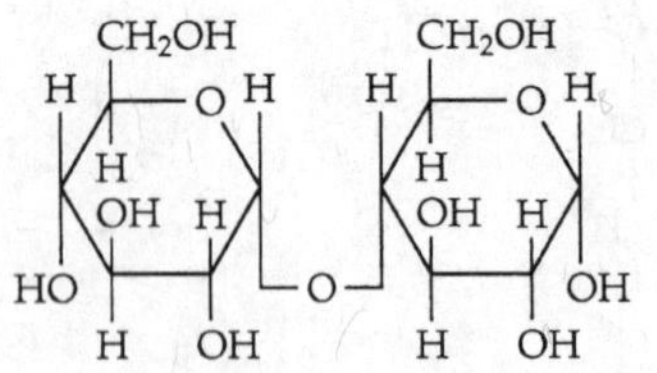

α-linkage 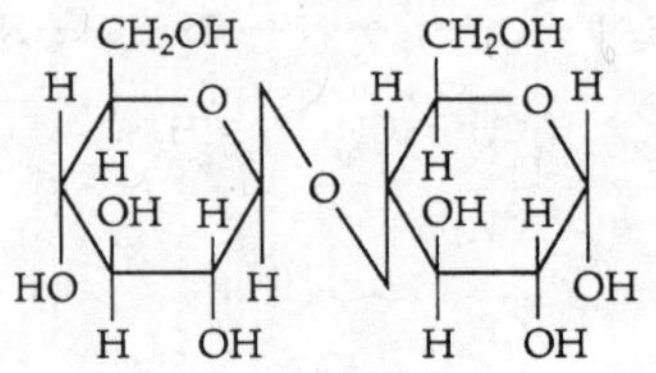   β-linkage

25.65   (a)   In the linear form of galactose, the aldehydic carbon is C1. Carbon atoms 2, 3, 4, and 5 are chiral because they each carry four different groups. Carbon 6 is not chiral because it contains two H atoms.

   (b)   The structure is best deduced by comparing galactose with glucose, and inverting the configurations at the appropriate carbon atoms. Recall from Solution 25.64 that both the β-form (shown here) and the α-form (OH on carbon 1 on the opposite side of ring as the CH_2OH on carbon 5) are possible.

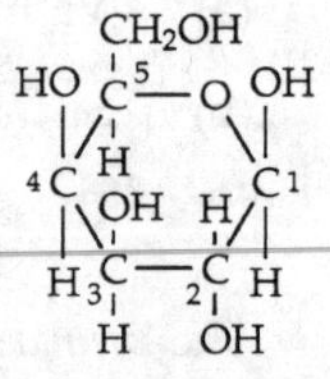

galactose

25.66   (a)   In the linear form of mannose, the aldehydic carbon is C1. Carbon atoms 2, 3, 4, and 5 are chiral because they each carry four different groups. Carbon 6 is not chiral because it contains two H atoms.

   (b)   Both the α (left) and β (right) forms are possible.

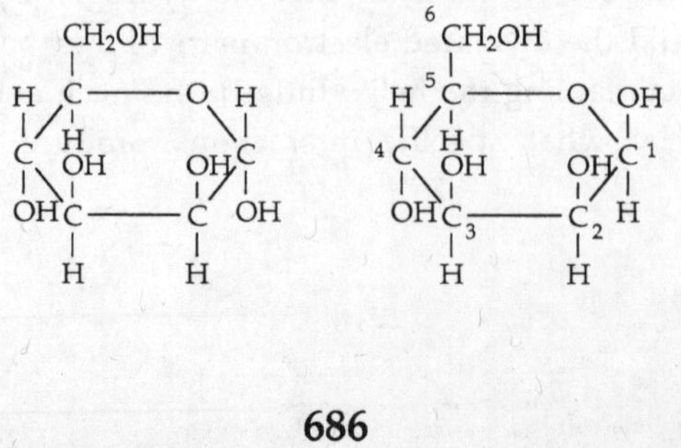

25.67   The empirical formula of glycogen is $C_6H_{10}O_5$. The six-membered ring form of glucose is the unit that forms the basis of glycogen. The monomeric glucose units are joined by α linkages.

25.68   The empirical formula of cellulose is $C_6H_{10}O_5$. As in glycogen, the six-membered ring form of glucose forms the monomer unit that is the basis of the polymer cellulose. In cellulose, glucose monomer units are joined by β linkages.

Nucleic Acids

25.69   A *nucleotide* consists of a nitrogen-containing aromatic compound, a sugar in the furanose (five-membered) ring form, and a phosphoric acid group. The structure of deoxycytidine monophosphate is shown at right.

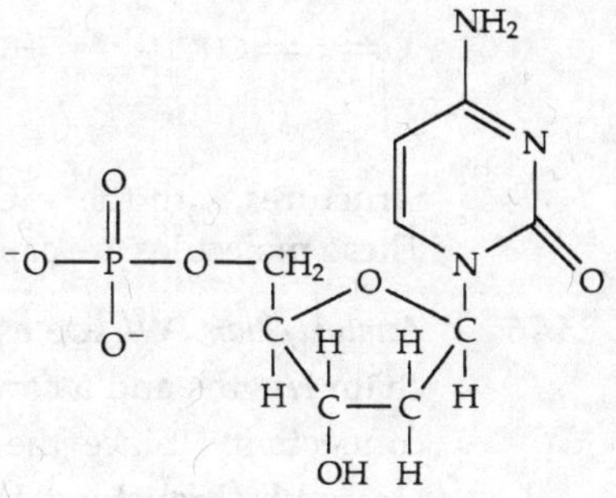

25.70

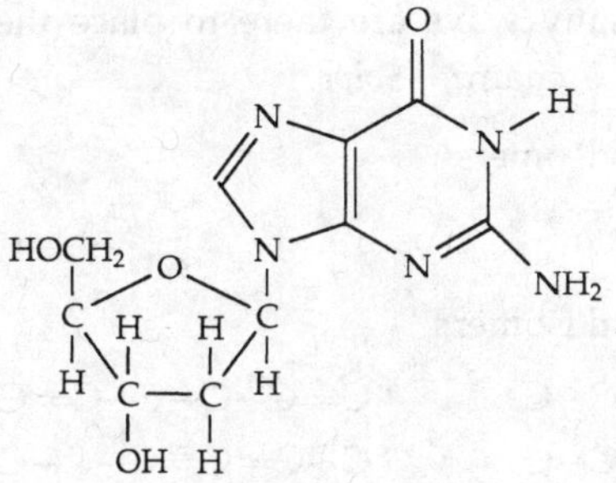

25.71 $C_4H_7O_3CH_2OH + HPO_4^{2-} \rightarrow C_4H_7O_3CH_2-O-PO_3^{2-} + H_2O$

25.72

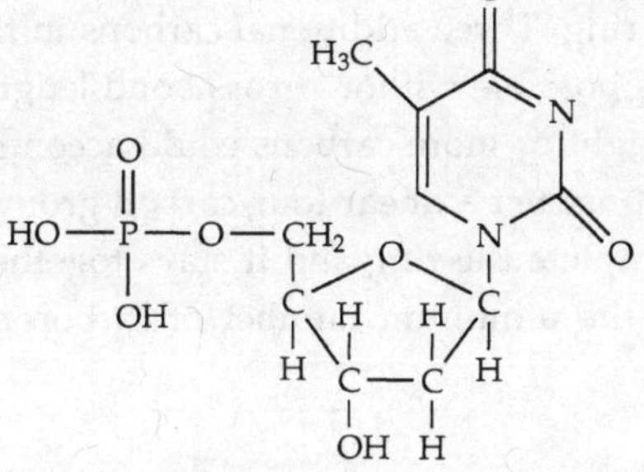

25.73   In the helical structure for DNA, the strands of the polynucleotides are held together by hydrogen-bonding interactions between particular pairs of bases. It happens that adenine and thymine form an especially effective base pair, and that guanine and cytosine are similarly related. Thus, each adenine has a thymine as its opposite number in the other strand, and each guanine has a cytosine as its opposite number. In the overall analysis of the double strand, total adenine must then equal total thymine, and total guanine equals total cytosine.

25.74

```
—A—C—T—C—G—A—
  :   :   :   :   :   :
—T—G—A—G—C—T—  ◄——— complementary strand
```

Additional Exercises

25.75

```
       H  O                    H
       |  ||                    C
H₂C —— C —— C —— H    H —— C ⁄⁄ \ C —— OH    H —— C ≡ C —— CH₂OH
                                              
                                            O
                                            ||
O ═ C ═ CHCH₃      HC ≡ COCH₃                C
                                          ⁄     \
                                      H₂C —— CH₂
```

Structures with the —OH group attached to an alkene carbon atom are not included. These molecules are called "vinyl alcohols" and are not the major form at equilibrium.

25.76 *Analyze/Plan.* We are asked the number of structural isomers for two specified carbon chain lengths and a certain number of double bonds. Structural isomers have different connectivity. Since the chain length is specified, we can ignore structural isomers created by branching. We are not asked about geometrical isomers, so we ignore those as well. The resulting question is: How many ways are there to place the specified number of double bonds along the specified C chain? *Solve.*

5 C chain with one double bond: 2 structural isomers

$$C═C–C–C–C \qquad C–C═C–C–C$$

6 C chain with two double bonds: 6 structural isomers

$$C═C–C═C–C–C \qquad C═C–C–C═C–C \qquad C═C–C–C–C═C$$
$$C–C═C–C═C–C \qquad C═C═C–C–C–C \qquad C–C═C═C–C–C$$

25.77 Because of the strain in bond angles about the ring, cyclic alkynes with less than eight carbons are not stable. Alkyne carbon atoms preferentially have 180° bond angles; this requires a linear four-carbon group in the ring. Three additional carbons in the ring do not provide enough flexibility to make this possible without gross bond length or angle distortions. It is possible that a ring with eight or more carbons could accommodate an alkyne linkage. To test this with models, construct a linear four-carbon group and then add tetrahedral C-atoms until you can complete the ring and it stays together without intervention. This is a good indication of the minimum number of carbon atoms in a stable ring that contains an alkyne linkage.

25.78

```
  H          H               CH₃         H
   \        /                  \        /
    C ═══ C                      C ═══ C
   /        \                  /        \
 CH₃       CH₂CH₃            H          CH₂CH₃
      cis                         trans
```

Cyclopentene does not show cis-trans isomerism because the existence of the ring demands that the C—C bonds be cis to one another.

25.79 In alkanes, carbon forms only single, sigma bonds. Alkenes contain at least one C—C double bond, consisting of one sigma and one pi bond. Alkynes have a least one C—C triple bond, composed of one sigma and two pi bonds. In both alkenes and alkynes,

C atoms are involved in pi overlap. The question is, what feature of Si prevents it from forming double or triple bonds which involve pi overlap.

According to Table 8.5, the average C—C single bond length is 1.54 Å, C=C is 1.34 Å, and C≡C is 1.20 Å. These distances show that pi overlap requires substantially closer approach of the two bonded atoms than sigma overlap alone. The bonding atomic radius of Si is 1.11 Å, while that of C is 0.77 Å (Figure 7.6). The close approach of Si atoms that is required for pi overlap is not possible because of its large bonding atomic radius. Thus, silicon analogs of alkenes and alkynes that involve multiple bonds and pi overlap are virtually unknown. Silicon analogs of alkanes with exclusively sigma overlap are known; the average Si—Si single bond length is 2.22 Å.

25.80   The C—Cl bonds in the trans compound are pointing in exactly opposite directions. Thus, the C—Cl bond dipoles cancel (Section 9.3). This is not the case in the cis compound, as can be seen by drawing the structure:

25.81 H_2C=CH—CH_2OH

(Structures with the —OH group attached to an alkene carbon atom are not included. These molecules are called "vinyl alcohols" and are not the major form at equilibrium.)

25.82   One. One molecule of HBr would add to the C=C according to the reaction

Because the π electrons in the phenyl ring are delocalized, the group is particularly stable and resistant to addition reactions. The phenyl group could undergo substitution with HBr, but only with a catalyst or special conditions.

25.83   Two plausible decomposition reactions are:

(i) $CH_2Cl_2(l) \rightarrow C(s) + H_2(g) + Cl_2(g)$

(ii) $CH_2(NO_2)_2(l) \rightarrow N_2(g) + CO_2(g) + H_2O(g) + 1/2 O_2(g)$

Use bond dissociation energies (Table 8.4) to evaluate approximate ΔH values for each reaction.

(i) $\Delta H = 2D(C-H) + 2D(C-Cl) - D(H-H) - D(Cl-Cl)$

        $= 2(413) + 2(328) - 436 - 242 = +804$ kJ

(ii) $\Delta H = 2D(C-H) + 2D(C-N) + 2D(N=O) + 2D(N-O) - D(N\equiv N) - 2D(C=O)$

$$- 2D(O-H) - 1/2D(O=O)$$

$$= 2(413) + 2(293) + 2(607) + 2(201) - 941 - 2(799) - 2(463) - 1/2(495)$$

$$\Delta H = -685 \text{ kJ}$$

Clearly, the decomposition of $CH_2(NO_2)_2$ is thermodynamically favorable, while the decomposition of CH_2Cl_2 is not. In particular, this is because of the stability of N_2 and CO_2 relative to $CH_2(NO_2)_2$. For CH_2Cl_2, no oxygen atoms are available to form stable products such as CO_2 and H_2O.

25.84   (a) Ether, $C-O-C$; alkene, $-CH=CH_2$

(b) carboxylic acid, $-\overset{\displaystyle O}{\overset{\displaystyle \|}{C}}-OH$; ester, $CH_3\overset{\displaystyle O}{\overset{\displaystyle \|}{C}}-O-$; aromatic,

(c) ketone, $-\overset{\displaystyle O}{\overset{\displaystyle \|}{C}}-$; alkene, $-CH=CH-$; alcohol $-C-OH$

25.85   (a) $CH_3CH_2CH_2\overset{\displaystyle O}{\overset{\displaystyle \|}{C}}OH$ or $(CH_3)_2CH\overset{\displaystyle O}{\overset{\displaystyle \|}{C}}OH$

(c)

(b) $CH_3-\overset{\displaystyle OH}{\overset{\displaystyle |}{C}}H-\overset{\displaystyle OH}{\overset{\displaystyle |}{C}}H_2$ or $\overset{\displaystyle OH}{\overset{\displaystyle |}{C}}H_2-CH_2-\overset{\displaystyle OH}{\overset{\displaystyle |}{C}}H_2$

(d)

25.86   The difference between an alcoholic hydrogen and a carboxylic acid hydrogen is two-fold. First, the electronegative carbonyl oxygen in a carboxylic acid withdraws electron density from the $O-H$ bond, rendering the bond more polar and the H more ionizable. Second, the conjugate base of a carboxylic acid, carboxylate anion, exhibits resonance. This stabilizes the conjugate base and encourages ionization of the carboxylic acid. In an alcohol no electronegative atoms are bound to the carbon that holds the $-OH$ group, and the H is tightly bound to the O.

25.87   (a) $CH_3\overset{\displaystyle O}{\overset{\displaystyle \|}{C}}-OH$, C_6H_5OH (b) $C_6H_5\overset{\displaystyle O}{\overset{\displaystyle \|}{C}}-OH$, CH_3OH

25.88   In order for indole to be planar, the N atom must be sp^2 hybridized. The nonbonded electron pair on N is in a pure p orbital perpendicular to the plane of the molecule. The electrons that form the π bonds in the molecule are also in pure p orbitals perpendicular to the plane of the molecule. Thus, each of these p orbitals is in the correct orientation for π overlap; the delocalized π system extends over the entire molecule and includes the "nonbonded" electron pair on N. The reason that indole is such a weak base (H^+ acceptor) is that the nonbonded electron pair is delocalized and a H^+ ion does not feel the attraction of a full localized electron pair.

25.89   (a)   None

(b)   The carbon bearing the secondary —OH has four different groups attached, and is thus chiral.

(c)   The carbon bearing the —NH_2 group and the carbon bearing the CH_3 group are both chiral.

25.90   (a)

$$H_3\overset{+}{N}CHC\overset{O}{\underset{\overset{|}{HC(CH_3)_2}}{\overset{||}{|}}}NCH_2C\overset{O}{\underset{\overset{|}{\underset{COO^-}{CH_2}}}{\overset{H}{\underset{}{}}\overset{O}{||}}}NCHCO^-$$

(b)

$$H_3\overset{+}{N}\overset{H}{\underset{\overset{|}{\underset{}{CH_2}}}{\overset{|}{C}}}\overset{O}{\underset{}{C}}NCHC\overset{O}{\underset{\overset{|}{CH_2OH}}{\overset{||}{}}}NCHCO^-$$

25.91   Glu-Cys-Gly is the only possible order. Glutamic acid has two carboxyl groups that can form a peptide bond with cysteine, so there are two possible structures.

$$H_3\overset{+}{N}CHC\underset{\overset{|}{\underset{COO^-}{(CH_2)_2}}}{\overset{O}{||}}NCHC\underset{\overset{|}{\underset{SH}{CH_2}}}{\overset{H\ O}{|\ ||}}NCH_2CO^-$$ or $$H_3\overset{+}{N}CHCH_2CH_2C\underset{\overset{|}{\underset{}{COO^-}}}{\overset{O}{||}}NCHC\underset{\overset{|}{\underset{SH}{CH_2}}}{\overset{H\ O}{|\ ||}}NCH_2CO^-$$

25.92   Starch, glycogen, and cellulose are all biopolymers built by linking glucose monomers. Starch and glycogen have alpha (α) glucose linkages, where the bridging O atom is on the opposite side of the ring as the CH_2OH group. The smallest repeating unit in starch and glycogen is a single glucose unit. Starch and glycogen can have branched structures, while cellulose is always linear.

Cellulose has beta (β) glucose linkages, where the bridging O atom is on the same side of one of the rings as the CH_2OH group and on the opposite side of the CH_2OH group on the second ring. The geometry of the β linkage requires that the two linked glucose units have different orientations and that the smallest repeating unit in cellulose is two glucose units with a β linkage.

The molecular weight of a polymer is an indication of the number of monomer units present. Starch, glycogen, and cellulose all have a range of molecular weights. Glycogen has the widest range of molecular weights, 5,000–5,000,000 amu, and is potentially the largest polymer. Cellulose is intermediate in size with an average molar mass of 500,000 amu.

Starch and cellulose are produced in plants, while glycogen is produced in animals and serves as an energy storage mechanism.

25.93   Both glucose and fructose contain six C atoms, so both are hexoses. Glucose contains an aldehyde group at C1, so it is an aldohexose. Fructose has a ketone at C2, so it is a ketohexose.

25.94

```
—G—G—T—A—C—T——
  ⋮   ⋮   ⋮   ⋮   ⋮   ⋮
—C—C—A—T—G—A——  ←— complementary strand
```

Integrative Exercises

25.95 CH_3CH_2OH CH_3-O-CH_3
         ethanol   dimethyl ether

Ethanol contains —O—H bonds which form strong intermolecular hydrogen bonds, while dimethyl ether experiences only weak dipole-dipole and dispersion forces.

difluoromethane   tetrafluoromethane

CH_2F_2 is a polar molecule, while CF_4 is nonpolar. CH_2F_2 experiences dipole-dipole and dispersion forces, while CF_4 experiences only dispersion forces.

In both cases, stronger intermolecular forces lead to the higher boiling point.

25.96   Determine the empirical formula of the unknown compound and its oxidation product. Use chemical properties to propose possible structures.

$$68.1\,g\,C \times \frac{1\,mol\,C}{12.01\,g\,C} = 5.6703;\ 5.6703/1.1375 = 4.98 \approx 5$$

$$13.7\,g\,H \times \frac{1\,mol\,H}{1.008\,g\,H} = 13.5913;\ 13.5913/1.1375 = 11.95 \approx 12$$

$$18.2\,g\,P \times \frac{1\,mol\,O}{16.00\,g\,O} = 1.1375;\ 1.1375/1.1375 = 1$$

The empirical formula of the unknown is $C_5H_{12}O$.

$$69.7\,g\,C \times \frac{1\,mol\,C}{12.01\,g\,C} = 5.8035;\ 5.8035/1.1625 = 4.99 \approx 5$$

$$11.7\,g\,H \times \frac{1\,mol\,H}{1.008\,g\,H} = 11.6071;\ 11.6071/1.1625 = 9.99 \approx 10$$

$$18.6\,g\,O \times \frac{1\,mol\,O}{16.00\,g\,O} = 1.1625;\ 1.1625/1.1625 = 1$$

The empirical formula of the oxidation product is $C_5H_{10}O$.

The compound is clearly an alcohol. Its slight solubility in water is consistent with the properties expected of a secondary alcohol with a five-carbon chain. The fact that oxidation results in a ketone, rather than an aldehyde or a carboxylic acid, tells us that it is a secondary alcohol. Some reasonable structures for the unknown secondary alcohol are:

$$CH_3CHCH_2CH_2CH_3 \quad CH_3CHCHCH_2CH_3 \quad CH_3CHCH(CH_3)_2$$
$$\qquad | \qquad\qquad\qquad | \qquad\qquad\qquad |$$
$$\quad OH \qquad\qquad\qquad OH \qquad\qquad\qquad OH$$

25.97 Determine the empirical formula, molar mass, and thus molecular formula of the compound. Confirm with physical data.

$$66.7\,g\,C \times \frac{1\,mol\,C}{12.01\,g\,C} = 5.554 \ \ mol\,C;\ 5.554/1.388 = 4$$

$$11.2\,g\,H \times \frac{1\,mol\,H}{1.008\,g\,H} = 11.11 \ \ mol\,H;\ 11.11/1.388 = 8$$

$$22.2\,g\,O \times \frac{1\,mol\,O}{16.00\,g\,O} = 1.388 \ \ mol\,O;\ 1.388/1.388 = 1$$

The empirical formula is C_4H_8O. Using Equation 10.11 (MM = molar mass):

$$MM = \frac{(2.28\,g\,/\,L)(0.08206\,L\bullet atm\,/\,mol\bullet K)(373K)}{0.970\,atm} = 71.9\,g\,/\,mol$$

The formula weight of C_4H_8O is 72, so the molecular formula is also C_4H_8O. Since the compound has a carbonyl group and cannot be oxidized to an acid, the only possibility is 2-butanone.

$$\overset{\displaystyle O}{\overset{\displaystyle \|}{CH_3CCH_2CH_3}}$$

The boiling point of 2-butanone is 79.6°C, confirming the identification.

25.98 Determine the empirical formula, molar mass, and thus molecular formula of the compound. Confirm with physical data.

$$85.7\,g\,C \times \frac{1\,mol\,C}{12.01\,g\,C} = 7.136\,mol\,C;\ 7.136/7.136 = 1$$

$$14.3\,g\,H \times \frac{1\,mol\,H}{1.008\,g\,H} = 14.19\,mol\,H;\ 14.19/7.136 \approx 2$$

Empirical formula is CH_2. Using Equation 10.11 (MM = molar mass):

$$MM = \frac{(2.21\,g\,/\,L)(0.08206\,L\bullet atm\,/\,mol\bullet K)(373K)}{(735\,/\,760)\,atm} = 69.9\,g\,/\,mol$$

The molecular formula is thus C_5H_{10}. The absence of reaction with aqueous Br_2 indicates that the compound is not an alkene, so the compound is probably the cycloalkane cyclopentane. According to the *Handbook of Chemistry and Physics*, the boiling point of cyclopentane is 49°C at 760 torr. This confirms the identity of the unknown.

25.99 The reaction is: $2NH_2CH_2COOH(aq) \rightarrow NH_2CH_2CONHCH_2COOH(aq) + H_2O(l)$

$\Delta G° = (-488) + (-237.13) - 2(-369) = 12.87 = 13\,kJ$

25.100　(a)　A = adenosine = $C_{10}H_{12}O_3N_5$

$$\left[A-O-\underset{\underset{O}{|}}{\overset{\overset{O}{\|}}{P}}-O-\underset{\underset{O}{|}}{\overset{\overset{O}{\|}}{P}}-O-\underset{\underset{O}{|}}{\overset{\overset{O}{\|}}{P}}-O \right]^{4-} + H_2O \longrightarrow$$

$$\left[A-O-\underset{\underset{O}{|}}{\overset{\overset{O}{\|}}{P}}-O-\underset{\underset{O}{|}}{\overset{\overset{O}{\|}}{P}}-OH \right]^{2-} + HPO_4^{2-}$$

$$[A-P_3O_{10}]^{4-} + H_2O \longrightarrow [A-P_2O_6(OH)]^{2-} + HPO_4^{2-}$$

(The placement of the H^+ in these reactions is somewhat arbitrary; H^+ is attracted to the strongest base, but the equilibria are complex.)

(b)　If the hydrolysis reaction is spontaneous, the sign of ΔG must be negative.

(c)　Adenosine monophosphate (AMP) + inorganic phosphate

$$\left[A-O-\underset{\underset{O}{|}}{\overset{\overset{O}{\|}}{P}}-OH \right]^{-} + HPO_4^{2-}$$

(The placement of the H^+ in these reactions is somewhat arbitrary; H^+ is attracted to the strongest base, but the equilibria are complex.)

25.101　(a)　At low pH, the amine   At high pH, the amine
　　　　　　and carboxyl groups   and carboxyl groups
　　　　　　are protonated.   are deprotonated.

$$\overset{+}{H_3N}-\underset{\underset{CH_3}{|}}{\overset{\overset{H}{|}}{C}}-\overset{\overset{O}{\|}}{C}-OH \qquad NH_2-\underset{\underset{CH_3}{|}}{\overset{\overset{H}{|}}{C}}-\overset{\overset{O}{\|}}{C}-O^-$$

(b)　$$CH_3\overset{\overset{O}{\|}}{C}-OH(aq) \longrightarrow CH_3\overset{\overset{O}{\|}}{C}-O^-(aq) + H^+(aq)$$

$K_a = 1.8 \times 10^{-5}$, $pK_a = -\log(1.8 \times 10^{-5}) = 4.74$

The conjugate acid of NH_3 is NH_4^+.

$NH_4^+(aq) \rightleftharpoons NH_3(aq) + H^+(aq)$

$K_a = K_w / K_b = 1 \times 10^{-14} / 1.8 \times 10^{-5} = 5.55 \times 10^{-10} = 5.6 \times 10^{-10}$

$pK_a = -\log(5.55 \times 10^{-10}) = 9.26$

In general, a $-COOH$ group is a stronger acid than a $-NH_3^+$ group. The lower pK_a value for amino acids is for the ionization (deprotonation) of the $-COOH$ group and the higher pK_a is for the deprotonation of the $-NH_3^+$ group.

25.102　(a)　Because the native form is most stable, it has a lower, more negative free energy than the denatured form. Another way to say this is that ΔG for the process of denaturing the protein is positive.

(b)　ΔS is negative in going from the denatured form to the folded (native) form; the native protein is more ordered.

(c)　The four S—S linkages are strong covalent links holding the chain in place in the folded structure. A folded structure without these links would be less stable (higher G) and have more motional freedom (more positive entropy).

(d)　After reduction, the eight S—H groups will form hydrogen-bond-like interations with acceptors along the protein backbone, but these will be weaker and less specifically located than the S—S covalent bonds of the native protein. Overall, the tertiary structure of the reduced protein will be looser and less compact due to the loss of the S—S linkages.

25.103　$AMPOH^-(aq) \rightleftharpoons AMPO^{2-}(aq) + H^+(aq)$

$pK_a = 7.21; K_a = 10^{-pK_a} = 6.17 \times 10^{-8} = 6.2 \times 10^{-8}$

$K_a = \dfrac{[AMPO^{2-}][H^+]}{[AMPOH^-]} = 6.2 \times 10^{-8}$. When pH = 7.40, $[H^+] = 3.98 \times 10^{-8} = 4 \times 10^{-8}$.

Then $\dfrac{[AMPOH^-]}{[AMPO^{2-}]} = 3.98 \times 10^{-8} / 6.17 \times 10^{-8} = 0.6457 = 0.6$